Calculus

AND ITS APPLICATIONS
Expanded Version

Marvin L. Bittinger
Indiana University Purdue University Indianapolis

David J. Ellenbogen
Community College of Vermont

Scott A. Surgent
Arizona State University

PEARSON

Boston   Columbus   Indianapolis   New York   San Francisco
Amsterdam   Cape Town   Dubai   London   Madrid   Milan   Munich   Paris   Montréal   Toronto
Delhi   Mexico City   São Paulo   Sydney   Hong Kong   Seoul   Singapore   Taipei   Tokyo

Editor in Chief	Deirdre Lynch
Executive Editor	Jennifer Crum
Senior Content Editor	Rachel S. Reeve
Program Manager	Tatiana Anacki
Editorial Assistant	Joanne Wendelken
Managing Editor	Karen Wernholm
Senior Production Supervisor	Ron Hampton
Associate Design Director	Andrea Nix
Senior Designer/Cover Design	Barbara Atkinson
Digital Assets Manager	Marianne Groth
Media Producer	Nicholas Sweeney, Jonathan Wooding
Software Development	Mary Durnwald, TestGen; Bob Carroll, Kristina Evans MathXL
Executive Marketing Manager	Jeff Weidenaar
Marketing Assistant	Caitlin Crain, Brooke Smith
Senior Technology Specialist	Joe Vetere
Rights and Permissions Advisor	Pam Foley
Manufacturing Buyer	Debbie Rossi, Carol Melville
Text Design, Art Editing, Photo Research	Geri Davis/The Davis Group, Inc.
Production Coordination	Jane Hoover/Lifland et al., Bookmakers
Composition	PreMediaGlobal
Illustrations	Network Graphics, Inc., and William Melvin
Cover Image	Shutter stock/luri

Photo Credits

For permission to use copyrighted material, grateful acknowledgment is made to the copyright holders on page 842, which is hereby made part of this copyright page.

Many of the designations used by manufacturers and sellers to distinguish their products are claimed as trademarks. Where those designations appear in this book, and Pearson was aware of a trademark claim, the designations have been printed in initial caps or all caps.

Library of Congress Cataloging-in-Publication Data
Bittinger, Marvin L.
  [Calculus and its applications (expanded version)]
  Calculus and its applications: expanded version: media update/Marvin L. Bittinger, Indiana University Purdue University Indianapolis, David J. Ellenbogen, Community College of Vermont, Scott A. Surgent, Arizona State University.—1st edition.
    pages cm
Includes index.
ISBN 978-0-13-412258-8—ISBN 0-13-412258-5
1. Calculus—Textbooks.   I. Ellenbogen, David.   II. Surgent, Scott Adam.   III. Title.
QA303.2.B4667 2016
515—dc23   2014041546

1 2 3 4 5 6 7 8 9 10—RRD—16 15

www.pearsonhighered.com

ISBN-10: 0-13-412258-5
ISBN-13: 978-0-13-412258-8

To: Elaine, Victoria, and Beth

Contents

Supplementary On-line Chapters

Available electronically within MyMathLab or in print via a custom version of the text. Contact your local Pearson representative for more details.

Preface

Calculus and Its Applications, Expanded Version, is derived from the most student-oriented applied calculus text on the market: *Calculus and Its Applications*, Tenth Edition. The authors believe that appealing to students' intuition and speaking in a direct, down-to-earth manner make this text accessible to any student possessing the prerequisite math skills. By presenting more topics in a conceptual and often visual manner and adding student self-assessment and teaching aids, this text truly addresses students' needs. However, the authors recognize that it is not enough for a text to be accessible—it must also provide students with motivation to learn. Tapping into areas of student interest, the authors provide an abundant supply of examples and exercises rich in real-world data from business, economics, environmental studies, health care, and the life sciences. Relevant examples cover applications ranging from the distribution of wealth to the growth of membership in Facebook. Found in every chapter, realistic applications draw students into the discipline and help them to generalize the material and apply it to new and novel situations. To further spark student interest, hundreds of meticulously drawn graphs and illustrations appear throughout the text, making it a favorite among students who are visual learners.

Calculus and Its Applications, Expanded Version, covers enough content to support two semesters of applied calculus easily. Topics included in this expanded version are trigonometric functions, additional coverage of differential equations, sequences and series, and probability. Material on systems and matrices, linear programming, and discrete probability is also available in print via custom editions (contact your Pearson representative for details) or within MyMathLab. A course in intermediate algebra is assumed to be a prerequisite. The Prerequisite Skills Diagnostic Test that follows this preface is a tool for gauging students' preparedness. Appendix A: Review of Basic Algebra, together with Chapter R: Functions, Graphs, and Models, should provide a sufficient foundation to unify the diverse backgrounds of most students.

Our Approach

Intuitive Presentation

Although the word *intuitive* has many meanings and interpretations, its use here means "experience based, without proof." Throughout the text, when a concept is discussed, its presentation is designed so that the students' learning process is based on their earlier mathematical experience. This is illustrated by the following situations.

- Before the formal definition of *continuity* is presented, an informal explanation is given, complete with graphs that make use of student intuition about ways in which a function could be discontinuous (see pp. 113–114).
- The definition of *derivative*, in Chapter 1 (see p. 135), is presented after the discussion of average rates of change. This presentation is more accessible and realistic than the strictly geometric idea of slope.
- When maximization problems involving volume are introduced (see p. 264), a function is derived that is to be maximized. Instead of forging ahead with the standard calculus solution, the student is first asked to stop, make a table of function values, graph the function, and then estimate the maximum value. This experience provides students with more insight into the problem. They recognize not only that different dimensions yield different volumes, but also that the dimensions yielding the maximum volume may be conjectured or estimated as a result of the calculations.
- Relative maxima and minima (Sections 2.1 and 2.2) and absolute maxima and minima (Section 2.4) are covered in separate sections in Chapter 2, so that students gradually build

up an understanding of these topics as they consider graphing using calculus concepts (see pp. 198–234 and 250–262).

- The explanation underlying the definition of the number e is presented in Chapter 3 both graphically and through a discussion of continuously compounded interest (see pp. 345–347).
- Chapter 9 starts off with a sequence, something students have seen many times before, even if they don't know it by name. The text stresses that a sequence is a function, with the special restriction that the inputs are integers, allowing students to relate sequences to something familiar—functions. The chapter's discussion of arithmetic and geometric sequences then ties them to linear and exponential functions, strengthening the connection to what students already know.
- Probability is also something that all students have been exposed to. They intuitively understand that a coin lands heads up half the time, that rolling two dice is more likely to yield 7 than 12, and that certain real-life events are more likely than others. Chapter 10 uses this kind of common knowledge to ease students into an understanding of formal probability and its rules.

Strong Algebra Review

One of the most critical factors underlying success in this course sequence is a strong foundation in algebra skills. We recognize that students start the first course with varying degrees of skills, so we have included multiple opportunities to help students target their weak areas and remediate or refresh the needed skills.

- **Prerequisite Skills Diagnostic Test (Part A).** This portion of the diagnostic test assesses skills refreshed in Appendix A: Review of Basic Algebra. Answers to the questions reference specific examples within the appendix.
- **Appendix A: Review of Basic Algebra.** This 11-page appendix provides examples on topics such as exponents, equations and inequalities, and applied problems. It ends with an exercise set, for which answers are provided at the back of the book so that students can check their understanding.
- **Prerequisite Skills Diagnostic Test (Part B).** This portion of the diagnostic test assesses skills that are reviewed in Chapter R, and the answers reference specific sections in that chapter. Some instructors may choose to cover these topics thoroughly in class, making this assessment less critical. Other instructors may use all or portions of this test to determine whether there is a need to spend time remediating before moving on with Chapter 1.
- **Chapter R.** This chapter covers basic concepts related to functions, graphing, and modeling. It is an optional chapter based on the prerequisite skills students have.
- **"Getting Ready for Calculus" in MyMathLab.** This optional but valuable material is included in MyMathLab just prior to appropriate chapters and as a separate chapter. It can be used to assess algebra weaknesses in students individually or as a class, so that the instructor can remediate accordingly or instruct students to remediate on their own. An individualized study plan in conjunction with such assessment allows MyMathLab to generate homework specific to each student's individual needs. Together, this assessment, remediation, and practice constitute a very powerful resource for instructors.
- **Basic Skills Videos:** Basic skills tutorial videos have been added within specific exercises to address prerequisite skills for that exercise. These videos augment existing example videos (that address the content of that specific exercise) by focusing on the prerequisite skills needed for that exercise.

Applications

Relevant and factual applications drawn from a broad spectrum of fields are integrated throughout the text as applied examples and exercises and are also featured in separate application sections. We use real data as often as possible to illustrate for students the relevance of the applications. In addition, each chapter opener in this text includes an application that serves as a preview of what students will learn in the chapter.

The applications in the exercise sets are grouped under headings that identify them as reflecting real-life situations: Business and Economics, Life and Physical Sciences, Social Sciences, and General Interest. This organization allows the instructor to gear the assigned exercises to a particular student and also allows the student to know whether a particular exercise applies to his or her major.

MathTalk Videos have been added to MyMathLab to help motivate students by pointing out relevant connections to their majors—especially business. The videos feature Andrea Young from Ripon College (WI), a dynamic math professor (and actor!). The videos can be used as lecture starters or as part of homework assignments (in regular, online, or flipped classes).

Furthermore, the Index of Applications at the back of the book provides students and instructors with a comprehensive list of the many different fields considered throughout the text.

Approach to Technology

This text emphasizes mathematical modeling, utilizing the advantages of technology as appropriate. Though the use of technology is optional, its use meshes well with the text's more intuitive approach to applied calculus. For example, the use of the graphing calculator in modeling, as an optional topic, is introduced in Section R.6 and then reinforced many times throughout the text.

Technology Connections

Technology Connections are included throughout the text to illustrate the use of technology. Whenever appropriate, art that simulates graphs or tables generated by a graphing calculator are included as well. The text also includes discussion of the smartphone applications Graphicus, iPlot, and Quick Graph to take advantage of technology to which many students have access.

There are four types of Technology Connections for students and instructors to use for exploring key ideas.

- **Lesson/Teaching.** These provide students with an example, followed by exercises to work within the lesson.
- **Checking.** These tell the students how to verify a solution within an example by using a graphing calculator.
- **Exploratory/Investigation.** These provide questions to guide students through an investigation.
- **Technology Connection Exercises.** Most exercise sets contain technology-based exercises identified with either an icon or the heading "Technology Connection." These exercises also appear in the Chapter Review Exercises and the Chapter Tests. The Printable Test Forms include technology-based exercises as well.

Use of Art and Color

One of our hallmarks is the pervasive use of color as a pedagogical tool. Color is used in a methodical and precise manner so that it enhances the readability of the text for students and instructors. When two curves are graphed using the same set of axes, one is usually red and the other blue with the red graph being the curve of major importance. This is exemplified in the graphs from Chapter R (pp. 54 and 82) below. Note that the equation labels are the same color as the curves. When the instructions say "Graph," the dots match the color of the curve.

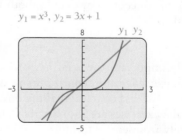

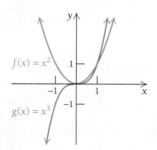

The following figure from Chapter 1 (p. 134) shows the use of colors to distinguish between secant and tangent lines. Throughout the text, blue is used for secant lines and red for tangent lines.

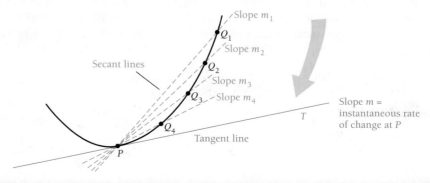

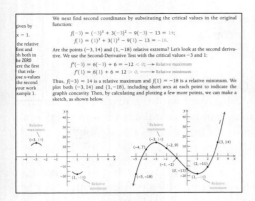

In the text from Chapter 2 (p. 219) shown at the left, the color red denotes substitution in equations and blue highlights the corresponding outputs, including maximum and minimum values. The specific use of color is carried out in the figure that follows. Note that when dots are used for emphasis other than just merely plotting, they are black.

Beginning with the discussion of integration in Chapter 4, the color amber is used to highlight areas in graphs. The figure to the left below, from Chapter 4 (p. 427), illustrates the use of blue and red for the curves and labels and amber for the area.

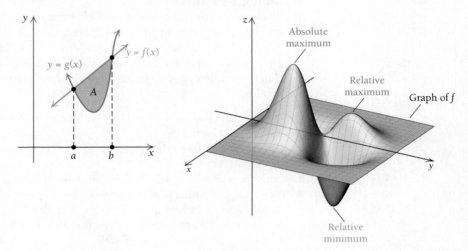

Throughout the text, the three-dimensional art has been carefully rendered to make it easier for students to visualize complex graphs, such as the one above (p. 543).

Interactive Figures

We have gone beyond the pages of this book to take advantage of students' ability to learn visually. Within MyMathLab, interactive figures are provided. These can be used by instructors in presentations or in assessing students' understanding, as well as by students for independent exploration of concepts. The easily manipulated figures take advantage of students' intuition and extend their visual understanding of concepts. For details about these figures, see the description on page xvii or online in MyMathLab.

Accuracy

We know how vitally important the accuracy of a textbook is to both students and instructors. To that end, we have exceeded the typical pursuit of accuracy. We went to great pains to ensure that the examples are clear and concise, that the direction lines and problem-solving processes are consistent, and that the exercises are supported by complementary examples in the section and are gradated appropriately from easier to more challenging. The accuracy checking process for this text involved a total of four proofreaders at two different stages in the process and independent reviews of the examples and exercises by multiple accuracy checkers. Additional proofreading and accuracy reviews by professors who teach this course were conducted so that subjective improvements and refinements could be made. And, lastly, a thorough cross-check of the solutions in the instructor and student solution manuals with the answers in the text constituted a final audit for consistency and accuracy. Although some of the material in this text is newly written, it fulfills the unusually high expectations associated with the Bittinger name and with *Calculus and Its Applications*, Tenth Edition.

Pedagogy of *Calculus and Its Applications,* Expanded Version

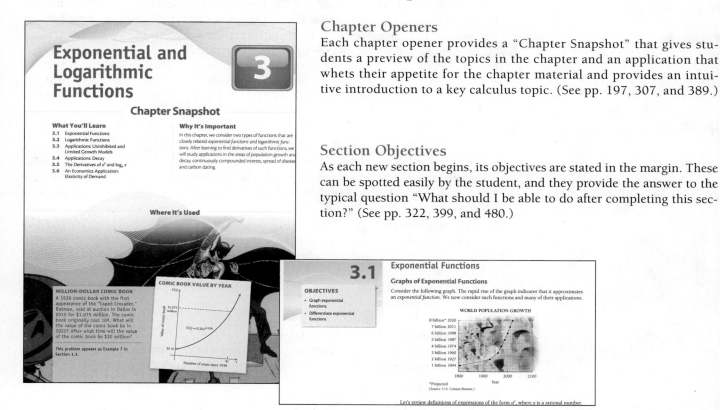

Chapter Openers

Each chapter opener provides a "Chapter Snapshot" that gives students a preview of the topics in the chapter and an application that whets their appetite for the chapter material and provides an intuitive introduction to a key calculus topic. (See pp. 197, 307, and 389.)

Section Objectives

As each new section begins, its objectives are stated in the margin. These can be spotted easily by the student, and they provide the answer to the typical question "What should I be able to do after completing this section?" (See pp. 322, 399, and 480.)

Technology Connections

The text allows the instructor to incorporate graphing calculators, spreadsheets, and smart phone applications into classes. All use of technology is clearly labeled so that it can be included or omitted as desired. (See pp. 54–56 and 209–212.)

Teaching Tips

The Annotated Instructor's Edition provides tips for instructors who are new to teaching this course as a way to help them avoid common missteps often made by students.

Quick Check Exercises

Giving students the opportunity to check their understanding of a new concept or skill is vital to their learning and their confidence. Quick Check exercises follow and mirror selected examples in the text, allowing students to both practice and assess the skills they are learning. Instructors may include these as part of a lecture as a means of gauging skills and gaining immediate feedback. Answers to the Quick Check exercises are provided at the end of each section following the exercise set. (See pp. 236, 331, and 412.)

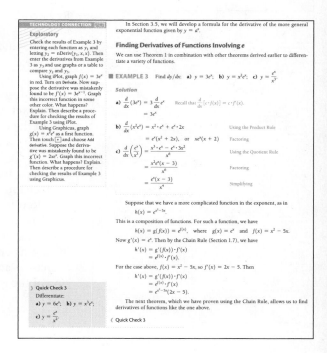

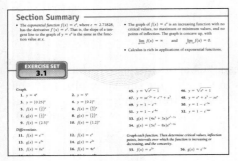

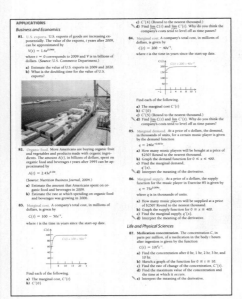

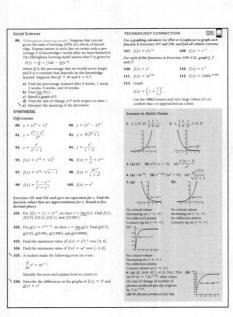

Section Summary

To assist students in identifying the key topics for each section, a Section Summary precedes every exercise set. Key concepts and definitions are presented in bulleted list format to help focus students' attention on the most important ideas presented in the section. (See pp. 106, 246, and 360.)

Variety of Exercises

There are over 5000 exercises in this text. All exercise sets are enhanced by the inclusion of real-world applications, detailed art pieces, and illustrative graphs.

Applications

A section of applied problems is included in nearly every exercise set. The problems are grouped under headings that identify them as business and economics, life and physical sciences, social sciences, or general interest. Each problem is accompanied by a brief description of its subject matter (see pp. 155–157, 347–351, and 397–398).

Thinking and Writing Exercises

Identified by a ✎, these exercises ask students to explain mathematical concepts in their own words, thereby strengthening their understanding (see pp. 143, 249, and 422).

Synthesis Exercises

Synthesis exercises are included in every exercise set, including the Chapter Review Exercises and Chapter Tests. They require students to go beyond the immediate objectives of the section or chapter and are designed to both challenge students and make them think about what they are learning (see pp. 176, 276, and 364).

Technology Connection Exercises

These exercises appear in the Technology Connections (see pp. 29, 141, and 327) and in the exercise sets (see pp. 120, 249, and 425). They allow students to solve problems or check solutions using a graphing calculator or smart phone.

Concept Reinforcement Exercises

Each chapter closes with a set of Chapter Review Exercises, which includes Concept Reinforcement exercises at the beginning. The exercises are confidence builders for students who have completed their study of the chapter. Presented in matching, true/false, or fill-in-the-blank format, these exercises can also be used in class as oral exercises. As with all review exercises, each concept reinforcement exercise is accompanied by a bracketed section reference to indicate where discussion of the concept appears in the chapter. (See pp. 301, 382, and 466.)

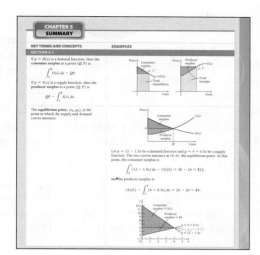

User-Friendly Chapter Summaries

Chapter summaries are formatted in a tabular style that makes it easy for students to distill key ideas. Each chapter summary presents a section-by-section list of key definitions, concepts, and theorems, with examples for further clarification. (See pp. 185, 295, and 378.)

Chapter Reviews and Tests

At the end of each chapter are review exercises and a test. The Chapter Review Exercises, which include bracketed references to the sections in which the related course content first appears, provide comprehensive coverage of each chapter's material (see pp. 190–192).

The Chapter Test includes synthesis and technology questions (see pp. 192–193). There is also a Cumulative Review at the end of the text that can serve as a practice final examination. The answers, including section references, to the chapter tests and the Cumulative Review are at the back of the book. Additional forms of each of the chapter tests and the final examination, accompanied by answer keys and ready for classroom use, appear in the Printable Test Forms.

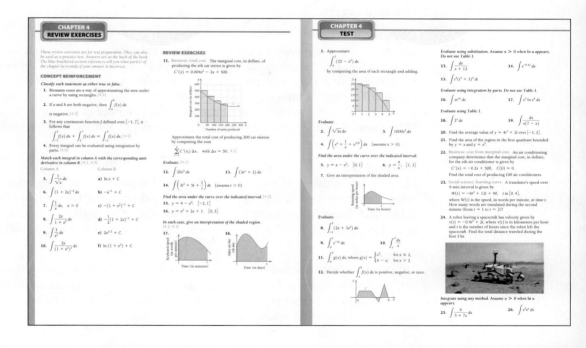

Extended Technology Applications

Extended Technology Applications at the end of each chapter use real applications and real data. They require a step-by-step analysis that encourages group work. More challenging in nature, the exercises in these features involve a variety of high-level technology uses such as the use of regression to create models on a graphing calculator.

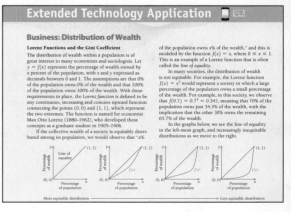

Approach to Content

Chapter 1

Chapter 1 starts with an intuitive discussion of limits, stressing various methods that are used to determine a limit, including numerical (tabular), algebraic simplification, and graphical. We do not feel it is necessary or appropriate for the intended audience of this text to introduce the more rigorous epsilon-delta theory of limits. Once limits are established, we immediately introduce rates of change, stressing applications, and from there, move to instantaneous rates of change and the derivative. The remainder of Chapter 1 is devoted to techniques and applications of differentiation.

Chapter 2

We use differentiation as the focus of Chapter 2, in which we discuss the behavior of functions and the techniques used to graph functions. We discuss polynomial and rational functions and incorporate applications and technology connections as often as possible. We strive to help the student understand not only *how* derivatives can be useful, but also *why*.

Chapter 3

We introduce exponential and logarithmic functions and their derivatives and applications in Chapter 3. We feel that effective coverage of these classes of functions deserves a separate chapter, from which the student should gain the ability to differentiate most such functions and understand their applications.

Chapter 4

We introduce integration in Chapter 4, beginning the chapter with a section on antidifferentiation. We feel this "mechanical" skill serves as a way to segue from differentiation to integration by stressing that the latter process is the reverse of the former. Section 4.2 discusses the geometrical concept of integration using Riemann sums, and Section 4.3 ties the first two sections together with a discussion of the Fundamental Theorem of Calculus. We stress the applications of integration and devote the remaining sections of Chapter 4 to the techniques of antidifferentiation.

Chapter 5

Chapter 5 showcases the variety of methods and applications of integration: its applications to economics are presented in Sections 5.1 and 5.2, and improper integrals are discussed in Section 5.3. Numerical integration, using Riemann sums, the Trapezoidal Rule, and Simpson's Rule, is presented in Section 5.4, followed by a discussion of volumes of rotated solids in Section 5.5.

Chapter 6

Multivariable calculus is covered in Chapter 6, with differentiation and applications discussed in the first four sections and integration in the latter two. This chapter can be studied first in a second-semester course. At that point, basic calculus skills can be assumed and other topics of mathematics can be explored on a much deeper level.

Chapter 7

Trigonometry is discussed in Chapter 7. The first section is a review of the main points of trigonometry, and instructors may cover as much or as little of this as they believe their students need. Differentiation is discussed in Section 7.2, integration in Section 7.3, and inverse trigonometric functions in Section 7.4. All the sections in this chapter include applications that we feel demonstrate the utility of trigonometry in business and the life sciences.

We know that not all schools cover trigonometry. However, we chose to place this text's coverage of the topic before later chapters so that the material can be used at the instructor's discretion. Chapter 7 can be skipped in a course that does not cover trigonometry with no disruption to the flow of other topics, and the Trigonometry Connections in the following chapters are set apart for easy integration or omission.

Chapter 8

Chapter 8 covers differential equations. We focus on ordinary differential equations and the techniques used to solve certain equations, as well as applications that grow from these forms. Differential equations involving trigonometry are segregated in the final section, allowing instructors who choose to cover trigonometry to easily find this material and those who choose not to cover trigonometry to simply skip it.

Chapter 9

Chapter 9 focuses on sequences and series. The general arithmetic and geometric forms are discussed in the first two sections. After that, the chapter allows for two paths: instructors who wish to focus on financial applications involving sequences and series can proceed to Sections 9.3 and 9.4, while those who desire a deeper discussion of power series and Taylor series can move on to Sections 9.5 and 9.6. As in Chapter 8, the material involving trigonometry is deliberately segregated in the final section for instructors' convenience.

Chapter 10

Probability and probability distributions are covered in Chapter 10. A review of sets is provided in Section 10.1, and, again, instructors can choose how much, if any, of this material to cover, based on their students' assumed knowledge. Basic probability is presented in Section 10.2. We do not cover combinatorics or specialized probability techniques such as the binomial model, the hypergeometric model, and Bayes Rule. Instead of focusing on the myriad ways to calculate probabilities, we choose instead to lay the foundation for discussing discrete probability distributions in Section 10.3. This, then, segues nicely into Section 10.4, where continuous probability distributions are discussed and integration is used to study them. Section 10.5 caps this chapter with further discussion of the normal distribution, perhaps the most well-known of the continuous probability distributions.

Trigonometry Connections

The topic of trigonometry is strategically placed in the middle of the text (Chapter 7) rather than at the end, as so many books do, so those who cover it can naturally expand on topics such as higher-order differential equations and Taylor series. Because many courses in applied calculus do not have time to cover trigonometry, Chapter 7 and the Trigonometry Connections subsections in subsequent chapters can be skipped without causing any disruption to the flow of content. Sections that contain a Trigonometry Connection subsection also include specifically labeled exercises that assess this content.

Additional Chapters Online

Two chapters are available online within MyMathLab or in print via a custom version of the text. Chapters include: Chapter 11: Systems and Matrices, and Chapter 12: Discrete Probability. See the Table of Contents at the front of this text or contact your local Pearson representative for more details.

Applications

For most instructors, the ultimate goal is for students to be able to apply what they learn in this course to everyday scenarios. This ability motivates learning and brings student understanding to a higher level. To further this goal, we have included almost 1200 applications in the examples and exercises in the text. These will motivate students to apply what they're learning to their future careers.

Annotated Instructor's Edition

An Annotated Instructor's Edition is included in the long list of instructor resources. Located in the margins in the AIE are Teaching Tips, which are ideal for new or less experienced instructors. In addition, answers to exercises are provided on the same page, making it easier than ever to check student work.

Supplements

Student Supplements	Instructor Supplements

Student's Solutions Manual
(ISBN: 0-321-84417-3 | 978-0-321-84417-0)
- Provides detailed solutions to all odd-numbered exercises, with the exception of the Thinking and Writing exercises

Graphing Calculator Manual (downloadable)
- Provides instructions and keystroke operations for the TI-83/84 Plus, and TI-84 Plus with new operating system, featuring MathPrint™.
- Includes worked-out examples taken directly from the text
- Available in MyMathLab

Video Lectures with optional captioning (online)
- Complete set of digitized videos for student use at home or on campus
- Available in MyMathLab

Supplementary Chapters
- Two online chapters on systems and matrices, and discrete probability
- Available within MyMathLab
- Can also be customized as part of the text through Pearson Custom Learning or through a Pearson representative

All of the student supplements listed above are included in MyMathLab.

Annotated Instructor's Edition
(ISBN: 0-13-411120-6 | 978-0-13-411120-9)
- Includes numerous Teaching Tips
- Includes all of the answers, usually on the same page as the exercises, for quick reference

Instructor's Solutions Manual (downloadable)
- Provides complete solutions to all text exercises
- Available to qualified instructors through the Pearson Instructor Resource Center, www.pearsonhighered.com/irc, and MyMathLab

Printable Test Forms (downloadable)
- Contains several alternative tests per chapter
- Contains several comprehensive final exams
- Includes answer keys
- Available to qualified instructors through the Pearson Instructor Resource Center, www.pearsonhighered.com/irc, and MyMathLab

TestGen® (downloadable)
- Enables instructors to build, edit, print, and administer tests using a computerized bank of questions developed to cover all the objectives of the text
- Algorithmically based, allowing instructors to create multiple but equivalent versions of the same question with the click of a button
- Allows instructors to modify test bank questions or add new questions
- Can be downloaded from www.pearsoned.com/testgen

PowerPoint Lecture Presentation (downloadable)
- Classroom presentation software oriented specifically to the text's topic sequence
- Available to qualified instructors through the Pearson Instructor Resource Center, www.pearsonhighered.com/irc, and MyMathLab

Media Supplements

MyMathLab® Online Course (access code required)

MyMathLab® is a text-specific, easily customizable online course that integrates interactive multimedia instruction with textbook content.

MyMathLab delivers **proven results** in helping individual students succeed.

- MyMathLab has a consistently positive impact on the quality of learning in higher education math instruction. MyMathLab can be successfully implemented in any environment—lab-based, hybrid, fully online, or traditional—and demonstrates the quantifiable difference that integrated usage has on student retention, subsequent success, and overall achievement.
- MyMathLab's comprehensive online gradebook automatically tracks students' results on tests, quizzes, homework, and in the study plan. Instructors can use the gradebook to quickly intervene if students

have trouble or to provide positive feedback on a job well done. The data within MyMathLab are easily exported to a variety of spreadsheet programs, such as Microsoft Excel. Instructors can determine which points of data to export, and then analyze the results to evaluate success.

MyMathLab provides **engaging experiences** that personalize, stimulate, and measure learning for each student.

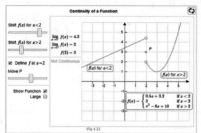

- **Exercises** in MyMathLab are correlated to the exercises in the textbook, and they regenerate algorithmically to give students unlimited opportunity for practice and mastery. The software offers immediate, helpful feedback when students enter incorrect answers.
- **Multimedia learning aids** that accompany the exercises include guided solutions, sample problems, animations, video lectures, podcasts, and eText clips for extra help at point of use.
  - **Interactive figures** included within MyMathLab serve as both teaching and learning tools. They can be used by instructors during lectures to illustrate some of the more difficult and visually challenging calculus topics. Used in this manner, the figures engage students more fully and save time otherwise spent rendering figures by hand. Instructors may also choose to assign the questions that accompany the figures, which lead students to discover key concepts. The interactive figures are also available to students, who may explore them on their own as a way to better visualize the concepts being presented.
  - **MathTalk Videos** help motivate students by pointing out relevant connections to their majors—especially business. The videos feature Andrea Young from Ripon College (WI), a dynamic math professor (and actor!). The videos can be used as lecture starters or as part of homework assignments (in regular, online or flipped classes).

And, MyMathLab comes from a **trusted partner** with educational expertise and an eye on the future.

- Knowing that you are using a Pearson product means knowing that you are using quality content: eTexts are accurate and assessment tools work. Whether you are just getting started with MyMathLab or have a question along the way, we're here to help you learn about our technologies and how to incorporate them into your course.

To learn more about how MyMathLab combines proven learning applications with powerful assessment, visit www.mymathlab.com or contact your Pearson representative.

MyMathLab® Integrated Review Course (access code required)

The Integrated Review course has all of the MyMathLab features mentioned above but makes it easier for instructors to address gaps in prerequisite skills. In this MyMathLab course, the Skills Check Quizzes are pre-assigned and coupled with Personalized Homework. The latter provides help for students who did not show mastery of prerequisite skills on the quiz. This allows the instructor to focus on course content while MyMathLab addresses the skills gaps that often impede student progress. Ask your Pearson representative for details or to see a copy of this course.

MathXL® Online Course (access code required)

MathXL® is the homework and assessment engine that runs MyMathLab. (MyMathLab is MathXL plus a learning management system.) With MathXL, instructors can

- create, edit, and assign online homework and tests using algorithmically generated exercises correlated at the objective level to the textbook;
- create and assign their own online exercises and import TestGen tests for added flexibility; and
- maintain records of all student work tracked in MathXL's online gradebook.

With MathXL, students can

- take chapter tests in MathXL and receive personalized study plans and/or personalized homework assignments based on their test results;
- use the study plan and/or the homework to link directly to tutorial exercises for the objectives they need to study; and
- access supplemental animations and video clips directly from selected exercises.

MathXL is available to qualified adopters. For more information, visit www.mathxl.com, or contact your Pearson representative.

Acknowledgments

As authors, we have taken many steps to ensure the accuracy of this text. Many devoted individuals comprised the team that was responsible for monitoring the revision and production process in a manner that makes this a work of which we can all be proud. We are thankful for our publishing team at Pearson, as well as all of the Pearson representatives who share our book with educators across the country. Many thanks to Michelle Christian, who was instrumental in getting Scott Surgent's first book printed and in bringing him to the attention of the Pearson team.

We would like to thank Jane Hoover for her many helpful suggestions, proofreading, and checking of art. Jane's attention to detail and pleasant demeanor made our work as low in stress as humanly possible, given the demands of the production process.

We also wish to thank Michelle Beecher Lanosga for her incredibly helpful data research. Her efforts make the real-world problems in this text as up-to-date as possible, given the production deadlines we faced. Geri Davis deserves credit for both the attractive design of the text and the coordination of the many illustrations, photos, and graphs. She is always a distinct pleasure to work with and sets the standard by which all other art editors are measured.

We are very grateful for Mary Ann Teel's contributions to this edition: her thoughtful comments while reviewing draft chapters, her careful reading of the exposition for accuracy and consistency, and her work on the testing manual. Many thanks to Lisa Grilli and Donna Krichiver for providing helpful teaching tips for the Annotated Instructor's Edition. We greatly appreciate Dave Dubriske's work on the solutions manuals and Steve Ouellette's work on Appendix D and the *Graphing Calculator Manual*. Many thanks also to John Morin, Thomas Wegleitner, Lauri Semarne, Patricia Nelson, Deanna Raymond, and Doug Ewert for their careful checking of the manuscript and typeset pages. Thank you to Douglas Williams for his help with generating some of the 3D images in Chapter 6. In addition, thank you to Hugh Cornell of the University of North Florida, Jerry DeGroot of Purdue University North Central, Jigarkumar S. Patel of the University of Texas at Dallas, Curtis Paul of Moorpark College, G. Brock Williams of Texas Tech University, and Mary Jane Sterling and Tiffany Troutman of Bradley University, who accuracy checked and provided insight for this text that only those who teach the course can provide.

Finally, the following reviewers provided thoughtful and insightful comments that helped immeasurably with the development of this expanded version (these reviewers are identified with an asterisk) and with the revision of the Tenth Edition.

*Jay Abramson, *Arizona State University*
*Mohammed Ali, *Prince George's Community College*
 Nilay Tanik Argon, *The University of North Carolina at Chapel Hill*
*Anthony Barcellos, *American River College*
*Melkana Brakalova, *Fordham University*
*Naala Brewer, *Arizona State University*
*Adena Calden, *University of Massachusetts*
*Debra S. Carney, *University of Denver*
*Nelson Castaneda, *Central Connecticut State University*
*Martha Morrow Chalhoub, *Collin College*
*K. Joseph Chen, *Purdue University*
*Mei Qin Chen, *The Citadel*
*Woonjung Choi, *Arizona State University*
*Hugh Cornell, *University of North Florida*
 Rakissa Cribari, Ph.D., *University of Colorado–Denver*
*Jerry DeGroot, *Purdue University North Central*
 Samantha C. Fay, *Jefferson College*
*Daria Filippova, *Bowling Green State University*

Burt K. Fischer, CPA
*Igor Fulman, *Arizona State University*
Lewis A. Germann, *Troy University*
John R. Griggs, *North Carolina State University*
Lisa Grilli, *Northern Illinois University*
*Steven Hair, *Penn State University*
Mary Beth Headlee, *Manatee Community College*
*Glenn Jablonski, *Triton College*
Darin Kapanjie, *Temple University*
*Karla Karstens, *University of Vermont*
*Theresa Killebrew, *Mesa Community College*
*Donna S. Krichiver, *Johnson County Community College*
*Donna LaLonde, *Washburn University*
Rebecca E. Lynn, *Colorado State University*
*Joseph Mayne, *Loyola University of Chicago*
*Gail Nord, *Gonzaga University*
*Richard O'Beirne, *George Mason University*
*Jigarkumar S. Patel, *The University of Texas at Dallas*
*Curtis Paul, *Moorpark College*
Shahla Peterman, *University of Missouri–St. Louis*
*Timothy Pilachowski, *University of Maryland, College Park*
Mohammed Rajah, *MiraCosta College*
*Yvonne M. Sandoval, *Pima Community College, West Campus*
*Andrew Schwartz, *Southeast Missouri State University*
Charlie Snygg, *Pikes Peak Community College*
*Mary Jane Sterling, *Bradley University*
Scott R. Sykes, *University of West Georgia*
Mary Ann Teel, *University of North Texas*
Bruce Thomas, *Kennesaw State University*
*Tiffany Troutman, *Bradley University*
*Ani P. Velo, *University of San Diego*
*Dennis Walsh, *Middle Tennessee State University*
Patrick Ward, *Illinois Central College*
*G. Brock Williams, *Texas Tech University*
*Jane-Marie Wright, *Suffolk Community College, Ammerman Campus*
*David Zeigler, *California State University, Sacramento*

Prerequisite Skills Diagnostic Test

········ To the Student and the Instructor ·······················

To the Student and the Instructor

Part A of this diagnostic test covers basic algebra concepts, such as properties of exponents, multiplying and factoring polynomials, equation solving, and applied problems. Part B covers topics discussed in Chapter R, such as graphs, slope, equations of lines, and functions, most of which come from a course in intermediate or college algebra. This diagnostic test does not cover regression, though it is considered in Chapter R and used throughout the text. This test can be used to assess student needs for this course. Students who miss most of the questions in part A should study Appendix A before moving to Chapter R. Those who miss most of the questions in part B should study Chapter R. Students who miss just a few questions might study the related topics in either Appendix A or Chapter R before continuing with the calculus chapters.

Part A: Answers and locations of worked-out solutions appear on p. A-46.

Express each of the following without an exponent.

1. 4^3 **2.** $(-2)^5$ **3.** $\left(\frac{1}{2}\right)^3$ **4.** $(-2x)^1$ **5.** e^0

Express each of the following without a negative exponent.

6. x^{-5} **7.** $\left(\frac{1}{4}\right)^{-2}$ **8.** t^{-1}

Multiply. Express each answer without a negative exponent.

9. $x^5 \cdot x^6$ **10.** $x^{-5} \cdot x^6$ **11.** $2x^{-3} \cdot 5x^{-4}$

Divide. Express each answer without a negative exponent.

12. $\dfrac{a^3}{a^2}$ **13.** $\dfrac{e^3}{e^{-4}}$

Simplify. Express each answer without a negative exponent.

14. $(x^{-2})^3$ **15.** $(2x^4y^{-5}z^3)^{-3}$

Multiply.

16. $3(x - 5)$ **17.** $(x - 5)(x + 3)$ **18.** $(a + b)(a + b)$

19. $(2x - t)^2$ **20.** $(3c + d)(3c - d)$

Factor.

21. $2xh + h^2$ **22.** $x^2 - 6xy + 9y^2$ **23.** $x^2 - 5x - 14$

24. $6x^2 + 7x - 5$ **25.** $x^3 - 7x^2 - 4x + 28$

Solve.

26. $-\frac{5}{6}x + 10 = \frac{1}{2}x + 2$ **27.** $3x(x - 2)(5x + 4) = 0$

28. $4x^3 = x$ **29.** $\dfrac{2x}{x - 3} - \dfrac{6}{x} = \dfrac{18}{x^2 - 3x}$

30. $17 - 8x \geq 5x - 4$

31. After a 5% gain in weight, a grizzly bear weighs 693 lb. What was the bear's original weight?

32. Raggs, Ltd., a clothing firm, determines that its total revenue, in dollars, from the sale of x suits is given by $200x + 50$. Determine the number of suits the firm must sell to ensure that its total revenue will be more than $70,050.

Part B: Answers and locations of worked-out solutions appear on pp. A-46 and A-47.

Graph.

1. $y = 2x + 1$ **2.** $3x + 5y = 10$

3. $y = x^2 - 1$ **4.** $x = y^2$

5. A function f is given by $f(x) = 3x^2 - 2x + 8$. Find each of the following: $f(0), f(-5),$ and $f(7a)$.

6. A function f is given by $f(x) = x - x^2$. Find and simplify $\dfrac{f(x + h) - f(x)}{h}$, for $h \neq 0$.

7. Graph the function f defined as follows:
$$f(x) = \begin{cases} 4, & \text{for } x \leq 0, \\ 3 - x^2, & \text{for } 0 < x \leq 2, \\ 2x - 6, & \text{for } x > 2. \end{cases}$$

8. Write interval notation for $\{x | -4 < x < 5\}$.

9. Find the domain: $f(x) = \dfrac{3}{2x - 5}$.

10. Find the slope and y-intercept of $2x - 4y - 7 = 0$.

11. Find an equation of the line that has slope 3 and contains the point $(-1, -5)$.

12. Find the slope of the line containing the points $(-2, 6)$ and $(-4, 9)$.

Graph.

13. $f(x) = x^2 - 2x - 3$ **14.** $f(x) = x^3$

15. $f(x) = \dfrac{1}{x}$ **16.** $f(x) = |x|$

17. $f(x) = -\sqrt{x}$

18. Suppose that $1000 is invested at 5%, compounded annually. How much is the investment worth at the end of 2 yr?

Functions, Graphs, and Models

R

Chapter Snapshot

What You'll Learn

R.1 Graphs and Equations
R.2 Functions and Models
R.3 Finding Domain and Range
R.4 Slope and Linear Functions
R.5 Nonlinear Functions and Models
R.6 Mathematical Modeling and Curve Fitting

Why It's Important

This chapter introduces functions and covers their graphs, notation, and applications. Also presented are many topics that we will consider often throughout the text: supply and demand, total cost, total revenue, total profit, the concept of a mathematical model, and curve fitting.

Skills in using a graphing calculator are also introduced in optional Technology Connections. Details on keystrokes are given in the *Graphing Calculator Manual* (GCM).

Part A of the diagnostic test (p. xxi), on basic algebra concepts, allows students to determine whether they need to review Appendix A (p. 819) before studying this chapter. Part B, on college algebra topics, assesses the need to study this chapter before moving on to the calculus chapters.

Where It's Used

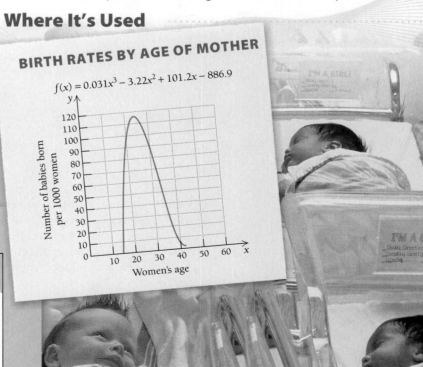

BIRTH RATES

What is the average number of live births per 1000 women age 20?

This problem appears as an example in a Technology Connection in Section R.6.

BIRTH RATES BY AGE OF MOTHER

$$f(x) = 0.031x^3 - 3.22x^2 + 101.2x - 886.9$$

(vertical axis) Number of babies born per 1000 women

(horizontal axis) Women's age

BIRTH RATES FOR WOMEN OF SELECTED AGES

AGE, x	AVERAGE NUMBER OF LIVE BIRTHS PER 1000 WOMEN
16	34
18.5	86.5
22	111.1
27	113.9
32	84.5
37	35.4
42	6.8

(Source: *Centers for Disease Control and Prevention.*)

R.1

OBJECTIVES

- Graph equations.
- Use graphs as mathematical models to make predictions.
- Carry out calculations involving compound interest.

Graphs and Equations

What Is Calculus?

What is calculus? This is a common question at the start of a course like this. Let's consider a simplified answer for now.

Consider a protein energy drink box, as shown below, at left. The following is a typical problem from an algebra course. Try to solve it. (If you need some algebra review, refer to Appendix A at the end of the book.)

Algebra Problem

The sum of the height, width, and length of a box is 207 mm. If the height is three times the width and the length is 7 mm more than the width, find the dimensions of the box.

The box has a width of 40 mm, a length of 47 mm, and a height of 120 mm.*

The following is a calculus problem that a manufacturer of boxes might need to solve.

Calculus Problem

A protein energy drink box is to hold 200 cm^3 (6.75 fl oz) of protein energy drink. If the height of the box must be twice the width, what dimensions will minimize the surface area of the box?

Height

Length

Width

Dimensions that assume that the height is twice the width			Total Surface Area, $2wh + 2lh + 2wl$
Width, w	Height, h	Length, l	
5 cm	10 cm	4 cm	220 cm^2
4 cm	8 cm	6.25 cm	214 cm^2 ←——— Smallest
3 cm	6 cm	11.$\overline{1}$ cm	236 cm^2
2 cm	4 cm	25 cm	316 cm^2

We selected combinations for which $h = 2w$ and the product $w \cdot h \cdot l$ is 200.

One way to solve this problem might be to choose several sets of dimensions for a 200-cm^3 box that is twice as tall as it is wide, compute the resulting areas, and determine which is the least. If you have access to spreadsheet software, you might create a spreadsheet and expand the table at left. We let w = width, h = height, l = length, and A = surface area. Then the surface area A is given by

$$A = 2wh + 2lh + 2wl, \quad \text{with } h = 2w.$$

From the data in the table, we might conclude that the smallest surface area is 214 cm^2. But how can we be certain that there are no other dimensions that yield a smaller area? We need the tools of calculus to answer this. We will study such maximum–minimum problems in more detail in Chapter 2.

Other topics we will consider in calculus are the slope of a curve at a point, rates of change, area under a curve, accumulations of quantities, and some statistical applications.

---

*To find this, let w = width, h = height, and l = length. Then $h = 3w$ and $l = w + 7$, so $w + 3w + w + 7 = 207$. This yields $w = 40$, and thus $h = 120$ and $l = 47$.

Graphs

The study of graphs is an essential aspect of calculus. A graph offers the opportunity to visualize relationships. For instance, the graph below shows how life expectancy has changed over time in the United States. One topic that we consider later in calculus is how a change on one axis affects the change on another.

ESTIMATED LIFE EXPECTANCY OF U.S. NEWBORNS BY YEAR OF BIRTH, 1929–2010

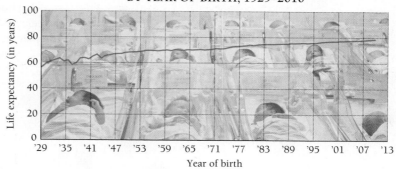

(*Source*: U.S. National Center for Health Statistics.)

Ordered Pairs and Graphs

Each point in a plane corresponds to an ordered pair of numbers. Note in the figure at the right that the point corresponding to the pair $(2, 5)$ is different from the point corresponding to the pair $(5, 2)$. This is why we call a pair like $(2, 5)$ an **ordered pair**. The first number is called the **first coordinate** of the point, and the second number is called the **second coordinate**. Together these are the *coordinates of the point*. The vertical line is often called the *y-axis*, and the horizontal line is often called the *x-axis*.

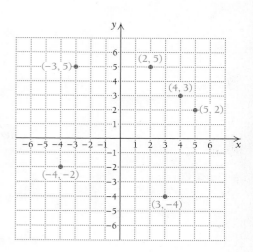

Graphs of Equations

A **solution** of an equation in two variables is an ordered pair of numbers that, when substituted for the variables, forms a true sentence. If not directed otherwise, we usually take the variables in *alphabetical* order. For example, $(-1, 2)$ is a solution of the equation $3x^2 + y = 5$, because when we substitute -1 for x and 2 for y, we get a true sentence:

$$3x^2 + y = 5$$
$$3(-1)^2 + 2 \stackrel{?}{=} 5$$
$$3 + 2 \stackrel{?}{=} 5$$
$$5 = 5. \quad \text{TRUE}$$

DEFINITION

The **graph** of an equation is a drawing that represents all ordered pairs that are solutions of the equation.

We obtain the graph of an equation by plotting enough ordered pairs (that are solutions) to see a pattern. The graph could be a line, a curve (or curves), or some other configuration.

■ **EXAMPLE 1**   Graph: $y = 2x + 1$.

Solution   We first find some ordered pairs that are solutions and arrange them in a table. To find an ordered pair, we can choose any number for x and then determine y. For example, if we choose -2 for x and substitute in $y = 2x + 1$, we find that $y = 2(-2) + 1 = -4 + 1 = -3$. Thus, $(-2, -3)$ is a solution. We select both negative numbers and positive numbers, as well as 0, for x. If a number takes us off the graph paper, we usually omit the pair from the graph.

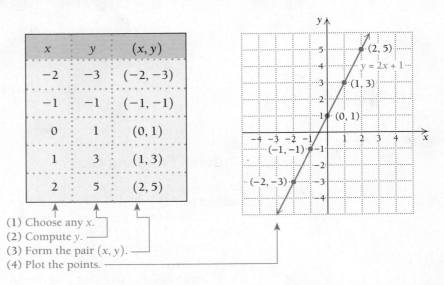

x	y	(x, y)
-2	-3	$(-2, -3)$
-1	-1	$(-1, -1)$
0	1	$(0, 1)$
1	3	$(1, 3)$
2	5	$(2, 5)$

(1) Choose any x.
(2) Compute y.
(3) Form the pair (x, y).
(4) Plot the points.

After we plot the points, we look for a pattern in the graph. If we had enough points, they would suggest a solid line. We draw the line with a straightedge and label it $y = 2x + 1$.

〉 **Quick Check 1**

Graph: $y = 3 - x$.

〈 Now try Quick Check 1

■ **EXAMPLE 2**   Graph: $3x + 5y = 10$.

Solution   We could choose x-values, substitute, and solve for y-values, but we first solve for y to ease the calculations.*

$$3x + 5y = 10$$
$$3x + 5y - 3x = 10 - 3x \qquad \text{Subtracting } 3x \text{ from both sides}$$
$$5y = 10 - 3x \qquad \text{Simplifying}$$
$$\tfrac{1}{5} \cdot 5y = \tfrac{1}{5} \cdot (10 - 3x) \qquad \text{Multiplying both sides by } \tfrac{1}{5}, \text{ or dividing both sides by 5}$$
$$y = \tfrac{1}{5} \cdot (10) - \tfrac{1}{5} \cdot (3x) \qquad \text{Using the distributive law}$$
$$= 2 - \tfrac{3}{5}x \qquad \text{Simplifying}$$
$$= -\tfrac{3}{5}x + 2$$

---

*Be sure to consult Appendix A, as needed, for a review of algebra.

Next we use $y = -\frac{3}{5}x + 2$ to find three ordered pairs, choosing multiples of 5 for x to avoid fractions.

x	y	(x, y)
0	2	$(0, 2)$
5	-1	$(5, -1)$
-5	5	$(-5, 5)$

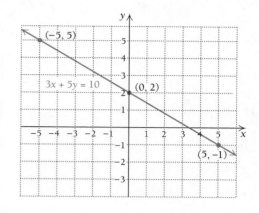

> **Quick Check 2**
>
> Graph: $3x - 5y = 10$.

We plot the points, draw the line, and label the graph as shown.

❮ Now try Quick Check 2

Examples 1 and 2 show graphs of linear equations. Such graphs are considered in greater detail in Section R.4.

■ **EXAMPLE 3**   Graph: $y = x^2 - 1$.

Solution

x	y	(x, y)
-2	3	$(-2, 3)$
-1	0	$(-1, 0)$
0	-1	$(0, -1)$
1	0	$(1, 0)$
2	3	$(2, 3)$

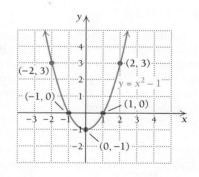

This time the pattern of the points is a curve called a *parabola*. We plot enough points to see a pattern and draw the graph.

> **Quick Check 3**
>
> Graph: $y = 2 - x^2$.

❮ Now try Quick Check 3

■ **EXAMPLE 4**   Graph: $x = y^2$.

Solution   In this case, x is expressed in terms of the variable y. Thus, we first choose numbers for y and then compute x.

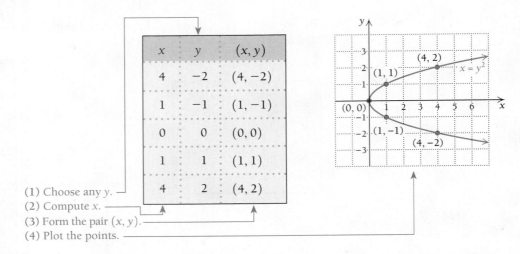

x	y	(x, y)
4	-2	$(4, -2)$
1	-1	$(1, -1)$
0	0	$(0, 0)$
1	1	$(1, 1)$
4	2	$(4, 2)$

(1) Choose any y.
(2) Compute x.
(3) Form the pair (x, y).
(4) Plot the points.

We plot these points, keeping in mind that x is still the first coordinate and y the second. We look for a pattern and complete the graph, usually connecting the points.

❯ **Quick Check 4**

Graph: $x = 1 + y^2$.

❮ Now try Quick Check 4

TECHNOLOGY CONNECTION

Introduction to the Use of a Graphing Calculator: Windows and Graphs

Viewing Windows

In this first of the optional Technology Connections, we begin to create graphs using a graphing calculator. Most of the coverage will refer to a TI-84 Plus or TI-83 Plus graphing calculator but in a somewhat generic manner, discussing features common to most graphing calculators. Although some keystrokes will be listed, exact keystrokes can be found in the owner's manual for your calculator or in the *Graphing Calculator Manual* (GCM) that accompanies this text.

The **viewing window** is a feature common to all graphing calculators. This is the rectangular screen in which a graph appears. Windows are described by four numbers, [**L, R, B, T**], which represent the **L**eft and **R**ight endpoints of the x-axis and the **B**ottom and **T**op endpoints of the y-axis. A WINDOW feature can be used to set these dimensions. Below is a window setting of $[-20, 20, -5, 5]$ with axis scaling denoted as Xscl = 5 and Yscl = 1, which means that there are 5 units between tick marks extending from -20 to 20 on the x-axis and 1 unit between tick marks extending from -5 to 5 on the y-axis.

Scales should be chosen with care, since tick marks become blurred and indistinguishable when too many appear. On most graphing calculators, a setting of $[-10, 10, -10, 10]$, Xscl = 1, Yscl = 1, Xres = 1 is considered **standard**.

Graphs are made up of black rectangular dots called **pixels**. The setting Xres allows users to set pixel resolution at 1 through 8 for graphs of equations. At Xres = 1, equations are evaluated and graphed at each pixel on the x-axis. At Xres = 8, equations are evaluated and graphed at every eighth pixel on the x-axis. The resolution is better for smaller Xres values than for larger values.

Graphs

Let's use a graphing calculator to graph the equation $y = x^3 - 5x + 1$. The equation can be entered using the notation Y=X^3−5X+1. We obtain the following graph in the standard viewing window.

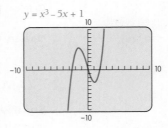

$y = x^3 - 5x + 1$

It is often necessary to change viewing windows in order to best reveal the curvature of a graph. For example,

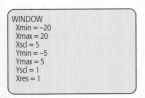

```
WINDOW
  Xmin = −20
  Xmax = 20
  Xscl = 5
  Ymin = −5
  Ymax = 5
  Yscl = 1
  Xres = 1
```

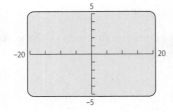

(continued)

each of the following is a graph of $y = 3x^5 - 20x^3$, but with a different viewing window. Which do you think best displays the curvature of the graph?

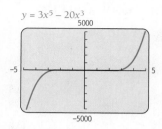

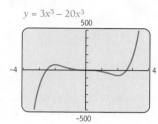

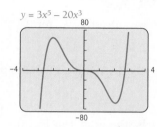

In general, choosing a window that best reveals a graph's characteristics involves some trial and error and, in some cases, some knowledge about the shape of the graph. We will learn more about the shape of graphs as we continue through the text.

To graph an equation like $3x + 5y = 10$, most calculators require that the equation be solved for y. Thus, we must rewrite and enter the equation as

$$y = \frac{(-3x + 10)}{5}, \quad \text{or} \quad y = \left(-\frac{3}{5}\right)x + 2.$$

(See Example 2.) Its graph is shown below in the standard window.

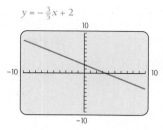

To graph an equation like $x = y^2$, we solve for y and get $y = \sqrt{x}$ or $y = -\sqrt{x}$, which can be written as $y = \pm\sqrt{x}$. We then graph the individual equations $y_1 = \sqrt{x}$ and $y_2 = -\sqrt{x}$.

EXERCISES
Graph each of the following equations. Select the standard window, $[-10, 10, -10, 10]$, with axis scaling Xscl $= 1$ and Yscl $= 1$.

1. $y = x + 3$ **2.** $y = x - 5$

3. $y = 2x - 1$ **4.** $y = 3x + 1$

5. $y = -\frac{2}{3}x + 4$ **6.** $y = -\frac{4}{5}x + 3$

7. $2x - 3y = 18$ **8.** $5y + 3x = 4$

9. $y = x^2$ **10.** $y = (x + 4)^2$

11. $y = 8 - x^2$ **12.** $y = 4 - 3x - x^2$

13. $y + 10 = 5x^2 - 3x$ **14.** $y - 2 = x^3$

15. $y = x^3 - 7x - 2$ **16.** $y = x^4 - 3x^2 + x$

17. $y = |x|$ (On most calculators, this is entered as $y = \text{abs}(x)$.)

18. $y = |x - 5|$

19. $y = |x| - 5$

20. $y = 9 - |x|$

Mathematical Models

When a real-world situation can be described in mathematical language, the description is a **mathematical model**. For example, the natural numbers constitute a mathematical model for situations in which counting is essential. Situations in which algebra can be brought to bear often require the use of functions as models. See Example 5, which follows.

Mathematical models are abstracted from real-world situations. The mathematical model may give results that allow us to predict what will happen in the real-world situation. If the predictions are inaccurate or the results of experimentation do not conform to the model, the model must be changed or discarded.

Mathematical modeling is often an ongoing process. For example, finding a mathematical model that will provide an accurate prediction of population growth is not a simple task. Any population model that one might devise would need to be reshaped as further information is acquired.

CREATING A
MATHEMATICAL MODEL

1. Recognize a
real-world problem.

2. Collect data.

3. Analyze the data.

4. Construct a model.

5. Test and refine the model.

6. Explain and predict.

■ **EXAMPLE 5** The graph below shows participation by females in high school athletics from 2002 to 2011.

FEMALES IN HIGH SCHOOL ATHLETICS

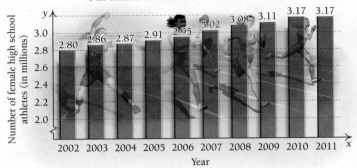

(*Source*: National Federation of State High School Associations.)

Use the model $N = 0.044t + 2.80$, where t is the number of years after 2002 and N is the number of participants, in millions, to predict the number of female high school athletes in 2014.

Solution Since 2014 is 12 years after 2002, we substitute 12 for t:

$$N = 0.044t + 2.80 = 0.044(12) + 2.80 = 3.328.$$

According to this model, in 2014, approximately 3.33 million females will participate in high school athletics.

As is the case with many kinds of models, the model in Example 5 is not perfect. For example, for $t = 1$, we get $N = 2.844$, a number slightly different from the 2.86 in the original data. But, for purposes of estimating, the model is adequate. The cubic model $N = -0.0006x^3 + 0.0089x^2 + 0.013x + 2.81$ also fits the data, at least in the short term: For $t = 1$, we get $N \approx 2.83$, close to the original data value. But for $t = 12, N \approx 3.21$, quite different from the prediction in Example 5. The difficulty with a cubic model here is that, eventually, its predictions become undependable. For example, the model in Example 5 predicts that there will be 3.59 million female high school athletes in 2020, but the cubic model predicts only 2.43 million. We always have to subject our models to careful scrutiny.

One important model that is extremely precise involves **compound interest**. Suppose that we invest P dollars at interest rate i, expressed as a decimal and compounded annually. The amount A_1 in the account at the end of the first year is given by

$$A_1 = P + Pi \qquad \text{The original amount invested, } P, \text{ is called the principal.}$$
$$= P(1 + i) = Pr,$$

where, for convenience, we let

$$r = 1 + i.$$

Going into the second year, we have Pr dollars, so by the end of the second year, we will have the amount A_2 given by

$$A_2 = A_1 \cdot r = (Pr)r$$
$$= Pr^2 = P(1 + i)^2.$$

Going into the third year, we have Pr^2 dollars, so by the end of the third year, we will have the amount A_3 given by

$$A_3 = A_2 \cdot r = (Pr^2)r$$
$$= Pr^3 = P(1 + i)^3.$$

In general, we have the following theorem.

THEOREM 1

If an amount P is invested at interest rate i, expressed as a decimal and compounded annually, in t years it will grow to the amount A given by

$$A = P(1 + i)^t.$$

■ **EXAMPLE 6** **Business: Compound Interest.** Suppose that $1000 is invested in Fibonacci Investment Fund at 5%, compounded annually. How much is in the account at the end of 2 yr?

Solution We substitute 1000 for P, 0.05 for i, and 2 for t into the equation $A = P(1 + i)^t$ and get

$$
\begin{aligned}
A &= 1000(1 + 0.05)^2 && \text{Adding terms in parentheses} \\
&= 1000(1.05)^2 \\
&= 1000(1.1025) && \text{Squaring} \\
&= \$1102.50. && \text{Multiplying}
\end{aligned}
$$

There is $1102.50 in the account after 2 yr.

⟨ Now try Quick Check 5

> **Quick Check 5**
>
> **Business.** Repeat Example 6 for an interest rate of 6%.

For interest that is compounded quarterly (four times per year), we can find a formula like the one above, as illustrated in the following diagram.

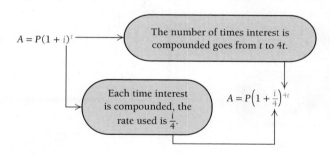

$A = P(1 + i)^t$ → The number of times interest is compounded goes from t to $4t$.

Each time interest is compounded, the rate used is $\frac{i}{4}$.

$A = P\left(1 + \frac{i}{4}\right)^{4t}$

"Compounded quarterly" means that the interest is divided by 4 and compounded four times per year. In general, the following theorem applies.

THEOREM 2

If a principal P is invested at interest rate i, expressed as a decimal and compounded n times a year, in t years it will grow to an amount A given by

$$A = P\left(1 + \frac{i}{n}\right)^{nt}.$$

■ **EXAMPLE 7** **Business: Compound Interest.** Suppose that $1000 is invested in Wellington Investment Fund at 5%, compounded quarterly. How much is in the account at the end of 3 yr?

Solution   We use the equation $A = P(1 + i/n)^{nt}$, substituting 1000 for P, 0.05 for i, 4 for n (compounding quarterly), and 3 for t. Then we get

$$A = 1000\left(1 + \frac{0.05}{4}\right)^{4\cdot 3}$$
$$= 1000(1 + 0.0125)^{12}$$
$$= 1000(1.0125)^{12}$$
$$= 1000(1.160754518) \qquad \text{Using a calculator to approximate } (1.0125)^{12}$$
$$= 1160.754518$$
$$\approx \$1160.75. \qquad \text{The symbol} \approx \text{means "approximately equal to."}$$

There is $1160.75 in the account after 3 yr.

⟨ Now try Quick Check 6

> **Quick Check 6**
>
> **Business: Compound Interest.** Repeat Example 7 for an interest rate of 6%.

A calculator with a ⟨yˣ⟩ or ⟨∧⟩ key and a ten-digit readout was used to find $(1.02)^{12}$ in Example 7. The number of places on a calculator may affect the accuracy of the answer. Thus, you may occasionally find that your answers do not agree with those at the back of the book, which were found on a calculator with a ten-digit readout. In general, when using a calculator, do all computations and round only at the end, as in Example 7. Usually, your answer will agree to at least four digits. It is usually wise to consult with your instructor on the accuracy required.

Section Summary

- Most graphs can be created by plotting points and looking for patterns. A graphing calculator can create graphs rapidly.

- Mathematical equations can serve as models of many kinds of applications.

EXERCISE SET
R.1

Exercises designated by the symbol ✎ are Thinking and Writing Exercises. They should be answered using one or two English sentences. Because answers to many such exercises will vary, solutions are not given at the back of the book.

Graph. (Unless directed otherwise, assume that "Graph" means "Graph by hand.")

1. $y = x + 4$

2. $y = x - 1$

3. $y = -3x$

4. $y = -\frac{1}{4}x$

5. $y = \frac{2}{3}x - 4$

6. $y = -\frac{5}{3}x + 3$

7. $x + y = 5$

8. $x - y = 4$

9. $8y - 2x = 4$

10. $6x + 3y = -9$

11. $5x - 6y = 12$

12. $2x + 5y = 10$

13. $y = x^2 - 5$

14. $y = x^2 - 3$

15. $x = 2 - y^2$

16. $x = y^2 + 2$

17. $y = |x|$

18. $y = |4 - x|$

19. $y = 7 - x^2$

20. $y = 5 - x^2$

21. $y + 1 = x^3$

22. $y - 7 = x^3$

APPLICATIONS

23. Running records. According to at least one study, the world record in any running race can be modeled by a linear equation. In particular, the world record R, in minutes, for the mile run in year x can be modeled by

$$R = -0.00582x + 15.3476.$$

Use this model to estimate the world records for the mile run in 1954, 2008, and 2012. Round your answers to the nearest hundredth of a minute.

24. Medicine. Ibuprofen is a medication used to relieve pain. The function

$$A = 0.5t^4 + 3.45t^3 - 96.65t^2 + 347.7t, \; 0 \le t \le 6,$$

can be used to estimate the number of milligrams, *A*, of ibuprofen in the bloodstream *t* hours after 400 mg of the medication has been swallowed. (*Source:* Based on data from Dr. P. Carey, Burlington, VT.) How many milligrams of ibuprofen are in the bloodstream 2 hr after 400 mg has been swallowed?

25. Snowboarding in the half-pipe. Shaun White, "The Flying Tomato," won a gold medal in the 2010 Winter Olympics for snowboarding in the half-pipe. He soared an unprecedented 25 ft above the edge of the half-pipe. His speed $v(t)$, in miles per hour, upon reentering the pipe can be approximated by $v(t) = 10.9t$, where *t* is the number of seconds for which he was airborne. White was airborne for 2.5 sec. (*Source:* "White Rides to Repeat in Halfpipe, Lago Takes Bronze," Associated Press, 2/18/2010.) How fast was he going when he reentered the half-pipe?

26. Skateboard bomb drop. The distance $s(t)$, in feet, traveled by a body falling freely from rest in *t* seconds is approximated by

$$s(t) = 16t^2.$$

On April 6, 2006, pro skateboarder Danny Way smashed the world record for the "bomb drop" by free-falling 28 ft from the Fender Stratocaster guitar atop the Hard Rock Hotel & Casino in Las Vegas onto a ramp below. (*Source:* www.skateboardingmagazine.com.) How long did it take until he hit the ramp?

27. Hearing-impaired Americans. The number *N*, in millions, of hearing-impaired Americans of age *x* can be approximated by the graph that follows. (*Source:* American Speech-Language Hearing Association.)

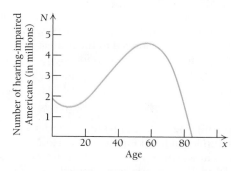

Use the graph to answer the following.

a) Approximate the number of hearing-impaired Americans of ages 20, 40, 50, and 60.

b) For what ages is the number of hearing-impaired Americans approximately 4 million?

c) Examine the graph and try to determine the age at which the greatest number of Americans is hearing-impaired.

d) What difficulty do you have in making this determination?

28. Life science: incidence of breast cancer. The following graph approximates the incidence of breast cancer *y*, per 100,000 women, as a function of age *x*, where *x* represents ages 25 to 102.

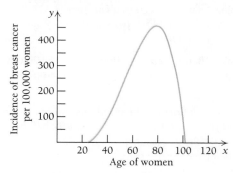

(*Source:* National Cancer Institute.)

a) What is the incidence of breast cancer in 40-yr-old women?

b) For what ages is the incidence of breast cancer about 400 per 100,000 women?

c) Examine the graph and try to determine the age at which the largest incidence of breast cancer occurs.

d) What difficulty do you have making this determination?

29. Compound interest. An investor purchases a $100,000 certificate of deposit from Newton Bank, at 2.8%. How much is the investment worth (rounded to the nearest cent) at the end of 1 yr, if interest is compounded:

a) annually?   **b)** semiannually?
c) quarterly?   **d)** daily (use 365 days
e) hourly?   for 1 yr)?

30. Compound interest. An investor purchases a $300,000 certificate of deposit from Descartes Bank, at 2.2%. How much is the investment worth (rounded to the nearest cent) at the end of 1 yr, if interest is compounded:

a) annually?   **b)** semiannually?
c) quarterly?   **d)** daily (use 365 days
e) hourly?   for 1 yr)?

31. Compound interest. An investor deposits $30,000 in Godel Municipal Bond Funds, at 4%. How much is the investment worth (rounded to the nearest cent) at the end of 1 yr, if interest is compounded:

a) annually? **b)** semiannually?
c) quarterly? **d)** daily (use 365 days
e) hourly? for 1 yr)?

32. Compound interest. An investor deposits $1000 in Wiles Municipal Bond Funds, at 5%. How much is the investment worth (rounded to the nearest cent) at the end of 1 yr, if interest is compounded:

a) annually? **b)** semiannually?
c) quarterly? **d)** daily (use 365 days
e) hourly? for 1 yr)?

*Determining monthly payments on a loan. **If P dollars are borrowed, the monthly payment M, made at the end of each month for n months, is given by***

$$M = P \frac{\frac{i}{12}\left(1 + \frac{i}{12}\right)^n}{\left(1 + \frac{i}{12}\right)^n - 1},$$

where i is the annual interest rate and n is the total number of monthly payments.

33. Fermat's Last Bank makes a car loan of $18,000, at 6.4% interest and with a loan period of 3 yr. What is the monthly payment?

34. At Haken Bank, Ken Appel takes out a $100,000 mortgage at an interest rate of 4.8% for a loan period of 30 yr. What is the monthly payment?

*Annuities. **If P dollars are invested annually in an annuity (investment fund), after n years, the annuity will be worth***

$$W = P\left[\frac{(1 + i)^n - 1}{i}\right],$$

where i is the interest rate, compounded annually.

35. You invest $3000 annually in an annuity from Mersenne Fund Annuities that earns 6.57% interest. How much is the investment worth after 18 yr? Round to the nearest cent.

36. Suppose that you establish an annuity that earns $7\frac{1}{4}\%$ interest, and you want it to be worth $50,000 in 20 yr. How much will you need to invest annually to achieve this goal?

37. Deer population in Maine. The deer population in Maine from 1986 to 2006 is approximated in the graph below.

DEER POPULATION IN MAINE

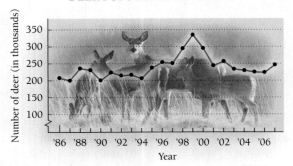

(*Source*: State of Maine, Department of Inland Fisheries and Wildlife.)

a) In what years was the deer population in Maine at or above 250,000?
b) In what years was the deer population at 200,000?
c) In what year was the deer population highest?
d) In what years was the deer population lowest?

38. Speculate as to why the deer population in Maine was so high in the years 1997–2001.

SYNTHESIS

*Retirement account. **Sally makes deposits into a retirement account every year from the age of 30 until she retires at age 65.***

39. a) If Sally deposits $1200 per year and the account earns interest at a rate of 8% per year, compounded annually, how much does she have in the account when she retires? (*Hint:* Use the annuity formula for Exercises 35 and 36.)

b) How much of that total amount is from Sally's deposits? How much is interest?

40. a) Sally plans to take regular monthly distributions from her retirement account from the time she retires until she is 80 years old, when the account will have a value of $0. How much should she take each month? Assume the interest rate is 8% per year, compounded monthly. (*Hint:* Use the formula for Exercises 33 and 34 that calculates the monthly payments on a loan.)

b) What is the total of the payments she will receive? How much of the total will be her own money (see part b of Exercise 39), and how much will be interest?

TECHNOLOGY CONNECTION

The Technology Connection heading indicates exercises designed to provide practice using a graphing calculator.

Graph.

41. $y = x - 150$ **42.** $y = 25 - |x|$

43. $y = x^3 + 2x^2 - 4x - 13$ **44.** $y = \sqrt{23 - 7x}$

45. $9.6x + 4.2y = -100$ **46.** $y = -2.3x^2 + 4.8x - 9$

47. $x = 4 + y^2$ **48.** $x = 8 - y^2$

Answers to Quick Checks

1. $y = 3 - x$

2. $3x - 5y = 10$

3. $y = 2 - x^2$

4. $x = 1 + y^2$

5. There is $1123.60 in the account after 2 yr.
6. There is $1195.62 in the account after 3 yr.

Functions and Models

Identifying Functions

The idea of a *function* is one of the most important concepts in mathematics. Put simply, a function is a special kind of correspondence between two sets. Let's look at the following.

> To each letter on a telephone keypad there corresponds a number.
>
> To each model of cell phone in a store there corresponds its price.
>
> To each real number there corresponds the cube of that number.

In each of these examples, the first set is called the *domain* and the second set is called the *range*. Given a member of the domain, there is *exactly one* member of the range to which it corresponds. This type of correspondence is called a *function*.

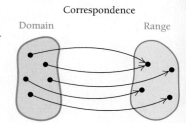

Correspondence

DEFINITION

A **function** is a correspondence between a first set, called the **domain**, and a second set, called the **range**, such that each member of the domain corresponds to *exactly one* member of the range.

■ **EXAMPLE 1**   Determine whether or not each correspondence is a function.

a) *Cumulative number of iPhones sold*

Domain	Range
2006 ⟶	0
2007 ⟶	1,389,000
2008 ⟶	11,627,000
2009 ⟶	20,371,000

(*Source:* Apple Inc.)

b) *Squaring*

Domain	Range
3 ⟶	9
4 ⟶	16
5	
−5 ⟶	25

c) *Baseball teams*

Domain	Range
Arizona ⟶	Diamondbacks
Chicago ⟶	Cubs
	White Sox
Baltimore ⟶	Orioles

d) *Baseball teams*

Domain	Range
Diamondbacks ⟶	Arizona
Cubs ⟶	Chicago
White Sox	
Orioles ⟶	Baltimore

Solution

a) The correspondence *is* a function because each member of the domain corresponds (is matched) to only one member of the range.

b) The correspondence *is* a function because each member of the domain corresponds to only one member of the range, even though two members of the domain correspond to 25.

c) The correspondence *is not* a function because one member of the domain, Chicago, corresponds to two members of the range, the Cubs and the White Sox.

d) The correspondence *is* a function because each member of the domain corresponds to only one member of the range, even though two members of the domain correspond to Chicago.

■ EXAMPLE 2   Determine whether or not each correspondence is a function.

Domain	Correspondence	Range
a) A family	Each person's weight	A set of positive numbers
b) The integers $\{\ldots, -3, -2, -1, 0, 1, 2, 3, \ldots\}$	Each number's square	A set of nonnegative integers: $\{0, 1, 4, 9, 16, 25, \ldots\}$
c) The set of all states	Each state's members of the U.S. Senate	The set of all U.S. senators

Solution

a) The correspondence is a function because each person has *only one* weight.

b) The correspondence is a function because each integer has *only one* square.

c) The correspondence is *not* a function because each state has *two* U.S. Senators.

Consistent with the definition on p. 13, we will regard a function as a set of ordered pairs, such that no two pairs have the same first coordinate paired with different second coordinates. The domain is the set of all first coordinates, and the range is the set of all second coordinates. Function names are usually represented by lowercase letters. Thus, if f represents the function in Example 1(b), we have

$$f = \{(3, 9), (4, 16), (5, 25), (-5, 25)\}$$

and

$$\text{Domain of } f = \{3, 4, 5, -5\}; \quad \text{Range of } f = \{9, 16, 25\}.$$

Finding Function Values

Most functions considered in mathematics are described by an equation like $y = 2x + 3$ or $y = 4 - x^2$. To graph the function given by $y = 2x + 3$, we find ordered pairs by performing calculations for selected x values.

for $x = 4, y = 2x + 3 = 2 \cdot 4 + 3 = 11;$ The graph includes $(4, 11)$.

for $x = -5, y = 2x + 3 = 2 \cdot (-5) + 3 = -7;$ The graph includes $(-5, -7)$.

for $x = 0, y = 2x + 3 = 2 \cdot 0 + 3 = 3;$ and so on.   The graph includes $(0, 3)$.

For $y = 2x + 3$, the **inputs** (members of the domain) are the values of x substituted into the equation. The **outputs** (members of the range) are the resulting values of y. If we call the function f, we can use x to represent an arbitrary input and $f(x)$, read "f of x" or "f at x" or "the value of f at x," to represent the corresponding output. In this notation, the function given by $y = 2x + 3$ is written as $f(x) = 2x + 3$, and the calculations above can be written more concisely as

$$f(4) = 2 \cdot 4 + 3 = 11;$$
$$f(-5) = 2 \cdot (-5) + 3 = -7;$$
$$f(0) = 2 \cdot 0 + 3 = 3; \text{ and so on.}$$

Thus, instead of writing "when $x = 4$, the value of y is 11," we can simply write "$f(4) = 11$," which is most commonly read as "f of 4 is 11."

It helps to think of a function as a machine. Think of $f(4) = 11$ as the result of putting a member of the domain (an input), 4, into the machine. The machine knows the correspondence $f(x) = 2x + 3$, computes $2 \cdot 4 + 3$, and produces a member of the range (the output), 11.

Function: $f(x) = 2x + 3$	
Input	Output
4	11
-5	-7
0	3
t	$2t + 3$
$a + h$	$2(a + h) + 3$

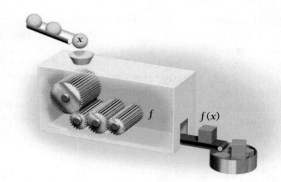

Remember that $f(x)$ ***does not mean*** "f times x" and should never be read that way.

■ **EXAMPLE 3**   The squaring function f is given by

$$f(x) = x^2.$$

Find $f(-3), f(1), f(k), f(\sqrt{k}), f(1 + t),$ and $f(x + h)$.

Solution   We have

$$f(-3) = (-3)^2 = 9;$$
$$f(1) = 1^2 = 1;$$
$$f(k) = k^2;$$
$$f(\sqrt{k}) = (\sqrt{k})^2 = k;$$
$$f(1 + t) = (1 + t)^2 = 1 + 2t + t^2;$$
$$f(x + h) = (x + h)^2 = x^2 + 2xh + h^2.$$

For a review of algebra, see Appendix A on p. 819.

To find $f(x + h)$, remember what the function does: It squares the input. Thus, $f(x + h) = (x + h)^2 = x^2 + 2xh + h^2$. This amounts to replacing x on both sides of $f(x) = x^2$ with $x + h$.

❮ Quick Check 1

❭ **Quick Check 1**

A function f is given by $f(x) = 3x + 5$. Find $f(4), f(-5), f(0), f(a),$ and $f(a + h)$.

■ **EXAMPLE 4**   A function f is given by $f(x) = 3x^2 - 2x + 8$. Find $f(0), f(-5),$ and $f(7a)$.

Solution   One way to find function values when a formula is given is to think of the formula with blanks, or placeholders, as follows:

$$f(\quad) = 3\,\boxed{}^{\,2} - 2\,\boxed{} + 8.$$

To find an output for a given input, we think: "Whatever goes in the blank on the left goes in the blank(s) on the right."

$$f(0) = 3 \cdot 0^2 - 2 \cdot 0 + 8 = 8$$
$$f(-5) = 3(-5)^2 - 2 \cdot (-5) + 8 = 3 \cdot 25 + 10 + 8 = 75 + 10 + 8 = 93$$
$$f(7a) = 3(7a)^2 - 2(7a) + 8 = 3 \cdot 49a^2 - 14a + 8 = 147a^2 - 14a + 8$$

❮ Quick Check 2

❭ **Quick Check 2**

A function f is given by $f(x) = 3x^2 + 2x - 7$. Find $f(4), f(-5), f(0), f(a),$ and $f(5a)$.

The TABLE Feature

The TABLE feature is one way to find ordered pairs of inputs and outputs of functions. To see how, consider the function given by $f(x) = x^3 - 5x + 1$. We enter it as $y_1 = x^3 - 5x + 1$. To use the TABLE feature, we access the TABLE SETUP screen and enter the x-value at which the table will start and an increment for the x-value. For this equation, let's set TblStart $= 0.3$ and ΔTbl $= 1$. (Other values can be chosen.) This means that the table's x-values will start at 0.3 and increase by 1.

```
TABLE  SETUP
  TblStart = .3
  ΔTbl = 1 ■
 Indpnt: Auto  Ask
 Depend: Auto  Ask
```

We next set Indpnt and Depend to Auto and then press TABLE. The result is shown below.

X	Y1
.3	−.473
1.3	−3.303
2.3	1.667
3.31	20.437
4.3	59.007
5.3	123.38
6.3	219.55
X = .3	

The arrow keys, ⌃ and ⌄, allow us to scroll up and down the table and extend it to other values not initially shown.

X	Y1
12.3	1800.4
13.3	2287.1
14.3	2853.7
15.3	3506.1
16.3	4250.2
17.3	5092.2
18.3	6038
X = 18.3	

EXERCISES

Use the function given by $f(x) = x^3 - 5x + 1$ for Exercises 1 and 2.

1. Use the TABLE feature to construct a table starting with $x = 10$ and ΔTbl $= 5$. Find the value of y when x is 10. Then find the value of y when x is 35.

2. Adjust the table settings to Indpnt: Ask. How does the table change? Enter a number of your choice and see what happens. Use this setting to find the value of y when x is 28.

■ **EXAMPLE 5**   A function f subtracts the square of an input from the input:

$$f(x) = x - x^2.$$

Find $f(4), f(x + h)$, and $\dfrac{f(x + h) - f(x)}{h}$.

Solution   We have

$$f(4) = 4 - 4^2 = 4 - 16 = -12;$$
$$f(x + h) = (x + h) - (x + h)^2$$
$$= x + h - (x^2 + 2xh + h^2)$$
$$= x + h - x^2 - 2xh - h^2 \qquad \text{Squaring the binomial}$$
$$\frac{f(x + h) - f(x)}{h} = \frac{x + h - x^2 - 2xh - h^2 - (x - x^2)}{h} = \frac{h - 2xh - h^2}{h}$$
$$= \frac{h(1 - 2x - h)}{h}$$
$$= 1 - 2x - h, \quad \text{for } h \neq 0.$$

❭ **Quick Check 3**

A function f is given by $f(x) = 2x - x^2$. Find $f(4), f(x + h)$, and $\dfrac{f(x + h) - f(x)}{h}$.

❮ Quick Check 3

Graphs of Functions

Consider again the squaring function. The input 3 is associated with the output 9. The input–output pair $(3, 9)$ is one point on the *graph* of this function.

> **DEFINITION**
>
> The **graph** of a function f is a draw-
> ing that represents all the input–output
> pairs $(x, f(x))$. In cases where the func-
> tion is given by an equation, the graph
> of the function is the graph of the equa-
> tion $y = f(x)$.

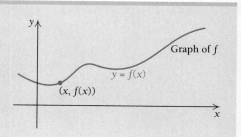

It is customary to locate input values (the domain) on the horizontal axis and output values (the range) on the vertical axis.

■ **EXAMPLE 6**   Graph: $f(x) = x^2 - 1$.

Solution

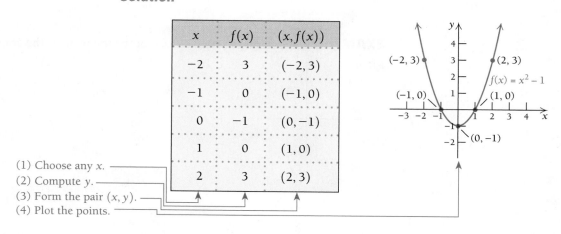

x	$f(x)$	$(x, f(x))$
-2	3	$(-2, 3)$
-1	0	$(-1, 0)$
0	-1	$(0, -1)$
1	0	$(1, 0)$
2	3	$(2, 3)$

(1) Choose any x.
(2) Compute y.
(3) Form the pair (x, y).
(4) Plot the points.

We plot the input–output pairs from the table and, in this case, draw a curve to complete
the graph.

> **Quick Check 4**
>
> Graph: $f(x) = 2 - x^2$.

❰ Quick Check 4

TECHNOLOGY CONNECTION

Graphs and Function Values

We discussed graphing equations in the Technology
Connection of Section R.1. Graphing a function makes
use of the same procedure. We just change the "$f(x) = $"
notation to "$y = $." Thus, to graph $f(x) = 2x^2 + x$,
we key in $y_1 = 2x^2 + x$.

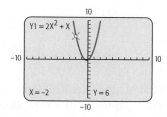

There are several ways in which to find function values.
One is to use the TABLE feature, as previously discussed.

Another way is to use the TRACE feature. To do so, graph the
function, press TRACE, and either move the cursor or enter any
x-value that is in the window. The corresponding y-value appears automatically. Function values can also be found using
the VALUE or Y-VARS feature. Consult an owner's manual or the
GCM for details.

EXERCISES

1. Graph $f(x) = x^2 + 3x - 4$. Then find $f(-5), f(-4.7)$,
$f(11)$, and $f(2/3)$. (*Hint:* To find $f(11)$, be sure that the
window dimensions for the x-values include $x = 11$.)

2. Graph $f(x) = 3.7 - x^2$. Then find $f(-5), f(-4.7)$,
$f(11)$, and $f(2/3)$.

3. Graph $f(x) = 4 - 1.2x - 3.4x^2$. Then find $f(-5)$,
$f(-4.7), f(11)$, and $f(2/3)$.

The Vertical-Line Test

Let's now determine how we can look at a graph and decide whether it is a graph of a function. In the graph at the right, note that the input x_1 has *two* outputs. Since a function has exactly *one* output for every input, this fact means that the graph does not represent a function. It also means that a vertical line could intersect the graph in more than one place.

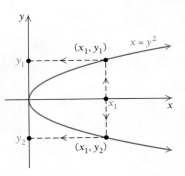

The Vertical-Line Test

A graph represents a function if it is impossible to draw a vertical line that intersects the graph more than once.

■ **EXAMPLE 7**   Determine whether each of the following is the graph of a function.

a)

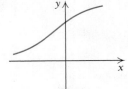

b)

c)

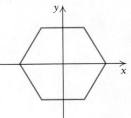

d)

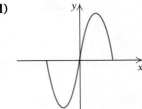

e)

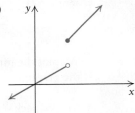

f)

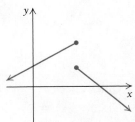

Solution

a) The graph is that of a function. It is impossible to draw a vertical line that intersects the graph more than once.

b) The graph is not that of a function. A vertical line (in fact, many) can intersect the graph more than once.

c) The graph is not that of a function.

d) The graph is that of a function.

e) The graph is that of a function.

f) The graph is not that of a function.

Functions Defined Piecewise

Sometimes functions are defined *piecewise*. That is, there are different output formulas for different parts of the domain, as in parts (e) and (f) of Example 7. To graph a piecewise-defined function, we usually work from left to right, paying special attention to the correspondence specified for the *x*-values on each part of the horizontal axis.

■ **EXAMPLE 8**   Graph the function defined as follows: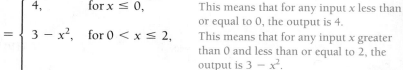

$$f(x) = \begin{cases} 4, & \text{for } x \le 0, \\ 3 - x^2, & \text{for } 0 < x \le 2, \\ 2x - 6, & \text{for } x > 2. \end{cases}$$

This means that for any input x less than or equal to 0, the output is 4.

This means that for any input x greater than 0 and less than or equal to 2, the output is $3 - x^2$.

This means that for any input x greater than 2, the output is $2x - 6$.

Solution   Working from left to right along the x-axis, we note that for any x-values less than or equal to 0, the graph is the horizontal line $y = 4$. Note that for $f(x) = 4$,

$$f(-2) = 4;$$
$$f(-1) = 4;$$
and $$f(0) = 4.$$

The solid dot indicates that $(0, 4)$ is part of the graph.

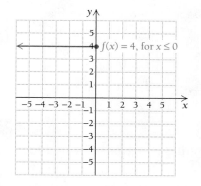

Next, observe that for x-values greater than 0 but not greater than 2, the graph is a portion of the parabola given by $y = 3 - x^2$. Note that for $f(x) = 3 - x^2$,

$$f(0.5) = 3 - 0.5^2 = 2.75;$$
$$f(1) = 2;$$
and $$f(2) = -1.$$

The open dot at $(0, 3)$ indicates that that point is *not* part of the graph.

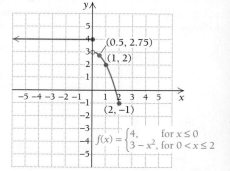

Finally, note that for x-values greater than 2, the graph is the line $y = 2x - 6$. Note that for $f(x) = 2x - 6$,

$$f(2.5) = 2 \cdot 2.5 - 6 = -1;$$
$$f(4) = 2;$$
and $$f(5) = 4.$$

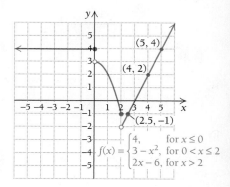

> **Quick Check 5**
>
> Graph the function defined as follows:
>
> $$f(x) = \begin{cases} 4, & \text{for } x \le 0, \\ 4 - x^2, & \text{for } 0 < x \le 2, \\ 2x - 6, & \text{for } x > 2. \end{cases}$$

❮ Quick Check 5

■ **EXAMPLE 9**   Graph the function defined as follows:

$$g(x) = \begin{cases} 3, & \text{for } x = 1, \\ -x + 2, & \text{for } x \ne 1. \end{cases}$$

Solution   The function is defined such that $g(1) = 3$ and for all other x-values (that is, for $x \ne 1$), we have $g(x) = -x + 2$. Thus, to graph this function, we graph the line given by $y = -x + 2$, but with an open dot at the point above $x = 1$. To complete the graph, we plot the point $(1, 3)$ since $g(1) = 3$.

x	$g(x)$	$(x, g(x))$
-3	$-(-3) + 2$	$(-3, 5)$
0	$-0 + 2$	$(0, 2)$
1	3	$(1, 3)$
2	$-2 + 2$	$(2, 0)$
3	$-3 + 2$	$(3, -1)$

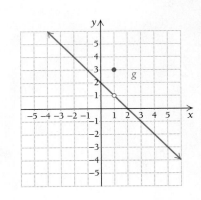

> **Quick Check 6**
>
> Graph the function defined as follows:
>
> $$f(x) = \begin{cases} 1, & \text{for } x = -2, \\ 2 - x, & \text{for } x \neq -2. \end{cases}$$

❰ Quick Check 6

TECHNOLOGY CONNECTION

Graphing Functions Defined Piecewise

Graphing functions defined piecewise generally involves the use of inequality symbols, which are often accessed using the TEST menu. The function in Example 8 is entered as follows:

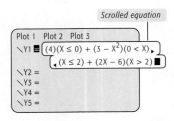

The graph is shown below.

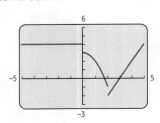

Note that most graphing calculators will not display solid or open dots.

EXERCISES

Graph.

1. $f(x) = \begin{cases} -x - 2, & \text{for } x < -2, \\ 4 - x^2, & \text{for } -2 \leq x < 2, \\ x + 3, & \text{for } x \geq 2 \end{cases}$

2. $f(x) = \begin{cases} x^2 - 2, & \text{for } x \leq 3, \\ 1, & \text{for } x > 3 \end{cases}$

3. $f(x) = \begin{cases} x + 3, & \text{for } x \leq -2, \\ 1, & \text{for } -2 < x \leq 3, \\ x^2 - 10, & \text{for } x > 3 \end{cases}$

Some Final Remarks

We sometimes use the terminology *y is a function of x*. This means that x is an input and y is an output. It also means that x is the **independent variable** because it represents inputs and y is the **dependent variable** because it represents outputs. We may refer to "a function $y = x^2$" without naming it using a letter f. We may also simply refer to x^2 (alone) as a function.

In calculus we will study how the outputs of a function change when the inputs change.

Section Summary

- *Functions* are a key concept in mathematics.

- The essential trait of a function is that to each number in the *domain* there corresponds one and only one number in the *range*.

EXERCISE SET
R.2

Note: A review of algebra can be found in Appendix A on p. 819.

Determine whether each correspondence is a function.

1.
Domain Range
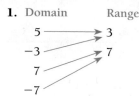

2.
Domain Range

3.
Domain Range
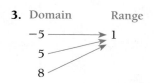

4.
Domain Range
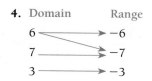

5. Sandwich prices.

DOMAIN	RANGE
Hamburger	$0.89
Cheeseburger	$0.95
Filet-O-Fish®	$3.00
Quarter Pounder® with cheese	$3.20
Big N' Tasty® with cheese	
Big Mac®	$3.20
Crispy Chicken	$3.40
Chicken McGrill®	$2.89
Double Quarter Pounder® with cheese	$3.80

(*Source*: www.mcdonalds.com.)

6. Sandwich calorie content.

DOMAIN	RANGE
Hamburger	250
Cheeseburger	300
Quarter Pounder®	410
Double Cheeseburger®	440
Filet-O-Fish®	380
Big N' Tasty®	460
McRib®	500
Big Mac®	540
Double Quarter Pounder® with cheese	740

(*Source*: www.mcdonalds.com.)

Determine whether each of the following is a function.

	Domain	Correspondence	Range
7.	A set of iPods	Each iPod's memory in gigabytes	A set of numbers
8.	A set of iPods	Each iPod's owner	A set of people
9.	A set of iPods	The number of songs on each iPod	A set of numbers
10.	A set of iPods	The number of Avril Lavigne songs on each iPod	A set of numbers
11.	The set of all real numbers	Square each number and then add 8.	The set of all positive numbers greater than or equal to 8
12.	The set of all real numbers	Raise each number to the fourth power.	The set of all nonnegative numbers
13.	A set of females	Each person's biological mother	A set of females
14.	A set of males	Each person's biological father	A set of males
15.	A set of avenues	An intersecting road	A set of cross streets
16.	A set of textbooks	An even-numbered page in each book	A set of pages
17.	A set of shapes	The area of each shape	A set of area measurements
18.	A set of shapes	The perimeter of each shape	A set of length measurements

19. A function f is given by

$$f(x) = 4x - 3.$$

This function takes a number x, multiplies it by 4, and subtracts 3.

a) Complete this table.

x	5.1	5.01	5.001	5
$f(x)$				

b) Find $f(4), f(3), f(-2), f(k), f(1 + t)$, and $f(x + h)$.

20. A function f is given by

$$f(x) = 3x + 2.$$

This function takes a number x, multiplies it by 3, and adds 2.

a) Complete this table.

x	4.1	4.01	4.001	4
$f(x)$				

b) Find $f(5), f(-1), f(k), f(1 + t)$, and $f(x + h)$.

21. A function g is given by

$$g(x) = x^2 - 3.$$

This function takes a number x, squares it, and subtracts 3. Find $g(-1), g(0), g(1), g(5), g(u), g(a + h)$, and

$$\frac{g(a + h) - g(a)}{h}.$$

22. A function g is given by

$$g(x) = x^2 + 4.$$

This function takes a number x, squares it, and adds 4. Find $g(-3), g(0), g(-1), g(7), g(v), g(a + h)$, and

$$\frac{g(a + h) - g(a)}{h}.$$

23. A function f is given by

$$f(x) = \frac{1}{(x + 3)^2}.$$

This function takes a number x, adds 3, squares the result, and takes the reciprocal of that result.

a) Find $f(4), f(-3), f(0), f(a), f(t + 4), f(x + h)$, and

$$\frac{f(x + h) - f(x)}{h}.$$ If an output is undefined, state that fact.

b) Note that f could also be given by

$$f(x) = \frac{1}{x^2 + 6x + 9}.$$

Explain what this does to an input number x.

24. A function f is given by

$$f(x) = \frac{1}{(x - 5)^2}.$$

This function takes a number x, subtracts 5 from it, squares the result, and takes the reciprocal of the square.

a) Find $f(3), f(-1), f(5), f(k), f(t - 1), f(t - 4)$, and $f(x + h)$. If an output is undefined, state that fact.

b) Note that f could also be given by

$$f(x) = \frac{1}{x^2 - 10x + 25}.$$

Explain what this does to an input number x.

Graph each function.

25. $f(x) = 2x - 5$

26. $f(x) = 3x - 1$

27. $g(x) = -4x$

28. $g(x) = -2x$

29. $f(x) = x^2 - 2$

30. $f(x) = x^2 + 4$

31. $f(x) = 6 - x^2$

32. $g(x) = -x^2 + 1$

33. $g(x) = x^3$

34. $g(x) = \frac{1}{2}x^3$

Use the vertical-line test to determine whether each graph is that of a function. (In Exercises 43–46, the vertical dashed lines are not part of the graph.)

35.

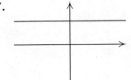

36.

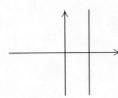

37.

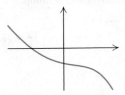

38.

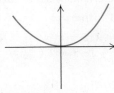

39.

40.

41.

42.

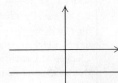

43.

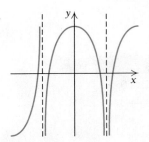

44.

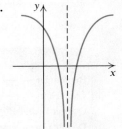

45.

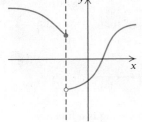

46.

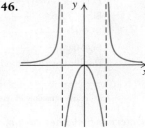

47. a) Graph $x = y^2 - 2$.
 b) Is this a function?

48. a) Graph $x = y^2 - 3$.
 b) Is this a function?

49. For $f(x) = x^2 - 3x$, find $\dfrac{f(x+h) - f(x)}{h}$.

50. For $f(x) = x^2 + 4x$, find $\dfrac{f(x+h) - f(x)}{h}$.

For Exercises 51–54, consider the function f given by

$$f(x) = \begin{cases} -2x + 1, & \text{for } x < 0, \\ 17, & \text{for } x = 0, \\ x^2 - 3, & \text{for } 0 < x < 4, \\ \frac{1}{2}x + 1, & \text{for } x \geq 4. \end{cases}$$

51. Find $f(-1)$ and $f(1)$.

52. Find $f(-3)$ and $f(3)$.

53. Find $f(0)$ and $f(10)$.

54. Find $f(-5)$ and $f(5)$.

Graph.

55. $f(x) = \begin{cases} 1, & \text{for } x < 0, \\ -1, & \text{for } x \geq 0 \end{cases}$

56. $f(x) = \begin{cases} 2, & \text{for } x \leq 3, \\ -2, & \text{for } x > 3 \end{cases}$

57. $f(x) = \begin{cases} 6, & \text{for } x = -2, \\ x^2, & \text{for } x \neq -2 \end{cases}$

58. $f(x) = \begin{cases} 5, & \text{for } x = 1, \\ x^3, & \text{for } x \neq 1 \end{cases}$

59. $g(x) = \begin{cases} -x, & \text{for } x < 0, \\ 4, & \text{for } x = 0, \\ x + 2, & \text{for } x > 0 \end{cases}$

60. $g(x) = \begin{cases} 2x - 3, & \text{for } x < 1, \\ 5, & \text{for } x = 1, \\ x - 2, & \text{for } x > 1 \end{cases}$

61. $g(x) = \begin{cases} \frac{1}{2}x - 1, & \text{for } x < 2, \\ -4, & \text{for } x = 2, \\ x - 3, & \text{for } x > 2 \end{cases}$

62. $g(x) = \begin{cases} x^2, & \text{for } x < 0, \\ -3, & \text{for } x = 0, \\ -2x + 3, & \text{for } x > 0 \end{cases}$

63. $f(x) = \begin{cases} -7, & \text{for } x = 2, \\ x^2 - 3, & \text{for } x \neq 2 \end{cases}$

64. $f(x) = \begin{cases} -6, & \text{for } x = -3, \\ -x^2 + 5, & \text{for } x \neq -3 \end{cases}$

Compound interest. *The amount of money, $A(t)$, in a savings account that pays 6% interest, compounded quarterly for t years, with an initial investment of P dollars, is given by*

$$A(t) = P\left(1 + \frac{0.06}{4}\right)^{4t}.$$

65. If $500 is invested at 6%, compounded quarterly, how much will the investment be worth after 2 yr?

66. If $800 is invested at 6%, compounded quarterly, how much will the investment be worth after 3 yr?

Chemotherapy. *In computing the dosage for chemotherapy, a patient's body surface area is needed. A good approximation of a person's surface area s, in square meters (m^2), is given by the formula*

$$s = \sqrt{\frac{hw}{3600}},$$

where w is the patient's weight in kilograms (kg) and h is the patient's height in centimeters (cm). (Source: U.S. Oncology.) Use the preceding information for Exercises 67 and 68. Round your answers to the nearest thousandth.

67. Assume that a patient's height is 170 cm. Find the patient's approximate surface area assuming that:

　a) The patient's weight is 70 kg.
　b) The patient's weight is 100 kg.
　c) The patient's weight is 50 kg.

68. Assume that a patient's weight is 70 kg. Approximate the patient's surface area assuming that:

　a) The patient's height is 150 cm.
　b) The patient's height is 180 cm.

69. *Scaling stress factors.* In psychology a process called *scaling* is used to attach numerical ratings to a group of life experiences. In the table below, various events have been rated on a scale from 1 to 100 according to their stress levels.

Event	Scale of Impact
Death of spouse	100
Divorce	73
Jail term	63
Marriage	50
Lost job	47
Pregnancy	40
Death of close friend	37
Loan over $10,000	31
Child leaving home	29
Change in schools	20
Loan less than $10,000	17
Christmas	12

(*Source:* Thomas H. Holmes, University of Washington School of Medicine.)

　a) Does the table represent a function? Why or why not?
　b) What are the inputs? What are the outputs?

SYNTHESIS

Solve for y in terms of x. Decide whether the resulting equation represents a function.

70. $2x + y - 16 = 4 - 3y + 2x$

71. $2y^2 + 3x = 4x + 5$

72. $(4y^{2/3})^3 = 64x$　　　**73.** $(3y^{3/2})^2 = 72x$

74. Explain why the vertical-line test works.

75. Is 4 in the domain of f in Exercises 51–54? Explain why or why not.

TECHNOLOGY CONNECTION ·················

In Exercises 76 and 77, use the TABLE *feature to construct a table for the function under the given conditions.*

76. $f(x) = x^3 + 2x^2 - 4x - 13$; TblStart $= -3$; ΔTbl $= 2$

77. $f(x) = \dfrac{3}{x^2 - 4}$; TblStart $= -3$; ΔTbl $= 1$

78. A function f is given by
$$f(x) = |x - 2| + |x + 1| - 5.$$
Find $f(-3), f(-2), f(0)$, and $f(4)$.

79. Graph the function in each of Exercises 76–78.

80. Use the TRACE feature to find several ordered-pair solutions of the function $f(x) = \sqrt{10 - x^2}$.

Answers to Quick Checks

1. $17, -10, 5, 3a - 5, 3a + 3h - 5$,
2. $49, 58, -7, 3a^2 + 2a - 7, 75a^2 + 10a - 7$
3. $-8, 2x + 2h - x^2 - 2xh - h^2, 2 - 2x - h$
4. $f(x) = 2 - x^2$

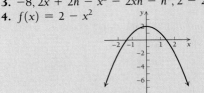

5.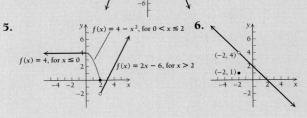

6.

R.3

Finding Domain and Range

Set Notation

A **set** is a collection of objects. The set we consider most in calculus is the set of **real numbers**, $\mathbb{R}$. There is a real number for every point on the number line.

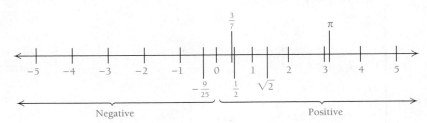

The set consisting of $-\frac{9}{25}$, 0, and $\sqrt{2}$ can be written $\left\{-\frac{9}{25}, 0, \sqrt{2}\right\}$. This method of describing sets is known as the **roster method**. It lists every member of the set. We describe larger sets using **set-builder notation**, which specifies conditions under which an object is in the set. For example, the set of all real numbers less than 4 can be described as follows in set-builder notation:

$$\{x \mid x \text{ is a real number less than } 4\} \text{ or } \{x \mid x < 4\}.$$

The set of
all x
such that
x is a real number less than 4.

Interval Notation

We can also describe sets using **interval notation**. If a and b are real numbers, with $a < b$, we define the interval (a, b) as the set of all numbers between but not including a and b, that is, the set of all x for which $a < x < b$. Thus,

$$(a, b) = \{x \mid a < x < b\}.$$

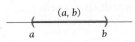

The points a and b are the **endpoints** of the interval. The parentheses indicate that the endpoints are *not* included in the interval.

The interval $[a, b]$ is defined as the set of all x for which $a \leq x \leq b$. Thus,

$$[a, b] = \{x \mid a \leq x \leq b\}.$$

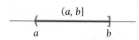

The brackets indicate that the endpoints *are* included in the interval.*

Be careful not to confuse the *interval* (a, b) with the *ordered pair* (a, b) used to represent a point in the plane, as in Section R.1. The context in which the notation appears usually makes the meaning clear.

Intervals like $(-2, 3)$, in which neither endpoint is included, are called **open intervals**; intervals like $[-2, 3]$, which include both endpoints, are said to be **closed intervals**. Thus, $[a, b]$ is read "the closed interval a, b" and (a, b) is read "the open interval a, b."

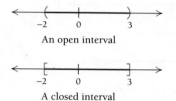

An open interval

A closed interval

Some intervals are **half-open** and include one endpoint but not the other:

$$(a, b] = \{x \mid a < x \leq b\}. \quad \text{The graph excludes } a \text{ and includes } b.$$

$$[a, b) = \{x \mid a \leq x < b\}. \quad \text{The graph includes } a \text{ and excludes } b.$$

Some intervals extend without bound in one or both directions. We use the symbols ∞, read "infinity," and $-\infty$, read "negative infinity," to name these intervals. The notation $(5, \infty)$ represents the set of all numbers greater than 5. That is,

$$(5, \infty) = \{x \mid x > 5\}.$$

Similarly, the notation $(-\infty, 5)$ represents the set of all numbers less than 5. That is,

$$(-\infty, 5) = \{x \mid x < 5\}.$$

The notations $[5, \infty)$ and $(-\infty, 5]$ are used when we want to include the endpoints. The interval $(-\infty, \infty)$ names the set of all real numbers.

$$(-\infty, \infty) = \{x \mid x \text{ is a real number}\}$$

Interval notation is summarized in the following table. Note that the symbols ∞ and $-\infty$ always have a parenthesis next to them; neither of these represents a real number.

*Some books use the representations ⊶⊷ and ╾┼─┼╼ instead of, respectively, ╾(───)╼ and ╾[───]╼.

Intervals: Notation and Graphs

Interval Notation	Set Notation	Graph
(a, b)	$\{x \mid a < x < b\}$	
$[a, b]$	$\{x \mid a \le x \le b\}$	
$[a, b)$	$\{x \mid a \le x < b\}$	
$(a, b]$	$\{x \mid a < x \le b\}$	
(a, ∞)	$\{x \mid x > a\}$	
$[a, \infty)$	$\{x \mid x \ge a\}$	
$(-\infty, b)$	$\{x \mid x < b\}$	
$(-\infty, b]$	$\{x \mid x \le b\}$	
$(-\infty, \infty)$	$\{x \mid x \text{ is a real number}\}$	

■ **EXAMPLE 1**   Write interval notation for each set or graph:

a) $\{x \mid -4 < x < 5\}$

b) $\{x \mid x \ge -2\}$

c)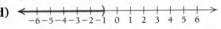

d)

> **Quick Check 1**
>
> Write interval notation for each set:
>
> **a)** $\{x \mid -2 \le x \le 5\}$;
>
> **b)** $\{x \mid -2 \le x < 5\}$;
>
> **c)** $\{x \mid -2 < x \le 5\}$;
>
> **d)** $\{x \mid -2 < x < 5\}$.

Solution

a) $\{x \mid -4 < x < 5\} = (-4, 5)$

b) $\{x \mid x \ge -2\} = [-2, \infty)$

c) $(-2, 4]$

d) $(-\infty, -1)$

❬ Quick Check 1

Finding Domain and Range

Recall that when a set of ordered pairs is such that no two different pairs share a common first coordinate, we have a function. The **domain** is the set of all first coordinates, and the **range** is the set of all second coordinates.

■ **EXAMPLE 2**   For the function f whose graph is shown to the right, determine the domain and the range.

Solution   This function consists of just four ordered pairs and can be written as

$$\{(-3, 1), (1, -2), (3, 0), (4, 5)\}.$$

We can determine the domain and the range by reading the x- and the y-values directly from the graph.

The domain is the set of all first coordinates, $\{-3, 1, 3, 4\}$. The range is the set of all second coordinates, $\{1, -2, 0, 5\}$.

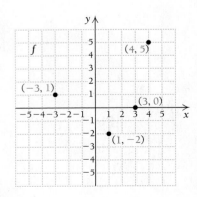

EXAMPLE 3 For the function f whose graph is shown to the right, determine each of the following.

a) The number in the range that is paired with 1 (from the domain). That is, find $f(1)$.

b) The domain of f

c) The number(s) in the domain that is (are) paired with 1 (from the range). That is, find all x-values for which $f(x) = 1$.

d) The range of f

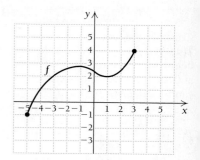

Solution

a) To determine which number in the range is paired with 1 in the domain, we locate 1 on the horizontal axis. Next, we find the point on the graph of f for which 1 is the first coordinate. From that point, we look to the vertical axis to find the corresponding y-coordinate, 2. The input 1 has the output 2—that is, $f(1) = 2$.

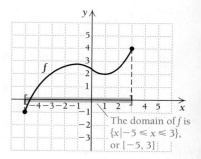

b) The domain of the function is the set of all x-values, or inputs, of the points on the graph. These extend from -5 to 3 and can be viewed as the curve's shadow, or *projection,* onto the x-axis. Thus, the domain is the set $\{x \mid -5 \le x \le 3\}$, or, in interval notation, $[-5, 3]$.

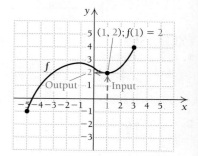

c) To determine which number(s) in the domain is (are) paired with 1 in the range, we locate 1 on the vertical axis (see the graph to the right). From there, we look left and right to the graph of f to find any points for which 1 is the second coordinate. One such point exists: $(-4, 1)$. For this function, we note that $x = -4$ is the only member of the domain paired with the range value of 1. For other functions, there might be more than one member of the domain paired with a member of the range.

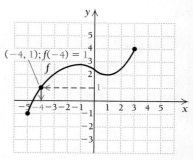

d) The range of the function is the set of all y-values, or outputs, of the points on the graph. These extend from -1 to 4 and can be viewed as the curve's shadow, or projection, onto the y-axis. Thus, the range is the set $\{y \mid -1 \le y \le 4\}$, or, in interval notation, $[-1, 4]$.

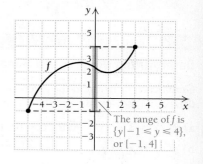

〉 **Quick Check 2**

For the function f whose graph follows, determine each of the following: $f(-1), f(1)$, the domain, and the range.

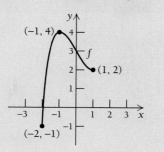

〈 Quick Check 2

Consider the function given by $f(x) = 1/(x - 3)$. The table below was obtained with ΔTbl set at 0.25. Note that the calculator cannot calculate $f(3)$.

ΔTbl = 0.25

X	Y1
2	−1
2.25	−1.333
2.5	−2
2.75	−4
3	ERR:
3.25	4
3.5	2
3.75	1.3333

X = 2

EXERCISES

1. Make a table for $f(x) = 1/(x^2 - 4)$ from $x = -3$ to $x = 0$ and with ΔTbl set at 0.5.

2. Create a table for the function given in Example 5.

> **Quick Check 3**
>
> Find the domain of each function. Express your answers in interval notation.
>
> **a)** $f(x) = \dfrac{5}{x - 8}$
>
> **b)** $f(x) = x^3 + |2x|$
>
> **c)** $f(x) = \sqrt{2x - 8}$

When a function is given by an equation or formula, the domain is understood to be the largest set of real numbers (inputs) for which function values (outputs) can be calculated. That is, the domain is the set of all allowable inputs into the formula. To find the domain, think, "For what input values does the function have an output?"

■ **EXAMPLE 4** Find the domain: $f(x) = |x|$.

Solution We ask, "What can we substitute?" Is there any number x for which we cannot calculate $|x|$? The answer is no. Thus, the domain of f is the set of all real numbers.

■ **EXAMPLE 5** Find the domain: $f(x) = \dfrac{3}{2x - 5}$.

Solution We recall that a denominator cannot equal zero and ask, "What can we substitute?" Is there any number x for which we cannot calculate $3/(2x - 5)$? Since $3/(2x - 5)$ cannot be calculated when the denominator $2x - 5$ is 0, we solve the following equation to find those real numbers that must be excluded from the domain of f:

$$2x - 5 = 0 \quad \text{Setting the denominator equal to 0}$$
$$2x = 5 \quad \text{Adding 5 to both sides}$$
$$x = \tfrac{5}{2}. \quad \text{Dividing both sides by 2}$$

Thus, $\tfrac{5}{2}$ is not in the domain, whereas all other real numbers are. We say that f is *not defined at* $\tfrac{5}{2}$, or $f\left(\tfrac{5}{2}\right)$ *does not exist.*

The domain of f is $\left\{x \mid x \text{ is a real number } and\ x \neq \tfrac{5}{2}\right\}$, or, in interval notation, $\left(-\infty, \tfrac{5}{2}\right) \cup \left(\tfrac{5}{2}, \infty\right)$. The symbol $\cup$ indicates the *union* of two sets and means that all elements in both sets are included in the domain.

■ **EXAMPLE 6** Find the domain: $f(x) = \sqrt{4 + 3x}$.

Solution We ask, "What can we substitute?" Is there any number x for which we cannot calculate $\sqrt{4 + 3x}$? We recall that radicands in even roots cannot be negative. Since $\sqrt{4 + 3x}$ is not a real number when the radicand $4 + 3x$ is negative, the domain is all real numbers for which $4 + 3x \geq 0$. We find the domain by solving the inequality. (See Appendix A for a review of inequality solving.)

$$4 + 3x \geq 0$$
$$3x \geq -4 \quad \text{Simplifying}$$
$$x \geq -\tfrac{4}{3} \quad \text{Dividing both sides by 3}$$

The domain is $\left[-\tfrac{4}{3}, \infty\right)$.

❰ Quick Check 3

Determining Domain and Range

Graph each function in the given viewing window. Then determine the domain and the range.

a) $f(x) = 3 - |x|, [-10, 10, -10, 10]$

b) $f(x) = x^3 - x, [-3, 3, -4, 4]$

c) $f(x) = \dfrac{12}{x}$, or $12x^{-1}, [-14, 14, -14, 14]$

d) $f(x) = x^4 - 2x^2 - 3, [-4, 4, -6, 6]$

(continued)

We have the following.

a) $y = 3 - |x|$

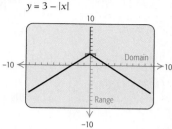

Domain = $\mathbb{R}$ (the real numbers), or $(-\infty, \infty)$
Range = $(-\infty, 3]$

b) $y = x^3 - x$

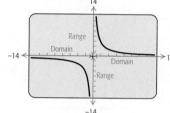

Domain = $\mathbb{R}$
Range = $\mathbb{R}$

c) $y = \dfrac{12}{x}$, or $12x^{-1}$

Xscl = 2, Yscl = 2

The number 0 is excluded as an input.
Domain = $\{x \mid x \text{ is a real number } and \, x \neq 0\}$, or
$(-\infty, 0) \cup (0, \infty)$;
Range = $\{y \mid y \text{ is a real number } and \, y \neq 0\}$, or
$(-\infty, 0) \cup (0, \infty)$

d) $y = x^4 - 2x^2 - 3$

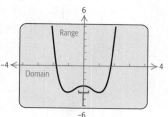

Domain = $\mathbb{R}$
Range = $[-4, \infty)$

We can confirm our results using the TRACE feature, moving the cursor along the curve or entering any *x*-value in which we have interest. We can also use the TABLE feature. In Example (d), it might not appear as though the domain is all real numbers because the graph seems "thin," but careful examination of the formula shows that we can indeed substitute any real number.

EXERCISES
Graph each function in the given viewing window. Then determine the domain and the range.

1. $f(x) = |x| - 4$, $[-10, 10, -10, 10]$

2. $f(x) = 2 + 3x - x^3$, $[-5, 5, -5, 5]$

3. $f(x) = \dfrac{-3}{x}$, or $-3x^{-1}$, $[-20, 20, -20, 20]$

4. $f(x) = x^4 - 2x^2 - 7$, $[-4, 4, -9, 9]$

5. $f(x) = \sqrt{x + 4}$, $[-8, 8, -8, 8]$

6. $f(x) = \sqrt{9 - x^2}$, $[-5, 5, -5, 5]$

7. $f(x) = -\sqrt{9 - x^2}$, $[-5, 5, -5, 5]$

8. $f(x) = x^3 - 5x^2 + x - 4$, $[-10, 10, -20, 10]$

Domains and Ranges in Applications

The domain and the range of a function given by a formula are sometimes affected by the context of an application. Let's look again at Exercise 65 in Section R.2.

■ **EXAMPLE 7**   **Business: Compound Interest.**   Suppose that $500 is invested at 6%, compounded quarterly for *t* years. From Theorem 2 in Section R.1, we know that the amount in the account is given by

$$A(t) = 500\left(1 + \frac{0.06}{4}\right)^{4t}$$
$$= 500(1.015)^{4t}.$$

The amount *A* is a function of the number of years for which the money is invested. Determine the domain.

Solution   We can substitute any real number for *t* into the formula, but a negative number of years is not meaningful. The context of the application excludes negative numbers. Thus, the domain is the set of all nonnegative numbers, $[0, \infty)$.

■ **EXAMPLE 8   Cellphone Calling Plans.**   Recently, Sprint® offered a cellphone calling plan in which a customer's monthly bill could be modeled by the graph below. Find the range of the function shown.

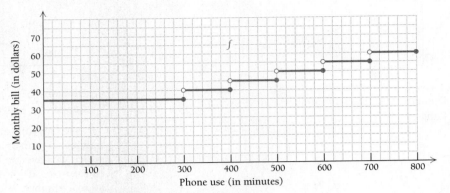

(*Source: The New York Times.* Bill does not include taxes and fees.)

Solution   The range is the set of all outputs—in this case, the different monthly bill amounts—shown in the graph. We see that only six different outputs are used, highlighted in blue. Thus, the range of the function shown is $\{35, 40, 45, 50, 55, 60\}$.

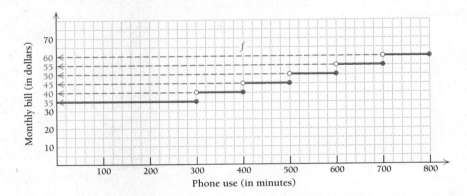

We will continue to determine the domain and the range of a function as we progress through this book.

Section Summary

The following is a review of the function concepts considered in Sections R.1–R.3.

Function Concepts

- Formula for f: $f(x) = x^2 - 7$
- For every input of f, there is exactly one output.
- For the input 1, -6 is the output.
- $f(1) = -6$
- $(1, -6)$ is on the graph.
- Domain = the set of all inputs
       = the set of all real numbers, $\mathbb{R}$
- Range = the set of all outputs
       = $[-7, \infty)$

Graph

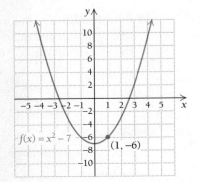

EXERCISE SET
R.3

In Exercises 1–10, write interval notation for each graph.

1.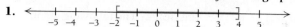

2.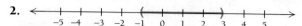

3.

4.

5.

6.

7.

8.

9.

10.

Write interval notation for each of the following. Then graph the interval on a number line.

11. The set of all numbers x such that $-2 \le x \le 2$

12. The set of all numbers x such that $-5 < x < 5$

13. $\{x \mid -4 \le x < -1\}$

14. $\{x \mid 6 < x \le 20\}$

15. $\{x \mid x \le -2\}$

16. $\{x \mid x > -3\}$

17. $\{x \mid -2 < x \le 3\}$

18. $\{x \mid -10 \le x < 4\}$

19. $\{x \mid x < 12.5\}$

20. $\{x \mid x \ge 12.5\}$

In Exercises 21–32, the graph is that of a function. Determine for each one (a) $f(1)$; (b) the domain; (c) all x-values such that $f(x) = 2$; and (d) the range.

21.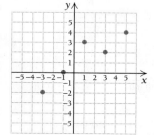

22.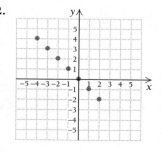

23.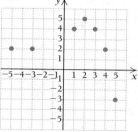

24.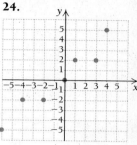

25.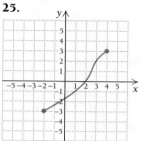

26.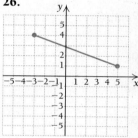

27.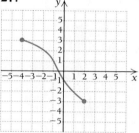

28.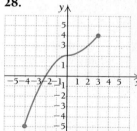

29.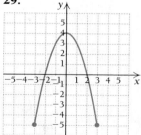

30.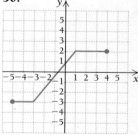

31.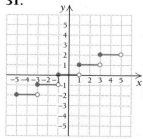

32.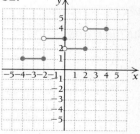

Find the domain of each function given below.

33. $f(x) = \dfrac{6}{2 - x}$

34. $f(x) = \dfrac{2}{x + 3}$

35. $f(x) = \sqrt{2x}$

36. $f(x) = \sqrt{x - 2}$

37. $f(x) = x^2 - 2x + 3$

38. $f(x) = x^2 + 3$

39. $f(x) = \dfrac{x - 2}{6x - 12}$

40. $f(x) = \dfrac{8}{3x - 6}$

41. $f(x) = |x - 4|$

42. $f(x) = |x| - 4$

43. $f(x) = \dfrac{3x - 1}{7 - 2x}$

44. $f(x) = \dfrac{2x - 1}{9 - 2x}$

45. $g(x) = \sqrt{4 + 5x}$

46. $g(x) = \sqrt{2 - 3x}$

47. $g(x) = x^2 - 2x + 1$

48. $g(x) = 4x^3 + 5x^2 - 2x$

49. $g(x) = \dfrac{2x}{x^2 - 25}$ (*Hint:* Factor the denominator.)

50. $g(x) = \dfrac{x - 1}{x^2 - 36}$ (*Hint:* Factor the denominator.)

51. $g(x) = |x| + 1$

52. $g(x) = |x + 7|$

53. $g(x) = \dfrac{2x - 6}{x^2 - 6x + 5}$ (*Hint:* Factor the denominator.)

54. $g(x) = \dfrac{3x - 10}{x^2 - 4x - 5}$ (*Hint:* Factor the denominator.)

55. For the function f whose graph is shown to the right, find all x-values for which $f(x) \le 0$.

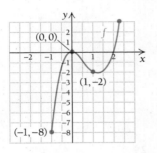

56. For the function g whose graph is shown to the right, find all x-values for which $g(x) = 1$.

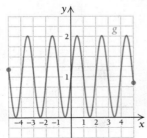

APPLICATIONS

Business and Economics

57. Compound interest. Suppose that $5000 is invested at 8% interest, compounded semiannually, for t years.

a) The amount A in the account is a function of time. Find an equation for this function.

b) Determine the domain of the function in part (a).

58. Compound interest. Suppose that $3000 is borrowed as a college loan, at 5% interest, compounded daily, for t years.

a) The amount A that is owed is a function of time. Find an equation for this function.

b) Determine the domain of the function in part (a).

Life and Physical Sciences

59. Hearing-impaired Americans. The following graph (considered in Exercise Set R.1) approximates the number N, in millions, of hearing-impaired Americans as a function of age x. (*Source:* American Speech-Language Hearing Association.) The equation for this graph is the function given by

$$N(x) = -0.00006x^3 + 0.006x^2 - 0.1x + 1.9.$$

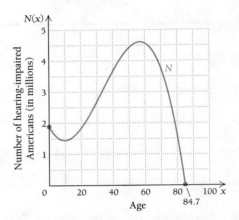

a) Use the graph to determine the domain of N.

b) Use the graph to determine the range of N.

c) If you were marketing a new type of hearing aid, at what age group (expressed as a 10-yr interval) would you target advertisements? Why?

60. Incidence of breast cancer. The following graph (considered in Exercise 28 of Exercise Set R.1 without an equation) approximates the incidence of breast cancer I, per 100,000 women, as a function of age x. The equation for this graph is the function given by

$$I(x) = -0.0000554x^4 + 0.0067x^3 - 0.0997x^2 - 0.84x - 0.25.$$

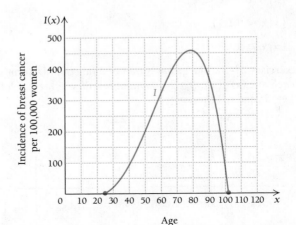

(*Source:* Based on data from the National Cancer Institute.)

a) Use the graph to determine the domain of *I*.
b) Use the graph to determine the range of *I*.
c) What 10-yr age interval sees the greatest increase in the incidence of breast cancer? Explain how you determined this.

61. Lung cancer. The following graph approximates the incidence of lung and bronchus cancer *L*, per 100,000 males, as a function of *t*, the number of years since 1940. The equation for this graph is the function given by

$$L(t) = -0.00054t^3 + 0.02917t^2 + 1.2329t + 8.$$

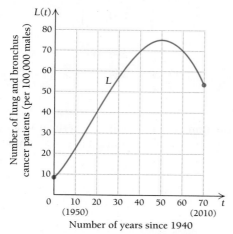

(*Source:* Based on data from the American Cancer Society Surveillance Research, 2005.)

a) Use the graph to estimate the domain of *L*.
b) Use the graph to estimate the range of *L*.

62. See Exercise 61.

a) Use the graph to approximate all the *x*-values (ages) where the cancer rate is 50 per 100,000.
b) Use the graph to approximate all the *x*-values (ages) where the cancer rate is 70 per 100,000.
c) Use the formula to approximate the lung and bronchus cancer rate in 2010.

SYNTHESIS

63. For a given function, $f(2) = -5$. Give as many interpretations of this fact as you can.

64. Explain how it is possible for the domain and the range of a function to be the same set.

65. Give an example of a function for which the number 3 is not in the domain, and explain why it is not.

TECHNOLOGY CONNECTION

66. Determine the range of each of the functions in Exercises 33, 35, 39, 40, and 47.

67. Determine the range of each of the functions in Exercises 34, 36, 48, 51, and 54.

Answers to Quick Checks

1. (a) $[-2, 5]$, **(b)** $[-2, 5)$, **(c)** $(-2, 5]$, **(d)** $(-2, 5)$
2. $f(-1) = 4, f(1) = 2$; domain is $[-2, 1]$, and range is $[-1, 4]$
3. (a) $\{x \mid x \text{ is a real number and } x \neq 8\}$ **(b)** $\mathbb{R}$ **(c)** $[4, \infty)$

R.4 Slope and Linear Functions

Horizontal and Vertical Lines

Let's consider graphs of equations $y = c$ and $x = a$, where c and a are real numbers.

OBJECTIVES

- Graph equations of the types $y = f(x) = c$ and $x = a$.
- Graph linear functions.
- Find an equation of a line when given the slope and one point on the line and when given two points on the line.
- Solve applied problems involving slope and linear functions.

■ **EXAMPLE 1**

a) Graph $y = 4$.
b) Decide whether the graph represents a function.

Solution

a) The graph consists of all ordered pairs whose second coordinate is 4. To see how a pair such as $(-2, 4)$ could be a solution of $y = 4$, we can consider the equation above in the form

$$y = 0x + 4.$$

Then $(-2, 4)$ is a solution because

$$0(-2) + 4 = 4$$

is true.

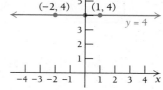

b) The vertical-line test holds. Thus, the graph represents a function.

■ **EXAMPLE 2**

a) Graph $x = -3$.

b) Decide whether the graph represents a function.

Solution

a) The graph consists of all ordered pairs whose first coordinate is -3. To see how a pair such as $(-3, 4)$ could be a solution of $x = -3$, we can consider the equation in the form

$$x + 0y = -3.$$

Then $(-3, 4)$ is a solution because

$$(-3) + 0(4) = -3$$

is true.

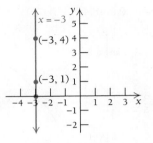

> **Quick Check 1**
>
> Graph each equation:
>
> a) $x = 4$;
>
> b) $y = -3$.

b) This graph does not represent a function because it fails the vertical-line test. The line itself meets the graph more than once—in fact, infinitely many times.

❮ Quick Check 1

In general, we have the following.

THEOREM 3

The graph of $y = c$, or $f(x) = c$, a horizontal line, is the graph of a function. Such a function is referred to as a **constant function**. The graph of $x = a$ is a vertical line, and $x = a$ is not a function.

TECHNOLOGY CONNECTION

Visualizing Slope

Exploratory: Squaring a Viewing Window

The standard $[-10, 10, -10, 10]$ viewing window shown below is not scaled identically on both axes. Note that the intervals on the y-axis are about two-thirds the length of those on the x-axis.

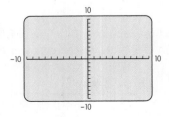

If we change the dimensions of the window to $[-6, 6, -4, 4]$, we get a graph for which the units are visually about the same on both axes.

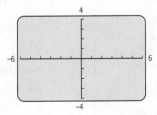

Creating such a window is called **squaring the window**. On many calculators, this is accomplished automatically by selecting the ZSquare option of the ZOOM menu.

(continued)

Each of the following is a graph of $y = 2x - 3$, but with different viewing windows. When the window is square, as shown in the last graph, we get the most accurate representation of the *slope* of the line.

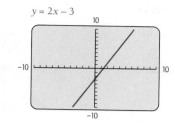

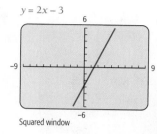

Squared window

EXERCISES

Use a squared viewing window for each of these exercises.

1. Graph $y = x + 1, y = 2x + 1, y = 3x + 1$, and $y = 10x + 1$. What do you think the graph of $y = 247x + 1$ will look like?

2. Graph $y = x, y = \frac{7}{8}x, y = 0.47x$, and $y = \frac{2}{31}x$. What do you think the graph of $y = 0.000018x$ will look like?

3. Graph $y = -x, y = -2x, y = -5x$, and $y = -10x$. What do you think the graph of $y = -247x$ will look like?

4. Graph $y = -x - 1, y = -\frac{3}{4}x - 1, y = -0.38x - 1$, and $y = -\frac{5}{32}x - 1$. What do you think the graph of $y = -0.000043x - 1$ will look like?

The Equation $y = mx$

Consider the following table of numbers and look for a pattern.

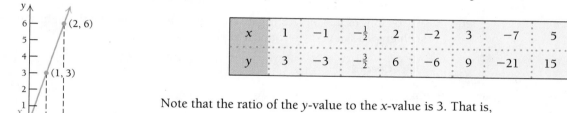

x	1	-1	$-\frac{1}{2}$	2	-2	3	-7	5
y	3	-3	$-\frac{3}{2}$	6	-6	9	-21	15

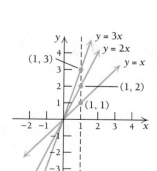

Note that the ratio of the y-value to the x-value is 3. That is,

$$\frac{y}{x} = 3, \quad \text{or} \quad y = 3x.$$

Ordered pairs from the table can be used to graph the equation $y = 3x$ (see the figure at the left). Note that this is a function.

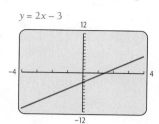

THEOREM 4

The graph of the function given by

$$y = mx \quad \text{or} \quad f(x) = mx$$

is the straight line through the origin $(0, 0)$ and the point $(1, m)$. The constant m is called the **slope** of the line.

Various graphs of $y = mx$ for positive values of m are shown to the left. Note that such graphs slant up from left to right. A line with large positive slope rises faster than a line with smaller positive slope.

When $m = 0$, $y = 0x$, or $y = 0$. On the left below is a graph of $y = 0$. Note that this is both the x-axis and a horizontal line.

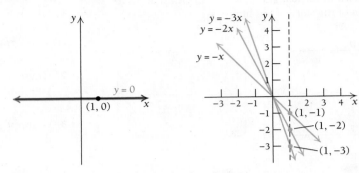

The rows of flowers form lines of equal slope.

Graphs of $y = mx$ for negative values of m are shown on the right above. Note that such graphs slant down from left to right.

❮ Quick Check 2

❯ **Quick Check 2**

Graph each equation:

a) $y = \dfrac{1}{2}x$;

b) $y = -\dfrac{1}{2}x$.

Direct Variation

There are many applications involving equations like $y = mx$, where m is some positive number. In such situations, we say that we have **direct variation**, and m (the slope) is called the **variation constant**, or **constant of proportionality**. Generally, only positive values of x and y are considered.

> **DEFINITION**
>
> The variable y **varies directly** as x if there is some positive constant m such that $y = mx$. We also say that y is **directly proportional** to x.

■ **EXAMPLE 3** **Life Science: Weight on Earth and the Moon.** The weight M, in pounds, of an object on the moon is directly proportional to the weight E of that object on Earth. An astronaut who weighs 180 lb on Earth will weigh 28.8 lb on the moon.

a) Find an equation of variation.

b) An astronaut weighs 120 lb on Earth. How much will the astronaut weigh on the moon?

Solution

a) The equation has the form $M = mE$. To find m, we substitute:

$$M = mE$$
$$28.8 = m \cdot 180$$
$$\frac{28.8}{180} = m$$
$$0.16 = m.$$

Thus, $M = 0.16E$ is the equation of variation.

b) To find the weight on the moon of an astronaut who weighs 120 lb on Earth, we substitute 120 for E in the equation of variation,

$$M = 0.16 \cdot 120, \qquad \text{Substituting 120 for } E$$

and get

$$M = 19.2.$$

Thus, an astronaut who weighs 120 lb on Earth weighs 19.2 lb on the moon.

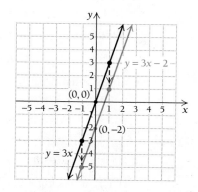

The Equation $y = mx + b$

Compare the graphs of the equations

$$y = 3x \quad \text{and} \quad y = 3x - 2$$

(see the following figure). Note that the graph of $y = 3x - 2$ is a shift 2 units down of the graph of $y = 3x$, and that $y = 3x - 2$ has y-intercept $(0, -2)$. Both graphs represent functions.

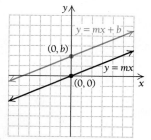

DEFINITION

A **linear function** is given by

$$y = mx + b \quad \text{or} \quad f(x) = mx + b$$

and has a graph that is the straight line parallel to the graph of $y = mx$ and crossing the y-axis at $(0, b)$. The point $(0, b)$ is called the **y-intercept**. (See the figure at the left.)

As before, the constant m is the slope of the line. When $m = 0$, $y = 0x + b = b$, and we have a constant function (see Theorem 3 at the beginning of this section). The graph of such a function is a horizontal line.

The Slope–Intercept Equation

Every nonvertical line l is uniquely determined by its slope m and its y-intercept $(0, b)$. In other words, the slope describes the "slant" of the line, and the y-intercept locates the point at which the line crosses the y-axis. Thus, we have the following definition.

DEFINITION

$y = mx + b$ is called the **slope–intercept equation** of a line.

■ **EXAMPLE 4**   Find the slope and the y-intercept of the graph of $2x - 4y - 7 = 0$.

Solution   We solve for y:

$$2x - 4y - 7 = 0$$

$$4y = 2x - 7 \qquad \text{Adding 4y to both sides}$$

$$y = \frac{2}{4}x - \frac{7}{4} \qquad \text{Dividing both sides by 4}$$

Slope: $\tfrac{1}{2}$ y-intercept: $\left(0, -\tfrac{7}{4}\right)$

> **Quick Check 3**
>
> Find the slope and the y-intercept of the graph of $3x - 6y - 7 = 0$.

❮ Quick Check 3

The Point–Slope Equation

Suppose that we know the slope of a line and some point on the line other than the y-intercept. We can still find an equation of the line.

■ **EXAMPLE 5**   Find an equation of the line with slope 3 containing the point $(-1, -5)$.

Solution   The slope is given as $m = 3$. From the slope–intercept equation, we have

$$y = 3x + b, \tag{1}$$

so we must determine b. Since $(-1, -5)$ is on the line, we substitute -5 for y and -1 for x:

$$-5 = 3(-1) + b$$

$$-5 = -3 + b,$$

so $-2 = b$

Then, replacing b in equation (1) with -2, we get $y = 3x - 2$.

More generally, if a point (x_1, y_1) is on the line given by

$$y = mx + b, \tag{2}$$

it must follow that

$$y_1 = mx_1 + b. \tag{3}$$

Subtracting the left and right sides of equation (3) from the left and right sides, respectively, of equation (2), we have

$$y - y_1 = (mx + b) - (mx_1 + b)$$
$$= mx + b - mx_1 - b \qquad \text{Multiplying by } -1$$
$$= mx - mx_1 \qquad \text{Combining like terms}$$
$$= m(x - x_1). \qquad \text{Factoring}$$

DEFINITION

$y - y_1 = m(x - x_1)$ is called the **point–slope equation** of a line. The point is (x_1, y_1), and the slope is m.

This definition allows us to write an equation of a line given its slope and the coordinates of *any* point on the line.

■ **EXAMPLE 6**   Find an equation of the line with slope $\frac{2}{3}$ containing the point $(-1, -5)$.

Solution   Substituting in

$$y - y_1 = m(x - x_1),$$

we get

$$y - (-5) = \tfrac{2}{3}[x - (-1)]$$
$$y + 5 = \tfrac{2}{3}(x + 1)$$
$$y + 5 = \tfrac{2}{3}x + \tfrac{2}{3} \qquad \text{Multiplying by } \tfrac{2}{3}$$
$$y = \tfrac{2}{3}x + \tfrac{2}{3} - 5 \qquad \text{Subtracting 5}$$
$$y = \tfrac{2}{3}x + \tfrac{2}{3} - \tfrac{15}{3}$$
$$y = \tfrac{2}{3}x - \tfrac{13}{3}. \qquad \text{Combining like terms}$$

〉 **Quick Check 4**

Find the equation of the line with slope $-\frac{2}{3}$ containing the point $(-3, 6)$.

〈 Quick Check 4

Which lines have the same slope?

Computing Slope

We now determine a method of computing the slope of a line when we know the coordinates of two of its points. Suppose that (x_1, y_1) and (x_2, y_2) are the coordinates of two different points, P_1 and P_2, respectively, on a line that is not vertical. Consider a right triangle with legs parallel to the axes, as shown in the following figure.

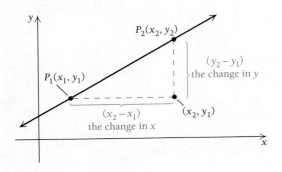

Note that the change in y is $y_2 - y_1$ and the change in x is $x_2 - x_1$. The ratio of these changes is the slope. To see this, consider the point–slope equation,

$$y - y_1 = m(x - x_1).$$

Since (x_2, y_2) is on the line, it must follow that

$$y_2 - y_1 = m(x_2 - x_1). \qquad \text{Substituting}$$

Since the line is not vertical, the two x-coordinates must be different; thus, $x_2 - x_1$ is nonzero, and we can divide by it to get the following theorem.

THEOREM 5

The slope of a line containing points (x_1, y_1) and (x_2, y_2) is

$$m = \frac{y_2 - y_1}{x_2 - x_1} = \frac{\text{change in } y}{\text{change in } x}.$$

■ **EXAMPLE 7**   Find the slope of the line containing the points $(-2, 6)$ and $(-4, 9)$.

Solution   We have

$$m = \frac{y_2 - y_1}{x_2 - x_1} = \frac{6 - 9}{-2 - (-4)} \qquad \text{We treated } (-2, 6) \text{ as } P_2 \text{ and } (-4, 9) \text{ as } P_1.$$

$$= \frac{-3}{2} = -\frac{3}{2}.$$

Note that it does not matter which point is taken first, so long as we subtract the coordinates in the same order. In this example, we can also find m as follows:

> **Quick Check 5**
>
> Find the slope of the line containing the points $(2, 3)$ and $(1, -4)$.

$$m = \frac{9 - 6}{-4 - (-2)} = \frac{3}{-2} = -\frac{3}{2}. \qquad \text{Here, } (-4, 9) \text{ serves as } P_2, \text{ and} (-2, 6) \text{ serves as } P_1.$$

❮ Quick Check 5

If a line is horizontal, the change in y for any two points is 0. Thus, a horizontal line has slope 0. If a line is vertical, the change in x for any two points is 0. Thus, the slope is *not defined* because we cannot divide by 0. A vertical line has undefined slope. Thus, "0 slope" and "undefined slope" are two very different concepts.

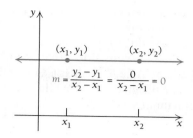

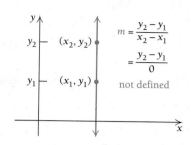

At top, two coordinate graphs illustrating slope formulas:

Left graph: points (x_1, y_1) and (x_2, y_2) on a horizontal line,
$$m = \frac{y_2 - y_1}{x_2 - x_1} = \frac{0}{x_2 - x_1} = 0$$

Right graph: points (x_2, y_2) and (x_1, y_1) on a vertical line,
$$m = \frac{y_2 - y_1}{x_2 - x_1} = \frac{y_2 - y_1}{0}$$
not defined

Applications of Slope

Slope has many real-world applications. For example, numbers like 2%, 3%, and 6% are often used to represent the *grade* of a road, a measure of how steep a road on a hill is. A 3% grade $\left(3\% = \frac{3}{100}\right)$ means that for every horizontal distance of 100 ft, the road rises 3 ft. In architecture, the *pitch* of a roof is a measure of how steeply it is angled—a steep pitch sheds more snow than a shallow pitch. Wheelchair-ramp design also involves slope: Building codes rarely allow the steepness of a wheelchair ramp to exceed $\frac{1}{12}$.

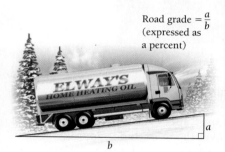

Road grade $= \dfrac{a}{b}$
(expressed as a percent)

24 ft   2 ft

Ski trail difficulty ratings, or *gradients*, are yet another application of slope. The following table presents examples.

Ski Trail Difficulty Ratings in North America

Trail Rating	Symbol	Level of Difficulty	Description
Green Circle	●	Easiest	A Green Circle trail is the easiest. These trails are generally wide and groomed, typically with slope gradients ranging from 6% to 25% (a 100% slope is a 45° angle).
Blue Square	■	Intermediate	A Blue Square trail is of intermediate difficulty. These trails have gradients ranging from 25% to 40%. They are usually groomed and are usually among the most heavily used.
Black Diamond	◆	Difficult	Black Diamond trails tend to be steep (typically 40% and up), may or may not be groomed, and are among the most difficult.

There are even more difficult ski trails. The rating of a trail is done at the discretion of the ski resort operators. There are a number of iPod and iPhone apps that skiers can use to estimate difficulty ratings. To estimate a gradient, hold your arm parallel to the ground out from your side—that is a 0% gradient. Hold it at a 45° angle—that is a 100% gradient. A 22.5° angle is a 41% gradient. And, surprisingly, an angle of only 3.5° constitutes a 6% gradient. What do you think the slope is of the steep road at the top of the mountain in the photo?

Frenchman Mountain near Las Vegas, Nevada.

Slope can also be considered as an **average rate of change**.

■ **EXAMPLE 8**   **Life Science: Amount Spent on Cancer Research.**   The amount spent on cancer research has increased steadily over the years and is approximated in the following graph. Find the average rate of change of the amount spent on research.

CANCER RESEARCH

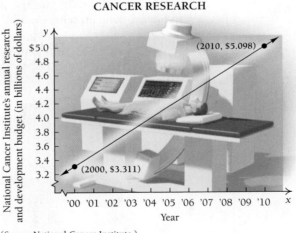

(*Source*: National Cancer Institute.)

Solution   First, we determine the coordinates of two points on the graph. In this case, they are given as (2000, \$3.311) and (2010, \$5.098). Then we compute the slope, or rate of change, as follows:

$$\text{Slope} = \text{average rate of change} = \frac{\text{change in } y}{\text{change in } x}$$

$$= \frac{\$5.098 - \$3.311}{2010 - 2000} = \frac{1.787}{10} \approx \$0.1787 \text{ billion/yr.}$$

Applications of Linear Functions

Many applications are modeled by linear functions.

■ EXAMPLE 9   **Business: Total Cost.**   Raggs, Ltd., a clothing firm, has **fixed costs** of $10,000 per year. These costs, such as rent, maintenance, and so on, must be paid no matter how much the company produces. To produce x units of a certain kind of suit, it costs $20 per suit (unit) in addition to the fixed costs. That is, the **variable costs** for producing x of these suits are $20x$ dollars. These costs are due to the amount produced and stem from items such as material, wages, fuel, and so on. The **total cost** $C(x)$ of producing x suits in a year is given by a function C:

$$C(x) = (\text{Variable costs}) + (\text{Fixed costs}) = 20x + 10{,}000.$$

a) Graph the variable-cost, the fixed-cost, and the total-cost functions.

b) What is the total cost of producing 100 suits? 400 suits?

Solution

a) The variable-cost and fixed-cost functions appear in the graph on the left below. The total-cost function is shown in the graph on the right. From a practical standpoint, the domains of these functions are nonnegative integers 0, 1, 2, 3, and so on, since it does not make sense to make either a negative number or a fractional number of suits. It is common practice to draw the graphs as though the domains were the entire set of nonnegative real numbers.

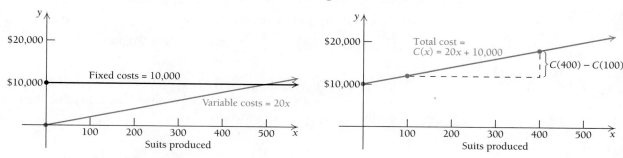

b) The total cost of producing 100 suits is

$$C(100) = 20 \cdot 100 + 10{,}000 = \$12{,}000.$$

The total cost of producing 400 suits is

$$C(400) = 20 \cdot 400 + 10{,}000$$
$$= \$18{,}000.$$

■ EXAMPLE 10   **Business: Profit-and-Loss Analysis.**   When a business sells an item, it receives the *price* paid by the consumer (this is normally greater than the *cost* to the business of producing the item).

a) The **total revenue** that a business receives is the product of the number of items sold and the price paid per item. Thus, if Raggs, Ltd., sells x suits at $80 per suit, the total revenue $R(x)$, in dollars, is given by

$$R(x) = \text{Unit price} \cdot \text{Quantity sold} = 80x.$$

If $C(x) = 20x + 10{,}000$ (see Example 9), graph R and C using the same set of axes.

b) The **total profit** that a business receives is the amount left after all costs have been subtracted from the total revenue. Thus, if $P(x)$ represents the total profit when x items are produced and sold, we have

$$P(x) = (\text{Total revenue}) - (\text{Total costs}) = R(x) - C(x).$$

Determine $P(x)$ and draw its graph using the same set of axes as was used for the graph in part (a).

c) The company will *break even* at that value of x for which $P(x) = 0$ (that is, no profit and no loss). This is the point at which $R(x) = C(x)$. Find the **break-even value** of x.

Solution

a) The graphs of $R(x) = 80x$ and $C(x) = 20x + 10,000$ are shown below. When $C(x)$ is above $R(x)$, a loss will occur. This is shown by the region shaded red. When $R(x)$ is above $C(x)$, a gain will occur. This is shown by the region shaded gray.

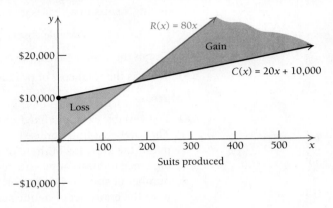

b) To find P, the profit function, we have

$$P(x) = R(x) - C(x) = 80x - (20x + 10,000)$$
$$= 60x - 10,000.$$

The graph of $P(x)$ is shown by the heavy line. The red portion of the line shows a "negative" profit, or loss. The black portion of the heavy line shows a "positive" profit, or gain.

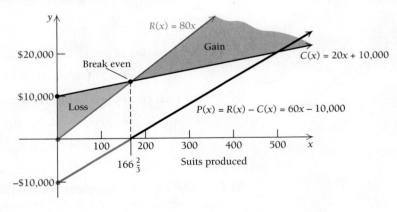

c) To find the break-even value, we solve $R(x) = C(x)$:

$$R(x) = C(x)$$
$$80x = 20x + 10,000$$
$$60x = 10,000$$
$$x = 166\tfrac{2}{3}.$$

How do we interpret the fractional answer, since it is not possible to produce $\tfrac{2}{3}$ of a suit? We simply round up to 167. Estimates of break-even values are usually sufficient since companies want to operate well away from break-even values in order to maximize profit.

> **Quick Check 6**
>
> **Business.** Suppose that in Examples 9 and 10 fixed costs are increased to $20,000. Find:
>
> **a)** the total-cost, total-revenue, and total-profit functions;
>
> **b)** the break-even value.

❮ Quick Check 6

Section Summary

- Graphs of functions that are straight lines (*linear functions*) are characterized by an equation of the type

$f(x) = mx + b$, where m is the slope and $(0, b)$ is the *y-intercept*, the point at which the graph crosses the *y*-axis.

EXERCISE SET
R.4

Graph.

1. $x = 3$
2. $x = 5$
3. $y = -2$
4. $y = -4$
5. $x = -4.5$
6. $x = -1.5$
7. $y = 3.75$
8. $y = 2.25$

Graph. List the slope and y-intercept.

9. $y = -2x$
10. $y = -3x$
11. $f(x) = 0.5x$
12. $f(x) = -0.5x$
13. $y = 3x - 4$
14. $y = 2x - 5$
15. $g(x) = -x + 3$
16. $g(x) = x - 2.5$
17. $y = 7$
18. $y = -5$

Find the slope and y-intercept.

19. $y - 3x = 6$
20. $y - 4x = 1$
21. $2x + y - 3 = 0$
22. $2x - y + 3 = 0$
23. $2x + 2y + 8 = 0$
24. $3x - 3y + 6 = 0$
25. $x = 3y + 7$
26. $x = -4y + 3$

Find an equation of the line:

27. with $m = -5$, containing $(-2, -3)$.
28. with $m = 7$, containing $(1, 7)$.
29. with $m = -2$, containing $(2, 3)$.
30. with $m = -3$, containing $(5, -2)$.
31. with slope 2, containing $(3, 0)$.
32. with slope -5, containing $(5, 0)$.
33. with *y*-intercept $(0, -6)$ and slope $\frac{1}{2}$.
34. with *y*-intercept $(0, 7)$ and slope $\frac{4}{3}$.
35. with slope 0, containing $(2, 3)$.
36. with slope 0, containing $(4, 8)$.

Find the slope of the line containing the given pair of points. If a slope is undefined, state that fact.

37. $(5, -3)$ and $(-2, 1)$
38. $(-2, 1)$ and $(6, 3)$
39. $(2, -3)$ and $(-1, -4)$
40. $(-3, -5)$ and $(1, -6)$
41. $(3, -7)$ and $(3, -9)$
42. $(-4, 2)$ and $(-4, 10)$
43. $\left(\frac{4}{5}, -3\right)$ and $\left(\frac{1}{2}, \frac{2}{5}\right)$
44. $\left(-\frac{3}{16}, -\frac{1}{2}\right)$ and $\left(\frac{5}{8}, -\frac{3}{4}\right)$
45. $(2, 3)$ and $(-1, 3)$
46. $\left(-6, \frac{1}{2}\right)$ and $\left(-7, \frac{1}{2}\right)$
47. $(x, 3x)$ and $(x + h, 3(x + h))$
48. $(x, 4x)$ and $(x + h, 4(x + h))$
49. $(x, 2x + 3)$ and $(x + h, 2(x + h) + 3)$
50. $(x, 3x - 1)$ and $(x + h, 3(x + h) - 1)$

51–60. *Find an equation of the line containing the pair of points in each of Exercises 37–46.*

61. Find the slope of the skateboard ramp.

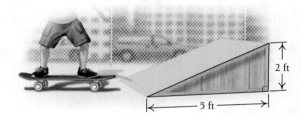

62. Find the slope (or grade) of the treadmill.

63. Find the slope (or head) of the river. Express the answer as a percentage.

APPLICATIONS

Business and Economics

64. Highway tolls. It has been suggested that since heavier vehicles are responsible for more of the wear and tear on highways, drivers should pay tolls in direct proportion to the weight of their vehicles. Suppose that a Toyota Camry weighing 3350 lb was charged $2.70 for traveling an 80-mile stretch of highway.

   a) Find an equation of variation that expresses the amount of the toll T as a function of the vehicle's weight w.

   b) What would the toll be if a 3700-lb Jeep Cherokee drove the same stretch of highway?

65. Inkjet cartridges. A registrar's office finds that the number of inkjet cartridges, I, required each year for its copiers and printers varies directly with the number of students enrolled, s.

   a) Find an equation of variation that expresses I as a function of s, if the office requires 16 cartridges when 2800 students enroll.

   b) How many cartridges would be required if 3100 students enrolled?

66. Profit-and-loss analysis. Boxowitz, Inc., a computer firm, is planning to sell a new graphing calculator. For the first year, the fixed costs for setting up the new production line are $100,000. The variable costs for producing each calculator are estimated at $20. The sales department projects that 150,000 calculators can be sold during the first year at a price of $45 each.

   a) Find and graph $C(x)$, the total cost of producing x calculators.

   b) Using the same axes as in part (a), find and graph $R(x)$, the total revenue from the sale of x calculators.

   c) Using the same axes as in part (a), find and graph $P(x)$, the total profit from the production and sale of x calculators.

   d) What profit or loss will the firm realize if the expected sale of 150,000 calculators occurs?

   e) How many calculators must the firm sell in order to break even?

67. Profit-and-loss analysis. Red Tide is planning a new line of skis. For the first year, the fixed costs for setting up production are $45,000. The variable costs for producing each pair of skis are estimated at $80, and the selling price will be $255 per pair. It is projected that 3000 pairs will sell the first year.

   a) Find and graph $C(x)$, the total cost of producing x pairs of skis.

   b) Find and graph $R(x)$, the total revenue from the sale of x pairs of skis. Use the same axes as in part (a).

   c) Using the same axes as in part (a), find and graph $P(x)$, the total profit from the production and sale of x pairs of skis.

   d) What profit or loss will the company realize if the expected sale of 3000 pairs occurs?

   e) How many pairs must the company sell in order to break even?

68. Straight-line depreciation. Quick Copy buys an office machine for $5200 on January 1 of a given year. The machine is expected to last for 8 yr, at the end of which time its *salvage value* will be $1100. If the company figures the decline in value to be the same each year, then the *book value*, $V(t)$, after t years, $0 \le t \le 8$, is given by

$$V(t) = C - t\left(\frac{C - S}{N}\right),$$

where C is the original cost of the item, N is the number of years of expected life, and S is the salvage value.

   a) Find the linear function for the straight-line depreciation of the office machine.

   b) Find the book value after 0 yr, 1 yr, 2 yr, 3 yr, 4 yr, 7 yr, and 8 yr.

69. Profit-and-loss analysis. Jimmy decides to mow lawns to earn money. The initial cost of his lawnmower is $250. Gasoline and maintenance costs are $4 per lawn.

   a) Formulate a function $C(x)$ for the total cost of mowing x lawns.

   b) Jimmy determines that the total-profit function for the lawnmowing business is given by $P(x) = 9x - 250$. Find a function for the total revenue from mowing x lawns. How much does Jimmy charge per lawn?

   c) How many lawns must Jimmy mow before he begins making a profit?

70. Straight-line depreciation. (See Exercise 68.) A business tenant spends $40 per square foot on improvements to a 25,000-ft^2 office space. Under IRS guidelines for straight-line depreciation, these improvements will depreciate completely—that is, have zero salvage value—after 39 yr. Find the depreciated value of the improvements after 10 yr.

71. Book value. (See Exercise 68.) The Video Wizard buys a new computer system for $60,000 and projects that its book value will be $2000 after 5 yr. Using straight-line depreciation, find the book value after 3 yr.

72. Book value. Tyline Electric uses the function $B(t) = -700t + 3500$ to find the book value, $B(t)$, in dollars, of a photocopier t years after its purchase.

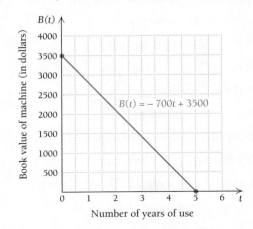

   a) What do the numbers -700 and 3500 signify?

   b) How long will it take the copier to depreciate completely?

   c) What is the domain of B? Explain.

General Interest

73. Stair requirements. A North Carolina state law requires that stairs have minimum treads of 9 in. and maximum risers of 8.25 in. (*Source*: North Carolina Office of the State Fire Marshal.) See the illustration below. According to this law, what is the maximum grade of stairs in North Carolina?

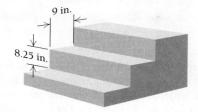

74. Health insurance premiums. Find the average rate of change in the annual premium for a family's health insurance.

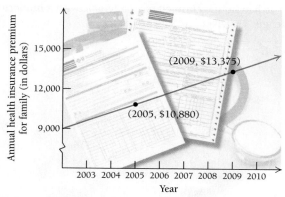

(*Source*: The Kaiser Family Foundation; Health Research and Education Trust.)

75. Health insurance premiums. Find the average rate of change in the annual premium for a single person.

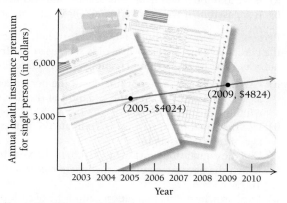

(*Source*: The Kaiser Family Foundation; Health Research and Education Trust.)

76. Two-year college tuitions. Find the average rate of change of the tuition and fees at public two-year colleges.

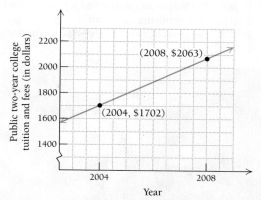

(*Source*: U.S. National Center for Education Statistics, *Digest of Education Statistics*, annual.)

77. Wedding cost. Find the average rate of change of the cost of a formal wedding.

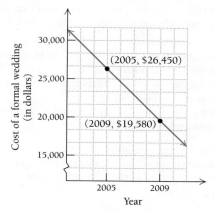

(*Source*: The Fairchild Bridal Group.)

78. Energy conservation. The R-factor of home insulation is directly proportional to its thickness T.

a) Find an equation of variation if $R = 12.51$ when $T = 3$ in.

b) What is the R-factor for insulation that is 6 in. thick?

79. Nerve impulse speed. Impulses in nerve fibers travel at a speed of 293 ft/sec. The distance D, in feet, traveled in t sec is given by $D = 293t$. How long would it take an impulse to travel from the brain to the toes of a person who is 6 ft tall?

80. Muscle weight. The weight M of the muscles in a human is directly proportional to the person's body weight W.

a) It is known that a person who weighs 200 lb has 80 lb of muscles. Find an equation of variation expressing M as a function of W.

b) Express the variation constant as a percent, and interpret the resulting equation.

c) What is the muscle weight of a person who weighs 120 lb?

Muscle weight is directly proportional to body weight.

81. Brain weight. The weight B of a human's brain is directly proportional to a person's body weight W.

a) It is known that a person who weighs 120 lb has a brain that weighs 3 lb. Find an equation of variation expressing B as a function of W.

b) Express the variation constant as a percent and interpret the resulting equation.

c) What is the weight of the brain of a person who weighs 160 lb?

82. Stopping distance on glare ice. The stopping distance (at some fixed speed) of regular tires on glare ice is given by a linear function of the air temperature F,

$$D(F) = 2F + 115,$$

where $D(F)$ is the stopping distance, in feet, when the air temperature is F, in degrees Fahrenheit.

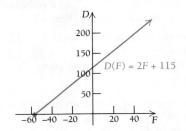

a) Find $D(0°)$, $D(-20°)$, $D(10°)$, and $D(32°)$.

b) Explain why the domain should be restricted to the interval $[-57.5°, 32°]$.

83. Reaction time. While driving a car, you see a child suddenly crossing the street. Your brain registers the emergency and sends a signal to your foot to hit the brake. The car travels a distance D, in feet, during this time, where D is a function of the speed r, in miles per hour, that the car is traveling when you see the child. That reaction distance is a linear function given by

$$D(r) = \frac{11r + 5}{10}.$$

a) Find $D(5)$, $D(10)$, $D(20)$, $D(50)$, and $D(65)$.

b) Graph $D(r)$.

c) What is the domain of the function? Explain.

84. Estimating heights. An anthropologist can use certain linear functions to estimate the height of a male or female, given the length of certain bones. The *humerus* is the bone from the elbow to the shoulder. Let $x =$ the length of the humerus, in centimeters. Then the height, in centimeters, of a male with a humerus of length x is given by

$$M(x) = 2.89x + 70.64.$$

The height, in centimeters, of a female with a humerus of length x is given by

$$F(x) = 2.75x + 71.48.$$

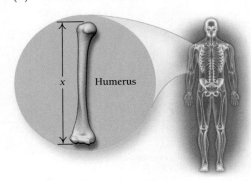

A 26-cm humerus was uncovered in some ruins.

a) If we assume it was from a male, how tall was he?

b) If we assume it was from a female, how tall was she?

85. Percentage of young adults using the Internet. In 2000, the percentage of 18- to 29-year-olds who used the Internet was 72%. In 2009, that percentage had risen to 92%.

a) Use the year as the x-coordinate and the percentage as the y-coordinate. Find the equation of the line that contains the data points.

b) Use the equation in part (a) to estimate the percentage of Internet users in 2010.

c) Use the equation in part (a) to estimate the year in which the percentage of Internet users will reach 100%.

d) Explain why a linear equation cannot be used for years after the year found in part (c).

86. Manatee population. In January 2005, 3143 manatees were counted in an aerial survey of Florida. In January 2009, 3802 manatees were counted. (*Source*: Florida Fish and Wildlife Conservation Commission.)

a) Using the year as the x-coordinate and the number of manatees as the y-coordinate, find an equation of the line that contains the two data points.

b) Use the equation in part (a) to estimate the number of manatees counted in January 2011.

c) The actual number counted in January 2011 was 4834. Does the equation found in part (a) give an accurate representation of the number of manatees counted each year?

87. Urban population. The population of Woodland is P. After a growth of 2%, its new population is N.

a) Assuming that N is directly proportional to P, find an equation of variation.

b) Find N when $P = 200,000$.

c) Find P when $N = 367,200$.

88. Median age of women at first marriage. In general, people in our society are marrying at a later age. The median age, $A(t)$, of women at first marriage can be approximated by the linear function

$$A(t) = 0.08t + 19.7,$$

where t is the number of years after 1950. Thus, $A(0)$ is the median age of women at first marriage in 1950, $A(50)$ is the median age in 2000, and so on.

a) Find $A(0), A(1), A(10), A(30)$, and $A(50)$.

b) What was the median age of women at first marriage in 2008?

c) Graph $A(t)$.

SYNTHESIS

89. Explain and compare the situations in which you would use the slope–intercept equation rather than the point–slope equation.

90. Discuss and relate the concepts of fixed cost, total cost, total revenue, and total profit.

91. Business: daily sales. Match each sentence below with the most appropriate of the following graphs (I, II, III, or IV).

a) After January 1, daily sales continued to rise, but at a slower rate.

b) After January 1, sales decreased faster than they ever grew.

c) The rate of growth in daily sales doubled after January 1.

d) After January 1, daily sales decreased at half the rate that they grew in December.

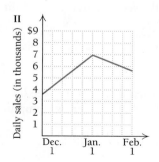

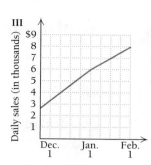

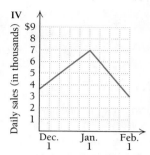

92. Business: depreciation. A large crane is being depreciated according to the model $V(t) = 900 - 60t$, where $V(t)$ is measured in thousands of dollars and t is the number of years since 2005. If the crane is to be depreciated until its value is $0, what is the domain of the depreciation model?

TECHNOLOGY CONNECTION

93. Graph some of the total-revenue, total-cost, and total-profit functions in this exercise set using the same set of axes. Identify regions of profit and loss.

Answers to Quick Checks

1.

2.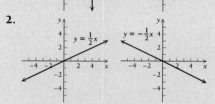

3. $m = \dfrac{1}{2}$, y-intercept: $\left(0, -\dfrac{7}{6}\right)$

4. $y = -\dfrac{2}{3}x + 4$ **5.** 7

6. (a) $C(x) = 20x + 20,000; R(x) = 80x;$ $P(x) = R(x) - C(x) = 60x - 20,000$ **(b)** 333 suits

R.5

Nonlinear Functions and Models

There are many functions that have graphs that are not lines. In this section, we study some of these **nonlinear functions** that we will frequently encounter throughout this course.

Quadratic Functions

DEFINITION

A **quadratic function** f is given by

$$f(x) = ax^2 + bx + c, \quad \text{where } a \neq 0.$$

We have already used quadratic functions—for example, $f(x) = x^2$ and $g(x) = x^2 - 1$. We can create hand-drawn graphs of quadratic functions using the following information.

The graph of a quadratic function $f(x) = ax^2 + bx + c$ is called a **parabola.**

a) It is always a cup-shaped curve, like those in Examples 1 and 2 that follow.

b) It opens upward if $a > 0$ or opens downward if $a < 0$.

c) It has a turning point, or **vertex**, whose first coordinate is

$$x = -\frac{b}{2a}.$$

d) The vertical line $x = -b/(2a)$ (not part of the graph) is the line of symmetry.

■ **EXAMPLE 1**   Graph: $f(x) = x^2 - 2x - 3$.

Solution   Note that for $f(x) = 1x^2 - 2x - 3$, we have $a = 1, b = -2$, and $c = -3$. Since $a > 0$, the graph opens upward. Let's next find the vertex, or turning point. The x-coordinate of the vertex is

$$x = -\frac{b}{2a}$$

$$= -\frac{-2}{2(1)} = 1.$$

Substituting 1 for x, we find the second coordinate of the vertex, $f(1)$:

$$f(1) = 1^2 - 2(1) - 3$$
$$= 1 - 2 - 3$$
$$= -4.$$

The vertex is $(1, -4)$. The vertical line $x = 1$ is the line of symmetry of the graph. We choose some x-values on each side of the vertex, compute y-values, plot the points, and graph the parabola.

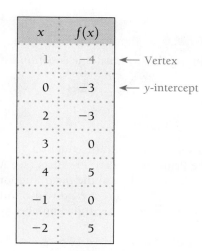

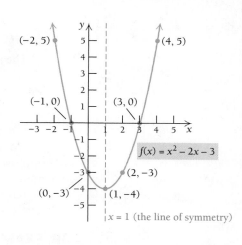

x	$f(x)$
1	-4
0	-3
2	-3
3	0
4	5
-1	0
-2	5

■ **EXAMPLE 2** Graph: $f(x) = -2x^2 + 10x - 7$.

Solution We first note that $a = -2$, and since $a < 0$, the graph will open downward. Let's next find the vertex, or turning point. The x-coordinate of the vertex is

$$x = -\frac{b}{2a}$$

$$= -\frac{10}{2(-2)} = \frac{5}{2}.$$

Substituting $\frac{5}{2}$ for x in the equation, we find the second coordinate of the vertex:

$$y = f\left(\tfrac{5}{2}\right) = -2\left(\tfrac{5}{2}\right)^2 + 10\left(\tfrac{5}{2}\right) - 7$$
$$= -2\left(\tfrac{25}{4}\right) + 25 - 7$$
$$= \tfrac{11}{2}.$$

The vertex is $\left(\frac{5}{2}, \frac{11}{2}\right)$, and the line of symmetry is $x = \frac{5}{2}$. We choose some x-values on each side of the vertex, compute y-values, plot the points, and graph the parabola:

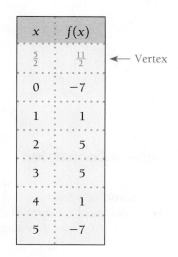

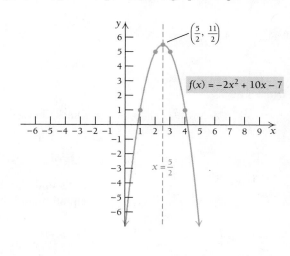

x	$f(x)$
$\frac{5}{2}$	$\frac{11}{2}$
0	-7
1	1
2	5
3	5
4	1
5	-7

⟨ Quick Check 1

TECHNOLOGY CONNECTION

EXERCISES

Using the procedure of Examples 1 and 2, graph each of the following by hand, using the TABLE feature to create an input–output table for each function. Then press GRAPH to check your sketch.

1. $f(x) = x^2 - 6x + 4$

2. $f(x) = -2x^2 + 4x + 1$

⟩ **Quick Check 1**

Graph each function:

a) $f(x) = x^2 + 2x - 3$;

b) $f(x) = -2x^2 - 10x - 5$.

First coordinates of points at which a quadratic function intersects the x-axis (x-intercepts), if they exist, can be found by solving the quadratic equation, $ax^2 + bx + c = 0$. If real-number solutions exist, they can be found using the *quadratic formula*. See Appendix A at the end of the book for additional review of this important result.

THEOREM 6 **The Quadratic Formula**

The solutions of any quadratic equation $ax^2 + bx + c = 0, a \neq 0$, are given by

$$x = \frac{-b \pm \sqrt{b^2 - 4ac}}{2a}.$$

When solving a quadratic equation, $ax^2 + bx + c = 0, a \neq 0$, first try to factor and use the Principle of Zero Products (see Appendix A). When factoring is not possible or seems difficult, use the quadratic formula. It will always give the solutions. When $b^2 - 4ac < 0$, there are no real-number solutions and thus no x-intercepts. There are solutions in an expanded number system called the *complex numbers*.

■ **EXAMPLE 3** Solve: $3x^2 - 4x = 2$.

Solution We first find the standard form $ax^2 + bx + c = 0$, and then determine a, b, and c:

$$3x^2 - 4x - 2 = 0,$$
$$a = 3, \quad b = -4, \quad c = -2.$$

We then use the quadratic formula:

$$\begin{aligned}
x &= \frac{-b \pm \sqrt{b^2 - 4ac}}{2a} \\
&= \frac{-(-4) \pm \sqrt{(-4)^2 - 4(3)(-2)}}{2 \cdot 3} \quad \text{Substituting} \\
&= \frac{4 \pm \sqrt{16 + 24}}{6} = \frac{4 \pm \sqrt{40}}{6} \quad \text{Simplifying} \\
&= \frac{4 \pm \sqrt{4 \cdot 10}}{6} = \frac{4 \pm 2\sqrt{10}}{6} \\
&= \frac{2(2 \pm \sqrt{10})}{2 \cdot 3} \\
&= \frac{2 \pm \sqrt{10}}{3}.
\end{aligned}$$

The solutions are $(2 + \sqrt{10})/3$ and $(2 - \sqrt{10})/3$, or approximately 1.721 and -0.387.

〉 **Quick Check 2**

Solve: $3x^2 + 2x = 7$.

❮ Quick Check 2

Algebraic–Graphical Connection

Let's make an algebraic–graphical connection between the solutions of a quadratic equation and the x-intercepts of a quadratic function.

We just graphed equations of the form $f(x) = ax^2 + bx + c, a \neq 0$. Let's look at the graph of $f(x) = x^2 + 6x + 8$ and its x-intercepts, which follows.

EXERCISE

1. a) Below is the graph of
$$f(x) = x^2 - 6x + 8.$$

Using *only* the graph, find the solutions of $x^2 - 6x + 8 = 0$.

$y = x^2 - 6x + 8$

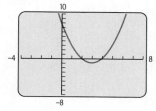

b) Using only a graph, find the solutions of $x^2 + 3x - 10 = 0$.

c) Use the TABLE feature to check your answers to parts (a) and (b).

c) Use a graph and a table to check the solutions of $x^2 + 6x + 8 = 0$, solved algebraically on the right.

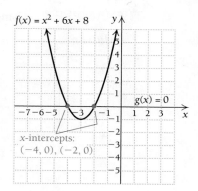

The **x-intercepts**, $(-4, 0)$ and $(-2, 0)$, are the points at which the graph crosses the x-axis. These pairs are also the points of intersection of the graphs of $f(x) = x^2 + 6x + 8$ and $g(x) = 0$ (the x-axis). The x-values, -4 and -2, can be found by solving $f(x) = g(x)$:

$$x^2 + 6x + 8 = 0$$
$$(x + 4)(x + 2) = 0 \qquad \text{Factoring; there is no need for the quadratic formula here.}$$
$$x + 4 = 0 \quad or \quad x + 2 = 0 \qquad \text{Principle of Zero Products}$$
$$x = -4 \quad or \quad x = -2.$$

The solutions of $x^2 + 6x + 8 = 0$ are -4 and -2, which are the first coordinates of the x-intercepts, $(-4, 0)$ and $(-2, 0)$, of the graph of $f(x) = x^2 + 6x + 8$. A brief review of factoring can be found in Appendix A at the end of the book.

Polynomial Functions

Linear and quadratic functions are part of a general class of *polynomial functions*.

DEFINITION

A **polynomial function** f is given by

$$f(x) = a_n x^n + a_{n-1} x^{n-1} + \cdots + a_2 x^2 + a_1 x^1 + a_0,$$

where n is a nonnegative integer and $a_n, a_{n-1}, \ldots, a_1, a_0$ are real numbers, called the **coefficients**.

The following are examples of polynomial functions:

$$f(x) = -5, \qquad \text{(A constant function)}$$
$$f(x) = 4x + 3, \qquad \text{(A linear function)}$$
$$f(x) = -x^2 + 2x + 3, \qquad \text{(A quadratic function)}$$
$$f(x) = 2x^3 - 4x^2 + x + 1. \qquad \text{(A cubic, or third-degree, function)}$$

In general, creating graphs of polynomial functions other than linear and quadratic functions is difficult without a calculator. We use calculus to sketch such graphs in Chapter 2. Some **power functions**, of the form

$$f(x) = ax^n,$$

are relatively easy to graph.

■ **EXAMPLE 4** Using the same set of axes, graph $f(x) = x^2$ and $g(x) = x^3$.

Solution We set up a table of values, plot the points, and then draw the graphs.

x	x^2	x^3
-2	4	-8
-1	1	-1
$-\frac{1}{2}$	$\frac{1}{4}$	$-\frac{1}{8}$
0	0	0
$\frac{1}{2}$	$\frac{1}{4}$	$\frac{1}{8}$
1	1	1
2	4	8

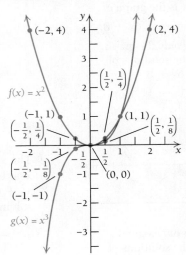

> **Quick Check 3**
>
> Graph each function using the same set of axes:
> $f(x) = 4 - x^2$ and
> $g(x) = x^3 - 1$.

❰ Quick Check 3

TECHNOLOGY CONNECTION 〰

Solving Polynomial Equations

The INTERSECT Feature

Consider solving the equation

$$x^3 = 3x + 1.$$

Doing so amounts to finding the x-coordinates of the point(s) of intersection of the graphs of

$$f(x) = x^3 \quad \text{and} \quad g(x) = 3x + 1.$$

We enter the functions as

$$y_1 = x^3 \quad \text{and} \quad y_2 = 3x + 1$$

and then graph. We use a $[-3, 3, -5, 8]$ window to see the curvature and possible points of intersection.

$$y_1 = x^3, \, y_2 = 3x + 1$$

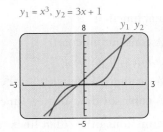

There appear to be at least three points of intersection. Using the INTERSECT feature in the CALC menu, we see that the point of intersection on the left is about $(-1.53, -3.60)$.

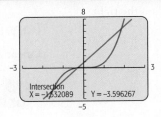

In a similar manner, we find the other points of intersection to be about $(-0.35, -0.04)$ and $(1.88, 6.64)$. The solutions of $x^3 = 3x + 1$ are the x-coordinates of these points, approximately

$$-1.53, -0.35, \text{ and } 1.88.$$

The ZERO Feature

The ZERO, or ROOT, feature can be used to solve an equation. The word "zero" in this context refers to an input, or x-value, for which the output of a function is 0. That is, c is a **zero** of the function f if $f(c) = 0$.

To use such a feature requires a 0 on one side of the equation. Thus, to solve $x^3 = 3x + 1$, we obtain $x^3 - 3x - 1 = 0$ by subtracting $3x + 1$ from both sides. Graphing $y = x^3 - 3x - 1$ and using the ZERO feature to find the zero on the left, we view a screen like the following.

(continued)

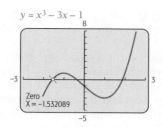

$y = x^3 - 3x - 1$

Zero
X = -1.532089

We see that $x^3 - 3x - 1 = 0$ when $x \approx -1.53$, so -1.53 is an approximate solution of the equation $x^3 = 3x + 1$. Proceeding in a similar manner, we can approximate the other solutions as -0.35 and 1.88. Note that the points of intersection of the graphs of f and g have the same x-values as the zeros of $x^3 - 3x - 1$.

EXERCISES

Using the INTERSECT feature, solve each equation.

1. $x^2 = 10 - 3x$

2. $2x + 24 = x^2$

3. $x^3 = 3x - 2$

4. $x^4 - 2x^2 = 0$

Using the ZERO feature, solve each equation.

5. $0.4x^2 = 280x$
(*Hint:* Use $[-200, 800, -100{,}000, 200{,}000]$.)

6. $\frac{1}{3}x^3 - \frac{1}{2}x^2 = 2x - 1$

7. $x^2 = 0.1x^4 + 0.4$

8. $0 = 2x^4 - 4x^2 + 2$

Find the zeros of each function.

9. $f(x) = 3x^2 - 4x - 2$

10. $f(x) = -x^3 + 6x^2 + 5$

11. $g(x) = x^4 + x^3 - 4x^2 - 2x + 4$

12. $g(x) = -x^4 + x^3 + 11x^2 - 9x - 18$

TECHNOLOGY CONNECTION

Apps for the iPhone and iPod Touch

The advent of the iPhone and other sophisticated mobile phones has made available many inexpensive mathematics applications. Two useful apps for the iPhone and iPod Touch are iPlot and Graphicus (which can be purchased at the iTunes Store). Each has more visually appealing displays than standard graphing calculators, but each also has limited capability. For example, neither iPlot nor Graphicus can do regression, as described in Section R.6.

iPlot

Among the features of this app is the ability to graph most of the functions we encounter in this book. There is a Zoom feature, a Trace feature that can be used to find function values, roots (zeros), and points of intersection of graphs, and a separate Root feature for finding zeros. Let's consider the function $f(x) = 2x^3 - x^4$ as an example.

To graph and find roots, first open the iPlot app. You will see a screen like that in Fig. 1. Notice the four icons at the bottom. The Functions icon is highlighted. Press ⊞ in the upper right; then enter the function as 2*x^3-x^4 (Fig. 2). Slide to the bottom and change the graph color if desired. Next, press Done in the upper right, followed by the Plot icon to obtain the graph (Fig. 3).

Below the graph are buttons for various options. To find a root of the function, press Root (firmly) so that it becomes highlighted. Then move the cursor close to a potential root (Fig. 4). Sometimes it is difficult to position the cursor directly on the root. But if you press the Apply icon in the

FIGURE 1   **FIGURE 2**

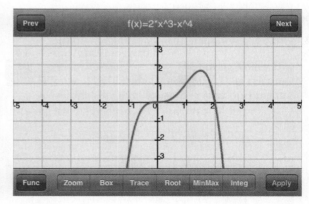

FIGURE 3

(continued)

Apps for the iPhone and iPod Touch (*continued*)

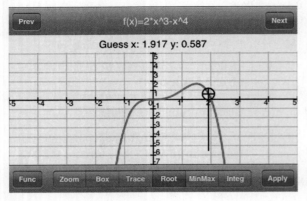

FIGURE 4

lower right once you are close, the cursor will jump to the answer, in this case, 2 (Fig. 5).

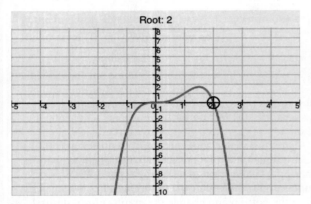

FIGURE 5

You will likely find it necessary to experiment with iPlot. Press some of the options buttons to explore other features. Use the Settings icon to change the window settings and make other modifications to the appearance. Missing symbols can be found by clicking on #+= at the left of the keypad.

Piecewise-defined functions can be entered, but an inappropriate vertical line may show up on the graph. The graph of a function like $f(x) = \dfrac{x^2 - 9}{x + 3}$ will not show the hole at $x = -3$. iPlot can graph more than one function on the same set of axes. Try graphing $g(x) = x^3 - 1$ along with $f(x) = 2x^3 - x^4$, and then pressing Trace to find points of intersection.

Sometimes iPlot will crash or lock up, giving an error message like "Unexpected End of Formula." If this occurs, go to the line where the function formula occurs, then press Edit function at the top to delete the function and enter it again.

For more information, consult the iPlot page at the iTunes Store or visit www.posimotion.com.

EXERCISES

Using iPlot, repeat Exercises 1–12 in the Technology Connection on page 55.

Graphicus

Graphicus is also very appealing visually and has the ability to graph most of the functions we encounter in this book. It can find roots and intersections, and it excels at many aspects of calculus, as we'll see later in this book. Let's look again at the function $f(x) = 2x^3 - x^4$.

To graph and find roots, first open Graphicus. Touch the blank rectangle at the top of the screen and enter the function as y(x)=2x^3-x^4. Press + in the upper right, and you will see the graph in Fig. 6. Notice the seven icons at the bottom. The one at the far left is for zooming. The fourth icon from the left is for finding roots. Touch it and note how quickly the roots of the function are highlighted (Fig. 7). Touch the symbol marking each root, and a box appears with its value; see Figs. 8 and 9.

Piecewise-defined functions cannot be entered. The graph of a function with a hole, such as $f(x) = \dfrac{x^2 - 9}{x + 3}$, will not show the hole at $x = -3$. Graphicus can graph more than one function on the same set of axes. Graph $g(x) = x^3 - 1$ along with $f(x) = 2x^3 - x^4$, and then press the fourth icon to find the points of intersection. Press some of the other icons to explore the uses of Graphicus.

For more information, consult the Graphicus page at the iTunes Store or visit www.facebook.com/pages/Graphicus/189699869029.

EXERCISES

Using Graphicus, repeat Exercises 1–12 in the Technology Connection on page 55.

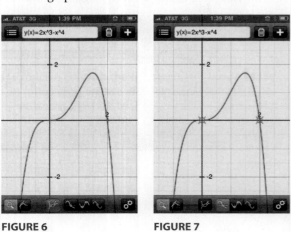

FIGURE 6   **FIGURE 7**

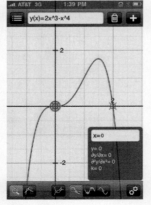

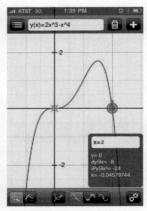

FIGURE 8   **FIGURE 9**

Rational Functions

> **DEFINITION**
>
> Functions given by the quotient, or ratio, of two polynomials are called **rational functions.**

The following are examples of rational functions:

$$f(x) = \frac{x^2 - 9}{x - 3}, \qquad h(x) = \frac{x - 3}{x^2 - x - 2},$$

$$g(x) = \frac{3x^2 - 4x}{2x + 10}, \qquad k(x) = \frac{x^3 - 2x + 7}{1} = x^3 - 2x + 7.$$

Note that as the function k illustrates, every polynomial function is also a rational function.

The domain of a rational function is restricted to those input values that do not result in division by zero. Thus, for f above, the domain consists of all real numbers except 3. To determine the domain of h, we set the denominator equal to 0 and solve:

$$x^2 - x - 2 = 0$$
$$(x + 1)(x - 2) = 0$$
$$x = -1 \quad or \quad x = 2.$$

Therefore, -1 and 2 are not in the domain. The domain of h consists of all real numbers except -1 and 2. The numbers -1 and 2 "split," or "separate," the intervals in the domain.

The graphing of most rational functions is rather complicated and is best dealt with using the tools of calculus that we will develop in Chapters 1 and 2. For now we will focus on graphs that are fairly basic and leave the more complicated graphs for Chapter 2.

■ **EXAMPLE 5**   Graph: $f(x) = \dfrac{x^2 - 9}{x - 3}.$

Solution   This particular function can be simplified before we graph it. We do so by factoring the numerator and removing a factor of 1 as follows:

$$f(x) = \frac{x^2 - 9}{x - 3} \qquad \text{Note that 3 is not in the domain of } f.$$

$$= \frac{(x - 3)(x + 3)}{x - 3}$$

$$= \frac{x - 3}{x - 3} \cdot \frac{x + 3}{1}$$

$$= x + 3, \quad x \neq 3. \qquad \text{We must specify } x \neq 3.$$

This simplification assumes that x is not 3. By writing $x \neq 3$, we indicate that for any x-value other than 3, the equation $f(x) = x + 3$ is used:

$$f(x) = x + 3, \quad x \neq 3.$$

If $y_1 = (x^2 - 9)/(x - 3)$ and $y_2 = x + 3$ are both graphed, the two graphs appear indistinguishable. To see this, graph both lines and use the arrow keys, ⌃ and ⌄, and the TRACE feature to move the cursor from line to line.

EXERCISES

1. Compare the results of using TRACE and entering the value 3.

2. Use the TABLE feature with TblStart set at −1 and Δ Tbl set at 1. How do y_1 and y_2 differ in the resulting table of values?

To find function values, we substitute any value for x other than 3. We make calculations as in the following table and draw the graph. The open circle at $(3, 6)$ indicates that this point is not part of the graph.

x	$f(x)$
−3	0
−2	1
−1	2
0	3
1	4
2	5
4	7

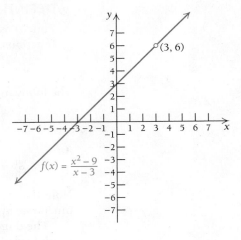

One important class of rational functions is given by $f(x) = k/x$, where k is a constant.

■ **EXAMPLE 6**   Graph: $f(x) = 1/x$.

Solution   We make a table of values, plot the points, and then draw the graph.

x	$f(x)$
−3	$-\frac{1}{3}$
−2	$-\frac{1}{2}$
−1	−1
$-\frac{1}{2}$	−2
$-\frac{1}{4}$	−4
$\frac{1}{4}$	4
$\frac{1}{2}$	2
1	1
2	$\frac{1}{2}$
3	$\frac{1}{3}$

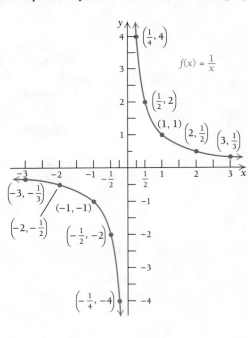

❭ **Quick Check 4**

Graph each function:

a) $f(x) = \dfrac{x^2 - 9}{x + 3}$;

b) $f(x) = -\dfrac{1}{x}$.

❬ Quick Check 4

Graphs of Rational Functions

Consider two graphs of the function given by

$$f(x) = \frac{2x + 1}{x - 3}.$$

CONNECTED mode:

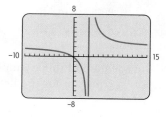

DOT mode:

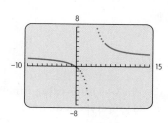

Here use of CONNECTED mode can lead to an *incorrect* graph. Because, in CONNECTED mode, points are joined with line segments, both branches of the graph are connected, making it appear as though the vertical line $x = 3$ is part of the graph.

On the other hand, in DOT mode, the calculator simply plots dots representing coordinates of points. When graphing rational functions, it is usually best to use DOT mode.

EXERCISES
Graph each of the following using DOT mode.

1. $f(x) = \dfrac{4}{x - 2}$ **2.** $f(x) = \dfrac{x}{x + 2}$

3. $f(x) = \dfrac{x^2 - 1}{x^2 + x - 6}$ **4.** $f(x) = \dfrac{x^2 - 4}{x - 1}$

5. $f(x) = \dfrac{10}{x^2 + 4}$ **6.** $f(x) = \dfrac{8}{x^2 - 4}$

7. $f(x) = \dfrac{2x + 3}{3x^2 + 7x - 6}$ **8.** $f(x) = \dfrac{2x^3}{x^2 + 1}$

In Example 6, note that 0 is not in the domain of f because it would yield a denominator of zero. The function is decreasing over the intervals $(-\infty, 0)$ and $(0, \infty)$. The function $f(x) = 1/x$ is an example of **inverse variation**.

DEFINITION

y **varies inversely** as *x* if there is some positive number *k* such that $y = k/x$. We also say that *y* is **inversely proportional** to *x*.

■ **EXAMPLE 7** **Business: Stocks and Gold.** Certain economists theorize that stock prices are inversely proportional to the price of gold. That is, when the price of gold goes up, the prices of stocks go down; and when the price of gold goes down, the prices of stocks go up. Let's assume that the Dow Jones Industrial Average D, an index of the overall prices of stocks, is inversely proportional to the price of gold G, in dollars per ounce. One day the Dow Jones was 10,619.70 and the price of gold was $1129.60 per ounce. What will the Dow Jones Industrial Average be if the price of gold rises to $1400?

Solution We assume that $D = k/G$, so $10{,}619.7 = k/1129.6$ and $k = 11{,}996{,}013.12$. Thus,

$$D = \frac{11{,}996{,}013.12}{G}.$$

We substitute 1400 for G and compute D:

$$D = \frac{11{,}996{,}013.12}{1400} \approx 8568.6.$$

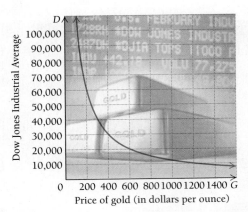

Dow Jones Industrial Average

Price of gold (in dollars per ounce)

Warning!   Do not put too much "stock" in the equation of this example. It is meant only to give an idea of economic relationships. An equation for predicting the stock market accurately has not been found!

〈 Quick Check 5

Absolute-Value Functions

The absolute value of a number is its distance from 0 on the number line. We denote the absolute value of a number x as $|x|$. The absolute-value function, given by $f(x) = |x|$, is very important in calculus, and its graph has a distinctive V shape.

■ **EXAMPLE 8**   Graph: $f(x) = |x|$.

Solution   We make a table of values, plot the points, and then draw the graph.

x	$f(x)$
-3	3
-2	2
-1	1
0	0
1	1
2	2
3	3

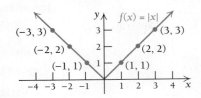

We can think of this function as being defined piecewise by considering the definition of absolute value:

$$f(x) = |x| = \begin{cases} x, & \text{if } x \geq 0, \\ -x, & \text{if } x < 0. \end{cases}$$

Square-Root Functions

The following is an example of a square-root function and its graph.

■ **EXAMPLE 9**   Graph: $f(x) = -\sqrt{x}$.

Solution   The domain of this function is the set of all nonnegative numbers—the interval $[0, \infty)$. You can find approximate values of square roots on your calculator. We set up a table of values, plot the points, and then draw the graph.

x	0	1	2	3	4	5
$f(x) = -\sqrt{x}$	0	−1	−1.4	−1.7	−2	−2.2

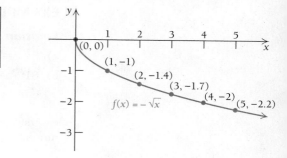

> **Quick Check 6**
>
> Graph each function:
> **a)** $f(x) = |x + 2|$;
> **b)** $f(x) = \sqrt{x} + 4$.

❬ Quick Check 6

Power Functions with Rational Exponents

We are motivated to define rational exponents so that the following laws of exponents still hold (also see Appendix A):

For any nonzero real number a and any integers n and m,

$$a^n \cdot a^m = a^{n+m}; \quad \frac{a^n}{a^m} = a^{n-m}; \quad (a^n)^m = a^{n \cdot m}; \quad a^{-m} = \frac{1}{a^m}.$$

This suggests that $a^{1/2}$ be defined so that $(a^{1/2})^2 = a^{(1/2) \cdot 2} = a^1$. Thus, we define $a^{1/2}$ as $\sqrt{a}$. Similarly, in order to have $(a^{1/3})^3 = a^{(1/3) \cdot 3} = a^1$, we define $a^{1/3}$ as $\sqrt[3]{a}$. In general,

$$a^{1/n} = \sqrt[n]{a}, \quad \text{provided } \sqrt[n]{a} \text{ is defined.}$$

Again, for the laws of exponents to hold, we have, assuming that $\sqrt[n]{a}$ exists,

$$a^{m/n} = (a^m)^{1/n} = \sqrt[n]{a^m} = (\sqrt[n]{a})^m,$$

and $a^{-m/n}$ is defined by

$$a^{-m/n} = \frac{1}{a^{m/n}} = \frac{1}{\sqrt[n]{a^m}}.$$

■ **EXAMPLE 10**   Rewrite each of the following as an equivalent expression with rational exponents:

a) $\sqrt[4]{x}$

b) $\sqrt[3]{r^2}$

c) $\sqrt{x^{10}}$,   for $x \geq 0$

d) $\dfrac{1}{\sqrt[3]{b^5}}$

Solution

a) $\sqrt[4]{x} = x^{1/4}$

b) $\sqrt[3]{r^2} = r^{2/3}$

c) $\sqrt{x^{10}} = x^{10/2} = x^5$, $x \geq 0$

d) $\dfrac{1}{\sqrt[3]{b^5}} = \dfrac{1}{b^{5/3}} = b^{-5/3}$

■ **EXAMPLE 11**   Rewrite each of the following as an equivalent expression using radical notation: **a)** $x^{1/3}$; **b)** $t^{6/7}$; **c)** $x^{-2/3}$; **d)** $r^{-1/4}$.

Solution

a) $x^{1/3} = \sqrt[3]{x}$ **b)** $t^{6/7} = \sqrt[7]{t^6}$

c) $x^{-2/3} = \dfrac{1}{x^{2/3}} = \dfrac{1}{\sqrt[3]{x^2}}$ **d)** $r^{-1/4} = \dfrac{1}{r^{1/4}} = \dfrac{1}{\sqrt[4]{r}}$

■ **EXAMPLE 12**   Simplify: **a)** $8^{5/3}$; **b)** $81^{3/4}$.

Solution

a) $8^{5/3} = (8^{1/3})^5 = (\sqrt[3]{8})^5 = 2^5 = 32$

b) $81^{3/4} = (81^{1/4})^3 = (\sqrt[4]{81})^3 = 3^3 = 27$

Because even roots (square roots, fourth roots, sixth roots, and so on) of negative numbers are not real numbers, the domain of a radical function may have restrictions.

■ **EXAMPLE 13**   Find the domain of the function given by
$$f(x) = \sqrt[4]{2x - 10}.$$

Solution   For $f(x)$ to be a real number, $2x - 10$ cannot be negative. Thus, to find the domain of f, we solve the inequality $2x - 10 \geq 0$:

$$2x - 10 \geq 0$$
$$2x \geq 10 \qquad \text{Adding 10 to both sides}$$
$$x \geq 5. \qquad \text{Dividing both sides by 2}$$

The domain of f is $\{x \mid x \geq 5\}$, or, in interval notation, $[5, \infty)$.

> **Quick Check 7**
>
> Find the domain of the function given by
> $$f(x) = \sqrt{x + 3}.$$

❰ Quick Check 7

Power functions of the form $f(x) = ax^k$, with k a fraction, occur in many applications.

■ **EXAMPLE 14**   **Life Science: Home Range.**   The *home range* of an animal is defined as the region to which the animal confines its movements. It has been shown that for carnivorous (meat-eating) mammals the area of that region can be approximated by the function

$$H(w) = 0.11w^{1.36},$$

where w is the mass of the animal, in grams, and $H(w)$ is the area of the home range, in hectares. Graph the function. (*Source: Based on information in Emlen, J. M., Ecology: An Evolutionary Approach, p. 200 (Reading, MA: Addison-Wesley, 1973), and Harestad, A. S., and Bunnel, F. L., "Home Range and Body Weight—A Reevaluation," Ecology, Vol. 60, No. 2 (April, 1979), pp. 389–402.*)

Solution   We can approximate function values using a power key, usually labeled [⌃] or [yˣ]. Note that $w^{1.36} = w^{136/100} = \sqrt[100]{w^{136}}$.

A lynx in its territorial area.

w	0	700	1400	2100	2800	3500
$H(w)$	0	814.2	2089.9	3627.5	5364.5	7266.5

The graph is shown below. Note that the function values increase from left to right. As body weight increases, the area over which the animal moves increases.

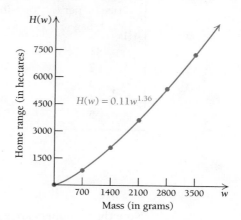

Supply and Demand Functions

Supply and demand in economics are modeled by increasing and decreasing functions.

Demand Functions

The table and graph below show the relationship between the price x per bag of sugar and the quantity q of 5-lb bags that consumers will demand at that price.

Demand Schedule

Price, x, per 5-lb Bag	Quantity, q, of 5-lb Bags (in millions)
$5	4
4	5
3	7
2	10
1	15

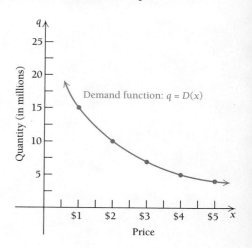

Note that the quantity consumers demand is inversely proportional to the price. As the price goes up, the quantity demanded goes down.

Supply Functions

The next table and graph show the relationship between the price x per bag of sugar and the quantity q of 5-lb bags that sellers are willing to supply, or sell, at that price.

Supply Schedule

Price, x, per 5-lb Bag	Quantity, q, of 5-lb Bags (in millions)
$1	0
2	10
3	16
4	20
5	22

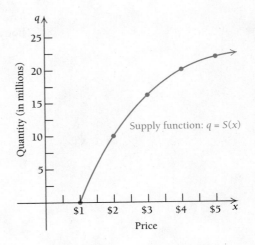

Note that suppliers are willing to supply greater quantities at higher prices than they are at lower prices.

Let's now look at these curves together. As price increases, supply increases and demand decreases; and as price decreases, demand increases but supply decreases. The point of intersection (x_E, q_E) is called the **equilibrium point**. The equilibrium price x_E (in this case, $2 per bag) corresponds to an equilibrium quantity q_E (in this case, 10 million bags). Sellers are willing to sell 10 million bags at $2/bag, and consumers are willing to buy 10 million bags at that price. The situation is analogous to a buyer and seller haggling over the sale of an item. The equilibrium point, or selling price, is what they finally agree on.

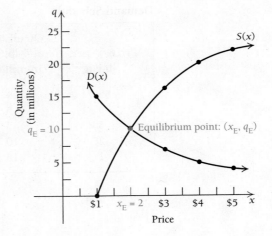

■ **EXAMPLE 15**   **Economics: Equilibrium Point.**   Find the equilibrium point for the demand and supply functions for the Ultra-Fine coffee maker. Here q represents the number of coffee makers produced, in hundreds, and x is the price, in dollars.

Demand: $q = 50 - \dfrac{1}{4}x$

Supply: $q = x - 25$

Solution   To find the equilibrium point, the quantity demanded must match the quantity produced:

$$50 - \frac{1}{4}x = x - 25$$

$$50 + 25 = x + \frac{1}{4}x \qquad \text{Adding } 25 + \frac{1}{4}x \text{ to each side}$$

EXERCISE

1. Use the INTERSECT feature to find the equilibrium point for the following demand and supply functions.

Demand: $q = 1123.6 - 61.4x$

Supply: $q = 201.8 + 4.6x$

$$75 = \frac{5}{4}x$$

$$75 \cdot \frac{4}{5} = x \qquad \text{Multiplying both sides by } \frac{4}{5}$$

$$60 = x.$$

>) **Quick Check 8**
>
> **Economics: Equilibrium Point.** Repeat Example 15 for the following functions:
>
> Demand: $\quad q = 70 - \frac{1}{5}x$
>
> Supply: $\quad q = x - 20$

Thus, $x_E = 60$. To find q_E, we substitute x_E into either function. We select the supply function:

$$q_E = x_E - 25 = 60 - 25 = 35.$$

Thus, the equilibrium quantity is 3500 units, and the equilibrium point is ($60, 3500$).

(Quick Check 8

Section Summary

- Many types of functions have graphs that are not straight lines; among these are *quadratic functions, polynomial functions, power functions, rational functions, absolute-value functions,* and *square-root functions.*

- Demand is modeled by a decreasing function. Supply is modeled by an increasing function. The point of intersection of graphs of demand and supply functions for the same product is called the *equilibrium point.*

EXERCISE SET
R.5

Graph each pair of equations on one set of axes.

1. $y = \frac{1}{2}x^2$ and $y = -\frac{1}{2}x^2$

2. $y = \frac{1}{4}x^2$ and $y = -\frac{1}{4}x^2$

3. $y = x^2$ and $y = x^2 - 1$

4. $y = x^2$ and $y = x^2 - 3$

5. $y = -2x^2$ and $y = -2x^2 + 1$

6. $y = -3x^2$ and $y = -3x^2 + 2$

7. $y = |x|$ and $y = |x - 3|$

8. $y = |x|$ and $y = |x - 1|$

9. $y = x^3$ and $y = x^3 + 2$

10. $y = x^3$ and $y = x^3 + 1$

11. $y = \sqrt{x}$ and $y = \sqrt{x - 1}$

12. $y = \sqrt{x}$ and $y = \sqrt{x - 2}$

For each of the following, state whether the graph of the function is a parabola. If the graph is a parabola, find the parabola's vertex.

13. $f(x) = x^2 + 4x - 7$

14. $f(x) = x^3 - 2x + 3$

15. $g(x) = 2x^4 - 4x^2 - 3$

16. $g(x) = 3x^2 - 6x$

Graph.

17. $y = x^2 - 4x + 3$

18. $y = x^2 - 6x + 5$

19. $y = -x^2 + 2x - 1$

20. $y = -x^2 - x + 6$

21. $f(x) = 2x^2 - 6x + 1$

22. $f(x) = 3x^2 - 6x + 4$

23. $g(x) = -3x^2 - 4x + 5$

24. $g(x) = -2x^2 - 3x + 7$

25. $y = \frac{2}{x}$

26. $y = \frac{3}{x}$

27. $y = -\frac{2}{x}$

28. $y = \frac{-3}{x}$

29. $y = \frac{1}{x^2}$

30. $y = \frac{1}{x - 1}$

31. $y = \sqrt[3]{x}$

32. $y = \frac{1}{|x|}$

33. $f(x) = \frac{x^2 + 5x + 6}{x + 3}$

34. $g(x) = \frac{x^2 + 7x + 10}{x + 2}$

35. $f(x) = \frac{x^2 - 1}{x - 1}$

36. $g(x) = \frac{x^2 - 25}{x - 5}$

Solve.

37. $x^2 - 2x = 2$

38. $x^2 - 2x + 1 = 5$

39. $x^2 + 6x = 1$

40. $x^2 + 4x = 3$

41. $4x^2 = 4x + 1$

42. $-4x^2 = 4x - 1$

43. $3y^2 + 8y + 2 = 0$

44. $2p^2 - 5p = 1$

45. $x + 7 + \dfrac{9}{x} = 0$ (*Hint:* Multiply both sides by x.)

46. $1 - \dfrac{1}{w} = \dfrac{1}{w^2}$

Rewrite each of the following as an equivalent expression with rational exponents.

47. $\sqrt{x^3}$

48. $\sqrt{x^5}$

49. $\sqrt[5]{a^3}$

50. $\sqrt[4]{b^2}, \quad b \geq 0$

51. $\sqrt[7]{t}$

52. $\sqrt[8]{c}$

53. $\sqrt[4]{x^{12}}, \quad x \geq 0$

54. $\sqrt[3]{t^6}$

55. $\dfrac{1}{\sqrt{t^5}}$

56. $\dfrac{1}{\sqrt{m^4}}$

57. $\dfrac{1}{\sqrt{x^2 + 7}}$

58. $\sqrt{x^3 + 4}$

Rewrite each of the following as an equivalent expression using radical notation.

59. $x^{1/5}$

60. $t^{1/7}$

61. $y^{2/3}$

62. $t^{2/5}$

63. $t^{-2/5}$

64. $y^{-2/3}$

65. $b^{-1/3}$

66. $b^{-1/5}$

67. $e^{-17/6}$

68. $m^{-19/6}$

69. $(x^2 - 3)^{-1/2}$

70. $(y^2 + 7)^{-1/4}$

71. $\dfrac{1}{t^{2/3}}$

72. $\dfrac{1}{w^{-4/5}}$

Simplify.

73. $9^{3/2}$ **74.** $16^{5/2}$ **75.** $64^{2/3}$

76. $8^{2/3}$ **77.** $16^{3/4}$ **78.** $25^{5/2}$

Determine the domain of each function.

79. $f(x) = \dfrac{x^2 - 25}{x - 5}$

80. $f(x) = \dfrac{x^2 - 4}{x + 2}$

81. $f(x) = \dfrac{x^3}{x^2 - 5x + 6}$

82. $f(x) = \dfrac{x^4 + 7}{x^2 + 6x + 5}$

83. $f(x) = \sqrt{5x + 4}$

84. $f(x) = \sqrt{2x - 6}$

85. $f(x) = \sqrt[4]{7 - x}$

86. $f(x) = \sqrt[5]{5 - x}$

APPLICATIONS

Business and Economics

Find the equilibrium point for each pair of demand and supply functions.

87. Demand: $q = 1000 - 10x$;   Supply: $q = 250 + 5x$

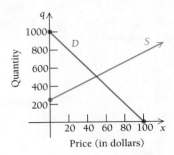

88. Demand: $q = 8800 - 30x$;   Supply: $q = 7000 + 15x$

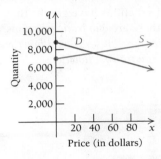

89. Demand: $q = \dfrac{5}{x}$;   Supply: $q = \dfrac{x}{5}$

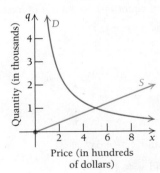

90. Demand: $q = \dfrac{4}{x}$;   Supply: $q = \dfrac{x}{4}$

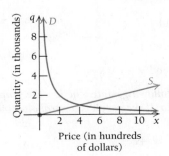

91. Demand: $q = (x - 3)^2$; Supply: $q = x^2 + 2x + 1$
(assume $x \leq 3$)

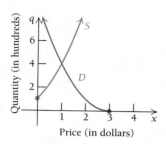

92. Demand: $q = (x - 4)^2$; Supply: $q = x^2 + 2x + 6$
(assume $x \leq 4$)

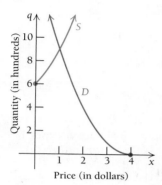

93. Demand: $q = 5 - x$; Supply: $q = \sqrt{x + 7}$

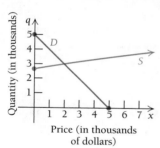

94. Demand: $q = 7 - x$; Supply: $q = 2\sqrt{x + 1}$

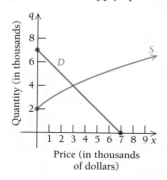

95. Stock prices and prime rate. It is theorized that the price per share of a stock is inversely proportional to the prime (interest) rate. In January 2010, the price per share S of Apple Inc. stock was $205.93 and the prime rate R was 3.25%. The prime rate rose to 4.75% in March 2010. (*Source*: finance.yahoo.com and Federal Reserve Board.) What was the price per share in March 2010 if the assumption of inverse proportionality is correct?

96. Demand. The quantity sold x of a plasma television is inversely proportional to the price p. If 85,000 plasma TVs sold for $2900 each, how many will be sold if the price is $850 each?

Life and Physical Sciences

97. Radar range. The function given by

$$R(x) = 11.74x^{0.25}$$

can be used to approximate the maximum range, $R(x)$, in miles, of ARSR-3 surveillance radar with a peak power of x watts.

  a) Determine the maximum radar range when the peak power is 40,000 watts, 50,000 watts, and 60,000 watts.
  b) Graph the function.

98. Home range. Refer to Example 14. The home range, in hectares, of an omnivorous mammal (one that eats both plants and meat) of mass w grams is given by

$$H(w) = 0.059w^{0.92}.$$

(*Source*: Harestad, A. S., and Bunnel, F. L., "Home Range and Body Weight—A Reevaluation," *Ecology*, Vol. 60, No. 2 (April, 1979), pp. 389–402.) Complete the table of approximate function values and graph the function.

w	0	1000	2000	3000	4000	5000	6000	7000
$H(w)$	0	34.0						

99. Life science: pollution control. Pollution control has become a very important concern in all countries. If controls are not put in place, it has been predicted that the function

$$P = 1000t^{5/4} + 14,000$$

will describe the average pollution, in pollutant particles per cubic centimeter of air, in most cities at time t, in years, where $t = 0$ corresponds to 1970 and $t = 35$ corresponds to 2005.

  a) Predict the pollution in 2005, 2008, and 2014.
  b) Graph the function over the interval $[0, 50]$.

100. Surface area and mass. The surface area of a person whose mass is 75 kg can be approximated by the function

$$f(h) = 0.144h^{1/2},$$

where $f(h)$ is measured in square meters and h is the person's height in centimeters. (*Source*: U.S. Oncology.)

  a) Find the approximate surface area of a person whose mass is 75 kg and whose height is 180 cm.
  b) Find the approximate surface area of a person whose mass is 75 kg and whose height is 170 cm.
  c) Graph the function f for $0 \leq h \leq 200$.

SYNTHESIS

101. Zipf's Law. According to Zipf's Law, the number of cities N with a population greater than S is inversely proportional to S. In 2008, there were 52 U.S. cities with a population greater than 350,000. Estimate the number of U.S. cities with a population between 350,000 and 500,000; between 300,000 and 600,000.

102. At most, how many *y*-intercepts can a function have? Explain.

103. Explain the difference between a rational function and a polynomial function. Is every polynomial function a rational function? Why or why not?

TECHNOLOGY CONNECTION

Use the ZERO *feature or the* INTERSECT *feature to approximate the zeros of each function to three decimal places.*

104. $f(x) = x^3 - x$
  (Also, use algebra to find the zeros of this function.)

105. $f(x) = 2x^3 - x^2 - 14x - 10$

106. $f(x) = \frac{1}{2}(|x - 4| + |x - 7|) - 4$

107. $f(x) = x^4 + 4x^3 - 36x^2 - 160x + 300$

108. $f(x) = \sqrt{7 - x^2} - 1$

109. $f(x) = |x + 1| + |x - 2| - 5$

110. $f(x) = |x + 1| + |x - 2|$

111. $f(x) = |x + 1| + |x - 2| - 3$

112. $f(x) = x^8 + 8x^7 - 28x^6 - 56x^5 + 70x^4 + 56x^3 - 28x^2 - 8x + 1$

113. Find the equilibrium point for the following demand and supply functions, where *q* is the quantity, in thousands of units, and *x* is the price per unit, in dollars.

  Demand: $q = 83 - x$

  Supply: $q = \dfrac{x^2}{576} - 1.9$

Answers to Quick Checks

1. (a) 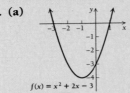 **(b)**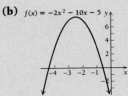

2. $\dfrac{-1 \pm \sqrt{22}}{3}$, or 1.230 and −1.897

3.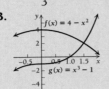

4. (a) 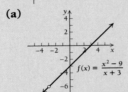 **(b)**

5. Dow Jones Industrial Average increases; $D = 11{,}996.01$.

6. (a)   **(b)**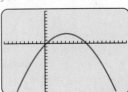

7. $\{x \mid x \geq -3\}$ **8.** Equilibrium point is ($75, 5500); price is $75 and quantity is 5500.

R.6 Mathematical Modeling and Curve Fitting

Fitting Functions to Data

OBJECTIVE

- Use curve fitting to find a mathematical model for a set of data and use the model to make predictions.

We have developed a library of functions that can serve as models for many applications. Although others will be introduced later, let's look at those that we have considered. (Cubic and quartic functions are covered in detail in Chapter 2, but we show them for reference.) We will not consider rational functions in this section.

Linear function:
$f(x) = mx + b$

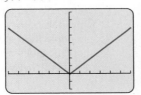

Quadratic function:
$f(x) = ax^2 + bx + c, \ a > 0$

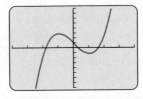

Quadratic function:
$f(x) = ax^2 + bx + c, \ a < 0$

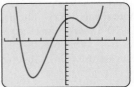

Absolute-value function:
$f(x) = |x|$

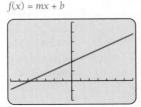

Cubic function:
$f(x) = ax^3 + bx^2 + cx + d, \ a > 0$

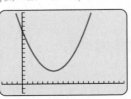

Quartic function:
$f(x) = ax^4 + bx^3 + cx^2 + dx + e, \ a > 0$

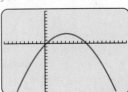

Now let's consider some real-world data. How can we decide which, if any, type of function might fit the data? One simple way is to examine a graph of the data called a **scatterplot**. Then we look for a pattern resembling one of the graphs on p. 68. For example, data might be modeled by a linear function if the graph resembles a straight line. The data might be modeled by a quadratic function if the graph rises and then falls, or falls and then rises, in a curved manner resembling a parabola.

Let's now use our library of functions to see which, if any, might fit certain data sets.

■ **EXAMPLE 1** **Choosing Models.** For the scatterplots and graphs below, determine which, if any, of the following functions might be used as a model for the data.

Linear, $f(x) = mx + b$

Quadratic, $f(x) = ax^2 + bx + c, a > 0$

Quadratic, $f(x) = ax^2 + bx + c, a < 0$

Polynomial, neither quadratic nor linear

a) 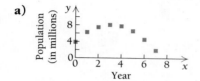 **b)** 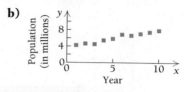 **c)**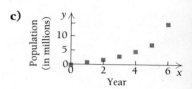

d)

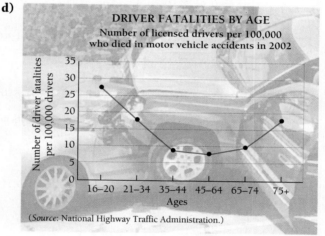

DRIVER FATALITIES BY AGE
Number of licensed drivers per 100,000 who died in motor vehicle accidents in 2002

(*Source*: National Highway Traffic Administration.)

e)

ACTUAL AND PROJECTED ANNUAL
SUIT SALES FOR RAGGS LTD.

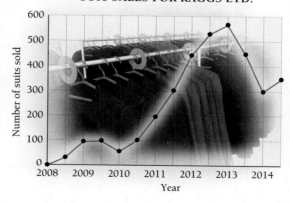

Solution

a)

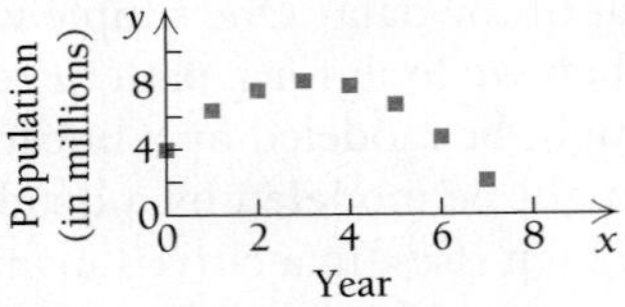
Population (in millions)
Year

The data rise and then fall in a curved manner fitting a quadratic function,

$$f(x) = ax^2 + bx + c, a < 0.$$

b)

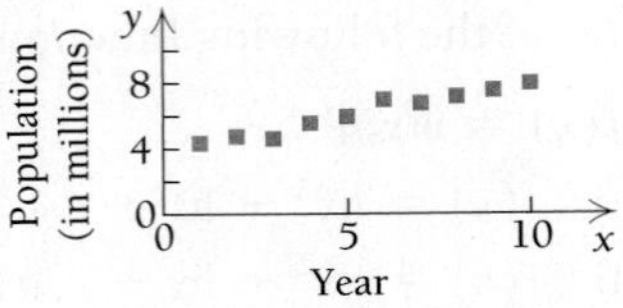
Population (in millions)
Year

The data seem to fit a linear function,

$$f(x) = mx + b.$$

c)

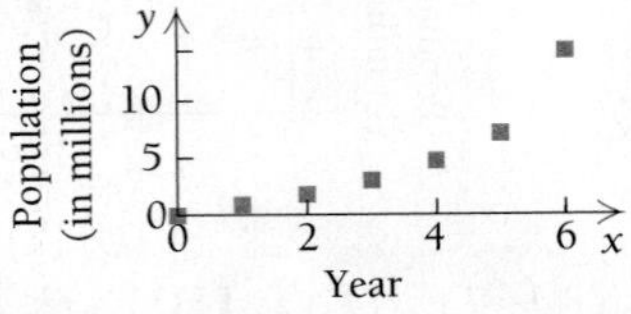
Population (in millions)
Year

The data rise in a manner fitting the right-hand side of a quadratic function,

$$f(x) = ax^2 + bx + c, a > 0.$$

d)

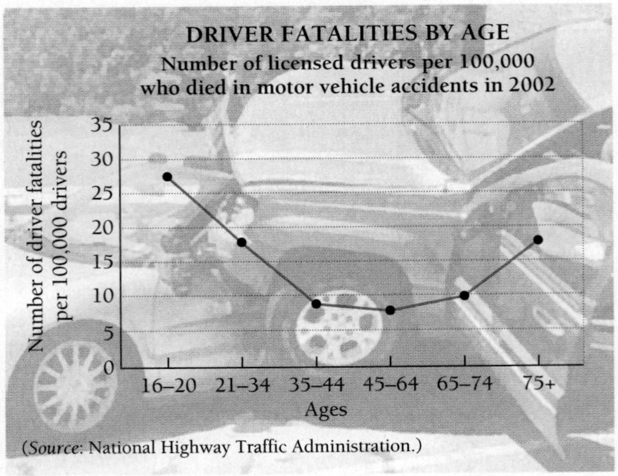
DRIVER FATALITIES BY AGE
Number of licensed drivers per 100,000
who died in motor vehicle accidents in 2002
Number of driver fatalities per 100,000 drivers
16–20
21–34
35–44
45–64
65–74
75+
Ages
(Source: National Highway Traffic Administration.)

The data fall and then rise in a curved manner fitting a quadratic function,

$$f(x) = ax^2 + bx + c, a > 0.$$

e)

**ACTUAL AND PROJECTED ANNUAL
SUIT SALES FOR RAGGS LTD.**

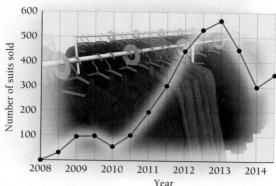

The data rise and fall more than once, so they do not fit a linear or quadratic function but might fit a polynomial function that is neither quadratic nor linear.

It is sometimes possible to find a mathematical model by graphing a set of data as a scatterplot, inspecting the graph to see if a known type of function seems to fit, and then using the data points to derive the equation of a specific function.

■ **EXAMPLE 2**   **Business: High-Speed Internet Subscribers.**   The following table shows the number of U.S. households with high-speed Internet and a scatterplot of the data. It appears that the data can be represented or modeled by a linear function.

Year, x	Number of Households with High-Speed Internet (in millions), y	Scatterplot
2005, 0	25.4	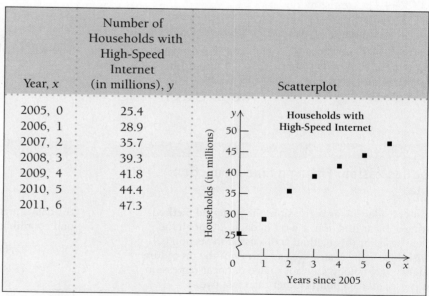
2006, 1	28.9	
2007, 2	35.7	
2008, 3	39.3	
2009, 4	41.8	
2010, 5	44.4	
2011, 6	47.3	

(*Source*: National Cable & Telecommunications Association.)

a) Find a linear function that fits the data.

b) Use the model to predict the number of households with high-speed Internet in 2014.

Solution

a) We can choose any two of the data points to determine an equation. Let's use $(2, 35.7)$ and $(6, 47.3)$.

We first determine the slope of the line:

$$m = \frac{47.3 - 35.7}{6 - 2} = \frac{11.6}{4} = 2.9.$$

Then we substitute 2.9 for m and either of the points $(2, 35.7)$ or $(6, 47.3)$ for (x_1, y_1) in the point–slope equation. Using $(6, 47.3)$, we get

$$y - 47.3 = 2.9(x - 6),$$

which simplifies to

$$y = 2.9x + 29.9,$$

where x is the number of years after 2005 and y is in millions. We can then graph the linear equation $y = 2.9x + 29.9$ on the scatterplot to see how it fits the data.

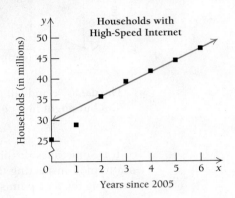

> **) Quick Check 1**
>
> **Business.** Repeat Example 2 using the data points $(1, 28.9)$ and $(5, 44.4)$. Are the model and the result different from those in Example 2? (This illustrates that a model and the predictions it makes are dependent on the pair of data points used.)

b) We can predict the number of households with high-speed Internet in 2014 by substituting 9 for x in the model $(2014 - 2005 = 9)$:

$$
\begin{aligned}
y &= 2.9x + 29.9 && \text{Model} \\
&= 2.9(9) + 29.9 && \text{Substituting} \\
&\approx 56.
\end{aligned}
$$

We can then predict that there will be about 56 million households with high-speed Internet in 2014.

⟨ Quick Check 1

TECHNOLOGY CONNECTION

Linear Regression: Fitting a Linear Function to Data

We now consider **linear regression**, the preferred method for fitting a linear function to a set of data. Although the complete basis for this method is discussed in Section 6.4, we consider it here because we can carry out the procedure easily using technology. One advantage of linear regression is that it uses *all* data points rather than just two.

EXAMPLE   **Business: High-Speed Internet Subscribers.** Consider the data in Example 2.

a) Find the equation of the regression line for the given data. Then graph the regression line with the graph.

b) Use the model to predict the number of households with high-speed Internet in 2014. Compare your answer to that found in Example 2.

Solution

a) To fit a linear function to the data using regression, we select the EDIT option of the STAT menu. We then enter

the data, entering the first coordinate for L1 and the second coordinate for L2.

$$(0, 25.4), (1, 28.9), (2, 35.7), (3, 39.3),$$
$$(4, 41.8), (5, 44.4), (6, 47.3)$$

L1	L2	L3
0	25.4	-----
1	28.9	
2	35.7	
3	39.3	
4	41.8	
5	44.4	
6	47.3	
L2(7) = 47.3		

L1	L2	L3 2
1	28.9	-----
2	35.7	
3	39.3	
4	41.8	
5	44.4	
6	47.3	
L2(7) = 47.3		

To view the data points, we turn on PLOT1 by pressing ⌃ **ENTER** from y_1 at the Y= screen and then pressing **GRAPH**. To set the window, we could select the ZoomStat option of ZOOM. Instead, we select a $[-1, 20, 20, 70]$ window.

(continued)

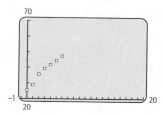

To find the line that best fits the data, we press **STAT**, select CALC, then LinReg(ax+b), and then press **ENTER**.

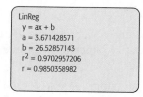

LinReg
y = ax + b
a = 3.671428571
b = 26.52857143
r² = 0.9702957206
r = 0.9850358982

The equation of the regression is approximately $y = 3.67x + 26.5$. We enter this as y_1 and press **GRAPH**.

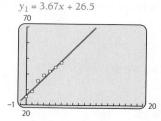

$y_1 = 3.67x + 26.5$

b) To find, or predict, the number of U.S. households with high-speed Internet in 2014, we press **2ND**, **CALC**, **1**, and enter 9.

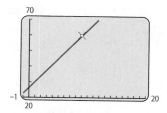

Note that the result found using the regression equation, 59.53 million, is more than the 56 million found in Example 2. Statisticians might give more credence to the regression value because more data points were used to construct the equation.

EXERCISE

1. Study time and test scores. The data in the following table relate study time and test scores.

Study Time (in hours)	Test Grade (in percent)
7	83
8	85
9	88
10	91
11	?

a) Fit a regression line to the data. Then make a scatterplot of the data and graph the regression line with the scatterplot.
b) Use the linear model to predict the test score received when one has studied for 11 hr.
c) Discuss the appropriateness of a linear model of these data.

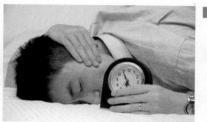

■ **EXAMPLE 3** Life Science: Hours of Sleep and Death Rate. In a study by Dr. Harold J. Morowitz of Yale University, data were gathered that showed the relationship between the death rate of men and the average number of hours per day that the men slept. These data are listed in the following table.

Average Number of Hours of Sleep, x	Death Rate per 100,000 Males, y
5	1121
6	805
7	626
8	813
9	967

(*Source*: Morowitz, Harold J., "Hiding in the Hammond Report," Hospital Practice.)

a) Make a scatterplot of the data, and determine whether the data seem to fit a quadratic function.

b) Find a quadratic function that fits the data.

c) Use the model to find the death rate for males who sleep 2 hr, 8 hr, and 10 hr.

Solution

a) The scatterplot is shown to the left. Note that the rate drops and then rises, which suggests that a quadratic function might fit the data.

Possible quadratic function that "fits" data

Death rate per 100,000 males

Average number of hours of sleep

b) We consider the quadratic model,

$$y = ax^2 + bx + c. \tag{1}$$

To derive the constants (or parameters) a, b, and c, we use the three data points $(5, 1121)$, $(7, 626)$, and $(9, 967)$. Since these points are to be solutions of equation (1), it follows that

$$1121 = a \cdot 5^2 + b \cdot 5 + c, \quad \text{or} \quad 1121 = 25a + 5b + c,$$
$$626 = a \cdot 7^2 + b \cdot 7 + c, \quad \text{or} \quad 626 = 49a + 7b + c,$$
$$967 = a \cdot 9^2 + b \cdot 9 + c, \quad \text{or} \quad 967 = 81a + 9b + c.$$

We solve this system of three equations in three variables using procedures of algebra and get

$$a = 104.5, \quad b = -1501.5, \quad \text{and} \quad c = 6016.$$

Substituting these values into equation (1), we get the function given by

$$y = 104.5x^2 - 1501.5x + 6016.$$

c) The death rate for males who sleep 2 hr is given by

$$y = 104.5(2)^2 - 1501.5(2) + 6016 = 3431.$$

The death rate for males who sleep 8 hr is given by

$$y = 104.5(8)^2 - 1501.5(8) + 6016 = 692.$$

The death rate for males who sleep 10 hr is given by

$$y = 104.5(10)^2 - 1501.5(10) + 6016 = 1451.$$

❰ Quick Check 2

> **Quick Check 2**
>
> **Life Science.** See the data on live births in the following Technology Connection. Use the data points $(16, 34)$, $(27, 113.9)$, and $(37, 35.4)$ to find a quadratic function that fits the data. Use the model to predict the average number of live births to women age 20.

TECHNOLOGY CONNECTION

Mathematical Modeling Using Regression: Fitting Quadratic and Other Polynomial Functions to Data

Regression can be extended to quadratic, cubic, and quartic polynomial functions.

EXAMPLE **Life Science: Live Births to Women of Age x.** The chart to the right relates the average number of live births to women of a particular age.

a) Fit a quadratic function to the data using REGRESSION. Then make a scatterplot of the data and graph the quadratic function with the scatterplot.

b) Fit a cubic function to the data using REGRESSION. Then make a scatterplot of the data and graph the cubic function with the scatterplot.

c) Which function seems to fit the data better?

Age, x	Average Number of Live Births per 1000 women
16	34
18.5	86.5
22	111.1
27	113.9
32	84.5
37	35.4
42	6.8

(*Source*: Centers for Disease Control and Prevention.)

(*continued*)

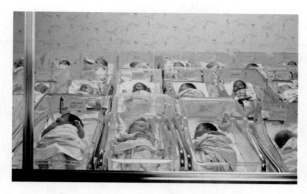

d) Use the function from part (c) to estimate the average number of live births to women of ages 20 and 30.

Solution   We proceed as follows.

a) To fit a quadratic function using REGRESSION, the procedure is similar to what is outlined in the preceding Technology Connection on linear regression. We enter the data but select QuadReg instead of LinReg(ax+b). For the graph below, we used ZoomStat to set the window. We wrote y_1 using approximations of a, b, and c from the QuadReg screen.

$$y_1 = -0.49x^2 + 25.95x - 238.49$$

 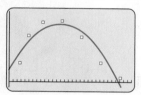

b) To fit a cubic function, we select CubicReg and obtain the following:

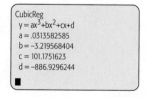

A convenient way to copy a regression function, in full detail, onto the Y= screen is to press $\boxed{\text{Y=}}$, move the cursor to Y2 (or wherever the function is to appear), press **VARS**, and select STATISTICS and then EQ and RegEq.

$$y_2 = 0.03135825845952x^3 - 3.2195684044498x^2 +$$
$$101.17516232219x - 886.92962438781$$

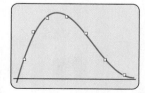

c) The graph of the cubic function seems to fit closer to the data points. Thus we choose it as a model.

d) We press **2ND** $\boxed{\text{QUIT}}$ to leave the graph screen. Pressing **VARS** and selecting Y-VARS and then FUNCTION and Y2, we have Y2(20) ≈ 99.6 and Y2(30) ≈ 97.4 as shown.

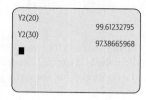

The TRACE feature can also be used. Thus, the average number of live births is 99.6 per 1000 women age 20 and 97.4 per 1000 women age 30.

EXERCISES

1. Life science: live births.
   a) Use the REGRESSION feature to fit a quartic equation to the live-birth data. Make a scatterplot of the data. Then graph the quartic function with the scatterplot. Decide whether the quartic function gives a better fit than either the quadratic or the cubic function.
   b) Explain why the domain of the cubic live-birth function should probably be restricted to the interval $[15, 45]$.

2. Business: median household income by age.

Age, x	Median Income in 2003
19.5	$27,053
29.5	44,779
39.5	55,044
49.5	60,242
59.5	49,215
65	23,787

(*Source:* Based on data in the *Statistical Abstract of the United States, 2005.*)

   a) Make a scatterplot of the data and fit a quadratic function to the data using QuadReg. Then graph the quadratic function with the scatterplot.
   b) Fit a cubic function to the data using CubicReg. Then graph the cubic function with the scatterplot.
   c) Fit a quartic function to the data using QuadReg. Then graph the quartic function with the scatterplot.
   d) Which of the quadratic, cubic, or quartic functions seems to best fit the data?
   e) Use the function from part (d) to estimate the median household income of people age 25; of people age 45.

3. Life science: hours of sleep and death rate. Repeat Example 3 using quadratic regression to fit a function to the data.

EXERCISE SET
R.6

Choosing models. *For the scatterplots and graphs in Exercises 1–9, determine which, if any, of the following functions might be used as a model for the data:*

Linear, $f(x) = mx + b$

Quadratic, $f(x) = ax^2 + bx + c, a > 0$

Quadratic, $f(x) = ax^2 + bx + c, a < 0$

Polynomial, neither quadratic nor linear

1.
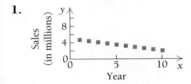

2.
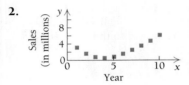

3.
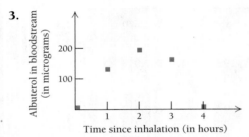

Business and Economics

4.
U.S. TRADE DEFICIT WITH JAPAN
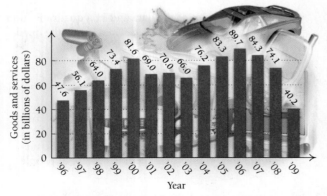

(*Source*: U.S. Census Bureau, Foreign Trade Statistics.)

5.
NBA AVERAGE SALARY
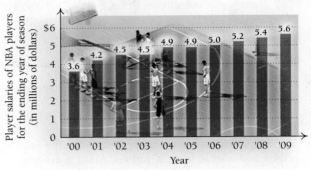

(*Source*: *The Compendium of Professional Basketball*, by Robert D. Bradley, Xavier Press, 2010.)

6.
MEDIAN INCOME
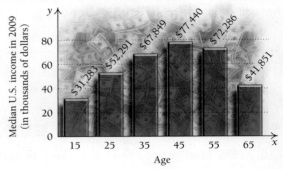

(*Source*: U.S. Census Bureau.)

7.
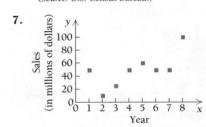

8.
AVERAGE MONTHLY TEMPERATURE IN DALLAS
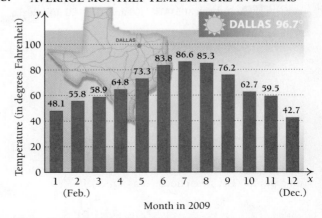

(*Source*: National Oceanic and Atmospheric Administration, National Weather Service Weather Forcast Office, Dallas/Ft. Worth, TX.)

9. **PRIME INTEREST RATE (on first day of the month)**

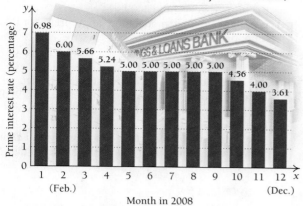

(*Source*: Federal Reserve Board.)

APPLICATIONS

Business and Economics

10. Prime interest rate.

a) For the prime interest rate data in Exercise 9, find a linear function that fits the data using the values given for January 2008 and December 2008.

b) Use the linear function to estimate the prime rate in June 2009.

11. Average salary in NBA. Use the data from the bar graph in Exercise 5.

a) Find a linear function that fits the data using the average salaries given for the years 2000 and 2009. Use 0 for 2000 and 9 for 2009.

b) Use the linear function to predict average salaries in 2012 and 2020.

c) In what year will the average salary reach 9.0 million?

Life and Physical Sciences

12. Absorption of an asthma medication. Use the data from Exercise 3.

a) Find a quadratic function that fits the data using the data points $(0, 0)$, $(2, 200)$, and $(3, 167)$.

b) Use the function to estimate the amount of albuterol in the bloodstream after 4 hr.

c) Does it make sense to use this function for $t = 6$? Why or why not?

13. Braking distance.

a) Find a quadratic function that fits the following data.

Travel Speed (mph)	Braking Distance (ft)
20	25
40	105
60	300

(*Source*: New Jersey Department of Law and Public Safety.)

b) Use the function to estimate the braking distance of a car that travels at 50 mph.

c) Does it make sense to use this function when speeds are less than 15 mph? Why or why not?

14. Daytime accidents.

a) Find a quadratic function that fits the following data.

Travel Speed (in km/h)	Number of Daytime Accidents (for every 200 million km driven)
60	100
80	130
100	200

b) Use the function to estimate the number of daytime accidents that occur at 50 km/h for every 200 million km driven.

15. High blood pressure in women.

a) Choose two points from the following data and find a linear function that fits the data.

Age of Female	Percentage of Females with High Blood Pressure
30	1.4
40	8.5
50	19.5
60	31.9
70	53.0

(*Source:* Based on data from Health United States 2005, CDC/NCHS.)

b) Graph the scatterplot and the function on the same set of axes.

c) Use the function to estimate the percentage of 55-yr-old women with high blood pressure.

16. High blood pressure in men.

a) Choose two points from the following data and find a linear function that fits the data.

Age of Male	Percentage of Males with High Blood Pressure
30	7.3
40	12.1
50	20.4
60	24.8
70	34.9

(*Source:* Based on data from Health United States 2005, CDC/NCHS.)

b) Graph the scatterplot and the function on the same set of axes.

c) Use the function to estimate the percentage of 55-yr-old men with high blood pressure.

SYNTHESIS

17. Suppose that you have just 3 or 4 data points. Why might it make better sense to use a linear function rather than a quadratic or cubic function that fits these data more closely?

18. When modeling the number of hours of daylight for the dates April 22 to August 22, which would be a better choice: a linear function or a quadratic function? Explain.

19. Explain the restrictions that should be placed on the domain of the quadratic function found in Exercise 12 and why such restrictions are needed.

20. Explain the restrictions that should be placed on the domain of the quadratic function found in Exercise 13 and why such restrictions are needed.

TECHNOLOGY CONNECTION

21. Business: prime interest rate.

   a) Use regression to fit a linear function to the data in Exercise 9.

   b) Use the function to estimate the prime rate in June 2009.

   c) Compare your answers to those found in Exercise 10. Which is more accurate?

   d) Fit a cubic function to the data and use it to estimate the prime rate in June 2009.

   e) Is a linear or cubic model more appropriate for this set of data? Explain.

22. Business: trade deficit with Japan.

   a) Use regression to fit a cubic function to the data in Exercise 4. Let x be the number of years after 1996.

   b) Use the function to estimate the trade deficit with Japan in 2012.

   c) Why might a linear function be a more logical choice than a cubic function for modeling this set of data?

Answers to Quick Checks

1. $y = 3.875x + 25.025$; 59.9 million households with high-speed Internet subscribers. The model is different, and the prediction is higher.

2. $y = -0.7197x^2 + 38.2106x - 393.1272$; 83.2 live births per 1000 women age 20.

KEY TERMS AND CONCEPTS	EXAMPLES

SECTION R.1

To **graph** an ordered pair (x, y), move x units horizontally, then y units vertically, depending on whether the coordinates are positive, negative, or zero.

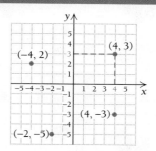

A **solution** of an equation in two variables is an ordered pair of numbers that, when substituted for the variables, forms a true equation. The **graph** of an equation is a drawing that represents all ordered pairs that are solutions of the equation.

This is the graph of the equation $y = x^4 - 2x^2$.

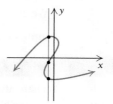

SECTION R.2

The mathematics used to represent the essential features of an applied problem comprise a **mathematical model**.

Business. The equation $A = \$1000\left(1 + \dfrac{i}{4}\right)^{7 \times 4}$ can be used to estimate the amount A, in dollars, to which $1000 will grow in 7 years, at interest rate i, compounded quarterly.

A **function** is a correspondence between a first set, called a **domain**, and a second set, called a **range**, such that each member of the domain corresponds to *exactly one* member of the range.

A graph represents a function if it is impossible to draw a vertical line that crosses the graph more than once.

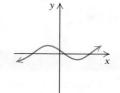

Not a function.   A function.

Function notation permits us to easily determine what member of the range, **output**, is paired with a member of the domain, **input**.

The function given by $f(x) = x^2 - 5x - 8$ allows us to determine what member of the range is paired with the number 2 of the domain:

$$f(x) = x^2 - 5x - 8,$$
$$f(2) = 2^2 - 5(2) - 8$$
$$= 4 - 10 - 8$$
$$= -14$$

(continued)

KEY TERMS AND CONCEPTS	EXAMPLES

SECTION R.2 (*continued*)

In calculus, it is important to be able to simplify an expression like

$$\frac{f(x + h) - f(x)}{h}.$$

For $f(x) = x^2 - 5x - 8$,

$$\frac{f(x + h) - f(x)}{h} = \frac{[(x + h)^2 - 5(x + h) - 8] - f(x)}{h}$$

$$= \frac{x^2 + 2xh + h^2 - 5x - 5h - 8 - [x^2 - 5x - 8]}{h}$$

$$= \frac{x^2 + 2xh + h^2 - 5x - 5h - 8 - x^2 + 5x + 8}{h}$$

$$= \frac{2xh + h^2 - 5h}{h}$$

$$= \frac{h(2x + h - 5)}{h}$$

$$= 2x + h - 5, \ x \neq 0$$

A function that is defined **piecewise** specifies different rules for parts of the domain.

The graph of

$$g(x) = \begin{cases} \dfrac{1}{3}x + 3, & \text{for } x < 3, \\ -x, & \text{for } x \geq 3, \end{cases}$$

is

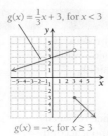

$g(x) = \frac{1}{3}x + 3$, for $x < 3$

$g(x) = -x$, for $x \geq 3$

We graph $\dfrac{1}{3}x + 3$ *only* for inputs x less than 3. We graph $-x$ *only* for inputs x greater than or equal to 3.

SECTION R.3

The domain of any function is a **set**, or collection of objects. In this book, the domain is usually a set of real numbers.

There are three ways of naming sets:

Roster Notation: $\{-2, 5, 9, \pi\}$

Set-Builder Notation: $\{x \mid x \text{ is a real number and } x \geq 3\}$

Interval Notation:

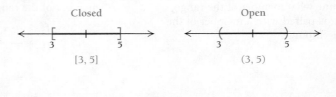

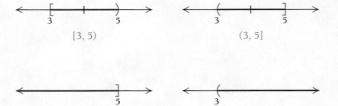

KEY TERMS AND CONCEPTS	**EXAMPLES**

For a function given by an equation or formula, the domain is the largest set of real numbers (inputs) for which function values (outputs) can be calculated.

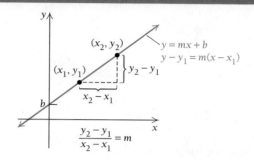

Function

$$f(x) = \frac{8x}{x^2 - 9} = \frac{8x}{(x - 3)(x + 3)}$$

$$g(x) = \sqrt{10 - 5x}$$

$$h(x) = |x - 3| + 5x^2$$

Domain

All real numbers except -3 and 3.

All real numbers x for which $10 - 5x \geq 0$ or $10 \geq 5x$ or $2 \geq x$. The domain is $(-\infty, 2]$.

All real numbers, $\mathbb{R}$.

SECTION R.4

The **slope m** of the line containing the points (x_1, y_1) and (x_2, y_2) is given by $m = \dfrac{y_2 - y_1}{x_2 - x_1}$ and can be regarded as an **average rate of change**. The graph of an equation $f(x) = mx + b$, called the **slope–intercept form**, is a line with slope m and **y-intercept** $(0, b)$.

The graph of an equation $y - y_1 = m(x - x_1)$, called the **point–slope form**, is a line with slope m passing through the point (x_1, y_1).

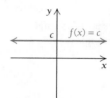

The graph of a **constant function**, given by $f(x) = c$, is a horizontal line.

The graph of an equation of the form $x = a$ is a vertical line and is not a function.

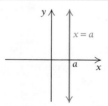

SECTION R.5

The graph of a **quadratic function** $f(x) = ax^2 + bx + c$ is a **parabola** with a **vertex** at $\left(-\dfrac{b}{2a}, f\left(-\dfrac{b}{2a}\right)\right)$. The graph opens **upward** if $a > 0$ and **downward** if $a < 0$. It is an example of a nonlinear function or model.

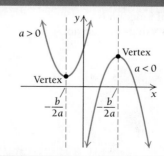

(continued)

KEY TERMS AND CONCEPTS	EXAMPLES

SECTION R.5 *(continued)*

The *x*-intercepts of the graph of a quadratic equation, $f(x) = ax^2 + bx + c$, are the solutions of $ax^2 + bx + c = 0$ and are given by the quadratic formula:

$$x = \frac{-b \pm \sqrt{b^2 - 4ac}}{2a}.$$

The *x*-intercepts of $f(x) = x^2 - 6x - 10$ are found by solving $x^2 - 6x - 10 = 0$. They are

$$\frac{-b + \sqrt{b^2 - 4ac}}{2a} = \frac{-(-6) + \sqrt{(-6)^2 - 4(1)(-10)}}{2(1)} = 3 + \sqrt{19}, \text{ and}$$

$$\frac{-b - \sqrt{b^2 - 4ac}}{2a} = \frac{-(-6) - \sqrt{(-6)^2 - 4(1)(-10)}}{2(1)} = 3 - \sqrt{19}.$$

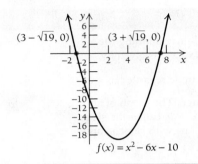

A **polynomial function** is a function like those shown at the right.

$f(x) = -8$, constant,
$g(x) = 3x - 7$, linear,
$h(x) = x^2 - 6x - 10$, quadratic,
$p(x) = 2x^3 - 7x + 0.23x - 10$, cubic.

Power functions are polynomial functions of the form $f(x) = ax^n$, where n is a positive integer.

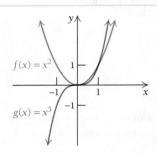

A **rational function** is a function given by the quotient of two polynomials.

$$g(x) = \frac{x^2 - 9}{x + 3}, \quad h(x) = \frac{x^3 - 4x^2 + x - 5}{x^2 - 6x - 10},$$

$$f(x) = \frac{x^3 - 4x^2 + x - 5}{1} = x^3 - 4x^2 + x - 5.$$

The domain of a rational function is restricted to those input values that do not result in division by zero.

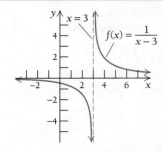

The domain is the set of all real numbers except 3, or $(-\infty, 3) \cup (3, \infty)$.

KEY TERMS AND CONCEPTS	**EXAMPLES**
An **absolute-value function** is described using absolute value symbols.	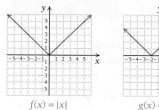 $f(x) = \lvert x \rvert$ $g(x) = \lvert x - 2 \rvert$
A **square-root function** is a function described using a square root symbol.	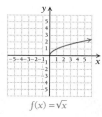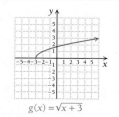 $f(x) = \sqrt{x}$ $g(x) = \sqrt{x + 3}$
The variable y **varies directly** as x if there exists some positive constant m for which $y = mx$. We say that m is the **variation constant**.	$$y = 3.14x, \qquad A = 1.08t, \qquad I = kR, \qquad y = \frac{1}{3}x$$
Total profit is the difference between **total revenue** and **total cost**. The value x at which $P(x) = 0$ is the **break-even value**.	For $P(x) = R(x) - C(x)$, $R(x) = 84x$, and $C(x) = 28x + 56{,}000$, the break-even value is found as follows: $$\begin{aligned} R(x) &= C(x) \\ 84x &= 28x + 56{,}000 \\ 56x &= 56{,}000 \\ x &= 10{,}000. \end{aligned}$$
The point at which consumer's demand and producer's supply is the same is called an **equilibrium point**.	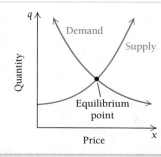

(*continued*)

KEY TERMS AND CONCEPTS

EXAMPLES

SECTION R.6

When real-world data are available, we can form a **scatterplot** and decide which, if any, of the graphs in this chapter best models the situation. Then we can use an appropriate model to make predictions.

Business.

Average Gasoline Prices, 2003–2012

Year, x	Price per Gallon
2003, 0	1.51
2004, 1	2.08
2005, 2	2.16
2006, 3	2.94
2007, 4	3.20
2008, 5	4.03
2009, 6	2.57
2010, 7	3.18
2011, 8	3.67
2012, 9	3.80

(*Source*: U.S. Department of Energy.)

Linear Model   *Quadratic Model*

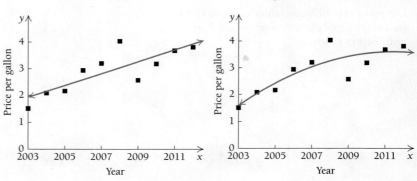

Cubic Model

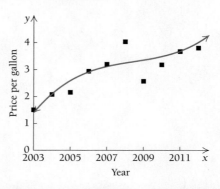

These review exercises are for test preparation. They can also be used as a practice test. Answers are at the back of the book. The blue bracketed section references tell you what part(s) of the chapter to restudy if your answer is incorrect.

CONCEPT REINFORCEMENT

For each equation in column A, select the most appropriate graph in column B. [R.1, R.4, R.5]

Column A | Column B

1. $y = |x|$

a)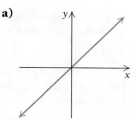

2. $f(x) = x^2 - 1$

b)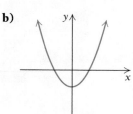

3. $y = -2x - 1$

c)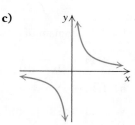

4. $y = x$

d)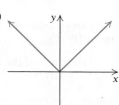

5. $g(x) = \sqrt{x}$

e)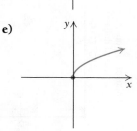

6. $f(x) = \dfrac{1}{x}$

f)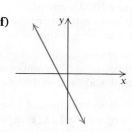

In Exercises 7–14, classify each statement as either true or false.

7. The graph of an equation is a drawing that represents all ordered pairs that are solutions of the equation. [R.1]

8. If $f(-3) = 5$ and $f(3) = 5$, then f cannot be a function. [R.2]

9. The notation $(3, 7)$ can represent a point or an interval. [R.3]

10. An equation of the form $y - y_1 = m(x - x_1)$ has a graph that is a line of slope m passing through (x_1, y_1). [R.4]

11. The graph of an equation of the form $f(x) = ax^2 + bx + c$ has its vertex at $x = b/(2a)$. [R.5]

12. A scatterplot is a random collection of points near a line. [R.6]

13. Unless stated otherwise, the domain of a polynomial function is the set of all real numbers. [R.5]

14. The graph of a constant function has a slope of 0. [R.4]

REVIEW EXERCISES

15. Life science: babies born to women of age x. The following graph relates the number of babies born per 1000 women to the women's age. [R.1, R.3]

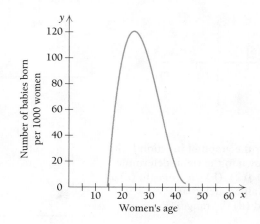

Use the graph to answer the following.

a) **What is the incidence of babies born to women of age 35?**

b) **For what ages are approximately 100 babies born per every 1000 women?**

c) **Make an estimate of the domain of the function, and explain why it should be so.**

16. Business: compound interest. Suppose that $1100 is invested at 5%, compounded semiannually. How much is in the account at the end of 4 yr? [R.1]

17. Finance: compound interest. Suppose that $4000 is borrowed at 12%, compounded annually. How much is owed at the end of 2 yr? [R.1]

18. Is the following correspondence a function? Why or why not? [R.2]

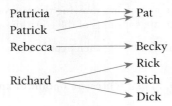

19. A function is given by $f(x) = -x^2 + x$. Find each of the following. [R.2]

   a) $f(3)$ **b)** $f(-5)$
   c) $f(a)$ **d)** $f(x + h)$

Graph. [R.5]

20. $y = |x + 1|$ 21. $f(x) = (x - 2)^2$

22. $f(x) = \dfrac{x^2 - 16}{x + 4}$ 23. $g(x) = \sqrt{x + 1}$

Use the vertical-line test to determine whether each of the following is the graph of a function. [R.2]

24. 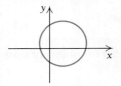   25.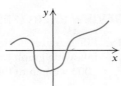

26. 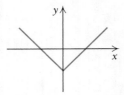   27.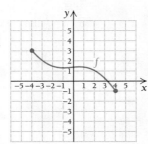

28. For the graph of function f shown to the right, determine **(a)** $f(2)$; **(b)** the domain; **(c)** all x-values such that $f(x) = 2$; and **(d)** the range. [R.3]

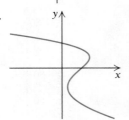

29. Consider the function given by

$$f(x) = \begin{cases} -x^2 + 2, & \text{for } x < 1, \\ 4, & \text{for } 1 \le x < 2, \\ \frac{1}{2}x, & \text{for } x \ge 2. \end{cases}$$

   a) Find $f(-1), f(1.5)$, and $f(6)$. [R.2]
   b) Graph the function. [R.2]

30. Write interval notation for each graph. [R.3]

   a)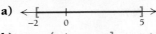

   b)

   c)

31. Write interval notation for each of the following. Then graph the interval on a number line. [R.3]

   a) $\{x | -4 \le x < 5\}$ **b)** $\{x | x > 2\}$

32. For the function graphed below, determine **(a)** $f(-3)$; **(b)** the domain; **(c)** all x-values for which $f(x) = 4$; **(d)** the range. [R.3]

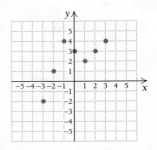

33. Find the domain of f. [R.3, R.5]

   a) $f(x) = \dfrac{7}{2x - 10}$

   b) $f(x) = \sqrt{x + 6}$

34. What are the slope and the y-intercept of $y = -3x + 2$? [R.4]

35. Find an equation of the line with slope $\frac{1}{4}$, containing the point $(8, -5)$. [R.4]

36. Find the slope of the line containing the points $(2, -5)$ and $(-3, 10)$. [R.4]

Find the average rate of change. [R.4]

37.

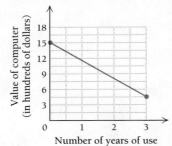

38.

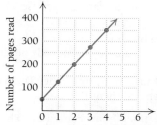

Number of days spent reading

39. Business: shipping charges. The amount A that Pet-Treats-to-U charges for shipping is directly proportional to the value V of the item(s) being shipped. If the business charges \$2.10 to ship a \$60 gift basket, find an equation of variation expressing A as a function of V. [R.4]

40. Business: profit-and-loss analysis. The band Soul Purpose has fixed costs of \$4000 for producing a new CD. Thereafter, the variable costs are \$0.50 per CD, and the CD will sell for \$10. [R.4]

 a) Find and graph $C(x)$, the total cost of producing x CDs.
 b) Find and graph $R(x)$, the total revenue from the sale of x CDs. Use the same axes as in part (a).
 c) Find and graph $P(x)$, the total profit from the production and sale of x CDs. Use the same axes as in part (b).
 d) How many CDs must the band sell in order to break even?

41. Graph each pair of equations on one set of axes. [R.5]
 a) $y = \sqrt{x}$ and $y = \sqrt{x} - 3$
 b) $y = x^3$ and $y = (x - 1)^3$

42. Graph each of the following. If the graph is a parabola, identify the vertex. [R.5]
 a) $f(x) = x^2 - 6x + 8$
 b) $g(x) = \sqrt[3]{x} + 2$
 c) $y = -\dfrac{1}{x}$
 d) $y = \dfrac{x^2 + x - 6}{x - 2}$

43. Solve each of the following. [R.5]
 a) $5 + x^2 = 4x + 2$
 b) $2x^2 = 4x + 3$

44. Rewrite each of the following as an equivalent expression with a rational exponent. [R.5]
 a) $\sqrt[5]{x^4}$
 b) $\sqrt{t^8}$
 c) $\dfrac{1}{\sqrt{3m^2}}$
 d) $\dfrac{1}{\sqrt{x^2 - 9}}$

45. Rewrite each of the following as an equivalent expression using radical notation. [R.5]
 a) $x^{2/5}$
 b) $m^{-3/5}$
 c) $(x^2 - 5)^{1/2}$
 d) $\dfrac{1}{t^{-1/3}}$

46. Determine the domain of the function given by $f(x) = \sqrt[8]{2x - 9}$. [R.5]

47. Economics: equilibrium point. Find the equilibrium point for the given demand and supply functions. [R.5]

 Demand: $q = (x - 7)^2$
 Supply: $q = x^2 + x + 4$ (assume $x \le 7$)

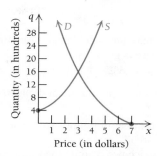

48. Trail maintenance. The amount of time required to maintain a section of the Appalachian Trail varies inversely as the number of volunteers working. If a particular section of trail can be cleared in 4 hr by 9 volunteers, how long would it take 11 volunteers to clear the same section? [R.5]

49. Life science: maximum heart rate. A person exercising should not exceed a maximum heart rate, which depends on his or her gender, age, and resting heart rate. The following table shows data relating resting heart rate and maximum heart rate for a 20-yr-old woman. [R.6]

Resting Heart Rate, r (in beats per minute)	Maximum Heart Rate, M (in beats per minute)
50	170
60	172
70	174
80	176

(*Source*: American Heart Association.)

 a) Using the data points $(50, 170)$ and $(80, 176)$, find a linear function that fits the data.
 b) Graph the scatterplot and the function on the same set of axes.
 c) Use the function to predict the maximum heart rate of a woman whose resting heart rate is 67.

50. Business: ticket profits. The Spring Valley Drama Troupe is performing a new play. Data relating the daily profit P to the number of days after opening night are given below. [R.6]

Days, x	0	9	18	27	36	45
Profit, P (in dollars)	870	548	-100	-100	510	872

 a) Make a scatterplot of the data.
 b) Decide whether the data seem to fit a quadratic function.
 c) Using the data points $(0, 870)$, $(18, -100)$, and $(45, 872)$, find a quadratic function that fits the data.

d) Use the function to estimate the profit made on the 30th day.

e) Make an estimate of the domain of this function. Explain its restrictions.

SYNTHESIS

51. *Economics: demand.* The demand function for Clifton Cheddar Cheese is given by

Demand: $q = 800 - x^3$, $0 \le x \le 9.28$,

where x is the price per pound and q is in thousands of pounds.

a) Find the number of pounds sold when the price per pound is \$6.50.

b) Find the price per pound when 720,000 lb are sold.

TECHNOLOGY CONNECTION ················

Graph the function and find the zeros, the domain, and the range. [R.5]

52. $f(x) = x^3 - 9x^2 + 27x + 50$

53. $f(x) = \sqrt[3]{|4 - x^2|} + 1$

54. Approximate the point(s) of intersection of the graphs of the two functions in Exercises 52 and 53. [R.5]

55. *Life science: maximum heart rate.* Use the data in Exercise 49. [R.6]

a) Use regression to fit a linear function to the data.

b) Use the linear function to predict the maximum heart rate of a woman whose resting heart rate is 67.

c) Compare your answer to that found in Exercise 49. Are the answers equally reliable? Why or why not?

56. *Business: ticket profits.* Use the data in Exercise 50. [R.6]

a) Use regression to fit a quadratic function to the data.

b) Use the function to estimate the profit made on the 30th day.

c) What factors might cause the Spring Valley Drama Troupe's profit to drop and then rise?

57. *Social sciences: time spent on home computer.* The data in the table below relate the average number of minutes spent per month on a home computer, A, to a person's age, x. [R.6]

a) Use regression to fit linear, quadratic, cubic, and quartic functions to the data.

b) Make a scatterplot of the data and graph each function on the scatterplot.

c) Which function fits the data best? Why?

Age (in years)	Average Use (in minutes per month)
6.5	363
14.5	645
21	1377
29.5	1727
39.5	1696
49.5	2052
55	2299

(*Source*: Media Matrix; The PC Meter Company.)

CHAPTER R
TEST

1. *Business: compound interest.* A person made an investment at 6.5% compounded annually. It has grown to \$798.75 in 1 yr. How much was originally invested?

2. A function is given by $f(x) = -x^2 + 5$. Find:

a) $f(-3)$;

b) $f(a + h)$.

3. What are the slope and the y-intercept of $y = \frac{4}{5}x - \frac{2}{3}$?

4. Find an equation of the line with slope $\frac{1}{4}$, containing the point $(-3, 7)$.

5. Find the slope of the line containing the points $(-9, 2)$ and $(3, -4)$.

Find the average rate of change.

6.

Value of color copier (in hundreds of dollars) vs. Number of years of use

7.

Weight gained (in pounds) vs. Number of bags of feed used

8. *Life science: body fluids.* The weight F of fluids in a human is directly proportional to body weight W. It is known that a person who weighs 180 lb has 120 lb of fluids. Find an equation of variation expressing F as a function of W.

9. *Business: profit-and-loss analysis.* A printing shop has fixed costs of \$8000 for producing a newly designed note card. Thereafter, the variable costs are \$0.08 per card. The revenue from each card is expected to be \$0.50.
   a) Formulate a function $C(x)$ for the total cost of producing x cards.
   b) Formulate a function $R(x)$ for the total revenue from the sale of x cards.
   c) Formulate a function $P(x)$ for the total profit from the production and sale of x cards.
   d) How many cards must the company sell in order to break even?

10. *Economics: equilibrium point.* Find the equilibrium point for these demand and supply functions:

   Demand: $q = (x - 8)^2$, $0 \le x \le 8$,
   Supply: $q = x^2 + x + 13$,

   given that x is the unit price, in dollars, and q is the quantity demanded or supplied, in thousands.

Use the vertical-line test to determine whether each of the following is the graph of a function.

11.

12.

13. For the following graph of a quadratic function f, determine **(a)** $f(1)$; **(b)** the domain; **(c)** all x-values such that $f(x) = 4$; and **(d)** the range.

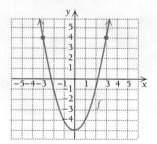

14. Graph: $f(x) = 8/x$.

15. Convert to a rational exponent: $1/\sqrt{t}$.

16. Convert to radical notation: $t^{-3/5}$.

17. Graph: $f(x) = \dfrac{x^2 - 1}{x + 1}$.

Determine the domain of each function.

18. $f(x) = \dfrac{x^2 + 20}{x^2 + 5x - 14}$ **19.** $f(x) = \dfrac{x}{\sqrt{3x + 6}}$

20. Write interval notation for the following graph.

21. Graph:

$$f(x) = \begin{cases} x^2 + 2, & \text{for } x \ge 0, \\ x^2 - 2, & \text{for } x < 0. \end{cases}$$

22. *Nutrition.* As people age, their daily caloric needs change. The following table shows data for physically active females, relating age to number of calories needed daily.

Age	Number of Calories Needed Daily
6	1800
11	2200
16	2400
24	2400
41	2200

(*Source:* Based on data from U.S. Department of Agriculture.)

   a) Make a scatterplot of the data.
   b) Do the data appear to fit a quadratic function?
   c) Using the data points $(6, 1800)$, $(16, 2400)$, and $(41, 2200)$, find a quadratic function that fits the data.
   d) Use the function from part (c) to predict the number of calories needed daily by a physically active 30-yr-old woman.
   e) Estimate the domain of the function from part (a). Explain its restrictions.

SYNTHESIS

23. Simplify: $(64^{4/3})^{-1/2}$.

24. Find the domain and the zeros of the function given by
$$f(x) = (5 - 3x)^{1/4} - 7.$$

25. Write an equation that has exactly three solutions: -3, 1, and 4. Answers will vary.

26. A function's average rate of change over the interval $[1, 5]$ is $-\frac{3}{7}$. If $f(1) = 9$, find $f(5)$.

TECHNOLOGY CONNECTION

27. Graph the function and find the zeros and the domain and the range:
$$f(x) = \sqrt[3]{|9 - x^2|} - 1.$$

28. *Nutrition.* Use the data in Exercise 22.
   a) Use REGRESSION to fit a quadratic function to the data.
   b) Use the function from part (a) to predict the number of calories needed daily by a physically active 30-yr-old woman.
   c) Compare your answer from part (b) with that from part (d) of Exercise 22. Which answer do you feel is more accurate? Why?

Average Price of a Movie Ticket

Extended Technology Applications occur at the end of each chapter. They are designed to consider certain applications in greater depth, make use of calculator skills, and allow for possible group or collaborative learning.

Have you noticed that the price of a movie ticket seems to increase? The table and graph that follow show the average price of a movie ticket for the years 1950 to 2010.

How much did you pay the last time you went to the movies in the evening (not a matinee)? The average prices in the table may seem low, but they reflect discounts for matinees and children's and senior citizens' tickets. Let's use our skills with REGRESSION to analyze the data.

YEAR, t	AVERAGE TICKET PRICE, P(t)
1950, 0	$ 0.46
1955, 5	0.58
1960, 10	0.76
1965, 15	1.01
1970, 20	1.55
1975, 25	2.05
1980, 30	2.69
1985, 35	3.55
1990, 40	4.23
1995, 45	4.35
2000, 50	5.39
2005, 55	6.41
2010, 60	8.01

(*Source*: Motion Picture Association of America.)

EXERCISES

1. **a)** Using REGRESSION, find a linear function that fits the data.
   b) Graph the linear function.
   c) Use the linear function to predict the average price of a movie ticket in 2012 and in 2020. Do these estimates appear reasonable?
   d) Use the function to predict when the average price of a ticket will reach $20. Does this estimate seem reasonable?

2. **a)** Using REGRESSION, find a quadratic function,
   $$y = ax^2 + bx + c,$$
   that fits the data.
   b) Graph the quadratic function.
   c) Use the quadratic function to predict the average price of a movie ticket in 2012 and in 2020. Do these estimates appear reasonable?
   d) Use the function to predict when the average price of a ticket will reach $20. Does this estimate seem reasonable?

3. **a)** Using REGRESSION, find a cubic function,
   $$y = ax^3 + bx^2 + cx + d,$$
   that fits the data.
   b) Graph the cubic function.
   c) Use the cubic function to predict the average price of a movie ticket in 2012 and in 2020. Do these estimates appear reasonable?
   d) Use the function to predict when the average price of a ticket will reach $20. Does this estimate seem reasonable?

4. **a)** Using REGRESSION, find a quartic function,
   $$y = ax^4 + bx^3 + cx^2 + dx + e,$$
   that fits the data.
   b) Graph the quartic function.
   c) Use the quartic function to predict the average price of a movie ticket in 2012 and in 2020. Do these estimates appear reasonable?
   d) Use the function to predict when the average price of a ticket will reach $20. Does this estimate seem reasonable?

5. You are a research statistician assigned the task of making an accurate prediction of movie ticket prices.
   a) Why might you not use the linear function?
   b) Why might you use the quadratic function rather than the linear function?
   c) Examine the graphs and equations of the four functions and the estimates they provided. One choice the research statistician makes is not to use a higher-order polynomial function when a lower-order one works just as well. Under this criterion, why would a quadratic work just as well as a cubic? Look at the coefficients of the leading terms.

There are yet other procedures a statistician uses to choose predicting functions but they are beyond the scope of this text.

Differentiation

Chapter Snapshot

What You'll Learn

Why It's Important

With this chapter, we begin our study of calculus. The first concepts we consider are *limits* and *continuity*. We apply those concepts to establishing the first of the two main building blocks of calculus: differentiation.

Differentiation is a process that takes a formula for a function and derives a formula for another function, called a *derivative*, that allows us to find the slope of a tangent line to a curve at a point. A derivative also represents an instantaneous rate of change. Throughout the chapter, we will learn various techniques for finding derivatives.

Where It's Used

MARKET SATURATION

A new product is placed on the market and becomes very popular. How can the derivative help us understand its rate of sales and the phenomenon of market saturation?

This problem appears as Example 8 in Section 1.7.

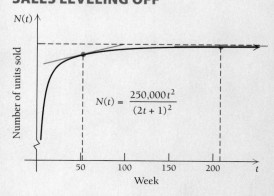

SALES LEVELING OFF

$$N(t) = \frac{250{,}000\,t^2}{(2t+1)^2}$$

Number of units sold — $N(t)$

Week — t

Limits: A Numerical and Graphical Approach

OBJECTIVE

- Find limits of functions, if they exist, using numerical or graphical methods.

In this section, we discuss the concept of a *limit*. The discussion is intuitive—that is, relying on prior experience and lacking formal proof.

Suppose a football team has the ball on its own 10-yard line. Then, because of a penalty, the referee moves the ball back half the distance to the goal line; the ball is now on the 5-yard line. If the team commits the same infraction again, the ball will again be moved half the distance to the goal line; now it's on the 2.5-yard line. If this kind of penalty were repeated over and over again, the ball would move steadily *closer* to the goal line but never actually be placed *on* the goal line. We would say that the *limit* of the distance between the ball and the goal line is zero.

Limits

One important aspect of the study of calculus is the analysis of how function values (outputs) change as input values change. Basic to this study is the notion of a limit. Suppose a function f is given and suppose the x-values (the inputs) get closer and closer to some number a. If the corresponding outputs—the values of $f(x)$—get closer and closer to another number, then that number is called the *limit* of $f(x)$ as x approaches a.

For example, let $f(x) = 2x + 3$ and select x-values that get closer and closer to 4. In the table and graph below, we see that as the input values approach 4 from the left (that is, are less than 4), the output values approach 11, and as the input values approach 4 from the right (that is, are greater than 4), the output values also approach 11. Thus, we say:

As x approaches 4 from either side, the function $f(x) = 2x + 3$ approaches 11.

Limit Numerically

Limit Graphically

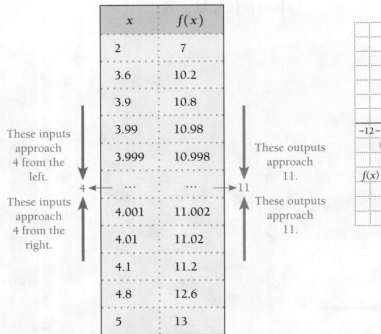

x	$f(x)$
2	7
3.6	10.2
3.9	10.8
3.99	10.98
3.999	10.998
...	...
4.001	11.002
4.01	11.02
4.1	11.2
4.8	12.6
5	13

These inputs approach 4 from the left.

These inputs approach 4 from the right.

These outputs approach 11.

These outputs approach 11.

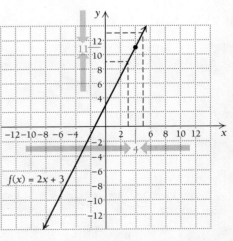

$f(x) = 2x + 3$

An arrow, →, is often used to stand for the words "approaches from either side." Thus, the statement above can be written:

$$\text{As } x \to 4, \quad 2x + 3 \to 11.$$

The number 11 is said to be the *limit* of $2x + 3$ as x approaches 4 from either side. We can abbreviate this statement as follows:

$$\lim_{x \to 4} (2x + 3) = 11.$$

This is read: "The limit, as x approaches 4, of $2x + 3$ is 11."

DEFINITION

As x approaches a, the **limit** of $f(x)$ is L, written

$$\lim_{x \to a} f(x) = L,$$

if all values of $f(x)$ are close to L for values of x that are sufficiently close, but not equal, to a. The limit L must be a unique real number.

When we write $\lim_{x \to a} f(x)$, we are indicating that x is approaching a from both sides. If we want to be specific about the side from which the x-values approach the value a, we use the notation

$$\lim_{x \to a^-} f(x) \text{ to indicate the limit from the left (that is, where } x < a),$$

or

$$\lim_{x \to a^+} f(x) \text{ to indicate the limit from the right (that is, where } x > a).$$

These are called *left-hand* and *right-hand limits*, respectively. In order for a limit to exist, both the left-hand and right-hand limits must exist and be the same. This leads to the following theorem.

THEOREM

As x approaches a, the limit of $f(x)$ is L if the limit from the left exists and the limit from the right exists and both limits are L. That is,

$$\text{if} \quad \lim_{x \to a^+} f(x) = \lim_{x \to a^-} f(x) = L, \quad \text{then} \quad \lim_{x \to a} f(x) = L.$$

The converse of this theorem is also true: if $\lim_{x \to a} f(x) = L$, then it is assumed that both the left-hand limit, $\lim_{x \to a^-} f(x)$, and the right-hand limit, $\lim_{x \to a^+} f(x)$, exist and are equal to L.

TECHNOLOGY CONNECTION

Finding Limits Using the TABLE and TRACE Features

Consider the function given by $f(x) = 3x - 1$. Let's use the TABLE feature to complete the following table. Note that the inputs do not have the same increment from one to the next, but do approach 6 from either the left or the right. We use TblSet and select Indpnt and Ask mode. Then we enter the inputs shown and use the corresponding outputs to complete the table.

(continued)

Finding Limits Using the TABLE and TRACE Features (*continued*)

$$f(x) = 3x - 1$$

x	$f(x)$
5	14
5.8	16.4
5.9	16.7
5.99	16.97
5.999	16.997
6 ←	→ ?
6.001	17.003
6.01	17.03
6.1	17.3
6.4	18.2
7	20

Now we set the table in Auto mode and starting (TblStart) with a number near 6, we make tables for some increments (ΔTbl) like 0.1, 0.01, −0.1, −0.01, and so on, to determine $\lim\limits_{x \to 6} f(x)$.

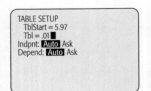

As an alternative, graphical approach, let's use the TRACE feature with the graph of f. We move the cursor from left to right so that the x-coordinate approaches 6 from the left, changing the window as needed, to see what happens. For example, let's use $[5.3, 6.4, 14, 18]$ and move the cursor from right to left so that the x-coordinate approaches 6 from the right. In general, the TRACE feature is not an efficient way to find limits, but it will help you

to visualize the limit process in this early stage of your learning.

Using the TABLE and TRACE features, let's complete the following:

$$\lim_{x \to 6^+} f(x) = \boxed{17} \text{ and } \lim_{x \to 6} f(x) = \boxed{17}.$$

Thus,

$$\lim_{x \to 6} f(x) = \boxed{17}.$$

EXERCISES

Consider $f(x) = 3x - 1$. Use the TABLE and TRACE features, making up your own tables, to find each of the following.

1. $\lim\limits_{x \to 2} f(x)$ **2.** $\lim\limits_{x \to -1} f(x)$

Consider $g(x) = x^3 - 2x - 2$ for Exercises 3–5.

3. Complete the following table.

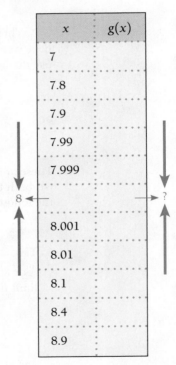

x	$g(x)$
7	
7.8	
7.9	
7.99	
7.999	
8 ←	→ ?
8.001	
8.01	
8.1	
8.4	
8.9	

Use the TABLE and TRACE features to find each of the following.

4. $\lim\limits_{x \to 8} g(x)$ **5.** $\lim\limits_{x \to -1} g(x)$

Refer again to the function $f(x) = 2x + 3$. We showed that as x approaches 4, the function values approach 11, summarized as $\lim\limits_{x \to 4} (2x + 3) = 11$. You may have thought "Why not just substitute 4 into the function to get 11?" You are partially correct to make this observation (and in the next section, we see that we can use such shortcuts in certain cases), but keep in mind that we are curious about the behavior of the function $f(x) = 2x + 3$ for values of x close to 4, not necessarily at 4 itself. It may help to summarize what we know already:

- At $x = 4$, the function value is 11, and this is visualized as the point $(4, 11)$ on the graph of f.

- For values of x close to 4, the values of $f(x)$ are correspondingly near 11; this is the *limit* we have been discussing.

The limit can help us understand the behavior of some functions a little more clearly. Consider the following example.

■ **EXAMPLE 1**   Let $f(x) = \dfrac{x^2 - 1}{x - 1}$.

a) What is $f(1)$?   **b)** What is the limit of $f(x)$ as x approaches 1?

Solution

a) There is no answer, since we get a 0 in the denominator:

$$f(1) = \frac{(1)^2 - 1}{(1) - 1} = \frac{0}{0}$$

Thus, $f(1)$ does not exist. There is no point on the graph of $f(x)$ shown below for $x = 1$.

Limit Numerically

$x \to 1^-$ $(x < 1)$	$f(x)$
0.9	1.9
0.99	1.99
0.999	1.999

$x \to 1^+$ $(x > 1)$	$f(x)$
1.1	2.1
1.01	2.01
1.001	2.001

Limit Graphically

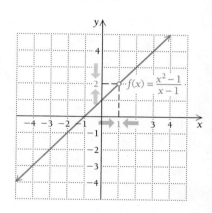

$f(x) = \dfrac{x^2 - 1}{x - 1}$

〉 Quick Check 1

Let

$$f(x) = \frac{x^2 - 9}{x - 3}.$$

(See Example 5 in Section R.5.)

a) What is $f(3)$?

b) What is the limit of f as x approaches 3?

b) We select x-values close to 1 on either side (see the table above), and we see that the function values get closer and closer to 2. Thus, the limit of $f(x)$ as x approaches 1 is 2: $\lim\limits_{x \to 1} f(x) = 2$.

The graph has a "hole" at the point $(1, 2)$. Thus, even though the function is not defined at $x = 1$, the limit *does* exist as $x \to 1$.

〈 Now try Quick Check 1

Limits are also useful when discussing the behavior of piecewise-defined functions, as the following example illustrates.

■ **EXAMPLE 2**   Consider the function H given by

$$H(x) = \begin{cases} 2x + 2, & \text{for } x < 1, \\ 2x - 4, & \text{for } x \geq 1. \end{cases}$$

Graph the function and find each of the following limits, if they exist. When necessary, state that the limit does not exist.

a) $\lim_{x \to 1} H(x)$ **b)** $\lim_{x \to -3} H(x)$

Solution   We check the limits from the left and from the right both numerically, with an input–output table, and graphically.

a)   *Limit Numerically*   *Limit Graphically*

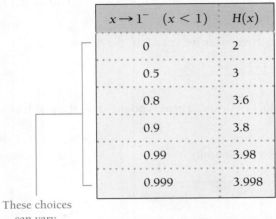

$x \to 1^-$ $(x < 1)$	$H(x)$
0	2
0.5	3
0.8	3.6
0.9	3.8
0.99	3.98
0.999	3.998

These choices can vary.

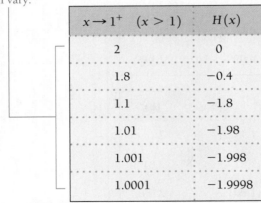

$x \to 1^+$ $(x > 1)$	$H(x)$
2	0
1.8	−0.4
1.1	−1.8
1.01	−1.98
1.001	−1.998
1.0001	−1.9998

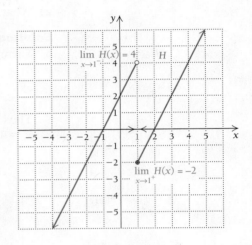

As inputs x approach 1 from the left, outputs $H(x)$ approach 4. Thus, the limit from the left is 4. That is,

$$\lim_{x \to 1^-} H(x) = 4.$$

But as inputs x approach 1 from the right, outputs $H(x)$ approach −2. Thus, the limit from the right is −2. That is,

$$\lim_{x \to 1^+} H(x) = -2.$$

Since the limit from the left, 4, is not the same as the limit from the right, 2, we say that

$$\lim_{x \to 1} H(x) \ does \ not \ exist.$$

We note in passing that $H(1) = -2$. In this example, the function value exists for $x = 1$, but the limit as x approaches 1 does not exist.

Exploratory

Check the results of Examples 1 and 2 using the TABLE feature. See Section R.2 to recall how to graph functions defined piecewise.

Note: Each Technology Connection labeled "Exploratory" is designed to lead you through a discovery process. Because of this, answers are not provided.

b)

Limit Numerically		Limit Graphically

$x \to -3^-$ ($x < -3$)	$H(x)$
-4	-6
-3.5	-5
-3.1	-4.2
-3.01	-4.02
-3.001	-4.002

$x \to -3^+$ ($x > -3$)	$H(x)$
-2	-2
-2.5	-3
-2.9	-3.8
-2.99	-3.98
-2.999	-3.998

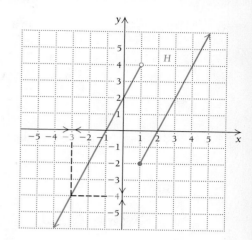

As inputs x approach -3 from the left, outputs $H(x)$ approach -4, so the limit from the left is -4. That is,

$$\lim_{x \to -3^-} H(x) = -4.$$

As inputs x approach -3 from the right, outputs $H(x)$ approach -4, so the limit from the right is -4. That is,

$$\lim_{x \to -3^+} H(x) = -4.$$

Since the limits from the left and from the right exist and are the same, we have

$$\lim_{x \to -3} H(x) = -4.$$

❬ Now try Quick Check 2

⟩ **Quick Check 2**

Let

$$k(x) = \begin{cases} -x + 4, & \text{for } x \leq 3, \\ 2x + 1, & \text{for } x > 3. \end{cases}$$

Find these limits:

a) $\lim\limits_{x \to 3^-} k(x)$, $\lim\limits_{x \to 3^+} k(x)$, and $\lim\limits_{x \to 3} k(x)$;

b) $\lim\limits_{x \to 1} k(x)$.

The limit at a number a *does not depend* on the function value at a or even on whether that function value, $f(a)$, exists. That is, whether or not a limit exists at a has *nothing* to do with the function value $f(a)$.

The "Wall" Method

As an alternative approach for Example 2, we can draw a "wall" at $x = 1$, as shown in blue on the graph to the left on the next page. We then follow the curve from left to right with a pencil until we hit the wall and mark the location with an ✕, assuming it can be determined. Then we follow the curve from right to left until we hit the wall and mark that location with an ×. If the locations are the same, as in the graph to the right below, a limit exists. Thus, for Example 2,

$$\lim_{x \to 1} H(x) \text{ does not exist,} \quad \text{and} \quad \lim_{x \to -3} H(x) = -4.$$

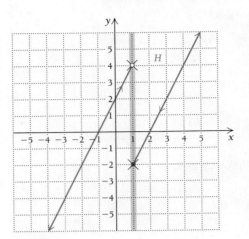 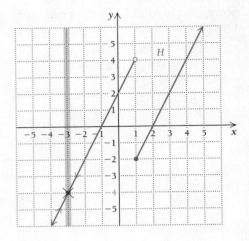

■ **EXAMPLE 3**   Consider the function defined as follows:

$$G(x) = \begin{cases} 5, & \text{for } x = 1, \\ x + 1, & \text{for } x \neq 1. \end{cases}$$

Graph the function, and find each of the following limits, if they exist. If necessary, state that the limit does not exist.

a) $\lim_{x \to 1} G(x)$ **b)** $\lim_{x \to -2} G(x)$

Solution   The graph of G follows.

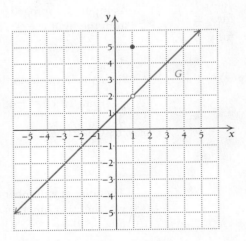

a) As inputs x approach 1 from the left, outputs $G(x)$ approach 2, so the limit from the left is 2. As inputs x approach 1 from the right, outputs $G(x)$ also approach 2, so the limit from the right is 2. Since the limit from the left, 2, is the same as the limit from the right, 2, we have

$$\lim_{x \to 1} G(x) = 2.$$

Note that the limit, 2, is not the same as the function value at 1, which is $G(1) = 5$.

Limit Numerically

$x \to 1^-$ $(x < 1)$	$G(x)$
0	1
0.5	1.5
0.9	1.9
0.99	1.99

$$\lim_{x \to 1} G(x) = 2$$

$x \to 1^+$ $(x > 1)$	$G(x)$
1.5	2.5
1.1	2.1
1.01	2.01
1.001	2.001

Limit Graphically

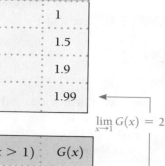

b) Using the same approach as in part (a), we have

$$\lim_{x \to -2} G(x) = -1.$$

Note that in this case, the limit, -1, is the same as the function value at -2, which is $G(-2) = -1$.

Limit Numerically

$x \to -2^-$ $(x < -2)$	$G(x)$
-3	-2
-2.5	-1.5
-2.1	-1.1
-2.01	-1.01

$$\lim_{x \to -2} G(x) = -1$$

$x \to -2^+$ $(x > -2)$	$G(x)$
-1.5	-0.5
-1.9	-0.9
-1.99	-0.99
-1.999	-0.999

Limit Graphically

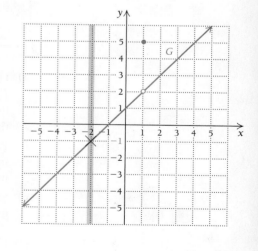

〉 **Quick Check 3**

Calculate the following limits based on the graph of f.

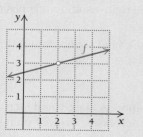

a) $\displaystyle\lim_{x \to 2^-} f(x)$

b) $\displaystyle\lim_{x \to 2^+} f(x)$

c) $\displaystyle\lim_{x \to 2} f(x)$

〈 Now try Quick Check 3

Limits Involving Infinity

Limits also help us understand the role of infinity with respect to some functions. Consider the following example.

■ EXAMPLE 4 Let $f(x) = \dfrac{1}{x}$.

a) Find $\lim\limits_{x \to 0^-} f(x)$.

b) Find $\lim\limits_{x \to 0^+} f(x)$.

c) Use the information from parts (a) and (b) to form a conclusion about $\lim\limits_{x \to 0} f(x)$.

Solution We note first that $f(0)$ does not exist: there is no point on the graph that corresponds to $x = 0$.

Limit Numerically

$x \to 0^-$ $(x < 0)$	$f(x)$
-0.1	-10
-0.01	-100
-0.001	$-1,000$
-0.0001	$-10,000$

$x \to 0^+$ $(x > 0)$	$f(x)$
0.1	10
0.01	100
0.001	$1,000$
0.0001	$10,000$

Limit Graphically

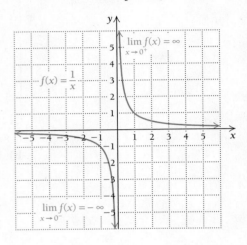

a) The table and graph show that as x approaches 0 from the left, the corresponding $f(x)$ values are decreasing without bound. We conclude that the left-hand limit is negative infinity; that is, $\lim\limits_{x \to 0^-} f(x) = -\infty$. We describe the notion of "infinity" by the symbol ∞. The symbol ∞ *does not represent* a real number.

b) The table and graph show that as x approaches 0 from the right, the $f(x)$ values increase without bound toward positive infinity. The right-hand limit is positive infinity; that is, $\lim\limits_{x \to 0^+} f(x) = \infty$.

c) Since the left-hand and right-hand limits do not match (and are not finite), $\lim\limits_{x \to 0} f(x)$ does not exist.

Sometimes we need to determine limits when the inputs get larger and larger without bound, that is, as the inputs approach infinity. In such cases, we are finding *limits at infinity*. Such a limit is expressed as

$$\lim_{x \to \infty} f(x) \quad \text{or} \quad \lim_{x \to -\infty} f(x).$$

These limits are approached from one side only: from the left if approaching positive infinity or from the right if approaching negative infinity.

■ **EXAMPLE 5**   Let $f(x) = \dfrac{1}{x}$. Find $\lim\limits_{x \to \infty} f(x)$ and $\lim\limits_{x \to -\infty} f(x)$.

Solution   The table shows that as x gets larger and larger in the positive direction, the values for $f(x)$ approach 0. Thus, $\lim\limits_{x \to \infty} f(x) = 0$. As x decreases in the negative direction, we get the same value for the limit: $\lim\limits_{x \to -\infty} f(x) = 0$.

Limit Numerically   *Limit Graphically*

$x \to \infty$	$f(x)$
10	0.1
100	0.01
1,000	0.001
10,000	0.0001

$$\lim_{x \to \infty} f(x) = 0$$

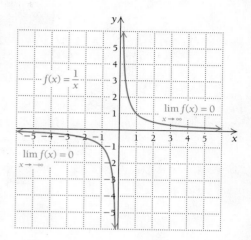

$x \to -\infty$	$f(x)$
−10	−0.1
−100	−0.01
−1,000	−0.001
−10,000	−0.0001

$$\lim_{x \to -\infty} f(x) = 0$$

■ **EXAMPLE 6**   Consider the function f given by

$$f(x) = \frac{1}{x - 2} + 3.$$

Graph the function, and find each of the following limits, if they exist. If necessary, state that the limit does not exist.

a) $\lim\limits_{x \to 3} f(x)$ **b)** $\lim\limits_{x \to 2} f(x)$

Solution   The graph of $f(x)$ is shown to the right. Note that it is the same as the graph of $f(x) = \dfrac{1}{x}$ but shifted 2 units to the right and 3 units up.

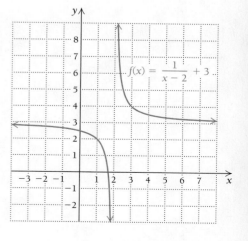

a) As inputs x approach 3 from the left, outputs $f(x)$ approach 4, so the limit from the left is 4. As inputs x approach 3 from the right, outputs $f(x)$ also approach 4. Since the limit from the left, 4, is the same as the limit from the right, we have

$$\lim_{x \to 3} f(x) = 4.$$

Limit Numerically

$x \to 3^-$ $(x < 3)$	$f(x)$
2.1	13
2.5	5
2.9	$4.\overline{1}$
2.99	$4.\overline{01}$

$$\lim_{x \to 3} f(x) = 4$$

$x \to 3^+$ $(x > 3)$	$f(x)$
3.5	$3.\overline{6}$
3.2	$3.8\overline{3}$
3.1	$3.\overline{90}$
3.01	$3.\overline{9900}$

Limit Graphically

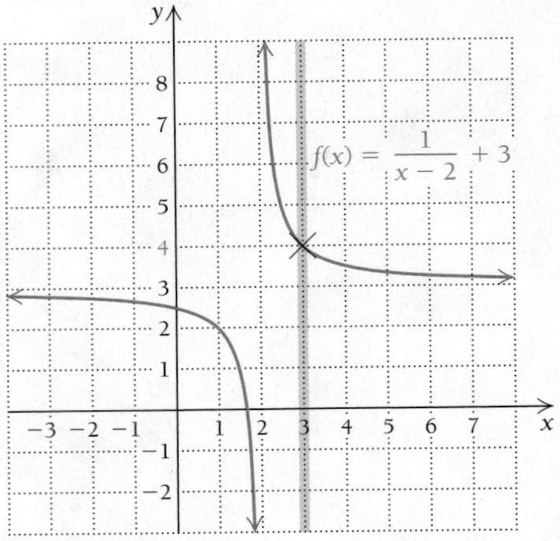

$$f(x) = \frac{1}{x-2} + 3$$

b) As inputs x approach 2 from the left, outputs $f(x)$ become more and more negative, without bound. These numbers do not approach any real number, although it might be said that the limit from the left is negative infinity, $-\infty$. That is,

$$\lim_{x \to 2^-} f(x) = -\infty.$$

As inputs x approach 2 from the right, outputs $f(x)$ become larger and larger, without bound. These numbers do not approach any real number, although it might be said that the limit from the right is infinity, ∞. That is,

$$\lim_{x \to 2^+} f(x) = \infty.$$

Because the left-sided limit differs from the right-sided limit,

$$\lim_{x \to 2} f(x) \text{ does not exist.}$$

Limit Numerically

$x \to 2^-$ $(x < 2)$	$f(x)$
1.5	1
1.9	-7
1.99	-97
1.999	-997

$\lim\limits_{x \to 2} f(x)$ does not exist.

$x \to 2^+$ $(x > 2)$	$f(x)$
2.5	5
2.1	13
2.01	103
2.001	1003

Limit Graphically

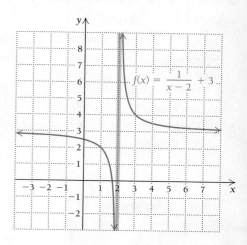

$f(x) = \dfrac{1}{x - 2} + 3$

■ **EXAMPLE 7** Consider again the function in Example 6, given by

$$f(x) = \frac{1}{x - 2} + 3.$$

Find $\lim\limits_{x \to \infty} f(x)$.

Solution As inputs x get larger and larger, outputs $f(x)$ get closer and closer to 3. We have

$$\lim_{x \to \infty} f(x) = 3.$$

Limit Numerically

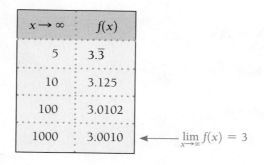

$x \to \infty$	$f(x)$
5	$3.\overline{3}$
10	3.125
100	3.0102
1000	3.0010

$\lim\limits_{x \to \infty} f(x) = 3$

Limit Graphically

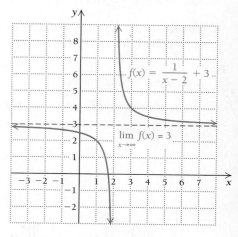

$f(x) = \dfrac{1}{x - 2} + 3$

$\lim\limits_{x \to \infty} f(x) = 3$

〉 **Quick Check 4**

Let $h(x) = \dfrac{1}{1 - x} + 6$. Find these limits:

a) $\lim\limits_{x \to 1} h(x)$; **b)** $\lim\limits_{x \to 2} h(x)$;

c) $\lim\limits_{x \to \infty} h(x)$.

〈 Now try Quick Check 4

Section Summary

- The *limit* of a function f, as x approaches a, is written $\lim_{x \to a} f(x) = L$. This means that as the values of x approach a, the corresponding values of $f(x)$ approach L. The value L must be a unique, finite number.
- A *left-hand limit* is written $\lim_{x \to a^-} f(x)$. The values of x are approaching a from the left, that is, $x < a$.
- A *right-hand limit* is written $\lim_{x \to a^+} f(x)$. The values of x are approaching a from the right, that is, $x > a$.

- If the left-hand and right-hand limits (as x approaches a) are *not* equal, the limit does *not* exist. On the other hand, if the left-hand and right-hand limits are equal and not infinite, the limit does exist.
- A limit $\lim_{x \to a} f(x)$ may exist even though the function value $f(a)$ does not. (See Example 1.)
- A limit $\lim_{x \to a} f(x)$ may exist and be different from the function value $f(a)$. (See Example 3b.)
- Graphs and tables are useful tools in determining limits.

EXERCISE SET
1.1

Complete each of the following statements.

1. As x approaches 3, the value of $2x + 5$ approaches _____.

2. As x approaches -4, the value of $3x + 7$ approaches _____.

3. As x approaches _____, the value of $-3x$ approaches 6.

4. As x approaches _____, the value of $x - 2$ approaches 5.

5. The notation $\lim_{x \to 4} f(x)$ is read _____.

6. The notation $\lim_{x \to 1} g(x)$ is read _____.

7. The notation $\lim_{x \to 5^-} F(x)$ is read _____.

8. The notation $\lim_{x \to 4^+} G(x)$ is read _____.

9. The notation _____ is read "the limit, as x approaches 2 from the right."

10. The notation _____ is read "the limit, as x approaches 3 from the left."

For Exercises 11–18, consider the function f given by

$$f(x) = \begin{cases} x - 2, & \text{for } x \leq 3, \\ x - 1, & \text{for } x > 3. \end{cases}$$

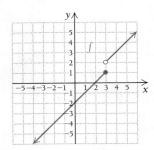

When necessary, state that the limit does not exist.

11. Find $\lim_{x \to 3^+} f(x)$.

12. Find $\lim_{x \to 3^-} f(x)$.

13. Find $\lim_{x \to -1^-} f(x)$.

14. Find $\lim_{x \to -1^+} f(x)$.

15. Find $\lim_{x \to 3} f(x)$.

16. Find $\lim_{x \to -1} f(x)$.

17. Find $\lim_{x \to 4} f(x)$.

18. Find $\lim_{x \to 2} f(x)$.

For Exercises 19–26, consider the function g given by

$$g(x) = \begin{cases} x + 6, & \text{for } x < -2, \\ -\tfrac{1}{2}x + 1, & \text{for } x \geq -2. \end{cases}$$

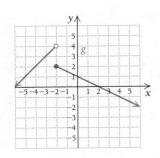

If a limit does not exist, state that fact.

19. $\lim_{x \to -2^-} g(x)$

20. $\lim_{x \to -2^+} g(x)$

21. $\lim_{x \to 4^+} g(x)$

22. $\lim_{x \to 4} g(x)$

23. $\lim_{x \to 4} g(x)$

24. $\lim_{x \to -2} g(x)$

25. $\lim_{x \to 2} g(x)$

26. $\lim_{x \to 4} g(x)$

For Exercises 27–34, use the following graph of F to find each limit. When necessary, state that the limit does not exist.

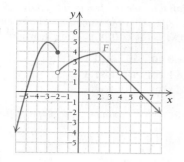

27. $\lim_{x \to -3} F(x)$

28. $\lim_{x \to 2} F(x)$

29. $\lim_{x \to -2} F(x)$

30. $\lim_{x \to -5} F(x)$

31. $\lim_{x \to 4} F(x)$

32. $\lim_{x \to 6} F(x)$

33. $\lim_{x \to -2^+} F(x)$

34. $\lim_{x \to -2^-} F(x)$

For Exercises 35–42, use the following graph of G to find each limit. When necessary, state that the limit does not exist.

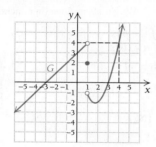

35. $\lim_{x \to -2} G(x)$

36. $\lim_{x \to 0} G(x)$

37. $\lim_{x \to 1^-} G(x)$

38. $\lim_{x \to 1^+} G(x)$

39. $\lim_{x \to 1} G(x)$

40. $\lim_{x \to 3^-} G(x)$

41. $\lim_{x \to 3^+} G(x)$

42. $\lim_{x \to 3} G(x)$

For Exercises 43–52, use the following graph of H to find each limit. When necessary, state that the limit does not exist.

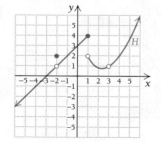

43. $\lim_{x \to -3} H(x)$

44. $\lim_{x \to -2^-} H(x)$

45. $\lim_{x \to -2^+} H(x)$

46. $\lim_{x \to -2} H(x)$

47. $\lim_{x \to 1^-} H(x)$

48. $\lim_{x \to 1^+} H(x)$

49. $\lim_{x \to 1} H(x)$

50. $\lim_{x \to 3^-} H(x)$

51. $\lim_{x \to 3^+} H(x)$

52. $\lim_{x \to 3} H(x)$

For Exercises 53–62, use the following graph of f to find each limit. When necessary, state that the limit does not exist.

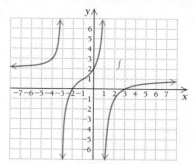

53. $\lim_{x \to -1} f(x)$

54. $\lim_{x \to 2} f(x)$

55. $\lim_{x \to -3} f(x)$

56. $\lim_{x \to 0} f(x)$

57. $\lim_{x \to 3} f(x)$

58. $\lim_{x \to 1} f(x)$

59. $\lim_{x \to -4} f(x)$

60. $\lim_{x \to -2} f(x)$

61. $\lim_{x \to \infty} f(x)$

62. $\lim_{x \to -\infty} f(x)$

For Exercises 63–80, graph each function and then find the specified limits. When necessary, state that the limit does not exist.

63. $f(x) = |x|$;   find $\lim_{x \to 0} f(x)$ and $\lim_{x \to -2} f(x)$.

64. $f(x) = x^2$;   find $\lim_{x \to -1} f(x)$ and $\lim_{x \to 0} f(x)$.

65. $g(x) = x^2 - 5$;   find $\lim_{x \to 0} g(x)$ and $\lim_{x \to -1} g(x)$.

66. $g(x) = |x| + 1$;   find $\lim_{x \to -3} g(x)$ and $\lim_{x \to 0} g(x)$.

67. $F(x) = \dfrac{1}{x - 3}$;   find $\lim_{x \to 3} F(x)$ and $\lim_{x \to 4} F(x)$.

68. $G(x) = \dfrac{1}{x + 2}$;   find $\lim_{x \to -1} G(x)$ and $\lim_{x \to -2} G(x)$.

69. $f(x) = \dfrac{1}{x} - 2$;   find $\lim_{x \to \infty} f(x)$ and $\lim_{x \to 0} f(x)$.

70. $f(x) = \dfrac{1}{x} + 3$;   find $\lim_{x \to \infty} f(x)$ and $\lim_{x \to 0} f(x)$.

71. $g(x) = \dfrac{1}{x + 2} + 4$;   find $\lim_{x \to \infty} g(x)$ and $\lim_{x \to -2} g(x)$.

72. $g(x) = \dfrac{1}{x - 3} + 2$;   find $\lim_{x \to \infty} g(x)$ and $\lim_{x \to 3} g(x)$.

73. $F(x) = \begin{cases} 2x + 1, & \text{for } x < 1, \\ x, & \text{for } x \geq 1. \end{cases}$
Find $\lim_{x \to 1^-} F(x)$, $\lim_{x \to 1^+} F(x)$, and $\lim_{x \to 1} F(x)$.

74. $G(x) = \begin{cases} -x + 3, & \text{for } x < 2, \\ x + 1, & \text{for } x \geq 2. \end{cases}$
Find $\lim_{x \to 2^-} G(x)$, $\lim_{x \to 2^+} G(x)$, and $\lim_{x \to 2} G(x)$.

75. $g(x) = \begin{cases} -x + 4, & \text{for } x < 3, \\ x - 3, & \text{for } x > 3. \end{cases}$
Find $\lim_{x \to 3^-} g(x)$, $\lim_{x \to 3^+} g(x)$, and $\lim_{x \to 3} g(x)$.

76. $f(x) = \begin{cases} 3x - 4, & \text{for } x < 1, \\ x - 2, & \text{for } x > 1. \end{cases}$

Find $\lim_{x \to 1^-} f(x)$, $\lim_{x \to 1^+} f(x)$, and $\lim_{x \to 1} f(x)$.

77. $G(x) = \begin{cases} x^2, & \text{for } x < -1, \\ x + 2, & \text{for } x > -1. \end{cases}$ Find $\lim_{x \to -1} G(x)$.

78. $F(x) = \begin{cases} -2x - 3, & \text{for } x < -1, \\ x^3, & \text{for } x > -1. \end{cases}$ Find $\lim_{x \to -1} F(x)$.

79. $H(x) = \begin{cases} x + 1, & \text{for } x < 0, \\ 2, & \text{for } 0 \le x < 1, \\ 3 - x, & \text{for } x \ge 1. \end{cases}$

Find $\lim_{x \to 0} H(x)$ and $\lim_{x \to 1} H(x)$.

80. $G(x) = \begin{cases} 2 + x, & \text{for } x \le -1, \\ x^2, & \text{for } -1 < x < 3, \\ 9, & \text{for } x \ge 3. \end{cases}$

Find $\lim_{x \to -1} G(x)$ and $\lim_{x \to 3} G(x)$.

APPLICATIONS

Business and Economics

Taxicab fares. **In New York City, taxicabs charge passengers $2.50 for entering a cab and then $0.40 for each one-fifth of a mile (or fraction thereof) traveled. (There are additional charges for slow traffic and idle times, but these are not considered in this problem.) If x represents the distance traveled in miles, then C(x) is the cost of the taxi fare, where**

$$C(x) = \$2.50, \quad \text{if} \quad x = 0,$$
$$C(x) = \$2.90, \quad \text{if} \quad 0 < x \le 0.2,$$
$$C(x) = \$3.30, \quad \text{if} \quad 0.2 < x \le 0.4,$$
$$C(x) = \$3.70, \quad \text{if} \quad 0.4 < x \le 0.6,$$

and so on. The graph of C is shown below. (Source: New York City Taxi and Limousine Commission.)

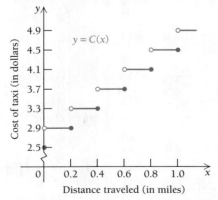

Using the graph of the taxicab fare function, find each of the following limits, if it exists.

81. $\lim_{x \to 0.25^-} C(x)$, $\lim_{x \to 0.25^+} C(x)$, $\lim_{x \to 0.25} C(x)$

82. $\lim_{x \to 0.2^-} C(x)$, $\lim_{x \to 0.2^+} C(x)$, $\lim_{x \to 0.2} C(x)$

83. $\lim_{x \to 0.6^-} C(x)$, $\lim_{x \to 0.6^+} C(x)$, $\lim_{x \to 0.6} C(x)$

The postage function. **The cost of sending a large envelope via U.S. first-class mail is $0.90 for the first ounce and $0.20 for each additional ounce (or fraction thereof). (Source: www.usps.com.) If x represents the weight of a large envelope, in ounces, then p(x) is the cost of mailing it, where**

$$p(x) = \$0.90, \quad \text{if} \quad 0 < x \le 1,$$
$$p(x) = \$1.10, \quad \text{if} \quad 1 < x \le 2,$$
$$p(x) = \$1.30, \quad \text{if} \quad 2 < x \le 3,$$

and so on, up through 13 ounces. The graph of p is shown below.

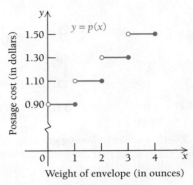

Using the graph of the postage function, find each of the following limits, if it exists.

84. $\lim_{x \to 1^-} p(x)$, $\lim_{x \to 1^+} p(x)$, $\lim_{x \to 1} p(x)$

85. $\lim_{x \to 2^-} p(x)$, $\lim_{x \to 2^+} p(x)$, $\lim_{x \to 2} p(x)$

86. $\lim_{x \to 2.6^-} p(x)$, $\lim_{x \to 2.6^+} p(x)$, $\lim_{x \to 2.6} p(x)$

87. $\lim_{x \to 3} p(x)$

88. $\lim_{x \to 3.4} p(x)$

Natural Sciences

Population growth. **In a certain habitat, the deer population (in hundreds) as a function of time (in years) is given in the graph of p below.**

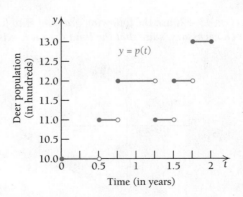

Use the graph for Exercises 89–91.

89. Find $\lim_{t \to 1.5^-} p(t)$, $\lim_{t \to 1.5^+} p(t)$, and $\lim_{t \to 1.5} p(t)$.

90. Find $\lim_{t \to 1.75^-} p(t)$, $\lim_{t \to 1.75^+} p(t)$, and $\lim_{t \to 1.75} p(t)$.

91. Explain what event(s) might account for the points at which no limit exists.

Population growth. *The population of bears in a certain region is given in the graph of p below. Time, t, is measured in months.*

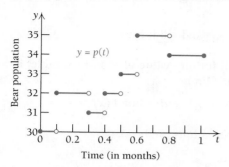

Time (in months)

Use the graph for Exercises 92–94.

92. Find $\lim\limits_{t\to 0.6^-} p(t)$, $\lim\limits_{t\to 0.6^+} p(t)$, and $\lim\limits_{t\to 0.6} p(t)$.

93. Find $\lim\limits_{t\to 0.8^-} p(t)$, $\lim\limits_{t\to 0.8^+} p(t)$, and $\lim\limits_{t\to 0.8} p(t)$.

94. Explain what event(s) might account for the points at which no limit exists.

SYNTHESIS

In Exercises 95–97, fill in each blank so that $\lim\limits_{x\to 2} f(x)$ exists.

95. $f(x) = \begin{cases} \frac{1}{2}x + \underline{}, & \text{for } x < 2, \\ -x + 6, & \text{for } x > 2, \end{cases}$

96. $f(x) = \begin{cases} -\frac{1}{2}x + 1, & \text{for } x < 2, \\ \frac{3}{2}x + \underline{}, & \text{for } x > 2, \end{cases}$

97. $f(x) = \begin{cases} x^2 - 9, & \text{for } x < 2, \\ -x^2 + \underline{}, & \text{for } x > 2, \end{cases}$

TECHNOLOGY CONNECTION

98. Graph the function f given by

$$f(x) = \begin{cases} -3, & \text{for } x = -2, \\ x^2, & \text{for } x \neq -2. \end{cases}$$

Use the GRAPH and TRACE features to find each of the following limits. When necessary, state that the limit does not exist.

a) $\lim\limits_{x\to -2^+} f(x)$

b) $\lim\limits_{x\to -2^-} f(x)$

c) $\lim\limits_{x\to -2} f(x)$

d) $\lim\limits_{x\to 2^+} f(x)$

e) $\lim\limits_{x\to 2} f(x)$

f) Does $\lim\limits_{x\to -2} f(x) = f(-2)$?

g) Does $\lim\limits_{x\to 2} f(x) = f(2)$?

In Exercises 99–101, use the GRAPH *and* TRACE *features to find each limit. When necessary, state that the limit does not exist.*

99. For $f(x) = \begin{cases} x^2 - 2, & \text{for } x < 0, \\ 2 - x^2, & \text{for } x \geq 0, \end{cases}$
find $\lim\limits_{x\to 0} f(x)$ and $\lim\limits_{x\to -2} f(x)$.

100. For $g(x) = \dfrac{20x^2}{x^3 + 2x^2 + 5x}$,
find $\lim\limits_{x\to \infty} g(x)$ and $\lim\limits_{x\to -\infty} g(x)$.

101. For $f(x) = \dfrac{1}{x^2 - 4x - 5}$,
find $\lim\limits_{x\to -1} f(x)$ and $\lim\limits_{x\to 5} f(x)$.

Answers to Quick Checks

1. **(a)** $f(3)$ does not exist; **(b)** 6
2. **(a)** 1, 7, does not exist; **(b)** 3
3. **(a)** 3; **(b)** 3; **(c)** 3
4. **(a)** Limit does not exist; **(b)** 5; **(c)** 6

1.2 Algebraic Limits and Continuity

Using numerical and graphical methods for finding limits can be time-consuming. In this section, we develop methods to more quickly evaluate limits for a wide variety of functions. We then use limits to study *continuity*, a concept of great importance in calculus.

OBJECTIVES

• Develop and use the Limit Properties to calculate limits.

• Determine whether a function is continuous at a point.

Algebraic Limits

Consider the functions given by $f(x) = x$, $g(x) = 3$, and $F(x) = x + 3$, displayed in the following graphs. Note that function F is the sum of functions f and g.

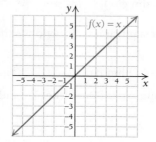

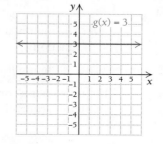

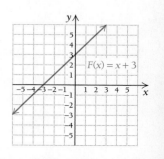

Suppose we are interested in the limits of $f(x)$, $g(x)$, and $F(x)$ as x approaches 2. In Section 1.1, we learned numerical and graphical techniques that can be used to show that

$$\lim_{x \to 2} f(x) = 2, \quad \lim_{x \to 2} g(x) = 3, \quad \text{and} \quad \lim_{x \to 2} F(x) = 5.$$

These techniques work equally well for any value of a. For example, if we choose $a = -1$, we can compute the following limits:

$$\lim_{x \to -1} f(x) = -1, \quad \lim_{x \to -1} g(x) = 3, \quad \text{and} \quad \lim_{x \to -1} F(x) = 2.$$

From these results, the following observations can be made:

1. For any real number a, $\lim_{x \to a} x = a$.
2. For any real number a, $\lim_{x \to a} 3 = 3$.

Recalling that $F(x) = f(x) + g(x)$, we make this reasonable conclusion:

3. For any real number a, $\lim_{x \to a} (x + 3) = a + 3$.

We determined the limits of these functions by observing basic behaviors and making reasonable generalizations. For what other functions can limits (as $x \to a$) be found by simply evaluating the function at $x = a$? The following list summarizes common limit properties that allow us to calculate limits much more efficiently.

Limit Properties

If $\lim_{x \to a} f(x) = L$ and $\lim_{x \to a} g(x) = M$ and c is any constant, then we have the following limit properties.

L1. The limit of a constant is the constant:

$$\lim_{x \to a} c = c.$$

L2. The limit of a power is the power of that limit, and the limit of a root is the root of that limit (assuming n is a positive integer):

$$\lim_{x \to a} [f(x)]^n = \left[\lim_{x \to a} f(x) \right]^n = L^n,$$

$$\lim_{x \to a} \sqrt[n]{f(x)} = \sqrt[n]{\lim_{x \to a} f(x)} = \sqrt[n]{L}.$$

In the case of the root, we must have $L \geq 0$ if n is even.

L3. The limit of a sum or a difference is the sum or the difference of the limits:

$$\lim_{x \to a} [f(x) \pm g(x)] = \lim_{x \to a} f(x) \pm \lim_{x \to a} g(x) = L \pm M.$$

L4. The limit of a product is the product of the limits:

$$\lim_{x \to a} [f(x) \cdot g(x)] = \left[\lim_{x \to a} f(x) \right] \cdot \left[\lim_{x \to a} g(x) \right] = L \cdot M.$$

L5. The limit of a quotient is the quotient of the limits:

$$\lim_{x \to a} \frac{f(x)}{g(x)} = \frac{\lim_{x \to a} f(x)}{\lim_{x \to a} g(x)} = \frac{L}{M}, \quad \text{assuming } M \neq 0.$$

L6. The limit of a constant times a function is the constant times the limit of the function:

$$\lim_{x \to a} c \cdot f(x) = c \cdot \lim_{x \to a} f(x) = cL.$$

Property L6 is a combination of Properties L1 and L4 but is stated here for emphasis as it is used frequently.

■ **EXAMPLE 1** Use the Limit Properties to find $\lim\limits_{x\to4} (x^2 - 3x + 7)$.

Solution We know that $\lim\limits_{x\to4} x$ is 4. By Limit Property L4,

$$\lim_{x\to4} x^2 = \lim_{x\to4} x \cdot \lim_{x\to4} x = 4 \cdot 4 = 16.$$

By Limit Property L6,

$$\lim_{x\to4} (-3x) = -3 \cdot \lim_{x\to4} x = -3 \cdot 4 = -12.$$

By Limit Property L1,

$$\lim_{x\to4} 7 = 7.$$

Combining the above results using Limit Property L3, we have

$$\lim_{x\to4} (x^2 - 3x + 7) = 16 - 12 + 7 = 11.$$

The result of Example 1 is extended in the following theorem.

THEOREM ON LIMITS OF RATIONAL FUNCTIONS

For any rational function F, with a in the domain of F,

$$\lim_{x\to a} F(x) = F(a).$$

Rational functions are a family of common functions, including all polynomial functions (which include constant functions and linear functions) and ratios composed of such functions (see Section R.5). Thus, the Limit Properties allow us to evaluate limits of rational functions very quickly without the need for tables or graphs, as illustrated by the following examples.

■ **EXAMPLE 2** Find $\lim\limits_{x\to2} (x^4 - 5x^3 + x^2 - 7)$.

Solution It follows from the Theorem on Limits of Rational Functions that we can find the limit by substitution:

$$\begin{aligned}
\lim_{x\to2} (x^4 - 5x^3 + x^2 - 7) &= 2^4 - 5 \cdot 2^3 + 2^2 - 7 \\
&= 16 - 40 + 4 - 7 \\
&= -27.
\end{aligned}$$

> **Quick Check 1**
>
> Find these limits and note the Limit Property you use at each step.
>
> **a)** $\lim\limits_{x\to1} 2x^3 + 3x^2 - 6$
>
> **b)** $\lim\limits_{x\to4} \dfrac{2x^2 + 5x - 1}{3x - 2}$
>
> **c)** $\lim\limits_{x\to-2} \sqrt{1 + 3x^2}$

■ **EXAMPLE 3** Find $\lim\limits_{x\to0} \sqrt{x^2 - 3x + 2}$.

Solution The Theorem on Limits of Rational Functions and Limit Property L2 tell us that we can substitute to find the limit:

$$\begin{aligned}
\lim_{x\to0} \sqrt{x^2 - 3x + 2} &= \sqrt{0^2 - 3 \cdot 0 + 2} \\
&= \sqrt{2}.
\end{aligned}$$

❰ Quick Check 1

In the following example, the function is rational, but a is not in the domain of the function, since it would result in a zero in the denominator. Nevertheless, we can still find the limit.

■ **EXAMPLE 4**   Let $r(x) = \dfrac{x^2 - x - 12}{x + 3}$. Find $\lim\limits_{x \to -3} r(x)$.

Solution   We note that $r(-3)$ does not exist, since substituting -3 for x would give zero in the denominator. Since it is impossible to determine this limit by direct evaluation (the assumption for Limit Property L5 is not met), we use a table and a graph. Although $x \neq -3$, x can be as close to -3 as we wish:

Limit Numerically   *Limit Graphically*

$x \to -3^-$ $(x < -3)$	$r(x)$
-3.1	-7.1
-3.01	-7.01
-3.001	-7.001

$x \to -3^+$ $(x > -3)$	$r(x)$
-2.9	-6.9
-2.99	-6.99
-2.999	-6.999

$$\lim_{x \to -3} r(x) = -7$$

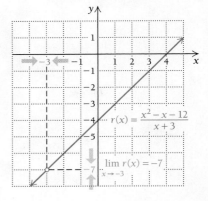

Both the table and the graph show that the limit is

$$\lim_{x \to -3} \left(\frac{x^2 - x - 12}{x + 3} \right) = -7.$$

Limit Algebraically.   The function is simplified by factoring the numerator and noting that the factor $x + 3$ is present in both numerator and denominator. As long as $x \neq -3$, the expression can be simplified:

$$\frac{x^2 - x - 12}{x + 3} = \frac{(x + 3)(x - 4)}{(x + 3)} = x - 4, \qquad \text{for } x \neq -3.$$

We then evaluate the limit on the simplified form:

$$\lim_{x \to -3} (x - 4) = (-3) - 4 = -7.$$

As we saw above, the graph of $r(x)$ is a line with a "hole" at the point $(-3, -7)$. Even though $r(-3)$ does not exist, the limit *does* exist since we are only concerned about the behavior of $r(x)$ for x-values *close* to -3. The decision to simplify was made by noting that $[(-3)^2 - (-3) - 12]/[(-3) + 3] = 0/0$. This *indeterminate form* indicates that the polynomials in the numerator and the denominator share a common factor, in this case, $x + 3$. The $0/0$ form is a hint that a limit may exist. Look for ways to simplify the function algebraically, or use a table and a graph to determine the limit.

❭ **Quick Check 2**

Using a table, a graph, and algebra, find

$$\lim_{x \to 2} \frac{x^2 + 4x - 12}{x^2 - 4}.$$

❬ Quick Check 2

A common error in determining limits is to assume that all limits can be found by direct evaluation. A student will often attempt to calculate the limit of a function like the one in Example 4, get a zero in the denominator, and then make the erroneous assumption that the limit does not exist. Remember, finding a limit involves analyzing the behavior of the function when x is *close* to the a-value, not necessarily at the a-value. As Example 4 illustrated, the function may not be defined at a certain a-value, but its limit may still be determined.

Limits like that in Example 4 that cannot be determined by direct evaluation can be found by another method, called l'Hôpital's Rule, which uses calculus and derivatives. This method is discussed in Appendix B (p. 830).

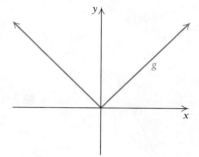
EXERCISES

Find each limit, if it exists, using the TABLE feature.

1. $\lim\limits_{x \to -2} (x^4 - 5x^3 + x^2 - 7)$

2. $\lim\limits_{x \to 1} \sqrt{x^2 + 3x + 4}$

3. $\lim\limits_{x \to 5} \dfrac{x - 5}{x^2 - 6x + 5}$

4. $\lim\limits_{x \to 3} \dfrac{x - 3}{x^2 - 9}$

■ **EXAMPLE 5**   Find $\lim\limits_{h \to 0} (3x^2 + 3xh + h^2)$.

Solution   We treat x as a constant since we are interested only in the way in which the expression varies when h approaches 0. We use the Limit Properties to find that

$$\lim_{h \to 0} (3x^2 + 3xh + h^2) = 3x^2 + 3x(0) + 0^2$$
$$= 3x^2.$$

Continuity

The following are the graphs of functions that are *continuous* over the whole real number line, that is, over $(-\infty, \infty)$.

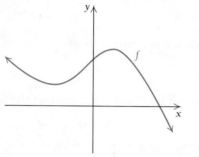

Note that there are no "jumps" or "holes" in the graphs. For now, we use an intuitive definition of continuity, which we will soon refine. We say that a function is **continuous over**, or **on, some interval** of the real number line if its graph can be traced without lifting the pencil from the paper. If there is any point in an interval where a "jump" or a "hole" occurs, then we say that the function is *not continuous* over that interval. The graphs of functions F, G, and H, which follow, show that these functions are *not* continuous over the whole real line.

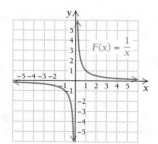

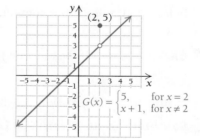

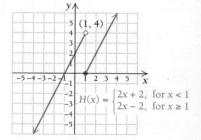

In each case, the graph *cannot* be traced without lifting the pencil from the paper. However, each case represents a different situation:

- F is not continuous over $(-\infty, \infty)$ because the point $x = 0$ is not part of the domain. Thus, there is no point to trace at $x = 0$. Note that F is continuous over the intervals $(-\infty, 0)$ and $(0, \infty)$.

- G is not continuous over $(-\infty, \infty)$ because it is not continuous at $x = 2$. To see this, trace the graph of G starting to the left of $x = 2$. As x approaches 2 from either side, $G(x)$ approaches 3. However, *at* $x = 2$, $G(x)$ *jumps* up to 5. Note that G is continuous over $(-\infty, 2)$ and $(2, \infty)$.

- H is not continuous over $(-\infty, \infty)$ because it is not continuous at $x = 1$. To see this, trace the graph of H starting to the left of $x = 1$. As x approaches 1 from the left, $H(x)$ approaches 4. However, as x approaches 1 from the right, $H(x)$ is close to 0. Note that H is continuous over $(-\infty, 1)$ and $(1, \infty)$.

A continuous curve.

Each of the above graphs has a *point of discontinuity*. The graph of F is discontinuous at 0, because $F(0)$ does not exist; the graph of G is discontinuous at 2, because $\lim\limits_{x \to 2} G(x) \neq G(2)$; and the graph of H is discontinuous at 1, because $\lim\limits_{x \to 1} H(x)$ does not exist.

DEFINITION

A function f is **continuous** at $x = a$ if:

a) $f(a)$ exists,   (The output at a exists.)
b) $\lim\limits_{x \to a} f(x)$ exists,   (The limit as $x \to a$ exists.)

and

c) $\lim\limits_{x \to a} f(x) = f(a)$.   (The limit is the same as the output.)

A function is **continuous over an interval I** if it is continuous at each point in I.

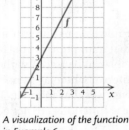

A visualization of the function in Example 6

■ **EXAMPLE 6**   Determine whether the function given by

$$f(x) = 2x + 3$$

is continuous at $x = 4$.

Solution   This function is continuous at $x = 4$ because:

a) $f(4)$ exists, $(f(4) = 11)$
b) $\lim\limits_{x \to 4} f(x)$ exists, $(\lim\limits_{x \to 4} f(x) = 11$ was found on pp. 94–95.)

and

c) $\lim\limits_{x \to 4} f(x) = 11 = f(4)$.

In fact, $f(x) = 2x + 3$ is continuous at any point on the real number line.

A visualization of the function in Example 7

■ **EXAMPLE 7**   Is the function f given by

$$f(x) = x^2 - 5$$

continuous at $x = 3$? Why or why not?

Solution   By the Theorem on Limits of Rational Functions, we have

$$\lim\limits_{x \to 3} f(x) = 3^2 - 5 = 9 - 5 = 4.$$

Since

$$f(3) = 3^2 - 5 = 4,$$

we have

$$\lim\limits_{x \to 3} f(x) = f(3).$$

Thus, f is continuous at $x = 3$. This function is also continuous over all real x.

A visualization of the function in Example 8

■ **EXAMPLE 8**   Is the function g, given by

$$g(x) = \begin{cases} \frac{1}{2}x + 3, & \text{for } x < -2, \\ x - 1, & \text{for } x \geq -2, \end{cases}$$

continuous at $x = -2$? Why or why not?

〉 Quick Check 3

Let

$$g(x) = \begin{cases} 3x - 5, & \text{for } x < 2, \\ 2x + 1, & \text{for } x \geq 2. \end{cases}$$

Is g continuous at $x = 2$?
Why or why not?

Solution To find out if g is continuous at -2, we must determine whether $\lim\limits_{x \to -2} g(x) = g(-2)$. Thus, we first note that $g(-2) = -2 - 1 = -3$. To find $\lim\limits_{x \to -2} g(x)$, we look at left- and right-hand limits:

$$\lim\limits_{x \to -2^-} g(x) = \frac{1}{2}(-2) + 3 = -1 + 3 = 2; \ \lim\limits_{x \to -2^+} g(x) = -2 - 1 = -3.$$

Since $\lim\limits_{x \to -2^-} g(x) \neq \lim\limits_{x \to -2^+} g(x)$, we see that $\lim\limits_{x \to -2} g(x)$ does not exist. Thus, g is not continuous at -2. It is continuous at all other x-values.

〈 Quick Check 3

■ EXAMPLE 9 Is the function F, given by

$$F(x) = \begin{cases} \dfrac{x^2 - 16}{x - 4}, & \text{for } x \neq 4, \\ 7, & \text{for } x = 4, \end{cases}$$

continuous at $x = 4$? Why or why not?

Solution For F to be continuous at 4, we must have $\lim\limits_{x \to 4} F(x) = F(4)$. Note that $F(4) = 7$. To find $\lim\limits_{x \to 4} F(x)$, we note that, for $x \neq 4$,

$$\frac{x^2 - 16}{x - 4} = \frac{(x - 4)(x + 4)}{x - 4} = x + 4.$$

Thus

$$\lim\limits_{x \to 4} F(x) = 4 + 4 = 8.$$

We see that F is *not* continuous at $x = 4$ since

$$\lim\limits_{x \to 4} F(x) \neq F(4).$$

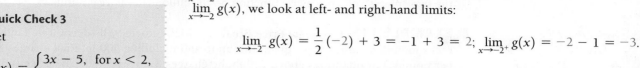

A visualization of the function in Example 9

The Limit Properties and the Theorem on Limits of Rational Functions can be used to show that if $f(x)$ and $g(x)$ are two arbitrary polynomial functions, then f and g are both continuous. Furthermore, $f + g$, $f - g$, $f \cdot g$, and, assuming $g(x) \neq 0$, f/g are also continuous. We can also use the Limit Properties to show that for n, an integer greater than 1, $\sqrt[n]{f(x)}$ is continuous, provided that $f(x) \geq 0$ when n is even.

A visualization of the function in Example 10

■ EXAMPLE 10 Is the function G, given by

$$G(x) = \begin{cases} -x + 3, & \text{for } x \leq 2, \\ x^2 - 3, & \text{for } x > 2, \end{cases}$$

continuous for all x? Why or why not?

Solution For G to be continuous, it must be continuous for all real numbers. Since $y = -x + 3$ is continuous on $(-\infty, 2]$ and $y = x^2 - 3$ is continuous on $(2, \infty)$, we need only to determine whether $\lim\limits_{x \to 2} G(x) = G(2)$, that is, whether G is continuous at $x = 2$:

$$G(2) = -2 + 3 = 1;$$

$$\lim\limits_{x \to 2^-} G(x) = -2 + 3 = 1 \quad \text{and} \quad \lim\limits_{x \to 2^+} G(x) = (2)^2 - 3 = 4 - 3 = 1,$$

so $\lim\limits_{x \to 2} G(x) = 1.$

Since $\lim\limits_{x \to 2} G(x) = G(2)$, we have shown that G is continuous at $x = 2$, and we can conclude that G is continuous at all x.

〈 Quick Check 4

〉 Quick Check 4

a) Let

$$h(x) = \begin{cases} \dfrac{x^2 - 9}{x - 3}, & \text{for } x \neq 3, \\ 7, & \text{for } x = 3. \end{cases}$$

Is h continuous at $x = 3$?
Why or why not?

b) Let

$$p(x) = \begin{cases} \dfrac{x^2 - 25}{x - 5}, & \text{for } x \neq 5, \\ c, & \text{for } x = 5. \end{cases}$$

Determine c such that p is continuous at $x = 5$.

Example 10 showed that a piecewise-defined function may be continuous for all real numbers. In Examples 11 and 12, we explore a situation where it is *preferred* that the two "pieces" of a piecewise function be continuous for all real numbers.

■ **EXAMPLE 11**   **Business: Price Breaks.**   Rick's Rocks sells decorative landscape rocks in bulk quantities. For quantities up to and including 500 lb, Rick charges $2.50 per pound. For quantities above 500 lb, he charges $2 per pound. The price function can be stated as a piecewise function:

$$p(x) = \begin{cases} 2.50x, & \text{for } 0 < x \le 500, \\ 2x, & \text{for } x > 500. \end{cases}$$

where p is the price in dollars and x is the quantity in pounds. Is the price function $p(x)$ continuous at $x = 500$? Why or why not?

Solution   The graph of $p(x)$ follows.

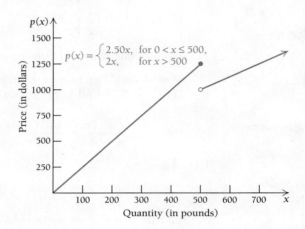

As x approaches 500 from the left, we have $\lim\limits_{x \to 500^-} p(x) = 1250$, and when x approaches 500 from the right, we have $\lim\limits_{x \to 500^+} p(x) = 1000$. Since the left-hand and right-hand limits are not equal, the limit $\lim\limits_{x \to 500} p(x)$ does not exist. Thus, the function is *not* continuous at $x = 500$. This graph literally shows a price "break."

For the record, the function value at $x = 500$ is $p(500) = 1250$, but this fact plays no role with regard to whether or not the limit exists.

■ **EXAMPLE 12**   **Business.**   Rick, of Rick's Rocks in Example 11, realizes that his customers are taking advantage of him: for example, they pay less for 550 lb of rocks than they would for 500 lb of rocks. For Rick, this means lost revenue, so he decides to add a quantity discount surcharge for quantities above 500 lb. If k represents this surcharge, the price function becomes:

$$p(x) = \begin{cases} 2.50x, & \text{for } 0 < x \le 500, \\ 2x + k, & \text{for } x > 500. \end{cases}$$

Find k such that the function is continuous at $x = 500$.

Solution   If p is continuous at $x = 500$, its limit must exist there as well. We must therefore have

$$\lim_{x \to 500^-} p(x) = \lim_{x \to 500^+} p(x).$$

This means that the left-hand and right-hand limits must be equal so that the two pieces of the graph actually connect (as in Example 10). We know from Example 11 that the left-hand limit is 1250, so we want the right-hand limit to be 1250 as well.

We set the right-hand limit equal to 1250:

$$\lim_{x \to 500^+} (2x + k) = 1250.$$

We allow x to approach 500 from the right. By Limit Property L3, we have

$$2(500) + k = 1250.$$

This simplifies to $1000 + k = 1250$, which gives $k = 250$. Thus, if $k = 250$, the function will be continuous at $x = 500$.

> **Quick Check 5**
>
> A reservoir is empty at time $t = 0$ minutes. It fills at a rate of 3 gallons of water per minute for 30 minutes. At 30 minutes, the reservoir is no longer being filled and a valve is opened, allowing water to escape at a rate of 4 gallons per minute. The volume v after t minutes is given by the function
>
> $$v(t) = \begin{cases} 3t, & \text{for } 0 \le t \le 30, \\ k - 4t, & \text{for } t > 30. \end{cases}$$
>
> Determine k such that the volume function v is continuous at $t = 30$. Explain why this must be true.

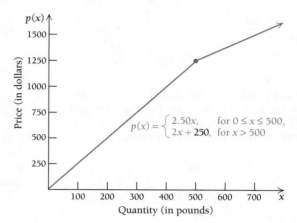

$$p(x) = \begin{cases} 2.50x, & \text{for } 0 \le x \le 500, \\ 2x + 250, & \text{for } x > 500 \end{cases}$$

This is a fair compromise: the customers still get a cheaper rate per pound once x is above 500 pounds, but Rick is no longer losing money as he was with his previous pricing function.

❮ Quick Check 5

Section Summary

- For a rational function for which a is in the domain, the limit as x approaches a can be found by direct evaluation of the function at a.
- If direct evaluation leads to the *indeterminate form* $0/0$, the limit may still exist: algebraic simplification and/or a table and graph are used to find the limit.
- Informally, a function is *continuous* if its graph can be sketched without lifting the pencil off the paper.

- Formally, a function is continuous at $x = a$ if
  (1) the function value $f(a)$ exists,
  (2) the limit as x approaches a exists, and
  (3) the function value and the limit are equal.
  This can be summarized as $\lim_{x \to a} f(x) = f(a)$.
- If any part of the continuity definition fails, then the function is discontinuous at $x = a$.

EXERCISE SET
1.2

Classify each statement as either true or false.

1. $\lim_{x \to 3} 7 = 7$

2. If $\lim_{x \to 2} f(x) = 9$, then $\lim_{x \to 2} \sqrt{f(x)} = 3$.

3. If $\lim_{x \to 1} g(x) = 5$, then $\lim_{x \to 1} [g(x)]^2 = 10$.

4. If $\lim_{x \to 4} F(x) = 7$, then $\lim_{x \to 4} [c \cdot F(x)] = 7c$.

5. If f is continuous at $x = 2$, then $f(2)$ must exist.

6. If g is discontinuous at $x = 3$, then $g(3)$ must not exist.

7. If $\lim_{x \to 4} F(x)$ exists, then F must be continuous at $x = 4$.

8. If $\lim_{x \to 7} G(x)$ equals $G(7)$, then G must be continuous at $x = 7$.

Use the Theorem on Limits of Rational Functions to find the following limits. When necessary, state that the limit does not exist.

9. $\lim\limits_{x \to 1} (3x + 2)$

10. $\lim\limits_{x \to 2} (4x - 5)$

11. $\lim\limits_{x \to -1} (x^2 - 4)$

12. $\lim\limits_{x \to -2} (x^2 + 3)$

13. $\lim\limits_{x \to 3} (x^2 - 4x + 7)$

14. $\lim\limits_{x \to 5} (x^2 - 6x + 9)$

15. $\lim\limits_{x \to 2} (2x^4 - 3x^3 + 4x - 1)$

16. $\lim\limits_{x \to -1} (3x^5 + 4x^4 - 3x + 6)$

17. $\lim\limits_{x \to 3} \dfrac{x^2 - 8}{x - 2}$

18. $\lim\limits_{x \to 3} \dfrac{x^2 - 25}{x^2 - 5}$

For Exercises 19–30, the initial substitution of $x = a$ yields the form 0/0. Look for ways to simplify the function algebraically, or use a table and/or a graph to determine the limit. When necessary, state that the limit does not exist.

19. $\lim\limits_{x \to 3} \dfrac{x^2 - 9}{x - 3}$

20. $\lim\limits_{x \to 5} \dfrac{x^2 - 25}{x - 5}$

21. $\lim\limits_{x \to 1} \dfrac{x^2 + 5x - 6}{x^2 - 1}$

22. $\lim\limits_{x \to -2} \dfrac{x^2 - 2x - 8}{x^2 - 4}$

23. $\lim\limits_{x \to 2} \dfrac{3x^2 + x - 14}{x^2 - 4}$

24. $\lim\limits_{x \to -3} \dfrac{2x^2 - x - 21}{9 - x^2}$

25. $\lim\limits_{x \to 1} \dfrac{x^3 - 1}{x - 1}$

26. $\lim\limits_{x \to 2} \dfrac{x^3 - 8}{2 - x}$

27. $\lim\limits_{x \to 25} \dfrac{\sqrt{x} - 5}{x - 25}$

28. $\lim\limits_{x \to 9} \dfrac{9 - x}{\sqrt{x} - 3}$

29. $\lim\limits_{x \to -1} \dfrac{x^2 + 5x + 4}{x^2 + 2x + 1}$

30. $\lim\limits_{x \to 2} \dfrac{x^2 + 3x - 10}{x^2 - 4x + 4}$

Use the Limit Properties to find the following limits. If a limit does not exist, state that fact.

31. $\lim\limits_{x \to 4} \sqrt{x^2 - 9}$

32. $\lim\limits_{x \to 5} \sqrt{x^2 - 16}$

33. $\lim\limits_{x \to 2} \sqrt{x^2 - 9}$

34. $\lim\limits_{x \to 3} \sqrt{x^2 - 16}$

35. $\lim\limits_{x \to 3^+} \sqrt{x^2 - 9}$

36. $\lim\limits_{x \to -4^-} \sqrt{x^2 - 16}$

Determine whether each of the functions shown in Exercises 37–41 is continuous over the interval $(-6, 6)$.

37.

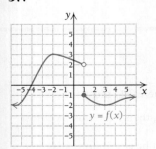

38.

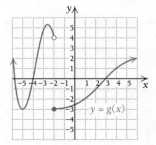

39.

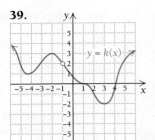

40.

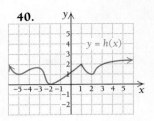

41.

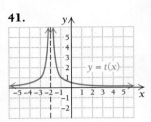

Use the graphs and functions in Exercises 37–41 to answer each of the following. If an expression does not exist, state that fact.

42. **a)** Find $\lim\limits_{x \to 1^+} f(x)$, $\lim\limits_{x \to 1^-} f(x)$, and $\lim\limits_{x \to 1} f(x)$.
   b) Find $f(1)$.
   c) Is f continuous at $x = 1$? Why or why not?
   d) Find $\lim\limits_{x \to -2} f(x)$.
   e) Find $f(-2)$.
   f) Is f continuous at $x = -2$? Why or why not?

43. **a)** Find $\lim\limits_{x \to 1^+} g(x)$, $\lim\limits_{x \to 1^-} g(x)$, and $\lim\limits_{x \to 1} g(x)$.
   b) Find $g(1)$.
   c) Is g continuous at $x = 1$? Why or why not?
   d) Find $\lim\limits_{x \to -2} g(x)$.
   e) Find $g(-2)$.
   f) Is g continuous at $x = -2$? Why or why not?

44. **a)** Find $\lim\limits_{x \to -1} k(x)$.
   b) Find $k(-1)$.
   c) Is k continuous at $x = -1$? Why or why not?
   d) Find $\lim\limits_{x \to 3} k(x)$.
   e) Find $k(3)$.
   f) Is k continuous at $x = 3$? Why or why not?

45. **a)** Find $\lim\limits_{x \to 1} h(x)$.
   b) Find $h(1)$.
   c) Is h continuous at $x = 1$? Why or why not?
   d) Find $\lim\limits_{x \to -2} h(x)$.
   e) Find $h(-2)$.
   f) Is h continuous at $x = -2$? Why or why not?

46. **a)** Find $\lim\limits_{x \to 1} t(x)$.
   b) Find $t(1)$.
   c) Is t continuous at $x = 1$? Why or why not?
   d) Find $\lim\limits_{x \to -2} t(x)$.
   e) Find $t(-2)$.
   f) Is t continuous at $x = -2$? Why or why not?

In Exercises 47 and 48, use the graphs to find the limits and answer the related questions.

47.

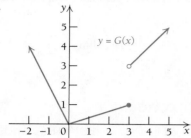

a) Find $\lim_{x \to 3^+} G(x)$.
b) Find $\lim_{x \to 3^-} G(x)$.
c) Find $\lim_{x \to 3} G(x)$.
d) Find $G(3)$.
e) Is G continuous at $x = 3$? Why or why not?
f) Is G continuous at $x = 0$? Why or why not?
g) Is G continuous at $x = 2.9$? Why or why not?

48. Consider the function

$$C(x) = \begin{cases} -1, & \text{for } x < 2, \\ 1, & \text{for } x \ge 2. \end{cases}$$

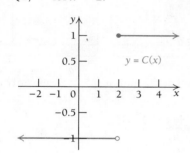

a) Find $\lim_{x \to 2^+} C(x)$.
b) Find $\lim_{x \to 2^-} C(x)$.
c) Find $\lim_{x \to 2} C(x)$.
d) Find $C(2)$.
e) Is C continuous at $x = 2$? Why or why not?
f) Is C continuous at $x = 1.95$? Why or why not?

49. Is the function given by $f(x) = 3x - 2$ continuous at $x = 5$? Why or why not?

50. Is the function given by $g(x) = x^2 - 3x$ continuous at $x = 4$? Why or why not?

51. Is the function given by $G(x) = \dfrac{1}{x}$ continuous at $x = 0$? Why or why not?

52. Is the function given by $F(x) = \sqrt{x}$ continuous at $x = -1$? Why or why not?

53. Is the function given by

$$g(x) = \begin{cases} \frac{1}{3}x + 4, & \text{for } x \le 3, \\ 2x - 1, & \text{for } x > 3, \end{cases}$$

continuous at $x = 3$? Why or why not?

54. Is the function given by

$$f(x) = \begin{cases} \frac{1}{2}x + 1, & \text{for } x < 4, \\ -x + 7, & \text{for } x \ge 4, \end{cases}$$

continuous at $x = 4$? Why or why not?

55. Is the function given by

$$F(x) = \begin{cases} \frac{1}{3}x + 4, & \text{for } x \le 3, \\ 2x - 5, & \text{for } x > 3, \end{cases}$$

continuous at $x = 3$? Why or why not?

56. Is the function given by

$$G(x) = \begin{cases} \frac{1}{2}x + 1, & \text{for } x < 4, \\ -x + 5, & \text{for } x > 4, \end{cases}$$

continuous at $x = 4$? Why or why not?

57. Is the function given by

$$f(x) = \begin{cases} \frac{1}{3}x + 4, & \text{for } x < 3, \\ 2x - 1, & \text{for } x \ge 3, \end{cases}$$

continuous at $x = 3$? Why or why not?

58. Is the function given by

$$g(x) = \begin{cases} \frac{1}{2}x + 1, & \text{for } x < 4, \\ -x + 7, & \text{for } x > 4, \end{cases}$$

continuous at $x = 4$? Why or why not?

59. Is the function given by

$$G(x) = \begin{cases} \dfrac{x^2 - 4}{x - 2}, & \text{for } x \ne 2, \\ 5, & \text{for } x = 2, \end{cases}$$

continuous at $x = 2$? Why or why not?

60. Is the function given by

$$F(x) = \begin{cases} \dfrac{x^2 - 1}{x - 1}, & \text{for } x \ne 1, \\ 4, & \text{for } x = 1, \end{cases}$$

continuous at $x = 1$? Why or why not?

61. Is the function given by

$$f(x) = \begin{cases} \dfrac{x^2 - 4x - 5}{x - 5}, & \text{for } x < 5, \\ x + 1, & \text{for } x \ge 5, \end{cases}$$

continuous at $x = 5$? Why or why not?

62. Is the function given by

$$G(x) = \begin{cases} \dfrac{x^2 - 3x - 4}{x - 4}, & \text{for } x < 4, \\ 2x - 3, & \text{for } x \ge 4, \end{cases}$$

continuous at $x = 4$? Why or why not?

63. Is the function given by $g(x) = \dfrac{1}{x^2 - 7x + 10}$ continuous at $x = 5$? Why or why not?

64. Is the function given by $f(x) = \dfrac{1}{x^2 - 6x + 8}$ continuous at $x = 3$? Why or why not?

65. Is the function given by $F(x) = \dfrac{1}{x^2 - 7x + 10}$ continuous at $x = 4$? Why or why not?

66. Is the function given by $G(x) = \dfrac{1}{x^2 - 6x + 8}$ continuous at $x = 2$? Why or why not?

67. Is the function given by $g(x) = x^2 - 3x + 2$ continuous over the interval $(-4, 4)$? Why or why not?

68. Is the function given by $F(x) = x^2 - 5x + 6$ continuous over the interval $(-5, 5)$? Why or why not?

69. Is the function given by $f(x) = \dfrac{1}{x} + 3$ continuous over the interval $(-7, 7)$? Why or why not?

70. Is the function given by $G(x) = \dfrac{1}{x - 1}$ continuous over the interval $(0, \infty)$? Why or why not?

71. Is the function given by $g(x) = 4x^3 - 6x$ continuous on $\mathbb{R}$?

72. Is the function given by $F(x) = \dfrac{3}{x - 5}$ continuous on $\mathbb{R}$?

APPLICATIONS

Business and Economics

73. The Candy Factory sells candy by the pound, charging \$1.50 per pound for quantities up to and including 20 pounds. Above 20 pounds, the Candy Factory charges \$1.25 per pound for the entire quantity, plus a quantity surcharge k. If x represents the number of pounds, the price function is

$$p(x) = \begin{cases} 1.50x, & \text{for } x \le 20, \\ 1.25x + k, & \text{for } x > 20. \end{cases}$$

a) Find k such that the price function p is continuous at $x = 20$.

b) Explain why it is preferable to have continuity at $x = 20$.

Life and Physical Sciences

74. A lab technician controls the temperature T inside a kiln. From an initial temperature of 0 degrees Celsius (°C), he allows the kiln to increase by 2°C per minute for the next 60 min. After the 60th minute, he allows the kiln to cool at the rate of 3°C per minute. The temperature function T is defined by

$$T(t) = \begin{cases} 2t, & \text{for } t \le 60, \\ k - 3t, & \text{for } t > 60. \end{cases}$$

a) Find k such that T is continuous at $t = 60$.

b) Explain why T must be continuous at $t = 60$ min.

SYNTHESIS

Find each limit, if it exists. If a limit does not exist, state that fact.

75. $\displaystyle\lim_{x \to 0} \frac{|x|}{x}$

76. $\displaystyle\lim_{x \to -2} \frac{x^3 + 8}{x^2 - 4}$

TECHNOLOGY CONNECTION

In Section 1.1, we discussed how to use the TABLE feature to find limits. Consider

$$\lim_{x \to 0} \frac{\sqrt{1 + x} - 1}{x}.$$

Input–output tables for this function are shown below. The table on the left uses TblStart $= -1$ and ΔTbl $= 0.5$. By using smaller and smaller step values and beginning closer to 0, we can refine the table and obtain a better estimate of the limit. On the right is an input–output table with TblStart $= -0.03$ and ΔTbl $= 0.01$.

X	Y1		X	Y1
-1	1		-0.03	0.50381
-0.5	0.58579		-0.02	0.50253
0	ERROR		-0.01	0.50126
0.5	0.44949		0	ERROR
1	0.41421		0.01	0.49876
1.5	0.38743		0.02	0.49752
2	0.36603		0.03	0.49631
X = -1			X = -0.03	

It appears that the limit is 0.5. We can verify this by graphing

$$y = \frac{\sqrt{1 + x} - 1}{x}$$

and tracing the curve near $x = 0$, zooming in on that portion of the curve.

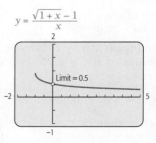

We see that

$$\lim_{x \to 0} \frac{\sqrt{1 + x} - 1}{x} = 0.5.$$

To verify this algebraically, multiply $\dfrac{\sqrt{1 + x} - 1}{x}$ by 1, using $\dfrac{\sqrt{1 + x} + 1}{\sqrt{1 + x} + 1}$. Then simplify the result and find the limit.

In Exercises 77–84, find each limit. Use the TABLE feature and start with $\Delta Tbl = 0.1$. Then use 0.01, 0.001, and 0.0001. When you think you know the limit, graph and use the TRACE feature to verify your assertion. Then try to verify it algebraically.

77. $\displaystyle\lim_{a \to -2} \frac{a^2 - 4}{\sqrt{a^2 + 5} - 3}$

78. $\displaystyle\lim_{x \to 1} \frac{\sqrt{x} - 1}{x - 1}$

79. $\displaystyle\lim_{x \to 0} \frac{\sqrt{3 - x} - \sqrt{3}}{x}$

80. $\displaystyle\lim_{x \to 0} \frac{\sqrt{4 + x} - \sqrt{4 - x}}{x}$

81. $\displaystyle\lim_{x \to 1} \frac{x - \sqrt[4]{x}}{x - 1}$

82. $\displaystyle\lim_{x \to 0} \frac{\sqrt{7 + 2x} - \sqrt{7}}{x}$

83. $\displaystyle\lim_{x \to 4} \frac{2 - \sqrt{x}}{4 - x}$

84. $\displaystyle\lim_{x \to 0} \frac{7 - \sqrt{49 - x^2}}{x}$

Answers to Quick Checks

1. (a) -1; L1, L2, L3, L6
(b) $\dfrac{51}{10}$, or 5.1; L1, L2, L3, L5, L6
(c) $\sqrt{13}$; L1, L2, L3, L6
2. 2 **3.** No, the limit as $x \to 2$ does not exist.
4. (a) No, the limit as $x \to 3$ is 6, but $h(3) = 7$, so the limit does not equal the function value. **(b)** $c = 10$
5. $k = 210$; the amount of water (in gallons) is continuous as a function of time.

1.3

Average Rates of Change

OBJECTIVES

- Compute an average rate of change.
- Find a simplified difference quotient.

Let's say that a car travels 110 mi in 2 hr. Its *average rate of change* (*speed*) is 110 mi/2 hr, or 55 mi/hr (55 mph). Suppose that you are on a freeway and you begin accelerating. Glancing at the speedometer, you see that at that instant your *instantaneous rate of change* is 55 mph. These are two quite different concepts. The first you are probably familiar with. The second involves ideas of limits and calculus. To understand instantaneous rate of change, we first need to develop a solid understanding of average rate of change.

The following graph shows the total production of suits by Raggs, Ltd., during one morning of work. Industrial psychologists have found curves like this typical of the production of factory workers.

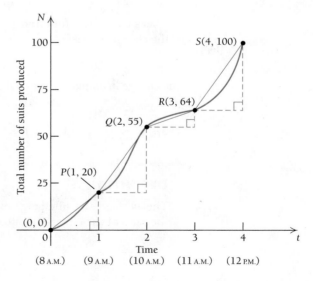

■ **EXAMPLE 1** **Business: Production.** What was the number of suits produced at Raggs, Ltd., from 9 A.M. to 10 A.M.?

Solution At 9 A.M., 20 suits had been produced. At 10 A.M., 55 suits had been produced. In the hour from 9 A.M. to 10 A.M., the number of suits produced was

$$55 \text{ suits} - 20 \text{ suits}, \quad \text{or} \quad 35 \text{ suits}.$$

Note that 35 is the slope of the line from P to Q.

EXAMPLE 2 **Business: Average Rate of Change.** What was the average number of suits produced per hour from 9 A.M. to 11 A.M.?

Solution We have

$$\frac{64 \text{ suits} \ - \ 20 \text{ suits}}{11 \text{ A.M.} \ - \ 9 \text{ A.M.}} = \frac{44 \text{ suits}}{2 \text{ hr}}$$

$$= 22 \frac{\text{suits}}{\text{hr}}.$$

Note that 22 is the slope of the line from P to R.

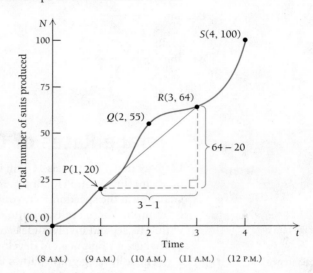

Quick Check 1

State the average rate of change for each situation in a short sentence. Be sure to include units.

a) It rained 4 inches over a period of 8 hours.

b) Your car travels 250 miles on 20 gallons of gas.

c) At 2 P.M., the temperature was 82 degrees. At 5 P.M., the temperature was 76 degrees.

❮ Quick Check 1

Let's consider a function $y = f(x)$ and two inputs x_1 and x_2. The *change in input*, or the *change in x*, is

$$x_2 - x_1.$$

The *change in output*, or the *change in y*, is

$$y_2 - y_1,$$

where $y_1 = f(x_1)$ and $y_2 = f(x_2)$.

DEFINITION

The **average rate of change of *y* with respect to *x***, as x changes from x_1 to x_2, is the ratio of the change in output to the change in input:

$$\frac{y_2 - y_1}{x_2 - x_1}, \quad \text{where } x_2 \neq x_1.$$

If we look at a graph of the function, we see that

$$\frac{y_2 - y_1}{x_2 - x_1} = \frac{f(x_2) - f(x_1)}{x_2 - x_1},$$

which is both the average rate of change and the slope of the line from $P(x_1, y_1)$ to $Q(x_2, y_2)$.* The line passing through P and Q, denoted $\overleftrightarrow{PQ}$, is called a **secant line**.

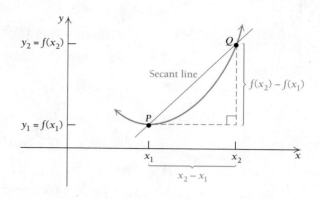

The slope of the secant line is interpreted as the average rate of change of f from x_1 to x_2.

■ **EXAMPLE 3**   For $y = f(x) = x^2$, find the average rate of change as:

a) x changes from 1 to 3.

b) x changes from 1 to 2.

c) x changes from 2 to 3.

Solution   The following graph is not necessary to the computations but gives us a look at two of the secant lines whose slopes are being computed.

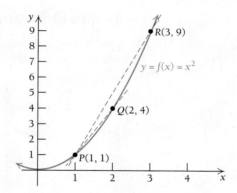

a) When $x_1 = 1$,

$$y_1 = f(x_1) = f(1) = 1^2 = 1;$$

and when $x_2 = 3$,

$$y_2 = f(x_2) = f(3) = 3^2 = 9.$$

The average rate of change is

$$\frac{y_2 - y_1}{x_2 - x_1} = \frac{f(x_2) - f(x_1)}{x_2 - x_1}$$

$$= \frac{9 - 1}{3 - 1}$$

$$= \frac{8}{2} = 4.$$

---

*The notation $P(x_1, y_1)$ simply means that point P has coordinates (x_1, y_1).

b) When $x_1 = 1$,
$$y_1 = f(x_1) = f(1) = 1^2 = 1;$$
and when $x_2 = 2$,
$$y_2 = f(x_2) = f(2) = 2^2 = 4.$$
The average rate of change is
$$\frac{4 - 1}{2 - 1} = \frac{3}{1} = 3.$$

c) When $x_1 = 2$,
$$y_1 = f(x_1) = f(2) = 2^2 = 4;$$
and when $x_2 = 3$,
$$y_2 = f(x_2) = f(3) = 3^2 = 9.$$
The average rate of change is
$$\frac{9 - 4}{3 - 2} = \frac{5}{1} = 5.$$

⟨ Quick Check 2

> **Quick Check 2**
>
> For $f(x) = x^3$, find the average rate of change between:
>
> **a)** $x = 1$ and $x = 4$;
>
> **b)** $x = 1$ and $x = 2$;
>
> **c)** $x = 1$ and $x = 1.2$.

For a linear function, the average rate of change is the same for any choice of x_1 and x_2. As we saw in Example 3, a function that is not linear has average rates of change that vary with the choice of x_1 and x_2.

Difference Quotients as Average Rates of Change

We now develop a notation for average rates of change that does not require subscripts. Instead of x_1, we will write simply x; in place of x_2, we will write $x + h$.

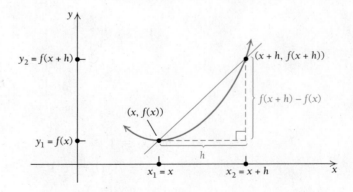

It may help to think of the h as the horizontal distance between the inputs x_1 and x_2. That is, to get from x_1, or x, to x_2, we move a distance h. Thus, $x_2 = x + h$. Then the average rate of change, also called a *difference quotient,* is given by

$$\frac{y_2 - y_1}{x_2 - x_1} = \frac{f(x_2) - f(x_1)}{x_2 - x_1} = \frac{f(x + h) - f(x)}{(x + h) - x} = \frac{f(x + h) - f(x)}{h}.$$

EXERCISES

Use a calculator to show that $f(x + h) \neq f(x) + f(h)$ for each of the following functions.

1. $f(x) = x^4 + x^2$; let $x = 6$ and $h = 0.02$.

2. $f(x) = x^3 - 2x^2 + 4$; let $x = 6$ and $h = 0.1$.

DEFINITION

The average rate of change of f with respect to x is also called the **difference quotient**. It is given by

$$\frac{f(x + h) - f(x)}{h}, \quad \text{where } h \neq 0.$$

The difference quotient is equal to the slope of the secant line from $(x, f(x))$ to $(x + h, f(x + h))$.

Keep in mind that, in general, $f(x + h) \neq f(x) + f(h)$. (You can check this using $f(x) = x^2$, as in the Technology Connection at the left.)

■ **EXAMPLE 4**   For $f(x) = x^2$, find the difference quotient when:

a) $x = 5$ and $h = 3$.

b) $x = 5$ and $h = 0.1$.

c) $x = 5$ and $h = 0.01$.

Solution

a) We substitute $x = 5$ and $h = 3$ into the formula:

$$\frac{f(x + h) - f(x)}{h} = \frac{f(5 + 3) - f(5)}{3} = \frac{f(8) - f(5)}{3}.$$

Now $f(8) = 8^2 = 64$ and $f(5) = 5^2 = 25$, and we have

$$\frac{f(8) - f(5)}{3} = \frac{64 - 25}{3} = \frac{39}{3} = 13.$$

The difference quotient is 13. It is also the slope of the line from $(5, 25)$ to $(8, 64)$.

b) We substitute $x = 5$ and $h = 0.1$ into the formula:

$$\frac{f(x + h) - f(x)}{h} = \frac{f(5 + 0.1) - f(5)}{0.1} = \frac{f(5.1) - f(5)}{0.1}.$$

Now $f(5.1) = (5.1)^2 = 26.01$ and $f(5) = 25$, and we have

$$\frac{f(5.1) - f(5)}{0.1} = \frac{26.01 - 25}{0.1} = \frac{1.01}{0.1} = 10.1.$$

c) We substitute $x = 5$ and $h = 0.01$ into the formula:

$$\frac{f(x + h) - f(x)}{h} = \frac{f(5 + 0.01) - f(5)}{0.01} = \frac{f(5.01) - f(5)}{0.01}.$$

Now $f(5.01) = (5.01)^2 = 25.1001$ and $f(5) = 25$, and we have

$$\frac{f(5.01) - f(5)}{0.01} = \frac{25.1001 - 25}{0.01} = \frac{0.1001}{0.01} = 10.01.$$

Note the trend in the average rate of change as h gets closer to 0.

For the function in Example 4, let's find a form of the difference quotient that will allow for more efficient computations.

■ **EXAMPLE 5**   For $f(x) = x^2$, find a simplified form of the difference quotient. Then find the value of the difference quotient when $x = 5$ and $h = 0.1$ and when $x = 5$ and $h = 0.01$.

Solution   We have

$$f(x) = x^2,$$

so

$$f(x + h) = (x + h)^2 = x^2 + 2xh + h^2. \quad \text{Multiplying } (x + h)(x + h)$$

Then

$$f(x + h) - f(x) = (x^2 + 2xh + h^2) - x^2 = 2xh + h^2. \quad \begin{array}{l} \text{The } x^2 \text{ terms} \\ \text{sum to 0.} \end{array}$$

Thus,

$$\frac{f(x + h) - f(x)}{h} = \frac{2xh + h^2}{h} = \frac{h(2x + h)}{h} = 2x + h, \quad h \neq 0. \quad h/h = 1$$

This is a simplified form of this difference quotient. It is important to note that any difference quotient is defined only when $h \neq 0$. The above simplification is valid only for nonzero values of h.

When $x = 5$ and $h = 0.1$,

$$\frac{f(x + h) - f(x)}{h} = 2x + h = 2 \cdot 5 + 0.1 = 10 + 0.1 = 10.1.$$

When $x = 5$ and $h = 0.01$,

$$\frac{f(x + h) - f(x)}{h} = 2x + h = 2 \cdot 5 + 0.01 = 10.01.$$

Although the expression $2x + h$ is valid only when $h \neq 0$, there is nothing stopping us from allowing h to get closer and closer to 0. Perhaps you can sense what the value of the difference quotient would be in this example if you allowed h to get close to 0 as a limit.

Compare the results of Example 4(b) and 4(c) and Example 5. In general, computations are easier when a simplified form of a difference quotient is found before any specific calculations are performed.

■ **EXAMPLE 6**   For $f(x) = x^3$, find a simplified form of the difference quotient.

Solution   For $f(x) = x^3$,

$$f(x + h) = (x + h)^3 = x^3 + 3x^2h + 3xh^2 + h^3.$$

(This is shown in Appendix A at the end of this book.) Then

$$f(x + h) - f(x) = (x^3 + 3x^2h + 3xh^2 + h^3) - x^3 = 3x^2h + 3xh^2 + h^3.$$

Thus,

$$\frac{f(x + h) - f(x)}{h} = \frac{3x^2h + 3xh^2 + h^3}{h} \quad \text{It is understood that } h \neq 0.$$

$$= \frac{h(3x^2 + 3xh + h^2)}{h} \quad \text{Factoring out } h; \quad h/h = 1$$

$$= 3x^2 + 3xh + h^2, \quad h \neq 0.$$

Again, this is true *only* for $h \neq 0$.

❭ **Quick Check 3**

Use the result of Example 6 to calculate the slope of the secant line (average rate of change) at $x = 2$, for $h = 0.1$, $h = 0.01$, and $h = 0.001$.

❬ Quick Check 3

The next two examples illustrate the development of the difference quotients for a simple rational function (Example 7) and for the square-root function (Example 8). These two forms are very common in calculus.

■ **EXAMPLE 7** For $f(x) = 1/x$, find a simplified form of the difference quotient.

Solution For $f(x) = 1/x$,

$$f(x + h) = \frac{1}{x + h}.$$

Then

$$f(x + h) - f(x) = \frac{1}{x + h} - \frac{1}{x}$$

$$= \frac{1}{x + h} \cdot \frac{x}{x} - \frac{1}{x} \cdot \frac{x + h}{x + h} \quad \text{Here we are multiplying by 1 to get a common denominator.}$$

$$= \frac{x - (x + h)}{x(x + h)}$$

$$= \frac{x - x - h}{x(x + h)}$$

$$= \frac{-h}{x(x + h)}.$$

Thus,

$$\frac{f(x + h) - f(x)}{h} = \frac{\dfrac{-h}{x(x + h)}}{h}$$

$$= \frac{-h}{x(x + h)} \cdot \frac{1}{h} = \frac{-1}{x(x + h)}, \quad h \neq 0.$$

This is true *only* for $h \neq 0$.

■ **EXAMPLE 8** For $f(x) = \sqrt{x}$, find a simplified form of the difference quotient.

Solution For $f(x) = \sqrt{x}$, we have $f(x + h) = \sqrt{x + h}$, so the difference quotient is

$$\frac{f(x + h) - f(x)}{h} = \frac{\sqrt{x + h} - \sqrt{x}}{h}.$$

Algebraic simplification of this difference quotient leads to

$$\frac{f(x + h) - f(x)}{h} = \frac{1}{\sqrt{x + h} + \sqrt{x}}, \quad h \neq 0.$$

Demonstration of this simplification is left as Exercise 54. You should note this simplified difference quotient as it will be seen again in Section 1.5.

In all of the above cases where we simplified a difference quotient using algebra, we ended up with two variables: x and h, where h cannot be 0. Although h cannot be exactly 0, we may let h be as close to 0 as we desire. This is a limit! In the next section, we will take that final step: allowing h to approach 0 as a limit.

Section Summary

- An *average rate of change* is the slope of a line between two points. If the two points are (x_1, y_1) and (x_2, y_2), then the average rate of change is $\dfrac{y_2 - y_1}{x_2 - x_1}$.

- If the two points are solutions to a single function, an equivalent form of the slope formula is $\dfrac{f(x + h) - f(x)}{h}$, where h is the horizontal difference between the two

x-values. This is called the *difference quotient*. The line connecting these two points is called a *secant line*.

- The difference quotient is the same as the slope formula. Both give the slope of the line between two points.

- The difference quotient gives the *average rate of change* between two points on a graph, represented by the secant line.

- It is preferable to simplify a difference quotient algebraically before evaluating it for particular values of x and h.

EXERCISE SET
1.3

For each function in Exercises 1–16, (a) find a simplified form of the difference quotient and then (b) complete the following table.

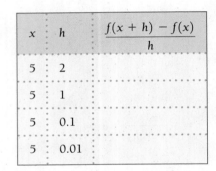

x	h	$\dfrac{f(x + h) - f(x)}{h}$
5	2	
5	1	
5	0.1	
5	0.01	

1. $f(x) = 4x^2$ **2.** $f(x) = 5x^2$

3. $f(x) = -4x^2$ **4.** $f(x) = -5x^2$

5. $f(x) = x^2 + x$ **6.** $f(x) = x^2 - x$

7. $f(x) = \dfrac{2}{x}$ **8.** $f(x) = \dfrac{9}{x}$

9. $f(x) = -2x + 5$ **10.** $f(x) = 2x + 3$

11. $f(x) = 1 - x^3$ **12.** $f(x) = 12x^3$

13. $f(x) = x^2 - 3x$ **14.** $f(x) = x^2 - 4x$

15. $f(x) = x^2 + 4x - 3$ **16.** $f(x) = x^2 - 3x + 5$

APPLICATIONS

Business and Economics

For Exercises 17–24, use each graph to estimate the average rate of change of the percentage of new employees in that type

of employment from 2000 to 2005, from 2005 to 2009, and from 2000 to 2009. (Source: Bureau of Labor Statistics.)

17. Total employment

18. Construction.

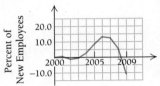

19. Professional services.

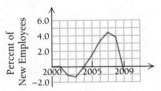

20. Health care.

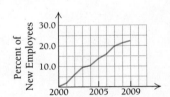

21. Education.

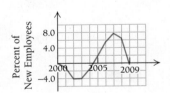

22. Government.

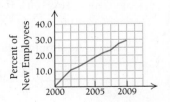

23. Mining and logging.

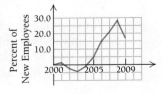

24. Manufacturing.

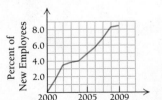
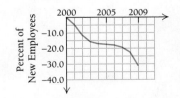

25. Use the following graph to find the average rate of change in U.S. energy consumption from 1970 to 1980, from 1980 to 1990, and from 2000 to 2010.

U.S. ENERGY CONSUMPTION

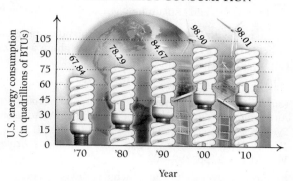

(*Source*: U.S. Energy Information Administration.)

26. Use the following graph to find the average rate of change of the U.S. trade deficit with Japan from 1990 to 1995, from 1995 to 2000, and from 2000 to 2010.

U.S. TRADE DEFICIT WITH JAPAN

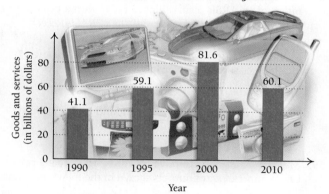

(*Source*: U.S. Census Bureau, *Statistical Abstract of the United States, 2011*.)

27. Utility. Utility is a type of function that occurs in economics. When a consumer receives x units of a certain product, a certain amount of pleasure, or utility U, is derived. The following is a graph of a typical utility function.

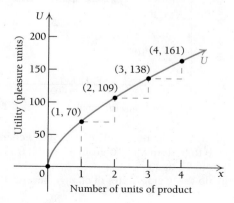

a) Find the average rate of change of U as x changes from 0 to 1; from 1 to 2; from 2 to 3; from 3 to 4.
b) Why do you think the average rates of change are decreasing as x increases?

28. Advertising results. The following graph shows a typical response to advertising. After an amount a is spent on advertising, the company sells $N(a)$ units of a product.

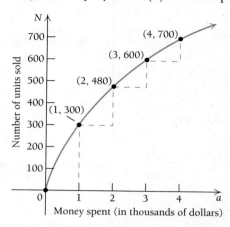

a) Find the average rate of change of N as a changes from 0 to 1; from 1 to 2; from 2 to 3; from 3 to 4.
b) Why do you think the average rates of change are decreasing as x increases?

29. Baseball ticket prices. Based on data from Major League Baseball, the average price of a ticket to a major league game can be approximated by

$$p(x) = 0.03x^2 + 0.56x + 8.63,$$

where x is the number of years after 1991 and $p(x)$ is in dollars. (*Source*: Based on data from www.teammarketing.com.)

a) Find $p(4)$.
b) Find $p(17)$.
c) Find $p(17) - p(4)$.
d) Find $\dfrac{p(17) - p(4)}{17 - 4}$, and interpret this result.

30. Compound interest. The amount of money, $A(t)$, in a savings account that pays 6% interest, compounded quarterly for t years, when an initial investment of $2000 is made, is given by

$$A(t) = 2000(1.015)^{4t}.$$

a) Find $A(3)$.
b) Find $A(5)$.
c) Find $A(5) - A(3)$.
d) Find $\dfrac{A(5) - A(3)}{5 - 3}$, and interpret this result.

31. Credit card debt. When a balance of $5000 is owed on a credit card and interest is being charged at a rate of 14% per year, the total amount owed after t years, $A(t)$, is given by

$$A(t) = 5000(1.14)^t.$$

Find $\dfrac{A(3) - A(2)}{3 - 2}$, and interpret this result.

32. Credit card debt. When a balance of $3000 is owed on a credit card and interest is charged at a rate of 17% per year, the total amount owed after t years, $A(t)$, is given by

$$A(t) = 3000(1.17)^t.$$

Find $\dfrac{A(4) - A(3)}{4 - 3}$, and interpret this result.

33. Total cost. Suppose that Sport Stylz Inc. determines that the cost, in dollars, of producing x cellphone-sunglasses is given by

$$C(x) = -0.05x^2 + 50x.$$

Find $\dfrac{C(301) - C(300)}{301 - 300}$, and interpret the significance of this result to the company.

34. Total revenue. Suppose that Sports Stylz Inc. determines that the revenue, in dollars, from the sale of x cellphone-sunglasses is given by

$$R(x) = -0.01x^2 + 1000x.$$

Find $\dfrac{R(301) - R(300)}{301 - 300}$, and interpret the significance of this result to the company.

Life and Physical Sciences

35. Growth of a baby. The median weights of babies at age t months are graphed below.

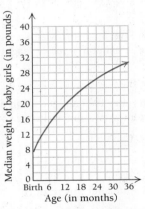 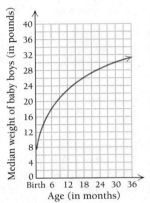

(*Source:* Developed by the National Center for Health Statistics in collaboration with the National Center for Chronic Disease Prevention and Health Promotion, 2000.)

Use the graph of girls' median weight to estimate:

a) The average growth rate of a girl during her first 12 months. (Your answer should be in pounds per month.)

b) The average growth rate of a girl during her second 12 months.

c) The average growth rate of a girl during her first 24 months.

d) Based on your answers in parts (a)–(c) and the graph, estimate the growth rate of a typical 12-month-old girl. Use a straightedge.

e) When does the graph indicate that a baby girl's growth rate is greatest?

36. Growth of a baby. Use the graph of boys' median weight in Exercise 35 to estimate:

a) The average growth rate of a boy during his first 15 months. (Your answer should be in pounds per month.)

b) The average growth rate of a boy during his second 15 months. (Your answer should be in pounds per month.)

c) The average growth rate of a boy during his first 30 months. (Your answer should be in pounds per month.)

d) Based on your answers in parts (a)–(c) and the graph, estimate the growth rate of a typical boy at exactly 15 months, and explain how you arrived at this figure.

37. Home range. It has been shown that the home range, in hectares, of a carnivorous mammal weighing w grams can be approximated by

$$H(w) = 0.11w^{1.36}.$$

(*Source:* Based on information in Emlen, J. M., *Ecology: An Evolutionary Approach*, p. 200, Reading, MA: Addison-Wesley, 1973; and Harestad, A. S., and Bunnel, F. L., "Home Range and Body Weight—A Reevaluation," *Ecology*, Vol. 60, No. 2, pp. 389–402.)

a) Find the average rate at which a carnivorous mammal's home range increases as the animal's weight grows from 500 g to 700 g.

b) Find $\dfrac{H(300) - H(200)}{300 - 200}$, and interpret this result.

38. Radar range. The function given by $R(x) = 11.74x^{1/4}$ can be used to approximate the maximum range $R(x)$, in miles, of an ARSR-3 surveillance radar with a peak power of x watts (W). (*Source: Introduction to RADAR Techniques*, Federal Aviation Administration, 1988.)

a) Find the rate at which the maximum radar range changes as peak power increases from 40,000 W to 60,000 W.

b) Find $\dfrac{R(60,000) - R(50,000)}{60,000 - 50,000}$, and interpret this result.

39. Memory. The total number of words, $M(t)$, that a person can memorize in t minutes is shown in the following graph.

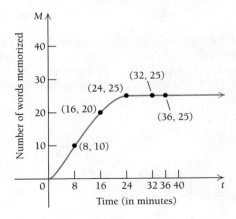

a) Find the average rate of change of M as t changes from 0 to 8; from 8 to 16; from 16 to 24; from 24 to 32; from 32 to 36.

b) Why do the average rates of change become 0 after 24 min?

40. Average velocity. Suppose that in t hours, a truck travels $s(t)$ miles, where

$$s(t) = 10t^2.$$

a) Find $s(5) - s(2)$. What does this represent?

b) Find the average rate of change of distance with respect to time as t changes from $t_1 = 2$ to $t_2 = 5$. This is known as **average velocity**, or **speed**.

41. *Average velocity.* In t seconds, an object dropped from a certain height will fall $s(t)$ feet, where

$$s(t) = 16t^2.$$

a) Find $s(5) - s(3)$.

b) What is the average rate of change of distance with respect to time during the period from 3 to 5 sec? This is also *average velocity*.

42. *Gas mileage.* At the beginning of a trip, the odometer on a car reads 30,680, and the car has a full tank of gas. At the end of the trip, the odometer reads 31,077. It takes 13.5 gal of gas to refill the tank.

a) What is the average rate at which the car was traveling, in miles per gallon?

b) What is the average rate of gas consumption in gallons per mile?

Social Sciences

43. *Population growth.* The two curves below describe the numbers of people in two countries at time t, in years.

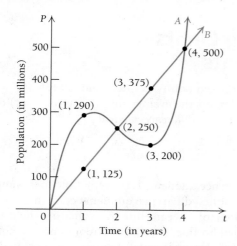

a) Find the average rate of change of each population with respect to time t as t changes from 0 to 4. This is often called the **average growth rate**.

b) If the calculation in part (a) were the only one made, would we detect the fact that the populations were growing differently? Explain.

c) Find the average rates of change of each population as t changes from 0 to 1; from 1 to 2; from 2 to 3; from 3 to 4.

d) For which population does the statement "the population grew consistently at a rate of 125 million per year" convey accurate information? Why?

SYNTHESIS

44. *Business: comparing rates of change.* The following two graphs show the number of federally insured banks and the Nasdaq Composite Stock Index over a 6-month period.

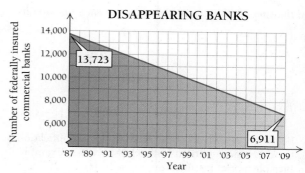

(*Source:* Federal Deposit Insurance Corp.)

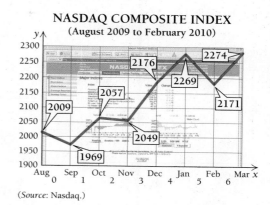

(*Source:* Nasdaq.)

Explain the difference between these graphs in as many ways as you can. Be sure to mention average rates of change.

45. *Rising cost of college.* Like the cost of most things, the cost of a college education has gone up over the past 35 years. The graphs below display the yearly costs of 4-year colleges in 2008 dollars—indicating that the costs prior to 2008 have been adjusted for inflation.

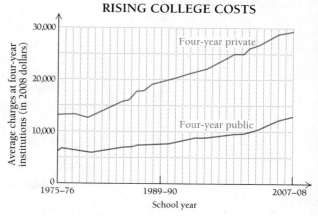

(*Source:* National Center for Educational Statistics, *Annual Digest of Educational Statistics,* 2008.)

a) In what school year did the cost of a private 4-year college increase the most?

b) In what school year(s) did the cost of a public 4-year college increase the most?

c) Assuming an annual inflation rate of 3%, calculate the cost of a year at a public and at a private 4-year college in 1975. Express the costs in 1975 dollars.

Find the simplified difference quotient for each function listed.

46. $f(x) = mx + b$

47. $f(x) = ax^2 + bx + c$

48. $f(x) = ax^3 + bx^2$

49. $f(x) = x^4$

50. $f(x) = x^5$

51. $f(x) = ax^5 + bx^4$

52. $f(x) = \dfrac{1}{x^2}$

53. $f(x) = \dfrac{1}{1 - x}$

54. Below are shown all of the steps in the simplification of the difference quotient for $f(x) = \sqrt{x}$ (see Example 8). Provide a brief justification for each step.

$$\frac{f(x + h) - f(x)}{h} = \frac{\sqrt{x + h} - \sqrt{x}}{h}$$

a) $= \dfrac{\sqrt{x + h} - \sqrt{x}}{h} \cdot \left(\dfrac{\sqrt{x + h} + \sqrt{x}}{\sqrt{x + h} + \sqrt{x}} \right)$

b) $= \dfrac{x + h + \sqrt{x}\sqrt{x + h} - \sqrt{x}\sqrt{x + h} - x}{h(\sqrt{x + h} + \sqrt{x})}$

c) $= \dfrac{x + h - x}{h(\sqrt{x + h} + \sqrt{x})}$

d) $= \dfrac{h}{h(\sqrt{x + h} + \sqrt{x})}$

e) $= \dfrac{1}{\sqrt{x + h} + \sqrt{x}}$

For Exercises 55 and 56, find the simplified difference quotient.

55. $f(x) = \sqrt{2x + 1}$

56. $f(x) = \dfrac{1}{\sqrt{x}}$

Answers to Quick Checks

1. (a) It rained 1/2 inch per hour.   **(b)** Your car gets 12.5 miles per gallon.   **(c)** The temperature dropped 2 degrees per hour.   **2. (a)** 21;   **(b)** 7;   **(c)** 3.64
3. 12.61, 12.0601, 12.006001

1.4 Differentiation Using Limits of Difference Quotients

OBJECTIVES

- Find derivatives and values of derivatives.
- Find equations of tangent lines.

The *slope of the secant line* connecting two points $(x, f(x))$ and $(x + h, f(x + h))$ on the graph of a function $y = f(x)$ represents the *average rate of change* of $f(x)$ over the interval $[x, x + h]$. This rate is given by the *difference quotient*

$$\frac{f(x + h) - f(x)}{h}, \qquad h \neq 0.$$

In Section 1.3, we worked out difference quotients for several functions, simplifying them as much as possible. These simplified difference quotients contain the variables x and h, and we can calculate the slope of a secant line by evaluating the difference quotient for a given x-value and a given h-value. Recall that h represents the horizontal distance between x and $x + h$. Although h cannot equal zero, we can consider the case where h approaches zero as a limit. In this section, we explore this possibility.

Tangent Lines

A line that touches a circle at exactly one point is called a *tangent line*. The word "tangent" derives from the Latin *tangentem*, meaning "touch," whereas the word "secant" derives from the Latin *secantem*, meaning "cut." In the figure below, the secant line cuts through the circle, while the tangent line touches, but does not cut through, the circle.

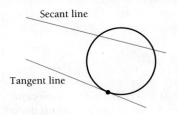

Secant line

Tangent line

These notions can be extended to any smooth curve: a tangent line touches a curve at a single point only, in the same way as the tangent line touches the circle in the figure at the bottom of p. 132. In Fig. 1, the line L touches the curve exactly once in the small interval containing P, the *point of tangency*. We are not concerned with the behavior of the line far from the point of tangency. We see that L does pass through the curve elsewhere, but it is still considered a tangent line to the curve at point P.

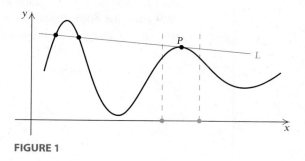

FIGURE 1

In Figure 2, line M crosses the curve only at point P, but is not tangent to the curve; it does not touch the curve in the desired manner.

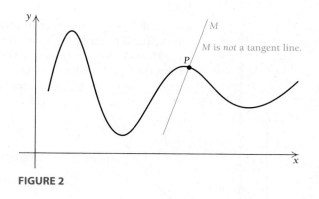

FIGURE 2

In Figure 3, all of the lines except for L_1 and L_2 are tangent lines.

The power lines run tangent to the tower brackets at the turn.

Exploratory

Graph $y_1 = 3x^5 - 20x^3$ with the viewing window $[-3, 3, -80, 80]$, with Xscl $= 1$ and Yscl $= 10$. Then also graph the lines $y_2 = -7x - 10$, $y_3 = -30x + 13$, and $y_4 = -45x + 28$. Which line appears to be tangent to the graph of y_1 at $(1, -17)$? If necessary, zoom in near $(1, -17)$ to refine your guess.

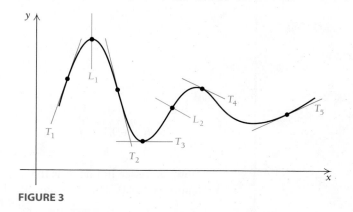

FIGURE 3

One more observation: if a curve is smooth (has no corners), then each point on the curve will have a *unique* tangent line; that is, exactly one tangent line is possible at any given point.

Differentiation Using Limits

We now define *tangent line* so that it makes sense for *any* curve. To do this, we use the notion of limit.

To obtain the line tangent to the curve at point P, consider secant lines through P and neighboring points Q_1, Q_2, and so on. As the Q's approach P, the secant lines approach line T. Each secant line has a slope. The slopes m_1, m_2, m_3, and so on, of the secant lines approach the slope m of line T. We *define* line T as the **tangent line**, the line that contains point P and has slope m, where m is the limit of the slopes of the secant lines as the points Q approach P.

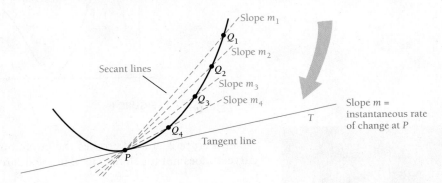

Think of the sequence of secant lines as an animation: as the points Q move closer to the fixed point P, the resulting secant lines "lie down" on the tangent line.

How might we calculate the limit m? Suppose that P has coordinates $(x, f(x))$. Then the first coordinate of Q is x plus some number h, or $x + h$. The coordinates of Q are $(x + h, f(x + h))$, as shown in Fig. 4.

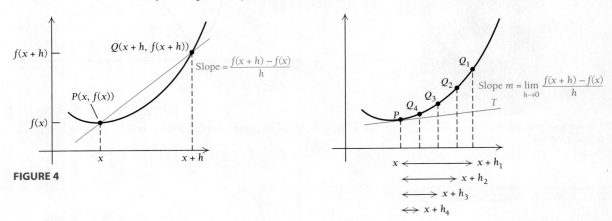

FIGURE 4

FIGURE 5

From Section 1.3, we know that the slope of the secant line $\overleftrightarrow{PQ}$ is given by

$$\frac{f(x + h) - f(x)}{h}.$$

Now, as we see in Fig. 5, as the Q's approach P, the values of $x + h$ approach x. That is, h approaches 0. Thus, we have the following.

The slope of the tangent line at $(x, f(x)) = m = \lim_{h \to 0} \frac{f(x + h) - f(x)}{h}$.

This limit is also the **instantaneous rate of change** of $f(x)$ at x.

The formal definition of the *derivative of a function f* can now be given. We will designate the derivative at x as $f'(x)$, rather than m. The notation $f'(x)$ is read "the derivative of f at x," "f prime at x," or "f prime of x."

> ### DEFINITION
>
> For a function $y = f(x)$, its **derivative** at x is the function f' defined by
>
> $$f'(x) = \lim_{h \to 0} \frac{f(x + h) - f(x)}{h},$$
>
> provided that the limit exists. If $f'(x)$ exists, then we say that f is **differentiable** at x. We sometimes call f' the **derived function**.

Let's now calculate some formulas for derivatives. That is, given a formula for a function f, we will attempt to find a formula for f'.

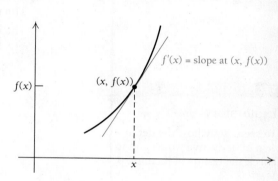

There are three steps in calculating a derivative.

1. Write the difference quotient, $(f(x + h) - f(x))/h$.
2. Simplify the difference quotient.
3. Find the limit as h approaches 0.

■ **EXAMPLE 1**   For $f(x) = x^2$, find $f'(x)$. Then find $f'(-3)$ and $f'(4)$.

Solution   We have

1. $\dfrac{f(x + h) - f(x)}{h} = \dfrac{(x + h)^2 - x^2}{h}$

2. $\dfrac{f(x + h) - f(x)}{h} = \dfrac{x^2 + 2xh + h^2 - x^2}{h}$ Evaluating $f(x + h)$ and $f(x)$

$\qquad\qquad\qquad = \dfrac{2xh + h^2}{h} = \dfrac{h(2x + h)}{h}$ $\Big\}$ Simplifying

$\qquad\qquad\qquad = 2x + h, \quad h \neq 0$

3. We want to find

$$\lim_{h \to 0} \frac{f(x + h) - f(x)}{h} = \lim_{h \to 0} (2x + h).$$

Recall that in Section 1.3, we calculated the slope of a secant line by evaluating the difference quotient at a particular x-value and a series of h-values that grew closer to zero: $h = 0.1$, $h = 0.01$, and so on. If we allow h to approach 0 as close as we desire, we have the derivative. That is, for

$$\lim_{h \to 0} (2x + h) = 2x,$$

we have

$$f'(x) = 2x.$$

Using the fact that $f'(x) = 2x$, it follows that

$$f'(-3) = 2 \cdot (-3) = -6, \quad \text{and} \quad f'(4) = 2 \cdot 4 = 8.$$

This tells us that at $x = -3$, the curve has a tangent line whose slope is

$$f'(-3) = -6,$$

and at $x = 4$, the tangent line has slope

$$f'(4) = 8.$$

We can also say:

- The tangent line to the curve at the point $(-3, 9)$ has slope -6.
- The tangent line to the curve at the point $(4, 16)$ has slope 8.
- The instantaneous rate of change at $x = -3$ is -6.
- The instantaneous rate of change at $x = 4$ is 8.

TECHNOLOGY CONNECTION

Exploratory

To see exactly how the definition of derivative works and to check the results of Example 1, let $y_1 = ((-3 + x)^2 - (-3)^2)/x$ and note that we are using x in place of h. Use the TABLE feature and enter smaller and smaller values for x. Then repeat the procedure for $y_2 = ((4 + x)^2 - 4^2)/x$. How do these results confirm those in Example 1?

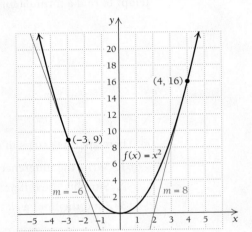

■ **EXAMPLE 2**   For $f(x) = x^3$, find $f'(x)$. Then find $f'(-1)$ and $f'(1.5)$.

Solution

1. We have

$$\frac{f(x + h) - f(x)}{h} = \frac{(x + h)^3 - x^3}{h}.$$

2. In Example 6 of Section 1.3 (on p. 126), we showed how this difference quotient can be simplified to

$$\frac{f(x + h) - f(x)}{h} = 3x^2 + 3xh + h^2, \quad h \neq 0.$$

3. We then have

$$f'(x) = \lim_{h \to 0} \frac{f(x + h) - f(x)}{h} = \lim_{h \to 0} (3x^2 + 3xh + h^2) = 3x^2.$$

Thus,

$$f'(-1) = 3(-1)^2 = 3 \quad \text{and} \quad f'(1.5) = 3(1.5)^2 = 6.75.$$

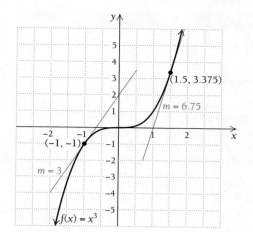

〈 Quick Check 1

> **Quick Check 1**
>
> Use the results from Examples 1 and 2 to find the derivative of $f(x) = x^3 + x^2$, and then calculate $f'(-2)$ and $f'(4)$. Interpret these results.

A common error in simplifying a difference quotient is to write $f(x) + h$, which is incorrect! The expression $x + h$ must remain a unit when substituting for x in the function. For example, if $f(x) = x^2$, we write $f(x + h) = (x + h)^2$. Once we have set up the expression for the difference quotient correctly, we can *then* simplify it using normal algebraic techniques.

■ **EXAMPLE 3**   For $f(x) = 3x - 4$, find $f'(x)$ and $f'(2)$.

Solution   We follow the three steps given above.

1. $\dfrac{f(x + h) - f(x)}{h} = \dfrac{3(x + h) - 4 - (3x - 4)}{h}$ The parentheses are important.

2. $\dfrac{f(x + h) - f(x)}{h} = \dfrac{3x + 3h - 4 - 3x + 4}{h}$ Using the distributive law

    $= \dfrac{3h}{h} = 3, \quad h \neq 0;$ Simplifying

3. $\lim\limits_{h \to 0} \dfrac{f(x + h) - f(x)}{h} = \lim\limits_{h \to 0} 3 = 3$, since 3 is a constant.

    Thus, if $f(x) = 3x - 4$, then $f'(x) = 3$ and $f'(2) = 3$.

The result of Example 3 suggests that, for a straight line, the slope of a tangent line is the slope of the straight line itself. That is, a general formula for the derivative of a linear function

$$f(x) = mx + b$$

is $f'(x) = m.$

The formula can be verified in a manner similar to that used in Example 3.

Examples 1–3 and Example 4, which follows, involve a somewhat lengthy process, but in Section 1.5 we will develop some faster techniques. It is very important in this section, however, to fully understand the concept of a derivative.

■ **EXAMPLE 4**   For $f(x) = \dfrac{1}{x}$:

a) Find $f'(x)$.   **b)** Find $f'(2)$.

c) Find an equation of the tangent line to the curve at $x = 2$.

Curves and Tangent Lines

EXERCISES

1. For $f(x) = 3/x$, find $f'(x)$, $f'(-2)$, and $f'(-\frac{1}{2})$.

2. Find an equation of the tangent line to the graph of $f(x) = 3/x$ at $(-2, -\frac{1}{2})$ and an equation of the tangent line to the curve at $(-\frac{1}{2}, -6)$. Then graph the curve $f(x) = 3/x$ and both tangent lines. Use ZOOM to view the graphs near the points of tangency.

3. To check your equations of tangent lines, first graph $f(x) = 3/x$. Select Tangent from the DRAW menu, and enter first $x = -2$ and then $x = -\frac{1}{2}$.

Solution

a) 1. We have

$$\frac{f(x + h) - f(x)}{h} = \frac{[1/(x + h)] - (1/x)}{h}.$$

2. In Example 7 of Section 1.3 (on p. 127), we showed that this difference quotient simplifies to

$$\frac{f(x + h) - f(x)}{h} = \frac{-1}{x(x + h)}, \quad h \neq 0.$$

3. We want to find

$$\lim_{h \to 0} \frac{f(x + h) - f(x)}{h} = \lim_{h \to 0} \frac{-1}{x(x + h)}.$$

As $h \to 0$, we have $x + h \to x$. Thus,

$$f'(x) = \lim_{h \to 0} \frac{-1}{x(x + h)} = \frac{-1}{x \cdot x} = \frac{-1}{x^2}.$$

b) Since $f'(x) = -1/x^2$, we have

$$f'(2) = \frac{-1}{2^2} = -\frac{1}{4}. \qquad \text{This is the slope of the tangent line at } x = 2.$$

c) We can find an equation of the tangent line at $x = 2$ if we know the line's slope and a point that is on the line. In part (b), we found that the slope at $x = 2$ is $-\frac{1}{4}$. To find a point on the line, we compute $f(2)$:

$$f(2) = \frac{1}{2}. \qquad \begin{array}{l}\text{CAUTION! Be careful to use } f \text{ when computing} \\ y\text{-values and } f' \text{ when computing slope.}\end{array}$$

We have

Point: $(2, \frac{1}{2})$,   This is $(x_1, f(x_1))$.
Slope: $-\frac{1}{4}$.   This is $f'(x_1)$.

We substitute into the point–slope equation (see Section R.4):

$$y - y_1 = m(x - x_1)$$

$$y - \tfrac{1}{2} = -\tfrac{1}{4}(x - 2)$$

$$y = -\tfrac{1}{4}x + \tfrac{1}{2} + \tfrac{1}{2} \qquad \left.\begin{array}{l} \\ \end{array}\right\} \begin{array}{l}\text{Rewriting in} \\ \text{slope–intercept form}\end{array}$$

$$= -\tfrac{1}{4}x + 1.$$

The equation of the tangent line to the curve at $x = 2$ is

$$y = -\tfrac{1}{4}x + 1.$$

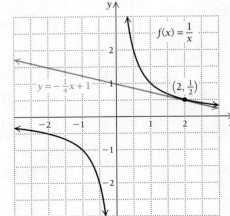

> **Quick Check 2**
>
> Repeat Example 4a for $f(x) = -\dfrac{2}{x}$. What are the similarities in your method?

⟨ Quick Check 2

In Example 4, note that since $f(0)$ does not exist for $f(x) = 1/x$, we cannot evaluate the difference quotient

$$\frac{f(0 + h) - f(0)}{h}.$$

Thus, $f'(0)$ does not exist. We say that "f is not differentiable at 0."

When a function is not defined at a point, it is not differentiable at that point. In general, if a function is discontinuous at a point, it is not differentiable at that point.

Sometimes a function f is continuous at a point, but its derivative f' is not defined at this point. The function $f(x) = |x|$ is an example. It is continuous at $x = 0$ since it meets all the requirements for continuity there. But what about a tangent line at this point?

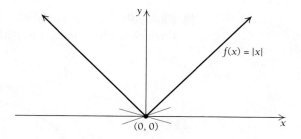

$$f(x) = |x|$$

$(0, 0)$

Suppose that we try to draw a tangent line at $(0, 0)$. A function like this with a corner (not smooth) would seem to have many tangent lines at $(0, 0)$, and thus many slopes. The derivative at such a point would not be unique. Let's try to calculate the derivative at 0.

Since

$$f(x) = |x| = \begin{cases} x, & \text{for } x \geq 0, \\ -x, & \text{for } x < 0, \end{cases}$$

it follows from our earlier work with lines that

$$f'(x) = \begin{cases} 1, & \text{for } x > 0, \\ -1, & \text{for } x < 0. \end{cases}$$

Then, since

$$\lim_{x \to 0^+} f'(x) \neq \lim_{x \to 0^-} f'(x),$$

it follows that $f'(0)$ does not exist.

In general, if a function has a "corner," it will not have a derivative at that point. The following graphs show examples of "corners."

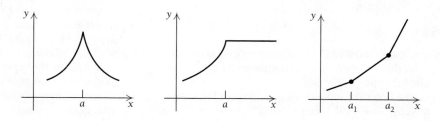

A function will also fail to be differentiable at a point if it has a vertical tangent at that point. For example, the function shown to the right has a vertical tangent at point *a*. Recall that since the slope of a vertical line is undefined, there is no derivative at such a point.

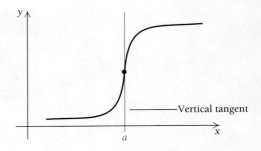

The function given by $f(x) = |x|$ illustrates the fact that although a function may be continuous at each point in an interval *I*, it may not be differentiable at each point in *I*. That is, continuity does not imply differentiability. However, differentiability *does* guarantee continuity. That is, if $f'(a)$ exists, then *f* is continuous at *a*. The function $f(x) = x^2$ is an example of a function that is differentiable over the interval $(-\infty, \infty)$ and is therefore continuous everywhere. Thus, when a function is differentiable over an interval, it is not just continuous, but is also *smooth* in the sense that there are no "corners" in its graph.

■ **EXAMPLE 5**   Below is the graph of a function $y = t(x)$. List the points in the graph at which the function *t* is not differentiable.

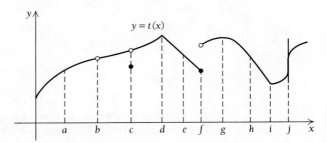

Solution   A function is not differentiable at a point if there is (1) a discontinuity, (2) a corner, or (3) a vertical tangent at that point. Therefore, the function $y = t(x)$ is not differentiable at $x = b, x = c$, and $x = f$ since the function is discontinuous at these points; it is not differentiable at $x = d$ and $x = i$ since there are corners at these points; and it is not differentiable at $x = j$ as there is a vertical tangent line at this point (the slope is undefined). The function is differentiable at $x = a, x = e, x = g$, and $x = h$.

❭ **Quick Check 3**

Where is $f(x) = |x + 6|$ not differentiable? Why?

❬ Quick Check 3

TECHNOLOGY CONNECTION

Calculus Apps for the iPhone or iPod Touch

A Technology Connection in Section R.5 introduced two apps for the iPhone and iPod Touch: iPlot and Graphicus (both available from the iTunes Store). Here we begin to cover uses of Graphicus in calculus.

Graphicus

This app can graph most of the functions we encounter in this book and draw tangent lines at points on a curve. Let's consider the function $f(x) = x^4 - 2x^2$.

To graph a function and examine its tangent lines, first open Graphicus. Touch the blank rectangle at the top of the screen and enter the function as y(x)=x^4-2x^2. Press ＋ at the upper right. You will see the graph, as shown in Fig. 1.

Notice the seven icons at the bottom of the screen. The first one is for zooming. Touch it to zoom in and zoom out. Touch the screen with two fingers and move vertically or horizontally to adjust the windows and scaling. The second icon is for visualizing tangent lines. Touch it and then touch the graph at a point, and you will see a highlighted point on the graph and a tangent line to the curve at that point (Fig. 2). Move your finger left to right along the graph and you will see highlighted various other points on the curve and tangent lines at those points (Figs. 3 and 4). Think about the slopes of the tangent lines as you move from left to right. When are they positive, negative, or 0? Note where the slope changes from positive to 0 to negative.

(continued)

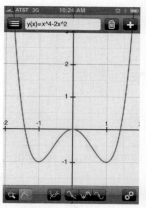

FIGURE 1

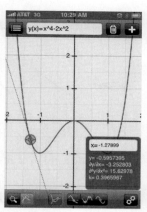

FIGURE 2

FIGURE 3

FIGURE 4

EXERCISES

Use Graphicus to graph each function. You may need to use the zoom function to alter the window. Visualize tangent lines, noting where they are positive, negative, and 0. Then try to determine where the slope changes from positive to 0 to negative.

1. $f(x) = 2x^3 - x^4$

2. $f(x) = x(200 - x)$

3. $f(x) = x^3 - 6x^2$

4. $f(x) = -4.32 + 1.44x + 3x^2 - x^3$

5. $g(x) = x\sqrt{4 - x^2}$

6. $g(x) = \dfrac{4x}{x^2 + 1}$

7. $f(x) = \dfrac{x^2 - 3x}{x - 1}$

8. $f(x) = |x + 2| - 3$. What happens to the tangent line at the point $(-2, -3)$?

Section Summary

- A *tangent line* is a line that touches a (smooth) curve at a single point, the *point of tangency*. See Fig. 3 (on p. 133) for examples of tangent lines (and lines that are not considered tangent lines).
- The *derivative* of a function $f(x)$ is defined by

$$f'(x) = \lim_{h \to 0} \frac{f(x + h) - f(x)}{h}.$$

- The *slope* of the tangent line to the graph of $y = f(x)$ at $x = a$ is the value of the derivative at $x = a$; that is, the slope of the tangent line at $x = a$ is $f'(a)$.
- Slopes of tangent lines are interpreted as *instantaneous rates of change*.

- The equation of a tangent line at $x = a$ is found by simplifying $y - f(a) = f'(a)(x - a)$.
- If a function is differentiable at a point $x = a$, then it is *continuous* at $x = a$. That is, differentiability implies continuity.
- However, continuity at a point $x = a$ does *not* imply differentiability at $x = a$. A good example is the absolute-value function, $f(x) = |x|$, or any function whose graph has a corner. Continuity alone is not sufficient to guarantee differentiability.
- A function is not differentiable at a point $x = a$ if:
  (1) there is a discontinuity at $x = a$,
  (2) there is a corner at $x = a$, or
  (3) there is a vertical tangent at $x = a$.

EXERCISE SET
1.4

In Exercises 1–16:

a) *Graph the function.*

b) *Draw tangent lines to the graph at points whose x-coordinates are* −2, 0, *and* 1.

c) *Find* $f'(x)$ *by determining* $\lim\limits_{h \to 0} \dfrac{f(x + h) - f(x)}{h}$.

d) *Find* $f'(-2)$, $f'(0)$, *and* $f'(1)$. *These slopes should match those of the lines you drew in part (b).*

1. $f(x) = \frac{3}{2}x^2$

2. $f(x) = \frac{1}{2}x^2$

3. $f(x) = -2x^2$

4. $f(x) = -3x^2$

5. $f(x) = x^3$

6. $f(x) = -x^3$

7. $f(x) = 2x + 3$

8. $f(x) = -2x + 5$

9. $f(x) = \frac{1}{2}x - 3$

10. $f(x) = \frac{3}{4}x - 2$

11. $f(x) = x^2 + x$

12. $f(x) = x^2 - x$

13. $f(x) = 2x^2 + 3x - 2$

14. $f(x) = 5x^2 - 2x + 7$

15. $f(x) = \dfrac{1}{x}$

16. $f(x) = \dfrac{2}{x}$

17. Find an equation of the tangent line to the graph of $f(x) = x^2$ at **(a)** $(3, 9)$; **(b)** $(-1, 1)$; **(c)** $(10, 100)$. See Example 1.

18. Find an equation of the tangent line to the graph of $f(x) = x^3$ at **(a)** $(-2, -8)$; **(b)** $(0, 0)$; **(c)** $(4, 64)$. See Example 2.

19. Find an equation of the tangent line to the graph of $f(x) = 2/x$ at **(a)** $(1, 2)$; **(b)** $(-1, -2)$; **(c)** $(100, 0.02)$. See Exercise 16.

20. Find an equation of the tangent line to the graph of $f(x) = -1/x$ at **(a)** $(-1, 1)$; **(b)** $(2, -\frac{1}{2})$; **(c)** $(-5, \frac{1}{5})$.

21. Find an equation of the tangent line to the graph of $f(x) = 4 - x^2$ at **(a)** $(-1, 3)$; **(b)** $(0, 4)$; **(c)** $(5, -21)$.

22. Find an equation of the tangent line to the graph of $f(x) = x^2 - 2x$ at **(a)** $(-2, 8)$; **(b)** $(1, -1)$; **(c)** $(4, 8)$.

23. Find $f'(x)$ for $f(x) = mx + b$.

24. Find $f'(x)$ for $f(x) = ax^2 + bx$.

For Exercises 25–28, list the points in the graph at which each function is not differentiable.

25.

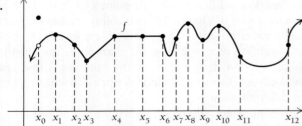

26.

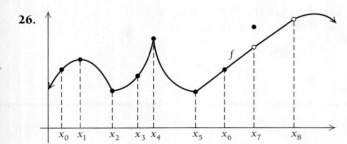

27.

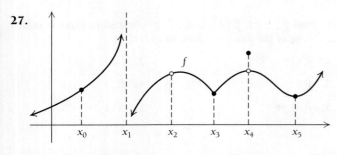

28.

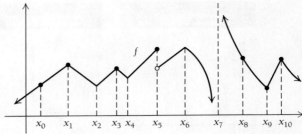

29. Draw a graph that is continuous, but not differentiable, at $x = 3$.

30. Draw a graph that has a horizontal tangent line at $x = 5$.

31. Draw a graph that is differentiable and has horizontal tangent lines at $x = 0$, $x = 2$, and $x = 4$.

32. Draw a graph that has horizontal tangent lines at $x = 2$ and $x = 5$ and is continuous, but not differentiable, at $x = 3$.

33. Draw a graph that is smooth, but not differentiable, at $x = 1$.

34. Draw a graph that is smooth for all x, but not differentiable at $x = -1$ and $x = 2$.

APPLICATIONS

Business and Economics

35. *The postage function.* Consider the postage function defined in Exercises 84–88 of Exercise Set 1.1, on p. 108. At what values in the domain is the function not differentiable?

36. *The taxicab fare function.* Consider the taxicab fare function defined in Exercises 81–83 of Exercise Set 1.1, on p. 108. At what values is the function not differentiable?

37. *Baseball ticket prices.* Consider the model for average Major League Baseball ticket prices in Exercise 29 of Exercise Set 1.3, on p. 129. At what values is the function not differentiable?

The values of the Dow Jones Industrial Average for the week of January 4–11, 2010, are graphed below, where x is the day of the month.

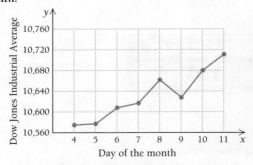

(*Source:* www.moneycentral.msn.com.)

38. On what day did the Dow Jones Industrial Average show the greatest rate of increase? On what day did it show the greatest rate of decrease? Give the rates.

39. Is the function differentiable at the given x-values? Why or why not?

SYNTHESIS

40. Which of the lines in the following graph appear to be tangent lines? Try to explain why or why not.

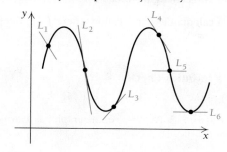

41. On the following graph, use a blue colored pencil to draw each secant line from point P to the points Q. Then use a red colored pencil to draw a tangent line to the curve at P. Describe what happens.

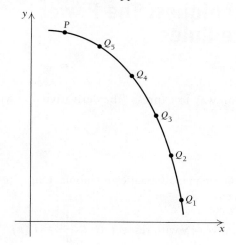

For Exercises 42–48, find $f'(x)$ for the given function.

42. $f(x) = x^4$ (See Exercise 49 in Section 1.3.)

43. $f(x) = \dfrac{1}{1 - x}$ (See Exercise 53 in Section 1.3.)

44. $f(x) = x^5$ (See Exercise 50 in Section 1.3.)

45. $f(x) = \dfrac{1}{x^2}$ (See Exercise 52 in Section 1.3.)

46. $f(x) = \sqrt{x}$ (See Example 8 in Section 1.3.)

47. $f(x) = \sqrt{2x + 1}$ (See Exercise 55 in Section 1.3.)

48. $f(x) = \dfrac{1}{\sqrt{x}}$ (See Exercise 56 in Section 1.3.)

49. Consider the function f given by
$$f(x) = \frac{x^2 - 9}{x + 3}.$$

a) For what x-value(s) is this function not differentiable?
b) Describe the simplest way to find $f'(4)$.

50. Consider the function g given by
$$g(x) = \frac{x^2 + x}{2x}.$$

a) For what x-value(s) is this function not differentiable?
b) What is $g'(3)$? Describe the simplest way to determine this.

51. Consider the function h given by
$$h(x) = |x - 3| + 2.$$

a) For what x-value(s) is the function not differentiable?
b) Evaluate $h'(0)$, $h'(1)$, $h'(4)$, and $h'(10)$. Is there a shortcut you can use to find these slopes?

52. Consider the function k given by
$$k(x) = 2|x + 5|.$$

a) For what x-value(s) is this function not differentiable?
b) Evaluate $k'(-10)$, $k'(-7)$, $k'(-2)$, and $k'(0)$. Is there a shortcut you can use to find these slopes?

53. Let $f(x) = \dfrac{x^2 + 4x + 3}{x + 1}$. A student recognizes that this function can be simplified as follows:
$$f(x) = \frac{x^2 + 4x + 3}{x + 1} = \frac{(x + 1)(x + 3)}{x + 1} = x + 3.$$

Since $y = x + 3$ is a line with slope 1, the student makes the following conclusions: $f'(-2) = 1$, $f'(-1) = 1$, $f'(0) = 1$, $f'(1) = 1$. Where did the student make an error?

54. Let $g(x) = \sqrt[3]{x}$. A student graphed this function, and the graph appeared to be smooth and continuous for all real numbers x. The student concluded that $g(x)$ is differentiable for all x, which is false. Identify the error, and explain why the conclusion is false. What is the correct conclusion regarding the differentiability of $g(x)$?

55. Let F be a piecewise-defined function given by
$$F(x) = \begin{cases} x^2 + 1, & \text{for } x \le 2, \\ 2x + 1, & \text{for } x > 2. \end{cases}$$

a) Verify that F is continuous at $x = 2$.
b) Is F differentiable at $x = 2$? Explain why or why not.

56. Let G be a piecewise-defined function given by
$$G(x) = \begin{cases} x^3, & \text{for } x \le 1, \\ 3x - 2, & \text{for } x > 1. \end{cases}$$

a) Verify that G is continuous at $x = 1$.
b) Is G differentiable at $x = 1$? Explain why or why not.

57. Let H be a piecewise-defined function given by
$$H(x) = \begin{cases} 2x^2 - x, & \text{for } x \le 3, \\ mx + b, & \text{for } x > 3. \end{cases}$$

Determine the values of m and b that make H differentiable at $x = 3$.

58–63. Use a calculator to check your answers to Exercises 17–22.

64. Business: growth of an investment. A company determines that the value of an investment is V, in millions of dollars, after time t, in years, where V is given by

$$V(t) = 5t^3 - 30t^2 + 45t + 5\sqrt{t}.$$

Note: Calculators often use only the variables y and x, so you may need to change the variables when entering this function.

a) Graph V over the interval $[0, 5]$.
b) Find the equation of the secant line passing through the points $(1, V(1))$ and $(5, V(5))$. Then graph this secant line using the same axes as in part (a).
c) Find the average rate of change of the investment between year 1 and year 5.

d) Repeat parts (b) and (c) for the following pairs of points: $(1, V(1))$ and $(4, V(4))$; $(1, V(1))$ and $(3, V(3))$; $(1, V(1))$ and $(1.5, V(1.5))$.
e) What *appears* to be the slope of the tangent line to the graph at the point $(1, V(1))$?
f) Approximate the rate at which the value of the investment is changing after 1 yr.

65. Use a calculator to determine where $f'(x)$ does not exist, if $f(x) = \sqrt[3]{x} - 5$.

Answers to Quick Checks

1. $f'(x) = 3x^2 + 2x; f'(-2) = 8, f'(4) = 56$

2. $f'(x) = \dfrac{2}{x^2}$ **3.** At $x = -6$, the graph has a corner.

1.5   Differentiation Techniques: The Power and Sum–Difference Rules

OBJECTIVES

- Differentiate using the Power Rule or the Sum–Difference Rule.
- Differentiate a constant or a constant times a function.
- Determine points at which a tangent line has a specified slope.

Leibniz Notation

Let y be a function of x. A common way to express "the derivative of y with respect to x" is the notation

$$\frac{dy}{dx}.$$

This notation was invented by the German mathematician Leibniz. Using this notation, we can write the following sentence:

If $y = f(x)$, then the derivative of y with respect to x is $\dfrac{dy}{dx} = f'(x).$

In practice, we often use *prime notation*, such as y' or $f'(x)$, to represent a derivative when there is no confusion as to which variables are involved. The dy/dx notation is a little more formal than the prime notation but has the same meaning. We will use both types of notation often.

When we wish to evaluate a derivative at a number, we write

$$\frac{dy}{dx}\bigg|_{x=2} = f'(2).$$

The vertical line is interpreted as "evaluated at," so the above expression is read as "the derivative of y with respect to x evaluated at $x = 2$ is the value $f'(2)$."

We can also write

$$\frac{d}{dx}f(x).$$

This is identical in meaning to dy/dx and is another way to denote the derivative of the function. When placed next to a function, d/dx is treated as a command to find the function's derivative. Using functions from previous sections, we can write

$$\frac{d}{dx}x^2 = 2x, \qquad \frac{d}{dx}x^3 = 3x^2, \qquad \frac{d}{dx}\left(\frac{1}{x}\right) = -\frac{1}{x^2}, \qquad \text{and so on.}$$

Historical Note: The German mathematician and philosopher Gottfried Wilhelm von Leibniz (1646–1716) and the English mathematician, philosopher, and physicist Sir Isaac Newton (1642–1727) are both credited with the invention of calculus, though each performed his work independently. Newton used the dot notation $\dot{y}$ for dy/dt, where y is a function of time; this notation is still used, though it is not as common as Leibniz notation.

We will use all of these derivative forms often, and with practice, their use will become natural. In Chapter 2, we will discuss specific meanings of dy and dx.

The Power Rule

In Section 1.4, we calculated the derivative for some simple power functions. Look at the following table and see if you can identify a pattern:

Function	Derivative	
x^2	$2x^1$	Example 1 in Section 1.4
x^3	$3x^2$	Example 2 in Section 1.4
x^4	$4x^3$	Exercise 42 in Section 1.4
$\dfrac{1}{x} = x^{-1}$	$-1 \cdot x^{-2} = \dfrac{-1}{x^2}$	Example 4 in Section 1.4
$\dfrac{1}{x^2} = x^{-2}$	$-2 \cdot x^{-3} = \dfrac{-2}{x^3}$	Exercise 45 in Section 1.4
$\sqrt{x} = x^{1/2}$	$\dfrac{1}{2}x^{-1/2} = \dfrac{1}{2\sqrt{x}}$	Exercise 46 in Section 1.4

The pattern can be described as follows: "To find the derivative of a power function, bring the exponent to the front of the variable as a coefficient and reduce the exponent by 1."

1. Write the exponent as the coefficient.

$$\frac{d}{dx}x^k = k \cdot x^{k-1}$$

2. Subtract 1 from the exponent.

This rule is summarized as the following theorem.

THEOREM 1 **The Power Rule**

For any real number k, if $y = x^k$, then

$$\frac{d}{dx}x^k = k \cdot x^{k-1}.$$

We proved this theorem for the cases where $k = 2$, 3, and -1 in Examples 1, 2, and 4 in Section 1.4 and for other cases as exercises at the end of that section. The proof of this theorem for the case where k is any positive integer is very elegant.

Proof. Let $f(x) = x^k$. We need to find the expanded form for $f(x + h) = (x + h)^k$ so that we can set up the difference quotient. When $(x + h)^k$ is multiplied (expanded), a pattern becomes evident, as the following shows:

$$(x + h)^1 = x + h,$$
$$(x + h)^2 = x^2 + 2xh + h^2,$$
$$(x + h)^3 = x^3 + 3x^2h + 3xh^2 + h^3,$$
$$(x + h)^4 = x^4 + 4x^3h + 6x^2h^2 + 4xh^3 + h^4.$$

The first term is x^k, and the second term is $kx^{k-1}h$. The terms in the shaded triangle all contain h to the power of 2 or greater. Calling these "shaded terms," we can summarize the above expansion as follows:

$$(x + h)^k = x^k + kx^{k-1}h + (\textit{shaded terms}).$$

We now substitute for $(x + h)^k$ in the difference quotient:

$$\frac{f(x + h) - f(x)}{h} = \frac{(x + h)^k - x^k}{h} = \frac{x^k + kx^{k-1}h + (shaded\ terms) - x^k}{h}.$$

The x^k terms in the numerator sum to 0, and h is factored out. The "shaded terms" now contain h to the power of 1 or greater, and we refer to them as the "reduced shaded terms." The h's in the numerator and the denominator cancel, and we have

$$\frac{h[kx^{k-1} + (reduced\ shaded\ terms)]}{h} = kx^{k-1} + (reduced\ shaded\ terms).$$

When we take the limit as $h \to 0$, the "reduced shaded terms" become 0:

$$f'(x) = \lim_{h \to 0} kx^{k-1} + (reduced\ shaded\ terms) = kx^{k-1}. \quad \blacksquare$$

Although we have proved the Power Rule only for the case where k is a positive integer, it is valid for all real numbers k. However, a complete proof of this fact is outside the scope of this book.

■ **EXAMPLE 1**   Differentiate each of the following:

a) $y = x^5$;   b) $y = x$;   c) $y = x^{-4}$.

Solution

a) $\dfrac{d}{dx}x^5 = 5 \cdot x^{5-1} = 5x^4$ Using the Power Rule

b) $\dfrac{d}{dx}x = 1 \cdot x^{1-1} = 1 \cdot x^0 = 1$

c) $\dfrac{d}{dx}x^{-4} = -4 \cdot x^{-4-1} = -4x^{-5}$,   or $-4 \cdot \dfrac{1}{x^5}$,   or $-\dfrac{4}{x^5}$

❰ Quick Check 1

> **Quick Check 1**
>
> **a)** Differentiate:
>   (i) $y = x^{15}$; (ii) $y = x^{-7}$.
>
> **b)** Explain why $\dfrac{d}{dx}(\pi^2) = 0$,
>
>   not 2π.

The Power Rule also allows us to differentiate expressions with rational exponents.

■ **EXAMPLE 2**   Differentiate:

a) $y = \sqrt[5]{x}$;   b) $y = x^{0.7}$.

Solution

a) $\dfrac{d}{dx}\sqrt[5]{x} = \dfrac{d}{dx}x^{1/5} = \dfrac{1}{5} \cdot x^{(1/5)-1}$

$= \dfrac{1}{5}x^{-4/5}$,   or $\dfrac{1}{5} \cdot \dfrac{1}{x^{4/5}}$,   or $\dfrac{1}{5} \cdot \dfrac{1}{\sqrt[5]{x^4}}$,   or $\dfrac{1}{5\sqrt[5]{x^4}}$

b) $\dfrac{d}{dx}x^{0.7} = 0.7x^{(0.7)-1} = 0.7x^{-0.3}$

❰ Quick Check 2

> **Quick Check 2**
>
> Differentiate:
> **a)** $y = \sqrt[4]{x}$;
> **b)** $y = x^{-1.25}$.

Numerical Differentiation and Tangent Lines

Consider $f(x) = x\sqrt{4 - x^2}$, graphed below.

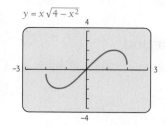

To find the value of dy/dx at a point, we select dy/dx from the CALC menu.

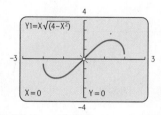

Next we key in the desired *x*-value or use the arrow keys to move the cursor to the desired point. We then press **ENTER** to obtain the value of the derivative at the given *x*-value.

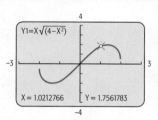

We can also use the Tangent feature from the DRAW menu to draw the tangent line at the point where the derivative was found. Both the line and its equation will appear on the calculator screen.

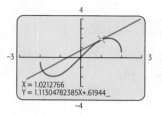

EXERCISES

For each of the following functions, use dy/dx to find the derivative, and then draw the tangent line at the given point. When selecting the viewing window, be sure to include the specified *x*-values.

1. $f(x) = x(200 - x)$;
$x = 24, \ x = 138, x = 150, x = 190$

2. $f(x) = x^3 - 6x^2$;
$x = -2, x = 0, x = 2, x = 4, x = 6.3$

3. $f(x) = -4.32 + 1.44x + 3x^2 - x^3$;
$x = -0.5, x = 0.5, x = 2.1$

In Section 1.3, we found the simplified difference quotient for two common functions: $f(x) = 1/x$ and $f(x) = \sqrt{x}$. By taking the limit of each difference quotient as $h \to 0$, we find the derivative of the function as follows:

• For $f(x) = 1/x$, we have

$$f'(x) = \lim_{h \to 0} \frac{-1}{x(x + h)} = -\frac{1}{x^2}. \quad \text{(See Example 7 in Section 1.3.)}$$

• For $f(x) = \sqrt{x}$, we have

$$f'(x) = \lim_{h \to 0} \left(\frac{1}{\sqrt{x} + \sqrt{x + h}} \right) = \frac{1}{2\sqrt{x}}. \quad \text{(See Example 8 in Section 1.3.)}$$

These two functions are very common in calculus, so it may be helpful to memorize their derivative forms. The Power Rule can be used to confirm these results.

• For $f(x) = 1/x$, we rewrite the function as a power: $f(x) = x^{-1}$. The Power Rule then gives

$$f'(x) = (-1)x^{(-1)-1} = (-1)x^{-2} = -\frac{1}{x^2}.$$

- For $f(x) = \sqrt{x}$, we rewrite the function as a power: $f(x) = x^{1/2}$. The Power Rule then gives

$$f'(x) = \frac{1}{2} x^{(1/2)-1} = \frac{1}{2} x^{-1/2} = \frac{1}{2x^{1/2}} = \frac{1}{2\sqrt{x}}.$$

TECHNOLOGY CONNECTION

Exploratory

Graph the constant function $y = -3$. Then find the derivative of this function at $x = -6$, $x = 0$, and $x = 8$. What do you conclude about the derivative of a constant function?

The Derivative of a Constant Function

Consider the constant function given by $F(x) = c$. Note that the slope at each point on its graph is 0.

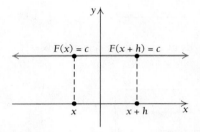

This suggests the following theorem.

THEOREM 2

The derivative of a constant function is 0. That is, $\dfrac{d}{dx} c = 0$.

Proof. Let F be the function given by $F(x) = c$. Then

$$\frac{F(x+h) - F(x)}{h} = \frac{c - c}{h}$$

$$= \frac{0}{h} = 0.$$

The difference quotient for this function is always 0. Thus, as h approaches 0, the limit of the difference quotient is 0, so $F'(x) = 0$. ■

The Derivative of a Constant Times a Function

Now let's consider differentiating functions such as

$$f(x) = 5x^2 \quad \text{and} \quad g(x) = -7x^4.$$

Note that we already know how to differentiate x^2 and x^4. Let's look for a pattern in the results of Section 1.4 and its exercise set.

Function	Derivative
$5x^2$	$10x$
$3x^{-1}$	$-3x^{-2}$
$\frac{3}{2}x^2$	$3x$
$1 \cdot x^3$	$3x^2$

Perhaps you have discovered the following theorem.

THEOREM 3

The derivative of a constant times a function is the constant times the derivative of the function. Using derivative notation, we can write this as

$$\frac{d}{dx}[c \cdot f(x)] = c \cdot \frac{d}{dx}f(x).$$

Proof. Let $F(x) = cf(x)$. Then

$$F'(x) = \lim_{h \to 0} \frac{F(x + h) - F(x)}{h} \qquad \text{Using the definition of a derivative}$$

$$= \lim_{h \to 0} \frac{cf(x + h) - cf(x)}{h} \qquad \text{Substituting}$$

$$= c \cdot \lim_{h \to 0} \left[\frac{f(x + h) - f(x)}{h}\right] \qquad \text{Factoring and using the Limit Properties}$$

$$= c \cdot f'(x). \qquad \text{Using the definition of a derivative} \quad \blacksquare$$

Combining this rule with the Power Rule allows us to find many derivatives.

■ EXAMPLE 3 Find each of the following derivatives:

a) $\dfrac{d}{dx} 7x^4$;

b) $\dfrac{d}{dx}(-9x)$;

c) $\dfrac{d}{dx}\left(\dfrac{1}{5x^2}\right)$.

Solution

a) $\dfrac{d}{dx}7x^4 = 7\dfrac{d}{dx}x^4 = 7 \cdot 4 \cdot x^{4-1} = 28x^3$ \qquad With practice, this may be done in one step.

b) $\dfrac{d}{dx}(-9x) = -9\dfrac{d}{dx}x = -9 \cdot 1 = -9$

c) $\dfrac{d}{dx}\left(\dfrac{1}{5x^2}\right) = \dfrac{d}{dx}\left(\dfrac{1}{5}x^{-2}\right) = \dfrac{1}{5} \cdot \dfrac{d}{dx}x^{-2}$

$$= \frac{1}{5}(-2)x^{-2-1}$$

$$= -\frac{2}{5}x^{-3}, \quad \text{or} \quad -\frac{2}{5x^3}$$

❮ Quick Check 3

> **Quick Check 3**
>
> Differentiate each of the following:
>
> **a)** $y = 10x^9$;
>
> **b)** $y = \pi x^3$;
>
> **c)** $y = \dfrac{2}{3x^4}$.

A common mistake is to write an expression such as $1/(5x^2)$ as $(5x)^{-2}$, which is incorrect. The exponent 2 applies only to the x; the 5 is part of the coefficient $\frac{1}{5}$. Carefully examine Example 3c, noting how the final answer is simplified.

Recall from Section 1.4 that a function's derivative at x is also its instantaneous rate of change at x.

■ EXAMPLE 4 **Life Science: Volume of a Tumor.** The volume V of a spherical tumor can be approximated by

$$V(r) = \tfrac{4}{3}\pi r^3,$$

where r is the radius of the tumor, in centimeters.

a) Find the rate of change of the volume with respect to the radius.

b) Find the rate of change of the volume at $r = 1.2$ cm.

Solution

a) $\dfrac{dV}{dr} = V'(r) = 3 \cdot \dfrac{4}{3} \cdot \pi r^2 = 4\pi r^2$

(This expression turns out to be equal to the tumor's surface area.)

b) $V'(1.2) = 4\pi(1.2)^2 = 5.76\pi \approx 18\dfrac{\text{cm}^3}{\text{cm}} = 18 \text{ cm}^2$

When the radius is 1.2 cm, the volume is changing at the rate of 18 cm^3 for every change of 1 cm in the radius.

The Derivative of a Sum or a Difference

In Exercise 11 of Exercise Set 1.4, you found that the derivative of

$$f(x) = x^2 + x$$

is $$f'(x) = 2x + 1.$$

Note that the derivative of x^2 is $2x$, the derivative of x is 1, and the sum of these derivatives is $f'(x)$. This illustrates the following.

THEOREM 4   **The Sum–Difference Rule**

Sum. The derivative of a sum is the sum of the derivatives:

$$\frac{d}{dx}[f(x) + g(x)] = \frac{d}{dx}f(x) + \frac{d}{dx}g(x).$$

Difference. The derivative of a difference is the difference of the derivatives:

$$\frac{d}{dx}[f(x) - g(x)] = \frac{d}{dx}f(x) - \frac{d}{dx}g(x).$$

Proof. The proof of the Sum Rule relies on the fact that the limit of a sum is the sum of the limits. Let $F(x) = f(x) + g(x)$. Then

$$\lim_{h \to 0} \frac{F(x + h) - F(x)}{h} = \lim_{h \to 0} \frac{[f(x + h) + g(x + h)] - [f(x) + g(x)]}{h}$$

$$= \lim_{h \to 0} \left[\frac{f(x + h) - f(x)}{h} + \frac{g(x + h) - g(x)}{h} \right] = f'(x) + g'(x).$$

To prove the Difference Rule, we note that

$$\frac{d}{dx}(f(x) - g(x)) = \frac{d}{dx}(f(x) + (-1)g(x)) = \frac{d}{dx}f(x) + \frac{d}{dx}(-1)g(x)$$

$$= f'(x) + (-1)\frac{d}{dx}g(x) = f'(x) - g'(x). \quad \blacksquare$$

Any function that is a sum or difference of several terms can be differentiated term by term.

■ **EXAMPLE 5** Find each of the following derivatives:

a) $\dfrac{d}{dx}(5x^3 - 7)$;

b) $\dfrac{d}{dx}\left(24x - \sqrt{x} + \dfrac{5}{x}\right)$.

Solution

a) $\dfrac{d}{dx}(5x^3 - 7) = \dfrac{d}{dx}(5x^3) - \dfrac{d}{dx}(7)$

$\qquad\qquad\qquad = 5\dfrac{d}{dx}x^3 - 0$

$\qquad\qquad\qquad = 5 \cdot 3x^2$

$\qquad\qquad\qquad = 15x^2$

b) $\dfrac{d}{dx}\left(24x - \sqrt{x} + \dfrac{5}{x}\right) = \dfrac{d}{dx}(24x) - \dfrac{d}{dx}(\sqrt{x}) + \dfrac{d}{dx}\left(\dfrac{5}{x}\right)$

$\qquad\qquad\qquad = 24 \cdot \dfrac{d}{dx}x - \dfrac{d}{dx}x^{1/2} + 5 \cdot \dfrac{d}{dx}x^{-1}$

$\qquad\qquad\qquad = 24 \cdot 1 - \dfrac{1}{2}x^{(1/2)-1} + 5(-1)x^{-1-1}$

$\qquad\qquad\qquad = 24 - \dfrac{1}{2}x^{-1/2} - 5x^{-2}$

$\qquad\qquad\qquad = 24 - \dfrac{1}{2\sqrt{x}} - \dfrac{5}{x^2}$

> **Quick Check 4**
>
> Differentiate:
>
> $y = 3x^5 + 2\sqrt[3]{x} + \dfrac{1}{3x^2} + \sqrt{5}.$

❮ Quick Check 4

A word of caution! The derivative of

$f(x) + c,$

a function plus a constant, is just the derivative of the function,

$f'(x).$

The derivative of

$c \cdot f(x),$

a function times a constant, is the constant times the derivative

$c \cdot f'(x).$

That is, the constant is retained for a product, but not for a sum.

Slopes of Tangent Lines

It is important to be able to determine points at which the tangent line to a curve has a certain slope, that is, points at which the derivative attains a certain value.

■ **EXAMPLE 6**   Find the points on the graph of $f(x) = -x^3 + 6x^2$ at which the tangent line is horizontal.

Solution   The derivative is used to find the slope of a tangent line, and a horizontal tangent line has slope 0. Therefore, we are seeking all x for which $f'(x) = 0$:

$$f'(x) = 0 \qquad \text{Setting the derivative equal to 0}$$

$$\frac{d}{dx}(-x^3 + 6x^2) = 0$$

$$-3x^2 + 12x = 0. \qquad \text{Differentiating}$$

We factor and solve:

$$-3x(x - 4) = 0$$
$$-3x = 0 \quad or \quad x - 4 = 0$$
$$x = 0 \quad or \qquad x = 4.$$

We are to find the points *on the graph*, so we must determine the second coordinates from the original equation, $f(x) = -x^3 + 6x^2$.

$$f(0) = -0^3 + 6 \cdot 0^2 = 0.$$
$$f(4) = -4^3 + 6 \cdot 4^2 = -64 + 96 = 32.$$

Thus, the points we are seeking are $(0, 0)$ and $(4, 32)$, as shown on the graph.

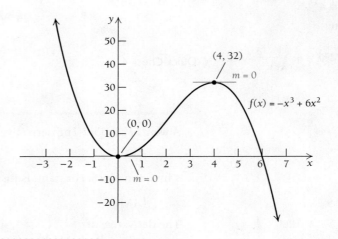

■ **EXAMPLE 7**   Find the points on the graph of $f(x) = -x^3 + 6x^2$ at which the tangent line has slope 9.

Solution   We want to find values of x for which $f'(x) = 9$. That is, we want to find x such that

$$-3x^2 + 12x = 9. \qquad \text{As in Example 6, note that } \frac{d}{dx}(-x^3 + 6x^2) = -3x^2 + 12x.$$

To solve, we add -9 on both sides and get

$$-3x^2 + 12x - 9 = 0.$$

We then multiply both sides of the equation by $-\frac{1}{3}$, giving

$$x^2 - 4x + 3 = 0,$$

which is factored as follows:

$$(x - 3)(x - 1) = 0.$$

EXERCISE

1. Graph $y = \frac{1}{3}x^3 - 2x^2 + 4x$, and draw tangent lines at various points. Estimate points at which the tangent line is horizontal. Then use calculus, as in Examples 6 and 7, to find the exact results.

⟩ **Quick Check 5**

For the function in Example 7, find the x-values for which $f'(x) = -15$.

We have two solutions: $x = 1$ or $x = 3$. We need the actual coordinates: when $x = 1$, we have $f(1) = -(1)^3 + 6(1)^2 = 5$. Therefore, at the point $(1, 5)$ on the graph of $f(x)$, the tangent line has a slope of 9. In a similar way, we can state that the tangent line at the point $(3, 27)$ has a slope of 9 as well. All of this is illustrated in the following graph.

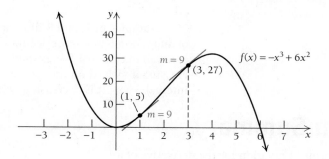

《 Quick Check 5

Analyzing a Function by Its Derivative

Some functions are always increasing or always decreasing. For example, the function $f(x) = x^3 + 2x$ is always increasing. That is, at no time does the graph of this function run "downhill" or lie flat. It is steadily increasing: all tangent lines have positive slopes. How can we use the function's derivative to demonstrate this fact?

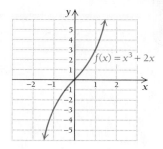

The derivative of this function is $dy/dx = f'(x) = 3x^2 + 2$. For any x-value, x^2 will be nonnegative; the expression $3x^2 + 2$ is thus positive for all x. It is impossible to set this derivative equal to any negative quantity and solve for x (try it). The graph of the function shows the always increasing trend that the derivative proves has to be true.

■ **EXAMPLE 8** Let $f(x) = -x^3 - 5x + 1$. Is this function always increasing or always decreasing? Use its derivative to support your conjecture.

Solution The graph of f is shown at the right.

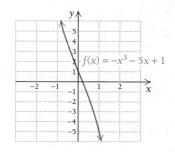

Based on the graph alone, the function appears to be always decreasing, but how do we know we aren't missing something, since we are looking at only a small portion of the graph? The graph alone is not enough to "prove" our observation. We need to use the derivative:

$$f'(x) = -3x^2 - 5.$$

Since x^2 is always 0 or positive, $-3x^2$ is always negative or 0. Subtracting 5 from $-3x^2$ will always give a negative result. Therefore, the derivative is always negative for all real numbers x. This means all tangent lines to this graph have a negative ("downhill") slope. Thus, the graph is always decreasing.

Examples 6, 7, and 8 illustrate ways to use the derivative to analyze the behavior of a function much more accurately than can be done by observation alone. In fact, our eyes can deceive us! For example, the graph of $f(x) = x^3 - x^2$ appears to be always increasing if viewed on the standard window of the TI-83. However, it does have a small interval where it is decreasing, which will be shown in Exercise 133.

Section Summary

- Common forms of notation for the derivative of a function are y', $f'(x)$, $\dfrac{dy}{dx}$, and $\dfrac{d}{dx}f(x)$.

- The *Power Rule* for differentiation is $\dfrac{d}{dx}[x^k] = kx^{k-1}$, for all real numbers k.

- The derivative of a constant is zero: $\dfrac{d}{dx}c = 0$.

- The derivative of a constant times a function is the constant times the derivative of the function:
$$\frac{d}{dx}[c \cdot f(x)] = c \cdot \frac{d}{dx}f(x).$$

- The derivative of a sum (or difference) is the sum (or difference) of the derivatives of the terms:
$$\frac{d}{dx}[f(x) \pm g(x)] = \frac{d}{dx}f(x) \pm \frac{d}{dx}g(x).$$

EXERCISE SET
1.5

Find $\dfrac{dy}{dx}$.

1. $y = x^7$

2. $y = x^8$

3. $y = -3x$

4. $y = -0.5x$

5. $y = 12$

6. $y = 7$

7. $y = 2x^{15}$

8. $y = 3x^{10}$

9. $y = x^{-6}$

10. $y = x^{-8}$

11. $y = 4x^{-2}$

12. $y = 3x^{-5}$

13. $y = x^3 + 3x^2$

14. $y = x^4 - 7x$

15. $y = 8\sqrt{x}$

16. $y = 4\sqrt{x}$

17. $y = x^{0.9}$

18. $y = x^{0.7}$

19. $y = \frac{1}{2}x^{4/5}$

20. $y = -4.8x^{1/3}$

21. $y = \dfrac{7}{x^3}$

22. $y = \dfrac{6}{x^4}$

23. $y = \dfrac{4x}{5}$

24. $y = \dfrac{3x}{4}$

Find each derivative.

25. $\dfrac{d}{dx}\left(\sqrt[4]{x} - \dfrac{3}{x}\right)$

26. $\dfrac{d}{dx}\left(\sqrt[5]{x} - \dfrac{2}{x}\right)$

27. $\dfrac{d}{dx}\left(\sqrt{x} - \dfrac{2}{\sqrt{x}}\right)$

28. $\dfrac{d}{dx}\left(\sqrt[3]{x} + \dfrac{4}{\sqrt{x}}\right)$

29. $\dfrac{d}{dx}(-2\sqrt[3]{x^5})$

30. $\dfrac{d}{dx}(-\sqrt[4]{x^3})$

31. $\dfrac{d}{dx}(5x^2 - 7x + 3)$

32. $\dfrac{d}{dx}(6x^2 - 5x + 9)$

Find $f'(x)$.

33. $f(x) = 0.6x^{1.5}$

34. $f(x) = 0.3x^{1.2}$

35. $f(x) = \dfrac{2x}{3}$

36. $f(x) = \dfrac{3x}{4}$

37. $f(x) = \dfrac{4}{7x^3}$

38. $f(x) = \dfrac{2}{5x^6}$

39. $f(x) = \dfrac{5}{x} - x^{2/3}$

40. $f(x) = \dfrac{4}{x} - x^{3/5}$

41. $f(x) = 4x - 7$

42. $f(x) = 7x - 14$

43. $f(x) = \dfrac{x^{4/3}}{4}$

44. $f(x) = \dfrac{x^{3/2}}{3}$

45. $f(x) = -0.01x^2 - 0.5x + 70$

46. $f(x) = -0.01x^2 + 0.4x + 50$

Find y'.

47. $y = 3x^{-2/3} + x^{3/4} + x^{6/5} + \dfrac{8}{x^3}$

48. $y = x^{-3/4} - 3x^{2/3} + x^{5/4} + \dfrac{2}{x^4}$

49. $y = \dfrac{2}{x} - \dfrac{x}{2}$ **50.** $y = \dfrac{x}{7} + \dfrac{7}{x}$

51. If $f(x) = x^2 + 4x - 5$, find $f'(10)$.

52. If $f(x) = \sqrt{x}$, find $f'(4)$.

53. If $y = \dfrac{4}{x^2}$, find $\dfrac{dy}{dx}\Big|_{x=-2}$

54. If $y = x + \dfrac{2}{x^3}$, find $\dfrac{dy}{dx}\Big|_{x=1}$

55. If $y = x^3 + 2x - 5$, find $\dfrac{dy}{dx}\Big|_{x=-2}$

56. If $y = \sqrt[3]{x} + \sqrt{x}$, find $\dfrac{dy}{dx}\Big|_{x=64}$

57. If $y = \dfrac{1}{3x^4}$, find $\dfrac{dy}{dx}\Big|_{x=-1}$

58. If $y = \dfrac{2}{5x^3}$, find $\dfrac{dy}{dx}\Big|_{x=4}$

59. Find an equation (in $y = mx + b$ form) of the tangent line to the graph of $f(x) = x^3 - 2x + 1$

  a) at $(2, 5)$;   **b)** at $(-1, 2)$;   **c)** at $(0, 1)$.

60. Find an equation of the tangent line to the graph of $f(x) = x^2 - \sqrt{x}$

  a) at $(1, 0)$;   **b)** at $(4, 14)$;   **c)** at $(9, 78)$.

61. Find an equation of the tangent line to the graph of $f(x) = \dfrac{1}{x^2}$

  a) at $(1, 1)$;   **b)** at $(3, \tfrac{1}{9})$;   **c)** at $(-2, \tfrac{1}{4})$.

62. Find the equation of the tangent line to the graph of $g(x) = \sqrt[3]{x^2}$

  a) at $(-1, 1)$;   **b)** at $(1, 1)$;   **c)** at $(8, 4)$.

For each function, find the points on the graph at which the tangent line is horizontal. If none exist, state that fact.

63. $y = x^2 - 3$ **64.** $y = -x^2 + 4$

65. $y = -x^3 + 1$ **66.** $y = x^3 - 2$

67. $y = 3x^2 - 5x + 4$ **68.** $y = 5x^2 - 3x + 8$

69. $y = -0.01x^2 - 0.5x + 70$

70. $y = -0.01x^2 + 0.4x + 50$

71. $y = 2x + 4$ **72.** $y = -2x + 5$

73. $y = 4$ **74.** $y = -3$

75. $y = -x^3 + x^2 + 5x - 1$

76. $y = -\tfrac{1}{3}x^3 + 6x^2 - 11x - 50$

77. $y = \tfrac{1}{3}x^3 - 3x + 2$ **78.** $y = x^3 - 6x + 1$

79. $f(x) = \tfrac{1}{3}x^3 + \tfrac{1}{2}x^2 - 2$

80. $f(x) = \tfrac{1}{3}x^3 - 3x^2 + 9x - 9$

For each function, find the points on the graph at which the tangent line has slope 1.

81. $y = 20x - x^2$ **82.** $y = 6x - x^2$

83. $y = -0.025x^2 + 4x$ **84.** $y = -0.01x^2 + 2x$

85. $y = \tfrac{1}{3}x^3 + 2x^2 + 2x$

86. $y = \tfrac{1}{3}x^3 - x^2 - 4x + 1$

APPLICATIONS

Life Sciences

87. Healing wound. The circular area A, in square centimeters, of a healing wound is approximated by
$$A(r) = 3.14r^2,$$
where r is the wound's radius, in centimeters.
  a) Find the rate of change of the area with respect to the radius.
  b) Explain the meaning of your answer to part (a).

88. Healing wound. The circumference C, in centimeters, of a healing wound is approximated by
$$C(r) = 6.28r,$$
where r is the wound's radius, in centimeters.
  a) Find the rate of change of the circumference with respect to the radius.
  b) Explain the meaning of your answer to part (a).

89. Growth of a baby. The median weight of a boy whose age is between 0 and 36 months can be approximated by the function
$$w(t) = 8.15 + 1.82t - 0.0596t^2 + 0.000758t^3,$$
where t is measured in months and w is measured in pounds.

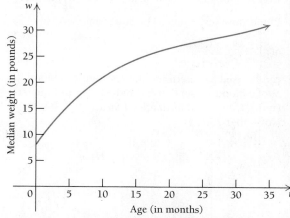

(*Source*: Centers for Disease Control. Developed by the National Center for Health Statistics in collaboration with the National Center for Chronic Disease Prevention and Health Promotion, 2000.)

Use this approximation to find the following for a boy with median weight:
  a) The rate of change of weight with respect to time.
  b) The weight of the baby at age 10 months.
  c) The rate of change of the baby's weight with respect to time at age 10 months.

90. Temperature during an illness. The temperature T of a person during an illness is given by

$$T(t) = -0.1t^2 + 1.2t + 98.6,$$

where T is the temperature, in degrees Fahrenheit, at time t, in days.

a) Find the rate of change of the temperature with respect to time.

b) Find the temperature at $t = 1.5$ days.

c) Find the rate of change at $t = 1.5$ days.

91. Heart rate. The equation

$$R(v) = \frac{6000}{v}$$

can be used to determine the heart rate, R, of a person whose heart pumps 6000 milliliters (mL) of blood per minute and v milliliters of blood per beat. (*Source: Mathematics Teacher*, Vol. 99, No. 4, November 2005.)

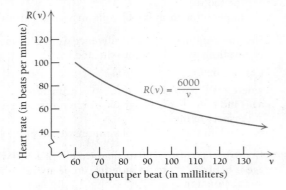

a) Find the rate of change of heart rate with respect to v, the output per beat.

b) Find the heart rate at $v = 80$ mL per beat.

c) Find the rate of change at $v = 80$ mL per beat.

92. Blood flow resistance. The equation

$$S(r) = \frac{1}{r^4}$$

can be used to determine the resistance to blood flow, S, of a blood vessel that has radius r, in millimeters (mm). (*Source: Mathematics Teacher*, Vol. 99, No. 4, November 2005.)

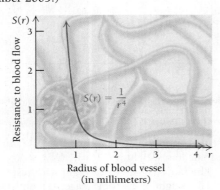

a) Find the rate of change of resistance with respect to r, the radius of the blood vessel.

b) Find the resistance at $r = 1.2$ mm.

c) Find the rate of change of S with respect to r when $r = 0.8$ mm.

Social Sciences

93. Population growth rate. The population of a city grows from an initial size of 100,000 to a size P given by

$$P(t) = 100,000 + 2000t^2,$$

where t is in years.

a) Find the **growth rate**, dP/dt.

b) Find the population after 10 yr.

c) Find the growth rate at $t = 10$.

d) Explain the meaning of your answer to part (c).

94. Median age of women at first marriage. The median age of women at first marriage can be approximated by the linear function

$$A(t) = 0.08t + 19.7,$$

where $A(t)$ is the median age of women marrying for the first time at t years after 1950.

a) Find the rate of change of the median age A with respect to time t.

b) Explain the meaning of your answer to part (a).

General Interest

95. View to the horizon. The view V, or distance in miles, that one can see to the horizon from a height h, in feet, is given by

$$V = 1.22\sqrt{h}.$$

a) Find the rate of change of V with respect to h.

b) How far can one see to the horizon from an airplane window at a height of 40,000 ft?

c) Find the rate of change at $h = 40,000$.

d) Explain the meaning of your answers to parts (a) and (c).

96. Baseball ticket prices. The average price, in dollars, of a ticket for a Major League baseball game x years after 1990 can be estimated by

$$p(x) = 9.41 - 0.19x + 0.09x^2.$$

a) Find the rate of change of the average ticket price with respect to the year, dp/dx.

b) What is the average ticket price in 2010?

c) What is the rate of change of the average ticket price in 2010?

SYNTHESIS

For each function, find the interval(s) for which $f'(x)$ is positive.

97. $f(x) = x^2 - 4x + 1$

98. $f(x) = x^2 + 7x + 2$

99. $f(x) = \frac{1}{3}x^3 - x^2 - 3x + 5$

100. Find the points on the graph of
$$y = x^4 - \frac{4}{3}x^2 - 4$$
at which the tangent line is horizontal.

101. Find the points on the graph of
$$y = 2x^6 - x^4 - 2$$
at which the tangent line is horizontal.

Use the derivative to help show whether each function is always increasing, always decreasing, or neither.

102. $f(x) = x^5 + x^3$

103. $f(x) = x^3 + 2x$

104. $f(x) = \frac{1}{x}, \quad x \neq 0$

105. $f(x) = \sqrt{x}, \quad x \geq 0$

106. The function $f(x) = x^3 + ax$ is always increasing if $a > 0$, but not if $a < 0$. Use the derivative of f to explain why this observation is true.

Find dy/dx. Each function can be differentiated using the rules developed in this section, but some algebra may be required beforehand.

107. $y = (x + 3)(x - 2)$

108. $y = (x - 1)(x + 1)$

109. $y = \dfrac{x^5 - x^3}{x^2}$

110. $y = \dfrac{5x^2 - 8x + 3}{8}$

111. $y = \dfrac{x^5 + x}{x^2}$

112. $y = \dfrac{x^5 - 3x^4 + 2x + 4}{x^2}$

113. $y = (-4x)^3$

114. $y = \sqrt{7x}$

115. $y = \sqrt[3]{8x}$

116. $y = (x - 3)^2$

117. $y = \left(\sqrt{x} - \dfrac{1}{\sqrt{x}}\right)^2$

118. $y = (\sqrt{x} + \sqrt[3]{x})^2$

119. $y = (x + 1)^3$

120. Use Theorem 1 to prove that the derivative of 1 is 0.

121. When might Leibniz notation be more convenient than function notation?

122. Write a short biographical paper on Leibniz and/or Newton. Emphasize the contributions each man made to many areas of science and society.

TECHNOLOGY CONNECTION

Graph each of the following. Then estimate the x-values at which tangent lines are horizontal.

123. $f(x) = x^4 - 3x^2 + 1$

124. $f(x) = 1.6x^3 - 2.3x - 3.7$

125. $f(x) = 10.2x^4 - 6.9x^3$

126. $f(x) = \dfrac{5x^2 + 8x - 3}{3x^2 + 2}$

For each of the following, graph f and f' and then determine $f'(1)$. For Exercises 131 and 132, use nDeriv on the TI-83.

127. $f(x) = 20x^3 - 3x^5$

128. $f(x) = x^4 - 3x^2 + 1$

129. $f(x) = x^3 - 2x - 2$

130. $f(x) = x^4 - x^3$

131. $f(x) = \dfrac{4x}{x^2 + 1}$

132. $f(x) = \dfrac{5x^2 + 8x - 3}{3x^2 + 2}$

133. The function $f(x) = x^3 - x^2$ (mentioned after Example 8) appears to be always increasing, or possibly flat, on the default viewing window of the TI-83.

a) Graph the function in the default window; then zoom in until you see a small interval in which f is decreasing.

b) Use the derivative to determine the point(s) at which the graph has horizontal tangent lines.

c) Use your result from part (b) to infer the interval for which f is decreasing. Does this agree with your calculator's image of the graph?

d) Is it possible there are other intervals for which f is decreasing? Explain why or why not.

Answers to Quick Checks

1. (a) (i) $y' = 15x^{14}$, (ii) $y' = -7x^{-8}$;
(b) because π^2 is a constant

2. (a) $y' = \frac{1}{4}x^{-3/4} = \dfrac{1}{4\sqrt[4]{x^3}}$; (b) $y' = -1.25x^{-2.25}$

3. (a) $y' = 90x^8$; (b) $y' = 3\pi x^2$; (c) $y' = -\dfrac{8}{3x^5}$

4. $y' = 15x^4 + \dfrac{2}{3\sqrt[3]{x^2}} - \dfrac{2}{3x^3}$

5. $x = -1$ and $x = 5$

1.6

Differentiation Techniques: The Product and Quotient Rules

The Product Rule

A function can be written as the product of two other functions. For example, the function $F(x) = x^3 \cdot x^4$ can be viewed as the product of the two functions $f(x) = x^3$ and $g(x) = x^4$, yielding $F(x) = f(x) \cdot g(x)$. Is the derivative of $F(x)$ the product of the derivatives of its factors, $f(x)$ and $g(x)$? The answer is no. To see this, note that the product of x^3 and x^4 is x^7, and the derivative of this product is $7x^6$. However, the derivatives of the two functions are $3x^2$ and $4x^3$, and the product of these derivatives is $12x^5$. This example shows that, in general, *the derivative of a product is **not** the product of the derivatives*. The following is a rule for finding the derivative of a product.

THEOREM 5   The Product Rule

Let $F(x) = f(x) \cdot g(x)$. Then

$$F'(x) = \frac{d}{dx}[f(x) \cdot g(x)] = f(x) \cdot \left[\frac{d}{dx}g(x)\right] + g(x) \cdot \left[\frac{d}{dx}f(x)\right].$$

The derivative of a product is the first factor times the derivative of the second factor, plus the second factor times the derivative of the first factor.

The proof of the Product Rule is outlined in Exercise 123 at the end of this section.
   Let's check the Product Rule for $x^2 \cdot x^5$. There are five steps:

$$\frac{d}{dx}(x^2 \cdot x^5)$$

$$= x^2 \cdot 5x^4 + x^5 \cdot 2x \quad ⑤$$
$$= 5x^6 + 2x^6$$
$$= 7x^6$$

1. Write down the first factor.
2. Multiply it by the derivative of the second factor.
3. Write down the second factor.
4. Multiply it by the derivative of the first factor.
5. Add the result of steps (1) and (2) to the result of steps (3) and (4).

Usually we try to write the results in simplified form. In Examples 1 and 2, we do not simplify in order to better emphasize the steps being performed.

■ **EXAMPLE 1**   Find $\dfrac{d}{dx}[(x^4 - 2x^3 - 7)(3x^2 - 5x)]$. Do not simplify.

Solution   We let $f(x) = x^4 - 2x^3 - 7$ and $g(x) = 3x^2 - 5x$. We differentiate each of these, obtaining $f'(x) = 4x^3 - 6x^2$ and $g'(x) = 6x - 5$. By the Product Rule, the derivative of the given function is then

$$\qquad\qquad f(x) \quad\cdot\quad g(x) \qquad\qquad\qquad f(x) \quad\cdot\quad g'(x) \quad+\quad g(x) \quad\cdot\quad f'(x)$$

$$\frac{d}{dx}[(x^4 - 2x^3 - 7)(3x^2 - 5x)] = (x^4 - 2x^3 - 7)(6x - 5) + (3x^2 - 5x)(4x^3 - 6x^2)$$

In this example, we could have first multiplied the polynomials and then differentiated. Both methods give the same solution after simplification.

It makes no difference which factor of the given function is called $f(x)$ and which is called $g(x)$. We usually let the first function listed be $f(x)$ and the second function be $g(x)$, but if we switch the names, the process still gives the same answer. Try it by repeating Example 1 with $g(x) = x^4 - 2x^3 - 7$ and $f(x) = 3x^2 - 5x$.

❭ **Quick Check 1**

Use the Product Rule to differentiate each of the following functions. Do not simplify.

a)
$$y = (2x^5 + x - 1)(3x - 2)$$
b) $y = (\sqrt{x} + 1)(\sqrt[5]{x} - x)$

■ **EXAMPLE 2** For $F(x) = (x^2 + 4x - 11)(7x^3 - \sqrt{x})$, find $F'(x)$. Do not simplify.

Solution We rewrite this as

$$F(x) = (x^2 + 4x - 11)(7x^3 - x^{1/2}).$$

Then, using the Product Rule, we have

$$F'(x) = (x^2 + 4x - 11)(21x^2 - \tfrac{1}{2}x^{-1/2}) + (7x^3 - x^{1/2})(2x + 4).$$

❬ **Quick Check 1**

The Quotient Rule

The derivative of a quotient is *not* the quotient of the derivatives. To see why, consider x^5 and x^2. The quotient x^5/x^2 is x^3, and the derivative of this quotient is $3x^2$. The individual derivatives are $5x^4$ and $2x$, and the quotient of these derivatives, $5x^4/(2x)$, is $(5/2)x^3$, which is not $3x^2$.

The rule for differentiating quotients is as follows.

> **THEOREM 6**   **The Quotient Rule**
>
> If $Q(x) = \dfrac{N(x)}{D(x)}$,   then $Q'(x) = \dfrac{D(x) \cdot N'(x) - N(x) \cdot D'(x)}{[D(x)]^2}$.
>
> The derivative of a quotient is the denominator times the derivative of the numerator, minus the numerator times the derivative of the denominator, all divided by the square of the denominator.
>
> (If we think of the function in the numerator as the first function and the function in the denominator as the second function, then we can reword the Quotient Rule as "the derivative of a quotient is the second function times the derivative of the first function minus the first function times the derivative of the second function, all divided by the square of the second function.")

A proof of this result is outlined in Exercise 101 of Section 1.7 (on p. 176).

The Quotient Rule is illustrated below.

$$\frac{d}{dx}\left[\frac{N(x)}{D(x)}\right] = \frac{D(x) \cdot N'(x) - N(x) \cdot D'(x)}{[D(x)]^2}$$

There are six steps:

1. Write the denominator.
2. Multiply the denominator by the derivative of the numerator.
3. Write a minus sign.
4. Write the numerator.
5. Multiply it by the derivative of the denominator.
6. Divide by the square of the denominator.

■ **EXAMPLE 3**   For $Q(x) = x^5/x^2$, find $Q'(x)$.

Solution   We have already seen that $x^5/x^2 = x^3$ and $\dfrac{d}{dx}x^3 = 3x^2$, but we wish to practice using the Quotient Rule. We have $D(x) = x^2$ and $N(x) = x^5$:

$$\overbrace{D(x) \cdot N'(x) - N(x) \cdot D'(x)}$$

$$Q'(x) = \frac{x^2 \;\cdot\; 5x^4 \;-\; x^5 \;\cdot\; 2x}{(x^2)^2}$$

$$\underbrace{[D(x)]^2}$$

$$= \frac{5x^6 - 2x^6}{x^4} = \frac{3x^6}{x^4} = 3x^2. \quad \text{This checks with the result above.}$$

■ **EXAMPLE 4**   Differentiate: $f(x) = \dfrac{1 + x^2}{x^3}$.

Solution

$$\begin{aligned}
f'(x) &= \frac{x^3 \cdot 2x - (1 + x^2) \cdot 3x^2}{(x^3)^2} && \text{Using the Quotient Rule} \\[2mm]
&= \frac{2x^4 - 3x^2 - 3x^4}{x^6} = \frac{-x^4 - 3x^2}{x^6} && \\[2mm]
&= \frac{x^2(-x^2 - 3)}{x^2 \cdot x^4} && \text{Factoring} \\[2mm]
&= \frac{-x^2 - 3}{x^4} && \text{Removing a factor equal to 1: } \frac{x^2}{x^2} = 1
\end{aligned}$$

■ **EXAMPLE 5**   Differentiate: $f(x) = \dfrac{x^2 - 3x}{x - 1}$.

Solution   We have

$$\begin{aligned}
f'(x) &= \frac{(x - 1)(2x - 3) - (x^2 - 3x) \cdot 1}{(x - 1)^2} && \text{Using the Quotient Rule} \\[2mm]
&= \frac{2x^2 - 5x + 3 - x^2 + 3x}{(x - 1)^2} && \text{Using the distributive law} \\[2mm]
&= \frac{x^2 - 2x + 3}{(x - 1)^2}. && \text{Simplifying}
\end{aligned}$$

It is not necessary to multiply out $(x - 1)^2$.

❮ Quick Check 2

❯ **Quick Check 2**

a) Differentiate: $f(x) = \dfrac{1 - 3x}{x^2 + 2}$. Simplify your result.

b) Show that

$$\frac{d}{dx}\left[\frac{ax + 1}{bx + 1}\right] = \frac{a - b}{(bx + 1)^2}.$$

TECHNOLOGY CONNECTION

Checking Derivatives Graphically

To check Example 5, we first enter the function:

$$y_1 = \frac{x^2 - 3x}{x - 1}.$$

Then we enter the possible derivative:

$$y_2 = \frac{x^2 - 2x + 3}{(x - 1)^2}.$$

For the third function, we enter

$$y_3 = \text{nDeriv}(y_1, x, x).$$

Next, we deselect y_1 and graph y_2 and y_3. We use different graph styles and the Sequential mode to see each graph as it appears on the screen.

$$y_2 = \frac{x^2 - 2x + 3}{(x - 1)^2}, \quad y_3 = \text{nDeriv}(y_1, x, x)$$

Since the graphs appear to coincide, it appears that $y_2 = y_3$ and we have a check. This is considered a partial check, however, because the graphs might not coincide at a point not in the viewing window.

We can also use a table to check that $y_2 = y_3$.

X	Y2	Y3
5.97	1.081	1.081
5.98	1.0806	1.0806
5.99	1.0803	1.0803
6	1.08	1.08
6.01	1.0797	1.0797
6.02	1.0794	1.0794
6.03	1.079	1.079
X = 5.97		

You should verify that had we miscalculated the derivative as, say, $y_2 = (x^2 - 2x - 8)/(x - 1)^2$, neither the tables nor the graphs of y_2 and y_3 would agree.

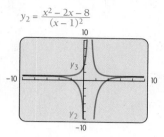

$$y_2 = \frac{x^2 - 2x - 8}{(x - 1)^2}$$

EXERCISES

1. For the function

$$f(x) = \frac{x^2 - 4x}{x + 2},$$

use graphs and tables to determine which of the following seems to be the correct derivative.

a) $f'(x) = \dfrac{-x^2 - 4x - 8}{(x + 2)^2}$

b) $f'(x) = \dfrac{x^2 - 4x + 8}{(x + 2)^2}$

c) $f'(x) = \dfrac{x^2 + 4x - 8}{(x + 2)^2}$

2-5. Check the results of Examples 1–4 in this section.

Application of the Quotient Rule

The total cost, total revenue, and total profit functions, discussed in Section R.4, pertain to the accumulated cost, revenue, and profit when x items are produced. Because of economies of scale and other factors, it is common for the cost, revenue (price), and profit for, say, the 10th item to differ from those for the 1000th item. For this reason, a business is often interested in the *average* cost, revenue, and profit associated with the production and sale of x items.

DEFINITION

If $C(x)$ is the cost of producing x items, then the **average cost** of producing x items is $\dfrac{C(x)}{x}$;

if $R(x)$ is the revenue from the sale of x items, then the **average revenue** from selling x items is $\dfrac{R(x)}{x}$;

if $P(x)$ is the profit from the sale of x items, then the **average profit** from selling x items is $\dfrac{P(x)}{x}$.

■ **EXAMPLE 6**   **Business.** Paulsen's Greenhouse finds that the cost, in dollars, of growing x hundred geraniums is modeled by

$$C(x) = 200 + 100\sqrt[4]{x}.$$

If the revenue from the sale of x hundred geraniums is modeled by

$$R(x) = 120 + 90\sqrt{x},$$

find each of the following.

a) The average cost, the average revenue, and the average profit when x hundred geraniums are grown and sold.

b) The rate at which average profit is changing when 300 geraniums are being grown and sold.

Solution

a) We let A_C, A_R, and A_P represent average cost, average revenue, and average profit, respectively. Then

$$A_C(x) = \frac{C(x)}{x} = \frac{200 + 100\sqrt[4]{x}}{x};$$

$$A_R(x) = \frac{R(x)}{x} = \frac{120 + 90\sqrt{x}}{x};$$

$$A_P(x) = \frac{P(x)}{x} = \frac{R(x) - C(x)}{x} = \frac{-80 + 90\sqrt{x} - 100\sqrt[4]{x}}{x}.$$

b) To find the rate at which average profit is changing when 300 geraniums are being grown, we calculate $A_P{}'(3)$ (remember that x is in hundreds):

$$A_P{}'(x) = \frac{d}{dx}\left[\frac{-80 + 90x^{1/2} - 100x^{1/4}}{x}\right]$$

$$= \frac{x(\frac{1}{2}\cdot 90x^{1/2-1} - \frac{1}{4}\cdot 100x^{1/4-1}) - (-80 + 90x^{1/2} - 100x^{1/4})\cdot 1}{x^2}$$

$$= \frac{45x^{1/2} - 25x^{1/4} + 80 - 90x^{1/2} + 100x^{1/4}}{x^2} = \frac{75x^{1/4} - 45x^{1/2} + 80}{x^2};$$

$$A_P{}'(3) = \frac{75\sqrt[4]{3} - 45\sqrt{3} + 80}{3^2} \approx 11.20.$$

When 300 geraniums are being grown, the average profit is increasing by $11.20 per hundred plants, or about 11.2 cents per plant.

Using Y-VARS

One way to save keystrokes on most calculators is to use the Y-VARS option on the VARS menu.

To check Example 6, we let $y_1 = 200 + 100x^{0.25}$ and $y_2 = 120 + 90x^{0.5}$. To express the profit function as y_3, we press ⌐ Y= ⌐ and move the cursor to enter y_3. Next we press **VARS** and select Y-VARS and then FUNCTION. From the FUNCTION menu we select Y2, which then appears on the Y= screen. After pressing ⌐ − ⌐, we repeat the procedure to get Y1 on the Y= screen.

```
Plot 1   Plot 2   Plot 3
\Y1 = 200 + 100X^0.25
\Y2 = 120 + 90X^0.5
\Y3 = Y2 − Y1
\Y4 =
\Y5 =
\Y6 =
```

EXERCISES

1. Use the Y-VARS option to enter $y_4 = y_1/x$, $y_5 = y_2/x$, and $y_6 = y_3/x$, and explain what each of the functions represents.

2. Use nDeriv from the MATH menu or dy/dx from the CALC menu to check part (b) of Example 6.

Section Summary

- The *Product Rule* is

$$\frac{d}{dx}[f(x) \cdot g(x)] = f(x) \cdot \frac{d}{dx}[g(x)] + g(x) \cdot \frac{d}{dx}[f(x)].$$

- The *Quotient Rule* is

$$\frac{d}{dx}\left[\frac{f(x)}{g(x)}\right] = \frac{g(x) \cdot \frac{d}{dx}[f(x)] - f(x) \cdot \frac{d}{dx}[g(x)]}{[g(x)]^2}.$$

- Be careful to note the order in which you write out the factors when using the Quotient Rule. Because the Quotient Rule involves subtraction and division, the order in which you perform the operations is important.

EXERCISE SET
1.6

Differentiate two ways: first, by using the Product Rule; then, by multiplying the expressions before differentiating. Compare your results as a check.

1. $y = x^5 \cdot x^6$

2. $y = x^9 \cdot x^4$

3. $f(x) = (2x + 5)(3x - 4)$

4. $g(x) = (3x - 2)(4x + 1)$

5. $G(x) = 4x^2(x^3 + 5x)$

6. $F(x) = 3x^4(x^2 - 4x)$

7. $y = (3\sqrt{x} + 2)x^2$

8. $y = (4\sqrt{x} + 3)x^3$

9. $g(x) = (4x - 3)(2x^2 + 3x + 5)$

10. $f(x) = (2x + 5)(3x^2 - 4x + 1)$

11. $F(t) = (\sqrt{t} + 2)(3t - 4\sqrt{t} + 7)$

12. $G(t) = (2t + 3\sqrt{t} + 5)(\sqrt{t} + 4)$

Differentiate two ways: first, by using the Quotient Rule; then, by dividing the expressions before differentiating. Compare your results as a check.

13. $y = \dfrac{x^7}{x^3}$

14. $y = \dfrac{x^6}{x^4}$

15. $f(x) = \dfrac{2x^5 + x^2}{x}$

16. $g(x) = \dfrac{3x^7 - x^3}{x}$

17. $G(x) = \dfrac{8x^3 - 1}{2x - 1}$

18. $F(x) = \dfrac{x^3 + 27}{x + 3}$

19. $y = \dfrac{t^2 - 16}{t + 4}$

20. $y = \dfrac{t^2 - 25}{t - 5}$

Differentiate each function.

21. $f(x) = (3x^2 - 2x + 5)(4x^2 + 3x - 1)$

22. $g(x) = (5x^2 + 4x - 3)(2x^2 - 3x + 1)$

23. $y = \dfrac{5x^2 - 1}{2x^3 + 3}$

24. $y = \dfrac{3x^4 + 2x}{x^3 - 1}$

25. $G(x) = (8x + \sqrt{x})(5x^2 + 3)$

26. $F(x) = (-3x^2 + 4x)(7\sqrt{x} + 1)$

27. $g(t) = \dfrac{t}{3 - t} + 5t^3$

28. $f(t) = \dfrac{t}{5 + 2t} - 2t^4$

29. $F(x) = (x + 3)^2$

[*Hint:* $(x + 3)^2 = (x + 3)(x + 3)$.]

30. $G(x) = (5x - 4)^2$

31. $y = (x^3 - 4x)^2$

32. $y = (3x^2 - 4x + 5)^2$

33. $g(x) = 5x^{-3}(x^4 - 5x^3 + 10x - 2)$

34. $f(x) = 6x^{-4}(6x^3 + 10x^2 - 8x + 3)$

35. $F(t) = \left(t + \dfrac{2}{t}\right)(t^2 - 3)$

36. $G(t) = (3t^5 - t^2)\left(t - \dfrac{5}{t}\right)$

37. $y = \dfrac{x^2 + 1}{x^3 - 1} - 5x^2$

38. $y = \dfrac{x^3 - 1}{x^2 + 1} + 4x^3$

39. $y = \dfrac{\sqrt[3]{x} - 7}{\sqrt{x} + 3}$

40. $y = \dfrac{\sqrt{x} + 4}{\sqrt[3]{x} - 5}$

41. $f(x) = \dfrac{x}{x^{-1} + 1}$

42. $f(x) = \dfrac{x^{-1}}{x + x^{-1}}$

43. $F(t) = \dfrac{1}{t - 4}$

44. $G(t) = \dfrac{1}{t + 2}$

45. $f(x) = \dfrac{3x^2 + 2x}{x^2 + 1}$

46. $f(x) = \dfrac{3x^2 - 5x}{x^2 - 1}$

47. $g(t) = \dfrac{-t^2 + 3t + 5}{t^2 - 2t + 4}$

48. $f(t) = \dfrac{3t^2 + 2t - 1}{-t^2 + 4t + 1}$

49–96. Use a graphing calculator to check the results of Exercises 1–48.

97. Find an equation of the tangent line to the graph of $y = 8/(x^2 + 4)$ at (a) $(0, 2)$; (b) $(-2, 1)$.

98. Find an equation of the tangent line to the graph of $y = \sqrt{x}/(x + 1)$ at (a) $x = 1$; (b) $x = \frac{1}{4}$.

99. Find an equation of the tangent line to the graph of $y = x^2 + 3/(x - 1)$ at (a) $x = 2$; (b) $x = 3$.

100. Find an equation of the tangent line to the graph of $y = 4x/(1 + x^2)$ at (a) $(0, 0)$; (b) $(-1, -2)$.

APPLICATIONS

Business and Economics

101. Average cost. Summertime Fabrics finds that the cost, in dollars, of producing x jackets is given by $C(x) = 950 + 15\sqrt{x}$. Find the rate at which the average cost is changing when 400 jackets have been produced.

102. Average cost. Tongue-Tied Sauces, Inc., finds that the cost, in dollars, of producing x bottles of barbecue sauce is given by $C(x) = 375 + 0.75x^{3/4}$. Find the rate at which the average cost is changing when 81 bottles of barbecue sauce have been produced.

103. Average revenue. Summertime Fabrics finds that the revenue, in dollars, from the sale of x jackets is given by $R(x) = 85\sqrt{x}$. Find the rate at which average revenue is changing when 400 jackets have been produced.

104. Average revenue. Tongue-Tied Sauces, Inc., finds that the revenue, in dollars, from the sale of x bottles of barbecue sauce is given by $R(x) = 7.5x^{0.7}$. Find the rate at which average revenue is changing when 81 bottles of barbecue sauce have been produced.

105. Average profit. Use the information in Exercises 101 and 103 to determine the rate at which Summertime Fabrics' average profit per jacket is changing when 400 jackets have been produced and sold.

106. Average profit. Use the information in Exercises 102 and 104 to determine the rate at which Tongue-Tied Sauces' average profit per bottle of barbecue sauce is changing when 81 bottles have been produced and sold.

107. Average profit. Sparkle Pottery has determined that the cost, in dollars, of producing x vases is given by

$$C(x) = 4300 + 2.1x^{0.6}.$$

If the revenue from the sale of x vases is given by $R(x) = 65x^{0.9}$, find the rate at which the average profit per vase is changing when 50 vases have been made and sold.

108. Average profit. Cruzin' Boards has found that the cost, in dollars, of producing x skateboards is given by

$$C(x) = 900 + 18x^{0.7}.$$

If the revenue from the sale of x skateboards is given by $R(x) = 75x^{0.8}$, find the rate at which the average profit per skateboard is changing when 20 skateboards have been built and sold.

109. Gross domestic product. The U.S. gross domestic product (in billions of dollars) can be approximated using the function

$$P(t) = 567 + t(36t^{0.6} - 104),$$

where t is the number of the years since 1960.

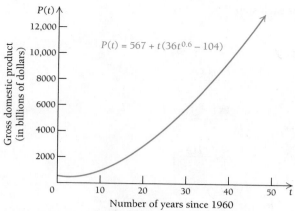

$P(t) = 567 + t(36t^{0.6} - 104)$

Gross domestic product (in billions of dollars)

Number of years since 1960

(*Source*: U.S. Bureau of Economic Analysis.)

a) Find $P'(t)$.
b) Find $P'(45)$.
c) In words, explain what $P'(45)$ means.

Social Sciences

110. Population growth. The population P, in thousands, of a small city is given by

$$P(t) = \frac{500t}{2t^2 + 9},$$

where t is the time, in years.

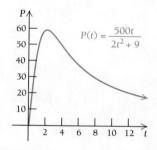

$P(t) = \dfrac{500t}{2t^2 + 9}$

a) Find the growth rate.
b) Find the population after 12 yr.
c) Find the growth rate at $t = 12$ yr.

Life and Physical Sciences

111. Temperature during an illness. The temperature T of a person during an illness is given by

$$T(t) = \frac{4t}{t^2 + 1} + 98.6,$$

where T is the temperature, in degrees Fahrenheit, at time t, in hours.

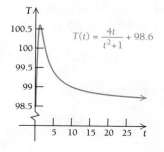

$T(t) = \dfrac{4t}{t^2+1} + 98.6$

a) Find the rate of change of the temperature with respect to time.
b) Find the temperature at $t = 2$ hr.
c) Find the rate of change of the temperature at $t = 2$ hr.

SYNTHESIS

Differentiate each function.

112. $f(x) = \dfrac{7 - \dfrac{3}{2x}}{\dfrac{4}{x^2} + 5}$ (*Hint*: Simplify before differentiating.)

113. $y(t) = 5t(t - 1)(2t + 3)$

114. $f(x) = x(3x^3 + 6x - 2)(3x^4 + 7)$

115. $g(x) = (x^3 - 8) \cdot \dfrac{x^2 + 1}{x^2 - 1}$

116. $f(t) = (t^5 + 3) \cdot \dfrac{t^3 - 1}{t^3 + 1}$

117. $f(x) = \dfrac{(x - 1)(x^2 + x + 1)}{x^4 - 3x^3 - 5}$

118. Let $f(x) = \dfrac{x}{x + 1}$ and $g(x) = \dfrac{-1}{x + 1}$.
  a) Compute $f'(x)$.
  b) Compute $g'(x)$.
  c) What can you conclude about f and g on the basis of your results from parts (a) and (b)?

119. Let $f(x) = \dfrac{x^2}{x^2 - 1}$ and $g(x) = \dfrac{1}{x^2 - 1}$.
  a) Compute $f'(x)$.
  b) Compute $g'(x)$.
  c) What can you conclude about the graphs of f and g on the basis of your results from parts (a) and (b)?

120. Write a rule for finding the derivative of $f(x) \cdot g(x) \cdot h(x)$. Describe the rule in words.

121. Is the derivative of the reciprocal of $f(x)$ the reciprocal of the derivative of $f'(x)$? Why or why not?

122. Sensitivity. The reaction R of the body to a dose Q of medication is often represented by the general function

$$R(Q) = Q^2\left(\frac{k}{2} - \frac{Q}{3}\right),$$

where k is a constant and R is in millimeters of mercury (mmHg) if the reaction is a change in blood pressure or in degrees Fahrenheit (°F) if the reaction is a change in temperature. The rate of change dR/dQ is defined to be the body's *sensitivity* to the medication.

a) Find a formula for the sensitivity.
b) Explain the meaning of your answer to part (a).

123. A proof of the Product Rule appears below. Provide a justification for each step.

a) $\dfrac{d}{dx}[f(x) \cdot g(x)] = \lim\limits_{h \to 0} \dfrac{f(x+h)g(x+h) - f(x)g(x)}{h}$

b) $= \lim\limits_{h \to 0} \dfrac{f(x+h)g(x+h) - f(x+h)g(x) + f(x+h)g(x) - f(x)g(x)}{h}$

c) $= \lim\limits_{h \to 0} \dfrac{f(x+h)g(x+h) - f(x+h)g(x)}{h} + \lim\limits_{h \to 0} \dfrac{f(x+h)g(x) - f(x)g(x)}{h}$

d) $= \lim\limits_{h \to 0} \left[f(x+h) \cdot \dfrac{g(x+h) - g(x)}{h} \right] + \lim\limits_{h \to 0} \left[g(x) \cdot \dfrac{f(x+h) - f(x)}{h} \right]$

e) $= f(x) \cdot \lim\limits_{h \to 0} \dfrac{g(x+h) - g(x)}{h} + g(x) \cdot \lim\limits_{h \to 0} \dfrac{f(x+h) - f(x)}{h}$

f) $= f(x) \cdot g'(x) + g(x) \cdot f'(x)$

g) $= f(x) \cdot \left[\dfrac{d}{dx}g(x) \right] + g(x) \cdot \left[\dfrac{d}{dx}f(x) \right]$

TECHNOLOGY CONNECTION · · · · · · · · · · · ·

124. Business. Refer to Exercises 102, 104, and 106. At what rate is Tongue-Tied Sauces' profit changing at the break-even point? At what rate is the average profit per bottle of barbecue sauce changing at that point?

125. Business. Refer to Exercises 101, 103, and 105. At what rate is Summertime Fabrics' profit changing at the break-even point? At what rate is the average profit per jacket changing at that point?

For the function in each of Exercises 126–131, graph f and f'. Then estimate points at which the tangent line to f is horizontal. If no such point exists, state that fact.

126. $f(x) = x^2(x - 2)(x + 2)$

127. $f(x) = \left(x + \dfrac{2}{x} \right)(x^2 - 3)$

128. $f(x) = \dfrac{x^3 - 1}{x^2 + 1}$ **129.** $f(x) = \dfrac{0.3x}{0.04 + x^2}$

130. $f(x) = \dfrac{0.01x^2}{x^4 + 0.0256}$ **131.** $f(x) = \dfrac{4x}{x^2 + 1}$

132. Use a graph to decide which of the following seems to be the correct derivative of the function in Exercise 131.

$y_1 = \dfrac{2}{x}$

$y_2 = \dfrac{4 - 4x}{x^2 + 1}$

$y_3 = \dfrac{4 - 4x^2}{(x^2 + 1)^2}$

$y_4 = \dfrac{4x^2 - 4}{(x^2 + 1)^2}$

Answers to Quick Checks

1. (a) $y' = (2x^5 + x - 1)(3) + (3x - 2)(10x^4 + 1)$

(b) $y' = (\sqrt{x} + 1)\left(\dfrac{1}{5\sqrt[5]{x^4}} - 1 \right) + (\sqrt[5]{x} - x)\left(\dfrac{1}{2\sqrt{x}} \right)$

2. (a) $y' = \dfrac{3x^2 - 2x - 6}{(x^2 + 2)^2}$

(b) $\dfrac{(bx + 1)(a) - (ax + 1)(b)}{(bx + 1)^2} = \dfrac{abx + a - abx - b}{(bx + 1)^2}$

$= \dfrac{a - b}{(bx + 1)^2}$

1.7   The Chain Rule

The Extended Power Rule

OBJECTIVES

- Find the composition of two functions.
- Differentiate using the Extended Power Rule or the Chain Rule.

Some functions are considered simple. *Simple* is a subjective concept, and it may be best to illustrate what it means with examples. The following functions are considered simple and are similar to those we have seen in Sections 1.5 and 1.6:

$$f(x) = 2x, \qquad g(x) = 3x^2 - 5x, \qquad h(x) = 2\sqrt{x}, \qquad j(x) = \dfrac{2x - 1}{x^2 - 3}.$$

On the other hand, these functions are not considered simple:

$$f(x) = (x^3 + 2x)^5, \qquad g(x) = \sqrt{2x + 5}, \qquad h(x) = \left(\dfrac{3x - 7}{4x^2 + 1} \right)^3.$$

In these cases, we see that the variable x is part of one or more expressions that are raised to some power. How can we use the concepts of differentiation from Sections 1.5 and 1.6 to differentiate functions of this form? In this section, we will introduce and discuss the *Chain Rule,* but we begin our discussion with a special case of the Chain Rule called the *Extended Power Rule.*

The Extended Power Rule

The function $y = 1 + x^2$ is considered simple, and its derivative can be found directly from the Power Rule. However, if we nest this function in some manner, for example, $y = (1 + x^2)^3$, we now have a more complicated form. How do we determine the derivative? We might guess the following:

$$\frac{d}{dx}[(1 + x^2)^3] \stackrel{?}{=} 3(1 + x^2)^2. \qquad \text{Remember, this is a guess.}$$

To check this, we expand the function $y = (1 + x^2)^3$:

$$\begin{aligned} y &= (1 + x^2)^3 \\ &= (1 + x^2) \cdot (1 + x^2) \cdot (1 + x^2) \\ &= (1 + 2x^2 + x^4) \cdot (1 + x^2) \qquad \text{Multiplying the first two factors} \\ &= 1 + 3x^2 + 3x^4 + x^6. \qquad \text{Multiplying by the third factor} \end{aligned}$$

Taking the derivative of this function, we have

$$y' = 6x + 12x^3 + 6x^5.$$

Now we can factor out $6x$:

$$y' = 6x(1 + 2x^2 + x^4).$$

We rewrite $6x$ as $3 \cdot 2x$ and factor the expression within the parentheses:

$$y' = 3(1 + x^2)^2 \cdot 2x$$

Thus, it seems our original *guess* was close: it lacked only the extra factor, $2x$, which is the derivative of the expression inside the parentheses. The correct derivative of $y = (1 + x^2)^3$ is $y' = 3(1 + x^2)^2 \cdot 2x$, which suggests a general pattern for differentiating functions of this form, in which an expression is raised to a power k.

THEOREM 7   The Extended Power Rule

Suppose that $g(x)$ is a differentiable function of x. Then, for any real number k,

$$\frac{d}{dx}[g(x)]^k = k[g(x)]^{k-1} \cdot \frac{d}{dx}g(x).$$

The Extended Power Rule allows us to differentiate functions such as $y = (1 + x^2)^{89}$ without having to expand the expression $1 + x^2$ to the 89th power (very time-consuming) and functions such as $y = (1 + x^2)^{1/3}$, for which "expanding" to the $\frac{1}{3}$ power is impractical.

Let's differentiate $(1 + x^3)^5$. There are three steps to carry out.

1. Mentally block out the "inside" function, $1 + x^3$. $\qquad (1 + x^3)^5$
2. Differentiate the "outside" function, $(1 + x^3)^5$. $\qquad 5(1 + x^3)^4$
3. Multiply by the derivative of the "inside" function. $\qquad 5(1 + x^3)^4 \cdot 3x^2$
$$= 15x^2(1 + x^3)^4 \qquad \text{Simplified}$$

Step (3) is quite commonly overlooked. Do not forget it!

Exploratory

One way to check your differentiation of y_1 is to enter your derivative as y_2 and see if the graph of y_2 coincides with the graph of $y_3 = \text{nDeriv}(y_1, x, x)$. Use this approach to check Example 1. Be sure to use the Sequential mode and different graph styles for the two curves.

The Extended Power Rule is best illustrated by examples. Carefully examine each of the following examples, noting the three-step process for applying the rule.

■ **EXAMPLE 1**   Differentiate: $f(x) = (1 + x^3)^{1/2}$.

Solution

$$\frac{d}{dx}(1 + x^3)^{1/2} = \frac{1}{2}(1 + x^3)^{1/2 - 1} \cdot 3x^2$$

$$= \frac{3x^2}{2}(1 + x^3)^{-1/2}$$

$$= \frac{3x^2}{2\sqrt{1 + x^3}}$$

■ **EXAMPLE 2**   Differentiate: $y = (1 - x^2)^3 + (5 + 4x)^2$.

Solution   Here we combine the Sum–Difference Rule and the Extended Power Rule:

$$\frac{dy}{dx} = 3(1 - x^2)^2(-2x) + 2(5 + 4x)^1 \cdot 4.$$

We differentiate each term using the Extended Power Rule.

> **Quick Check 1**
>
> **a)** Use the Extended Power Rule to differentiate $y = (x^4 + 2x^2 + 1)^3$.
>
> **b)** Explain why
> $$\frac{d}{dx}\left[(x^2 + 4x + 1)^4\right] = 4(x^2 + 4x + 1)^3 \cdot 2x + 4$$
> is incorrect.

Since $dy/dx = 3(1 - x^2)^2(-2x) + 2(5 + 4x) \cdot 4$, it follows that

$$\frac{dy}{dx} = -6x(1 - x^2)^2 + 8(5 + 4x)$$

$$= -6x(1 - 2x^2 + x^4) + 40 + 32x$$

$$= -6x + 12x^3 - 6x^5 + 40 + 32x$$

$$= 40 + 26x + 12x^3 - 6x^5.$$

❮ Quick Check 1

■ **EXAMPLE 3**   Differentiate: $f(x) = (3x - 5)^4(7 - x)^{10}$.

Solution   Here we combine the Product Rule and the Extended Power Rule:

$$f'(x) = (3x - 5)^4 \cdot 10(7 - x)^9(-1) + (7 - x)^{10} 4(3x - 5)^3(3)$$

$$= -10(3x - 5)^4(7 - x)^9 + (7 - x)^{10} 12(3x - 5)^3$$

$$= 2(3x - 5)^3(7 - x)^9[-5(3x - 5) + 6(7 - x)]$$

We factor out $2(3x - 5)^3(7 - x)^9$.

$$= 2(3x - 5)^3(7 - x)^9(-15x + 25 + 42 - 6x)$$

$$= 2(3x - 5)^3(7 - x)^9(67 - 21x).$$

■ **EXAMPLE 4**   Differentiate: $f(x) = \sqrt[4]{\dfrac{x + 3}{x - 2}}$.

Solution   Here we use the Quotient Rule to differentiate the inside function:

$$\frac{d}{dx}\sqrt[4]{\frac{x + 3}{x - 2}} = \frac{d}{dx}\left(\frac{x + 3}{x - 2}\right)^{1/4} = \frac{1}{4}\left(\frac{x + 3}{x - 2}\right)^{1/4 - 1}\left[\frac{(x - 2)1 - 1(x + 3)}{(x - 2)^2}\right]$$

$$= \frac{1}{4}\left(\frac{x + 3}{x - 2}\right)^{-3/4}\left[\frac{x - 2 - x - 3}{(x - 2)^2}\right]$$

> **Quick Check 2**
>
> Differentiate:
> $$f(x) = \frac{(2x^2 - 1)}{(3x^4 + 2)^2}.$$

$$= \frac{1}{4}\left(\frac{x + 3}{x - 2}\right)^{-3/4}\left[\frac{-5}{(x - 2)^2}\right], \quad \text{or} \quad \frac{-5}{4(x + 3)^{3/4}(x - 2)^{5/4}}.$$

❮ Quick Check 2

Composition of Functions and the Chain Rule

Before discussing the Chain Rule, let's consider *composition of functions*.

One author of this text exercises three times a week at a local YMCA. When he recently bought a pair of running shoes, he found a label on which the numbers at the bottom indicate equivalent shoe sizes in five countries.

Author Marv Bittinger and his size-$11\frac{1}{2}$ running shoes

This label suggests that there are functions that convert one country's shoe sizes to those used in another country. There is, indeed, a function g that gives a correspondence between shoe sizes in the United States and those in France:

$$g(x) = \frac{4x + 92}{3},$$

where x is the U.S. size and $g(x)$ is the French size. Thus, a U.S. size $11\frac{1}{2}$ corresponds to a French size

$$g\left(11\tfrac{1}{2}\right) = \frac{4 \cdot 11\frac{1}{2} + 92}{3}, \quad \text{or } 46.$$

There is also a function f that gives a correspondence between shoe sizes in France and those in Japan. The function is given by

$$f(x) = \frac{15x - 100}{2},$$

where x is the French size and $f(x)$ is the corresponding Japanese size. Thus, a French size 46 corresponds to a Japanese size

$$f(46) = \frac{15 \cdot 46 - 100}{2}, \quad \text{or } 295.$$

It seems reasonable to conclude that a shoe size of $11\frac{1}{2}$ in the United States corresponds to a size of 295 in Japan and that some function h describes this correspondence. Can we find a formula for h?

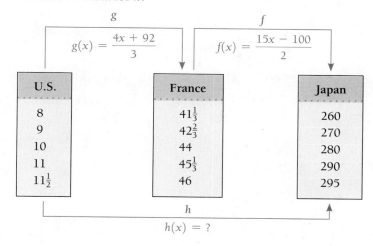

A shoe size x in the United States corresponds to a shoe size $g(x)$ in France, where

$$g(x) = \frac{4x + 92}{3}.$$

Thus, $(4x + 92)/3$ represents a shoe size in France. If we replace x in $f(x)$ with $(4x + 92)/3$, we can find the corresponding shoe size in Japan:

$$f(g(x)) = \frac{15\left(\dfrac{4x + 92}{3}\right) - 100}{2}$$

$$= \frac{5(4x + 92) - 100}{2} = \frac{20x + 460 - 100}{2}$$

$$= \frac{20x + 360}{2} = 10x + 180.$$

This gives a formula for h: $h(x) = 10x + 180$. As a check, a shoe size of $11\frac{1}{2}$ in the United States corresponds to a shoe size of $h\left(11\frac{1}{2}\right) = 10\left(11\frac{1}{2}\right) + 180 = 295$ in Japan. The function h is the *composition* of f and g, symbolized by $f \circ g$ and read as "f composed with g," or simply "f circle g."

DEFINITION

The **composed** function $f \circ g$, the **composition** of f and g, is defined as

$$(f \circ g)(x) = f(g(x)).$$

We can visualize the composition of functions as shown below.

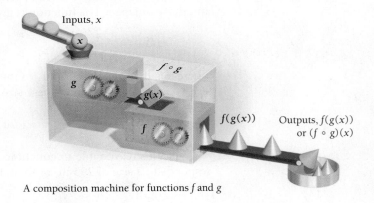

A composition machine for functions f and g

To find $(f \circ g)(x)$, we substitute $g(x)$ for x in $f(x)$. The function $g(x)$ is nested within $f(x)$.

■ **EXAMPLE 5**   For $f(x) = x^3$ and $g(x) = 1 + x^2$, find $(f \circ g)(x)$ and $(g \circ f)(x)$.

Solution   Consider each function separately:

$$f(x) = x^3 \qquad \text{This function cubes each input.}$$

and

$$g(x) = 1 + x^2. \qquad \text{This function adds 1 to the square of each input.}$$

a) The function $f \circ g$ first does what g does (adds 1 to the square) and then does what f does (cubes). We find $f(g(x))$ by substituting $g(x)$ for x:

$$(f \circ g)(x) = f(g(x)) = f(1 + x^2) \qquad \text{Using } g(x) \text{ as an input}$$
$$= (1 + x^2)^3$$
$$= 1 + 3x^2 + 3x^4 + x^6.$$

b) The function $g \circ f$ first does what f does (cubes) and then does what g does (adds 1 to the square). We find $g(f(x))$ by substituting $f(x)$ for x:

$$(g \circ f)(x) = g(f(x)) = g(x^3) \qquad \text{Using } f(x) \text{ as an input}$$
$$= 1 + (x^3)^2 = 1 + x^6.$$

> **Quick Check 3**
>
> For the functions in Example 6, find:
> **a)** $(f \circ f)(x)$;
> **b)** $(g \circ g)(x)$.

■ **EXAMPLE 6**   For $f(x) = \sqrt{x}$ and $g(x) = x - 1$, find $(f \circ g)(x)$ and $(g \circ f)(x)$.

Solution

$$(f \circ g)(x) = f(g(x)) = f(x - 1) = \sqrt{x - 1}$$
$$(g \circ f)(x) = g(f(x)) = g(\sqrt{x}) = \sqrt{x} - 1$$

❰ Quick Check 3

Keep in mind that, in general, $(f \circ g)(x) \neq (g \circ f)(x)$. We see this fact demonstrated in Examples 5 and 6.

How do we differentiate a composition of functions? The following theorem tells us.

THEOREM 8   The Chain Rule

The derivative of the composition $f \circ g$ is given by

$$\frac{d}{dx}\big[(f \circ g)(x)\big] = \frac{d}{dx}\big[f(g(x))\big] = f'(g(x)) \cdot g'(x).$$

As we noted earlier, the Extended Power Rule is a special case of the Chain Rule. Consider $f(x) = x^k$. For any other function $g(x)$, we have $(f \circ g)(x) = [g(x)]^k$, and the derivative of the composition is

$$\frac{d}{dx}[g(x)]^k = k[g(x)]^{k-1} \cdot g'(x).$$

The Chain Rule often appears in another form. Suppose that $y = f(u)$ and $u = g(x)$. Then

$$\frac{dy}{dx} = \frac{dy}{du} \cdot \frac{du}{dx}.$$

To better understand the Chain Rule, suppose that a video game manufacturer wished to determine its rate of profit, in *dollars* per *minute*. One way to find this rate would be to multiply the rate of profit, in *dollars* per *item*, by the production rate, in *items* per *minute*. That is,

$$\begin{bmatrix} \text{Change in profits} \\ \text{with respect to time} \end{bmatrix} = \begin{bmatrix} \text{Change in profits with respect} \\ \text{to number of games produced} \end{bmatrix} \cdot \begin{bmatrix} \text{Change in number of games} \\ \text{produced with respect to time.} \end{bmatrix}$$

■ **EXAMPLE 7**   For $y = 2 + \sqrt{u}$ and $u = x^3 + 1$, find dy/du, du/dx, and dy/dx.

Solution   First we find dy/du and du/dx:

$$\frac{dy}{du} = \frac{1}{2}u^{-1/2} \quad \text{and} \quad \frac{du}{dx} = 3x^2.$$

Then

$$\frac{dy}{dx} = \frac{dy}{du} \cdot \frac{du}{dx}$$

$$= \frac{1}{2\sqrt{u}} \cdot 3x^2$$

$$= \frac{3x^2}{2\sqrt{x^3 + 1}}. \quad \text{Substituting } x^3 + 1 \text{ for } u$$

❰ Quick Check 4

> **Quick Check 4**
>
> If $y = u^2 + u$ and
> $u = x^2 + x$, find $\dfrac{dy}{dx}$.

■ **EXAMPLE 8**   **Business.**   A new product is placed on the market and becomes very popular. Its quantity sold N is given as a function of time t, where t is measured in weeks:

$$N(t) = \frac{250{,}000t^2}{(2t + 1)^2}, \quad t > 0.$$

Differentiate this function. Then use the derivative to evaluate $N'(52)$ and $N'(208)$, and interpret these results.

Solution   To determine $N'(t)$, we use the Quotient Rule along with the Extended Power Rule:

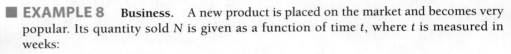

$$N'(t) = \frac{d}{dt}\left[\frac{250{,}000t^2}{(2t + 1)^2}\right] = \frac{(2t + 1)^2 \cdot \frac{d}{dt}\left[250{,}000t^2\right] - 250{,}000t^2 \cdot \frac{d}{dt}\left[(2t + 1)^2\right]}{\left[(2t + 1)^2\right]^2}$$

$$= \frac{(2t + 1)^2 \cdot (500{,}000t) - 250{,}000t^2 \cdot 2(2t + 1)^1 \cdot 2}{(2t + 1)^4} \quad \begin{array}{l}\text{The Extended}\\ \text{Power Rule is}\\ \text{used here.}\end{array}$$

$$= \frac{(2t + 1)^2(500{,}000t) - 1{,}000{,}000t^2(2t + 1)}{(2t + 1)^4}$$

$$= \frac{500{,}000t(2t + 1)\left[(2t + 1) - 2t\right]}{(2t + 1)^4}. \quad \begin{array}{l}\text{The expression } 500{,}000t(2t + 1)\\ \text{is factored out in the numerator;}\\ \text{the } 2t \text{ terms inside the square}\\ \text{brackets sum to } 0.\end{array}$$

Therefore, $N'(t) = \dfrac{500{,}000t}{(2t + 1)^3}.$ $\begin{array}{l}\text{The factor } (2t + 1) \text{ in the numerator}\\ \text{cancels } (2t + 1) \text{ in the denominator.}\end{array}$

We evaluate $N'(t)$ at $t = 52$:

$$N'(52) = \frac{500{,}000(52)}{(2(52) + 1)^3} \approx 22.5.$$

Thus, after 52 weeks (1 yr), the quantity sold is increasing by about 22.5 units per week.

For $t = 208$ weeks (4 yr), we get

$$N'(208) = \frac{500{,}000(208)}{(2(208) + 1)^3} \approx 1.4.$$

After 4 yr, the quantity sold is increasing at about 1.4 units per week. What is happening here? Consider the graph of $N(t)$:

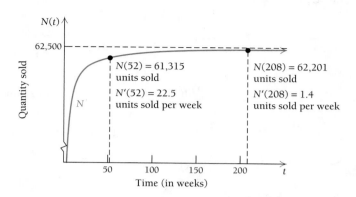

We see that the slopes of the tangent lines, representing the change in numbers of units sold per week, are leveling off as t increases. Perhaps the market is becoming saturated with this product: while sales continue to increase, the rate of the sales increase per week is leveling off.

Section Summary

- The *Extended Power Rule* tells us that if $y = [f(x)]^k$, then

$$y' = \frac{d}{dx}[f(x)]^k = k[f(x)]^{k-1} \cdot f'(x).$$

- The *composition* of $f(x)$ with $g(x)$ is written $(f \circ g)(x)$ and is defined as $(f \circ g)(x) = f(g(x))$.
- In general, $(f \circ g)(x) \neq (g \circ f)(x)$.

- The *Chain Rule* is used to differentiate a composition of functions. If

$$F(x) = (f \circ g)(x) = f(g(x)),$$

then

$$F'(x) = \frac{d}{dx}[(f \circ g)(x)] = f'(g(x)) \cdot g'(x).$$

EXERCISE SET
1.7

Differentiate each function.

1. $y = (2x + 1)^2$ ⎱ Check by expanding and
2. $y = (3 - 2x)^2$ ⎰ then differentiating.

3. $y = (7 - x)^{55}$

4. $y = (8 - x)^{100}$

5. $y = \sqrt{1 + 8x}$

6. $y = \sqrt{1 - x}$

7. $y = \sqrt{3x^2 - 4}$

8. $y = \sqrt{4x^2 + 1}$

9. $y = (8x^2 - 6)^{-40}$

10. $y = (4x^2 + 1)^{-50}$

11. $y = (x - 4)^8(2x + 3)^6$

12. $y = (x + 5)^7(4x - 1)^{10}$

13. $y = \dfrac{1}{(3x + 8)^2}$

14. $y = \dfrac{1}{(4x + 5)^2}$

15. $y = \dfrac{4x^2}{(7 - 5x)^3}$

16. $y = \dfrac{7x^3}{(4 - 9x)^5}$

17. $f(x) = (1 + x^3)^3 - (2 + x^8)^4$

18. $f(x) = (3 + x^3)^5 - (1 + x^7)^4$

19. $f(x) = x^2 + (200 - x)^2$

20. $f(x) = x^2 + (100 - x)^2$

21. $g(x) = \sqrt{x} + (x - 3)^3$

22. $G(x) = \sqrt[3]{2x - 1} + (4 - x)^2$

23. $f(x) = -5x(2x - 3)^4$

24. $f(x) = -3x(5x + 4)^6$

25. $g(x) = (3x - 1)^7(2x + 1)^5$

26. $F(x) = (5x + 2)^4(2x - 3)^8$

27. $f(x) = x^2\sqrt{4x - 1}$

28. $f(x) = x^3\sqrt{5x + 2}$

29. $G(x) = \sqrt[3]{x^5 + 6x}$

30. $F(x) = \sqrt[4]{x^2 - 5x + 2}$

31. $f(x) = \left(\dfrac{3x - 1}{5x + 2}\right)^4$ **32.** $f(x) = \left(\dfrac{2x}{x^2 + 1}\right)^3$

33. $g(x) = \sqrt{\dfrac{4 - x}{3 + x}}$ **34.** $g(x) = \sqrt{\dfrac{3 + 2x}{5 - x}}$

35. $f(x) = (2x^3 - 3x^2 + 4x + 1)^{100}$

36. $f(x) = (7x^4 + 6x^3 - x)^{204}$

37. $g(x) = \left(\dfrac{2x + 3}{5x - 1}\right)^{-4}$ **38.** $h(x) = \left(\dfrac{1 - 3x}{2 - 7x}\right)^{-5}$

39. $f(x) = \sqrt{\dfrac{x^2 + x}{x^2 - x}}$ **40.** $f(x) = \sqrt[3]{\dfrac{4 - x^3}{x - x^2}}$

41. $f(x) = \dfrac{(2x + 3)^4}{(3x - 2)^5}$ **42.** $f(x) = \dfrac{(5x - 4)^7}{(6x + 1)^3}$

43. $f(x) = 12(2x + 1)^{2/3}(3x - 4)^{5/4}$

44. $y = 6\sqrt[3]{x^2 + x}(x^4 - 6x)^3$

Find $\dfrac{dy}{du}, \dfrac{du}{dx}, $ *and* $\dfrac{dy}{dx}.$

45. $y = \sqrt{u}$ and $u = x^2 - 1$

46. $y = \dfrac{15}{u^3}$ and $u = 2x + 1$

47. $y = u^{50}$ and $u = 4x^3 - 2x^2$

48. $y = \dfrac{u + 1}{u - 1}$ and $u = 1 + \sqrt{x}$

49. $y = u(u + 1)$ and $u = x^3 - 2x$

50. $y = (u + 1)(u - 1)$ and $u = x^3 + 1$

Find $\dfrac{dy}{dx}$ *for each pair of functions.*

51. $y = 5u^2 + 3u$ and $u = x^3 + 1$

52. $y = u^3 - 7u^2$ and $u = x^2 + 3$

53. $y = \sqrt[3]{2u + 5}$ and $u = x^2 - x$

54. $y = \sqrt{7 - 3u}$ and $u = x^2 - 9$

55. Find $\dfrac{dy}{dt}$ if $y = \dfrac{1}{u^2 + u}$ and $u = 5 + 3t.$

56. Find $\dfrac{dy}{dt}$ if $y = \dfrac{1}{3u^5 - 7}$ and $u = 7t^2 + 1.$

57. Find an equation for the tangent line to the graph of $y = \sqrt{x^2 + 3x}$ at the point $(1, 2).$

58. Find an equation for the tangent line to the graph of $y = (x^3 - 4x)^{10}$ at the point $(2, 0).$

59. Find an equation for the tangent line to the graph of $y = x\sqrt{2x + 3}$ at the point $(3, 9).$

60. Find an equation for the tangent line to the graph of $y = \left(\dfrac{2x + 3}{x - 1}\right)^3$ at the point $(2, 343).$

61. Consider

$$f(x) = \dfrac{x^2}{(1 + x)^5}.$$

   a) Find $f'(x)$ using the Quotient Rule and the Extended Power Rule.

   b) Note that $f(x) = x^2(1 + x)^{-5}.$ Find $f'(x)$ using the Product Rule and the Extended Power Rule.

   c) Compare your answers to parts (a) and (b).

62. Consider

$$g(x) = \left(\dfrac{6x + 1}{2x - 5}\right)^2.$$

   a) Find $g'(x)$ using the Extended Power Rule.

   b) Note that

$$g(x) = \dfrac{36x^2 + 12x + 1}{4x^2 - 20x + 25}.$$

   Find $g'(x)$ using the Quotient Rule.

   c) Compare your answers to parts (a) and (b). Which approach was easier, and why?

In Exercises 63–66, find $f(x)$ and $g(x)$ such that $h(x) = (f \circ g)(x).$ Answers may vary.

63. $h(x) = (3x^2 - 7)^5$ **64.** $h(x) = \dfrac{1}{\sqrt{7x + 2}}$

65. $h(x) = \dfrac{x^3 + 1}{x^3 - 1}$ **66.** $h(x) = (\sqrt{x} + 5)^4$

Do Exercises 67–70 in two ways. First, use the Chain Rule to find the answer. Next, check your answer by finding $f(g(x)),$ taking the derivative, and substituting.

67. $f(u) = u^3, \quad g(x) = u = 2x^4 + 1$
   Find $(f \circ g)'(-1).$

68. $f(u) = \dfrac{u + 1}{u - 1}, \quad g(x) = u = \sqrt{x}$
   Find $(f \circ g)'(4).$

69. $f(u) = \sqrt[3]{u}, \quad g(x) = u = 1 + 3x^2$
   Find $(f \circ g)'(2).$

70. $f(u) = 2u^5, \quad g(x) = u = \dfrac{3 - x}{4 + x}$
   Find $(f \circ g)'(-10).$

For Exercises 71–74, use the Chain Rule to differentiate each function. You may need to apply the rule more than once.

71. $f(x) = (2x^3 + (4x - 5)^2)^6$

72. $f(x) = (-x^5 + 4x + \sqrt{2x + 1})^3$

73. $f(x) = \sqrt{x^2 + \sqrt{1 - 3x}}$

74. $f(x) = \sqrt[3]{2x + (x^2 + x)^4}$

APPLICATIONS

Business and Economics

75. Total revenue. A total-revenue function is given by

$$R(x) = 1000\sqrt{x^2 - 0.1x},$$

where $R(x)$ is the total revenue, in thousands of dollars, from the sale of x items. Find the rate at which total revenue is changing when 20 items have been sold.

76. Total cost. A total-cost function is given by

$$C(x) = 2000(x^2 + 2)^{1/3} + 700,$$

where $C(x)$ is the total cost, in thousands of dollars, of producing x items. Find the rate at which total cost is changing when 20 items have been produced.

77. Total profit. Use the total-cost and total-revenue functions in Exercises 75 and 76 to find the rate at which total profit is changing when x items have been produced and sold.

78. Total cost. A company determines that its total cost, in thousands of dollars, for producing x items is

$$C(x) = \sqrt{5x^2 + 60},$$

and it plans to boost production t months from now according to the function

$$x(t) = 20t + 40.$$

How fast will costs be rising 4 months from now?

79. Consumer credit. The total outstanding consumer credit of the United States (in billions of dollars) can be modeled by the function

$$C(x) = 0.21x^4 - 5.92x^3 + 50.53x^2 - 18.92x + 1114.93,$$

where x is the number of years since 1995.

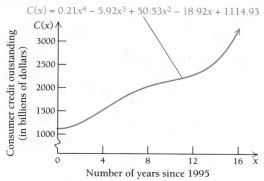

$C(x) = 0.21x^4 - 5.92x^3 + 50.53x^2 - 18.92x + 1114.93$

(*Source*: Federal Reserve Board.)

a) Find dC/dx.
b) Interpret the meaning of dC/dx.
c) Using this model, estimate how quickly outstanding consumer credit was rising in 2010.

80. Utility. Utility is a type of function that occurs in economics. When a consumer receives x units of a product, a certain amount of pleasure, or utility, U, is derived. Suppose that the utility related to the number of tickets x for a ride at a county fair is

$$U(x) = 80\sqrt{\frac{2x + 1}{3x + 4}}.$$

Find the rate at which the utility changes with respect to the number of tickets bought.

81. Compound interest. If $1000 is invested at interest rate i, compounded annually, in 3 yr it will grow to an amount A given by (see Section R.1)

$$A = \$1000(1 + i)^3.$$

a) Find the rate of change, dA/di.
b) Interpret the meaning of dA/di.

82. Compound interest. If $1000 is invested at interest rate i, compounded quarterly, in 5 yr it will grow to an amount, A, given by

$$A = \$1000\left(1 + \frac{i}{4}\right)^{20}.$$

a) Find the rate of change, dA/di.
b) Interpret the meaning of dA/di.

83. Consumer demand. Suppose that the demand function for a product is given by

$$D(p) = \frac{80,000}{p},$$

and that price p is a function of time given by $p = 1.6t + 9$, where t is in days.

a) Find the demand as a function of time t.
b) Find the rate of change of the quantity demanded when $t = 100$ days.

84. Business profit. A company is selling laptop computers. It determines that its total profit, in dollars, is given by

$$P(x) = 0.08x^2 + 80x,$$

where x is the number of units produced and sold. Suppose that x is a function of time, in months, where $x = 5t + 1$.

a) Find the total profit as a function of time t.
b) Find the rate of change of total profit when $t = 48$ months.

Life and Physical Sciences

85. Chemotherapy. The dosage for Carboplatin chemotherapy drugs depends on several parameters of the particular drug as well as the age, weight, and sex of the patient. For female patients, the formulas giving the dosage for such drugs are

$$D = 0.85A(c + 25) \quad \text{and} \quad c = (140 - y)\frac{w}{72x},$$

where A and x depend on which drug is used, D is the dosage in milligrams (mg), c is called the creatine clearance, y is the patient's age in years, and w is the patient's weight in kilograms (kg). (*Source: U.S. Oncology.*)

a) Suppose that a patient is a 45-year-old woman and the drug has parameters $A = 5$ and $x = 0.6$. Use this information to write formulas for D and c that give D as a function of c and c as a function of w.

b) Use your formulas from part (a) to compute dD/dc.

c) Use your formulas from part (a) to compute dc/dw.

d) Compute dD/dw.

e) Interpret the meaning of the derivative dD/dw.

SYNTHESIS

*If $f(x)$ is a function, then $(f \circ f)(x) = f(f(x))$ is the composition of f with itself. This is called an **iterated function**, and the composition can be repeated many times. For example, $(f \circ f \circ f)(x) = f(f(f(x)))$. Iterated functions are very useful in many areas, including finance (compound interest is a simple case) and the sciences (in weather forecasting, for example). For each function, use the Chain Rule to find the derivative.*

86. If $f(x) = x^2 + 1$, find $\dfrac{d}{dx}[(f \circ f)(x)]$.

87. If $f(x) = x + \sqrt{x}$, find $\dfrac{d}{dx}[(f \circ f)(x)]$.

88. If $f(x) = x^2 + 1$, find $\dfrac{d}{dx}[(f \circ f \circ f)(x)]$

89. If $f(x) = \sqrt[3]{x}$, find $\dfrac{d}{dx}[(f \circ f \circ f)(x)]$.

Do you see a shortcut?

Differentiate.

90. $y = \sqrt{(2x - 3)^2 + 1}$

91. $y = \sqrt[3]{x^3 + 6x + 1} \cdot x^5$

92. $s = \sqrt[4]{t^4 + 3t^2 + 8} \cdot 3t$

93. $y = \left(\dfrac{x}{\sqrt{x - 1}}\right)^3$

94. $y = (x\sqrt{1 + x^2})^3$

95. $y = \dfrac{\sqrt{1 - x^2}}{1 - x}$

96. $w = \dfrac{u}{\sqrt{1 + u^2}}$

97. $y = \left(\dfrac{x^2 - x - 1}{x^2 + 1}\right)^3$

98. $g(x) = \sqrt{\dfrac{x^2 - 4x}{2x + 1}}$

99. $f(t) = \sqrt{3t + \sqrt{t}}$

100. $F(x) = [6x(3 - x)^5 + 2]^4$

101. The following is the beginning of an alternative proof of the Quotient Rule that uses the Product Rule and the Power Rule. Complete the proof, giving reasons for each step.

Proof. Let

$$Q(x) = \frac{N(x)}{D(x)}.$$

Then

$$Q(x) = N(x) \cdot [D(x)]^{-1}.$$

Therefore, ...

102. The Extended Power Rule (for positive integer powers) can be verified using the Product Rule. For example, if $y = [f(x)]^2$, then the Product Rule is applied by recognizing that $[f(x)]^2 = [f(x)] \cdot [f(x)]$. Therefore,

$$\frac{d}{dx}([f(x)] \cdot [f(x)]) = f(x) \cdot f'(x) + f'(x) \cdot f(x)$$

$$= 2f(x) \cdot f'(x).$$

a) Use the Product Rule to show that $\dfrac{d}{dx}[f(x)]^3 = 3[f(x)]^2 \cdot f'(x)$. [*Hint:* $[f(x)]^3 = [f(x)]^2 \cdot f(x)$.]

b) Use the Product Rule to show that $\dfrac{d}{dx}[f(x)]^4 = 4[f(x)]^3 \cdot f'(x)$.

TECHNOLOGY CONNECTION

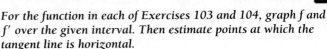

For the function in each of Exercises 103 and 104, graph f and f' over the given interval. Then estimate points at which the tangent line is horizontal.

103. $f(x) = 1.68x\sqrt{9.2 - x^2}; [-3, 3]$

104. $f(x) = \sqrt{6x^3 - 3x^2 - 48x + 45}; [-5, 5]$

Find the derivative of each of the following functions analytically. Then use a calculator to check the results.

105. $f(x) = x\sqrt{4 - x^2}$

106. $g(x) = \dfrac{4x}{\sqrt{x - 10}}$

107. $f(x) = (\sqrt{2x - 1} + x^3)^5$

Answers to Quick Checks

1. (a) $y' = 3(x^4 + 2x^2 + 1)^2(4x^3 + 4x)$
(b) The result lacks parentheses around $2x + 4$. It should be written: $y' = 4(x^2 + 4x + 1)^3(2x + 4)$.
2. $y' = \dfrac{-36x^5 + 24x^3 + 8x}{(3x^4 + 2)^3}$
3. (a) $f(f(x)) = \sqrt{\sqrt{x}} = \sqrt[4]{x}$;
(b) $g(g(x)) = (x - 1) - 1 = x - 2$
4. $\dfrac{dy}{dx} = \dfrac{dy}{du} \cdot \dfrac{du}{dx} = (2u + 1)(2x + 1) =$
$(2(x^2 + x) + 1)(2x + 1) = (2x^2 + 2x + 1)(2x + 1)$

Higher-Order Derivatives

Consider the function given by

$$y = f(x) = x^5 - 3x^4 + x.$$

Its derivative f' is given by

$$y' = f'(x) = 5x^4 - 12x^3 + 1.$$

The derivative function f' can also be differentiated. We can think of the derivative of f' as the rate of change of the slope of the tangent lines of f. It can also be regarded as the rate at which $f'(x)$ is changing. We use the notation f'' for the derivative $(f')'$. That is,

$$f''(x) = \frac{d}{dx}f'(x).$$

We call f'' the *second derivative* of f. For $f(x) = x^5 - 3x^4 + x$, the second derivative is given by

$$y'' = f''(x) = 20x^3 - 36x^2.$$

Continuing in this manner, we have

$$f'''(x) = 60x^2 - 72x, \qquad \text{The third derivative of } f$$
$$f''''(x) = 120x - 72, \qquad \text{The fourth derivative of } f$$
$$f'''''(x) = 120. \qquad \text{The fifth derivative of } f$$

When notation like $f'''(x)$ gets lengthy, we abbreviate it using a number or n in parentheses. Thus, $f^{(n)}(x)$ is the nth derivative. For the function above,

$$f^{(4)}(x) = 120x - 72,$$
$$f^{(5)}(x) = 120,$$
$$f^{(6)}(x) = 0, \quad \text{and}$$
$$f^{(n)}(x) = 0, \quad \text{for any integer } n \geq 6.$$

Leibniz notation for the second derivative of a function given by $y = f(x)$ is

$$\frac{d^2y}{dx^2}, \quad \text{or} \quad \frac{d}{dx}\left(\frac{dy}{dx}\right),$$

read "the second derivative of y with respect to x." The 2's in this notation are *not* exponents. If $y = x^5 - 3x^4 + x$, then

$$\frac{d^2y}{dx^2} = 20x^3 - 36x^2.$$

Leibniz notation for the third derivative is d^3y/dx^3; for the fourth derivative, d^4y/dx^4; and so on:

$$\frac{d^3y}{dx^3} = 60x^2 - 72x,$$

$$\frac{d^4y}{dx^4} = 120x - 72,$$

$$\frac{d^5y}{dx^5} = 120.$$

■ **EXAMPLE 1**   For $y = 1/x$, find d^2y/dx^2.

Solution   We have $y = x^{-1}$, so

$$\frac{dy}{dx} = -1 \cdot x^{-1-1} = -x^{-2}, \quad \text{or} \quad -\frac{1}{x^2}.$$

Then

$$\frac{d^2y}{dx^2} = (-2)(-1)x^{-2-1} = 2x^{-3}, \quad \text{or} \quad \frac{2}{x^3}.$$

■ **EXAMPLE 2**   For $y = (x^2 + 10x)^{20}$, find y' and y''.

Solution   To find y', we use the Extended Power Rule:

$$
\begin{aligned}
y' &= 20(x^2 + 10x)^{19}(2x + 10) \\
&= 20(x^2 + 10x)^{19} \cdot 2(x + 5) \qquad \text{Factor out a 2.} \\
&= 40(x^2 + 10x)^{19}(x + 5). \qquad 20 \times 2 = 40
\end{aligned}
$$

To find y'', we use the Product Rule and the Extended Power Rule:

$$
\begin{aligned}
y'' &= 40(x^2 + 10x)^{19}(1) + (x + 5)19 \cdot 40(x^2 + 10x)^{18}(2x + 10) \\
&= 40(x^2 + 10x)^{19} + 760(x + 5)(x^2 + 10x)^{18} 2(x + 5) \qquad 19 \times 40 = 760 \\
&= 40(x^2 + 10x)^{19} + 1520(x + 5)^2(x^2 + 10x)^{18} \qquad 760 \times 2 = 1520 \\
&= 40(x^2 + 10x)^{18}[(x^2 + 10x) + 38(x + 5)^2] \qquad \text{Factoring} \\
&= 40(x^2 + 10x)^{18}[x^2 + 10x + 38(x^2 + 10x + 25)] \\
&= 40(x^2 + 10x)^{18}(39x^2 + 390x + 950).
\end{aligned}
$$

❮ Quick Check 1

> **Quick Check 1**
>
> **a)** Find y'':
>
> (i) $y = -6x^4 + 3x^2$;
>
> (ii) $y = \dfrac{2}{x^3}$;
>
> (iii) $y = (3x^2 + 1)^2$.
>
> **b)** Find $\dfrac{d^4}{dx^4}\left[\dfrac{1}{x}\right]$.

Velocity and Acceleration

We have already seen that a function's derivative represents its instantaneous rate of change. When the function relates a change in distance to a change in time, the instantaneous rate of change is called *speed*, or *velocity*.* The letter v is generally used to stand for velocity.

DEFINITION

The **velocity** of an object that is $s(t)$ units from a starting point at time t is given by

$$\text{Velocity} = v(t) = s'(t) = \lim_{h \to 0} \frac{s(t + h) - s(t)}{h}.$$

---

*In this text, the words "speed" and "velocity" are used interchangeably. In physics and engineering, this is not done, since velocity requires direction and speed does not.

■ **EXAMPLE 3** **Physical Science: Velocity.** Suppose that an object travels so that its distance s, in miles, from its starting point is a function of time t, in hours, as follows:

$$s(t) = 10t^2.$$

a) Find the average velocity between the times $t = 2$ hr and $t = 5$ hr.

b) Find the (instantaneous) velocity when $t = 4$ hr.

Solution

a) From $t = 2$ hr to $t = 5$ hr, we have

$$\frac{\text{Difference in miles}}{\text{Difference in hours}} = \frac{s(5) - s(2)}{3}$$

$$= \frac{10 \cdot 5^2 \text{ mi} - 10 \cdot 2^2 \text{ mi}}{3 \text{ hr}} = 70 \frac{\text{mi}}{\text{hr}}.$$

b) The instantaneous velocity is given by

$$\lim_{h \to 0} \frac{s(t + h) - s(t)}{h} = s'(t).$$

We know how to find this limit quickly from the special techniques learned in Section 1.5. Thus, $s'(t) = 20t$, and

$$s'(4) = 20 \cdot 4 = 80 \frac{\text{mi}}{\text{hr}}.$$

Often velocity itself is a function of time. When a jet takes off or a vehicle comes to a sudden stop, the change in velocity is easily felt by passengers. The rate at which velocity changes is called *acceleration*. Suppose that Car A reaches a speed of 65 mi/hr in 8.4 sec and Car B reaches a speed of 65 mi/hr in 8 sec; then B has a *faster acceleration* than A. We generally use the letter a for acceleration. It is useful to think of acceleration as the rate at which velocity is changing.

DEFINITION

Acceleration $= a(t) = v'(t) = s''(t)$.

■ **EXAMPLE 4** **Physical Science: Distance, Velocity, and Acceleration.** For $s(t) = 10t^2$, find $v(t)$ and $a(t)$, where s is the distance from the starting point, in miles, and t is in hours. Then find the distance, velocity, and acceleration when $t = 4$ hr.

Solution We have $s(t) = 10t^2$. Thus,

$$v(t) = s'(t) = 20t \quad \text{and} \quad a(t) = v'(t) = s''(t) = 20.$$

It follows that

$$s(4) = 10(4)^2 = 160 \text{ mi},$$
$$v(4) = 20(4) = 80 \text{ mi/hr},$$

and $\quad a(4) = 20 \text{ mi/hr}^2.$

If this distance function applies to motion of a vehicle, then at time $t = 4$ hr, the vehicle has traveled 160 mi, the velocity is 80 mi/hr, and the acceleration is 20 miles per hour per hour, which we abbreviate as 20 mi/hr^2.

Note from Example 4 that since acceleration represents the *rate* at which velocity is changing, the units in which it is measured involve a unit of time squared:

$$\frac{\text{Change in velocity}}{\text{Change in time}} = \frac{\text{mi/hr}}{\text{hr}} = \frac{\text{mi}}{\text{hr}} \cdot \frac{1}{\text{hr}} = \text{mi/hr}^2.$$

■ **EXAMPLE 5** **Free Fall.** When an object is dropped, the distance it falls in t seconds, assuming that air resistance is negligible, is given by

$$s(t) = 4.905t^2,$$

where $s(t)$ is in meters (m). If a stone is dropped from a cliff, find each of the following, assuming that air resistance is negligible: **(a)** how far it has traveled 5 sec after being dropped, **(b)** how fast it is traveling 5 sec after being dropped, and **(c)** the stone's acceleration after it has been falling for 5 sec.

Solution

a) After 5 sec, the stone has traveled

$$s(5) = 4.905(5)^2 = 4.905(25) = 122.625 \text{ m}.$$

b) The speed at which the stone falls is given by

$$v(t) = s'(t) = 9.81t.$$

Thus,

$$v(5) = 9.81 \cdot 5 = 49.05 \text{ m/sec}.$$

c) The stone's acceleration after t sec is constant:

$$a(t) = v'(t) = s''(t) = 9.81 \text{ m/sec}^2.$$

Thus, $s''(5) = 9.81 \text{ m/sec}^2.$

❬ Quick Check 2

> **Quick Check 2**
>
> A pebble is dropped from a hot-air balloon. Find how far it has fallen, how fast it is falling, and its acceleration after 3.5 sec. Let $s(t) = 16t^2$, where t is in seconds and s is in feet.

In Example 8 in Section 1.7, $N(t)$ represented the quantity sold N of a product after t weeks on the market. Its first derivative was always positive (always increasing) indicating that sales were always increasing. But sales were leveling off toward zero. In the following example, we see how the second derivative can help us understand this observation.

■ **EXAMPLE 6** **Business.** In Example 8 in Section 1.7, the function $N(t) = \dfrac{250,000t^2}{(2t + 1)^2}$, $t > 0$, represented the quantity sold N of a product after t weeks on the market. Its derivative is

$$N'(t) = \frac{500,000t}{(2t + 1)^3}.$$

Recall that $N'(t)$ represented the *rate of change* in number of units sold per week. Find $N''(t)$; then use it to calculate $N''(52)$ and $N''(208)$ and interpret these results.

Solution We use the Quotient Rule along with the Extended Power Rule:

$$\frac{d}{dt}[N'(t)] = \frac{(2t + 1)^3 \cdot \dfrac{d}{dt}[500,000t] - (500,000t) \cdot \dfrac{d}{dt}[(2t + 1)^3]}{[(2t + 1)^3]^2} \qquad \text{Extended Power Rule}$$

$$= \frac{(2t + 1)^3(500,000) - (500,000t)(3)(2t + 1)^2 \cdot 2}{[(2t + 1)^3]^2}$$

$$= \frac{(2t + 1)^2[(2t + 1)(500,000) - 6(500,000t)]}{(2t + 1)^6}. \qquad \text{Factoring out } (2t + 1)^2$$

After simplification, we have

$$N''(t) = \frac{-2{,}000{,}000t + 500{,}000}{(2t + 1)^4}.$$

At $t = 52$, we have

$$N''(52) = \frac{-2{,}000{,}000(52) + 500{,}000}{[2(52) + 1]^4} \approx -0.852.$$

Thus, after 52 weeks (1 yr), the rate of the rate of sales is decreasing at -0.852 units per week per week. In other words, although sales are increasing during the 52nd week (remember, $N'(52) > 0$), the rate at which sales are increasing is decreasing. To put it in most basic terms: sales are increasing but not as fast as before.

This is more evident when $t = 208$:

$$N''(208) = \frac{-2{,}000{,}000(208) + 500{,}000}{[2(208) + 1]^4} \approx -0.014.$$

After 208 weeks (4 yr), sales have nearly leveled off. Remember, sales are increasing at the 208th week (recall that $N'(208) > 0$), but since the market is nearly saturated with this product, the rate at which sales are increasing has slowed to near 0.

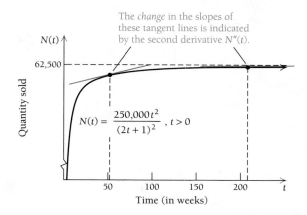

There are many important real-world applications that make use of the second derivative. In Chapter 2, we will examine a number of these, including applications in the fields of economics, health care, and the natural and physical sciences.

Section Summary

- The *second derivative* is the derivative of the first derivative of a function. In symbols, $f''(x) = [f'(x)]'$.
- The second derivative describes the rate of change of the rate of change. In other words, it describes the rate of change of the first derivative.

- A real-life example of a second derivative is *acceleration*. If $s(t)$ represents distance as a function of time of a moving object, then $v(t) = s'(t)$ describes the speed (velocity) of the object. Any change in the speed of the object is the acceleration: $a(t) = v'(t) = s''(t)$.
- The common notation for the nth derivative of a function is $f^{(n)}(x)$ or $\dfrac{d^n}{dx^n} f(x)$.

TECHNOLOGY CONNECTION   ⌒

Exploratory

Many calculators have a tangent-drawing feature. This feature can be used to explore the behavior of the second derivative. Graph the function $f(x) = x^2$ in the standard window. Select DRAW and then select Tangent. Choose an x-value by typing it in or using the arrow keys to trace along the graph. Press **ENTER**, and a tangent line will be drawn at the selected x-value. In the lower-left corner of the screen, the equation of the tangent line is given. The slope of the line is the coefficient of x. You can create a table of slope values at various x-values, and from this table, infer how the second derivative helps describe the shape of a graph. (This activity can be easily adapted for Graphicus and iPlot.)

EXERCISES

1. Let $f(x) = x^2$. Use the tangent-drawing feature to complete the following table:

x	Slope at x
-4	
-2	
.0	
.2	
.4	

2. The x-values in the table are increasing. What is the corresponding behavior of the slopes (increasing or decreasing)?

3. How is the graph "turning"? What conclusion can you make about the behavior of the slopes as x increases in value?

4. Make a table of slopes for $f(x) = -x^2$. Analyze their behavior relative to the "turning" of this function's graph.

5. What general conclusion can you make about the second derivative and the "turning" of a graph?

This "turning" behavior of a graph is known as *concavity*, and it is a very useful concept in the analysis of functions. We will develop more about the second derivative and concavity in Chapter 2.

EXERCISE SET
1.8

Find d^2y/dx^2.

1. $y = x^5 + 9$

2. $y = x^4 - 7$

3. $y = 2x^4 - 5x$

4. $y = 5x^3 + 4x$

5. $y = 4x^2 + 3x - 1$

6. $y = 4x^2 - 5x + 7$

7. $y = 7x + 2$

8. $y = 6x - 3$

9. $y = \dfrac{1}{x^2}$

10. $y = \dfrac{1}{x^3}$

11. $y = \sqrt{x}$

12. $y = \sqrt[4]{x}$

Find $f''(x)$.

13. $f(x) = x^4 + \dfrac{3}{x}$

14. $f(x) = x^3 - \dfrac{5}{x}$

15. $f(x) = x^{1/5}$

16. $f(x) = x^{1/3}$

17. $f(x) = 4x^{-3}$

18. $f(x) = 2x^{-2}$

19. $f(x) = (x^2 + 3x)^7$

20. $f(x) = (x^3 + 2x)^6$

21. $f(x) = (2x^2 - 3x + 1)^{10}$

22. $f(x) = (3x^2 + 2x + 1)^5$

23. $f(x) = \sqrt[4]{(x^2 + 1)^3}$

24. $f(x) = \sqrt[3]{(x^2 - 1)^2}$

Find y''.

25. $y = x^{2/3} + 4x$

26. $y = x^{3/2} - 5x$

27. $y = (x^3 - x)^{3/4}$

28. $y = (x^4 + x)^{2/3}$

29. $y = 2x^{5/4} + x^{1/2}$

30. $y = 3x^{4/3} - x^{1/2}$

31. $y = \dfrac{2}{x^3} + \dfrac{1}{x^2}$

32. $y = \dfrac{3}{x^4} - \dfrac{1}{x}$

33. $y = (x^2 + 3)(4x - 1)$

34. $y = (x^3 - 2)(5x + 1)$

35. $y = \dfrac{3x + 1}{2x - 3}$

36. $y = \dfrac{2x + 3}{5x - 1}$

37. For $y = x^4$, find d^4y/dx^4.

38. For $y = x^5$, find d^4y/dx^4.

39. For $y = x^6 - x^3 + 2x$, find d^5y/dx^5.

40. For $y = x^7 - 8x^2 + 2$, find d^6y/dx^6.

41. For $f(x) = x^{-2} - x^{1/2}$, find $f^{(4)}(x)$.

42. For $f(x) = x^{-3} + 2x^{1/3}$, find $f^{(5)}(x)$.

43. For $g(x) = x^4 - 3x^3 - 7x^2 - 6x + 9$, find $g^{(6)}(x)$.

44. For $g(x) = 6x^5 + 2x^4 - 4x^3 + 7x^2 - 8x + 3$, find $g^{(7)}(x)$.

APPLICATIONS

Life and Physical Sciences

45. Given

$$s(t) = t^3 + t,$$

where s is in feet and t is in seconds, find each of the following.

a) $v(t)$ **b)** $a(t)$
c) The velocity and acceleration when $t = 4$ sec

46. Given

$$s(t) = -10t^2 + 2t + 5,$$

where s is in meters and t is in seconds, find each of the following.

a) $v(t)$
b) $a(t)$
c) The velocity and acceleration when $t = 1$ sec

47. Given

$$s(t) = 3t + 10,$$

where s is in miles and t is in hours, find each of the following.

a) $v(t)$ **b)** $a(t)$
c) The velocity and acceleration when $t = 2$ hr
d) When the distance function is given by a linear function, we have *uniform motion*. What does uniform motion mean in terms of velocity and acceleration?

48. Given

$$s(t) = t^2 - \frac{1}{2}t + 3,$$

where s is in meters and t is in seconds, find each of the following.

a) $v(t)$ **b)** $a(t)$
c) The velocity and acceleration when $t = 1$ sec

49. Free fall. When an object is dropped, the distance it falls in t seconds, assuming that air resistance is negligible, is given by

$$s(t) = 16t^2,$$

where $s(t)$ is in feet. Suppose that a medic's reflex hammer is dropped from a hovering helicopter. Find **(a)** how far the hammer falls in 3 sec, **(b)** how fast the hammer is traveling 3 sec after being dropped, and **(c)** the hammer's acceleration after it has been falling for 3 sec.

50. Free fall. (See Exercise 49.) Suppose a worker drops a bolt from a scaffold high above a work site. Assuming that air resistance is negligible, find **(a)** how far the bolt falls in 2 sec; **(b)** how fast the bolt is traveling 2 sec after being dropped, and **(c)** the bolt's acceleration after it has been falling for 2 sec.

51. Free fall. Find the velocity and acceleration of the stone in Example 5 after it has been falling for 2 sec.

52. Free fall. Find the velocity and acceleration of the stone in Example 5 after it has been falling for 3 sec.

53. The following graph describes an airplane's distance from its last point of rest.

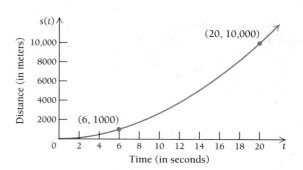

a) Is the plane's velocity greater at $t = 6$ sec or $t = 20$ sec? How can you tell?
b) Is the plane's acceleration positive or negative? How can you tell?

54. The following graph describes a bicycle racer's distance from a roadside television camera.

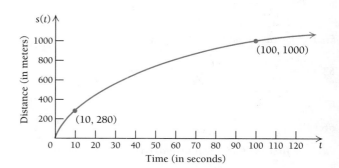

a) When is the bicyclist's velocity the greatest? How can you tell?
b) Is the bicyclist's acceleration positive or negative? How can you tell?

55. Sales. A company determines that monthly sales S, in thousands of dollars, after t months of marketing a product is given by

$$S(t) = 2t^3 - 40t^2 + 220t + 160.$$

a) Find $S'(1)$, $S'(2)$, and $S'(4)$.
b) Find $S''(1)$, $S''(2)$, and $S''(4)$.
c) Interpret the meaning of your answers to parts (a) and (b).

56. Sales. A business discovers that the number of items sold t days after launching a new sales promotion is given by

$$N(t) = 2t^3 - 3t^2 + 2t.$$

a) Find $N'(1)$, $N'(2)$, and $N'(4)$.
b) Find $N''(1)$, $N''(2)$, and $N''(4)$.
c) Interpret the meaning of your answers to parts (a) and (b).

57. Population. The function $p(t) = \dfrac{2000t}{4t + 75}$ models the population p of deer in an area after t months.

a) Find $p'(10)$, $p'(50)$, and $p'(100)$.
b) Find $p''(10)$, $p''(50)$, and $p''(100)$.
c) Interpret the meaning of your answers to parts (a) and (b). What is happening to this population of deer in the long term?

58. Medicine. A medication is injected into the bloodstream, where it is quickly metabolized. The percent concentration p of the medication after t minutes in the bloodstream is modeled by the function $p(t) = \dfrac{2.5t}{t^2 + 1}$.

a) Find $p'(0.5)$, $p'(1)$, $p'(5)$, and $p'(30)$.
b) Find $p''(0.5)$, $p''(1)$, $p''(5)$, and $p''(30)$.
c) Interpret the meaning of your answers to parts (a) and (b). What is happening to the concentration of medication in the bloodstream in the long term?

SYNTHESIS

Find y''' for each function.

59. $y = \dfrac{1}{1 - x}$

60. $y = x\sqrt{1 + x^2}$

61. $y = \dfrac{1}{\sqrt{2x + 1}}$

62. $y = \dfrac{3x - 1}{2x + 3}$

Find y'' for each function.

63. $y = \dfrac{\sqrt{x} + 1}{\sqrt{x} - 1}$

64. $y = \dfrac{x}{\sqrt{x - 1}}$

65. For $y = x^k$, find d^5y/dx^5.

66. For $y = ax^3 + bx^2 + cx + d$, find d^3y/dx^3.

Find the first through the fourth derivatives. Be sure to simplify at each stage before continuing.

67. $f(x) = \dfrac{x - 1}{x + 2}$

68. $f(x) = \dfrac{x + 3}{x - 2}$

69. Baseball. A baseball is dropping from a height of 180 ft. For how many seconds must it fall to reach a speed of 50 mi/hr? (*Hint:* See Exercise 49.)

70. Free fall. All free-fall distance functions follow this form on Earth: $s(t) = 4.905t^2$, where t is in seconds and s is in meters. The second derivative always has the same value. What does that value represent?

71. Free fall. On the moon, all free-fall distance functions are of the form $s(t) = 0.81t^2$, where t is in seconds and s is in meters. An object is dropped from a height of 200 meters above the moon. After $t = 2$ sec,

a) How far has the object fallen?
b) How fast is it traveling?
c) What is its acceleration?
d) Explain the meaning of the second derivative of this free-fall function.

72. Hang time. On Earth, an object will have traveled 4.905 m after 1 sec of free fall. Thus, by symmetry, a jumper requires 1 sec to leap 4.905 m high, then another second to come back to the ground. Assume a jumper starts on level ground. Explain why it is impossible for a human being, even Michael Jordan, to stay in the air for (have a "hang time" of) 2 sec. Can a human have a hang time of 1.5 sec? 1 sec? What do you think is the longest possible hang time achievable by humans jumping from level ground?

73. Free fall. Skateboarder Danny Way free-fell 28 ft from the Fender Stratocaster Guitar atop the Hard Rock Hotel & Casino in Las Vegas onto a ramp below. The distance $s(t)$, in feet, traveled by a body falling freely from rest in t seconds is approximated by $s(t) = 16t^2$. Estimate Way's velocity at the moment he touched down onto the ramp. (*Note:* You will need the result from Exercise 26 in Section R.1.)

TECHNOLOGY CONNECTION · · · · · · · · · ·

For the distance function in each of Exercises 74–77, graph s, v, and a over the given interval. Then use the graphs to determine the point(s) at which the velocity will switch from increasing to decreasing or from decreasing to increasing.

74. $s(t) = 0.1t^4 - t^2 + 0.4;$ $[-5, 5]$

75. $s(t) = -t^3 + 3t;$ $[-3, 3]$

76. $s(t) = t^4 + t^3 - 4t^2 - 2t + 4;$ $[-3, 3]$

77. $s(t) = t^3 - 3t^2 + 2;$ $[-2, 4]$

Answers to Quick Checks

1. (a) (i) $y'' = -72x^2 + 6$,   (ii) $y'' = \dfrac{24}{x^5}$,

(iii) $y'' = 108x^2 + 12$;   **(b)** $y^{(4)} = \dfrac{24}{x^5}$

2. Distance $= s(3.5) = 196$ ft;
velocity $= s'(3.5) = 112$ ft/sec;
acceleration $= s''(3.5) = 32$ ft/sec^2.

KEY TERMS AND CONCEPTS

EXAMPLES

SECTION 1.1

As x approaches (but is not equal to) a, the **limit** of $f(x)$ is L, written as

$$\lim_{x \to a} f(x) = L.$$

$$\lim_{x \to 4} (2x + 3) = 11 \qquad\qquad \lim_{x \to 1} \frac{x^2 - 1}{x - 1} = 2$$

Limit Numerically

	x	$2x + 3$
$x < 4$	3.9	10.8
	3.99	10.98
	3.999	10.998
$x > 4$	4.1	11.2
	4.01	11.02
	4.001	11.002

Limit Graphically

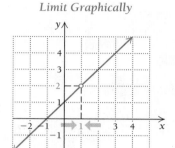

If x approaches a from the left ($x < a$), we have a **left-hand limit**, written as

$$\lim_{x \to a^-} f(x).$$

If x approaches a from the right ($x > a$), we have a **right-hand limit**, written as

$$\lim_{x \to a^+} f(x).$$

If the left-hand and right-hand limits are equal, then the limit as x approaches a exists.

If the left-hand and right-hand limits are *not* equal, then the limit as x approaches a does *not* exist.

Consider the function G given by

$$G(x) = \begin{cases} 4 - x, & \text{for } x < 3, \\ \sqrt{x - 2} + 1, & \text{for } x \geq 3. \end{cases}$$

Graph the function and find each limit, if it exists.

a) $\lim_{x \to 1} G(x)$ \qquad\qquad **b)** $\lim_{x \to 3} G(x)$

We check the limits from the left and from the right, both numerically and graphically.

a) *Limit Numerically*

$x \to 1^- \ (x < 1)$	$G(x)$
0.9	3.1
0.99	3.01
0.999	3.001

$x \to 1^+ \ (x > 1)$	$G(x)$
1.1	2.9
1.01	2.99
1.001	2.999

Limit Graphically

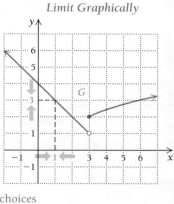

These choices can vary.

Both the tables and the graph show that as x gets closer to 1, the outputs $G(x)$ get closer to 3. Thus, $\lim_{x \to 1} G(x) = 3$.

(continued)

KEY TERMS AND CONCEPTS	**EXAMPLES**

SECTION 1.1 (*continued*)

b)   *Limit Numerically*   *Limit Graphically*

$x \to 3^-$ $(x < 3)$	$G(x)$
2.5	1.5
2.9	1.1
2.99	1.01
2.999	1.001

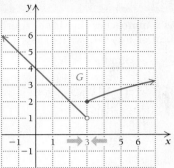

$x \to 3^+$ $(x > 3)$	$G(x)$
3.5	2.2247
3.1	2.0488
3.01	2.0050
3.001	2.0005

$\lim\limits_{x \to 3} G(x)$ does not exist.

Both the tables and the graph indicate that $\lim\limits_{x \to 3^-} G(x) \neq \lim\limits_{x \to 3^+} G(x)$. Since the left-hand and right-hand limits differ, $\lim\limits_{x \to 3} G(x)$ does not exist.

SECTION 1.2

For any rational function F (see Section R.5) with a in its domain, we have

$$\lim_{x \to a} F(x) = F(a).$$

Let $f(x) = 2x^2 + 3x - 1$, and let $a = 2$. We have

$$\lim_{x \to 2} f(x) = f(2) = 2(2)^2 + 3(2) - 1 = 13.$$

Let $g(x) = \dfrac{x^2 - 16}{x + 4}$, and let $a = 6$. We have

$$\lim_{x \to 6} g(x) = \frac{(6)^2 - 16}{(6) + 4} = \frac{20}{10} = 2.$$

A function f is **continuous** at $x = a$ if the following three conditions are met:

1. $f(a)$ exists. (The output at a exists.)
2. $\lim\limits_{x \to a} f(x)$ exists. (The limit as $x \to a$ exists.)
3. $\lim\limits_{x \to a} f(x) = f(a)$. (The limit is the same as the output.)

If any one of these conditions is not fulfilled, the function is **discontinuous** at $x = a$.

Is the function g given by $g(x) = \dfrac{x^2 - 3x - 4}{x + 1}$ continuous over $[-3, 3]$?

For g to be continuous over $[-3, 3]$, it must be continuous at each point in $[-3, 3]$. Note that

$$g(x) = \frac{x^2 - 3x - 4}{x + 1}$$
$$= \frac{(x + 1)(x - 4)}{x + 1}$$
$$= x - 4, \quad \text{provided } x \neq -1.$$

Since -1 is not in the domain of g, it follows that $g(-1)$ does not exist. Thus, g is not continuous over $[-3, 3]$.

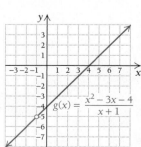

KEY TERMS AND CONCEPTS	EXAMPLES

SECTION 1.3

The **average rate of change** of y with respect to x between two points (x_1, y_1) and (x_2, y_2) is the slope of the line connecting the points:

$$\frac{y_2 - y_1}{x_2 - x_1}.$$

Business. At 1 P.M., a bookstore had revenue of \$570 for the day, and at 4 P.M., it had revenue of \$900 for the day. Therefore, the average rate of revenue with respect to time is $\frac{900 - 570}{4 - 1} = \frac{330}{3} = 110$, or \$110 dollars per hour for the period of time between 1 P.M. and 4 P.M.

Let f be a function. Any line connecting two points on the graph of f is called a **secant line**. Its slope is the average rate of change of f, which is given by the **difference quotient**:

$$\frac{f(x + h) - f(x)}{h},$$

where h is the difference between the two input x-values.

Let $f(x) = 3x^2$. Then

$$f(x + h) = 3(x + h)^2 = 3x^2 + 6xh + 3h^2.$$

The difference quotient for this function simplifies to

$$\frac{f(x + h) - f(x)}{h} = \frac{3x^2 + 6xh + 3h^2 - 3x^2}{h} = 6x + 3h.$$

Therefore, if $x = 2$ and $h = 0.05$, the slope of the secant line is $6(2) + 3(0.05) = 12.15$.

SECTION 1.4

The **derivative** of a function f is defined by

$$f'(x) = \lim_{h \to 0} \frac{f(x + h) - f(x)}{h}.$$

The derivative gives the slope of the tangent line to f at $x = a$, and that slope is interpreted as the **instantaneous rate of change**. The process of finding a derivative is called **differentiation**.

Let $f(x) = 3x^2$. Its simplified difference quotient is $6x + 3h$. Therefore, the derivative is

$$f'(x) = \lim_{h \to 0} (6x + 3h) = 6x.$$

The slope of the tangent line at $x = 2$ is

$$f'(2) = 6(2) = 12.$$

For $f(x) = -x^2 + 5$, find $f'(x)$ and $f'(2)$.

We have

$$\frac{f(x + h) - f(x)}{h} = \frac{-(x + h)^2 + 5 - (-x^2 + 5)}{h}$$

$$= \frac{-(x^2 + 2xh + h^2) + 5 + x^2 - 5}{h}$$

$$= \frac{-2xh - h^2}{h}$$

$$= -2x - h, \quad h \neq 0.$$

Since $f'(x) = \lim\limits_{h \to 0} \dfrac{f(x + h) - f(x)}{h} = \lim\limits_{h \to 0} (-2x - h),$

we have

$$f'(x) = -2x.$$

It follows that $f'(2) = -2 \cdot 2 = -4$.

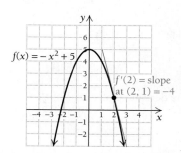

(continued)

KEY TERMS AND CONCEPTS	EXAMPLES

SECTION 1.4 (*continued*)

Continuity:

1. If a function f is differentiable at $x = a$, then it is continuous at $x = a$. (Differentiability implies continuity.)

2. Continuity of a function f at $x = a$ does *not* necessarily mean that f is differentiable at $x = a$. Any function whose graph has a corner is continuous but not differentiable at the corner.

3. If a function f is discontinuous at $x = a$, then it is not differentiable at $x = a$.

1. Let $f(x) = 3x^2$. Since we know the derivative is $f'(x) = 6x$ and the derivative at $x = 2$ exists, we can conclude that $f(x)$ is continuous at $x = 2$.

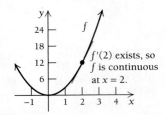

2. The absolute-value function is continuous at $x = 0$ but not differentiable at $x = 0$, since there is a corner at $x = 0$.

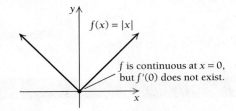

3. The function $g(x) = \dfrac{x^2 - 16}{x + 4}$ is discontinuous at $x = -4$; therefore, the derivative $g'(x)$ is not defined at $x = -4$. (Note that the derivative is defined at other values of x.)

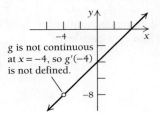

SECTION 1.5

If $y = f(x)$, the derivative in Leibniz notation is written $\dfrac{dy}{dx}$ or $\dfrac{d}{dx} f(x)$. Each has the same meaning as $f'(x)$.

Let $y = x^3$. Then, in Leibniz notation, $\dfrac{dy}{dx} = 3x^2$.

The **Power Rule:**

For any real number k,

$$\frac{d}{dx}x^k = k \cdot x^{k-1}.$$

$$\frac{d}{dx}x^7 = 7x^6$$

$$\frac{d}{dx}\sqrt{x} = \frac{d}{dx}x^{1/2} = \frac{1}{2}x^{-1/2} = \frac{1}{2\sqrt{x}}$$

$$\frac{d}{dx}\left(\frac{1}{x}\right) = \frac{d}{dx}x^{-1} = -1 \cdot x^{-2} = -\frac{1}{x^2}$$

The derivative of a constant is

$$\frac{d}{dx}c = 0.$$

$$\frac{d}{dx}34 = 0$$

$$\frac{d}{dx}\sqrt{2} = 0$$

KEY TERMS AND CONCEPTS	EXAMPLES
The derivative of a constant times a function is $$\frac{d}{dx}[c \cdot f(x)] = c \cdot \frac{d}{dx}f(x).$$	$$\frac{d}{dx}3x^8 = 3 \cdot \frac{d}{dx}x^8 = 3 \cdot 8x^7 = 24x^7$$ $$\frac{d}{dx}\left(\frac{2}{3x^2}\right) = \frac{2}{3} \cdot \frac{d}{dx}\left(\frac{1}{x^2}\right) = \frac{2}{3} \cdot (-2x^{-3}) = -\frac{4}{3x^3}$$
The **Sum–Difference Rule:** $$\frac{d}{dx}\left[f(x) \pm g(x)\right] = \frac{d}{dx}f(x) \pm \frac{d}{dx}g(x)$$	$$\frac{d}{dx}(x^7 + 3x) = 7x^6 + 3$$ $$\frac{d}{dx}(5x - x^4) = 5 - 4x^3$$

SECTION 1.6

The **Product Rule:** $$\frac{d}{dx}[f(x) \cdot g(x)]$$ $$= f(x) \cdot g'(x) + g(x) \cdot f'(x).$$	$$\frac{d}{dx}\left[(2x + 3)\sqrt{x}\right] = (2x + 3)\frac{1}{2}x^{-1/2} + \sqrt{x} \cdot 2$$ $$= \frac{2x + 3}{2\sqrt{x}} + 2\sqrt{x}$$
The **Quotient Rule:** $$\frac{d}{dx}\left[\frac{f(x)}{g(x)}\right] = \frac{g(x) \cdot f'(x) - f(x) \cdot g'(x)}{[g(x)]^2}.$$	$$\frac{d}{dx}\left(\frac{3x - 1}{2x + 5}\right) = \frac{(2x + 5)3 - (3x - 1)2}{(2x + 5)^2} = \frac{17}{(2x + 5)^2}$$

SECTION 1.7

The **Extended Power Rule:** $$\frac{d}{dx}[g(x)]^k = k[g(x)]^{k-1} \cdot \frac{d}{dx}g(x).$$	If $f(x) = (2x^5 + 4x)^7$, then $$f'(x) = 7(2x^5 + 4x)^6(10x^4 + 4).$$ The parentheses are required if the derivative of the "inside" function consists of more than one term.
The **Chain Rule:** $$\frac{d}{dx}\left[(f \circ g)(x)\right] = \frac{d}{dx}\left[f(g(x))\right]$$ $$= f'(g(x)) \cdot g'(x).$$	Let $y = u^3$, where $u = 2x^4 + 7$. Find $\frac{dy}{dx}$. We have $\frac{du}{dx} = 8x^3$ and $\frac{dy}{du} = 3u^2$. Since $\frac{dy}{dx} = \frac{dy}{du} \cdot \frac{du}{dx}$, $$\frac{dy}{dx} = 3u^2 \cdot 8x^3 = 24x^3(2x^4 + 7)^2.$$ Note that the Extended Power Rule is a special case of the Chain Rule.

SECTION 1.8

The **second derivative** is the derivative of the first derivative. $$\frac{d}{dx}[f'(x)] = [f'(x)]' = f''(x).$$ The second derivative describes the rate of change of the derivative.	Let $f(x) = 5x^7 + 20x$. Then $$\frac{d}{dx}f(x) = f'(x) = 35x^6 + 20,$$ and therefore, $$\frac{d^2}{dx^2}f(x) = f''(x) = 210x^5.$$

(continued)

KEY TERMS AND CONCEPTS	EXAMPLES
SECTION 1.8 (*continued*)	

Higher-order derivatives include second, third, fourth, and so on, derivatives of a function. The ***n*th derivative** of a function is written as

$$\frac{d^n y}{dx^n} = f^{(n)}(x).$$

For $y = 5x^7 + 20x$, we have the following:

$$\frac{d^3 y}{dx^3} = f^{(3)}(x) = 1050x^4,$$

$$\frac{d^4 y}{dx^4} = f^{(4)}(x) = 4200x^3,$$

$$\frac{d^5 y}{dx^5} = f^{(5)}(x) = 12{,}600x^2, \text{ and so on.}$$

A real-life application of the second derivative is **acceleration**. If $s(t)$ represents distance as a function of time, then velocity is $v(t) = s'(t)$ and acceleration is the change in velocity: $a(t) = v'(t) = s''(t)$.

Physical Sciences.   A particle moves according to the distance function $s(t) = 5t^3$, where t is in seconds and s in feet. Therefore, its velocity function is $v(t) = s'(t) = 15t^2$ and its acceleration function is $a(t) = v'(t) = s''(t) = 30t$. At $t = 2$ sec, the particle is $s(2) = 40$ ft from the starting point, traveling at $v(2) = 60$ ft/sec and accelerating at $a(2) = 60$ ft/sec^2 (it's speeding up).

CHAPTER 1
REVIEW EXERCISES

These review exercises are for test preparation. They can also be used as a practice test. Answers are at the back of the book. The blue bracketed section references tell you what part(s) of the chapter to restudy if your answer is incorrect.

CONCEPT REINFORCEMENT

Classify each statement as either true or false.

1. If $\lim\limits_{x \to 5} f(x)$ exists, then $f(5)$ must exist. [1.1]

2. If $\lim\limits_{x \to 2} f(x) = L$, then $L = f(2)$. [1.1]

3. If f is continuous at $x = 3$, then $\lim\limits_{x \to 3} f(x) = f(3)$. [1.2]

4. A function's average rate of change over the interval $[2, 8]$ is the same as its instantaneous rate of change at $x = 5$. [1.3, 1.4]

5. A function's derivative at a point, if it exists, can be found as the limit of a difference quotient. [1.4]

6. For $f'(5)$ to exist, f must be continuous at 5. [1.4]

7. If f is continuous at 5, then $f'(5)$ must exist. [1.4]

8. The acceleration function is the derivative of the velocity function. [1.8]

Match each function in column A with the rule in column B that would be the most appropriate to use for differentiating the function. [1.5, 1.6]

Column A

9. $f(x) = x^7$

10. $g(x) = x + 9$

11. $F(x) = (5x - 3)^4$

12. $G(x) = \dfrac{2x + 1}{3x - 4}$

13. $H(x) = f(x) \cdot g(x)$

14. $f(x) = 2x - 7$

Column B

a) Extended Power Rule

b) Product Rule

c) Sum Rule

d) Difference Rule

e) Power Rule

f) Quotient Rule

REVIEW EXERCISES

For Exercises 15–17, consider

$$\lim_{x \to -7} f(x), \quad \text{where } f(x) = \frac{x^2 + 4x - 21}{x + 7}.$$

15. Limit numerically. [1.1]

    a) Complete the following input–output tables.

$x \to -7^-$	$f(x)$
-8	
-7.5	
-7.1	
-7.01	
-7.001	
-7.0001	

$x \to -7^+$	$f(x)$
-6	
-6.5	
-6.9	
-6.99	
-6.999	
-6.9999	

    b) Find $\lim_{x \to -7^-} f(x)$, $\lim_{x \to -7^+} f(x)$, and $\lim_{x \to -7} f(x)$, if each exists.

16. Limit graphically. Graph the function, and use the graph to find the limit. [1.1]

17. Limit algebraically. Find the limit algebraically. Show your work. [1.2]

Find each limit, if it exists. If a limit does not exist, state that fact. [1.1, 1.2]

18. $\lim_{x \to -2} \dfrac{8}{x}$

19. $\lim_{x \to 1} (4x^3 - x^2 + 7x)$

20. $\lim_{x \to -7} \dfrac{x^2 + 2x - 35}{x + 7}$

21. $\lim_{x \to \infty} \dfrac{1}{x} + 3$ [1.1]

From the graphs in Exercises 22 and 23, determine whether each function is continuous and explain why or why not. [1.2]

22.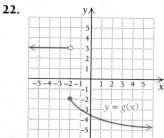

23.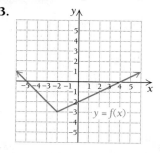

For the function graphed in Exercise 22, answer the following.

24. Find $\lim_{x \to 1} g(x)$. [1.2]

25. Find $g(1)$. [1.2]

26. Is g continuous at 1? Why or why not? [1.2]

27. Find $\lim_{x \to -2} g(x)$. [1.2]

28. Find $g(-2)$. [1.2]

29. Is g continuous at -2? Why or why not? [1.2]

30. For $f(x) = x^3 + x^2 - 2x$, find the average rate of change as x changes from -1 to 2. [1.3]

31. Find a simplified difference quotient for $g(x) = -3x + 2$. [1.3]

32. Find a simplified difference quotient for $f(x) = 2x^2 - 3$. [1.3]

33. Find an equation of the tangent line to the graph of $y = x^2 + 3x$ at the point $(-1, -2)$. [1.4]

34. Find the points on the graph of $y = -x^2 + 8x - 11$ at which the tangent line is horizontal. [1.5]

35. Find the points on the graph of $y = 5x^2 - 49x + 12$ at which the tangent line has slope 1. [1.5]

Find dy/dx.

36. $y = 9x^5$ [1.5]

37. $y = 8\sqrt[3]{x}$ [1.5]

38. $y = \dfrac{-3}{x^8}$ [1.5]

39. $y = 15x^{2/5}$ [1.5]

40. $y = 0.1x^7 - 3x^4 - x^3 + 6$ [1.5]

Differentiate.

41. $f(x) = \dfrac{5}{12}x^6 + 8x^4 - 2x$ [1.5]

42. $y = \dfrac{x^3 + x}{x}$ [1.5, 1.6]

43. $y = \dfrac{x^2 + 8}{8 - x}$ [1.6]

44. $g(x) = (5 - x)^2(2x - 1)^5$ [1.6]

45. $f(x) = (x^5 - 3)^7$ [1.7]

46. $f(x) = x^2(4x + 2)^{3/4}$ [1.7]

47. For $y = x^3 - \dfrac{2}{x}$, find $\dfrac{d^4y}{dx^4}$. [1.8]

48. For $y = \dfrac{3}{42}x^7 - 10x^3 + 13x^2 + 28x - 2$, find y''. [1.8]

49. For $s(t) = t + t^4$, with t in seconds and $s(t)$ in feet, find each of the following. [1.8]

    a) $v(t)$

    b) $a(t)$

    c) The velocity and the acceleration when $t = 2$ sec

50. Business: average revenue, cost, and profit. Given revenue and cost functions $R(x) = 40x$ and $C(x) = 5\sqrt{x} + 100$, find each of the following. Assume $R(x)$ and $C(x)$ are in dollars and x is the number of items produced. [1.6]

    a) The average cost, the average revenue, and the average profit when x items are produced and sold

    b) The rate at which average cost is changing when 9 items are produced

51. Social science: growth rate. The population of a city grows from an initial size of 10,000 to a size P, given by $P = 10{,}000 + 50t^2$, where t is in years. [1.5]

a) Find the growth rate.

b) Find the number of people in the city after 20 yr (at $t = 20$).

c) Find the growth rate at $t = 20$.

52. Find $(f \circ g)(x)$ and $(g \circ f)(x)$, given that $f(x) = x^2 + 5$ and $g(x) = 1 - 2x$. [1.7]

SYNTHESIS

53. Differentiate $y = \dfrac{x\sqrt{1 + 3x}}{1 + x^3}$. [1.7]

TECHNOLOGY CONNECTION

Create an input–output table that includes each of the following limits. Start with ΔTbl $= 0.1$ and then go to 0.01, 0.001, and 0.0001. When you think you know the limit, graph the functions, and use the TRACE feature to verify your assertion.

54. $\displaystyle\lim_{x \to 1} \dfrac{2 - \sqrt{x + 3}}{x - 1}$ [1.1, 1.5]

55. $\displaystyle\lim_{x \to 11} \dfrac{\sqrt{x - 2} - 3}{x - 11}$ [1.1, 1.5]

56. Graph f and f' over the given interval. Then estimate points at which the tangent line to f is horizontal. [1.5]

$$f(x) = 3.8x^5 - 18.6x^3; \quad [-3, 3]$$

CHAPTER 1
TEST

For Exercises 1–3, consider

$$\lim_{x \to 6} f(x), \text{ where } f(x) = \dfrac{x^2 - 36}{x - 6}.$$

1. Numerical limits.

a) Complete the following input–output tables.

$x \to 6^-$	$f(x)$		$x \to 6^+$	$f(x)$
5			7	
5.7			6.5	
5.9			6.1	
5.99			6.01	
5.999			6.001	
5.9999			6.0001	

b) Find $\displaystyle\lim_{x \to 6^-} f(x)$, $\displaystyle\lim_{x \to 6^+} f(x)$, and $\displaystyle\lim_{x \to 6} f(x)$, if each exists.

2. Graphical limits. Graph the function, and use the graph to find the limit.

3. Algebraic limits. Find the limit algebraically. Show your work.

Graphical limits. *Consider the following graph of function f for Exercises 4–11.*

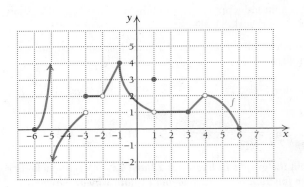

Find each limit, if it exists. If a limit does not exist, state that fact.

4. $\displaystyle\lim_{x \to -5} f(x)$

5. $\displaystyle\lim_{x \to -4} f(x)$

6. $\displaystyle\lim_{x \to -3} f(x)$

7. $\displaystyle\lim_{x \to -2} f(x)$

8. $\displaystyle\lim_{x \to -1} f(x)$

9. $\displaystyle\lim_{x \to 1} f(x)$

10. $\displaystyle\lim_{x \to 2} f(x)$

11. $\displaystyle\lim_{x \to 3} f(x)$

Determine whether each function is continuous. If a function is not continuous, state why.

12.

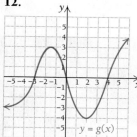

13.

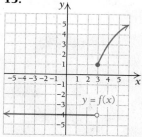

Consider the function shown in Exercise 13.

14. Find $\lim\limits_{x \to 3} f(x)$.

15. Find $f(3)$.

16. Is f continuous at 3?

17. Find $\lim\limits_{x \to 4} f(x)$.

18. Find $f(4)$.

19. Is f continuous at 4?

Find each limit, if it exists. If a limit does not exist, state why.

20. $\lim\limits_{x \to 1} (3x^4 - 2x^2 + 5)$

21. $\lim\limits_{x \to 2^+} \dfrac{x - 2}{x(x^2 - 4)}$

22. $\lim\limits_{x \to 0} \dfrac{7}{x}$

23. Find a simplified difference quotient for
$f(x) = 2x^2 + 3x - 9$.

24. Find an equation of the tangent line to the graph of $y = x + (4/x)$ at the point $(4, 5)$.

25. Find the point(s) on the graph of $y = x^3 - 3x^2$ at which the tangent line is horizontal.

Find dy/dx.

26. $y = x^{23}$

27. $y = 4\sqrt[3]{x} + 5\sqrt{x}$

28. $y = \dfrac{-10}{x}$

29. $y = x^{5/4}$

30. $y = -0.5x^2 + 0.61x + 90$

Differentiate.

31. $y = \dfrac{1}{3}x^3 - x^2 + 2x + 4$

32. $y = \dfrac{3x - 4}{x^3}$

33. $f(x) = \dfrac{x}{5 - x}$

34. $f(x) = (x + 3)^4(7 - x)^5$

35. $y = (x^5 - 4x^3 + x)^{-5}$

36. $f(x) = x\sqrt{x^2 + 5}$

37. For $y = x^4 - 3x^2$, find $\dfrac{d^3y}{dx^3}$.

38. *Business: average revenue, cost, and profit.* Given revenue and cost functions

$$R(x) = 50x \quad \text{and} \quad C(x) = x^{2/3} + 750,$$

where x is the number of items produced and $R(x)$ and $C(x)$ are in dollars, find the following:

a) The average revenue, the average cost, and the average profit when x items are produced

b) The rate at which average cost is changing when 8 items are produced

39. *Social sciences: memory.* In a certain memory experiment, a person is able to memorize M words after t minutes, where $M = -0.001t^3 + 0.1t^2$.

a) Find the rate of change of the number of words memorized with respect to time.

b) How many words are memorized during the first 10 min (at $t = 10$)?

c) At what rate are words being memorized after 10 min (at $t = 10$)?

40. Find $(f \circ g)(x)$ and $(g \circ f)(x)$ for $f(x) = x^2 - x$ and $g(x) = 2x^3$.

SYNTHESIS

41. Differentiate $y = \sqrt{(1 - 3x)^{2/3}(1 + 3x)^{1/3}}$.

42. Find $\lim\limits_{x \to 3} \dfrac{x^3 - 27}{x - 3}$.

TECHNOLOGY CONNECTION

43. Graph f and f' over the interval $[0, 5]$. Then estimate points at which the tangent line to f is horizontal.

$$f(x) = 5x^3 - 30x^2 + 45x + 5\sqrt{x}; \quad [0, 5]$$

44. Find the following limit by creating a table of values:

$$\lim\limits_{x \to 0} \dfrac{\sqrt{5x + 25} - 5}{x}.$$

Start with ΔTbl $= 0.1$ and then go to 0.01 and 0.001. When you think you know the limit, graph

$$y = \dfrac{\sqrt{5x + 25} - 5}{x},$$

and use the TRACE feature to verify your assertion.

Path of a Baseball: The Tale of the Tape

Have you ever watched a baseball game and seen a home run ball hit an obstruction after it has cleared the fence? Suppose the ball hits a sign at a location that is 60 ft above the ground at a distance of 400 ft from home plate. An announcer or a message on the scoreboard might proclaim, "According to the tale of the tape, the ball would have traveled 442 ft." How is such a calculation made? The answer is related to the curve formed by the path of a baseball.

Whatever the path of a well-hit baseball is, it is *not* the graph of a parabola,

$$f(x) = ax^2 + bx + c.$$

A well-hit baseball follows the path of a "skewed" parabola, as shown at the lower right. One reason that the ball's flight is not parabolic is that a well-hit ball has backspin. This fact, combined with the frictional effect of the ball's stitches with the air, skews the path of the ball in the direction of its landing.

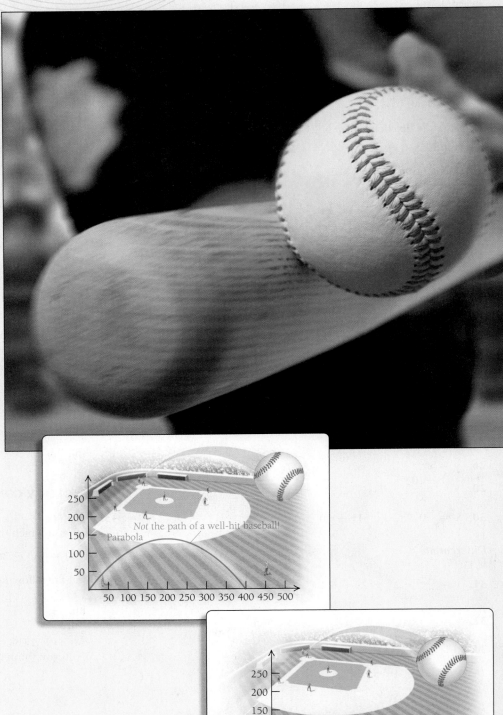

Let's see if we can model the path of a baseball. Consider the following data.

HORIZONTAL DISTANCE, x (IN FEET)	VERTICAL DISTANCE, y (IN FEET)
0	4.5
50	43
100	82
200	130
285	142
300	134
360	100
400	60

Assume for the given data that $(0, 4.5)$ is the point at home plate at which the ball is hit, roughly 4.5 ft off the ground. Also, assume that the ball has hit a billboard 60 ft above the ground and 400 ft from home plate.

EXERCISES

1. Plot the points and connect them with line segments. This can be done on many calculators by pressing STAT PLOT, turning on PLOT, and selecting the appropriate type.

2. **a)** Use REGRESSION to find a cubic function
$$y = ax^3 + bx^2 + cx + d$$
that fits the data.
   b) Graph the function over the interval $[0, 500]$.
   c) Does the function closely model the given data?
   d) Predict the horizontal distance from home plate at which the ball would have hit the ground had it not hit the billboard.

e) Find the rate of change of the ball's height with respect to its horizontal distance from home plate.
f) Find the point(s) at which the graph has a horizontal tangent line. Explain the significance of the point(s).

3. **a)** Use REGRESSION to find a quartic function
$$y = ax^4 + bx^3 + cx^2 + dx + e$$
that fits the data.
   b) Graph the function over the interval $[0, 500]$.
   c) Does the function closely model the given data?
   d) Predict the horizontal distance from home plate at which the ball would have hit the ground had it not hit the billboard.
   e) Find the rate of change of the ball's height with respect to its horizontal distance from home plate.
   f) Find the point(s) at which the graph has a horizontal tangent line. Explain the significance of the point(s).

4. **a)** Although most calculators cannot fit such a function to the data, assume that the equation
$$y = 0.0015x\sqrt{202,500 - x^2}$$
has been found using some type of curve-fitting technique. Graph the function over the interval $[0, 500]$.
   b) Predict the horizontal distance from home plate at which the ball would have hit the ground had it not hit the billboard.
   c) Find the rate of change of the ball's height with respect to its horizontal distance from home plate.
   d) Find the point(s) at which the graph has a horizontal tangent line. Explain the significance of the point(s).

5. Compare the answers in Exercises 2(d), 3(d), and 4(b). Discuss the relative merits of using the quartic model in Exercise 3 with the model in Exercise 4 to make the prediction.

Tale of the tape.   Actually, scoreboard operators in the major leagues use different models to predict the distance that a home run ball would have traveled. The models are linear and are related to the trajectory of the ball, that is, how high the ball is hit. See the following graph.

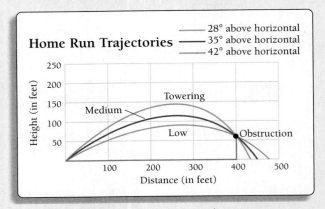

Suppose that a ball hits an obstruction d feet horizontally from home plate at a height of H feet. Then the estimated horizontal distance D that the ball would have traveled, depending on its trajectory type, is

Low trajectory:	$D = 1.1H + d,$
Medium trajectory:	$D = 0.7H + d,$
Towering trajectory:	$D = 0.5H + d.$

6. For a ball striking an obstacle at $d = 400$ ft and $H = 60$ ft, estimate how far the ball would have traveled if it were following a low trajectory, or a medium trajectory, or a towering trajectory.

7. In 1953, Hall-of-Famer Mickey Mantle hit a towering home run in old Griffith Stadium in Washington, D.C., that hit an obstruction 60 ft high and 460 ft from home plate. Reporters asserted at the time that the ball would have traveled 565 ft. Is this estimate valid?

8. Use the appropriate formula to estimate the distance D for each the following famous long home runs.

a) Ted Williams (Boston Red Sox, June 9, 1946): Purportedly the longest home run ball ever hit to right field at Boston's Fenway Park, Williams's ball landed in the stands 502 feet from home plate, 30 feet above the ground. Assume a medium trajectory.

b) Reggie Jackson (Oakland Athletics, July 13, 1971): Jackson's mighty blast hit an electrical transformer on top of the right-field roof at old Tiger Stadium in the 1971 All-Star Game. The transformer was 380 feet from home plate, 100 feet up. Assume a towering trajectory. Jackson's home run was reported to still be on the upward arc when it hit the transformer.

c) Richie Sexson (Arizona Diamondbacks, April 26, 2004): Sexson hit a drive that caromed off the center-field scoreboard at Bank One Ballpark in Phoenix. The scoreboard is 414 feet from home plate and 75 feet high. Assume a medium trajectory.

The reported distances these balls would have traveled are 527 feet for Williams's home run, 530 feet for Jackson's, and 469 feet for Sexson's (*Source:* www.hittrackeronline.com). How close are your estimates?

The Babe Would Be Proud

Many thanks to Robert K. Adair, professor of physics at Yale University, for many of the ideas presented in this application.

Applications of Differentiation

2

Chapter Snapshot

What You'll Learn

2.1 Using First Derivatives to Find Maximum and Minimum Values and Sketch Graphs

2.2 Using Second Derivatives to Find Maximum and Minimum Values and Sketch Graphs

2.3 Graph Sketching: Asymptotes and Rational Functions

2.4 Using Derivatives to Find Absolute Maximum and Minimum Values

2.5 Maximum–Minimum Problems; Business and Economics Applications

2.6 Marginals and Differentials

2.7 Implicit Differentiation and Related Rates

Why It's Important

In this chapter, we explore many applications of differentiation. We learn to find maximum and minimum values of functions, and that skill allows us to solve many kinds of problems in which we need to find the largest and/or smallest value in a real-world situation. We also apply our differentiation skills to graphing and to approximating function values.

Where It's Used

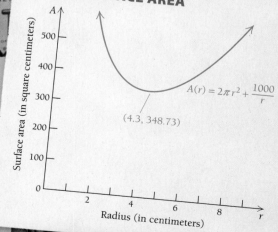

MINIMIZING MATERIAL USED

Minimizing the amount of material used is a common goal in manufacturing, as it reduces overall costs as well as increases efficiency. For example, cylindrical food cans come in a variety of sizes. Suppose a can is to have a volume of 500 milliliters. Are there optimal dimensions for the can's height and radius that will minimize the material needed to produce each can? Can you see how minimizing the material used per can translates into minimized costs and conservation of resources?

This problem appears as Example 3 in Section 2.5.

RELATING THE RADIUS OF A CAN TO ITS SURFACE AREA

$$A(r) = 2\pi r^2 + \frac{1000}{r}$$

(4.3, 348.73)

Surface area (in square centimeters)

Radius (in centimeters)

2.1

Using First Derivatives to Find Maximum and Minimum Values and Sketch Graphs

The graph below shows a typical life cycle of a retail product and is similar to graphs we will consider in this chapter. Note that the number of items sold varies with respect to time. Sales begin at a small level and increase to a point of maximum sales, after which they taper off to a low level, where the decline is probably due to the effect of new competitive products. The company then rejuvenates the product by making improvements. Think about versions of certain products: televisions can be traditional, flat-screen, or high-definition; music recordings have been produced as phonograph (vinyl) records, audiotapes, compact discs, and MP3 files. Where might each of these products be in a typical product life cycle? Does the curve seem appropriate for each product?

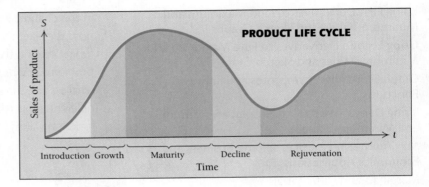

Finding the largest and smallest values of a function—that is, the maximum and minimum values—has extensive applications. The first and second derivatives of a function are calculus tools that provide information we can use in graphing functions and finding minimum and maximum values. Throughout this section we will assume, unless otherwise noted, that all functions are continuous. However, continuity of a function does not guarantee that its first and second derivatives are continuous.

Increasing and Decreasing Functions

If the graph of a function rises from left to right over an interval I, the function is said to be **increasing** on, or over, I.

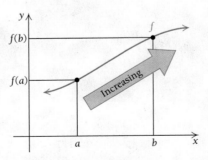

f is an increasing function over I:
for all a, b in I, if $a < b$, then $f(a) < f(b)$.

Exploratory

Graph the function

$$y = -\tfrac{1}{3}x^3 + 6x^2 - 11x - 50$$

and its derivative

$$y' = -x^2 + 12x - 11$$

using the window $[-10, 25, -100, 150]$, with Xscl = 5 and Yscl = 25. Then TRACE from left to right along each graph. As you move the cursor from left to right, note that the x-coordinate always increases. If a function is increasing over an interval, the y-coordinate will increase as well. If a function is decreasing over an interval, the y-coordinate will decrease.

Over what intervals is the function increasing?
Over what intervals is the function decreasing?
Over what intervals is the derivative positive?
Over what intervals is the derivative negative?

What rules can you propose relating the sign of y' to the behavior of y?

If the graph drops from left to right, the function is said to be **decreasing** on, or over, I.

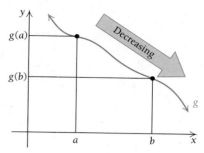

g is a decreasing function over I:
for all a, b in I, if $a < b$, then $g(a) > g(b)$.

We can describe these phenomena mathematically as follows.

DEFINITIONS

A function f is **increasing** over I if, for every a and b in I,

$$\text{if } a < b, \quad \text{then } f(a) < f(b).$$

(If the input a is less than the input b, then the output for a is less than the output for b.)

A function f is **decreasing** over I if, for every a and b in I,

$$\text{if } a < b, \quad \text{then } f(a) > f(b).$$

(If the input a is less than the input b, then the output for a is greater than the output for b.)

The above definitions can be restated in terms of secant lines. If a graph is increasing over an interval I, then, for all a and b in I such that $a < b$, the slope of the secant line between $x = a$ and $x = b$ is positive. Similarly, if a graph is decreasing over an interval I, then, for all a and b in I such that $a < b$, the slope of the secant line between $x = a$ and $x = b$ is negative:

Increasing: $\dfrac{f(b) - f(a)}{b - a} > 0.$ Decreasing: $\dfrac{f(b) - f(a)}{b - a} < 0.$

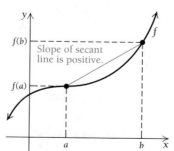
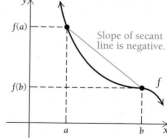

The following theorem shows how we can use the derivative (the slope of a tangent line) to determine whether a function is increasing or decreasing.

THEOREM 1

If $f'(x) > 0$ for all x in an open interval I, then f is increasing over I.
If $f'(x) < 0$ for all x in an open interval I, then f is decreasing over I.

Theorem 1 is illustrated in the following graph.

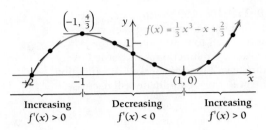

f is increasing over the intervals $(-\infty, -1)$ and $(1, \infty)$; slopes of tangent lines are positive.
f is decreasing over the interval $(-1, 1)$; slopes of tangent lines are negative.

For determining increasing or decreasing behavior using a derivative, the interval I is an open interval; that is, it does not include its endpoints. Note how the intervals on which f is increasing and decreasing are written in the preceding graph: $x = -1$ and $x = 1$ are not included in any interval over which the function is increasing or decreasing. These values are examples of *critical values*.

Critical Values

Consider the graph of a continuous function f in Fig. 1.

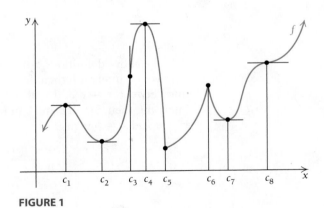

FIGURE 1

Note the following:

1. $f'(c) = 0$ at $x = c_1, c_2, c_4, c_7,$ and c_8. That is, the tangent line to the graph is horizontal for these values.
2. $f'(c)$ does not exist at $x = c_3, c_5,$ and c_6. The tangent line is vertical at c_3, and there are corner points at both c_5 and c_6. (See also the discussion at the end of Section 1.4.)

DEFINITION

A **critical value** of a function f is any number c in the domain of f for which the tangent line at $(c, f(c))$ is horizontal or for which the derivative does not exist. That is, c is a critical value if $f(c)$ exists and

$$f'(c) = 0 \quad \text{or} \quad f'(c) \text{ does not exist.}$$

Thus, in the graph of f in Fig. 1:

1. $c_1, c_2, c_4, c_7,$ and c_8 are critical values because $f'(c) = 0$ for each value.
2. $c_3, c_5,$ and c_6 are critical values because $f'(c)$ does not exist for each value.

Also note that a continuous function can change from increasing to decreasing or from decreasing to increasing *only* at a critical value. In the graph in Fig. 1, $c_1, c_2, c_4, c_5, c_6,$ and c_7 separate the intervals over which the function changes from increasing to decreasing or from decreasing to increasing. Although c_3 and c_8 are critical values, they do not separate intervals over which the function changes from increasing to decreasing or from decreasing to increasing.

Finding Relative Maximum and Minimum Values

Now consider the graph in Fig. 2. Note the "peaks" and "valleys" at the interior points $c_1, c_2,$ and c_3.

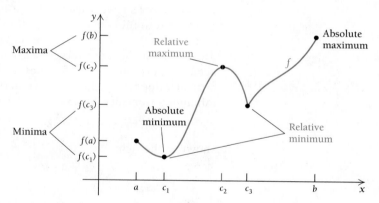

FIGURE 2

Here $f(c_2)$ is an example of a **relative maximum** (plural: **maxima**). Each of $f(c_1)$ and $f(c_3)$ is called a **relative minimum** (plural: **minima**). The terms **local maximum** and **local minimum** are also used.

DEFINITIONS

Let I be the domain of f.

$f(c)$ is a **relative minimum** if there exists within I an open interval I_1 containing c such that $f(c) \leq f(x)$, for all x in I_1;

and

$f(c)$ is a **relative maximum** if there exists within I an open interval I_2 containing c such that $f(c) \geq f(x)$, for all x in I_2.

A relative maximum can be thought of loosely as the second coordinate of a "peak" that may or may not be the highest point over all of I. Similarly, a relative minimum can be thought of as the second coordinate of a "valley" that may or may not be the lowest point on I. The second coordinates of the points that are the highest and the lowest on the interval are, respectively, the **absolute maximum** and the **absolute minimum**. For now, we focus on finding relative maximum or minimum values, collectively referred to as **relative extrema** (singular: **extremum**).

Look again at the graph in Fig. 2. The *x*-values at which a continuous function has relative extrema are those values for which the derivative is 0 or for which the derivative does not exist—the critical values.

THEOREM 2

If a function *f* has a relative extreme value $f(c)$ on an open interval, then *c* is a critical value, so

$$f'(c) = 0 \quad \text{or} \quad f'(c) \text{ does not exist.}$$

A *relative extreme point,* $(c, f(c))$, is higher or lower than all other points over some open interval containing *c*. A relative minimum point will have a *y*-value that is lower than that of points both to the left and to the right of it, and, similarly, a relative maximum point will have a *y*-value that is higher than that of points to the left and right of it. Thus, relative extrema cannot be located at the endpoints of a closed interval, since an endpoint lacks "both sides" with which to make the necessary comparisons. However, as we will see in Section 2.4, endpoints *can* be absolute extrema. Note that the right endpoint of the curve in Fig. 2 is the absolute maximum point.

Theorem 2 is very useful, but it is important to understand it precisely. What it says is that to find relative extrema, we need only consider those inputs for which the derivative is 0 or for which it does not exist. We can think of a critical value as a *candi-date* for a value where a relative extremum *might* occur. That is, Theorem 2 does not say that every critical value will yield a relative maximum or minimum. Consider, for example, the graph of

$$f(x) = (x - 1)^3 + 2,$$

shown at the right. Note that

$$f'(x) = 3(x - 1)^2,$$

and

$$f'(1) = 3(1 - 1)^2 = 0.$$

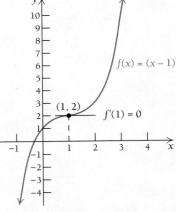

The function has $c = 1$ as a critical value, but has no relative maximum or minimum at that value.

Theorem 2 does say that if a relative maximum or minimum occurs, then the first coordinate of that extremum will be a critical value. How can we tell when the existence of a critical value leads us to a relative extremum? The following graph leads us to a test.

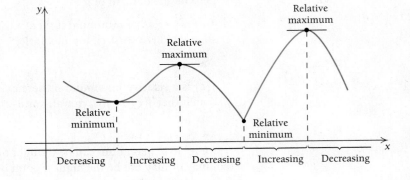

Note that at a critical value where there is a relative minimum, the function is decreasing to the left of the critical value and increasing to the right. At a critical value where there is a relative maximum, the function is increasing to the left of the critical value and decreasing to the right. In both cases, the derivative changes signs on either side of the critical value.

Graph over the interval (a, b)	$f(c)$	Sign of $f'(x)$ for x in (a, c)	Sign of $f'(x)$ for x in (c, b)	Increasing or decreasing
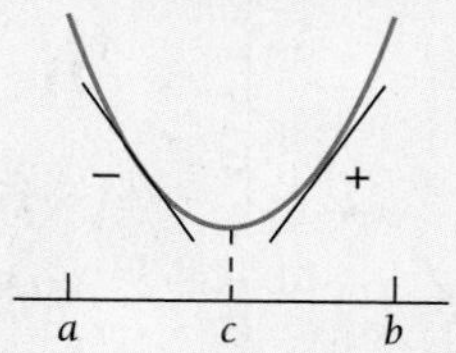	Relative minimum	−	+	Decreasing on (a, c); increasing on (c, b)
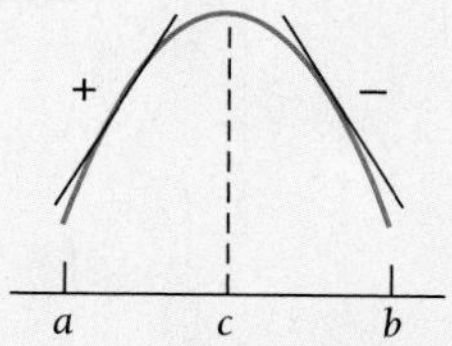	Relative maximum	+	−	Increasing on (a, c); decreasing on (c, b)
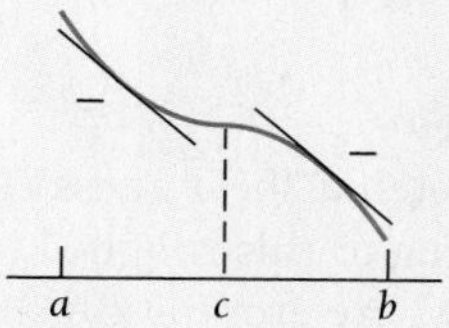	No relative maxima or minima	−	−	Decreasing on (a, b)
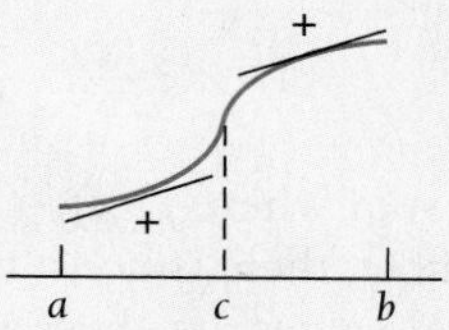	No relative maxima or minima	+	+	Increasing on (a, b)

Derivatives tell us when a function is increasing or decreasing. This leads us to the First-Derivative Test.

THEOREM 3   **The First-Derivative Test for Relative Extrema**

For any continuous function f that has exactly one critical value c in an open interval (a, b):

F1. f has a relative minimum at c if $f'(x) < 0$ on (a, c) and $f'(x) > 0$ on (c, b). That is, f is decreasing to the left of c and increasing to the right of c.

F2. f has a relative maximum at c if $f'(x) > 0$ on (a, c) and $f'(x) < 0$ on (c, b). That is, f is increasing to the left of c and decreasing to the right of c.

F3. f has neither a relative maximum nor a relative minimum at c if $f'(x)$ has the same sign on (a, c) as on (c, b).

Now let's see how we can use the First-Derivative Test to find relative extrema and create accurate graphs.

■ **EXAMPLE 1**   Graph the function f given by

$$f(x) = 2x^3 - 3x^2 - 12x + 12,$$

and find the relative extrema.

Solution   Suppose that we are trying to graph this function but don't know any calculus. What can we do? We could plot several points to determine in which direction the graph seems to be turning. Let's pick some x-values and see what happens.

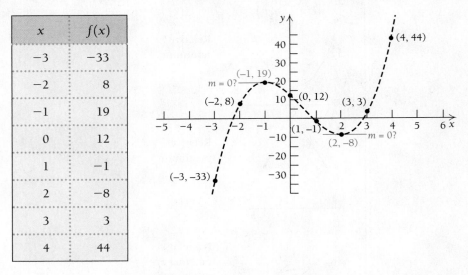

x	$f(x)$
-3	-33
-2	8
-1	19
0	12
1	-1
2	-8
3	3
4	44

We plot the points and use them to sketch a "best guess" of the graph, shown as the dashed line in the figure above. According to this rough sketch, it appears that the graph has a tangent line with slope 0 somewhere around $x = -1$ and $x = 2$. But how do we know for sure? We use calculus to support our observations. We begin by finding a general expression for the derivative:

$$f'(x) = 6x^2 - 6x - 12.$$

We next determine where $f'(x)$ does not exist or where $f'(x) = 0$. Since we can evaluate $f'(x) = 6x^2 - 6x - 12$ for any real number, there is no value for which $f'(x)$ does not exist. So the only possibilities for critical values are those where $f'(x) = 0$, locations at which there are horizontal tangents. To find such values, we solve $f'(x) = 0$:

$$6x^2 - 6x - 12 = 0$$
$$x^2 - x - 2 = 0 \qquad \text{Dividing both sides by 6}$$
$$(x + 1)(x - 2) = 0 \qquad \text{Factoring}$$
$$x + 1 = 0 \quad \text{or} \quad x - 2 = 0 \qquad \text{Using the Principle of Zero Products}$$
$$x = -1 \quad \text{or} \qquad x = 2.$$

The critical values are -1 and 2. Since it is at these values that a relative maximum or minimum might exist, we examine the intervals on each side of the critical values: A is $(-\infty, -1)$, B is $(-1, 2)$, and C is $(2, \infty)$, as shown below.

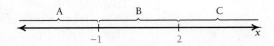

Next, we analyze the sign of the derivative on each interval. If $f'(x)$ is positive for one value in the interval, then it will be positive for all values in the interval. Similarly, if it is negative for one value, it will be negative for all values in the interval. Thus, we choose a test value in each interval and make a substitution. The test values we choose are -2, 0, and 4.

A: Test -2, $f'(-2) = 6(-2)^2 - 6(-2) - 12$
$$= 24 + 12 - 12 = 24 > 0;$$

B: Test 0, $f'(0) = 6(0)^2 - 6(0) - 12 = -12 < 0;$

C: Test 4, $f'(4) = 6(4)^2 - 6(4) - 12$
$$= 96 - 24 - 12 = 60 > 0.$$

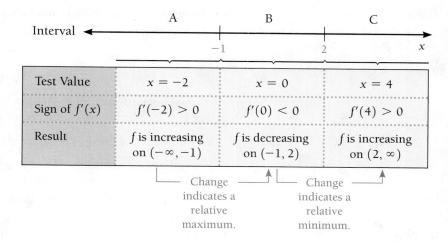

	A	B	C
Test Value	$x = -2$	$x = 0$	$x = 4$
Sign of $f'(x)$	$f'(-2) > 0$	$f'(0) < 0$	$f'(4) > 0$
Result	f is increasing on $(-\infty, -1)$	f is decreasing on $(-1, 2)$	f is increasing on $(2, \infty)$

Change indicates a relative maximum.   Change indicates a relative minimum.

Therefore, by the First-Derivative Test,

f has a relative maximum at $x = -1$ given by

$$f(-1) = 2(-1)^3 - 3(-1)^2 - 12(-1) + 12$$ *Substituting into the original function*

$$= 19$$ This is a relative maximum.

and f has a relative minimum at $x = 2$ given by

$$f(2) = 2(2)^3 - 3(2)^2 - 12(2) + 12 = -8.$$ This is a relative minimum.

Thus, there is a relative maximum at $(-1, 19)$ and a relative minimum at $(2, -8)$, as we suspected from the sketch of the graph.

The information we have obtained from the first derivative can be very useful in sketching a graph of the function. We know that this polynomial is continuous, and we know where the function is increasing, where it is decreasing, and where it has relative extrema. We complete the graph by using a calculator to generate some additional function values. The graph of the function, shown below in red, has been scaled to clearly show its curving nature.

 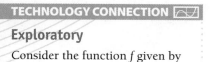

Exploratory

Consider the function f given by

$$f(x) = x^3 - 3x + 2.$$

Graph both f and f' using the same set of axes. Examine the graphs using the TABLE and TRACE features. Where do you think the relative extrema of $f(x)$ occur? Where is the derivative equal to 0? Where does $f(x)$ have critical values?

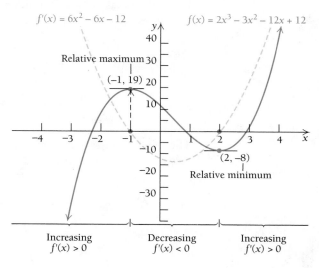

For reference, the graph of the derivative is shown in blue. Note that $f'(x) = 0$ where $f(x)$ has relative extrema. We summarize the behavior of this function as follows, by noting where it is increasing or decreasing, and by characterizing its critical points:

- The function f is increasing over the interval $(-\infty, -1)$.
- The function f has a relative maximum at the point $(-1, 19)$.
- The function f is decreasing over the interval $(-1, 2)$.
- The function f has a relative minimum at the point $(2, -8)$.
- The function f is increasing over the interval $(2, \infty)$.

❮ Quick Check 1

> **Quick Check 1**
>
> Graph the function g given by $g(x) = x^3 - 27x - 6$, and find the relative extrema.

Interval notation and point notation look alike. Be clear when stating your answers whether you are identifying an interval or a point.

To use the first derivative for graphing a function f:

1. Find all critical values by determining where $f'(x)$ is 0 and where $f'(x)$ is undefined (but $f(x)$ is defined). Find $f(x)$ for each critical value.

2. Use the critical values to divide the x-axis into intervals and choose a test value in each interval.

3. Find the sign of $f'(x)$ for each test value chosen in step 2, and use this information to determine where $f(x)$ is increasing or decreasing and to classify any extrema as relative maxima or minima.

4. Plot some additional points and sketch the graph.

The *derivative* f' is used to find the critical values of f. The test values are substituted into the *derivative* f', and the function values are found using the *original* function f.

■ **EXAMPLE 2** Find the relative extrema of the function f given by

$$f(x) = 2x^3 - x^4.$$

Then sketch the graph.

Solution First, we must determine the critical values. To do so, we find $f'(x)$:

$$f'(x) = 6x^2 - 4x^3.$$

Next, we find where $f'(x)$ does not exist or where $f'(x) = 0$. Since $f'(x) = 6x^2 - 4x^3$ is a polynomial, it exists for all real numbers x. Therefore, the only candidates for critical values are where $f'(x) = 0$, that is, where the tangent line is horizontal:

$$\begin{aligned}
6x^2 - 4x^3 &= 0 && \text{Setting } f'(x) \text{ equal to 0} \\
2x^2(3 - 2x) &= 0 && \text{Factoring} \\
2x^2 = 0 \quad &\text{or} \quad 3 - 2x = 0 \\
x^2 = 0 \quad &\text{or} \qquad\quad 3 = 2x \\
x = 0 \quad &\text{or} \qquad\quad x = \tfrac{3}{2}.
\end{aligned}$$

The critical values are 0 and $\tfrac{3}{2}$. We use these values to divide the x-axis into three intervals as shown below: A is $(-\infty, 0)$; B is $(0, \tfrac{3}{2})$; and C is $(\tfrac{3}{2}, \infty)$.

Note that $f\left(\tfrac{3}{2}\right) = 2\left(\tfrac{3}{2}\right)^3 - \left(\tfrac{3}{2}\right)^4 = \tfrac{27}{16}$ and $f(0) = 2 \cdot 0^3 - 0^4 = 0$ are possible extrema.

We now determine the sign of the derivative on each interval by choosing a test value in each interval and substituting. We generally choose test values for which it is easy to compute $f'(x)$.

A: Test -1, $f'(-1) = 6(-1)^2 - 4(-1)^3$
$= 6 + 4 = 10 > 0;$

B: Test 1, $f'(1) = 6(1)^2 - 4(1)^3$
$= 6 - 4 = 2 > 0;$

C: Test 2, $f'(2) = 6(2)^2 - 4(2)^3$
$= 24 - 32 = -8 < 0.$

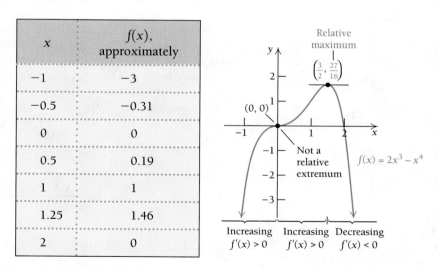

Test Value	$x = -1$	$x = 1$	$x = 2$
Sign of $f'(x)$	$f'(-1) > 0$	$f'(1) > 0$	$f'(2) < 0$
Result	f is increasing on $(-\infty, 0)$	f is increasing on $\left(0, \frac{3}{2}\right)$	f is decreasing on $\left(\frac{3}{2}, \infty\right)$

— No change — — Change — indicates a relative maximum.

Therefore, by the First-Derivative Test, f has no extremum at $x = 0$ (since $f(x)$ is increasing on both sides of 0) and has a relative maximum at $x = \frac{3}{2}$. Thus, $f\left(\frac{3}{2}\right)$, or $\frac{27}{16}$, is a relative maximum.

We use the information obtained to sketch the graph below. Other function values are listed in the table.

x	$f(x)$, approximately
-1	-3
-0.5	-0.31
0	0
0.5	0.19
1	1
1.25	1.46
2	0

We summarize the behavior of f:

- The function f is increasing over the interval $(-\infty, 0)$.
- The function f has a critical point at $(0, 0)$, which is neither a minimum nor a maximum.

- The function f is increasing over the interval $\left(0, \frac{3}{2}\right)$.
- The function f has relative maximum at the point $\left(\frac{3}{2}, \frac{27}{16}\right)$.
- The function f is decreasing over the interval $\left(\frac{3}{2}, \infty\right)$.

Since f is increasing over the intervals $(-\infty, 0)$ and $\left(0, \frac{3}{2}\right)$, we can say that f is increasing over $\left(-\infty, \frac{3}{2}\right)$ despite the fact that $f'(0) = 0$ within this interval. In this case, we can observe that any secant line connecting two points within this interval will have a positive slope.

❱ **Quick Check 2**

Find the relative extrema of the function h given by $h(x) = x^4 - \frac{8}{3}x^3$. Then sketch the graph.

❰ Quick Check 2

◼ **EXAMPLE 3**   Find the relative extrema of the function f given by

$$f(x) = (x - 2)^{2/3} + 1.$$

Then sketch the graph.

Solution   First, we determine the critical values. To do so, we find $f'(x)$:

$$f'(x) = \frac{2}{3}(x - 2)^{-1/3}$$

$$= \frac{2}{3\sqrt[3]{x - 2}}.$$

Next, we find where $f'(x)$ does not exist or where $f'(x) = 0$. Note that $f'(x)$ does not exist at 2, although $f(x)$ does. Thus, 2 is a critical value. Since the only way for a fraction to be 0 is if its numerator is 0, we see that $f'(x) = 0$ has no solution. Thus, 2 is the only critical value. We use 2 to divide the x-axis into the intervals A, which is $(-\infty, 2)$, and B, which is $(2, \infty)$. Note that $f(2) = (2 - 2)^{2/3} + 1 = 1$.

To determine the sign of the derivative, we choose a test value in each interval and substitute each value into the derivative. We choose test values 0 and 3. It is not necessary to find an exact value of the derivative; we need only determine the sign. Sometimes we can do this by just examining the formula for the derivative:

A: Test 0, $f'(0) = \dfrac{2}{3\sqrt[3]{0 - 2}} < 0$;

B: Test 3, $f'(3) = \dfrac{2}{3\sqrt[3]{3 - 2}} > 0.$

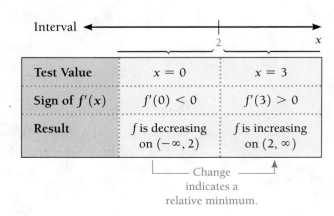

	Test Value	$x = 0$	$x = 3$
	Sign of $f'(x)$	$f'(0) < 0$	$f'(3) > 0$
	Result	f is decreasing on $(-\infty, 2)$	f is increasing on $(2, \infty)$

Change indicates a relative minimum.

Since we have a change from decreasing to increasing, we conclude from the First-Derivative Test that a relative minimum occurs at $(2, f(2))$, or $(2, 1)$. The graph has *no* tangent line at $(2, 1)$ since $f'(2)$ does not exist.

We use the information obtained to sketch the graph. Other function values are listed in the table.

x	$f(x)$, approximately
-1	3.08
-0.5	2.84
0	2.59
0.5	2.31
1	2
1.5	1.63
2	1
2.5	1.63
3	2
3.5	2.31
4	2.59

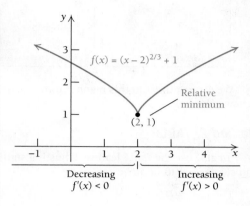

We summarize the behavior of f:

> **Quick Check 3**
>
> Find the relative extrema of the function g given by $g(x) = 3 - x^{1/3}$. Then sketch the graph.

- The function f is decreasing over the interval $(-\infty, 2)$.
- The function f has a relative minimum at the point $(2, 1)$.
- The function f is increasing over the interval $(2, \infty)$.

❮ Quick Check 3

TECHNOLOGY CONNECTION

Finding Relative Extrema

To explore some methods for approximating relative extrema, let's find the relative extrema of

$$f(x) = -0.4x^3 + 6.2x^2 - 11.3x - 54.8.$$

We first graph the function, using a window that reveals the curvature.

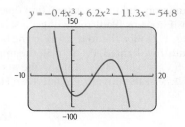

Method 1: TRACE

Beginning with the window shown at left, we press **TRACE** and move the cursor along the curve, noting where relative extrema might occur.

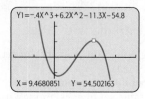

A relative maximum seems to be about $y = 54.5$ at $x = 9.47$. We can refine the approximation by zooming in to obtain the following window. We press **TRACE** and move

(continued)

Finding Relative Extrema (*continued*)

the cursor along the curve, again noting where the y-value is largest. The approximation is about $y = 54.61$ at $x = 9.31$.

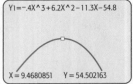 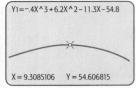

We can continue in this manner until the desired accuracy is achieved.

Method 2: TABLE

We can also use the TABLE feature, adjusting starting points and step values to improve accuracy:

$$\text{TblStart} = 9.3 \quad \Delta\text{Tbl} = .01$$

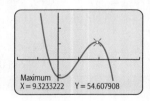

The approximation seems to be nearly $y = 54.61$ at an x-value between 9.32 and 9.33. We could next set up a new table showing function values between $f(9.32)$ and $f(9.33)$ to refine the approximation.

Method 3: MAXIMUM, MINIMUM

Using the MAXIMUM option from the CALC menu, we find that a relative maximum of about 54.61 occurs at $x \approx 9.32$.

Method 4: fMax *or* fMin

This feature calculates a relative maximum or minimum value over any specified closed interval. We see from the initial graph that a relative maximum occurs in the interval $[-10, 20]$. Using the fMax option from the MATH menu, we see that a relative maximum occurs on $[-10, 20]$ when $x \approx 9.32$.

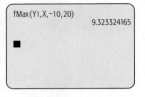

To obtain the maximum value, we evaluate the function at the given x-value, obtaining the following.

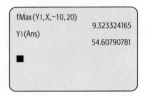

The approximation is about $y = 54.61$ at $x = 9.32$.

Using any of these methods, we find the relative minimum to be about $y = -60.30$ at $x = 1.01$.

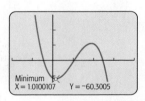

EXERCISE

1. Using one of the methods just described, approximate the relative extrema of the function in Example 1.

TECHNOLOGY CONNECTION

Finding Relative Extrema with iPlot

We can use iPlot to graph a function and its derivative and then find relative extrema.

iPlot has the capability of graphing a function and its derivative on the same set of axes, though it does not give a formula for the derivative but merely draws the graph. As an example, let's consider the function given by $f(x) = x^3 - 3x + 4$.

To graph a function and its derivative, first open the iPlot app on your iPhone or iPad. You will get a screen like the one in Fig. 1. Notice the four icons at the bottom. The Functions icon is highlighted. Press $+$ in the upper right; then enter $f(x) = x^3 - 3x + 4$ using the notation x^3–3*x+4. Press Done at the upper right and then Plot at the lower right (Fig. 2).

(*continued*)

FIGURE 1

FIGURE 2

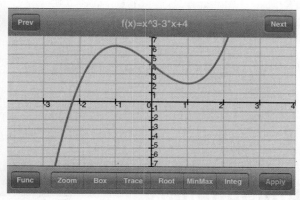

FIGURE 3

FIGURE 4

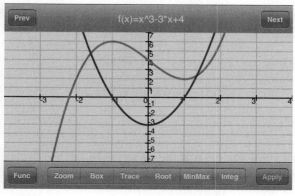

FIGURE 5

The graph of $f(x) = x^3 - 3x + 4$ is shown in red in Fig. 3. To graph the derivative of f, first click on the Functions icon again, and then press $\boxed{+}$. You will get the screen shown in Fig. 4.

Next, slide the Derivate button to the left. ("Derivate" means "Differentiate.") Then enter the same function as before, x^3–3*x+4, and press Done. D(x^3–3*x+4) will appear in the second line. Press Plot, and you will see both functions plotted, as shown in Fig. 5. Look over the two graphs, and use Trace to find various function values. Press Prev to jump between the function and its derivative. Look for x-values where the derivative is 0. What happens at these values of

the original function? Examining the graphs in this way reveals that the graph of $f(x) = x^3 - 3x + 4$ has a relative maximum point at $(-1, 6)$ and a relative minimum point at $(1, 2)$.

iPlot has an additional feature that allows us to be more certain about these relative extrema. Go back to the original plot of $f(x) = x^3 - 3x + 4$ (Fig. 3), and press Settings. Change the window to $[-3, 3, 12, -10]$ to better see the graph. Press the MinMax button at the bottom. Touch the screen as closely as possible to what might be a relative extremum. See Figs. 6 and 7 for the relative maximum. The relative minimum can be found similarly.

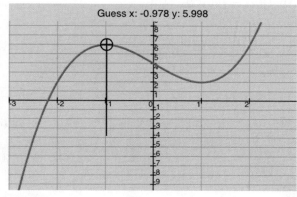

FIGURE 6

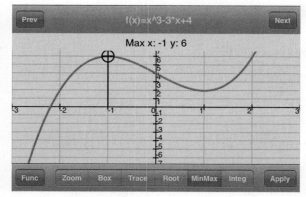

FIGURE 7

(continued)

EXERCISES

For each function, use iPlot to create the graph and find the derivative. Then explore each graph to look for possible relative extrema. Use MinMax to determine the relative extrema.

1. $f(x) = 2x^3 - x^4$

2. $f(x) = x(200 - x)$

3. $f(x) = x^3 - 6x^2$

4. $f(x) = -4.32 + 1.44x + 3x^2 - x^3$

5. $g(x) = x\sqrt{4 - x^2}$

6. $g(x) = \dfrac{4x}{x^2 + 1}$

7. $f(x) = \dfrac{x^2 - 3x}{x - 1}$

8. $f(x) = |x + 2| - 3$

Section Summary

- A function f is *increasing* over an interval I if, for all a and b in I such that $a < b$, $f(a) < f(b)$. Equivalently, the slope of the secant line connecting a and b is positive:
$$\frac{f(b) - f(a)}{b - a} > 0.$$

- A function f is *decreasing* over an interval I if, for all a and b in I such that $a < b$, $f(a) > f(b)$. Equivalently, the slope of the secant line connecting a and b is negative:
$$\frac{f(b) - f(a)}{b - a} < 0.$$

- Using the first derivative, a function is *increasing* over an open interval I if, for all x in I, the slope of the tangent line at x is positive; that is, $f'(x) > 0$. Similarly, a function is *decreasing* over an open interval I if, for all x in I, the slope of the tangent line is negative; that is, $f'(x) < 0$.

- A *critical value* is a number c in the domain of f such that $f'(c) = 0$ or $f'(c)$ does not exist. The point $(c, f(c))$ is called a *critical point*.
- A relative maximum point is higher than all other points in some interval containing it. Similarly, a relative minimum point is lower than all other points in some interval containing it. The y-value of such a point is called a *relative maximum* (or *minimum*) *value* of the function.
- Minimum and maximum points are collectively called *extrema*.
- Critical values are candidates for possible relative extrema. The *First-Derivative Test* is used to classify a critical value as a relative minimum, a relative maximum, or neither.

EXERCISE SET
2.1

Find the relative extrema of each function, if they exist. List each extremum along with the x-value at which it occurs. Then sketch a graph of the function.

1. $f(x) = x^2 + 4x + 5$

2. $f(x) = x^2 + 6x - 3$

3. $f(x) = 5 - x - x^2$

4. $f(x) = 2 - 3x - 2x^2$

5. $g(x) = 1 + 6x + 3x^2$

6. $F(x) = 0.5x^2 + 2x - 11$

7. $G(x) = x^3 - x^2 - x + 2$

8. $g(x) = x^3 + \frac{1}{2}x^2 - 2x + 5$

9. $f(x) = x^3 - 3x + 6$

10. $f(x) = x^3 - 3x^2$

11. $f(x) = 3x^2 + 2x^3$

12. $f(x) = x^3 + 3x$

13. $g(x) = 2x^3 - 16$

14. $F(x) = 1 - x^3$

15. $G(x) = x^3 - 6x^2 + 10$

16. $f(x) = 12 + 9x - 3x^2 - x^3$

17. $g(x) = x^3 - x^4$

18. $f(x) = x^4 - 2x^3$

19. $f(x) = \frac{1}{3}x^3 - 2x^2 + 4x - 1$

20. $F(x) = -\frac{1}{3}x^3 + 3x^2 - 9x + 2$

21. $g(x) = 2x^4 - 20x^2 + 18$

22. $f(x) = 3x^4 - 15x^2 + 12$

23. $F(x) = \sqrt[3]{x - 1}$

24. $G(x) = \sqrt[3]{x + 2}$

25. $f(x) = 1 - x^{2/3}$

26. $f(x) = (x + 3)^{2/3} - 5$

27. $G(x) = \dfrac{-8}{x^2 + 1}$

28. $F(x) = \dfrac{5}{x^2 + 1}$

29. $g(x) = \dfrac{4x}{x^2 + 1}$

30. $g(x) = \dfrac{x^2}{x^2 + 1}$

31. $f(x) = \sqrt[3]{x}$

32. $f(x) = (x + 1)^{1/3}$

33. $g(x) = \sqrt{x^2 + 2x + 5}$

34. $F(x) = \dfrac{1}{\sqrt{x^2 + 1}}$

35–68. Check the results of Exercises 1–34 using a calculator.

For Exercises 69–84, draw a graph to match the description given. Answers will vary.

69. $f(x)$ is increasing over $(-\infty, 2)$ and decreasing over $(2, \infty)$.

70. $g(x)$ is decreasing over $(-\infty, -3)$ and increasing over $(-3, \infty)$.

71. $G(x)$ is decreasing over $(-\infty, 4)$ and $(9, \infty)$ and increasing over $(4, 9)$.

72. $F(x)$ is increasing over $(-\infty, 5)$ and $(12, \infty)$ and decreasing over $(5, 12)$.

73. $g(x)$ has a positive derivative over $(-\infty, -3)$ and a negative derivative over $(-3, \infty)$.

74. $f(x)$ has a negative derivative over $(-\infty, 1)$ and a positive derivative over $(1, \infty)$.

75. $F(x)$ has a negative derivative over $(-\infty, 2)$ and $(5, 9)$ and a positive derivative over $(2, 5)$ and $(9, \infty)$.

76. $G(x)$ has a positive derivative over $(-\infty, -2)$ and $(4, 7)$ and a negative derivative over $(-2, 4)$ and $(7, \infty)$.

77. $f(x)$ has a positive derivative over $(-\infty, 3)$ and $(3, 9)$, a negative derivative over $(9, \infty)$, and a derivative equal to 0 at $x = 3$.

78. $g(x)$ has a negative derivative over $(-\infty, 5)$ and $(5, 8)$, a positive derivative over $(8, \infty)$, and a derivative equal to 0 at $x = 5$.

79. $F(x)$ has a negative derivative over $(-\infty, -1)$ and a positive derivative over $(-1, \infty)$, and $F'(-1)$ does not exist.

80. $G(x)$ has a positive derivative over $(-\infty, 0)$ and $(3, \infty)$ and a negative derivative over $(0, 3)$, but neither $G'(0)$ nor $G'(3)$ exists.

81. $f(x)$ has a negative derivative over $(-\infty, -2)$ and $(1, \infty)$ and a positive derivative over $(-2, 1)$, and $f'(-2) = 0$, but $f'(1)$ does not exist.

82. $g(x)$ has a positive derivative over $(-\infty, -3)$ and $(0, 3)$, a negative derivative over $(-3, 0)$ and $(3, \infty)$, and a derivative equal to 0 at $x = -3$ and $x = 3$, but $g'(0)$ does not exist.

83. $H(x)$ is increasing over $(-\infty, \infty)$, but the derivative does not exist at $x = 1$.

84. $K(x)$ is decreasing over $(-\infty, \infty)$, but the derivative does not exist at $x = 0$ and $x = 2$.

85. Consider this graph.

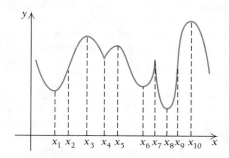

Explain the idea of a critical value. Then determine which x-values are critical values, and state why.

86. Consider this graph.

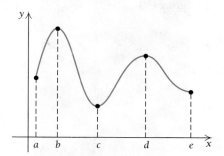

Using the graph and the intervals noted, explain how to relate the concept of the function being increasing or decreasing to the first derivative.

APPLICATIONS

Business and Economics

87. *Employment.* According to the U.S. Bureau of Labor Statistics, the number of professional services employees fluctuated during the period 2000–2009, as modeled by

$$E(t) = -28.31t^3 + 381.86t^2 - 1162.07t + 16{,}905.87,$$

where t is the number of years since 2000 ($t = 0$ corresponds to 2000) and E is thousands of employees. (*Source*: www.data.bls.gov.) Find the relative extrema of this function, and sketch the graph. Interpret the meaning of the relative extrema.

88. *Advertising.* Brody Electronics estimates that it will sell N units of a new toy after spending a thousands of dollars on advertising, where

$$N(a) = -a^2 + 300a + 6, \quad 0 \leq a \leq 300.$$

Find the relative extrema and sketch a graph of the function.

Life and Physical Sciences

89. *Temperature during an illness.* The temperature of a person during an intestinal illness is given by

$$T(t) = -0.1t^2 + 1.2t + 98.6, \quad 0 \leq t \leq 12,$$

where T is the temperature (°F) at time t, in days. Find the relative extrema and sketch a graph of the function.

90. Solar eclipse. On January 15, 2010, the longest annular solar eclipse until 3040 occurred over Africa and the Indian Ocean (in an annular eclipse, the sun is partially obscured by the moon and looks like a ring). The path of the full eclipse on the earth's surface is modeled by

$$f(x) = 0.0125x^2 - 1.157x + 22.864, \quad 15 < x < 90,$$

where x is the number of degrees of longitude east of the prime meridian and $f(x)$ is the number of degrees of latitude north (positive) or south (negative) of the equator. (*Source*: NASA.) Find the longitude and latitude of the southernmost point at which the full eclipse could be viewed.

SYNTHESIS

In Exercises 91–96, the graph of a derivative f' is shown. Use the information in each graph to determine where f is increasing or decreasing and the x-values of any extrema. Then sketch a possible graph of f.

91.

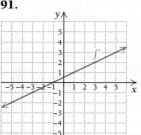

92.

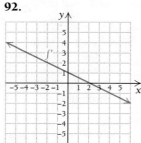

93.

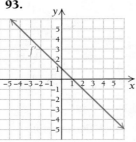

94.

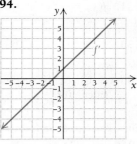

95.

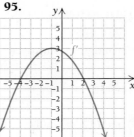

96.

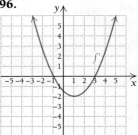

TECHNOLOGY CONNECTION

Graph each function. Then estimate any relative extrema.

97. $f(x) = -x^6 - 4x^5 + 54x^4 + 160x^3 - 641x^2 - 828x + 1200$

98. $f(x) = x^4 + 4x^3 - 36x^2 - 160x + 400$

99. $f(x) = \sqrt[3]{|4 - x^2|} + 1$ **100.** $f(x) = x\sqrt{9 - x^2}$

Use your calculator's absolute-value feature to graph the following functions and determine relative extrema and intervals over which the function is increasing or decreasing. State the x-values at which the derivative does not exist.

101. $f(x) = |x - 2|$

102. $f(x) = |2x - 5|$

103. $f(x) = |x^2 - 1|$

104. $f(x) = |x^2 - 3x + 2|$

105. $f(x) = |9 - x^2|$

106. $f(x) = |-x^2 + 4x - 4|$

107. $f(x) = |x^3 - 1|$

108. $f(x) = |x^4 - 2x^2|$

Life science: caloric intake and life expectancy. *The data in the following table give, for various countries, daily caloric intake, projected life expectancy, and infant mortality. Use the data for Exercises 109 and 110.*

Country	Daily Caloric Intake	Life Expectancy at Birth (in years)	Infant Mortality (number of deaths before age 1 per 1000 births)
Argentina	3004	77	13
Australia	3057	82	5
Bolivia	2175	67	46
Canada	3557	81	5
Dominican Republic	2298	74	30
Germany	3491	79	4
Haiti	1835	61	62
Mexico	3265	76	17
United States	3826	78	6
Venezuela	2453	74	17

(Source: *U.N. FAO Statistical Yearbook, 2009.*)

109. Life expectancy and daily caloric intake.

  a) Use the regression procedures of Section R.6 to fit a cubic function $y = f(x)$ to the data in the table, where x is daily caloric intake and y is life expectancy. Then fit a quartic function and decide which fits best. Explain.

  b) What is the domain of the function?

  c) Does the function have any relative extrema? Explain.

110. Infant mortality and daily caloric intake.

  a) Use the regression procedures of Section R.6 to fit a cubic function $y = f(x)$ to the data in the table, where x is daily caloric intake and y is infant mortality. Then fit a quartic function and decide which fits best. Explain.

  b) What is the domain of the function?

  c) Does the function have any relative extrema? Explain.

111. Describe a procedure that can be used to select an appropriate viewing window for the functions given in (a) Exercises 1–16 and (b) Exercises 97–100.

Answers to Quick Checks

1. Relative maximum at $(-3, 48)$, relative minimum at $(3, -60)$

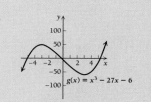

2. Relative minimum at $\left(2, -\frac{16}{3}\right)$

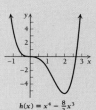

3. There are no extrema.

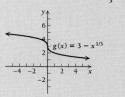

Using Second Derivatives to Find Maximum and Minimum Values and Sketch Graphs

The "turning" behavior of a graph is called its *concavity*. The second derivative plays a pivotal role in analyzing the concavity of a function's graph.

OBJECTIVES

- Classify the relative extrema of a function using the Second-Derivative Test.
- Sketch the graph of a continuous function.

Concavity: Increasing and Decreasing Derivatives

The graphs of two functions are shown below. The graph of f is turning up and the graph of g is turning down. Let's see if we can relate these observations to the functions' derivatives.

Consider first the graph of f. Take a ruler, or straightedge, and draw tangent lines as you move along the curve from left to right. What happens to the slopes of the tangent lines? Do the same for the graph of g. Look at the curvature and decide whether you see a pattern.

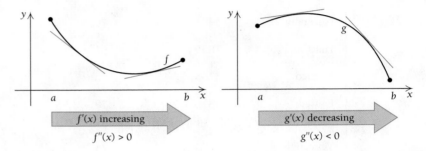

For the graph of f, the slopes of the tangent lines are increasing. That is, f' is increasing over the interval. This can be determined by noting that $f''(x)$ is positive, since the relationship between f' and f'' is like the relationship between f and f'. Note also that all the tangent lines for f are below the graph. For the graph of g, the slopes are decreasing. This can be determined by noting that g' is decreasing whenever $g''(x)$ is negative. For g, all tangent lines are above the graph.

DEFINITION

Suppose that f is a function whose derivative f' exists at every point in an open interval I. Then

f is **concave up** on I if f' is increasing over I.

f is **concave down** on I if f' is decreasing over I.

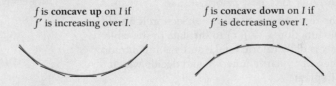

The following theorem states how the concavity of a function's graph and the second derivative of the function are related.

THEOREM 4   A Test for Concavity

1. If $f''(x) > 0$ on an interval I, then the graph of f is concave up.
    (f' is increasing, so f is turning up on I.)
2. If $f''(x) < 0$ on an interval I, then the graph of f is concave down.
    (f' is decreasing, so f is turning down on I.)

Keep in mind that a function can be decreasing and concave up, decreasing and concave down, increasing and concave up, or increasing and concave down. That is, *concavity* and *increasing/decreasing* are independent concepts. It is the increasing or decreasing aspect of the *derivative* that tells us about the function's concavity.

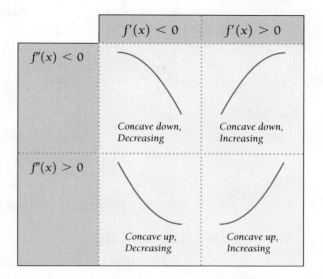

TECHNOLOGY CONNECTION

Exploratory

Graph the function

$$f(x) = -\tfrac{1}{3}x^3 + 6x^2 - 11x - 50$$

and its second derivative,

$$f''(x) = -2x + 12,$$

using the viewing window $[-10, 25, -100, 150]$, with Xscl = 5 and Yscl = 25.

Over what intervals is the graph of f concave up?
Over what intervals is the graph of f concave down?
Over what intervals is the graph of f'' positive?
Over what intervals is the graph of f'' negative?

What can you conjecture?

Now graph the first derivative

$$f'(x) = -x^2 + 12x - 11$$

and the second derivative

$$f''(x) = -2x + 12$$

using the viewing window $[-10, 25, -200, 50]$, with Xscl = 5 and Yscl = 25.

Over what intervals is the first derivative f' increasing?
Over what intervals is the first derivative f' decreasing?
Over what intervals is the graph of f'' positive?
Over what intervals is the graph of f'' negative?

What can you conjecture?

Classifying Relative Extrema Using Second Derivatives

Let's see how we can use second derivatives to determine whether a function has a relative extremum on an open interval.

The following graphs show both types of concavity at a critical value (where $f'(c) = 0$). When the second derivative is positive (graph is concave up) at the critical value, the critical point is a relative minimum point, and when the second derivative is negative (graph is concave down), the critical point is a relative maximum point.

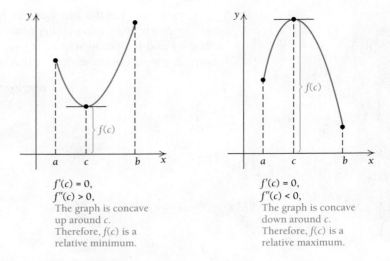

$f'(c) = 0,$
$f''(c) > 0,$
The graph is concave
up around c.
Therefore, $f(c)$ is a
relative minimum.

$f'(c) = 0,$
$f''(c) < 0,$
The graph is concave
down around c.
Therefore, $f(c)$ is a
relative maximum.

This analysis is summarized in Theorem 5:

THEOREM 5 **The Second-Derivative Test for Relative Extrema**

Suppose that f is differentiable for every x in an open interval (a, b) and that there is a critical value c in (a, b) for which $f'(c) = 0$. Then:

  1. $f(c)$ is a relative minimum if $f''(c) > 0$.
  2. $f(c)$ is a relative maximum if $f''(c) < 0$.

For $f''(c) = 0$, the First-Derivative Test can be used to determine whether $f(x)$ is a relative extremum.

Consider the following graphs. In each one, f' and f'' are both 0 at $c = 2$, but the first function has an extremum and the second function does not. When c is a critical value and $f''(c) = 0$, an extremum may or may not exist at c. Note too that if $f'(c)$ does not exist and c is a critical value, then $f''(c)$ also does not exist. Again, an approach other than the Second-Derivative Test must be used to determine whether $f(c)$ is an extremum.

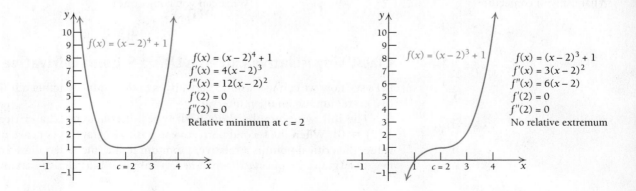

$f(x) = (x - 2)^4 + 1$
$f'(x) = 4(x - 2)^3$
$f''(x) = 12(x - 2)^2$
$f'(2) = 0$
$f''(2) = 0$
Relative minimum at $c = 2$

$f(x) = (x - 2)^3 + 1$
$f'(x) = 3(x - 2)^2$
$f''(x) = 6(x - 2)$
$f'(2) = 0$
$f''(2) = 0$
No relative extremum

The second derivative is used to help identify extrema and determine the overall behavior of a graph, as we see in the following examples.

■ **EXAMPLE 1** Find the relative extrema of the function f given by

$$f(x) = x^3 + 3x^2 - 9x - 13,$$

and sketch the graph.

Solution To find any critical values, we determine $f'(x)$. To determine whether any critical values lead to extrema, we also find $f''(x)$:

$$f'(x) = 3x^2 + 6x - 9,$$
$$f''(x) = 6x + 6.$$

Then we solve $f'(x) = 0$:

$$3x^2 + 6x - 9 = 0$$
$$x^2 + 2x - 3 = 0 \qquad \text{Dividing both sides by 3}$$
$$(x + 3)(x - 1) = 0 \qquad \text{Factoring}$$
$$x + 3 = 0 \quad \text{or} \quad x - 1 = 0 \qquad \text{Using the Principle of Zero Products}$$
$$x = -3 \quad \text{or} \qquad x = 1.$$

We next find second coordinates by substituting the critical values in the original function:

$$f(-3) = (-3)^3 + 3(-3)^2 - 9(-3) - 13 = 14;$$
$$f(1) = (1)^3 + 3(1)^2 - 9(1) - 13 = -18.$$

Are the points $(-3, 14)$ and $(1, -18)$ relative extrema? Let's look at the second derivative. We use the Second-Derivative Test with the critical values -3 and 1:

$$f''(-3) = 6(-3) + 6 = -12 < 0; \longrightarrow \text{Relative maximum}$$
$$f''(1) = 6(1) + 6 = 12 > 0. \longrightarrow \text{Relative minimum}$$

Thus, $f(-3) = 14$ is a relative maximum and $f(1) = -18$ is a relative minimum. We plot both $(-3, 14)$ and $(1, -18)$, including short arcs at each point to indicate the graph's concavity. Then, by calculating and plotting a few more points, we can make a sketch, as shown below.

Exploratory

Consider the function given by

$$y = x^3 - 3x^2 - 9x - 1.$$

Use a graph to estimate the relative extrema. Then find the first and second derivatives. Graph both in the same window. Use the ZERO feature to determine where the first derivative is zero. Verify that relative extrema occur at those x-values by checking the sign of the second derivative. Then check your work using the approach of Example 1.

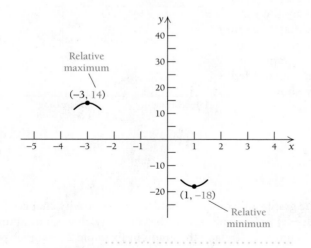

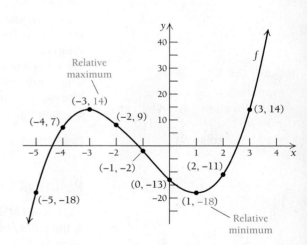

■ **EXAMPLE 2** Find the relative extrema of the function f given by

$$f(x) = 3x^5 - 20x^3,$$

and sketch the graph.

Solution We find both the first and second derivatives:

$$f'(x) = 15x^4 - 60x^2,$$
$$f''(x) = 60x^3 - 120x.$$

Then we solve $f'(x) = 0$ to find any critical values:

$$15x^4 - 60x^2 = 0$$
$$15x^2(x^2 - 4) = 0$$
$$15x^2(x + 2)(x - 2) = 0 \qquad \text{Factoring}$$
$$15x^2 = 0 \quad \text{or} \quad x + 2 = 0 \quad \text{or} \quad x - 2 = 0 \qquad \text{Using the Principle of Zero Products}$$
$$x = 0 \quad \text{or} \qquad x = -2 \quad \text{or} \qquad x = 2.$$

We next find second coordinates by substituting in the original function:

$$f(-2) = 3(-2)^5 - 20(-2)^3 = 64;$$
$$f(0) = 3(0)^5 - 20(0)^3 = 0.$$
$$f(2) = 3(2)^5 - 20(2)^3 = -64;$$

All three of these y-values are candidates for relative extrema.

We now use the Second-Derivative Test with the numbers -2, 2, and 0:

$$f''(-2) = 60(-2)^3 - 120(-2) = -240 < 0; \longrightarrow \text{Relative maximum}$$
$$f''(0) = 60(0)^3 - 120(0) = 0. \longrightarrow \text{The Second-Derivative Test fails. Use the First-Derivative Test.}$$
$$f''(2) = 60(2)^3 - 120(2) = 240 > 0; \longrightarrow \text{Relative minimum}$$

Thus, $f(-2) = 64$ is a relative maximum, and $f(2) = -64$ is a relative minimum. Since $f'(-1) < 0$ and $f'(1) < 0$, we know that f is decreasing on both $(-2, 0)$ and $(0, 2)$. Thus, we know by the First-Derivative Test that f has no relative extremum at $(0, 0)$. We complete the graph, plotting other points as needed. The extrema are shown in the graph at right.

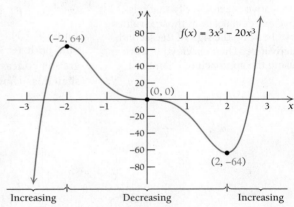

❮ Quick Check 1

> **Quick Check 1**
>
> Find the relative extrema of the function g given by $g(x) = 10x^3 - 6x^5$, and sketch the graph.

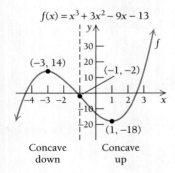

Points of Inflection

Look again at the graphs in Examples 1 and 2. The concavity changes from down to up at the point $(-1, -2)$ in Example 1, and the concavity changes from up to down at the point $(0, 0)$ in Example 2. (In fact, the graph in Example 2 has other points where the concavity changes direction. This is addressed in Example 3.)

A **point of inflection**, or an **inflection point**, is a point across which the direction of concavity changes. For example, in Figs. 1–3, point P is an inflection point. The figures display the sign of $f''(x)$ to indicate the concavity on either side of P.

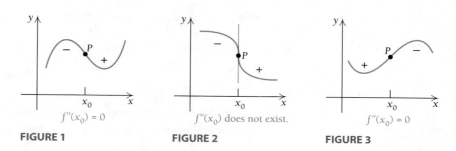

FIGURE 1   **FIGURE 2**   **FIGURE 3**

As we move to the right along each curve, the concavity changes at P. Since, as we move through P, the sign of $f''(x)$ changes, either the value of $f''(x_0)$ at P must be 0, as in Figs. 1 and 3, or $f''(x_0)$ must not exist, as in Fig. 2.

THEOREM 6   Finding Points of Inflection

If a function f has a point of inflection, it must occur at a point x_0, where

$$f''(x_0) = 0 \quad \text{or} \quad f''(x_0) \text{ does not exist.}$$

The converse of Theorem 6 is not necessarily true. That is, if $f''(x_0)$ is 0 or does not exist, then there is not necessarily a point of inflection at x_0. There must be a change in the direction of concavity on either side of x_0 for $(x_0, f(x_0))$ to be a point of inflection. For example, for $f(x) = (x - 2)^4 + 1$ (see the graph on p. 218), we have $f''(2) = 0$, but $(2, f(2))$ is not a point of inflection since the graph of f is concave up to the left and to the right of $x = 2$.

To find candidates for points of inflection, we look for numbers x_0 for which $f''(x_0) = 0$ or for which $f''(x_0)$ does not exist. Then, if $f''(x)$ changes sign as x moves through x_0 (see Figs. 1–3), we have a point of inflection at x_0.

Theorem 6, about points of inflection, is completely analogous to Theorem 2 about relative extrema. Theorem 2 tells us that relative extrema occur when $f'(x) = 0$ or $f'(x)$ does not exist. Theorem 6 tells us that points of inflection occur when $f''(x) = 0$ or $f''(x)$ does not exist.

■ **EXAMPLE 3**   Use the second derivative to determine the point(s) of inflection for the function in Example 2.

Solution   The function is $f(x) = 3x^5 - 20x^3$, and its second derivative is $f''(x) = 60x^3 - 120x$. We set the second derivative equal to 0 and solve for x:

$$60x^3 - 120x = 0$$
$$60x(x^2 - 2) = 0 \qquad \text{Factoring out } 60x$$
$$60x(x + \sqrt{2})(x - \sqrt{2}) = 0 \qquad \text{Factoring } x^2 - 2 \text{ as a difference of squares}$$
$$x = 0 \quad \text{or} \quad x = -\sqrt{2} \quad \text{or} \quad x = \sqrt{2}. \qquad \text{Using the Principle of Zero Products}$$

Next, we check the sign of $f''(x)$ over the intervals bounded by these three x-values. We are looking for a change in sign from one interval to the next:

Interval		$-\sqrt{2}$		0		$\sqrt{2}$		x
Test Value	$x = -2$	$x = -1$	$x = 1$	$x = 2$
Sign of $f''(x)$	$f''(-2) < 0$	$f''(-1) > 0$	$f''(1) < 0$	$f''(2) > 0$
Result	Concave down	Concave up	Concave down	Concave up

The graph changes from concave down to concave up at $x = -\sqrt{2}$, from concave up to concave down at $x = 0$, and from concave down to concave up at $x = \sqrt{2}$. Therefore, $\left(-\sqrt{2}, f\left(-\sqrt{2}\right)\right), (0, f(0))$, and $\left(\sqrt{2}, f\left(\sqrt{2}\right)\right)$ are points of inflection. Since

$$f\left(-\sqrt{2}\right) = 3\left(-\sqrt{2}\right)^5 - 20\left(-\sqrt{2}\right)^3 = 28\sqrt{2},$$
$$f(0) = 3(0)^5 - 20(0)^3 = 0,$$

and $\qquad f\left(\sqrt{2}\right) = 3\left(\sqrt{2}\right)^5 - 20\left(\sqrt{2}\right)^3 = -28\sqrt{2},$

these points are $\left(-\sqrt{2}, 28\sqrt{2}\right), (0, 0)$, and $\left(\sqrt{2}, -28\sqrt{2}\right)$, shown in the graph below.

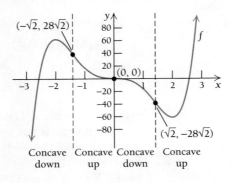

> **Quick Check 2**
>
> Determine the points of inflection for the function given by $g(x) = 10x^3 - 6x^5$.

❮ Quick Check 2

Curve Sketching

The first and second derivatives enhance our ability to sketch curves. We use the following strategy:

Strategy for Sketching Graphs*

a) *Derivatives and domain.* Find $f'(x)$ and $f''(x)$. Note the domain of f.

b) *Critical values of f.* Find the critical values by solving $f'(x) = 0$ and finding where $f'(x)$ does not exist. These numbers yield candidates for relative maxima or minima. Find the function values at these points.

c) *Increasing and/or decreasing; relative extrema.* Substitute each critical value, x_0, from step (b) into $f''(x)$. If $f''(x_0) < 0$, then $f(x_0)$ is a relative maximum and f is increasing to the left of x_0 and decreasing to the right. If $f''(x_0) > 0$, then $f(x_0)$ is a relative minimum and f is decreasing to the left of x_0 and increasing to the right.

d) *Inflection points.* Determine candidates for inflection points by finding where $f''(x) = 0$ or where $f''(x)$ does not exist. Find the function values at these points.

e) *Concavity.* Use the candidates for inflection points from step (d) to define intervals. Substitute test values into $f''(x)$ to determine where the graph is concave up ($f''(x) > 0$) and where it is concave down ($f''(x) < 0$).

f) *Sketch the graph.* Sketch the graph using the information from steps (a)–(e), calculating and plotting extra points as needed.

*This strategy is refined further, for rational functions, in Section 2.3.

The examples that follow apply this step-by-step strategy to sketch the graphs of several functions.

■ **EXAMPLE 4**   Find the relative extrema of the function f given by

$$f(x) = x^3 - 3x + 2,$$

and sketch the graph.

Solution

a) *Derivatives and domain.* Find $f'(x)$ and $f''(x)$:

$$f'(x) = 3x^2 - 3,$$
$$f''(x) = 6x.$$

The domain of f (and of any polynomial function) is $(-\infty, \infty)$, or the set of all real numbers, which is also written as $\mathbb{R}$.

b) *Critical values of f.* Find the critical values by determining where $f'(x)$ does not exist and by solving $f'(x) = 0$. We know that $f'(x) = 3x^2 - 3$ exists for all values of x, so the only critical values are where $f'(x)$ is 0:

$$3x^2 - 3 = 0 \qquad \text{Setting } f'(x) \text{ equal to 0}$$
$$3x^2 = 3$$
$$x^2 = 1$$
$$x = \pm 1.$$

We have $f(-1) = 4$ and $f(1) = 0$, so $(-1, 4)$ and $(1, 0)$ are on the graph.

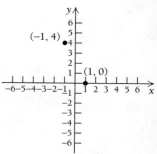

c) *Increasing and/or decreasing; relative extrema.* Substitute the critical values into $f''(x)$:

$$f''(-1) = 6(-1) = -6 < 0,$$

so $f(-1) = 4$ is a relative maximum, with f increasing on $(-\infty, -1)$ and decreasing on $(-1, 1)$.

$$f''(1) = 6 \cdot 1 = 6 > 0,$$

so $f(1) = 0$ is a relative minimum, with f decreasing on $(-1, 1)$ and increasing on $(1, \infty)$.

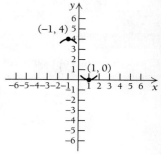

d) *Inflection points.* Find possible inflection points by finding where $f''(x)$ does not exist and by solving $f''(x) = 0$. We know that $f''(x) = 6x$ exists for all values of x, so we try to solve $f''(x) = 0$:

$$6x = 0 \qquad \text{Setting } f''(x) \text{ equal to 0}$$
$$x = 0. \qquad \text{Dividing both sides by 6}$$

We have $f(0) = 2$, which gives us another point, $(0, 2)$, that lies on the graph.

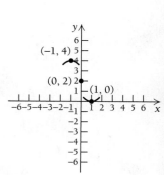

e) *Concavity.* Find the intervals on which f is concave up or concave down, using the point $(0, 2)$ from step (d). From step (c), we can conclude that f is concave down over the interval $(-\infty, 0)$ and concave up over $(0, \infty)$.

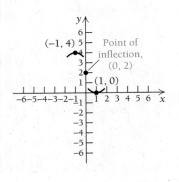

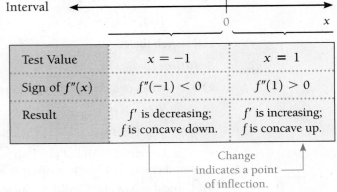

Interval	$x = -1$	$x = 1$
Test Value	$x = -1$	$x = 1$
Sign of $f''(x)$	$f''(-1) < 0$	$f''(1) > 0$
Result	f' is decreasing; f is concave down.	f' is increasing; f is concave up.

Change indicates a point of inflection.

f) *Sketch the graph.* Sketch the graph using the information in steps (a)–(e). Calculate some extra function values if desired. The graph follows.

x	$f(x)$
-3	-16
-2	0
-1	4
0	2
1	0
2	4
3	20

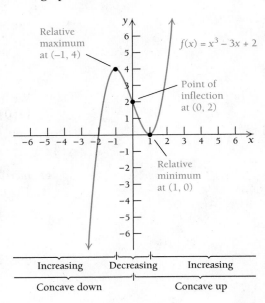

TECHNOLOGY CONNECTION

Check the results of Example 4 using a calculator.

■ **EXAMPLE 5**   Find the relative maxima and minima of the function f given by

$$f(x) = x^4 - 2x^2,$$

and sketch the graph.

Solution

a) *Derivatives and domain.* Find $f'(x)$ and $f''(x)$:

$$f'(x) = 4x^3 - 4x,$$
$$f''(x) = 12x^2 - 4.$$

The domain of f is $\mathbb{R}$.

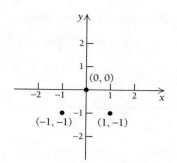

b) *Critical values.* Since $f'(x) = 4x^3 - 4x$ exists for all values of x, the only critical values are where $f'(x) = 0$:

$$4x^3 - 4x = 0 \qquad \text{Setting } f'(x) \text{ equal to } 0$$
$$4x(x^2 - 1) = 0$$
$$4x = 0 \quad \text{or} \quad x^2 - 1 = 0$$
$$x = 0 \quad \text{or} \qquad x^2 = 1$$
$$x = \pm 1.$$

We have $f(0) = 0, f(-1) = -1$, and $f(1) = -1$, which gives the points $(0, 0)$, $(-1, -1)$, and $(1, -1)$ on the graph.

c) *Increasing and/or decreasing; relative extrema.* Substitute the critical values into $f''(x)$:

$$f''(0) = 12 \cdot 0^2 - 4 = -4 < 0,$$

so $f(0) = 0$ is a relative maximum, with f increasing on $(-1, 0)$ and decreasing on $(0, 1)$.

$$f''(-1) = 12(-1)^2 - 4 = 8 > 0,$$

so $f(-1) = -1$ is a relative minimum, with f decreasing on $(-\infty, -1)$ and increasing on $(-1, 0)$.

$$f''(1) = 12 \cdot 1^2 - 4 = 8 > 0,$$

so $f(1) = -1$ is also a relative minimum, with f decreasing on $(0, 1)$ and increasing on $(1, \infty)$.

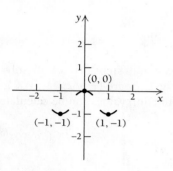

d) *Inflection points.* Find where $f''(x)$ does not exist and where $f''(x) = 0$. Since $f''(x)$ exists for all real numbers, we just solve $f''(x) = 0$:

$$12x^2 - 4 = 0 \qquad \text{Setting } f''(x) \text{ equal to } 0$$
$$4(3x^2 - 1) = 0$$
$$3x^2 - 1 = 0$$
$$3x^2 = 1$$
$$x^2 = \frac{1}{3}$$
$$x = \pm\sqrt{\frac{1}{3}}$$
$$= \pm\frac{1}{\sqrt{3}}.$$

We have

$$f\left(\frac{1}{\sqrt{3}}\right) = \left(\frac{1}{\sqrt{3}}\right)^4 - 2\left(\frac{1}{\sqrt{3}}\right)^2$$
$$= \frac{1}{9} - \frac{2}{3} = -\frac{5}{9}$$

and

$$f\left(-\frac{1}{\sqrt{3}}\right) = -\frac{5}{9}.$$

These values give

$$\left(-\frac{1}{\sqrt{3}}, -\frac{5}{9}\right) \text{ and } \left(\frac{1}{\sqrt{3}}, -\frac{5}{9}\right)$$

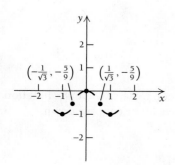

as possible inflection points.

e) *Concavity.* Find the intervals on which f is concave up or concave down, using the points $\left(-\dfrac{1}{\sqrt{3}}, -\dfrac{5}{9}\right)$ and $\left(\dfrac{1}{\sqrt{3}}, -\dfrac{5}{9}\right)$, from step (d). From step (c), we can conclude that f is concave up over the intervals $\left(-\infty, -1/\sqrt{3}\right)$ and $\left(1/\sqrt{3}, \infty\right)$ and concave down over the interval $\left(-1/\sqrt{3}, 1/\sqrt{3}\right)$.

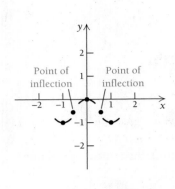

Point of inflection   Point of inflection

Interval

$-1/\sqrt{3}$ $-1/\sqrt{3}$ x

Test Value	$x = -1$	$x = 0$	$x = 1$
Sign of $f''(x)$	$f''(-1) > 0$	$f''(0) < 0$	$f''(1) > 0$
Result	f' is increasing; f is concave up.	f' is decreasing; f is concave down.	f' is increasing; f is concave up.

Change indicates a point of inflection.

Change indicates a point of inflection.

f) *Sketch the graph.* Sketch the graph using the information in steps (a)–(e). By solving $x^4 - 2x^2 = 0$, we can find the x-intercepts easily. They are $\left(-\sqrt{2}, 0\right)$, $(0, 0)$, and $\left(\sqrt{2}, 0\right)$. This also aids with graphing. Extra function values can be calculated if desired. The graph is shown below.

x	$f(x)$, approximately
-2	8
-1.5	0.56
-1	-1
-0.5	-0.44
0	0
0.5	-0.44
1	-1
1.5	0.56
2	8

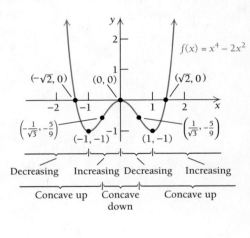

$f(x) = x^4 - 2x^2$

$(-\sqrt{2}, 0)$ $(0, 0)$ $(\sqrt{2}, 0)$

$\left(-\frac{1}{\sqrt{3}}, -\frac{5}{9}\right)$ $(-1, -1)$ $(1, -1)$ $\left(\frac{1}{\sqrt{3}}, -\frac{5}{9}\right)$

Decreasing   Increasing   Decreasing   Increasing

Concave up   Concave down   Concave up

TECHNOLOGY CONNECTION

EXERCISE

1. Consider

$$f(x) = x^3(x - 2)^3.$$

How many relative extrema do you anticipate finding? Where do you think they will be?

Graph f, f', and f'' using $[-1, 3, -2, 6]$ as a viewing window. Estimate the relative extrema and the inflection points of f. Then check your work using the analytic methods of Examples 4 and 5.

》 Quick Check 3

Find the relative maxima and minima of the function f given by $f(x) = 1 + 8x^2 - x^4$, and sketch the graph.

《 Quick Check 3

■ EXAMPLE 6 Graph the function f given by

$$f(x) = (2x - 5)^{1/3} + 1.$$

List the coordinates of any extrema and points of inflection. State where the function is increasing or decreasing, as well as where it is concave up or concave down.

Solution

a) *Derivatives and domain.* Find $f'(x)$ and $f''(x)$:

$$f'(x) = \tfrac{1}{3}(2x - 5)^{-2/3} \cdot 2 = \tfrac{2}{3}(2x - 5)^{-2/3}, \text{ or } \frac{2}{3(2x - 5)^{2/3}};$$

$$f''(x) = -\tfrac{4}{9}(2x - 5)^{-5/3} \cdot 2 = -\tfrac{8}{9}(2x - 5)^{-5/3}, \text{ or } \frac{-8}{9(2x - 5)^{5/3}}.$$

The domain of f is $\mathbb{R}$.

b) *Critical values.* Since

$$f'(x) = \frac{2}{3(2x - 5)^{2/3}}$$

is never 0 (a fraction equals 0 only when its numerator is 0), the only critical value is when $f'(x)$ does not exist. The only time $f'(x)$ does not exist is when its denominator is 0:

$$3(2x - 5)^{2/3} = 0$$
$$(2x - 5)^{2/3} = 0 \qquad \text{Dividing both sides by 3}$$
$$(2x - 5)^2 = 0 \qquad \text{Cubing both sides}$$
$$2x - 5 = 0$$
$$2x = 5$$
$$x = \tfrac{5}{2}$$

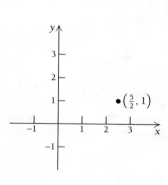

We now have $f\left(\tfrac{5}{2}\right) = \left(2 \cdot \tfrac{5}{2} - 5\right)^{1/3} + 1 = 0 + 1 = 1$, so the point $\left(\tfrac{5}{2}, 1\right)$ is on the graph.

c) *Increasing and/or decreasing; relative extrema.* Substitute the critical value into $f''(x)$:

$$f''\left(\tfrac{5}{2}\right) = \frac{-8}{9\left(2 \cdot \tfrac{5}{2} - 5\right)^{5/3}} = \frac{-8}{9 \cdot 0} = \frac{-8}{0}.$$

Since $f''\left(\tfrac{5}{2}\right)$ does not exist, the Second-Derivative Test cannot be used at $x = \tfrac{5}{2}$. Instead, we use the First-Derivative Test, selecting 2 and 3 as test values on either side of $\tfrac{5}{2}$:

$$f'(2) = \frac{2}{3(2 \cdot 2 - 5)^{2/3}} = \frac{2}{3(-1)^{2/3}} = \frac{2}{3 \cdot 1} = \frac{2}{3},$$

and $f'(3) = \dfrac{2}{3(2 \cdot 3 - 5)^{2/3}} = \dfrac{2}{3 \cdot 1^{2/3}} = \dfrac{2}{3 \cdot 1} = \dfrac{2}{3}.$

Since $f'(x) > 0$ on either side of $x = \tfrac{5}{2}$, we know that f is increasing on both $\left(-\infty, \tfrac{5}{2}\right)$ and $\left(\tfrac{5}{2}, \infty\right)$; thus, $f\left(\tfrac{5}{2}\right) = 1$ is not an extremum.

d) *Inflection points.* Find where $f''(x)$ does not exist and where $f''(x) = 0$. Since $f''(x)$ is never 0 (why?), we only need to find where $f''(x)$ does not exist. Since $f''(x)$ cannot exist where $f'(x)$ does not exist, we know from step (b) that a possible inflection point is $\left(\tfrac{5}{2}, 1\right)$.

e) *Concavity.* We check the concavity on either side of $x = \tfrac{5}{2}$. We choose $x = 2$ and $x = 3$ as our test values.

$$f''(2) = \frac{-8}{9(2 \cdot 2 - 5)^{5/3}} = \frac{-8}{9(-1)} > 0,$$

so f is concave up on $\left(-\infty, \frac{5}{2}\right)$.

$$f''(3) = \frac{-8}{9(2 \cdot 3 - 5)^{5/3}} = \frac{-8}{9 \cdot 1} < 0,$$

so f is concave down on $\left(\frac{5}{2}, \infty\right)$.

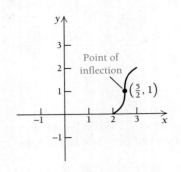

Point of inflection
$\left(\frac{5}{2}, 1\right)$

Interval

Test Value	$x = 2$	$x = 3$
Sign of $f''(x)$	$f''(2) > 0$	$f''(3) < 0$
Result	f' is increasing; f is concave up.	f' is decreasing; f is concave down.

Change indicates a point of inflection.

f) *Sketch the graph.* Sketch the graph using the information in steps (a)–(e). By solving $(2x - 5)^{1/3} + 1 = 0$, we can find the x-intercept—it is $(2, 0)$. Extra function values can be calculated, if desired. The graph is shown below.

x	$f(x)$, approximately
0	-0.71
1	-0.44
2	0
$\frac{5}{2}$	1
3	2
4	2.44
5	2.71

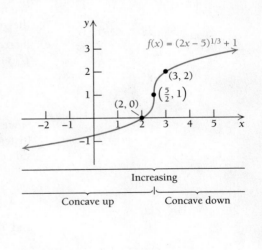

The following figures illustrate some information concerning the function in Example 1 that can be found from the first and second derivatives of f. The relative extrema are shown in Figs. 4 and 5. In Fig. 5, we see that the x-coordinates of the x-intercepts of f' are the critical values of f. Note that the intervals over which f is increasing or decreasing are those intervals for which f' is positive or negative, respectively.

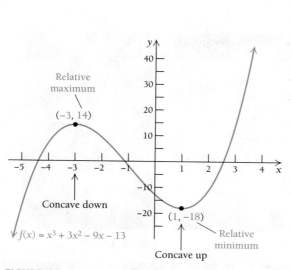

FIGURE 4

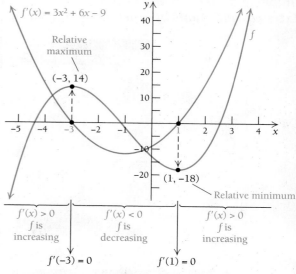

FIGURE 5

In Fig. 6, the intervals over which f' is increasing or decreasing are, respectively, those intervals over which f'' is positive or negative. And finally, in Fig. 7, we note that when $f''(x) < 0$, the graph of f is concave down, and when $f''(x) > 0$, the graph of f is concave up.

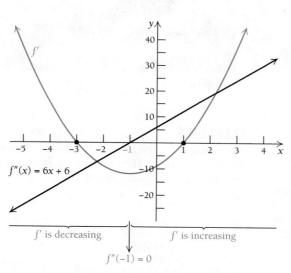

FIGURE 6

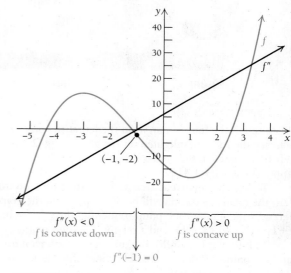

FIGURE 7

Using Graphicus to Find Roots, Extrema, and Inflection Points

Graphicus has the capability of finding roots, relative extrema, and points of inflection. Let's consider the function of Example 5, given by $f(x) = x^4 - 2x^2$.

Graphing a Function and Finding Its Roots, Relative Extrema, and Points of Inflection

After opening Graphicus, touch the blank rectangle at the top of the screen and enter the function as y(x) = x^4-2x^2. Press + in the upper right. You will see the graph (Fig. 1).

(continued)

Graphing a Function and Finding Its Roots, Relative Extrema, and Points of Inflection (continued)

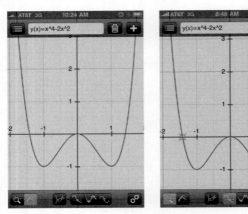

FIGURE 1   FIGURE 2   FIGURE 3

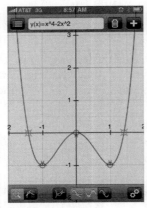

FIGURE 4   FIGURE 5   FIGURE 6   FIGURE 7

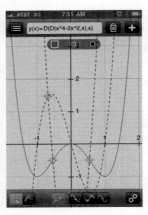

Then touch the fourth icon from the left at the bottom, and you will see the roots highlighted on the graph (Fig. 2). Touch the left-hand root symbol, and a value for one root, in this case, an approximation, −1.414214, is displayed (Fig. 3).

To find the relative extrema, touch the fifth icon from the left; the relative extrema will be highlighted on the graph in a different color. Touch each point to identify the relative minima at $(-1, -1)$, and $(1, -1)$, and a relative maximum at $(0, 0)$ (Fig. 4).

To find points of inflection, touch the sixth icon from the left, and the points of inflection will be highlighted as shown in Fig. 5. Touch these points and the approximations $(-0.577, -0.556)$ and $(0.577, -0.556)$ will be displayed.

Graphing a Function and Its Derivatives

Graphicus can graph derivatives. Go back to the original graph of f (Fig. 1). Touch ⊞; then press Add derivative, and you will see the original graph and the graph of the first derivative as a dashed line of a different color (Fig. 6). Touch ⊞ and Add derivative again, and you will see the graph of the original function, with dashed graphs of the first and second derivatives in different colors (Fig. 7). You can toggle between the derivatives by pressing the colored squares above the graphs.

EXERCISES

Use Graphicus to graph each function and its first and second derivatives. Then find approximations for roots, relative extrema, and points of inflection.

1. $f(x) = 2x^3 - x^4$

2. $f(x) = x(200 - x)$

3. $f(x) = x^3 - 6x^2$

4. $f(x) = -4.32 + 1.44x + 3x^2 - x^3$

5. $g(x) = x\sqrt{4 - x^2}$

6. $g(x) = \dfrac{4x}{x^2 + 1}$

7. $f(x) = \dfrac{x^2 - 3x}{x - 1}$

8. $f(x) = |x + 2| - 3$

EXERCISES

Graph the following:

$$f(x) = 3x^5 - 5x^3, \quad f'(x) = 15x^4 - 15x^2,$$

and

$$f''(x) = 60x^3 - 30x,$$

using the window $[-3, 3, -10, 10]$.

1. From the graph of f', estimate the critical values of f.

2. From the graph of f'', estimate the x-values of any inflection points of f.

Section Summary

- The second derivative f'' determines the *concavity* of the graph of function f.
- If $f''(x) > 0$ for all x in an open interval I, then the graph of f is *concave up* over I.
- If $f''(x) < 0$ for all x in an open interval I, then the graph of f is *concave down* over I.
- If c is a critical value and $f''(c) > 0$, then $f(c)$ is a relative minimum.
- If c is a critical value and $f''(c) < 0$, then $f(c)$ is a relative maximum.

- If c is a critical value and $f''(c) = 0$, the First-Derivative Test must be used to classify $f(c)$.
- If $f''(x_0) = 0$ or $f''(x_0)$ does not exist, and there is a change in concavity to the left and to the right of x_0, then the point $(x_0, f(x_0))$ is called a *point of inflection*.
- Finding relative extrema, intervals over which a function is increasing or decreasing, intervals of upward or downward concavity, and points of inflection is all part of a strategy for accurate curve sketching.

EXERCISE SET
2.2

For each function, find all relative extrema and classify each as a maximum or minimum. Use the Second-Derivative Test where possible.

1. $f(x) = 5 - x^2$

2. $f(x) = 4 - x^2$

3. $f(x) = x^2 - x$

4. $f(x) = x^2 + x - 1$

5. $f(x) = -5x^2 + 8x - 7$

6. $f(x) = -4x^2 + 3x - 1$

7. $f(x) = 8x^3 - 6x + 1$

8. $f(x) = x^3 - 12x - 1$

Sketch the graph of each function. List the coordinates of where extrema or points of inflection occur. State where the function is increasing or decreasing, as well as where it is concave up or concave down.

9. $f(x) = x^3 - 12x$

10. $f(x) = x^3 - 27x$

11. $f(x) = 3x^3 - 36x - 3$

12. $f(x) = 2x^3 - 3x^2 - 36x + 28$

13. $f(x) = \frac{8}{3}x^3 - 2x + \frac{1}{3}$

14. $f(x) = 80 - 9x^2 - x^3$

15. $f(x) = -x^3 + 3x^2 - 4$

16. $f(x) = -x^3 + 3x - 2$

17. $f(x) = 3x^4 - 16x^3 + 18x^2$
(Round results to three decimal places.)

18. $f(x) = 3x^4 + 4x^3 - 12x^2 + 5$
(Round results to three decimal places.)

19. $f(x) = x^4 - 6x^2$

20. $f(x) = 2x^2 - x^4$

21. $f(x) = x^3 - 2x^2 - 4x + 3$

22. $f(x) = x^3 - 6x^2 + 9x + 1$

23. $f(x) = 3x^4 + 4x^3$

24. $f(x) = x^4 - 2x^3$

25. $f(x) = x^3 - 6x^2 - 135x$

26. $f(x) = x^3 - 3x^2 - 144x - 140$

27. $f(x) = x^4 - 4x^3 + 10$

28. $f(x) = \frac{4}{3}x^3 - 2x^2 + x$

29. $f(x) = x^3 - 6x^2 + 12x - 6$

30. $f(x) = x^3 + 3x + 1$

31. $f(x) = 5x^3 - 3x^5$

32. $f(x) = 20x^3 - 3x^5$
(Round results to three decimal places.)

33. $f(x) = x^2(3 - x)^2$
(Round results to three decimal places.)

34. $f(x) = x^2(1 - x)^2$
(Round results to three decimal places.)

35. $f(x) = (x + 1)^{2/3}$

36. $f(x) = (x - 1)^{2/3}$

37. $f(x) = (x - 3)^{1/3} - 1$

38. $f(x) = (x - 2)^{1/3} + 3$

39. $f(x) = -2(x - 4)^{2/3} + 5$

40. $f(x) = -3(x - 2)^{2/3} + 3$

41. $f(x) = x\sqrt{4 - x^2}$ **42.** $f(x) = -x\sqrt{1 - x^2}$

43. $f(x) = \dfrac{x}{x^2 + 1}$ **44.** $f(x) = \dfrac{8x}{x^2 + 1}$

45. $f(x) = \dfrac{3}{x^2 + 1}$ **46.** $f(x) = \dfrac{-4}{x^2 + 1}$

For Exercises 47–56, sketch a graph that possesses the characteristics listed. Answers may vary.

47. f is increasing and concave up on $(-\infty, 4)$,
f is increasing and concave down on $(4, \infty)$.

48. f is decreasing and concave up on $(-\infty, 2)$,
f is decreasing and concave down on $(2, \infty)$.

49. f is increasing and concave down on $(-\infty, 1)$,
f is increasing and concave up on $(1, \infty)$.

50. f is decreasing and concave down on $(-\infty, 3)$,
f is decreasing and concave up on $(3, \infty)$.

51. f is concave down at $(1, 5)$, concave up at $(7, -2)$, and has an inflection point at $(4, 1)$.

52. f is concave up at $(1, -3)$, concave down at $(8, 7)$, and has an inflection point at $(5, 4)$.

53. $f'(-1) = 0, f''(-1) > 0, f(-1) = -5$;
$f'(7) = 0, f''(7) < 0, f(7) = 10; f''(3) = 0$, and $f(3) = 2$

54. $f'(-3) = 0, f''(-3) < 0, f(-3) = 8; f'(9) = 0$,
$f''(9) > 0, f(9) = -6; f''(2) = 0$, and $f(2) = 1$

55. $f'(-1) = 0, f''(-1) > 0, f(-1) = -2$;
$f'(1) = 0, f''(1) > 0, f(1) = -2; f'(0) = 0$,
$f''(0) < 0$, and $f(0) = 0$

56. $f'(0) = 0, f''(0) < 0, f(0) = 5; f'(2) = 0, f''(2) > 0$,
$f(2) = 2; f'(4) = 0, f''(4) < 0$, and $f(4) = 3$

57–102. Check the results of Exercises 1–46 with a graphing calculator.

APPLICATIONS

Business and Economics

Total revenue, cost, and profit.   *Using the same set of axes, sketch the graphs of the total-revenue, total-cost, and total-profit functions.*

103. $R(x) = 50x - 0.5x^2$, $C(x) = 4x + 10$

104. $R(x) = 50x - 0.5x^2$, $C(x) = 10x + 3$

105. Small business.   The percentage of the U.S. national income generated by nonfarm proprietors may be modeled by the function

$$p(x) = \frac{13x^3 - 240x^2 - 2460x + 585,000}{75,000},$$

where x is the number of years since 1970. Sketch the graph of this function for $0 \le x \le 40$.

106. Labor force.   The percentage of the U.S. civilian labor force aged 45–54 may be modeled by the function

$$f(x) = 0.025x^2 - 0.71x + 20.44,$$

where x is the number of years after 1970. Sketch the graph of this function for $0 \le x \le 30$.

Life and Physical Sciences

107. Coughing velocity.   A person coughs when a foreign object is in the windpipe. The velocity of the cough depends on the size of the object. Suppose a person has a windpipe with a 20-mm radius. If a foreign object has a radius r, in millimeters, then the velocity V, in millimeters per second, needed to remove the object by a cough is given by

$$V(r) = k(20r^2 - r^3),   0 \le r \le 20,$$

where k is some positive constant. For what size object is the maximum velocity required to remove the object?

108. New York temperatures.   The average temperature in New York can be approximated by the function

$$T(x) = 43.5 - 18.4x + 8.57x^2 - 0.996x^3 + 0.0338x^4,$$

where T represents the temperature, in degrees Fahrenheit, $x = 1$ represents the middle of January, $x = 2$ represents the middle of February, and so on. (*Source:* www.WorldClimate.com.)

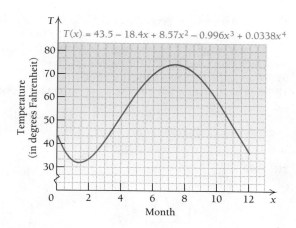

$T(x) = 43.5 - 18.4x + 8.57x^2 - 0.996x^3 + 0.0338x^4$

Month

a) Based on the graph, when would you expect the highest temperature to occur in New York?

b) Based on the graph, when would you expect the lowest temperature to occur?

c) Use the Second-Derivative Test to estimate the points of inflection for the function $T(x)$. What is the significance of these points?

109. Hours of daylight. The number of hours of daylight in Chicago is represented in the graph below. On what dates is the number of hours of daylight changing most rapidly? How can you tell?

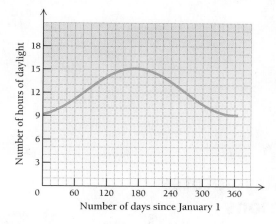

Number of days since January 1

(*Source*: Astronomical Applications Dept., U.S. Naval Observatory.)

SYNTHESIS

In each of Exercises 110 and 111, determine which graph is the derivative of the other and explain why.

110.

111.

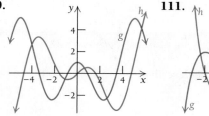
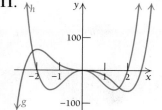

112. Social sciences: three aspects of love. Researchers at Yale University have suggested that the following graphs may represent three different aspects of love.

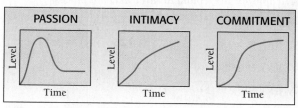

(*Source*: From "A Triangular Theory of Love," by R. J. Sternberg, 1986, *Psychological Review*, 93(2), 119–135. Copyright 1986 by the American Psychological Association, Inc. Reprinted by permission.)

Analyze each of these graphs in terms of the concepts you have learned: relative extrema, concavity, increasing, decreasing, and so on. Do you agree with the researchers regarding the shapes of these graphs? Why or why not?

113. Use calculus to prove that the relative minimum or maximum for any function f for which

$$f(x) = ax^2 + bx + c, a \neq 0,$$

occurs at $x = -b/(2a)$.

114. Use calculus to prove that the point of inflection for any function g given by

$$g(x) = ax^3 + bx^2 + cx + d, \qquad a \neq 0,$$

occurs at $x = -b/(3a)$.

For Exercises 115–121, assume the function f is differentiable over the interval $(-\infty, \infty)$; that is, it is smooth and continuous for all real numbers x and has no corners or vertical tangents. Classify each of the following statements as either true or false. If you choose false, explain why.

115. If f has exactly two critical values at $x = a$ and $x = b$, where $a < b$, then there must exist exactly one point of inflection at $x = c$ such that $a < c < b$. In other words, exactly one point of inflection must exist between any two critical points.

116. If f has exactly two critical values at $x = a$ and $x = b$, where $a < b$, then there must exist at least one point of inflection at $x = c$ such that $a < c < b$. In other words, at least one point of inflection must exist between any two critical points.

117. The function f can have no extrema but can have at least one point of inflection.

118. If the function f has two points of inflection, then there must be a critical value located between those points of inflection.

119. The function f can have a point of inflection at a critical value.

120. The function f can have a point of inflection at an extreme value.

121. The function f can have exactly one extreme value but no points of inflection.

TECHNOLOGY CONNECTION · · · · · · · · · · · · · · ·

Graph each function. Then estimate any relative extrema. Where appropriate, round to three decimal places.

122. $f(x) = 3x^{2/3} - 2x$ **123.** $f(x) = 4x - 6x^{2/3}$

124. $f(x) = x^2(x - 2)^3$ **125.** $f(x) = x^2(1 - x)^3$

126. $f(x) = x - \sqrt{x}$

127. $f(x) = (x - 1)^{2/3} - (x + 1)^{2/3}$

128. Social sciences: time spent on home computer. The following data relate the average number of minutes spent per month on a home computer to a person's age.

Age (in years)	Average Use (in minutes per month)
6.5	363
14.5	645
21	1377
29.5	1727
39.5	1696
49.5	2052
55 and up	2299

(Source: *Media Matrix; The PC Meter Company.*)

a) Use the regression procedures of Section R.6 to fit linear, cubic, and quartic functions $y = f(x)$ to the data, where x is age and y is average use per month. Decide which function best fits the data. Explain.

b) What is the domain of the function?

c) Does the function have any relative extrema? Explain.

Answers to Quick Checks

1. Relative minimum: -4 at $x = -1$; relative maximum: 4 at $x = 1$

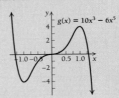

2. $(0, 0)$, $\left(\frac{\sqrt{2}}{2}, 2.475\right)$, and $\left(-\frac{\sqrt{2}}{2}, -2.475\right)$

3. Relative maxima: 17 at $x = -2$ and $x = 2$; relative minimum: 1 at $x = 0$

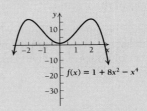

2.3 Graph Sketching: Asymptotes and Rational Functions

OBJECTIVES

- Find limits involving infinity.
- Determine the asymptotes of a function's graph.
- Graph rational functions.

Thus far we have considered a strategy for graphing a continuous function using the tools of calculus. We now want to consider some discontinuous functions, most of which are rational functions. Our graphing skills must now allow for discontinuities as well as certain lines called *asymptotes*.

Let's review the definition of a rational function.

Rational Functions

> **DEFINITION**
>
> A **rational function** is a function f that can be described by
>
> $$f(x) = \frac{P(x)}{Q(x)},$$
>
> where $P(x)$ and $Q(x)$ are polynomials, with $Q(x)$ not the zero polynomial. The domain of f consists of all inputs x for which $Q(x) \neq 0$.

Polynomials are themselves a special kind of rational function, since $Q(x)$ can be 1. Here we are considering graphs of rational functions in which the denominator is not a constant. Before we do so, however, we need to reconsider limits.

Vertical and Horizontal Asymptotes

Figure 1 shows the graph of the rational function

$$f(x) = \frac{x^2 - 1}{x^2 + x - 6} = \frac{(x-1)(x+1)}{(x-2)(x+3)}.$$

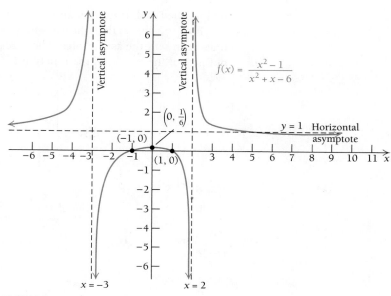

FIGURE 1

Note that as x gets closer to 2 from the left, the function values get smaller and smaller negatively, approaching $-\infty$. As x gets closer to 2 from the right, the function values get larger and larger positively. Thus,

$$\lim_{x \to 2^-} f(x) = -\infty \quad \text{and} \quad \lim_{x \to 2^+} f(x) = \infty.$$

For this graph, we can think of the line $x = 2$ as a "limiting line" called a *vertical asymptote*. Similarly, the line $x = -3$ is another vertical asymptote.

DEFINITION

The line $x = a$ is a **vertical asymptote** if any of the following limit statements is true:

$$\lim_{x \to a} f(x) = \infty, \quad \lim_{x \to a} f(x) = -\infty, \quad \lim_{x \to a^+} f(x) = \infty, \quad \text{or} \quad \lim_{x \to a^+} f(x) = -\infty.$$

The graph of a rational function *never* crosses a vertical asymptote. If the expression that defines the rational function f is simplified, meaning that it has no common factor other than -1 or 1, then if a is an input that makes the denominator 0, the line $x = a$ is a vertical asymptote.

For example,

$$f(x) = \frac{x^2 - 9}{x - 3} = \frac{(x - 3)(x + 3)}{x - 3}$$

does not have a vertical asymptote at $x = 3$, even though 3 is an input that makes the denominator 0. This is because when $(x^2 - 9)/(x - 3)$ is simplified, it has $x - 3$ as a common factor of the numerator and the denominator. In contrast,

$$g(x) = \frac{x^2 - 4}{x^2 + x - 12} = \frac{(x + 2)(x - 2)}{(x - 3)(x + 4)}$$

is simplified and has $x = 3$ and $x = -4$ as vertical asymptotes.

Figure 2 shows the four ways in which a vertical asymptote can occur. The dashed lines represent the asymptotes. They are sketched in for visual assistance only; they are not part of the graphs of the functions.

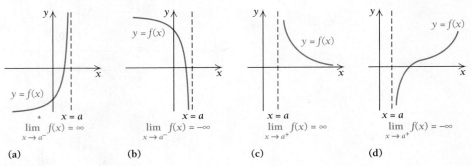

FIGURE 2

> **Quick Check 1**
>
> Determine the vertical asymptotes:
>
> $$f(x) = \frac{1}{x(x^2 - 16)}.$$

■ **EXAMPLE 1**   Determine the vertical asymptotes: $f(x) = \dfrac{3x - 2}{x(x - 5)(x + 3)}.$

Solution   The expression is in simplified form. The vertical asymptotes are the lines $x = 0$, $x = 5$, and $x = -3$.

❬ Quick Check 1

■ **EXAMPLE 2**   Determine the vertical asymptotes of the function given by

$$f(x) = \frac{x^2 - 2x}{x^3 - x}.$$

Solution   We write the expression in simplified form:

$$f(x) = \frac{x^2 - 2x}{x^3 - x} = \frac{x(x - 2)}{x(x - 1)(x + 1)}$$

$$= \frac{x - 2}{(x - 1)(x + 1)}, \quad x \neq 0.$$

The expression is now in simplified form. The vertical asymptotes are the lines $x = -1$ and $x = 1$.

> **Quick Check 2**
>
> For the function in Example 2, explain why $x = 0$ does not correspond to a vertical asymptote. What kind of discontinuity occurs at $x = 0$?

❬ Quick Check 2

Asymptotes

Our discussion here allows us to attach the term "vertical asymptote" to those mysterious vertical lines that appear when graphing rational functions in the **CONNECTED** mode. For example, consider the graph of $f(x) = 8/(x^2 - 4)$, using the window $[-6, 6, -8, 8]$. Vertical asymptotes occur at $x = -2$ and $x = 2$. These lines are not part of the graph.

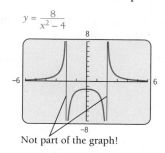

Not part of the graph!

EXERCISES

Graph each of the following in both **DOT** and **CONNECTED** modes. Try to locate the vertical asymptotes visually. Then verify your results using the method of Examples 1 and 2. You may need to try different viewing windows.

1. $f(x) = \dfrac{x^2 + 7x + 10}{x^2 + 3x - 28}$ **2.** $f(x) = \dfrac{x^2 + 5}{x^3 - x^2 - 6x}$

Look again at the graph in Fig. 1. Note that function values get closer and closer to 1 as x approaches $-\infty$, meaning that $f(x) \to 1$ as $x \to -\infty$. Also, function values get closer and closer to 1 as x approaches ∞, meaning that $f(x) \to 1$ as $x \to \infty$. Thus,

$$\lim_{x \to -\infty} f(x) = 1 \quad \text{and} \quad \lim_{x \to \infty} f(x) = 1.$$

The line $y = 1$ is called a *horizontal asymptote*.

DEFINITION

The line $y = b$ is a **horizontal asymptote** if either or both of the following limit statements is true:

$$\lim_{x \to -\infty} f(x) = b \quad \text{or} \quad \lim_{x \to \infty} f(x) = b.$$

The graph of a rational function may or may not cross a horizontal asymptote. Horizontal asymptotes occur when the degree of the numerator is less than or equal to the degree of the denominator. (The degree of a polynomial in one variable is the highest power of that variable.)

In Figs. 3–5, we see three ways in which horizontal asymptotes can occur.

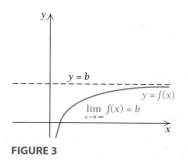

FIGURE 3

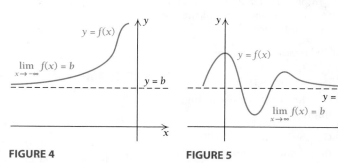

FIGURE 4   **FIGURE 5**

Horizontal asymptotes are found by determining the limit of a rational function as inputs approach $-\infty$ or ∞.

■ **EXAMPLE 3** Determine the horizontal asymptote of the function given by

$$f(x) = \frac{3x - 4}{x}.$$

Solution To find the horizontal asymptote, we consider

$$\lim_{x \to \infty} f(x) = \lim_{x \to \infty} \frac{3x - 4}{x}.$$

One way to find such a limit is to use an input–output table, as follows, using progressively larger x-values.

Inputs, x	1	10	50	100	2000
Outputs, $\dfrac{3x - 4}{x}$	-1	2.6	2.92	2.96	2.998

EXERCISES

1. Verify the limit

$$\lim_{x\to\infty} \frac{3x-4}{x} = 3$$

by using the TABLE feature with larger and larger x-values.

X	Y1	
50	2.92	
150	2.9733	
250	2.984	
350	2.9886	
450	2.9911	
550	2.9927	
650	2.9938	
X = 50		

X	Y1	
500	2.992	
1500	2.9973	
2500	2.9984	
3500	2.9989	
4500	2.9991	
5500	2.9993	
6500	2.9994	
X = 500		

2. Graph the function

$$f(x) = \frac{3x-4}{x}$$

in DOT mode. Then use TRACE, moving the cursor along the graph from left to right, and observe the behavior of the y-coordinates.

For Exercises 3 and 4, consider $\lim_{x\to\infty} \dfrac{2x+5}{x}$.

3. Use the TABLE feature to find the limit.

4. Graph the function in DOT mode, and use TRACE to find the limit.

> **Quick Check 3**
>
> Determine the horizontal asymptote of the function given by
>
> $$f(x) = \frac{(2x-1)(x+1)}{(3x+2)(5x+6)}.$$

As the inputs get larger and larger without bound, the outputs get closer and closer to 3. Thus,

$$\lim_{x\to\infty} \frac{3x-4}{x} = 3.$$

Another way to find this limit is to use some algebra and the fact that

as $x \to \infty$, we have $\dfrac{1}{x} \to 0$, and more generally, $\dfrac{b}{ax^n} \to 0$,

for any positive integer n and any constants a and b, $a \neq 0$. We multiply by 1, using $(1/x) \div (1/x)$. This amounts to dividing both the numerator and the denominator by x:

$$\lim_{x\to\infty} \frac{3x-4}{x} = \lim_{x\to\infty} \frac{3x-4}{x} \cdot \frac{(1/x)}{(1/x)}$$

$$= \lim_{x\to\infty} \frac{\dfrac{3x}{x} - \dfrac{4}{x}}{\dfrac{x}{x}}$$

$$= \lim_{x\to\infty} \frac{3 - \dfrac{4}{x}}{1}$$

$$= \lim_{x\to\infty} \left(3 - \frac{4}{x}\right)$$

$$= 3 - 0 = 3.$$

In a similar manner, it can be shown that

$$\lim_{x\to-\infty} f(x) = 3.$$

The horizontal asymptote is the line $y = 3$.

■ **EXAMPLE 4**   Determine the horizontal asymptote of the function given by

$$f(x) = \frac{3x^2 + 2x - 4}{2x^2 - x + 1}.$$

Solution   As in Example 3, the degree of the numerator is the same as the degree of the denominator. Let's adapt the algebraic approach used in that example.

To do so, we divide the numerator and the denominator by x^2 and find the limit as $|x|$ gets larger and larger:

$$f(x) = \frac{3x^2 + 2x - 4}{2x^2 - x + 1} = \frac{3 + \dfrac{2}{x} - \dfrac{4}{x^2}}{2 - \dfrac{1}{x} + \dfrac{1}{x^2}}.$$

As $|x|$ gets very large, the numerator approaches 3 and the denominator approaches 2. Therefore, the value of the function gets very close to $\frac{3}{2}$. Thus,

$$\lim_{x\to-\infty} f(x) = \frac{3}{2} \quad \text{and} \quad \lim_{x\to\infty} f(x) = \frac{3}{2}.$$

The line $y = \frac{3}{2}$ is a horizontal asymptote.

❮ Quick Check 3

To see why a horizontal asymptote is determined by the leading terms in the expression's numerator and denominator, consider the terms $3x^2$, $2x$, and -4 in Example 4. If $y_1 = 3x^2$ and $y_2 = 3x^2 + 2x - 4$, a table reveals that as x gets larger, the difference between y_2 and y_1 becomes less significant. This is because $3x^2$ grows far more rapidly than does $2x - 4$.

X	Y1	Y2
20	1200	1236
120	43200	43436
220	145200	145636
320	307200	307836
420	529200	530036
520	811200	812236
620	1.15 E6	1.15 E6

X = 20

EXERCISES

1. Let $y_1 = 2x^2$ and $y_2 = 2x^2 - x + 1$. Use a table to show that as x gets large, $y_1 \approx y_2$.

2. Let $y_1 = (3x^2)/(2x^2)$ and $y_2 = (3x^2 + 2x - 1)/(2x^2 - x + 1)$. Show that for large x, we have $y_1 \approx y_2$.

EXERCISES

Graph each of the following. Try to locate the horizontal asymptotes using the TABLE and TRACE features. Verify your results using the methods of Examples 3–5.

1. $f(x) = \dfrac{x^2 + 5}{x^3 - x^2 - 6x}$

2. $f(x) = \dfrac{9x^4 - 7x^2 - 9}{3x^4 + 7x^2 + 9}$

3. $f(x) = \dfrac{135x^5 - x^2}{x^7}$

4. $f(x) = \dfrac{3x^2 - 4x + 3}{6x^2 + 2x - 5}$

Examples 3 and 4 lead to the following result.

> When the degree of the numerator is the same as the degree of the denominator, the line $y = a/b$ is a horizontal asymptote, where a is the leading coefficient of the numerator and b is the leading coefficient of the denominator.

■ **EXAMPLE 5** Determine the horizontal asymptote:

$$f(x) = \frac{2x + 3}{x^3 - 2x^2 + 4}.$$

Solution Since the degree of the numerator is less than the degree of the denominator, there is a horizontal asymptote. To identify that asymptote, we divide both the numerator and denominator by the highest power of x in the denominator, just as in Examples 3 and 4, and find the limits as $|x| \to \infty$:

$$f(x) = \frac{2x + 3}{x^3 - 2x^2 + 4} = \frac{\dfrac{2}{x^2} + \dfrac{3}{x^3}}{1 - \dfrac{2}{x} + \dfrac{4}{x^3}}.$$

As x gets smaller and smaller negatively, $|x|$ gets larger and larger. Similarly, as x gets larger and larger positively, $|x|$ gets larger and larger. Thus, as $|x|$ becomes very large, every expression with a denominator that is a power of x gets ever closer to 0. Thus, the numerator of $f(x)$ approaches 0 as its denominator approaches 1; hence, the entire expression takes on values ever closer to 0. That is, for $x \to -\infty$ or $x \to \infty$, we have

$$f(x) \approx \frac{0 + 0}{1 - 0 + 0},$$

so

$$\lim_{x \to -\infty} f(x) = 0 \quad \text{and} \quad \lim_{x \to \infty} f(x) = 0,$$

and the x-axis, the line $y = 0$, is a horizontal asymptote.

> When the degree of the numerator is less than the degree of the denominator, the x-axis, or the line $y = 0$, is a horizontal asymptote.

Slant Asymptotes

Some asymptotes are neither vertical nor horizontal. For example, in the graph of

$$f(x) = \frac{x^2 - 4}{x - 1},$$

shown at right, as $|x|$ gets larger and larger, the curve gets closer and closer to $y = x + 1$. The line $y = x + 1$ is called a *slant asymptote*, or *oblique asymptote*. In Example 6, we will see how the line $y = x + 1$ was determined.

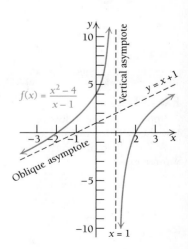

DEFINITION

A linear asymptote that is neither vertical nor horizontal is called a **slant asymptote**, or an **oblique asymptote**. For any rational function of the form $f(x) = p(x)/q(x)$, a slant asymptote occurs when the degree of $p(x)$ is exactly 1 more than the degree of $q(x)$. A graph can cross a slant asymptote.

How can we find a slant asymptote? One way is by division.

■ **EXAMPLE 6** Find the slant asymptote:

$$f(x) = \frac{x^2 - 4}{x - 1}.$$

TECHNOLOGY CONNECTION 〰️

EXERCISES

Graph each of the following. Try to visually locate the slant asymptotes. Then use the method in Example 6 to find each slant asymptote and graph it along with the original function.

1. $f(x) = \dfrac{3x^2 - 7x + 8}{x - 2}$

2. $f(x) = \dfrac{5x^3 + 2x + 1}{x^2 - 4}$

Solution When we divide the numerator by the denominator, we obtain a quotient of $x + 1$ and a remainder of -3:

$$\begin{array}{r} x + 1 \\ x - 1 \overline{)\, x^2 - 4} \\ \underline{x^2 - x } \\ x - 4 \\ \underline{x - 1} \\ -3 \end{array} \qquad f(x) = \frac{x^2 - 4}{x - 1} = (x + 1) + \frac{-3}{x - 1}.$$

Now we can see that when $|x|$ gets very large, $-3/(x - 1)$ approaches 0. Thus, for very large $|x|$, the expression $x + 1$ is the dominant part of

$$(x + 1) + \frac{-3}{x - 1}.$$

Thus, $y = x + 1$ is a slant asymptote.

❮ Quick Check 4

〉 **Quick Check 4**

Find the slant asymptote:

$$g(x) = \frac{2x^2 + x - 1}{x - 3}.$$

Intercepts

If they exist, the **x-intercepts** of a function occur at those values of x for which $y = f(x) = 0$, and they give us points at which the graph crosses the x-axis. If it exists, the **y-intercept** of a function occurs at the value of y for which $x = 0$, and it gives us the point at which the graph crosses the y-axis.

■ **EXAMPLE 7** Find the intercepts of the function given by

$$f(x) = \frac{x^3 - x^2 - 6x}{x^2 - 3x + 2}.$$

TECHNOLOGY CONNECTION 〰️

EXERCISES

Graph each of the following. Use the ZERO feature and a table in ASK mode to find the x- and y-intercepts.

1. $f(x) = \dfrac{x(x - 3)(x + 5)}{(x + 2)(x - 4)}$

2. $f(x) = \dfrac{x^3 + 2x^2 - 3x}{x^2 + 5}$

Solution We factor the numerator and the denominator:

$$f(x) = \frac{x(x + 2)(x - 3)}{(x - 1)(x - 2)}.$$

To find the x-intercepts, we solve the equation $f(x) = 0$. Such values occur when the numerator is 0 and the denominator is not. Thus, we solve the equation

$$x(x + 2)(x - 3) = 0.$$

The x-values that make the numerator 0 are $0, -2$, and 3. Since none of these make the denominator 0, they yield the x-intercepts $(0, 0)$, $(-2, 0)$, and $(3, 0)$.

To find the y-intercept, we let $x = 0$:

$$f(0) = \frac{0^3 - 0^2 - 6(0)}{0^2 - 3(0) + 2} = 0.$$

In this case, the y-intercept is also an x-intercept, $(0, 0)$.

⟨ Quick Check 5

> **Quick Check 5**

Find the intercepts of the function given by

$$h(x) = \frac{x^3 - x}{x^2 - 4}.$$

Sketching Graphs

We can now refine our analytic strategy for graphing.

Strategy for Sketching Graphs

a) *Intercepts.* Find the x-intercept(s) and the y-intercept of the graph.

b) *Asymptotes.* Find any vertical, horizontal, or slant asymptotes.

c) *Derivatives and domain.* Find $f'(x)$ and $f''(x)$. Find the domain of f.

d) *Critical values of f.* Find any inputs for which $f'(x)$ is not defined or for which $f'(x) = 0$.

e) *Increasing and/or decreasing; relative extrema.* Substitute each critical value, x_0, from step (d) into $f''(x)$. If $f''(x_0) < 0$, then x_0 yields a relative maximum and f is increasing to the left of x_0 and decreasing to the right. If $f''(x_0) > 0$, then x_0 yields a relative minimum and f is decreasing to the left of x_0 and increasing to the right. On intervals where no critical value exists, use f' and test values to find where f is increasing or decreasing.

f) *Inflection points.* Determine candidates for inflection points by finding x-values for which $f''(x)$ does not exist or for which $f''(x) = 0$. Find the function values at these points. If a function value $f(x)$ does not exist, then the function does not have an inflection point at x.

g) *Concavity.* Use the values from step (f) as endpoints of intervals. Determine the concavity over each interval by checking to see where f' is increasing—that is, where $f''(x) > 0$—and where f' is decreasing—that is, where $f''(x) < 0$. Do this by substituting a test value from each interval into $f''(x)$. Use the results of step (d).

h) *Sketch the graph.* Use the information from steps (a)–(g) to sketch the graph, plotting extra points as needed.

■ **EXAMPLE 8**   Sketch the graph of $f(x) = \dfrac{8}{x^2 - 4}$.

Solution

a) *Intercepts.* The x-intercepts occur at values for which the numerator is 0 but the denominator is not. Since in this case the numerator is the constant 8, there are no x-intercepts. To find the y-intercept, we compute $f(0)$:

$$f(0) = \frac{8}{0^2 - 4} = \frac{8}{-4} = -2.$$

This gives us one point on the graph, $(0, -2)$.

b) *Asymptotes.*

Vertical: The denominator, $x^2 - 4 = (x + 2)(x - 2)$, is 0 for x-values of -2 and 2. Thus, the graph has the lines $x = -2$ and $x = 2$ as vertical asymptotes. We draw them using dashed lines (they are *not* part of the actual graph, just guidelines).

Horizontal: The degree of the numerator is less than the degree of the denominator, so the *x*-axis, $y = 0$, is the horizontal asymptote.

Slant: There is no slant asymptote since the degree of the numerator is not 1 more than the degree of the denominator.

c) *Derivatives and domain.* We find $f'(x)$ and $f''(x)$ using the Quotient Rule:

$$f'(x) = \frac{-16x}{(x^2 - 4)^2} \quad \text{and} \quad f''(x) = \frac{16(3x^2 + 4)}{(x^2 - 4)^3}.$$

The domain of f is $(-\infty, -2) \cup (-2, 2) \cup (2, \infty)$ as determined in step (b).

d) *Critical values of f.* We look for values of *x* for which $f'(x) = 0$ or for which $f'(x)$ does not exist. From step (c), we see that $f'(x) = 0$ for values of *x* for which $-16x = 0$, but the denominator is not 0. The only such number is 0 itself. The derivative $f'(x)$ does not exist at -2 and 2, but neither value is in the domain of f. Thus, the only critical value is 0.

e) *Increasing and/or decreasing; relative extrema.* We use the undefined values and the critical values to determine the intervals over which f is increasing and the intervals over which f is decreasing. The values to consider are -2, 0, and 2.
Since

$$f''(0) = \frac{16(3 \cdot 0^2 + 4)}{(0^2 - 4)^3} = \frac{64}{-64} < 0,$$

we know that a relative maximum exists at $(0, f(0))$, or $(0, -2)$. Thus, f is increasing on the interval $(-2, 0)$ and decreasing on $(0, 2)$.
Since $f''(x)$ does not exist for the *x*-values -2 and 2, we use $f'(x)$ and test values to see if f is increasing or decreasing on $(-\infty, -2)$ and $(2, \infty)$:

$$\text{Test } -3, \quad f'(-3) = \frac{-16(-3)}{[(-3)^2 - 4]^2} = \frac{48}{25} > 0, \text{ so } f \text{ is increasing on } (-\infty, -2);$$

$$\text{Test } 3, \quad f'(3) = \frac{-16(3)}{[(3)^2 - 4]^2} = \frac{-48}{25} < 0, \text{ so } f \text{ is decreasing on } (2, \infty).$$

f) *Inflection points.* We determine candidates for inflection points by finding where $f''(x)$ does not exist and where $f''(x) = 0$. The only values for which $f''(x)$ does not exist are where $x^2 - 4 = 0$, or -2 and 2. Neither value is in the domain of f, so we focus solely on where $f''(x) = 0$, or

$$16(3x^2 + 4) = 0.$$

Since $16(3x^2 + 4) > 0$ for all real numbers *x*, there are no points of inflection.

g) *Concavity.* Since no values were found in step (f), the only place where concavity could change is on either side of the vertical asymptotes, $x = -2$ and $x = 2$. To determine the concavity, we check to see where $f''(x)$ is positive or negative. The numbers -2 and 2 divide the *x*-axis into three intervals. We choose test values in each interval and make a substitution into f'':

$$\text{Test } -3, \quad f''(-3) = \frac{16[3(-3)^2 + 4]}{[(-3)^2 - 4]^3} > 0;$$

$$\text{Test } 0, \quad f''(0) = \frac{16[3(0)^2 + 4]}{[(0)^2 - 4]^3} < 0; \qquad \text{We already knew this from step (e).}$$

$$\text{Test } 3, \quad f''(3) = \frac{16[3(3)^2 + 4]}{[(3)^2 - 4]^3} > 0.$$

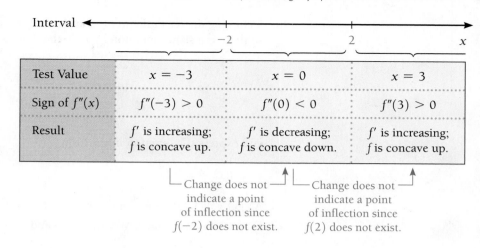

Interval			
	-2	2	x
Test Value	$x = -3$	$x = 0$	$x = 3$
Sign of $f''(x)$	$f''(-3) > 0$	$f''(0) < 0$	$f''(3) > 0$
Result	f' is increasing; f is concave up.	f' is decreasing; f is concave down.	f' is increasing; f is concave up.

Change does not indicate a point of inflection since $f(-2)$ does not exist.

Change does not indicate a point of inflection since $f(2)$ does not exist.

The function is concave up over the intervals $(-\infty, -2)$ and $(2, \infty)$. The function is concave down over the interval $(-2, 2)$.

h) *Sketch the graph.* We sketch the graph using the information in the following table, plotting extra points as needed. The graph is shown below.

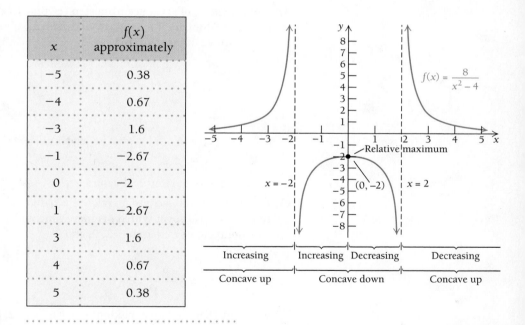

x	$f(x)$ approximately
-5	0.38
-4	0.67
-3	1.6
-1	-2.67
0	-2
1	-2.67
3	1.6
4	0.67
5	0.38

$f(x) = \dfrac{8}{x^2 - 4}$

Relative maximum

$(0, -2)$

$x = -2$ $\qquad$ $x = 2$

Increasing $\quad$ Increasing Decreasing $\quad$ Decreasing

Concave up $\quad$ Concave down $\quad$ Concave up

■ **EXAMPLE 9** Sketch the graph of the function given by $f(x) = \dfrac{x^2 + 4}{x}$.

Solution

a) *Intercepts.* The equation $f(x) = 0$ has no real-number solution. Thus, there are no x-intercepts. The number 0 is not in the domain of the function. Thus, there is no y-intercept.

b) *Asymptotes.*

Vertical: Since replacing x with 0 makes the denominator 0, the line $x = 0$ is a vertical asymptote.

Horizontal: The degree of the numerator is greater than the degree of the denominator, so there is no horizontal asymptote.

Slant: The degree of the numerator is 1 greater than the degree of the denominator, so there is a slant asymptote. We do the division

$$
\begin{array}{r}
x \\
x\overline{)x^2 + 4} \\
\underline{x^2 } \\
4
\end{array}
$$

and express the function in the form

$$f(x) = x + \frac{4}{x}.$$

As $|x|$ gets larger, $4/x$ approaches 0, so the line $y = x$ is a slant asymptote.

c) *Derivatives and domain.* We find $f'(x)$ and $f''(x)$:

$$f'(x) = 1 - 4x^{-2} = 1 - \frac{4}{x^2};$$

$$f''(x) = 8x^{-3} = \frac{8}{x^3}.$$

The domain of f is $(-\infty, 0)\cup(0, \infty)$, or all real numbers except 0.

d) *Critical values of f.* We see from step (c) that $f'(x)$ is undefined at $x = 0$, but 0 is not in the domain of f. Thus, to find critical values, we solve $f'(x) = 0$, looking for solutions other than 0:

$$1 - \frac{4}{x^2} = 0 \qquad \text{Setting } f'(x) \text{ equal to } 0$$

$$1 = \frac{4}{x^2}$$

$$x^2 = 4 \qquad \text{Multiplying both sides by } x^2$$

$$x = \pm 2.$$

Thus, -2 and 2 are critical values.

e) *Increasing and/or decreasing; relative extrema.* We use the points found in step (d) to find intervals over which f is increasing and intervals over which f is decreasing. The points to consider are -2, 0, and 2.

   Since

$$f''(-2) = \frac{8}{(-2)^3} = -1 < 0,$$

we know that a relative maximum exists at $(-2, f(-2))$, or $(-2, -4)$. Thus, f is increasing on $(-\infty, -2)$ and decreasing on $(-2, 0)$.

   Since

$$f''(2) = \frac{8}{(2)^3} = 1 > 0,$$

we know that a relative minimum exists at $(2, f(2))$, or $(2, 4)$. Thus, f is decreasing on $(0, 2)$ and increasing on $(2, \infty)$.

f) *Inflection points.* We determine candidates for inflection points by finding where $f''(x)$ does not exist or where $f''(x) = 0$. The only value for which $f''(x)$ does not exist is 0, but 0 is not in the domain of f. Thus, the only place an inflection point could occur is where $f''(x) = 0$:

$$\frac{8}{x^3} = 0.$$

But this equation has no solution. Thus, there are no points of inflection.

g) *Concavity.* Since no values were found in step (f), the only place where concavity could change would be on either side of the vertical asymptote $x = 0$. In step (e), we used the Second-Derivative Test to determine relative extrema. From that work, we know that f is concave down over the interval $(-\infty, 0)$ and concave up over $(0, \infty)$.

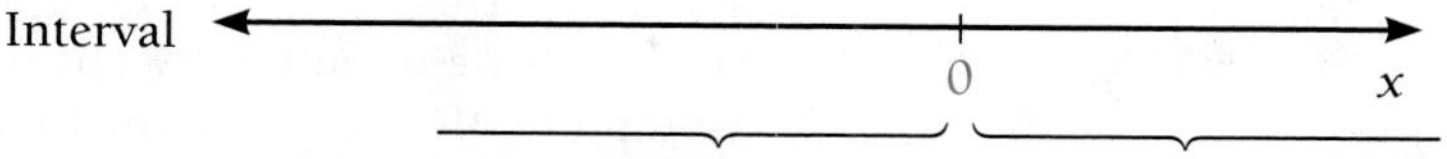

Interval		
Test Value	$x = -2$	$x = 2$
Sign of $f''(x)$	$f''(-2) < 0$	$f''(2) > 0$
Result	f' is decreasing; f is concave down.	f' is increasing; f is concave up.

Change does not indicate a point of inflection since $f(0)$ does not exist.

h) *Sketch the graph.* We sketch the graph using the preceding information and additional computed values of f, as needed. The graph follows.

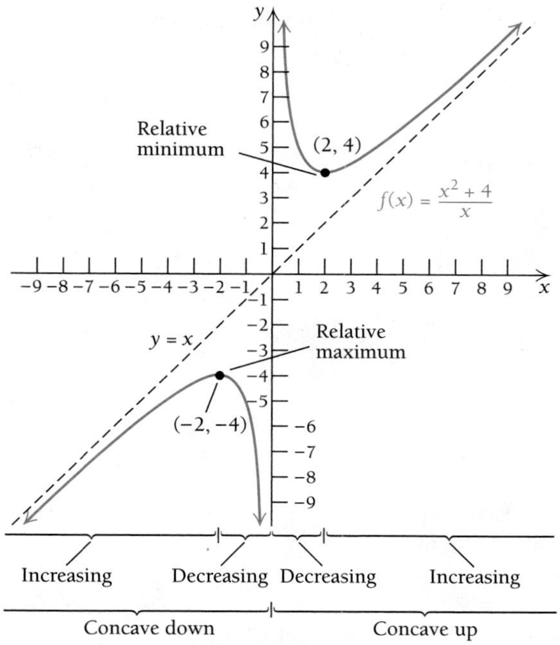

x	$f(x)$, approximately
-6	-6.67
-5	-5.8
-4	-5
-3	-4.3
-2	-4
-1	-5
-0.5	-8.5
0.5	8.5
1	5
2	4
3	4.3
4	5
5	5.8
6	6.67

> **Quick Check 6**
>
> Sketch the graph of the function given by
>
> $$f(x) = \frac{x^2 - 9}{x - 1}.$$

❮ Quick Check 6

We can apply our analytic strategy for graphing to "building" a rational function that meets certain initial conditions.

■ **EXAMPLE 10**   Determine a rational function f (in lowest terms) whose graph has vertical asymptotes at $x = -5$ and $x = 2$ and a horizontal asymptote at $y = 2$ and for which $f(1) = 3$.

Solution   We know that the graph of f has vertical asymptotes at $x = -5$ and $x = 2$, so we can conclude that the denominator must contain the factors $x + 5$ and $x - 2$. Writing these as a product, we see that the denominator will have degree 2 (although we do not carry out the multiplication). Since the graph has a horizontal asymptote, the function must have a polynomial of degree 2 in the numerator, and the leading coefficients must form a ratio of 2. Therefore, a reasonable first guess for f is given by

$$f(x) = \frac{2x^2}{(x + 5)(x - 2)}. \qquad \text{This is a first guess.}$$

However, this does not satisfy the requirement that $f(1) = 3$. We can add a constant in the numerator and then use the fact that $f(1) = 3$ to solve for this constant:

$$f(x) = \frac{2x^2 + B}{(x + 5)(x - 2)}$$

$$3 = \frac{2(1)^2 + B}{(1 + 5)(1 - 2)} \qquad \text{Setting } x = 1 \text{ and } f(x) = 3$$

$$3 = \frac{2 + B}{-6}$$

Multiplying both sides by -6, we have $-18 = 2 + B$. Therefore, $B = -20$. The rational function is given by

$$f(x) = \frac{2x^2 - 20}{(x + 5)(x - 2)}.$$

A sketch of its graph serves as a visual check that all of the initial conditions are met.

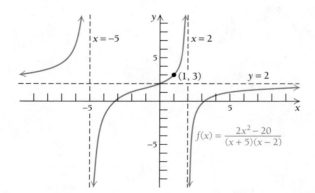

> **Quick Check 7**
>
> Determine a rational function g that has vertical asymptotes at $x = -2$ and $x = 2$ and a horizontal asymptote at $y = 3$, and for which $g(1) = -3$.

❮ Quick Check 7

Section Summary

- A line $x = a$ is a *vertical asymptote* if $\lim_{x \to a^-} f(x) = \pm \infty$ or $\lim_{x \to a^+} f(x) = \pm \infty$.
- A line $y = b$ is a *horizontal asymptote* if $\lim_{x \to \infty} f(x) = b$ or $\lim_{x \to -\infty} f(x) = b$.
- A graph may cross a horizontal asymptote but never a vertical asymptote.

- A *slant asymptote* occurs when the degree of the numerator is 1 greater than the degree of the denominator. Long division of polynomials can be used to determine the equation of the slant asymptote.
- Vertical, horizontal, and slant asymptotes can be used as guides for accurate curve sketching. Asymptotes are not a part of a graph but are visual guides only.

EXERCISE SET
2.3

Determine the vertical asymptote(s) of each function. If none exists, state that fact.

1. $f(x) = \dfrac{2x - 3}{x - 5}$

2. $f(x) = \dfrac{x + 4}{x - 2}$

3. $f(x) = \dfrac{3x}{x^2 - 9}$

4. $f(x) = \dfrac{5x}{x^2 - 25}$

5. $f(x) = \dfrac{x + 2}{x^3 - 6x^2 + 8x}$

6. $f(x) = \dfrac{x + 3}{x^3 - x}$

7. $f(x) = \dfrac{x + 6}{x^2 + 7x + 6}$

8. $f(x) = \dfrac{x + 2}{x^2 + 6x + 8}$

9. $f(x) = \dfrac{6}{x^2 + 36}$

10. $f(x) = \dfrac{7}{x^2 + 49}$

Determine the horizontal asymptote of each function. If none exists, state that fact.

11. $f(x) = \dfrac{6x}{8x + 3}$

12. $f(x) = \dfrac{3x^2}{6x^2 + x}$

13. $f(x) = \dfrac{4x}{x^2 - 3x}$

14. $f(x) = \dfrac{2x}{3x^3 - x^2}$

15. $f(x) = 5 - \dfrac{3}{x}$

16. $f(x) = 4 + \dfrac{2}{x}$

17. $f(x) = \dfrac{8x^4 - 5x^2}{2x^3 + x^2}$

18. $f(x) = \dfrac{6x^3 + 4x}{3x^2 - x}$

19. $f(x) = \dfrac{6x^4 + 4x^2 - 7}{2x^5 - x + 3}$

20. $f(x) = \dfrac{4x^3 - 3x + 2}{x^3 + 2x - 4}$

21. $f(x) = \dfrac{2x^3 - 4x + 1}{4x^3 + 2x - 3}$

22. $f(x) = \dfrac{5x^4 - 2x^3 + x}{x^5 - x^3 + 8}$

Sketch the graph of each function. Indicate where each function is increasing or decreasing, where any relative extrema occur, where asymptotes occur, where the graph is concave up or concave down, where any points of inflection occur, and where any intercepts occur.

23. $f(x) = -\dfrac{5}{x}$

24. $f(x) = \dfrac{4}{x}$

25. $f(x) = \dfrac{1}{x - 5}$

26. $f(x) = \dfrac{-2}{x - 5}$

27. $f(x) = \dfrac{1}{x + 2}$

28. $f(x) = \dfrac{1}{x - 3}$

29. $f(x) = \dfrac{-3}{x - 3}$

30. $f(x) = \dfrac{-2}{x + 5}$

31. $f(x) = \dfrac{3x - 1}{x}$

32. $f(x) = \dfrac{2x + 1}{x}$

33. $f(x) = x + \dfrac{2}{x}$

34. $f(x) = x + \dfrac{9}{x}$

35. $f(x) = \dfrac{-1}{x^2}$

36. $f(x) = \dfrac{2}{x^2}$

37. $f(x) = \dfrac{x}{x + 2}$

38. $f(x) = \dfrac{x}{x - 3}$

39. $f(x) = \dfrac{-1}{x^2 + 2}$

40. $f(x) = \dfrac{1}{x^2 + 3}$

41. $f(x) = \dfrac{x + 3}{x^2 - 9}$ (*Hint: Simplify.*)

42. $f(x) = \dfrac{x - 1}{x^2 - 1}$

43. $f(x) = \dfrac{x - 1}{x + 2}$

44. $f(x) = \dfrac{x - 2}{x + 1}$

45. $f(x) = \dfrac{x^2 - 4}{x + 3}$

46. $f(x) = \dfrac{x^2 - 9}{x + 1}$

47. $f(x) = \dfrac{x + 1}{x^2 - 2x - 3}$

48. $f(x) = \dfrac{x - 3}{x^2 + 2x - 15}$

49. $f(x) = \dfrac{2x^2}{x^2 - 16}$

50. $f(x) = \dfrac{x^2 + x - 2}{2x^2 - 2}$

51. $f(x) = \dfrac{1}{x^2 - 1}$

52. $f(x) = \dfrac{10}{x^2 + 4}$

53. $f(x) = \dfrac{x^2 + 1}{x}$

54. $f(x) = \dfrac{x^3}{x^2 - 1}$

55. $f(x) = \dfrac{x^2 - 9}{x - 3}$

56. $f(x) = \dfrac{x^2 - 16}{x + 4}$

In Exercises 57–62, determine a rational function that meets the given conditions, and sketch its graph.

57. The function f has a vertical asymptote at $x = 2$, a horizontal asymptote at $y = -2$, and $f(0) = 0$.

58. The function f has a vertical asymptote at $x = 0$, a horizontal asymptote at $y = 3$, and $f(1) = 2$.

59. The function g has vertical asymptotes at $x = -1$ and $x = 1$, a horizontal asymptote at $y = 1$, and $g(0) = 2$.

60. The function g has vertical asymptotes at $x = -2$ and $x = 0$, a horizontal asymptote at $y = -3$, and $g(1) = 4$.

61. The function h has vertical asymptotes at $x = -3$ and $x = 2$, a horizontal asymptote at $y = 0$, and $h(1) = 2$.

62. The function h has vertical asymptotes at $x = -\frac{1}{2}$ and $x = \frac{1}{2}$, a horizontal asymptote at $y = 0$, and $h(0) = -3$.

APPLICATIONS

Business and Economics

63. *Depreciation.* Suppose that the value V of the inventory at Fido's Pet Supply decreases, or depreciates, with time t, in months, where

$$V(t) = 50 - \frac{25t^2}{(t + 2)^2}.$$

a) Find $V(0)$, $V(5)$, $V(10)$, and $V(70)$.
b) Find the maximum value of the inventory over the interval $[0, \infty)$.
c) Sketch a graph of V.
d) Does there seem to be a value below which $V(t)$ will never fall? Explain.

64. *Average cost.* The total-cost function for Acme, Inc., to produce x units of a product is given by

$$C(x) = 3x^2 + 80.$$

a) The *average cost* is given by $A(x) = C(x)/x$. Find $A(x)$.
b) Graph the average cost.
c) Find the slant asymptote for the graph of $y = A(x)$, and interpret its significance.

65. *Cost of pollution control.* Cities and companies find that the cost of pollution control increases along with the percentage of pollutants to be removed in a situation. Suppose that the cost C, in dollars, of removing $p\%$ of the pollutants from a chemical spill is given by

$$C(p) = \frac{48{,}000}{100 - p}.$$

a) Find $C(0)$, $C(20)$, $C(80)$, and $C(90)$.
b) Find the domain of C.
c) Sketch a graph of C.
d) Can the company or city afford to remove 100% of the pollutants due to this spill? Explain.

66. *Total cost and revenue.* The total cost and total revenue, in dollars, from producing x couches are given by

$$C(x) = 5000 + 600x \quad \text{and} \quad R(x) = -\tfrac{1}{2}x^2 + 1000x.$$

a) Find the total-profit function, $P(x)$.
b) The *average profit* is given by $A(x) = P(x)/x$. Find $A(x)$.
c) Graph the average profit.
d) Find the slant asymptote for the graph of $y = A(x)$.

67. *Purchasing power.* Since 1970, the purchasing power of the dollar, as measured by consumer prices, can be modeled by the function

$$P(x) = \frac{2.632}{1 + 0.116x},$$

where x is the number of years since 1970. (*Source*: U.S. Bureau of Economic Analysis.)

a) Find $P(10)$, $P(20)$, and $P(40)$.

b) When was the purchasing power 0.50?
c) Find $\lim\limits_{x \to \infty} P(x)$.

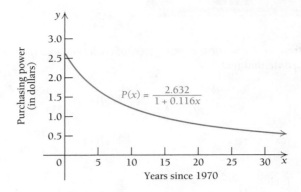

Life and Physical Sciences

68. *Medication in the bloodstream.* After an injection, the amount of a medication A, in cubic centimeters (cc), in the bloodstream decreases with time t, in hours. Suppose that under certain conditions A is given by

$$A(t) = \frac{A_0}{t^2 + 1},$$

where A_0 is the initial amount of the medication. Assume that an initial amount of 100 cc is injected.
a) Find $A(0)$, $A(1)$, $A(2)$, $A(7)$, and $A(10)$.
b) Find the maximum amount of medication in the bloodstream over the interval $[0, \infty)$.
c) Sketch a graph of the function.
d) According to this function, does the medication ever completely leave the bloodstream? Explain your answer.

General Interest

69. *Baseball: earned-run average.* A pitcher's *earned-run average* (the average number of runs given up every 9 innings, or 1 game) is given by

$$E = 9 \cdot \frac{r}{n},$$

where r is the number of earned runs allowed in n innings. Suppose that we fix the number of earned runs allowed at 4 and let n vary. We get a function given by

$$E(n) = 9 \cdot \frac{4}{n}.$$

a) Complete the following table, rounding to two decimal places:

Innings Pitched, n	9	6	3	1	$\frac{2}{3}$	$\frac{1}{3}$
Earned-Run Average, E						

b) The number of innings pitched n is equivalent to the number of outs that a pitcher is able to get while pitching, divided by 3. For example, if the pitcher gets just 1 out, he is credited with pitching $\frac{1}{3}$ of an inning. Find $\lim\limits_{n\to 0} E(n)$. Under what circumstances might this limit be plausible?

c) Suppose a pitcher gives up 4 earned runs over two complete games, or 18 innings. Calculate the pitcher's earned-run average, and interpret this result.

While pitching for the St. Louis Cardinals in 1968, Bob Gibson had an earned-run average of 1.12, a record low.

SYNTHESIS

70. Explain why a vertical asymptote is only a guide and is not part of the graph of a function.

71. Using graphs and limits, explain the idea of an asymptote to the graph of a function. Describe three types of asymptotes.

Find each limit, if it exists.

72. $\lim\limits_{x\to -\infty} \dfrac{-3x^2 + 5}{2 - x}$

73. $\lim\limits_{x\to 0} \dfrac{|x|}{x}$

74. $\lim\limits_{x\to -2} \dfrac{x^3 + 8}{x^2 - 4}$

75. $\lim\limits_{x\to \infty} \dfrac{-6x^3 + 7x}{2x^2 - 3x - 10}$

76. $\lim\limits_{x\to -\infty} \dfrac{-6x^3 + 7x}{2x^2 - 3x - 10}$

77. $\lim\limits_{x\to 1} \dfrac{x^3 - 1}{x^2 - 1}$

78. $\lim\limits_{x\to -\infty} \dfrac{7x^5 + x - 9}{6x + x^3}$

79. $\lim\limits_{x\to -\infty} \dfrac{2x^4 + x}{x + 1}$

TECHNOLOGY CONNECTION

Graph each function using a calculator, iPlot, or Graphicus.

80. $f(x) = x^2 + \dfrac{1}{x^2}$

81. $f(x) = \dfrac{x}{\sqrt{x^2 + 1}}$

82. $f(x) = \dfrac{x^3 + 4x^2 + x - 6}{x^2 - x - 2}$

83. $f(x) = \dfrac{x^3 + 2x^2 - 15x}{x^2 - 5x - 14}$

84. $f(x) = \dfrac{x^3 + 2x^2 - 3x}{x^2 - 25}$

85. $f(x) = \left| \dfrac{1}{x} - 2 \right|$

86. Graph the function

$$f(x) = \dfrac{x^2 - 3}{2x - 4}.$$

a) Find all the x-intercepts.
b) Find the y-intercept.
c) Find all the asymptotes.

87. Graph the function given by

$$f(x) = \dfrac{\sqrt{x^2 + 3x + 2}}{x - 3}.$$

a) Estimate $\lim\limits_{x\to \infty} f(x)$ and $\lim\limits_{x\to -\infty} f(x)$ using the graph and input–output tables as needed to refine your estimates.
b) Describe the outputs of the function over the interval $(-2, -1)$.
c) What appears to be the domain of the function? Explain.
d) Find $\lim\limits_{x\to -2^-} f(x)$ and $\lim\limits_{x\to -1^+} f(x)$.

88. Not all asymptotes are linear. Use long division to find an equation for the nonlinear asymptote that is approached by the graph of

$$f(x) = \dfrac{x^5 + x - 9}{x^3 + 6x}.$$

Then graph the function and its asymptote.

89. Refer to Fig. 1 on p. 235. The function is given by

$$f(x) = \dfrac{x^2 - 1}{x^2 + x - 6}.$$

a) Inspect the graph and estimate the coordinates of any extrema.
b) Find f' and use it to determine the critical values. (*Hint*: you will need the quadratic formula.) Round the x-values to the nearest hundredth.
c) Graph this function in the window $[0, 0.2, 0.16, 0.17]$. Use TRACE or MAXIMUM to confirm your results from part (b).
d) Graph this function in the window $[9.8, 10, 0.9519, 0.95195]$. Use TRACE or MINIMUM to confirm your results from part (b).
e) How close were your estimates of part (a)? Would you have been able to identify the relative minimum point without calculus techniques?

Answers to Quick Checks

1. The lines $x = 0$, $x = 4$, and $x = -4$ are vertical asymptotes.
2. The line $x = 0$ is not a vertical asymptote because $\lim\limits_{x\to 0} f(x) = 2$. There is a deleted point discontinuity at $x = 0$.
3. The line $y = \frac{2}{15}$ is a horizontal asymptote.
4. The line $y = 2x + 7$ is a slant asymptote.
5. x-intercepts: $(1, 0)$, $(-1, 0)$, $(0, 0)$; y-intercept: $(0, 0)$
6.

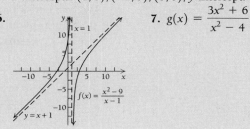

7. $g(x) = \dfrac{3x^2 + 6}{x^2 - 4}$

Using Derivatives to Find Absolute Maximum and Minimum Values

An extremum may be at the highest or lowest point for a function's entire graph, in which case it is called an *absolute extremum*. For example, the parabola given by $f(x) = x^2$ has a relative minimum at $(0, 0)$. This is also the lowest point for the *entire* graph of f, so it is also called the *absolute minimum*. Relative extrema are useful for graph sketching and understanding the behavior of a function. In many applications, however, we are more concerned with absolute extrema.

Absolute Maximum and Minimum Values

A relative minimum may or may not be an absolute minimum, meaning the smallest value of the function over its entire domain. Similarly, a relative maximum may or may not be an absolute maximum, meaning the greatest value of a function over its entire domain.

The function in the following graph has relative minima at interior points c_1 and c_3 of the closed interval $[a, b]$.

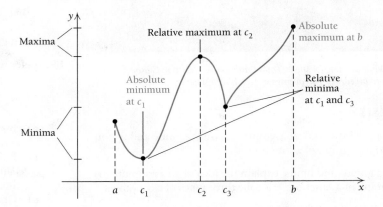

The relative minimum at c_1 is also the absolute minimum. On the other hand, the relative maximum at c_2 is *not* the absolute maximum. The absolute maximum occurs at the endpoint b.

> **DEFINITION**
>
> Suppose that f is a function with domain I.
>
> $f(c)$ is an **absolute minimum** if $f(c) \leq f(x)$ for all x in I.
>
> $f(c)$ is an **absolute maximum** if $f(c) \geq f(x)$ for all x in I.

Finding Absolute Maximum and Minimum Values over Closed Intervals

We first consider a continuous function for which the domain is a closed interval. Look at the graphs in Figs. 1 and 2 and try to determine where the absolute maxima and minima (extrema) occur for each interval.

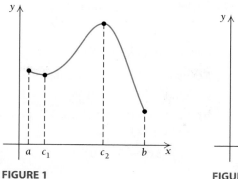

FIGURE 1

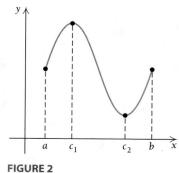

FIGURE 2

Note that each of the functions does indeed have an absolute maximum value and an absolute minimum value. This leads us to the following theorem.

THEOREM   **The Extreme-Value Theorem**

A continuous function f defined over a closed interval $[a, b]$ must have an absolute maximum value and an absolute minimum value over $[a, b]$.

Look again at the graphs in Figs. 1 and 2 and consider the critical values and the endpoints. In Fig. 1, the graph starts at $f(a)$ and falls to $f(c_1)$. Then it rises from $f(c_1)$ to $f(c_2)$. From there, it falls to $f(b)$. In Fig. 2, the graph starts at $f(a)$ and rises to $f(c_1)$. Then it falls from $f(c_1)$ to $f(c_2)$. From there, it rises to $f(b)$. It seems reasonable that whatever the maximum and minimum values are, they occur among the function values $f(a)$, $f(c_1)$, $f(c_2)$, and $f(b)$. This leads us to a procedure for determining *absolute extrema*.

THEOREM 8   **Maximum–Minimum Principle 1**

Suppose that f is a continuous function defined over a closed interval $[a, b]$. To find the absolute maximum and minimum values over $[a, b]$:

a) First find $f'(x)$.

b) Then determine all critical values in $[a, b]$. That is, find all c in $[a, b]$ for which

$$f'(c) = 0 \quad \text{or} \quad f'(c) \text{ does not exist.}$$

c) List the values from step (b) and the endpoints of the interval:

$$a, c_1, c_2, \ldots, c_n, b.$$

d) Evaluate $f(x)$ for each value in step (c):

$$f(a), f(c_1), f(c_2), \ldots, f(c_n), f(b).$$

The largest of these is the **absolute maximum of f over** $[a, b]$. The smallest of these is the **absolute minimum of f over** $[a, b]$.

A reminder: endpoints of a closed interval can be absolute extrema but *not* relative extrema.

■ **EXAMPLE 1**   Find the absolute maximum and minimum values of

$$f(x) = x^3 - 3x + 2$$

over the interval $\left[-2, \frac{3}{2}\right]$.

Solution   Keep in mind that we are considering only the interval $\left[-2, \frac{3}{2}\right]$.

a) Find $f'(x)$: $f'(x) = 3x^2 - 3$.

b) Find the critical values. The derivative exists for all real numbers. Thus, we merely solve $f'(x) = 0$:

$$3x^2 - 3 = 0$$
$$3x^2 = 3$$
$$x^2 = 1$$
$$x = \pm 1.$$

c) List the critical values and the endpoints: $-2, -1, 1,$ and $\frac{3}{2}$.

d) Evaluate f for each value in step (c):

$$f(-2) = (-2)^3 - 3(-2) + 2 = -8 + 6 + 2 = 0; \longrightarrow \text{Minimum}$$
$$f(-1) = (-1)^3 - 3(-1) + 2 = -1 + 3 + 2 = 4; \longrightarrow \text{Maximum}$$
$$f(1) = (1)^3 - 3(1) + 2 = 1 - 3 + 2 = 0; \longrightarrow \text{Minimum}$$
$$f\left(\tfrac{3}{2}\right) = \left(\tfrac{3}{2}\right)^3 - 3\left(\tfrac{3}{2}\right) + 2 = \tfrac{27}{8} - \tfrac{9}{2} + 2 = \tfrac{7}{8}$$

The largest of these values, 4, is the maximum. It occurs at $x = -1$. The smallest of these values is 0. It occurs twice: at $x = -2$ and $x = 1$. Thus, over the interval $\left[-2, \frac{3}{2}\right]$, the

$$\text{absolute maximum} = 4 \text{ at } x = -1$$

and the

$$\text{absolute minimum} = 0 \text{ at } x = -2 \text{ and } x = 1.$$

A visualization of Example 1

Note that an absolute maximum or minimum value can occur at more than one point.

TECHNOLOGY CONNECTION

Finding Absolute Extrema

To find the absolute extrema of Example 1, we can use any of the methods described in the Technology Connection on pp. 209–210. In this case, we adapt Methods 3 and 4.

Method 3

Method 3 is selected because there are relative extrema in the interval $\left[-2, \frac{3}{2}\right]$. This method gives us approximations for the relative extrema.

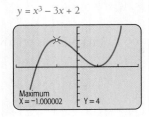

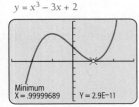

Next, we check function values at these x-values and at the endpoints, using Maximum–Minimum Principle 1 to determine the absolute maximum and minimum values over $\left[-2, \frac{3}{2}\right]$.

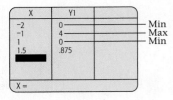

Method 4

Example 2 considers the same function as in Example 1, but over a different interval. Because there are no relative extrema, we can use fMax and fMin features from the MATH menu. The minimum and maximum values occur at the endpoints, as the following graphs show.

(continued)

```
fMin(Y1,X,−3,−1.5)
                     −2.999994692
Y1(Ans)
                     −15.99987261
```

```
fMax(Y1,X,−3,−1.5)
                     −1.500005458
Y1(Ans)
                     3.124979532
```

EXERCISE

1. Use a graph to estimate the absolute maximum and minimum values of $f(x) = x^3 - x^2 - x + 2$, first over the interval $[-2, 1]$ and then over the interval $[-1, 2]$. Then check your work using the methods of Examples 1 and 2.

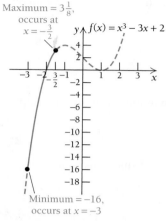

A *visualization of Example 2*

⟩ **Quick Check 1**

Find the absolute maximum and minimum values of the function given in Example 2 over the interval $[0, 3]$.

■ **EXAMPLE 2**   Find the absolute maximum and minimum values of

$$f(x) = x^3 - 3x + 2$$

over the interval $\left[-3, -\frac{3}{2}\right]$.

Solution   As in Example 1, the derivative is 0 at -1 and 1. But neither -1 nor 1 is in the interval $\left[-3, -\frac{3}{2}\right]$, so there are no critical values in this interval. Thus, the maximum and minimum values occur at the endpoints:

$$f(-3) = (-3)^3 - 3(-3) + 2$$
$$= -27 + 9 + 2 = -16; \longrightarrow \text{Minimum}$$
$$f\left(-\frac{3}{2}\right) = \left(-\frac{3}{2}\right)^3 - 3\left(-\frac{3}{2}\right) + 2$$
$$= -\frac{27}{8} + \frac{9}{2} + 2 = \frac{25}{8} = 3\frac{1}{8}. \longrightarrow \text{Maximum}$$

Thus, the absolute maximum over the interval $\left[-3, -\frac{3}{2}\right]$, is $3\frac{1}{8}$, which occurs at $x = -\frac{3}{2}$, and the absolute minimum over $\left[-3, -\frac{3}{2}\right]$ is -16, which occurs at $x = -3$.

⟨ Quick Check 1

Finding Absolute Maximum and Minimum Values over Other Intervals

When there is only one critical value c in I, we may not need to check endpoint values to determine whether the function has an absolute maximum or minimum value at that point.

> **THEOREM 9**   **Maximum–Minimum Principle 2**
>
> Suppose that f is a function such that $f'(x)$ exists for every x in an interval I and that there is *exactly one* (critical) value c in I, for which $f'(c) = 0$. Then
>
> $f(c)$ is the absolute maximum value over I if $f''(c) < 0$
>
> or
>
> $f(c)$ is the absolute minimum value over I if $f''(c) > 0$.

Theorem 9 holds no matter what the interval I is—whether open, closed, or infinite in length. If $f''(c) = 0$, either we must use Maximum–Minimum Principle 1 or we must know more about the behavior of the function over the given interval.

TECHNOLOGY CONNECTION 〰️

Finding Absolute Extrema

Let's do Example 3 graphically, by adapting Methods 1 and 2 of the Technology Connection on pp. 209–210. Strictly speaking, we cannot use the fMin or fMax options of the MATH menu or the MAXIMUM or MINIMUM options from the CALC menu since we do not have a closed interval.

Methods 1 and 2

We create a graph, examine its shape, and use TRACE and/or TABLE. This procedure leads us to see that there is indeed no absolute minimum. We do find an absolute maximum: $f(x) = 4$ at $x = 2$.

EXERCISE

1. Use a graph to estimate the absolute maximum and minimum values of $f(x) = x^2 - 4x$. Then check your work using the method of Example 3.

■ **EXAMPLE 3** Find the absolute maximum and minimum values of

$$f(x) = 4x - x^2.$$

Solution When no interval is specified, we consider the entire domain of the function. In this case, the domain is the set of all real numbers.

a) Find $f'(x)$:

$$f'(x) = 4 - 2x.$$

b) Find the critical values. The derivative exists for all real numbers. Thus, we merely solve $f'(x) = 0$:

$$4 - 2x = 0$$
$$-2x = -4$$
$$x = 2.$$

c) Since there is only one critical value, we can apply Maximum–Minimum Principle 2 using the second derivative:

$$f''(x) = -2.$$

The second derivative is constant. Thus, $f''(2) = -2$, and since this is negative, we have the absolute maximum:

$$f(2) = 4 \cdot 2 - 2^2,$$
$$= 8 - 4 = 4 \text{ at } x = 2.$$

The function has no minimum, as the graph, shown below, indicates.

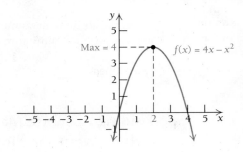

■ **EXAMPLE 4** Find the absolute maximum and minimum values of $f(x) = 4x - x^2$ over the interval $[1, 4]$.

Solution By the reasoning in Example 3, we know that the absolute maximum of f on $(-\infty, \infty)$ is $f(2)$, or 4. Since 2 is in the interval $[1, 4]$, we know that the absolute maximum of f over $[1, 4]$ will occur at 2. To find the absolute minimum, we need to check the endpoints:

$$f(1) = 4 \cdot 1 - 1^2 = 3$$

and

$$f(4) = 4 \cdot 4 - 4^2 = 0.$$

We see from the graph that the minimum is 0. It occurs at $x = 4$. Thus, the

$$\text{absolute maximum} = 4 \text{ at } x = 2,$$

and the

$$\text{absolute minimum} = 0 \text{ at } x = 4.$$

> **Quick Check 2**

Find the absolute maximum and minimum values of $f(x) = x^2 - 10x$ over each interval:

a) $[0, 6]$; **b)** $[4, 10]$.

❮ Quick Check 2

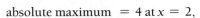

A Strategy for Finding Absolute Maximum and Minimum Values

The following general strategy can be used when finding absolute maximum and minimum values of continuous functions.

A Strategy for Finding Absolute Maximum and Minimum Values

To find absolute maximum and minimum values of a continuous function over an interval:

a) Find $f'(x)$.

b) Find the critical values.

c) If the interval is closed and there is more than one critical value, use Maximum–Minimum Principle 1.

d) If the interval is closed and there is exactly one critical value, use either Maximum–Minimum Principle 1 or Maximum–Minimum Principle 2. If it is easy to find $f''(x)$, use Maximum–Minimum Principle 2.

e) If the interval is not closed, such as $(-\infty, \infty)$, $(0, \infty)$, or (a, b), and the function has only one critical value, use Maximum–Minimum Principle 2. In such a case, if the function has a maximum, it will have no minimum; and if it has a minimum, it will have no maximum.

Finding absolute maximum and minimum values when more than one critical value occurs in an interval that is not closed, such as any of those listed in step (e) above, requires a detailed graph or techniques beyond the scope of this book.

■ **EXAMPLE 5** Find the absolute maximum and minimum values of

$$f(x) = (x - 2)^3 + 1.$$

Solution

a) Find $f'(x)$.

$$f'(x) = 3(x - 2)^2.$$

b) Find the critical values. The derivative exists for all real numbers. Thus, we solve $f'(x) = 0$:

$$3(x - 2)^2 = 0$$
$$(x - 2)^2 = 0$$
$$x - 2 = 0$$
$$x = 2.$$

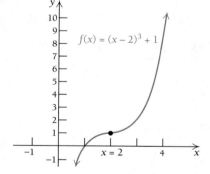

A visualization of Example 5

c) Since there is only one critical value and there are no endpoints, we can try to apply Maximum–Minimum Principle 2 using the second derivative:

$$f''(x) = 6(x - 2).$$

We have

$$f''(2) = 6(2 - 2) = 0,$$

so Maximum–Minimum Principle 2 does not apply. We cannot use Maximum–Minimum Principle 1 because there are no endpoints. But note that $f'(x) = 3(x - 2)^2$ is never negative. Thus, $f(x)$ is increasing everywhere except at $x = 2$, so there is no maximum and no minimum. For $x < 2$, say $x = 1$, we have $f''(1) = -6 < 0$. For $x > 2$, say $x = 3$, we have $f''(3) = 6 > 0$. Thus, at $x = 2$, the function has a *point of inflection*.

> **Quick Check 3**
>
> Let $f(x) = x^n$, where n is a positive odd integer. Explain why functions of this form never have an absolute minimum or maximum.

❮ Quick Check 3

■ **EXAMPLE 6**   Find the absolute maximum and minimum values of

$$f(x) = 5x + \frac{35}{x}$$

over the interval $(0, \infty)$.

Solution

a) Find $f'(x)$. We first express $f(x)$ as

$$f(x) = 5x + 35x^{-1}.$$

Then

$$f'(x) = 5 - 35x^{-2}$$

$$= 5 - \frac{35}{x^2}.$$

b) Find the critical values. Since $f'(x)$ exists for all values of x in $(0, \infty)$, the only critical values are those for which $f'(x) = 0$:

$$5 - \frac{35}{x^2} = 0$$

$$5 = \frac{35}{x^2}$$

$$5x^2 = 35 \qquad \text{Multiplying both sides by } x^2, \text{ since } x \neq 0$$

$$x^2 = 7$$

$$x = \pm\sqrt{7} \approx \pm 2.646.$$

c) The interval is not closed and is $(0, \infty)$. The only critical value is $\sqrt{7}$. Therefore, we can apply Maximum–Minimum Principle 2 using the second derivative,

$$f''(x) = 70x^{-3} = \frac{70}{x^3},$$

to determine whether we have a maximum or a minimum. Since

$$f''(\sqrt{7}) = \frac{70}{(\sqrt{7})^3} > 0,$$

an absolute minimum occurs at $x = \sqrt{7}$:

$$\text{Absolute minimum} = f(\sqrt{7})$$

$$= 5 \cdot \sqrt{7} + \frac{35}{\sqrt{7}}$$

$$= 5\sqrt{7} + \frac{35}{\sqrt{7}} \cdot \frac{\sqrt{7}}{\sqrt{7}}$$

$$= 5\sqrt{7} + \frac{35\sqrt{7}}{7}$$

$$= 5\sqrt{7} + 5\sqrt{7}$$

$$= 10\sqrt{7} \approx 26.458 \quad \text{at } x = \sqrt{7}.$$

TECHNOLOGY CONNECTION

Finding Absolute Extrema

Let's do Example 6 using MAXIMUM and MINIMUM from the CALC menu. The shape of the graph leads us to see that there is no absolute maximum, but there is an absolute minimum.

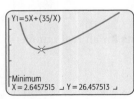

[0, 10, 0, 50]

Note that

$$\sqrt{7} \approx 2.6458 \quad \text{and} \quad 10\sqrt{7} \approx 26.458,$$

which confirms the analytic solution.

EXERCISE

1. Use a graph to estimate the absolute maximum and minimum values of $f(x) = 10x + 1/x$ over the interval $(0, \infty)$. Then check your work using the analytic method of Example 6.

The function has no maximum value, which can happen since the interval $(0, \infty)$ is *not* closed.

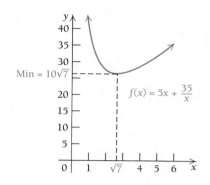

> **Quick Check 4**
>
> Find the absolute maximum and minimum values of
>
> $$g(x) = \frac{2x^2 + 18}{x}$$
>
> over the interval $(0, \infty)$.

《 Quick Check 4

Section Summary

- An *absolute minimum* of a function f is a value $f(c)$ such that $f(c) \le f(x)$ for all x in the domain of f.
- An *absolute maximum* of a function f is a value $f(c)$ such that $f(c) \ge f(x)$ for all x in the domain of f.
- If the domain of f is a closed interval and f is continuous over that domain, then the *Extreme–Value Theorem* guarantees the existence of both an absolute minimum and an absolute maximum.

- Endpoints of a closed interval may be absolute extrema, but not relative extrema.
- If there is exactly one critical value c such that $f'(c) = 0$ in the domain of f, then *Maximum–Minimum Principle 2* may be used. Otherwise, *Maximum–Minimum Principle 1* has to be used.

EXERCISE SET
2.4

1. Fuel economy. According to the U.S. Department of Energy, a vehicle's fuel economy, in miles per gallon (mpg), decreases rapidly for speeds over 60 mph.

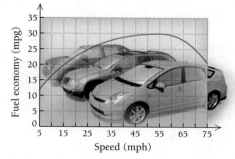

(*Sources*: U.S. Dept. of Energy and a study by West, B.H., McGill, R.N., Hodgson, J.W., Sluder, S.S., and Smith, D.E., Oak Ridge National Laboratory, 1999.)

a) Estimate the speed at which the absolute maximum gasoline mileage is obtained.

b) Estimate the speed at which the absolute minimum gasoline mileage is obtained.

c) What is the mileage obtained at 70 mph?

2. Fuel economy. Using the graph in Exercise 1, estimate the absolute maximum and the absolute minimum fuel economy over the interval $[30, 70]$.

Find the absolute maximum and minimum values of each function over the indicated interval, and indicate the x-values at which they occur.

3. $f(x) = 5 + x - x^2$; $[0, 2]$

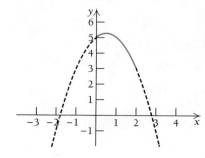

4. $f(x) = 4 + x - x^2$; $[0, 2]$

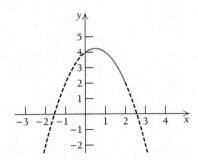

5. $f(x) = x^3 - x^2 - x + 2$; $[-1, 2]$

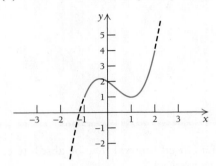

6. $f(x) = x^3 - \frac{1}{2}x^2 - 2x + 5$; $[-2, 1]$

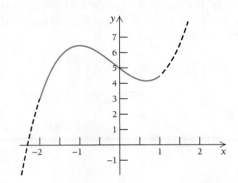

7. $f(x) = x^3 - x^2 - x + 3$; $[-1, 0]$

8. $f(x) = x^3 + \frac{1}{2}x^2 - 2x + 4$; $[-2, 0]$

9. $f(x) = 5x - 7$; $[-2, 3]$

10. $f(x) = 2x + 4$; $[-1, 1]$

11. $f(x) = 7 - 4x$; $[-2, 5]$

12. $f(x) = -2 - 3x$; $[-10, 10]$

13. $f(x) = -5$; $[-1, 1]$

14. $g(x) = 24$; $[4, 13]$

15. $f(x) = x^2 - 6x - 3$; $[-1, 5]$

16. $f(x) = x^2 - 4x + 5$; $[-1, 3]$

17. $f(x) = 3 - 2x - 5x^2$; $[-3, 3]$

18. $f(x) = 1 + 6x - 3x^2$; $[0, 4]$

19. $f(x) = x^3 - 3x^2$; $[0, 5]$

20. $f(x) = x^3 - 3x + 6$; $[-1, 3]$

21. $f(x) = x^3 - 3x$; $[-5, 1]$

22. $f(x) = 3x^2 - 2x^3$; $[-5, 1]$

23. $f(x) = 1 - x^3$; $[-8, 8]$

24. $f(x) = 2x^3$; $[-10, 10]$

25. $f(x) = 12 + 9x - 3x^2 - x^3$; $[-3, 1]$

26. $f(x) = x^3 - 6x^2 + 10$; $[0, 4]$

27. $f(x) = x^4 - 2x^3$; $[-2, 2]$

28. $f(x) = x^3 - x^4$; $[-1, 1]$

29. $f(x) = x^4 - 2x^2 + 5$; $[-2, 2]$

30. $f(x) = x^4 - 8x^2 + 3$; $[-3, 3]$

31. $f(x) = (x + 3)^{2/3} - 5$; $[-4, 5]$

32. $f(x) = 1 - x^{2/3}$; $[-8, 8]$

33. $f(x) = x + \dfrac{1}{x}$; $[1, 20]$

34. $f(x) = x + \dfrac{4}{x}$; $[-8, -1]$

35. $f(x) = \dfrac{x^2}{x^2 + 1}$; $[-2, 2]$

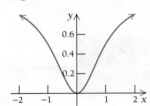

36. $f(x) = \dfrac{4x}{x^2 + 1}$; $[-3, 3]$

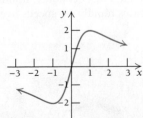

37. $f(x) = (x + 1)^{1/3}$; $[-2, 26]$

38. $f(x) = \sqrt[3]{x}$; $[8, 64]$

39–48. Check Exercises 3, 5, 9, 13, 19, 23, 33, 35, 37, and 38 with a graphing calculator.

Find the absolute maximum and minimum values of each function, if they exist, over the indicated interval. Also indicate the x-value at which each extremum occurs. When no interval is specified, use the real line, $(-\infty, \infty)$.

49. $f(x) = 12x - x^2$

50. $f(x) = 30x - x^2$

51. $f(x) = 2x^2 - 40x + 270$

52. $f(x) = 2x^2 - 20x + 340$

53. $f(x) = x - \frac{4}{3}x^3$; $(0, \infty)$

54. $f(x) = 16x - \frac{4}{3}x^3$; $(0, \infty)$

55. $f(x) = x(60 - x)$

56. $f(x) = x(25 - x)$

57. $f(x) = \frac{1}{3}x^3 - 3x$; $[-2, 2]$

58. $f(x) = \frac{1}{3}x^3 - 5x$; $[-3, 3]$

59. $f(x) = -0.001x^2 + 4.8x - 60$

60. $f(x) = -0.01x^2 + 1.4x - 30$

61. $f(x) = -\frac{1}{3}x^3 + 6x^2 - 11x - 50$; $(0, 3)$

62. $f(x) = -x^3 + x^2 + 5x - 1$; $(0, \infty)$

63. $f(x) = 15x^2 - \frac{1}{2}x^3$; $[0, 30]$

64. $f(x) = 4x^2 - \frac{1}{2}x^3$; $[0, 8]$

65. $f(x) = 2x + \dfrac{72}{x}$; $(0, \infty)$

66. $f(x) = x + \dfrac{3600}{x}$; $(0, \infty)$

67. $f(x) = x^2 + \dfrac{432}{x}$; $(0, \infty)$

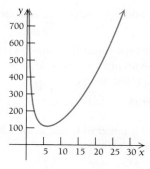

68. $f(x) = x^2 + \dfrac{250}{x}$; $(0, \infty)$

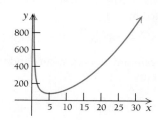

69. $f(x) = 2x^4 - x$; $[-1, 1]$

70. $f(x) = 2x^4 + x$; $[-1, 1]$

71. $f(x) = \sqrt[3]{x}$; $[0, 8]$

72. $f(x) = \sqrt{x}$; $[0, 4]$

73. $f(x) = (x + 1)^3$

74. $f(x) = (x - 1)^3$

75. $f(x) = 2x - 3$; $[-1, 1]$

76. $f(x) = 9 - 5x$; $[-10, 10]$

77. $f(x) = 2x - 3$; $[-1, 5)$

78. $f(x) = 9 - 5x$; $[-2, 3)$

79. $f(x) = x^{2/3}$; $[-1, 1]$

80. $g(x) = x^{2/3}$

81. $f(x) = \frac{1}{3}x^3 - x + \frac{2}{3}$

82. $f(x) = \frac{1}{3}x^3 - \frac{1}{2}x^2 - 2x + 1$

83. $f(x) = \frac{1}{3}x^3 - 2x^2 + x$; $[0, 4]$

84. $g(x) = \frac{1}{3}x^3 + 2x^2 + x$; $[-4, 0]$

85. $t(x) = x^4 - 2x^2$

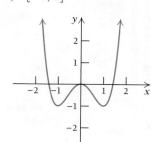

86. $f(x) = 2x^4 - 4x^2 + 2$

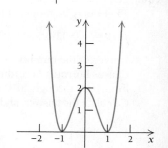

87–96. Check Exercises 49, 51, 53, 57, 61, 65, 67, 69, 73, and 85 with a graphing calculator.

APPLICATIONS

Business and Economics

97. **Monthly productivity.** An employee's monthly productivity M, in number of units produced, is found to be a function of t, the number of years of service. For a certain product, a productivity function is given by

$$M(t) = -2t^2 + 100t + 180, \quad 0 \le t \le 40.$$

Find the maximum productivity and the year in which it is achieved.

98. Advertising. Sound Software estimates that it will sell N units of a program after spending *a* dollars on advertising, where

$$N(a) = -a^2 + 300a + 6, \quad 0 \le a \le 300,$$

and *a* is in thousands of dollars. Find the maximum number of units that can be sold and the amount that must be spent on advertising in order to achieve that maximum.

99. Small business. The percentage of the U.S. national income generated by nonfarm proprietors may be modeled by

$$p(x) = \frac{13x^3 - 240x^2 - 2460x + 585,000}{75,000},$$

where *x* is the number of years since 1970. (*Source*: U.S. Census Bureau.) According to this model, in what year from 1970 through 2000 was this percentage a minimum? Calculate the answer, and then check it on the graph.

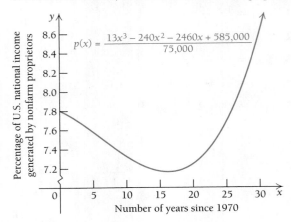

100. The percentage of the U.S. civilian labor force aged 35–44 may be modeled by

$$f(x) = -0.029x^2 + 0.928x + 19.103,$$

where *x* is the number of years since 1980. (*Source*: U.S. Census Bureau.) According to this model, in what year from 1980 through 2010 was this percentage a maximum? Calculate the answer, and then check it on the graph.

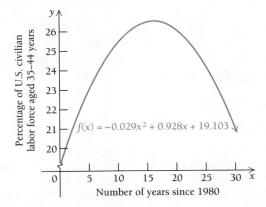

101. Worldwide oil production. One model of worldwide oil production is the function given by

$$P(t) = 0.000008533t^4 - 0.001685t^3 + 0.090t^2 - 0.687t + 4.00, \quad 0 \le t \le 90,$$

where $P(t)$ is the number of barrels, in billions, produced in a year, *t* years after 1950. (*Source*: *Beyond Oil*, by Kenneth S. Deffeyes, p. xii, Hill and Wang, New York, 2005.) According to this model, in what year did worldwide oil production achieve an absolute maximum? What was that maximum? (*Hint*: Do not solve $P'(t) = 0$ algebraically.)

102. Maximizing profit. Corner Stone Electronics determines that its total weekly profit, in dollars, from the production and sale of *x* amplifiers is given by

$$P(x) = \frac{1500}{x^2 - 6x + 10}.$$

Find the number of amplifiers, *x*, for which the total weekly profit is a maximum.

Maximizing profit. *The total-cost and total-revenue functions for producing x items are*

$$C(x) = 5000 + 600x \quad and \quad R(x) = -\frac{1}{2}x^2 + 1000x,$$

where $0 \le x \le 600$. Use these functions for Exercises 103 and 104.

103. a) Find the total-profit function $P(x)$.
  b) Find the number of items, *x*, for which the total profit is a maximum.

104. a) The *average profit* is given by $A(x) = P(x)/x$. Find $A(x)$.
  b) Find the number of items, *x*, for which the average profit is a maximum.

Life and Physical Sciences

105. Blood pressure. For a dosage of *x* cubic centimeters (cc) of a certain drug, the resulting blood pressure B is approximated by

$$B(x) = 305x^2 - 1830x^3, \quad 0 \le x \le 0.16.$$

Find the maximum blood pressure and the dosage at which it occurs.

SYNTHESIS

106. Explain the usefulness of the second derivative in finding the absolute extrema of a function.

For Exercises 107–110, find the absolute maximum and minimum values of each function, and sketch the graph.

107. $f(x) = \begin{cases} 2x + 1 & \text{for } -3 \le x \le 1, \\ 4 - x^2, & \text{for } 1 < x \le 2 \end{cases}$

108. $g(x) = \begin{cases} x^2, & \text{for } -2 \le x \le 0, \\ 5x, & \text{for } 0 < x \le 2 \end{cases}$

109. $h(x) = \begin{cases} 1 - x^2, & \text{for } -4 \le x < 0, \\ 1 - x, & \text{for } 0 \le x < 1, \\ x - 1, & \text{for } 1 \le x \le 2 \end{cases}$

110. $F(x) = \begin{cases} x^2 + 4, & \text{for } -2 \le x < 0, \\ 4 - x, & \text{for } 0 \le x < 3, \\ \sqrt{x - 2}, & \text{for } 3 \le x \le 67 \end{cases}$

111. Consider the piecewise-defined function f defined by:

$$f(x) = \begin{cases} x^2 + 2, & \text{for } -2 \le x \le 0, \\ 2, & \text{for } 0 < x < 4, \\ x - 2, & \text{for } 4 \le x \le 6. \end{cases}$$

   a) Sketch its graph.
   b) Identify the absolute maximum.
   c) How would you describe the absolute minimum?

112. *Physical science: dry lake elevation.* Dry lakes are common in the Western deserts of the United States. These beds of ancient lakes are notable for having perfectly flat terrain. Rogers Dry Lake in California has been used as a landing site for space shuttle missions in recent years. The graph shows the elevation E, in feet, as a function of the distance x, in miles, from a point west ($x = 0$) of Rogers Dry Lake to a point east of the dry lake. (*Source*: www.mytopo.com.)

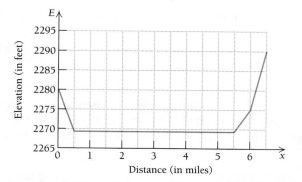

   a) What is the maximum elevation?
   b) How would you describe the minimum elevation?

Find the absolute maximum and minimum values of the function, if they exist, over the indicated interval.

113. $g(x) = x\sqrt{x + 3};$ $[-3, 3]$

114. $h(x) = x\sqrt{1 - x};$ $[0, 1]$

115. *Business: total cost.* Certain costs in a business environment can be separated into two components: those that increase with volume and those that decrease with volume. For example, customer service becomes more expensive as its quality increases, but part of the increased cost is offset by fewer customer complaints. A firm has determined that its cost of service, $C(x)$, in thousands of dollars, is modeled by

$$C(x) = (2x + 4) + \left(\frac{2}{x - 6}\right), \quad x > 6,$$

where x represents the number of "quality units." Find the number of "quality units" that the firm should use in order to minimize its total cost of service.

116. Let

$$y = (x - a)^2 + (x - b)^2.$$

For what value of x is y a minimum?

117. Explain the usefulness of the first derivative in finding the absolute extrema of a function.

TECHNOLOGY CONNECTION

118. *Business: worldwide oil production.* Refer to Exercise 101. In what year was worldwide oil production increasing most rapidly and at what rate was it increasing?

119. *Business: U.S. oil production.* One model of oil production in the United States is given by

$$P(t) = 0.0000000219t^4 - 0.0000167t^3 + 0.00155t^2 + 0.002t + 0.22, \quad 0 \le t \le 110,$$

where $P(t)$ is the number of barrels of oil, in billions, produced in a year, t years after 1910. (*Source: Beyond Oil*, by Kenneth S. Deffeyes, p. 41, Hill and Wang, New York, 2005.)

   a) According to this model, what is the absolute maximum amount of oil produced in the United States and in what year did that production occur?
   b) According to this model, at what rate was United States oil production declining in 2004 and in 2010?

Graph each function over the given interval. Visually estimate where absolute maximum and minimum values occur. Then use the TABLE *feature to refine your estimate.*

120. $f(x) = x^{2/3}(x - 5);$ $[1, 4]$

121. $f(x) = \frac{3}{4}(x^2 - 1)^{2/3};$ $\left[\frac{1}{2}, \infty\right)$

122. $f(x) = x\left(\frac{x}{2} - 5\right)^4;$ $\mathbb{R}$

123. Life and physical sciences: contractions during pregnancy. The following table and graph give the pressure of a pregnant woman's contractions as a function of time.

Time, t (in minutes)	Pressure (in millimeters of mercury)
0	10
1	8
2	9.5
3	15
4	12
5	14
6	14.5

Use a calculator that has the REGRESSION option.

a) Fit a linear equation to the data. Predict the pressure of the contractions after 7 min.

b) Fit a quartic polynomial to the data. Predict the pressure of the contractions after 7 min. Find the smallest contraction over the interval $[0, 10]$.

2.5 Maximum–Minimum Problems; Business and Economics Applications

OBJECTIVE

• Solve maximum–minimum problems using calculus.

An important use of calculus is the solving of maximum–minimum problems, that is, finding the absolute maximum or minimum value of some varying quantity Q and the point at which that maximum or minimum occurs.

■ **EXAMPLE 1** **Maximizing Area.** A hobby store has 20 ft of fencing to fence off a rectangular area for an electric train in one corner of its display room. The two sides up against the wall require no fence. What dimensions of the rectangle will maximize the area? What is the maximum area?

Solution At first glance, we might think that it does not matter what dimensions we use: They will all yield the same area. This is not the case. Let's first make a drawing and express the area in terms of one variable. If we let $x =$ the length, in feet, of one side and $y =$ the length, in feet, of the other, then, since the sum of the lengths must be 20 ft, we have

$$x + y = 20 \quad \text{and} \quad y = 20 - x.$$

Thus, the area is given by

$$\begin{aligned} A &= xy \\ &= x(20 - x) \\ &= 20x - x^2. \end{aligned}$$

EXERCISES

1. Complete this table, using a calculator as needed.

x	$y = 20 - x$	$A = x(20 - x)$
0		
4		
6.5		
8		
10		
12		
13.2		
20		

2. Graph $A(x) = x(20 - x)$ over the interval $[0, 20]$.

3. Estimate the maximum value, and state where it occurs.

> **Quick Check 1**
>
> Repeat Example 1 starting with 50 ft of fencing, and again starting with 100 ft of fencing. Do you detect a pattern? If you had n feet of fencing, what would be the dimensions of the maximum area (in terms of n)?

We are trying to find the maximum value of

$$A(x) = 20x - x^2$$

over the interval $(0, 20)$. We consider the interval $(0, 20)$ because x is a length and cannot be negative or 0. Since there is only 20 ft of fencing, x cannot be greater than 20. Also, x cannot be 20 because then the length of y would be 0.

a) We first find $A'(x)$: $A'(x) = 20 - 2x$.

b) This derivative exists for all values of x in $(0, 20)$. Thus, the only critical values are where

$$A'(x) = 20 - 2x = 0$$
$$-2x = -20$$
$$x = 10.$$

Since there is only one critical value, we can use the second derivative to determine whether we have a maximum. Note that

$$A''(x) = -2,$$

which is a constant. Thus, $A''(10)$ is negative, so $A(10)$ is a maximum. Now

$$A(10) = 10(20 - 10)$$
$$= 10 \cdot 10$$
$$= 100.$$

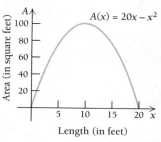

Thus, the maximum area of 100 ft^2 is obtained using 10 ft for the length of one side and $20 - 10$, or 10 ft for the other. Note that $A(5) = 75$, $A(16) = 64$, and $A(12) = 96$; so length does affect area.

⟨ Quick Check 1

Here is a general strategy for solving maximum–minimum problems. Although it may not guarantee success, it should certainly improve your chances.

A Strategy for Solving Maximum–Minimum Problems

1. Read the problem carefully. If relevant, make a drawing.

2. Make a list of appropriate variables and constants, noting what varies, what stays fixed, and what units are used. Label the measurements on your drawing, if one exists.

3. Translate the problem to an equation involving a quantity Q to be maximized or minimized. Try to represent Q in terms of the variables of step 2.

4. Try to express Q as a function of one variable. Use the procedures developed in Sections 2.1–2.4 to determine the maximum or minimum values and the points at which they occur.

■ **EXAMPLE 2**   **Maximizing Volume.**   From a thin piece of cardboard 8 in. by 8 in., square corners are cut out so that the sides can be folded up to make a box. What dimensions will yield a box of maximum volume? What is the maximum volume?

Solution   We might again think at first that it does not matter what the dimensions are, but our experience with Example 1 suggests otherwise. We make a drawing in which x is the length, in inches, of each square to be cut. It is important to note that since the original square is 8 in. by 8 in., after the smaller squares are removed, the lengths of the sides of the box will be $(8 - 2x)$ in. by $(8 - 2x)$ in.

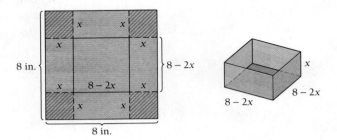

EXERCISES

1. Complete this table to help visualize Example 2.

x	$8 - 2x$	$4x^3 - 32x^2 + 64x$
0		
0.5		
1.0		
1.5		
2.0		
2.5		
3.0		
3.5		
4.0		

2. Graph $V(x) = 4x^3 - 32x^2 + 64x$ over the interval $(0, 4)$.

3. Estimate a maximum value, and state where it occurs.

After the four small squares are removed and the sides are folded up, the volume V of the resulting box is

$$V = l \cdot w \cdot h = (8 - 2x) \cdot (8 - 2x) \cdot x,$$

or $V(x) = (64 - 32x + 4x^2)x = 4x^3 - 32x^2 + 64x.$

Since $8 - 2x > 0$, this means that $x < 4$. Thus, we need to maximize

$$V(x) = 4x^3 - 32x^2 + 64x \quad \text{over the interval } (0, 4).$$

To do so, we first find $V'(x)$:

$$V'(x) = 12x^2 - 64x + 64.$$

Since $V'(x)$ exists for all x in the interval $(0, 4)$, we can set it equal to 0 to find the critical values:

$$V'(x) = 12x^2 - 64x + 64 = 0$$
$$4(3x^2 - 16x + 16) = 0$$
$$4(3x - 4)(x - 4) = 0$$
$$3x - 4 = 0 \quad \text{or} \quad x - 4 = 0$$
$$3x = 4 \quad \text{or} \qquad x = 4$$
$$x = \tfrac{4}{3} \quad \text{or} \qquad x = 4.$$

The only critical value in $(0, 4)$ is $\tfrac{4}{3}$. Thus, we can use the second derivative,

$$V''(x) = 24x - 64,$$

to determine whether we have a maximum. Since

$$V''\left(\tfrac{4}{3}\right) = 24 \cdot \tfrac{4}{3} - 64 = 32 - 64 < 0,$$

we know that $V\left(\tfrac{4}{3}\right)$ is a maximum.

Thus, to maximize the box's volume, small squares with edges measuring $\frac{4}{3}$ in., or $1\frac{1}{3}$ in., should be cut from each corner of the original 8 in. by 8 in. piece of cardboard. When the sides are folded up, the resulting box will have sides of length

$$8 - 2x = 8 - 2 \cdot \frac{4}{3} = \frac{24}{3} - \frac{8}{3} = \frac{16}{3} = 5\frac{1}{3} \text{ in.}$$

and a height of $1\frac{1}{3}$ in. The maximum volume is

$$V\left(\frac{4}{3}\right) = 4\left(\frac{4}{3}\right)^3 - 32\left(\frac{4}{3}\right)^2 + 64\left(\frac{4}{3}\right) = \frac{1024}{27} = 37\frac{25}{27} \text{ in}^3.$$

❮ Quick Check 2

❯ **Quick Check 2**

Repeat Example 2 starting with a sheet of cardboard measuring 8.5 in. by 11 in. (the size of a typical sheet of paper). Will this box hold 1 liter (L) of liquid? (*Hint:* $1\text{ L} = 1000\text{ cm}^3$ and $1\text{ in}^3 = 16.38\text{ cm}^3$.)

In manufacturing, minimizing the amount of material used is always preferred, both from a cost standpoint and in terms of efficiency.

■ **EXAMPLE 3** **Minimizing Material: Surface Area.** A manufacturer of food-storage containers makes a cylindrical can with a volume of 500 milliliters (mL; $1\text{ mL} = 1\text{ cm}^3$). What dimensions (height and radius) will minimize the material needed to produce each can, that is, minimize the surface area?

Solution We let $h = $ height of the can and $r = $ radius, both measured in centimeters. The formula for volume of a cylinder is

$$V = \pi r^2 h.$$

Since we know the volume is 500 cm^3, this formula allows us to relate h and r, expressing one in terms of the other. It is easier to solve for h in terms of r:

$$\pi r^2 h = 500$$
$$h = \frac{500}{\pi r^2}.$$

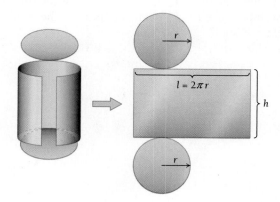

The can is composed of two circular ends, each with an area equal to πr^2, and a side wall that, when laid out flat, is a rectangle with a height h and a length the same as the circumference of the circular ends, or $2\pi r$. Thus, the area of this rectangle is $2\pi rh$.

The total surface area A is the sum of the areas of the two circular ends and the side wall:

$$A = 2(\pi r^2) + (2\pi rh)$$

$$= 2\pi r^2 + 2\pi r\left(\frac{500}{\pi r^2}\right). \qquad \text{Substituting for } h.$$

Simplifying, we have area A as a function of radius r:

$$A(r) = 2\pi r^2 + \frac{1000}{r}.$$

The nature of this problem situation requires that $r > 0$. We differentiate:

$$A'(r) = 4\pi r - \frac{1000}{r^2} \qquad \text{Note that } \frac{d}{dr}\left(\frac{1000}{r}\right) = \frac{d}{dr}\left(1000r^{-1}\right) = -1000r^{-2} = -\frac{1000}{r^2}.$$

We set the derivative equal to 0 and solve for r to determine the critical values:

$$4\pi r - \frac{1000}{r^2} = 0$$

$$4\pi r = \frac{1000}{r^2}$$

$$4\pi r^3 = 1000$$

$$r^3 = \frac{1000}{4\pi} = \frac{250}{\pi}$$

$$r = \sqrt[3]{\frac{250}{\pi}} \approx 4.3 \text{ cm.}$$

The critical value $r \approx 4.3$ is the only critical value in the interval $(0, \infty)$. The second derivative is

$$A''(r) = 4\pi + \frac{2000}{r^3}.$$

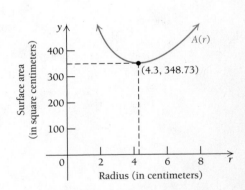

Evaluating $A''(x)$ at the critical value, we get a positive value:

$$A''(4.3) = 4\pi + \frac{2000}{(4.3)^3} > 0.$$

⟩ **Quick Check 3**

Repeat Example 3 for a cylindrical can with a volume of 1000 cm³. What do you notice about the relationship between the cylinder's radius and height? Repeat the example again for any other volume. Does the relationship between radius and height still hold? State this relationship.

The graph is concave up at the critical value, so the critical value indicates a minimum point. Thus, the radius should be $\sqrt[3]{\dfrac{250}{\pi}}$, or approximately 4.3 cm. The height is approximately $h = \dfrac{500}{\pi(4.3)^2} \approx 8.6$ cm, and the minimum total surface area is approximately 348.73 cm². Assuming that the material used for the side and the ends costs the same, minimizing the surface area will also minimize the cost to produce each can.

⟨ Quick Check 3

■ EXAMPLE 4 **Business: Maximizing Revenue.** A stereo manufacturer determines that in order to sell x units of a new stereo, the price per unit, in dollars, must be

$$p(x) = 1000 - x.$$

The manufacturer also determines that the total cost of producing x units is given by

$$C(x) = 3000 + 20x.$$

a) Find the total revenue $R(x)$.

b) Find the total profit $P(x)$.

c) How many units must the company produce and sell in order to maximize profit?

d) What is the maximum profit?

e) What price per unit must be charged in order to make this maximum profit?

Solution

a) $R(x)$ = Total revenue
$$= \text{(Number of units)} \cdot \text{(Price per unit)}$$
$$= \qquad x \qquad \cdot \qquad p$$
$$= x(1000 - x) = 1000x - x^2$$

b) $P(x)$ = Total revenue − Total cost
$$= R(x) - C(x)$$
$$= (1000x - x^2) - (3000 + 20x)$$
$$= -x^2 + 980x - 3000$$

c) To find the maximum value of $P(x)$, we first find $P'(x)$:

$$P'(x) = -2x + 980.$$

This is defined for all real numbers, so the only critical values will come from solving $P'(x) = 0$:

$$P'(x) = -2x + 980 = 0$$
$$-2x = -980$$
$$x = 490.$$

There is only one critical value. We can therefore try to use the second derivative to determine whether we have an absolute maximum. Note that

$$P''(x) = -2, \text{ a constant.}$$

Thus, $P''(490)$ is negative, and so profit is maximized when 490 units are produced and sold.

d) The maximum profit is given by

$$P(490) = -(490)^2 + 980 \cdot 490 - 3000$$
$$= \$237{,}100.$$

Thus, the stereo manufacturer makes a maximum profit of \$237,100 by producing and selling 490 stereos.

e) The price per unit needed to make the maximum profit is

$$p = 1000 - 490 = \$510.$$

❰ Quick Check 4

> **Quick Check 4**
>
> Repeat Example 4 with the price function
> $$p(x) = 1750 - 2x$$
> and the cost function
> $$C(x) = 2250 + 15x.$$
> Round your answers when necessary.

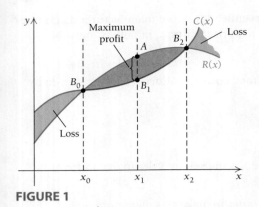

FIGURE 1

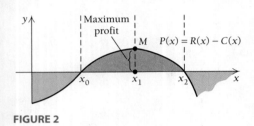

FIGURE 2

Let's take a general look at the total-profit function and its related functions. Figure 1 shows an example of total-cost and total-revenue functions. We can estimate what the maximum profit might be by looking for the widest gap between $R(x)$ and $C(x)$, when $R(x) > C(x)$. Points B_0 and B_2 are break-even points.

Figure 2 shows the related total-profit function. Note that when production is too low ($< x_0$), there is a loss, perhaps due to high fixed or initial costs and low revenue. When production is too high ($> x_2$), there is also a loss, perhaps due to the increased cost of overtime pay or expansion.

The business operates at a profit everywhere between x_0 and x_2. Note that maximum profit occurs at a critical value x_1 of $P(x)$. If we assume that $P'(x)$ exists for all x in some interval, usually $[0, \infty)$, this critical value occurs at some number x such that

$$P'(x) = 0 \quad \text{and} \quad P''(x) < 0.$$

Since $P(x) = R(x) - C(x)$, it follows that

$$P'(x) = R'(x) - C'(x) \quad \text{and} \quad P''(x) = R''(x) - C''(x).$$

Thus, the maximum profit occurs at some number x such that

$$P'(x) = R'(x) - C'(x) = 0 \quad \text{and} \quad P''(x) = R''(x) - C''(x) < 0,$$

or

$$R'(x) = C'(x) \quad \text{and} \quad R''(x) < C''(x).$$

In summary, we have the following theorem.

THEOREM 10

Maximum profit occurs at those x-values for which

$$R'(x) = C'(x) \quad \text{and} \quad R''(x) < C''(x).*$$

You can check that the results in parts (c) and (d) of Example 4 can be easily found using Theorem 10.

■ **EXAMPLE 5** **Business: Determining a Ticket Price.** Promoters of international fund-raising concerts must walk a fine line between profit and loss, especially when determining the price to charge for admission to closed-circuit TV showings in local theaters. By keeping records, a theater determines that at an admission price of $26, it averages 1000 people in attendance. For every drop in price of $1, it gains 50 customers. Each customer spends an average of $4 on concessions. What admission price should the theater charge in order to maximize total revenue?

Solution Let x be the number of dollars by which the price of $26 should be decreased. (If x is negative, the price is increased.) We first express the total revenue R as a function of x. Note that the increase in ticket sales is $50x$ when the price drops x dollars:

$$\begin{aligned}
R(x) &= (\text{Revenue from tickets}) + (\text{Revenue from concessions}) \\
&= (\text{Number of people}) \cdot (\text{Ticket price}) + (\text{Number of people}) \cdot 4 \\
&= (1000 + 50x)(26 - x) + (1000 + 50x) \cdot 4 \\
&= 26{,}000 - 1000x + 1300x - 50x^2 + 4000 + 200x,
\end{aligned}$$

or

$$R(x) = -50x^2 + 500x + 30{,}000.$$

---

*In Section 2.6, the concepts of *marginal revenue* and *marginal cost* are introduced, allowing $R'(x) = C'(x)$ to be regarded as Marginal revenue = Marginal cost.

To find x such that $R(x)$ is a maximum, we first find $R'(x)$:

$$R'(x) = -100x + 500.$$

This derivative exists for all real numbers x. Thus, the only critical values are where $R'(x) = 0$; so we solve that equation:

$$-100x + 500 = 0$$
$$-100x = -500$$
$$x = 5 \qquad \text{This corresponds to lowering the price by \$5.}$$

Since this is the only critical value, we can use the second derivative,

$$R''(x) = -100,$$

to determine whether we have a maximum. Since $R''(5)$ is negative, $R(5)$ is a maximum. Therefore, in order to maximize revenue, the theater should charge

$$\$26 - \$5, \quad \text{or} \quad \$21 \text{ per ticket.}$$

《 Quick Check 5

> **〉Quick Check 5**
>
> A baseball team charges $30 per ticket and averages 20,000 people in attendance per game. Each person spends an average of $8 on concessions. For every drop of $1 in the ticket price, the attendance rises by 800 people. What ticket price should the team charge to maximize total revenue?

Minimizing Inventory Costs

A retail business outlet needs to be concerned about inventory costs. Suppose, for example, that an appliance store sells 2500 television sets per year. It *could* operate by ordering all the sets at once. But then the owners would face the carrying costs (insurance, building space, and so on) of storing them all. Thus, they might make several, say 5, smaller orders, so that the largest number they would ever have to store is 500. However, each time they reorder, there are costs for paperwork, delivery charges, labor, and so on. It seems, therefore, that there must be some balance between carrying costs and reorder costs. Let's see how calculus can help determine what that balance might be. We are trying to minimize the following function:

Total inventory costs = (Yearly carrying costs) + (Yearly reorder costs).

The *lot size x* is the largest number ordered each reordering period. If x units are ordered each period, then during that time somewhere between 0 and x units are in stock. To have a representative expression for the amount in stock at any one time in the period, we can use the average, $x/2$. This represents the average amount held in stock over the course of each time period.

Refer to the graphs shown below and on the next page. If the lot size is 2500, then during the period between orders, there are somewhere between 0 and 2500 units in stock. On average, there are $2500/2$, or 1250 units in stock. If the lot size is 1250, then during the period between orders, there are somewhere between 0 and 1250 units in stock. On average, there are $1250/2$, or 625 units in stock. In general, if the lot size is x, the average inventory is $x/2$.

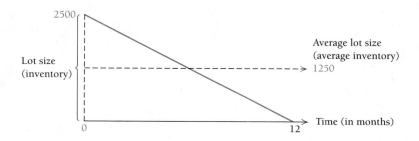

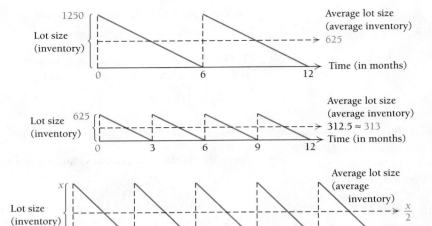

■ **EXAMPLE 6** **Business: Minimizing Inventory Costs.** A retail appliance store sells 2500 television sets per year. It costs $10 to store one set for a year. To reorder, there is a fixed cost of $20, plus a fee of $9 per set. How many times per year should the store reorder, and in what lot size, to minimize inventory costs?

Solution Let x = the lot size. Inventory costs are given by

$$C(x) = (\text{Yearly carrying costs}) + (\text{Yearly reorder costs}).$$

We consider each component of inventory costs separately.

a) *Yearly carrying costs.* The average amount held in stock is $x/2$, and it costs $10 per set for storage. Thus,

$$\text{Yearly carrying costs} = \left(\begin{array}{c}\text{Yearly cost}\\\text{per item}\end{array}\right) \cdot \left(\begin{array}{c}\text{Average number}\\\text{of items}\end{array}\right)$$

$$= 10 \cdot \frac{x}{2}.$$

b) *Yearly reorder costs.* We know that x is the lot size, and we let N be the number of reorders each year. Then $Nx = 2500$, and $N = 2500/x$. Thus,

$$\text{Yearly reorder costs} = \left(\begin{array}{c}\text{Cost of each}\\\text{order}\end{array}\right) \cdot \left(\begin{array}{c}\text{Number of}\\\text{reorders}\end{array}\right)$$

$$= (20 + 9x)\frac{2500}{x}.$$

c) Thus, we have

$$C(x) = 10 \cdot \frac{x}{2} + (20 + 9x)\frac{2500}{x}$$

$$= 5x + \frac{50,000}{x} + 22,500 = 5x + 50,000x^{-1} + 22,500.$$

d) To find a minimum value of C over $[1, 2500]$, we first find $C'(x)$:

$$C'(x) = 5 - \frac{50,000}{x^2}.$$

e) $C'(x)$ exists for all x in $[1, 2500]$, so the only critical values are those x-values such that $C'(x) = 0$. We solve $C'(x) = 0$:

$$5 - \frac{50{,}000}{x^2} = 0$$

$$5 = \frac{50{,}000}{x^2}$$

$$5x^2 = 50{,}000$$

$$x^2 = 10{,}000$$

$$x = \pm 100.$$

Since there is only one critical value in $[1, 2500]$, that is, $x = 100$, we can use the second derivative to see whether it yields a maximum or a minimum:

$$C''(x) = \frac{100{,}000}{x^3}.$$

$C''(x)$ is positive for all x in $[1, 2500]$, so we have a minimum at $x = 100$. Thus, to minimize inventory costs, the store should order $2500/100$, or 25, times per year. The lot size is 100 sets.

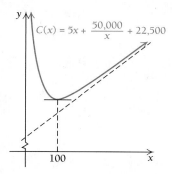

Exploratory

Many calculators can make tables and/or spreadsheets of function values. In reference to Example 6, without using calculus, one might make an estimate of the lot size that will minimize total inventory costs by using a table like the one below. Complete the table, and estimate the solution of Example 6.

EXERCISES

1. Graph $C(x)$ over the interval $[1, 2500]$.

2. Graphically estimate the minimum value, and note where it occurs. Does the table confirm the graph?

Lot Size, x	Number of Reorders, $\frac{2500}{x}$	Average Inventory, $\frac{x}{2}$	Carrying Costs, $10 \cdot \frac{x}{2}$	Cost of Each Order, $20 + 9x$	Reorder Costs, $(20 + 9x)\frac{2500}{x}$	Total Inventory Costs, $C(x) = 10 \cdot \frac{x}{2} + (20 + 9x)\frac{2500}{x}$
2500	1	1250	$12,500	$22,520	$22,520	$35,020
1250	2	625	6,250	11,270	22,540	
500	5	250	2,500	4,520		
250	10	125				
167	15	84				
125	20					
100	25					
90	28					
50	50					

What happens in problems like Example 6 if the answer is not a whole number? For those cases, we consider the two whole numbers closest to the answer and substitute them into $C(x)$. The value that yields the smaller $C(x)$ is the lot size.

■ **EXAMPLE 7**   **Business: Minimizing Inventory Costs.**   Reconsider Example 6, but change the $10 storage cost to $20. How many times per year should the store reorder television sets, and in what lot size, in order to minimize inventory costs?

Solution   Comparing this situation with that in Example 6, we find that the inventory cost function becomes

$$C(x) = 20 \cdot \frac{x}{2} + (20 + 9x)\frac{2500}{x}$$

$$= 10x + \frac{50,000}{x} + 22,500 = 10x + 50,000x^{-1} + 22,500.$$

Then we find $C'(x)$, set it equal to 0, and solve for x:

$$C'(x) = 10 - \frac{50,000}{x^2} = 0$$

$$10 = \frac{50,000}{x^2}$$

$$10x^2 = 50,000$$

$$x^2 = 5000$$

$$x = \sqrt{5000}$$

$$\approx 70.7.$$

It is impossible to reorder 70.7 sets each time, so we consider the two numbers closest to 70.7, which are 70 and 71. Since

$$C(70) \approx \$23,914.29 \quad \text{and} \quad C(71) \approx \$23,914.23,$$

it follows that the lot size that will minimize cost is 71, although the difference, $0.06, is not much. (*Note*: Such a procedure will not work for all functions but will work for the type we are considering here.) The number of times an order should be placed is $2500/71$ with a remainder of 15, indicating that 35 orders should be placed. Of those, $35 - 15 = 20$ will be for 71 items and 15 will be for 72 items.

〉 **Quick Check 6**

Repeat Example 7 with a storage cost of $30 per set and assuming that the store sells 3000 sets per year.

〈 Quick Check 6

The lot size that minimizes total inventory costs is often referred to as the *economic ordering quantity*. Three assumptions are made in using the preceding method to determine the economic ordering quantity. First, the demand for the product is the same year round. For television sets, this may be reasonable, but for seasonal items such as clothing or skis, this assumption is unrealistic. Second, the time between the placing of an order and its receipt remains consistent throughout the year. Finally, the various costs involved, such as storage, shipping charges, and so on, do not vary. This assumption may not be reasonable in a time of inflation, although variation in these costs can be allowed for by anticipating what they might be and using average costs. Regardless, the model described above is useful, allowing us to analyze a seemingly difficult problem using calculus.

Section Summary

- In many real-life applications, we wish to determine the minimum or maximum value of some function modeling a situation.

- Identify a realistic interval for the domain of the input variable. If it is a closed interval, its endpoints should be considered as possible critical values.

EXERCISE SET
2.5

1. Of all numbers whose sum is 50, find the two that have the maximum product. That is, maximize $Q = xy$, where $x + y = 50$.

2. Of all numbers whose sum is 70, find the two that have the maximum product. That is, maximize $Q = xy$, where $x + y = 70$.

3. In Exercise 1, can there be a minimum product? Why or why not?

4. In Exercise 2, can there be a minimum product? Why or why not?

5. Of all numbers whose difference is 4, find the two that have the minimum product.

6. Of all numbers whose difference is 6, find the two that have the minimum product.

7. Maximize $Q = xy^2$, where x and y are positive numbers such that $x + y^2 = 1$.

8. Maximize $Q = xy^2$, where x and y are positive numbers such that $x + y^2 = 4$.

9. Minimize $Q = 2x^2 + 3y^2$, where $x + y = 5$.

10. Minimize $Q = x^2 + 2y^2$, where $x + y = 3$.

11. Maximize $Q = xy$, where x and y are positive numbers such that $\frac{4}{3}x^2 + y = 16$.

12. Maximize $Q = xy$, where x and y are positive numbers such that $x + \frac{4}{3}y^2 = 1$.

13. Maximizing area. A lifeguard needs to rope off a rectangular swimming area in front of Long Lake Beach, using 180 yd of rope and floats. What dimensions of the rectangle will maximize the area? What is the maximum area? (Note that the shoreline is one side of the rectangle.)

14. Maximizing area. A rancher wants to enclose two rectangular areas near a river, one for sheep and one for cattle. There are 240 yd of fencing available. What is the largest total area that can be enclosed?

15. Maximizing area. A carpenter is building a rectangular shed with a fixed perimeter of 54 ft. What are the dimensions of the largest shed that can be built? What is its area?

16. Maximizing area. Of all rectangles that have a perimeter of 42 ft, find the dimensions of the one with the largest area. What is its area?

17. Maximizing volume. From a 50-cm-by-50-cm sheet of aluminum, square corners are cut out so that the sides can be folded up to make a box. What dimensions will yield a box of maximum volume? What is the maximum volume?

18. Maximizing volume. From a thin piece of cardboard 20 in. by 20 in., square corners are cut out so that the sides can be folded up to make a box. What dimensions will yield a box of maximum volume? What is the maximum volume?

19. Minimizing surface area. Drum Tight Containers is designing an open-top, square-based, rectangular box that will have a volume of 62.5 in³. What dimensions will minimize surface area? What is the minimum surface area?

20. Minimizing surface area. A soup company is constructing an open-top, square-based, rectangular metal tank that will have a volume of 32 ft³. What dimensions will minimize surface area? What is the minimum surface area?

21. Minimizing surface area. Open Air Waste Management is designing a rectangular construction dumpster that will be twice as long as it is wide and must hold 12 yd^3 of debris. Find the dimensions of the dumpster that will minimize its surface area.

22. Minimizing surface area. Ever Green Gardening is designing a rectangular compost container that will be twice as tall as it is wide and must hold 18 ft^3 of composted food scraps. Find the dimensions of the compost container with minimal surface area (include the bottom and top).

APPLICATIONS

Business and Economics

Maximizing profit. *Find the maximum profit and the number of units that must be produced and sold in order to yield the maximum profit. Assume that revenue, R(x), and cost, C(x), are in dollars for Exercises 23–26.*

23. $R(x) = 50x - 0.5x^2$, $C(x) = 4x + 10$

24. $R(x) = 50x - 0.5x^2$, $C(x) = 10x + 3$

25. $R(x) = 2x$, $C(x) = 0.01x^2 + 0.6x + 30$

26. $R(x) = 5x$, $C(x) = 0.001x^2 + 1.2x + 60$

27. $R(x) = 9x - 2x^2$, $C(x) = x^3 - 3x^2 + 4x + 1$; assume that $R(x)$ and $C(x)$ are in thousands of dollars, and x is in thousands of units.

28. $R(x) = 100x - x^2$,
$C(x) = \frac{1}{3}x^3 - 6x^2 + 89x + 100$;
assume that $R(x)$ and $C(x)$ are in thousands of dollars, and x is in thousands of units.

29. Maximizing profit. Raggs, Ltd., a clothing firm, determines that in order to sell x suits, the price per suit must be

$$p = 150 - 0.5x.$$

It also determines that the total cost of producing x suits is given by

$$C(x) = 4000 + 0.25x^2.$$

a) Find the total revenue, $R(x)$.
b) Find the total profit, $P(x)$.
c) How many suits must the company produce and sell in order to maximize profit?
d) What is the maximum profit?
e) What price per suit must be charged in order to maximize profit?

30. Maximizing profit. Riverside Appliances is marketing a new refrigerator. It determines that in order to sell x refrigerators, the price per refrigerator must be

$$p = 280 - 0.4x.$$

It also determines that the total cost of producing x refrigerators is given by

$$C(x) = 5000 + 0.6x^2.$$

a) Find the total revenue, $R(x)$.
b) Find the total profit, $P(x)$.
c) How many refrigerators must the company produce and sell in order to maximize profit?
d) What is the maximum profit?
e) What price per refrigerator must be charged in order to maximize profit?

31. Maximizing revenue. A university is trying to determine what price to charge for tickets to football games. At a price of $18 per ticket, attendance averages 40,000 people per game. Every decrease of $3 adds 10,000 people to the average number. Every person at the game spends an average of $4.50 on concessions. What price per ticket should be charged in order to maximize revenue? How many people will attend at that price?

32. Maximizing profit. Gritz-Charlston is a 300-unit luxury hotel. All rooms are occupied when the hotel charges $80 per day for a room. For every increase of x dollars in the daily room rate, there are x rooms vacant. Each occupied room costs $22 per day to service and maintain. What should the hotel charge per day in order to maximize profit?

33. Maximizing yield. An apple farm yields an average of 30 bushels of apples per tree when 20 trees are planted on an acre of ground. Each time 1 more tree is planted per acre, the yield decreases by 1 bushel (bu) per tree as a result of crowding. How many trees should be planted on an acre in order to get the highest yield?

34. Nitrogen prices. During 2001, nitrogen prices fell by 41%. Over the same period of time, nitrogen demand went up by 12%. (*Source: Chemical Week.*)

a) Assuming a linear change in demand, find the demand function, $q(x)$, by finding the equation of the line that passes through the points $(1, 1)$ and $(0.59, 1.12)$. Here x is the price as a fraction of the January 2001 price, and $q(x)$ is the demand as a fraction of the demand in January.

b) As a percentage of the January 2001 price, what should the price of nitrogen be to maximize revenue?

35. Vanity license plates. According to a pricing model, increasing the fee for vanity license plates by $1 decreases the percentage of a state's population that will request them by 0.04%. (*Source: E. D. Craft, "The demand for vanity (plates): Elasticities, net revenue maximization, and deadweight loss," Contemporary Economic Policy, Vol. 20, 133–144 (2002).*)

a) Recently, the fee for vanity license plates in Maryland was $25, and the percentage of the state's population that had vanity plates was 2.13%. Use this information to construct the demand function, $q(x)$, for the percentage of Maryland's population that will request vanity license plates for a fee of x dollars.

b) Find the fee, x, that will maximize revenue from vanity plates.

36. Maximizing revenue. When a theater owner charges $5 for admission, there is an average attendance of 180 people. For every $0.10 increase in admission, there is a loss of 1 customer from the average number. What admission should be charged in order to maximize revenue?

37. Minimizing costs. A rectangular box with a volume of 320 ft^3 is to be constructed with a square base and top. The cost per square foot for the bottom is 15¢, for the top is 10¢, and for the sides is 2.5¢. What dimensions will minimize the cost?

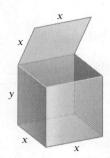

38. Maximizing area. A publisher decides that each page in a new book must have an area of 73.125 in^2, a 0.75-in. margin at the top and at the bottom of each page, and a 0.5-in. margin on each of the sides. What should the outside dimensions of each page be so that the printed area is a maximum?

39. Minimizing inventory costs. A sporting goods store sells 100 pool tables per year. It costs $20 to store one pool table for a year. To reorder, there is a fixed cost of $40 per shipment plus $16 for each pool table. How many times per year should the store order pool tables, and in what lot size, in order to minimize inventory costs?

40. Minimizing inventory costs. A pro shop in a bowling center sells 200 bowling balls per year. It costs $4 to store one bowling ball for a year. To reorder, there is a fixed cost of $1, plus $0.50 for each bowling ball. How many times per year should the shop order bowling balls, and in what lot size, in order to minimize inventory costs?

41. Minimizing inventory costs. A retail outlet for Boxowitz Calculators sells 720 calculators per year. It costs $2 to store one calculator for a year. To reorder, there is a fixed cost of $5, plus $2.50 for each calculator. How many times per year should the store order calculators, and in what lot size, in order to minimize inventory costs?

42. Minimizing inventory costs. Bon Temps Surf and Scuba Shop sells 360 surfboards per year. It costs $8 to store one surfboard for a year. Each reorder costs $10, plus an additional $5 for each surfboard ordered. How many times per year should the store order surfboards, and in what lot size, in order to minimize inventory costs?

43. Minimizing inventory costs. Repeat Exercise 41 using the same data, but assume yearly sales of 256 calculators with the fixed cost of each reorder set at $4.

44. Minimizing inventory costs. Repeat Exercise 42 using the same data, but change the reorder costs from an additional $5 per surfboard to $6 per surfboard.

45. Minimizing surface area. A closed-top cylindrical container is to have a volume of 250 in^2. What dimensions (radius and height) will minimize the surface area?

46. Minimizing surface area. An open-top cylindrical container is to have a volume of 400 cm^2. What dimensions (radius and height) will minimize the surface area?

47. Minimizing cost. Assume that the costs of the materials for making the cylindrical container described in Exercise 45 are $0.005/in^2 for the circular base and top and $0.003/in^2 for the wall. What dimensions will minimize the cost of materials?

48. Minimizing cost. Assume that the costs of the materials for making the cylindrical container described in Exercise 46 are $0.0015/cm^2 for the base and $0.008/cm^2 for the wall. What dimensions will minimize the cost of materials?

General Interest

49. Maximizing volume. The postal service places a limit of 84 in. on the combined length and girth of (distance around) a package to be sent parcel post. What dimensions of a rectangular box with square cross-section will contain the largest volume that can be mailed? (*Hint:* There are two different girths.)

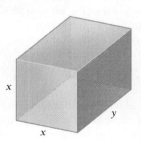

50. Minimizing cost. A rectangular play area is to be fenced off in a person's yard and is to contain 48 yd^2. The next-door neighbor agrees to pay half the cost of the fence on the side of the play area that lies along the property line. What dimensions will minimize the cost of the fence?

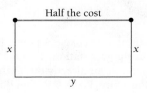

51. Maximizing light. A Norman window is a rectangle with a semicircle on top. Suppose that the perimeter of a particular Norman window is to be 24 ft. What should its dimensions be in order to allow the maximum amount of light to enter through the window?

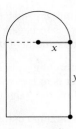

52. Maximizing light. Repeat Exercise 51, but assume that the semicircle is to be stained glass, which transmits only half as much light as clear glass does.

SYNTHESIS

53. For what positive number is the sum of its reciprocal and five times its square a minimum?

54. For what positive number is the sum of its reciprocal and four times its square a minimum?

55. Business: maximizing profit. The amount of money that customers deposit in a bank in savings accounts is directly proportional to the interest rate that the bank pays on that money. Suppose that a bank was able to turn around and loan out all the money deposited in its savings accounts at an interest rate of 18%. What interest rate should it pay on its savings accounts in order to maximize profit?

56. A 24-in. piece of wire is cut in two pieces. One piece is used to form a circle and the other to form a square. How should the wire be cut so that the sum of the areas is a minimum? A maximum?

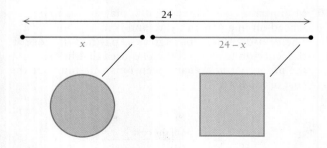

57. Business: minimizing costs. A power line is to be constructed from a power station at point A to an island at point C, which is 1 mi directly out in the water from a point B on the shore. Point B is 4 mi downshore from the power station at A. It costs $5000 per mile to lay the power line under water and $3000 per mile to lay the line under ground. At what point S downshore from A should

the line come to the shore in order to minimize cost? Note that S could very well be B or A. (*Hint*: The length of CS is $\sqrt{1 + x^2}$.)

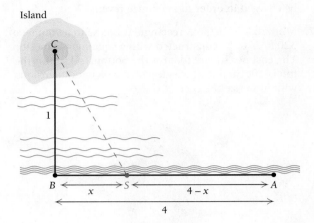

58. Life science: flights of homing pigeons. It is known that homing pigeons tend to avoid flying over water in the daytime, perhaps because the downdrafts of air over water make flying difficult. Suppose that a homing pigeon is released on an island at point C, which is 3 mi directly out in the water from a point B on shore. Point B is 8 mi downshore from the pigeon's home loft at point A. Assume that a pigeon flying over water uses energy at a rate 1.28 times the rate over land. Toward what point S downshore from A should the pigeon fly in order to minimize the total energy required to get to the home loft at A? Assume that

Total energy =
(Energy rate over water) · (Distance over water)
+ (Energy rate over land) · (Distance over land).

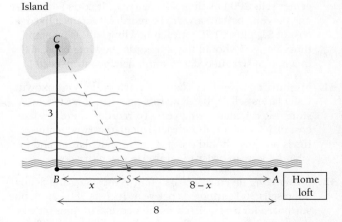

59. Business: minimizing distance. A road is to be built between two cities C_1 and C_2, which are on opposite sides of a river of uniform width r. C_1 is a units from the river, and C_2 is b units from the river, with $a \leq b$. A bridge will carry the traffic across the river. Where should the bridge be located in order to minimize the total distance between the cities? Give a general solution using the constants a, b, p, and r as shown in the figure.

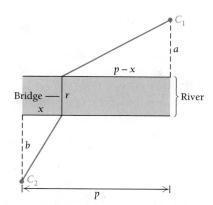

60. Business: minimizing cost. The total cost, in dollars, of producing x units of a certain product is given by

$$C(x) = 8x + 20 + \frac{x^3}{100}.$$

a) Find the average cost, $A(x) = C(x)/x$.
b) Find $C'(x)$ and $A'(x)$.
c) Find the minimum of $A(x)$ and the value x_0 at which it occurs. Find $C'(x_0)$.
d) Compare $A(x_0)$ and $C'(x_0)$.

61. Business: minimizing cost. Consider

$$A(x) = C(x)/x.$$

a) Find $A'(x)$ in terms of $C'(x)$ and $C(x)$.
b) Show that if $A(x)$ has a minimum, then it will occur at that value of x_0 for which

$$C'(x_0) = A(x_0)$$

$$= \frac{C(x_0)}{x_0}.$$

This result shows that if average cost can be minimized, such a minimum will occur when marginal cost equals average cost.

62. Minimize $Q = x^3 + 2y^3$, where x and y are positive numbers, such that $x + y = 1$.

63. Minimize $Q = 3x + y^3$, where $x^2 + y^2 = 2$.

64. Business: minimizing inventory costs—a general solution. A store sells Q units of a product per year. It costs a dollars to store one unit for a year. To reorder, there is a fixed cost of b dollars, plus c dollars for each unit. How many times per year should the store reorder, and in what lot size, in order to minimize inventory costs?

65. Business: minimizing inventory costs. Use the general solution found in Exercise 64 to find how many times per year a store should reorder, and in what lot size, when $Q = 2500$, $a = \$10$, $b = \$20$, and $c = \$9$.

Answers to Quick Checks

1. With 50 ft of fencing, the dimensions are 25 ft by 25 ft (625 ft² area); with 100 ft of fencing, they are 50 ft by 50 ft (2500 ft² area); in general, n feet of fencing gives $n/2$ ft by $n/2$ ft ($n^2/4$ ft² area).
2. The dimensions are approximately 1.585 in. by 5.33 in. by 7.83 in.; the volume is 66.15 in³, or 1083.5 cm³, slightly more than 1 L.
3. $r \approx 5.42$ cm, $h \approx 10.84$ cm, surface area ≈ 553.58 cm²; the relationship is $h = 2r$ (height equals diameter).
4. (a) $R(x) = 1750x - 2x^2$
(b) $P(x) = -2x^2 + 1735x - 2250$ **(c)** $x = 434$ units
(d) Maximum profit $= \$374,028$
(e) Price per unit $= \$882.00$
5. $\$23.50$ **6.** $x \approx 63$; the store should place 8 orders for 63 sets and 39 orders for 64 sets.

Marginals and Differentials

2.6

OBJECTIVES

- Find marginal cost, revenue, and profit.
- Find Δy and dy.
- Use differentials for approximations.

In this section, we consider ways of using calculus to make linear approximations. Suppose, for example, that a company is considering an increase in production. Usually the company wants at least an approximation of what the resulting changes in *cost*, *revenue*, and *profit* will be.

Marginal Cost, Revenue, and Profit

Suppose that a band is producing its own CD and considering an increase in monthly production from 12 cartons to 13. To estimate the resulting increase in cost, it would be reasonable to find the rate at which cost is increasing when 12 cartons are produced and add that to the cost of producing 12 cartons. That is,

$$C(13) \approx C(12) + C'(12).$$

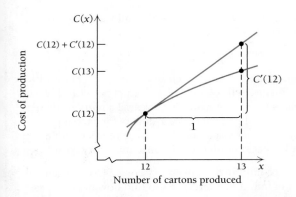

Cost of production

$C(x)$

$C(12) + C'(12)$

$C(13)$

$C'(12)$

$C(12)$

1

12   13 x

Number of cartons produced

The number $C'(12)$ is called the *marginal cost at 12*. Remember that $C'(12)$ is the slope of the tangent line at the point $(12, C(12))$. If, for example, this slope is $\frac{3}{4}$, we can regard it as a vertical change of 3 with a horizontal change of 4, or a vertical change of $\frac{3}{4}$ with a horizontal change of 1. It is this latter interpretation that we use for estimating. Graphically, this interpretation can be viewed as shown at the left. Note in the figure that $C'(12)$ is slightly more than the difference between $C(13)$ and $C(12)$, or $C(13) - C(12)$. For other curves, $C'(12)$ may be slightly less than $C(13) - C(12)$. Almost always, however, it is simpler to compute $C'(12)$ than it is to compute $C(13) - C(12)$.

Generalizing, we have the following.

DEFINITIONS

Let $C(x)$, $R(x)$, and $P(x)$ represent, respectively, the total cost, revenue, and profit from the production and sale of x items.

The **marginal cost** * at x, given by $C'(x)$, is the approximate cost of the $(x + 1)$st item:

$$C'(x) \approx C(x + 1) - C(x), \text{ or } C(x + 1) \approx C(x) + C'(x).$$

The **marginal revenue** at x, given by $R'(x)$, is the approximate revenue from the $(x + 1)$st item:

$$R'(x) \approx R(x + 1) - R(x), \text{ or } R(x + 1) \approx R(x) + R'(x).$$

The **marginal profit** at x, given by $P'(x)$, is the approximate profit from the $(x + 1)$st item:

$$P'(x) \approx P(x + 1) - P(x), \text{ or } P(x + 1) \approx P(x) + P'(x).$$

You can confirm that $P'(x) = R'(x) - C'(x)$.

■ **EXAMPLE 1**   **Business: Marginal Cost, Revenue, and Profit.**   Given

$$C(x) = 62x^2 + 27{,}500 \quad \text{and}$$
$$R(x) = x^3 - 12x^2 + 40x + 10,$$

find each of the following.

a) Total profit, $P(x)$

b) Total cost, revenue, and profit from the production and sale of 50 units of the product

c) The marginal cost, revenue, and profit when 50 units are produced and sold

Solution

a) Total profit $= P(x) = R(x) - C(x)$
$$= x^3 - 12x^2 + 40x + 10 - (62x^2 + 27{,}500)$$
$$= x^3 - 74x^2 + 40x - 27{,}490$$

---

*The term "marginal" comes from the Marginalist School of Economic Thought, which originated in Austria for the purpose of applying mathematics and statistics to the study of economics.

Business: Marginal Revenue, Cost, and Profit

EXERCISE

1. Using the viewing window $[0, 100, 0, 2000]$, graph these total-revenue and total-cost functions:

$$R(x) = 50x - 0.5x^2$$

and

$$C(x) = 10x + 3.$$

Then find $P(x)$ and graph it using the same viewing window. Find $R'(x)$, $C'(x)$, and $P'(x)$, and graph them using $[0, 60, 0, 60]$. Then find $R(40)$, $C(40)$, $P(40)$, $R'(40)$, $C'(40)$, and $P'(40)$. Which marginal function is constant?

b) $C(50) = 62 \cdot 50^2 + 27{,}500 = \$182{,}500$ (the total cost of producing the first 50 units);
$R(50) = 50^3 - 12 \cdot 50^2 + 40 \cdot 50 + 10 = \$97{,}010$ (the total revenue from the sale of the first 50 units);

$$
\begin{aligned}
P(50) &= R(50) - C(50) \\
&= \$97{,}010 - \$182{,}500 \qquad \text{We could also use } P(x) \text{ from part (a).} \\
&= -\$85{,}490 \qquad\qquad \text{There is a } \textit{loss} \text{ of } \$85{,}490 \text{ when 50 units are} \\
&\qquad\qquad\qquad\qquad\qquad \text{produced and sold.}
\end{aligned}
$$

c) $C'(x) = 124x$, so $C'(50) = 124 \cdot 50 = \$6200$. Once 50 units have been made, the approximate cost of the 51st unit (marginal cost) is $\$6200$.

$$R'(x) = 3x^2 - 24x + 40, \text{ so } R'(50) = 3 \cdot 50^2 - 24 \cdot 50 + 40 = \$6340.$$

Once 50 units have been sold, the approximate revenue from the 51st unit (marginal revenue) is $\$6340$.

$$P'(x) = 3x^2 - 148x + 40, \text{ so } P'(50) = 3 \cdot 50^2 - 148 \cdot 50 + 40 = \$140.$$

Once 50 units have been produced and sold, the approximate profit from the sale of the 51st item (marginal profit) is $\$140$.

Often, in business, formulas for $C(x)$, $R(x)$, and $P(x)$ are not known, but information may exist about the cost, revenue, and profit trends at a particular value $x = a$. For example, $C(a)$ and $C'(a)$ may be known, allowing a reasonable prediction to be made about $C(a + 1)$. In a similar manner, predictions can be made for $R(a + 1)$ and $P(a + 1)$. In Example 1, formulas *do* exist, so it is possible to see how accurate our predictions were. We check $C(51) - C(50)$ and leave the checks of $R(51) - R(50)$ and $P(51) - P(50)$ for you (see the Technology Connection below, at left):

$$
\begin{aligned}
C(51) - C(50) &= 62 \cdot 51^2 + 27{,}500 - (62 \cdot 50^2 + 27{,}500) \\
&= 6262,
\end{aligned}
$$

whereas $C'(50) = 6200.$

In this case, $C'(50)$ provides an approximation of $C(51) - C(50)$ that is within 1% of the actual value.

Note that marginal cost is different from *average* cost:

$$
\text{Average cost per unit for 50 units} = \frac{C(50)}{50} \xleftarrow{\text{Total cost of 50 units}}
$$
$$\xleftarrow{\text{The number of units, 50}}$$
$$= \frac{182{,}500}{50} = \$3650,$$

whereas

$$\text{Marginal cost when 50 units are produced} = \$6200$$
$$\approx \text{cost of the 51st unit.}$$

Differentials and Delta Notation

Just as the marginal cost $C'(x_0)$ can be used to estimate $C(x_0 + 1)$, the value of the derivative of any continuous function, $f'(x_0)$, can be used to estimate values of $f(x)$ for x-values near x_0. Before we do so, however, we need to develop some notation.

Recall the difference quotient

$$\frac{f(x + h) - f(x)}{h},$$

To check the accuracy of $R'(50)$ as an estimate of $R(51) - R(50)$, let $y_1 = x^3 - 12x^2 + 40x + 10$, $y_2 = y_1(x + 1) - y_1(x)$, and $y_3 = \text{nDeriv}(y_1, x, x)$. By using **TABLE** with Indpnt: Ask, we can display a table in which y_2 (the difference between $y_1(x + 1)$ and $y_1(x)$) can be compared with $y_1'(x)$.

X	Y2	Y3
40	3989	3880
48	5933	5800
50	6479	6340
▉		

X =

EXERCISE

1. Create a table to check the effectiveness of using $P'(50)$ to approximate $P(51) - P(50)$.

illustrated in the graph at the right. The difference quotient is used to define the derivative of a function at x. The number h is considered to be a *change* in x. Another notation for such a change is Δx, read "delta x" and called **delta notation**. The expression Δx is *not* the product of Δ and x; it is a new type of variable that represents the *change* in the value of x from a *first* value to a *second*. Thus,

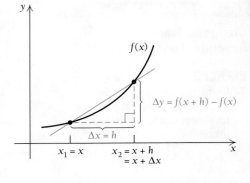

$$\Delta x = (x + h) - x = h.$$

If subscripts are used for the first and second values of x, we have

$$\Delta x = x_2 - x_1, \quad \text{or} \quad x_2 = x_1 + \Delta x.$$

Note that the value of Δx can be positive or negative. For example,

$$\text{if } x_1 = 4 \text{ and } \Delta x = 0.7, \text{ then } x_2 = 4.7,$$

and   if $x_1 = 4$ and $\Delta x = -0.7$, then $x_2 = 3.3$.

We generally omit the subscripts and use x and $x + \Delta x$. Now suppose that we have a function given by $y = f(x)$. A change in x from x to $x + \Delta x$ yields a change in y from $f(x)$ to $f(x + \Delta x)$. The change in y is given by

$$\Delta y = f(x + \Delta x) - f(x).$$

■ **EXAMPLE 2**   For $y = x^2, x = 4$, and $\Delta x = 0.1$, find Δy.

Solution   We have

$$\Delta y = (4 + 0.1)^2 - 4^2$$
$$= (4.1)^2 - 4^2 = 16.81 - 16 = 0.81.$$

■ **EXAMPLE 3**   For $y = x^3, x = 2$, and $\Delta x = -0.1$, find Δy.

Solution   We have

$$\Delta y = [2 + (-0.1)]^3 - 2^3$$
$$= (1.9)^3 - 2^3 = 6.859 - 8 = -1.141.$$

❭ **Quick Check 1**

For $y = 2x^4 + x, x = 2$, and $\Delta x = -0.05$, find Δy.

❬ Quick Check 1

Let's now use calculus to predict function values. If delta notation is used, the difference quotient

$$\frac{f(x + h) - f(x)}{h}$$

becomes

$$\frac{f(x + \Delta x) - f(x)}{\Delta x} = \frac{\Delta y}{\Delta x}.$$

We can then express the derivative as

$$\frac{dy}{dx} = \lim_{\Delta x \to 0} \frac{\Delta y}{\Delta x}.$$

Note that the delta notation resembles Leibniz notation (see Section 1.5).

For values of Δx close to 0, we have the approximation

$$\frac{dy}{dx} \approx \frac{\Delta y}{\Delta x}, \quad \text{or} \quad f'(x) \approx \frac{\Delta y}{\Delta x}.$$

Multiplying both sides of the second expression by Δx gives us

$$\Delta y \approx f'(x)\, \Delta x.$$

We can see this in the graph at the right.

From this graph, it seems reasonable to assume that, for small values of Δx, the y-values on the tangent line can be used to estimate function values on the curve.

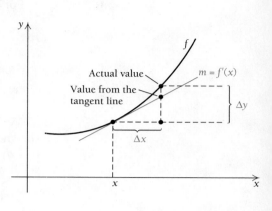

For f, a continuous, differentiable function, and small Δx,

$$f'(x) \approx \frac{\Delta y}{\Delta x} \quad \text{and} \quad \Delta y \approx f'(x) \cdot \Delta x.$$

Let's illustrate this idea by considering the square-root function, $f(x) = \sqrt{x}$. We know how to approximate $\sqrt{27}$ using a calculator. But suppose we didn't. We could begin with $\sqrt{25}$ and use as a change in input $\Delta x = 2$. We would use the corresponding change in y, that is, $\Delta y \approx f'(x)\Delta x$, to estimate $\sqrt{27}$.

■ **EXAMPLE 4** Approximate $\sqrt{27}$ using $\Delta y \approx f'(x)\Delta x$.

Solution We first think of the number closest to 27 that is a perfect square. This is 25. What we will do is approximate how $y = \sqrt{x}$ changes when x changes from 25 to 27.

From the box above, we have

$$\left.\begin{array}{l} \Delta y \approx f'(x) \cdot \Delta x \\ = \tfrac{1}{2}x^{-1/2} \cdot \Delta x \end{array}\right\} \text{Using } y = \sqrt{x} = x^{1/2} \text{ as } f(x)$$

We are interested in Δy as x changes from 25 to 27, so

$$\Delta y \approx \frac{1}{2\sqrt{25}} \cdot 2 \qquad \text{Replacing } x \text{ with 25 and } \Delta x \text{ with 2}$$

$$= \frac{1}{5} = 0.2.$$

We can now approximate $\sqrt{27}$:

$$\sqrt{27} = \sqrt{25} + \Delta y$$
$$= 5 + \Delta y$$
$$\approx 5 + 0.2$$
$$\approx 5.2.$$

To five decimal places, $\sqrt{27} = 5.19615$. Thus, our approximation is fairly accurate.

❮ Quick Check 2

❯ **Quick Check 2**

Approximate $\sqrt{98}$ using $\Delta y \approx f'(x)\,\Delta x$. To five decimal places, $\sqrt{98} = 9.89949$. How close is your approximation?

Up to now, we have not defined the symbols dy and dx as separate entities, but have treated dy/dx as one symbol. We now define dy and dx. These symbols are called **differentials**.

DEFINITION

For $y = f(x)$, we define

dx, called the **differential of x**, by $dx = \Delta x$

and

dy, called the **differential of y**, by $dy = f'(x)\, dx$.

We can illustrate dx and dy as shown at the right. Note that $dx = \Delta x$, but $dy \neq \Delta y$, though $dy \approx \Delta y$, for small values of dx.

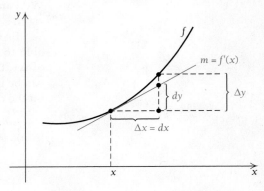

■ **EXAMPLE 5**   For $y = x(4 - x)^3$:

a) Find dy.

b) Find dy when $x = 5$ and $dx = 0.01$.

c) Compare dy to Δy.

Solution

a) First, we find dy/dx:

$$\frac{dy}{dx} = x\big[3(4 - x)^2(-1)\big] + (4 - x)^3 \qquad \text{Using the Product and Chain Rules}$$

$$= -3x(4 - x)^2 + (4 - x)^3$$
$$= (4 - x)^2\big[-3x + (4 - x)\big] \qquad \text{Factoring out } (4 - x)^2$$
$$= (4 - x)^2\big[-4x + 4\big]$$
$$= -4(4 - x)^2(x - 1). \qquad \text{Factoring out } -4$$

Then we solve for dy:

$$dy = -4(4 - x)^2(x - 1)\, dx.$$

b) When $x = 5$ and $dx = 0.01$,

$$dy = -4(4 - 5)^2(5 - 1)(0.01) = -4(-1)^2(4)(0.01) = -0.16.$$

c) The value $dy = -0.16$ is the approximate change in y between $x_1 = 5$ and $x_2 = 5.01$ (that is, $x_2 = x_1 + dx = 5 + 0.01$). The actual change in y is determined by evaluating the function for x_2 and x_1 and subtracting:

$$\Delta y = \big[5.01(4 - 5.01)^3\big] - \big[5(4 - 5)^3\big]$$
$$= \big[5.01(-1.01)^3\big] - \big[5(-1)^3\big]$$
$$= \big[5.01(-1.030301)\big] - \big[5(-1)\big]$$
$$= -0.16180801.$$

We see that the approximation dy and the actual change Δy are reasonably close. It is easier to calculate the approximation since that involves fewer steps, but the trade-off is some loss in accuracy. As long as dx is small, this loss in accuracy is acceptable for many applications.

Differentials are often used in applications involving measurements and tolerance. When we measure an object, we accept that our measurements are not exact, and we allow for a small tolerance in our measurements. If x represents a measurement (a length, a weight, a volume, etc.), then dx represents the tolerance. Even a small tolerance for the input can have a significant effect on the output, as the following example shows.

■ **EXAMPLE 6**   **Business: Cost and Tolerance.**   In preparation for laying new tile, Michelle measures the floor of a large conference room and finds it to be square, measuring 100 ft by 100 ft. Suppose her measurements are accurate to ± 6 in. (the tolerance).

a) Use a differential to estimate the difference in area (dA) due to the tolerance.

b) Compare the result from part (a) with the actual difference in area (ΔA).

c) If each tile covers 1 ft^2 and a box of 12 tiles costs \$24, how much extra cost should Michelle allot for the potential overage in floor area?

Solution

a) The floor is a square, with a presumed measurement of 100 ft per side and a tolerance of ± 6 in. $= \pm 0.5$ ft. The area A in square feet (ft^2) for a square of side length x ft is

$$A(x) = x^2.$$

The derivative is $dA/dx = 2x$, and solving for dA gives the differential of A:

$$dA = 2x\,dx.$$

To find dA, we substitute $x = 100$ and $dx = \pm 0.5$:

$$dA = 2(100)(\pm 0.5) = \pm 100.$$

The value of dA is interpreted as the approximate difference in area due to the inexactness in measuring. Therefore, if Michelle's measurements are off by half a foot, the total area can differ by approximately ± 100 ft^2. A small "error" in measurement can lead to quite a large difference in the resulting area.

b) The actual difference in area (ΔA) is calculated directly. We set $x_1 = 100$ ft, the presumed length measurement and let x_2 represent the length plus or minus the tolerance.

If the true length is at the low end, we have $x_2 = 99.5$ ft, that is, 100 ft minus the tolerance of 0.5 ft. The floor's area is then $99.5^2 = 9900.25$ ft^2. The actual difference in area is

$$\Delta A = A(x_2) - A(x_1)$$
$$= A(99.5) - A(100)$$
$$= 99.5^2 - 100^2$$
$$= 9900.25 - 10{,}000$$
$$= -99.75 \text{ ft}^2.$$

Thus, the actual difference in area is $\Delta A = -99.75$ ft^2, which compares well with the approximate value of $dA = -100$ ft^2.

If the true length is at the high end, we have $x_2 = 100.5$ ft. The floor's area is then $100.5^2 = 10{,}100.25$ ft^2. The actual difference in area is

$$\Delta A = A(x_2) - A(x_1)$$
$$= A(100.5) - A(100)$$
$$= 100.5^2 - 100^2$$
$$= 10{,}100.25 - 10{,}000$$
$$= 100.25 \text{ ft}^2.$$

In this case, the actual difference in area is $\Delta A = 100.25$ ft^2, which again compares well with the approximate value of $dA = +100$ ft^2.

c) The tiles (each measuring 1 ft^2) come 12 to a box. Thus, if the room were exactly 100 ft by 100 ft (an area of 10,000 ft^2), Michelle would need $10{,}000/12 = 833.33 \ldots$, or 834 boxes to cover the floor. To take into account the possibility that the room is larger by 100 ft^2, she needs a total of $10{,}100/12 = 841.67 \ldots$, or 842 boxes of tiles. Therefore, she should buy 8 extra boxes of tiles, meaning an extra cost of $(8)(24) = \$192$.

《 Quick Check 3

> **Quick Check 3**
>
> The four walls of a room measure 10 ft by 10 ft each, with a tolerance of ± 0.25 ft.
>
> **a)** Calculate the approximate difference in area, dA, for the four walls.
>
> **b)** Workers will be texturing the four walls using "knockdown" spray. Each bottle of knockdown spray costs $9 and covers 12 ft^2. How much extra cost for knockdown spray should the workers allot for the potential overage in wall area?

We see that there is an advantage to using a differential to calculate an approximate difference in an output variable. There is less actual calculating, and the result is often quite accurate. Compare the arithmetic steps needed in parts (a) and (b) of Example 6. Even though dA is an approximation, it is accurate enough for Michelle's needs: it is sufficient for her to know that the area can be off by as much as "about" 100 ft^2.

Historically, differentials were quite valuable when used to make approximations. However, with the advent of computers and graphing calculators, such use has diminished considerably. The use of marginals remains important in the study of business and economics.

Section Summary

- If $C(x)$ represents the cost for producing x items, then *marginal cost* $C'(x)$ is its derivative, and $C'(x) \approx C(x + 1) - C(x)$. Thus, the cost to produce the $(x + 1)$st item can be approximated by $C(x + 1) \approx C(x) + C'(x)$.
- If $R(x)$ represents the revenue from selling x items, then *marginal revenue* $R'(x)$ is its derivative, and $R'(x) \approx R(x + 1) - R(x)$. Thus, the revenue from the $(x + 1)$st item can be approximated by $R(x + 1) \approx R(x) + R'(x)$.
- If $P(x)$ represents profit from selling x items, then *marginal profit* $P'(x)$ is its derivative, and $P'(x) \approx P(x + 1) - P(x)$. Thus, the profit from the $(x + 1)$st item can be approximated by $P(x + 1) \approx P(x) + P'(x)$.

- In general, profit = revenue − cost, or $P(x) = R(x) - C(x)$.
- In *delta notation*, $\Delta x = (x + h) - x = h$, and $\Delta y = f(x + h) - f(x)$. For small values of Δx, we have $\dfrac{\Delta y}{\Delta x} \approx f'(x)$, which is equivalent to $\Delta y \approx f'(x)\,\Delta x$.

- The *differential* of x is $dx = \Delta x$. Since $\dfrac{dy}{dx} = f'(x)$, we have $dy = f'(x)\,dx$. In general, $dy \approx \Delta y$, and the approximation can be very close for sufficiently small dx.

EXERCISE SET
2.6

APPLICATIONS

Business and Economics

1. Marginal revenue, cost, and profit. Let $R(x)$, $C(x)$, and $P(x)$ be, respectively, the revenue, cost, and profit, in dollars, from the production and sale of x items. If

$$R(x) = 5x \quad \text{and} \quad C(x) = 0.001x^2 + 1.2x + 60,$$

find each of the following.

a) $P(x)$
b) $R(100)$, $C(100)$, and $P(100)$
c) $R'(x)$, $C'(x)$, and $P'(x)$
d) $R'(100)$, $C'(100)$, and $P'(100)$
e) Describe in words the meaning of each quantity in parts (b) and (d).

2. Marginal revenue, cost, and profit. Let $R(x)$, $C(x)$, and $P(x)$ be, respectively, the revenue, cost, and profit, in dollars, from the production and sale of x items. If

$$R(x) = 50x - 0.5x^2 \text{ and } C(x) = 4x + 10,$$

find each of the following.

a) $P(x)$
b) $R(20)$, $C(20)$, and $P(20)$
c) $R'(x)$, $C'(x)$, and $P'(x)$
d) $R'(20)$, $C'(20)$, and $P'(20)$

3. Marginal cost. Suppose that the monthly cost, in dollars, of producing x chairs is

$$C(x) = 0.001x^3 + 0.07x^2 + 19x + 700,$$

and currently 25 chairs are produced monthly.

a) What is the current monthly cost?
b) What would be the additional cost of increasing production to 26 chairs monthly?
c) What is the marginal cost when $x = 25$?
d) Use marginal cost to estimate the difference in cost between producing 25 and 27 chairs per month.
e) Use the answer from part (d) to predict $C(27)$.

4. Marginal cost. Suppose that the daily cost, in dollars, of producing x radios is

$$C(x) = 0.002x^3 + 0.1x^2 + 42x + 300,$$

and currently 40 radios are produced daily.

a) What is the current daily cost?
b) What would be the additional daily cost of increasing production to 41 radios daily?
c) What is the marginal cost when $x = 40$?
d) Use marginal cost to estimate the daily cost of increasing production to 42 radios daily.

5. Marginal revenue. Pierce Manufacturing determines that the daily revenue, in dollars, from the sale of x lawn chairs is

$$R(x) = 0.005x^3 + 0.01x^2 + 0.5x.$$

Currently, Pierce sells 70 lawn chairs daily.

a) What is the current daily revenue?
b) How much would revenue increase if 73 lawn chairs were sold each day?
c) What is the marginal revenue when 70 lawn chairs are sold daily?
d) Use the answer from part (c) to estimate $R(71)$, $R(72)$, and $R(73)$.

6. Marginal profit. For Sunshine Motors, the weekly profit, in dollars, of selling x cars is

$$P(x) = -0.006x^3 - 0.2x^2 + 900x - 1200,$$

and currently 60 cars are sold weekly.

a) What is the current weekly profit?
b) How much profit would be lost if the dealership were able to sell only 59 cars weekly?
c) What is the marginal profit when $x = 60$?
d) Use marginal profit to estimate the weekly profit if sales increase to 61 cars weekly.

7. Marginal profit. Crawford Computing finds that its weekly profit, in dollars, from the production and sale of x laptop computers is

$$P(x) = -0.004x^3 - 0.3x^2 + 600x - 800.$$

Currently Crawford builds and sells 9 laptops weekly.

a) What is the current weekly profit?
b) How much profit would be lost if production and sales dropped to 8 laptops weekly?
c) What is the marginal profit when $x = 9$?
d) Use the answers from parts (a)–(c) to estimate the profit resulting from the production and sale of 10 laptops weekly.

8. Marginal revenue. Solano Carriers finds that its monthly revenue, in dollars, from the sale of x carry-on suitcases is

$$R(x) = 0.007x^3 - 0.5x^2 + 150x.$$

Currently Solano is selling 26 carry-on suitcases monthly.

a) What is the current monthly revenue?
b) How much would revenue increase if sales increased from 26 to 28 suitcases?
c) What is the marginal revenue when 26 suitcases are sold?
d) Use the answers from parts (a)–(c) to estimate the revenue resulting from selling 27 suitcases per month.

9. Sales. Let $N(x)$ be the number of computers sold annually when the price is x dollars per computer. Explain in words what occurs if $N(1000) = 500{,}000$ and $N'(1000) = -100$.

10. Sales. Estimate the number of computers sold in Exercise 9 if the price is raised to $1025.

For Exercises 11–16, assume that C(x) and R(x) are in dollars and x is the number of units produced and sold.

11. For the total-cost function
    $$C(x) = 0.01x^2 + 0.6x + 30,$$
    find ΔC and $C'(x)$ when $x = 70$ and $\Delta x = 1$.

12. For the total-cost function
    $$C(x) = 0.01x^2 + 1.6x + 100,$$
    find ΔC and $C'(x)$ when $x = 80$ and $\Delta x = 1$.

13. For the total-revenue function
    $$R(x) = 2x,$$
    find ΔR and $R'(x)$ when $x = 70$ and $\Delta x = 1$.

14. For the total-revenue function
    $$R(x) = 3x,$$
    find ΔR and $R'(x)$ when $x = 80$ and $\Delta x = 1$.

15. **a)** Using $C(x)$ from Exercise 11 and $R(x)$ from Exercise 13, find the total profit, $P(x)$.
    b) Find ΔP and $P'(x)$ when $x = 70$ and $\Delta x = 1$.

16. **a)** Using $C(x)$ from Exercise 12 and $R(x)$ from Exercise 14, find the total profit, $P(x)$.
    b) Find ΔP and $P'(x)$ when $x = 80$ and $\Delta x = 1$.

17. **Marginal supply.** The supply, S, of a new rollerball pen is given by
    $$S = 0.007p^3 - 0.5p^2 + 150p,$$
    where p is the price in dollars.
    a) Find the rate of change of quantity with respect to price, dS/dp.
    b) How many units will producers want to supply when the price is $25 per unit?
    c) Find the rate of change at $p = 25$, and interpret this result.
    d) Would you expect dS/dp to be positive or negative? Why?

18. **Marginal productivity.** An employee's monthly productivity, M, in number of units produced, is found to be a function of the number of years of service, t. For a certain product, the productivity function is given by
    $$M(t) = -2t^2 + 100t + 180.$$
    a) Find the productivity of an employee after 5 yr, 10 yr, 25 yr, and 45 yr of service.
    b) Find the marginal productivity.
    c) Find the marginal productivity at $t = 5$; $t = 10$; $t = 25$; $t = 45$; and interpret the results.
    d) Explain how the employee's marginal productivity might be related to experience and to age.

19. **Average cost.** The average cost for a company to produce x units of a product is given by the function
    $$A(x) = \frac{13x + 100}{x}.$$
    Use $A'(x)$ to estimate the change in average cost as production goes from 100 units to 101 units.

20. **Supply.** A supply function for a certain product is given by
    $$S(p) = 0.08p^3 + 2p^2 + 10p + 11,$$
    where $S(p)$ is the number of items produced when the price is p dollars. Use $S'(p)$ to estimate how many more units a producer will supply when the price changes from $18.00 per unit to $18.20 per unit.

21. **Gross domestic product.** The U.S. gross domestic product, in billions of current dollars, may be modeled by the function
    $$P(x) = 567 + x(36x^{0.6} - 104),$$
    where x is the number of years since 1960. (*Source*: U.S. Bureau for Economic Analysis.) Use $P'(x)$ to estimate how much the gross domestic product increased from 2009 to 2010.

22. **Advertising.** Norris Inc. finds that it sells N units of a product after spending x thousands of dollars on advertising, where
    $$N(x) = -x^2 + 300x + 6.$$
    Use $N'(x)$ to estimate how many more units Norris will sell by increasing its advertising expenditure from $100,000 to $101,000.

Marginal tax rate. **Businesses and individuals are frequently concerned about their marginal tax rate, or the rate at which the next dollar earned is taxed. In progressive taxation, the 80,001st dollar earned is taxed at a higher rate than the 25,001st dollar earned and at a lower rate than the 140,001st dollar earned. Use the graph below, showing the marginal tax rate for 2005, to answer Exercises 23–26.**

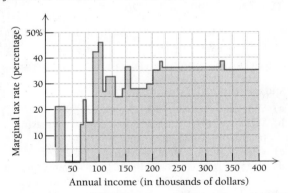

(*Source*: "Towards Fundamental Tax Reform" by Alan Auerbach and Kevin Hassett, *New York Times*, 5/5/05, p. C2.)

23. Was the taxation in 2005 progressive? Why or why not?

24. Marcy and Tyrone work for the same marketing agency. Because she is not yet a partner, Marcy's year-end income is approximately $95,000; Tyrone's year-end income is approximately $150,000. Suppose that one of them is to receive another $5000 in income for the year. Which one would keep more of that $5000 after taxes? Why?

25. Alan earns $25,000 per year and is considering a second job that would earn him another $2000 annually. At what rate will his tax liability (the amount he must pay in taxes) change if he takes the extra job? Express your answer in tax dollars paid per dollar earned.

26. Iris earns $50,000 per year and is considering extra work that would earn her an extra $3000 annually. At what rate will her tax liability grow if she takes the extra work (see Exercise 25)?

Find Δy and $f'(x)\Delta x$. Round to four and two decimal places, respectively.

27. For $y = f(x) = x^2$, $x = 2$, and $\Delta x = 0.01$

28. For $y = f(x) = x^3$, $x = 2$, and $\Delta x = 0.01$

29. For $y = f(x) = x + x^2$, $x = 3$, and $\Delta x = 0.04$

30. For $y = f(x) = x - x^2$, $x = 3$, and $\Delta x = 0.02$

31. For $y = f(x) = 1/x^2$, $x = 1$, and $\Delta x = 0.5$

32. For $y = f(x) = 1/x$, $x = 1$, and $\Delta x = 0.2$

33. For $y = f(x) = 3x - 1$, $x = 4$, and $\Delta x = 2$

34. For $y = f(x) = 2x - 3$, $x = 8$, and $\Delta x = 0.5$

Use $\Delta y \approx f'(x)\Delta x$ to find a decimal approximation of each radical expression. Round to three decimal places.

35. $\sqrt{26}$

36. $\sqrt{10}$

37. $\sqrt{102}$

38. $\sqrt{103}$

39. $\sqrt[3]{1005}$

40. $\sqrt[3]{28}$

Find dy.

41. $y = \sqrt{x + 1}$

42. $y = \sqrt{3x - 2}$

43. $y = (2x^3 + 1)^{3/2}$

44. $y = x^3(2x + 5)^2$

45. $y = \sqrt[5]{x + 27}$

46. $y = \dfrac{x^3 + x + 2}{x^2 + 3}$

47. $y = x^4 - 2x^3 + 5x^2 + 3x - 4$

48. $y = (7 - x)^8$

49. In Exercise 47, find dy when $x = 2$ and $dx = 0.1$.

50. In Exercise 48, find dy when $x = 1$ and $dx = 0.01$.

51. For $y = (3x - 10)^5$, find dy when $x = 4$ and $dx = 0.03$.

52. For $y = x^5 - 2x^3 - 7x$, find dy when $x = 3$ and $dx = 0.02$.

53. For $f(x) = x^4 - x^2 + 8$, use a differential to approximate $f(5.1)$.

54. For $f(x) = x^3 - 5x + 9$, use a differential to approximate $f(3.2)$.

SYNTHESIS

Life and Physical Sciences

55. Body surface area. Certain chemotherapy dosages depend on a patient's surface area. According to the Gehan and George model,

$$S = 0.02235h^{0.42246}w^{0.51456},$$

where h is the patient's height in centimeters, w is his or her weight in kilograms, and S is the approximation to his or her surface area in square meters. (*Source*: www.halls.md.)

Joanne is 160 cm tall and weighs 60 kg. Use a differential to estimate how much her surface area changes after her weight decreases by 1 kg.

56. Healing wound. The circular area of a healing wound is given by $A = \pi r^2$, where r is the radius, in centimeters. By approximately how much does the area decrease when the radius is decreased from 2 cm to 1.9 cm? Use 3.14 for π.

57. Medical dosage. The function

$$N(t) = \frac{0.8t + 1000}{5t + 4}$$

gives the bodily concentration $N(t)$, in parts per million, of a dosage of medication after time t, in hours. Use differentials to determine whether the concentration changes more from 1.0 hr to 1.1 hr or from 2.8 hr to 2.9 hr.

General Interest

58. Major League ticket prices. The average ticket price of a major league baseball game can be modeled by the function

$$p(x) = 0.09x^2 - 0.19x + 9.41,$$

where x is the number of years after 1990. (*Source*: Major League Baseball.) Use differentials to predict whether ticket price will increase more between 2010 and 2012 or between 2030 and 2031.

59. Suppose that a rope surrounds the earth at the equator. The rope is lengthened by 10 ft. By about how much is the rope raised above the earth?

Business and Economics

60. Marginal average cost. In Section 1.6, we defined the average cost of producing x units of a product in terms of the total cost $C(x)$ by $A(x) = C(x)/x$. Find a general expression for *marginal average cost, $A'(x)$*.

61. Cost and tolerance. A painting firm contracts to paint the exterior of a large water tank in the shape of a half-dome (a hemisphere). The radius of the tank is measured to be 100 ft with a tolerance of ± 6 in. (± 0.5 ft). (The formula for the surface area of a hemisphere is $A = 2\pi r^2$; use 3.14 as an approximation for π.) Each can of paint costs $30 and covers 300 ft^2.

  a) Calculate dA, the approximate difference in the surface area due to the tolerance.

  b) Assuming the painters cannot bring partial cans of paint to the job, how many extra cans should they bring to cover the extra area they may encounter?

  c) How much extra should the painters plan to spend on paint to account for the possible extra area?

62. Strategic oil supply. The U.S. Strategic Petroleum Reserve (SPR) stores petroleum in large spherical caverns built into salt deposits along the Gulf of Mexico. (*Source*: U.S. Department of Energy.) These caverns can be enlarged by filling the void with water, which dissolves the surrounding salt, and then pumping brine out. Suppose a cavern has a radius of 400 ft, which engineers want to enlarge by 5 ft. Use a differential to estimate how much volume will be added to form the enlarged cavern. (The formula for the volume of a sphere is $V = \frac{4}{3}\pi r^3$; use 3.14 as an approximation for π.)

Marginal revenue. *In each of Exercises 63–67, a demand function, $p = D(x)$, expresses price, in dollars, as a function of the number of items produced and sold. Find the marginal revenue.*

63. $p = 100 - \sqrt{x}$

64. $p = 400 - x$

65. $p = 500 - x$

66. $p = \dfrac{4000}{x} + 3$

67. $p = \dfrac{3000}{x} + 5$

68. Look up "differential" in a book or Web site devoted to math history. In a short paragraph, describe your findings.

69. Explain the uses of the differential.

Answers to Quick Checks

1. $\Delta y = -3.1319875$ **2.** $\sqrt{98} \approx \frac{99}{10}$, or 9.9.
This is within 0.001 of the actual value of $\sqrt{98}$.
3. **(a)** $\pm 20 \, \text{ft}^2$ **(b)** $\$18$ (for 2 extra bottles)

2.7 Implicit Differentiation and Related Rates*

OBJECTIVES

• Differentiate implicitly.

• Solve related-rate problems.

We often write a function with the output variable (usually y) isolated on one side of the equation. For example, if we write $y = x^3$, we have expressed y as an *explicit* function of x. Sometimes, with an equation like $y^3 + x^2 y^5 - x^4 = 27$, it may be cumbersome or nearly impossible to isolate the output variable; in such a case, we have an implicit relationship between the variables x and y. Then, we can find the derivative of y with respect to x using a process called *implicit differentiation*.

Implicit Differentiation

Consider the equation

$$y^3 = x.$$

This equation *implies* that y is a function of x, for if we solve for y, we get

$$y = \sqrt[3]{x}$$
$$= x^{1/3}.$$

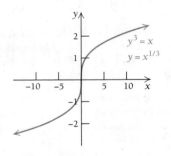

We know from our earlier work that

$$\frac{dy}{dx} = \frac{1}{3} x^{-2/3}. \tag{1}$$

A method known as **implicit differentiation** allows us to find dy/dx *without* solving for y. To do so, we use the Chain Rule, treating y as a function of x, and differentiate both sides of

$$y^3 = x$$

with respect to x:

$$\frac{d}{dx} y^3 = \frac{d}{dx} x.$$

The derivative on the left side is found using the Extended Power Rule:

$$3y^2 \frac{dy}{dx} = 1. \qquad \text{\small Remembering that the derivative of } y \text{ with respect} \\ \text{\small to } x \text{ is written } dy/dx$$

Finally, we solve for dy/dx by dividing both sides by $3y^2$:

$$\frac{dy}{dx} = \frac{1}{3y^2}, \quad \text{or} \quad \frac{1}{3} y^{-2}.$$

We can show that this indeed gives us the same answer as equation (1) by replacing y with $x^{1/3}$:

$$\frac{dy}{dx} = \frac{1}{3} y^{-2} = \frac{1}{3} (x^{1/3})^{-2} = \frac{1}{3} x^{-2/3}.$$

---

*This section can be omitted without loss of continuity.

Often, it is difficult or impossible to solve for y and to express dy/dx solely in terms of x. For example, the equation

$$y^3 + x^2y^5 - x^4 = 27$$

determines y as a function of x, but it would be difficult to solve for y. We can nevertheless find a formula for the derivative of y *without* solving for y. To do so usually involves computing $\dfrac{d}{dx}y^n$ for various integers n, and hence involves the Extended Power Rule in the form

$$\frac{d}{dx}y^n = ny^{n-1} \cdot \frac{dy}{dx}.$$

■ **EXAMPLE 1**   For $y^3 + x^2y^5 - x^4 = 27$:

a) Find dy/dx using implicit differentiation.

b) Find the slope of the tangent line to the curve at the point $(0, 3)$.

Solution

a) We differentiate the term x^2y^5 using the Product Rule. Because y is a function of x, it is critical that dy/dx is included as a factor in the result any time a term involving y is differentiated. When an expression involving just x is differentiated, there is no factor dy/dx.

$$\frac{d}{dx}(y^3 + x^2y^5 - x^4) = \frac{d}{dx}(27) \qquad \begin{array}{l}\text{Differentiating both sides}\\ \text{with respect to } x\end{array}$$

$$\frac{d}{dx}y^3 + \frac{d}{dx}x^2y^5 - \frac{d}{dx}x^4 = 0$$

$$3y^2 \cdot \frac{dy}{dx} + x^2 \cdot 5y^4 \cdot \frac{dy}{dx} + y^5 \cdot 2x - 4x^3 = 0. \qquad \begin{array}{l}\text{Using the Extended Power}\\ \text{Rule and the Product Rule}\end{array}$$

We next isolate those terms with dy/dx as a factor on one side:

$$3y^2 \cdot \frac{dy}{dx} + 5x^2y^4 \cdot \frac{dy}{dx} = 4x^3 - 2xy^5 \qquad \text{Adding } 4x^3 - 2xy^5 \text{ to both sides}$$

$$(3y^2 + 5x^2y^4)\frac{dy}{dx} = 4x^3 - 2xy^5 \qquad \text{Factoring out } dy/dx$$

$$\frac{dy}{dx} = \frac{4x^3 - 2xy^5}{3y^2 + 5x^2y^4}. \qquad \begin{array}{l}\text{Solving for } dy/dx \text{ and leaving the}\\ \text{answer in terms of } x \text{ and } y\end{array}$$

b) To find the slope of the tangent line to the curve at $(0, 3)$, we replace x with 0 and y with 3:

$$\frac{dy}{dx} = \frac{4 \cdot 0^3 - 2 \cdot 0 \cdot 3^5}{3 \cdot 3^2 + 5 \cdot 0^2 \cdot 3^4} = 0.$$

Exploratory

Graphicus can be used to graph equations that relate x and y implicitly. Press $\boxed{+}$ and then $\boxed{\text{g(x,y)=0}}$, and enter $y^3 + x^2y^3 - x^4 = 27$ as

$$\text{y^3+x^2(y^3)-x^4=27}$$

EXERCISES

Graph each equation.

1. $y^3 + x^2y^3 - x^4 = 27$

2. $y^2x + 2x^3y^3 = y + 1$

> **Quick Check 1**

For $y^2x + 2x^3y^3 = y + 1$, find dy/dx using implicit differentiation.

❮ Quick Check 1

It is not uncommon for the expression for dy/dx to contain *both* variables x and y. When using the derivative to calculate a slope, we must evaluate it at both the x-value and the y-value of the point of tangency.

The steps in Example 1 are typical of those used when differentiating implicitly.

To differentiate implicitly:

a) Differentiate both sides of the equation with respect to x (or whatever variable you are differentiating with respect to).

b) Apply the rules for differentiation (the Power, Product, Quotient, and Chain Rules) as necessary. Any time an expression involving y is differentiated, dy/dx will be a factor in the result.

c) Isolate all terms with dy/dx as a factor on one side of the equation.

d) If necessary, factor out dy/dx.

e) If necessary, divide both sides of the equation to isolate dy/dx.

The demand function for a product (see Section R.5) is often given implicitly.

■ **EXAMPLE 2**   For the demand equation $x = \sqrt{200 - p^3}$, differentiate implicitly to find dp/dx.

Solution

$$\frac{d}{dx}x = \frac{d}{dx}\sqrt{200 - p^3}$$

$$1 = \frac{1}{2}(200 - p^3)^{-1/2} \cdot (-3p^2) \cdot \frac{dp}{dx} \qquad \text{Using the Extended Power Rule twice}$$

$$1 = \frac{-3p^2}{2\sqrt{200 - p^3}} \cdot \frac{dp}{dx}$$

$$\frac{2\sqrt{200 - p^3}}{-3p^2} = \frac{dp}{dx}$$

Related Rates

Suppose that y is a function of x, say

$$y = f(x),$$

and x is a function of time, t. Since y depends on x and x depends on t, it follows that y depends on t. The Chain Rule gives the following:

$$\frac{dy}{dt} = \frac{dy}{dx} \cdot \frac{dx}{dt}.$$

Thus, the rate of change of y is *related* to the rate of change of x. Let's see how this comes up in problems. It helps to keep in mind that any variable can be thought of as a function of time t, even though a specific expression in terms of t may not be given or its rate of change with respect to t may be 0.

■ **EXAMPLE 3**   **Business: Service Area.**   A restaurant supplier services the restaurants in a circular area in such a way that the radius r is increasing at the rate of 2 mi per year at the moment when $r = 5$ mi. At that moment, how fast is the area increasing?

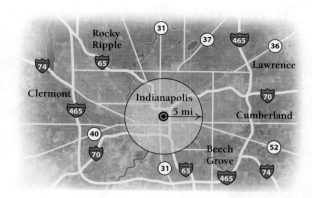

Solution The area A and the radius r are always related by the equation for the area of a circle:

$$A = \pi r^2.$$

We take the derivative of both sides with respect to t:

$$\frac{dA}{dt} = 2\pi r \cdot \frac{dr}{dt}. \qquad \text{The factor } dr/dt \text{ results from the Chain Rule and the fact that } r \text{ is assumed to be a function of } t.$$

At the moment in question, $dr/dt = 2$ mi/yr and $r = 5$ mi, so

$$\frac{dA}{dt} = 2\pi(5\text{mi})\left(2\,\frac{\text{mi}}{\text{yr}}\right)$$

$$= 20\pi\,\frac{\text{mi}^2}{\text{yr}} \approx 63 \text{ square miles per year.}$$

❰ Quick Check 2

❱ **Quick Check 2**

A spherical balloon is deflating, losing 20 cm³ of air per minute. At the moment when the radius of the balloon is 8 cm, how fast is the radius decreasing? (*Hint:* $V = \frac{4}{3}\pi r^3$.)

■ **EXAMPLE 4** **Business: Rates of Change of Revenue, Cost, and Profit.** For Luce Landscaping, the total revenue from the yard maintenance of x homes is given by

$$R(x) = 1000x - x^2,$$

and the total cost is given by

$$C(x) = 3000 + 20x.$$

Suppose that Luce is adding 10 homes per day at the moment when the 400th customer is signed. At that moment, what is the rate of change of (a) total revenue, (b) total cost, and (c) total profit?

Solution

a) $\dfrac{dR}{dt} = 1000 \cdot \dfrac{dx}{dt} - 2x \cdot \dfrac{dx}{dt}$ \qquad Differentiating both sides with respect to time

$\qquad = 1000 \cdot 10 - 2(400)10$ \qquad Substituting 10 for dx/dt and 400 for x

$\qquad = \$2000$ per day

b) $\dfrac{dC}{dt} = 20 \cdot \dfrac{dx}{dt}$ \qquad Differentiating both sides with respect to time

$\qquad = 20(10)$

$\qquad = \$200$ per day

c) Since $P = R - C$,

$$\frac{dP}{dt} = \frac{dR}{dt} - \frac{dC}{dt}$$

$\qquad = \$2000$ per day $- \$200$ per day

$\qquad = \$1800$ per day.

Section Summary

- If variables x and y are related to one another by an equation but neither variable is isolated on one side of the equation, we say that x and y have an implicit relationship. To find dy/dx without solving such an equation for y, we use *implicit differentiation*.
- Whenever we implicitly differentiate y with respect to x, the factor dy/dx will appear as a result of the Chain Rule.

- To determine the slope of a tangent line at a point on the graph of an implicit relationship, we may need to evaluate the derivative by inserting both the x-value and the y-value of the point of tangency.

EXERCISE SET
2.7

Differentiate implicitly to find dy/dx. Then find the slope of the curve at the given point.

1. $x^3 + 2y^3 = 6$; $(2, -1)$

2. $3x^3 - y^2 = 8$; $(2, 4)$

3. $2x^2 - 3y^3 = 5$; $(-2, 1)$

4. $2x^3 + 4y^2 = -12$; $(-2, -1)$

5. $x^2 - y^2 = 1$; $(\sqrt{3}, \sqrt{2})$

6. $x^2 + y^2 = 1$; $\left(\dfrac{1}{2}, \dfrac{\sqrt{3}}{2}\right)$

7. $3x^2y^4 = 12$; $(2, -1)$

8. $2x^3y^2 = -18$; $(-1, 3)$

9. $x^3 - x^2y^2 = -9$; $(3, -2)$

10. $x^4 - x^2y^3 = 12$; $(-2, 1)$

11. $xy - x + 2y = 3$; $\left(-5, \dfrac{2}{3}\right)$

12. $xy + y^2 - 2x = 0$; $(1, -2)$

13. $x^2y - 2x^3 - y^3 + 1 = 0$; $(2, -3)$

14. $4x^3 - y^4 - 3y + 5x + 1 = 0$; $(1, -2)$

Differentiate implicitly to find dy/dx.

15. $2xy + 3 = 0$

16. $x^2 + 2xy = 3y^2$

17. $x^2 - y^2 = 16$

18. $x^2 + y^2 = 25$

19. $y^5 = x^3$

20. $y^3 = x^5$

21. $x^2y^3 + x^3y^4 = 11$

22. $x^3y^2 + x^5y^3 = -19$

For each demand equation in Exercises 23–30, differentiate implicitly to find dp/dx.

23. $p^3 + p - 3x = 50$

24. $p^2 + p + 2x = 40$

25. $xp^3 = 24$

26. $x^3p^2 = 108$

27. $\dfrac{xp}{x + p} = 2$ (*Hint:* Clear the fraction first.)

28. $\dfrac{x^2p + xp + 1}{2x + p} = 1$ (*Hint:* Clear the fraction first.)

29. $(p + 4)(x + 3) = 48$

30. $1000 - 300p + 25p^2 = x$

31. Two variable quantities A and B are found to be related by the equation

$$A^3 + B^3 = 9.$$

What is the rate of change dA/dt at the moment when $A = 2$ and $dB/dt = 3$?

32. Two nonnegative variable quantities G and H are found to be related by the equation

$$G^2 + H^2 = 25.$$

What is the rate of change dH/dt when $dG/dt = 3$ and $G = 0$? $G = 1$? $G = 3$?

APPLICATIONS

Business and Economics

Rates of change of total revenue, cost, and profit. **In Exercises 33–36, find the rates of change of total revenue, cost, and profit with respect to time. Assume that R(x) and C(x) are in dollars.**

33. $R(x) = 50x - 0.5x^2$,
    $C(x) = 4x + 10$,
    when $x = 30$ and $dx/dt = 20$ units per day

34. $R(x) = 50x - 0.5x^2$,
    $C(x) = 10x + 3$,
    when $x = 10$ and $dx/dt = 5$ units per day

35. $R(x) = 2x$,
    $C(x) = 0.01x^2 + 0.6x + 30$,
    when $x = 20$ and $dx/dt = 8$ units per day

36. $R(x) = 280x - 0.4x^2$,
    $C(x) = 5000 + 0.6x^2$,
    when $x = 200$ and $dx/dt = 300$ units per day

37. Change of sales. Suppose that the price p, in dollars, and number of sales, x, of a certain item follow the equation

$$5p + 4x + 2px = 60.$$

Suppose also that p and x are both functions of time, measured in days. Find the rate at which x is changing when

$$x = 3, p = 5, and \frac{dp}{dt} = 1.5.$$

38. Change of revenue. For x and p as described in Exercise 37, find the rate at which the total revenue $R = xp$ is changing when

$$x = 3, p = 5, and \frac{dp}{dt} = 1.5.$$

Life and Natural Sciences

39. Rate of change of the Arctic ice cap. In a trend that scientists attribute, at least in part, to global warming, the floating cap of sea ice on the Arctic Ocean has been shrinking since 1950. The ice cap always shrinks in summer and grows in winter. Average minimum size of the ice cap, in square miles, can be approximated by

$$A = \pi r^2.$$

In 2012, the radius of the ice cap was approximately 648 mi and was shrinking at a rate of approximately 4.4 mi/yr. (*Source*: National Snow and Ice Data Center, Sept. 19, 2012.) How fast was the area changing at that time?

40. Rate of change of a healing wound. The area of a healing wound is given by

$$A = \pi r^2.$$

The radius is decreasing at the rate of 1 millimeter per day (-1 mm/day) at the moment when $r = 25$ mm. How fast is the area decreasing at that moment?

41. Body surface area. Certain chemotherapy dosages depend on a patient's surface area. According to the Mosteller model,

$$S = \frac{\sqrt{hw}}{60},$$

where h is the patient's height in centimeters, w is the patient's weight in kilograms, and S is the approximation to the patient's surface area in square meters. (*Source*: www.halls.md.) Assume that Tom's height is a constant 165 cm, but he is on a diet. If he loses 2 kg per month, how fast is his surface area decreasing at the instant he weighs 70 kg?

Poiseuille's Law. *The flow of blood in a blood vessel is faster toward the center of the vessel and slower toward the outside. The speed of the blood V, in millimeters per second (mm/sec), is given by*

$$V = \frac{p}{4Lv}(R^2 - r^2),$$

where R is the radius of the blood vessel, r is the distance of the blood from the center of the vessel, and p, L, and v are physical constants related to pressure, length, and viscosity of the blood vessels, respectively. Assume that dV/dt is measured in millimeters per second squared (mm/sec^2). Use this formula for Exercises 42 and 43.

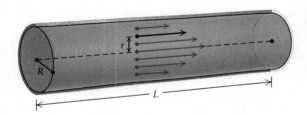

42. Assume that r is a constant as well as p, L, and v.

a) Find the rate of change dV/dt in terms of R and dR/dt when $L = 80$ mm, $p = 500$, and $v = 0.003$.

b) A person goes out into the cold to shovel snow. Cold air has the effect of contracting blood vessels far from the heart. Suppose that a blood vessel contracts at a rate of

$$\frac{dR}{dt} = -0.0002 \text{ mm/sec}$$

at a place in the blood vessel where the radius $R = 0.075$ mm. Find the rate of change, dV/dt, at that location.

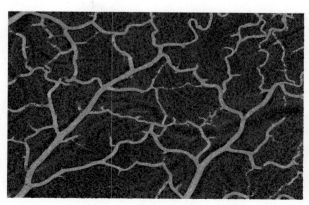

The flow of blood in a blood vessel can be modeled by Poiseuille's Law.

43. Assume that r is a constant as well as p, L, and v.

   a) Find the rate of change dV/dt in terms of R and dR/dt when $L = 70$ mm, $p = 400$, and $v = 0.003$.

   b) When shoveling snow in cold air, a person with a history of heart trouble can develop angina (chest pains) due to contracting blood vessels. To counteract this, he or she may take a nitroglycerin tablet, which dilates the blood vessels. Suppose that after a nitroglycerin tablet is taken, a blood vessel dilates at a rate of

$$\frac{dR}{dt} = 0.00015 \text{ mm/sec}$$

at a place in the blood vessel where the radius $R = 0.1$ mm. Find the rate of change, dV/dt.

General Interest

44. Two cars start from the same point at the same time. One travels north at 25 mph, and the other travels east at 60 mph. How fast is the distance between them increasing at the end of 1 hr? (*Hint:* $D^2 = x^2 + y^2$. To find D after 1 hr, solve $D^2 = 25^2 + 60^2$.)

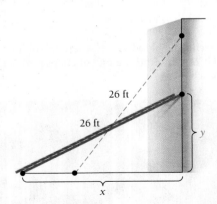

45. A ladder 26 ft long leans against a vertical wall. If the lower end is being moved away from the wall at the rate of 5 ft/s, how fast is the height of the top changing (this will be a negative rate) when the lower end is 10 ft from the wall?

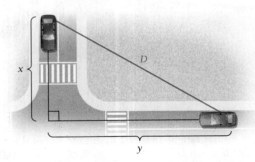

46. An inner city revitalization zone is a rectangle that is twice as long as it is wide. A diagonal through the region is growing at a rate of 90 m per year at a time when the region is 440 m wide. How fast is the area changing at that point in time?

47. The volume of a cantaloupe is given by

$$V = \tfrac{4}{3}\pi r^3.$$

The radius is growing at the rate of 0.7 cm/week, at a time when the radius is 7.5 cm. How fast is the volume changing at that moment?

SYNTHESIS

Differentiate implicitly to find dy/dx.

48. $\sqrt{x} + \sqrt{y} = 1$

49. $\dfrac{1}{x^2} + \dfrac{1}{y^2} = 5$

50. $y^3 = \dfrac{x - 1}{x + 1}$

51. $y^2 = \dfrac{x^2 - 1}{x^2 + 1}$

52. $x^{3/2} + y^{2/3} = 1$

53. $(x - y)^3 + (x + y)^3 = x^5 + y^5$

Differentiate implicitly to find d²y/dx².

54. $xy + x - 2y = 4$

55. $y^2 - xy + x^2 = 5$

56. $x^2 - y^2 = 5$

57. $x^3 - y^3 = 8$

58. Explain the usefulness of implicit differentiation.

59. Look up the word "implicit" in a dictionary. Explain how that definition can be related to the concept of a function that is defined "implicitly."

TECHNOLOGY CONNECTION · · · · · · · · · ·

Graph each of the following equations. Equations must be solved for y before they can be entered into most calculators. Graphicus does not require that equations be solved for y.

60. $x^2 + y^2 = 4$
   Note: You will probably need to sketch the graph in two parts: $y = \sqrt{4 - x^2}$ and $y = -\sqrt{4 - x^2}$. Then graph the tangent line to the graph at the point $(-1, \sqrt{3})$.

61. $x^4 = y^2 + x^6$
   Then graph the tangent line to the graph at the point $(-0.8, 0.384)$.

62. $y^4 = y^2 - x^2$

63. $x^3 = y^2(2 - x)$

64. $y^2 = x^3$

Answers to Quick Checks
1. $\dfrac{dy}{dx} = \dfrac{y^2 + 6x^2y^3}{1 - 2xy - 6x^3y^2}$ **2.** Approximately -0.025 cm/min

KEY TERMS AND CONCEPTS

EXAMPLES

SECTION 2.1

A function is **increasing** over an open interval I if, for all x in I, $f'(x) > 0$.

A function is **decreasing** over an open interval I if, for all x in I, $f'(x) < 0$.

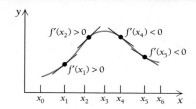

f is increasing over the interval (x_0, x_3) and decreasing over (x_3, x_6).

If f is a continuous function, then a **critical value** is any number c for which $f'(c) = 0$ or $f'(c)$ does not exist.

If $f'(c)$ does not exist, then the graph of f may have a corner or a vertical tangent at $(c, f(c))$.

The ordered pair $(c, f(c))$ is called a *critical point*.

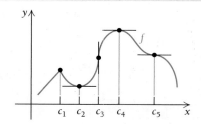

The values c_1, c_2, c_3, c_4, and c_5 are critical values of f.

- $f'(c_1)$ does not exist (corner).
- $f'(c_2) = 0$.
- $f'(c_3)$ does not exist (vertical tangent).
- $f'(c_4) = 0$.
- $f'(c_5) = 0$.

If f is a continuous function, then a **relative extremum** (**maximum** or **minimum**) always occurs at a critical value.

The converse is not true: a critical value may not correspond to an extremum.

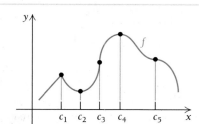

- The critical point $(c_1, f(c_1))$ is a relative maximum.
- The critical point $(c_2, f(c_2))$ is a relative minimum.
- The critical point $(c_3, f(c_3))$ is neither a relative maximum nor a relative minimum.
- The critical point $(c_4, f(c_4))$ is a relative maximum.
- The critical point $(c_5, f(c_5))$ is neither a relative maximum nor a relative minimum.

(continued)

KEY TERMS AND CONCEPTS	EXAMPLES

SECTION 2.1 *(continued)*

The **First-Derivative Test** allows us to classify a critical value as a relative maximum, a relative minimum, or neither.

Find the relative extrema of the function given by

$$f(x) = \frac{1}{3}x^3 - \frac{1}{2}x^2 - 20x + 7.$$

Critical values occur where $f'(x) = 0$ or $f'(x)$ does not exist. The derivative $f'(x) = x^2 - x - 20$ exists for all real numbers, so the only critical values occur when $f'(x) = 0$. Setting $x^2 - x - 20 = 0$ and solving, we have $x = -4$ and $x = 5$ as the critical values.

To apply the First-Derivative Test, we check the sign of $f'(x)$ to the left and the right of each critical value, using test values:

Test Value	$x = -5$	$x = 0$	$x = 6$
Sign of $f'(x)$	$f'(-5) > 0$	$f'(0) < 0$	$f'(6) > 0$
Result	f increasing	f decreasing	f increasing

Therefore, there is a relative maximum at $x = -4$: $f(-4) = 57\frac{2}{3}$. There is a relative minimum at $x = 5$: $f(5) = -63\frac{5}{6}$.

SECTION 2.2

If the graph of f is smooth and continuous, then the **second derivative**, $f''(x)$, determines the **concavity** of the graph.

If $f''(x) > 0$ for all x in an open interval I, then the graph of f is **concave up** over I.

If $f''(x) < 0$ for all x in an open interval I, then the graph of f is **concave down** over I.

A **point of inflection** occurs at $(x_0, f(x_0))$ if $f''(x_0) = 0$ and there is a change in concavity on either side of x_0.

The function given by

$$f(x) = \frac{1}{3}x^3 - \frac{1}{2}x^2 - 20x + 7$$

has the second derivative $f''(x) = 2x - 1$. Setting the second derivative equal to 0, we have $x_0 = \frac{1}{2}$. Using test values, we can check the concavity on either side of $x_0 = \frac{1}{2}$:

Test Value	$x = 0$	$x = 1$
Sign of $f''(x)$	$f''(0) < 0$	$f''(1) > 0$
Result	f is concave down	f is concave up

Therefore, the function is concave down over the interval $\left(-\infty, \frac{1}{2}\right)$ and concave up over the interval $\left(\frac{1}{2}, \infty\right)$. Since there is a change in concavity on either side of $x_0 = \frac{1}{2}$, we also conclude that the point $\left(\frac{1}{2}, -3\frac{1}{12}\right)$ is a point of inflection, where $f\left(\frac{1}{2}\right) = -3\frac{1}{12}$.

The **Second-Derivative Test** can also be used to classify relative extrema:

If $f'(c) = 0$ and $f''(c) > 0$, then $f(c)$ is a relative minimum.

If $f'(c) = 0$ and $f''(c) < 0$, then $f(c)$ is a relative maximum.

If $f'(c) = 0$ and $f''(c) = 0$, then the First-Derivative Test must be used to classify $f(c)$.

For the function

$$f(x) = \frac{1}{3}x^3 - \frac{1}{2}x^2 - 20x + 7,$$

evaluating the second derivative, $f''(x) = 2x - 1$, at the critical values yields the following conclusions:

- At $x = -4$, we have $f''(-4) < 0$. Since $f'(-4) = 0$ and the graph is concave down, we conclude that there is a relative maximum at $x = -4$.
- At $x = 5$, we have $f''(5) > 0$. Since $f'(5) = 0$ and the graph is concave up, we conclude that there is a relative minimum at $x = 5$.

KEY TERMS AND CONCEPTS	EXAMPLES

SECTION 2.3

A line $x = a$ is a **vertical asymptote** if

$$\lim_{x \to a} f(x) = \infty,$$

$$\lim_{x \to a} f(x) = -\infty,$$

$$\lim_{x \to a^+} f(x) = \infty,$$

or

$$\lim_{x \to a^+} f(x) = -\infty.$$

The graph of a rational function never crosses a vertical asymptote

Consider the function given by

$$f(x) = \frac{x^2 - 1}{x^2 + x - 6}.$$

Factoring, we have

$$f(x) = \frac{(x + 1)(x - 1)}{(x + 3)(x - 2)}.$$

This expression is simplified. Therefore, $x = -3$ and $x = 2$ are vertical asymptotes since

$$\lim_{x \to -3^-} f(x) = \infty \quad \text{and} \quad \lim_{x \to -3^+} f(x) = -\infty$$

and

$$\lim_{x \to 2^-} f(x) = -\infty \quad \text{and} \quad \lim_{x \to 2^+} f(x) = \infty.$$

A line $y = b$ is a **horizontal asymptote** if

$$\lim_{x \to -\infty} f(x) = b$$

or

$$\lim_{x \to \infty} f(x) = b.$$

The graph of a function can cross a horizontal asymptote. An asymptote is usually sketched as a dashed line; it is not part of the graph itself.

Also, $y = 1$ is a horizontal asymptote since
$$\lim_{x \to \infty} f(x) = 1 \quad \text{and} \quad \lim_{x \to -\infty} f(x) = 1.$$

For a rational function of the form $f(x) = p(x)/q(x)$, a **slant asymptote** occurs if the degree of the numerator is 1 greater than the degree of the denominator.

Let $f(x) = \frac{x^2 + 1}{x + 3}$. Long division yields $f(x) = x - 3 + \frac{10}{x + 3}$. As $x \to \infty$ or $x \to -\infty$, the remainder $\frac{10}{x + 3} \to 0$. Therefore, the slant asymptote is $y = x - 3$.

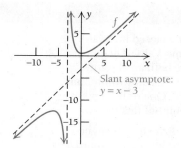

Slant asymptote: $y = x - 3$

Asymptotes, extrema, x- and y-intercepts, points of inflection, concavity, and intervals of increasing or decreasing are all used in the strategy for accurate graph sketching.

Consider the function given by $f(x) = \frac{1}{3}x^3 - \frac{1}{2}x^2 - 20x + 7$.

- f has a relative maximum point at $\left(-4, 57\frac{2}{3}\right)$ and a relative minimum point at $\left(5, -63\frac{5}{6}\right)$.
- f has a point of inflection at $\left(\frac{1}{2}, -3\frac{1}{12}\right)$.

(continued)

| KEY TERMS AND CONCEPTS | EXAMPLES |

SECTION 2.3 *(continued)*

- f is increasing over the interval $(-\infty, -4)$ and over the interval $(5, \infty)$, decreasing over the interval $(-4, 5)$, concave down over the interval $\left(-\infty, \frac{1}{2}\right)$, and concave up over the interval $\left(\frac{1}{2}, \infty\right)$.
- f has a y-intercept at $(0, 7)$.

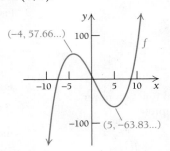

Consider the function given by $f(x) = \dfrac{x^2 - 1}{x^2 + x - 6}$.

- f has vertical asymptotes $x = -3$ and $x = 2$.
- f has a horizontal asymptote given by $y = 1$.
- f has a y-intercept at $\left(0, \frac{1}{6}\right)$.
- f has x-intercepts at $(-1, 0)$ and $(1, 0)$.

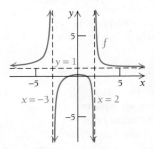

SECTION 2.4

If f is continuous over a closed interval $[a, b]$, then the **Extreme-Value Theorem** tells us that f will have both an absolute maximum value and an absolute minimum value over this interval. One or both points may occur at an endpoint of this interval.

Maximum–Minimum Principle 1 can be used to determine these absolute extrema: We find all critical values $c_1, c_2, c_3, \ldots, c_n$, in $[a, b]$, then evaluate

$$f(a), f(c_1), f(c_2), f(c_3), \ldots, f(c_n), f(b).$$

The largest of these is the **absolute maximum**, and the smallest is the **absolute minimum**.

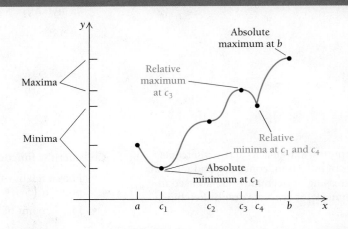

KEY TERMS AND CONCEPTS	EXAMPLES

If f is differentiable for all x in an interval I, and there is exactly one critical value c in I such that $f'(c) = 0$, then, according to **Maximum–Minimum Principle 2**, $f(c)$ is an absolute minimum if $f''(c) > 0$ or an absolute maximum if $f''(c) < 0$.

Let $f(x) = x + \dfrac{2}{x}$, for $x > 0$. The derivative is $f'(x) = 1 - \dfrac{2}{x^2}$. We solve for the critical value:

$$1 - \frac{2}{x^2} = 0$$
$$x^2 = 2$$
$$x = \pm\sqrt{2}.$$

The only critical value over the interval where $x > 0$ is $x = \sqrt{2}$. The second derivative is $f''(x) = \dfrac{4}{x^3}$. We see that $f''(\sqrt{2}) = \dfrac{4}{(\sqrt{2})^3} > 0$. Therefore, $(\sqrt{2}, f(\sqrt{2}))$ is an absolute minimum.

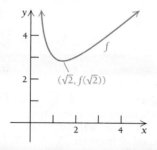

SECTION 2.5

Many real-world applications involve maximum–minimum problems.

See Examples 1–7 in Section 2.5 and the problem-solving strategy on p. 263.

SECTION 2.6

Marginal cost, **marginal revenue**, and **marginal profit** are estimates of the cost, revenue, and profit for the $(x + 1)$st item produced:

- $C'(x) \approx C(x + 1) - C(x)$,
  so $C(x + 1) \approx C(x) + C'(x)$.
- $R'(x) \approx R(x + 1) - R(x)$,
  so $R(x + 1) \approx R(x) + R'(x)$.
- $P'(x) \approx P(x + 1) - P(x)$,
  so $P(x + 1) \approx P(x) + P'(x)$.

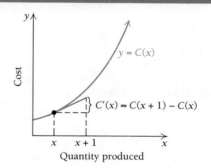

Delta notation represents the change in a variable:

$$\Delta x = x_2 - x_1$$

and

$$\Delta y = f(x_2) - f(x_1).$$

If $x_2 = x_1 + h$, then $\Delta x = h$.

If Δx is small, then the derivative can be used to approximate Δy:

$$\Delta y \approx f'(x) \cdot \Delta x.$$

For $f(x) = x^2$, let $x_1 = 2$ and $x_2 = 2.1$. Then, $\Delta x = 2.1 - 2 = 0.1$. Since $f(x_1) = f(2) = 4$ and $f(x_2) = f(2.1) = 4.41$, we have

$$\Delta y = f(x_2) - f(x_1) = 4.41 - 4 = 0.41.$$

Since $\Delta x = 0.1$ is small, we can approximate Δy by

$$\Delta y \approx f'(x)\,\Delta x.$$

The derivative is $f'(x) = 2x$. Therefore,

$$\Delta y \approx f'(2) \cdot (0.1) = 2(2) \cdot (0.1) = 0.4.$$

This approximation, 0.4, is very close to the actual difference, 0.41.

(continued)

KEY TERMS AND CONCEPTS	**EXAMPLES**

SECTION 2.6 (*continued*)

Differentials allow us to approximate changes in the output variable y given a change in the input variable x:

$$dx = \Delta x$$

and

$$dy = f'(x)\, dx.$$

If Δx is small, then $dy \approx \Delta y$.

In practice, it is often simpler to calculate dy, and it will be very close to the true value of Δy.

For $y = \sqrt[3]{x}$, find dy when $x = 27$ and $dx = 2$.

Note that $\dfrac{dy}{dx} = \dfrac{1}{3\sqrt[3]{x^2}}$. Thus, $dy = \dfrac{1}{3\sqrt[3]{x^2}}\, dx$. Evaluating, we have

$$dy = \frac{1}{3\sqrt[3]{(27)^2}}\,(2) = \frac{2}{27} \approx 0.074.$$

This result can be used to approximate the value of $\sqrt[3]{29}$, using the fact that $\sqrt[3]{29} \approx \sqrt[3]{27} + dy$:

$$\sqrt[3]{29} = \sqrt[3]{27} + dy \approx 3 + \frac{2}{27} \approx 3.074.$$

Thus, the approximation $\sqrt[3]{29} \approx 3.074$ is very close to the actual value, $\sqrt[3]{29} = 3.07231\ldots.$

SECTION 2.7

If an equation has variables x and y and y is not isolated on one side of the equation, the derivative dy/dx can be found without solving for y by the method of **implicit differentiation**.

Find $\dfrac{dy}{dx}$ if $y^5 = x^3 + 7$.

We differentiate both sides with respect to x, then solve for $\dfrac{dy}{dx}$.

$$\frac{d}{dx} y^5 = \frac{d}{dx} x^3 + \frac{d}{dx} 7$$

$$5y^4 \frac{dy}{dx} = 3x^2$$

$$\frac{dy}{dx} = \frac{3x^2}{5y^4}.$$

A **related rate** occurs when the rate of change of one variable (with respect to time) can be calculated in terms of the rate of change (with respect to time) of another variable of which it is a function.

A cube of ice is melting, losing 30 cm^3 of its volume (V) per minute. When the side length (x) of the cube is 20 cm, how fast is the side length decreasing?

Since $V = x^3$ and both V and x are changing with time, we differentiate each variable with respect to time:

$$\frac{dV}{dt} = 3x^2 \frac{dx}{dt}.$$

We have $x = 20$ and $\dfrac{dV}{dt} = -30$. Evaluating gives

$$-30 = 3(20)^2 \frac{dx}{dt}$$

or

$$\frac{dx}{dt} = -\frac{30}{3(20)^2} = -0.025\text{ cm/min.}$$

These review exercises are for test preparation. They can also be used as a practice test. Answers are at the back of the book. The blue bracketed section references tell you what part(s) of the chapter to restudy if your answer is incorrect.

CONCEPT REINFORCEMENT

Match each description in column A with the most appropriate graph in column B. [2.1–2.4]

Column A   Column B

1. A function with a relative maximum but no absolute extrema

a)
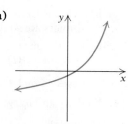

2. A function with both a vertical asymptote and a horizontal asymptote

b)
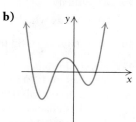

3. A function that is concave up and decreasing

c)
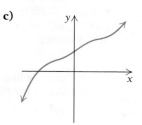

4. A function that is concave up and increasing

d)
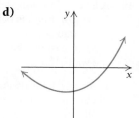

5. A function with three critical values

e)
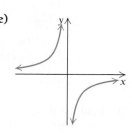

6. A function with one critical value and a second derivative that is always positive

f)
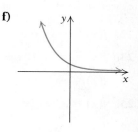

7. A function with a first derivative that is always positive

g)
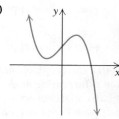

In Exercises 8–13, classify each statement as either true or false.

8. Every continuous function has at least one critical value. [2.1]

9. If a continuous function $y = f(x)$ has extrema, they will occur where $f'(x) = 0$. [2.1]

10. If $f'(c) = 0$ and $f''(c) > 0$, then $f(c)$ is a relative minimum. [2.2]

11. If $f'(c) = 0$ and $f''(c) = 0$, then $f(c)$ cannot be a relative minimum. [2.2]

12. If the graph of $f(x) = P(x)/Q(x)$ has a horizontal asymptote, then the degree of the polynomial $P(x)$ must be the same as that of the polynomial $Q(x)$. [2.3]

13. Absolute extrema of a continuous function f always occur at the endpoints of a closed interval. [2.4]

REVIEW EXERCISES

For each function given, find any extrema, along with the x-value at which they occur. Then sketch a graph of the function. [2.1]

14. $f(x) = 4 - 3x - x^2$ **15.** $f(x) = x^4 - 2x^2 + 3$

16. $f(x) = \dfrac{-8x}{x^2 + 1}$ **17.** $f(x) = 4 + (x - 1)^3$

18. $f(x) = x^3 + x^2 - x + 3$ **19.** $f(x) = 3x^{2/3}$

20. $f(x) = 2x^3 - 3x^2 - 12x + 10$

21. $f(x) = x^3 - 3x + 2$

Sketch the graph of each function. List any minimum or maximum values and where they occur, as well as any points of inflection. State where the function is increasing or decreasing, as well as where it is concave up or concave down. [2.2]

22. $f(x) = \dfrac{1}{3}x^3 + 3x^2 + 9x + 2$

23. $f(x) = x^2 - 10x + 8$

24. $f(x) = 4x^3 - 6x^2 - 24x + 5$

25. $f(x) = x^4 - 2x^2$

26. $f(x) = 3x^4 + 2x^3 - 3x^2 + 1$ (Round to three decimal places where appropriate.)

27. $f(x) = \frac{1}{5}x^5 + \frac{3}{4}x^4 - \frac{4}{3}x^3 + 8$ (Round to three decimal places where appropriate.)

Sketch the graph of each function. Indicate where each function is increasing or decreasing, the coordinates at which relative extrema occur, where any asymptotes occur, where the graph is concave up or concave down, and where any intercepts occur. [2.3]

28. $f(x) = \dfrac{2x + 5}{x + 1}$ **29.** $f(x) = \dfrac{x}{x - 2}$

30. $f(x) = \dfrac{5}{x^2 - 16}$ **31.** $f(x) = -\dfrac{x + 1}{x^2 - x - 2}$

32. $f(x) = \dfrac{x^2 - 2x + 2}{x - 1}$ **33.** $f(x) = \dfrac{x^2 + 3}{x}$

Find the absolute maximum and minimum values of each function, if they exist, over the indicated interval. Indicate the x-value at which each extremum occurs. Where no interval is specified, use the real line. [2.4]

34. $f(x) = x^4 - 2x^2 + 3$; $[0, 3]$

35. $f(x) = 8x^2 - x^3$; $[-1, 8]$

36. $f(x) = x + \dfrac{50}{x}$; $(0, \infty)$

37. $f(x) = x^4 - 2x^2 + 1$

38. Of all numbers whose sum is 60, find the two that have the maximum product. [2.5]

39. Find the minimum value of $Q = x^2 - 2y^2$, where $x - 2y = 1$. [2.5]

40. Business: maximizing profit. If
$$R(x) = 52x - 0.5x^2 \quad \text{and} \quad C(x) = 22x - 1,$$
find the maximum profit and the number of units that must be produced and sold in order to yield this maximum profit. Assume that $R(x)$ and $C(x)$ are in dollars. [2.5]

41. Business: minimizing cost. A rectangular box with a square base and a cover is to have a volume of 2500 ft³. If the cost per square foot for the bottom is $2, for the top is $3, and for the sides is $1, what should the dimensions be in order to minimize the cost? [2.5]

42. Business: minimizing inventory cost. A store in California sells 360 hybrid bicycles per year. It costs $8 to store one bicycle for a year. To reorder, there is a fixed cost of $10, plus $2 for each bicycle. How many times per year should the store order bicycles, and in what lot size, in order to minimize inventory costs? [2.5]

43. Business: marginal revenue. Crane Foods determines that its daily revenue, $R(x)$, in dollars, from the sale of x frozen dinners is
$$R(x) = 4x^{3/4}.$$

a) What is Crane's daily revenue when 81 frozen dinners are sold?

b) What is Crane's marginal revenue when 81 frozen dinners are sold?

c) Use the answers from parts (a) and (b) to estimate R(82). [2.6]

For Exercises 44 and 45, $y = f(x) = 2x^3 + x$. [2.6]

44. Find Δy and dy, given that $x = 1$ and $\Delta x = -0.05$.

45. a) Find dy.

b) Find dy when $x = -2$ and $dx = 0.01$.

46. Approximate $\sqrt{83}$ using $\Delta y \approx f'(x)\,\Delta x$. [2.6]

47. Physical science: waste storage. The Waste Isolation Pilot Plant (WIPP) in New Mexico consists of large rooms carved into a salt deposit and is used for long-term storage of radioactive waste. (*Source*: www.wipp.energy.gov.) A new storage room in the shape of a cube with an edge length of 200 ft is to be carved into the salt. Use a differential to estimate the potential difference in the volume of this room if the edge measurements have a tolerance of ±2 ft. [2.6]

48. Differentiate the following implicitly to find dy/dx. Then find the slope of the curve at the given point. [2.7]
$$2x^3 + 2y^3 = -9xy; \quad (-1, -2)$$

49. A ladder 25 ft long leans against a vertical wall. If the lower end is being moved away from the wall at the rate of 6 ft/sec, how fast is the height of the top decreasing when the lower end is 7 ft from the wall? [2.7]

50. Business: total revenue, cost, and profit. Find the rates of change, with respect to time, of total revenue, cost, and profit for
$$R(x) = 120x - 0.5x^2 \quad \text{and} \quad C(x) = 15x + 6,$$
when $x = 100$ and $dx/dt = 30$ units per day. Assume that $R(x)$ and $C(x)$ are in dollars. [2.7]

SYNTHESIS

51. Find the absolute maximum and minimum values, if they exist, over the indicated interval. [2.4]
$$f(x) = (x - 3)^{2/5}; \quad (-\infty, \infty)$$

52. Find the absolute maximum and minimum values of the piecewise-defined function given by

$$f(x) = \begin{cases} 2 - x^2, & \text{for } -2 \le x \le 1, \\ 3x - 2, & \text{for } 1 < x < 2, \\ (x - 4)^2, & \text{for } 2 \le x \le 6. \end{cases} \ [2.4]$$

53. Differentiate implicitly to find dy/dx:

$$(x - y)^4 + (x + y)^4 = x^6 + y^6. \ [2.7]$$

54. Find the relative maxima and minima of

$$y = x^4 - 8x^3 - 270x^2. \ [2.1 \text{ and } 2.2]$$

55. Determine a rational function f whose graph has a vertical asymptote at $x = -2$ and a horizontal asymptote at $y = 3$ and includes the point $(1, 2)$. [2.4]

TECHNOLOGY CONNECTION

Use a calculator to estimate the relative extrema of each function. [2.1 and 2.2]

56. $f(x) = 3.8x^5 - 18.6x^3$

57. $f(x) = \sqrt[3]{|9 - x^2|} - 1$

58. Life and physical sciences: incidence of breast cancer.   The following table provides data relating the incidence of breast cancer per 100,000 women of various ages.

a) Use REGRESSION to fit linear, quadratic, cubic, and quartic functions to the data.

Age	Incidence per 100,000
0	0
27	10
32	25
37	60
42	125
47	187
52	224
57	270
62	340
67	408
72	437
77	475
82	460
87	420

(Source: *National Cancer Institute.*)

b) Which function best fits the data?

c) Determine the domain of the function on the basis of the function and the problem situation, and explain.

d) Determine the maximum value of the function on the domain. At what age is the incidence of breast cancer the greatest? [2.1 and 2.2]

Note: The function used in Exercise 28 of Section R.1 was found in this manner.

CHAPTER 2
TEST

Find all relative minimum or maximum values as well as the x-values at which they occur. State where each function is increasing or decreasing. Then sketch a graph of the function.

1. $f(x) = x^2 - 4x - 5$

2. $f(x) = 4 + 3x - x^3$

3. $f(x) = (x - 2)^{2/3} - 4$

4. $f(x) = \dfrac{16}{x^2 + 4}$

Sketch a graph of each function. List any extrema, and indicate any asymptotes or points of inflection.

5. $f(x) = x^3 + x^2 - x + 1$

6. $f(x) = 2x^4 - 4x^2 + 1$

7. $f(x) = (x - 2)^3 + 3$

8. $f(x) = x\sqrt{9 - x^2}$

9. $f(x) = \dfrac{2}{x - 1}$

10. $f(x) = \dfrac{-8}{x^2 - 4}$

11. $f(x) = \dfrac{x^2 - 1}{x}$

12. $f(x) = \dfrac{x - 3}{x + 2}$

Find the absolute maximum and minimum values, if they exist, of each function over the indicated interval. Where no interval is specified, use the real line.

13. $f(x) = x(6 - x)$

14. $f(x) = x^3 + x^2 - x + 1; \ \left[-2, \tfrac{1}{2}\right]$

15. $f(x) = -x^2 + 8.6x + 10$

16. $f(x) = -2x + 5; \ [-1, 1]$

17. $f(x) = -2x + 5$

18. $f(x) = 3x^2 - x - 1$

19. $f(x) = x^2 + \dfrac{128}{x}; \ (0, \infty)$

20. Of all numbers whose difference is 8, find the two that have the minimum product.

21. Minimize $Q = x^2 + y^2$, where $x - y = 10$.

22. Business: maximum profit. Find the maximum profit and the number of units, x, that must be produced and sold in order to yield the maximum profit. Assume that $R(x)$ and $C(x)$ are the revenue and cost, in dollars, when x units are produced:

$R(x) = x^2 + 110x + 60$,
$C(x) = 1.1x^2 + 10x + 80$.

23. Business: minimizing cost. From a thin piece of cardboard 60 in. by 60 in., square corners are cut out so that the sides can be folded up to make an open box. What dimensions will yield a box of maximum volume? What is the maximum volume?

24. Business: minimizing inventory costs. Ironside Sports sells 1225 tennis rackets per year. It costs $2 to store one tennis racket for a year. To reorder, there is a fixed cost of $1, plus $0.50 for each tennis racket. How many times per year should Ironside order tennis rackets, and in what lot size, in order to minimize inventory costs?

25. For $y = f(x) = x^2 - 3$, $x = 5$, and $\Delta x = 0.1$, find Δy and $f'(x) \, \Delta x$.

26. Approximate $\sqrt{50}$ using $\Delta y \approx f'(x) \, \Delta x$.

27. For $y = \sqrt{x^2 + 3}$:

a) Find dy.

b) Find dy when $x = 2$ and $dx = 0.01$.

28. Differentiate the following implicitly to find dy/dx. Then find the slope of the curve at $(1, 2)$:

$x^3 + y^3 = 9$.

29. A spherical balloon has a radius of 15 cm. Use a differential to find the approximate change in the volume of the balloon if the radius is increased or decreased by 0.5 cm. (The volume of a sphere is $V = \frac{4}{3}\pi r^3$. Use 3.14 for π.)

30. A pole 13 ft long leans against a vertical wall. If the lower end is moving away from the wall at the rate of 0.4 ft/sec, how fast is the upper end coming down when the lower end is 12 ft from the wall?

SYNTHESIS

31. Find the absolute maximum and minimum values of the following function, if they exist, over $[0, \infty)$:

$$f(x) = \frac{x^2}{1 + x^3}.$$

32. Business: minimizing average cost. The total cost in dollars of producing x units of a product is given by

$$C(x) = 100x + 100\sqrt{x} + \frac{\sqrt{x^3}}{100}.$$

How many units should be produced to minimize the average cost?

TECHNOLOGY CONNECTION

33. Use a calculator to estimate any extrema of this function:

$f(x) = 5x^3 - 30x^2 + 45x + 5\sqrt{x}$.

34. Use a calculator to estimate any extrema of this function:

$g(x) = x^5 - x^3$.

35. Business: advertising. The business of manufacturing and selling bowling balls is one of frequent changes. Companies introduce new models to the market about every 3 to 4 months. Typically, a new model is created because of advances in technology such as new surface stock or a new way to place weight blocks in a ball. To decide how to best use advertising dollars, companies track the sales in relation to the amount spent on advertising. Suppose that a company has the following data from past sales.

Amount Spent on Advertising (in thousands)	Number of Bowling Balls Sold, N
$ 0	8
50	13,115
100	19,780
150	22,612
200	20,083
250	12,430
300	4

a) Use REGRESSION to fit linear, quadratic, cubic, and quartic functions to the data.

b) Determine the domain of the function in part (a) that best fits the data and the problem situation. Justify your answer.

c) Determine the maximum value of the function on the domain. How much should the company spend on advertising its next new model in order to maximize the number of bowling balls sold?

Maximum Sustainable Harvest

In certain situations, biologists are able to determine what is called a **reproduction curve**. This is a function

$$y = f(P)$$

such that if P is the population after P years, then $f(P)$ is the population a year later, at time $t + 1$. Such a curve is shown below.

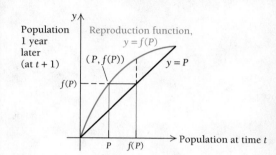

The line $y = P$ is significant because if it ever coincides with the curve $y = f(P)$, then we know that the population stays the same from year to year. Here the graph of f lies mostly above the line, indicating that the population is increasing.

Too many deer in a forest can deplete the food supply and eventually cause the population to decrease for lack of food. In such cases, often with some controversy, hunters are allowed to "harvest" some of the deer. Then with a greater food supply, the remaining deer population may prosper and increase.

We know that a population P will grow to a population $f(P)$ in a year. If this were a population of fur-bearing animals and the population were increasing, then hunters could "harvest" the amount

$$f(P) - P$$

each year without shrinking the initial population P. If the population were remaining the same or decreasing, then such a harvest would deplete the population.

Suppose that we want to know the value of P_0 that would allow the harvest to be the largest. If we could determine that P_0, we could let the population grow until it reached that level and then begin harvesting year after year the amount $f(P_0) - P_0$.

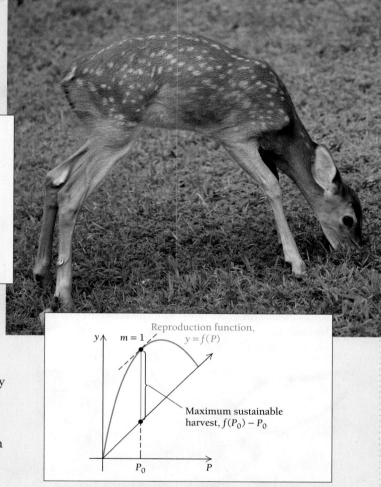

Let the harvest function H be given by

$$H(P) = f(P) - P.$$

Then $H'(P) = f'(P) - 1.$

Now, if we assume that $H'(P)$ exists for all values of P and that there is only one critical value, it follows that the *maximum sustainable harvest* occurs at that value P_0 such that

$$H'(P_0) = f'(P_0) - 1 = 0$$

and $H''(P_0) = f''(P_0) < 0.$

Or, equivalently, we have the following.

THEOREM

The **maximum sustainable harvest** occurs at P_0 such that

$$f'(P_0) = 1 \quad \text{and} \quad f''(P_0) < 0,$$

and is given by

$$H(P_0) = f(P_0) - P_0.$$

EXERCISES

For Exercises 1–3, do the following.

a) *Graph the reproduction curve, the line $y = P$, and the harvest function using the same viewing window.*
b) *Find the population at which the maximum sustainable harvest occurs. Use both a graphical solution and a calculus solution.*
c) *Find the maximum sustainable harvest.*

1. $f(P) = P(10 - P)$, where P is measured in thousands.
2. $f(P) = -0.025P^2 + 4P$, where P is measured in thousands. This is the reproduction curve in the Hudson Bay area for the snowshoe hare, a fur-bearing animal.

3. $f(P) = -0.01P^2 + 2P$, where P is measured in thousands. This is the reproduction curve in the Hudson Bay area for the lynx, a fur-bearing animal.

For Exercises 4 and 5, do the following.

a) *Graph the reproduction curve, the line $y = P$, and the harvest function using the same viewing window.*
b) *Graphically determine the population at which the maximum sustainable harvest occurs.*
c) *Find the maximum sustainable harvest.*

4. $f(P) = 40\sqrt{P}$, where P is measured in thousands. Assume that this is the reproduction curve for the brown trout population in a large lake.

5. $f(P) = 0.237P\sqrt{2000 - P^2}$, where P is measured in thousands.
6. The table below lists data regarding the reproduction of a certain animal.

a) Use REGRESSION to fit a cubic polynomial to these data.
b) Graph the reproduction curve, the line $y = P$, and the harvest function using the same viewing window.
c) Graphically determine the population at which the maximum sustainable harvest occurs.

POPULATION, P (in thousands)	POPULATION, $f(P)$, 1 YEAR LATER
10	9.7
20	23.1
30	37.4
40	46.2
50	42.6

Exponential and Logarithmic Functions

Chapter Snapshot

What You'll Learn

Why It's Important

In this chapter, we consider two types of functions that are closely related: *exponential functions* and *logarithmic functions*. After learning to find derivatives of such functions, we will study applications in the areas of population growth and decay, continuously compounded interest, spread of disease, and carbon dating.

Where It's Used

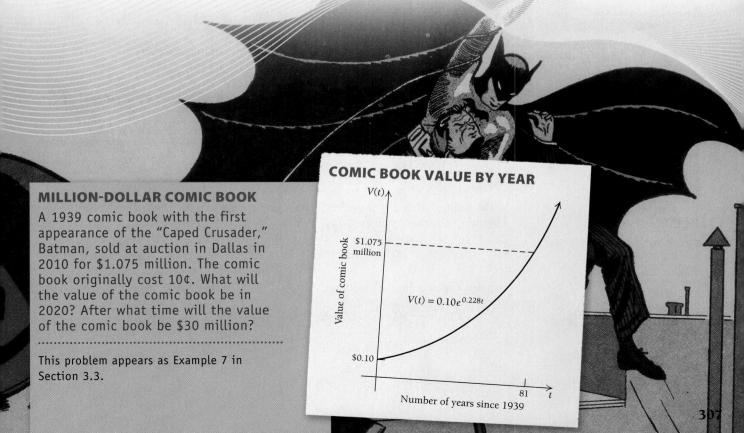

MILLION-DOLLAR COMIC BOOK

A 1939 comic book with the first appearance of the "Caped Crusader," Batman, sold at auction in Dallas in 2010 for $1.075 million. The comic book originally cost 10¢. What will the value of the comic book be in 2020? After what time will the value of the comic book be $30 million?

This problem appears as Example 7 in Section 3.3.

COMIC BOOK VALUE BY YEAR

$V(t)$

$1.075 million

$V(t) = 0.10e^{0.228t}$

$0.10

81 — t

Number of years since 1939

3.1 Exponential Functions

Graphs of Exponential Functions

Consider the following graph. The rapid rise of the graph indicates that it approximates an *exponential function*. We now consider such functions and many of their applications.

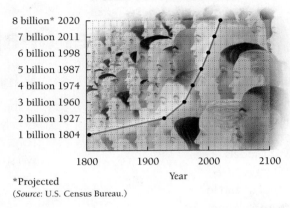

WORLD POPULATION GROWTH

8 billion* 2020
7 billion 2011
6 billion 1998
5 billion 1987
4 billion 1974
3 billion 1960
2 billion 1927
1 billion 1804

1800   1900   2000   2100
Year

*Projected
(*Source*: U.S. Census Bureau.)

Let's review definitions of expressions of the form a^x, where x is a rational number. For example,

$$a^{2.34} \quad \text{or} \quad a^{234/100}$$

means "raise a to the 234th power and then take the 100th root $\left(\sqrt[100]{a^{234}}\right)$."

What about expressions with irrational exponents, such as $2^{\sqrt{2}}, 2^{\pi}$, and $2^{-\sqrt{3}}$? An *irrational number* is a number named by an infinite, nonrepeating decimal. Let's consider 2^{π}. We know that π is irrational with an infinite, nonrepeating decimal expansion:

$$3.141592653\ldots.$$

This means that π is approached as a limit by the rational numbers

$$3, 3.1, 3.14, 3.141, 3.1415, \ldots,$$

so it seems reasonable that 2^{π} should be approached as a limit by the rational powers

$$2^3, 2^{3.1}, 2^{3.14}, 2^{3.141}, 2^{3.1415}, \ldots.$$

Estimating each power with a calculator, we get the following:

$$8, 8.574188, 8.815241, 8.821353, 8.824411, \ldots.$$

In general, a^x is approximated by the values of a^r for rational numbers r near x; a^x is the limit of a^r as r approaches x through rational values. Thus, for $a > 0$, the usual laws of exponents, such as

$$a^x \cdot a^y = a^{x+y}, \qquad a^x \div a^y = a^{x-y}, \qquad (a^x)^y = a^{xy}, \qquad \text{and} \qquad a^{-x} = \frac{1}{a^x},$$

can be applied to real number exponents. Moreover, the function so obtained, $f(x) = a^x$, is continuous.

DEFINITION

An **exponential function** f is given by

$$f(x) = a^x,$$

where x is any real number, $a > 0$, and $a \neq 1$. The number a is called the **base**.

The following are examples of exponential functions:

$$f(x) = 2^x, \qquad f(x) = \left(\tfrac{1}{2}\right)^x, \qquad f(x) = (0.4)^x.$$

Note that in contrast to power functions like $y = x^2$ and $y = x^3$, an exponential function has the variable in the exponent, not as the base. Exponential functions have countless applications. For now, however, let's consider their graphs.

■ **EXAMPLE 1** Graph: $y = f(x) = 2^x$.

Solution First, we find some function values. Note that 2^x is always positive:

$$x = 0, \quad y = 2^0 = 1;$$

$$x = \frac{1}{2}, \quad y = 2^{1/2} = \sqrt{2} \approx 1.4;$$

$$x = 1, \quad y = 2^1 = 2;$$

$$x = 2, \quad y = 2^2 = 4;$$

$$x = 3, \quad y = 2^3 = 8;$$

$$x = -1, \quad y = 2^{-1} = \frac{1}{2};$$

$$x = -2, \quad y = 2^{-2} = \frac{1}{2^2} = \frac{1}{4}.$$

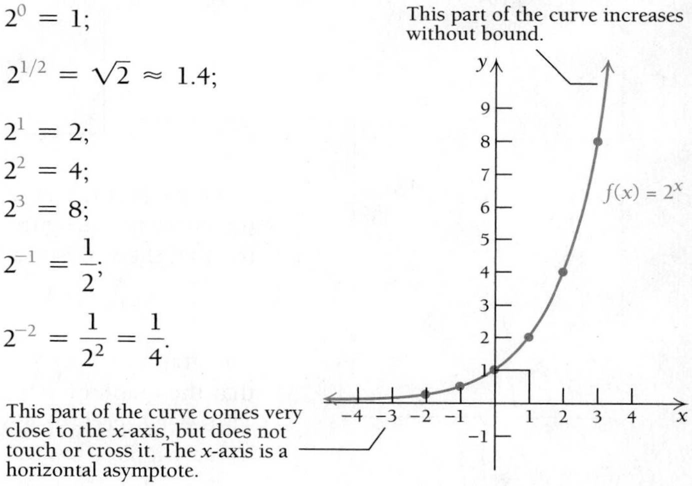

This part of the curve increases without bound.

$f(x) = 2^x$

This part of the curve comes very close to the x-axis, but does not touch or cross it. The x-axis is a horizontal asymptote.

x	0	$\frac{1}{2}$	1	2	3	-1	-2
$y = f(x) = 2^x$	1	1.4	2	4	8	$\frac{1}{2}$	$\frac{1}{4}$

Next, we plot the points and connect them with a smooth curve, as shown above. The graph is continuous, increasing without bound, and concave up. We see too that the x-axis is a horizontal asymptote (see Section 2.3), that is,

$$\lim_{x \to -\infty} f(x) = 0 \quad \text{and} \quad \lim_{x \to \infty} f(x) = \infty.$$

❮ Quick Check 1

> ❯ **Quick Check 1**
>
> For $f(x) = 3^x$, complete this table of function values.
>
> | x | $f(x) = 3^x$ |
> |---|---|
> | 0 | |
> | 1 | |
> | 2 | |
> | 3 | |
> | -1 | |
> | -2 | |
> | -3 | |
>
> Graph $f(x) = 3^x$.

■ **EXAMPLE 2** Graph: $y = g(x) = \left(\tfrac{1}{2}\right)^x$.

Solution First, we note that

$$y = g(x) = \left(\tfrac{1}{2}\right)^x$$
$$= (2^{-1})^x$$
$$= 2^{-x}.$$

This will ease our work in calculating function values:

$$x = 0, \quad y = 2^{-0} = 1;$$

$$x = \frac{1}{2}, \quad y = 2^{-1/2} = \frac{1}{2^{1/2}}$$

$$= \frac{1}{\sqrt{2}} \approx \frac{1}{1.4} \approx 0.7;$$

$$x = 1, \quad y = 2^{-1} = \frac{1}{2};$$

$$x = 2, \quad y = 2^{-2} = \frac{1}{4};$$

$$x = -1, \quad y = 2^{-(-1)} = 2;$$

$$x = -2, \quad y = 2^{-(-2)} = 4;$$

$$x = -3, \quad y = 2^{-(-3)} = 8.$$

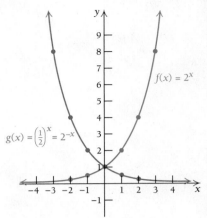

> **Quick Check 2**
>
> For $g(x) = \left(\frac{1}{3}\right)^x$, complete this table of function values.
>
> | x | $g(x) = \left(\frac{1}{3}\right)^x$ |
> |---|---|
> | 0 | |
> | 1 | |
> | 2 | |
> | 3 | |
> | −1 | |
> | −2 | |
> | −3 | |
>
> Graph $g(x) = \left(\frac{1}{3}\right)^x$.

x	0	$\frac{1}{2}$	1	2	−1	−2	−3
$y = f(x) = \left(\frac{1}{2}\right)^x$	1	0.7	$\frac{1}{2}$	$\frac{1}{4}$	2	4	8

Next, we plot these points and connect them with a smooth curve, as shown by the red curve in the figure. The graph is continuous, decreasing, and concave up. We see too that the x-axis is a horizontal asymptote, that is,

$$\lim_{x \to \infty} g(x) = 0 \quad \text{and} \quad \lim_{x \to -\infty} g(x) = \infty.$$

The graph of $f(x) = 2^x$ of Example 1 is shown as a blue curve, for comparison. Note that the graph of $y = g(x)$ is the reflection of the graph of $y = f(x)$ across the y-axis. Thus, we *expect* the graphs of $y = \left(\frac{1}{2}\right)^x = 2^{-x}$ and $y = 2^x$ to be symmetric with respect to the y-axis.

❬ Quick Check 2

TECHNOLOGY CONNECTION

Exploratory Exercise: Growth

Take a sheet of $8\frac{1}{2}$-in.-by-11-in. paper and cut it into two equal pieces. Then cut these again to obtain four equal pieces. Then cut these to get eight equal pieces, and so on, performing five cutting steps.

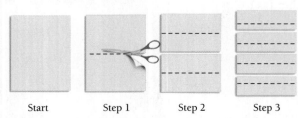

	t	$0.004 \cdot 2^t$
Start	0	$0.004 \cdot 2^0$, or 0.004
Step 1	1	$0.004 \cdot 2^1$, or 0.008
Step 2	2	$0.004 \cdot 2^2$, or 0.016
Step 3	3	
Step 4	4	
Step 5	5	

a) Place all the pieces in a stack and measure the thickness.

b) A piece of paper is typically 0.004 in. thick. Check the measurement in part (a) by completing the table.

c) Graph the function $f(t) = 0.004(2)^t$.

d) Compute the thickness of the stack (in miles) after 25 steps.

Exploratory

Use a calculator or iPlot or Graphicus to check the graphs of the functions in Examples 1 and 2. Then graph

$$f(x) = 3^x \quad \text{and}$$
$$g(x) = \left(\tfrac{1}{3}\right)^x.$$

and look for patterns.

1. The function given by $f(x) = a^x$, with $a > 1$, is a positive, increasing, continuous function. As x gets smaller, a^x approaches 0. The graph is concave up, and the x-axis is the horizontal asymptote.

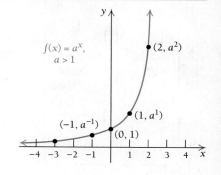

2. The function given by $f(x) = a^x$, with $0 < a < 1$, is a positive, decreasing, continuous function. As x gets larger, a^x approaches 0. The graph is concave up, and the x-axis is the horizontal asymptote.

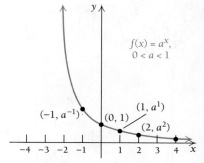

For $a = 1$, we have $f(x) = a^x = 1^x = 1$; so, in this case, f is a constant function. This is why we do not allow 1 to be the base of an exponential function.

The Number e and the Derivative of e^x

Let's consider finding the derivative of the exponential function

$$f(x) = a^x.$$

The derivative is given by

$$f'(x) = \lim_{h \to 0} \frac{f(x+h) - f(x)}{h} \qquad \text{Definition of the derivative}$$

$$= \lim_{h \to 0} \frac{a^{x+h} - a^x}{h} \qquad \text{Substituting } a^{x+h} \text{ for } f(x+h) \text{ and } a^x \text{ for } f(x)$$

$$= \lim_{h \to 0} \frac{a^x \cdot a^h - a^x \cdot 1}{h} \qquad \text{Using the laws for exponents}$$

$$= \lim_{h \to 0} \left(a^x \cdot \frac{a^h - 1}{h} \right) \qquad \text{Factoring}$$

$$= a^x \cdot \lim_{h \to 0} \frac{a^h - 1}{h}. \qquad a^x \text{ is constant with respect to } h, \text{ and the limit of a constant times a function is the constant times the limit of that function. (See Limit Property L6 in Section 1.2.)}$$

We get

$$f'(x) = a^x \cdot \lim_{h \to 0} \frac{a^h - 1}{h}.$$

In particular, for $g(x) = 2^x$,

$$g'(x) = 2^x \cdot \lim_{h \to 0} \frac{2^h - 1}{h}. \qquad \text{Substituting 2 for } a$$

h	$\dfrac{2^h - 1}{h}$
0.5	0.8284
0.25	0.7568
0.175	0.7369
0.0625	0.7084
0.03125	0.7007
0.00111	0.6934
0.000001	0.6931

Note that the limit does not depend on the value of x at which we are evaluating the derivative. In order for $g'(x)$ to exist, we must determine whether

$$\lim_{h \to 0} \frac{2^h - 1}{h} \text{ exists.}$$

Let's investigate this question.

We choose a sequence of numbers h approaching 0 and compute $(2^h - 1)/h$, listing the results in a table, as shown at left. It seems reasonable to assume that $(2^h - 1)/h$ has a limit as h approaches 0 and that its approximate value is 0.7; thus,

$$g'(x) \approx (0.7)2^x.$$

In other words, the derivative is a constant times 2^x. Similarly, for $t(x) = 3^x$,

$$t'(x) = 3^x \cdot \lim_{h \to 0} \frac{3^h - 1}{h}. \qquad \text{Substituting 3 for } a$$

h	$\dfrac{3^h - 1}{h}$
0.5	1.4641
0.25	1.2643
0.175	1.2113
0.0625	1.1372
0.03125	1.1177
0.00111	1.0993
0.000001	1.0986

Again, we can find an approximation for the limit that does not depend on the value of x at which we are evaluating the derivative. Consider the table at left. Again, it seems reasonable to conclude that $(3^h - 1)/h$ has a limit as h approaches 0. This time, the approximate value is 1.1; thus,

$$t'(x) \approx (1.1)3^x.$$

In other words, the derivative is a constant times 3^x.

Let's now analyze what we have done. We proved that

$$\text{if } f(x) = a^x, \quad \text{then } f'(x) = a^x \cdot \lim_{h \to 0} \frac{a^h - 1}{h}.$$

Consider $\lim\limits_{h \to 0} \dfrac{a^h - 1}{h}$.

1. For $a = 2$,

$$\lim_{h \to 0} \frac{a^h - 1}{h} = \lim_{h \to 0} \frac{2^h - 1}{h} \approx 0.6931.$$

2. For $a = 3$,

$$\lim_{h \to 0} \frac{a^h - 1}{h} = \lim_{h \to 0} \frac{3^h - 1}{h} \approx 1.0986.$$

It seems reasonable to conclude that, for some choice of a between 2 and 3, we have

$$\lim_{h \to 0} \frac{a^h - 1}{h} = 1.$$

To find that a, it suffices to look for a value such that

$$\frac{a^h - 1}{h} = 1. \tag{1}$$

Multiplying both sides by h and then adding 1 to both sides, we have

$$a^h = 1 + h.$$

Raising both sides to the power $1/h$, we have

$$a = (1 + h)^{1/h}. \tag{2}$$

Since equations (1) and (2) are equivalent, it follows that

$$\lim_{h \to 0} \frac{a^h - 1}{h} = \lim_{h \to 0} 1 \quad \text{and} \quad \lim_{h \to 0} a = \lim_{h \to 0} (1 + h)^{1/h}$$

h	$(1 + h)^{1/h}$
0.5	2.25
0.1	2.5937
0.01	2.7048
0.001	2.7169
−0.01	2.732
−0.001	2.7196
0.0001	2.7181

are also equivalent. Thus,

$$\lim_{h \to 0} \frac{a^h - 1}{h} = 1 \quad \text{and} \quad a = \lim_{h \to 0} (1 + h)^{1/h}$$

are equivalent. We conclude that, for there to be a number a for which $\lim\limits_{h \to 0} \dfrac{a^h - 1}{h} = 1$, we must have $a = \lim\limits_{h \to 0} (1 + h)^{1/h}$. This last equation gives us the special number we are searching for.* The number is named e, in honor of Leonhard Euler (pronounced "Oiler"), the great Swiss mathematician (1707–1783) who did groundbreaking work with it.

DEFINITION

$$e = \lim_{h \to 0} (1 + h)^{1/h} \approx 2.718281828459$$

We call e the *natural base*.

It follows that, for the exponential function $f(x) = e^x$,

$$f'(x) = e^x \cdot \lim_{h \to 0} \frac{e^h - 1}{h}$$
$$= e^x \cdot 1$$
$$= e^x.$$

That is, the derivative of e^x is e^x.

We have shown that if $f(x) = e^x$, it follows that $f'(x) = e^x$.

THEOREM 1

The derivative of the function f given by $f(x) = e^x$ is itself:

$$f'(x) = f(x), \qquad \text{or} \qquad \frac{d}{dx} e^x = e^x.$$

Theorem 1 says that for the function $f(x) = e^x$, the derivative at x (the slope of the tangent line) is the same as the function value at x. That is, on the graph of $y = e^x$, at the point $(0, 1)$, the slope is $m = 1$; at the point $(1, e)$, the slope is $m = e$; at the point $(2, e^2)$, the slope is $m = e^2$, and so on. The function $y = e^x$ is the *only* exponential function for which this correlation between the function and its derivative is true.

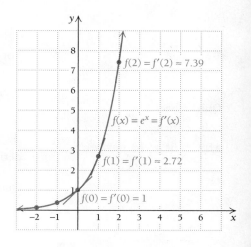

---

* This derivation is based on one presented in Appendix 4 of *e: The Story of a Number*, by Eli Maor (Princeton University Press, 1998).

Exploratory

Check the results of Example 3 by entering each function as y_1 and letting $y_2 = \text{nDeriv}(y_1, x, x)$. Then enter the derivatives from Example 3 as y_3 and use graphs or a table to compare y_2 and y_3.

Using iPlot, graph $f(x) = 3e^x$ in red. Turn on Derivate. Now suppose the derivative was mistakenly found to be $f'(x) = 3e^{x-1}$. Graph this incorrect function in some other color. What happens? Explain. Then describe a procedure for checking the results of Example 3 using iPlot.

Using Graphicus, graph $g(x) = x^2 e^x$ as a first function. Then touch $+$ and choose Add derivative. Suppose the derivative was mistakenly found to be $g'(x) = 2xe^x$. Graph this incorrect function. What happens? Explain. Then describe a procedure for checking the results of Example 3 using Graphicus.

In Section 3.5, we will develop a formula for the derivative of the more general exponential function given by $y = a^x$.

Finding Derivatives of Functions Involving e

We can use Theorem 1 in combination with other theorems derived earlier to differentiate a variety of functions.

■ **EXAMPLE 3** Find dy/dx: **a)** $y = 3e^x$; **b)** $y = x^2 e^x$; **c)** $y = \dfrac{e^x}{x^3}$.

Solution

a) $\dfrac{d}{dx}(3e^x) = 3\dfrac{d}{dx}e^x$ Recall that $\dfrac{d}{dx}[c \cdot f(x)] = c \cdot f'(x)$.

                 $= 3e^x$

b) $\dfrac{d}{dx}(x^2 e^x) = x^2 \cdot e^x + e^x \cdot 2x$ Using the Product Rule

               $= e^x(x^2 + 2x)$,   or $xe^x(x + 2)$ Factoring

c) $\dfrac{d}{dx}\left(\dfrac{e^x}{x^3}\right) = \dfrac{x^3 \cdot e^x - e^x \cdot 3x^2}{x^6}$ Using the Quotient Rule

            $= \dfrac{x^2 e^x(x - 3)}{x^6}$ Factoring

            $= \dfrac{e^x(x - 3)}{x^4}$ Simplifying

Suppose that we have a more complicated function in the exponent, as in

$$h(x) = e^{x^2 - 5x}.$$

This is a composition of functions. For such a function, we have

$$h(x) = g(f(x)) = e^{f(x)}, \quad \text{where} \quad g(x) = e^x \quad \text{and} \quad f(x) = x^2 - 5x.$$

Now $g'(x) = e^x$. Then by the Chain Rule (Section 1.7), we have

$$h'(x) = g'(f(x)) \cdot f'(x)$$
$$= e^{f(x)} \cdot f'(x).$$

For the case above, $f(x) = x^2 - 5x$, so $f'(x) = 2x - 5$. Then

$$h'(x) = g'(f(x)) \cdot f'(x)$$
$$= e^{f(x)} \cdot f'(x)$$
$$= e^{x^2 - 5x}(2x - 5).$$

The next theorem, which we have proven using the Chain Rule, allows us to find derivatives of functions like the one above.

> **Quick Check 3**
>
> Differentiate:
>
> **a)** $y = 6e^x$; **b)** $y = x^3 e^x$;
>
> **c)** $y = \dfrac{e^x}{x^2}$.

❮ Quick Check 3

THEOREM 2

The derivative of e to some power is the product of e to that power and the derivative of the power:

$$\frac{d}{dx} e^{f(x)} = e^{f(x)} \cdot f'(x)$$

or

$$\frac{d}{dx} e^u = e^u \cdot \frac{du}{dx}.$$

The following gives us a way to remember this rule.

$$h(x) = e^{x^2 - 5x}$$

$\downarrow$ Rewrite the original function.

$$h'(x) = e^{x^2 - 5x}(2x - 5)$$ Multiply by the derivative of the exponent.

■ EXAMPLE 4   Differentiate each of the following with respect to x:

a) $y = e^{8x}$;   **b)** $y = e^{-x^2 + 4x - 7}$;   **c)** $e^{\sqrt{x^2 - 3}}$.

Solution

a) $\dfrac{d}{dx} e^{8x} = e^{8x} \cdot 8$, or $8e^{8x}$

b) $\dfrac{d}{dx} e^{-x^2 + 4x - 7} = e^{-x^2 + 4x - 7}(-2x + 4)$,   or $-2(x - 2)e^{-x^2 + 4x - 7}$

c) $\dfrac{d}{dx} e^{\sqrt{x^2 - 3}} = \dfrac{d}{dx} e^{(x^2 - 3)^{1/2}}$

$\qquad = e^{(x^2 - 3)^{1/2}} \cdot \frac{1}{2}(x^2 - 3)^{-1/2} \cdot 2x$ Using the Chain Rule twice

$\qquad = e^{\sqrt{x^2 - 3}} \cdot x \cdot (x^2 - 3)^{-1/2}$

$\qquad = \dfrac{e^{\sqrt{x^2 - 3}} \cdot x}{\sqrt{x^2 - 3}}$,   or $\dfrac{xe^{\sqrt{x^2 - 3}}}{\sqrt{x^2 - 3}}$

❮ Quick Check 4

> **Quick Check 4**
>
> Differentiate:
>
> **a)** $f(x) = e^{-4x}$;
> **b)** $g(x) = e^{x^3 + 8x}$;
> **c)** $h(x) = e^{\sqrt{x^2 + 5}}$.

Graphs of e^x, e^{-x}, and $1 - e^{-kx}$

Now that we know how to find the derivative of $f(x) = e^x$, let's look at the graph of $f(x) = e^x$ from the standpoint of calculus concepts and the curve-sketching techniques discussed in Section 2.2.

■ **EXAMPLE 5**   Graph: $f(x) = e^x$. Analyze the graph using calculus.

Solution   We simply find some function values using a calculator, plot the points, and sketch the graph as shown below.

x	$f(x)$
-2	0.135
-1	0.368
0	1
1	2.718
2	7.389

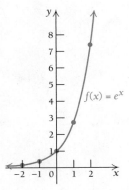

We analyze the graph using calculus as follows.

a) *Derivatives.* Since $f(x) = e^x$, it follows that $f'(x) = e^x$, so $f''(x) = e^x$.

b) *Critical values of f.* Since $f'(x) = e^x > 0$ for all real numbers x, we know that the derivative exists for all real numbers and there is no solution of the equation $f'(x) = 0$. There are no critical values and therefore no maximum or minimum values.

c) *Increasing.* We have $f'(x) = e^x > 0$ for all real numbers x, so the function f is increasing over the entire real line, $(-\infty, \infty)$.

d) *Inflection points.* We have $f''(x) = e^x > 0$ for all real numbers x, so the equation $f''(x) = 0$ has no solution and there are no points of inflection.

e) *Concavity.* Since $f''(x) = e^x > 0$ for all real numbers x, the function f' is increasing and the graph is concave up over the entire real line.

■ **EXAMPLE 6**   Graph: $g(x) = e^{-x}$. Analyze the graph using calculus.

Solution   First, we find some function values, plot the points, and sketch the graph.

x	$g(x)$
-2	7.389
-1	2.718
0	1
1	0.368
2	0.135

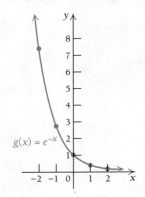

We can then analyze the graph using calculus as follows.

a) *Derivatives.* Since $g(x) = e^{-x}$, we have

$$g'(x) = e^{-x}(-1) = -e^{-x},$$

so

$$g''(x) = -e^{-x}(-1) = e^{-x}.$$

b) *Critical values of g.* Since $e^{-x} = 1/e^x > 0$, we have $g'(x) = -e^{-x} < 0$ for all real numbers x. Thus, the derivative exists for all real numbers, and the equation $g'(x) = 0$ has no solution. There are no critical values and therefore no maximum or minimum values.

c) *Decreasing.* Since the derivative $g'(x) = -e^{-x} < 0$ for all real numbers x, the function g is decreasing over the entire real line.

d) *Inflection points.* We have $g''(x) = e^{-x} > 0$, so the equation $g''(x) = 0$ has no solution and there are no points of inflection.

e) *Concavity.* We also know that since $g''(x) = e^{-x} > 0$ for all real numbers x, the function g' is increasing and the graph is concave up over the entire real line.

Functions of the type $f(x) = 1 - e^{-kx}$, with $x \geq 0$, have important applications.

■ **EXAMPLE 7**   Graph: $h(x) = 1 - e^{-2x}$, with $x \geq 0$. Analyze the graph using calculus.

Solution   First, we find some function values, plot the points, and sketch the graph.

x	$h(x)$
0	0
0.5	0.63212
1	0.86466
2	0.98168
3	0.99752
4	0.99966
5	0.99995

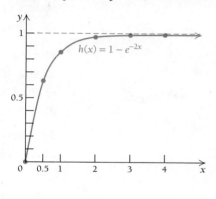

We can analyze the graph using calculus as follows.

a) *Derivatives.* Since $h(x) = 1 - e^{-2x}$,

$$h'(x) = -e^{-2x}(-2) = 2e^{-2x}$$

and

$$h''(x) = 2e^{-2x}(-2) = -4e^{-2x}.$$

b) *Critical values.* Since $e^{-2x} = 1/e^{2x} > 0$, we have $2e^{-2x} > 0$. Thus, $h'(x) = 0$ has no solution, and since $h'(x)$ exists for all $x > 0$, it follows that there are no critical values on the interval $(0, \infty)$

c) *Increasing.* Since $2e^{-2x} > 0$ for all real numbers x, we know that h is increasing over the interval $[0, \infty)$.

d) *Inflection points.* Since $h''(x) = -4e^{-2x} < 0$, we know that the equation $h''(x) = 0$ has no solution; thus, there are no points of inflection.

e) *Concavity.* Since $h''(x) = -4e^{-2x} < 0$, we know that h' is decreasing and the graph is concave down over the interval $(0, \infty)$.

〉 Quick Check 5

Graph each function. Then determine critical values, intervals over which the function is increasing or decreasing, inflection points, and the concavity.

a) $f(x) = 2e^{-x}$;

b) $g(x) = 2e^x$;

c) $h(x) = 1 - e^{-x}$.

In general, for $k > 0$, the graph of $h(x) = 1 - e^{-kx}$ is increasing, which we expect since $h'(x) = ke^{-kx}$ is always positive. Note that $h(x)$ approaches 1 as x approaches ∞.

A word of caution! Functions of the type a^x (for example, 2^x, 3^x, and e^x) are different from functions of the type x^a (for example, $x^2, x^3, x^{1/2}$). For a^x, the variable is in the exponent. For x^a, the variable is in the base. The derivative of a^x is not xa^{x-1}. In particular, we have the following:

$$\frac{d}{dx} e^x \neq xe^{x-1}, \quad \text{but} \quad \frac{d}{dx} e^x = e^x.$$

《 Quick Check 5

■ EXAMPLE 8 **Business: Worker Efficiency.** It is reasonable for a manufacturer to expect the daily output of a new worker to start out slow and continue to increase over time, but then tend to level off, never exceeding a certain amount. A firm manufactures 5G smart phones and determines that after working t days, the efficiency, in number of phones produced per day, of most workers can be modeled by the function

$$N(t) = 80 - 70e^{-0.13t}.$$

a) Find $N(0), N(1), N(5), N(10), N(20),$ and $N(30)$.

b) Graph $N(t)$.

c) Find $N'(t)$ and interpret this derivative in terms of rate of change.

d) What number of phones seems to determine where worker efficiency levels off?

Solution

a) We make a table of input–output values.

t	0	1	5	10	20	30
$N(t)$	10	18.5	43.5	60.9	74.8	78.6

b) Using these values and/or a graphing calculator, we obtain the graph.

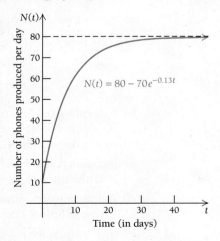

c) $N'(t) = -70e^{-0.13t}(-0.13) = 9.1e^{-0.13t}$; after t days, the rate of change of number of phones produced per day is given by $9.1e^{-0.13t}$.

d) Examining the graph and expanding the table of function values, it seems that worker efficiency levels off at no more than 80 phones produced per day.

〉 Quick Check 6

Business. Repeat Example 8 for the efficiency function

$$N(t) = 80 - 60e^{-0.12t}.$$

《 Quick Check 6

Section Summary

- The *exponential function* $f(x) = e^x$, where $e \approx 2.71828$, has the derivative $f'(x) = e^x$. That is, the slope of a tangent line to the graph of $y = e^x$ is the same as the function value at x.

- The graph of $f(x) = e^x$ is an increasing function with no critical values, no maximum or minimum values, and no points of inflection. The graph is concave up, with

$$\lim_{x \to \infty} f(x) = \infty \qquad \text{and} \qquad \lim_{x \to -\infty} f(x) = 0.$$

- Calculus is rich in applications of exponential functions.

EXERCISE SET
3.1

Graph.

1. $y = 4^x$

2. $y = 5^x$

3. $y = (0.25)^x$

4. $y = (0.2)^x$

5. $f(x) = \left(\frac{3}{2}\right)^x$

6. $f(x) = \left(\frac{4}{3}\right)^x$

7. $g(x) = \left(\frac{2}{3}\right)^x$

8. $g(x) = \left(\frac{3}{4}\right)^x$

9. $f(x) = (2.5)^x$

10. $f(x) = (1.2)^x$

Differentiate.

11. $f(x) = e^{-x}$

12. $f(x) = e^x$

13. $g(x) = e^{3x}$

14. $g(x) = e^{2x}$

15. $f(x) = 6e^x$

16. $f(x) = 4e^x$

17. $F(x) = e^{-7x}$

18. $F(x) = e^{-4x}$

19. $G(x) = 2e^{4x}$

20. $g(x) = 3e^{5x}$

21. $f(x) = -3e^{-x}$

22. $G(x) = -7e^{-x}$

23. $g(x) = \frac{1}{2}e^{-5x}$

24. $f(x) = \frac{1}{3}e^{-4x}$

25. $F(x) = -\frac{2}{3}e^{x^2}$

26. $g(x) = -\frac{4}{5}e^{x^3}$

27. $G(x) = 7 + 3e^{5x}$

28. $F(x) = 4 - e^{2x}$

29. $f(x) = x^5 - 2e^{6x}$

30. $G(x) = x^3 - 5e^{2x}$

31. $g(x) = x^5 e^{2x}$

32. $f(x) = x^7 e^{4x}$

33. $F(x) = \dfrac{e^{2x}}{x^4}$

34. $g(x) = \dfrac{e^{3x}}{x^6}$

35. $f(x) = (x^2 + 3x - 9)e^x$

36. $f(x) = (x^2 - 2x + 2)e^x$

37. $f(x) = \dfrac{e^x}{x^4}$

38. $f(x) = \dfrac{e^x}{x^5}$

39. $f(x) = e^{-x^2 + 7x}$

40. $f(x) = e^{-x^2 + 8x}$

41. $f(x) = e^{-x^2/2}$

42. $f(x) = e^{x^2/2}$

43. $y = e^{\sqrt{x-7}}$

44. $y = e^{\sqrt{x-4}}$

45. $y = \sqrt{e^x - 1}$

46. $y = \sqrt{e^x + 1}$

47. $y = xe^{-2x} + e^{-x} + x^3$

48. $y = e^x + x^3 - xe^x$

49. $y = 1 - e^{-x}$

50. $y = 1 - e^{-3x}$

51. $y = 1 - e^{-kx}$

52. $y = 1 - e^{-mx}$

53. $g(x) = (4x^2 + 3x)e^{x^2 - 7x}$

54. $g(x) = (5x^2 - 8x)e^{x^2 - 4x}$

Graph each function. Then determine critical values, inflection points, intervals over which the function is increasing or decreasing, and the concavity.

55. $f(x) = e^{2x}$

56. $g(x) = e^{-2x}$

57. $g(x) = e^{(1/2)x}$

58. $f(x) = e^{(1/3)x}$

59. $f(x) = \frac{1}{2}e^{-x}$

60. $g(x) = \frac{1}{3}e^{-x}$

61. $F(x) = -e^{(1/3)x}$

62. $G(x) = -e^{(1/2)x}$

63. $g(x) = 2(1 - e^{-x})$, for $x \ge 0$

64. $f(x) = 3 - e^{-x}$, for $x \ge 0$

65–74. For each function given in Exercises 55–64, graph the function and its first and second derivatives using a calculator, iPlot, or Graphicus.

75. Find the slope of the line tangent to the graph of $f(x) = e^x$ at the point $(0, 1)$.

76. Find the slope of the line tangent to the graph of $f(x) = 2e^{-3x}$ at the point $(0, 2)$.

77. Find an equation of the line tangent to the graph of $G(x) = e^{-x}$ at the point $(0, 1)$.

78. Find an equation of the line tangent to the graph of $f(x) = e^{2x}$ at the point $(0, 1)$.

79. and 80. For each of Exercises 77 and 78, graph the function and the tangent line using a calculator, iPlot, or Graphicus.

APPLICATIONS

Business and Economics

81. U.S. exports. U.S. exports of goods are increasing exponentially. The value of the exports, t years after 2009, can be approximated by

$$V(t) = 1.6e^{0.046t},$$

where $t = 0$ corresponds to 2009 and V is in billions of dollars. (*Source*: U.S. Commerce Department.)

a) Estimate the value of U.S. exports in 2009 and 2020.
b) What is the doubling time for the value of U.S. exports?

82. Organic food. More Americans are buying organic fruit and vegetables and products made with organic ingredients. The amount $A(t)$, in billions of dollars, spent on organic food and beverages t years after 1995 can be approximated by

$$A(t) = 2.43e^{0.18t}.$$

(*Source*: *Nutrition Business Journal*, 2004.)

a) Estimate the amount that Americans spent on organic food and beverages in 2009.
b) Estimate the rate at which spending on organic food and beverages was growing in 2006.

83. Marginal cost. A company's total cost, in millions of dollars, is given by

$$C(t) = 100 - 50e^{-t},$$

where t is the time in years since the start-up date.

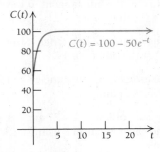

Find each of the following.

a) The marginal cost, $C'(t)$
b) $C'(0)$

c) $C'(4)$ (Round to the nearest thousand.)
d) Find $\lim_{t \to \infty} C(t)$ and $\lim_{t \to \infty} C'(t)$. Why do you think the company's costs tend to level off as time passes?

84. Marginal cost. A company's total cost, in millions of dollars, is given by

$$C(t) = 200 - 40e^{-t},$$

where t is the time in years since the start-up date.

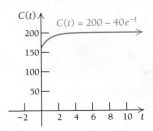

Find each of the following.

a) The marginal cost $C'(t)$
b) $C'(0)$
c) $C'(5)$ (Round to the nearest thousand.)
d) Find $\lim_{t \to \infty} C(t)$ and $\lim_{t \to \infty} C'(t)$. Why do you think the company's costs tend to level off as time passes?

85. Marginal demand. At a price of x dollars, the demand, in thousands of units, for a certain music player is given by the demand function

$$q = 240e^{-0.003x}.$$

a) How many music players will be bought at a price of $250? Round to the nearest thousand.
b) Graph the demand function for $0 \le x \le 400$.
c) Find the marginal demand, $q'(x)$.
d) Interpret the meaning of the derivative.

86. Marginal supply. At a price of x dollars, the supply function for the music player in Exercise 85 is given by

$$q = 75e^{0.004x},$$

where q is in thousands of units.

a) How many music players will be supplied at a price of $250? Round to the nearest thousand.
b) Graph the supply function for $0 \le x \le 400$.
c) Find the marginal supply, $q'(x)$.
d) Interpret the meaning of the derivative.

Life and Physical Sciences

87. Medication concentration. The concentration C, in parts per million, of a medication in the body t hours after ingestion is given by the function

$$C(t) = 10t^2e^{-t}.$$

a) Find the concentration after 0 hr, 1 hr, 2 hr, 3 hr, and 10 hr.
b) Sketch a graph of the function for $0 \le t \le 10$.
c) Find the rate of change of the concentration, $C'(t)$.
d) Find the maximum value of the concentration and the time at which it occurs.
e) Interpret the meaning of the derivative.

Social Sciences

88. *Ebbinghaus learning model.* Suppose that you are given the task of learning 100% of a block of knowledge. Human nature is such that we retain only a percentage P of knowledge t weeks after we have learned it. The *Ebbinghaus learning model* asserts that P is given by

$$P(t) = Q + (100 - Q)e^{-kt},$$

where Q is the percentage that we would never forget and k is a constant that depends on the knowledge learned. Suppose that $Q = 40$ and $k = 0.7$.

a) Find the percentage retained after 0 weeks, 1 week, 2 weeks, 6 weeks, and 10 weeks.
b) Find $\lim_{t \to \infty} P(t)$.
c) Sketch a graph of P.
d) Find the rate of change of P with respect to time t.
e) Interpret the meaning of the derivative.

SYNTHESIS

Differentiate.

89. $y = (e^{3x} + 1)^5$

90. $y = (e^{x^2} - 2)^4$

91. $y = \dfrac{e^{3t} - e^{7t}}{e^{4t}}$

92. $y = \sqrt[3]{e^{3t} + t}$

93. $y = \dfrac{e^x}{x^2 + 1}$

94. $y = \dfrac{e^x}{1 - e^x}$

95. $f(x) = e^{\sqrt{x}} + \sqrt{e^x}$

96. $f(x) = \dfrac{1}{e^x} + e^{1/x}$

97. $f(x) = e^{x/2} \cdot \sqrt{x - 1}$

98. $f(x) = \dfrac{xe^{-x}}{1 + x^2}$

99. $f(x) = \dfrac{e^x - e^{-x}}{e^x + e^{-x}}$

100. $f(x) = e^{e^x}$

Exercises 101 and 102 each give an expression for e. Find the function values that are approximations for e. Round to five decimal places.

101. For $f(t) = (1 + t)^{1/t}$, we have $e = \lim_{t \to 0} f(t)$. Find $f(1)$, $f(0.5), f(0.2), f(0.1)$, and $f(0.001)$.

102. For $g(t) = t^{1/(t-1)}$, we have $e = \lim_{t \to 1} g(t)$. Find $g(0.5)$, $g(0.9), g(0.99), g(0.999)$, and $g(0.9998)$.

103. Find the maximum value of $f(x) = x^2 e^{-x}$ over $[0, 4]$.

104. Find the minimum value of $f(x) = xe^x$ over $[-2, 0]$.

105. A student made the following error on a test:

$$\frac{d}{dx} e^x = xe^{x-1}.$$

Identify the error and explain how to correct it.

106. Describe the differences in the graphs of $f(x) = 3^x$ and $g(x) = x^3$.

Use a graphing calculator (or iPlot or Graphicus) to graph each function in Exercises 107 and 108, and find all relative extrema.

107. $f(x) = x^2 e^{-x}$

108. $f(x) = e^{-x^2}$

For each of the functions in Exercises 109–112, graph f, f', and f''.

109. $f(x) = e^x$

110. $f(x) = e^{-x}$

111. $f(x) = 2e^{0.3x}$

112. $f(x) = 1000e^{-0.08x}$

113. Graph

$$f(x) = \left(1 + \frac{1}{x}\right)^x.$$

Use the TABLE feature and very large values of x to confirm that e is approached as a limit.

Answers to Quick Checks

1. $1, 3, 9, 27, \dfrac{1}{3}, \dfrac{1}{9}, \dfrac{1}{27}$

2. $1, \dfrac{1}{3}, \dfrac{1}{9}, \dfrac{1}{27}, 3, 9, 27$

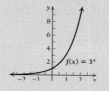

3. **(a)** $6e^x$; **(b)** $x^2 e^x (x + 3)$; **(c)** $\dfrac{e^x (x - 2)}{x^3}$

4. **(a)** $-4e^{-4x}$; **(b)** $e^{x^3 + 8x} (3x^2 + 8)$; **(c)** $\dfrac{xe^{\sqrt{x^2 + 5}}}{\sqrt{x^2 + 5}}$

5. **(a)** **(b)**

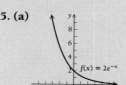

No critical values
Decreasing on $(-\infty, \infty)$
No inflection points
Concave up on $(-\infty, \infty)$

No critical values
Increasing on $(-\infty, \infty)$
No inflection points
Concave up on $(-\infty, \infty)$

(c)

No critical values
Increasing on $(-\infty, \infty)$
No inflection points
Concave down on $(-\infty, \infty)$

6. **(a)** $20, 26.8, 47.1, 61.9, 74.6, 78.4$ **(b)**
(c) $N'(t) = 7.2e^{-0.12t}$; after t days, the rate of change of number of phones produced per day is given by $7.2e^{-0.12t}$.
(d) 80 phones produced per day

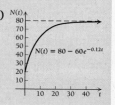

3.2 Logarithmic Functions

Logarithmic Functions and Their Graphs

Suppose that we want to solve the equation

$$10^y = 1000.$$

We are trying to find the power of 10 that will give 1000. Since $10^3 = 1000$, the answer is 3. The number 3 is called "the logarithm, base 10, of 1000."

DEFINITION

A **logarithm** is defined as follows:

$$\log_a x = y \qquad \text{means} \qquad a^y = x, \quad a > 0, a \neq 1.$$

The number $\log_a x$ is the power y to which we raise a to get x. The number a is called the *logarithmic base*. We read $\log_a x$ as "the logarithm, base a, of x."

For logarithms with base 10, $\log_{10} x$ is the power y such that $10^y = x$. Therefore, a logarithm can be thought of as an exponent. We can convert from a logarithmic equation to an exponential equation, and conversely, as follows.

Logarithmic Equation	Exponential Equation
$\log_a M = N$	$a^N = M$
$\log_{10} 100 = 2$	$10^2 = 100$
$\log_5 \frac{1}{25} = -2$	$5^{-2} = \frac{1}{25}$
$\log_{49} 7 = \frac{1}{2}$	$49^{1/2} = 7$

In order to graph a logarithmic equation, we can graph its equivalent exponential equation.

■ **EXAMPLE 1**   Graph: $y = \log_2 x$.

Solution   We first write the equivalent exponential equation:

$$2^y = x.$$

We select values for y and find the corresponding values of 2^y. Then we plot points, remembering that x is still the first coordinate, and connect the points with a smooth curve.

x, or 2^y	y
1	0
2	1
4	2
8	3
$\frac{1}{2}$	−1
$\frac{1}{4}$	−2

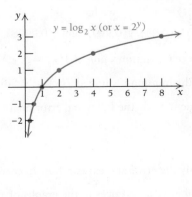

① Select y.
② Compute x.

Graphing Logarithmic Functions

To graph $y = \log_2 x$, we first graph $y_1 = 2^x$. We next select the DrawInv option from the DRAW menu and then the Y-VARS option from the VARS menu, followed by ⓵, ⓵, and **ENTER** to draw the inverse of y_1. Both graphs are drawn together.

$y_1 = 2^x, \ \ y_2 = \log_2 x$

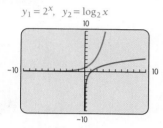

To use Graphicus to graph $y = \log_2 x$, press ⊞ and New x(y). Then enter 2^y as 2^y and press Done. The screen will display the graph of the equation $x = 2^y$, which by the definition of logarithms is also the graph of $y = \log_2 x$.

EXERCISES

Graph.

1. $y = \log_3 x$

2. $y = \log_5 x$

3. $f(x) = \log_e x$

4. $f(x) = \log_{10} x$

The graphs of $f(x) = 2^x$ and $g(x) = \log_2 x$ are shown below on the same set of axes. Note that we can obtain the graph of g by reflecting the graph of f across the line $y = x$. Functions whose graphs can be obtained in this manner are known as *inverses* of each other.

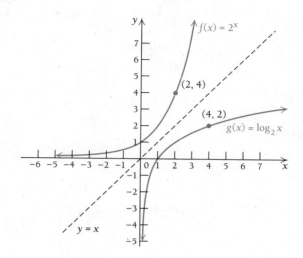

Although we do not develop inverses in detail here, it is important to note that they "undo" each other. For example,

$$f(3) = 2^3 = 8, \qquad \text{The input 3 gives the output 8.}$$

and $\quad g(8) = \log_2 8 = 3. \qquad$ The input 8 gets us back to 3.

Basic Properties of Logarithms

The following are some basic properties of logarithms. The proofs of P1–P3 follow from properties of exponents and are outlined in Exercises 107–109 at the end of this section. Properties P4–P6 follow directly from the definition of a logarithm, and a proof of P7 is outlined in Exercise 110.

THEOREM 3   Properties of Logarithms

For any positive numbers M, N, a, and b, with $a, b \neq 1$, and any real number k:

P1. $\log_a (MN) = \log_a M + \log_a N$

P2. $\log_a \dfrac{M}{N} = \log_a M - \log_a N$

P3. $\log_a (M^k) = k \cdot \log_a M$

P4. $\log_a a = 1$

P5. $\log_a (a^k) = k$

P6. $\log_a 1 = 0$

P7. $\log_b M = \dfrac{\log_a M}{\log_a b} \qquad$ (The change-of-base formula)

Let's illustrate these properties.

■ **EXAMPLE 2**   Given

$$\log_a 2 = 0.301 \quad \text{and} \quad \log_a 3 = 0.477,$$

find each of the following: **a)** $\log_a 6$;   **b)** $\log_a \frac{2}{3}$;   **c)** $\log_a 81$;

d) $\log_a \frac{1}{3}$;   **e)** $\log_a \sqrt{a}$;   **f)** $\log_a (2a)$;   **g)** $\dfrac{\log_a 3}{\log_a 2}$;   **h)** $\log_a 5$.

Solution

a) $\log_a 6 = \log_a (2 \cdot 3)$

$\qquad\quad = \log_a 2 + \log_a 3 \qquad$ By P1

$\qquad\quad = 0.301 + 0.477$

$\qquad\quad = 0.778$

b) $\log_a \frac{2}{3} = \log_a 2 - \log_a 3 \qquad$ By P2

$\qquad\quad = 0.301 - 0.477$

$\qquad\quad = -0.176$

c) $\log_a 81 = \log_a 3^4$

$\qquad\quad = 4 \log_a 3 \qquad$ By P3

$\qquad\quad = 4(0.477)$

$\qquad\quad = 1.908$

d) $\log_a \frac{1}{3} = \log_a 1 - \log_a 3 \qquad$ By P2

$\qquad\quad = 0 - 0.477 \qquad$ By P6

$\qquad\quad = -0.477$

e) $\log_a \sqrt{a} = \log_a (a^{1/2}) = \frac{1}{2} \qquad$ By P5

f) $\log_a (2a) = \log_a 2 + \log_a a \qquad$ By P1

$\qquad\quad = 0.301 + 1 \qquad$ By P4

$\qquad\quad = 1.301$

g) $\dfrac{\log_a 3}{\log_a 2} = \dfrac{0.477}{0.301} \approx 1.58$

We simply divided and used none of the properties.

h) There is no way to find $\log_a 5$ using the properties of logarithms ($\log_a 5 \neq \log_a 2 + \log_a 3$).

❬ Quick Check 1

> **Quick Check 1**
>
> Given $\log_b 2 = 0.356$ and $\log_b 5 = 0.827$, find each of the following:
>
> **a)** $\log_b 10$;   **b)** $\log_b \frac{2}{5}$;
>
> **c)** $\log_b \frac{5}{2}$;   **d)** $\log_b 16$;
>
> **e)** $\log_b 5b$;   **f)** $\log_b \sqrt{b}$.

Common Logarithms

The number $\log_{10} x$ is the **common logarithm** of x and is abbreviated $\log x$; that is:

DEFINITION

For any positive number x, $\quad \log x = \log_{10} x.$

Thus, when we write "$\log x$" with no base indicated, base 10 is understood. Note the following comparison of common logarithms and powers of 10.

$$1000 = 10^3$$
$$100 = 10^2$$
$$10 = 10^1$$
$$1 = 10^0$$
$$0.1 = 10^{-1}$$
$$0.01 = 10^{-2}$$
$$0.001 = 10^{-3}$$

The common logarithms at the right follow from the powers at the left.

$$\log 1000 = 3$$
$$\log 100 = 2$$
$$\log 10 = 1$$
$$\log 1 = 0$$
$$\log 0.1 = -1$$
$$\log 0.01 = -2$$
$$\log 0.001 = -3$$

Since $\log 100 = 2$ and $\log 1000 = 3$, it seems reasonable that $\log 500$ is somewhere between 2 and 3. Tables of logarithms were originally used for such approximations, but since the advent of the calculator, such tables are rarely used. Using a calculator with a **LOG** key, we find that $\log 500 \approx 2.6990$.

Before calculators became so readily available, common logarithms were used extensively to do certain computations. In fact, computation is the reason logarithms were developed. Since standard notation for numbers is based on 10, it was logical to use base-10, or common, logarithms for computations. Today, computations with common logarithms are mainly of historical interest; the logarithmic functions, base e, are far more important.

Natural Logarithms

The number e, which is approximately 2.718282, was developed in Section 3.1, and has extensive application in many fields. The number $\log_e x$ is the **natural logarithm** of x and is abbreviated $\ln x$.

DEFINITION

For any positive number x, $\quad \ln x = \log_e x$.

The following basic properties of natural logarithms parallel those given earlier for logarithms in general.

TECHNOLOGY CONNECTION

To enter $y = \log_{10} x$, the key labeled **LOG** can be used.

EXERCISES

1. Graph $f(x) = 10^x$, $y = x$, and $g(x) = \log_{10} x$ using the same set of axes. Then find $f(3)$, $f(0.699)$, $g(5)$, and $g(1000)$.

2. Use the **LOG** key and P7 of Theorem 4 to graph $y = \log_2 x$.

THEOREM 4   **Properties of Natural Logarithms**

P1. $\ln (MN) = \ln M + \ln N$ **P5.** $\ln (e^k) = k$

P2. $\ln \dfrac{M}{N} = \ln M - \ln N$ **P6.** $\ln 1 = 0$

P3. $\ln (a^k) = k \cdot \ln a$ **P7.** $\log_b M = \dfrac{\ln M}{\ln b}$ and $\ln M = \dfrac{\log M}{\log e}$

P4. $\ln e = 1$

Let's illustrate the properties of Theorem 4.

■ **EXAMPLE 3**   Given

$$\ln 2 = 0.6931 \quad \text{and} \quad \ln 3 = 1.0986,$$

find each of the following: **a)** $\ln 6$;   **b)** $\ln 81$;   **c)** $\ln \frac{1}{3}$;   **d)** $\ln (2e^5)$;   **e)** $\log_2 3$.

Solution

a) $\ln 6 = \ln (2 \cdot 3) = \ln 2 + \ln 3$ By P1
$$ = 0.6931 + 1.0986$$
$$ = 1.7917$$

b) $\ln 81 = \ln (3^4)$
$$ = 4 \ln 3 \quad \text{By P3}$$
$$ = 4(1.0986)$$
$$ = 4.3944$$

c) $\ln \frac{1}{3} = \ln 1 - \ln 3$ By P2
$$\phantom{\ln \tfrac{1}{3}} = 0 - 1.0986 \quad \text{By P6}$$
$$\phantom{\ln \tfrac{1}{3}} = -1.0986$$

d) $\ln (2e^5) = \ln 2 + \ln (e^5)$ By P1
$$ = 0.6931 + 5 \quad \text{By P5}$$
$$ = 5.6931$$

e) $\log_2 3 = \dfrac{\ln 3}{\ln 2} = \dfrac{1.0986}{0.6931} \approx 1.5851$ By P7

❬ Quick Check 2

> ❭ **Quick Check 2**
>
> Given $\ln 2 = 0.6931$ and $\ln 5 = 1.6094$, find each of the following:
>
> **a)** $\ln 10$;   **b)** $\ln \frac{5}{2}$;
> **c)** $\ln \frac{2}{5}$;   **d)** $\ln 32$;
> **e)** $\ln 5e^2$;   **f)** $\log_5 2$.

Finding Natural Logarithms Using a Calculator

You should have a calculator with an **LN** key. You can find natural logarithms directly using this key.

■ **EXAMPLE 4**   Approximate each of the following to six decimal places:

a) $\ln 5.24$;   **b)** $\ln 0.001278$.

Solution   We use a calculator with an **LN** key.

a) $\ln 5.24 \approx 1.656321$ **b)** $\ln 0.001278 \approx -6.662459$

Exponential Equations

If an equation contains a variable in an exponent, the equation is **exponential**. We can use logarithms to manipulate or solve exponential equations.

■ **EXAMPLE 5**   Solve $e^t = 40$ for t.

Solution   We have

$$\ln e^t = \ln 40 \quad \text{Taking the natural logarithm on both sides}$$
$$t = \ln 40 \quad \text{By P5; remember that } \ln e^t \text{ means } \log_e e^t.$$
$$t \approx 3.688879 \quad \text{Using a calculator}$$
$$t \approx 3.7$$

Note that this is an approximation for t even though an equals sign is often used.

■ **EXAMPLE 6**   Solve $e^{-0.04t} = 0.05$ for t.

Solution   We have

$$\ln e^{-0.04t} = \ln 0.05 \qquad \text{Taking the natural logarithm on both sides}$$
$$-0.04t = \ln 0.05 \qquad \text{By P5}$$
$$t = \frac{\ln 0.05}{-0.04}$$
$$t \approx \frac{-2.995732}{-0.04} \qquad \text{Using a calculator}$$
$$t \approx 75.$$

In Example 6, we rounded $\ln 0.05$ to -2.995732 in an intermediate step. When using a calculator, you should find

$$\frac{\ln 0.05}{-0.04}$$

by keying in

pressing **ENTER**, and rounding at the end. Answers at the back of this book have been found in this manner. Remember, the number of places in a table or on a calculator affects the accuracy of the answer. Usually, your answer should agree with that in the Answers section to at least three digits.

> **Quick Check 3**
>
> Solve each equation for t:
>
> **a)** $e^t = 80$;
>
> **b)** $e^{-0.08t} = 0.25$.

❮ Quick Check 3

TECHNOLOGY CONNECTION

Solving Exponential Equations

Let's solve the equation of Example 5, $e^t = 40$, graphically.

Method 1: The INTERSECT Feature

We change the variable to x and consider the system of equations $y_1 = e^x$ and $y_2 = 40$. We graph the equations in the window $[-1, 8, -10, 70]$ to see the curvature and point of intersection.

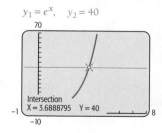

$y_1 = e^x$, $y_2 = 40$

Then we use the INTERSECT option from the CALC menu to find the point of intersection, about $(3.7, 40)$. The x-coordinate, 3.7, is the solution of $e^t = 40$.

Method 2: The ZERO Feature

We change the variable to x and get a 0 on one side of the equation: $e^x - 40 = 0$. Then we graph $y = e^x - 40$ in the window $[-1, 8, -10, 10]$.

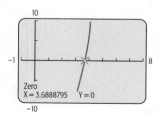

Using the ZERO option from the CALC menu, we see that the x-intercept is about $(3.7, 0)$, so 3.7 is the solution of $e^t = 40$.

EXERCISES

Solve graphically using a calculator, iPlot, or Graphicus.

1. $e^t = 1000$ **2.** $e^{-x} = 60$

3. $e^{-0.04t} = 0.05$ **4.** $e^{0.23x} = 41{,}378$

5. $15e^{0.2x} = 34{,}785.13$

Graphs of Natural Logarithmic Functions

There are two ways in which we might graph $y = f(x) = \ln x$. One is to graph the equivalent equation $x = e^y$ by selecting values for y and calculating the corresponding values of e^y. We then plot points, remembering that x is still the first coordinate.

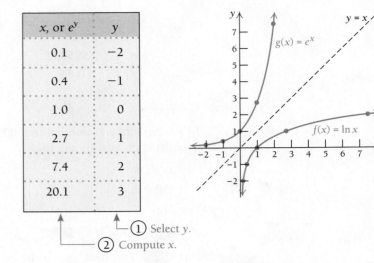

x, or e^y	y
0.1	−2
0.4	−1
1.0	0
2.7	1
7.4	2
20.1	3

① Select y.
② Compute x.

The graph above shows the graph of $g(x) = e^x$ for comparison. Note again that the functions are inverses of each other. That is, the graph of $y = \ln x$, or $x = e^y$, is a reflection, or mirror image, across the line $y = x$ of the graph of $y = e^x$. Any ordered pair (a, b) on the graph of g yields an ordered pair (b, a) on f. Note too that $\lim_{x \to 0^+} \ln x = -\infty$ and the y-axis is a vertical asymptote.

The second method of graphing $y = \ln x$ is to use a calculator to find function values. For example, when $x = 2$, then $y = \ln 2 \approx 0.6931 \approx 0.7$. This gives the pair $(2, 0.7)$ shown on the graph.

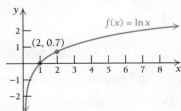

The following properties can be observed from the graph.

THEOREM 5

$\ln x$ exists only for positive numbers x. The domain is $(0, \infty)$.

$\ln x < 0$ for $0 < x < 1$.

$\ln x = 0$ when $x = 1$.

$\ln x > 0$ for $x > 1$.

The function given by $f(x) = \ln x$ is always increasing. The range is the entire real line, $(-\infty, \infty)$, or the set of real numbers, $\mathbb{R}$.

Derivatives of Natural Logarithmic Functions

Let's find the derivative of

$$f(x) = \ln x. \tag{1}$$

We first write its equivalent exponential equation:

$$e^{f(x)} = x. \qquad \ln x = \log_e x = f(x), \text{ so } e^{f(x)} = x, \qquad (2)$$
$$\text{by the definition of logarithms.}$$

Now we differentiate on both sides of this equation:

$$\frac{d}{dx} e^{f(x)} = \frac{d}{dx} x$$

$$e^{f(x)} \cdot f'(x) = 1 \qquad \text{By the Chain Rule}$$

$$x \cdot f'(x) = 1 \qquad \text{Substituting } x \text{ for } e^{f(x)} \text{ from equation (2)}$$

$$f'(x) = \frac{1}{x}.$$

Thus, we have the following.

TECHNOLOGY CONNECTION

Exploratory

Use Graphicus to graph $y = \ln x$. Then touch $+$ and choose Add derivative. Using the tangent line feature, move along the curve, noting the x-values, the y-values, and the values of dy/dx, to verify that $\dfrac{dy}{dx} = \dfrac{1}{x}$.

> ### THEOREM 6
>
> For any positive number x,
>
> $$\frac{d}{dx} \ln x = \frac{1}{x}.$$

To visualize the meaning of Theorem 6, look back at the graph of $f(x) = \ln x$ on p. 328. Take a small ruler or a credit card and place it as if its edge were a tangent line. Start on the left and move the ruler or card along the curve toward the right, noting how the tangent lines flatten out. Think about the slopes of these tangent lines. The slopes approach 0 as a limit, though they never actually become 0. This is consistent with the formula

$$\frac{d}{dx} \ln x = \frac{1}{x}, \quad \text{because } \lim_{x \to \infty} \frac{1}{x} = 0.$$

Theorem 6 asserts that to find the slope of the tangent line at x for the function $f(x) = \ln x$, we need only take the reciprocal of x. This is true only for positive values of x, since $\ln x$ is defined only for positive numbers. (For negative numbers x, this derivative formula becomes

$$\frac{d}{dx} \ln |x| = \frac{1}{x},$$

but we will seldom consider such a case in this text.)

Let's find some derivatives.

■ **EXAMPLE 7** Differentiate:

a) $y = 3 \ln x$; **b)** $y = x^2 \ln x + 5x$; **c)** $y = \dfrac{\ln x}{x^3}$.

Solution

a) $\dfrac{d}{dx}(3 \ln x) = 3 \dfrac{d}{dx} \ln x$

$$= \frac{3}{x}$$

b) $\dfrac{d}{dx}(x^2 \ln x + 5x) = x^2 \cdot \dfrac{1}{x} + \ln x \cdot 2x + 5 \qquad$ Using the Product Rule on $x^2 \ln x$

$$= x + 2x \cdot \ln x + 5 \qquad \text{Simplifying}$$

c) $\dfrac{d}{dx} \dfrac{\ln x}{x^3} = \dfrac{x^3 \cdot (1/x) - (\ln x)(3x^2)}{x^6}$ By the Quotient Rule

$= \dfrac{x^2 - 3x^2 \ln x}{x^6}$

$= \dfrac{x^2(1 - 3 \ln x)}{x^6}$ Factoring

$= \dfrac{1 - 3 \ln x}{x^4}$ Simplifying

> **Quick Check 4**
>
> Differentiate:
>
> **a)** $y = 5 \ln x$;
>
> **b)** $y = x^3 \ln x + 4x$;
>
> **c)** $y = \dfrac{\ln x}{x^2}$.

❰ Quick Check 4

Suppose that we want to differentiate a more complicated function that is of the form $h(x) = \ln f(x)$, such as

$$h(x) = \ln (x^2 - 8x).$$

This can be regarded as

$$h(x) = g(f(x)), \qquad \text{where} \quad g(x) = \ln x \quad \text{and} \quad f(x) = x^2 - 8x.$$

Now $g'(x) = 1/x$, so by the Chain Rule (Section 1.7), we have

$$h'(x) = g'(f(x)) \cdot f'(x)$$

$$= \dfrac{1}{f(x)} \cdot f'(x).$$

For the above case, $f(x) = x^2 - 8x$, so $f'(x) = 2x - 8$. Then

$$h'(x) = \dfrac{1}{x^2 - 8x} \cdot (2x - 8) = \dfrac{2x - 8}{x^2 - 8x}.$$

The following rule, which we have proven using the Chain Rule, allows us to find derivatives of functions like the one above.

THEOREM 7

The derivative of the natural logarithm of a function is the derivative of the function divided by the function:

$$\dfrac{d}{dx} \ln f(x) = \dfrac{1}{f(x)} \cdot f'(x) = \dfrac{f'(x)}{f(x)},$$

or

$$\dfrac{d}{dx} \ln u = \dfrac{1}{u} \cdot \dfrac{du}{dx}.$$

The following gives us a way of remembering this rule.

$$h(x) = \ln \underbrace{(x^2 - 8x)}$$

① Differentiate the "inside" function.

$$h'(x) = \dfrac{2x - 8}{x^2 - 8x}$$

② Divide by the "inside" function.

■ **EXAMPLE 8** Differentiate:

a) $y = \ln(3x)$; **b)** $y = \ln(x^2 - 5)$; **c)** $f(x) = \ln(\ln x)$; **d)** $f(x) = \ln\left(\dfrac{x^3 + 4}{x}\right)$.

Solution

a) If $y = \ln(3x)$, then

$$\frac{dy}{dx} = \frac{3}{3x} = \frac{1}{x}.$$

Note that we could have done this using the fact that $\ln(MN) = \ln M + \ln N$:

$$\ln(3x) = \ln 3 + \ln x;$$

then, since $\ln 3$ is a constant, we have

$$\frac{d}{dx}\ln(3x) = \frac{d}{dx}\ln 3 + \frac{d}{dx}\ln x = 0 + \frac{1}{x} = \frac{1}{x}.$$

b) If $y = \ln(x^2 - 5)$, then

$$\frac{dy}{dx} = \frac{2x}{x^2 - 5}.$$

c) If $f(x) = \ln(\ln x)$, then

$$f'(x) = \frac{1}{\ln x} \cdot \frac{d}{dx}\ln x = \frac{1}{\ln x} \cdot \frac{1}{x} = \frac{1}{x \ln x}.$$

d) If $f(x) = \ln\left(\dfrac{x^3 + 4}{x}\right)$, then, since $\ln\dfrac{M}{N} = \ln M - \ln N$, we have

$$f'(x) = \frac{d}{dx}\left[\ln(x^3 + 4) - \ln x\right] \qquad \text{By P2; this avoids use of the Quotient Rule.}$$

$$= \frac{3x^2}{x^3 + 4} - \frac{1}{x}$$

$$= \frac{3x^2}{x^3 + 4} \cdot \frac{x}{x} - \frac{1}{x} \cdot \frac{x^3 + 4}{x^3 + 4} \qquad \text{Finding a common denominator}$$

$$\left.\begin{array}{l} = \dfrac{(3x^2)x - (x^3 + 4)}{x(x^3 + 4)} \\[2mm] = \dfrac{3x^3 - x^3 - 4}{x(x^3 + 4)} = \dfrac{2x^3 - 4}{x(x^3 + 4)}. \end{array}\right\} \quad \text{Simplifying}$$

❬ **Quick Check 5**

❭ **Quick Check 5**

Differentiate:

a) $y = \ln 5x$;

b) $y = \ln(3x^2 + 4)$;

c) $y = \ln(\ln 5x)$;

d) $y = \ln\left(\dfrac{x^5 - 2}{x}\right)$.

TECHNOLOGY CONNECTION 〰

Exploratory

To check part (a) of Example 8, we let $y_1 = \ln(3x)$, $y_2 = \mathrm{nDeriv}(y_1, x, x)$, and $y_3 = 1/x$. Either GRAPH or TABLE can then be used to show that $y_2 = y_3$. Use this approach to check parts (b), (c), and (d) of Example 8. (Or use Graphicus to check those results.)

Applications

■ **EXAMPLE 9** **Social Science: Forgetting.** In a psychological experiment, students were shown a set of nonsense syllables, such as POK, RIZ, DEQ, and so on, and asked to recall them every minute thereafter. The percentage $R(t)$ who retained the syllables after t minutes was found to be given by the logarithmic learning model

$$R(t) = 80 - 27 \ln t, \quad \text{for} \quad t \geq 1.$$

Exploratory

Graph $y = 80 - 27 \ln x$, from Example 9, using the viewing window $[1, 14, -1, 100]$. Trace along the graph. Describe the meaning of each coordinate in an ordered pair.

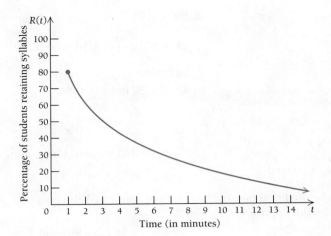

a) What percentage of students retained the syllables after 1 min?

b) Find $R'(2)$, and explain what it represents.

Solution

a) $R(1) = 80 - 27 \cdot \ln 1 = 80 - 27 \cdot 0 = 80\%$

b) $\dfrac{d}{dt}(80 - 27 \ln t) = 0 - 27 \cdot \dfrac{1}{t} = -\dfrac{27}{t}$,

so $R'(2) = -\dfrac{27}{2} = -13.5$.

This result indicates that 2 min after students have been shown the syllables, the percentage of them who remember the syllables is shrinking at the rate of 13.5% per minute.

▪ **EXAMPLE 10** **Business: An Advertising Model.** A company begins a radio advertising campaign in New York City to market a new product. The percentage of the "target market" that buys a product is normally a function of the duration of the advertising campaign. The radio station estimates this percentage, as a decimal, by using $f(t) = 1 - e^{-0.04t}$ for this type of product, where t is the number of days of the campaign. The target market is approximately 1,000,000 people and the price per unit is $0.50. If the campaign costs $1000 per day, how long should it last in order to maximize profit?

Solution
Modeling the percentage of the target market that buys the product, expressed as a decimal, by using $f(t) = 1 - e^{-0.04t}$ is justified if we graph f. The function increases from 0 (0%) toward 1 (100%). The longer the advertising campaign, the larger the percentage of the market that has bought the product.

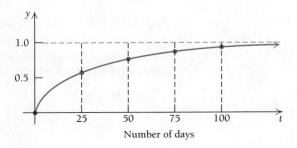

The total-profit function, here expressed in terms of time t, is given by

Profit = Revenue − Cost

$$P(t) = R(t) - C(t).$$

We find $R(t)$ and $C(t)$:

$$R(t) = (\text{Percentage buying}) \cdot (\text{Target market}) \cdot (\text{Price per unit})$$
$$= (1 - e^{-0.04t})(1,000,000)(0.5) = 500,000 - 500,000e^{-0.04t},$$

and

$$C(t) = (\text{Advertising costs per day}) \cdot (\text{Number of days}) = 1000t.$$

Next, we find $P(t)$ and take its derivative:

$$P(t) = R(t) - C(t)$$
$$= 500,000 - 500,000e^{-0.04t} - 1000t,$$
$$P'(t) = -500,000e^{-0.04t}(-0.04) - 1000$$
$$= 20,000e^{-0.04t} - 1000.$$

We then set the first derivative equal to 0 and solve:

$$20,000e^{-0.04t} - 1000 = 0$$
$$20,000e^{-0.04t} = 1000$$
$$e^{-0.04t} = \frac{1000}{20,000} = 0.05$$
$$\ln e^{-0.04t} = \ln 0.05$$
$$-0.04t = \ln 0.05$$
$$t = \frac{\ln 0.05}{-0.04}$$
$$t \approx 75.$$

We have only one critical value, so we can use the second derivative to determine whether we have a maximum:

$$P''(t) = 20,000e^{-0.04t}(-0.04)$$
$$= -800e^{-0.04t}.$$

Since exponential functions are positive, $e^{-0.04t} > 0$ for all numbers t. Thus, since $-800e^{-0.04t} < 0$ for all t, we have $P''(75) < 0$, and we have a maximum.

The advertising campaign should run for 75 days in order to maximize profit.

⟨ Quick Check 6

P graph with $P(t) = 500,000 - 500,000e^{-0.04t} - 1000t$

TECHNOLOGY CONNECTION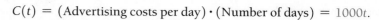

Exploratory

Graph the function P in Example 10, and verify that there is maximum profit if the length of the advertising campaign is about 75 days.

❭ **Quick Check 6**

Business: An Advertising Model. Repeat Example 10 using the function

$$f(t) = 1 - e^{-0.08t}$$

and assuming that the campaign costs $2000 per day.

Section Summary

- A logarithmic function $y = \log_a x$ is defined by $a^y = x$, for $a > 0$ and $a \neq 1$.
- The common logarithmic function g is defined by $g(x) = \log_{10} x = \log x$, for $x > 0$.
- The natural logarithm function f is defined by $f(x) = \log_e x = \ln x$, where $e \approx 2.71828$. The derivative of f is $f'(x) = \dfrac{1}{x}$, for $x > 0$. The slope of a tangent line to the graph of f at x is found by taking the reciprocal of the input x.

- The graph of $f(x) = \ln x$ is an increasing function with no critical values, no maximum or minimum values, and no points of inflection. The domain is $(0, \infty)$. The range is $(-\infty, \infty)$, or $\mathbb{R}$. The graph is concave down, with
$$\lim_{x \to \infty} f(x) = \infty \quad \text{and} \quad \lim_{x \to 0} f(x) = -\infty.$$
- Properties of logarithms are described in Theorems 3 and 4.
- Calculus is rich in applications of natural logarithmic functions.

EXERCISE SET
3.2

Write an equivalent exponential equation.

1. $\log_2 8 = 3$

2. $\log_3 81 = 4$

3. $\log_8 2 = \frac{1}{3}$

4. $\log_{27} 3 = \frac{1}{3}$

5. $\log_a K = J$

6. $\log_a J = K$

7. $-\log_{10} h = p$

8. $-\log_b V = w$

Write an equivalent logarithmic equation.

9. $e^M = b$

10. $e^t = p$

11. $10^2 = 100$

12. $10^3 = 1000$

13. $10^{-1} = 0.1$

14. $10^{-2} = 0.01$

15. $M^p = V$

16. $Q^n = T$

Given $\log_b 3 = 1.099$ and $\log_b 5 = 1.609$, find each value.

17. $\log_b \frac{5}{3}$

18. $\log_b \frac{1}{5}$

19. $\log_b 15$

20. $\log_b \sqrt{b^3}$

21. $\log_b (5b)$

22. $\log_b 75$

Given $\ln 4 = 1.3863$ and $\ln 5 = 1.6094$, find each value. Do not use a calculator.

23. $\ln 20$

24. $\ln 80$

25. $\ln \frac{5}{4}$

26. $\ln \frac{1}{5}$

27. $\ln (5e)$

28. $\ln (4e)$

29. $\ln \sqrt{e^6}$

30. $\ln \sqrt{e^8}$

31. $\ln \frac{1}{4}$

32. $\ln \frac{4}{5}$

33. $\ln \left(\frac{e}{5} \right)$

34. $\ln \left(\frac{4}{e} \right)$

Find each logarithm. Round to six decimal places.

35. $\ln 5894$

36. $\ln 99,999$

37. $\ln 0.0182$

38. $\ln 0.00087$

39. $\ln 8100$

40. $\ln 0.011$

Solve for t.

41. $e^t = 80$

42. $e^t = 10$

43. $e^{2t} = 1000$

44. $e^{3t} = 900$

45. $e^{-t} = 0.1$

46. $e^{-t} = 0.01$

47. $e^{-0.02t} = 0.06$

48. $e^{0.07t} = 2$

Differentiate.

49. $y = -8 \ln x$

50. $y = -9 \ln x$

51. $y = x^4 \ln x - \frac{1}{2} x^2$

52. $y = x^6 \ln x - \frac{1}{4} x^4$

53. $f(x) = \ln (6x)$

54. $f(x) = \ln (9x)$

55. $g(x) = x^2 \ln (7x)$

56. $g(x) = x^5 \ln (3x)$

57. $y = \dfrac{\ln x}{x^4}$

58. $y = \dfrac{\ln x}{x^5}$

59. $y = \ln \dfrac{x^2}{4}$ $\left(\textit{Hint: } \ln \dfrac{A}{B} = \ln A - \ln B. \right)$

60. $y = \ln \dfrac{x^4}{2}$

61. $y = \ln (3x^2 + 2x - 1)$

62. $y = \ln (7x^2 + 5x + 2)$

63. $f(x) = \ln \left(\dfrac{x^2 - 7}{x} \right)$

64. $f(x) = \ln \left(\dfrac{x^2 + 5}{x} \right)$

65. $g(x) = e^x \ln x^2$

66. $g(x) = e^{2x} \ln x$

67. $f(x) = \ln (e^x + 1)$

68. $f(x) = \ln (e^x - 2)$

69. $g(x) = (\ln x)^4$ (*Hint:* Use the Extended Power Rule.)

70. $g(x) = (\ln x)^3$

71. $f(x) = \ln (\ln (8x))$

72. $f(x) = \ln (\ln (3x))$

73. $g(x) = \ln (5x) \cdot \ln (3x)$

74. $g(x) = \ln (2x) \cdot \ln (7x)$

75. Find the equation of the line tangent to the graph of $y = (x^2 - x) \ln (6x)$ at $x = 2$.

76. Find the equation of the line tangent to the graph of $y = e^{3x} \cdot \ln (4x)$ at $x = 1$.

77. Find the equation of the line tangent to the graph of $y = (\ln x)^2$ at $x = 3$.

78. Find the equation of the line tangent to the graph of $y = \ln (4x^2 - 7)$ at $x = 2$.

APPLICATIONS
Business and Economics

79. Advertising. A model for consumers' response to advertising is given by

$$N(a) = 2000 + 500 \ln a, \quad a \geq 1,$$

where $N(a)$ is the number of units sold and a is the amount spent on advertising, in thousands of dollars.

a) How many units were sold after spending $1000 on advertising?

b) Find $N'(a)$ and $N'(10)$.

c) Find the maximum and minimum values, if they exist.

d) Find $\lim\limits_{a \to \infty} N'(a)$. Discuss whether it makes sense to continue to spend more and more dollars on advertising.

80. Advertising. A model for consumers' response to advertising is given by

$$N(a) = 1000 + 200 \ln a, \quad a \geq 1,$$

where $N(a)$ is the number of units sold and a is the amount spent on advertising, in thousands of dollars.

a) How many units were sold after spending $1000 on advertising?

b) Find $N'(a)$ and $N'(10)$.

c) Find the maximum and minimum values of N, if they exist.

d) Find $N'(a)$. Discuss $\lim\limits_{a \to \infty} N'(a)$. Does it make sense to spend more and more dollars on advertising? Why or why not?

81. An advertising model. Solve Example 10 given that the advertising campaign costs $2000 per day.

82. An advertising model. Solve Example 10 given that the advertising campaign costs $4000 per day.

83. Growth of a stock. The value, $V(t)$, in dollars, of a stock t months after it is purchased is modeled by

$$V(t) = 58(1 - e^{-1.1t}) + 20.$$

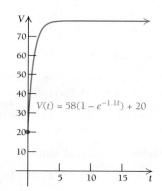

a) Find $V(1)$ and $V(12)$.

b) Find $V'(t)$.

c) After how many months will the value of the stock first reach $75?

d) Find $\lim\limits_{t \to \infty} V(t)$. Discuss the value of the stock over a long period of time. Is this trend typical?

84. Marginal revenue. The demand for a new computer game can be modeled by

$$p(x) = 53.5 - 8 \ln x,$$

where $p(x)$ is the price consumers will pay, in dollars, and x is the number of games sold, in thousands. Recall that total revenue is given by $R(x) = x \cdot p(x)$.

a) Find $R(x)$.

b) Find the marginal revenue, $R'(x)$.

c) Is there any price at which revenue will be maximized? Why or why not?

85. Marginal profit. The profit, in thousands of dollars, from the sale of x thousand mechanical pencils, can be estimated by

$$P(x) = 2x - 0.3x \ln x.$$

a) Find the marginal profit, $P'(x)$.

b) Find $P'(150)$, and explain what this number represents.

c) How many thousands of mechanical pencils should be sold to maximize profit?

Life and Physical Sciences

86. Acceptance of a new medicine. The percentage P of doctors who prescribe a certain new medicine is

$$P(t) = 100(1 - e^{-0.2t}),$$

where t is the time, in months.

a) Find $P(1)$ and $P(6)$.

b) Find $P'(t)$.

c) How many months will it take for 90% of doctors to prescribe the new medicine?

d) Find $\lim\limits_{t \to \infty} P(t)$, and discuss its meaning.

Social Sciences

87. Forgetting. Students in a botany class took a final exam. They took equivalent forms of the exam at monthly intervals thereafter. After t months, the average score $S(t)$, as a percentage, was found to be

$$S(t) = 68 - 20 \ln (t + 1), \quad t \geq 0.$$

a) What was the average score when the students initially took the test?

b) What was the average score after 4 months?

c) What was the average score after 24 months?

d) What percentage of their original answers did the students retain after 2 years (24 months)?

e) Find $S'(t)$.

f) Find the maximum value, if one exists.

g) Find $\lim\limits_{t \to \infty} S(t)$, and discuss its meaning.

88. Forgetting. Students in a zoology class took a final exam. They took equivalent forms of the exam at monthly intervals thereafter. After t months, the average score $S(t)$, as a percentage, was found to be given by

$$S(t) = 78 - 15 \ln (t + 1), \quad t \geq 0.$$

a) What was the average score when they initially took the test, $t = 0$?

b) What was the average score after 4 months?

c) What was the average score after 24 months?

d) What percentage of their original answers did the students retain after 2 years (24 months)?

e) Find $S'(t)$.

f) Find the maximum and minimum values, if they exist.

g) Find $\lim\limits_{t \to \infty} S(t)$ and discuss its meaning.

89. Walking speed. Bornstein and Bornstein found in a study that the average walking speed v, in feet per second, of a person living in a city of population p, in thousands, is

$$v(p) = 0.37 \ln p + 0.05.$$

(*Source:* M. H. Bornstein and H. G. Bornstein, "The Pace of Life," *Nature*, Vol. 259, pp. 557–559 (1976).)

a) The population of Seattle is 571,000 ($p = 571$). What is the average walking speed of a person living in Seattle?

b) The population of New York is 8,100,000. What is the average walking speed of a person living in New York?

c) Find $v'(p)$.

d) Interpret $v'(p)$ found in part (c).

90. Hullian learning model. A keyboarder learns to type W words per minute after t weeks of practice, where W is given by

$$W(t) = 100(1 - e^{-0.3t}).$$

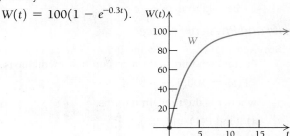

a) Find $W(1)$ and $W(8)$.
b) Find $W'(t)$.
c) After how many weeks will the keyboarder's speed be 95 words per minute?
d) Find $\lim_{t \to \infty} W(t)$, and discuss its meaning.

SYNTHESIS

91. Solve $P = P_0 e^{kt}$ for t.

Differentiate.

92. $f(x) = \ln (x^3 + 1)^5$

93. $f(t) = \ln (t^2 - t)^7$

94. $g(x) = [\ln (x + 5)]^4$

95. $f(x) = \ln [\ln (\ln(3x))]$

96. $f(t) = \ln [(t^3 + 3)(t^2 - 1)]$

97. $f(t) = \ln \dfrac{1 - t}{1 + t}$

98. $y = \ln \dfrac{x^5}{(8x + 5)^2}$

99. $f(x) = \log_5 x$

100. $f(x) = \log_7 x$

101. $y = \ln \sqrt{5 + x^2}$

102. $f(t) = \dfrac{\ln t^2}{t^2}$

103. $f(x) = \dfrac{1}{5} x^5 \left(\ln x - \dfrac{1}{5} \right)$

104. $y = \dfrac{x^{n+1}}{n + 1} \left(\ln x - \dfrac{1}{n + 1} \right)$

105. $f(x) = \ln \dfrac{1 + \sqrt{x}}{1 - \sqrt{x}}$

106. $f(x) = \ln (\ln x)^3$

To prove Properties P1, P2, P3, and P7 of Theorem 3, let $X = \log_a M$ and $Y = \log_a N$, and give reasons for the steps listed in Exercises 107–110.

107. Proof of P1 of Theorem 3.
$M = a^X$ and $N = a^Y$, _____
so $MN = a^X \cdot a^Y = a^{X+Y}$. _____
Thus, $\log_a (MN) = X + Y$ _____
$= \log_a M + \log_a N.$ _____

108. Proof of P2 of Theorem 3.
$M = a^X$ and $N = a^Y$, _____
so $\dfrac{M}{N} = \dfrac{a^X}{a^Y} = a^{X-Y}.$ _____
Thus, $\log_a \dfrac{M}{N} = X - Y$ _____
$= \log_a M - \log_a N.$ _____

109. Proof of P3 of Theorem 3.
$M = a^X$, _____
so $M^k = (a^X)^k$ _____
$= a^{Xk}.$ _____
Thus, $\log_a M^k = Xk$ _____
$= k \cdot \log_a M.$ _____

110. Proof of P7 of Theorem 3.
Let $\log_b M = R.$ _____
Then $b^R = M,$ _____
and $\log_a (b^R) = \log_a M.$ _____
Thus, $R \cdot \log_a b = \log_a M,$ _____
and $R = \dfrac{\log_a M}{\log_a b}.$ _____
It follows that
$\log_b M = \dfrac{\log_a M}{\log_a b}.$ _____

111. Find $\lim_{h \to 0} \dfrac{\ln (1 + h)}{h}$.

112. For any $k > 0$, $\ln (kx) = \ln k + \ln x$. Use this fact to show graphically why

$$\dfrac{d}{dx} \ln (kx) = \dfrac{d}{dx} \ln x = \dfrac{1}{x}.$$

TECHNOLOGY CONNECTION

113. Use natural logarithms to determine which is larger, e^π or π^e. (*Hint:* $y = \ln x$ is an increasing function.)

114. Find $\sqrt[x]{e}$. Compare it to other expressions of the type $\sqrt[x]{x}$, with $x > 0$. What can you conclude?

Use input–output tables to find each limit.

115. $\lim_{x \to 1} \ln x$

116. $\lim_{x \to \infty} \ln x$

Graph each function f and its derivative f'. Use a graphing calculator, iPlot, or Graphicus.

117. $f(x) = \ln x$

118. $f(x) = x \ln x$

119. $f(x) = x^2 \ln x$

120. $f(x) = \dfrac{\ln x}{x^2}$

Find the minimum value of each function. Use a graphing calculator, iPlot, or Graphicus.

121. $f(x) = x \ln x$

122. $f(x) = x^2 \ln x$

Answers to Quick Checks
1. (a) 1.183; **(b)** −0.471; **(c)** 0.471; **(d)** 1.424; **(e)** 1.827; **(f)** $\frac{1}{2}$ **2. (a)** 2.3025; **(b)** 0.9163; **(c)** −0.9163; **(d)** 3.4655; **(e)** 3.6094; **(f)** 0.4307 **3. (a)** $t \approx 4.3820$; **(b)** $t \approx 17.329$ **4. (a)** $\dfrac{5}{x}$; **(b)** $x^2 + 3x^2 \ln x + 4$; **(c)** $\dfrac{1 - 2 \ln x}{x^3}$ **5. (a)** $\dfrac{1}{x}$; **(b)** $\dfrac{6x}{3x^2 + 4}$; **(c)** $\dfrac{1}{x(\ln 5x)}$; **(d)** $\dfrac{4x^5 + 2}{x(x^5 - 2)}$ **6.** The advertising campaign should run for about 37 days to maximize profit.

3.3 Applications: Uninhibited and Limited Growth Models

Exponential Growth

Consider the function

$$f(x) = 2e^{3x}.$$

Differentiating, we get

$$f'(x) = 2e^{3x} \cdot 3$$
$$= f(x) \cdot 3.$$

Graphically, this says that the derivative, or slope of the tangent line, is simply the constant 3 times the function value.

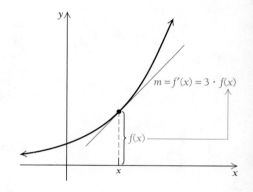

> **Quick Check 1**
>
> Differentiate $f(x) = 5e^{4x}$. Then express $f'(x)$ in terms of $f(x)$.

❰ Quick Check 1

Although we do not prove it here, the exponential function $f(x) = ce^{kx}$ is the only function for which the derivative is a constant times the function itself.

THEOREM 8

A function $y = f(x)$ satisfies the equation

$$\frac{dy}{dx} = ky \qquad \text{or} \qquad f'(x) = k \cdot f(x)$$

if and only if

$$y = ce^{kx} \qquad \text{or} \qquad f(x) = ce^{kx}$$

for some constant c.

■ **EXAMPLE 1** Find the general form of the function that satisfies the equation

$$\frac{dA}{dt} = 5A.$$

Solution The function is $A = ce^{5t}$, or $A(t) = ce^{5t}$, where c is an arbitrary constant. As a check, note that

$$A'(t) = ce^{5t} \cdot 5 = 5 \cdot A(t).$$

■ **EXAMPLE 2** Find the general form of the function that satisfies the equation

$$\frac{dP}{dt} = kP.$$

❯ **Quick Check 2**

Find the general form of the function that satisfies the equation

$$\frac{dN}{dt} = kN.$$

Solution The function is $P = ce^{kt}$, or $P(t) = ce^{kt}$, where c is an arbitrary constant. *Check*:

$$\frac{dP}{dt} = ce^{kt} \cdot k = kP.$$

❮ Quick Check 2

Whereas the solution of an algebraic equation is a number, the solutions of the equations in Examples 1 and 2 are functions. For example, the solution of $2x + 5 = 11$ is the number 3, and the solution of the equation $dP/dt = kP$ is the function $P(t) = ce^{kt}$. An equation like $dP/dt = kP$, which includes a derivative and which has a function as a solution, is called a *differential equation*.

■ **EXAMPLE 3** Solve the differential equation

$$f'(z) = k \cdot f(z).$$

Solution The solution is $f(z) = ce^{kz}$. *Check*: $f'(z) = ce^{kz} \cdot k = f(z) \cdot k$.

We will discuss differential equations in more depth in Chapter 8.

Uninhibited Population Growth

The equation

$$\frac{dP}{dt} = kP \quad \text{or} \quad P'(t) = kP(t), \quad \text{with } k > 0,$$

is the basic model of uninhibited (unrestrained) population growth, whether the population is comprised of humans, bacteria in a culture, or dollars invested with interest compounded continuously. In the absence of inhibiting or stimulating factors, a population normally reproduces at a rate proportional to its size, and this is exactly what $dP/dt = kP$ says. The only function that satisfies this differential equation is given by

$$P(t) = ce^{kt},$$

where t is time and k is the rate expressed in decimal notation. Note that

$$P(0) = ce^{k \cdot 0} = ce^0 = c \cdot 1 = c,$$

so c represents the initial population, which we denote P_0:

$$P(t) = P_0 e^{kt}.$$

The graph of $P(t) = P_0 e^{kt}$, for $k > 0$, shows how uninhibited growth produces a "population explosion."

What will the world population be in 2020?

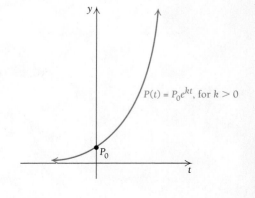

$P(t) = P_0 e^{kt}$, for $k > 0$

The constant k is called the **rate of exponential growth**, or simply the **exponential growth rate**. This is not the rate of change of the population size, which varies according to

$$\frac{dP}{dt} = kP,$$

but the constant by which P must be multiplied in order to get the instantaneous rate of change at any point in time. It is similar to the daily interest rate paid by a bank. If the daily interest rate is $0.07/365$, then any given balance P is growing at the rate of $0.07/365 \cdot P$ dollars per day. Because of the compounding, after 1 year, the interest earned will exceed 7% of P. When interest is compounded continuously, the interest rate is a true exponential growth rate. A detailed explanation of this is presented at the end of this section.

■ **EXAMPLE 4** **Business: Interest Compounded Continuously.** Suppose that an amount P_0, in dollars, is invested in the Von Neumann Hi-Yield Fund, with interest compounded continuously at 7% per year. That is, the balance P grows at the rate given by

$$\frac{dP}{dt} = 0.07P.$$

a) Find the function that satisfies the equation. Write it in terms of P_0 and 0.07.

b) Suppose that $100 is invested. What is the balance after 1 yr?

c) In what period of time will an investment of $100 double itself?

Solution

a) $P(t) = P_0 e^{0.07t}$ Note that $P(0) = P_0$.

b) $P(1) = 100 e^{0.07(1)} = 100 e^{0.07}$

$$\approx 100(1.072508) \qquad \text{It is best to skip this step when using a calculator.}$$

$$\approx \$107.25$$

c) We are looking for a number T such that $P(T) = \$200$. The number T is called the **doubling time**. To find T, we solve the equation

$$200 = 100 e^{0.07 \cdot T}$$
$$2 = e^{0.07T}.$$

We use natural logarithms to solve this equation:

$$\ln 2 = \ln e^{0.07T} \qquad \text{Finding the natural logarithm of both sides}$$
$$\ln 2 = 0.07T \qquad \text{By P5: } \ln e^k = k$$
$$\frac{\ln 2}{0.07} = T$$
$$9.9 \approx T.$$

Thus, $100 will double itself in approximately 9.9 yr.

> **Quick Check 3**
>
> Business: Interest Compounded Continuously. Repeat Example 4 for interest compounded continuously at 4% per year.

❰ Quick Check 3

To find a general expression relating the exponential growth rate k and the doubling time T, we solve the following:

$$2P_0 = P_0 e^{kT}$$
$$2 = e^{kT} \qquad \text{Dividing by } P_0$$
$$\ln 2 = \ln e^{kT}$$
$$\ln 2 = kT.$$

Note that this relationship between k and T does not depend on P_0. We now have the following theorem.

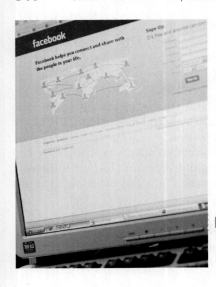

THEOREM 9

The *exponential growth rate k* and the *doubling time T* are related by

$$kT = \ln 2 \approx 0.693147,$$

or

$$k = \frac{\ln 2}{T} \approx \frac{0.693147}{T},$$

and

$$T = \frac{\ln 2}{k} \approx \frac{0.693147}{k}.$$

■ **EXAMPLE 5**   **Business: Facebook Membership.**   The social-networking Web site Facebook connects people with other members they designate as friends. Membership in Facebook has been doubling every 6 months. What is the exponential growth rate of Facebook membership, as a percentage?

Solution   We have

$$k = \frac{\ln 2}{T} \approx \frac{0.693147}{6 \text{ months}} \qquad \text{If possible, enter the calculation as } (\ln 2)/6 \text{ without approximating the logarithmic value.}$$

$$\approx 0.116 \cdot \frac{1}{\text{month}}.$$

The exponential growth rate of Facebook membership is 11.6% per month.

❰ Quick Check 4

> **Quick Check 4**
>
> Business: Internet Use. Worldwide use of the Internet is increasing at an exponential rate, with traffic doubling every 100 days. What is the exponential growth rate of Internet use?

The Rule of 70

The relationship between doubling time *T* and interest rate *k* is the basis of a rule often used in business, called the **Rule of 70**. To estimate how long it takes to double your money, divide 70 by the rate of return:

$$T = \frac{\ln 2}{k} \approx \frac{0.693147}{k} = \frac{100}{100} \cdot \frac{0.693147}{k}$$

$$\approx \frac{69.3147}{100k} \approx \frac{70}{100k}. \qquad \text{Remember that } k \text{ is the interest rate written as a decimal.}$$

■ **EXAMPLE 6**   **Life Science: World Population Growth.**   The world population was approximately 6.0400 billion at the beginning of 2000. It has been estimated that the population is growing exponentially at the rate of 0.016, or 1.6% per year. (How was this estimate determined? The answer is in the model we develop in the following Technology Connection.) Thus,

$$\frac{dP}{dt} = 0.016P,$$

where *t* is the time, in years, after 2000. (*Source*: U.S. Census Bureau.)

a) Find the function that satisfies the equation. Assume that $P_0 = 6.0400$ and $k = 0.016$.

b) Estimate the world population at the beginning of 2020 ($t = 20$).

c) After what period of time will the population be double that in 2000?

) **Quick Check 5**

Life Science: Population Growth in China. In 2006, the population of China was 1.314 billion, and the exponential growth rate was 0.6% per year. Thus,

$$\frac{dP}{dt} = 0.006P,$$

where t is the time, in years, after 2006. (*Source: Time Almanac, 2007.*)

a) Find the function that satisfies the equation. Assume that $P_0 = 1.314$ and $k = 0.006$.

b) Estimate the population of China at the beginning of 2020.

c) After what period of time will the population be double that in 2006?

Solution

a) $P(t) = 6.0400e^{0.016t}$

b) $P(20) = 6.0400e^{0.016(20)} = 6.0400e^{0.32} \approx 8.3179$ billion

c) $T = \dfrac{\ln 2}{k} = \dfrac{\ln 2}{0.016} = 43.3$ yr

Thus, according to this model, the population in 2000 will double itself by 2043. (No wonder environmentalists are alarmed!)

Under ideal conditions, the growth rate of this rapidly growing population of rabbits might be 11.7% per day. When will this population of rabbits double?

⟨ Quick Check 5

Exponential Models Using Regression

Projecting World Population Growth

The table below shows data regarding world population growth. A graph illustrating these data, along with the projected population in 2020, appeared in Section 3.1.

Year	World Population (in billions)
1927	2
1960	3
1974	4
1987	5
1998	6
2011	7

How was the population projected for 2020? The graph shows a rapidly growing population that can be modeled with an exponential function. We carry out the regression procedure very much as we did in Section R.6, but here we choose ExpReg rather than LinReg.

```
EDIT CALC TESTS
7↑QuartReg
8:LinReg(a+bx)
9:LnReg
0⎕ExpReg
A:PwrReg
B:Logistic
C:SinReg
```

```
ExpReg
y = a*b^x
a = 2.168521ᴇ−13
b = 1.015594968

■
```

Note that this gives us an exponential model of the type $y = a \cdot b^x$, where y is the population, in billions, in year x.

$$y = (2.168521 \cdot 10^{-13})(1.015594968)^x. \qquad (1)$$

The base here is not e, but we can make a conversion to an exponential function, base e, using the fact that $b = e^{\ln b}$ and then multiplying exponents:

$$b^x = (1.015594968)^x = (e^{\ln 1.015594968})^x$$
$$= e^{(\ln 1.015594968)x}$$
$$\approx e^{0.0154746161x}.$$

We can now write equation (1) as

$$y = (2.168521 \cdot 10^{-13})e^{0.0154746161x}. \qquad (2)$$

The advantage of this form is that we see the growth rate. Here the world population growth rate is about 0.015, or 1.5%. To find world population in 2012, we can substitute 2012 for x in either equation (1) or (2). We choose equation (2):

$$y = (2.168521 \cdot 10^{-13})e^{0.0154746161(2012)}$$
$$\approx 7.209 \text{ billion}.$$

EXERCISES

Use equation (1) or equation (2) to estimate world population in each year.

1. 2020   **2.** 2050   **3.** 2060   **4.** 2080

(continued)

Exponential Models Using Regression (*continued*)

Projecting College Costs

For Exercises 5 and 6, use the data regarding projected college costs (tuition and room and board) listed in the table below.

School Year, x	Costs of Attending a Public 4-year College or University (2006–2007 dollars)
1998–1999, 0	9,959
1999–2000, 1	9,978
2000–2001, 2	10,089
2001–2002, 3	10,535
2002–2003, 4	10,971
2003–2004, 5	11,709
2004–2005, 6	12,168
2005–2006, 7	12,421
2006–2007, 8	12,797
2007–2008, 9	12,944

(Source: National Center for Education Statistics, *Annual Digest of Education Statistics*: 2008.)

EXERCISES

5. Use REGRESSION to fit an exponential function $y = a \cdot b^x$ to the data. Let 1998–1999 be represented by $x = 0$ and let $y =$ the cost, in dollars. Then convert that formula to an exponential function, base e, and determine the exponential growth rate.

6. Use either of the exponential functions found in Exercise 5 to estimate college costs in the 2014–2015, 2017–2018, and 2039–2040 school years.

In the preceding Technology Connection, we used *regression* to create an exponential model. There is another way to create such a model if regression is not an option. As shown in Example 7, two representative data points are sufficient to determine P_0 and k in $P(t) = P_0 e^{kt}$.

■ **EXAMPLE 7**   **Business: Batman Comic Book.**   A 1939 comic book with the first appearance of the "Caped Crusader," Batman, sold at auction in Dallas in 2010 for a record $1.075 million. The comic book originally cost 10¢ (or $0.10). Using two representative data points (0, $0.10) and (71, $1,075,000), we can find an exponential function that models the increasing value of the comic book. The modeling assumption is that the value V of the comic book has grown exponentially, as given by

$$\frac{dV}{dt} = kV.$$

(*Source*: Heritage Auction Galleries.)

a) Find the function that satisfies this equation. Assume that $V_0 =$ $0.10.

b) Estimate the value of the comic book in 2020.

c) What is the doubling time for the value of the comic book?

d) In what year will the value of the comic book be $30 million, assuming there is no change in the growth rate?

Solution

a) Because of the modeling assumption, we have $V(t) = V_0 e^{kt}$. Since $V_0 =$ $0.10, it follows that

$$V(t) = 0.10e^{kt}.$$

We have made use of the data point (0, $0.10). Next, we use the data point (71, $1,075,000) to determine k. We solve

$$V(t) = 0.10e^{kt}, \quad \text{or} \quad 1{,}075{,}000 = 0.10^{k(71)}$$

for k, using natural logarithms:

$$1{,}075{,}000 = 0.10e^{k(71)} = 0.10e^{71k}$$

$$\frac{1{,}075{,}000}{0.10} = e^{71k} \qquad \text{Dividing to simplify}$$

$$10{,}750{,}000 = e^{71k}$$

$$\ln 10{,}750{,}000 = \ln e^{71k} \qquad \text{Finding the logarithm of both sides}$$

$$\ln 10{,}750{,}000 = 71k \qquad \text{By P5: } \ln e^k = k$$

$$\frac{16.190416}{71} \approx k \qquad \text{Skip this step when using a calculator.}$$

$$0.228 \approx k. \qquad \text{Rounding to the nearest thousandth}$$

The desired function is $V(t) = 0.10e^{0.228t}$, where V is in dollars and t is the number of years since 1939.

b) To estimate the value of the comic book in 2020, which is $2020 - 1939 = 81$ years after 1939, we substitute 81 for t in the equation:

$$V(t) = 0.10e^{0.228t}$$

$$V(81) = 0.10e^{0.228(81)} \approx \$10{,}484{,}567.$$

Not a bad resale value for a 10¢ comic book, presuming someone will pay the price!

c) The doubling time T is given by

$$T = \frac{\ln 2}{k} = \frac{\ln 2}{0.228} \approx 3.04 \text{ yr.}$$

d) We substitute $30,000,000 for $V(t)$ and solve for t:

$$V(t) = 0.10e^{0.228t}$$

$$30{,}000{,}000 = 0.10e^{0.228t}$$

$$\frac{30{,}000{,}000}{0.10} = e^{0.228t}$$

$$300{,}000{,}000 = e^{0.228t}$$

$$\ln 300{,}000{,}000 = \ln e^{0.228t}$$

$$\ln 300{,}000{,}000 = 0.228t$$

$$\frac{\ln 300{,}000{,}000}{0.228} = t$$

$$\frac{19.519293}{0.228} \approx t \qquad \text{Skip this step when using a calculator.}$$

$$86 \approx t. \qquad \text{Rounding to the nearest year}$$

We add 86 to 1939 to get 2025 as the year in which the value of the comic book will reach $30 million.

Note that in part (a) of this example, we find $\ln 10{,}750{,}000$ and divide by 71, obtaining approximately 0.228. We then use that value for k in part (b). Answers are found this way in the exercises. You may note some variation in the last one or two decimal places of your answers if you round as you go.

❰ Quick Check 6

❱ **Quick Check 6**

Business: Batman Comic Book. In Example 7, the consigner had bought the comic book in the late 1960s for $100. (*Source*: Heritage Auction Galleries.) Assume that the year of purchase was 1969 and that the value V of the comic book has since grown exponentially, as given by

$$\frac{dV}{dt} = kV,$$

where t is the number of years since 1969.

a) Use the data points (0, $100) and (41, $1,075,000) to find the function that satisfies the equation.

b) Estimate the value of the comic book in 2020, and compare your answer to that of Example 7.

c) What is the doubling time for the value of the comic book?

d) In what year will the value of the comic book be $30 million? Compare your answer to that of Example 7.

Models of Limited Growth

The growth model $P(t) = P_0e^{kt}$ has many applications to unlimited population growth, as we have seen in this section. However, there are often factors that prevent a population from exceeding some limiting value L—perhaps a limitation on food, living space, or other natural resources. One model of such growth is

$$P(t) = \frac{L}{1 + be^{-kt}}, \quad \text{for } k > 0,$$

which is called the *logistic equation,* or *logistic function.*

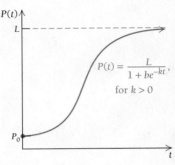

■ **EXAMPLE 8**   **Business: Satellite Radio Subscribers.**   Satellite radio companies provide subscribers with clear signals of hundreds of radio stations, including music, talk, and sports. The provider XM started up in 2001, followed by Sirius in 2002. Both companies did well, experiencing what seemed like exponential growth, but the slowing of this growth led Sirius to buy out XM in 2008, forming Sirius XM. The combined number of subscribers N, in millions, after time t, in years since 2000, with $t = 1$ corresponding to 2001, can be modeled by the logistic equation

$$N(t) = \frac{19.362}{1 + 295.393e^{-1.11t}}.$$

(*Source*: Sirius XM Radio, Inc.)

a) Find the combined number of subscribers after 1 yr (in 2001), 3 yr, 5 yr, and 8 yr.

b) Find the rate at which the number of subscribers was growing after 8 yr.

c) Graph the equation.

d) Explain why an uninhibited growth model is inappropriate but a logistic equation is appropriate to model this growth.

Solution

a) We use a calculator to find the function values:

$$N(1) = 0.197 \text{ million,}$$
$$N(3) = 1.673 \text{ million,}$$
$$N(5) = 9.013 \text{ million,}$$
$$N(8) = 18.598 \text{ million.}$$

After 1 yr, there were about 197,000 subscribers.
After 3 yr, there were about 1,673,000 subscribers.
After 5 yr, there were about 9,013,000 subscribers.
After 8 yr, there were about 18,598,000 subscribers.

b) We find the rate of change using the Quotient Rule:

$$N(t) = \frac{19.362}{1 + 295.393e^{-1.11t}},$$

$$N'(t) = \frac{(1 + 295.393e^{-1.11t}) \cdot 0 - 19.362(295.393e^{-1.11t})(-1.11)}{(1 + 295.393e^{-1.11t})^2}$$

$$= \frac{6348.533e^{-1.11t}}{(1 + 295.393e^{-1.11t})^2}.$$

) Quick Check 7

Life Science: Spread of an Epidemic. In a town whose population is 3500, an epidemic of a disease occurs. The number of people N infected t days after the disease first appears is given by

$$N(t) = \frac{3500}{1 + 19.9e^{-0.6t}}.$$

a) How many people are initially infected with the disease ($t = 0$)?

b) Find the number infected after 2 days, 5 days, 8 days, 12 days, and 18 days.

c) Graph the equation.

d) Find the rate at which the disease is spreading after 16 days.

e) Using this model, determine whether all 3500 residents will ever be infected.

Next, we use a calculator to evaluate the derivative at $t = 8$:

$$N'(8) = 0.815.$$

After 8 yr, the number of subscribers was growing at a rate of 0.815 million, or 815,000, per year.

c) The graph follows.

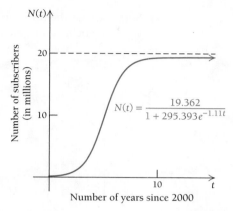

d) An uninhibited growth model is inappropriate because as more and more subscribers are added, the population contains fewer who have not subscribed, perhaps because of the cost or a lack of awareness or interest, or simply because the population is finite. The logistic equation, graphed in part (c), displays the rapid rise in the number of subscribers over the early years of the business as well as the slower growth in later years. It would appear that the limiting value of subscribers is between 19 and 20 million.

《 Quick Check 7

Another model of limited growth is provided by the function

$$P(t) = L(1 - e^{-kt}), \text{ for } k > 0,$$

which is shown graphed below. This function also increases over the entire interval $[0, \infty)$, but increases most rapidly at the beginning, unlike the logistic equation.

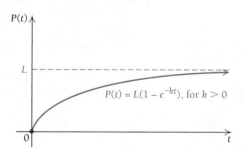

Models of limited growth are discussed in more detail in Section 8.3.

Business Application: An Alternative Derivation of e and $P(t) = P_0e^{kt}$

The number e can also be found using the *compound-interest formula* (which was developed in Chapter R),

$$A = P\left(1 + \frac{i}{n}\right)^{nt},$$

where A is the amount that an initial investment P will be worth after t years at interest rate i, expressed as a decimal, compounded n times per year.

Suppose that \$1 is invested at 100% interest ($i = 100\% = 1$) for 1 yr (though obviously no bank would pay this). The formula becomes

$$A = \left(1 + \frac{1}{n}\right)^n.$$

Suppose that the number of compounding periods, n, increases indefinitely. Let's investigate the behavior of the function. We obtain the following table of values and graph.

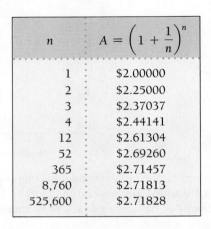

n	$A = \left(1 + \dfrac{1}{n}\right)^n$
1	\$2.00000
2	\$2.25000
3	\$2.37037
4	\$2.44141
12	\$2.61304
52	\$2.69260
365	\$2.71457
8,760	\$2.71813
525,600	\$2.71828

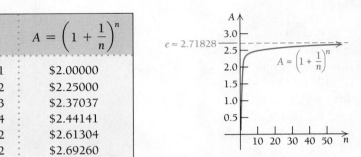

If interest is compounded *continuously*, we have $A = \lim\limits_{n \to \infty} \left(1 + \dfrac{1}{n}\right)^n$, or equivalently,

$$A = \lim_{n \to \infty} \left(1 + \frac{1}{1/h}\right)^{1/h}, \quad \text{or} \quad A = \lim_{h \to 0} (1 + h)^{1/h}.$$

Recall from Section 3.1 that $\lim\limits_{h \to 0} (1 + h)^{1/h} = e$. Thus,

$$\lim_{n \to \infty} \left(1 + \frac{1}{n}\right)^n = e.$$

This result is confirmed by the graph and table above. For \$1, invested at an interest rate of 100% with increasingly frequent compounding periods, the greatest value it could grow to in 1 yr is about \$2.7183.

To develop the formula

$$P(t) = P_0 e^{kt},$$

we again start with the compound-interest formula,

$$A = P\left(1 + \frac{i}{n}\right)^{nt},$$

and assume that interest will be compounded continuously. Let $P = P_0$ and $i = k$ to obtain

$$P(t) = P_0\left(1 + \frac{k}{n}\right)^{nt}.$$

We are interested in what happens as n approaches ∞. To find this limit, we first let

$$\frac{k}{n} = \frac{1}{q}, \quad \text{so that} \quad qk = n.$$

TECHNOLOGY CONNECTION 📉

Exploratory

Graph

$$y = \left(1 + \frac{1}{x}\right)^x$$

using the viewing window $[0, 5000, 0, 5]$, with Xscl = 1000 and Yscl = 1. Trace along the graph. Why does the graph appear to be horizontal? As you trace to the right, note the value of the y-coordinate. Is it approaching a constant? What seems to be its limiting value?

Note that since k is a positive constant, as n gets large, so must q. Thus,

$$P(t) = \lim_{n \to \infty} \left[P_0 \left(1 + \frac{k}{n}\right)^{nt} \right]$$ Letting the number of compounding periods become infinite

$$= P_0 \lim_{q \to \infty} \left[\left(1 + \frac{1}{q}\right)^{qkt} \right]$$ The limit of a constant times a function is the constant times the limit. We also substitute $1/q$ for k/n and qk for n. Also, $q \to \infty$ because $n \to \infty$.

$$= P_0 \left[\lim_{q \to \infty} \left(1 + \frac{1}{q}\right)^q \right]^{kt}$$ The limit of a power is the power of the limit: a form of Limit Property L2 in Section 1.2.

$$= P_0 [e]^{kt}.$$

Section Summary

- Uninhibited growth can be modeled by a *differential equation* of the type $\dfrac{dP}{dt} = kP$, whose solutions are $P(t) = P_0 e^{kt}$.

- The *rate of exponential growth k* and the *doubling time T* are related by the equation $T = \dfrac{\ln 2}{k}$, or $k = \dfrac{\ln 2}{T}$.

- Certain kinds of limited growth can be modeled by equations such as $P(t) = \dfrac{L}{1 + be^{-kt}}$ and $P(t) = L(1 - e^{-kt})$, for $k > 0$.

EXERCISE SET
3.3

1. Find the general form of f if $f'(x) = 4f(x)$.

2. Find the general form of g if $g'(x) = 6g(x)$.

3. Find the general form of the function that satisfies $dA/dt = -9A$.

4. Find the general form of the function that satisfies $dP/dt = -3P(t)$.

5. Find the general form of the function that satisfies $dQ/dt = kQ$.

6. Find the general form of the function that satisfies $dR/dt = kR$.

APPLICATIONS

Business and Economics

7. U.S. patents. The number of applications for patents, N, grew dramatically in recent years, with growth averaging about 4.6% per year. That is,

$$N'(t) = 0.046N(t).$$

(*Source*: New York Times, 11/13/05, p. C1.)

a) Find the function that satisfies this equation. Assume that $t = 0$ corresponds to 1980, when approximately 112,000 patent applications were received.
b) Estimate the number of patent applications in 2020.
c) Estimate the doubling time for $N(t)$.

8. Franchise expansion. Pete Zah's, Inc., is selling franchises for pizza shops throughout the country. The marketing manager estimates that the number of franchises, N, will increase at the rate of 10% per year, that is,

$$\frac{dN}{dt} = 0.10N.$$

a) Find the function that satisfies this equation. Assume that the number of franchises at $t = 0$ is 50.
b) How many franchises will there be in 20 yr?
c) In what period of time will the initial number of 50 franchises double?

9. Compound interest. Suppose that P_0 is invested in the Mandelbrot Bond Fund for which interest is compounded continuously at 5.9% per year. That is, the balance P grows at the rate given by

$$\frac{dP}{dt} = 0.059P.$$

a) Find the function that satisfies the equation. Write it in terms of P_0 and 0.059.

b) Suppose that $1000 is invested. What is the balance after 1 yr? After 2 yr?

c) When will an investment of $1000 double itself?

10. Compound interest. Suppose that P_0 is invested in a savings account for which interest is compounded continuously at 4.3% per year. That is, the balance P grows at the rate given by

$$\frac{dP}{dt} = 0.043P.$$

a) Find the function that satisfies the equation. Write it in terms of P_0 and 0.043.

b) Suppose that $20,000 is invested. What is the balance after 1 yr? After 2 yr?

c) When will an investment of $20,000 double itself?

11. Bottled water sales. Since 2000, sales of bottled water have increased at the rate of approximately 9.3% per year. That is, the volume of bottled water sold, G, in billions of gallons, t years after 2000 is growing at the rate given by

$$\frac{dG}{dt} = 0.093G.$$

(*Source:* The Beverage Marketing Corporation.)

a) Find the function that satisfies the equation, given that approximately 4.7 billion gallons of bottled water were sold in 2000.

b) Predict the number of gallons of water sold in 2025.

c) What is the doubling time for $G(t)$?

12. Annual net sales. Green Mountain Coffee Roasters produces many varieties of flavored coffees, teas, and K-cups. The net sales S of the company have grown exponentially at the rate of 36.1% per year, and the growth can be approximated by

$$\frac{dS}{dt} = 0.361S,$$

where t is the number of years since 2004. (*Source*: Green Mountain Coffee Roasters financial statements.)

a) Find the function that satisfies the equation, given that net sales in 2004 ($t = 0$) were approximately $120,400.

b) Estimate net sales in 2006, 2008, and 2015.

c) What is the doubling time for $S(t)$?

13. Annual interest rate. Euler Bank advertises that it compounds interest continuously and that it will double your money in 15 yr. What is its annual interest rate?

14. Annual interest rate. Hardy Bank advertises that it compounds interest continuously and that it will double your money in 12 yr. What is its annual interest rate?

15. Oil demand. The growth rate of the demand for oil in the United States is 10% per year. When will the demand be double that of 2006?

16. Coal demand. The growth rate of the demand for coal in the world is 4% per year. When will the demand be double that of 2006?

Interest compounded continuously. *For Exercises 17–20, complete the following.*

Initial Investment at $t = 0$, P_0	Interest Rate, k	Doubling Time, T (in years)	Amount after 5 yr
17. $75,000	6.2%		
18. $5,000			$7,130.90
19.	8.4%		$11,414.71
20.		11	$17,539.32

21. Art masterpieces. In 2004, a collector paid $104,168,000 for Pablo Picasso's "Garcon à la Pipe." The same painting sold for $30,000 in 1950. (*Source*: BBC News, 5/6/04.)

Boy with a Pipe (1905), Pablo Picasso. © 2011 Picasso Estate/ARS

a) Find the exponential growth rate k, to three decimal places, and determine the exponential growth function V, for which $V(t)$ is the painting's value, in dollars, t years after 1950.

b) Predict the value of the painting in 2015.

c) What is the doubling time for the value of the painting?

d) How long after 1950 will the value of the painting be $1 billion?

22. Per capita income. In 2007, U.S. per capita personal income I was $46,459. In 2011, it was $47,153. (*Source*: data.worldbank.org.) Assume that the growth of U.S. per capita personal income follows an exponential model.

a) Letting $t = 0$ be 2007, write the function.

b) Predict what U.S. per capita income will be in 2020.

c) In what year will U.S. per capita income be double that of 2007?

23. Federal receipts. In 1990, U.S. federal receipts (money taken in), *E*, were $1.031 billion. In 2009, federal receipts were $2.523 billion. (*Source:* National Center for Education Statistics.) Assume that the growth of federal receipts follows an exponential model and use 1990 as the base year ($t = 0$).

  a) Find the value of *k* to six decimal places, and write the function, with $E(t)$ in billions of dollars.
  b) Estimate federal receipts in 2015.
  c) When will federal receipts be $10 billion?

24. Consumer price index. The *consumer price index* compares the costs, *c*, of goods and services over various years, where 1983 is used as a base ($t = 0$). The same goods and services that cost $100 in 1983 cost $226 in 2012. (*Source:* Bureau of Labor Statistics.) Assuming an exponential model:

  a) Write the function, rounding *k* to five decimal places.
  b) Estimate what the goods and services costing $100 in 1983 will cost in 2020.
  c) In what year did the same goods and services cost twice the 1983 price?

Sales of paper shredders. *Data in the following bar graph show paper shredder sales in recent years. Use these data for Exercises 25 and 26.*

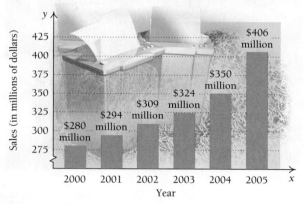

SHREDDER BOOM:
ESTIMATED PAPER SHREDDER SALES

(*Source:* www.sfgate.com.)

25. a) Use REGRESSION to fit an exponential function $y = a \cdot b^x$ to the data. Let *y* be in millions of dollars. Then convert that formula to an exponential function, base *e*, where *x* is the number of years after 1990, and determine the exponential growth rate. (See the Technology Connection on p. 341.)
  b) Estimate the total sales of paper shredders in 2007 and in 2012.
  c) After what amount of time will sales be $500 million?
  d) What is the doubling time for sales of shredders?

26. a) To find an exponential function, base *e*, that fits the data, find *k* using the points (10, 280) and (15, 406). Then write the function. (Assume that *x* is the number of years after 1990.)
  b) Estimate the total sales of paper shredders in 2007 and in 2012.
  c) After what amount of time will total sales be $500 million?

  d) What is the doubling time for sales of shredders?
  e) Compare your answers to parts (a)–(d) with those from Exercise 25. Decide which exponential function seems better to you, and explain why.

27. Value of Manhattan Island. Peter Minuit of the Dutch West India Company purchased Manhattan Island from the natives living there in 1626 for $24 worth of merchandise. Assuming an exponential rate of inflation of 5%, how much will Manhattan be worth in 2020?

28. Total revenue. Intel, a computer chip manufacturer, reported $1265 million in total revenue in 1986. In 2005, the total revenue was $38.8 billion. (*Source*: U.S. Securities and Exchange Commission.) Assuming an exponential model, find the growth rate *k*, to four decimal places, and write the revenue function *R*, with $R(t)$ in billions of dollars. Then predict the company's total revenue for 2012.

29. The U.S. Forever Stamp. On May 12, 2008, the U.S. Postal Service reissued the Forever Stamp (which features an image of the Liberty Bell). The Forever Stamp is always valid as first-class postage on standard envelopes weighing 1 ounce or less, regardless of any subsequent increases in the first-class rate. (*Source*: U.S. Postal Service.)

  a) The cost of a first-class postage stamp was 4¢ in 1962 and 45¢ in 2011. This increase represents exponential growth. Write the function *S* for the cost of a stamp *t* years after 1962 ($t = 0$).
  b) What was the growth rate in the cost?
  c) Predict the cost of a first-class postage stamp in 2013, 2016, and 2019.
  d) An advertising firm spent $4500 on 10,000 first-class postage stamps in 2011. Knowing it will need 10,000 first-class stamps in each of the years 2011–2021, it decides at the beginning of 2011 to try to save money by spending $4500 on 10,000 Forever Stamps, but also buying enough of the stamps to cover the years 2012 through 2021. Assuming there is a postage increase in each of the years 2013, 2016, and 2019 to the cost predicted in part (c), how much money will the firm save by buying Forever Stamps?
  e) Discuss the pros and cons of the purchase decision described in part (d).

30. Average salary of Major League baseball players. In 1970, the average salary of Major League baseball players was $29,303. In 2005, the average salary was $2,632,655. (*Source: Baseball Almanac.*) Assuming exponential growth occurred, what was the growth rate to the nearest hundredth of a percent? What will the average salary be in 2015? In 2020?

31. Effect of advertising. Suppose that SpryBorg Inc. introduces a new computer game in Houston using television advertisements. Surveys show that $P\%$ of the target audience buy the game after x ads are broadcast, satisfying

$$P(x) = \frac{100}{1 + 49e^{-0.13x}}.$$

a) What percentage buy the game without seeing a TV ad ($x = 0$)?

b) What percentage buy the game after the ad is run 5 times? 10 times? 20 times? 30 times? 50 times? 60 times?

c) Find the rate of change, $P'(x)$.

d) Sketch a graph of the function.

32. Cost of a Hershey bar. The cost of a Hershey bar was $0.05 in 1962 and $0.75 in 2010 (in a supermarket, not in a movie theater).

a) Find an exponential function that fits the data.

b) Predict the cost of a Hershey bar in 2015 and 2025.

33. Superman comic book. Three days before the sale of the Batman comic book (Example 7) in 2010, a 1938 comic book with the first appearance of Superman sold at auction in Dallas for a record $1.0 million. The comic book originally cost 10¢ ($0.10). (*Source*: Heritage Auction Galleries.) Using two representative data points, (0, $0.10) and (72, $1,000,000), we can approximate the data with an exponential function. The modeling assumption is that the value V of the comic book has grown exponentially, as given by

$$\frac{dV}{dt} = kV.$$

(In the summer of 2010, a family in the southern United States was facing foreclosure on their mortgage and loss of their home. Then, as they were packing, an amazing twist of fate occurred; they came across some old comic books in the basement and one of them was this first Superman comic. They sold it and saved their house.)

a) Find the function that satisfies this equation. Assume that $V_0 = \$0.10$.

b) Estimate the value of the comic book in 2020.

c) What is the doubling time for the value of the comic book?

d) After what time will the value of the comic book be $30 million, assuming there is no change in the growth rate?

34. Batman comic book. Refer to Example 7. In what year will the value of the comic book be $5 million?

35. Batman comic book. Refer to Example 7. In what year will the value of the comic book be $10 million?

Life and Physical Sciences

Population growth. *For Exercises 36–40, complete the following.*

	Population	Exponential Growth Rate, k	Doubling Time, T (in years)
36.	Mexico	3.5%/yr	
37.	Europe		69.31
38.	Oil reserves		6.931
39.	Coal reserves		17.3
40.	Alaska	2.794%/yr	

41. Yellowstone grizzly bears. In 1972, the population of grizzly bears in Yellowstone National Park had shrunk to approximately 190. In 2005, the number of Yellowstone grizzlies had grown to about 610. (*Source: New York Times*, 9/26/05.) Find an exponential function that fits the data, and then predict Yellowstone's grizzly bear population in 2016. Round k to three decimal places.

42. Bicentennial growth of the United States. The population of the United States in 1776 was about 2,508,000. In the country's bicentennial year, the population was about 216,000,000.

a) Assuming an exponential model, what was the growth rate of the United States through its bicentennial year?

b) Is exponential growth a reasonable assumption? Explain.

43. Limited population growth. A ship carrying 1000 passengers has the misfortune to be wrecked on a small island from which the passengers are never rescued. The natural resources of the island restrict the growth of the population to a *limiting value* of 5780, to which the population gets closer and closer but which it never reaches. The population of the island after time t, in years, is approximated by the logistic equation

$$P(t) = \frac{5780}{1 + 4.78e^{-0.4t}}.$$

a) Find the population after 0 yr, 1 yr, 2 yr, 5 yr, 10 yr, and 20 yr.

b) Find the rate of change, $P'(t)$.

c) Sketch a graph of the function.

This island may have achieved its limited population growth.

44. Limited population growth. A lake is stocked with 400 rainbow trout. The size of the lake, the availability of food, and the number of other fish restrict growth in the lake to a *limiting value* of 2500. (See Exercise 43.) The population of trout in the lake after time t, in months, is approximated by

$$P(t) = \frac{2500}{1 + 5.25e^{-0.32t}}.$$

a) Find the population after 0 months, 1 month, 5 months, 10 months, 15 months, and 20 months.

b) Find the rate of change, $P'(t)$.

c) Sketch a graph of the function.

Social Sciences

45. Women college graduates. The number of women graduating from 4-yr colleges in the United States grew from 1930, when 48,869 women earned a bachelor's degree, to 2005, when approximately 832,000 women received such a degree. (*Source:* National Center for Education Statistics.) Find an exponential function that fits the data, and the exponential growth rate, rounded to the nearest hundredth of a percent.

46. Hullian learning model. The Hullian learning model asserts that the probability p of mastering a task after t learning trials is approximated by

$$p(t) = 1 - e^{-kt},$$

where k is a constant that depends on the task to be learned. Suppose that a new dance is taught to an aerobics class. For this particular dance, the constant $k = 0.28$.

a) What is the probability of mastering the dance's steps in 1 trial? 2 trials? 5 trials? 11 trials? 16 trials? 20 trials?

b) Find the rate of change, $p'(t)$.

c) Sketch a graph of the function.

47. Diffusion of information. Pharmaceutical firms invest significantly in testing new medications. After a drug is approved by the Federal Drug Administration, it still takes time for physicians to fully accept and start prescribing the medication. The acceptance by physicians approaches a *limiting value* of 100%, or 1, after time t, in months. Suppose that the percentage P of physicians prescribing a new cancer medication after t months is approximated by

$$P(t) = 100(1 - e^{-0.4t}).$$

a) What percentage of doctors are prescribing the medication after 0 months? 1 month? 2 months? 3 months? 5 months? 12 months? 16 months?

b) Find $P'(7)$, and interpret its meaning.

c) Sketch a graph of the function.

48. Spread of infection. Spread by skin-to-skin contact or via shared towels or clothing, methicillin-resistant *Staphylococcus aureus* (MRSA) can easily infect growing numbers of students at a university. Left unchecked, the number of cases of MRSA on a university campus t weeks after the first 9 cases occur can be modeled by

$$N(t) = \frac{568.803}{1 + 62.200e^{-0.092t}}.$$

(*Source:* Vermont Department of Health, Epidemiology Division.)

a) Find the number of infected students beyond the first 9 cases after 3 weeks, 40 weeks, and 80 weeks.

b) Find the rate at which the disease is spreading after 20 weeks.

c) Explain why an unrestricted growth model is inappropriate but a logistic equation is appropriate for this situation. Then use a calculator to graph the equation.

49. Spread of a rumor. The rumor "People who study math all get scholarships" spreads across a college campus. Data in the following table show the number of students N who have heard the rumor after time t, in days.

a) Use REGRESSION to fit a logistic equation,

$$N(t) = \frac{c}{1 + ae^{-bt}},$$

to the data.

b) Estimate the limiting value of the function. At most, how many students will hear the rumor?

Time, t (in days)	Number, N, Who Have Heard the Rumor
1	1
2	2
3	4
4	7
5	12
6	18
7	24
8	26
9	28
10	28
11	29
12	30

c) Graph the function.

d) Find the rate of change, $N'(t)$.

e) Find $\lim\limits_{t \to \infty} N'(t)$, and explain its meaning.

SYNTHESIS

We have now studied models for linear, quadratic, exponential, and logistic growth. In the real world, understanding which is the most appropriate type of model for a given situation is an important skill. For each situation in Exercises 50–56, identify the most appropriate type of model and explain why you chose that model. List any restrictions you would place on the domain of the function.

50. The growth in value of a U.S. savings bond

51. The growth in the length of Zachary's hair following a haircut

52. The growth in sales of cellphones

53. The drop and rise of a lake's water level during and after a drought

54. The rapidly growing sales of organic foods

55. The number of manufacturing jobs that have left the United States since 1995

56. The life expectancy of the average American

57. Find an expression relating the exponential growth rate k and the *quadrupling time* T_4.

58. Find an expression relating the exponential growth rate k and the *tripling time* T_3.

59. A quantity Q_1 grows exponentially with a doubling time of 1 yr. A quantity Q_2 grows exponentially with a doubling time of 2 yr. If the initial amounts of Q_1 and Q_2 are the same, how long will it take for Q_1 to be twice the size of Q_2?

60. To what exponential growth rate per hour does a growth rate of 100% per day correspond?

Business: effective annual yield. Suppose that $100 is invested at 7%, compounded continuously, for 1 yr. We know from Example 4 that the ending balance will be $107.25. This would also be the ending balance if $100 were invested at 7.25%, compounded once a year (simple interest). The rate of 7.25% is called the effective annual yield. In general, if P_0 is invested at interest rate k, compounded continuously, then the effective annual yield is that number Y satisfying $P_0(1 + Y) = P_0e^k$. Then, $1 + Y = e^k$, or

$$\text{Effective annual yield} = Y = e^k - 1.$$

61. An amount is invested at 7.3% per year compounded continuously. What is the effective annual yield?

62. An amount is invested at 8% per year compounded continuously. What is the effective annual yield?

63. The effective annual yield on an investment compounded continuously is 9.42%. At what rate was it invested?

64. The effective annual yield on an investment compounded continuously is 6.61%. At what rate was it invested?

65. To show that the exponential growth rate can be determined using any two points, let

$$y_1 = Ce^{kt_1}, \quad \text{and} \quad y_2 = Ce^{kt_2}.$$

Solve this system of equations for k to show that k can be calculated directly, using (t_1, y_1) and (t_2, y_2).

66. Complete the table below, which relates growth rate k and doubling time T.

Growth Rate, k (per year)	1%	2%		14%
Doubling Time, T (in years)			15	10

Graph $T = (\ln 2)/k$. Is this a linear relationship? Explain.

67. Describe the differences in the graphs of an exponential function and a logistic function.

68. Explain how the Rule of 70 could be useful to someone studying inflation.

69. Business: total revenue. The revenue of Red Rocks, Inc., in millions of dollars, is given by the function

$$R(t) = \frac{4000}{1 + 1999e^{-0.5t}},$$

where t is measured in years.

a) What is $R(0)$, and what does it represent?

b) Find $\lim\limits_{t \to \infty} R(t)$. Call this value $R_{\max}$, and explain what it means.

c) Find the value of t (to the nearest integer) for which $R(t) = 0.99R_{\max}$.

Answers to Quick Checks

1. $f'(x) = 20e^{4x}; f'(x) = 4f(x)$

2. $N(t) = ce^{kt}$, where c is an arbitrary constant

3. (a) $P(t) = P_0e^{0.04t}$; **(b)** $104.08; **(c)** 17.3 yr

4. 0.69% per day

5. (a) $P(t) = 1.314e^{0.006t}$; **(b)** 1.429 billion; **(c)** 115.5 yr

6. (a) $V(t) = 100e^{0.226t}$, which is almost the same as the growth rate found in Example 7; **(b)** $10,131,604, which is quite close to the estimate found in Example 7; **(c)** 3.07 yr; **(d)** after 55.8 yr, or in 2025, again about the same as was found in Example 7

7. (a) 167; **(b)** 500, 1758, 3007, 3449, 3499

(c)

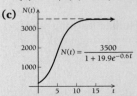

$$N(t) = \frac{3500}{1 + 19.9e^{-0.6t}}$$

(d) After 16 days, the number of people infected is growing at the rate of about 2.8 people per day. **(e)** According to the model, $\lim\limits_{t \to \infty} N(t) = 3500$, so virtually all of the town's residents will be infected.

Applications: Decay

In the equation of population growth, $dP/dt = kP$, the constant k is actually given by

$$k = (\text{Birth rate}) - (\text{Death rate}).$$

Thus, a population "grows" only when the *birth rate* is greater than the *death rate*. When the birth rate is less than the death rate, k will be negative, and the population will be decreasing, or "decaying," at a rate proportional to its size. For convenience in our computations, we will express such a negative value as $-k$, where $k > 0$. The equation

$$\frac{dP}{dt} = -kP, \quad \text{where } k > 0,$$

shows P to be *decreasing* as a function of time, and the solution

$$P(t) = P_0e^{-kt}$$

shows it to be decreasing exponentially. This is called **exponential decay**. The amount present initially at $t = 0$ is again P_0.

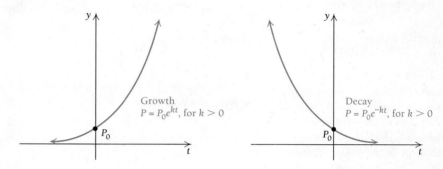

Radioactive Decay

Radioactive elements decay exponentially; that is, they disintegrate at a rate that is proportional to the amount present.

EXAMPLE 1 **Life Science: Decay.** Strontium-90 has a decay rate of 2.8% per year. The rate of change of an amount N of this radioactive isotope is given by

$$\frac{dN}{dt} = -0.028N.$$

a) Find the function that satisfies the equation. Let N_0 represent the amount present at $t = 0$.

b) Suppose that 1000 grams (g) of strontium-90 is present at $t = 0$. How much will remain after 70 yr?

c) After how long will half of the 1000 g remain?

Solution

a) $N(t) = N_0e^{-0.028t}$

b) $N(70) = 1000e^{-0.028(70)}$

$$= 1000e^{-1.96}$$

$$\approx 1000(0.1408584209)$$

$$\approx 140.8584209.$$

After 70 yr, about 140.9 g of the strontium-90 remains.

❯ Quick Check 1

Life Science: Decay. Xenon-133 has a decay rate of 14% per day. The rate of change of an amount N of this radioactive isotope is given by

$$\frac{dN}{dt} = -0.14N.$$

a) Find the function that satisfies the equation. Let N_0 represent the amount present at $t = 0$.

b) Suppose 1000 g of xenon-133 is present at $t = 0$. How much will remain after 10 days?

c) After how long will half of the 1000 g remain?

How can scientists determine that the remains of an animal or plant have lost 30% of the carbon-14? The assumption is that the percentage of carbon-14 in the atmosphere and in living plants and animals is the same. When a plant or animal dies, the amount of carbon-14 decays exponentially. The scientist burns the remains and uses a Geiger counter to determine the percentage of carbon-14 in the smoke. The amount by which this varies from the percentage in the atmosphere indicates how much carbon-14 has been lost through decay.

The process of carbon-14 dating was developed by the American chemist Willard F. Libby in 1952. It is known that the radioactivity of a living plant measures 16 disintegrations per gram per minute. Since the half-life of carbon-14 is 5730 years, a dead plant with an activity of 8 disintegrations per gram per minute is 5730 years old, one with an activity of 4 disintegrations per gram per minute is about 11,500 years old, and so on. Carbon-14 dating can be used to determine the age of organic objects from 30,000 to 40,000 years old. Beyond such an age, it is too difficult to measure the radioactivity, and other methods are used.

c) We are asking, "At what time T will $N(T)$ be half of N_0, or $\frac{1}{2} \cdot 1000$?" The number T is called the **half-life**. To find T, we solve the equation

$$
\begin{aligned}
500 &= 1000e^{-0.028T} & &\text{We use 500 because } 500 = \tfrac{1}{2} \cdot 1000. \\
\tfrac{1}{2} &= e^{-0.028T} & &\text{Dividing both sides by 1000} \\
\ln \tfrac{1}{2} &= \ln e^{-0.028T} & &\text{Taking the natural logarithm of both sides} \\
\ln 1 - \ln 2 &= -0.028T & &\text{Using the properties of logarithms} \\
0 - \ln 2 &= -0.028T & & \\
\frac{-\ln 2}{-0.028} &= T & &\text{Dividing both sides by } -0.028 \\
\frac{\ln 2}{0.028} &= T & & \\
\frac{0.693147}{0.028} &\approx T & & \\
25 &\approx T. & &
\end{aligned}
$$

Thus, the half-life of strontium-90 is about 25 yr.

❮ Quick Check 1

We can find a general expression relating the decay rate k and the half-life T by solving the equation

$$
\begin{aligned}
\tfrac{1}{2}P_0 &= P_0 e^{-kT} \\
\tfrac{1}{2} &= e^{-kT} \\
\ln \tfrac{1}{2} &= \ln e^{-kT} \\
\ln 1 - \ln 2 &= -kT \\
0 - \ln 2 &= -kT \\
-\ln 2 &= -kT \\
\ln 2 &= kT.
\end{aligned}
$$

Again, we have the following.

THEOREM 10

The *decay rate*, k, and the *half-life*, T, are related by

$$kT = \ln 2 = 0.693147,$$

or

$$k = \frac{\ln 2}{T} \quad \text{and} \quad T = \frac{\ln 2}{k}.$$

Thus, the half-life, T, depends only on the decay rate, k. In particular, it is independent of the initial population size.

The effect of half-life is shown in the radioactive decay curve below. Note that the exponential function gets close to, but never reaches, 0 as t gets larger. Thus, in theory, a radioactive substance never completely decays.

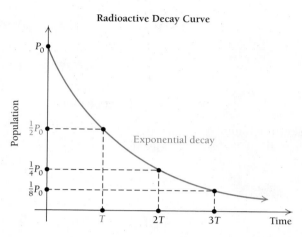

Radioactive Decay Curve

EXAMPLE 2 **Life Science: Half-life.** Plutonium-239, a common product of a functioning nuclear reactor, can be deadly to people exposed to it. Its decay rate is about 0.0028% per year. What is its half-life?

Solution We have

$$T = \frac{\ln 2}{k}$$

$$= \frac{\ln 2}{0.000028} \qquad \text{Converting the percentage to decimal notation}$$

$$\approx 24{,}755.$$

Thus, the half-life of plutonium-239 is about 24,755 yr.

⟨ Quick Check 2

> **Quick Check 2**
>
> **Life Science: Half-life.**
>
> **a)** The decay rate of cesium-137 is 2.3% per year. What is its half-life?
>
> **b)** The half-life of barium-140 is 13 days. What is its decay rate?

EXAMPLE 3 **Life Science: Carbon Dating.** The radioactive element carbon-14 has a half-life of 5730 yr. The percentage of carbon-14 present in the remains of plants and animals can be used to determine age. Archaeologists found that the linen wrapping from one of the Dead Sea Scrolls had lost 22.3% of its carbon-14. How old was the linen wrapping?

Solution Our plan is to find the exponential equation of the form $N(t) = N_0 e^{-kt}$, replace $N(t)$ with $(1 - 0.223)N_0$, and solve for t. First, however, we must find the decay rate, k:

$$k = \frac{\ln 2}{T} = \frac{0.693147}{5730} \approx 0.00012097, \quad \text{or} \quad 0.012097\% \text{ per year.}$$

Thus, the amount $N(t)$ that remains from an initial amount N_0 after t years is given by:

$$N(t) = N_0 e^{-0.00012097t}. \qquad \text{Remember: } k \text{ is positive, so } -k \text{ is negative.}$$

(*Note:* This equation can be used for all subsequent carbon-dating problems.)

In 1947, a Bedouin youth looking for a stray goat climbed into a cave at Kirbet Qumran on the shores of the Dead Sea near Jericho and came upon earthenware jars containing an incalculable treasure of ancient manuscripts, which concern the Jewish books of the Bible. Shown here are fragments of those so-called Dead Sea Scrolls, a portion of some 600 or so texts found so far. Officials date them before A.D. 70, making them the oldest biblical manuscripts by 1000 years.

If the linen wrapping of a Dead Sea Scroll lost 22.3% of its carbon-14 from an initial amount P_0, then $77.7\% \cdot P_0$ remains. To find the age t of the wrapping, we solve the following equation for t:

$$77.7\% \, N_0 = N_0 e^{-0.00012097t}$$
$$0.777 = e^{-0.00012097t}$$
$$\ln 0.777 = \ln e^{-0.00012097t}$$
$$\ln 0.777 = -0.00012097t$$
$$\frac{\ln 0.777}{-0.00012097} = t$$
$$2086 \approx t.$$

> **Quick Check 3**
>
> **Life Science: Carbon Dating.**
> How old is a skeleton found at an archaeological site if tests show that it has lost 60% of its carbon-14?

Thus, the linen wrapping of the Dead Sea Scroll is about 2086 yr old.

❮ Quick Check 3

A Business Application: Present Value

A representative of a financial institution is often asked to solve a problem like the following.

■ **EXAMPLE 4**   **Business: Present Value.**   Following the birth of their granddaughter, two grandparents want to make an initial investment of P_0 that will grow to $10,000 by the child's 20th birthday. Interest is compounded continuously at 6%. What should the initial investment be?

Solution   Using the equation $P = P_0 e^{kt}$, we find P_0 such that

$$10,000 = P_0 e^{0.06 \cdot 20},$$

or

$$10,000 = P_0 e^{1.2}.$$

Now

$$\frac{10,000}{e^{1.2}} = P_0,$$

or

$$10,000 e^{-1.2} = P_0,$$

and, using a calculator, we have

$$P_0 = 10,000 e^{-1.2}$$
$$\approx \$3011.94.$$

Thus, the grandparents must deposit $3011.94, which will grow to $10,000 by the child's 20th birthday.

Economists call $3011.94 the *present value* of $10,000 due 20 yr from now at 6%, compounded continuously. The process of computing present value is called **discounting**. Another way to pose this problem is to ask "What must I invest now, at 6%, compounded continuously, in order to have $10,000 in 20 years?" The answer is $3011.94, and it is the present value of $10,000.

Computing present value can be interpreted as exponential decay from the future back to the present.

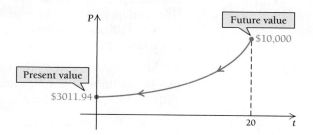

〉 Quick Check 4

Business: Present Value.
Repeat Example 4 for an interest rate of 4%.

❰ Quick Check 4

In general, the present value P_0 of an amount P due t years later is found by solving the following equation for P_0:

$$P_0 e^{kt} = P$$

$$P_0 = \frac{P}{e^{kt}} = P e^{-kt}.$$

THEOREM 11

The **present value** P_0 of an amount P due t years later, at interest rate k, compounded continuously, is given by

$$P_0 = P e^{-kt}.$$

Newton's Law of Cooling

Consider the following situation. A hot cup of soup, at a temperature of 200°, is placed in a 70° room.* The temperature of the soup decreases over time t, in minutes, according to the model known as **Newton's Law of Cooling**.

Newton's Law of Cooling

The temperature T of a cooling object drops at a rate that is proportional to the difference $T - C$, where C is the constant temperature of the surrounding medium. Thus,

$$\frac{dT}{dt} = -k(T - C). \tag{1}$$

The function that satisfies equation (1) is

$$T = T(t) = ae^{-kt} + C. \tag{2}$$

To check that $T(t) = ae^{-kt} + C$ is the solution, find dT/dt and substitute dT/dt and $T(t)$ into equation (1). This check is left to the student.

---

*Assume throughout this section that all temperatures are in degrees Fahrenheit unless noted otherwise.

■ **EXAMPLE 5** **Life Science: Scalding Coffee.** McDivett's Pie Shoppes, a national restaurant firm, finds that the temperature of its freshly brewed coffee is 130°. The company fears that if customers spill hot coffee on themselves, lawsuits might result. Room temperature in the restaurants is generally 72°. The temperature of the coffee cools to 120° after 4.3 min. The company determines that it is safer to serve the coffee at a temperature of 105°. How long does it take a cup of coffee to cool to 105°?

Solution Note that C, the surrounding air temperature, is 72°. To find the value of a in equation (2) for Newton's Law of Cooling, we observe that at $t = 0$, we have $T(0) = 130°$. We solve for a as follows:

$$130 = ae^{-k \cdot 0} + 72$$
$$130 = a + 72$$
$$58 = a. \qquad \text{Note that } a = 130 - 72, \text{ the difference between the original temperatures.}$$

Next, we find k using the fact that $T(4.3) = 120$:

$$120 = 58e^{-k \cdot (4.3)} + 72$$
$$48 = 58e^{-4.3k}$$
$$\frac{48}{58} = e^{-4.3k}$$
$$\ln \frac{48}{58} = \ln e^{-4.3k}$$
$$-0.1892420 \approx -4.3k$$
$$k \approx 0.044.$$

We now have $T(t) = 58e^{-0.044t} + 72$. To see how long it will take the coffee to cool to 105°, we set $T(t) = 105$ and solve for t:

$$105 = 58e^{-0.044t} + 72$$
$$33 = 58e^{-0.044t}$$
$$\frac{33}{58} = e^{-0.044t}$$
$$\ln \frac{33}{58} = \ln e^{-0.044t}$$
$$-0.5639354 \approx -0.044t$$
$$t \approx 12.8 \text{ min.}$$

Thus, to cool to 105°, the coffee should be allowed to cool for about 13 min.

❰ Quick Check 5

❱ **Quick Check 5**

Life Science: Scalding Coffee. Repeat Example 5, but assume that the coffee is sold in an ice cream shop, where the room temperature is 70°, and it cools to a temperature of 120° in 4 min.

The graph of $T(t) = ae^{-kt} + C$ shows that $\lim_{t \to \infty} T(t) = C$. The temperature of the object decreases toward the temperature of the surrounding medium.

Mathematically, this model tells us that the object's temperature never quite reaches C. In practice, the temperature of the cooling object will get so close to that of the surrounding medium that no device could detect a difference. Let's now see how Newton's Law of Cooling can be used in solving a crime.

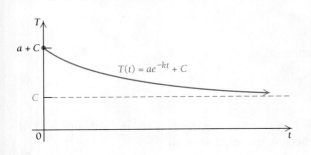

■ **EXAMPLE 6** Forensics: When Was the Murder Committed? A body is found slumped over a desk in a study. A coroner arrives at noon, immediately takes the temperature of the body, and finds it to be 94.6°. She waits 1 hr, takes the temperature again, and finds it to be 93.4°. She also notes that the temperature of the room is 70°. When was the murder committed?

Solution Note that C, the surrounding air temperature, is 70°. To find a in $T(t) = ae^{-kt} + C$, we assume that the temperature of the body was normal when the murder occurred. Thus, $T = 98.6°$ at $t = 0$:

$$98.6 = ae^{-k \cdot 0} + 70,$$
$$a = 28.6.$$

This gives $T(t) = 28.6e^{-kt} + 70$.

To find the number of hours N since the murder was committed, we must first determine k. From the two temperature readings the coroner made, we have

$$94.6 = 28.6e^{-kN} + 70, \quad \text{or} \quad 24.6 = 28.6e^{-kN}; \tag{3}$$
$$93.4 = 28.6e^{-k(N+1)} + 70, \quad \text{or} \quad 23.4 = 28.6e^{-k(N+1)}. \tag{4}$$

Dividing equation (3) by equation (4), we get

$$\frac{24.6}{23.4} = \frac{28.6e^{-kN}}{28.6e^{-k(N+1)}}$$
$$= e^{-kN + k(N+1)}$$
$$= e^{-kN + kN + k} = e^k.$$

We solve this equation for k:

$$\ln \frac{24.6}{23.4} = \ln e^k \qquad \text{Taking the natural logarithm on both sides}$$
$$0.05 \approx k.$$

Next, we substitute back into equation (3) and solve for N:

$$24.6 = 28.6e^{-0.05N}$$
$$\frac{24.6}{28.6} = e^{-0.05N}$$
$$\ln \frac{24.6}{28.6} = \ln e^{-0.05N}$$
$$-0.150660 \approx -0.05N$$
$$3 \approx N.$$

Since the coroner arrived at noon, or 12 o'clock, the murder occurred at about 9:00 A.M.

❬ Quick Check 6

> **Quick Check 6**
>
> **Forensics.** Repeat Example 6, assuming that the coroner arrives at 2 A.M., immediately takes the temperature of the body, and finds it to be 92.8°. She waits 1 hr, takes the temperature again, and finds it to be 90.6°. She also notes that the temperature of the room is 72°. When was the murder committed?

Section Summary

- Several types of functions are additional candidates for curve fitting and applications:

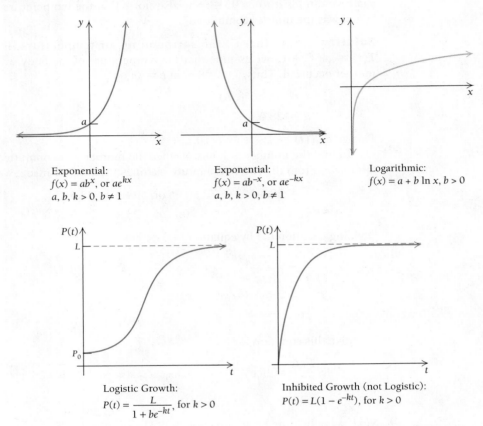

Exponential:
$f(x) = ab^x$, or ae^{kx}
$a, b, k > 0, b \neq 1$

Exponential:
$f(x) = ab^{-x}$, or ae^{-kx}
$a, b, k > 0, b \neq 1$

Logarithmic:
$f(x) = a + b \ln x, b > 0$

Logistic Growth:
$P(t) = \dfrac{L}{1 + be^{-kt}}$, for $k > 0$

Inhibited Growth (not Logistic):
$P(t) = L(1 - e^{-kt})$, for $k > 0$

When we analyze a set of data, we can consider these models, as well as linear, quadratic, polynomial, and rational functions.

EXERCISE SET
3.4

APPLICATIONS

Life and Physical Sciences

1. Radioactive decay. Iodine-131 has a decay rate of 9.6% per day. The rate of change of an amount N of iodine-131 is given by

$$\frac{dN}{dt} = -0.096\,N,$$

where t is the number of days since the decay began.

a) Let N_0 represent the amount of iodine-131 present at $t = 0$. Find the exponential function that models the situation.

b) Suppose that 500 g of iodine-131 is present at $t = 0$. How much will remain after 4 days?

c) After how many days will half of the 500 g of iodine-131 remain?

2. Radioactive decay. Carbon-14 has a decay rate of 0.012097% per year. The rate of change of an amount N of carbon-14 is given by

$$\frac{dN}{dt} = -0.00012097N,$$

where t is the number of years since the decay began.

a) Let N_0 represent the amount of carbon-14 present at $t = 0$. Find the exponential function that models the situation.

b) Suppose 200 g of carbon-14 is present at $t = 0$. How much will remain after 800 yr?

c) After how many years will half of the 200 g of carbon-14 remain?

3. Chemistry. Substance A decomposes at a rate proportional to the amount of A present.

  a) Write an equation relating A to the amount left of an initial amount A_0 after time t.

  b) It is found that 10 lb of A will reduce to 5 lb in 3.3 hr. After how long will there be only 1 lb left?

4. Chemistry. Substance A decomposes at a rate proportional to the amount of A present.

  a) Write an equation relating A to the amount left of an initial amount A_0 after time t.

  b) It is found that 8 g of A will reduce to 4 g in 3 hr. After how long will there be only 1 g left?

Radioactive decay. *For Exercises 5–8, complete the following.*

Radioactive Substance	Decay Rate, k	Half-life, T
5. Polonium-218		3 min
6. Radium-226		1600 yr
7. Lead-210	3.15%/yr	
8. Strontium-90	2.77%/yr	

9. Half-life. Of an initial amount of 1000 g of lead-210, how much will remain after 100 yr? See Exercise 7 for the value of k.

10. Half-life. Of an initial amount of 1000 g of polonium-218, how much will remain after 20 min? See Exercise 5 for the value of k.

11. Carbon dating. How old is an ivory tusk that has lost 40% of its carbon-14?

12. Carbon dating. How old is a piece of wood that has lost 90% of its carbon-14?

13. Cancer treatment. Iodine-125 is often used to treat cancer and has a half-life of 60.1 days. In a sample, the amount of iodine-125 decreased by 25% while in storage. How long was the sample sitting on the shelf?

14. Carbon dating. How old is a Chinese artifact that has lost 60% of its carbon-14?

15. Carbon dating. Recently, while digging in Chaco Canyon, New Mexico, archaeologists found corn pollen that had lost 38.1% of its carbon-14. The age of this corn pollen was evidence that Indians had been cultivating crops in the Southwest centuries earlier than scientists had thought. (*Source: American Anthropologist.*) What was the age of the pollen?

Chaco Canyon, New Mexico

Business and Economics

16. Present value. Following the birth of a child, a parent wants to make an initial investment P_0 that will grow to $30,000 by the child's 20th birthday. Interest is compounded continuously at 6%. What should the initial investment be?

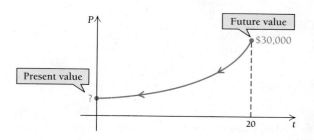

17. Present value. Following the birth of a child, a parent wants to make an initial investment P_0 that will grow to $40,000 by the child's 20th birthday. Interest is compounded continuously at 5.3%. What should the initial investment be?

18. Present value. A homeowner wants to have $15,000 available in 5 yr to pay for new siding. Interest is 4.3%, compounded continuously. How much money should be invested?

19. Sports salaries. An athlete signs a contract that guarantees a $9-million salary 6 yr from now. Assuming that money can be invested at 5.7%, with interest compounded continuously, what is the present value of that year's salary?

20. Actors' salaries. An actor signs a film contract that will pay $12 million when the film is completed 3 yr from now. Assuming that money can be invested at 6.2%, with interest compounded continuously, what is the present value of that payment?

21. Estate planning. A person has a trust fund that will yield $80,000 in 13 yr. A CPA is preparing a financial statement for this client and wants to take into account the present value of the trust fund in computing the client's net worth. Interest is compounded continuously at 4.8%. What is the present value of the trust fund?

22. Supply and demand. The supply and demand for stereos produced by a sound company are given by

$$S(x) = \ln x \quad \text{and} \quad D(x) = \ln \frac{163{,}000}{x},$$

where $S(x)$ is the number of stereos that the company is willing to sell at price x and $D(x)$ is the quantity that the public is willing to buy at price x. Find the equilibrium point. (See Section R.5.)

23. Salvage value. A business estimates that the salvage value $V(t)$, in dollars, of a piece of machinery after t years is given by

$$V(t) = 40{,}000e^{-t}.$$

  a) What did the machinery cost initially?

  b) What is the salvage value after 2 yr?

c) Find the rate of change of the salvage value, and explain its meaning.

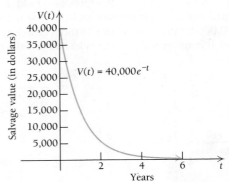

24. Salvage value. A company tracks the value of a particular photocopier over a period of years. The data in the table below show the value of the copier at time t, in years, after the date of purchase.

Time, t (in years)	Salvage Value
0	$34,000
1	22,791
2	15,277
3	10,241
4	6,865
5	4,600
6	3,084

(*Source: International Data Corporation.*)

a) Use REGRESSION to fit an exponential function $y = a \cdot b^x$ to the data. Then convert that formula to $V(t) = V_0 e^{-kt}$, where V_0 is the value when the copier is purchased and t is the time, in years, from the date of purchase. (See the Technology Connection on p. 341.)

b) Estimate the salvage value of the copier after 7 yr; 10 yr.

c) After what amount of time will the salvage value be $1000?

d) After how long will the copier be worth half of its original value?

e) Find the rate of change of the salvage value, and interpret its meaning.

25. Actuarial science. An actuary works for an insurance company and calculates insurance premiums. Given an actual mortality rate (probability of death) for a given age, actuaries sometimes need to project future expected mortality rates of people of that age. An example of a formula that is used to project future mortality rates is

$$Q(t) = (Q_0 - 0.00055)e^{0.163t} + 0.00055,$$

where t is the number of years into the future and Q_0 is the mortality rate when $t = 0$.

a) Suppose the initial actual mortality rate of a group of females aged 25 is 0.014 (14 deaths per 1000). What is the future expected mortality rate of this group of females 3, 5, and 10 yr in the future?

b) Sketch the graph of the mortality function $Q(t)$ for the group in part (a) for $0 \le t \le 10$.

26. Actuarial science. Use the formula from Exercise 25.

a) Suppose the initial actual mortality rate of a group of males aged 25 is 0.023 (23 deaths per 1000). What is the future expected mortality rate of this group of males 3, 5, and 10 yr in the future?

b) Sketch the graph of the mortality function $Q(t)$ for the group in part (a) for $0 \le t \le 10$.

c) What is the ratio of the mortality rate for 25-year-old males 10 yr in the future to that for 25-year-old females 10 yr in the future (Exercise 25a)?

27. U.S. farms. The number N of farms in the United States has declined continually since 1950. In 1950, there were 5,650,000 farms, and in 2005, that number had decreased to 2,100,990. (*Sources*: U.S. Department of Agriculture; National Agricultural Statistics Service.)

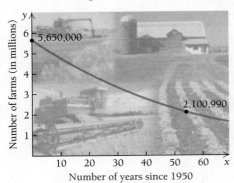

Assuming the number of farms decreased according to the exponential decay model:

a) Find the value of k, and write an exponential function that describes the number of farms after time t, where t is the number of years since 1950.

b) Estimate the number of farms in 2009 and in 2015.

c) At this decay rate, when will only 1,000,000 farms remain?

Social Sciences

28. Forgetting. In an art history class, students took a final exam. They were subsequently retested with an equivalent test at monthly intervals. Their average retest scores t months later are given in the following table.

Time, t (in months)	Score, y
1	84.9%
2	84.6%
3	84.4%
4	84.2%
5	84.1%
6	83.9%

a) Use REGRESSION to fit a logarithmic function $y = a + b \ln x$ to the data.

b) Use the function to predict the average test score after 8 months, 10 months, 24 months, and 36 months.

c) After how long will the test scores fall below 82%?

d) Find the rate of change of the scores, and interpret its meaning.

29. Decline in beef consumption. The annual consumption of beef per person was about 64.6 lb in 2000 and about 61.2 lb in 2008. Assuming that $B(t)$, the annual beef consumption t years after 2000, is decreasing according to the exponential decay model:

a) Find the value of k, and write the equation.
b) Estimate the consumption of beef in 2015.
c) In what year (theoretically) will the consumption of beef be 20 lb per person?

30. Population decrease of Russia. The population of Russia dropped from 150 million in 1995 to 138 million in 2012. (*Source*: CIA–*The World Factbook*.) Assume that $P(t)$, the population, in millions, t years after 1995, is decreasing according to the exponential decay model.

a) Find the value of k, and write the equation.
b) Estimate the population of Russia in 2016.
c) When will the population of Russia be 100 million?

31. Population decrease of Ukraine. The population of Ukraine dropped from 51.9 million in 1995 to 44.9 million in 2012. (*Source*: CIA–*The World Factbook*.) Assume that $P(t)$, the population, in millions, t years after 1995, is decreasing according to the exponential decay model.

a) Find the value of k, and write the equation.
b) Estimate the population of Ukraine in 2015.
c) After how many years will the population of Ukraine be 1 million, according to this model?

Life and Natural Sciences

32. Cooling. After warming the water in a hot tub to $100°$, the heating element fails. The surrounding air temperature is $40°$, and in 5 min the water temperature drops to $95°$.

a) Find the value of the constant a in Newton's Law of Cooling.
b) Find the value of the constant k. Round to five decimal places.
c) What is the water temperature after 10 min?
d) How long does it take the water to cool to $41°$?
e) Find the rate of change of the water temperature, and interpret its meaning.

33. Cooling. The temperature in a whirlpool bath is $102°$, and the room temperature is $75°$. The water cools to $90°$ in 10 min.

a) Find the value of the constant a in Newton's Law of Cooling.
b) Find the value of the constant k. Round to five decimal places.
c) What is the water temperature after 20 min?
d) How long does it take the water to cool to $80°$?
e) Find the rate of change of the water temperature, and interpret its meaning.

34. Forensics. A coroner arrives at a murder scene at 2 A.M. He takes the temperature of the body and finds it to be $61.6°$. He waits 1 hr, takes the temperature again, and finds it to be $57.2°$. The body is in a meat freezer, where the temperature is $10°$. When was the murder committed?

35. Forensics. A coroner arrives at a murder scene at 11 P.M. She finds the temperature of the body to be $85.9°$. She waits 1 hr, takes the temperature again, and finds it to be $83.4°$. She notes that the room temperature is $60°$. When was the murder committed?

36. Prisoner-of-war protest. The initial weight of a prisoner of war is 140 lb. To protest the conditions of her imprisonment, she begins a fast. Her weight t days after her last meal is approximated by

$$W = 140e^{-0.009t}.$$

a) How much does the prisoner weigh after 25 days?
b) At what rate is the prisoner's weight changing after 25 days?

37. Political protest. A monk weighing 170 lb begins a fast to protest a war. His weight after t days is given by

$$W = 170e^{-0.008t}.$$

a) When the war ends 20 days later, how much does the monk weigh?
b) At what rate is the monk losing weight after 20 days (before any food is consumed)?

38. Atmospheric pressure. Atmospheric pressure P at altitude a is given by

$$P = P_0e^{-0.00005a},$$

where P_0 is the pressure at sea level. Assume that $P_0 = 14.7 \text{ lb/in}^2$ (pounds per square inch).

a) Find the pressure at an altitude of 1000 ft.
b) Find the pressure at an altitude of 20,000 ft.
c) At what altitude is the pressure 14.7 lb/in^2?
d) Find the rate of change of the pressure, and interpret its meaning.

39. Satellite power. The power supply of a satellite is a radioisotope (radioactive substance). The power output P, in watts (W), decreases at a rate proportional to the amount present; P is given by

$$P = 50e^{-0.004t},$$

where t is the time, in days.

a) How much power will be available after 375 days?
b) What is the half-life of the power supply?
c) The satellite's equipment cannot operate on fewer than 10 W of power. How long can the satellite stay in operation?
d) How much power did the satellite have to begin with?
e) Find the rate of change of the power output, and interpret its meaning.

40. Cases of tuberculosis. The number of cases N of tuberculosis in the United States has decreased continually since 1956, as shown in the following graph. In 1956 ($t = 0$), there were 69,895 cases. By 2006 ($t = 50$), this number had decreased by over 80%, to 13,767 cases.

CASES OF TUBERCULOSIS IN THE UNITED STATES

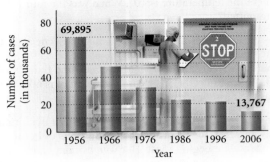

(*Source:* Centers for Disease Control and Prevention.)

a) Find the value of k, and write an exponential function that describes the number of tuberculosis cases after time t, where t is the number of years since 1956.
b) Estimate the number of cases in 2012 and in 2020.
c) At this decay rate, in what year will there be 5000 cases?

Modeling

For each of the scatterplots in Exercises 41–50, determine which, if any, of these functions might be used as a model for the data:

a) Quadratic: $f(x) = ax^2 + bx + c$
b) Polynomial, not quadratic
c) Exponential: $f(x) = ae^{kx}, k > 0$
d) Exponential: $f(x) = ae^{-kx}, k > 0$
e) Logarithmic: $f(x) = a + b \ln x$
f) Logistic: $f(x) = \dfrac{a}{1 + be^{-kx}}$

41.

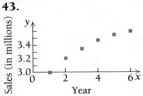

42.

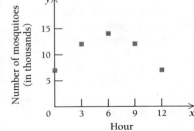

43.

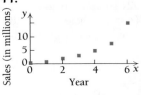

44.

45.

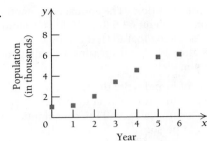

46.

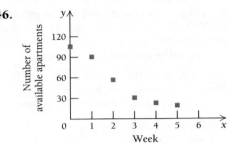

47.

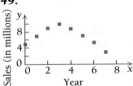

48.

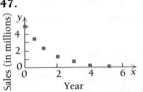

49.

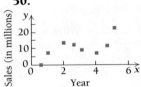

50.

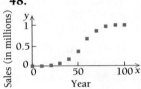

SYNTHESIS

51. Economics: supply and demand. The demand, $D(x)$, and supply, $S(x)$, functions for a certain type of multi-purpose printer are as follows:

$$D(x) = q = 480e^{-0.003x}$$

and

$$S(x) = q = 150e^{0.004x}.$$

Find the equilibrium point. Assume that x is the price in dollars.

The Beer–Lambert Law. **A beam of light enters a medium such as water or smoky air with initial intensity I_0. Its intensity is decreased depending on the thickness (or concentration) of the medium. The intensity I at a depth (or concentration) of x units is given by**

$$I = I_0 e^{-\mu x}.$$

The constant μ ("mu"), called the coefficient of absorption, varies with the medium. Use this law for Exercises 52 and 53.

52. Light through smog. Concentrations of particulates in the air due to pollution reduce sunlight. In a smoggy area, $\mu = 0.01$ and x is the concentration of particulates measured in micrograms per cubic meter (mcg/m³).

What change is more significant—dropping pollution levels from 100 mcg/m³ to 90 mcg/m³ or dropping them from 60 mcg/m³ to 50 mcg/m³? Why?

53. Light through sea water. Sea water has $\mu = 1.4$ and x is measured in meters. What would increase cloudiness more—dropping x from 2 m to 5 m or dropping x from 7 m to 10 m? Explain.

54. Newton's Law of Cooling. Consider the following exploratory situation. Fill a glass with hot tap water. Place a thermometer in the glass and measure the temperature. Check the temperature every 30 min thereafter. Plot your data on this graph, and connect the points with a smooth curve.

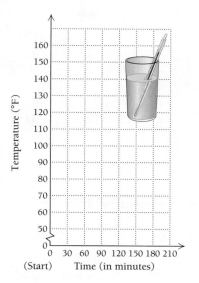

a) What was the temperature of the water when you began?

b) At what temperature does there seem to be a leveling off of the graph?

c) What is the difference between your answers to parts (a) and (b)?

d) How does the water temperature in part (b) compare with the room temperature?

e) Find an equation that fits the data. Use this equation to check values of other data points. How do they compare?

f) Is it ever "theoretically" possible for the temperature of the water to be the same as the room temperature? Explain.

g) Find the rate of change of the temperature, and interpret its meaning.

55. An interest rate decreases from 8% to 7.2%. Explain why this increases the present value of an amount due 10 yr later.

Answers to Quick Checks

1. **(a)** $N(t) = N_0 e^{-0.14t}$;   **(b)** 246.6 g;   **(c)** 4.95 days
2. **(a)** 30.1 yr;   **(b)** 5.3% per day   **3.** 7574 yr
4. $4493.29   **5.** 11.7 min   **6.** $N = 2.2$ hr, so the murder was committed 2 hr and 12 min before the coroner arrived, at about 11:48 P.M. on the previous day.

3.5

The Derivatives of a^x and $\log_a x$

The Derivative of a^x

To find the derivative of a^x, for any base a, we first express a^x as a power of e. To do this, we recall that $\log_b x$ is the power to which b is raised in order to get x. Thus,

$$b^{\log_b x} = x.$$

In particular, it follows that

$$e^{\log_e A} = A, \quad \text{or } e^{\ln A} = A.$$

If we replace A with a^x, we have

$$e^{\ln a^x} = a^x, \quad \text{or } a^x = e^{\ln a^x}. \tag{1}$$

To find the derivative of a^x, we differentiate both sides:

$$\frac{d}{dx} a^x = \frac{d}{dx} e^{\ln a^x}$$

$$= \frac{d}{dx} e^{x \ln a} \qquad \text{Using a property of logarithms}$$

$$= \frac{d}{dx} e^{(\ln a)x}$$

$$= e^{(\ln a)x} \cdot \ln a \qquad \text{Differentiating } e^{kx} \text{ with respect to } x$$

$$= e^{\ln a^x} \cdot \ln a \qquad \text{Using a property of logarithms}$$

$$= a^x \cdot \ln a. \qquad \text{Using equation (1)}$$

Thus, we have the following theorem.

THEOREM 12

$$\frac{d}{dx} a^x = (\ln a)a^x$$

■ **EXAMPLE 1** Differentiate: **a)** $y = 2^x$; **b)** $y = (1.4)^x$; **c)** $f(x) = 3^{2x}$.

Solution

a) $\dfrac{d}{dx} 2^x = (\ln 2)2^x$ \qquad Using Theorem 12

Note that $\ln 2 \approx 0.7$, so this equation verifies our earlier approximation of the derivative of 2^x as $(0.7)2^x$ in Section 3.1.

b) $\dfrac{d}{dx} (1.4)^x = (\ln 1.4)(1.4)^x$

c) Since $f(x) = 3^{2x}$ is of the form $f(x) = 3^{g(x)}$, the Chain Rule applies:

$$f'(x) = (\ln 3)3^{2x} \cdot \frac{d}{dx}(2x)$$

$$= \ln 3 \cdot 3^{2x} \cdot 2 = 2 \ln 3 \cdot 3^{2x}.$$

❬ Quick Check 1

Compare these formulas:

$$\frac{d}{dx} a^x = (\ln a)a^x \quad \text{and} \quad \frac{d}{dx} e^x = e^x.$$

The simplicity of the latter formula is a reason for the use of base e in calculus. The many applications of e in natural phenomena provide additional reasons.

One other result also follows from what we have done. If

$$f(x) = a^x,$$

we now know that

$$f'(x) = a^x(\ln a).$$

Alternatively, in Section 3.1, we showed that if $f(x) = a^x$, then

$$f'(x) = a^x \cdot \lim_{h \to 0} \frac{a^h - 1}{h}.$$

Thus,

$$a^x(\ln a) = a^x \cdot \lim_{h \to 0} \frac{a^h - 1}{h}.$$

Dividing both sides by a^x, we have the following.

THEOREM 13

$$\ln a = \lim_{h \to 0} \frac{a^h - 1}{h}$$

The Derivative of $\log_a x$

Just as the derivative of a^x is expressed in terms of $\ln a$, so too is the derivative of $\log_a x$. To find this derivative, we first express $\log_a x$ in terms of $\ln a$ using the change-of-base formula (P7 of Theorem 3) from Section 3.2:

$$\frac{d}{dx} \log_a x = \frac{d}{dx} \left(\frac{\log_e x}{\log_e a} \right) \qquad \text{Using the change-of-base formula}$$

$$= \frac{d}{dx} \left(\frac{\ln x}{\ln a} \right)$$

$$= \frac{1}{\ln a} \cdot \frac{d}{dx} (\ln x) \qquad \frac{1}{\ln a} \text{ is a constant.}$$

$$= \frac{1}{\ln a} \cdot \frac{1}{x}.$$

THEOREM 14

$$\frac{d}{dx} \log_a x = \frac{1}{\ln a} \cdot \frac{1}{x}$$

Comparing this equation with

$$\frac{d}{dx} \ln x = \frac{1}{x},$$

we see another reason for the use of base e in calculus: we avoid obtaining the constant $1/(\ln a)$ when taking the derivative.

■ **EXAMPLE 2**   Differentiate: **a)** $y = \log_8 x$;   **b)** $y = \log x$;
c) $f(x) = \log_3 (x^2 + 1)$;   **d)** $f(x) = x^3 \log_5 x$.

Solution

a) $\dfrac{d}{dx} \log_8 x = \dfrac{1}{\ln 8} \cdot \dfrac{1}{x}$ Using Theorem 14

b) $\dfrac{d}{dx} \log x = \log_{10} x$ $\log x$ means $\log_{10} x$.

$\qquad = \dfrac{1}{\ln 10} \cdot \dfrac{1}{x}$

c) Note that $f(x) = \log_3 (x^2 + 1)$ is of the form $f(x) = \log_3 (g(x))$, so the Chain Rule is required:

$$f'(x) = \frac{1}{\ln 3} \cdot \frac{1}{x^2 + 1} \cdot \frac{d}{dx}(x^2 + 1) \quad \text{Using the Chain Rule}$$

$$= \frac{1}{\ln 3} \cdot \frac{1}{x^2 + 1} \cdot 2x$$

$$= \frac{2x}{(\ln 3)(x^2 + 1)}.$$

d) Since $f(x) = x^3 \log_5 x$ is of the form $f(x) = g(x) \cdot h(x)$, the Product Rule is applied:

$$f'(x) = x^3 \cdot \frac{d}{dx} \log_5 x + \log_5 x \cdot 3x^2 \quad \text{Using the Product Rule}$$

$$= x^3 \cdot \frac{1}{\ln 5} \cdot \frac{1}{x} + \log_5 x \cdot 3x^2$$

$$= \frac{x^2}{\ln 5} + 3x^2 \log_5 x, \quad \text{or} \quad x^2 \left(\frac{1}{\ln 5} + 3 \log_5 x \right).$$

❰ *Quick Check 2*

TECHNOLOGY CONNECTION

Exploratory

Using the nDeriv feature, check the results of Examples 1 and 2 graphically. Then differentiate $y = \log_2 x$, and check the result with your calculator.

❭ **Quick Check 2**

Differentiate:

a) $y = \log_2 x$;
b) $f(x) = -7 \log x$;
c) $g(x) = x^6 \log x$;
d) $y = \log_8 (x^3 - 7)$.

Section Summary

- The following rules apply when we differentiate exponential and logarithmic functions whose bases are positive but not the number e:

$$\frac{d}{dx} a^x = (\ln a)a^x, \quad \text{and} \quad \frac{d}{dx} \log_a x = \frac{1}{\ln a} \cdot \frac{1}{x}.$$

EXERCISE SET
3.5

Differentiate.

1. $y = 7^x$

2. $y = 6^x$

3. $f(x) = 8^x$

4. $f(x) = 15^x$

5. $g(x) = x^3(5.4)^x$

6. $g(x) = x^5(3.7)^x$

7. $y = 7^{x^4+2}$

8. $y = 4^{x^2+5}$

9. $y = e^{8x}$

10. $y = e^{x^2}$

11. $f(x) = 3^{x^4+1}$

12. $f(x) = 12^{7x-4}$

13. $y = \log_4 x$

14. $y = \log_8 x$

15. $y = \log_{17} x$

16. $y = \log_{23} x$

17. $g(x) = \log_6 (5x + 1)$

18. $g(x) = \log_{32} (9x - 2)$

19. $F(x) = \log (6x - 7)$

20. $G(x) = \log (5x + 4)$

21. $y = \log_8 (x^3 + x)$

22. $y = \log_9 (x^4 - x)$

23. $f(x) = 4 \log_7 \left(\sqrt{x} - 2 \right)$

24. $g(x) = -\log_6 \left(\sqrt[3]{x} + 5 \right)$

25. $y = 6^x \cdot \log_7 x$

26. $y = 5^x \cdot \log_2 x$

27. $G(x) = (\log_{12} x)^5$

28. $F(x) = (\log_9 x)^7$

29. $g(x) = \dfrac{7^x}{4x + 1}$

30. $f(x) = \dfrac{6^x}{5x - 1}$

31. $y = 5^{2x^3 - 1} \cdot \log (6x + 5)$

32. $y = \log (7x + 3) \cdot 4^{2x^4 + 8}$

33. $F(x) = 7^x \cdot (\log_4 x)^9$

34. $G(x) = \log_9 x \cdot (4^x)^6$

35. $f(x) = (3x^5 + x)^5 \log_3 x$

36. $g(x) = \sqrt{x^3 - x} \, (\log_5 x)$

APPLICATIONS

Business and Economics

37. Double declining balance depreciation. An office machine is purchased for $5200. Under certain assumptions, its salvage value, V, in dollars, is depreciated according to a method called *double declining balance*, by basically 80% each year, and is given by

$$V(t) = 5200(0.80)^t,$$

where t is the time, in years, after purchase.

  a) Find $V'(t)$.

  b) Interpret the meaning of $V'(t)$.

38. Recycling aluminum cans. It is known that 45% of all aluminum cans distributed will be recycled each year. A beverage company uses 250,000 lb of aluminum cans. After recycling, the amount of aluminum, in pounds, still in use after t years is given by

$$N(t) = 250,000(0.45)^t.$$

(*Source*: The Container Recycling Institute.)

  a) Find $N'(t)$.

  b) Interpret the meaning of $N'(t)$.

39. Household liability. The total financial liability, in billions of dollars, of U.S. households can be modeled by the function

$$L(t) = 1547(1.083)^t,$$

where t is the number of years after 1980. The graph of this function follows.

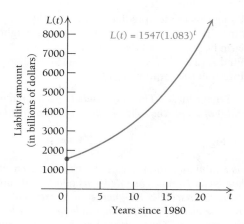

  a) Using this model, predict the total financial liability of U.S. households in 2012.

  b) Find $L'(25)$.

  c) Interpret the meaning of $L'(25)$.

40. Small business. The number of nonfarm proprietorships, in thousands, in the United States can be modeled by the function

$$N(t) = 8400 \ln t - 10{,}500.$$

where t is the number of years after 1970. The graph of this function is given below.

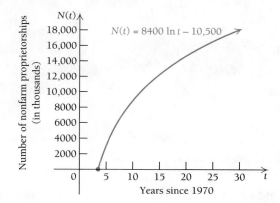

  a) Using this model, predict the number of nonfarm proprietorships in the United States in 2014.

  b) Find $N'(45)$.

  c) Interpret the meaning of $N'(45)$.

Life and Physical Sciences

41. Agriculture. Farmers wishing to avoid the use of genetically modified (GMO) seeds are increasingly concerned about inadvertently growing GMO plants as a result of pollen drifting from nearby farms. Assuming that these farmers raise their own seeds, the fractional portion of their crop that remains free of GMO plants t years later can be approximated by

$$P(t) = (0.98)^t.$$

a) Using this model, predict the fractional portion of the crop that will be GMO-free 10 yr after a neighboring farm begins to use GMO seeds.

b) Find $P'(15)$.

c) Interpret the meaning of $P'(15)$.

Earthquake magnitude. *The magnitude R (measured on the Richter scale) of an earthquake of intensity I is defined as*

$$R = \log \frac{I}{I_0},$$

where I_0 is a minimum intensity used for comparison. When one earthquake is 10 times as intense as another, its magnitude on the Richter scale is 1 higher. If one earthquake is 100 times as intense as another, its magnitude on the Richter scale is 2 higher, and so on. Thus, an earthquake whose magnitude is 6 on the Richter scale is 10 times as intense as an earthquake whose magnitude is 5. Earthquake intensities can be interpreted as multiples of the minimum intensity I_0. Use this information for Exercises 42 and 43.

42. On January 12, 2010, a devastating earthquake struck the Caribbean nation of Haiti. It had an intensity of $I_0 10^7$. What was its magnitude on the Richter scale?

This photograph shows part of the damage in Haiti due to the earthquake of January 2010.

43. On March 11, 2011, an earthquake more intense than the one in Haiti struck the nation of Japan. It had an intensity of $I_0 10^{9.0}$. What was its magnitude on the Richter scale?

The location of the March 2011 earthquake in Japan.

44. **Earthquake intensity.** The intensity of an earthquake is given by

$$I = I_0 10^R,$$

where R is the magnitude on the Richter scale and I_0 is the minimum intensity, at which $R = 0$, used for comparison.

a) Find I, in terms of I_0, for an earthquake of magnitude 7 on the Richter scale.

b) Find I, in terms of I_0, for an earthquake of magnitude 8 on the Richter scale.

c) Compare your answers to parts (a) and (b).

d) Find the rate of change dI/dR.

e) Interpret the meaning of dI/dR.

45. **Intensity of sound.** The intensity of a sound is given by

$$I = I_0 10^{0.1L},$$

where L is the loudness of the sound as measured in decibels and I_0 is the minimum intensity detectable by the human ear.

a) Find I, in terms of I_0, for the loudness of a power mower, which is 100 decibels.

b) Find I, in terms of I_0, for the loudness of a just audible sound, which is 10 decibels.

c) Compare your answers to parts (a) and (b).

d) Find the rate of change dI/dL.

e) Interpret the meaning of dI/dL.

46. **Earthquake magnitude.** The magnitude R (measured on the Richter scale) of an earthquake of intensity I is defined as

$$R = \log \frac{I}{I_0},$$

where I_0 is the minimum intensity (used for comparison). (The exponential form of this definition is given in Exercise 44.)

a) Find the rate of change dR/dI.

b) Interpret the meaning of dR/dI.

47. **Loudness of sound.** The loudness L of a sound of intensity I is defined as

$$L = 10 \log \frac{I}{I_0},$$

where I_0 is the minimum intensity detectable by the human ear and L is the loudness measured in decibels. (The exponential form of this definition is given in Exercise 45.)

a) Find the rate of change dL/dI.

b) Interpret the meaning of dL/dI.

48. **Response to drug dosage.** The response y to a dosage x of a drug is given by

$$y = m \log x + b,$$

where m and b are constants. The response may be hard to measure with a number. The patient might perspire more, have an increase in temperature, or faint.

a) Find the rate of change dy/dx.

b) Interpret the meaning of dy/dx.

SYNTHESIS

49. Find $\lim\limits_{h \to 0} \dfrac{3^h - 1}{h}$. (*Hint*: See p. 367.)

Use the Chain Rule, implicit differentiation, and other techniques to differentiate each function given in Exercises 50–57.

50. $f(x) = 3^{(2^x)}$

51. $y = 2^{x^4}$

52. $y = x^x$, for $x > 0$

53. $y = \log_3 (\log x)$

54. $f(x) = x^{e^x}$, for $x > 0$

55. $y = a^{f(x)}$

56. $y = \log_a f(x)$, for $f(x)$ positive

57. $y = [f(x)]^{g(x)}$, for $f(x)$ positive

58. In your own words, derive the formula for finding the derivative of $f(x) = a^x$.

59. In your own words, derive the formula for finding the derivative of $f(x) = \log_a x$.

Answers to Quick Checks

1. (a) $(\ln 5)5^x$; **(b)** $(\ln 4)4^x$; **(c)** $(\ln 4.3)(4.3)^x$

2. (a) $\dfrac{1}{\ln 2} \cdot \dfrac{1}{x}$; **(b)** $\dfrac{-7}{\ln 10} \cdot \dfrac{1}{x}$;

(c) $x^5 \left(\dfrac{1}{\ln 10} + 6 \log x \right)$; **(d)** $\dfrac{3x^2}{(\ln 8)(x^3 - 7)}$

3.6

An Economics Application: Elasticity of Demand

OBJECTIVES

- Find the elasticity of a demand function.
- Find the maximum of a total-revenue function.
- Characterize demand in terms of elasticity.

Retailers and manufacturers often need to know how a small change in price will affect the demand for a product. If a small increase in price produces no change in demand, a price increase may make sense; if a small increase in price creates a large drop in demand, the increase is probably ill advised. To measure the sensitivity of demand to a small percent increase in price, economists calculate the *elasticity of demand*.

Suppose that q represents a quantity of goods purchased and x is the price per unit of the goods. Recall that q and x are related by the demand function

$$q = D(x).$$

Suppose that there is a change Δx in the price per unit. The percent change in price is given by

$$\frac{\Delta x}{x} = \frac{\Delta x}{x} \cdot \frac{100}{100} = \frac{\Delta x \cdot 100}{x}\%.$$

A change in the price produces a change Δq in the quantity sold. The percent change in quantity is given by

$$\frac{\Delta q}{q} = \frac{\Delta q \cdot 100}{q}\%.$$

The ratio of the percent change in quantity to the percent change in price is

$$\frac{\Delta q / q}{\Delta x / x},$$

which can be expressed as

$$\frac{x}{q} \cdot \frac{\Delta q}{\Delta x}. \qquad (1)$$

Note that for differentiable functions,

$$\lim_{\Delta x \to 0} \frac{\Delta q}{\Delta x} = \frac{dq}{dx},$$

so the limit as Δx approaches 0 of the expression in equation (1) becomes

$$\lim_{\Delta x \to 0} \frac{x}{q} \cdot \frac{\Delta q}{\Delta x} = \frac{x}{q} \cdot \frac{dq}{dx} = \frac{x}{q} \cdot D'(x) = \frac{x}{D(x)} \cdot D'(x).$$

This result is the basis of the following definition.

DEFINITION

The **elasticity of demand** E is given as a function of price x by

$$E(x) = -\frac{x \cdot D'(x)}{D(x)}.$$

To understand the purpose of the negative sign in the preceding definition, note that the price, x, and the demand, $D(x)$, are both nonnegative. Since $D(x)$ is normally decreasing, $D'(x)$ is usually negative. By inserting a negative sign in the definition, economists make $E(x)$ nonnegative and easier to work with.

■ **EXAMPLE 1** **Economics: Demand for DVD Rentals.** Klix Video has found that demand for rentals of its DVDs is given by

$$q = D(x) = 120 - 20x,$$

where q is the number of DVDs rented per day at x dollars per rental. Find each of the following.

a) The quantity demanded when the price is $2 per rental

b) The elasticity as a function of x

c) The elasticity at $x = 2$ and at $x = 4$. Interpret the meaning of these values of the elasticity.

d) The value of x for which $E(x) = 1$. Interpret the meaning of this price.

e) The total-revenue function, $R(x) = x \cdot D(x)$

f) The price x at which total revenue is a maximum

Solution

a) For $x = 2$, we have $D(2) = 120 - 20(2) = 80$. Thus, 80 DVDs per day will be rented at a price of $2 per rental.

b) To find the elasticity, we first find the derivative $D'(x)$:

$$D'(x) = -20.$$

Then we substitute -20 for $D'(x)$ and $120 - 20x$ for $D(x)$ in the expression for elasticity:

$$E(x) = -\frac{x \cdot D'(x)}{D(x)} = -\frac{x \cdot (-20)}{120 - 20x} = \frac{20x}{120 - 20x} = \frac{x}{6 - x}.$$

c) $E(2) = \dfrac{2}{6 - 2} = \dfrac{1}{2}$

At $x = 2$, the elasticity is $\frac{1}{2}$, which is less than 1. Thus, the ratio of the percent change in quantity to the percent change in price is less than 1. A small percentage increase in price will cause an even smaller percentage decrease in the quantity sold.

$$E(4) = \frac{4}{6 - 4} = 2$$

At $x = 4$, the elasticity is 2, which is greater than 1. Thus, the ratio of the percent change in quantity to the percent change in price is greater than 1. A small percentage increase in price will cause a larger percentage decrease in the quantity sold.

Economics: Demand for In-Ear Radios

A company determines that the demand function for in-ear radios is

$$q = D(x) = 300 - x,$$

where q is the number sold per day when the price is x dollars per radio.

EXERCISES

1. Find the elasticity E and the total revenue R.

2. Using only the first quadrant, graph the demand, elasticity, and total-revenue functions on the same set of axes.

3. Find the price x for which the total revenue is a maximum. Use the method of Example 1. Check the answer graphically.

d) We set $E(x) = 1$ and solve for p:

$$\frac{x}{6 - x} = 1$$

$$x = 6 - x \qquad \text{We multiply both sides by } 6 - x, \text{ assuming that } x \neq 6.$$

$$2x = 6$$

$$x = 3.$$

Thus, when the price is $3 per rental, the ratio of the percent change in quantity to the percent change in price is 1.

e) Recall that the total revenue $R(x)$ is given by $x \cdot D(x)$. Then

$$R(x) = x \cdot D(x) = x(120 - 20x) = 120x - 20x^2.$$

f) To find the price x that maximizes total revenue, we find $R'(x)$:

$$R'(x) = 120 - 40x.$$

We see that $R'(x)$ exists for all x in the interval $[0, \infty)$. Thus, we solve:

$$R'(x) = 120 - 40x = 0$$

$$-40x = -120$$

$$x = 3.$$

Since there is only one critical value, we can try to use the second derivative to see if we have a maximum:

$$R''(x) = -40 < 0.$$

Thus, $R''(3)$ is negative, so $R(3)$ is a maximum. That is, total revenue is a maximum at $3 per rental.

Note in parts (d) and (f) of Example 1 that the value of x for which $E(x) = 1$ is the same as the value of x for which total revenue is a maximum. The following theorem states that this is always the case.

THEOREM 15

Total revenue is increasing at those x-values for which $E(x) < 1$.

Total revenue is decreasing at those x-values for which $E(x) > 1$.

Total revenue is maximized at the value(s) of x for which $E(x) = 1$.

Proof. We know that

$$R(x) = x \cdot D(x),$$

so $\qquad R'(x) = x \cdot D'(x) + D(x) \cdot 1 \qquad$ Using the Product Rule

$$= D(x)\left[\frac{x \cdot D'(x)}{D(x)} + 1\right] \qquad \text{Check this by multiplying.}$$

$$= D(x)[-E(x) + 1]$$

$$= D(x)[1 - E(x)].$$

(continued)

Since we can assume that $D(x) > 0$, it follows that $R'(x)$ is positive for $E(x) < 1$, is negative for $E(x) > 1$, and is 0 when $E(x) = 1$. Thus, total revenue is increasing for $E(x) < 1$, is decreasing for $E(x) > 1$, and is maximized when $E(x) = 1$. ∎

Elasticity and Revenue

For a particular value of the price x:

1. The demand is *inelastic* if $E(x) < 1$. An increase in price will bring an increase in revenue. If demand is inelastic, then revenue is increasing.

2. The demand has *unit elasticity* if $E(x) = 1$. The demand has unit elasticity when revenue is at a maximum.

3. The demand is *elastic* if $E(x) > 1$. An increase in price will bring a decrease in revenue. If demand is elastic, then revenue is decreasing.

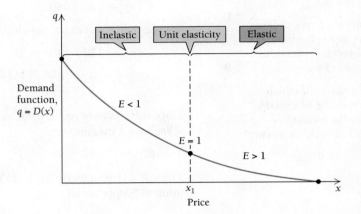

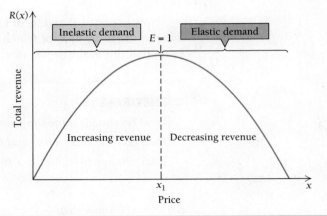

In summary, suppose that Klix Video in Example 1 raises the price per rental and that the total revenue increases. Then we say the demand is *inelastic*. If the total revenue decreases, we say the demand is *elastic*. Some price elasticities in the U.S. economy are listed in the following table.

Price Elasticities in the U.S. Economy

Industry	Elasticity
Elastic Demands	
Metals	1.52
Electrical engineering products	1.39
Mechanical engineering products	1.30
Furniture	1.26
Motor vehicles	1.14
Instrument engineering products	1.10
Professional services	1.09
Transportation services	1.03
Inelastic Demands	
Gas, electricity, and water	0.92
Oil	0.91
Chemicals	0.89
Beverages (all types)	0.78
Tobacco	0.61
Food	0.58
Banking and insurance services	0.56
Housing services	0.55
Clothing	0.49
Agricultural and fish products	0.42
Books, magazines, and newspapers	0.34
Coal	0.32

(*Source*: Ahsan Mansur and John Whalley, "Numerical specification of applied general equilibrium models: Estimation, calibration, and data." In H. E. Scarf and J. B. Shoven (eds.), *Applied General Equilibrium Analysis*. (New York: Cambridge University Press, 1984), p. 109.)

> **Quick Check 1**

Economics: Demand for DVD Rentals. Internet rentals affect Klix Video in such a way that the demand for rentals of its DVDs changes to

$$q = D(x) = 30 - 5x.$$

a) Find the quantity demanded when the price is $2 per rental, $3 per rental, and $5 per rental.

b) Find the elasticity of demand as a function of x.

c) Find the elasticity at $x = 2$, $x = 3$, and $x = 5$. Interpret the meaning of these values.

d) Find the value of x for which $E(x) = 1$. Interpret the meaning of this price.

e) Find the total-revenue function, $R(x) = x \cdot D(x)$.

f) Find the price x at which total revenue is a maximum.

❮ Quick Check 1

Section Summary

• The *elasticity of demand E* is given as a function of price x by

$$E(x) = -\frac{x \cdot D'(x)}{D(x)}.$$

Elasticity provides a means of evaluating the change in revenue that results from an increase in price.

For the demand function given in each of Exercises 1–12, find the following.

   a) *The elasticity*
   b) *The elasticity at the given price, stating whether the demand is elastic or inelastic*
   c) *The value(s) of x for which total revenue is a maximum (assume that x is in dollars)*

1. $q = D(x) = 400 - x$; $x = 125$

2. $q = D(x) = 500 - x$; $x = 38$

3. $q = D(x) = 200 - 4x$; $x = 46$

4. $q = D(x) = 500 - 2x$; $x = 57$

5. $q = D(x) = \dfrac{400}{x}$; $x = 50$

6. $q = D(x) = \dfrac{3000}{x}$; $x = 60$

7. $q = D(x) = \sqrt{600 - x}$; $x = 100$

8. $q = D(x) = \sqrt{300 - x}$; $x = 250$

9. $q = D(x) = 100e^{-0.25x}$; $x = 10$

10. $q = D(x) = 200e^{-0.05x}$; $x = 80$

11. $q = D(x) = \dfrac{100}{(x + 3)^2}$; $x = 1$

12. $q = D(x) = \dfrac{500}{(2x + 12)^2}$; $x = 8$

APPLICATIONS

Business and Economics

13. Demand for chocolate chip cookies. Good Times Bakers works out a demand function for its chocolate chip cookies and finds it to be

   $q = D(x) = 967 - 25x$,

   where q is the quantity of cookies sold when the price per cookie, in cents, is x.

   a) Find the elasticity.
   b) At what price is the elasticity of demand equal to 1?
   c) At what prices is the elasticity of demand elastic?
   d) At what prices is the elasticity of demand inelastic?
   e) At what price is the revenue a maximum?
   f) At a price of 20¢ per cookie, will a small increase in price cause the total revenue to increase or decrease?

14. Demand for oil. Suppose that you have been hired as an economic consultant concerning the world demand for oil. The demand function is

   $q = D(x) = 63{,}000 + 50x - 25x^2$, $0 \le x \le 50$,

where q is measured in millions of barrels of oil per day at a price of x dollars per barrel.

   a) Find the elasticity.
   b) Find the elasticity at a price of $10 per barrel, stating whether the demand is elastic or inelastic at that price.
   c) Find the elasticity at a price of $20 per barrel, stating whether the demand is elastic or inelastic at that price.
   d) Find the elasticity at a price of $30 per barrel, stating whether the demand is elastic or inelastic at that price.
   e) At what price is the revenue a maximum?
   f) What quantity of oil will be sold at the price that maximizes revenue? Compare the current world price to your answer.
   g) At a price of $30 per barrel, will a small increase in price cause the total revenue to increase or decrease?

15. Demand for computer games. High Wire Electronics determines the following demand function for a new game:

   $q = D(x) = \sqrt{200 - x^3}$,

   where q is the number of games sold per day when the price is x dollars per game.

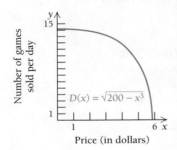

   a) Find the elasticity.
   b) Find the elasticity when $x = 3$.
   c) At $x = 3$, will a small increase in price cause the total revenue to increase or decrease?

16. Demand for tomato plants. Sunshine Gardens determines the following demand function during early summer for tomato plants:

   $q = D(x) = \dfrac{2x + 300}{10x + 11}$,

   where q is the number of plants sold per day when the price is x dollars per plant.

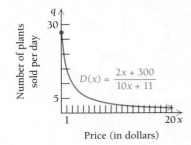

a) Find the elasticity.
b) Find the elasticity when $x = 3$.
c) At \$3 per plant, will a small increase in price cause the total revenue to increase or decrease?

SYNTHESIS

17. Economics: constant elasticity curve.

a) Find the elasticity of the demand function

$$q = D(x) = \frac{k}{x^n},$$

where k is a positive constant and n is an integer greater than 0.

b) Is the value of the elasticity dependent on the price per unit?

c) Does the total revenue have a maximum? When?

18. Economics: exponential demand curve.

a) Find the elasticity of the demand function

$$q = D(x) = Ae^{-kx},$$

where A and k are positive constants.

b) Is the value of the elasticity dependent on the price per unit?

c) Does the total revenue have a maximum? At what value of x?

19. Let

$$L(x) = \ln D(x).$$

Describe the elasticity in terms of $L'(x)$.

20. Explain in your own words the concept of elasticity and its usefulness to economists. Do some library or online research or consult an economist in order to determine when and how this concept was first developed.

21. Explain how the elasticity of demand for a product can be affected by the availability of substitutes for the product.

Answers to Quick Checks

1. (a) 20, 15, 5; **(b)** $E(x) = \dfrac{x}{6 - x}$;
(c) 0.5, 1, 5 (see Example 1 for the interpretations);
(d) \$3; **(e)** $R(x) = 30x - 5x^2$; **(f)** \$3

KEY TERMS AND CONCEPTS

EXAMPLES

SECTION 3.1

An **exponential function** f is a function of the form $f(x) = a^x$, where x is any real number and a, the **base**, is any positive number other than 1.

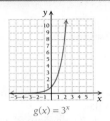

$f(x) = \left(\frac{1}{2}\right)^x \qquad g(x) = 3^x$

The natural base e is such that

$$\frac{d}{dx}e^x = e^x,$$

where $e \approx 2.718$ and $e = \lim_{h \to 0}(1 + h)^{1/h}$

and $\dfrac{d}{dx}e^{f(x)} = f'(x)e^x.$

To differentiate a function like $f(x) = (5x^2 + 3)e^{7x}$, both the Product Rule and the Chain Rule are needed:

$$\frac{d}{dx}\left[(5x^2 + 3)e^{7x}\right] = (5x^2 + 3)e^{7x} \cdot 7 + e^{7x} \cdot 10x.$$

SECTION 3.2

A **logarithmic function** g is any function of the form $g(x) = \log_a x$, where a, the **base**, is a positive number other than 1:

$$y = \log_a x \quad \text{means} \quad x = a^y.$$

Logarithmic functions are inverses of exponential functions.

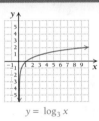

$y = \log_3 x$

Remember that $\log_b c$ *is the exponent to which* b *is raised to get* c. The notation for common logarithms and natural logarithms is

$$\log c = \log_{10} c \quad \text{and} \quad \ln c = \log_e c.$$

$\log 1000 = \log_{10} 10^3 = 3 \quad \text{and} \quad \ln e^{\sqrt{5}} = \log_e e^{\sqrt{5}} = \sqrt{5}$

The following important properties of logarithms allow us to manipulate expressions. We assume that M, N, and a are positive, with $a \neq 1$, and k is a real number:

$$\log_a (MN) = \log_a M + \log_a N$$

$$\log_a \frac{M}{N} = \log_a M - \log_a N$$

$$\log_a (M^k) = k \log_a M$$

$$\log_a a = 1$$

$$\log_a a^k = k$$

$$\log_a 1 = 0$$

$$\log_b M = \frac{\log_a M}{\log_a b}$$

$\log_2 5x = \log_2 5 + \log_2 x$

$\log_2 \dfrac{2}{Q} = \log_2 2 - \log_2 Q = 1 - \log_2 Q$

$\ln e^7 = 7$

$\log_a t = \dfrac{\ln t}{\ln a}$

$\log_{10} 1 = 0$

KEY TERMS AND CONCEPTS	EXAMPLES

Logarithms are used to solve certain exponential equations.

Solve: $5e^{2t} = 80$.

We have

$$5e^{2t} = 80$$
$$e^{2t} = 16 \quad \text{Dividing both sides by 5}$$
$$\ln e^{2t} = \ln 16 \quad \text{Taking the natural log of both sides}$$
$$2t = \ln 16 \quad \text{Using a property of logarithms}$$
$$t = \frac{\ln 16}{2}$$
$$t \approx 1.386$$

The derivative of the natural logarithm function of x, where x is any positive number, is the reciprocal of x:

$$\frac{d}{dx}\ln x = \frac{1}{x},$$

and

$$\frac{d}{dx}\ln f(x) = \frac{f'(x)}{f(x)}.$$

Differentiating a natural logarithmic function may require applying the Product, Quotient, and Chain Rules.

$$\frac{d}{dx}\left[\ln(5x) \cdot (x^3 - 7x)\right] = \ln(5x) \cdot (3x^2 - 7) + (x^3 - 7x) \cdot \frac{1}{5x} \cdot 5$$

$$= \ln(5x) \cdot (3x^2 - 7) + x \cdot (x^2 - 7) \cdot \frac{1}{5x} \cdot 5$$

$$= \ln(5x) \cdot (3x^2 - 7) + (x^2 - 7)$$

SECTION 3.3

Because exponential functions are the only functions for which the derivative (rate of change) is directly proportional to the function value at any point in time, they can be used to model many real-world situations involving uninhibited growth.

If $\frac{dP}{dt} = kP$, with $k > 0$, then $P(t) = P_0 e^{kt}$, where P_0 is the initial population at $t = 0$.

Business. The balance P in an account with Turing Mutual Funds grows at a rate given by

$$\frac{dP}{dt} = 0.04P,$$

where t is time, in years. Find the function that satisfies the equation (let $P(0) = P_0$). After what period of time will an initial investment, P_0, double itself?

The function is

$$P(t) = P_0 e^{0.04t}.$$

Check:

$$\frac{d}{dt}P_0 e^{0.04t} = P_0 e^{0.04t} \cdot 0.04$$
$$= 0.04 P_0 e^{0.04t}$$
$$= 0.04 P(t)$$

To find the time for the amount to double, we set $P(t) = 2P_0$ and solve for t:

$$2P_0 = P_0 e^{0.04t}$$
$$2 = e^{0.04t} \quad \text{Dividing both sides by } P_0$$
$$\ln 2 = \ln(e^{0.04t}) \quad \text{Taking the natural log of both sides}$$
$$\ln 2 = 0.04t$$
$$\frac{\ln 2}{0.04} = t, \quad \text{or} \quad t \approx 17.3 \text{ yr.}$$

(continued)

KEY TERMS AND CONCEPTS	EXAMPLES

SECTION 3.3 (continued)

The exponential **growth rate** k and the **doubling time** T are related by

$$k = \frac{\ln 2}{T} \quad \text{and} \quad T = \frac{\ln 2}{k}.$$

The **Rule of 70** expresses the doubling time for a sum of money in terms of the interest rate k expressed as a decimal:

$$T \approx \frac{70}{100k}.$$

If the exponential growth rate is 4% per year, then the doubling time is

$$T = \frac{\ln 2}{4\%} \approx \frac{0.693147}{0.04} \approx 17.3 \text{ yr,}$$

or using the Rule of 70, $T \approx \dfrac{70}{100(0.04)} \approx 17.5$ yr.

The doubling time for the number of downloads per day from iTunes is 614 days. The exponential growth rate is

$$k = \frac{\ln 2}{T} \approx \frac{0.693147}{614} \approx 0.0011289 = 0.11\% \text{ per day.}$$

Two models for limited or inhibited growth are

$$P(t) = \frac{L}{1 + be^{-kt}},$$

and

$$P(t) = L(1 - e^{-kt}),$$

where $k > 0$ and L is the limiting value.

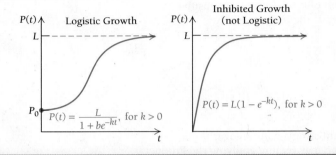

SECTION 3.4

Exponential growth is modeled by $P(t) = P_0 e^{kt}$, $k > 0$, and **exponential decay** is modeled by $P(t) = P_0 e^{-kt}$, $k > 0$.

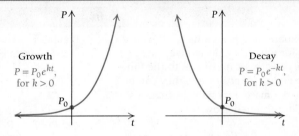

Exponential decay characterizes many real-world phenomena. One of the most common is *radioactive decay*.

Physical Science. Lead-210 has a decay rate of 3.15% per year. The rate of change of an amount N of lead-210 is given by

$$\frac{dN}{dt} = -0.0315N.$$

Find the function that satisfies the equation. How much of an 80-g sample of lead-210 will remain after 20 yr?

The function is $N(t) = N_0 e^{-0.0315t}$.

Check:

$$\frac{d}{dt}\left(N_0 e^{-0.0315t}\right) = N_0 e^{-0.0315t}(-0.0315)$$

$$= -0.0315 \cdot N(t).$$

Then the amount remaining after 20 yr is, $N(20) = 80e^{-0.0315(20)} \approx 42.6$ g.

KEY TERMS AND CONCEPTS	EXAMPLES

Half-life, T, and *decay rate, k,* are related by

$$k = \frac{\ln 2}{T} \quad \text{and} \quad T = \frac{\ln 2}{k}.$$

The decay equation, $P_0 = Pe^{-kt}$, can also be used to calculate present value.

The half-life of a radioactive isotope is 38 days. The decay rate is

$$T = \frac{\ln 2}{38} = \frac{0.693147}{38} \approx 0.0182 = 1.82\% \text{ per day.}$$

The **present value** P_0 of an amount P due t years later, at interest rate k, compounded continuously, is given by

$$P_0 = Pe^{-kt}.$$

The present value of $200,000 due 8 yr from now, at 4.6% interest, compounded continuously, is given by

$$P_0 = Pe^{-kt} = 200{,}000e^{-0.046 \cdot 8} = \$138{,}423.44.$$

SECTION 3.5

The following formulas can be used to differentiate exponential and logarithmic functions for any base a, other than e:

$$\frac{d}{dx}a^x = (\ln a)a^x \quad \text{and} \quad \frac{d}{dx}\log_a x = \frac{1}{\ln a} \cdot \frac{1}{x}.$$

$$\frac{d}{dx}7^{0.3x} = \ln 7 \cdot 7^{0.3x} \cdot 0.3$$

$$\frac{d}{dx}\log_6 x = \frac{1}{\ln 6} \cdot \frac{1}{x}$$

SECTION 3.6

The **elasticity of demand** E is a function of x, the price:

$$E(x) = -\frac{x \cdot D'(x)}{D(x)}.$$

When $E(x) > 1$, total revenue is decreasing; when $E(x) < 1$, total revenue is increasing; and
when $E(x) = 1$, total revenue is maximized.

Business. The Leslie Davis Band finds that demand for its CD at performances is given by

$$q = D(x) = 50 - 2x,$$

where x is the price, in dollars, of each CD sold and q is the number of CDs sold at a performance. Find the elasticity when the price is $10 per CD, and interpret the result. Then, find the price at which revenue is maximized.

Elasticity at x is given by

$$\begin{aligned} E(x) &= -\frac{x \cdot D'(x)}{D(x)} \\ &= -\frac{x(-2)}{50 - 2x} \\ &= -\frac{-2x}{50 - 2x} = \frac{x}{25 - x}. \end{aligned}$$

Thus,

$$E(10) = \frac{10}{25 - 10} = \frac{2}{3}.$$

Since $E(10)$ is less than 1, the demand for the CD is *inelastic*, and an increase in price will increase revenue.
Revenue is maximized when $E(x) = 1$:

$$\begin{aligned} \frac{x}{25 - x} &= 1 \\ x &= 25 - x \\ 2x &= 25 \\ x &= 12.5. \end{aligned}$$

At a price of $12.50 per CD, revenue will be maximized.

These review exercises are for test preparation. They can also be used as a practice test. Answers are at the back of the book. The blue bracketed section references tell you what part(s) of the chapter to restudy if your answer is incorrect.

CONCEPT REINFORCEMENT

Match each equation in column A with the most appropriate graph in column B. [3.1–3.4]

Column A   Column B

1. $P(t) = 50e^{0.03t}$ **a)**

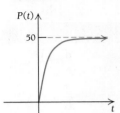

2. $P(t) = \dfrac{50}{1 + 2e^{-0.02t}}$ **b)**

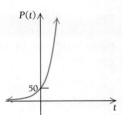

3. $P(t) = 50e^{-0.20t}$ **c)**

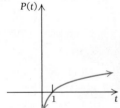

4. $P(t) = \ln t$ **d)**

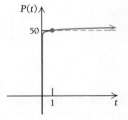

5. $P(t) = 50(1 - e^{-0.04t})$ **e)**

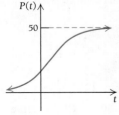

6. $P(t) = 50 + \ln t$ **f)**

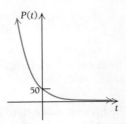

Classify each statement as either true or false.

7. The base a in the exponential function given by $f(x) = a^x$ must be greater than 1. [3.1]

8. The base a in the logarithmic function given by $g(x) = \log_a x$ must be greater than 0. [3.2]

9. If $f'(x) = c \cdot f(x)$ for $c \neq 0$ and $f(x) \neq 0$, then f must be an exponential function. [3.3]

10. With exponential growth, the doubling time depends on the size of the original population. [3.3]

11. A radioactive isotope's half-life determines the value of its decay constant. [3.4]

12. A radioactive isotope's half-life depends on how much of the substance is initially present. [3.4]

13. For any exponential function of the form $f(x) = a^x$, it follows that $f'(x) = \ln a \cdot a^x$. [3.5]

14. For any logarithmic function of the form $g(x) = \log_a x$, it follows that $g'(x) = \dfrac{1}{a} \cdot \dfrac{1}{x}$. [3.5]

15. Revenue is maximized when the elasticity of demand is 1. [3.6]

REVIEW EXERCISES

Differentiate each function.

16. $y = \ln x$ [3.2]

17. $y = e^x$ [3.1]

18. $y = \ln(x^4 + 5)$ [3.2]

19. $y = e^{2\sqrt{x}}$ [3.1]

20. $f(x) = \ln\sqrt{x}$ [3.2]

21. $f(x) = x^4 e^{3x}$ [3.1]

22. $f(x) = \dfrac{\ln x}{x^3}$ [3.2]

23. $f(x) = e^{x^2} \cdot \ln 4x$ [3.1, 3.2]

24. $f(x) = e^{4x} - \ln\dfrac{x}{4}$ [3.1, 3.2]

25. $g(x) = x^8 - 8\ln x$ [3.2]

26. $y = \dfrac{\ln e^x}{e^x}$ [3.1, 3.2]

27. $F(x) = 9^x$ [3.5]

28. $g(x) = \log_2 x$ [3.5]

29. $y = 3^x \cdot \log_4(2x + 1)$ [3.5]

Graph each function. [3.1]

30. $f(x) = 4^x$

31. $g(x) = \left(\tfrac{1}{3}\right)^x$

Given $\log_a 2 = 1.8301$ *and* $\log_a 7 = 5.0999,$
find each logarithm. [3.2]

32. $\log_a 14$ **33.** $\log_a \tfrac{2}{7}$ **34.** $\log_a 28$

35. $\log_a 3.5$ **36.** $\log_a \sqrt{7}$ **37.** $\log_a \tfrac{1}{4}$

38. Find the function that satisfies $dQ/dt = 7Q$, given that $Q(0) = 25$. [3.3]

39. Life science: population growth. The population of Boomtown doubled in 16 yr. What was the growth rate of the city? Round to the nearest tenth of a percent. [3.3]

40. Business: interest compounded continuously. Suppose that $8300 is invested in Noether Bond Fund, where the interest rate is 6.8%, compounded continuously. How long will it take for the $8300 to double itself? Round to the nearest tenth of a year. [3.3]

41. Business: cost of a prime-rib dinner. The average cost C of a prime-rib dinner was $15.81 in 1986. In 2010, it was $27.95. Assuming that the exponential growth model applies: [3.3]

a) Find the exponential growth rate to three decimal places, and write the function that models the situation.

b) What will the cost of such a dinner be in 2012? In 2020?

42. Business: franchise growth. A clothing firm is selling franchises throughout the United States and Canada. It is estimated that the number of franchises N will increase at the rate of 12% per year, that is,

$$\dfrac{dN}{dt} = 0.12N,$$

where t is the time, in years. [3.3]

a) Find the function that satisfies the equation, assuming that the number of franchises in 2007 ($t = 0$) is 60.

b) How many franchises will there be in 2013?

c) After how long will the number of franchises be 120? Round to the nearest tenth of a year.

43. Life science: decay rate. The decay rate of a certain radioactive substance is 13% per year. What is its half-life? Round to the nearest tenth of a year. [3.4]

44. Life science: half-life. The half-life of radon-222 is 3.8 days. What is its decay rate? Round to the nearest tenth of a percent. [3.4]

45. Life science: decay rate. A certain radioactive isotope has a decay rate of 7% per day, that is,

$$\dfrac{dA}{dt} = -0.07A,$$

where A is the amount of the isotope present at time t, in days. [3.4]

a) Find a function that satisfies the equation if the amount of the isotope present at $t = 0$ is 800 g.

b) After 20 days, how much of the 800 g will remain? Round to the nearest gram.

c) After how long will half of the original amount remain?

46. Social science: Hullian learning model. The probability p of mastering a certain assembly-line task after t learning trials is given by

$$p(t) = 1 - e^{-0.7t}.\ [3.3]$$

a) What is the probability of learning the task after 1 trial? 2 trials? 5 trials? 10 trials? 14 trials?

b) Find the rate of change, $p'(t)$.

c) Interpret the meaning of $p'(t)$.

d) Sketch a graph of the function.

47. Business: present value. Find the present value of $1,000,000 due 40 yr later at 4.2%, compounded continuously. [3.4]

48. Economics: elasticity of demand. Consider the demand function

$$q = D(x) = \dfrac{600}{(x + 4)^2}.\quad [3.6]$$

a) Find the elasticity.

b) Find the elasticity at $x = \$1$, stating whether the demand is elastic or inelastic.

c) Find the elasticity at $x = \$12$, stating whether the demand is elastic or inelastic.

d) At a price of $12, will a small increase in price cause the total revenue to increase or decrease?

e) Find the value of x for which the total revenue is a maximum.

SYNTHESIS

49. Differentiate: $y = \dfrac{e^{2x} + e^{-2x}}{e^{2x} - e^{-2x}}$. [3.1]

50. Find the minimum value of $f(x) = x^4 \ln(4x)$. [3.2]

TECHNOLOGY CONNECTION

51. Graph: $f(x) = \dfrac{e^{1/x}}{(1 + e^{1/x})2}$. [3.1]

52. Find $\displaystyle\lim_{x \to 0} \dfrac{e^{1/x}}{(1 + e^{1/x})^2}$. [3.1]

53. Business: shopping on the Internet. Online sales of all types of consumer products increased at an exponential rate in the last decade or so. Data in the following table show online retail sales, in billions of dollars. [3.3]

Years, t, after 1998	U.S. Online Retail Sales (in billions)
0	$ 4.9
1	14.7
2	28.0
3	34.3
4	44.7
5	55.7
6	69.2
7	86.3

(Source: U.S. Census Bureau.)

a) Use REGRESSION to fit an exponential function $y = a \cdot b^x$ to the data. Then convert that formula to an exponential function, base e, where t is the number of years after 1998, and determine the exponential growth rate.
b) Estimate online sales in 2010; in 2020.
c) After what amount of time will online sales be $400 billion?
d) What is the doubling time of online sales?

CHAPTER 3
TEST

Differentiate.

1. $y = 2e^{3x}$

2. $y = (\ln x)^4$

3. $f(x) = e^{-x^2}$

4. $f(x) = \ln \dfrac{x}{7}$

5. $f(x) = e^x - 5x^3$

6. $f(x) = 3e^x \ln x$

7. $y = 7^x + 3^x$

8. $y = \log_{14} x$

Given $\log_b 2 = 0.2560$ *and* $\log_b 9 = 0.8114$, *find each of the following.*

9. $\log_b 18$ 　　　**10.** $\log_b 4.5$ 　**11.** $\log_b 3$

12. Find the function that satisfies $dM/dt = 6M$, with $M(0) = 2$.

13. The doubling time for a certain bacteria population is 3 hr. What is the growth rate? Round to the nearest tenth of a percent.

APPLICATIONS

14. Business: interest compounded continuously. An investment is made at 6.931% per year, compounded continuously. What is the doubling time? Round to the nearest tenth of a year.

15. Business: cost of milk. The cost C of a gallon of milk was $3.22 in 2006. In 2010, it was $3.50. (*Source*: U.S. Department of Labor, Bureau of Labor Statistics.) Assuming that the exponential growth model applies:
a) Find the exponential growth rate to the nearest tenth of a percent, and write the equation.
b) Find the cost of a gallon of milk in 2012 and 2018.

16. Life science: drug dosage. A dose of a drug is injected into the body of a patient. The drug amount in the body decreases at the rate of 10% per hour, that is,

$$\frac{dA}{dt} = -0.1A,$$

where A is the amount in the body and t is the time, in hours.
a) A dose of 3 cubic centimeters (cc) is administered. Assuming $A_0 = 3$, find the function that satisfies the equation.
b) How much of the initial dose of 3 cc will remain after 10 hr?
c) After how long does half of the original dose remain?

17. Life science: decay rate. The decay rate of radium-226 is 4.209% per century. What is its half-life?

18. Life science: half-life. The half-life of bohrium-267 is 17 sec. What is its decay rate? Express the rate as a percentage rounded to four decimal places.

19. Business: effect of advertising. Twin City Roasters introduced a new coffee in a trial run. The firm advertised the coffee on television and found that the percentage P of people who bought the coffee after t ads had been run satisfied the function

$$P(t) = \frac{100}{1 + 24e^{-0.28t}}.$$

a) What percentage of people bought the coffee before seeing the ad ($t = 0$)?
b) What percentage bought the coffee after the ad had been run 1 time? 5 times? 10 times? 15 times? 20 times? 30 times? 35 times?
c) Find the rate of change, $P'(t)$.
d) Interpret the meaning of $P'(t)$.
e) Sketch a graph of the function.

20. In 2010, a professional athlete signed a contract paying him $13 million in 2016. Find the present value of that amount in 2010, assuming 4.3% interest, compounded continuously.

21. Economics: elasticity of demand. Consider the demand function

$$q = D(x) = 400e^{-0.2x}.$$

a) Find the elasticity.
b) Find the elasticity at $x = 3$, and state whether the demand is elastic or inelastic.
c) Find the elasticity at $x = 18$, and state whether the demand is elastic or inelastic.
d) At a price of $3, will a small increase in price cause the total revenue to increase or decrease?
e) Find the price for which the total revenue is a maximum.

SYNTHESIS

22. Differentiate: $y = x(\ln x)^2 - 2x \ln x + 2x$.

23. Find the maximum and minimum values of $f(x) = x^4 e^{-x}$ over $[0, 10]$.

TECHNOLOGY CONNECTION

24. Graph: $f(x) = \dfrac{e^x - e^{-x}}{e^x + e^{-x}}$.

25. Find $\lim\limits_{x \to 0} \dfrac{e^x - e^{-x}}{e^x + e^{-x}}$.

26. Business: average price of a television commercial. The cost of a 30-sec television commercial that runs during the Super Bowl was increasing exponentially from 1991 to 2012. Data in the table below show costs for those years.

Years, t, after 1990	Cost of commercial
1	$ 800,000
3	850,000
5	1,000,000
8	1,300,000
13	2,100,000
16	2,600,000
22	3,500,000

(Source: *National Football League.*)

a) Use REGRESSION to fit an exponential function $y = a \cdot b^x$ to the data. Then convert that formula to an exponential function, base e, where t is the number of years after 1990.

b) Estimate the cost of a commercial run during the Super Bowl in 2014 and 2017.
c) After what amount of time will the cost be $1 billion?
d) What is the doubling time of the cost of a commercial run during the Super Bowl?
e) The cost of a Super Bowl commercial in 2009 turned out to be $3 million, and in 2010 it dropped to about $2.8 million, possibly due to the decline in the world economy. Expand the table of costs, and make a scatterplot of the data. Does the cost still seem to follow an exponential function? Explain. What kind of curve seems to fit the data best? Fit that curve using REGRESSION, and predict the cost of a Super Bowl commercial in 2013 and in 2015. Compare your answers to those of part (b).

Extended Technology Application

The Business of Motion Picture Revenue and DVD Release

There has been increasing pressure by motion picture executives to narrow the gap between the theater release of a movie and the release to DVD. The executives want to reduce marketing expenses, adapt to audiences' increasing consumption of on-demand movies, and boost decreasing DVD sales. Theater owners, on the other hand, want to protect declining ticket sales; the number of people attending movies in the United States and Canada decreased from 1.57 billion in 2002 to 1.42 billion in 2009. The owners are fearful that if movie executives shorten the time between theater release and DVD release, the number of ticket buyers will drop even more, as many people will be willing to wait for the DVD.

The following table presents the number of days between theater release and DVD release for 10 movies. Note that average gap in time is about 4 months.

MOVIE	RELEASE DATES	NUMBER OF DAYS SEPARATING RELEASE DATES
The Girl with the Dragon Tattoo	Theater: Dec. 21, 2011 DVD: Mar. 20, 2012	90
The Artist	Theater: Nov. 25, 2011 DVD: June 26, 2012	214
X-men, First Class	Theater: June 3, 2011 DVD: Sept. 9, 2011	98
Avatar	Theater: Dec. 18, 2009 DVD: Apr. 22, 2010	126
The Blind Side	Theater: Nov. 27, 2009 DVD: Mar. 26, 2010	119
The Twilight Saga: New Moon	Theater: Nov. 20, 2009 DVD: Mar. 22, 2010	121
Precious	Theater: Nov. 6, 2009 DVD: Mar. 9, 2010	121
Julie & Julia	Theater: Aug. 7, 2009 DVD: Dec. 8, 2009	122
Slumdog Millionaire	Theater: Jan. 23, 2009 DVD: Mar. 31, 2009	67
Iron Man	Theater: May 2, 2008 DVD: Sept. 30, 2008	151

Average = 123 days, or about 4 months

Let's examine the data for *The Blind Side,* for which Sandra Bullock won the 2010 Oscar for best actress. The movie was based on a true story about a family that takes in a destitute young man and nurtures him into adulthood, when he becomes a player for the Baltimore Ravens in the National Football League. The following table presents weekly estimates of gross revenue, G, for the movie. The total revenue, R, is approximated by adding each week's gross revenue to the preceding week's total revenue. (Occasionally, other revenues, such as from permission fees, are added to box office revenue, so total revenue might be more than this sum.)

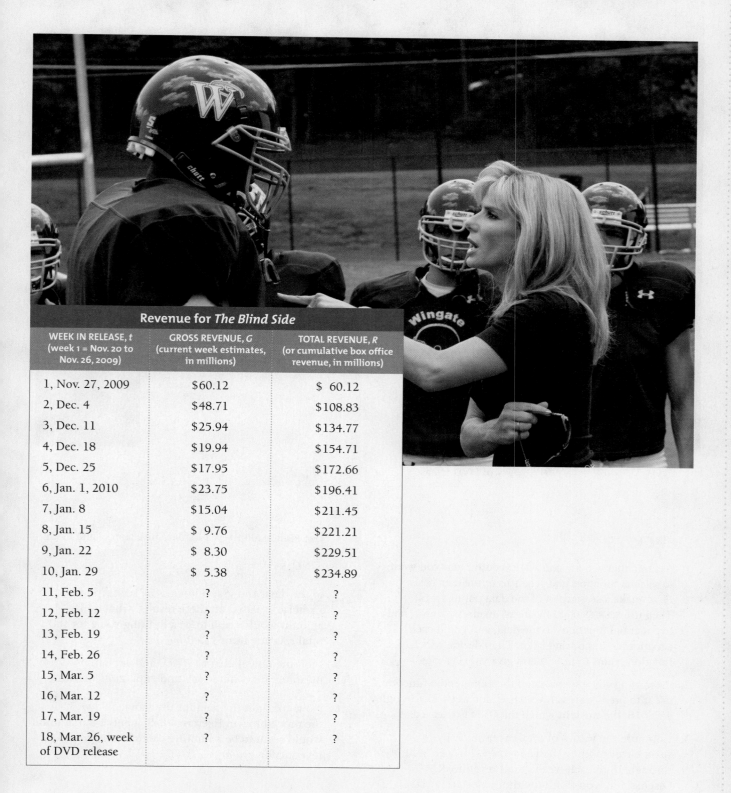

Revenue for *The Blind Side*		
WEEK IN RELEASE, t (week 1 = Nov. 20 to Nov. 26, 2009)	GROSS REVENUE, G (current week estimates, in millions)	TOTAL REVENUE, R (or cumulative box office revenue, in millions)
1, Nov. 27, 2009	$60.12	$ 60.12
2, Dec. 4	$48.71	$108.83
3, Dec. 11	$25.94	$134.77
4, Dec. 18	$19.94	$154.71
5, Dec. 25	$17.95	$172.66
6, Jan. 1, 2010	$23.75	$196.41
7, Jan. 8	$15.04	$211.45
8, Jan. 15	$ 9.76	$221.21
9, Jan. 22	$ 8.30	$229.51
10, Jan. 29	$ 5.38	$234.89
11, Feb. 5	?	?
12, Feb. 12	?	?
13, Feb. 19	?	?
14, Feb. 26	?	?
15, Mar. 5	?	?
16, Mar. 12	?	?
17, Mar. 19	?	?
18, Mar. 26, week of DVD release	?	?

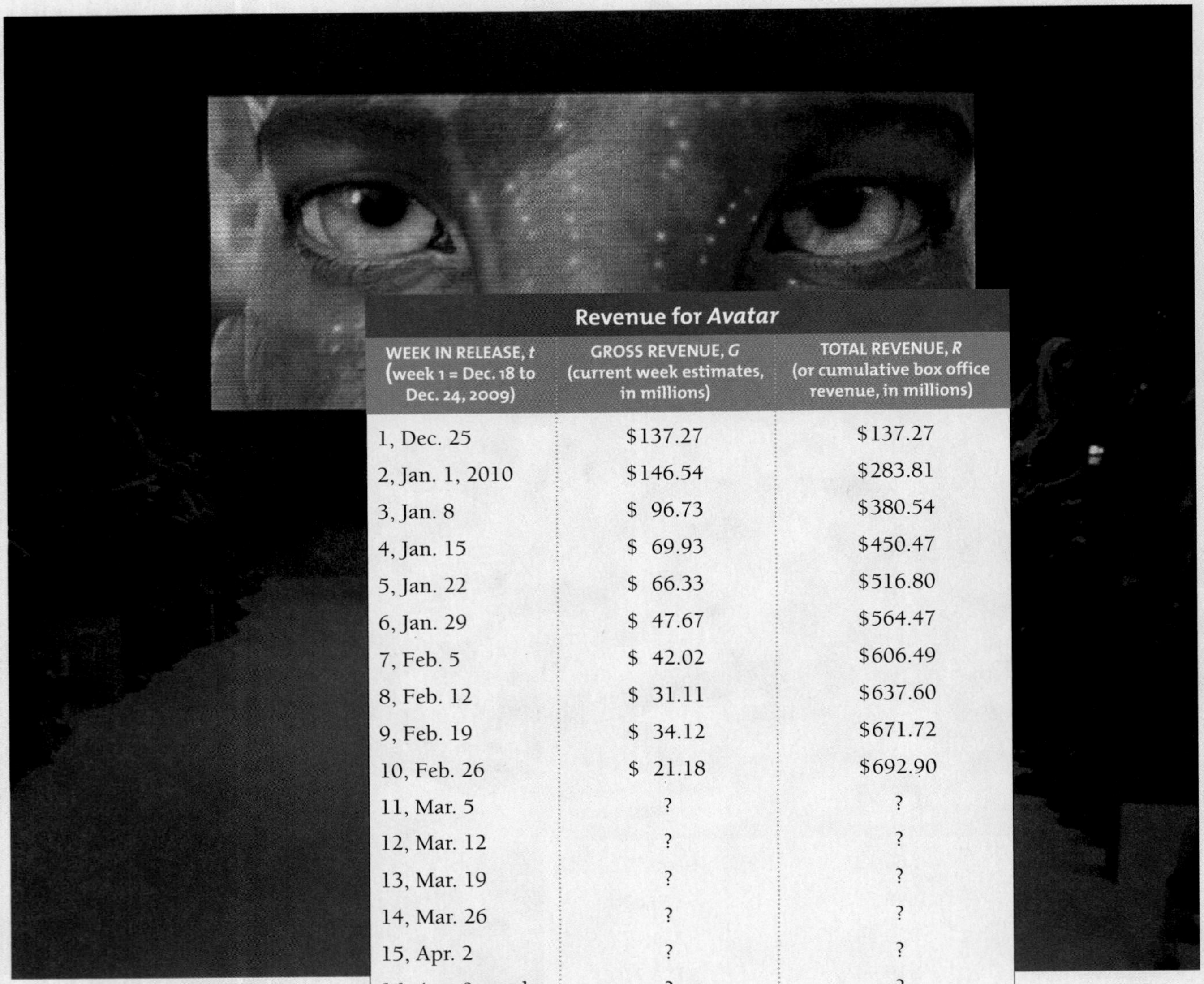

Revenue for *Avatar*		
WEEK IN RELEASE, *t* (week 1 = Dec. 18 to Dec. 24, 2009)	**GROSS REVENUE, *G*** (current week estimates, in millions)	**TOTAL REVENUE, *R*** (or cumulative box office revenue, in millions)
1, Dec. 25	$137.27	$137.27
2, Jan. 1, 2010	$146.54	$283.81
3, Jan. 8	$ 96.73	$380.54
4, Jan. 15	$ 69.93	$450.47
5, Jan. 22	$ 66.33	$516.80
6, Jan. 29	$ 47.67	$564.47
7, Feb. 5	$ 42.02	$606.49
8, Feb. 12	$ 31.11	$637.60
9, Feb. 19	$ 34.12	$671.72
10, Feb. 26	$ 21.18	$692.90
11, Mar. 5	?	?
12, Mar. 12	?	?
13, Mar. 19	?	?
14, Mar. 26	?	?
15, Apr. 2	?	?
16, Apr. 9, week of DVD release	?	?

EXERCISES

1. Assume that you are a movie executive and you want to select a function that seems to fit the data best. First, make a scatterplot of the data points (t, G). Then use REGRESSION to fit linear, quadratic, cubic, and exponential functions to the data, and graph each equation with the scatterplot. Then decide which function seems to fit best and give your reasons.

2. Presuming you have selected the exponential function, use it to predict gross revenue, G, for week 11 through week 18, the week in which the DVD is released.

3. Compute the values of total revenue, R, by successively adding the values of G for weeks 11 through 18 to each week's total revenue, R. Discuss why you think the time selected for DVD release is appropriate.

4. Use REGRESSION to fit a logistic function of the form
$$R(t) = \frac{c}{1 + ae^{-bt}}$$
to the data, and graph it with the scatterplot of Exercise 1. Based on these results, what dollar amount would seem to be a limiting value for the total revenue from *The Blind Side*?

5. Find the rate of change $R'(t)$, and explain its meaning. Find $\lim_{t \to \infty} R'(t)$, and explain its meaning.

6. Now, consider the data for the movie *Avatar*. Using the procedures in Exercises 1–5, what dollar amount would seem to be a limiting value on the gross revenue for *Avatar*?

Integration

4

Chapter Snapshot

What You'll Learn

Why It's Important

Is it possible to determine the distance a vehicle has traveled if we know its velocity function? Can we determine a company's total profit if we know its marginal-profit function? We can, using a process called *integration*, which is one of the two main branches of calculus, the other being differentiation. We will see that we can use integration to find the area under a curve, which has many practical applications in science, business, and statistics.

Where It's Used

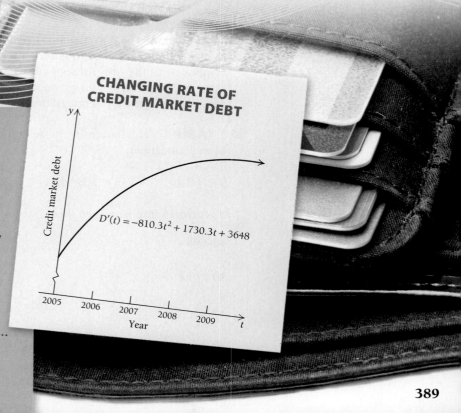

NATIONAL CREDIT MARKET DEBT

For the years 2005 through 2009, the annual rate of change in the national credit market debt, in billions of dollars per year, could be modeled by the function

$$D'(t) = -810.3t^2 + 1730.3t + 3648,$$

where t is the number of years since 2005. (***Source:*** Federal Reserve System.) Find the national credit market debt in 2009, given that $D(0) = 41,267$.

CHANGING RATE OF CREDIT MARKET DEBT

$$D'(t) = -810.3t^2 + 1730.3t + 3648$$

This problem appears as Exercise 60 in Exercise Set 4.1.

4.1

OBJECTIVES

- Find an antiderivative of a function.
- Evaluate indefinite integrals using basic rules of antidifferentiation.
- Use initial conditions to determine an antiderivative.

Antidifferentiation

Suppose we do the reverse of differentiation: given a function, we find another function whose derivative is the given function. This is called *antidifferentiation*, and it is a part of the larger process of *integration*. Integration, the main topic of this chapter, is the second main branch of calculus, the first being differentiation. We will see that integration can be used to find the area under a curve over a closed interval, which has many important applications.

Antidifferentiation is the process of differentiation in reverse. Given a function $f(x)$, we determine another function $F(x)$ such that the derivative of $F(x)$ is $f(x)$; that is, $\dfrac{d}{dx}F(x) = f(x)$.

For example, let $f(x) = 2x$. The function $F(x) = x^2$ is an antiderivative of $f(x)$ since $\dfrac{d}{dx}x^2 = 2x$. However, other functions also have a derivative of $2x$. For example, $y = x^2 + 1$, $y = x^2 - 10$, and $y = x^2 + 250$ also differentiate to $2x$; the x^2 term differentiates to $2x$, and the constant term differentiates to zero. Therefore, an antiderivative of $f(x) = 2x$ is any function that can be written in the form $F(x) = x^2 + C$, where C is a constant. This leads us to the following theorem.

THEOREM 1

The **antiderivative** of $f(x)$ is the set of functions $F(x) + C$ such that

$$\frac{d}{dx}[F(x) + C] = f(x).$$

The constant C is called the **constant of integration.**

Theorem 1 can be restated as follows: if two functions $F(x)$ and $G(x)$ have the same derivative $f(x)$, then $F(x)$ and $G(x)$ differ by at most a constant: $F(x) = G(x) + C$.

If $F(x)$ is an antiderivative of a function $f(x)$, we write

$$\int f(x)\, dx = F(x) + C.$$

This equation is read as "the antiderivative of $f(x)$, with respect to x, is the set of functions $F(x) + C$." The expression on the left side is called an **indefinite integral**. The symbol $\int$ is the *integral sign* and is a command for antidifferentiation. The function $f(x)$ is called the *integrand,* and the meaning of dx will be made clear when we develop the geometry of integration in Section 4.2.

■ **EXAMPLE 1** Determine these indefinite integrals. That is, find the antiderivative of each integrand:

a) $\int 8\, dx$;   **b)** $\int 3x^2\, dx$;   **c)** $\int e^x\, dx$;   **d)** $\int \dfrac{1}{x}\, dx\ (x > 0)$.

Solution You have seen these integrands before as derivatives of other functions.

a) $\int 8\, dx = 8x + C$ Check: $\dfrac{d}{dx}(8x + C) = 8$.

b) $\int 3x^2\, dx = x^3 + C$ Check: $\dfrac{d}{dx}(x^3 + C) = 3x^2$.

c) $\int e^x\, dx = e^x + C$ Check: $\dfrac{d}{dx}(e^x + C) = e^x$.

d) $\int \dfrac{1}{x}\, dx = \ln x + C$ Check: $\dfrac{d}{dx}(\ln x + C) = \dfrac{1}{x}$.

Always check each antiderivative you determine by differentiating it.

The results of Example 1 suggest several useful rules of antidifferentiation, which are summarized in Theorem 2.

THEOREM 2   Rules of Antidifferentiation

A1. Constant Rule:

$$\int k\,dx = kx + C.$$

A2. Power Rule (where $n \neq -1$):

$$\int x^n\,dx = \frac{1}{n+1}x^{n+1} + C, \qquad n \neq -1.$$

A3. Natural Logarithm Rule:

$$\int \frac{1}{x}\,dx = \ln x + C, \qquad x > 0.$$

A4. Exponential Rule (base e):

$$\int e^{ax}\,dx = \frac{1}{a}e^{ax} + C, \qquad a \neq 0.$$

Let's use these rules in the following examples.

EXAMPLE 2   Find the antiderivative of $f(x) = x^4$. That is, determine $\int x^4\,dx$.

Solution   We know that the derivative of a power function has an exponent decreased by 1, so we might guess that $F(x) = x^5 + C$ is an antiderivative of $f(x) = x^4$. However, $\frac{d}{dx}x^5 = 5x^4$, so our guess is not correct. It is close, however: including a coefficient of $\frac{1}{5}$ gives us the desired antiderivative:

$$\int x^4\,dx = \frac{1}{5}x^5 + C. \qquad \text{Check: } \frac{d}{dx}\left(\frac{1}{5}x^5 + C\right) = \frac{1}{5}(5x^4) = x^4.$$

Note that $\frac{1}{5}$ times 5 gives the coefficient 1.

Using the Power Rule of Antidifferentiation can be viewed as a two-step process:

$$\int x^n\,dx \overset{①}{=} \frac{1}{n+1}x^{n+1} + C$$
②

1. Raise the power by 1.
2. Divide the term by the new power.

EXAMPLE 3   Use the Power Rule of Antidifferentiation to determine these indefinite integrals:

a) $\int x^7\,dx$; **b)** $\int x^{99}\,dx$; **c)** $\int \sqrt{x}\,dx$; **d)** $\int \frac{1}{x^3}\,dx$.

Be sure to check each answer by differentiation.

Solution

a) $\int x^7\,dx = \frac{x^{7+1}}{7+1} + C = \frac{1}{8}x^8 + C \qquad \text{Check: } \frac{d}{dx}\left(\frac{1}{8}x^8 + C\right) = \frac{1}{8}(8x^7) = x^7.$

b) $\int x^{99}dx = \dfrac{x^{99+1}}{99+1} + C$

$\qquad = \dfrac{1}{100}x^{100} + C$ Check: $\dfrac{d}{dx}\left(\dfrac{1}{100}x^{100} + C\right) = \dfrac{1}{100}(100x^{99}) = x^{99}.$

c) We note that $\sqrt{x} = x^{1/2}$. Therefore,

$$\int \sqrt{x}\,dx = \int x^{1/2}\,dx = \dfrac{x^{(1/2)+1}}{\left(\frac{1}{2}\right)+1} + C = \dfrac{x^{3/2}}{\frac{3}{2}} + C$$

$$= \dfrac{2}{3}x^{3/2} + C.\quad \text{Check: } \dfrac{d}{dx}\left(\dfrac{2}{3}x^{3/2} + C\right) = \dfrac{2}{3}\left(\dfrac{3}{2}x^{1/2}\right) = x^{1/2} = \sqrt{x}.$$

》 Quick Check 1

Determine these indefinite integrals:

a) $\int x^{10}\,dx;$

b) $\int x^{200}\,dx;$

c) $\int \sqrt[6]{x}\,dx;$

d) $\int \dfrac{1}{x^4}\,dx.$

d) We note that $\dfrac{1}{x^3} = x^{-3}$. Therefore,

$$\int \dfrac{1}{x^3}\,dx = \int x^{-3}\,dx = \dfrac{x^{-3+1}}{-3+1} + C = -\dfrac{1}{2}x^{-2} + C$$

$$= -\dfrac{1}{2x^2} + C.\quad \text{Check: } \dfrac{d}{dx}\left(-\dfrac{1}{2}x^{-2} + C\right) = -\dfrac{1}{2}(-2x^{-3}) = x^{-3} = \dfrac{1}{x^3}.$$

《 Quick Check 1

The Power Rule of Antidifferentiation is valid for all real numbers n, except for $n = -1$. Attempting to use the Power Rule when $n = -1$ will result in a 0 in the denominator of the coefficient. However, as we saw in Example 1(d), if $n = -1$, we have $x^{-1} = \dfrac{1}{x}$, which is the derivative of the natural logarithm function, $y = \ln x$. Therefore,

$$\int \dfrac{1}{x}\,dx = \ln x + C, \text{ for } x > 0.$$

Caution! Note the key difference between the indefinite integrals $\int \dfrac{1}{x^3}\,dx$ and $\int \dfrac{1}{x}\,dx$. Although they look similar, the first of these integrals is determined by the Power Rule, while the second is determined by the Natural Logarithm Rule.

The exponential function $f(x) = e^x$ has the property that $\dfrac{d}{dx}e^x = e^x$; therefore, we can conclude that $\int e^x\,dx = e^x + C$. In Example 4, we explore the case of $f(x) = e^{ax}$.

■ **EXAMPLE 4**   Determine the indefinite integral $\int e^{4x}\,dx$.

Solution   Since we know that $\dfrac{d}{dx}e^x = e^x$, it is reasonable to make this initial guess:

$$\int e^{4x}\,dx = e^{4x} + C.$$

But this is (slightly) wrong, since $\dfrac{d}{dx}(e^{4x} + C) = 4e^{4x}$, with the coefficient 4 in the derivative resulting from application of the Chain Rule. We modify our guess by inserting $\frac{1}{4}$ to obtain the correct antiderivative:

$$\int e^{4x}\,dx = \dfrac{1}{4}e^{4x} + C.$$

》 Quick Check 2

Find each antiderivative:

a) $\int e^{-3x}\,dx;$

b) $\int e^{(1/2)x}\,dx.$

This checks: $\dfrac{d}{dx}\left(\dfrac{1}{4}e^{4x} + C\right) = \dfrac{1}{4}(4e^{4x}) = e^{4x}$; multiplying $\dfrac{1}{4}$ and 4 gives 1.

《 Quick Check 2

Two useful properties of antidifferentiation are presented in Theorem 3.

THEOREM 3   Properties of Antidifferentiation

P1. A constant factor can be moved to the front of an indefinite integral:

$$\int \big[c \cdot f(x)\big]\, dx = c \cdot \int f(x)\, dx.$$

P2. The antiderivative of a sum or a difference is the sum or the difference of the antiderivatives:

$$\int \big[f(x) \pm g(x)\big]\, dx = \int f(x)\, dx \pm \int g(x)\, dx.$$

In Example 5, we use the rules of antidifferentiation in conjunction with the properties of antidifferentiation. In part (b), we algebraically simplify the integrand before performing the antidifferentiation steps.

■ **EXAMPLE 5**   Determine these indefinite integrals. Assume $x > 0$.

a) $\int (3x^5 + 7x^2 + 8)\, dx;$ **b)** $\int \dfrac{4 + 3x + 2x^4}{x}\, dx.$

Solution

a) We antidifferentiate each term separately:

$$\int (3x^5 + 7x^2 + 8)\, dx = \int 3x^5\, dx + \int 7x^2\, dx + \int 8\, dx$$

By Antidifferentiation Property P2

$$= 3\big(\tfrac{1}{6}x^6\big) + 7\big(\tfrac{1}{3}x^3\big) + 8x$$

By Antidifferentiation Property P1 and Rules A1 and A2

$$= \tfrac{1}{2}x^6 + \tfrac{7}{3}x^3 + 8x + C.$$

Note the simplification of coefficients and the inclusion of just one constant of integration.

b) We algebraically simplify the integrand by noting that x is a common denominator and then reducing each ratio as much as possible:

$$\frac{4 + 3x + 2x^4}{x} = \frac{4}{x} + \frac{3x}{x} + \frac{2x^4}{x} = \frac{4}{x} + 3 + 2x^3 + C.$$

Therefore,

$$\int \frac{4 + 3x + 2x^4}{x}\, dx = \int \Big(\frac{4}{x} + 3 + 2x^3\Big)dx$$

$$= 4 \ln x + 3x + \tfrac{1}{2}x^4 + C.$$

By Antidifferentiation Properties P1 and P2 and Rules A1, A2, and A3

❮ Quick Check 3

> **Quick Check 3**
>
> Determine these indefinite integrals:
>
> **a)** $\int (2x^4 + 3x^3 - 7x^2 + x - 5)\, dx;$
>
> **b)** $\int (x - 5)^2\, dx;$
>
> **c)** $\int \dfrac{x^2 - 7x + 2}{x^2}\, dx.$

Initial Conditions

The constant of integration C may be of interest in some applications. In such cases, we may specify a point that is a solution of the antiderivative, thereby allowing us to solve for C. This point is called an **initial condition**.

■ **EXAMPLE 6**   Find a function f such that $f'(x) = 2x + 3$ and $f(1) = -2$.

Solution   The antiderivative of $f'(x) = 2x + 3$ is

$$f(x) = \int (2x + 3)\, dx = x^2 + 3x + C.$$

Since $f(1) = -2$, we let $x = 1$ and $f(1) = -2$, and solve for C:

$$-2 = (1)^2 + 3(1) + C.$$

Simplifying, we have $-2 = 4 + C$, which gives $C = -6$. Therefore, the specific antiderivative of $f'(x) = 2x + 3$ that satisfies the initial condition is

$$f(x) = x^2 + 3x - 6.$$

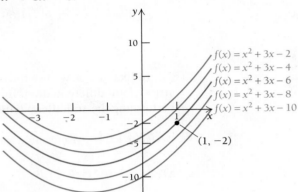

> **Quick Check 4**

Find an antiderivative of $g(x) = e^{2x}$ such that the point $(0, 3)$ is a solution of the antiderivative.

❮ Quick Check 4

The antiderivative of a function has many applications. For example, in Section 1.8, we saw that velocity is the derivative of a distance function, and, therefore, distance is the antiderivative of a velocity function. If information about the distance of an object at some time t is known, it provides us with an initial condition.

■ **EXAMPLE 7**   **Physical Sciences: Height of a Thrown Object.**   A rock is thrown directly upward with an initial velocity of 50 ft/sec from an initial height of 10 ft. The velocity of this rock is modeled by the function $v(t) = -32t + 50$, where t is in seconds, v is in feet per second, and $t = 0$ represents the moment the rock is released.

a) Determine a distance function h as a function of t (in this case, "distance" is the same as "height"). Be sure to consider the fact that at $t = 0$, the rock is 10 ft above the ground.

b) Determine the height and the velocity of the rock after 3 sec.

Solution

a) Since distance (height) is the antiderivative of velocity, we have the following:

$$h(t) = \int (-32t + 50)\, dt$$
$$= -16t^2 + 50t + C.$$

The constant of integration C can be determined since we know the initial height of the rock, which gives us the ordered pair $(0, 10)$ as an initial condition. We substitute 0 for t and 10 for $h(t)$, and solve for C:

$$10 = -16(0)^2 + 50(0) + C$$
$$10 = C.$$

Therefore, the distance function is $h(t) = -16t^2 + 50t + 10$.

b) To determine the height of the rock after 3 sec, we substitute 3 for t in our distance function:

$$h(3) = -16(3)^2 + 50(3) + 10 = 16 \text{ ft}.$$

The velocity after 3 sec is

$$v(3) = -32(3) + 50 = -46 \text{ ft/sec}.$$

Thus, the rock is 16 ft above the ground, but the negative velocity indicates that it is moving downward.

■ **EXAMPLE 8** **Life Sciences: Change in Population.** The rate of change of the population of Phoenix, Arizona, is modeled by the exponential function $P'(t) = 11.7e^{0.026t}$, where t is the number of years since 1960 and $P'(t)$ is in thousands of people per year. In 1980, Phoenix had a population of 790,000. (*Source:* U.S. Census Bureau.)

a) Find the population model $P(t)$.

b) Estimate the population of Phoenix in 2012.

Solution

a) We antidifferentiate the rate-of-change model:

$$P(t) = \int 11.7e^{0.026t} \, dt$$

$$= \frac{11.7}{0.026} e^{0.026t} + C \qquad \text{By Antidifferentiation Property P1 and Rule A4}$$

$$= 450e^{0.026t} + C$$

The population in 1980 is treated as the initial condition: $(20, 790)$. We make the substitutions and solve for C:

$$790 = 450e^{0.026(20)} + C$$

$$790 = 756.9 + C$$

$$C = 33.1.$$

Therefore, the population model is $P(t) = 450e^{0.026t} + 33.1$.

b) The year 2012 corresponds to $t = 52$, so we make the substitution:

$$P(52) = 450e^{0.026(52)} + 33.1$$

$$= 1772.$$

According to this model, the population of Phoenix in 2012 should be about 1,772,000. Given that Phoenix had a population of 1,567,000 in 2008, this prediction is reasonable.

❮ Quick Check 5

> **❯ Quick Check 5**
>
> A town's rate of population change is modeled by $P'(t) = 34t + 16$, where t is the number of years since 1990 and $P'(t)$ is in people per year.
>
> **a)** Find the population model for this town if it is known that in 2000, the town had a population of 2500.
>
> **b)** Forecast the town's population in 2015.

TECHNOLOGY CONNECTION

Antiderivatives and Area

A graphing calculator can calculate the area under the graph of a function. In the Y= window, enter the function $f(x) = 2x$, for $x \geq 0$, and graph it in $[0, 10, 0, 20]$. Press **2ND** and ⟨CALC⟩ and then select $\int f(x) \, dx$ from the list. For "Lower Limit," type in 0, and press **ENTER**. For "Upper Limit," let $x = 1$, and press **ENTER**. The calculator will shade in the region and report the area in the lower-left corner.

Do this for a series of x-values, and put the information in a table like that to the right.

Base, x	Height, $f(x)$	Area of region, $A(x)$
1	2	1
2	4	4
3		
4		
5		
6		
7		

(continued)

Antiderivatives and Area (*continued*)

EXERCISES

1. a) Fill in the entire table.
   b) If $x = 20$, what is the area under the graph of f?
   c) What is the relationship between the value of x in the first column and the area in the third column?
   d) Form your observation from part (c) into an area function $A(x)$.
   e) What is the relationship between the area function A from part (d) and the given function f?

2. Repeat parts (a) through (e) of Exercise 1 for $f(x) = 3$, and look for a pattern in an relationship between the area function A and the given function f.

3. Repeat parts (a) through (e) of Exercise 1 for $f(x) = 3x^2$, and look for a pattern in the relationship between the area function A and the given function f.

Section Summary

- The *antiderivative* of a function $f(x)$ is a set of functions $F(x) + C$ such that

$$\frac{d}{dx}[F(x) + C] = f(x),$$

  where the constant C is called the *constant of integration*.
- An antiderivative is denoted by an *indefinite integral* using the integral sign, $\int$. If $F(x)$ is an antiderivative of $f(x)$, we write

$$\int f(x)\, dx = F(x) + C.$$

  We check the correctness of an antiderivative we have found by differentiating it.

- The *Constant Rule of Antidifferentiation* is $\int k\, dx = kx + C$.
- The *Power Rule of Antidifferentiation* is

$$\int x^n\, dx = \frac{1}{n + 1}x^{n+1} + C, \quad \text{for } n \neq -1.$$

- The *Natural Logarithm Rule of Antidifferentiation* is

$$\int \frac{1}{x}\, dx = \ln x + C, \quad \text{for } x > 0.$$

- The *Exponential Rule* (base e) of *Antidifferentiation* is

$$\int e^{ax}\, dx = \frac{1}{a}e^{ax} + C, \quad \text{for } a \neq 0.$$

- An *initial condition* is an ordered pair that is a solution of a particular antiderivative of an integrand.

EXERCISE SET
4.1

Determine these indefinite integrals.

1. $\displaystyle\int x^6\, dx$

2. $\displaystyle\int x^7\, dx$

3. $\displaystyle\int 2\, dx$

4. $\displaystyle\int 4\, dx$

5. $\displaystyle\int x^{1/4}\, dx$

6. $\displaystyle\int x^{1/3}\, dx$

7. $\displaystyle\int (x^2 + x - 1)\, dx$

8. $\displaystyle\int (x^2 - x + 2)\, dx$

9. $\displaystyle\int (2t^2 + 5t - 3)\, dt$

10. $\displaystyle\int (3t^2 - 4t + 7)\, dt$

11. $\displaystyle\int \frac{1}{x^3}\, dx$

12. $\displaystyle\int \frac{1}{x^5}\, dx$

13. $\displaystyle\int \sqrt[3]{x}\, dx$

14. $\displaystyle\int \sqrt{x}\, dx$

15. $\displaystyle\int \sqrt{x^5}\, dx$

16. $\displaystyle\int \sqrt[3]{x^2}\, dx$

17. $\displaystyle\int \frac{dx}{x^4}$

18. $\displaystyle\int \frac{dx}{x^2}$

19. $\displaystyle\int \frac{1}{x}\, dx$

20. $\displaystyle\int \frac{2}{x}\, dx$

21. $\displaystyle\int \left(\frac{3}{x} + \frac{5}{x^2}\right) dx$

22. $\displaystyle\int \left(\frac{4}{x^3} + \frac{7}{x}\right) dx$

23. $\displaystyle\int \frac{-7}{\sqrt[3]{x^2}}\, dx$

24. $\displaystyle\int \frac{5}{\sqrt[4]{x^3}}\, dx$

25. $\displaystyle\int 2e^{2x}\, dx$

26. $\displaystyle\int 4e^{4x}\, dx$

27. $\displaystyle\int e^{3x}\, dx$

28. $\displaystyle\int e^{5x}\, dx$

29. $\displaystyle\int e^{7x}\, dx$

30. $\displaystyle\int e^{6x}\, dx$

31. $\displaystyle\int 5e^{3x}\, dx$

32. $\displaystyle\int 2e^{5x}\, dx$

33. $\displaystyle\int 6e^{8x}\,dx$

34. $\displaystyle\int 12e^{3x}\,dx$

35. $\displaystyle\int \frac{2}{3}e^{-9x}\,dx$

36. $\displaystyle\int \frac{4}{5}e^{-10x}\,dx$

37. $\displaystyle\int (5x^2 - 2e^{7x})\,dx$

38. $\displaystyle\int (2x^5 - 4e^{3x})\,dx$

39. $\displaystyle\int \left(x^2 - \frac{3}{2}\sqrt{x} + x^{-4/3}\right)dx$

40. $\displaystyle\int \left(x^4 + \frac{1}{8\sqrt{x}} - \frac{4}{5}x^{-2/5}\right)dx$

41. $\displaystyle\int (3x + 2)^2\,dx$ (*Hint:* Expand first.)

42. $\displaystyle\int (x + 4)^2\,dx$

43. $\displaystyle\int \left(\frac{3}{x} - 5e^{2x} + \sqrt{x^7}\right)dx$

44. $\displaystyle\int \left(2e^{6x} - \frac{3}{x} + \sqrt[3]{x^4}\right)dx$

45. $\displaystyle\int \left(\frac{7}{\sqrt{x}} - \frac{2}{3}e^{5x} - \frac{8}{x}\right)dx$

46. $\displaystyle\int \left(\frac{4}{\sqrt[5]{x}} + \frac{3}{4}e^{6x} - \frac{7}{x}\right)dx$

Find f such that:

47. $f'(x) = x - 3,\quad f(2) = 9$

48. $f'(x) = x - 5,\quad f(1) = 6$

49. $f'(x) = x^2 - 4,\quad f(0) = 7$

50. $f'(x) = x^2 + 1,\quad f(0) = 8$

51. $f'(x) = 5x^2 + 3x - 7,\quad f(0) = 9$

52. $f'(x) = 8x^2 + 4x - 2,\quad f(0) = 6$

53. $f'(x) = 3x^2 - 5x + 1,\quad f(1) = \frac{7}{2}$

54. $f'(x) = 6x^2 - 4x + 2,\quad f(1) = 9$

55. $f'(x) = 5e^{2x},\quad f(0) = \frac{1}{2}$ **56.** $f'(x) = 3e^{4x},\quad f(0) = \frac{7}{4}$

57. $f'(x) = \dfrac{4}{\sqrt{x}},\quad f(1) = -5$

58. $f'(x) = \dfrac{2}{\sqrt[3]{x}},\quad f(1) = 1$

APPLICATIONS

Business and Economics

Credit market debt. *From 2005 to 2009, the annual rate of change in the national credit market debt, in billions of dollars per year, could be modeled by the function*

$$D'(t) = -810.3t^2 + 1730.3t + 3648,$$

where t is the number of years since 2005. (Source: Federal Reserve System.) Use the preceding information for Exercises 59 and 60.

59. Find the national credit market debt, $D(t)$, during the years 2005 through 2009 given that $D(0) = 41{,}267$.

60. What was the national credit market debt in 2009, given that $D(0) = 41{,}267$?

61. Total cost from marginal cost. A company determines that the marginal cost, C', of producing the xth unit of a product is given by

$$C'(x) = x^3 - 2x.$$

Find the total-cost function, C, assuming that $C(x)$ is in dollars and that fixed costs are \$7000.

62. Total cost from marginal cost. A company determines that the marginal cost, C', of producing the xth unit of a product is given by

$$C'(x) = x^3 - x.$$

Find the total-cost function, C, assuming that $C(x)$ is in dollars and that fixed costs are \$6500.

63. Total revenue from marginal revenue. A company determines that the marginal revenue, R', in dollars, from selling the xth unit of a product is given by

$$R'(x) = x^2 - 3.$$

a) Find the total-revenue function, R, assuming that $R(0) = 0$.

b) Why is $R(0) = 0$ a reasonable assumption?

64. Total revenue from marginal revenue. A company determines that the marginal revenue, R', in dollars, from selling the xth unit of a product is given by

$$R'(x) = x^2 - 1.$$

a) Find the total-revenue function, R, assuming that $R(0) = 0$.

b) Why is $R(0) = 0$ a reasonable assumption?

65. Demand from marginal demand. A company finds that the rate at which the quantity of a product that consumers demand changes with respect to price is given by the marginal-demand function

$$D'(x) = -\frac{4000}{x^2},$$

where x is the price per unit, in dollars. Find the demand function if it is known that 1003 units of the product are demanded by consumers when the price is \$4 per unit.

66. Supply from marginal supply. A company finds that the rate at which a seller's quantity supplied changes with respect to price is given by the marginal-supply function

$$S'(x) = 0.24x^2 + 4x + 10,$$

where x is the price per unit, in dollars. Find the supply function if it is known that the seller will sell 121 units of the product when the price is $5 per unit.

67. Efficiency of a machine operator. The rate at which a machine operator's efficiency, E (expressed as a percentage), changes with respect to time t is given by

$$\frac{dE}{dt} = 30 - 10t,$$

where t is the number of hours the operator has been at work.

A machine operator's efficiency changes with respect to time.

a) Find $E(t)$, given that the operator's efficiency after working 2 hr is 72%; that is, $E(2) = 72$.

b) Use the answer to part (a) to find the operator's efficiency after 3 hr; after 5 hr.

68. Efficiency of a machine operator. The rate at which a machine operator's efficiency, E (expressed as a percentage), changes with respect to time t is given by

$$\frac{dE}{dt} = 40 - 10t,$$

where t is the number of hours the operator has been at work.

a) Find $E(t)$, given that the operator's efficiency after working 2 hr is 72%; that is, $E(2) = 72$.

b) Use the answer to part (a) to find the operator's efficiency after 4 hr; after 8 hr.

Social and Life Sciences

69. Spread of an influenza. During 18 weeks from November 2009 to February 2010, the rate at which the number of cases of swine flu changed could be approximated by

$$I'(t) = -6.34t + 141.6,$$

where I is the total number of people who have contracted swine flu and t is time measured in weeks. (*Source*: Centers for Disease Control and Prevention.)

a) Estimate $I(t)$, the total number who have contracted influenza by time t. Assume that $I(0) = 1408$.

b) Approximately how many people contracted influenza during the first 8 weeks?

c) Approximately how many people contracted influenza during the whole 18 weeks?

d) Approximately how many people per 100,000 contracted influenza during the last 7 of the 18 weeks?

70. Memory. In a memory experiment, the rate at which students memorize Spanish vocabulary is found to be given by

$$M'(t) = 0.2t - 0.003t^2,$$

where $M(t)$ is the number of words memorized in t minutes.

a) Find $M(t)$ if it is known that $M(0) = 0$.

b) How many words are memorized in 8 min?

Physical Sciences

71. Physics: height of a thrown baseball. A baseball is thrown directly upward with an initial velocity of 75 ft/sec from an initial height of 30 ft. The velocity of the baseball is given by the function

$$v(t) = -32t + 75,$$

where t is the number of seconds since the ball was released and v is in feet per second.

a) Find the function h that gives the height (in feet) of the baseball after t seconds, using the fact that at $t = 0$, the ball is 30 ft above the ground.

b) What are the height and the velocity of the baseball after 2 sec of flight?

c) After how many seconds does the ball reach its highest point? (*Hint*: The ball "stops" for a moment before starting its downward fall.)

d) How high does the ball get at its highest point?

e) After how many seconds will the ball hit the ground?

f) What is the ball's velocity at the moment it hits the ground?

General Interest

72. Population growth. The rates of change in population for two cities are as follows:

Alphaville: $P'(t) = 45,$
Betaburgh: $Q'(t) = 105e^{0.03t},$

where t is the number of years since 1990, and both P' and Q' are measured in people per year. In 1990, Alphaville had a population of 5000, and Betaburgh had a population of 3500.

a) Determine the population models for both cities.

b) What were the populations of Alphaville and Betaburgh, to the nearest hundred, in 2000?

c) Sketch the graph of each city's population model and estimate the year in which the two cities have the same population.

SYNTHESIS

Find f.

73. $f'(t) = \sqrt{t} + \dfrac{1}{\sqrt{t}}, \quad f(4) = 0$

74. $f'(t) = t^{\sqrt{3}}, \quad f(0) = 8$

Evaluate. Each of the following can be determined using the rules developed in this section, but some algebra may be required beforehand.

75. $\displaystyle\int (5t + 4)^2\, t^4\, dt$

76. $\displaystyle\int (x - 1)^2\, x^3\, dx$

77. $\displaystyle\int (1 - t)\sqrt{t}\, dt$

78. $\displaystyle\int \frac{(t + 3)^2}{\sqrt{t}}\, dt$

79. $\displaystyle\int \frac{x^4 - 6x^2 - 7}{x^3}\, dx$

80. $\displaystyle\int (t + 1)^3\, dt$

81. $\displaystyle\int \frac{1}{\ln 10}\frac{dx}{x}$

82. $\displaystyle\int be^{ax}\, dx$

83. $\displaystyle\int (3x - 5)(2x + 1)^2\, dx$

84. $\displaystyle\int \sqrt[3]{64x^4}\, dx$

85. $\displaystyle\int \frac{x^2 - 1}{x + 1}\, dx$

86. $\displaystyle\int \frac{t^3 + 8}{t + 2}\, dt$

87. On a test, a student makes this statement: "The function $f(x) = x^2$ has a unique antiderivative." Is this a true statement? Why or why not?

88. Describe the graphical interpretation of an antiderivative.

Antiderivatives as Areas

4.2

OBJECTIVES

- Find the area under a graph to solve real-world problems.
- Use rectangles to approximate the area under a graph.

Integral calculus is primarily concerned with the *area* below the graph of a function (specifically, the area between the graph of a function and the *x*-axis). There are many situations where the area can be interpreted in a meaningful way. In this section, we assume that all functions are nonnegative; that is, $f(x) \geq 0$. Consider the following examples.

■ **EXAMPLE 1**   **Physical Sciences: Distance as Area.**   A vehicle travels at 50 mi/hr for 2 hr. How far has the vehicle traveled?

Solution   The answer is 100 mi. We treat the vehicle's velocity as a function, $v(x) = 50$. We graph this function, sketch a vertical line at $x = 2$, and obtain a rectangle. This rectangle measures 2 units horizontally and 50 units vertically. Its area is the distance the vehicle has traveled:

$$2\text{ hr} \cdot \frac{50\text{ mi}}{1\text{ hr}} = 100\text{ mi.} \qquad \frac{2\text{ hr}}{1\text{ hr}} = 2$$

Note that the units of hours cancel.

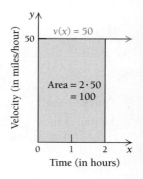

■ **EXAMPLE 2**   **Business: Total Cost as Area.**   Green Leaf Skateboards determines that for the first 50 skateboards produced, its cost is \$40 per skateboard. What is the total cost to produce 50 skateboards?

Solution   The marginal-cost function is $C'(x) = 40$, $0 \leq x \leq 50$. Its graph is a horizontal line. If we mark off 50 units along the *x*-axis, we get a rectangle, as in Example 1. The area of this rectangle is $40 \cdot 50 = 2000$. Therefore, the total cost to produce 50 skateboards is \$2000:

$$\left(50\text{ skateboards} \cdot 40\,\frac{\text{dollars}}{\text{skateboard}} = 2000\text{ dollars}\right).$$

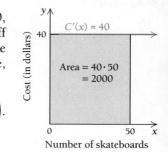

Geometry and Areas

For the time being, we will deal with linear functions. In these cases, we can use geometry to find the area formed by the graph of a function. The following two formulas, where $b =$ base and $h =$ height, will be useful.

Area of a rectangle: $A = bh$ Area of a triangle: $A = \frac{1}{2}bh$

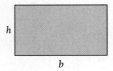

Examples 3 and 4 continue to explore the themes of Example 1 (distance as area) and Example 2 (total cost as area), respectively. In these cases, the graph of the function is linear with a nonzero slope.

■ **EXAMPLE 3**   **Physical Sciences: Distance as Area.**   The velocity of a moving object is given by the function $v(x) = 3x$, where x is in hours and v is in miles per hour. Use geometry to find the area under the graph, which is the distance the object has traveled:

a) during the first 3 hr $(0 \le x \le 3)$;

b) between the third hour and the fifth hour $(3 \le x \le 5)$.

Solution

a) The graph of the velocity function is shown at the right. We see the region corresponding to the time interval $0 \le x \le 3$ is a triangle with base 3 and height 9 (since $v(3) = 9$). Therefore, the area of this region is $A = \frac{1}{2}(3)(9) = \frac{27}{2} = 13.5$. The object traveled 13.5 mi during the first 3 hr.

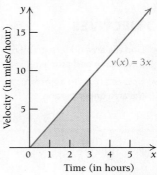

b) The region corresponding to the time interval $3 \le x \le 5$ is a trapezoid. It can be decomposed into a rectangle and a triangle as indicated in the figure to the right. The rectangle has base 2 and height 9, and thus an area $A = (2)(9) = 18$; the triangle has base 2 and height 6, for an area $A = \frac{1}{2}(2)(6) = 6$. Summing these, we get 24. Therefore, the object traveled 24 mi between the third hour and the fifth hour.

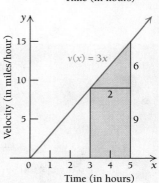

> **Quick Check 1**
>
> An object moves with a velocity of $v(t) = \frac{1}{2}t$, where t is in minutes and v is in feet per minute.
>
> **a)** How far does the object travel during the first 30 min?
>
> **b)** How far does the object travel between the first hour and the second hour?

❮ Quick Check 1

■ **EXAMPLE 4** **Business: Total Profit as Area.** Cousland, Inc., has a marginal-profit function modeled by the linear function $P'(x) = 0.15x$, where x is in months and P' is in thousands of dollars per month. Sketch this graph and use it to determine the total profit earned by Cousland, Inc., in a year ($0 \leq x \leq 12$).

Solution The graph of P' is shown below:

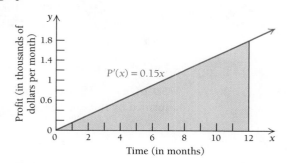

〉 Quick Check 2

Calculate the total profit of Cousland, Inc., between the fifth and the twelfth month.

For the 12-month period, the area is calculated by using the formula for a triangle: $A = \frac{1}{2}(12\text{ months})\left(1.8 \dfrac{\text{thousands of dollars}}{\text{month}}\right) = 10.8$ thousand dollars. Cousland, Inc., earned a total profit of $10,800 in a year.

❮ **Quick Check 2**

In each of the Examples 1 through 4, the function was a *rate* function; its output units formed a rate (miles per hour in Examples 1 and 3, dollars per skateboard in Example 2, thousands of dollars per month in Example 4). The units of the area were derived by multiplying input units by output units.

Riemann Summation

How does an antiderivative of a function translate into the area below that function's graph? The Technology Connection on p. 395 and Examples 1 through 4 in this section suggest a pattern:

- If $f(x) = k$, where k is a constant, its graph is a horizontal line of height k. The region under this graph over the interval $[0, x]$ is a rectangle, and its area is $A = k \cdot x$ (height times base).

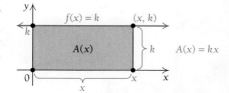

- If $f(x) = mx$, its graph is a line of slope m, passing through the origin. The region under this graph over an interval $[0, x]$ is a triangle, and its area is $A = \frac{1}{2}(x)(mx) = \frac{1}{2}mx^2$.

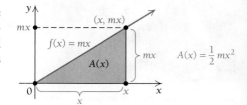

In these two cases, the area function is an antiderivative of the function that generated the graph. Is this always true? Is the formula for the area under the graph of any function that function's antiderivative? How do we handle curved graphs for which area formulas may not be known? We investigate these questions using geometry, in a procedure called *Riemann summation* (pronounced "Ree-mahn") in honor of the great German mathematician G. F. Bernhard Riemann (1826–1866).

Before we consider areas under curves, let's revisit Green Leaf Skateboards.

■ **EXAMPLE 5**   **Business: Total Cost.**   Green Leaf Skateboards has the following marginal-cost function for producing skateboards: For up to 50 skateboards, the cost is $40 per skateboard. For quantities from 51 through 125 skateboards, the cost drops to $30 per skateboard. After 125 skateboards, it drops to $25 per skateboard. If x represents the number of skateboards produced, the marginal-cost function C' is

$$C'(x) = \begin{cases} 40, & \text{for } 0 \le x \le 50, \\ 30, & \text{for } 50 < x \le 125, \\ 25, & \text{for } 125 < x \le 150. \end{cases}$$

Find the total cost to produce 150 skateboards.

Solution   We are extending Example 2. We calculate the areas of the rectangles formed by the horizontal lines of the graph of the marginal-cost function:

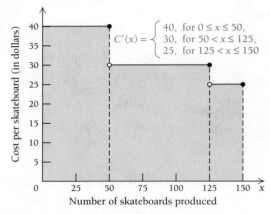

The total cost to produce 150 skateboards is found by summing those areas:

$$\text{Total cost} = (40)(50) + (30)(75) + (25)(25) = \$4875.$$

Example 5 illustrates the first steps of **Riemann summation**, a method that allows us to determine the area under curved graphs. We use rectangles to approximate the area under a curve given by $y = f(x)$, a continuous function, over an interval $[a, b]$. Riemann summation is accomplished with the use of *summation notation*, introduced below.

In the following figure, $[a, b]$ is divided into four subintervals, each having width $\Delta x = (b - a)/4$.

The area under a curve can be approximated by a sum of rectangular areas.

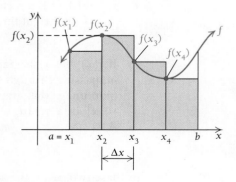

The heights of the rectangles shown are

$$f(x_1), \qquad f(x_2), \qquad f(x_3), \qquad \text{and} \qquad f(x_4).$$

The area of the region under the curve is approximately the sum of the areas of the four rectangles:

$$f(x_1)\,\Delta x + f(x_2)\,\Delta x + f(x_3)\,\Delta x + f(x_4)\,\Delta x.$$

We can denote this sum with **summation notation**, or **sigma notation**, which uses the Greek capital letter sigma, Σ:

$$\sum_{i=1}^{4} f(x_i)\, \Delta x.$$

This is read "the sum of the product $f(x_i)\, \Delta x$ from $i = 1$ to $i = 4$." To recover the original expression, we substitute the numbers 1 through 4 successively for i in $f(x_i)\, \Delta x$ and write plus signs between the results.

Before we continue, let's consider some examples involving summation notation.

■ **EXAMPLE 6** Write summation notation for $2 + 4 + 6 + 8 + 10$.

Solution Note that we are adding consecutive multiples of 2:

$$2 + 4 + 6 + 8 + 10 = \sum_{i=1}^{5} 2i.$$

❮ Quick Check 3

■ **EXAMPLE 7** Write summation notation for

$$g(x_1)\, \Delta x + g(x_2)\, \Delta x + \cdots + g(x_{19})\, \Delta x.$$

Solution

$$g(x_1)\, \Delta x + g(x_2)\, \Delta x + \cdots + g(x_{19})\, \Delta x = \sum_{i=1}^{19} g(x_i)\, \Delta x$$

■ **EXAMPLE 8** Express $\displaystyle\sum_{i=1}^{4} 3^i$ without using summation notation.

Solution

$$\sum_{i=1}^{4} 3^i = 3^1 + 3^2 + 3^3 + 3^4, \quad \text{or} \quad 120$$

❮ Quick Check 4

■ **EXAMPLE 9** Express $\displaystyle\sum_{i=1}^{30} h(x_i)\, \Delta x$ without using summation notation.

Solution

$$\sum_{i=1}^{30} h(x_i)\, \Delta x = h(x_1)\, \Delta x + h(x_2)\, \Delta x + \cdots + h(x_{30})\, \Delta x$$

Approximation of area by rectangles becomes more accurate as we use smaller subintervals and hence more rectangles, as shown in the following figures.

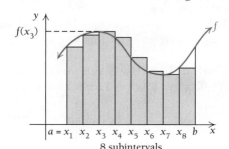

8 subintervals

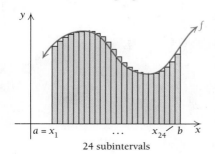

24 subintervals

> **Quick Check 3**
>
> Write summation notation for each expression:
> **a)** $5 + 10 + 15 + 20 + 25$;
> **b)** $33 + 44 + 55 + 66$.

> **Quick Check 4**
>
> Express $\displaystyle\sum_{i=1}^{6} (i^2 + i)$ without using summation notation.

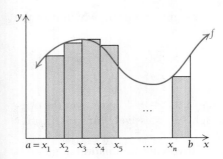

In general, suppose that the interval $[a, b]$ is divided into n equally sized subintervals, each of width $\Delta x = (b - a)/n$. We construct rectangles with heights

$$f(x_1), f(x_2), \dots, f(x_n).$$

The width of each rectangle is Δx, so the first rectangle has an area of $f(x_1)\Delta x$, the second rectangle has an area of $f(x_2)\Delta x$, and so on. The area of the region under the curve is approximated by the sum of the areas of the rectangles:

$$\sum_{i=1}^{n} f(x_i)\Delta x.$$

■ **EXAMPLE 10** Consider the graph of

$$f(x) = 600x - x^2$$

over the interval $[0, 600]$.

a) Approximate the area by dividing the interval into 6 subintervals.

b) Approximate the area by dividing the interval into 12 subintervals.

Solution

a) We divide $[0, 600]$ into 6 subintervals of size

$$\Delta x = \frac{600 - 0}{6} = 100,$$

with x_i ranging from $x_1 = 0$ to $x_6 = 500$. Thus, the area under the curve is approximately

$$\begin{aligned}
\sum_{i=1}^{6} f(x_i)\,\Delta x &= f(0) \cdot 100 + f(100) \cdot 100 + f(200) \cdot 100 \\
&\quad + f(300) \cdot 100 + f(400) \cdot 100 + f(500) \cdot 100 \\
&= 0 \cdot 100 + 50{,}000 \cdot 100 + 80{,}000 \cdot 100 \\
&\quad + 90{,}000 \cdot 100 + 80{,}000 \cdot 100 + 50{,}000 \cdot 100 \\
&= 35{,}000{,}000.
\end{aligned}$$

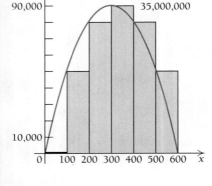

Total area = 35,000,000

b) We divide $[0, 600]$ into 12 subintervals of size $\Delta x = (600 - 0)/12 = 50$, with x_i ranging from $x_1 = 0$ to $x_{12} = 550$. Thus, we have another approximation of the area under the curve:

$$\begin{aligned}
\sum_{i=1}^{12} f(x_i)\,\Delta x &= f(0) \cdot 50 + f(50) \cdot 50 + f(100) \cdot 50 + f(150) \cdot 50 \\
&\quad + f(200) \cdot 50 + f(250) \cdot 50 + f(300) \cdot 50 + f(350) \cdot 50 \\
&\quad + f(400) \cdot 50 + f(450) \cdot 50 + f(500) \cdot 50 + f(550) \cdot 50 \\
&= 0 \cdot 50 + 27{,}500 \cdot 50 + 50{,}000 \cdot 50 + 67{,}500 \cdot 50 \\
&\quad + 80{,}000 \cdot 50 + 87{,}500 \cdot 50 + 90{,}000 \cdot 50 + 87{,}500 \cdot 50 \\
&\quad + 80{,}000 \cdot 50 + 67{,}500 \cdot 50 + 50{,}000 \cdot 50 + 27{,}500 \cdot 50 \\
&= 35{,}750{,}000.
\end{aligned}$$

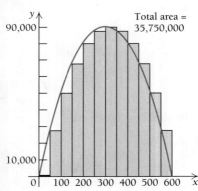

Total area = 35,750,000

Note that in Example 10 the approximation using $n = 12$ is closer to the exact value than the one using $n = 6$.

The sums used in Example 10 to approximate the area under a curve are called *Riemann sums*. Riemann sums can be calculated using any x-value within each subinterval.

■ **EXAMPLE 11** Use 5 subintervals to approximate the area under the graph of $f(x) = 0.1x^3 - 2.3x^2 + 12x + 25$ over the interval $[1, 16]$.

Solution We divide $[1, 16]$ into 5 subintervals of size $\Delta x = (16 - 1)/5 = 3$, with x_i ranging from $x_1 = 1$ to $x_5 = 13$. Although a drawing is not required, we can make one to help visualize the area.

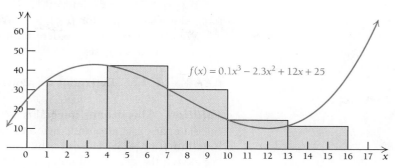

The area under the curve from 1 to 16 is approximately

$$\sum_{i=1}^{5} f(x_i)\,\Delta x = f(1)\cdot 3 + f(4)\cdot 3 + f(7)\cdot 3 + f(10)\cdot 3 + f(13)\cdot 3$$
$$= 34.8\cdot 3 + 42.6\cdot 3 + 30.6\cdot 3 + 15\cdot 3 + 12\cdot 3$$
$$= 405.$$

> **Quick Check 5**
>
> Use 6 subintervals to approximate the area under the graph of the function in Example 11 over the interval $[0, 12]$.

❮ Quick Check 5

Steps for the Process of Riemann Summation

1. Draw the graph of $f(x)$.
2. Subdivide the interval $[a, b]$ into n subintervals of equal width. Calculate the width of each rectangle by using the formula $\Delta x = \dfrac{b - a}{n}$.
3. Construct rectangles above the subintervals such that the top left corner of each rectangle touches the graph.
4. Determine the area of each rectangle.
5. Sum these areas to arrive at an approximation for the total area under the curve.

Definite Integrals

The key concept being developed in this section is that the more subintervals we use, the more accurate the approximation of area becomes. As the number of subdivisions n increases, the width of each rectangle Δx decreases. If n is allowed to approach infinity, then Δx approaches 0; these are limits, and the approximations of area become more and more exact to the true area under the graph. The *exact* area underneath the graph of a continuous function $y = f(x)$ over an interval $[a, b]$ is, by definition, given by a definite integral.

DEFINITION

Let $y = f(x)$ be continuous and nonnegative, $f(x) \geq 0$, over an interval $[a, b]$. A **definite integral** is the limit as $n \to \infty$ (equivalently, $\Delta x \to 0$) of the Riemann sum of the areas of rectangles under the graph of the function $y = f(x)$ over the interval $[a, b]$.

$$\text{Exact area} = \lim_{\Delta x \to 0} \sum_{i=1}^{n} f(x_i)\cdot\Delta x = \int_{a}^{b} f(x)\,dx.$$

Notice that the summation symbol becomes an integral sign (the elongated "s" is Leibniz notation representing "sum") and Δx becomes dx. The interval endpoints a and b are placed at the bottom right and top right, respectively, of the integral sign.

If $f(x) \geq 0$ over an interval $[a, b]$, *the definite integral represents area.* The definite integral is also defined for $f(x) < 0$. We will discuss its interpretation in Section 4.3.

We can use geometry to determine the value of some definite integrals, as the following example suggests.

■ **EXAMPLE 12**   Determine the value of $\displaystyle\int_{0}^{2} (3x + 2)\, dx$.

Solution   This definite integral is a command to calculate the exact area underneath the graph of the function $f(x) = 3x + 2$ over the interval $[0, 2]$. We sketch the graph and note that the region is a trapezoid. Thus, we can use geometry to determine this area (a Riemann sum is not needed here).

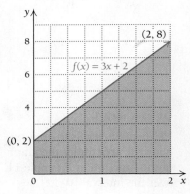

> **Quick Check 6**
>
> Use geometry to determine the values of these definite integrals:
>
> **a)** $\displaystyle\int_{0}^{3} (x + 1)\, dx$;
>
> **b)** $\displaystyle\int_{4}^{7} (15 - 2x)\, dx$.

Using a method similar to that in Example 3(b), we find that the area is 10. Therefore,

$$\int_{0}^{2} (3x + 2)\, dx = 10.$$

❮ Quick Check 6

Section Summary

- The area under a curve can often be interpreted in a meaningful way.
- The units of the area are found by multiplying the units of the input variable by the units of the output variable. It is crucial that the units are consistent.
- Geometry can be used to find areas of regions formed by graphs of linear functions.

- A *Riemann sum* uses rectangles to approximate the area under a curve. The more rectangles, the better the approximation.
- The *definite integral,* $\int_{a}^{b} f(x)\, dx$, is a representation of the exact area under the graph of a continuous function $y = f(x)$, where $f(x) \geq 0$, over an interval $[a, b]$.

EXERCISE SET
4.2

APPLICATIONS

Business and Economics

In Exercises 1–8, calculate total cost, disregarding any fixed costs.

1. Total cost from marginal cost. Redline Roasting has found that the cost, in dollars per pound, of the coffee it roasts is

$$C'(x) = -0.012x + 6.50, \quad \text{for } x \le 300,$$

where x is the number of pounds of coffee roasted. Find the total cost of roasting 200 lb of coffee.

2. Total cost from marginal cost. Sylvie's Old World Cheeses has found that the cost, in dollars per kilogram, of the cheese it produces is

$$C'(x) = -0.003x + 4.25, \quad \text{for } x \le 500,$$

where x is the number of kilograms of cheese produced. Find the total cost of producing 400 kg of cheese.

3. Total cost from marginal cost. Photos from Nature has found that the cost per card of producing x note cards is given by

$$C'(x) = -0.04x + 85, \quad \text{for } x \le 1000,$$

where $C'(x)$ is the cost, in cents, per card. Find the total cost of producing 650 cards.

4. Total cost from marginal cost. Cleo's Custom Fabrics has found that the cost per yard of producing x yards of a particular fabric is given by

$$C'(x) = -0.007x + 12, \quad \text{for } x \le 350,$$

where $C'(x)$ is the cost in dollars. Find the total cost of producing 200 yd of this material.

5. Total profit from marginal profit. A concert promoter sells x tickets and has a marginal-profit function given by

$$P'(x) = 2x - 1150,$$

where $P'(x)$ is in dollars per ticket. This means that the rate of change of total profit with respect to the number of tickets sold, x, is $P'(x)$. Find the total profit from the sale of the first 300 tickets.

6. Total profit from marginal profit. Poyse Inc. has a marginal-profit function given by

$$P'(x) = -2x + 80,$$

where $P'(x)$ is in dollars per unit. This means that the rate of change of total profit with respect to the number of units produced, x, is $P'(x)$. Find the total profit from the production and sale of the first 40 units.

7. Total cost from marginal costs. Raggs, Ltd., determines that its marginal cost, in dollars per dress, is given by

$$C'(x) = -\frac{2}{25}x + 50, \quad \text{for } x \le 450.$$

Find the total cost of producing the first 200 dresses.

8. Total cost from marginal cost. Using the information and answer from Exercise 7, find the cost of producing the 201st dress through the 400th dress.

9. Total cost from marginal cost. Ship Shape Woodworkers has found that the marginal cost of producing x feet of custom molding is given by

$$C'(x) = -0.00002x^2 - 0.04x + 45, \quad \text{for } x \le 800,$$

where $C'(x)$ is in cents. Approximate the total cost of manufacturing 800 ft of molding, using 5 subintervals over $[0, 800]$ and the left endpoint of each subinterval.

10. Total cost from marginal cost. Soulful Scents has found that the marginal cost of producing x ounces of a new fragrance is given by

$$C'(x) = 0.0005x^2 - 0.1x + 30, \quad \text{for } x \le 125,$$

where $C'(x)$ is in dollars. Use 5 subintervals over $[0, 100]$ and the left endpoint of each subinterval to approximate the total cost of producing 100 oz of the fragrance.

11. Total cost from marginal cost. Shelly's Roadside Fruit has found that the marginal cost of producing x pints of fresh-squeezed orange juice is given by

$$C'(x) = 0.000008x^2 - 0.004x + 2, \quad \text{for } x \le 350,$$

where $C'(x)$ is in dollars. Approximate the total cost of producing 270 pt of juice, using 3 subintervals over $[0, 270]$ and the left endpoint of each subinterval.

12. Total cost from marginal cost. Mangianello Paving, Inc., has found that the marginal cost, in dollars, of paving a road surface with asphalt is given by

$$C'(x) = \frac{1}{6}x^2 - 20x + 1800, \quad \text{for } x \le 80,$$

where x is measured in hundreds of feet. Use 4 subintervals over $[0, 40]$ and the left endpoint of each subinterval to approximate the total cost of paving 4000 ft of road surface.

In Exercises 13–18, write summation notation for each expression.

13. $3 + 6 + 9 + 12 + 15 + 18$

14. $5 + 10 + 15 + 20 + 25 + 30 + 35$

15. $f(x_1) + f(x_2) + f(x_3) + f(x_4)$

16. $g(x_1) + g(x_2) + g(x_3) + g(x_4) + g(x_5)$

17. $G(x_1) + G(x_2) + \cdots + G(x_{15})$

18. $F(x_1) + F(x_2) + \cdots + F(x_{17})$

19. Express $\displaystyle\sum_{i=1}^{4} 2^i$ without using summation notation.

20. Express $\sum_{i=0}^{5} (-2)^i$ without using summation notation.

21. Express $\sum_{i=1}^{5} f(x_i)$ without using summation notation.

22. Express $\sum_{i=1}^{4} g(x_i)$ without using summation notation.

23. a) Approximate the area under the following graph of $f(x) = \dfrac{1}{x^2}$ over the interval $[1, 7]$ by computing the area of each rectangle to four decimal places and then adding.

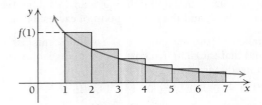

b) Approximate the area under the graph of $f(x) = \dfrac{1}{x^2}$ over the interval $[1, 7]$ by computing the area of each rectangle to four decimal places and then adding. Compare your answer to that for part (a).

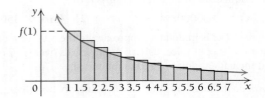

24. a) Approximate the area under the graph of $f(x) = x^2 + 1$ over the interval $[0, 5]$ by computing the area of each rectangle and then adding.

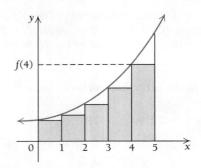

b) Approximate the area under the graph of $f(x) = x^2 + 1$ over the interval $[0, 5]$ by computing

the area of each rectangle and then adding. Compare your answer to that for part (a).

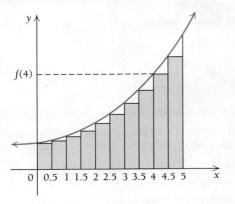

25. Total profit from marginal profit. Holcomb Hill Fitness has found that the marginal profit, $P'(x)$, in cents, is given by

$$P'(x) = -0.0006x^3 + 0.28x^2 + 55.6x, \quad \text{for } x \le 500,$$

where x is the number of members currently enrolled at the health club.

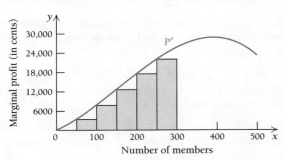

Approximate the total profit when 300 members are enrolled by computing the sum

$$\sum_{i=1}^{6} P'(x_i)\,\Delta x,$$

with $\Delta x = 50$.

26. Total cost from marginal cost. Raggs, Ltd., has found that the marginal cost, in dollars, for the xth jacket produced is given by

$$C'(x) = 0.0003x^2 - 0.2x + 50.$$

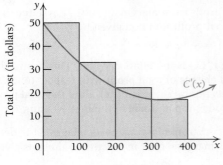

Approximate the total cost of producing 400 jackets by computing the sum

$$\sum_{i=1}^{4} C'(x_i)\, \Delta x,$$

with $\Delta x = 100$.

27. Approximate the area under the graph of

$$f(x) = 0.01x^4 - 1.44x^2 + 60$$

over the interval $[2, 10]$ by dividing the interval into 4 subintervals.

28. Approximate the area under the graph of

$$g(x) = -0.02x^4 + 0.28x^3 - 0.3x^2 + 20$$

over the interval $[3, 12]$ by dividing the interval into 4 subintervals.

29. Approximate the area under the graph of

$$F(x) = 0.2x^3 + 2x^2 - 0.2x - 2$$

over the interval $[-8, -3]$ using 5 subintervals.

30. Approximate the area under the graph of

$$G(x) = 0.1x^3 + 1.2x^2 - 0.4x - 4.8$$

over the interval $[-10, -4]$ using 6 subintervals.

In Exercises 31–39, use geometry to evaluate each definite integral.

31. $\displaystyle\int_{0}^{2} 2\, dx$

32. $\displaystyle\int_{0}^{5} 6\, dx$

33. $\displaystyle\int_{2}^{6} 3\, dx$

34. $\displaystyle\int_{-1}^{4} 4\, dx$

35. $\displaystyle\int_{0}^{3} x\, dx$

36. $\displaystyle\int_{0}^{5} 4x\, dx$

37. $\displaystyle\int_{0}^{10} \frac{1}{2}x\, dx$

38. $\displaystyle\int_{0}^{5} (2x + 5)\, dx$

39. $\displaystyle\int_{2}^{4} (10 - 2x)\, dx$

SYNTHESIS

40. Show that, for any function f defined for all x_i's, and any constant k, we have

$$\sum_{i=1}^{4} kf(x_i) = k\sum_{i=1}^{4} f(x_i).$$

Then show that, in general,

$$\sum_{i=1}^{n} kf(x_i) = k\sum_{i=1}^{n} f(x_i),$$

for any constant k and any function f defined for all x_i's.

41. Use the following graph of $y = f(x)$ to evaluate each definite integral.

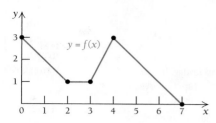

a) $\displaystyle\int_{0}^{2} f(x)\,dx$

b) $\displaystyle\int_{2}^{3} f(x)\,dx$

c) $\displaystyle\int_{3}^{4} f(x)\,dx$

d) $\displaystyle\int_{4}^{7} f(x)\,dx$

e) Use the results from parts (a)–(d) to evaluate

$$\int_{0}^{7} f(x)\,dx.$$

42. Use geometry and the following graph of $f(x) = \frac{1}{2}x$ to evaluate each definite integral.

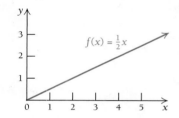

a) $\displaystyle\int_{0}^{1} f(x)\,dx$

b) $\displaystyle\int_{1}^{3} f(x)\,dx$

c) Find c such that $\displaystyle\int_{0}^{c} f(x)\,dx = 4$.

d) Find c such that $\displaystyle\int_{0}^{c} f(x)\,dx = 3$.

TECHNOLOGY CONNECTION

The exact area of a semicircle of radius r can be found using the formula $A = \frac{1}{2}\pi r^2$. Using this equation, compare the answers to Exercises 43 and 44 with the exact area. Note that most calculators do not show entire semicircles.

43. Approximate the area under the graph of $f(x) = \sqrt{25 - x^2}$ using 10 rectangles.

44. Approximate the area under the graph of $g(x) = \sqrt{49 - x^2}$ using 14 rectangles.

Answers to Quick Checks

1. (a) 225 ft; **(b)** 2700 ft   **2.** Total profit $= \$8925$

3. (a) $\displaystyle\sum_{i=1}^{5} 5i$; **(b)** $\displaystyle\sum_{i=3}^{6} 11i$ **4.** 112   **5.** 368   **6. (a)** 7.5;

(b) 12

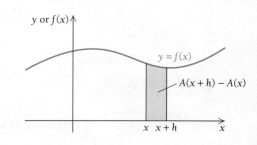

Area and Definite Integrals

4.3

In Sections 4.1 and 4.2, we considered the relationship between the area under the graph of a function f and the antiderivative of f. We have yet to establish the general rule that the antiderivative of a function f does in fact lead to the exact area under the graph of f. As we will see, we can use the antiderivative of a function to determine the exact area under the graph of the function. This process is called *integration*.

OBJECTIVES

- Find the area under the graph of a nonnegative function over a given closed interval.
- Evaluate a definite integral.
- Interpret an area below the horizontal axis.
- Solve applied problems involving definite integrals.

The Fundamental Theorem of Calculus

The area under the graph of a nonnegative continuous function f over an interval $[a, b]$ is determined as an area function A, which is an antiderivative of f; that is, $\dfrac{d}{dx} A(x) = f(x)$. We have established this fact for a few cases in which f was a constant or linear function by using geometry formulas for areas of a rectangle and a triangle. When the graph of f is a curve, we can approximate the area underneath the graph using a Riemann sum, which suggests a general method for calculating the area underneath the graph of *any* nonnegative continuous function f.

The following table summarizes some of the area functions we determined by geometry.

Function	Area Function	Text Reference
$f(x) = 2x$	$A(x) = x^2$	Technology Connection (p. 395)
$f(x) = 3$	$A(x) = 3x$	Technology Connection (p. 395)
$f(x) = 3x^2$	$A(x) = x^3$	Technology Connection (p. 395)
$f(x) = k$	$A(x) = kx$	Section 4.2
$f(x) = mx$	$A(x) = \frac{1}{2}mx^2$	Section 4.2

You may have noticed that each time the derivative of the area function, $A(x)$, is $f(x)$. Is this always the case?

We answer this by letting $A(x)$ represent the area under a nonnegative continuous function f over the interval $[0, x]$. To find $A'(x)$, we use the definition of the derivative:

$$A'(x) = \lim_{h \to 0} \frac{A(x + h) - A(x)}{h}.$$

Since $A(x + h)$ is the area under f over the interval $[0, x + h]$, it follows that $A(x + h) - A(x)$ is the area under f between x and $x + h$.

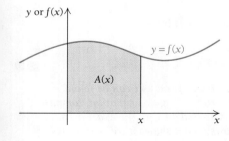

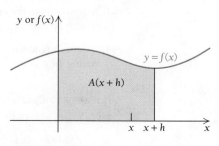

As h approaches zero, the area $A(x + h) - A(x)$ approaches the area of a rectangle with width h and height $f(x)$. That is,

$$A(x + h) - A(x) \approx h \cdot f(x).$$

Thus, $\dfrac{A(x + h) - A(x)}{h} \approx f(x),$ Dividing both sides by h

and $\lim\limits_{h \to 0} \dfrac{A(x + h) - A(x)}{h} = \lim\limits_{h \to 0} f(x) = f(x),$

which demonstrates that $A'(x) = f(x)$. We have proved the following remarkable result.

THEOREM 4

Let f be a nonnegative continuous function over an interval $[0, b]$, and let $A(x)$ be the area between the graph of f and the x-axis over the interval $[0, x]$, with $0 < x < b$. Then $A(x)$ is a differentiable function of x and $A'(x) = f(x)$.

Theorem 4 resolves the question posed earlier: Yes, the derivative of the area function is always the function under which the area is being calculated. However, Theorem 4 holds only for the interval $[0, x]$. How can we adapt Theorem 4 to apply when f is defined over any interval $[a, b]$? Referring to the graph below, we see that the area over $[a, b]$ is the same as the area over $[0, b]$ minus the area over $[0, a]$ or $A(b) - A(a)$.

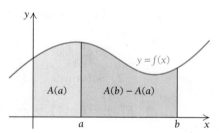

In Section 4.1, we also found that a function's antiderivatives can differ only in their constant terms. Thus, if $F(x)$ is another antiderivative of $f(x)$, then $A(x) = F(x) + C$, for some constant C, and

$$\text{Area} = A(b) - A(a) = F(b) + C - (F(a) + C) = F(b) - F(a).$$

This result tells us that as long as an area is computed by substituting an interval's endpoints into an antiderivative and then subtracting, *any* antiderivative—and any choice of C—can be used. It generally simplifies computations to choose 0 as the value of C.

■ EXAMPLE 1   Find the area under the graph of $f(x) = \frac{1}{5}x^2 + 3$ over the interval $[2, 5]$.

Solution   Although making a drawing is not required, doing so helps us visualize the problem.

Note that every antiderivative of $f(x) = \frac{1}{5}x^2 + 3$ is of the form

$$F(x) = \frac{1}{15}x^3 + 3x + C.$$

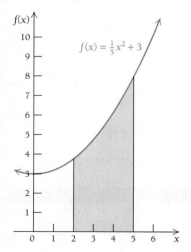

For simplicity, we set $C = 0$, so

$$\text{Area over } [2, 5] = F(5) - F(2)$$
$$= \frac{125}{15} + 15 - \left(\frac{8}{15} + 6\right)$$
$$= 16\frac{4}{5}.$$

Although it is possible to express the area under a curve as the limit of a Riemann sum, as we did in Section 4.2, it is usually much easier to work with antiderivatives.

To find the area under the graph of a nonnegative continuous function f over the interval $[a, b]$:

1. Find any antiderivative $F(x)$ of $f(x)$. Let $C = 0$ for simplicity.
2. Evaluate $F(x)$ at $x = b$ and $x = a$, and compute $F(b) - F(a)$. The result is the area under the graph over the interval $[a, b]$.

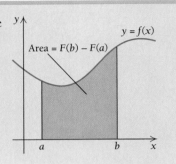

■ EXAMPLE 2   Find the area under the graph of $y = x^2 + 1$ over the interval $[-1, 2]$.

Solution   In this case, $f(x) = x^2 + 1$, with $a = -1$ and $b = 2$.

1. Find any antiderivative $F(x)$ of $f(x)$. We choose the simplest one:

$$F(x) = \frac{x^3}{3} + x.$$

2. Substitute 2 and -1, and find the difference $F(2) - F(-1)$:

$$F(2) - F(-1) = \left[\frac{2^3}{3} + 2\right] - \left[\frac{(-1)^3}{3} + (-1)\right]$$
$$= \frac{8}{3} + 2 - \left[\frac{-1}{3} - 1\right]$$
$$= \frac{8}{3} + 2 + \frac{1}{3} + 1$$
$$= 6.$$

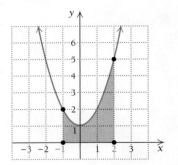

We can make a partial check by counting the squares and parts of squares shaded on the graph to the right.

❮ Quick Check 1

❯ Quick Check 1

Refer to the function and graph in Example 2.

a) Calculate the area over the interval $[0, 5]$.

b) Calculate the area over the interval $[-2, 2]$.

c) Can you suggest a shortcut for part (b)?

TECHNOLOGY CONNECTION

Using iPlot to Find the Area under a Graph

The iPlot app can be used on your iPhone or iPad to evaluate definite integrals and find the area under a continuous nonnegative function over a closed interval. Let's find the area under the graph of the function considered in Example 2, $f(x) = x^2 + 1$, over the interval $[-1, 2]$. After opening iPlot, touch the Functions icon at the bottom of the screen. Press $\boxed{+}$ in the upper right. Then enter the function

Using iPlot to Find the Area under a Graph (*continued*)

as X^2+1. Press Done in the upper right and then Plot to obtain the graph (Fig. 1).

To find the area under the curve over the interval $[-1, 2]$, press Integ (second from the right at the bottom) until it changes color. You may have to press it firmly. Then touch the screen, and you will see a tracing cursor (Fig. 2). Locate the cursor as close as you can to the lower bound of the interval. Then firmly press Apply (in the lower right corner). The screen will glow and then a display like that in Fig. 3 will appear. Touch the number at the top, and change it to -1. Then touch the upper bound number, and change it to 2; the screen will look like Fig. 4.

Next, press OK firmly. The complete curve with the shaded area and the answer, 6, will be displayed (Fig. 5).

Caution! This app seems to crash frequently. You may need to start over. Hopefully, an update will improve the app's stability.

iPlot will also evaluate definite integrals over intervals where the function is not nonnegative. For example, $\int_1^2 (x^2 - 3)\,dx = -\frac{2}{3} \approx -0.666667$, as seen in Fig. 6. As we shall see later in this section, since there is more area below the x-axis than above, the result is negative.

FIGURE 3   **FIGURE 4**

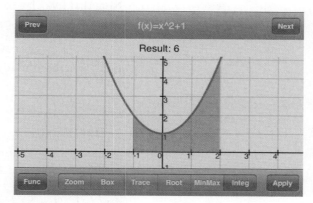

FIGURE 5

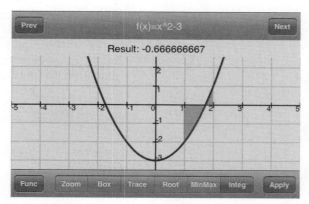

FIGURE 6

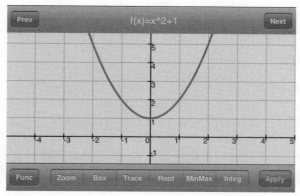

FIGURE 1

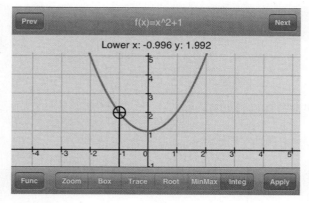

FIGURE 2

EXERCISES

Evaluate each definite integral.

1. $\displaystyle\int_{-2}^{2} (4 - x^2)\,dx$

2. $\displaystyle\int_{-1}^{0} (x^3 - 3x + 1)\,dx$

3. $\displaystyle\int_{1}^{6} \frac{\ln x}{x^2}\,dx$

4. $\displaystyle\int_{-8}^{2} \frac{4}{(1 + e^x)^2}\,dx$

5. $\displaystyle\int_{4}^{15} (0.002x^4 - 0.3x^2 + 4x - 7)\,dx$

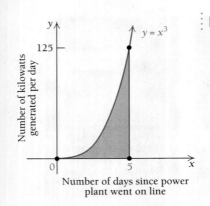

y = x³

Number of kilowatts generated per day

125

0   5

Number of days since power plant went on line

■ **EXAMPLE 3**   Let $y = x^3$ represent the number of kilowatts (kW) generated by a new power plant each day, x days after going on line. Find the area under the graph of $y = x^3$ over the interval $[0, 5]$ and interpret the significance of the area.

Solution   In this case, $f(x) = x^3, a = 0$, and $b = 5$.

1. Find any antiderivative $F(x)$ of $f(x)$. We choose the simplest one:

$$F(x) = \frac{x^4}{4}.$$

2. Substitute 5 and 0, and find the difference $F(5) - F(0)$:

$$F(5) - F(0) = \frac{5^4}{4} - \frac{0^4}{4} = \frac{625}{4} = 156\tfrac{1}{4}.$$

The area represents the total number of kilowatts generated during the first 5 days. Note that kW/day · days = kW.

The difference $F(b) - F(a)$ has the same value for all antiderivatives F of a function f whether the function is nonnegative or not. It is called the *definite integral* of f from a to b.

DEFINITION

Let f be any continuous function over the interval $[a, b]$ and F be any antiderivative of f. Then the **definite integral** of f from a to b is

$$\int_a^b f(x)\, dx = F(b) - F(a).$$

Evaluating definite integrals is called *integration*. The numbers a and b are known as the **limits of integration**. Note that this use of the word *limit* indicates an endpoint of an interval, not a value that is being approached, as you learned in Chapter 1.

■ **EXAMPLE 4**   Evaluate: $\int_a^b x^2\, dx.$

Solution   Using the antiderivative $F(x) = x^3/3$, we have

$$\int_a^b x^2\, dx = \frac{b^3}{3} - \frac{a^3}{3}.$$

It is convenient to use an intermediate notation:

$$\int_a^b f(x)\, dx = \left[F(x)\right]_a^b = F(b) - F(a),$$

where $F(x)$ is an antiderivative of $f(x)$.

■ **EXAMPLE 5**   Evaluate each of the following:

a) $\displaystyle\int_{-1}^4 (x^2 - x)\, dx;$ **b)** $\displaystyle\int_0^3 e^x\, dx;$ **c)** $\displaystyle\int_1^e \left(1 + 2x - \frac{1}{x}\right) dx$ (assume $x > 0$).

Solution

a) $\int_{-1}^{4} (x^2 - x)\, dx = \left[\dfrac{x^3}{3} - \dfrac{x^2}{2}\right]_{-1}^{4} = \left(\dfrac{4^3}{3} - \dfrac{4^2}{2}\right) - \left(\dfrac{(-1)^3}{3} - \dfrac{(-1)^2}{2}\right)$

$$= \left(\dfrac{64}{3} - \dfrac{16}{2}\right) - \left(\dfrac{-1}{3} - \dfrac{1}{2}\right)$$

$$= \dfrac{64}{3} - 8 + \dfrac{1}{3} + \dfrac{1}{2} = 14\tfrac{1}{6}$$

b) $\int_{0}^{3} e^x\, dx = \left[e^x\right]_0^3 = e^3 - e^0 = e^3 - 1$

〉 Quick Check 2

Quick Check 2

Evaluate each of the following:

a) $\int_{2}^{4} (2x^3 - 3x)\, dx;$

b) $\int_{0}^{\ln 4} 2e^x\, dx;$

c) $\int_{1}^{5} \dfrac{x-1}{x}\, dx.$

c) $\int_{1}^{e} \left(1 + 2x - \dfrac{1}{x}\right) dx = \left[x + x^2 - \ln x\right]_1^e$ We assume $x > 0$.

$$= (e + e^2 - \ln e) - (1 + 1^2 - \ln 1)$$

$$= (e + e^2 - 1) - (1 + 1 - 0)$$

$$= e + e^2 - 1 - 1 - 1$$

$$= e + e^2 - 3$$

《 Quick Check 2

The fact that we can express the integral of a function either as a limit of a sum or in terms of an antiderivative is so important that it has a name: the *Fundamental Theorem of Integral Calculus*.

The Fundamental Theorem of Integral Calculus

If a continuous function f has an antiderivative F over $[a, b]$, then

$$\lim_{n \to \infty} \sum_{i=1}^{n} f(x_i)\Delta x = \int_{a}^{b} f(x)\, dx = F(b) - F(a).$$

It is helpful to envision taking the limit as stretching the summation sign, Σ, into something resembling an S (the integral sign) and redefining Δx as dx. Because Δx is used in the limit, dx appears in the integral notation.

More on Area

When we evaluate the definite integral of a nonnegative function, we get the area under the graph over an interval.

■ EXAMPLE 6   Suppose that y is the profit per mile traveled and x is the number of miles traveled, in thousands. Find the area under $y = 1/x$ over the interval $[1, 4]$ and interpret the significance of this area.

Solution

$$\int_{1}^{4} \dfrac{dx}{x} = \left[\ln x\right]_1^4 = \ln 4 - \ln 1$$

$$= \ln 4 - 0 \approx 1.3863$$

Considering the units, (dollars/mile) · miles = dollars, we see that the area represents a to-tal profit of \$1386.30 when the miles traveled increase from 1000 to 4000 miles.

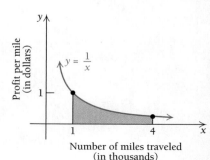

■ **EXAMPLE 7** Find the area under $y = 1/x^2$ over the interval $[1, b]$.

Solution

$$\int_1^b \frac{dx}{x^2} = \int_1^b x^{-2}\, dx$$

$$= \left[\frac{x^{-2+1}}{-2+1}\right]_1^b$$

$$= \left[\frac{x^{-1}}{-1}\right]_1^b = \left[-\frac{1}{x}\right]_1^b$$

$$= \left(-\frac{1}{b}\right) - \left(-\frac{1}{1}\right)$$

$$= 1 - \frac{1}{b}$$

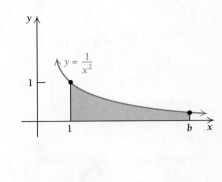

Now let's compare the definite integrals of the functions $y = x^2$ and $y = -x^2$:

$$\int_0^2 x^2\, dx = \left[\frac{x^3}{3}\right]_0^2 \qquad \int_0^2 -x^2\, dx = \left[-\frac{x^3}{3}\right]_0^2$$

$$= \frac{2^3}{3} - \frac{0^3}{3} = \frac{8}{3} \qquad\qquad = -\frac{2^3}{3} + \frac{0^3}{3} = -\frac{8}{3}$$

The graphs of the functions $y = x^2$ and $y = -x^2$ are reflections of each other across the x-axis. Thus, the shaded areas are the same, $\frac{8}{3}$. The evaluation procedure for $y = -x^2$ gave us $-\frac{8}{3}$. This illustrates that for negative-valued functions, the definite integral gives us the opposite of the area between the curve and the x-axis.

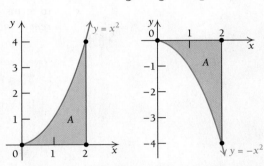

Now let's consider $f(x) = x^2 - 1$ over the interval $[0, 2]$. It has both positive and negative values. We apply the preceding evaluation procedure, even though the function values are not all nonnegative. We do so in two ways.

First, let's use the fact that for any a, b, c, if $a < b < c$, then

$$\int_a^c f(x)\, dx = \int_a^b f(x)\, dx + \int_b^c f(x)\, dx.$$

The area from a to b plus the area from b to c is the area from a to c.

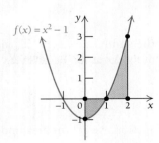

(We will consider this property of integrals again in Section 4.4.) Note that 1 is the x-intercept in $[0, 2]$.

$$\int_0^2 (x^2 - 1)\, dx = \int_0^1 (x^2 - 1)\, dx + \int_1^2 (x^2 - 1)\, dx$$

$$= \left[\frac{x^3}{3} - x\right]_0^1 + \left[\frac{x^3}{3} - x\right]_1^2$$

$$= \left[\left(\frac{1^3}{3} - 1\right) - \left(\frac{0^3}{3} - 0\right)\right] + \left[\left(\frac{2^3}{3} - 2\right) - \left(\frac{1^3}{3} - 1\right)\right]$$

$$= \underbrace{\left[\frac{1}{3} - 1\right]}_{} + \underbrace{\left[\frac{8}{3} - 2 - \frac{1}{3} + 1\right]}_{}$$

$$= -\frac{2}{3} + \frac{4}{3} = \frac{2}{3}.$$

This shows that the area above the x-axis exceeds the area below the x-axis by $\frac{2}{3}$ unit.

Now let's evaluate the original integral in another, more direct, way:

$$\int_0^2 (x^2 - 1)\, dx = \left[\frac{x^3}{3} - x \right]_0^2$$

$$= \left(\frac{2^3}{3} - 2 \right) - \left(\frac{0^3}{3} - 0 \right)$$

$$= \left(\frac{8}{3} - 2 \right) - 0 = \frac{2}{3}.$$

The definite integral of a continuous function over an interval is the sum of the areas above the x-axis minus the sum of the areas below the x-axis.

■ **EXAMPLE 8**   Consider $\int_{-1}^2 (-x^3 + 3x - 1)\, dx$. Predict the sign of the result by examining the graph, and then evaluate the integral.

Solution   From the graph, it appears that there is considerably more area below the x-axis than above. Thus, we expect that

$$\int_{-1}^2 (-x^3 + 3x - 1)\, dx < 0.$$

Evaluating the integral, we have

$$\int_{-1}^2 (-x^3 + 3x - 1)\, dx = \left[-\frac{x^4}{4} + \frac{3}{2}x^2 - x \right]_{-1}^2$$

$$= \left(-\frac{2^4}{4} + \frac{3}{2} \cdot 2^2 - 2 \right) - \left(-\frac{(-1)^4}{4} + \frac{3}{2}(-1)^2 - (-1) \right)$$

$$= (-4 + 6 - 2) - \left(-\frac{1}{4} + \frac{3}{2} + 1 \right) = 0 - 2\tfrac{1}{4}$$

$$= -2\tfrac{1}{4}.$$

As a partial check, we note that the result is negative, as predicted.

❭ **Quick Check 3**

Let $f(x) = x^4 - x^2$.

a) Predict the sign of the value of $\int_0^2 f(x)\, dx$ by examining the graph.

b) Evaluate this integral.

❬ Quick Check 3

The graph shows $y = -x^3 + 3x - 1$.

TECHNOLOGY CONNECTION

Approximating Definite Integrals

There are two methods for evaluating definite integrals with a calculator. Let's consider the function from Example 8: $f(x) = -x^3 + 3x - 1$.

Method 1: fnInt

First, we select fnInt from the MATH menu. Next, we enter the function, the variable, and the endpoints of the interval over which we are integrating. The calculator returns the same value for the definite integral as we found in Example 8.

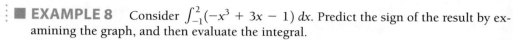

$$\int_{-1}^2 (-X^3 + 3X - 1)dX$$

-2.25

Method 2: $\int f(x)\, dx$

We first graph $y_1 = -x^3 + 3x - 1$. Then we select $\int f(x)dx$ from the CALC menu and enter the lower and upper limits of integration. The calculator shades the area and returns

(continued)

the same value for the definite integral as found in Example 8.

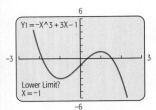

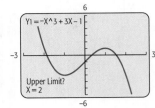

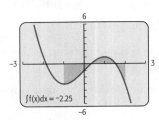

EXERCISES

Evaluate each definite integral.

1. $\displaystyle\int_{-1}^{2} (x^2 - 1)\, dx$

2. $\displaystyle\int_{-2}^{3} (x^3 - 3x + 1)\, dx$

3. $\displaystyle\int_{1}^{6} \frac{\ln x}{x^2}\, dx$

4. $\displaystyle\int_{-8}^{2} \frac{4}{(1 + e^x)^2}\, dx$

5. $\displaystyle\int_{-10}^{10} (0.002x^4 - 0.3x^2 + 4x - 7)\, dx$

Applications Involving Definite Integrals

Determining Total Profit

■ **EXAMPLE 9** **Business: Total Profit from Marginal Profit.** Northeast Airlines determines that the marginal profit resulting from the sale of x seats on a jet traveling from Atlanta to Kansas City, in hundreds of dollars, is given by

$$P'(x) = \sqrt{x} - 6.$$

Find the total profit when 60 seats are sold.

Solution We integrate to find $P(60)$:

$$P(60) = \int_{0}^{60} P'(x)\, dx$$

$$= \int_{0}^{60} \left(\sqrt{x} - 6\right) dx$$

$$= \left[\frac{2}{3} x^{3/2} - 6x\right]_{0}^{60}$$

$$\approx -50.1613. \qquad \text{Using a calculator}$$

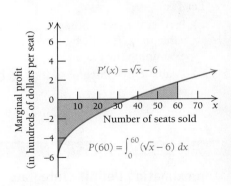

When 60 seats are sold, Northeast's profit is $-\$5016.13$. That is, the airline will lose $\$5016.13$ on the flight.

❮ Quick Check 4

> **Quick Check 4**
>
> **Business.** Referring to Example 9, find the total profit of Northeast Airlines when 140 seats are sold.

Finding Velocity and Distance from Acceleration

Recall that the position coordinate at time t of an object moving along a number line is $s(t)$. Then

$$s'(t) = v(t) = \text{the } \textbf{velocity} \text{ at time } t,$$

$$s''(t) = v'(t) = a(t) = \text{the } \textbf{acceleration} \text{ at time } t.$$

■ **EXAMPLE 10** **Physical Science: Distance.** Suppose that $v(t) = 5t^4$ and $s(0) = 9$. Find $s(t)$. Assume that $s(t)$ is in feet and $v(t)$ is in feet per second.

Solution We first find $s(t)$ by integrating:

$$s(t) = \int v(t)\, dt = \int 5t^4\, dt = t^5 + C.$$

Next we determine C by using the initial condition $s(0) = 9$, which is the starting position for s at time $t = 0$:

$$s(0) = 0^5 + C = 9$$
$$C = 9.$$

Thus, $s(t) = t^5 + 9$.

■ **EXAMPLE 11** **Physical Science: Distance.** Suppose that $a(t) = 12t^2 - 6$, with $v(0) =$ the initial velocity $= 5$, and $s(0) =$ the initial position $= 10$. Find $s(t)$, and graph $a(t), v(t),$ and $s(t)$.

Solution

1. We first find $v(t)$ by integrating $a(t)$:

$$v(t) = \int a(t)\, dt$$

$$= \int (12t^2 - 6)\, dt$$

$$= 4t^3 - 6t + C_1.$$

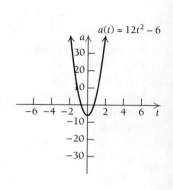

The condition $v(0) = 5$ allows us to find C_1:

$$v(0) = 4 \cdot 0^3 - 6 \cdot 0 + C_1 = 5$$
$$C_1 = 5.$$

Thus, $v(t) = 4t^3 - 6t + 5$.

2. Next we find $s(t)$ by integrating $v(t)$:

$$s(t) = \int v(t)\, dt$$

$$= \int (4t^3 - 6t + 5)\, dt$$

$$= t^4 - 3t^2 + 5t + C_2.$$

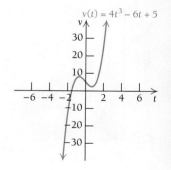

The condition $s(0) = 10$ allows us to find C_2:

$$s(0) = 0^4 - 3 \cdot 0^2 + 5 \cdot 0 + C_2 = 10$$
$$C_2 = 10.$$

Thus, $s(t) = t^4 - 3t^2 + 5t + 10$.

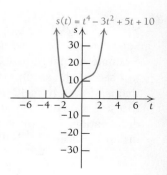

■ **EXAMPLE 12**   **Physical Science: Total Distance Traveled.**   A particle starts out from the origin. Its velocity, in miles per hour, is given by

$$v(t) = \sqrt{t} + t,$$

where t is the number of hours since the particle left the origin. How far does the particle travel during the second, third, and fourth hours (from $t = 1$ to $t = 4$)?

Solution   Recall that velocity, or speed, is the rate of change of distance with respect to time. In other words, velocity is the derivative of the distance function, and the distance function is an antiderivative of the velocity function. To find the total distance traveled from $t = 1$ to $t = 4$, we evaluate the integral

$$\int_1^4 \left(\sqrt{t} + t \right) dt.$$

We have

$$
\begin{aligned}
\int_1^4 \left(\sqrt{t} + t \right) dt &= \int_1^4 \left(t^{1/2} + t \right) dt \\
&= \left[\tfrac{2}{3} t^{3/2} + \tfrac{1}{2} t^2 \right]_1^4 \\
&= \tfrac{2}{3} \cdot 4^{3/2} + \tfrac{1}{2} \cdot 4^2 - \left(\tfrac{2}{3} \cdot 1^{3/2} + \tfrac{1}{2} \cdot 1^2 \right) \\
&= \tfrac{16}{3} + \tfrac{16}{2} - \tfrac{2}{3} - \tfrac{1}{2} \\
&= \tfrac{14}{3} + \tfrac{15}{2} \\
&= \tfrac{73}{6} \\
&= 12 \tfrac{1}{6} \text{ mi.}
\end{aligned}
$$

As a check, we can count shaded squares and parts of squares on the graph at the left. As another partial check, we can observe that the units used are $(\text{mi/hr}) \cdot \text{hr} = \text{mi}$.

⟨ Quick Check 5

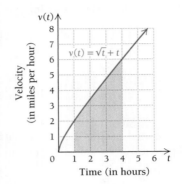

> **Quick Check 5**

Use the grid in the graph in Example 12 to support the claim that the particle travels between 6 and 7 mi during the fifth hour (from $t = 4$ to $t = 5$). Then calculate the actual distance traveled during that hour.

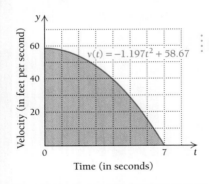

> **Quick Check 6**

Suppose that the driver in Example 13 braked to a stop in 7 sec but did so "linearly," that is, slowed from 40 mi/hr to a stop at a constant rate of deceleration. Find the braking distance in feet.

■ **EXAMPLE 13**   **Physical Science: Braking Distance.**   The driver of a vehicle traveling at 40 mi/hr (58.67 ft/sec) applies the brakes, softly at first, then harder, coming to a complete stop after 7 sec. The velocity as a function of time is modeled by the function $v(t) = -1.197t^2 + 58.67$, where v is in feet per second, t is in seconds, and $0 \le t \le 7$. How far did the vehicle travel while the driver was braking?

Solution   The distance traveled is given by the definite integral of $v(t)$:

$$
\begin{aligned}
\int_0^7 (-1.197t^2 + 58.67)\, dt &= \left[-\frac{1.197}{3} t^3 + 58.67t \right]_0^7 \\
&= -\frac{1.197}{3} (7)^3 + 58.67(7) - 0 \\
&= 273.83 \text{ ft.}
\end{aligned}
$$

This is nearly the length of a football field! In the graph of v, the shaded area represents the distance the vehicle traveled during the 7 sec.

⟨ Quick Check 6

Section Summary

- The exact area between the *x*-axis and the graph of the nonnegative continuous function $y = f(x)$ over the interval $[a, b]$ is found by evaluating the *definite integral*

$$\int_a^b f(x)\, dx = F(b) - F(a),$$

where F is an antiderivative of f.

- If a function has areas both below and above the *x*-axis, the definite integral gives the net total area, or the difference between the sum of the areas above the *x*-axis and the sum of the areas below the *x*-axis.

  ○ If there is more area above the *x*-axis than below, the definite integral will be positive.

  ○ If there is more area below the *x*-axis than above, the definite integral will be negative.

  ○ If the areas above and below the *x*-axis are the same, the definite integral will be 0.

EXERCISE SET 4.3

Find the area under the given curve over the indicated interval.

1. $y = 4;$ $[1, 3]$

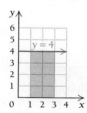

2. $y = 5;$ $[1, 3]$

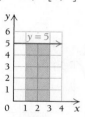

3. $y = 2x;$ $[1, 3]$

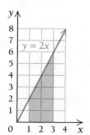

4. $y = x^2;$ $[0, 3]$

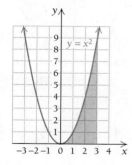

5. $y = x^2;$ $[0, 5]$

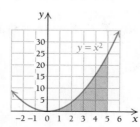

6. $y = x^3;$ $[0, 2]$

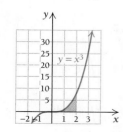

7. $y = x^3;$ $[0, 1]$

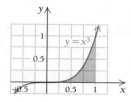

8. $y = 1 - x^2;$ $[-1, 1]$

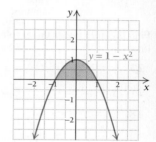

9. $y = 4 - x^2;$ $[-2, 2]$

10. $y = e^x;$ $[0, 2]$

11. $y = e^x;$ $[0, 3]$

12. $y = \dfrac{2}{x};$ $[1, 4]$

13. $y = \dfrac{3}{x};$ $[1, 6]$

14. $y = x^2 - 4x;$ $[-4, -2]$

In each of Exercises 15–24, explain what the shaded area represents.

15.

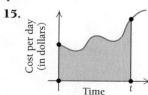

16.

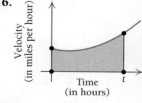

17.

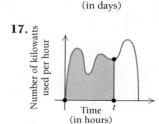

18.

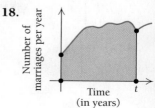

19.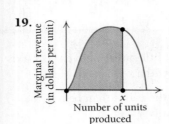

20.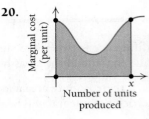

21.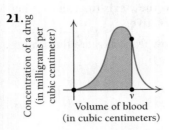

22.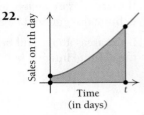

23.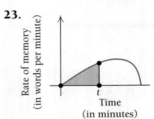

24.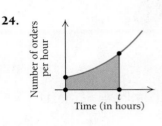

Find the area under the graph of each function over the given interval.

25. $y = x^3$; $[0, 2]$

26. $y = x^4$; $[0, 1]$

27. $y = x^2 + x + 1$; $[2, 3]$

28. $y = 2 - x - x^2$; $[-2, 1]$

29. $y = 5 - x^2$; $[-1, 2]$

30. $y = e^x$; $[-2, 3]$

31. $y = e^x$; $[-1, 5]$

32. $y = 2x + \dfrac{1}{x^2}$; $[1, 4]$

In Exercises 33 and 34, determine visually whether $\int_a^b f(x)\, dx$ is positive, negative, or zero, and express $\int_a^b f(x)\, dx$ in terms of the area A. Explain your result.

33. a)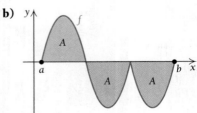

b)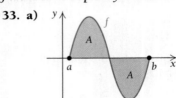

34. a)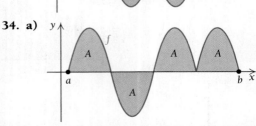

b)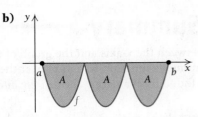

Evaluate. Then interpret the result in terms of the area above and/or below the x-axis.

35. $\displaystyle\int_0^{1.5} (x - x^2)\, dx$

36. $\displaystyle\int_0^2 (x^2 - x)\, dx$

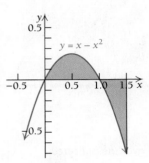

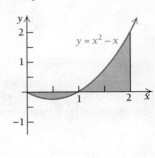

37. $\displaystyle\int_{-1}^1 (x^3 - 3x)\, dx$

38. $\displaystyle\int_0^b -2e^{3x}\, dx$

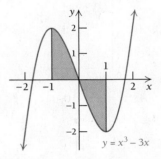

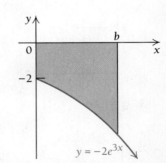

39–42. Check the results of each of Exercises 35–38 using a graphing calculator.

Evaluate.

43. $\displaystyle\int_1^3 (3t^2 + 7)\, dt$

44. $\displaystyle\int_1^2 (4t^3 - 1)\, dt$

45. $\displaystyle\int_1^4 \left(\sqrt{x} - 1\right) dx$

46. $\displaystyle\int_1^8 \left(\sqrt[3]{x} - 2\right) dx$

47. $\displaystyle\int_{-2}^5 (2x^2 - 3x + 7)\, dx$

48. $\displaystyle\int_{-2}^3 (-x^2 + 4x - 5)\, dx$

49. $\displaystyle\int_{-5}^2 e^t\, dt$

50. $\displaystyle\int_{-2}^3 e^{-t}\, dt$

51. $\displaystyle\int_a^b \tfrac{1}{2}x^2\, dx$

52. $\displaystyle\int_a^b \tfrac{1}{5}x^3\, dx$

53. $\displaystyle\int_a^b e^{2t}\, dt$

54. $\displaystyle\int_a^b -e^t\, dt$

55. $\int_1^e \left(x + \dfrac{1}{x} \right) dx$

56. $\int_1^e \left(x - \dfrac{1}{x} \right) dx$

57. $\int_0^2 \sqrt{2x}\, dx$ (*Hint:* Simplify first.)

58. $\int_0^{27} \sqrt{3x}\, dx$

APPLICATIONS

Business and Economics

59. Business: total profit. Pure Water Enterprises finds that the marginal profit, in dollars, from drilling a well that is x feet deep is given by

$$P'(x) = \sqrt[5]{x}.$$

Find the profit when a well 250 ft deep is drilled.

60. Business: total revenue. Sally's Sweets finds that the marginal revenue, in dollars, from the sale of x pounds of maple-coated pecans is given by

$$R'(x) = 6x^{-1/6}.$$

Find the revenue when 300 lb of maple-coated pecans are produced.

61. Business: increasing total cost. Kitchens-to-Please Contracting determines that the marginal cost, in dollars per foot, of installing x feet of kitchen countertop is given by

$$C'(x) = 8x^{-1/3}.$$

Find the cost of installing an extra 14 ft of countertop after 50 ft have already been ordered.

62. Business: increasing total profit. Laso Industries finds that the marginal profit, in dollars, from the sale of x digital control boards is given by

$$P'(x) = 2.6x^{0.1}.$$

A customer orders 1200 digital control boards and later increases the order to 1500. Find the extra profit resulting from the increase in order size.

63. Accumulated sales. A company estimates that its sales will grow continuously at a rate given by the function

$$S'(t) = 20e^t,$$

where $S'(t)$ is the rate at which sales are increasing, in dollars per day, on day t.

a) Find the accumulated sales for the first 5 days.
b) Find the sales from the 2nd day through the 5th day. (This is the integral from 1 to 5.)

64. Accumulated sales. Raggs, Ltd., estimates that its sales will grow continuously at a rate given by the function

$$S'(t) = 10e^t,$$

where $S'(t)$ is the rate at which sales are increasing, in dollars per day, on day t.

a) Find the accumulated sales for the first 5 days.
b) Find the sales from the 2nd day through the 5th day. (This is the integral from 1 to 5.)

Credit market debt. *The annual rate of change in the national credit market debt (in billions of dollars per year) can be modeled by the function*

$$D'(t) = 857.98 + 829.66t - 197.34t^2 + 15.36t^3,$$

where t is the number of years since 1995. (Source: Federal Reserve System.) Use the preceding information for Exercises 65 and 66.

65. By how much did the credit market debt increase between 1996 and 2000?

66. By how much did the credit market debt increase between 1999 and 2005?

Industrial learning curve. *A company is producing a new product. Due to the nature of the product, the time required to produce each unit decreases as workers become more familiar with the production procedure. It is determined that the function for the learning process is*

$$T(x) = 2 + 0.3\left(\dfrac{1}{x} \right),$$

where $T(x)$ is the time, in hours, required to produce the xth unit. Use this information for Exercises 67 and 68.

67. Find the total time required for a new worker to produce units 1 through 10; units 20 through 30.

68. Find the total time required for a new worker to produce units 1 through 20; units 20 through 40.

Social Sciences

Memorizing. *The rate of memorizing information initially increases. Eventually, however, a maximum rate is reached, after which it begins to decrease.*

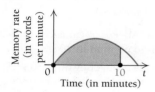

69. Suppose that in a memory experiment the rate of memorizing is given by

$$M'(t) = -0.009t^2 + 0.2t,$$

where $M'(t)$ is the memory rate, in words per minute. How many words are memorized in the first 10 min (from $t = 0$ to $t = 10$)?

70. Suppose that in another memory experiment the rate of memorizing is given by

$$M'(t) = -0.003t^2 + 0.2t,$$

where $M'(t)$ is the memory rate, in words per minute. How many words are memorized in the first 10 min (from $t = 0$ to $t = 10$)?

71. See Exercise 69. How many words are memorized during minutes 10–15?

72. See Exercise 70. How many words are memorized during minutes 10–17?

Life and Physical Sciences

Find $s(t)$.

73. $v(t) = 3t^2$, $s(0) = 4$

74. $v(t) = 2t$, $s(0) = 10$

Find $v(t)$.

75. $a(t) = 4t$, $v(0) = 20$

76. $a(t) = 6t$, $v(0) = 30$

Find $s(t)$.

77. $a(t) = -2t + 6$, with $v(0) = 6$ and $s(0) = 10$

78. $a(t) = -6t + 7$, with $v(0) = 10$ and $s(0) = 20$

79. Physics. A particle is released as part of an experiment. Its speed t seconds after release is given by $v(t) = -0.5t^2 + 10t$, where $v(t)$ is in meters per second.

   a) How far does the particle travel during the first 5 sec?
   b) How far does it travel during the second 5 sec?

80. Physics. A particle is released during an experiment. Its speed t minutes after release is given by $v(t) = -0.3t^2 + 9t$, where $v(t)$ is in kilometers per minute.

   a) How far does the particle travel during the first 10 min?
   b) How far does it travel during the second 10 min?

81. Distance and speed. A motorcycle accelerates at a constant rate from 0 mph ($v(0) = 0$) to 60 mph in 15 sec.

   a) How fast is it traveling after 15 sec?
   b) How far has it traveled after 15 sec? (*Hint*: Convert seconds to hours.)

82. Distance and speed. A car accelerates at a constant rate from 0 mph to 60 mph in 30 sec.

   a) How fast is it traveling after 30 sec?
   b) How far has it traveled after 30 sec?

83. Distance and speed. A bicyclist decelerates at a constant rate from 30 km/hr to a standstill in 45 sec.

   a) How fast is the bicyclist traveling after 20 sec?
   b) How far has the bicyclist traveled after 45 sec?

84. Distance and speed. A cheetah decelerates at a constant rate from 50 km/hr to a complete stop in 20 sec.

   a) How fast is the cheetah moving after 10 sec?
   b) How far has the cheetah traveled after 20 sec?

85. Distance. For a freely falling object, $a(t) = -32$ ft/sec^2, $v(0) = $ initial velocity $= v_0$ (in ft/sec), and $s(0) = $ initial height $= s_0$ (in ft). Find a general expression for $s(t)$ in terms of v_0 and s_0.

86. Time. A ball is thrown upward from a height of 10 ft, that is, $s(0) = 10$, at an initial velocity of 80 ft/sec, or $v(0) = 80$. How long will it take before the ball hits the ground? (See Exercise 85.)

87. Distance. A car accelerates at a constant rate from 0 to 60 mph in min. How far does the car travel during that time?

88. Distance. A motorcycle accelerates at a constant rate from 0 to 50 mph in 15 sec. How far does it travel during that time?

89. Physics. A particle starts out from the origin. Its velocity, in miles per hour, after t hours is given by

$$v(t) = 3t^2 + 2t.$$

How far does it travel from the 2nd hour through the 5th hour (from $t = 1$ to $t = 5$)?

90. Physics. A particle starts out from the origin. Its velocity, in miles per hour, after t hours is given by

$$v(t) = 4t^3 + 2t.$$

How far does it travel from the start through the 3rd hour (from $t = 0$ to $t = 3$)?

SYNTHESIS

91. Accumulated sales. Bluetape, Inc. estimates that its sales will grow continuously at a rate given by

$$S'(t) = 0.5e^t,$$

where $S'(t)$ is the rate at which sales are increasing, in dollars per day, on day t. On what day will accumulated sales first exceed $10,000?

92. Total pollution. A factory is polluting a lake in such a way that the rate of pollutants entering the lake at time t, in months, is given by

$$N'(t) = 280t^{3/2},$$

where N is the total number of pounds of pollutants in the lake at time t.

   a) How many pounds of pollutants enter the lake in 16 months?
   b) An environmental board tells the factory that it must begin cleanup procedures after 50,000 lb of pollutants have entered the lake. After what length of time will this occur?

Evaluate.

93. $\displaystyle\int_{2}^{3}\frac{x^2-1}{x-1}\,dx$

94. $\displaystyle\int_{1}^{5}\frac{x^5-x^{-1}}{x^2}\,dx$

95. $\displaystyle\int_{4}^{16}(x-1)\sqrt{x}\,dx$

96. $\displaystyle\int_{0}^{1}(x+2)^3\,dx$

97. $\displaystyle\int_{1}^{8}\frac{\sqrt[3]{x^2}-1}{\sqrt[3]{x}}\,dx$

98. $\displaystyle\int_{0}^{1}\frac{x^3+8}{x+2}\,dx$

99. $\displaystyle\int_{2}^{5}\left(t+\sqrt{3}\right)\left(t-\sqrt{3}\right)\,dt$

100. $\displaystyle\int_{0}^{1}(t+1)^3\,dt$

101. $\displaystyle\int_{1}^{3}\left(x-\frac{1}{x}\right)^2\,dx$

102. $\displaystyle\int_{1}^{3}\frac{t^5-t}{t^3}\,dt$

103. $\displaystyle\int_{4}^{9}\frac{t+1}{\sqrt{t}}\,dt$

Find the error in each of the following. Explain.

104. $\displaystyle\int_{1}^{2}(x^2+x+1)\,dx=\left[\frac{1}{3}x^3+\frac{1}{2}x^2+x\right]_{1}^{2}$

$$=\left(\frac{1}{3}\cdot 2^3+\frac{1}{2}\cdot 2^2+2\right)$$

$$=\frac{20}{3}$$

105. $\displaystyle\int_{1}^{2}(\ln x-e^x)\,dx=\left[\frac{1}{x}-e^x\right]_{1}^{2}$

$$=\left(\frac{1}{2}-e^2\right)-(1-e^1)$$

$$=e-e^2-\frac{1}{2}$$

 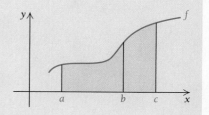

Evaluate.

106. $\displaystyle\int_{-1.2}^{6.3}(x^3-9x^2+27x+50)\,dx$

107. $\displaystyle\int_{-8}^{1.4}(x^4+4x^3-36x^2-160x+300)\,dx$

108. $\displaystyle\int_{-2}^{2}\sqrt{4-x^2}\,dx$

109. $\displaystyle\int_{-1}^{1}\left(3+\sqrt{1-x^2}\right)\,dx$

110. $\displaystyle\int_{0}^{8}x(x-5)^4\,dx$

111. $\displaystyle\int_{-2}^{2}x^{2/3}\left(\frac{5}{2}-x\right)\,dx$

112. $\displaystyle\int_{2}^{4}\frac{x^2-4}{x^2-3}\,dx$

113. $\displaystyle\int_{-10}^{10}\frac{8}{x^2+4}\,dx$

114. Prove that $\displaystyle\int_{a}^{b}f(x)\,dx=-\int_{b}^{a}f(x)\,dx$.

Answers to Quick Checks

1. (a) $46\frac{2}{3}$, or $\frac{140}{3}$; **(b)** $9\frac{1}{3}$, or $\frac{28}{3}$; **(c)** integrate from 0 to 2, then double the result. **2. (a)** 102; **(b)** 6; **(c)** $4-\ln 5\approx 2.39$ **3. (a)** Positive; **(b)** $\frac{56}{15}$ **4.** Approximately \$26,433.49 **5.** Approximately 6.62 mi **6.** Approximately 205.35 ft

---

4.4

Properties of Definite Integrals

The Additive Property of Definite Integrals

OBJECTIVES

- Use properties of definite integrals to find the area between curves.
- Solve applied problems involving definite integrals.
- Determine the average value of a function.

We have seen that the definite integral

$$\int_{a}^{c}f(x)\,dx$$

can be regarded as the area under the graph of $y=f(x)\geq 0$ over the interval $[a,c]$. Thus, if b is such that $a<b<c$, the above integral can be expressed as a sum. This **additive property of definite integrals** is stated in the following theorem.

THEOREM 5

For $a<b<c$,

$$\int_{a}^{c}f(x)\,dx=\int_{a}^{b}f(x)\,dx+\int_{b}^{c}f(x)\,dx.$$

For any number b between a and c, the integral from a to c is the integral from a to b plus the integral from b to c.

Theorem 5 is especially useful when a function is defined piecewise, in different ways over different subintervals.

■ **EXAMPLE 1** Find the area under the graph of $y = f(x)$ from -4 to 5, where

$$f(x) = \begin{cases} 9, & \text{for } x < 3, \\ x^2, & \text{for } x \geq 3. \end{cases}$$

Solution

$$\int_{-4}^{5} f(x)\,dx = \int_{-4}^{3} f(x)\,dx + \int_{3}^{5} f(x)\,dx$$

$$= \int_{-4}^{3} 9\,dx + \int_{3}^{5} x^2\,dx$$

$$= 9\big[x\big]_{-4}^{3} + \left[\frac{x^3}{3}\right]_{3}^{5}$$

$$= 9(3 - (-4)) + \left(\frac{5^3}{3} - \frac{3^3}{3}\right)$$

$$= 95\tfrac{2}{3}$$

❮ Quick Check 1

> **Quick Check 1**
>
> Find the area under the graph of $y = g(x)$ from -3 to 6, where
>
> $$g(x) = \begin{cases} x^2, & \text{for } x \leq 2, \\ 8 - x, & \text{for } x > 2. \end{cases}$$

■ **EXAMPLE 2** Evaluate each definite integral:

a) $\displaystyle\int_{-3}^{4} |x|\,dx;$ **b)** $\displaystyle\int_{0}^{3} |1 - x^2|\,dx.$

Solution

a) The absolute-value function $f(x) = |x|$ is defined piecewise as follows:

$$f(x) = |x| = \begin{cases} -x, & \text{for } x < 0, \\ x, & \text{for } x \geq 0. \end{cases}$$ See Example 8 in Section R.5.

Therefore,

$$\int_{-3}^{4} |x|\,dx = \int_{-3}^{0} (-x)\,dx + \int_{0}^{4} x\,dx$$

$$= \left[-\frac{x^2}{2}\right]_{-3}^{0} + \left[\frac{x^2}{2}\right]_{0}^{4}$$

$$= \left(-\frac{0^2}{2} - \left(-\frac{(-3)^2}{2}\right)\right) + \left(\frac{4^2}{2} - \frac{0^2}{2}\right)$$

$$= \frac{9}{2} + \frac{16}{2} = \frac{25}{2}.$$

As a check, this definite integral can also be evaluated using geometry, since the two regions are triangles.

b) The graph of $f(x) = |1 - x^2|$ is the graph of $y = 1 - x^2$, where any portion of the graph below the x-axis (that is, where $y < 0$) is reflected above the x-axis.

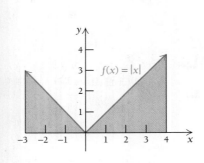

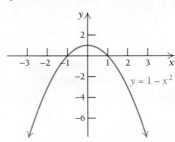

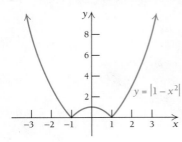

We see that y is negative when $x < -1$ or $x > 1$. Therefore, the function is defined piecewise:

$$f(x) = |1 - x^2| = \begin{cases} x^2 - 1, & \text{for } x < -1 \text{ or } x > 1, \\ 1 - x^2, & \text{for } -1 \le x \le 1. \end{cases}$$

See Exercises 101–108 in Section 2.1.

The definite integral of $f(x) = |1 - x^2|$ over the interval $[0, 3]$ is treated as the sum of two definite integrals, those of $y = 1 - x^2$ over the interval $[0, 1]$ and of $y = x^2 - 1$ over the interval $[1, 3]$:

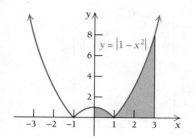

$$\int_0^3 |1 - x^2|\, dx = \int_0^1 (1 - x^2)\, dx + \int_1^3 (x^2 - 1)\, dx$$

$$= \left[x - \frac{1}{3}x^3 \right]_0^1 + \left[\frac{1}{3}x^3 - x \right]_1^3$$

$$= \left(1 - \frac{1}{3} \cdot 1^3 \right) - \left(0 - \frac{1}{3} \cdot 0^3 \right) + \left(\frac{1}{3} \cdot 3^3 - 3 \right) - \left(\frac{1}{3} \cdot 1^3 - 1 \right)$$

$$= \frac{22}{3}.$$

> **Quick Check 2**
>
> Evaluate $\int_0^7 |2x - 1|\, dx$.

⟨ Quick Check 2

The Area of a Region Bounded by Two Graphs

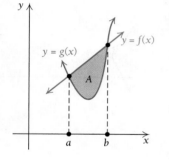

Suppose that we want to find the area of a region bounded by the graphs of two functions, $y = f(x)$ and $y = g(x)$, as shown at the left.

Note that the area of the desired region A is the area of A_2 minus that of A_1.

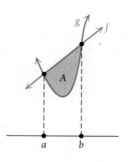 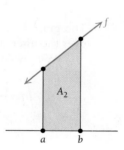 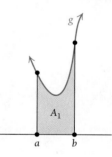

Thus,

$$A = \underbrace{\int_a^b f(x)\, dx}_{A_2} - \underbrace{\int_a^b g(x)\, dx}_{A_1},$$

or $A = \displaystyle\int_a^b [f(x) - g(x)]\, dx.$

In general, we have the following theorem.

THEOREM 6

Let f and g be continuous functions and suppose that $f(x) \geq g(x)$ over the interval $[a, b]$. Then the area of the region between the two curves, from $x = a$ to $x = b$, is

$$\int_a^b [f(x) - g(x)] \, dx.$$

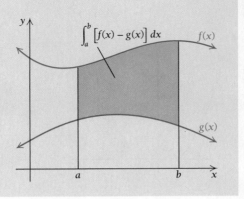

■ **EXAMPLE 3** Find the area of the region bounded by the graphs of $f(x) = 2x + 1$ and $g(x) = x^2 + 1$.

Solution First, we make a reasonably accurate sketch, as in the figure at the right, to determine which is the upper graph. To calculate the points of intersection, we set $f(x)$ equal to $g(x)$ and solve.

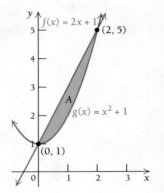

$$f(x) = g(x)$$
$$2x + 1 = x^2 + 1$$
$$0 = x^2 - 2x$$
$$0 = x(x - 2)$$
$$x = 0 \quad \text{or} \quad x = 2$$

The graphs intersect at $x = 0$ and $x = 2$. We see that, over the interval $[0, 2]$, f is the upper graph.

We now compute the area as follows:

$$\int_0^2 [(2x + 1) - (x^2 + 1)] \, dx = \int_0^2 (2x - x^2) \, dx$$

$$= \left[x^2 - \frac{x^3}{3} \right]_0^2$$

$$= \left(2^2 - \frac{2^3}{3} \right) - \left(0^2 - \frac{0^3}{3} \right)$$

$$= 4 - \frac{8}{3}$$

$$= \frac{4}{3}.$$

〉 **Quick Check 3**

Find the area of the region bounded by the graphs of $y = \sqrt{x}$ and $y = \frac{1}{3}x$.

❮ Quick Check 3

To find the area bounded by two graphs, such as $y_1 = -2x - 7$ and $y_2 = -x^2 - 4$, we graph each function and use INTERSECT from the CALC menu to determine the points of intersection.

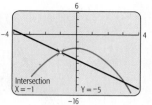

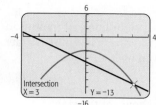

Next, we use fnInt from the MATH menu.

$$\int_{-1}^{3} (-X^2 - 4) - (-2X - 7)\,dX$$
$$10.66666667$$

The area bounded by the two curves is about 10.7.

EXERCISES

1. Find the area of the region in Example 3 using this approach.

2. Use iPlot to find the same area.

■ **EXAMPLE 4**   Find the area of the region bounded by

$$y = x^4 - 3x^3 - 4x^2 + 10, \qquad y = 40 - x^2, \qquad x = 1, \qquad \text{and} \qquad x = 3.$$

Solution   First, we make a reasonably accurate sketch, as in the figure below, to ensure that we have the correct configuration. Note that over $[1, 3]$, the upper graph is $y = 40 - x^2$. Thus, $40 - x^2 \geq x^4 - 3x^3 - 4x^2 + 10$ over $[1, 3]$.

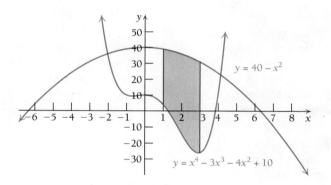

By subtracting a negative number (the area between the red curve and the *x*-axis), we are adding areas. The limits of integration are stated, so we can compute the area as follows:

$$\int_{1}^{3} \left[(40 - x^2) - (x^4 - 3x^3 - 4x^2 + 10) \right] dx$$

$$= \int_{1}^{3} (-x^4 + 3x^3 + 3x^2 + 30)\, dx$$

$$= \left[-\frac{x^5}{5} + \frac{3}{4}x^4 + x^3 + 30x \right]_{1}^{3}$$

$$= \left(-\frac{3^5}{5} + \frac{3}{4} \cdot 3^4 + 3^3 + 30 \cdot 3 \right) - \left(-\frac{1^5}{5} + \frac{3}{4} \cdot 1^4 + 1^3 + 30 \cdot 1 \right)$$

$$= 97.6.$$

An Environmental Application

■ **EXAMPLE 5**   **Life Science: Emission Control.**   A clever college student develops an engine that is believed to meet all state standards for emission control. The new engine's rate of emission is given by

$$E(t) = 2t^2,$$

where $E(t)$ is the emissions, in billions of pollution particulates per year, at time t, in years. The emission rate of a conventional engine is given by

$$C(t) = 9 + t^2.$$

The graphs of both curves are shown at the right.

a) At what point in time will the emission rates be the same?

b) What reduction in emissions results from using the student's engine?

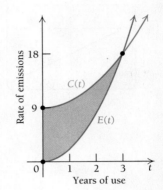

Solution

a) The rate of emission will be the same when $E(t) = C(t)$, or

$$2t^2 = 9 + t^2$$
$$t^2 - 9 = 0$$
$$(t - 3)(t + 3) = 0$$
$$t = 3 \quad \text{or} \quad t = -3.$$

Since negative time has no meaning in this problem, the emission rates will be the same when $t = 3$ yr.

b) The reduction in emissions is represented by the area of the shaded region in the figure above. It is the area between $C(t) = 9 + t^2$ and $E(t) = 2t^2$, from $t = 0$ to $t = 3$, and is computed as follows:

$$\int_0^3 \left[(9 + t^2) - 2t^2 \right] dt = \int_0^3 (9 - t^2)\, dt$$

$$= \left[9t - \frac{t^3}{3} \right]_0^3$$

$$= \left(9 \cdot 3 - \frac{3^3}{3} \right) - \left(9 \cdot 0 - \frac{0^3}{3} \right)$$

$$= 27 - 9$$

$$= 18 \text{ billion pollution particulates.}$$

❬ **Quick Check 4**

> ❭ **Quick Check 4**
>
> Two rockets are fired upward simultaneously. The first rocket's velocity is given by $v_1(t) = 4t$; the second rocket's velocity is given by $v_2(t) = \frac{1}{10}t^2$. In both cases, t is in seconds and velocity is in feet per second.
>
> **a)** When the two rockets' velocities are the same, how far ahead (in feet) is the first rocket?
>
> **b)** After how many seconds will the second rocket catch up to the first, and how far away will they be from the starting point?

Average Value of a Continuous Function

Another important use of the area under a curve is in finding the average value of a continuous function over a closed interval.

Suppose that

$$T = f(t)$$

is the temperature at time t recorded at a weather station on a certain day. The station uses a 24-hr clock, so the domain of the temperature function is the interval $[0, 24]$. The function is continuous, as shown in the following graph.

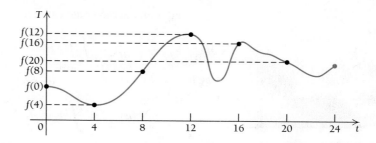

To find the average temperature for the day, we might take six temperature readings at 4-hr intervals, starting at midnight:

$$T_0 = f(0), \quad T_1 = f(4), \quad T_2 = f(8), \quad T_3 = f(12), \quad T_4 = f(16), \quad T_5 = f(20).$$

The average reading would then be the sum of these six readings divided by 6:

$$T_{av} = \frac{T_0 + T_1 + T_2 + T_3 + T_4 + T_5}{6}.$$

This computation of the average temperature has limitations. For example, suppose that it is a hot summer day, and at 2:00 in the afternoon (hour 14 on the 24-hr clock), there is a short thunderstorm that cools the air for an hour between our readings. This temporary dip would not show up in the average computed above.

What can we do? We could take 48 readings at half-hour intervals. This should give us a better result. In fact, the shorter the time between readings, the better the result should be. It seems reasonable that we might define the **average value** of T over $[0, 24]$ to be the limit, as n approaches ∞, of the average of n values:

$$\text{Average value } T = \lim_{n \to \infty} \left(\frac{1}{n} \sum_{i=1}^{n} T_i \right) = \lim_{n \to \infty} \left(\frac{1}{n} \sum_{i=1}^{n} f(t_i) \right).$$

Note that this is not too far from our definition of an integral. All we need is to get Δt, which is $(24 - 0)/n$, or $24/n$, into the summation. We accomplish this by multiplying by 1, writing 1 as $\frac{1}{\Delta t} \cdot \Delta t$:

$$\text{Average value of } T = \lim_{n \to \infty} \left(\frac{1}{\Delta t} \cdot \frac{1}{n} \sum_{i=1}^{n} f(t_i) \, \Delta t \right)$$

$$= \lim_{n \to \infty} \left(\frac{n}{24} \cdot \frac{1}{n} \sum_{i=1}^{n} f(t_i) \, \Delta t \right) \qquad \Delta t = \frac{24}{n}, \text{ so } \frac{1}{\Delta t} = \frac{n}{24}$$

$$= \frac{1}{24} \lim_{n \to \infty} \sum_{i=1}^{n} f(t_i) \, \Delta t$$

$$= \frac{1}{24} \int_0^{24} f(t) \, dt.$$

DEFINITION

Let f be a continuous function over a closed interval $[a, b]$. Its **average value**, y_{av}, over $[a, b]$ is given by

$$y_{av} = \frac{1}{b - a} \int_a^b f(x) \, dx.$$

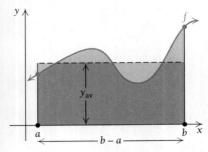

Let's consider average value in another way. If we multiply both sides of

$$y_{av} = \frac{1}{b-a}\int_a^b f(x)\,dx$$

by $b-a$, we get

$$(b-a)y_{av} = \int_a^b f(x)\,dx.$$

Now the expression on the left side is the area of a rectangle of length $b-a$ and height y_{av}. The area of such a rectangle is the same as the area under the graph of $y = f(x)$ over the interval $[a, b]$, as shown in the figure at the left.

TECHNOLOGY CONNECTION

EXERCISE

1. Graph $f(x) = x^4$. Compute the average value of the function over the interval $[0, 2]$, using the method of Example 6. Then use that value, y_{av}, and draw a graph of it as a horizontal line using the same set of axes. What does this line represent in comparison to the graph of $f(x) = x^4$ and its associated area?

■ **EXAMPLE 6** Find the average value of $f(x) = x^2$ over the interval $[0, 2]$.

Solution The average value is

$$\frac{1}{2-0}\int_0^2 x^2\,dx = \frac{1}{2}\left[\frac{x^3}{3}\right]_0^2$$

$$= \frac{1}{2}\left(\frac{2^3}{3} - \frac{0^3}{3}\right)$$

$$= \frac{1}{2}\cdot\frac{8}{3} = \frac{4}{3}, \quad \text{or } 1\frac{1}{3}.$$

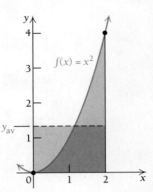

Note that although the values of $f(x)$ increase from 0 to 4 over $[0, 2]$, we do not expect the average value to be 2 (which is half of 4), because we see from the graph that $f(x)$ is less than 2 over more than half the interval.

■ **EXAMPLE 7** Rico's speed, in miles per hour, t minutes after entering the freeway, is given by

$$v(t) = -\frac{1}{200}t^3 + \frac{3}{20}t^2 - \frac{3}{8}t + 60, \quad t \le 30.$$

From 5 min after entering the freeway to 25 min after doing so, what was Rico's average speed? How far did he travel over that time interval?

Solution The average speed is

$$\frac{1}{25-5}\int_5^{25}\left(-\frac{1}{200}t^3 + \frac{3}{20}t^2 - \frac{3}{8}t + 60\right)dt$$

$$= \frac{1}{20}\left[-\frac{1}{800}t^4 + \frac{1}{20}t^3 - \frac{3}{16}t^2 + 60t\right]_5^{25}$$

$$= \frac{1}{20}\left[\left(-\frac{1}{800}\cdot25^4 + \frac{1}{20}\cdot25^3 - \frac{3}{16}\cdot25^2 + 60\cdot25\right)\right.$$

$$\left. -\left(-\frac{1}{800}\cdot5^4 + \frac{1}{20}\cdot5^3 - \frac{3}{16}\cdot5^2 + 60\cdot5\right)\right]$$

$$= \frac{1}{20}\left(\frac{53{,}625}{32} - \frac{9625}{32}\right) = 68\frac{3}{4} \text{ mph.}$$

To find how far Rico traveled over the time interval $[5, 25]$, we first note that t is given in minutes, not hours. Since 25 min − 5 min = 20 min is $\frac{1}{3}$ hr, the distance traveled over $[5, 25]$, is

$$\tfrac{1}{3}\cdot68\tfrac{3}{4} = 22\tfrac{11}{12} \text{ mi.}$$

❭ **Quick Check 5**

The temperature, in degrees Fahrenheit, in Minneapolis on a winter's day is modeled by the function

$$f(x) = -0.012x^3 + 0.38x^2 - 1.99x - 10.1,$$

where x is the number of hours from midnight ($0 \le x \le 24$). Find the average temperature in Minneapolis during this 24-hour period.

❬ Quick Check 5

Section Summary

- The *additive property of definite integrals* states that a definite integral can be expressed as the sum of two (or more) other definite integrals. If f is continuous on $[a, c]$ and we choose b such that $a < b < c$, then

$$\int_a^c f(x)\, dx = \int_a^b f(x)\, dx + \int_b^c f(x)\, dx.$$

- The area of a region bounded by the graphs of two functions, $f(x)$ and $g(x)$, where $f(x) \geq g(x)$ over an interval $[a, b]$, is

$$A = \int_a^b \left[f(x) - g(x) \right] dx.$$

- The *average value* of a continuous function f over an interval $[a, b]$ is

$$y_{av} = \frac{1}{b - a} \int_a^b f(x)\, dx.$$

EXERCISE SET 4.4

Find the area under the graph of f over the interval $[1, 5]$.

1. $f(x) = \begin{cases} 2x + 1, & \text{for } x \leq 3, \\ 10 - x, & \text{for } x > 3 \end{cases}$

2. $f(x) = \begin{cases} x + 5, & \text{for } x \leq 4, \\ 11 - \frac{1}{2}x, & \text{for } x > 4 \end{cases}$

Find the area under the graph of g over the interval $[-2, 3]$.

3. $g(x) = \begin{cases} x^2 + 4, & \text{for } x \leq 0, \\ 4 - x, & \text{for } x > 0 \end{cases}$

4. $g(x) = \begin{cases} -x^2 + 5, & \text{for } x \leq 0, \\ x + 5, & \text{for } x > 0 \end{cases}$

Find the area under the graph of f over the interval $[-6, 4]$.

5. $f(x) = \begin{cases} -x^2 - 6x + 7, & \text{for } x < 1, \\ \frac{3}{2}x - 1, & \text{for } x \geq 1 \end{cases}$

6. $f(x) = \begin{cases} -x - 1, & \text{for } x < -1, \\ -x^2 + 4x + 5, & \text{for } x \geq -1 \end{cases}$

Find the area represented by each definite integral.

7. $\int_0^4 |x - 3|\, dx$

8. $\int_{-1}^1 |3x - 2|\, dx$

9. $\int_0^2 |x^3 - 1|\, dx$

10. $\int_{-3}^4 |x^3|\, dx$

In Exercises 11–16, determine the x-values at which the graphs of f and g cross. If no such x-values exist, state that fact.

11. $f(x) = 9, \quad g(x) = x^2$

12. $f(x) = 8, \quad g(x) = \frac{1}{2}x^2$

13. $f(x) = 7, \quad g(x) = x^2 - 3x + 2$

14. $f(x) = -6, \quad g(x) = x^2 + 3x + 13$

15. $f(x) = x^2 - x - 5, \quad g(x) = x + 10$

16. $f(x) = x^2 - 7x + 20, \quad g(x) = 2x + 6$

Find the area of the shaded region.

17. $f(x) = 2x + x^2 - x^3, \quad g(x) = 0$

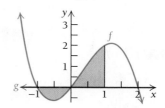

18. $f(x) = x^3 + 3x^2 - 9x - 12, \quad g(x) = 4x + 3$

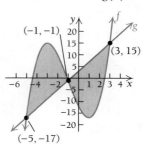

19. $f(x) = x^4 - 8x^3 + 18x^2, \quad g(x) = x + 28$

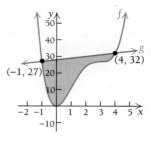

20. $f(x) = 4x - x^2$, $g(x) = x^2 - 6x + 8$

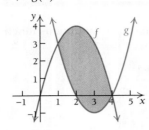

Find the area of the region bounded by the graphs of the given equations.

21. $y = x, y = x^3, x = 0, x = 1$

22. $y = x, y = x^4$

23. $y = x + 2, y = x^2$

24. $y = x^2 - 2x, y = x$

25. $y = 6x - x^2, y = x$

26. $y = x^2 - 6x, y = -x$

27. $y = 2x - x^2, y = -x$

28. $y = x^2, y = \sqrt{x}$

29. $y = x, y = \sqrt[4]{x}$

30. $y = 3, y = x, x = 0$

31. $y = 5, y = \sqrt{x}, x = 0$

32. $y = x^2, y = x^3$

33. $y = 4 - x^2, y = 4 - 4x$

34. $y = x^2 + 1, y = x^2, x = 1, x = 3$

35. $y = x^2 + 3, y = x^2, x = 1, x = 2$

36. $y = 2x^2 - x - 3, y = x^2 + x$

37. $y = 2x^2 - 6x + 5, y = x^2 + 6x - 15$

Find the average value over the given interval.

38. $y = 2x^3$; $[-1, 1]$

39. $y = 4 - x^2$; $[-2, 2]$

40. $y = e^x$; $[0, 1]$

41. $y = e^{-x}$; $[0, 1]$

42. $y = x^2 - x + 1$; $[0, 2]$

43. $f(x) = x^2 + x - 2$; $[0, 4]$

44. $f(x) = mx + 1$; $[0, 2]$

45. $f(x) = 4x + 5$; $[0, a]$

46. $f(x) = x^n, n \neq 0$; $[0, 1]$

47. $f(x) = x^n, n \neq 0$; $[1, 2]$

48. $f(x) = \dfrac{n}{x}$; $[1, 5]$

APPLICATIONS

Business and Economics

49. Total and average daily profit. Shylls, Inc., determines that its marginal revenue per day is given by

$$R'(t) = 100e^t, \quad R(0) = 0,$$

where $R(t)$ is the total accumulated revenue, in dollars, on the tth day. The company's marginal cost per day is given by

$$C'(t) = 100 - 0.2t, \quad C(0) = 0,$$

where $C(t)$ is the total accumulated cost, in dollars, on the tth day.

a) Find the total profit from $t = 0$ to $t = 10$ (the first 10 days). *Note:*

$$P(T) = R(T) - C(T) = \int_0^T \left[R'(t) - C'(t) \right] dt.$$

b) Find the average daily profit for the first 10 days (from $t = 0$ to $t = 10$).

50. Total and average daily profit. Great Green, Inc., determines that its marginal revenue per day is given by

$$R'(t) = 75e^t - 2t, \quad R(0) = 0,$$

where $R(t)$ is the total accumulated revenue, in dollars, on the tth day. The company's marginal cost per day is given by

$$C'(t) = 75 - 3t, \quad C(0) = 0,$$

where $C(t)$ is the total accumulated cost, in dollars, on the tth day.

a) Find the total profit from $t = 0$ to $t = 10$ (see Exercise 49).

b) Find the average daily profit for the first 10 days.

51. Accumulated sales. ProArt, Inc., determines that its weekly online sales, $S(t)$, in hundreds of dollars, t weeks after online sales began, can be estimated by

$$S(t) = 9e^t.$$

Find the average weekly sales for the first 5 weeks after online sales began.

52. Accumulated sales. Music Manager, Ltd., estimates that monthly revenue, $R(t)$, in thousands of dollars, attributable to its Web site t months after the Web site was launched, is given by

$$R(t) = 0.5e^t.$$

Find the average monthly revenue attributable to the Web site for its first 4 months of operation.

53. Refer to Exercise 51. Find ProArt's average weekly online sales for weeks 2 through 5 ($t = 1$ to $t = 5$).

54. Refer to Exercise 52. Find the average monthly revenue from Music Manager's Web site for months 3 through 5 ($t = 2$ to $t = 5$).

Social Sciences

55. Memorizing. In a memory experiment, Alice is able to memorize words at the rate given by

$$m'(t) = -0.009t^2 + 0.2t \quad \text{(words per minute)}.$$

In the same memory experiment, Ben is able to memorize words at the rate given by

$$M'(t) = -0.003t^2 + 0.2t \quad \text{(words per minute)}.$$

a) Who has the higher rate of memorization?

b) How many more words does that person memorize from $t = 0$ to $t = 10$ (during the first 10 min of the experiment)?

c) Over the first 10 min of the experiment, on average, how many words per minute did Alice memorize?

d) Over the first 10 min of the experiment, on average, how many words per minute did Ben memorize?

56. Results of studying. Antonio's score on a test is given by

$$s(t) = t^2, \quad 0 \le t \le 10,$$

where $s(t)$ is his score after t hours of studying. Bonnie's score on the same test is given by

$$S(t) = 10t, \quad 0 \le t \le 10,$$

where $S(t)$ is her score after t hours of studying.

a) For $0 < t < 10$, who will have the higher test score?

b) Find the average value of $s(t)$ over the interval $[7, 10]$, and explain what it represents.

c) Find the average value of $S(t)$ over the interval $[6, 10]$, and explain what it represents.

d) Assuming that both students have the same study habits and are equally likely to study for any number of hours, t, in $[0, 10]$, on average, how far apart will their test scores be?

57. Results of practice. A keyboarder's speed over a 5-min interval is given by

$$W(t) = -6t^2 + 12t + 90, \quad t \text{ in } [0, 5],$$

where $W(t)$ is the speed, in words per minute, at time t.

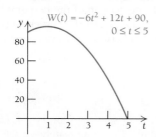

a) Find the speed at the beginning of the interval.

b) Find the maximum speed and when it occurs.

c) Find the average speed over the 5-min interval.

58. Average population. The population of the United States can be approximated by

$$P(t) = 282.3e^{0.01t},$$

where P is in millions and t is the number of years since 2000. (*Source:* Population Division, U.S. Census Bureau.) Find the average value of the population from 2001 to 2005.

Natural and Life Sciences

59. Average drug dose. The concentration, C, of phenyl-butazone, in micrograms per milliliter (μg/mL), in the plasma of a calf injected with this anti-inflammatory agent is given approximately by

$$C(t) = 42.03e^{-0.01050t},$$

where t is the number of hours after the injection and $0 \le t \le 120$. (*Source:* A. K. Arifah and P. Lees, "Pharma-codynamics and Pharmacokinetics of Phenylbutazone in

Calves," *Journal of Veterinary Pharmacology and Therapeutics,* Vol. 25, 299–309 (2002).)

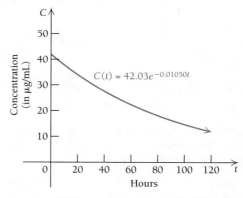

a) Given that this model is accurate for $0 \le t \le 120$, what is the initial dosage?

b) What is the average amount of phenylbutazone in the calf's body for the time between 10 and 120 hours?

60. New York temperature. For any date, the average temperature on that date in New York can be approximated by the function

$$T(x) = 43.5 - 18.4x + 8.57x^2 - 0.996x^3 + 0.0338x^4,$$

where T represents the temperature in degrees Fahrenheit, $x = 1$ represents the middle of January, $x = 2$ represents the middle of February, and so on. (*Source:* www.worldclimate.com.) Compute the average temperature in New York over the whole year to the nearest degree.

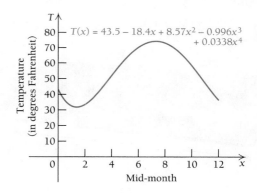

61. Outside temperature. The temperature over a 10-hr period is given by

$$f(t) = -t^2 + 5t + 40, \quad 0 \le t \le 10.$$

a) Find the average temperature.

b) Find the minimum temperature.

c) Find the maximum temperature.

62. Engine emissions. The emissions of an engine are given by

$$E(t) = 2t^2,$$

where $E(t)$ is the engine's rate of emission, in billions of pollution particulates per year, at time t, in years. Find the average emissions from $t = 1$ to $t = 5$.

SYNTHESIS

Find the area of the region bounded by the given graphs.

63. $y = x^2, y = x^{-2}, x = 5$

64. $y = e^x, y = e^{-x}, x = -2$

65. $y = x + 6, y = -2x, y = x^3$

66. $y = x^2, y = x^3, x = -1$

67. $x + 2y = 2, y - x = 1, 2x + y = 7$

68. Find the area bounded by $y = 3x^5 - 20x^3$, the x-axis, and the first coordinates of the relative maximum and minimum values of the function.

69. Find the area bounded by $y = x^3 - 3x + 2$, the x-axis, and the first coordinates of the relative maximum and minimum values of the function.

70. Life science: Poiseuille's Law. The flow of blood in a blood vessel is faster toward the center of the vessel and slower toward the outside. The speed of the blood is given by

$$V = \frac{p}{4Lv}(R^2 - r^2),$$

where R is the radius of the blood vessel, r is the distance of the blood from the center of the vessel, and p, v, and L are physical constants related to the pressure and viscosity of the blood and the length of the blood vessel. If R is constant, we can think of V as a function of r:

$$V(r) = \frac{p}{4Lv}(R^2 - r^2).$$

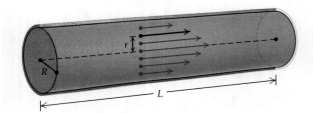

The total blood flow, Q, is given by

$$Q = \int_0^R 2\pi \cdot V(r) \cdot r \cdot dr.$$

Find Q.

71. Solve for K, given that

$$\int_1^2 \left[(3x^2 + 5x) - (3x + K)\right] dx = 6.$$

TECHNOLOGY CONNECTION

Find the area of the region enclosed by the given graphs.

72. $y = x^2 + 4x, y = \sqrt{16 - x^2}$

73. $y = x\sqrt{4 - x^2}, y = \dfrac{-4x}{x^2 + 1}, x = 0, x = 2$

74. $y = 2x^2 + x - 4, y = 1 - x + 8x^2 - 4x^4$

75. $y = \sqrt{1 - x^2}, y = 1 - x^2, x = -1, x = 1$

76. Consider the following functions:

$$f(x) = 3.8x^5 - 18.6x^3,$$
$$g(x) = 19x^4 - 55.8x^2.$$

a) Graph these functions in the window $[-3, 3, -80, 80]$, with Yscl $= 10$.

b) Estimate the first coordinates a, b, and c of the three points of intersection of the two graphs.

c) Find the area between the curves on the interval $[a, b]$.

d) Find the area between the curves on the interval $[b, c]$.

Answers to Quick Checks

1. $\frac{83}{3}$, or $27\frac{2}{3}$ **2.** $\frac{85}{2}$, or $42\frac{1}{2}$ **3.** $\frac{9}{2}$, or $4\frac{1}{2}$ **4. (a)** $1066\frac{2}{3}$ ft; **(b)** 60 sec, 7200 ft **5.** Approximately $-2.5°$F

4.5 Integration Techniques: Substitution

The following formulas provide a basis for an integration technique called **substitution**.

OBJECTIVES

- Evaluate integrals using substitution.

- Solve applied problems involving integration by substitution.

A. $\displaystyle\int u^r \, du = \frac{u^{r+1}}{r+1} + C,$ assuming $r \neq -1$

B. $\displaystyle\int e^u \, du = e^u + C$

C. $\displaystyle\int \frac{1}{u} \, du = \ln|u| + C;$ or $\displaystyle\int \frac{1}{u} \, du = \ln u + C,$ $u > 0$

(Unless noted otherwise, we will assume $u > 0$.)

In the above formulas, the variable u represents a more complicated expression in terms of x. First, consider the integral $\int x^7\, dx$. We can carry out this integration using the Power Rule of Antidifferentiation:

$$\int x^7\, dx = \frac{x^{7+1}}{7+1} + C = \frac{x^8}{8} + C, \quad \text{or} \quad \frac{1}{8}x^8 + C.$$

But, what about an integral like $\int (3x - 4)^7\, dx$, whose integrand is more complicated? Suppose we thought the antiderivative was

$$\frac{(3x - 4)^8}{8} + C.$$

If we do a check by differentiating, we get

$$8 \cdot \frac{1}{8} \cdot (3x - 4)^7 \cdot 3 \cdot dx.$$

This simplifies to

$$3(3x - 4)^7, \quad \text{which is } not \ (3x - 4)^7,$$

though it is off only by the constant factor 3. Instead, let's make this substitution:

$$u = 3x - 4.$$

Then $du/dx = 3$ and recalling our work on differentials (Section 2.6), we have

$$du = 3 \cdot dx, \quad \text{or} \quad \frac{du}{3} = dx.$$

With *substitution*, our original integral, $\int (3x - 4)^7\, dx$, takes the form

$$\int (3x - 4)^7\, dx = \int u^7 \cdot \frac{du}{3} \qquad \text{Substituting } u \text{ for } 3x - 4 \text{ and } \frac{du}{3} \text{ for } dx$$

$$= \frac{1}{3} \cdot \int u^7\, du \qquad \text{Factoring out the constant } \frac{1}{3}$$

$$= \frac{1}{3} \cdot \frac{u^8}{8} + C \qquad \text{By substitution formula A}$$

$$= \frac{1}{3 \cdot 8} \cdot (3x - 4)^8 + C = \frac{1}{24}(3x - 4)^8 + C.$$

We leave it to the student to check that this is indeed the antiderivative. Note how this procedure reverses the Chain Rule.

Recall the Leibniz notation, dy/dx, for a derivative. We gave specific definitions of the differentials dy and dx in Section 2.6. Recall that

$$\frac{dy}{dx} = f'(x) \qquad \text{and} \qquad dy = f'(x)\, dx.$$

We will make extensive use of this notation in this section.

■ **EXAMPLE 1**   For $y = f(x) = x^3$, find dy.

Solution   We have

$$\frac{dy}{dx} = f'(x) = 3x^2,$$

so $dy = f'(x)\, dx = 3x^2\, dx.$

■ **EXAMPLE 2** For $u = F(x) = x^{2/3}$, find du.

Solution We have

$$\frac{du}{dx} = F'(x) = \tfrac{2}{3}x^{-1/3},$$

so $\qquad du = F'(x)\,dx = \tfrac{2}{3}x^{-1/3}\,dx.$

. . . .

■ **EXAMPLE 3** For $u = g(x) = \ln x$, find du.

Solution We have

$$\frac{du}{dx} = g'(x) = \frac{1}{x},$$

so $\qquad du = g'(x)\,dx = \frac{1}{x}\,dx, \quad \text{or} \quad \frac{dx}{x}.$

. . . .

> **Quick Check 1**

Find each differential.
a) For $y = \sqrt{x}$, find dy.
b) For $u = x^2 - 3x$, find du.
c) For $y = \dfrac{1}{x^3}$, find dy.
d) For $u = 4x - 3$, find du.

■ **EXAMPLE 4** For $y = f(x) = e^{x^2}$, find dy.

Solution Using the Chain Rule, we have

$$\frac{dy}{dx} = f'(x) = e^{x^2} \cdot 2x,$$

so $\qquad dy = f'(x)\,dx = e^{x^2} \cdot 2x\,dx.$

❮ Quick Check 1

So far, the dx in

$$\int f(x)\,dx$$

has played no role in integration other than to indicate the variable of integration. Now it becomes convenient to make use of dx. Consider the integral

$$\int 2xe^{x^2}\,dx.$$

Finding an antiderivative may seem impossible. Yet, if we note that $2xe^{x^2} = e^{x^2} \cdot 2x$, we see in Example 4 that $f(x) = e^{x^2}$ is an antiderivative of $f'(x) = 2xe^{x^2}$. How might we find such an antiderivative directly? Suppose that we let $u = x^2$. Then

$$\frac{du}{dx} = 2x, \quad \text{and} \quad du = 2x\,dx.$$

If we substitute u for x^2 and du for $2x\,dx$, we have

$$\int 2xe^{x^2}\,dx = \int e^{x^2}\,2x\,dx = \int e^u\,du.$$

Since

$$\int e^u \, du = e^u + C,$$

it follows that

$$\int 2xe^{x^2} \, dx = \int e^u \, du$$
$$= e^u + C$$
$$= e^{x^2} + C.$$

In effect, we have used the Chain Rule in reverse. We can check the result by differentiating. The procedure is referred to as *substitution*, or *change of variable*. It can involve trial and error, but you will become more proficient the more you practice. If you try a substitution that doesn't result in an integrand that can be easily integrated, try another substitution. While there are many integrations that cannot be carried out using substitution, any integral that fits formula A, B, or C on p. 436 can be evaluated with this procedure.

■ **EXAMPLE 5** Evaluate: $\int 3x^2(x^3 + 1)^{10} \, dx$.

Solution Note that $3x^2$ is the derivative of x^3. Thus,

$$\int 3x^2(x^3 + 1)^{10} \, dx = \int (x^3 + 1)^{10} \, 3x^2 \, dx \qquad \underline{\text{Substitution}} \quad \begin{array}{l} u = x^3 + 1, \\ du = 3x^2 \, dx \end{array}$$

$$= \int u^{10} \, du$$

$$= \frac{u^{11}}{11} + C$$

$$= \tfrac{1}{11}(x^3 + 1)^{11} + C. \qquad \text{"Reversing" the substitution}$$

As a check, we differentiate:

$$\frac{d}{dx}\left[\tfrac{1}{11}(x^3 + 1)^{11} + C\right] = \tfrac{11}{11}(x^3 + 1)^{10} \cdot 3x^2 + 0$$
$$= (x^3 + 1)^{10} \cdot 3x^2$$
$$= 3x^2(x^3 + 1)^{10}.$$

> **Quick Check 2**

Evaluate: $\int 4x(2x^2 + 3)^3 \, dx$.

❰ Quick Check 2

■ **EXAMPLE 6** Evaluate: $\int \dfrac{2x \, dx}{1 + x^2}$.

Solution

$$\int \frac{2x \, dx}{1 + x^2} = \int \frac{du}{u} \qquad \underline{\text{Substitution}} \quad \begin{array}{l} u = 1 + x^2; \\ du = 2x \, dx. \end{array}$$

$$= \ln u + C \qquad \text{Remember: } \int \frac{du}{u} = \int \frac{1}{u} \cdot du.$$

$$= \ln (1 + x^2) + C$$

> **Quick Check 3**

Evaluate: $\int \dfrac{e^x}{1 + e^x} \, dx$.

❰ Quick Check 3

■ **EXAMPLE 7** Evaluate: $\int \dfrac{2x\,dx}{(1+x^2)^2}$.

Solution

$$\int \frac{2x\,dx}{(1+x^2)^2} = \int \frac{du}{u^2} \qquad \underline{\text{Substitution}} \quad \begin{array}{l} u = 1 + x^2, \\ du = 2x\,dx \end{array}$$

$$= \int u^{-2}\,du$$

$$= -u^{-1} + C$$

$$= -\frac{1}{u} + C$$

$$= -\frac{1}{1+x^2} + C \left.\vphantom{\begin{array}{c}a\\b\end{array}}\right\} \begin{array}{l} \text{Don't forget to reverse the} \\ \text{substitution after integrating.} \end{array}$$

> **Quick Check 4**

Evaluate: $\int \dfrac{6x^2}{\sqrt{3 + 2x^3}}\,dx$.

❰ Quick Check 4

■ **EXAMPLE 8** Evaluate: $\int \dfrac{\ln(3x)\,dx}{x}$.

Solution

$$\int \frac{\ln(3x)\,dx}{x} = \int u\,du \qquad \underline{\text{Substitution}} \quad \begin{array}{l} u = \ln(3x), \\ du = \dfrac{1}{x}\,dx \end{array}$$

$$= \frac{u^2}{2} + C$$

$$= \frac{(\ln(3x))^2}{2} + C$$

> **Quick Check 5**

Evaluate: $\int \dfrac{(\ln x)^2}{x}\,dx$.

❰ Quick Check 5

■ **EXAMPLE 9** Evaluate: $\int xe^{x^2}\,dx$.

Solution If we try $u = x^2$, we have $du = 2x\,dx$. We don't have $2x\,dx$ in $\int xe^{x^2}\,dx$. We do have $x\,dx$, so we need a factor of 2. To provide this factor, we multiply by 1, in the form $\frac{1}{2} \cdot 2$:

$$\int xe^{x^2}\,dx = \frac{1}{2}\int 2xe^{x^2}\,dx$$

$$= \frac{1}{2}\int e^{x^2}(2x\,dx) = \frac{1}{2}\int e^u\,du$$

$$= \frac{1}{2}e^u + C = \frac{1}{2}e^{x^2} + C.$$

> **Quick Check 6**

Evaluate: $\int x^2 e^{4x^3}\,dx$.

❰ Quick Check 6

With practice, you will be able to make certain substitutions mentally and just write down the answer. Example 10 illustrates one such case.

■ **EXAMPLE 10** Evaluate: $\int \dfrac{dx}{x + 3}$.

Solution

$$\int \frac{dx}{x + 3} = \int \frac{du}{u} \qquad \underline{\text{Substitution}} \quad \begin{array}{l} u = x + 3, \\ du = 1\,dx = dx \end{array}$$

$$= \ln u + C$$

$$= \ln (x + 3) + C \qquad \text{We assume } x + 3 > 0.$$

■ **EXAMPLE 11** Evaluate: $\int_0^1 5x\sqrt{x^2 + 3}\,dx$. Round to the nearest thousandth.

Solution We first find the indefinite integral and then evaluate that integral over $[0, 1]$:

$$\int 5x\sqrt{x^2 + 3}\,dx = 5\int x\sqrt{x^2 + 3}\,dx$$

$$= \frac{5}{2}\int 2x\sqrt{x^2 + 3}\,dx$$

$$= \frac{5}{2}\int \sqrt{x^2 + 3}\,2x\,dx$$

$$= \frac{5}{2}\int \sqrt{u}\,du \qquad \underline{\text{Substitution}} \quad \begin{array}{l} u = x^2 + 3, \\ du = 2x\,dx \end{array}$$

$$= \frac{5}{2}\int u^{1/2}\,du$$

$$= \frac{5}{2}\cdot\frac{2}{3}u^{3/2} + C$$

$$= \frac{5}{3}(x^2 + 3)^{3/2} + C. \qquad \begin{array}{l} \text{Reversing the substitution before} \\ \text{evaluating with the bounds} \end{array}$$

Using 0 for the constant C, we have

$$\int_0^1 5x\sqrt{x^2 + 3}\,dx = \left[\frac{5}{3}(x^2 + 3)^{3/2}\right]_0^1$$

$$= \frac{5}{3}\left[(x^2 + 3)^{3/2}\right]_0^1$$

$$= \frac{5}{3}\left[4^{3/2} - 3^{3/2}\right]$$

$$\approx 4.673.$$

❮ Quick Check 7

EXERCISE

1. Use a calculator to evaluate $\int_0^1 5x\sqrt{x^2 + 3}\,dx$.

❯ **Quick Check 7**

Evaluate:
$\int_0^2 (x + 1)(x^2 + 2x + 3)^4\,dx$.

In some cases, after a substitution is made, a further simplification can allow us to complete an integration:

■ **EXAMPLE 12** Evaluate: $\int \dfrac{x}{x+2}\, dx$. Assume that $x + 2 > 0$.

Solution We substitute $u = x + 2$ and $du = dx$. We observe that $x = u - 2$. The substitutions are made:

$$\int \frac{x}{x+2}\, dx = \int \frac{u-2}{u}\, du \qquad \underline{\text{Substitution}} \quad \boxed{\begin{aligned} u &= x + 2, \\ du &= dx \\ x &= u - 2 \end{aligned}}$$

$$= \int \left(1 - \frac{2}{u} \right) du \qquad \frac{u-2}{u} = \frac{u}{u} - \frac{2}{u} = 1 - \frac{2}{u}$$

$$= u - 2 \ln u + C_1$$

$$= x + 2 - 2 \ln (x + 2) + C_1 \qquad \text{Reversing the substitution}$$

$$= x - 2 \ln (x + 2) + C \qquad C = C_1 + 2$$

> **Quick Check 8**

Evaluate $\int \dfrac{x}{(x-1)^3}\, dx$ by letting $u = x - 1$.

❰ Quick Check 8

Strategy for Substitution

The following strategy may help in carrying out the procedure of substitution:

1. Decide which rule of antidifferentiation is appropriate.
    a) If you believe it is the Power Rule, let u be the base (see Examples 5, 7, 8, and 11).
    b) If you believe it is the Exponential Rule (base e), let u be the expression in the exponent (see Example 9).
    c) If you believe it is the Natural Logarithm Rule, let u be the denominator (see Examples 6 and 10).
2. Determine du.
3. Inspect the integrand to be sure the substitution accounts for all factors. You may need to insert constants (see Examples 9 and 11) or make an extra substitution (see Example 12).
4. Perform the antidifferentiation.
5. Reverse the substitution. If there are bounds, use them to evaluate the integral *after* the substitution has been reversed.
6. *Always check your answer by differentiation.*

Section Summary

- Integration by *substitution* is the reverse of applying the Chain Rule of Differentiation.
- The substitution is reversed after the integration has been performed.

- Results should be checked using differentiation.

**EXERCISE SET
4.5**

Evaluate. Assume $u > 0$ *when* $\ln u$ *appears. (Be sure to check by differentiating!)*

1. $\displaystyle\int (8 + x^3)^5\, 3x^2\, dx$

2. $\displaystyle\int (x^2 - 7)^6\, 2x\, dx$

3. $\displaystyle\int (x^2 - 6)^7 x\, dx$

4. $\displaystyle\int (x^3 + 1)^4 x^2\, dx$

5. $\displaystyle\int (3t^4 + 2)t^3\, dt$

6. $\displaystyle\int (2t^5 - 3)t^4\, dt$

7. $\displaystyle\int \frac{2}{1 + 2x}\, dx$

8. $\displaystyle\int \frac{5}{5x + 7}\, dx$

9. $\displaystyle\int (\ln x)^3 \frac{1}{x}\, dx$

10. $\displaystyle\int (\ln x)^7 \frac{1}{x}\, dx$

11. $\displaystyle\int e^{3x}\, dx$

12. $\displaystyle\int e^{7x}\, dx$

13. $\displaystyle\int e^{x/3}\, dx$

14. $\displaystyle\int e^{x/2}\, dx$

15. $\displaystyle\int x^4 e^{x^5}\, dx$

16. $\displaystyle\int x^3 e^{x^4}\, dx$

17. $\displaystyle\int t e^{-t^2}\, dt$

18. $\displaystyle\int t^2 e^{-t^3}\, dt$

19. $\displaystyle\int \frac{1}{5 + 2x}\, dx$

20. $\displaystyle\int \frac{1}{2 + 8x}\, dx$

21. $\displaystyle\int \frac{dx}{12 + 3x}$

22. $\displaystyle\int \frac{dx}{1 + 7x}$

23. $\displaystyle\int \frac{dx}{1 - x}$

24. $\displaystyle\int \frac{dx}{4 - x}$

25. $\displaystyle\int t(t^2 - 1)^5\, dt$

26. $\displaystyle\int t^2(t^3 - 1)^7\, dt$

27. $\displaystyle\int (x^4 + x^3 + x^2)^7(4x^3 + 3x^2 + 2x)\, dx$

28. $\displaystyle\int (x^3 - x^2 - x)^9(3x^2 - 2x - 1)\, dx$

29. $\displaystyle\int \frac{e^x\, dx}{4 + e^x}$

30. $\displaystyle\int \frac{e^t\, dt}{3 + e^t}$

31. $\displaystyle\int \frac{\ln x^2}{x}\, dx$ (*Hint:* Use the properties of logarithms.)

32. $\displaystyle\int \frac{(\ln x)^2}{x}\, dx$

33. $\displaystyle\int \frac{dx}{x \ln x}$

34. $\displaystyle\int \frac{dx}{x \ln x^2}$

35. $\displaystyle\int x\sqrt{ax^2 + b}\, dx$

36. $\displaystyle\int \sqrt{ax + b}\, dx$

37. $\displaystyle\int P_0 e^{kt}\, dt$

38. $\displaystyle\int b e^{ax}\, dx$

39. $\displaystyle\int \frac{x^3\, dx}{(2 - x^4)^7}$

40. $\displaystyle\int \frac{3x^2\, dx}{(1 + x^3)^5}$

41. $\displaystyle\int 12x\sqrt[5]{1 + 6x^2}\, dx$

42. $\displaystyle\int 5x\sqrt[4]{1 - x^2}\, dx$

Evaluate.

43. $\displaystyle\int_0^1 2x e^{x^2}\, dx$

44. $\displaystyle\int_0^1 3x^2 e^{x^3}\, dx$

45. $\displaystyle\int_0^1 x(x^2 + 1)^5\, dx$

46. $\displaystyle\int_1^2 x(x^2 - 1)^7\, dx$

47. $\displaystyle\int_0^4 \frac{dt}{1 + t}$

48. $\displaystyle\int_0^2 e^{4x}\, dx$

49. $\displaystyle\int_1^4 \frac{2x + 1}{x^2 + x - 1}\, dx$

50. $\displaystyle\int_1^3 \frac{2x + 3}{x^2 + 3x}\, dx$

51. $\displaystyle\int_0^b e^{-x}\, dx$

52. $\displaystyle\int_0^b 2e^{-2x}\, dx$

53. $\displaystyle\int_0^b m e^{-mx}\, dx$

54. $\displaystyle\int_0^b k e^{-kx}\, dx$

55. $\displaystyle\int_0^4 (x - 6)^2\, dx$

56. $\displaystyle\int_0^3 (x - 5)^2\, dx$

57. $\displaystyle\int_0^2 \frac{3x^2\, dx}{(1 + x^3)^5}$

58. $\displaystyle\int_{-1}^0 \frac{x^3\, dx}{(2 - x^4)^7}$

59. $\displaystyle\int_0^{\sqrt{7}} 7x\sqrt[3]{1 + x^2}\, dx$

60. $\displaystyle\int_0^1 12x\sqrt[5]{1 - x^2}\, dx$

61. Use a graphing calculator to check the results of any of Exercises 43–60.

Evaluate. Use the technique of Example 12.

62. $\displaystyle\int \frac{x}{x - 5}\, dx$

63. $\displaystyle\int \frac{3x}{2x + 1}\, dx$

64. $\displaystyle\int \frac{x}{1 - 4x}\, dx$

65. $\displaystyle\int \frac{x + 3}{x - 2}\, dx$ (*Hint:* $u = x - 2$.)

66. $\displaystyle\int \frac{2x + 3}{3x - 2}\, dx$

67. $\displaystyle\int x^2(x + 1)^{10}\, dx$ (*Hint:* $u = x + 1$.)

68. $\displaystyle\int x^3(x + 2)^7\, dx$

69. $\displaystyle\int x^2\sqrt{x - 2}\, dx$ (*Hint:* $u = x - 2$.)

70. $\displaystyle\int \frac{x}{\sqrt{x - 2}}\, dx$

APPLICATIONS
Business and Economics

71. Demand from marginal demand. A firm has the marginal-demand function

$$D'(x) = \frac{-2000x}{\sqrt{25 - x^2}}.$$

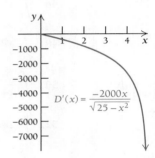

Find the demand function given that $D = 13{,}000$ when $x = \$3$ per unit.

72. Value of an investment. V. King Manufacturing buys a new machine for \$250,000. The marginal revenue from the sale of products produced by the machine after t years is given by

$$R'(t) = 4000t.$$

The salvage value of the machine, in dollars, after t years is given by

$$V(t) = 200{,}000 - 25{,}000e^{0.1t}.$$

The total profit from the machine, in dollars, after t years is given by

$$P(t) = \begin{pmatrix} \text{Revenue} \\ \text{from} \\ \text{sale of} \\ \text{product} \end{pmatrix} + \begin{pmatrix} \text{Revenue} \\ \text{from} \\ \text{sale of} \\ \text{machine} \end{pmatrix} - \begin{pmatrix} \text{Cost} \\ \text{of} \\ \text{machine} \end{pmatrix}.$$

The company knows that $R(0) = 0$.

a) Find $P(t)$.
b) Find $P(10)$.

73. Profit from marginal profit. A firm has the marginal-profit function

$$\frac{dP}{dx} = \frac{9000 - 3000x}{(x^2 - 6x + 10)^2}.$$

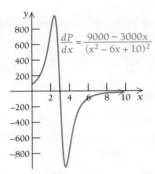

Find the total-profit function given that $P = \$1500$ at $x = 3$.

Social Sciences

74. Divorce rate. The divorce rate in the United States is approximated by

$$D(t) = 100{,}000e^{0.025t},$$

where $D(t)$ is the number of divorces occurring at time t and t is the number of years measured from 1900. That is, $t = 0$ corresponds to 1900, $t = 98\frac{9}{365}$ corresponds to January 9, 1998, and so on.

a) Find the total number of divorces from 1900 to 2005. Note that this is given by

$$\int_0^{105} D(t)\, dt.$$

b) Find the total number of divorces from 1980 to 2006. Note that this is given by

$$\int_{80}^{106} D(t)\, dt.$$

SYNTHESIS

Find the area of the shaded region.

75.

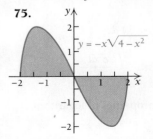

76.

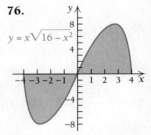

Evaluate. Assume $u > 0$ when $\ln u$ appears.

77. $\displaystyle\int \frac{dx}{ax + b}$

78. $\displaystyle\int 5x\sqrt{1 - 4x^2}\, dx$

79. $\displaystyle\int \frac{e^{\sqrt{t}}}{\sqrt{t}}\, dt$

80. $\displaystyle\int \frac{x^2}{e^{x^3}}\, dx$

81. $\int \frac{(\ln x)^{99}}{x}\, dx$

82. $\int \frac{e^{1/t}}{t^2}\, dt$

83. $\int (e^t + 2)e^t\, dt$

84. $\int \frac{dx}{x\,(\ln x)^4}$

85. $\int \frac{t^2}{\sqrt[4]{2 + t^3}}\, dt$

86. $\int x^2 \sqrt{x^3 + 1}\, dx$

87. $\int \frac{[(\ln x)^2 + 3(\ln x) + 4]}{x}\, dx$

88. $\int \frac{x - 3}{(x^2 - 6x)^{1/3}}\, dx$

89. $\int \frac{t^3 \ln (t^4 + 8)}{t^4 + 8}\, dt$

90. $\int \frac{t^2 + 2t}{(t + 1)^2}\, dt$

$\left(\text{Hint:} \dfrac{t^2 + 2t}{(t + 1)^2} = \dfrac{t^2 + 2t + 1 - 1}{t^2 + 2t + 1} = 1 - \dfrac{1}{(t + 1)^2}. \right)$

91. $\int \frac{x^2 + 6x}{(x + 3)^2}\, dx$ (Hint: See Exercise 90.)

92. $\int \frac{x + 3}{x + 1}\, dx$ $\left(\text{Hint:} \dfrac{x + 3}{x + 1} = 1 + \dfrac{2}{x + 1}. \right)$

93. $\int \frac{t - 5}{t - 4}\, dt$

94. $\int \frac{dx}{x(\ln x)^n}, \quad n \neq -1$

95. $\int \frac{dx}{e^x + 1}$ $\left(\text{Hint:} \dfrac{1}{e^x + 1} = \dfrac{e^{-x}}{1 + e^{-x}}. \right)$

96. $\int \frac{e^x - e^{-x}}{e^x + e^{-x}}\, dx$

97. $\int \frac{(\ln x)^n}{x}\, dx, \quad n \neq -1$

98. $\int \frac{dx}{x \ln x\,[\ln (\ln x)]}$

99. $\int \frac{e^{-mx}}{1 - ae^{-mx}}\, dx$

100. $\int 9x(7x^2 + 9)^n\, dx, \quad n \neq -1$

101. $\int 5x^2(2x^3 - 7)^n\, dx, \quad n \neq -1$

102. Determine whether the following is a theorem:

$$\int [2f(x)]\, dx = [f(x)]^2 + C.$$

Answers to Quick Checks

1. **(a)** $dy = \dfrac{1}{2\sqrt{x}}\, dx$; **(b)** $du = (2x - 3)\, dx$;

   (c) $dy = -\dfrac{3}{x^4}\, dx$; **(c)** $du = 4\, dx$

2. $\frac{1}{4}(2x^2 + 3)^4 + C$ **3.** $\ln (1 + e^x) + C$

4. $2\sqrt{3 + 2x^3} + C$ **5.** $\frac{1}{3}(\ln x)^3 + C$ **6.** $\frac{1}{12}e^{4x^3} + C$

7. 16,080.8 **8.** $-\dfrac{1}{x - 1} - \dfrac{1}{2(x - 1)^2} + C$

Integration Techniques: Integration by Parts

4.6

OBJECTIVES

- Evaluate integrals using the formula for integration by parts.
- Solve applied problems involving integration by parts.

Recall the Product Rule for differentiation:

$$\frac{d}{dx}(uv) = u\frac{dv}{dx} + v\frac{du}{dx}.$$

Integrating both sides with respect to x, we get

$$uv = \int u\frac{dv}{dx}\, dx + \int v\frac{du}{dx}\, dx$$

$$= \int u\, dv + \int v\, du.$$

Solving for $\int u\, dv$, we get the following theorem.

> **THEOREM 7**   **The Integration-by-Parts Formula**
>
> $$\int u \, dv = uv - \int v \, du$$

This equation can be used as a formula for integrating in certain situations—that is, situations in which an integrand is a product of two functions, and one of the functions can be integrated using the techniques we have already developed. For example,

$$\int xe^x \, dx$$

can be considered as

$$\int x(e^x \, dx) = \int u \, dv,$$

where we let

$$u = x \quad \text{and} \quad dv = e^x \, dx.$$

In this case, differentiating u gives

$$du = dx,$$

and integrating dv gives

$$v = e^x. \qquad \text{We select } C = 0 \text{ to obtain the simplest antiderivative.}$$

Then the Integration-by-Parts Formula gives us

$$\int \overset{u}{(x)} \overset{dv}{(e^x \, dx)} = \overset{u}{(x)} \overset{v}{(e^x)} - \int \overset{v}{(e^x)} \overset{du}{(dx)}$$

$$= xe^x - e^x + C.$$

This method of integrating is called **integration by parts**. As always, to check, we can simply differentiate. This check is left to the student.

Note that integration by parts, like substitution, is a trial-and-error process. In the preceding example, suppose that we had reversed the roles of x and e^x. We would have obtained

$$u = e^x, \qquad dv = x \, dx,$$

$$du = e^x \, dx, \qquad v = \frac{x^2}{2},$$

and

$$\int \overset{u}{(e^x)} \overset{dv}{(x \, dx)} = \overset{u}{(e^x)} \overset{v}{\left(\frac{x^2}{2}\right)} - \int \overset{v}{\left(\frac{x^2}{2}\right)} \overset{du}{(e^x \, dx)}.$$

Now the integrand on the right is more difficult to integrate than the one with which we began. When we can integrate *both* factors of an integrand, and thus have a choice as to how to apply the Integration-by-Parts Formula, it can happen that only one (or maybe none) of the possibilities will work.

Tips on Using Integration by Parts

1. If you have had no success using substitution, try integration by parts.
2. Use integration by parts when an integral is of the form

$$\int f(x)\, g(x)\, dx.$$

Match it with an integral of the form

$$\int u\, dv$$

by choosing a function to be $u = f(x)$, where $f(x)$ can be differentiated, and the remaining factor to be $dv = g(x)\, dx$, where $g(x)$ can be integrated.
3. Find du by differentiating and v by integrating.
4. If the resulting integral is more complicated than the original, make some other choice for u and dv.
5. To check your result, differentiate.

Let's consider some additional examples.

■ **EXAMPLE 1**   Evaluate: $\int \ln x\, dx$. Assume $x > 0$.

Solution   Note that $\int (dx/x) = \ln x + C$, but we do not yet know how to find $\int \ln x\, dx$ since we have not yet found a function whose derivative is $\ln x$. Since we can differentiate $\ln x$, we let

$$u = \ln x \quad \text{and} \quad dv = dx.$$

Then $du = \dfrac{1}{x}\, dx$ and $v = x$.

Using the Integration-by-Parts Formula gives

$$\int \overset{u}{(\ln x)}\overset{dv}{(dx)} = \overset{u}{(\ln x)}\overset{v}{x} - \int \overset{v}{x}\overset{du}{\left(\frac{1}{x}\, dx\right)}$$

$$= x \ln x - \int dx$$

$$= x \ln x - x + C.$$

> **Quick Check 1**
> Evaluate: $\int x e^{3x}\, dx$.

‹ Quick Check 1

■ **EXAMPLE 2**   Evaluate: $\int x \ln x\, dx$.

Solution   Let's examine several choices, as follows.
Attempt 1:   We let

$$u = 1 \quad \text{and} \quad dv = x \ln x\, dx.$$

This will not work because we do not as yet know how to integrate $dv = x \ln x\, dx$.

Attempt 2:   We let
$$u = x \ln x \qquad \text{and} \quad dv = dx.$$
Then $du = \left[x\left(\dfrac{1}{x}\right) + (\ln x)1 \right] dx$ and $v = x$

$$= (1 + \ln x)\, dx.$$

Using the Integration-by-Parts Formula, we have

$$\int \overset{u}{(x \ln x)}\, \overset{dv}{dx} = \overset{u}{(x \ln x)} \overset{v}{x} - \int \overset{v}{x}(\overset{du}{(1 + \ln x)\, dx})$$

$$= x^2 \ln x - \int (x + x \ln x)\, dx.$$

This integral seems more complicated than the original, but we will reconsider it in Example 6.

Attempt 3:   We let
$$u = \ln x \quad \text{and} \quad dv = x\, dx.$$
Then $du = \dfrac{1}{x}\, dx$ and $v = \dfrac{x^2}{2}.$

Using the Integration-by-Parts Formula, we have

$$\int \overset{u}{x} \overset{dv}{\ln x\, dx} = \overset{u}{\ln x} \cdot \overset{v}{\frac{x^2}{2}} - \int \overset{v}{\frac{x^2}{2}} \left(\overset{du}{\frac{1}{x}\, dx} \right) = \frac{x^2}{2} \ln x - \frac{1}{2} \int x\, dx$$

$$= \frac{x^2}{2} \ln x - \frac{x^2}{4} + C.$$

This choice of u and dv allows us to evaluate the integral.

■ EXAMPLE 3   Evaluate: $\int x \sqrt{5x + 1}\, dx.$

Solution   We let
$$u = x \quad \text{and} \quad dv = (5x + 1)^{1/2}\, dx.$$
Then $du = dx$ and $v = \frac{2}{15}(5x + 1)^{3/2}.$

Note that we have to use substitution in order to integrate dv:

$$\int (5x + 1)^{1/2}\, dx = \frac{1}{5} \int (5x + 1)^{1/2}\, 5\, dx = \frac{1}{5} \int w^{1/2}\, dw \quad \underline{\text{Substitution}} \quad \boxed{\begin{array}{l} w = 5x + 1, \\ dw = 5\, dx \end{array}}$$

$$v = \frac{1}{5} \cdot \frac{w^{1/2+1}}{\frac{1}{2} + 1} = \frac{2}{15} w^{3/2} = \frac{2}{15}(5x + 1)^{3/2}.$$

Using the Integration-by-Parts Formula gives us

$$\int \overset{u}{x} \Big(\overset{dv}{\sqrt{5x + 1}\, dx} \Big) = \overset{u}{x} \cdot \overset{v}{\tfrac{2}{15}(5x + 1)^{3/2}} - \int \overset{v}{\tfrac{2}{15}(5x + 1)^{3/2}} \overset{du}{dx}$$

$$= \tfrac{2}{15} x (5x + 1)^{3/2} - \tfrac{2}{15} \cdot \tfrac{2}{25}(5x + 1)^{5/2} + C$$

$$= \tfrac{2}{15} x (5x + 1)^{3/2} - \tfrac{4}{375}(5x + 1)^{5/2} + C.$$

This integral may also be evaluated using a substitution similar to that shown in Example 12 of Section 4.5. We revisit the evaluation of this integral in Exercise 43 at the end of this section.

⟩ **Quick Check 2**

Evaluate: $\int 2x\sqrt{3x-2}\,dx$.

⟨ Quick Check 2

■ **EXAMPLE 4**   Evaluate: $\int_1^2 \ln x \, dx$.

Solution   First, we find the indefinite integral (see Example 1). Next, we evaluate the definite integral:

$$\int_1^2 \ln x \, dx = \left[x \ln x - x \right]_1^2$$
$$= (2 \ln 2 - 2) - (1 \cdot \ln 1 - 1)$$
$$= 2 \ln 2 - 2 + 1$$
$$= 2 \ln 2 - 1 \approx 0.386.$$

Repeated Integration by Parts

In some cases, we may need to apply the Integration-by-Parts Formula more than once.

■ **EXAMPLE 5**   Evaluate $\int_0^7 x^2 e^{-x}\,dx$ to find the area of the shaded region shown to the left.

Solution   We first let

$$u = x^2 \qquad \text{and} \qquad dv = e^{-x}dx.$$

Then $du = 2x\,dx$ and $v = -e^{-x}$.

Using the Integration-by-Parts Formula gives

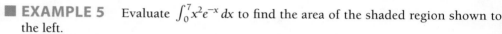

$$\int \overset{u}{x^2}(\overset{dv}{e^{-x}dx}) = \overset{u}{x^2}(\overset{v}{-e^{-x}}) - \int \overset{v}{-e^{-x}}(\overset{du}{2x\,dx}) \qquad (1)$$

$$= -x^2e^{-x} + \int 2xe^{-x}\,dx.$$

To evaluate the integral on the right, we can apply integration by parts again, as follows. We let

$$u = 2x \qquad \text{and} \qquad dv = e^{-x}dx.$$

Then $du = 2\,dx$ and $v = -e^{-x}$.

Using the Integration-by-Parts Formula once again, we get

$$\int \overset{u}{2x}(\overset{dv}{e^{-x}dx}) = \overset{u}{2x}(\overset{v}{-e^{-x}}) - \int \overset{v}{-e^{-x}}(\overset{du}{2\,dx})$$

$$= -2xe^{-x} - 2e^{-x} + C. \qquad (2)$$

When we substitute equation (2) into (1), the original integral becomes

$$\int x^2 e^{-x}\,dx = -x^2e^{-x} - 2xe^{-x} - 2e^{-x} + C$$
$$= -e^{-x}(x^2 + 2x + 2) + C. \qquad \text{Factoring simplifies the next step.}$$

TECHNOLOGY CONNECTION

EXERCISE

1. Use a calculator or iPlot to evaluate

$$\int_0^7 x^2 e^{-x}\,dx.$$

We now evaluate the definite integral:

$$\int_0^7 x^2 e^{-x}\, dx = \left[-e^{-x}(x^2 + 2x + 2)\right]_0^7$$

$$= \left[-e^{-7}(7^2 + 2(7) + 2)\right] - \left[-e^{-0}(0^2 + 2(0) + 2)\right]$$

$$= -65e^{-7} + 2 \approx 1.94.$$

》 **Quick Check 3**

Evaluate: $\displaystyle\int_0^3 \frac{x}{\sqrt{x+1}}\, dx.$

《 Quick Check 3

Recurring Integrals

Occasionally integration by parts yields an integral of the form $\int v\, du$ that is identical to the original integral. If we are alert and notice this when it occurs, we can find a solution of the original integral algebraically. Let's use this approach and reconsider Example 2.

■ **EXAMPLE 6**   Evaluate $\int x \ln x\, dx$ using the result of the second attempt in Example 2.

Solution   For the second attempt in Example 2, we let

$$u = x \ln x \qquad \text{and} \quad dv = dx,$$

so that

$$du = (1 + \ln x)\, dx \quad \text{and} \quad v = x.$$

Let's now work further with the result we abandoned earlier:

$$\int \overset{u}{(x \ln x)}\, \overset{dv}{dx} = \overset{u}{(x \ln x)}\overset{v}{x} - \int \overset{v}{x}(\overset{du}{(1 + \ln x)\, dx})$$

$$= x^2 \ln x - \int (x + x \ln x)\, dx$$

$$= x^2 \ln x - \int x\, dx - \int x \ln x\, dx \qquad \text{Substituting the integral of each term in the sum}$$

$$= x^2 \ln x - \tfrac{1}{2}x^2 - \int x \ln x\, dx \qquad \text{The original integral is duplicated.}$$

$$2\int x \ln x\, dx = x^2 \ln x - \tfrac{1}{2}x^2 \qquad \text{Adding } \int x \ln x\, dx \text{ to both sides}$$

$$\int x \ln x\, dx = \tfrac{1}{2}x^2 \ln x - \tfrac{1}{4}x^2 + C. \qquad \text{Dividing both sides by 2}$$

Tabular Integration by Parts

In situations like that in Example 5, we have an integral,

$$\int f(x)\, g(x)\, dx,$$

for which $f(x)$ can be repeatedly differentiated easily to a derivative that is eventually 0. The function $g(x)$ can also be repeatedly integrated easily. In such cases, we can use integration by parts more than once to evaluate the integral.

■ **EXAMPLE 7**   Evaluate: $\int x^3 e^x \, dx$.

Solution   We use integration by parts repeatedly, watching for patterns:

$$\int \overset{u}{x^3} \overset{dv}{e^x} \, dx = \overset{u}{x^3} \overset{v}{e^x} - \int \overset{v}{e^x} \overset{du}{3x^2} \, dx$$

$$= x^3 e^x - \int 3x^2 e^x \, dx. \qquad \text{This integral is simpler than the original.} \quad \textbf{(1)}$$

To solve $\int 3x^2 e^x \, dx$, we select $u = 3x^2$ and $dv = e^x \, dx$, so

$$du = 6x \, dx \quad \text{and} \quad v = e^x,$$

and

$$\int 3x^2 e^x \, dx = \overset{u}{3x^2} \overset{v}{e^x} - \int \overset{v}{e^x} \overset{du}{6x} \, dx$$

$$= 3x^2 e^x - \int 6x e^x \, dx. \qquad \text{This integral is the simplest so far.} \quad \textbf{(2)}$$

To solve $\int 6x e^x \, dx$, we select $u = 6x$ and $dv = e^x \, dx$, so

$$du = 6 \, dx \quad \text{and} \quad v = e^x,$$

and

$$\int 6x e^x \, dx = \overset{u}{6x} \overset{v}{e^x} - \int \overset{v}{e^x} \overset{du}{6} \, dx$$

$$= 6x e^x - 6 \int e^x \, dx$$

$$= 6x e^x - 6 e^x + C. \quad \textbf{(3)}$$

Combining equations (1), (2), and (3), we have

$$\int x^3 e^x \, dx = x^3 e^x - \left(3x^2 e^x - \int 6x e^x \, dx \right) \qquad \begin{array}{l}\text{Substituting equation (2) into} \\ \text{equation (1)}\end{array}$$

$$= x^3 e^x - 3x^2 e^x + \int 6x e^x \, dx$$

$$= x^3 e^x - 3x^2 e^x + 6x e^x - 6 e^x + C. \qquad \text{Substituting equation (3)}$$

As you can see, this approach can get complicated. Using **tabular integration**, as shown in the following table, can greatly simplify our work.

$f(x)$ and Repeated Derivatives	Sign of Product	$g(x)$ and Repeated Integrals
x^3	$(+)$	e^x
$3x^2$	$(-)$	e^x
$6x$	$(+)$	e^x
6	$(-)$	e^x
0		e^x

We then add products along the arrows, making the alternating sign changes, and obtain the correct result:

$$\int x^3 e^x \, dx = x^3 e^x - 3x^2 e^x + 6x e^x - 6 e^x + C.$$

> **Quick Check 4**
>
> Evaluate: $\int x^4 e^{2x} \, dx$.

❮ Quick Check 4

Section Summary

- The *Integration-by-Parts Formula* is the reverse of the Product Rule for differentiation:

$$\int u\, dv = uv - \int v\, du.$$

- The choices for u and dv should be such that the integral $\int v\, du$ is simpler than the original integral. If this does not turn out to be the case, other choices should be made.
- *Tabular integration* is useful in cases where repeated integration by parts is necessary.

EXERCISE SET
4.6

Evaluate using integration by parts or substitution. Check by differentiating.

1. $\int 4xe^{4x}\, dx$

2. $\int 3xe^{3x}\, dx$

3. $\int x^3(3x^2)\, dx$

4. $\int x^2(2x)\, dx$

5. $\int xe^{5x}\, dx$

6. $\int 2xe^{4x}\, dx$

7. $\int xe^{-2x}\, dx$

8. $\int xe^{-x}\, dx$

9. $\int x^2 \ln x\, dx$

10. $\int x^3 \ln x\, dx$

11. $\int x \ln \sqrt{x}\, dx$

12. $\int x^2 \ln x^3\, dx$

13. $\int \ln (x + 5)\, dx$

14. $\int \ln (x + 4)\, dx$

15. $\int (x + 2) \ln x\, dx$

16. $\int (x + 1) \ln x\, dx$

17. $\int (x - 1) \ln x\, dx$

18. $\int (x - 2) \ln x\, dx$

19. $\int x\sqrt{x + 2}\, dx$

20. $\int x\sqrt{x + 5}\, dx$

21. $\int x^3 \ln (2x)\, dx$

22. $\int x^2 \ln (5x)\, dx$

23. $\int x^2 e^x\, dx$

24. $\int (\ln x)^2\, dx$

25. $\int x^2 e^{2x}\, dx$

26. $\int x^{-5} \ln x\, dx$

27. $\int x^3 e^{-2x}\, dx$

28. $\int x^5 e^{4x}\, dx$

29. $\int (x^4 + 4)e^{3x}\, dx$

30. $\int (x^3 - x + 1)e^{-x}\, dx$

Evaluate using integration by parts.

31. $\int_1^2 x^2 \ln x\, dx$

32. $\int_1^2 x^3 \ln x\, dx$

33. $\int_2^6 \ln (x + 8)\, dx$

34. $\int_0^5 \ln (x + 7)\, dx$

35. $\int_0^1 xe^x\, dx$

36. $\int_0^1 (x^3 + 2x^2 + 3)e^{-2x}\, dx$

37. $\int_0^8 x\sqrt{x + 1}\, dx$

38. $\int_0^{\ln 3} x^2 e^{2x}\, dx$

39. Cost from marginal cost. A company determines that its marginal-cost function is given by

$$C'(x) = 4x\sqrt{x + 3}.$$

Find the total cost given that $C(13) = \$1126.40$.

40. Profit from marginal profit. A firm determines that its marginal-profit function is given by

$$P'(x) = 1000x^2 e^{-0.2x}.$$

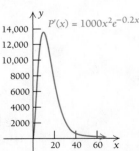

Find the total profit given that $P = -\$2000$ when $x = 0$.

Life and Physical Sciences

41. Electrical energy use. The rate at which electrical energy is used by the Ortiz family, in kilowatt-hours (kW-h) per day, is given by

$$K(t) = 10te^{-t},$$

where t is time, in hours. That is, t is in the interval $[0, 24]$.

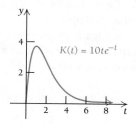

a) How many kilowatt-hours does the family use in the first T hours of a day ($t = 0$ to $t = T$)?
b) How many kilowatt-hours does the family use in the first 4 hours of the day?

42. Drug dosage. Suppose that an oral dose of a drug is taken. Over time, the drug is assimilated in the body and excreted through the urine. The total amount of the drug that has passed through the body in time T is given by

$$\int_0^T E(t)\, dt,$$

where E is the rate of excretion of the drug. A typical rate-of-excretion function is

$$E(t) = te^{-kt},$$

where $k > 0$ and t is the time, in hours.

a) Find a formula for

$$\int_0^T E(t)\, dt.$$

b) Find

$$\int_0^{10} E(t)\, dt, \quad \text{when } k = 0.2 \text{ mg/hr.}$$

SYNTHESIS

In Exercises 43 and 44, evaluate the given indefinite integral using substitution. Refer to Example 12 in Section 4.5 to review the technique.

43. Evaluate $\int x\sqrt{5x + 1}\, dx$ by letting $u = 5x + 1$ and $du = 5\, dx$ (so that $dx = \frac{1}{5}\, du$) and observing that $x = \dfrac{u - 1}{5}$. Compare your answer to that found in Example 3 of this section. Are they the same? (*Hint*: Simplify both forms of the answer into a common third form.)

44. Consider $\displaystyle\int \frac{x}{\sqrt{x - 3}}\, dx.$

a) Evaluate this integral using integration by parts.
b) Evaluate it using the substitution $u = x - 3$ and observing that $x = u + 3$.
c) Show algebraically that the answers from parts (a) and (b) are equivalent.

In Exercises 45 and 46, both substitution and integration by parts are used to determine the indefinite integral.

45. Evaluate $\int e^{\sqrt{x}}\, dx$ by letting $u = \sqrt{x}$. Note that $x = u^2$, so $dx = 2u\, du$. Make the substitutions and observe that the new integral (with variable u) can be evaluated using integration by parts.

46. Evaluate $\displaystyle\int \frac{1}{1 + \sqrt{x}}\, dx$ by letting $u = \sqrt{x}$ and following the procedure used in Exercise 45.

Evaluate using integration by parts.

47. $\displaystyle\int \sqrt{x} \ln x\, dx$

48. $\displaystyle\int \frac{te^t}{(t + 1)^2}\, dt$

49. $\displaystyle\int \frac{\ln x}{\sqrt{x}}\, dx$

50. $\displaystyle\int \frac{13t^2 - 48}{\sqrt[5]{4t + 7}}\, dt$

51. $\displaystyle\int (27x^3 + 83x - 2)\sqrt[6]{3x + 8}\, dx$

52. $\displaystyle\int x^2(\ln x)^2\, dx$

53. $\displaystyle\int x^n(\ln x)^2\, dx, \quad n \neq -1$

54. $\displaystyle\int x^n \ln x\, dx, \quad n \neq -1$

55. Verify that for any positive integer n,

$$\int x^n e^x\, dx = x^n e^x - n\int x^{n-1}e^x\, dx.$$

56. Verify that for any positive integer n,

$$\int (\ln x)^n\, dx = x(\ln x)^n - n\int (\ln x)^{n-1}\, dx.$$

57. Determine whether the following is a theorem:

$$\int f(x)g(x)\, dx = \int f(x)\, dx \cdot \int g(x)\, dx.$$

Explain.

58. Compare the procedures of differentiation and integration. Which seems to be the most complicated or difficult and why?

TECHNOLOGY CONNECTION

59. Use a graphing calculator to evaluate

$$\int_1^{10} x^5 \ln x\, dx.$$

Answers to Quick Checks

1. $\dfrac{x}{3}e^{3x} - \dfrac{1}{9}e^{3x} + C$

2. $\dfrac{4}{9}x(3x - 2)^{3/2} - \dfrac{4}{135}(3x - 2)^{5/2} + C$ **3.** $2\frac{2}{3}$

4. $e^{2x}\left(\dfrac{1}{2}x^4 - x^3 + \dfrac{3}{2}x^2 - \dfrac{3}{2}x + \dfrac{3}{4}\right) + C$

4.7

Integration Techniques: Tables

Tables of Integration Formulas

You have probably noticed that, generally speaking, integration is more challenging than differentiation. Because of this, integral formulas that are reasonable and/or important have been gathered into tables. Table 1, shown below and inside the back cover of this book, is a brief example of such a table. Entire books of integration formulas are available in libraries, and lengthy tables are also available online. Such tables are usually classified by the form of the integrand. The idea is to properly match the integral in question with a formula in the table. Sometimes some algebra or a technique such as substitution or integration by parts may be needed as well as a table.

TABLE 1   Integration Formulas

1. $\displaystyle\int x^n \, dx = \frac{x^{n+1}}{n+1} + C, \quad n \neq -1$

2. $\displaystyle\int \frac{dx}{x} = \ln x + C, \quad x > 0$

3. $\displaystyle\int u \, dv = uv - \int v \, du$

4. $\displaystyle\int e^x \, dx = e^x + C$

5. $\displaystyle\int e^{ax} \, dx = \frac{1}{a} \cdot e^{ax} + C$

6. $\displaystyle\int x e^{ax} \, dx = \frac{1}{a^2} \cdot e^{ax}(ax - 1) + C$

7. $\displaystyle\int x^n \, e^{ax} \, dx = \frac{x^n e^{ax}}{a} - \frac{n}{a}\int x^{n-1} e^{ax} \, dx + C$

8. $\displaystyle\int \ln x \, dx = x \ln x - x + C$

9. $\displaystyle\int (\ln x)^n \, dx = x(\ln x)^n - n\int (\ln x)^{n-1} \, dx + C, \quad n \neq -1$

10. $\displaystyle\int x^n \ln x \, dx = x^{n+1}\left[\frac{\ln x}{n+1} - \frac{1}{(n+1)^2}\right] + C, \quad n \neq -1$

11. $\displaystyle\int a^x \, dx = \frac{a^x}{\ln a} + C, \quad a > 0, a \neq 1$

12. $\displaystyle\int \frac{1}{\sqrt{x^2 + a^2}} \, dx = \ln \left| x + \sqrt{x^2 + a^2} \right| + C$

13. $\displaystyle\int \frac{1}{\sqrt{x^2 - a^2}} \, dx = \ln \left| x + \sqrt{x^2 - a^2} \right| + C$

14. $\displaystyle\int \frac{1}{x^2 - a^2} \, dx = \frac{1}{2a} \ln \left| \frac{x - a}{x + a} \right| + C$

15. $\displaystyle\int \frac{1}{a^2 - x^2} \, dx = \frac{1}{2a} \ln \left| \frac{a + x}{a - x} \right| + C$

16. $\displaystyle\int \frac{1}{x\sqrt{a^2 + x^2}} \, dx = -\frac{1}{a} \ln \left| \frac{a + \sqrt{a^2 + x^2}}{x} \right| + C$

(continued)

TABLE 1 *(continued)*

17. $\displaystyle\int \frac{1}{x\sqrt{a^2 - x^2}}\, dx = -\frac{1}{a}\ln\left|\frac{a + \sqrt{a^2 - x^2}}{x}\right| + C$

18. $\displaystyle\int \frac{x}{a + bx}\, dx = \frac{a}{b^2} + \frac{x}{b} - \frac{a}{b^2}\ln|a + bx| + C$

19. $\displaystyle\int \frac{x}{(a + bx)^2}\, dx = \frac{a}{b^2(a + bx)} + \frac{1}{b^2}\ln|a + bx| + C$

20. $\displaystyle\int \frac{1}{x(a + bx)}\, dx = \frac{1}{a}\ln\left|\frac{x}{a + bx}\right| + C$

21. $\displaystyle\int \frac{1}{x(a + bx)^2}\, dx = \frac{1}{a(a + bx)} + \frac{1}{a^2}\ln\left|\frac{x}{a + bx}\right| + C$

22. $\displaystyle\int \sqrt{x^2 \pm a^2}\, dx = \tfrac{1}{2}\left[x\sqrt{x^2 \pm a^2} \pm a^2\ln|x + \sqrt{x^2 \pm a^2}|\right] + C$

23. $\displaystyle\int x\sqrt{a + bx}\, dx = \frac{2}{15b^2}(3bx - 2a)(a + bx)^{3/2} + C$

24. $\displaystyle\int x^2\sqrt{a + bx}\, dx = \frac{2}{105b^3}(15b^2x^2 - 12abx + 8a^2)(a + bx)^{3/2} + C$

25. $\displaystyle\int \frac{x\, dx}{\sqrt{a + bx}} = \frac{2}{3b^2}(bx - 2a)\sqrt{a + bx} + C$

26. $\displaystyle\int \frac{x^2\, dx}{\sqrt{a + bx}} = \frac{2}{15b^3}(3b^2x^2 - 4abx + 8a^2)\sqrt{a + bx} + C$

■ **EXAMPLE 1** Evaluate: $\displaystyle\int \frac{dx}{x(3 - x)}$.

Solution The integral $\displaystyle\int \frac{dx}{x(3 - x)}$ fits *formula 20* in Table 1:

$$\int \frac{1}{x(a + bx)}\, dx = \frac{1}{a}\ln\left|\frac{x}{a + bx}\right| + C.$$

In the given integral, $a = 3$ and $b = -1$, so we have, by the formula,

$$\int \frac{dx}{x(3 - x)} = \frac{1}{3}\ln\left|\frac{x}{3 + (-1)x}\right| + C$$

$$= \frac{1}{3}\ln\left|\frac{x}{3 - x}\right| + C.$$

❯ **Quick Check 1**

Evaluate: $\displaystyle\int \frac{2x}{(3 - 5x)^2}\, dx$.

❮ Quick Check 1

■ **EXAMPLE 2** Evaluate: $\displaystyle\int \frac{5x}{7x - 8}\, dx$.

Solution We first factor 5 out of the integral. The integral then fits *formula 18* in Table 1:

$$\int \frac{x}{a + bx}\, dx = \frac{a}{b^2} + \frac{x}{b} - \frac{a}{b^2}\ln|a + bx| + C.$$

In the given integral, $a = -8$ and $b = 7$, so we have, by the formula,

$$\int \frac{5x}{7x - 8}\, dx = 5 \int \frac{x}{-8 + 7x}\, dx$$

$$= 5\left[\frac{-8}{7^2} + \frac{x}{7} - \frac{-8}{7^2}\ln|-8 + 7x|\right] + C$$

$$= 5\left[\frac{-8}{49} + \frac{x}{7} + \frac{8}{49}\ln|7x - 8|\right] + C$$

$$= -\frac{40}{49} + \frac{5x}{7} + \frac{40}{49}\ln|7x - 8| + C.$$

> **Quick Check 2**

Evaluate: $\displaystyle\int \frac{3}{2x(7 - 3x)^2}\, dx.$

❮ Quick Check 2

▪ **EXAMPLE 3**   Evaluate: $\int \sqrt{16x^2 + 3}\, dx.$

Solution   This integral *almost* fits *formula 22* in Table 1:

$$\int \sqrt{x^2 \pm a^2}\, dx = \tfrac{1}{2}\left[x\sqrt{x^2 \pm a^2} \pm a^2\ln\left|x + \sqrt{x^2 \pm a^2}\right|\right] + C.$$

But the coefficient of x^2 needs to be 1. To achieve this, we first factor out 16. Then we apply *formula 22*:

$$\int \sqrt{16x^2 + 3}\, dx = \int \sqrt{16\left(x^2 + \tfrac{3}{16}\right)}\, dx \qquad \text{Factoring}$$

$$= \int 4\sqrt{x^2 + \tfrac{3}{16}}\, dx \qquad \begin{array}{l}\text{Using the properties of}\\ \text{radicals; } \sqrt{16} = 4\end{array}$$

$$= 4\int \sqrt{x^2 + \tfrac{3}{16}}\, dx \qquad \text{We have } a^2 = \tfrac{3}{16} \text{ in formula 22.}$$

$$= 4 \cdot \tfrac{1}{2}\left[x\sqrt{x^2 + \tfrac{3}{16}} + \tfrac{3}{16}\ln\left|x + \sqrt{x^2 + \tfrac{3}{16}}\right|\right] + C$$

$$= 2\left[x\sqrt{x^2 + \tfrac{3}{16}} + \tfrac{3}{16}\ln\left|x + \sqrt{x^2 + \tfrac{3}{16}}\right|\right] + C.$$

In the given integral, $a^2 = 3/16$ and $a = \sqrt{3}/4$, though we did not need to use a in this form when applying the formula.

> **Quick Check 3**

Evaluate: $\int x^2\sqrt{8 + 3x}\, dx.$

❮ Quick Check 3

▪ **EXAMPLE 4**   Evaluate: $\displaystyle\int \frac{dx}{x^2 - 25}.$

Solution   This integral fits *formula 14* in Table 1:

$$\int \frac{1}{x^2 - a^2}\, dx = \frac{1}{2a}\ln\left|\frac{x - a}{x + a}\right| + C.$$

In the given integral, $a^2 = 25$, so $a = 5$. We have, by the formula,

$$\int \frac{dx}{x^2 - 25} = \frac{1}{10}\ln\left|\frac{x - 5}{x + 5}\right| + C.$$

> **Quick Check 4**

Evaluate: $\displaystyle\int \frac{4}{x^2 - 11}\, dx.$

❮ Quick Check 4

■ **EXAMPLE 5** Evaluate: $\int (\ln x)^3 \, dx$.

Solution This integral fits *formula 9* in Table 1:

$$\int (\ln x)^n \, dx = x(\ln x)^n - n \int (\ln x)^{n-1} \, dx + C, \quad n \neq -1.$$

We must apply the formula three times:

$$\int (\ln x)^3 \, dx = x(\ln x)^3 - 3 \int (\ln x)^2 \, dx + C \quad \text{Formula 9, with } n = 3$$

$$= x(\ln x)^3 - 3\left[x(\ln x)^2 - 2 \int \ln x \, dx \right] + C \quad \begin{array}{l}\text{Applying formula 9}\\ \text{again, with } n = 2\end{array}$$

$$= x(\ln x)^3 - 3\left[x(\ln x)^2 - 2\left(x \ln x - \int dx \right) \right] + C \quad \begin{array}{l}\text{Applying formula 9}\\ \text{for the third time,}\\ \text{with } n = 1\end{array}$$

$$= x(\ln x)^3 - 3x(\ln x)^2 + 6x \ln x - 6x + C.$$

❭ **Quick Check 5**

Evaluate: $\int x^4 \ln x \, dx$.

❬ **Quick Check 5**

The Web site www.integrals.com can be used to find integrals. If you have access to the Internet, use this Web site to check Examples 1–5 or to do the exercises in the following set.

Section Summary

- Tables of integrals or the Web site www.integrals.com can be used to evaluate many integrals.

- Some algebraic simplification of the integrand may be required before the correct integral form can be identified.

**EXERCISE SET
4.7**

Evaluate using Table 1.

1. $\int xe^{-3x} \, dx$

2. $\int 2xe^{3x} \, dx$

3. $\int 6^x \, dx$

4. $\int \dfrac{1}{\sqrt{x^2 - 9}} \, dx$

5. $\int \dfrac{1}{25 - x^2} \, dx$

6. $\int \dfrac{1}{x\sqrt{4 + x^2}} \, dx$

7. $\int \dfrac{x}{3 - x} \, dx$

8. $\int \dfrac{x}{(1 - x)^2} \, dx$

9. $\int \dfrac{1}{x(8 - x)^2} \, dx$

10. $\int \sqrt{x^2 + 9} \, dx$

11. $\int \ln (3x) \, dx$

12. $\int \ln \left(\dfrac{4}{5}x \right) \, dx$

13. $\int x^4 \ln x \, dx$

14. $\int x^3 e^{-2x} \, dx$

15. $\int x^3 \ln x \, dx$

16. $\int 5x^4 \ln x \, dx$

17. $\int \dfrac{dx}{\sqrt{x^2 + 7}}$

18. $\int \dfrac{3 \, dx}{x\sqrt{1 - x^2}}$

19. $\int \dfrac{10 \, dx}{x(5 - 7x)^2}$

20. $\int \dfrac{2}{5x(7x + 2)} \, dx$

21. $\int \dfrac{-5}{4x^2 - 1} \, dx$

22. $\int \sqrt{9t^2 - 1} \, dt$

23. $\int \sqrt{4m^2 + 16} \, dm$

24. $\int \dfrac{3 \ln x}{x^2} \, dx$

25. $\int \dfrac{-5 \ln x}{x^3} \, dx$

26. $\int (\ln x)^4 \, dx$

27. $\int \dfrac{e^x}{x^{-3}} \, dx$

28. $\int \dfrac{3}{\sqrt{4x^2 + 100}} \, dx$

29. $\int x\sqrt{1 + 2x} \, dx$

30. $\int x\sqrt{2 + 3x} \, dx$

APPLICATIONS

Business and Economics

31. Supply from marginal supply. A lawn machinery company introduces a new kind of lawn seeder. It finds that its marginal supply for the seeder satisfies the function

$$S'(x) = \frac{100x}{(20 - x)^2}, \quad 0 \le x \le 19,$$

where S is the quantity purchased when the price is x thousand dollars per seeder. Find the supply function, $S(x)$, given that the company will sell 2000 seeders when the price is 19 thousand dollars.

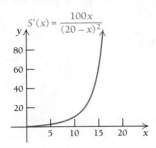

Social Sciences

32. Learning rate. The rate of change of the probability that an employee learns a task on a new assembly line is given by

$$p'(t) = \frac{1}{t(2 + t)^2},$$

where $p(t)$ is the probability of learning the task after t months. Find $p(t)$ given that $p = 0.8267$ when $t = 2$.

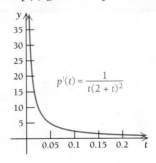

SYNTHESIS

Evaluate using Table 1 or the Web site www.integrals.com.

33. $\displaystyle\int \frac{8}{3x^2 - 2x}\, dx$

34. $\displaystyle\int \frac{x\, dx}{4x^2 - 12x + 9}$

35. $\displaystyle\int \frac{dx}{x^3 - 4x^2 + 4x}$

36. $\displaystyle\int e^x \sqrt{e^{2x} + 1}\, dx$

37. $\displaystyle\int \frac{-e^{-2x}\, dx}{9 - 6e^{-x} + e^{-2x}}$

38. $\displaystyle\int \frac{\sqrt{(\ln x)^2 + 49}}{2x}\, dx$

Answers to Quick Checks

1. Using formula 19: $\dfrac{6}{25(3 - 5x)} + \dfrac{2}{25} \ln |3 - 5x| + C$

2. Using formula 21: $\dfrac{3}{14(7 - 3x)} + \dfrac{3}{98} \ln \left| \dfrac{x}{7 - 3x} \right| + C$

3. Using formula 24:

$$\frac{2}{2835}(135x^2 - 288x + 512)(8 + 3x)^{3/2} + C$$

4. Using formula 14: $\dfrac{2}{11}\sqrt{11} \ln \left| \dfrac{(x - \sqrt{11})^2}{x^2 - 11} \right| + C$

5. Using formula 10: $\dfrac{x^5}{5} \ln x - \dfrac{x^5}{25} + C$

KEY TERMS AND CONCEPTS   **EXAMPLES**

SECTION 4.1

Antidifferentiation is the reverse of differentiation. A function F is an **antiderivative** of a function f if

$$\frac{d}{dx} F(x) = f(x).$$

Antiderivatives of a function f all differ by a constant C, called the **constant of integration**.

$F(x) = x^2$ is an antiderivative of $f(x) = 2x$ since $\frac{d}{dx}(x^2) = 2x$.

Antiderivatives of $f(x) = 2x$ have the form $x^2 + C$. These antiderivatives can also be expressed using the indefinite integral $\int 2x \, dx = x^2 + C$, where the function $f(x) = 2x$ is the integrand.

An **indefinite integral** of a function is symbolized by

$$\int f(x) \, dx = F(x) + C,$$

where $f(x)$ is called the **integrand**.

$G(x) = \ln x$ is an antiderivative of $g(x) = \frac{1}{x}$ since $\frac{d}{dx}(\ln x) = \frac{1}{x}$.

The corresponding indefinite integral is

$$\int \frac{1}{x} \, dx = \ln x + C.$$

We use four **rules of antidifferentiation**:

A1. $\int k \, dx = kx + C$

A2. $\int x^n \, dx = \frac{x^{n+1}}{n + 1} + C, \quad n \neq -1$

A3. $\int \frac{1}{x} \, dx = \ln x + C, \quad x > 0$

A4. $\int e^{ax} \, dx = \frac{1}{a} e^{ax} + C$

There are two common **properties of indefinite integrals**:

P1. $\int [c \cdot f(x)] \, dx = c \int f(x) \, dx$

P2. $\int [f(x) \pm g(x)] \, dx$
$\quad = \int f(x) \, dx \pm \int g(x) \, dx.$

- $\int 9 \, dx = 9x + C \qquad$ A1

- $\int x^6 \, dx = \frac{1}{7} x^7 + C \qquad$ A2

- $\int (3x^4 + 4x - 5) \, dx = \frac{3}{5} x^5 + 2x^2 - 5x + C \qquad$ A2, P1, P2

- $\int e^{3x} \, dx = \frac{1}{3} e^{3x} + C \qquad$ A4

- $\int \frac{5}{x} \, dx = 5 \ln x + C \qquad$ A3

- $\int \sqrt{x} \, dx = \int x^{1/2} \, dx = \frac{2}{3} x^{3/2} + C \qquad$ A2

- $\int \frac{1}{x^5} \, dx = \int x^{-5} \, dx = \frac{1}{-4} x^{-4} = -\frac{1}{4x^4} + C \qquad$ A2

- $\int \frac{x^2 - 1}{x} \, dx = \int \left(x - \frac{1}{x} \right) dx = \frac{1}{2} x^2 - \ln x + C \qquad$ A2, A3, P2

- $\int (x + 4)^2 \, dx = \int (x^2 + 8x + 16) \, dx$

$$= \frac{1}{3} x^3 + 4x^2 + 16x + C \qquad \text{A1, A2, P1, P2}$$

(continued)

KEY TERMS AND CONCEPTS	EXAMPLES

SECTION 4.1 (continued)

An **initial condition** is a point that is a solution of a particular antiderivative.

Find $\int (3x - 2)\, dx$ such that $(1, 4)$ is a solution of the antiderivative.

We antidifferentiate: $\int (3x - 2)\, dx = \dfrac{3}{2}x^2 - 2x + C$. Therefore, we

have $F(x) = \dfrac{3}{2}x^2 - 2x + C$. We are given $(1, 4)$ as an initial condition, so we substitute and solve for C:

$$4 = \frac{3}{2}(1)^2 - 2(1) + C$$

$$4 = \frac{3}{2} - 2 + C$$

$$4 = -\frac{1}{2} + C$$

$$C = \frac{9}{2}.$$

Therefore, the particular antiderivative that meets the initial condition is

$$F(x) = \frac{3}{2}x^2 - 2x + \frac{9}{2}.$$

SECTION 4.2

The **area under the graph of a function** can be interpreted in a meaningful way. The units of the area are determined by multiplying the units of the input variable by the units of the output variable.

Physical Science. A jogger runs according to a velocity function $v(t) = 6$, where t is in hours and v is in miles per hour. In 3 hr, the jogger will have run $3 \text{ hr} \cdot 6\, \dfrac{\text{mi}}{\text{hr}} = 18\text{ mi}$, which is the area under the line representing the velocity function.

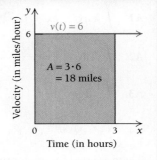

Common geometry formulas can sometimes be used to calculate the area.

Business. A company's marginal revenue is modeled by

$$R'(x) = 0.37x,$$

where x is the number of units sold and R' is in thousands of dollars per unit. The total revenue from selling 100 units is the area under the marginal revenue function.

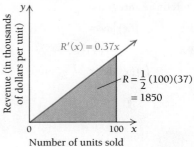

$$\text{Total revenue} = \frac{1}{2}(100 \text{ units})\left(37\, \frac{\text{thousands of dollars}}{\text{unit}}\right)$$

$$= 1850 \text{ thousand dollars, or } \$1{,}850{,}000.$$

KEY TERMS AND CONCEPTS	EXAMPLES

Riemann summation uses rectangles to approximate the area under a curve. The more subintervals (rectangles) used, the more accurate the approximation of the area.

If the number of subintervals is allowed to approach infinity, we have a **definite integral**, which represents the exact area under the graph of a continuous and nonnegative function $f(x) \geq 0$ over an interval $[a, b]$:

$$\text{Exact area} = \int_a^b f(x)\, dx.$$

Approximate the area under the graph of $f(x) = -\frac{1}{3}x^2 + 3x$ over the interval $[1, 9]$ using 4 subintervals.

Each subinterval will have width $\Delta x = \dfrac{9 - 1}{4} = 2$, with x_i ranging from $x_1 = 1$ to $x_4 = 7$. The area under the curve over $[1, 9]$ is approximated as follows:

$$\sum_{i=1}^{4} f(x_i) \cdot \Delta x = f(1) \cdot 2 + f(3) \cdot 2 + f(5) \cdot 2 + f(7) \cdot 2$$

$$= \frac{8}{3} \cdot 2 + 6 \cdot 2 + \frac{20}{3} \cdot 2 + \frac{14}{3} \cdot 2$$

$$= 40.$$

Thus, the area is approximately 40 square units.

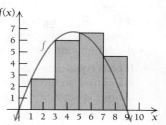

SECTION 4.3

The **Fundamental Theorem of Calculus** tells us that the exact area under a continuous function f over an interval $[a, b]$ is calculated directly using a definite integral:

$$\int_a^b f(x)\, dx = F(b) - F(a).$$

The function $F(x)$ is any antiderivative of $f(x)$ (we usually set the constant of integration equal to 0).

The exact area under the graph of $f(x) = -\frac{1}{3}x^2 + 3x$ over the interval $[1, 9]$ is

$$\int_1^9 \left(-\frac{1}{3}x^2 + 3x \right) dx = \left[-\frac{1}{9}x^3 + \frac{3}{2}x^2 \right]_1^9$$

$$= \left(-\frac{1}{9}(9)^3 + \frac{3}{2}(9)^2 \right) - \left(-\frac{1}{9}(1)^3 + \frac{3}{2}(1)^2 \right)$$

$$= \frac{352}{9} = 39\frac{1}{9}.$$

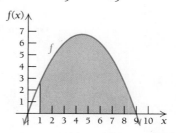

(*continued*)

KEY TERMS AND CONCEPTS	EXAMPLES

SECTION 4.3 (continued)

The Fundamental Theorem of Calculus is true for *all* continuous functions over an interval $[a, b]$.

The definite integral gives the *net* area between the graph of a continuous function f and the x-axis over an interval $[a, b]$:

- If f is negative over an interval $[a, b]$, then the definite integral will be negative.
- If f has more area above the x-axis than below it over an interval $[a, b]$, then the definite integral will be positive.
- If f has more area below the x-axis than above it over an interval $[a, b]$, then the definite integral will be negative.
- If f has equal areas above and below the x-axis over an interval $[a, b]$, then the definite integral will be zero.

Evaluate the definite integral of $f(x) = x^2 - 1$ over the interval $[-1, 1]$.

The function f is negative, therefore, the definite integral will be negative.

$$\int_{-1}^{1} (x^2 - 1) \, dx = \left[\frac{1}{3} x^3 - x \right]_{-1}^{1}$$

$$= \left(\frac{1}{3} (1)^3 - (1) \right) - \left(\frac{1}{3} (-1)^3 - (-1) \right)$$

$$= -\frac{4}{3}.$$

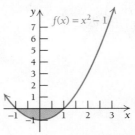

Area over [−1, 1]

Evaluate the definite integral of $f(x) = x^2 - 1$ over the interval $[0, 3]$.

The function f has more area above the x-axis than below. For the portion of the graph below the x-axis, we integrate from 0 to 1:

$$\int_{0}^{1} (x^2 - 1) \, dx = \left[\frac{1}{3} x^3 - x \right]_{0}^{1} = -\frac{2}{3}.$$

For the portion of the graph above the x-axis, we integrate from 1 to 3:

$$\int_{1}^{3} (x^2 - 1) \, dx = \left[\frac{1}{3} x^3 - x \right]_{1}^{3} = (9 - 3) - \left(\frac{1}{3} - 1 \right) = \frac{20}{3}.$$

We sum the two results:

$$\int_{0}^{3} (x^2 - 1) \, dx = \left[\frac{1}{3} x^3 - x \right]_{0}^{3} = -\frac{2}{3} + \frac{20}{3} = \frac{18}{3} = 6.$$

Thus, the net area is 6. We can also integrate from 0 to 3 directly:

$$\int_{0}^{3} (x^2 - 1) \, dx = \left[\frac{1}{3} x^3 - x \right]_{0}^{3} = (9 - 3) - (0) = 6.$$

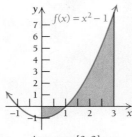

Area over [0, 3]

KEY TERMS AND CONCEPTS	**EXAMPLES**

The function $g(x) = x - 2$ over $[0, 4]$ has equal areas below and above the x-axis. Therefore,

$$\int_0^4 (x - 2)\, dx = \left[\frac{1}{2}x^2 - 2x\right]_0^4 = 8 - 8 = 0.$$

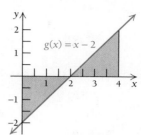

Area over [0, 4]

SECTION 4.4

There are many useful properties of definite integrals.

- **Additive property:** if $a < b < c$, we have

$$\int_a^c f(x)\, dx = \int_a^b f(x)\, dx + \int_b^c f(x)\, dx.$$

- **Area of a region bounded by two curves:** if $f(x) \geq g(x)$ over an interval $[a, b]$, then the area between the graphs of f and g from $x = a$ to $x = b$ is

$$A = \int_a^b \left[f(x) - g(x)\right] dx.$$

- **Average value:** the average value of a function f over an interval $[a, b]$ is given by

$$y_{\text{av}} = \frac{1}{b - a}\int_a^b f(x)\, dx.$$

The additive property is useful for piecewise-defined functions, which include the absolute-value function.

$$\int_{-2}^3 |x|\, dx = \int_{-2}^0 (-x)\, dx + \int_0^3 x\, dx$$

$$= 2 + \frac{9}{2}$$

$$= \frac{13}{2}.$$

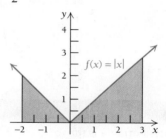

Let $f(x) = x^2$ and $g(x) = x + 2$. Setting these expressions equal to one another and solving for x, we find that the curves intersect when $x = -1$ and $x = 2$. Furthermore, we see that $g(x) = x + 2$ is the "top" function. Therefore, the area between these curves is

$$A = \int_{-1}^2 (x + 2 - x^2)\, dx = \left[\frac{1}{2}x^2 + 2x - \frac{1}{3}x^3\right]_{-1}^2$$

$$= \left(\frac{1}{2}(2)^2 + 2(2) - \frac{1}{3}(2)^3\right) - \left(\frac{1}{2}(-1)^2 + 2(-1) - \frac{1}{3}(-1)^3\right)$$

$$= \frac{9}{2}.$$

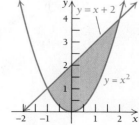

(*continued*)

KEY TERMS AND CONCEPTS	**EXAMPLES**

SECTION 4.4 (continued)

The average value of $y = x^3$ over the interval $[1, 4]$ is

$$y_{av} = \frac{1}{3} \int_1^4 x^3 \, dx = \left[\frac{1}{12} x^4 \right]_1^4 = 21\frac{1}{4}.$$

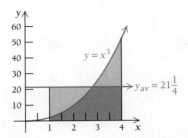

SECTION 4.5

Integration by **substitution** is the reverse of applying the Chain Rule. We choose u, determine du, and rewrite the integrand in terms of u. It may be necessary to multiply the integrand by a constant and the entire integral by its reciprocal to obtain the correct form. *Results should be checked using differentiation!*

Evaluate $\int 2x(x^2 + 1)^5 \, dx$.

We let $u = x^2 + 1$, so $du = 2x \, dx$. Therefore, we have

$$\int 2x(x^2 + 1)^5 \, dx = \int u^5 \, du = \frac{1}{6} u^6 + C = \frac{1}{6}(x^2 + 1)^6 + C.$$

Evaluate $\int \frac{1}{3x - 2} \, dx$.

We let $u = 3x - 2$, so $du = 3 \, dx$. We multiply the integrand by 3 and the integral by $\frac{1}{3}$ outside the integral:

$$\int \frac{1}{3x - 2} \, dx = \frac{1}{3} \int \frac{1}{3x - 2} 3 \, dx = \frac{1}{3} \int \frac{1}{u} \, du$$

$$= \frac{1}{3} \ln u + C = \frac{1}{3} \ln (3x - 2) + C, \text{ where } (3x - 2) > 0.$$

SECTION 4.6

The **Integration-by-Parts Formula** is the reverse of the product rule for derivatives:

$$\int u \, dv = uv - \int v \, du$$

Evaluate $\int x^3 \ln x \, dx$.

We let $u = \ln x$ and $dv = x^3 \, dx$. We have $du = \frac{1}{x} \, dx$ and $v = \frac{1}{4} x^4$.

Therefore,

$$\int x^3 \ln x \, dx = \frac{1}{4} x^4 \ln x - \int \left(\frac{1}{4} x^4 \right) \left(\frac{1}{x} \, dx \right)$$

$$= \frac{1}{4} x^4 \ln x - \frac{1}{4} \int x^3 \, dx$$

$$= \frac{1}{4} x^4 \ln x - \frac{1}{16} x^4 + C.$$

KEY TERMS AND CONCEPTS	**EXAMPLES**

Tabular integration by parts is useful when this formula has to be applied more than once to evaluate an integral.

Evaluate $\int x^3 e^{2x}\,dx$.

This will involve repeated integrations by parts. Since x^3 eventually differentiates to 0 and e^{2x} is easily integrable, we use tabular integration by parts:

$f(x)$ and Repeated Derivatives	Sign of Product	$g(x)$ and Repeated Integrals
x^3	$(+)$	e^{2x}
$3x^2$	$(-)$	$\frac{1}{2}e^{2x}$
$6x$	$(+)$	$\frac{1}{4}e^{2x}$
6	$(-)$	$\frac{1}{8}e^{2x}$
0		$\frac{1}{16}e^{2x}$

We multiply along the arrows, alternate signs, and simplify when possible. The antiderivative is

$$\int x^3 e^{2x}\,dx = \frac{1}{2}x^3 e^{2x} - \frac{3}{4}x^2 e^{2x} + \frac{6}{8}xe^{2x} - \frac{6}{16}e^{2x}$$

$$= e^{2x}\left(\frac{1}{2}x^3 - \frac{3}{4}x^2 + \frac{3}{4}x - \frac{3}{8}\right) + C.$$

SECTION 4.7

Integration tables show formulas for evaluating many general forms of integrals.

Evaluate $\int \dfrac{3}{\sqrt{x^2 + 64}}\,dx$.

We see that formula 12 from the table of integration formulas on pp. 454–455 is appropriate:

$$\int \frac{1}{\sqrt{x^2 + a^2}}\,dx = \ln\left|x + \sqrt{x^2 + a^2}\right| + C.$$

We bring the constant 3 to the front of the integral, and we note that $a^2 = 64$, so $a = 8$:

$$\int \frac{3}{\sqrt{x^2 + 64}}\,dx = 3\int \frac{1}{\sqrt{x^2 + 64}}\,dx = 3\ln\left|x + \sqrt{x^2 + 64}\right| + C.$$

These review exercises are for test preparation. They can also be used as a practice test. Answers are at the back of the book. The blue bracketed section references tell you what part(s) of the chapter to restudy if your answer is incorrect.

CONCEPT REINFORCEMENT

Classify each statement as either true or false.

1. Riemann sums are a way of approximating the area under a curve by using rectangles. [4.2]

2. If a and b are both negative, then $\int_a^b f(x)\, dx$ is negative. [4.3]

3. For any continuous function f defined over $[-1, 7]$, it follows that

$$\int_{-1}^2 f(x)\, dx + \int_2^7 f(x)\, dx = \int_{-1}^7 f(x)\, dx. \,[4.4]$$

4. Every integral can be evaluated using integration by parts. [4.6]

Match each integral in column A with the corresponding antiderivative in column B. [4.1, 4.5]

Column A   Column B

5. $\int \dfrac{1}{\sqrt{x}}\, dx$ **a)** $\ln x + C$

6. $\int (1 + 2x)^{-2}\, dx$ **b)** $-x^{-1} + C$

7. $\int \dfrac{1}{x}\, dx, \quad x > 0$ **c)** $-(1 + x^2)^{-1} + C$

8. $\int \dfrac{2x}{1 + x^2}\, dx$ **d)** $-\dfrac{1}{2}(1 + 2x)^{-1} + C$

9. $\int \dfrac{1}{x^2}\, dx$ **e)** $2x^{1/2} + C$

10. $\int \dfrac{2x}{(1 + x^2)^2}\, dx$ **f)** $\ln (1 + x^2) + C$

REVIEW EXERCISES

11. Business: total cost. The marginal cost, in dollars, of producing the xth car stereo is given by

$$C'(x) = 0.004x^2 - 2x + 500.$$

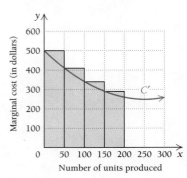

Approximate the total cost of producing 200 car stereos by computing the sum

$$\sum_{i=1}^4 C'(x_i)\, \Delta x, \quad \text{with } \Delta x = 50. \,[4.2]$$

Evaluate. [4.1]

12. $\int 20x^4\, dx$ **13.** $\int (3e^x + 2)\, dx$

14. $\int \left(3t^2 + 5t + \dfrac{1}{t} \right) dt$ (assume $t > 0$)

Find the area under the curve over the indicated interval. [4.3]

15. $y = 4 - x^2$: $[-2, 1]$

16. $y = x^2 + 2x + 1$: $[0, 3]$

In each case, give an interpretation of the shaded region. [4.2, 4.3]

17.

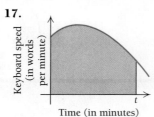

18.

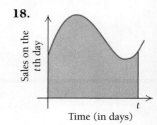

Evaluate. [4.3, 4.4]

19. $\displaystyle\int_a^b x^5 \, dx$

20. $\displaystyle\int_{-1}^1 (x^3 - x^4) \, dx$

21. $\displaystyle\int_0^1 (e^x + x) \, dx$

22. $\displaystyle\int_1^4 \frac{2}{x} \, dx$

23. $\displaystyle\int_{-2}^4 f(x) \, dx$, where $f(x) = \begin{cases} x + 2, & \text{for } x \le 0, \\ 2 - \frac{1}{2}\sqrt{x}, & \text{for } x > 0 \end{cases}$

Decide whether $\displaystyle\int_a^b f(x) \, dx$ is positive, negative, or zero. [4.3]

24.

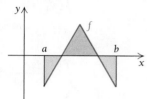

25.

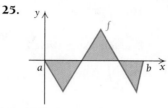

26.

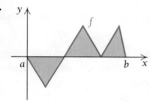

27. Find the area of the region bounded by
$y = 3x^2$ and $y = 9x$. [4.4]

Evaluate using substitution. Do not use Table 1. [4.5]

28. $\displaystyle\int x^3 e^{x^4} \, dx$

29. $\displaystyle\int \frac{24t^5}{4t^6 + 3} \, dt$

30. $\displaystyle\int \frac{\ln (4x)}{2x} \, dx$

31. $\displaystyle\int 2e^{-3x} \, dx$

Evaluate using integration by parts. Do not use Table 1. [4.6]

32. $\displaystyle\int 3x e^{3x} \, dx$

33. $\displaystyle\int \ln \sqrt[3]{x^2} \, dx$

34. $\displaystyle\int 3x^2 \ln x \, dx$

35. $\displaystyle\int x^4 e^{3x} \, dx$

Evaluate using Table 1. [4.7]

36. $\displaystyle\int \frac{1}{49 - x^2} \, dx$

37. $\displaystyle\int x^2 e^{5x} \, dx$

38. $\displaystyle\int \frac{x}{7x + 1} \, dx$

39. $\displaystyle\int \frac{dx}{\sqrt{x^2 - 36}}$

40. $\displaystyle\int x^6 \ln x \, dx$

41. $\displaystyle\int x e^{8x} \, dx$

42. Business: total cost. Refer to Exercise 11. Calculate the total cost of producing 200 car stereos. [4.4]

43. Find the average value of $y = x e^{-x}$ over $[0, 2]$. [4.4]

44. A particle starts out from the origin. Its velocity in mph after t hours is given by $v(t) = 3t^2 + 2t$. Find the distance that the particle travels during the first 4 hr (from $t = 0$ to $t = 4$). [4.3]

45. Business: total revenue. A company estimates that its revenue will grow continuously at a rate given by the function $S'(t) = 3e^{3t}$, where $S'(t)$ is the rate at which revenue is increasing on the tth day. Find the accumulated revenue for the first 4 days. [4.3]

Integrate using any method. [4.3–4.6]

46. $\displaystyle\int x^3 e^{0.1x} \, dx$

47. $\displaystyle\int \frac{12t^2}{4t^3 + 7} \, dt$

48. $\displaystyle\int \frac{x \, dx}{\sqrt{4 + 5x}}$

49. $\displaystyle\int 5x^4 e^{x^5} \, dx$

50. $\displaystyle\int \frac{dx}{x + 9}$ (assume $x > -9$)

51. $\displaystyle\int t^7 (t^8 + 3)^{11} \, dt$

52. $\displaystyle\int \ln (7x) \, dx$

53. $\displaystyle\int x \ln (8x) \, dx$

SYNTHESIS

Evaluate. [4.5–4.7]

54. $\displaystyle\int \frac{t^4 \ln (t^5 + 3)}{t^5 + 3} \, dt$

55. $\displaystyle\int \frac{dx}{e^x + 2}$

56. $\displaystyle\int \frac{\ln \sqrt{x}}{x} \, dx$

57. $\displaystyle\int x^{91} \ln x \, dx$

58. $\displaystyle\int \ln \left(\frac{x - 3}{x - 4} \right) \, dx$

59. $\displaystyle\int \frac{dx}{x (\ln x)^4}$

60. $\displaystyle\int x \sqrt[3]{x + 3} \, dx$

61. $\displaystyle\int \frac{x^2}{2x + 1} \, dx$

TECHNOLOGY CONNECTION

62. Use a graphing calculator to approximate the area between the following curves:
$$y = 2x^2 - 2x, \quad y = 12x^2 - 12x^3. \, [4.4]$$

1. Approximate

$$\int_0^5 (25 - x^2)\, dx$$

by computing the area of each rectangle and adding.

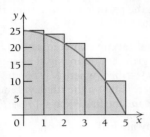

Evaluate.

2. $\int \sqrt{3x}\, dx$

3. $\int 1000x^5\, dx$

4. $\int \left(e^x + \dfrac{1}{x} + x^{3/8} \right) dx$ (assume $x > 0$)

Find the area under the curve over the indicated interval.

5. $y = x - x^2$; $[0, 1]$

6. $y = \dfrac{4}{x}$; $[1, 3]$

7. Give an interpretation of the shaded area.

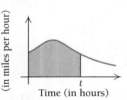

Evaluate.

8. $\int_{-1}^{2} (2x + 3x^2)\, dx$

9. $\int_0^1 e^{-2x}\, dx$

10. $\int_e^{e^2} \dfrac{dx}{x}$

11. $\int_0^5 g(x)\, dx$, where $g(x) = \begin{cases} x^2, & \text{for } x \le 2, \\ 6 - x, & \text{for } x > 2 \end{cases}$

12. Decide whether $\int_a^b f(x)\, dx$ is positive, negative, or zero.

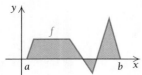

Evaluate using substitution. Assume $u > 0$ when $\ln u$ appears. Do not use Table 1.

13. $\int \dfrac{dx}{x + 12}$

14. $\int e^{-0.5x}\, dx$

15. $\int t^3 (t^4 + 3)^9\, dt$

Evaluate using integration by parts. Do not use Table 1.

16. $\int x e^{5x}\, dx$

17. $\int x^3 \ln x^4\, dx$

Evaluate using Table 1.

18. $\int 2^x\, dx$

19. $\int \dfrac{dx}{x(7 - x)}$

20. Find the average value of $y = 4t^3 + 2t$ over $[-1, 2]$.

21. Find the area of the region in the first quadrant bounded by $y = x$ and $y = x^5$.

22. Business: cost from marginal cost. An air conditioning company determines that the marginal cost, in dollars, for the xth air conditioner is given by

$$C'(x) = -0.2x + 500, \quad C(0) = 0.$$

Find the total cost of producing 100 air conditioners.

23. Social science: learning curve. A translator's speed over 4-min interval is given by

$$W(t) = -6t^2 + 12t + 90, \quad t \text{ in } [0, 4],$$

where $W(t)$ is the speed, in words per minute, at time t. How many words are translated during the second minute (from $t = 1$ to $t = 2$)?

24. A robot leaving a spacecraft has velocity given by $v(t) = -0.4t^2 + 2t$, where $v(t)$ is in kilometers per hour and t is the number of hours since the robot left the spacecraft. Find the total distance traveled during the first 3 hr.

Integrate using any method. Assume $u > 0$ when $\ln u$ appears.

25. $\int \dfrac{6}{5 + 7x}\, dx$

26. $\int x^5 e^x\, dx$

27. $\displaystyle\int x^5 e^{x^6}\,dx$

28. $\displaystyle\int \sqrt{x}\,\ln x\,dx$

29. $\displaystyle\int \frac{dx}{64 - x^2}$

30. $\displaystyle\int x^4 e^{-0.1x}\,dx$

31. $\displaystyle\int x\ln(13x)\,dx$

34. $\displaystyle\int \ln\!\left(\frac{x+3}{x+5}\right)dx$

35. $\displaystyle\int \frac{8x^3 + 10}{\sqrt[3]{5x - 4}}\,dx$

36. $\displaystyle\int \frac{x}{\sqrt{3x - 2}}\,dx$

37. $\displaystyle\int \frac{(x+4)^2}{x^2}\,dx$

38. Evaluate $\int 5^x\,dx$ without using Table 1.
(*Hint:* $5 = e^{\ln 5}$.)

SYNTHESIS

Evaluate using any method.

32. $\displaystyle\int x^3\sqrt{x^2 + 4}\,dx$

33. $\displaystyle\int \frac{\big[(\ln x)^3 - 4(\ln x)^2 + 5\big]}{x}\,dx$

TECHNOLOGY CONNECTION

39. Use a calculator to approximate the area between the following curves:

$$y = 3x - x^2, \quad y = 2x^3 - x^2 - 5x.$$

Extended Technology Application

Business: Distribution of Wealth

Lorenz Functions and the Gini Coefficient

The distribution of wealth within a population is of great interest to many economists and sociologists. Let $y = f(x)$ represent the percentage of wealth owned by x percent of the population, with x and y expressed as decimals between 0 and 1. The assumptions are that 0% of the population owns 0% of the wealth and that 100% of the population owns 100% of the wealth. With these requirements in place, the *Lorenz function* is defined to be any continuous, increasing and concave upward function connecting the points $(0, 0)$ and $(1, 1)$, which represent the two extremes. The function is named for economist Max Otto Lorenz (1880–1962), who developed these concepts as a graduate student in 1905–1906.

If the collective wealth of a society is equitably distributed among its population, we would observe that "x%

of the population owns x% of the wealth," and this is modeled by the function $f(x) = x$, where $0 \le x \le 1$. This is an example of a Lorenz function that is often called the *line of equality*.

In many societies, the distribution of wealth is not equitable. For example, the Lorenz function $f(x) = x^3$ would represent a society in which a large percentage of the population owns a small percentage of the wealth. For example, in this society, we observe that $f(0.7) = 0.7^3 = 0.343$, meaning that 70% of the population owns just 34.3% of the wealth, with the implication that the other 30% owns the remaining 65.7% of the wealth.

In the graphs below, we see the line of equality in the left-most graph, and increasingly inequitable distributions as we move to the right.

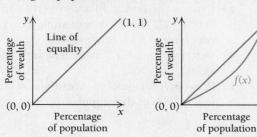

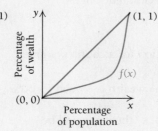

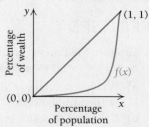

Most equitable distribution ————————————————————————————————→ Less equitable distribution

Note that the area between the line of equality and the graph of the Lorenz function $f(x)$ is small if the distribution of wealth is close to equitable and is large when the distribution is very unequitable. The *Gini coefficient* (named for the Italian statistician and demographer Corrado Gini, 1884–1965) is a measure of the difference between the actual distribution of wealth in a society and the ideal distribution represented by the line of equality. It is the ratio of the area between the line of equality and the graph of the Lorenz function to the area below the line of equality and above the *x*-axis. In the figure that follows, the Gini coefficient is represented by the formula

$$\text{Gini coefficient} = \frac{A}{A + B}.$$

The area A is found by calculating the area between two curves, $\int_0^1 (x - f(x))\,dx$, where x is the line of equality and $f(x)$ is the Lorenz function for a particular society. We observe that $A + B$ is a triangle with area $\frac{1}{2}(1)(1) = \frac{1}{2}$. Thus, the Gini coefficient can be written as an integral:

$$\text{Gini coefficient} = \frac{A}{A + B} = \frac{\displaystyle\int_0^1 (x - f(x))\,dx}{\left(\dfrac{1}{2}\right)}$$

$$= 2\int_0^1 (x - f(x))\,dx.$$

For the most equitable distribution of wealth, the Gini coefficient would be 0, since there would be no difference (area) between the graph of the Lorenz function and the line of equality; for the most inequitable distribution of wealth, the Gini coefficient would be 1. Often, the Gini coefficient is multiplied by 100 to give the *Gini index*: a Gini coefficient of 0.34 gives a Gini index of 34.

EXERCISES

1. Suppose the Lorenz function for a country is given by $f(x) = x^2, 0 \le x \le 1$.

a) What percentage of the wealth is owned by 60% of the population?

b) Calculate the Gini index (the value will be between 0 and 100).

2. Suppose the Lorenz function for a country is given by $f(x) = x^{3.5}, 0 \le x \le 1$.

a) What percentage of the wealth is owned by 60% of the population?

b) Calculate the Gini index.

Regression for Determining Lorenz Functions

If data exist on the distribution of wealth in a society, a Lorenz function can be determined using regression.

EXERCISES

3. The data in the table show the amount of wealth distributed within a population.

x	0.1	0.2	0.3	0.4	0.5	0.6	0.7	0.8	0.9
y	0.0178	0.06	0.122	0.201	0.297	0.409	0.536	0.677	0.833

Use regression to determine a power function that best fits these data. (*Note:* Entering the point $(0, 0)$ may cause an error message to appear. However, the point $(1, 1)$ should be entered along with the rest of the data.)

a) Express the Lorenz function in the form $f(x) = x^n$. The coefficient should be 1, so you may have to do some rounding.

b) Determine the Gini coefficient and the Gini index.

c) What percentage of the wealth is owned by the lowest 74% of this population?

4. A fast-food chain has many hundreds of franchises nationwide. Ideally, each franchise would generate equal amounts of revenue for the chain, but in reality, some perform better than others. An internal audit reveals the following results: the lowest 30% of the franchises account for just 6% of the total revenue, the lowest 50% account for 20% of the total revenue, and the lowest 70% account for 43.5% of the total revenue. (Assume that 100% of the franchises account for 100% of the total revenue.)

a) Use regression to determine a power function that models these data, and write the Lorenz function in the form $f(x) = x^n$. The coefficient should be 1, so you may have to do some rounding.

b) Determine the Gini coefficient and the Gini index.

c) What percentage of total revenue is generated by the lowest 45% of the franchises?

d) What percentage of total revenue is generated by the top 10% of the franchises?

Gini Coefficient as a Function of n

Functions of the form $f(x) = x^n$, $0 \leq x \leq 1$, where $n \geq 1$, meet the criteria for Lorenz functions. We can develop a function $G(n)$ that will allow us to calculate the Gini coefficient directly, given a value of n.

$$G(n) = 2\int_0^1 (x - x^n)\, dx$$
$$= \left[2\left(\frac{1}{2}x^2 - \frac{1}{n+1}x^{n+1} \right) \right]_0^1$$
$$= 2\left(\frac{1}{2} - \frac{1}{n+1} \right)$$
$$= 1 - \frac{2}{n+1}$$
$$= \frac{n-1}{n+1}.$$

EXERCISES

5. Verify your results for Exercises 3 and 4 using the function $G(n) = \dfrac{n-1}{n+1}$.

6. The United States had a Gini index of 45.0 in 2007. (*Source*: Department of Labor Statistics.) Express this as a decimal: $G = 0.45$.

a) Solve for n, and write the Lorenz function in the form $f(x) = x^n$.

b) According to this model, what percentage of the wealth was owned by the least wealthy 55% of U.S. citizens in 2007?

7. Canada's Gini index is usually 30.0.

a) Determine the Lorenz function.

b) What percentage of wealth is owned by the least wealthy 55% of the citizens in Canada?

Sometimes, raw data may not fit "neatly" into the $f(x) = x^n$ form, especially in cases where the distribution very heavily favors a small percentage of the population that holds most of the wealth. In these cases, an exponential function of the form $f(x) = a \cdot b^x$, $0 \leq x \leq 1$ may work better, as long as the value of a is extremely small.

8. In 2004, the distribution of net worth within the
United States was as given in the following table:

PERCENTAGE OF POPULATION	0.4	0.6	0.8	0.9	0.95	0.99	1
PERCENTAGE OF WEALTH	0.002	0.04	0.153	0.287	0.41	0.656	1

(*Source*: Prof. E. N. Wolff, Levy Institute of Economics at Bard College, 2007.)

a) According to the table, what percentage of net
worth was held by the top 1%? (*Hint*: What percentage did the other 99% hold?)

b) Use regression to fit an exponential function
$g(x) = a \cdot b^x$ to these data.

c) Determine the area between the line of equality
and the graph of $g(x)$ over the interval $[0, 1]$.
(*Hint*: Integrate $a \cdot b^x$ using formula 11 from
Table 1 in Section 4.7.)

d) Determine the Gini coefficient and the Gini
index. (*Note*: E. N. Wolff calculated the Gini
coefficient as 0.829.)

e) What percentage of the net worth was held
by the lowest 50% of the population?

f) What percentage of the net worth was held
by the top 15% of the population?

Applications of Integration

5

Chapter Snapshot

What You'll Learn

5.1 An Economics Application:
Consumer Surplus and Producer Surplus

5.2 Applications of Integrating Growth
and Decay Models

5.3 Improper Integrals

5.4 Numerical Integration

5.5 Volume

Why It's Important

In this chapter, we explore a wide variety of applications of integration to business and economics (consumer and producer surplus and income streams) and environmental science and finance (exponential growth and decay). We also see how to use integration to find volumes of solids and how to integrate using various numerical methods.

Where It's Used

BUNGEE JUMPING

Regina loves to go bungee jumping. The table shows the number of half-hours that Regina is willing to go bungee jumping at various prices. If Regina goes bungee jumping for 6 half-hours per month, what is her consumer surplus? At a price of $11.50 per half-hour, what is Regina's consumer surplus?

This problem appears as Exercise 21 in Exercise Set 5.1.

REGINA'S DEMAND DATA

TIME SPENT (in half-hours per month)	PRICE (per half-hour)
8	$ 2.50
7	5.00
6	7.50
5	10.00
4	12.50
3	15.00
2	17.50
1	20.00

<table>
<tr><td>

5.1

</td><td>

An Economics Application: Consumer Surplus and Producer Surplus

</td></tr>
</table>

OBJECTIVE

• Given demand and supply functions, find the consumer surplus and the producer surplus at the equilibrium point.

It has been convenient to think of demand and supply as quantities that are functions of price. For purposes of this section, we will find it convenient to think of them as prices that are functions of quantity: $p = D(x)$ and $p = S(x)$. Indeed, such an interpretation is common in economics. We can use integration to calculate quantities of interest to economists, such as *consumer surplus* and *producer surplus*.

The consumer's **demand curve** is the graph of $p = D(x)$, which shows the price per unit that the consumer is willing to pay for x units of a product. It is usually a decreasing function since the consumer expects to pay less per unit for large quantities of the product. The producer's **supply curve** is the graph of $p = S(x)$, which shows the price per unit the producer is willing to accept for selling x units. It is usually an increasing function since a higher price per unit is an incentive for the producer to make more units available for sale. The equilibrium point, (x_E, p_E), is the intersection of these two curves.

TECHNOLOGY CONNECTION

EXERCISE

1. Graph the demand and supply functions

$$D(x) = (x - 5)^2 \text{ and }$$
$$S(x) = x^2 + x + 3$$

using the viewing window $[0, 5, 0, 30]$, with Yscl $= 5$. Find the equilibrium point using the INTERSECT feature.

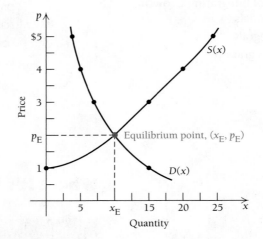

Utility is a function often considered in economics. When a consumer receives x units of a product, a certain amount of pleasure, or utility, U, is derived from them (see Exercise 27 in Exercise Set 1.3). For example, the number of movies that you see in a month gives you a certain utility. If you see four movies (unless they are not entertaining), you get more utility than if you see no movies. The same notion applies to having a meal in a restaurant or paying your heating bill to warm your home.

To help to explain the concepts of consumer surplus and producer surplus, we will consider the utility of seeing movies over a fixed amount of time, say, 1 month. We are also going to make the assumption that the movies seen are of about the same quality.

Samantha is a college student who likes movies. At a price of $10 per ticket, she will see no movies. At a price of $8.75 per ticket, she will see one movie per month, and at a price of $7.50 per ticket, she will see two movies per month. As the price per ticket decreases, Samantha tends to see more movies. As long as the number of movies (x) is small, Samantha's demand function for movies can be modeled by $p = 10 - 1.25x$. We want to examine the utility she receives from going to the movies.

At a ticket price of $8.75, Samantha sees one movie. Her total expenditure is $(1)\$8.75 = \$8.75$, as shown by the blue region in Fig. 1. However, the area under Samantha's demand curve over the interval $[0, 1]$ is $9.38 (rounded). This is what going to one movie per month is worth to Samantha—that is, what she is willing to pay. Since she spent $8.75, the difference in area, represented by the orange triangle, $\$9.38 - \$8.75 = \$0.63$, can be interpreted as the pleasure Samantha gets, but does not have to pay for, from the one movie. Economists define this amount as the *consumer surplus*. It is the extra utility that consumers enjoy when prices decrease as more units are purchased.

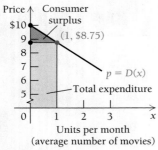

FIGURE 1

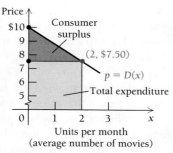

Price

$10
9
8
7
6
5

Consumer surplus

(2, $7.50)

$p = D(x)$

Total expenditure

0   1   2   3   *x*

Units per month
(average number of movies)

FIGURE 2

Suppose Samantha goes to two movies per month at $7.50 per ticket. Her total expenditure is $(2)\$7.50 = \$15.00$, which is represented by the blue region in Fig. 2. The area under Samantha's demand curve over the interval $[0, 2]$ is $17.50. Therefore, Samantha's consumer surplus is $2.50, which measures the pleasure Samantha received, but did not have to pay for, from the two movies.

Suppose that the graph of a demand function is a curve, as shown at the right.

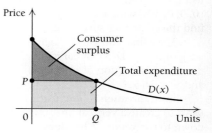

Price

Consumer surplus

Total expenditure

P

$D(x)$

0   Q Units

If Samantha goes to Q movies when the price is P, then her total expenditure is QP. The total area under the curve is the total utility, or the total enjoyment received, and is

$$\int_0^Q D(x)\, dx.$$

The *consumer surplus* is the total area under the curve minus the total expenditure. This surplus is the total utility minus the total cost and is given by

$$\int_0^Q D(x)\, dx - QP.$$

DEFINITION

Suppose that $p = D(x)$ describes the demand function for a commodity. Then the **consumer surplus** is defined for the point (Q, P) as

$$\int_0^Q D(x)\, dx - QP.$$

■ EXAMPLE 1   Find the consumer surplus for the demand function given by $D(x) = (x - 5)^2$ when $x = 3$.

Solution   When $x = 3$, we have $D(3) = (3 - 5)^2 = (-2)^2 = 4$. Then

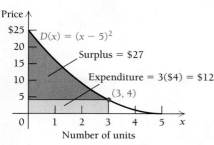

Price

$25
20
15
10
5

$D(x) = (x - 5)^2$

Surplus = $27

Expenditure = 3($4) = $12

(3, 4)

0   1   2   3   4   5   *x*

Number of units

$$\text{Consumer surplus} = \int_0^3 (x - 5)^2\, dx - 3 \cdot 4$$

$$= \int_0^3 (x^2 - 10x + 25)\, dx - 12$$

$$= \left[\frac{x^3}{3} - 5x^2 + 25x\right]_0^3 - 12$$

$$= \left[\left(\frac{3^3}{3} - 5(3)^2 + 25(3)\right) - \left(\frac{0^3}{3} - 5(0)^2 + 25(0)\right)\right] - 12$$

$$= (9 - 45 + 75) - 0 - 12$$

$$= \$27.$$

❯ Quick Check 1

Find the consumer surplus for the demand function given by $D(x) = x^2 - 6x + 16$ when $x = 1$.

❮ Quick Check 1

Exploratory

Graph $D(x) = (x - 5)^2$, the demand function in Example 1, using the viewing window $[0, 5, 0, 30]$, with Yscl = 5. To find the consumer surplus at $x = 3$, we first find $D(3)$. Then we graph $y = D(3)$. What is the point of intersection of $y = D(x)$ and $y = D(3)$?

From the intersection, use DRAW to create a vertical line down to the x-axis. What does the area of the resulting rectangle represent? What does the area above the horizontal line and below the curve represent?

Let's now look at a supply curve for a movie theater, as shown in Figs. 3 and 4. Suppose the movie theater will not sell tickets to a movie for any price at or below $4 (because this would not be enough to cover operating costs and return a profit), but will sell one ticket for one movie at $5.75 or two tickets for two movies at $7.50 each. For small numbers of movies (x), the theater's supply curve is modeled by $p = 4 + 1.75x$. The price $5.75 is within what Samantha is willing to pay for one movie, and the theater will take in a revenue of $(1)\$5.75 = \$5.75$ for selling Samantha one ticket for one movie. The area of the yellow region in Fig. 3 represents the total per-person cost to the theater for showing one movie, which is $4.88 (rounded). Since the theater takes in $5.75 for selling one ticket, the difference, $\$5.75 - \$4.88 = \$0.87$, represents the surplus over cost and is a contribution toward profit for the theater. Economists call this the *producer surplus*. It is the benefit a producer receives when supplying more units at a higher price than the price at which the producer expects to sell units. It is the extra revenue the producer receives as a result of not being forced to sell fewer units at a lower price.

At a price of $7.50, the theater will show Samantha 2 movies and collect total receipts of 2($7.50), or $15. The area of the yellow region in Fig. 4 represents the total cost to the theater of showing Samantha 2 movies, which is $11.50. The area of the green triangle is $\$15.00 - \$11.50 = \$3.50$ and is the producer's surplus. It is a contribution to the theater's profit.

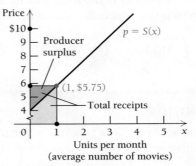

FIGURE 3   **FIGURE 4**

Suppose that the graph of the supply function is a curve, as shown at the right. If the theater shows Samantha Q movies when the price is P, the total receipts are QP. The *producer surplus* is the total receipts minus the area under the curve and is given by

$$QP - \int_0^Q S(x)\, dx.$$

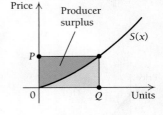

DEFINITION

Suppose that $p = S(x)$ is the supply function for a commodity. Then the **producer surplus** is defined for the point (Q, P) as

$$QP - \int_0^Q S(x)\, dx.$$

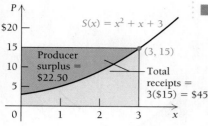

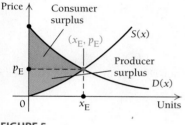

> **Quick Check 2**

Find the producer surplus for $S(x) = \frac{1}{3}x^2 + \frac{4}{3}x + 4$ when $x = 1$.

■ **EXAMPLE 2** Find the producer surplus for $S(x) = x^2 + x + 3$ when $x = 3$.

Solution When $x = 3$, $S(3) = 3^2 + 3 + 3 = 15$. Then

$$\text{Producer surplus} = 3 \cdot 15 - \int_0^3 (x^2 + x + 3)\, dx$$

$$= 45 - \left[\frac{x^3}{3} + \frac{x^2}{2} + 3x\right]_0^3$$

$$= 45 - \left[\left(\frac{3^3}{3} + \frac{3^2}{2} + 3(3)\right) - \left(\frac{0^3}{3} + \frac{0^2}{2} + 3(0)\right)\right]$$

$$= 45 - \left(9 + \frac{9}{2} + 9 - 0\right)$$

$$= \$22.50.$$

❮ Quick Check 2

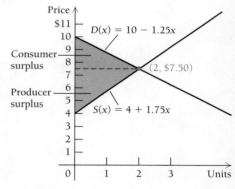

FIGURE 5

The **equilibrium point**, (x_E, p_E), in Fig. 5 is the point at which the supply and demand curves intersect. It is the point at which sellers and buyers come together and purchases and sales actually occur.

Let's reconsider the example involving Samantha and the movie theater. When the theater charged $5.75 for one ticket, Samantha saw one movie. Since seeing the movie was worth $9.38 to Samantha, she derived $9.38 − $5.75 = $3.63 in utility. To Samantha, this was a very good deal, since she paid much less than she was willing to pay. However, the theater lost potential revenue by "undercharging" Samantha.

We see in Fig. 6 that at a price of $7.50 per ticket, Samantha's demand curve and the theater's supply curve intersect. This point is advantageous for both Samantha and the theater, since Samantha is *willing* to see two movies at a price of $7.50 per ticket, while the theater can increase its surplus by selling the two tickets to Samantha. In other words, if the price per ticket is set too low, the theater will certainly sell tickets but will lose revenue it could be receiving if the price were set slightly higher, since Samantha (and the general population) are *willing* to pay more according to the demand curve. On the other extreme, if the theater sets the price too high, it simply will not sell enough tickets to make a profit. The $7.50 ticket price is the best "middle ground" for producer (the theater) and consumer (Samantha) alike.

FIGURE 6

■ EXAMPLE 3   Given

$$D(x) = (x - 5)^2 \quad \text{and} \quad S(x) = x^2 + x + 3,$$

find each of the following.

a) The equilibrium point

b) The consumer surplus at the equilibrium point

c) The producer surplus at the equilibrium point

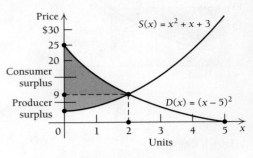

Solution

a) To find the equilibrium point, we set $D(x) = S(x)$ and solve:

$$(x - 5)^2 = x^2 + x + 3$$
$$x^2 - 10x + 25 = x^2 + x + 3$$
$$-10x + 25 = x + 3$$
$$22 = 11x$$
$$2 = x.$$

Thus, $x_E = 2$. To find p_E, we substitute x_E into either $D(x)$ or $S(x)$. If we choose $D(x)$, we have

$$
\begin{aligned}
p_E = D(x_E) &= D(2) \\
&= (2 - 5)^2 \\
&= (-3)^2 \\
&= \$9.
\end{aligned}
$$

Thus, the equilibrium point is $(2, \$9)$.

b) The consumer surplus at the equilibrium point is

$$\int_0^{x_E} D(x)\, dx - x_E p_E,$$

or

$$
\begin{aligned}
\int_0^2 (x - 5)^2\, dx - 2 \cdot 9 &= \left[\frac{(x - 5)^3}{3} \right]_0^2 - 18 \\
&= \left[\frac{(2 - 5)^3}{3} - \frac{(0 - 5)^3}{3} \right] - 18 \\
&= \frac{(-3)^3}{3} - \frac{(-5)^3}{3} - 18 = -\frac{27}{3} + \frac{125}{3} - \frac{54}{3} \\
&= \frac{44}{3} \approx \$14.67.
\end{aligned}
$$

c) The producer surplus at the equilibrium point is

$$x_E p_E - \int_0^{x_E} S(x)\, dx,$$

> **Quick Check 3**
>
> Given $D(x) = x^2 - 6x + 16$ and $S(x) = \frac{1}{3}x^2 + \frac{4}{3}x + 4$, find each of the following. Assume $x \le 5$.
>
> **a)** The equilibrium point
>
> **b)** The consumer surplus at the equilibrium point
>
> **c)** The producer surplus at the equilibrium point

or

$$2 \cdot 9 - \int_0^2 (x^2 + x + 3)\, dx = 2 \cdot 9 - \left[\frac{x^3}{3} + \frac{x^2}{2} + 3x \right]_0^2$$

$$= 18 - \left[\left(\frac{2^3}{3} + \frac{2^2}{2} + 3 \cdot 2 \right) - \left(\frac{0^3}{3} + \frac{0^2}{2} + 3 \cdot 0 \right) \right]$$

$$= 18 - \left(\frac{8}{3} + 2 + 6 \right)$$

$$= \frac{22}{3} \approx \$7.33.$$

⟨ Quick Check 3

Section Summary

- A *demand curve* is the graph of a function $p = D(x)$, which represents the unit price p a consumer is willing to pay for x items. It is usually a decreasing function.
- A *supply curve* is the graph of a function $p = S(x)$, which represents the unit price p a producer is willing to accept for x items. It is usually an increasing function.
- *Consumer surplus* at a point (Q, P) is defined as

$$\int_0^Q D(x)\, dx - QP.$$

- *Producer surplus* at a point (Q, P) is defined as

$$QP - \int_0^Q S(x)\, dx.$$

- The *equilibrium point*, (x_E, p_E), is the point at which the supply and demand curves intersect. The consumer surplus at the equilibrium point is

$$\int_0^{x_E} D(x) - x_E p_E.$$

The producer surplus at the equilibrium point is

$$x_E p_E - \int_0^{x_E} S(x)\, dx.$$

EXERCISE SET
5.1

In each of Exercises 1–14, $D(x)$ is the price, in dollars per unit, that consumers are willing to pay for x units of an item, and $S(x)$ is the price, in dollars per unit, that producers are willing to accept for x units. Find **(a)** *the equilibrium point,* **(b)** *the consumer surplus at the equilibrium point, and* **(c)** *the producer surplus at the equilibrium point.*

1. $D(x) = -\frac{5}{6}x + 9, \quad S(x) = \frac{1}{2}x + 1$

2. $D(x) = -3x + 7, \quad S(x) = 2x + 2$

3. $D(x) = (x - 4)^2, \quad S(x) = x^2 + 2x + 6$

4. $D(x) = (x - 3)^2, \quad S(x) = x^2 + 2x + 1$

5. $D(x) = (x - 6)^2, \quad S(x) = x^2$

6. $D(x) = (x - 8)^2, \quad S(x) = x^2$

7. $D(x) = 1000 - 10x, \quad S(x) = 250 + 5x$

8. $D(x) = 8800 - 30x, \quad S(x) = 7000 + 15x$

9. $D(x) = 5 - x,$ for $0 \le x \le 5; \quad S(x) = \sqrt{x + 7}$

10. $D(x) = 7 - x,$ for $0 \le x \le 7; \quad S(x) = 2\sqrt{x + 1}$

11. $D(x) = \dfrac{100}{\sqrt{x}}, \quad S(x) = \sqrt{x}$

12. $D(x) = \dfrac{1800}{\sqrt{x + 1}}, \quad S(x) = 2\sqrt{x + 1}$

13. $D(x) = (x - 4)^2, \quad S(x) = x^2 + 2x + 8$

14. $D(x) = 13 - x,$ for $0 \le x \le 13; \quad S(x) = \sqrt{x + 17}$

SYNTHESIS

For Exercises 15 and 16, follow the directions given for Exercises 1–14.

15. $D(x) = e^{-x + 4.5}, \quad S(x) = e^{x - 5.5}$

16. $D(x) = \sqrt{56 - x}, \quad S(x) = x$

17. Explain why both consumers and producers feel good when consumer and producer surpluses exist.

18. Do some research on consumer and producer surpluses in an economics book. Write a brief description.

TECHNOLOGY CONNECTION

For Exercises 19 and 20, graph each pair of demand and supply functions. Then:

a) *Find the equilibrium point using the* INTERSECT *feature or another feature that will allow you to find this point of intersection.*

b) *Graph* $y = D(x_E)$ *and determine the regions of both consumer and producer surpluses.*

c) *Find the consumer surplus.*

d) *Find the producer surplus.*

19. $D(x) = \dfrac{x + 8}{x + 1}$, $S(x) = \dfrac{x^2 + 4}{20}$

20. $D(x) = 15 - \frac{1}{3}x$, $S(x) = 2\sqrt[3]{x}$

21. Bungee jumping. Regina loves to go bungee jumping. The table shows the number of half-hours that Regina is willing to go bungee jumping at various prices.

Time Spent (in half-hours per month)	Price (per half-hour)
8	$ 2.50
7	5.00
6	7.50
5	10.00
4	12.50
3	15.00
2	17.50
1	20.00

a) Make a scatterplot of the data, and determine the type of function that you think fits best.

b) Fit that function to the data using REGRESSION.

c) If Regina goes bungee jumping for 6 half-hours per month, what is her consumer surplus?

d) At a price of $11.50 per half-hour, what is Regina's consumer surplus?

Answers to Quick Checks

1. $2.33   **2.** $0.89   **3. (a)** $(2, 8)$; **(b)** $6.67; **(c)** $4.44

5.2   Applications of Integrating Growth and Decay Models

OBJECTIVES

- Find the future value of an investment.
- Find the accumulated future value of a continuous income stream.
- Find the present value of an amount due in the future.
- Find the accumulated present value of an income stream.
- Calculate the total consumption of a natural resource.

Business and Economics Applications

We studied the exponential growth and decay models provided by the functions $P(t) = P_0 e^{kt}$ and $P(t) = P_0 e^{-kt}$ in Sections 3.3 and 3.4. Here we consider applications of the integrals of these functions. To ease our later work, let's find formulas for evaluating these integrals.

For the *growth* model, the formula is

$$\int_0^T P_0 e^{kt}\, dt = \left[\frac{P_0}{k} \cdot e^{kt} \right]_0^T \qquad \text{Using the substitution } u = e^{kt}$$

$$= \frac{P_0}{k} \left(e^{kT} - e^{k \cdot 0} \right) \qquad \text{Evaluating the integral}$$

$$= \frac{P_0}{k} \left(e^{kT} - 1 \right).$$

Similarly, for the *decay* model, the formula is $\displaystyle\int_0^T P_0 e^{-kt}\, dt = \frac{P_0}{k} \left(1 - e^{-kT} \right)$. Thus, we have the following integration formulas.

$$\text{Growth formula:} \int_0^T P_0 e^{kt}\, dt = \frac{P_0}{k}\left(e^{kT} - 1\right) \tag{1}$$

$$\text{Decay formula:} \int_0^T P_0 e^{-kt}\, dt = \frac{P_0}{k}\left(1 - e^{-kT}\right) \tag{2}$$

Now let's consider several applications of these formulas to business and economics.

Future Value

Recall the basic model for the growth of an amount of money, presented in the following definition.

> **DEFINITION**
>
> If P_0 is invested for t years at interest rate k, compounded continuously (Section 3.3), then
>
> $$P(t) = P_0 e^{kt}, \tag{3}$$
>
> where $P = P_0$ at $t = 0$. The value P is called the **future value** of P_0 dollars invested at interest rate k, compounded continuously, for t years.

■ **EXAMPLE 1** Business: Future Value of an Investment. Find the future value of $3650 invested for 3 yr at an interest rate of 5%, compounded continuously.

Solution Using equation (3) with $P_0 = 3650$, $k = 0.05$, and $t = 3$, we get

$$
\begin{aligned}
P(3) &= 3650 e^{0.05(3)} \\
&= 3650 e^{0.15} \\
&\approx 3650(1.161834) \\
&= \$4240.69.
\end{aligned}
$$

The future value of $3650 after 3 yr will be about $4240.69.

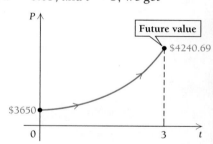

> **〉 Quick Check 1**
>
> **Business: Future Value of an Investment.** Find the future value of $10,000 invested for 3 yr at an interest rate of 6%, compounded continuously.

〈 Quick Check 1

Accumulated Future Value of a Continuous Income Stream

Let's consider a situation involving the accumulation of future values. The owner of a parking space near a convention center receives a yearly profit of $3650 at the end of each of 4 years; this is called an *income stream*. The owner invests the $3650 at 5% interest compounded continuously. When $3650 is received at the end of the first year, it is invested for $4 - 1$, or 3 yr. The future value is $4240.69, as we saw in Example 1. When $3650 is received at the end of the second year, it is invested for $4 - 2$, or 2 yr. The future value of this investment is $3650e^{0.05(4-2)}$, or $4,033.87. Note that this future value is less than $4240.69 because the time period is shorter. When $3650 is received at the end of the third year, it is invested for $4 - 3$, or 1 yr. That future value is $3650e^{0.05(4-3)}$, or $3837.14, smaller than each of the previous future values. When the last $3650 is received after the fourth year, it is reinvested for $4 - 4$, or 0 yr. This amount has no time to earn interest, so its future value is $3650. The *accumulated, or total, future value of the income stream* is the sum of the four future values:

$$
\begin{array}{llll}
\text{After 1st yr,} & t = 3: & \$3650 \longrightarrow 3650e^{0.05(3)} \longrightarrow & \$4{,}240.69 \\
\text{After 2nd yr,} & t = 2: & \$3650 \longrightarrow 3650e^{0.05(2)} \longrightarrow & \$4{,}033.87 \\
\text{After 3rd yr,} & t = 1: & \$3650 \longrightarrow 3650e^{0.05(1)} \longrightarrow & \$3{,}837.14 \\
\text{After 4th yr,} & t = 0: & \$3650 \longrightarrow 3650e^{0.05(0)} \longrightarrow & \$3{,}650.00 \\
\end{array}
$$

Total future value of the income stream = $15,761.70

Next, let's suppose that the owner of the parking space receives the profit at a rate of $3650 per year but in 365 payments of $10 per day. Each day, when the owner gets $10, it is invested at 5%, compounded continuously, and due at the end of the fourth year. The *first* day's investment grows to

$$
\frac{3650}{365} e^{0.05(4-1/365)} = 10e^{0.05(4-1/365)} \approx \$12.2124,
$$

since it will be invested for only 1 day, or $1/365$ yr, less than the full 4 yr. The value of the investment on the *second* day will grow to

$$
\frac{3650}{365} e^{0.05(4-2/365)} = 10e^{0.05(4-2/365)} \approx \$12.2107,
$$

at the end of the fourth year, and so on, for every day in the 4-yr period. Assuming that all deposits are made into the same account, the total of the future values is

$$
10e^{0.05(4-1/365)} + 10e^{0.05(4-2/365)} + \cdots + 10e^{0.05(2/365)} + 10e^{0.05(1/365)} + 10.
$$

Reversing the order of the terms in this sum, we have

$$
10 + 10e^{0.05(1/365)} + 10e^{0.05(2/365)} + \cdots + 10e^{0.05(4-2/365)} + 10e^{0.05(4-1/365)}.
$$

If we express 10 as $3650 \cdot \frac{1}{365}$ and let $\Delta t = \frac{1}{365}$, then we have a Riemann sum, with t in years, which can be approximated by the definite integral

$$
\int_0^4 3650e^{0.05(4-t)} \, dt = e^{0.2} \int_0^4 3650e^{-0.05t} \, dt. \tag{4}
$$

Let's further refine how the parking space owner receives profit. First, let's review the notion of *instantaneous rate of change*. The speedometer on a car provides an instantaneous speed, or rate of change. If the speedometer reads 58 mph, this means that at that *instant* the car's speed is 58 mph, and if the car continues at this speed for 1 hr, it will travel 58 mi.

Instead of receiving an income stream at the rate of $10 a day for 4 yr, suppose the owner could receive the money *continuously* at a rate of $3650 *per year* for 4 yr. This means that over the course of 4 yr, $3650 in profit will be received at a constant rate of $3650 per year in what is called a *continuous income stream, or flow*. If at each instant the money is invested at 5%, compounded continuously, then the *accumulated future value of the continuous income stream* is approximated by the definite integral in equation (4). Let's calculate that definite integral:

$$
e^{0.2} \int_0^4 3650e^{-0.05t} \, dt = -\frac{3650}{0.05} e^{0.2}(e^{-0.2} - 1) \qquad \text{Growth formula (1)}
$$

$$
\approx \$16{,}162.40 \qquad \text{Approximating using a calculator}
$$

Economists call $16,162.40 the **accumulated future value of a continuous income stream.**

DEFINITION   **Accumulated Future Value of a Continuous Income Stream**

Let $R(t)$ be a function that represents the rate, per year, of a continuous income stream, let k be the interest rate, compounded continuously, at which the continuous income stream is invested, and let T be the number of years for which the income stream is invested.

    Then the **accumulated future value of the continuous income stream** is given by

$$A = \int_0^T R(t)e^{k(T-t)}\, dt = e^{kT} \int_0^T R(t)\, e^{-kt}\, dt. \tag{5}$$

If $R(t)$ is a constant function, it can be factored out of the integral, and the formula becomes, after evaluating and simplifying,

$$A = \frac{R(t)}{k}\left(e^{kT} - 1\right). \tag{6}$$

If $R(t)$ is a nonconstant function, then equation (6) does not apply and the integral in equation (5) must be evaluated using some other technique such as integration by parts, tables, a graphing calculator, iPlot, or some other kind of software.

■ **EXAMPLE 2**   **Business: Insurance Settlement.**   A cardiac surgeon, Sarah Makahone, earns an income of $450,000 per year but is involved in an automobile accident that injures her legs in such a way that she can no longer stand up to perform heart surgery. In a legal settlement with an insurance company, Sarah is granted a continuous income stream of $225,000 per year for 20 yr, half her normal yearly income since she can practice other kinds of medicine while seated. Sarah invests the money at 3.2%, compounded continuously, in the Halmos Global Equities Fund. Find the accumulated future value of the continuous income stream.

Solution   This is an income stream flowing at a constant rate, so we can use equation (6), with $R(t) = \$225{,}000$, $k = 0.032$, and $T = 20$. We have

$$A = \frac{225{,}000}{0.032}\left(e^{0.032(20)} - 1\right) \approx \$6{,}303{,}381.18.$$

❰ Quick Check 2

> ❱ **Quick Check 2**
>
> **Business: Insurance Settlement.** Repeat Example 2 but assume that the insurance settlement is a continuous income stream of $125,000 per year for 25 yr and the money is invested at 5%, compounded continuously.

Present Value

We saw in Example 1 that the future value of $3650 invested for 3 yr at a continuously compounded interest rate of 5% is $4240.69. We call $3650 the **present value** of $4240.69 invested for 3 yr at interest rate 5%, compounded continuously. It answers this question: "What do we have to invest now at a certain interest to attain a certain future value?" (see Section 3.4).

    In general, the present value P_0 of an amount P invested at interest rate k and due t years later is found by solving the growth equation for P_0:

$$P_0 e^{kt} = P$$

$$P_0 = \frac{P}{e^{kt}} = Pe^{-kt}.$$

DEFINITION

The **present value, P_0,** of an amount P due t years later, at interest rate k, compounded continuously, is given by

$$P_0 = Pe^{-kt}.$$

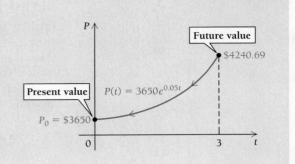

> **Quick Check 3**

Business: Finding the Present Value of a Trust. Mira Bell, following the birth of a grandchild, wants to set up a trust fund that will be worth $120,000 on the child's 18th birthday. Mira can get an interest rate of 5.6%, compounded continuously, for the time period. What amount will Mira have to deposit in the trust fund to achieve her goal?

■ **EXAMPLE 3** **Business: Finding the Present Value of a Trust.** In 10 years, Sam Bixby is going to receive $250,000 under the terms of a trust established by his uncle. If the money in the trust fund is invested at 4.8% interest, compounded continuously, what is the present value of Sam's legacy?

Solution Using the equation for present value given above, we have

$$P_0 = 250,000e^{-0.048(10)} \approx \$154,695.85.$$

❰ Quick Check 3

Accumulated Present Value of a Continuous Income Stream

To find the **accumulated present value of a continuous income stream,** when $R(t)$ is constant, we can work backward from equation (6):

$$A = \frac{R(t)}{k}(e^{kT} - 1).$$

We are looking for the principal B, the amount of a one-time deposit, at the interest rate k, that will yield the same accumulated value as the income stream. We choose B such that $Be^{kT} = A$ in equation (6). Then we solve for B:

$$Be^{kT} = \frac{R(t)}{k}(e^{kT} - 1)$$

$$\frac{Be^{kT}}{e^{kT}} = \frac{R(t)}{k}\left(\frac{e^{kT} - 1}{e^{kT}}\right) \qquad \text{Dividing by } e^{kT}$$

$$B = \frac{R(t)}{k}\left(\frac{e^{kT}}{e^{kT}} - \frac{1}{e^{kT}}\right)$$

$$B = \frac{R(t)}{k}(1 - e^{-kT}). \qquad \text{Simplifying}$$

> **DEFINITION** **Accumulated Present Value of a Continuous Income Stream**
>
> Let $R(t)$ be a function that represents the rate, per year, of a continuous income stream, let k be the interest rate, compounded continuously, at which the continuous income stream is invested, and let T be the number of years over which the income stream is received.
>
> If $R(t)$ is a constant function, then B, the **accumulated present value of the continuous income stream,** is given by
>
> $$B = \frac{R(t)}{k}\left(1 - e^{-kT}\right). \tag{7}$$
>
> If $R(t)$ is a nonconstant function, the accumulated present value of the continuous income stream is given by the following integral:
>
> $$B = \int_0^T R(t)e^{-kt}\, dt. \tag{8}$$

Accumulated present value is a useful tool in business decision making when evaluating a purchase, an investment, or a contract. It brings alternatives and allows for comparisons.

■ **EXAMPLE 4** **Business: Determining the Value of a Franchise.** Silver Spoon, Inc., operates frozen yogurt franchises. Chris Nelson, noting how much he enjoys the yogurt and yearning to be an entrepreneur, considers buying a franchise in his home town, Carmel, Indiana. As part of his decision to purchase, he wants to determine the accumulated present value of the income stream from the franchise over an 8-yr period. Silver Spoon tells Chris that he should expect a constant annual income stream given by

$$R_1(t) = \$275{,}000,$$

which Chris knows he can invest at an interest rate of 5%, compounded continuously.

However, Chris took a business calculus course like this one, and he does a linear regression on data from the annual reports of Silver Spoon, which indicates that there will be a nonconstant annual income stream of

$$R_2(t) = \$80{,}000t.$$

a) Evaluate the accumulated future value of the income stream at rate $R_1(t)$. Then evaluate the accumulated present value of the income stream, and interpret the results.

b) Evaluate the accumulated future value of the income stream at rate $R_2(t)$. Then evaluate the accumulated present value of the income stream, and interpret the results.

Round all answers to the nearest ten dollars.

Solution

a) Chris will have a constant income stream of $275,000 per year for 8 yr. Using equation (6), the accumulated *future* value is

$$A = \frac{R_1(t)}{k}\left(e^{kT} - 1\right) = \frac{275{,}000}{0.05}\left(e^{0.05(8)} - 1\right) \approx \$2{,}705{,}040.$$

This gives Chris a sense of the value of the franchise over the 8-yr period. The accumulated *present* value is found by using equation (7):

$$B = \frac{R_1(t)}{k}\left(1 - e^{-kT}\right) = \frac{275{,}000}{0.05}\left(1 - e^{-0.05(8)}\right) \approx \$1{,}813{,}240.$$

The first result tells us that if Chris were to buy the franchise now and invest the predicted income stream at 5%, compounded continuously, he would have $2,705,040 in 8 yr. The second result tells us that the first amount is worth $1,813,240 at the present.

b) With a nonconstant income stream, $R_2(t) = 80,000t$ per year, using equation (5), the accumulated *future* value is

$$e^{0.05(8)} \int_0^8 (80,000t)e^{-0.05t} = 80,000 \, e^{0.4} \int_0^8 te^{-0.05t} \, dt.$$

To evaluate this integral, we can use any of a variety of integration methods: integration by parts, tables, a graphing calculator, or iPlot. We use Formula 6, from Table 1 in Chapter 4 (p. 454), with $a = -0.05$ and $x = t$:

$$\int xe^{ax} \, dx = \frac{1}{a^2} \cdot e^{ax}(ax - 1) + C = \frac{1}{0.0025} \cdot e^{-0.05t}(-0.05t - 1)$$

$$= 400e^{-0.05t}(-0.05t - 1)$$

$$= -20te^{-0.05t} - 400e^{-0.05t} + C.$$

Then,

$$80,000 \, e^{0.4} \int_0^8 te^{-0.05t} \, dt = 80,000 \, e^{0.4}\big[(-20(8)e^{-0.05(8)} - 400 \, e^{-0.05(8)})$$

$$- (-20(0)e^{-0.05(0)} - 400e^{-0.05(0)})\big]$$

$$= 80,000 \, e^{0.4}\big[(-160e^{-0.4} - 400e^{-0.4}) - (-400)\big]$$

$$= 80,000 \, e^{0.4}\big[-560e^{-0.4} + 400\big] \approx \$2,938,390,$$

and

$$\int_0^8 (80,000t)e^{-0.05t} = 80,000 \int_0^8 te^{-0.05t} \, dt \approx \$1,969,660.$$

The first result tells us that if Chris were to buy the franchise now and invest the predicted income stream at 5%, compounded continuously, he would have $2,938,390 in 8 yr. The second result tells us that the first amount is worth $1,969,660 at the present.

Chris's computations yield a higher accumulated present value than that claimed by Silver Spoon, which gives him an indication that he is dealing with a reputable company.

❮ Quick Check 4

> **❯ Quick Check 4**
>
> **Business: Determining the Value of a Franchise.** Repeat Example 4, but with the following income streams:
>
> $R_1(t) = \$265,000,$
>
> $R_2(t) = 75,000t,$
>
> and an interest rate of 8%, compounded continuously.

■ **EXAMPLE 5   Business: Creating a College Trust.**   Emma and Jake Tuttle have a new grandchild, Erica. They want to create a college trust fund for her that will yield $100,000 by her 18th birthday.

a) What lump sum would they have to deposit now, in the Hilbert Prime Money Market Fund, at 6% interest, compounded continuously, to yield $100,000?

b) They discover that the required lump sum is more than they can afford at the time, so they decide to invest a constant stream of $R(t)$ dollars per year. Find $R(t)$ such that the accumulated future value of the continuous money stream is $100,000, assuming that the interest rate is 6%, compounded continuously.

Solution

a) The lump sum is the *present value* of $100,000, at 6% interest, compounded continuously, for 18 yr:

$$P_0 = Pe^{-kt} = 100,000e^{-0.06(18)} \approx \$33,959.55.$$

b) We want $R(t)$ such that

$$100,000 = \frac{R(t)}{0.06} \left(e^{0.06(18)} - 1 \right) \qquad \text{Using equation (4)}$$

$$0.06(100,000) = R(t)(e^{1.08} - 1)$$

$$\frac{6000}{(e^{1.08} - 1)} = R(t)$$

$$R(t) \approx \$3085.34.$$

A continuous money stream of $3085.34 per year, invested at 6%, compounded continuously for 18 yr, will yield a *future value* of $100,000.

⟩ Quick Check 5

Business: Creating a College Trust. Repeat Example 5 for a yield of $50,000 and an interest rate of 4%.

❮ Quick Check 5

■ **EXAMPLE 6** **Business: Contract Buyout.** A business executive is working under a contract that pays him $500,000 each year for 5 yr. After 2 yr, the company offers him a buyout of his contract. How much should the company offer him? Assume an annual percentage rate of 4.75%, compounded continuously.

Solution We can view the $500,000 as a continuous money stream. After 2 yr, the contract's accumulated future value, A_2, is

$$A_2 = \frac{500,000}{0.0475} \left(e^{0.0475(2)} - 1 \right) \approx \$1,049,040.58.$$

If the contract were allowed to run the full 5 yr, the accumulated future value, A_5, would be

$$A_5 = \frac{500,000}{0.0475} \left(e^{0.0475(5)} - 1 \right) \approx \$2,821,842.07.$$

The difference is

$$A_5 - A_2 = \$2,821,842.07 - \$1,049,040.58 = \$1,772,801.49.$$

Since the company is offering a lump sum payment to buy out the contract, the executive should expect an amount that, if allowed to grow at 4.75%, compounded continuously for the remaining 3 yr, would yield $1,772,801.49. That is, he should receive the present value of the difference, or

$$P_0 = 1,772,801.49e^{-0.0475(3)} = \$1,537,351.39.$$

⟩ Quick Check 6

Business: Contract Buyout. Repeat Example 6 for a $400,000 contract and an interest rate of 3.2%.

❮ Quick Check 6

Life and Physical Sciences: Consumption of Natural Resources

Another application of the integration of models of exponential growth uses

$$P(t) = P_0 e^{kt}$$

as a model of the demand for natural resources. Suppose that P_0 represents the annual amount of a natural resource (such as coal or oil) used at time $t = 0$ and that the growth rate for the use of this resource is k. Then, assuming exponential growth in demand (which is the case for the use of many resources), the amount used annually t years in the future is $P(t)$, given by

$$P(t) = P_0 e^{kt}.$$

The total amount used during an interval $[0, T]$ is then given by

$$\int_0^T P(t) \, dt = \int_0^T P_0 e^{kt} \, dt = \left[\frac{P_0}{k} e^{kt} \right]_0^T = \frac{P_0}{k} \left(e^{kT} - 1 \right).$$

Consumption of a Natural Resource

Suppose that $P(t)$ is the annual consumption of a natural resource in year t. If consumption of the resource is growing exponentially at growth rate k, then the total consumption of the resource after T years is given by

$$\int_0^T P_0 e^{kt}\, dt = \frac{P_0}{k}(e^{kT} - 1), \qquad (9)$$

where P_0 is the annual consumption at time $t = 0$.

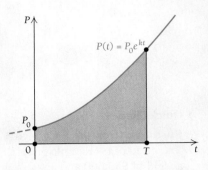

■ **EXAMPLE 7** **Physical Science: Gold Mining.** In 2009 ($t = 0$), world production of gold was 2572 metric tons, and it was growing exponentially at the rate of 9.2% per year. (*Source:* goldsheetlinks.com.) If the growth continues at this rate, how many tons of gold will be produced from 2009 to 2015?

Solution Using equation (9), we have

$$\int_0^6 2572 e^{0.092t}\, dt = \frac{2572}{0.092}(e^{0.092(6)} - 1)$$
$$= 27{,}956.5(e^{0.552} - 1)$$
$$\approx 20{,}596.2.$$

From 2009 to 2015, approximately 20,596 metric tons of gold will be produced.

> **Quick Check 7**
>
> **Life and Physical Science: Minerals from *Avatar*®.** The movie *Avatar* is set in the year 2154 on the moon Pandora, of the planet Polyphemus in the star system of Alpha Centauri. The conflict in the movie is centered around a precious but scarce mineral, Unobtanium.
>
> **a)** In 2010, the universe's production of Unobtanium was 6800 metric tons and it was being used at the rate of 0.8% per year. If Unobtanium continues to be used at this rate, how many tons of Unobtanium will be used between 2010 and 2024?
>
> **b)** In 2010, the universe's reserve of Unobtanium was 86,000 metric tons. Assuming that the growth rate of 0.8% per year continues and that no new reserves are discovered, when will the universe's reserve of Unobtanium be depleted?

■ **EXAMPLE 8** **Physical Science: Depletion of Gold Reserves.** The world reserves of gold in 2009 were estimated to be 47,000 metric tons. (*Source:* U.S. Geological Survey, U.S. Dept. of the Interior, January 2011.) Assuming that the growth rate for production given in Example 7 continues and that no new reserves are discovered, when will the world reserves of gold be depleted?

Solution Using equation (9), we want to find T such that

$$47{,}000 = \frac{2572}{0.092}(e^{0.092T} - 1).$$

We solve for T as follows:

$$47{,}000 = 27{,}956.5(e^{0.092T} - 1)$$

$$1.6812 \approx e^{0.092T} - 1 \qquad \text{Dividing both sides by 27,956.5}$$

$$2.6812 \approx e^{0.092T}$$

$$\ln 2.6812 \approx \ln e^{0.092T} \qquad \text{Taking the natural logarithm of each side}$$

$$\ln 2.6812 \approx 0.092T \qquad \text{Recall that } \ln e^k = k.$$

$$10.7 \approx T. \qquad \text{Dividing both sides by 0.092 and rounding}$$

Thus, assuming that world production of gold continues to increase at 9.2% per year and no new reserves are found, the world reserves of gold will be depleted 10.7 yr from 2009, or in 2019–2020.

❰ Quick Check 7

Section Summary

- The *future value* of an investment is given by $P = P_0 e^{kt}$, where P_0 dollars are invested for t years at interest rate k, compounded continuously.
- *The accumulated future value of a continuous income stream* is given by

$$A = e^{kT} \int_0^T R(t) e^{-kt}\, dt,$$

where $R(t)$ represents the rate of the continuous income stream, k is the interest rate, compounded continuously, at which the continuous income stream is invested, and T is the number of years for which the income stream is invested.

- If $R(t)$ is a constant function, then

$$A = \frac{R(t)}{k}(e^{kT} - 1).$$

- The *present value* is given by $P_0 = Pe^{-kt}$, where the amount P is due t years later and is invested at interest rate k, compounded continuously.
- The *accumulated present value of a continuous income stream* is given by

$$B = \int_0^T R(t) e^{-kt}\, dt,$$

where $R(t)$ represents the rate of the continuous income stream, k is the interest rate, compounded continuously, at which the continuous income stream is invested, and T is the number of years over which the income stream is received.

- If $R(t)$ is a constant function, then

$$B = \frac{R(t)}{k}(1 - e^{-kT}).$$

EXERCISE SET
5.2

For all the exercises in this exercise set, use a graphing calculator.

Find the future value P of each amount P_0 invested for time period t at interest rate k, compounded continuously.

1. $P_0 = \$100,000$, $t = 6$ yr, $k = 3\%$

2. $P_0 = \$55,000$, $t = 8$ yr, $k = 4\%$

3. $P_0 = \$140,000$, $t = 9$ yr, $k = 5.8\%$

4. $P_0 = \$88,000$, $t = 13$ yr, $k = 4.7\%$

Find the present value P_0 of each amount P due t years in the future and invested at interest rate k, compounded continuously.

5. $P = \$100,000$, $t = 6$ yr, $k = 3\%$

6. $P = \$100,000$, $t = 8$ yr, $k = 4\%$

7. $P = \$1,000,000$, $t = 25$ yr, $k = 7\%$

8. $P = \$2,000,000$, $t = 20$ yr, $k = 9\%$

Find the accumulated future value of each continuous income stream at rate R(t), for the given time T and interest rate k, compounded continuously. Round to the nearest \$10.

9. $R(t) = \$50,000$, $T = 22$ yr, $k = 7\%$

10. $R(t) = \$125,000$, $T = 20$ yr, $k = 6\%$

11. $R(t) = \$400,000$, $T = 20$ yr, $k = 8\%$

12. $R(t) = \$50,000$, $T = 22$ yr, $k = 7\%$

Find the accumulated present value of each continuous income stream at rate R(t), for the given time T and interest rate k, compounded continuously.

13. $R(t) = \$250,000$, $T = 18$ yr, $k = 4\%$

14. $R(t) = \$425,000$, $T = 15$ yr, $k = 7\%$

15. $R(t) = \$800,000$, $T = 20$ yr, $k = 8\%$

16. $R(t) = \$520,000$, $T = 25$ yr, $k = 6\%$

17. $R(t) = \$5200t$, $T = 18$ yr, $k = 7\%$

18. $R(t) = \$6400t$, $T = 20$ yr, $k = 4\%$

19. $R(t) = \$2000t + 7$, $T = 30$ yr, $k = 8\%$

20. $R(t) = t^2$, $T = 40$ yr, $k = 7\%$

APPLICATIONS

Business and Economics

21. Present value of a trust. In 18 yr, Maggie Oaks is to receive \$200,000 under the terms of a trust established by her grandparents. Assuming an interest rate of 5.8%, compounded continuously, what is the present value of Maggie's legacy?

22. Present value of a trust. In 16 yr, Claire Beasley is to receive \$180,000 under the terms of a trust established by her aunt. Assuming an interest rate of 6.2%, compounded continuously, what is the present value of Claire's legacy?

23. Salary value. At age 35, Rochelle earns her MBA and accepts a position as vice president of an asphalt company. Assume that she will retire at the age of 65, having received an annual salary of $95,000, and that the interest rate is 6%, compounded continuously.

a) What is the accumulated present value of her position?

b) What is the accumulated future value of her position?

24. Salary value. At age 25, Del earns his CPA and accepts a position in an accounting firm. Del plans to retire at the age of 65, having received an annual salary of $125,000. Assume an interest rate of 7%, compounded continuously.

a) What is the accumulated present value of his position?

b) What is the accumulated future value of his position?

25. Future value of an inheritance. Upon the death of his uncle, David receives an inheritance of $50,000, which he invests for 16 yr at 7.3%, compounded continuously. What is the future value of the inheritance?

26. Future value of an inheritance. Upon the death of his aunt, Burt receives an inheritance of $80,000, which he invests for 20 yr at 8.2%, compounded continuously. What is the future value of the inheritance?

27. Decision making. A group of entrepreneurs is considering the purchase of a fast-food franchise. Franchise A predicts that it will bring in a constant revenue stream of $80,000 per year for 10 yr. Franchise B predicts that it will bring in a constant revenue stream of $95,000 per year for 8 yr. Based on a comparison of accumulated present values, which franchise is the better buy, assuming the going interest rate is 6.1%, compounded continuously, and both franchises have the same purchase price?

28. Decision making. A group of entrepreneurs is considering the purchase of a fast-food franchise. Franchise A predicts that it will bring in a constant revenue stream of $120,000 per year for 10 yr. Franchise B predicts that it will bring in a constant revenue stream of $112,000 per year for 8 yr. Based on a comparison of accumulated present values, which franchise is the better buy, assuming the going interest rate is 7.4%, compounded continuously, and both franchises have the same purchase price?

29. Decision making. An athlete attains free agency and is looking for a new team. The Bronco Crunchers offer a salary of $100,000t$ for 8 yr. The Doppler Radars offer a salary of $83,000t$ for 9 yr.

a) Based on the accumulated present values of the salaries, which team has the better offer, assuming the going interest rate is 6%, compounded continuously?

b) What signing bonus should the team with the lower offer give to equalize the offers?

30. Capital outlay. A company determines that the rate of revenue coming in from a new machine is

$$R_1(t) = 8000 - 100t,$$

in dollars per year, for 8 yr, after which the machine will have to be replaced. The company also determines that a different brand of the machine will yield revenue at a rate of

$$R_2(t) = 7600 - 85t.$$

a) Find the accumulated present value of the income stream from each machine at an interest rate of 16%, compounded continuously.

b) Find the difference in the accumulated present values.

31. Trust fund. Bob and Ann MacKenzie have a new grandchild, Brenda. They want to create a trust fund for her that will yield $250,000 on her 24th birthday, when she might want to start her own business.

a) What lump sum would they have to deposit now at 5.8%, compounded continuously, to achieve $250,000?

b) The amount in part (a) is more than they can afford, so they decide to invest a constant money stream of $R(t)$ dollars per year. Find $R(t)$ such that the accumulated future value of the continuous money stream is $250,000, assuming an interest rate of 5.8%, compounded continuously.

32. Trust fund. Ted and Edith Markey have a new grandchild, Kurt. They want to create a trust fund for him that will yield $1,000,000 on his 22nd birthday so that he can start his own business when he is out of college.

a) What lump sum would they have to deposit now at 6.2%, compounded continuously, to achieve $1,000,000?

b) The amount in part (a) is more than they can afford, so they decide to invest a constant money stream of $R(t)$ dollars per year. Find $R(t)$ such that the accumulated future value of the continuous money stream is $1,000,000, assuming an interest rate of 6.2%, compounded continuously.

33. Early retirement. Lauren Johnson signs a 10-yr contract as a loan officer for a bank, at a salary of $84,000 per year. After 7 yr, the bank offers her early retirement. What is the least amount the bank should offer Lauren, given that the going interest rate is 7.4%, compounded continuously?

34. Early sports retirement. Tory Johnson signs a 10-yr contract to play for a football team at a salary of $5,000,000 per year. After 6 yr, his skills deteriorate, and the team offers to buy out the rest of his contract so they can drop his name from the roster. What is the least amount Tory should accept for the buyout, given that the going interest rate is 8.2%, compounded continuously?

35. Disability insurance settlement. A movie stuntman receives an annual salary of $180,000 per year, but becomes a quadriplegic after jumping from a cliff into water that is too shallow. He can never work again as a stuntman. Through a legal settlement with an insurance company, he is granted a continuous income stream of $120,000 per year for 20 yr. The stuntman invests the money at 8.2%, compounded continuously.

a) Find the accumulated future value of the continuous income stream. Round your answer to the nearest $10.

b) Thinking that he might not live 20 yr, the stuntman negotiates a flat sum payment from the insurance company, which is the accumulated present value of the continuous income steam. What is that amount? Round your answer to the nearest $10.

36. Disability insurance settlement. Dale is a furnace maintenance employee who receives an annual salary of $70,000 per year. He becomes partially paralyzed after falling through a ceiling while working on an attic air conditioner. Through a legal settlement with his employer's insurance company, he is granted a continuous income stream of $40,000 per year for 25 yr. Dale invests the money at 8%, compounded continuously.

a) Find the accumulated future value of the continuous income stream. Round your answer to the nearest $10.

b) Thinking that he might not live 25 yr more, Dale negotiates a flat sum payment from the insurance company, which is the accumulated present value of the continuous stream plus $100,000. What is that amount? Round your answer to the nearest $10.

37. Lottery winnings and risk analysis. Lucky Larry wins $1,000,000 in a state lottery. The standard way in which a state pays such lottery winnings is at a constant rate of $50,000 per year for 20 yr.

a) If Lucky invests each payment from the state at 7%, compounded continuously, what is the accumulated future value of the income stream? Round your answer to the nearest $10.

b) What is the accumulated present value of the income stream at 7%, compounded continuously? This amount represents what the state has to invest at the start of its lottery payments, assuming the 7% interest rate holds.

c) The risk for Lucky is that he doesn't know how long he will live or what the future interest rate will be; it might drop or rise, or it could vary considerably over 20 yr. This is the *risk* he assumes in accepting payments of $50,000 a year over 20 yr. Lucky has taken a course in business calculus so he is aware of the formulas for accumulated future value and present value. He calculates the accumulated present value of the income stream for interest rates of 4%, 6%, 8%, and 10%. What values does he obtain?

d) Lucky thinks "a bird in the hand (present value) is worth two in the bush (future value)" and decides to negotiate with the state for immediate payment of his lottery winnings. He asks the state for $600,000. They offer $400,000. Discuss the pros and cons of each amount. Lucky finally accepts $500,000. Is this a good decision?

38. Negotiating a sports contract. Gusto Stick is an excellent professional baseball player who has just become a free agent. His attorney begins negotiations with an interested team by

asking for a contract that provides Gusto with an income stream given by $R_1(t) = 800,000 + 340,000t$, over 10 yr, where t is in years. (Round all answers to the nearest $100.)

a) What is the accumulated future value of the offer, assuming an interest rate of 8%, compounded continuously?

b) What is the accumulated present value of the offer, assuming an interest rate of 8%, compounded continuously?

c) The team counters by offering an income stream given by $R_2(t) = 600,000 + 210,000t$. What is the accumulated present value of this counteroffer?

d) Gusto comes back with a demand for an income stream given by $R_3(t) = 1,000,000 + 250,000t$. What is the accumulated present value of this income stream?

e) Gusto signs a contract for the income stream in part (d) but decides to live on $500,000 each year, investing the rest at 8%, compounded continuously. What is the accumulated future value of the remaining income, assuming an interest rate of 8%, compounded continuously?

Life and Physical Sciences

39. Demand for natural gas. In 2010 ($t = 0$), the world consumption of natural gas was approximately 3.169 billion cubic meters and was growing exponentially at about 7.4% per year. (*Source:* www.bp.com.) If the demand continues to grow at this rate, how many cubic meters of natural gas will the world use from 2012 to 2025?

40. Demand for aluminum ore (bauxite). In 2010 ($t = 0$), bauxite production was approximately 209 million metric tons, and the demand was growing exponentially at a rate of 2.5% per year. (*Source:* U.S. Energy Information Administration.) If the demand continues to grow at this rate, how many metric tons of bauxite will the world use from 2010 to 2030?

41. Depletion of natural gas. The world reserves of natural gas were approximately 187.1 trillion cubic meters in 2010. (*Source:* www.bp.com.) Assuming the growth described in Exercise 39 continues and that no new reserves are found, when will the world reserves of natural gas be depleted?

42. Depletion of aluminum ore (bauxite). In 2010, the world reserves of bauxite were about 28 billion metric tons. (*Source:* U.S. Geological Survey summaries, Jan. 2011.) Assuming that the growth described in Exercise 40 continues and that no new reserves are discovered, when will the world reserves of bauxite be depleted?

43. Demand for and depletion of oil. Between 2006 and 2010, the annual world demand for oil was projected to increase from approximately 30.8 billion barrels to 34.5 billion barrels. (*Source:* U.S. Department of Energy and *Oil and Gas Journal*, Jan. 1, 2006.)

a) Assuming an exponential growth model, compute the growth rate of demand.
b) Predict the demand in 2015.
c) The world reserves of crude oil in 2006 were estimated at 1293 billion barrels. Assuming that no new oil is found, when will the reserves be depleted?

The model

$$\int_0^T Pe^{-kt}\,dt = \frac{P}{k}(1 - e^{-kT})$$

can be applied to calculate the buildup of a radioactive material that is being released into the atmosphere at a constant annual rate. Some of the material decays, but more continues to be released. The amount present at time T is given by the integral above, where P is the amount released per year and k is the half-life.

44. Radioactive buildup. Plutonium-239 has a decay rate of approximately 0.003% per year. Suppose that plutonium-239 is released into the atmosphere for 20 yr at a constant rate of 1 lb per year. How much plutonium-239 will be present in the atmosphere after 20 yr?

45. Radioactive buildup. Cesium-137 has a decay rate of 2.3% per year. Suppose cesium-137 is released into the atmosphere for 20 yr at at rate of 1 lb per year. How much cesium-137 will be present in the atmosphere after 20 yr?

SYNTHESIS

Capitalized cost. *The capitalized cost, c, of an asset over its lifetime is the total of the initial cost and the present value of all maintenance expenses that will occur in the future. It is computed with the formula*

$$c = c_0 + \int_0^L m(t)e^{-rt}\,dt,$$

where c_0 is the initial cost of the asset, L is the lifetime (in years), r is the interest rate (compounded continuously), and m(t) is the annual cost of maintenance. Find the capitalized cost under each set of assumptions.

46. $c_0 = \$500{,}000, r = 5\%, m(t) = \$20{,}000, L = 20$

47. $c_0 = \$400{,}000, r = 5.5\%, m(t) = \$10{,}000, L = 25$

48. $c_0 = \$600{,}000, r = 4\%,$
$m(t) = \$40{,}000 + \$1000e^{0.01t}, L = 40$

49. $c_0 = \$300{,}000, r = 5\%, m(t) = \$30{,}000 + \$500t,$
$L = 20$

50. Describe the idea of present value to a friend who is not a business major. Then describe accumulated present value.

51. Look up some data on rate of use and current world reserves of a natural resource not considered in this section. Predict when the world reserves for that resource will be depleted.

Answers to Quick Checks

1. $11,972.17   **2.** $6,225,857.39   **3.** $43,793.78
4. With R_1, accumulated future value is $2,969,590, and accumulated present value is $1,565,840. With R_2, accumulated future value is $3,005,640, and accumulated present value is $1,584,850.   **5.** $24,337.61, $1896.75
6. $1,219,822.71   **7.** **(a)** 100,736 metric tons;
(b) 12 yr, or by 2022

5.3

Improper Integrals

Let's try to find the area of the region under the graph of $y = 1/x^2$ over the interval $[1, \infty)$.

OBJECTIVES

- Determine whether an improper integral is convergent or divergent.

- Solve applied problems involving improper integrals.

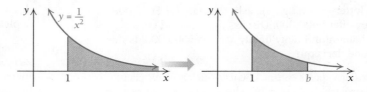

Note that this region is of infinite extent. We have not yet considered how to find the area of such a region. Let's find the area under the curve over the interval from

EXPLORATORY EXERCISES

1. Using a graphing calculator or iPlot, find

$$\int_1^{10} \frac{dx}{x^3},$$

$$\int_1^{100} \frac{dx}{x^3}, \quad \text{and}$$

$$\int_1^{1000} \frac{dx}{x^3}.$$

2. Predict the value of

$$\int_1^{\infty} \frac{dx}{x^3}.$$

1 to b, and then see what happens as b gets very large. The area under the graph over $[1, b]$ is

$$\int_1^b \frac{dx}{x^2} = \left[-\frac{1}{x}\right]_1^b$$

$$= \left(-\frac{1}{b}\right) - \left(-\frac{1}{1}\right)$$

$$= -\frac{1}{b} + 1$$

$$= 1 - \frac{1}{b}.$$

Then

$$\lim_{b \to \infty} (\text{area from 1 to } b) = \lim_{b \to \infty} \left(1 - \frac{1}{b}\right) = 1.$$

We *define* the area from 1 to infinity to be this limit. Here we have an example of an infinitely long region with a finite area.

Such areas may not always be finite. Let's try to find the area of the region under the graph of $y = 1/x$ over the interval $[1, \infty)$.

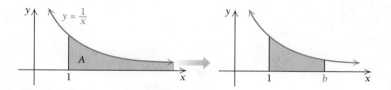

By definition, the area A from 1 to infinity is the limit as b approaches ∞ of the area from 1 to b, so

$$A = \lim_{b \to \infty} \int_1^b \frac{dx}{x} = \lim_{b \to \infty} \left[\ln x\right]_1^b$$

$$= \lim_{b \to \infty} (\ln b - \ln 1)$$

$$= \lim_{b \to \infty} \ln b.$$

In Section 3.2, we graphed $y = \ln x$ and saw that the function is always increasing. Therefore, the limit $\lim_{b \to \infty} \ln b$ does not exist and we have an infinitely long region with an infinite area.

Note that the graphs of $y = 1/x^2$ and $y = 1/x$ have similar shapes, but the region under one of them has a finite area and the region under the other does not.

An integral such as

$$\int_a^{\infty} f(x)\, dx,$$

with an upper limit of infinity, is an example of an **improper integral**. Its value is defined to be the following limit.

DEFINITION

$$\int_a^{\infty} f(x)\, dx = \lim_{b \to \infty} \int_a^b f(x)\, dx$$

If the limit exists, then we say that the improper integral **converges**, or is **convergent**. If the limit does not exist, then we say that the improper integral **diverges**, or is **divergent**. Thus,

$$\int_1^\infty \frac{dx}{x^2} = 1 \; converges, \quad \text{and} \quad \int_1^\infty \frac{dx}{x} \; diverges.$$

■ **EXAMPLE 1** Determine whether the following integral is convergent or divergent, and calculate its value if it is convergent:

$$\int_0^\infty 4e^{-2x} \, dx.$$

Solution We have

$$\int_0^\infty 4e^{-2x} \, dx = \lim_{b \to \infty} \int_0^b 4e^{-2x} \, dx$$

$$= \lim_{b \to \infty} \left[\frac{4}{-2} e^{-2x} \right]_0^b$$

$$= \lim_{b \to \infty} \left[-2e^{-2x} \right]_0^b$$

$$= \lim_{b \to \infty} \left[-2e^{-2b} - (-2e^{-2 \cdot 0}) \right]$$

$$= \lim_{b \to \infty} (-2e^{-2b} + 2)$$

$$= \lim_{b \to \infty} \left(2 - \frac{2}{e^{2b}} \right).$$

As b approaches ∞, we know that e^{2b} approaches ∞ (see the graphs of $y = a^x$ in Chapter 3), so

$$\frac{2}{e^{2b}} \to 0 \quad \text{and} \quad \left(2 - \frac{2}{e^{2b}} \right) \to 2.$$

Thus, $\displaystyle\int_0^\infty 4e^{-2x} \, dx = \lim_{b \to \infty} \left(2 - \frac{2}{e^{2b}} \right) = 2.$

The integral is convergent.

❮ Quick Check 1

> **Quick Check 1**
>
> Determine whether the following integral is convergent or divergent, and calculate its value if it is convergent:
>
> $$\int_2^\infty \frac{2}{x^3} \, dx.$$

Following are definitions of two other types of improper integrals.

DEFINITIONS

1. $\displaystyle\int_{-\infty}^b f(x) \, dx = \lim_{a \to -\infty} \int_a^b f(x) \, dx$

2. $\displaystyle\int_{-\infty}^\infty f(x) \, dx = \int_{-\infty}^c f(x) \, dx + \int_c^\infty f(x) \, dx,$

where c can be any real number.

In order for $\int_{-\infty}^\infty f(x) \, dx$ to converge, both integrals on the right in the second part of the definition must converge.

Applications of Improper Integrals

In Section 5.2, we learned that the accumulated present value of a continuous money flow (income stream) of P dollars per year, at a constant rate, from now until T years in the future can be found by integration:

$$\int_0^T Pe^{-kt}\, dt = \frac{P}{k}(1 - e^{-kT}),$$

where k is the interest rate and interest is compounded continuously. Suppose that the money flow is to continue perpetually (forever). Under this assumption, the accumulated present value of the money flow is

$$\int_0^\infty Pe^{-kt}\, dt = \lim_{T \to \infty} \int_0^T Pe^{-kt}\, dt$$

$$= \lim_{T \to \infty} \frac{P}{k}(1 - e^{-kT})$$

$$= \lim_{T \to \infty} \frac{P}{k}\left(1 - \frac{1}{e^{kT}}\right) = \frac{P}{k}.$$

THEOREM 1

The **accumulated present value** of a continuous money flow into an investment at the constant rate of P dollars per year perpetually is given by

$$\int_0^\infty Pe^{-kt}\, dt = \frac{P}{k},$$

where k is the interest rate and interest is compounded continuously.

〉 **Quick Check 2**

Find the accumulated present value of an investment for which there is a perpetual continuous money flow of $10,000 per year. Assume that the interest rate is 6%, compounded continuously.

■ **EXAMPLE 2**   **Business: Accumulated Present Value.**   Find the accumulated present value of an investment for which there is a perpetual continuous money flow of $2000 per year. Assume that the interest rate is 8%, compounded continuously.

Solution   The accumulated present value is 2000/0.08, or $25,000.

〈 Quick Check 2

When an amount P of radioactive material is being released into the atmosphere annually, the total amount that has been released at time T is given by

$$\int_0^T Pe^{-kt}\, dt = \frac{P}{k}(1 - e^{-kT}).$$

As T approaches ∞ (the radioactive material is released forever), the buildup of radioactive material approaches a limiting value P/k. It is no wonder that scientists and environmentalists are so concerned about radioactive waste. The radioactivity is "here to stay."

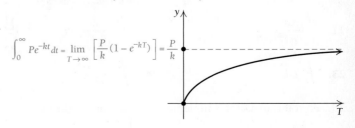

Section Summary

- An *improper integral* has infinity as one or both of its bounds and is evaluated using the limit:

$$\int_a^\infty f(x)\,dx = \lim_{b\to\infty} \int_a^b f(x)\,dx,$$

$$\int_{-\infty}^b f(x)\,dx = \lim_{a\to-\infty} \int_a^b f(x)\,dx,$$

and $\int_{-\infty}^\infty f(x)\,dx = \int_{-\infty}^c f(x)\,dx + \int_c^\infty f(x)\,dx,$

where c is any real number.

- The *accumulated present value* of a continuous money flow into an investment at the rate of P dollars per year perpetually is given by

$$\int_0^\infty Pe^{-kt}\,dt = \frac{P}{k},$$

where k is the interest rate compounded continuously.

EXERCISE SET
5.3

Determine whether each improper integral is convergent or divergent, and calculate its value if it is convergent.

1. $\displaystyle\int_2^\infty \frac{dx}{x^2}$

2. $\displaystyle\int_4^\infty \frac{dx}{x^2}$

3. $\displaystyle\int_3^\infty \frac{dx}{x}$

4. $\displaystyle\int_4^\infty \frac{dx}{x}$

5. $\displaystyle\int_0^\infty 3e^{-3x}\,dx$

6. $\displaystyle\int_0^\infty 4e^{-4x}\,dx$

7. $\displaystyle\int_1^\infty \frac{dx}{x^3}$

8. $\displaystyle\int_1^\infty \frac{dx}{x^4}$

9. $\displaystyle\int_0^\infty \frac{dx}{2+x}$

10. $\displaystyle\int_0^\infty \frac{4\,dx}{3+x}$

11. $\displaystyle\int_2^\infty 4x^{-2}\,dx$

12. $\displaystyle\int_2^\infty 7x^{-2}\,dx$

13. $\displaystyle\int_0^\infty e^x\,dx$

14. $\displaystyle\int_0^\infty e^{2x}\,dx$

15. $\displaystyle\int_3^\infty x^2\,dx$

16. $\displaystyle\int_5^\infty x^4\,dx$

17. $\displaystyle\int_0^\infty xe^x\,dx$

18. $\displaystyle\int_1^\infty \ln x\,dx$

19. $\displaystyle\int_0^\infty me^{-mx}\,dx,\ m>0$

20. $\displaystyle\int_0^\infty Qe^{-kt}\,dt,\ k>0$

21. $\displaystyle\int_\pi^\infty \frac{dt}{t^{1.001}}$

22. $\displaystyle\int_1^\infty \frac{2t}{t^2+1}\,dt$

23. $\displaystyle\int_{-\infty}^\infty t\,dt$

24. $\displaystyle\int_1^\infty \frac{3x^2}{(x^3+1)^2}\,dx$

25. Find the area, if it is finite, of the region under the graph of $y = 1/x^2$ over the interval $[2, \infty)$.

26. Find the area, if it is finite, of the region under the graph of $y = 1/x$ over the interval $[2, \infty)$.

27. Find the area, if it is finite, of the region bounded by $y = 2xe^{-x^2}, x = 0$, and $[0, \infty)$.

28. Find the area, if it is finite, of the region bounded by $y = 1/\sqrt{(3x-2)^3}, x = 6$, and $[6, \infty)$.

APPLICATIONS

Business and Economics

29. Accumulated present value. Find the accumulated present value of an investment for which there is a perpetual continuous money flow of $3600 per year at an interest rate of 7%, compounded continuously.

30. Accumulated present value. Find the accumulated present value of an investment for which there is a perpetual continuous money flow of $3500 per year at an interest rate of 6%, compounded continuously.

31. Total profit from marginal profit. A firm is able to determine that its marginal profit, in dollars, from producing x units of an item is given by

$$P'(x) = 200e^{-0.032x}.$$

Suppose that it were possible for the firm to make infinitely many units of this item. What would its total profit be?

32. Total profit from marginal profit. Find the total profit in Exercise 31 if

$$P'(x) = 200x^{-1.032}, \quad \text{where } x \geq 1.$$

33. Total cost from marginal cost. A company determines that its marginal cost, in dollars, for producing x units of a product is given by

$$C'(x) = 3600x^{-1.8}, \quad \text{where } x \geq 1.$$

Suppose that it were possible for the company to make infinitely many units of this product. What would the total cost be?

34. **Total production.** A firm determines that it can produce tires at a rate of
$$r(t) = 2000e^{-0.42t},$$
where t is the time, in years. Assuming that the firm endures forever (it never gets tired), how many tires can it make?

35. **Accumulated present value.** Find the accumulated present value of an investment for which there is a perpetual continuous money flow of $5000 per year, assuming continuously compounded interest at a rate of 8%.

36. **Accumulated present value.** Find the accumulated present value of an investment for which there is a perpetual continuous money flow of $2000e^{-0.01t}$ per year, assuming continuously compounded interest at a rate of 7%.

Capitalized cost. *The capitalized cost, c, of an asset for an unlimited lifetime is the total of the initial cost and the present value of all maintenance expenses that will occur in the future. It is computed by the formula*
$$c = c_0 + \int_0^{\infty} m(t)e^{-rt}\,dt,$$
where c_0 is the initial cost of the asset, r is the interest rate (compounded continuously), and $m(t)$ is the annual cost of maintenance. Find the capitalized cost under each set of assumptions.

37. $c_0 = \$500,000,$ $r = 5\%,$ $m(t) = \$20,000$

38. $c_0 = \$700,000,$ $r = 5\%,$ $m(t) = \$30,000$

Life and Physical Sciences

39. **Radioactive buildup.** Plutonium has a decay rate of 0.003% per year. Suppose that a nuclear accident causes plutonium to be released into the atmosphere perpetually at the rate of 1 lb each year. What is the limiting value of the radioactive buildup?

40. **Radioactive buildup.** Cesium-137 has a decay rate of 2.3% per year. Suppose that a nuclear accident causes cesium-137 to be released into the atmosphere perpetually at the rate of 1 lb each year. What is the limiting value of the radioactive buildup?

Radioactive implant treatments. *In the treatment of prostate cancer, radioactive implants are often used. The implants are left in the patient and never removed. The amount of energy that is transmitted to the body from the implant is measured in rem units and is given by*
$$E = \int_0^{a} P_0 e^{-kt}\,dt,$$
where k is the decay constant for the radioactive material, a is the number of years since the implant, and P_0 is the initial rate at which energy is transmitted. Use this information for Exercises 41 and 42.

41. Suppose that the treatment uses iodine-125, which has a half-life of 60.1 days.
   a) Find the decay rate, k, of iodine-125.
   b) How much energy (measured in rems) is transmitted in the first month if the initial rate of transmission is 10 rems per year?
   c) What is the total amount of energy that the implant will transmit to the body?

42. Suppose that the treatment uses palladium-103, which has a half-life of 16.99 days.
   a) Find the decay rate, k, of palladium-103.
   b) How much energy (measured in rems) is transmitted in the first month if the initial rate of transmission is 10 rems per year?
   c) What is the total amount of energy that the implant will transmit to the body?

SYNTHESIS

Determine whether each improper integral is convergent or divergent, and calculate its value if it is convergent.

43. $\displaystyle\int_0^{\infty} \frac{dx}{x^{2/3}}$ **44.** $\displaystyle\int_1^{\infty} \frac{dx}{\sqrt{x}}$

45. $\displaystyle\int_0^{\infty} \frac{dx}{(x+1)^{3/2}}$ **46.** $\displaystyle\int_{-\infty}^{0} e^{2x}\,dx$

47. $\displaystyle\int_0^{\infty} xe^{-x^2}\,dx$ **48.** $\displaystyle\int_{-\infty}^{\infty} xe^{-x^2}\,dx$

Life science: drug dosage. *Suppose that an oral dose of a drug is taken. Over time, the drug is assimilated in the body and excreted through the urine. The total amount of the drug that has passed through the body in time T is given by*
$$\int_0^{T} E(t)\,dt,$$
where $E(t)$ is the rate of excretion of the drug. A typical rate-of-excretion function is $E(t) = te^{-kt}$, where $k > 0$ and t is the time, in hours. Use this information for Exercises 49 and 50.

49. Find $\int_0^{\infty} E(t)\,dt$, and interpret the answer. That is, what does the integral represent?

50. A physician prescribes a dosage of 100 mg. Find k.

51. Consider the functions
$$y = \frac{1}{x^2} \quad \text{and} \quad y = \frac{1}{x}.$$
Suppose that you go to a paint store to buy paint to cover the region under each graph over the interval $[1, \infty)$. Discuss whether you could be successful and why or why not.

52. Suppose that you are the owner of a building that yields a continuous series of rental payments and you decide to sell the building. Explain how you would use the concept of the accumulated present value of a perpetual continuous money flow to determine a fair selling price.

TECHNOLOGY CONNECTION

53. Graph the function E and shade the area under the curve for each situation in Exercises 49 and 50.

Approximate each integral.

54. $\displaystyle\int_1^{\infty} \frac{4}{1+x^2}\,dx$ **55.** $\displaystyle\int_1^{\infty} \frac{6}{5+e^x}\,dx$

Numerical Integration

OBJECTIVES

- Solve problems using numerical integration methods such as Riemann sums, the Midpoint Rule, the Trapezoidal Rule, and Simpson's Rule.

- Find the percent error between an approximation and the actual value of a definite integral.

In Chapter 4, we considered many examples that required us to evaluate a definite integral. However, in many situations, it may be difficult or impossible to evaluate a definite integral using an antiderivative. To address this difficulty, we must develop ways to approximate the value of a definite integral using geometry—a process called *numerical integration*. Two common situations that can require numerical integration are the following:

- Finding an antiderivative is difficult or impossible. An example of this situation occurs in trying to evaluate $\int_a^b e^{-x^2}\, dx$, an integral used in statistics. There is no way to antidifferentiate $f(x) = e^{-x^2}$.

- The data may not be easily modeled by a continuous function. The data may consist of individual points, and so a continuous function that "fits" the data may not be easily found.

Let's consider a situation where the data consist of individual points. We will discover that it is not necessary to find a continuous function that models the data.

Riemann Sums

The most common geometric method for approximating area is to use rectangles, as we did in Section 4.2. We subdivide the interval $[a, b]$ into n subintervals, $[a, x_1], [x_1, x_2], [x_2, x_3]$, and so on, through $[x_{n-1}, b]$. Note that we can regard a as x_0 and b as x_n. If we require the subintervals to be equal in size, then each subinterval has a width of $\Delta x = \dfrac{b - a}{n}$. To illustrate, suppose the graph of $C(t)$ shown to the right is the rate of exertion (measured in calories per hour) after t 30-second intervals for a person exercising on a treadmill.

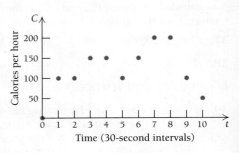

The total number of calories expended is approximated by the area under a graph of the data. Over each subinterval, a rectangle is drawn. One approach uses each data point as the top-left corner of such a rectangle. The sum of the areas of the rectangles is called a *Riemann sum*.

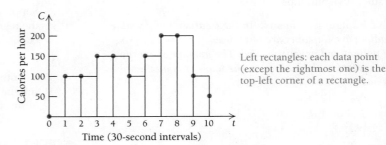

Left rectangles: each data point (except the rightmost one) is the top-left corner of a rectangle.

Since C is measured in calories per hour, we view each 30-second interval as $\frac{1}{120}$ of an hour. Thus, each rectangle has a width of $\frac{1}{120}$ and a height corresponding to the data point at the top-left corner. The sum of these *left-rectangles* is denoted as L_n, where n is the number of subintervals. We use a table to calculate the area of each rectangle, rounded to three decimal places. Note that the rightmost data point is not used.

Subinterval	Width	Height (x = left endpoint)	Area
$[0, 1]$	$\frac{1}{120}$ hr	$C(0) = 0$ cal/hr	$\frac{1}{120} \cdot 0 = 0$ cal
$[1, 2]$	$\frac{1}{120}$ hr	$C(1) = 100$ cal/hr	$\frac{1}{120} \cdot 100 \approx 0.833$ cal
$[2, 3]$	$\frac{1}{120}$ hr	$C(2) = 100$ cal/hr	$\frac{1}{120} \cdot 100 \approx 0.833$ cal
$[3, 4]$	$\frac{1}{120}$ hr	$C(3) = 150$ cal/hr	$\frac{1}{120} \cdot 150 = 1.250$ cal
$[4, 5]$	$\frac{1}{120}$ hr	$C(4) = 150$ cal/hr	$\frac{1}{120} \cdot 150 = 1.250$ cal
$[5, 6]$	$\frac{1}{120}$ hr	$C(5) = 100$ cal/hr	$\frac{1}{120} \cdot 100 \approx 0.833$ cal
$[6, 7]$	$\frac{1}{120}$ hr	$C(6) = 150$ cal/hr	$\frac{1}{120} \cdot 150 = 1.250$ cal
$[7, 8]$	$\frac{1}{120}$ hr	$C(7) = 200$ cal/hr	$\frac{1}{120} \cdot 200 \approx 1.667$ cal
$[8, 9]$	$\frac{1}{120}$ hr	$C(8) = 200$ cal/hr	$\frac{1}{120} \cdot 200 \approx 1.667$ cal
$[9, 10]$	$\frac{1}{120}$ hr	$C(9) = 100$ cal/hr	$\frac{1}{120} \cdot 100 \approx 0.833$ cal

The sum of the values in the last column is the total area of the 10 rectangles. This sum, L_{10}, approximates the total calories expended during 5 minutes on the treadmill. We have $L_{10} = 0 + 0.833 + 0.833 + 1.250 + 1.250 + 0.833 + 1.250 + 1.667 + 1.667 + 0.833 \approx 10.416$ calories.

We can also draw the rectangles so that each data point is the top-right corner of a rectangle. The sum of these *right-rectangles* is denoted as R_n, where n is the number of subintervals. The leftmost data point is not used.

We again use a table to find the area of each rectangle, rounded to three decimal places, and add the areas in the last column to approximate the total number of calories expended.

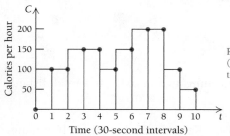

Right rectangles: each data point (except the leftmost one) is the top-right corner of a rectangle.

Subinterval	Width	Height (x = right endpoint)	Area
$[0, 1]$	$\frac{1}{120}$ hr	$C(1) = 100$ cal/hr	$\frac{1}{120} \cdot 100 \approx 0.833$ cal
$[1, 2]$	$\frac{1}{120}$ hr	$C(2) = 100$ cal/hr	$\frac{1}{120} \cdot 100 \approx 0.833$ cal
$[2, 3]$	$\frac{1}{120}$ hr	$C(3) = 150$ cal/hr	$\frac{1}{120} \cdot 150 = 1.250$ cal
$[3, 4]$	$\frac{1}{120}$ hr	$C(4) = 150$ cal/hr	$\frac{1}{120} \cdot 150 = 1.250$ cal
$[4, 5]$	$\frac{1}{120}$ hr	$C(5) = 100$ cal/hr	$\frac{1}{120} \cdot 100 \approx 0.833$ cal
$[5, 6]$	$\frac{1}{120}$ hr	$C(6) = 150$ cal/hr	$\frac{1}{120} \cdot 150 = 1.250$ cal
$[6, 7]$	$\frac{1}{120}$ hr	$C(7) = 200$ cal/hr	$\frac{1}{120} \cdot 200 \approx 1.667$ cal
$[7, 8]$	$\frac{1}{120}$ hr	$C(8) = 200$ cal/hr	$\frac{1}{120} \cdot 200 \approx 1.667$ cal
$[8, 9]$	$\frac{1}{120}$ hr	$C(9) = 100$ cal/hr	$\frac{1}{120} \cdot 100 \approx 0.833$ cal
$[9, 10]$	$\frac{1}{120}$ hr	$C(10) = 50$ cal/hr	$\frac{1}{120} \cdot 50 \approx 0.417$ cal

We have

$$R_{10} = 0.833 + 0.833 + 1.250 + 1.250 + 0.833 + 1.250 + 1.667$$
$$+ 1.667 + 0.833 + 0.417$$
$$\approx 10.833 \text{ calories.}$$

By averaging the two sums, L_{10} and R_{10}, we can conclude that the total number of calories expended is about

$$\frac{L_{10} + R_{10}}{2} = \frac{10.416 + 10.833}{2} = 10.625 \text{ calories.}$$

The Riemann sums method is summarized below.

DEFINITION **Riemann Sums**

Let the interval $[a, b]$ be subdivided into n equal subintervals, $[a, x_1], [x_1, x_2]$, $[x_2, x_3], \ldots, [x_{n-1}, b]$, each with width $\Delta x = \dfrac{b - a}{n}$. Assume f is defined for $a, x_1, x_2, \ldots, x_{n-1}, b$. The approximate value of $\displaystyle\int_a^b f(x)\,dx$ is given by L_n, where

$$L_n = \frac{b - a}{n} \cdot (f(a) + f(x_1) + f(x_2) + \cdots + f(x_{n-2}) + f(x_{n-1})).$$

The approximate value of $\displaystyle\int_a^b f(x)\,dx$ is also given by R_n, where

$$R_n = \frac{b - a}{n} \cdot (f(x_1) + f(x_2) + \cdots + f(x_{n-2}) + f(x_{n-1}) + f(b)).$$

In the following example, we use Riemann sums to approximate a total distance given velocity data for 1-minute intervals.

EXAMPLE 1 **Approximating Total Distance.** Ken rides his bicycle for 10 minutes. The table below shows Ken's speed, $v(t)$, in miles per hour, for each 1-minute interval of time. Find L_{10} to approximate the total distance Ken travels during the 10 minutes.

t (min)	0	1	2	3	4	5	6	7	8	9	10
$v(t)$, mi/hr	10	12	18	20	20	8	6	15	18	20	16

Solution We must have consistency of units, and each 1-minute interval of time is equivalent to $\frac{1}{60}$ of an hour. Since we are using left-rectangles, the rightmost data point will not be used. The following table summarizes the calculations:

Subinterval	Width	Height (x = left endpoint)	Area
$[0, 1]$	$\frac{1}{60}$ hr	$v(0) = 10$ mi/hr	$\frac{1}{60} \cdot 10 \approx 0.167$ mi
$[1, 2]$	$\frac{1}{60}$ hr	$v(1) = 12$ mi/hr	$\frac{1}{60} \cdot 12 = 0.2$ mi
$[2, 3]$	$\frac{1}{60}$ hr	$v(2) = 18$ mi/hr	$\frac{1}{60} \cdot 18 = 0.3$ mi
$[3, 4]$	$\frac{1}{60}$ hr	$v(3) = 20$ mi/hr	$\frac{1}{60} \cdot 20 \approx 0.333$ mi
$[4, 5]$	$\frac{1}{60}$ hr	$v(4) = 20$ mi/hr	$\frac{1}{60} \cdot 20 \approx 0.333$ mi
$[5, 6]$	$\frac{1}{60}$ hr	$v(5) = 8$ mi/hr	$\frac{1}{60} \cdot 8 \approx 0.133$ mi
$[6, 7]$	$\frac{1}{60}$ hr	$v(6) = 6$ mi/hr	$\frac{1}{60} \cdot 6 = 0.1$ mi
$[7, 8]$	$\frac{1}{60}$ hr	$v(7) = 15$ mi/hr	$\frac{1}{60} \cdot 15 = 0.25$ mi
$[8, 9]$	$\frac{1}{60}$ hr	$v(8) = 18$ mi/hr	$\frac{1}{60} \cdot 18 = 0.3$ mi
$[9, 10]$	$\frac{1}{60}$ hr	$v(9) = 20$ mi/hr	$\frac{1}{60} \cdot 20 \approx 0.333$ mi

> **Quick Check 1**
>
> Use the table in Example 1 to find R_{10}, an approximation of Ken's total distance traveled, and then find the average of L_{10} and R_{10}.

Thus, Ken travels approximately $L_{10} = 0.167 + 0.2 + 0.3 + 0.333 + 0.333 + 0.133 + 0.1 + 0.25 + 0.3 + 0.333 = 2.449$ miles.

❮ Quick Check 1

Many continuous functions have antiderivatives that are not easily found. For example, it is not possible to write the antiderivative of $f(x) = \sqrt{1 + x^3}$ using the basic set of elementary functions and arithmetic operations. The following example shows how we may approximate such a definite integral using Riemann sums.

■ **EXAMPLE 2**   Given the definite integral $\int_0^3 \sqrt{1 + x^3}\, dx$, find L_6 and R_6 and their average.

Solution   To approximate this definite integral using L_6, we graph $f(x) = \sqrt{1 + x^3}$ over $[0, 3]$ and draw six left-rectangles. Thus, $[0, 3]$ is subdivided into six equal subintervals of width 0.5. A table lists the areas of the rectangles, which are then added. To find the height of each rectangle, we evaluate $y = f(x)$ at the left endpoint of each subinterval and round the result to three decimal places.

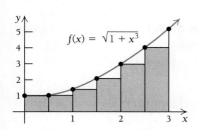

Subinterval	Width	Height (x = left endpoint)	Area
$[0, 0.5]$	0.5	$f(0) = 1$	$0.5 \cdot 1 = 0.500$
$[0.5, 1]$	0.5	$f(0.5) \approx 1.061$	$0.5 \cdot 1.061 \approx 0.531$
$[1, 1.5]$	0.5	$f(1) \approx 1.414$	$0.5 \cdot 1.414 \approx 0.707$
$[1.5, 2]$	0.5	$f(1.5) \approx 2.092$	$0.5 \cdot 2.092 \approx 1.046$
$[2, 2.5]$	0.5	$f(2) = 3$	$0.5 \cdot 3 = 1.500$
$[2.5, 3]$	0.5	$f(2.5) \approx 4.077$	$0.5 \cdot 4.077 \approx 2.039$

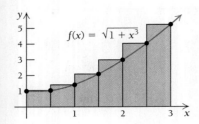

$f(x) = \sqrt{1 + x^3}$

Thus, we have

$$L_6 = 0.500 + 0.531 + 0.707 + 1.046 + 1.500 + 2.039 = 6.323.$$

To find R_6, the height of each rectangle is found by evaluating $y = f(x)$ at the right endpoint of each subinterval.

Subinterval	Width	Height ($x =$ right endpoint)	Area
$[0, 0.5]$	0.5	$f(0.5) \approx 1.061$	$0.5 \cdot 1.061 \approx 0.531$
$[0.5, 1]$	0.5	$f(1) \approx 1.414$	$0.5 \cdot 1.414 \approx 0.707$
$[1, 1.5]$	0.5	$f(1.5) \approx 2.092$	$0.5 \cdot 2.092 \approx 1.046$
$[1.5, 2]$	0.5	$f(2) = 3$	$0.5 \cdot 3 = 1.500$
$[2, 2.5]$	0.5	$f(2.5) \approx 4.077$	$0.5 \cdot 4.077 \approx 2.039$
$[2.5, 3]$	0.5	$f(3) \approx 5.292$	$0.5 \cdot 5.292 \approx 2.646$

We have $R_6 = 0.531 + 0.707 + 1.046 + 1.500 + 2.039 + 2.646 = 8.469$. We average the two sums to find a reasonable approximation of the definite integral:

$$\int_0^3 \sqrt{1 + x^3} \, dx \approx \frac{6.323 + 8.469}{2} = 7.396.$$

A calculator or algebra software shows that to three decimal places, the value of $\int_0^3 \sqrt{1 + x^3} \, dx$ is 7.341. As discussed in Chapter 4, we can find a more accurate approximation by subdividing the interval $[0, 3]$ into smaller subintervals.

❯ Quick Check 2

Approximate

$$\int_0^2 \ln(x + 1) \, dx$$

by finding L_6, R_6, and their average.

❮ Quick Check 2

Midpoint Rule

Instead of using the left or right endpoint of each subinterval to evaluate $y = f(x)$, we can evaluate the function at the *midpoint* of each subinterval. This is the basis of the *Midpoint Rule*, which gives an approximation M_n that is the sum of the areas of n rectangles whose heights are measured at the midpoints of the subintervals. The midpoint of each subinterval is the average of its left and right endpoints.

DEFINITION   Midpoint Rule

Let the interval $[a, b]$ be subdivided into n equal subintervals with width $\Delta x = \dfrac{b - a}{n}$, and assume that f is defined for $\dfrac{a + x_1}{2}, \dfrac{x_1 + x_2}{2}, \ldots, \dfrac{x_{n-1} + b}{2}$.

The approximate value of $\int_a^b f(x) \, dx$ is given by M_n, where

$$M_n = \frac{b - a}{n} \cdot \left(f\left(\frac{a + x_1}{2}\right) + f\left(\frac{x_1 + x_2}{2}\right) + f\left(\frac{x_2 + x_3}{2}\right) + \cdots \right.$$
$$\left. + f\left(\frac{x_{n-2} + x_{n-1}}{2}\right) + f\left(\frac{x_{n-1} + b}{2}\right) \right).$$

Let's find an approximation of $\int_0^3 \sqrt{1 + x^3} \, dx$ (from Example 2) using the Midpoint Rule.

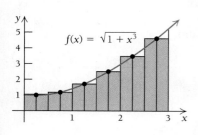

EXAMPLE 3 Use the Midpoint Rule to find M_6, an approximation of $\int_0^3 \sqrt{1 + x^3}\, dx$.

Solution We subdivide $[0, 3]$ into six subintervals, each with width 0.5, and then find the midpoint of each subinterval using the average of its two endpoints.

A table is used to find the area of each rectangle. To find the height of each rectangle, we evaluate $y = f(x)$ at the midpoint of each subinterval. Heights and areas are rounded to three decimal places.

Subinterval	Midpoint	Width	Height $(x = \text{midpoint})$	Area
$[0, 0.5]$	0.25	0.5	$f(0.25) \approx 1.008$	$0.5 \cdot 1.008 \approx 0.504$
$[0.5, 1]$	0.75	0.5	$f(0.75) \approx 1.192$	$0.5 \cdot 1.192 \approx 0.596$
$[1, 1.5]$	1.25	0.5	$f(1.25) \approx 1.718$	$0.5 \cdot 1.718 \approx 0.859$
$[1.5, 2]$	1.75	0.5	$f(1.75) \approx 2.522$	$0.5 \cdot 2.522 \approx 1.261$
$[2, 2.5]$	2.25	0.5	$f(2.25) \approx 3.520$	$0.5 \cdot 3.520 \approx 1.760$
$[2.5, 3]$	2.75	0.5	$f(2.75) \approx 4.669$	$0.5 \cdot 4.669 \approx 2.335$

Quick Check 3

Use the Midpoint Rule to find M_8, an approximation of $\int_0^2 \ln(x + 1)\, dx$.

Thus, we have $M_6 = 0.504 + 0.596 + 0.859 + 1.261 + 1.760 + 2.335 = 7.315$, and we conclude that $\int_0^3 \sqrt{1 + x^3}\, dx \approx 7.315$. This compares well with the average of L_6 and R_6 found in Example 2.

‹ Quick Check 3

The Trapezoidal Rule

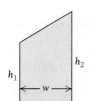

Instead of rectangles, we can use trapezoids to approximate the area under a curve. Recall that a *trapezoid* is a four-sided polygon of which two sides are parallel. If the parallel sides have lengths h_1 and h_2 and are w units apart, then the area of the trapezoid is $A = \left(\dfrac{h_1 + h_2}{2}\right) \cdot w$, where $\dfrac{h_1 + h_2}{2}$ is the average of the lengths h_1 and h_2.

Suppose we want to approximate the value of $\int_a^b f(x)\, dx$ using trapezoids, with $y = f(x)$ as shown below. Trapezoids are drawn over each subinterval with their parallel sides extending to the curve f. Assume that each trapezoid has width Δx.

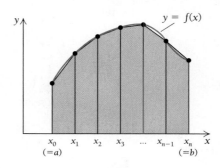

The first trapezoid has a width Δx and heights $f(x_0)$ and $f(x_1)$. Thus, the area of the first trapezoid is $\dfrac{f(x_0) + f(x_1)}{2} \cdot \Delta x$, the area of the second trapezoid is $\dfrac{f(x_1) + f(x_2)}{2} \cdot \Delta x$, and so on. Adding areas, we have

$$\int_a^b f(x)\, dx \approx \frac{f(x_0) + f(x_1)}{2} \cdot \Delta x + \frac{f(x_1) + f(x_2)}{2} \cdot \Delta x + \cdots + \frac{f(x_{n-1}) + f(x_n)}{2} \cdot \Delta x.$$

We now factor out Δx and express the fractions as a sum:

$$\int_a^b f(x)\, dx \approx \Delta x \cdot \left(\frac{f(x_0)}{2} + \frac{f(x_1)}{2} + \frac{f(x_1)}{2} + \frac{f(x_2)}{2} + \frac{f(x_2)}{2} + \cdots + \frac{f(x_{n-1})}{2} \right.$$
$$\left. + \frac{f(x_{n-1})}{2} + \frac{f(x_n)}{2} \right).$$

Note that $\dfrac{f(x_1)}{2} + \dfrac{f(x_1)}{2} = f(x_1), \dfrac{f(x_2)}{2} + \dfrac{f(x_2)}{2} = f(x_2)$, and so on. Only $\dfrac{f(x_0)}{2}$ and $\dfrac{f(x_n)}{2}$ cannot be combined. Thus, we have

$$\int_a^b f(x)\, dx \approx \Delta x \cdot \left(\frac{f(x_0)}{2} + f(x_1) + f(x_2) + \cdots + f(x_{n-1}) + \frac{f(x_n)}{2} \right).$$

This leads to the *Trapezoidal Rule*.

DEFINITION   **Trapezoidal Rule**

Let the interval $[a, b]$ be subdivided into n equal subintervals with width $\Delta x = \dfrac{b - a}{n}$, and assume that f is defined for $a, x_1, x_2, \ldots, x_{n-1}, b$. The approximate value of $\displaystyle\int_a^b f(x)\, dx$ is given by T_n, where

$$T_n = \frac{b - a}{n} \cdot \left(\frac{f(a)}{2} + f(x_1) + f(x_2) + \cdots + f(x_{n-1}) + \frac{f(b)}{2} \right).$$

■ **EXAMPLE 4**   Use the Trapezoidal Rule to find T_4, an approximation of $\int_1^3 \sqrt[3]{x + x^6}\, dx$.

Solution   The interval $[1, 3]$ is subdivided into four equal subintervals, $[1, 1.5]$, $[1.5, 2], [2, 2.5]$, and $[2.5, 3]$. Thus, $\Delta x = \dfrac{3 - 1}{4} = \dfrac{1}{2}$. Using a calculator, we have

$$f(1) = 1.260, \quad f(1.5) = 2.345, \quad f(2) = 4.041, \quad f(2.5) = 6.271, \quad f(3) = 9.012.$$

Therefore,

$$T_4 = \frac{1}{2} \cdot \left(\frac{f(1)}{2} + f(1.5) + f(2) + f(2.5) + \frac{f(3)}{2} \right)$$
$$\approx \frac{1}{2} \cdot \left(\frac{1.260}{2} + 2.345 + 4.041 + 6.271 + \frac{9.012}{2} \right)$$
$$= 8.897.$$

Thus, $\int_1^3 \sqrt[3]{x + x^6}\, dx \approx T_4 = 8.897$.

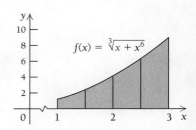

$f(x) = \sqrt[3]{x + x^6}$

> **Quick Check 4**
>
> Use the Trapezoidal Rule to find T_6, an approximation of $\int_0^3 \sqrt{1 + x^5}\, dx$

❰ Quick Check 4

Simpson's Rule

When working with Riemann sums, the Midpoint Rule, and the Trapezoidal Rule, we use line segments to approximate the curve of the graph of $y = f(x)$. However, we can instead approximate the curve using parabolas. This is the basis of *Simpson's Rule*, which gives an approximation denoted by S_n, where n is the number of subdivisions of the interval of integration. We will find that with Simpson's Rule, n must be even.

Consider $\int_a^b f(x)\, dx$ with the interval $[a, b]$ subdivided into two subintervals, $\left[a, \dfrac{a+b}{2}\right]$ and $\left[\dfrac{a+b}{2}, b\right]$, where $\dfrac{a+b}{2}$ is the midpoint of $[a, b]$. This provides three points: $(a, f(a))$, $\left(\dfrac{a+b}{2}, f\left(\dfrac{a+b}{2}\right)\right)$, and $(b, f(b))$. We can find a parabola that passes through these three points. This parabola is integrated and, after simplification, is used to approximate $\int_a^b f(x)\, dx$:

$$\int_a^b f(x)\, dx \approx \frac{b-a}{3n} \cdot \left(f(a) + 4f\left(\frac{a+b}{2}\right) + f(b)\right).$$

The proof of this formula is outlined in Exercise 73.

We now extend this rule to include more subdivisions. If n is even, then $n/2$ is a whole number and represents the number of parabolas. Again, $[a, b]$ is subdivided at $x_0, x_1, x_2, x_3, \ldots, x_{n-1}, x_n$, where $x_0 = a$ and $x_n = b$. A parabola is fitted to the first, second, and third points; another parabola is fitted to the third, fourth, and fifth points, and so on. This leads us to Simpson's Rule.

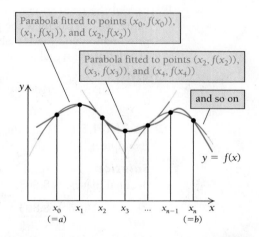

Parabola fitted to points $(x_0, f(x_0))$, $(x_1, f(x_1))$, and $(x_2, f(x_2))$

Parabola fitted to points $(x_2, f(x_2))$, $(x_3, f(x_3))$, and $(x_4, f(x_4))$

and so on

$y = f(x)$

DEFINITION Simpson's Rule

Let the interval $[a, b]$ be subdivided into an even number n of equal subintervals with width $\Delta x = \dfrac{b-a}{n}$, and assume that f is defined for $a, x_1, x_2, \ldots, x_{n-1}, b$.

The approximate value of $\displaystyle\int_a^b f(x)\, dx$ is given by S_n, where

$$S_n = \frac{b-a}{3n} \cdot \left(f(a) + 4f(x_1) + 2f(x_2) + 4f(x_3) + \cdots\right.$$
$$\left. + 2f(x_{n-2}) + 4f(x_{n-1}) + f(b)\right).$$

■ **EXAMPLE 5** Use Simpson's Rule to find S_2 and S_4, approximations of $\int_0^2 \sqrt{9 - x^2}\, dx$.

Solution For S_2, we subdivide $[0, 2]$ into two equal subintervals, $[0, 1]$ and $[1, 2]$. Note that $x_0 = 0, x_1 = 1$, and $x_2 = 2$. Thus, we have

$$S_2 = \frac{2 - 0}{3 \cdot 2} \cdot (f(0) + 4f(1) + f(2)) \qquad \text{Substituting}$$

$$\approx \tfrac{1}{3}(3 + 4 \cdot 2.828 + 2.236) \qquad \text{Calculating } f(0), f(1), \text{ and } f(2)$$

$$= 5.516 \text{ (rounded)}.$$

For S_4, we subdivide $[0, 2]$ into four equal subintervals, $[0, 0.5], [0.5, 1], [1, 1.5]$, and $[1.5, 2]$. We have $x_0 = 0, x_1 = 0.5, x_2 = 1, x_3 = 1.5$, and $x_4 = 2$. Thus,

$$S_4 = \frac{2 - 0}{3 \cdot 4} \cdot (f(0) + 4f(0.5) + 2f(1) + 4f(1.5) + f(2))$$

$$\approx \tfrac{1}{6} \cdot (3 + 4 \cdot 2.958 + 2 \cdot 2.828 + 4 \cdot 2.598 + 2.236)$$

$$= 5.519 \text{ (rounded)}.$$

> **Quick Check 5**
>
> Use Simpson's Rule to find S_2 and S_4, approximations of $\int_{-1}^1 e^{-x^2}\, dx$.

Algebra software shows that, to four decimal places, the value of $\int_0^2 \sqrt{9 - x^2}\, dx$ is 5.5198. Thus, both S_2 and S_4 are very accurate approximations of $\int_0^2 \sqrt{9 - x^2}\, dx$.

❮ Quick Check 5

Percent Error

All of the methods of numerical integration we have discussed provide an approximation of the true value of a definite integral. To compare such a result with the actual value, we can calculate the *percent error*, which is the percent difference between the approximation and the actual value of the definite integral. Assuming that the actual value of a definite integral is not zero, the percent error for any approximation is given by

$$\text{Percent error} = \frac{\text{approximation} - \text{actual value}}{\text{actual value}} \times 100.$$

■ **EXAMPLE 6** Find the percent error of T_4, the approximation of $\int_1^3 \sqrt[3]{x + x^6}\, dx$ in Example 4.

Solution We found $T_4 = 8.897$. The actual value of the definite integral, to four decimal places, is 8.803. Thus, the percent error is

$$\frac{8.897 - 8.803}{8.803} \times 100 = 1.07\%.$$

> **Quick Check 6**
>
> Find the percent error of S_2, the approximation of $\int_0^2 \sqrt{9 - x^2}\, dx$ in Example 5.

The approximation is 1.07% greater than the actual value of the definite integral.

❮ Quick Check 6

Section Summary

- The process of numerically approximating the value of the definite integral $\int_a^b f(x)\, dx$ using geometry is called *numerical integration*. We subdivide the interval $[a, b]$ into n subintervals, $[a, x_1], [x_1, x_2], \ldots, [x_{n-1}, b]$, regarding a as x_0 and b as x_n, and assume that f is defined for $a, x_1, x_2, \ldots, x_{n-1}, b$. If the subintervals are equal in size, then each subinterval has a width of $\Delta x = \frac{b - a}{n}$.

We may then use any of the following techniques.

- *Riemann sums*: Rectangles are drawn over each subinterval. With *left-rectangles*, the height of each rectangle is f evaluated at the left endpoint of the subinterval. The sum of the areas of all rectangles is denoted as L_n.

$$L_n = \frac{b - a}{n} \cdot (f(a) + f(x_1) + f(x_2) + \cdots + f(x_{n-2}) + f(x_{n-1})).$$

With *right-rectangles*, the height of each rectangle is f evaluated at the right endpoint of the subinterval. The sum of the areas of all rectangles is denoted as R_n.

$$R_n = \frac{b-a}{n} \cdot \big(f(x_1) + f(x_2) + \cdots + f(x_{n-2})$$
$$+ f(x_{n-1}) + f(b)\big).$$

We can average L_n and R_n (or use a larger value of n) for a better approximation of $\int_a^b f(x)\,dx$.

○ *Midpoint Rule*: Using rectangles, the height of each rectangle is f evaluated at the midpoint of the subinterval, assuming that f is defined for $\dfrac{a+x_1}{2}, \dfrac{x_1+x_2}{2}, \ldots,$ $\dfrac{x_{n-1}+b}{2}$. The sum of the areas of all rectangles is denoted as M_n.

$$M_n = \frac{b-a}{n} \cdot \left(f\left(\frac{a+x_1}{2}\right) + f\left(\frac{x_1+x_2}{2}\right) + \right.$$

$$\left. f\left(\frac{x_2+x_3}{2}\right) + \cdots + f\left(\frac{x_{n-2}+x_{n-1}}{2}\right) + f\left(\frac{x_{n-1}+b}{2}\right)\right).$$

○ *Trapezoidal Rule*: Trapezoids are drawn over each subinterval. The sum of the areas of all the trapezoids is denoted T_n, where

$$T_n = \frac{b-a}{n} \cdot \left(\frac{f(a)}{2} + f(x_1) + f(x_2) + \cdots \right.$$

$$\left. + f(x_{n-1}) + \frac{f(b)}{2}\right).$$

○ *Simpson's Rule*: The interval $[a, b]$ is divided into n subintervals, where n is even. The graph of f is then approximated by $n/2$ parabolas, and the area under f is approximated by S_n, where

$$S_n = \frac{b-a}{3n} \cdot \big(f(a) + 4f(x_1) + 2f(x_2) + 4f(x_3) + \cdots$$

$$+ 2f(x_{n-2}) + 4f(x_{n-1}) + f(b)\big).$$

- The *percent error* between an approximation and the actual value of a definite integral is given by

$$\text{Percent error} = \frac{\text{approximation} - \text{actual value}}{\text{actual value}} \times 100.$$

EXERCISE SET
5.4

In Exercises 1–10, find L_n, R_n, and their average for each definite integral using the indicated value of n. Give all answers to three decimal places.

1. $\displaystyle\int_0^4 (x^2 + 1)\,dx, n = 4$ **2.** $\displaystyle\int_2^6 (x^3 - 1)\,dx, n = 4$

3. $\displaystyle\int_0^3 (2x^3 - x)\,dx, n = 6$ **4.** $\displaystyle\int_{-1}^2 (x^2 + 2x)\,dx, n = 6$

5. $\displaystyle\int_2^4 \frac{1}{x}\,dx, n = 8$ **6.** $\displaystyle\int_3^8 \frac{1}{x-2}\,dx, n = 5$

7. $\displaystyle\int_0^3 e^{2x}\,dx, n = 6$ **8.** $\displaystyle\int_{-2}^0 e^{-x}\,dx, n = 4$

9. $\displaystyle\int_1^5 x(x^2 + 1)^2\,dx, n = 8$ **10.** $\displaystyle\int_0^5 x^2(x^3 + 2)^2\,dx, n = 10$

11–20. (a) Find M_n to three decimal places for each definite integral in Exercises 1–10, using the indicated value of n, and **(b)** find the percent error between M_n and the exact value of the definite integral.

In Exercises 21–28, use the Trapezoidal Rule to find T_n using the indicated value of n. Give all answers to three decimal places.

21. $\displaystyle\int_0^2 \sqrt{x^2 + 1}\,dx, n = 4$ **22.** $\displaystyle\int_1^4 \sqrt{x^2 + 3}\,dx, n = 6$

23. $\displaystyle\int_0^4 \frac{1}{x^2 + 1}\,dx, n = 8$ **24.** $\displaystyle\int_3^4 \frac{1}{x^2 - 1}\,dx, n = 3$

25. $\displaystyle\int_0^5 e^{\sqrt{x}}\,dx, n = 5$ **26.** $\displaystyle\int_1^3 e^{-\sqrt{x}}\,dx, n = 6$

27. $\displaystyle\int_0^4 \frac{1}{x^3 + 1}\,dx, n = 4$ **28.** $\displaystyle\int_1^6 \frac{x}{x^2 + 1}\,dx, n = 5$

In Exercises 29–36, use Simpson's Rule to find S_n using the indicated value of n. Give all answers to three decimal places.

29. $\displaystyle\int_2^4 \sqrt{x^2 - 1}\,dx, n = 4$ **30.** $\displaystyle\int_5^8 \sqrt{x^2 - 4}\,dx, n = 6$

31. $\displaystyle\int_{-1}^1 e^{-x^3}\,dx, n = 6$ **32.** $\displaystyle\int_0^1 e^{x^2}\,dx, n = 4$

33. $\displaystyle\int_1^3 \ln(x^2 + 1)\,dx, n = 6$ **34.** $\displaystyle\int_2^3 (1 + \ln x)^2\,dx, n = 4$

35. $\displaystyle\int_1^5 \frac{1}{\sqrt{x^2 + 1}}\,dx, n = 4$ **36.** $\displaystyle\int_{-1}^3 \frac{x}{\sqrt{x^3 + 2}}\,dx, n = 4$

TECHNOLOGY CONNECTION

37–52. Using algebra software or a graphing calculator, find, **(a)** the actual value of each definite integral in Exercises 21–36 to three decimal places and **(b)** the percent error between the approximation and the actual value of each integral.

APPLICATIONS

General Interest

53. Total distance. Moira drives her car for 8 minutes. Her speed, $v(t)$, in miles per hour, for 1-minute intervals, is shown below. Use L_8 and R_8 and their average to approximate the distance Moira travels to three decimal places. (*Hint:* 1 min $= \frac{1}{60}$ hr.)

t (min)	0	1	2	3	4	5	6	7	8
$v(t)$ (mi/hr)	0	25	30	35	30	22	20	10	10

54. Total distance. Walt goes for a 12-minute walk. His pedometer shows his speed, $v(t)$, in miles per hour, for 1-minute intervals. Use L_{12} and R_{12} and their average to approximate the distance Walt travels to three decimal places.

t (min)	0	1	2	3	4	5	6	7	8	9	10	11	12
$v(t)$ (mi/hr)	0	4	4	6	6	4	5	2	1	3	5	6	6

55. Cross-sectional area. A stream channel has the following profile, where the depths are in feet and measured at 2-foot intervals across the stream's width. Use the Trapezoidal Rule to approximate the cross-sectional area of the stream channel to two decimal places.

56. Surface area. The following diagram shows the distances across a pond, in feet, measured at 6-foot intervals. Use the Trapezoidal Rule to approximate the surface area of the pond to three decimal places.

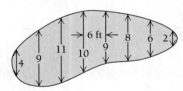

57. Area inside an ellipse. An ellipse is centered at the origin and has x-intercepts $(-4, 0)$ and $(4, 0)$ and y-intercepts $(0, -3)$ and $(0, 3)$. The portion of the ellipse in the first quadrant is given by $f(x) = \frac{3}{4}\sqrt{16 - x^2}$, for $0 \le x \le 4$.

  a) Approximate the area of the portion of the ellipse that is in the first quadrant by finding T_8 to three decimal places.

  b) Estimate the total area enclosed within the ellipse.

58. Area inside an ellipse. An ellipse is centered at the origin and has x-intercepts $(-5, 0)$ and $(5, 0)$ and y-intercepts $(0, -2)$ and $(0, 2)$. The portion of the ellipse in the first quadrant is given by $f(x) = \frac{2}{5}\sqrt{25 - x^2}$, for $0 \le x \le 5$.

  a) Approximate the area of the portion of the ellipse that is in the first quadrant by finding T_{10} to three decimal places.

  b) Estimate the total area enclosed within the ellipse.

For Exercises 59–62, the length of any differentiable curve $y = f(x)$ *over* $[a, b]$ *is given by* $\int_a^b \sqrt{1 + [f'(x)]^2}\, dx.$

59. Circumference of a circle. The portion of a unit circle that is in the first quadrant is given by $f(x) = \sqrt{1 - x^2}$, for $0 \le x \le 1$.

  a) Approximate the length of this curve, $\int_a^b \sqrt{1 + [f'(x)]^2}\, dx$, by finding M_6.

  b) Use a geometric formula to find the exact length of the curve.

  c) Find the percent error between the approximation from part (a) and the actual value from part (b).

60. Circumference of a circle. The portion of a circle of radius 2 that is in the first quadrant is given by $f(x) = \sqrt{4 - x^2}$, for $0 \le x \le 2$.

  a) Approximate the length of this curve, $\int_a^b \sqrt{1 + [f'(x)]^2}\, dx$, by finding M_8 to three decimal places.

  b) Use a geometric formula to find the exact length of the curve.

  c) Find the percent error between the approximation from part (a) and the actual value from part (b).

61. Length of a curve. Approximate the length of the curve $f(x) = e^x$ over $[-1, 2]$ by finding S_6 to three decimal places.

62. Length of a curve. Approximate the length of the curve $f(x) = 4 - x^2$ over $[-2, 2]$ by finding S_8 to three decimal places.

Business and Economics

63. Total cost. The back of a storage shed is in the shape shown below, where the heights (in feet) are given at 3-foot intervals. The cost to install metal sheeting over this surface is $6 per square foot. Approximately how much will it cost to cover the back of the shed with metal sheeting? Assume that the shed is symmetrical.

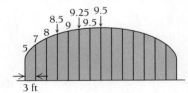

64. Total cost. The shape of a wall in a museum is shown below, where heights (in feet) are given at 6-foot intervals. The wall is to be paneled, and the cost to panel each square foot is $3.50. Find the approximate cost of paneling the wall.

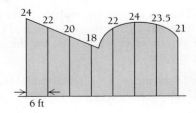

65. Total cost to maintain a green. The 17th green (called the "Island Green") at the Sawgrass TPC Golf Course in Ponte Vedra Beach, Florida, has the measurements shown below. (*Sources:* www.pgagolfd.com and www.yourgolftravel.com.)

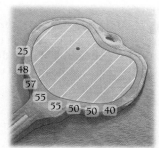

25
48
57
55
55 50 50 40

Lines spaced 10 feet apart

a) Use a numerical integration technique to approximate the area of the green.

b) If the cost to maintain the 17th green is $2.75 per square foot, what is the approximate total cost of maintaining the green?

66. Total cost to repave a parking lot. The back parking lot at Sun Devil Stadium in Tempe, Arizona, has the shape shown below, with lengths in feet, spaced every 50 feet. The average cost of repaving a parking lot using asphalt is about $3 per square foot. Find the approximate cost of repaving this parking lot using asphalt. (*Sources:* www.mapquest.com and www.durbinbrothers.net.)

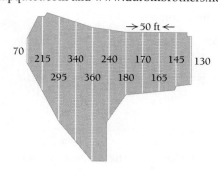

→ 50 ft ←

70
215 340 240 170 145 130
295 360 180 165

SYNTHESIS

67. Suppose $y = f(x)$ is differentiable, increasing, and concave up on $[a, b]$. Fill in each blank with $<$ or $>$.

a) L_n ____ $\int_a^b f(x)\,dx$ b) R_n ____ $\int_a^b f(x)\,dx$

c) T_n ____ $\int_a^b f(x)\,dx$

68. Suppose $y = f(x)$ is differentiable, decreasing, and concave up on $[a, b]$. Fill in each blank with $<$ or $>$.

a) L_n ____ $\int_a^b f(x)\,dx$ b) R_n ____ $\int_a^b f(x)\,dx$

c) T_n ____ $\int_a^b f(x)\,dx$

69. Suppose $y = f(x)$ is differentiable, increasing, and concave down on $[a, b]$. Fill in each blank with $<$ or $>$.

a) L_n ____ $\int_a^b f(x)\,dx$ b) R_n ____ $\int_a^b f(x)\,dx$

c) T_n ____ $\int_a^b f(x)\,dx$

70. Suppose $y = f(x)$ is differentiable, decreasing, and concave down on $[a, b]$. Fill in each blank with $<$ or $>$.

a) L_n ____ $\int_a^b f(x)\,dx$ b) R_n ____ $\int_a^b f(x)\,dx$

c) T_n ____ $\int_a^b f(x)\,dx$

71. Explain why the Trapezoidal Rule always gives an exact value for $\int_a^b (Cx + D)\,dx$.

72. Explain why Simpson's Rule always gives an exact value for $\int_a^b (Cx^2 + Dx + E)\,dx$.

73. The following is an outline of a proof of Simpson's Rule for the special case where $n = 2$ with f continuous over $[-1, 1]$. Here, $a = -1, b = 1$, and $\frac{a + b}{2} = 0$. We will show that

$$\int_{-1}^1 f(x)\,dx \approx \frac{1}{3} \cdot (f(-1) + 4f(0) + f(1)).$$

a) Let $(-1, y_1)$, $(0, y_2)$, and $(1, y_3)$ be three points, where $y_1 = f(-1), y_2 = f(0)$, and $y_3 = f(1)$, and let $y = Ax^2 + Bx + C$ be a parabola that passes through these three points. Find values for A, B and C in terms of y_1, y_2, and y_3. (*Hint:* Evaluate $y = Ax^2 + Bx + C$ at each point. You should be able to solve for C immediately. Then substitute and find expressions for A and B.)

b) Evaluate $\int_{-1}^1 (Ax^2 + Bx + C)\,dx$. Your answer will be in terms of y_1, y_2, and y_3. Write each term over a single common denominator.

74. Assume that a function f is continuous over the interval $[-1, 3]$. Using a method similar to that in Exercise 73, we can show that $\int_1^3 f(x)\,dx \approx \frac{1}{3} \cdot (f(1) + 4f(2) + f(3))$. Use the fact that $\int_{-1}^1 f(x)\,dx \approx \frac{1}{3} \cdot (f(-1) + 4f(0) + f(1))$ (from Exercise 73) to show that

$$\int_{-1}^3 f(x)\,dx \approx \frac{1}{3} \cdot (f(-1) + 4f(0) + 2f(1) + 4f(2) + f(3)),$$

where $n = 4$.

5.5

Volume

Consider the graph of $y = f(x)$ in Fig. 1. If the upper half-plane is rotated about the x-axis, then each point on the graph has a circular path, and the whole graph sweeps out a certain surface, called a *surface of revolution*.

The plane region bounded by the graph, the x-axis, $x = a$, and $x = b$ sweeps out a *solid of revolution*. To calculate the volume of this solid, we first approximate it as a finite sum of thin right circular cylinders, or disks (Fig. 2). We divide the interval $[a, b]$ into equal subintervals, each of length Δx. Thus, the height h of each disk is Δx (Fig. 3). The radius of each disk is $f(x_i)$, where x_i is the right-hand endpoint of the subinterval that determines that disk. If $f(x_i)$ is negative, we can use $|f(x_i)|$.

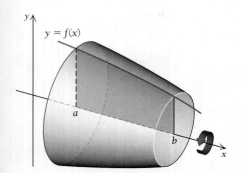

FIGURE 1

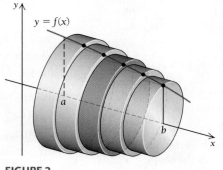

FIGURE 2

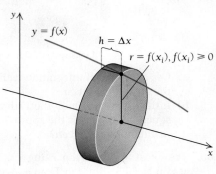

FIGURE 3

Since the volume of a right circular cylinder is given by

$$V = \pi r^2 h, \quad \text{or} \quad \text{Volume} = \text{area of the base} \cdot \text{height},$$

each of the approximating disks has the volume

$$\pi |f(x_i)|^2 \, \Delta x = \pi [f(x_i)]^2 \, \Delta x. \qquad \text{Squaring makes the use of absolute value unnecessary.}$$

The volume of the solid of revolution is approximated by the sum of the volumes of all the disks:

$$V \approx \sum_{i=1}^{n} \pi [f(x_i)]^2 \, \Delta x.$$

The actual volume is the limit as the thickness of the disks approaches zero, or the number of disks approaches infinity:

$$V = \lim_{n \to \infty} \sum_{i=1}^{n} \pi [f(x_i)]^2 \, \Delta x = \int_{a}^{b} \pi [f(x)]^2 \, dx.$$

(See Section 4.2.) That is, the volume is the value of the definite integral of the function $y = \pi [f(x)]^2$ from a to b.

THEOREM 2

For a continuous function f defined on $[a, b]$, the **volume, V, of the solid of revolution** obtained by rotating the area under the graph of f from a to b about the x-axis is given by

$$V = \int_{a}^{b} \pi [f(x)]^2 \, dx.$$

Explain how this could be interpreted as a solid of revolution.

> **Quick Check 1**
>
> Find the volume of the solid of revolution generated by rotating the region under the graph of $y = x^3$ from $x = 0$ to $x = 2$ about the x-axis.

■ **EXAMPLE 1**   Find the volume of the solid of revolution generated by rotating the region under the graph of $y = \sqrt{x}$ from $x = 0$ to $x = 1$ about the x-axis.

Solution

$$V = \int_0^1 \pi [f(x)]^2 \, dx$$

$$= \int_0^1 \pi [\sqrt{x}]^2 \, dx$$

$$= \int_0^1 \pi x \, dx$$

$$= \pi \left[\frac{x^2}{2} \right]_0^1$$

$$= \frac{\pi}{2} \left[x^2 \right]_0^1$$

$$= \frac{\pi}{2}(1^2 - 0^2) = \frac{\pi}{2}$$

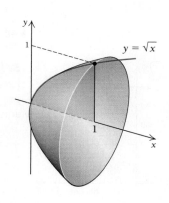

❰ Quick Check 1

■ **EXAMPLE 2**   Find the volume of the solid of revolution generated by rotating the region under the graph of

$$y = e^x$$

from $x = -1$ to $x = 2$ about the x-axis.

Solution

$$V = \int_{-1}^2 \pi [f(x)]^2 \, dx$$

$$= \int_{-1}^2 \pi [e^x]^2 \, dx$$

$$= \int_{-1}^2 \pi e^{2x} \, dx$$

$$= \left[\frac{\pi}{2} e^{2x} \right]_{-1}^2$$

$$= \frac{\pi}{2} \left[e^{2x} \right]_{-1}^2$$

$$= \frac{\pi}{2}(e^{2 \cdot 2} - e^{2(-1)})$$

$$= \frac{\pi}{2}(e^4 - e^{-2}) \approx 85.55$$

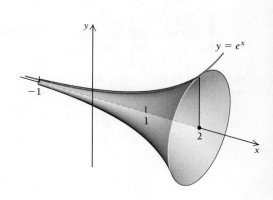

> **Quick Check 2**
>
> Find the volume of the solid of revolution generated by rotating the region under the graph of $y = \dfrac{1}{x}$ from $x = 1$ to $x = 3$ about the x-axis.

❰ Quick Check 2

■ EXAMPLE 3   Business: Water Storage.

A city's water storage tank is in the shape of the solid of revolution generated by rotating the region under the graph of

$$f(x) = 50\sqrt{1 - \frac{x^2}{40^2}}$$

from $x = -40$ ft to $x = 40$ ft about the x-axis. What is the volume of this tank?

Solution   The tank's shape is called an *oblate spheroid*: its vertical diameter (80 ft) is less than its horizontal diameter (100 ft). The graph of f is shown at left. Rotating the graph of f about the x-axis gives the shape of the tank, but it is standing on end. The actual tank has this shape turned on its side.

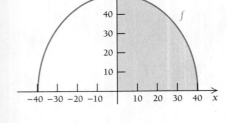

If we rotate the portion of the graph in the first quadrant, that is, from $x = 0$ to $x = 40$, we will get half of the solid. This has the advantage of using 0 as a bound of integration. We then multiply the result by 2 to determine the whole volume.

The volume for $0 \leq x \leq 40$ is

$$V = \int_0^{40} \pi \left[50\sqrt{1 - \frac{x^2}{40^2}} \right]^2 dx$$

$$= \int_0^{40} \pi \left[2500\left(1 - \frac{x^2}{1600}\right) \right] dx \qquad \text{$50^2 = 2500$, and the radical disappears due to squaring.}$$

$$= \int_0^{40} \pi \left[2500 - \frac{25}{16}x^2 \right] dx \qquad \text{Multiplying through by 2500}$$

$$= \pi \left[2500x - \frac{25}{48}x^3 \right]_0^{40} \qquad \text{Antidifferentiating}$$

$$= \frac{200{,}000}{3}\, \pi.$$

> ### Quick Check 3
>
> A tepee is a cone with a height of 15 ft at its center and a circular base with a radius of 8 ft. Determine the volume contained within this tepee. (*Hint*: Rotate the line $y = \frac{8}{15}x$ from $x = 0$ to $x = 15$ about the x-axis.)

Multiplying this result by 2 gives the tank's entire volume:

$$\frac{400{,}000}{3}\, \pi \approx 418{,}879\ \text{ft}^3.$$

Since 1 ft^3 holds 7.48 gal, this tank holds over 3.13 million gallons of water.

❮ Quick Check 3

Section Summary

• If a function f is continuous over an interval $[a, b]$, then the volume of the solid formed by rotating the area under the graph of f from a to b about the x-axis is given by

$$V = \int_a^b \pi [f(x)]^2\, dx.$$

EXERCISE SET
5.5

Find the volume generated by rotating about the x-axis the region bounded by the graphs of each set of equations.

1. $y = x, x = 0, x = 1$

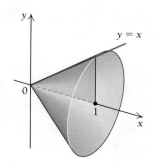

2. $y = x, x = 0, x = 2$

3. $y = \sqrt{x}, x = 1, x = 4$

4. $y = 2x, x = 1, x = 3$

5. $y = e^x, x = -2, x = 5$

6. $y = e^x, x = -3, x = 2$

7. $y = \dfrac{1}{x}, x = 1, x = 3$

8. $y = \dfrac{1}{x}, x = 1, x = 4$

9. $y = \dfrac{2}{\sqrt{x}}, x = 4, x = 9$

10. $y = \dfrac{1}{\sqrt{x}}, x = 1, x = 4$

11. $y = 4, x = 1, x = 3$

12. $y = 5, x = 1, x = 3$

13. $y = x^2, x = 0, x = 2$

14. $y = x + 1, x = -1, x = 2$

15. $y = \sqrt{1 + x}, x = 2, x = 10$

16. $y = 2\sqrt{x}, x = 1, x = 2$

17. $y = \sqrt{4 - x^2}, x = -2, x = 2$

18. $y = \sqrt{r^2 - x^2}, x = -r, x = r$ (assume $r > 0$)

APPLICATIONS

19. Cooling tower volume. Cooling towers at nuclear power plants have a "pinched" chimney shape (which promotes cooling within the tower) formed by rotating a hyperbola around an axis. The function

$$y = 50\sqrt{1 + \frac{x^2}{22,500}}, \quad \text{for} -250 \le x \le 150,$$

where x and y are in feet, describes the shape of such a tower (laying on its side). Determine the volume of the tower by rotating the region bounded by the graph of y about the x-axis. (*Hint:* See Example 3.)

20. Volume of a football. A regulation football used in the National Football League is 11 in. from tip to tip and 7 in. in diameter at its thickest (the regulations allow for slight variation in these dimensions). (*Source:* NFL.) The shape of a football can be modeled by the function

$$f(x) = -0.116x^2 + 3.5, \quad \text{for} -5.5 \le x \le 5.5,$$

where x is in inches. Find the volume of the football by rotating the region bounded by the graph of f about the x-axis.

SYNTHESIS

21. Graph $y = \sqrt{4 - x^2}$ and $y = \sqrt{r^2 - x^2}$, with $r > 0$, and explain how the results can be used to calculate the volume of a common shape. (See Exercises 17 and 18.)

22. Prove that the volume of a right circular cone of height h and radius r is $V = \frac{1}{3}\pi r^2 h$. (*Hint:* Rotate a line starting at the origin and ending at the point (h, r) about the x-axis.)

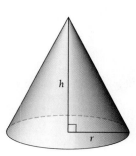

Find the volume generated by rotating about the x-axis the region bounded by the graphs of each set of equations.

23. $y = \sqrt{\ln x}, x = e, x = e^3$

24. $y = \sqrt{xe^{-x}}, x = 1, x = 2$

25. Consider the function $y = 1/x$ over the interval $[1, \infty)$. We showed in Section 5.3 that the area under the curve does not exist; that is,

$$\int_1^\infty \frac{1}{x}\, dx$$

diverges. Find the volume of the solid of revolution formed by rotating the region under the graph of $y = 1/x$ over the interval $[1, \infty)$ about the x-axis. That is, find

$$\int_1^\infty \pi \left[\frac{1}{x}\right]^2 dx.$$

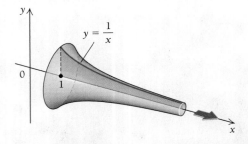

This solid is sometimes referred to as *Gabriel's horn*.

26. Paradox of Gabriel's horn or the infinite paint can. Though we cannot prove it here, the surface area of Gabriel's horn (see Exercise 25) is given by

$$S = \int_1^\infty \frac{2\pi}{x} \sqrt{1 + \frac{1}{x^4}}\, dx.$$

Show that the surface area of Gabriel's horn does not exist. The paradox is that the volume of the horn exists, but the surface area does not. This is like a can of paint that has a finite volume but, when full, does not hold enough paint to paint the outside of the can.

Answers to Quick Checks

1. $\dfrac{128\pi}{7}$ 2. $\dfrac{2\pi}{3}$ 3. 320π ft^3

KEY TERMS AND CONCEPTS

SECTION 5.1

If $p = D(x)$ is a demand function, then the **consumer surplus** at a point (Q, P) is

$$\int_0^Q D(x)\, dx - QP.$$

If $p = S(x)$ is a supply function, then the **producer surplus** at a point (Q, P) is

$$QP - \int_0^Q S(x)\, dx.$$

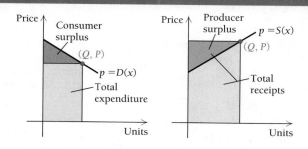

The **equilibrium point**, (x_E, p_E), is the point at which the supply and demand curves intersect.

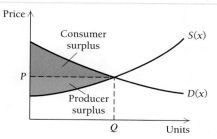

Let $p = 12 - 1.5x$ be a demand function and $p = 4 + 0.5x$ be a supply function. The two curves intersect at $(4, 6)$, the equilibrium point. At this point, the consumer surplus is

$$\int_0^4 (12 - 1.5x)\, dx - (4)(6) = 36 - 24 = \$12,$$

and the producer surplus is

$$(4)(6) - \int_0^4 (4 + 0.5x)\, dx = 24 - 20 = \$4.$$

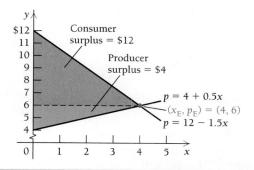

(*continued*)

KEY TERMS AND CONCEPTS	EXAMPLES

SECTION 5.2

The **future value** of P_0 dollars invested at an interest rate k for t years, compounded continuously, is given by $P = P_0 e^{kt}$.

The future value of $6000 invested at 6.75%, compounded continuously, for 5 yr is

$$P = 6000e^{0.0675(5)} = \$8408.64.$$

The amount P_0 is called the **present value**. If the future value P is known, then $P_0 = Pe^{-kt}$.

Sue wants to have $15,000 in 4 yr to make a down payment on a house. She opens a savings account that offers 4.5% interest, compounded continuously. The present value is the amount she needs to deposit now to have $15,000 in 4 yr:

$$P = 15{,}000e^{-0.045(4)} = \$12{,}529.05$$

The **accumulated future value of a continuous income stream** is given by

$$A = e^{kT}\int_0^T R(t)e^{-kt}\,dt,$$

where $R(t)$ is the rate of the continuous income stream, k is the interest rate, and T is the number of years. If $R(t)$ is a constant function, then

$$A = \frac{R(t)}{k}\left(e^{kT} - 1\right).$$

A baseball pitcher signs an $8,000,000 6-year contract and will be paid $1,150,000 per year. The money will be invested at 5%, compounded continuously, for the 6-yr term. The accumulated future value is

$$A = \frac{1{,}150{,}000}{0.05}\left(e^{0.05(6)} - 1\right) = \$8{,}046{,}752.57.$$

The **accumulated present value of a continuous income stream** is given by

$$B = \int_0^T R(t)e^{-kt}\,dt.$$

If $R(t)$ is a constant function, then

$$B = \frac{R(t)}{k}\left(1 - e^{-kT}\right).$$

If $R(t)$ is not a constant function, the definite integral must be evaluated with an appropriate integration technique.

The accumulated present value of the pitcher's contract is

$$B = \int_0^6 1{,}150{,}000e^{-0.05t}\,dt$$

$$= \frac{1{,}150{,}000}{0.05}\left(1 - e^{-0.05(6)}\right) = \$5{,}961{,}180.92.$$

Consumption of a natural resource can be modeled by

$$\int_0^T P_0 e^{kt}\,dt = \frac{P_0}{k}\left(e^{kt} - 1\right),$$

where $P(t) = P_0 e^{kt}$ is the annual consumption of the natural resource in year t and consumption is growing exponentially at a growth rate k.

Canada's diamond mines produce diamonds according to the model $P(t) = 2.5e^{0.272t}$, where $t = 0$ is 2000 and $P(t)$ is in millions of carats. (*Source*: USGS *Mineral Commodities Summaries*.) Using this model, we can forecast the total production of diamonds between 2000 and 2012:

$$\int_0^{12} 2.5e^{0.272t}\,dt = \frac{2.5}{0.272}\left(e^{0.272(12)} - 1\right) = 231.2 \text{ million carats.}$$

KEY TERMS AND CONCEPTS

EXAMPLES

SECTION 5.3

An integral with infinity as a bound is called an **improper integral**. All improper integrals are evaluated as limits:

$$\int_a^\infty f(x)\,dx = \lim_{b\to\infty} \int_a^b f(x)\,dx,$$

$$\int_{-\infty}^b f(x)\,dx = \lim_{a\to-\infty} \int_a^b f(x)\,dx.$$

If the limit exists, the improper integral is **convergent**. Otherwise, it is **divergent**.

$$\int_1^\infty \frac{1}{x^3}\,dx = \lim_{b\to\infty} \int_1^b \frac{1}{x^3}\,dx$$

$$= \lim_{b\to\infty}\left[-\frac{1}{2x^2}\right]_1^b$$

$$= \lim_{b\to\infty}\left[-\frac{1}{2(b)^2} - \left(-\frac{1}{2(1)^2}\right)\right]$$

$$= \frac{1}{2}$$

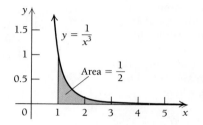

If both bounds are infinity, then the improper integral can be written as the sum of two integrals, where c is any real number:

$$\int_{-\infty}^\infty f(x)\,dx = \int_{-\infty}^c f(x)\,dx + \int_c^\infty f(x)\,dx.$$

$$\int_{-\infty}^0 e^{4x}\,dx = \lim_{a\to-\infty} \int_a^0 e^{4x}\,dx$$

$$= \lim_{a\to-\infty}\left[\tfrac{1}{4}e^{4x}\right]_a^0$$

$$= \lim_{a\to-\infty}\left[\tfrac{1}{4}\left(e^{4(0)} - e^{4(a)}\right)\right]$$

$$= \frac{1}{4}$$

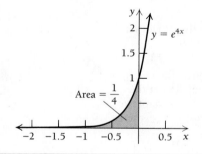

The **accumulated present value** of a continuous money flow into an investment at the rate of P dollars per year perpetually is

$$\int_0^\infty Pe^{-kt}\,dt = \frac{P}{k},$$

where k is the continuously compounded interest rate.

An investment of \$5000 per year perpetually at 7%, compounded continuously, has a present value of $\dfrac{5000}{0.07} = 71{,}428.57$.

(continued)

KEY TERMS AND CONCEPTS	EXAMPLES

SECTION 5.4

Numerical integration is a process that uses geometry to determine the approximate value of a definite integral, $\int_a^b f(x)\,dx$. The interval $[a, b]$ is first subdivided into n equal subintervals, $[a, x_1], [x_1, x_2], \ldots [x_{n-1}, b]$, each with width $\Delta x = \dfrac{b - a}{n}$.

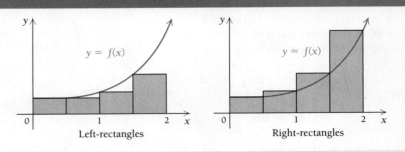

Left-rectangles   Right-rectangles

With the method of **Riemann sums**, rectangles drawn above each subinterval are used to approximate the area. With *left-rectangles*, the height of each rectangle is the value of $f(x)$ at the left endpoint of that subinterval. The sum of the areas of all n rectangles is denoted as L_n, where

$$L_n = \frac{b - a}{n} \cdot (f(a) + f(x_1) + f(x_2) + \cdots + f(x_{n-2}) + f(x_{n-1})).$$

With *right-rectangles*, the height of each rectangle is the value of $f(x)$ at the right endpoint of that subinterval. The sum of the areas of all n rectangles is denoted as R_n, where

$$R_n = \frac{b - a}{n} \cdot (f(x_1) + f(x_2) + \cdots + f(x_{n-2}) + f(x_{n-1}) + f(b)).$$

The average of L_n and R_n usually provides a better approximation of $\int_a^b f(x)\,dx$.

Let $\int_0^2 \sqrt{1 + x^4}\,dx$, with $n = 4$. Divide $[0, 2]$ into four subintervals, $[0, 0.5], [0.5, 1], [1, 1.5]$, and $[1.5, 2]$. Each subinterval has a width $\Delta x = \dfrac{2 - 0}{4} = \dfrac{1}{2}$. Therefore,

$$\begin{aligned} L_4 &= \Delta x \cdot (f(0) + f(0.5) + f(1) + f(1.5)) \\ &= \tfrac{1}{2} \cdot (1 + 1.031 + 1.414 + 2.462) \\ &= 2.954, \end{aligned}$$

and

$$\begin{aligned} R_4 &= \Delta x \cdot (f(0.5) + f(1) + f(1.5) + f(2)) \\ &= \tfrac{1}{2} \cdot (1.031 + 1.414 + 2.462 + 4.123) \\ &= 4.515. \end{aligned}$$

The average of L_4 and R_4, $\dfrac{2.954 + 4.515}{2} = 3.735$, is an approximate value of $\int_0^2 \sqrt{1 + x^4}\,dx$.

Midpoint Rule: Using rectangles, the height of each rectangle is the value of $f(x)$ at the midpoint of that subinterval. The sum of the areas of all n rectangles is denoted M_n, where

$$\begin{aligned} M_n = \frac{b - a}{n} \cdot \Bigl(& f\Bigl(\frac{a + x_1}{2}\Bigr) + f\Bigl(\frac{x_1 + x_2}{2}\Bigr) \\ & + f\Bigl(\frac{x_2 + x_3}{2}\Bigr) + \cdots + f\Bigl(\frac{x_{n-2} + x_{n-1}}{2}\Bigr) \\ & + f\Bigl(\frac{x_{n-1} + b}{2}\Bigr) \Bigr). \end{aligned}$$

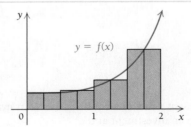

With $\int_0^2 \sqrt{1 + x^4}\,dx$ and $n = 4$, we have

$$\begin{aligned} M_4 &= \tfrac{1}{2} \cdot (f(0.25) + f(0.75) + f(1.25) + f(1.75)) \\ &= \tfrac{1}{2} \cdot (1.002 + 1.147 + 1.855 + 3.222) \\ &= 3.613. \end{aligned}$$

KEY TERMS AND CONCEPTS

EXAMPLES

Trapezoidal Rule: Trapezoids are drawn above each subinterval, and the definite integral is approximated by T_n, where

$$T_n = \frac{b-a}{n} \cdot \left(\frac{f(a)}{2} + f(x_1) + \right.$$

$$\left. f(x_2) + \cdots + f(x_{n-1}) + \frac{f(b)}{2} \right).$$

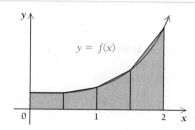

With $\int_0^2 \sqrt{1+x^4}\, dx$ and $n = 4$, we have

$$T_4 = \frac{1}{2} \cdot \left(\frac{f(0)}{2} + f(0.5) + f(1) + f(1.5) + \frac{f(2)}{2} \right)$$

$$= \frac{1}{2} \cdot \left(\frac{1}{2} + 1.031 + 1.414 + 2.462 + \frac{4.123}{2} \right)$$

$$= 3.734.$$

Simpson's Rule: The number of subintervals n is even, and $n/2$ parabolas are used to approximate the curve. The definite integral is approximated by S_n, where

$$S_n = \frac{b-a}{3n} \cdot (f(a) + 4f(x_1) + 2f(x_2)$$

$$+ 4f(x_3) + \cdots + 2f(x_{n-2}) + 4f(x_{n-1})$$

$$+ f(b)).$$

With $\int_0^2 \sqrt{1+x^4}\, dx$, and $n = 4$, we have

$$S_4 = \frac{1/2}{3} \cdot (f(0) + 4f(0.5) + 2f(1) + 4f(1.5) + f(2))$$

$$= \frac{1}{6} \cdot (1 + 4(1.031) + 2(1.414) + 4(2.462) + 4.123)$$

$$= 3.654.$$

The percent error between an approximation and the actual value of a definite integral is given by

Percent error =

$$\frac{\text{approximation} - \text{actual value}}{\text{actual value}} \times 100.$$

Algebra software shows that, to three decimal places, $\int_0^2 \sqrt{1+x^4}\, dx = 3.653$. Using the Midpoint Rule, $M_4 = 3.613$. Therefore, the percent error of this approximation is

$$\frac{3.613 - 3.653}{3.653} \times 100 \approx -1.09\%.$$

We obtained the approximation $S_4 = 3.654$ using Simpson's Rule. Its percent error is

$$\frac{3.654 - 3.653}{3.653} \times 100 \approx 0.03\%.$$

SECTION 5.5

If f is continuous over an interval $[a, b]$, the **volume of the solid of rotation** formed by rotating the area under the graph of f from a to b about the x-axis is given by

$$V = \int_a^b \pi [f(x)]^2 \, dx.$$

The volume of the solid formed by rotating the graph of $f(x) = \frac{1}{4} x^2$ from $x = -1$ to $x = 3$ about the x-axis is

$$V = \int_{-1}^3 \pi \left(\frac{1}{4} x^2 \right)^2 dx = \int_{-1}^3 \pi \left(\frac{1}{16} x^4 \right) dx = \frac{\pi}{16} \left[\frac{x^5}{5} \right]_{-1}^3$$

$$= \frac{\pi}{16} \left[\frac{243}{5} - \left(-\frac{1}{5} \right) \right] = \frac{244}{80} \pi = \frac{61}{20} \pi.$$

These review exercises are for test preparation. They can also be used as a practice test. Answers are at the back of the book. The blue bracketed section references tell you what part(s) of the chapter to restudy if your answer is incorrect.

CONCEPT REINFORCEMENT

Match each term in column A with the most appropriate graph in column B.

Column A   Column B

1. Consumer surplus [5.1]

   a)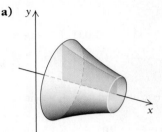

2. Producer surplus [5.1]

   b)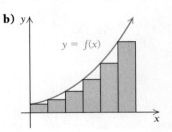
   $y = f(x)$

3. Midpoint Rule [5.4]

   c)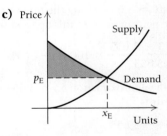
   Price, Supply, p_E, Demand, x_E, Units

4. Solid of revolution [5.5]

   d)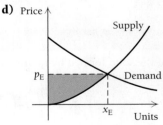
   Price, Supply, p_E, Demand, x_E, Units

5. Riemann sum with left-rectangles [5.4]

   e)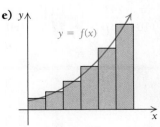
   $y = f(x)$

Classify each statement as either true or false.

6. The equilibrium point is the point at which the supply and demand curves intersect. [5.1]

7. The accumulated present value of an investment is the value of the investment as a tax-deductible present to a nonprofit charity. [5.2]

8. If an integral has $-\infty$ or ∞ as one of the limits of integration, it is an improper integral. [5.3]

9. An approximation of a definite integral by a Riemann sum using right-rectangles always gives an overestimate of the true area. [5.4]

10. Simpson's Rule requires that the interval $[a, b]$ be divided into an even number of subintervals. [5.4]

11. The Trapezoidal Rule always gives an exact answer if the integrand is linear. [5.4]

12. To find the volume of the solid of revolution obtained by rotating the graph of $y = f(x)$ about the x-axis, we must have $f(x) \geq 0$. [5.5]

REVIEW EXERCISES

Let $D(x) = (x - 6)^2$ be the price, in dollars per unit, that consumers are willing to pay for x units of an item, and $S(x) = x^2 + 12$ be the price, in dollars per unit, that producers are willing to accept for x units.

13. Find the equilibrium point. [5.1]

14. Find the consumer surplus at the equilibrium point. [5.1]

15. Find the producer surplus at the equilibrium point. [5.1]

16. Business: future value. Find the future value of $5000, at an annual percentage rate of 5.2%, compounded continuously, for 7 yr. [5.2]

17. Business: present value. Find the present value of $10,000 due in 5 yr, at an interest rate of 8.3%, compounded continuously. [5.2]

18. Business: future value of a continuous income stream. Find the accumulated future value of $2500 per year, at 6.25% compounded continuously, for 8 yr. [5.2]

19. Business: present accumulated value of a trust. The DeMars family welcomes a new baby, and the parents want to have $250,000 in 18 yr for their child's college education. Find the continuous money stream, at $R(t)$ dollars per year, that they need to invest at 5.75% compounded continuously, to generate $250,000. [5.2]

20. Business: early retirement. Cal Earl signs a 7-yr contract as a session drummer for a major recording company. His contract gives him a salary of $150,000

per year. After 3 yr, the company offers to buy out the remainder of his contract. What is the least amount Cal should accept, if the going interest rate is 6.15%, compounded continuously? [5.2]

21. Physical science: iron ore consumption. In 2010 $(t = 0)$, the world production of iron ore was estimated at 2.4 billion metric tons, and production was growing exponentially at the rate of 3% per year. (*Source:* U.S. Energy Information Administration.) If the production continues to grow at this rate, how much iron ore will be produced from 2010 to 2020? [5.2]

22. Physical science: depletion of iron ore. The world reserves of iron ore in 2010 were estimated to be 180 billion metric tons. (*Source:* U.S. Geological Survey.) Assuming that the growth rate in Exercise 21 continues and no new reserves are discovered, when will the world reserves of iron ore be depleted? [5.2]

Determine whether each improper integral is convergent or divergent, and calculate its value if it is convergent. [5.3]

23. $\int_1^\infty \frac{1}{x^2}\, dx$

24. $\int_1^\infty e^{4x}\, dx$

25. $\int_0^\infty e^{-2x}\, dx$

26. Approximate the value of $\int_0^2 \sqrt{2x^2 + x}\, dx$ by finding L_4, R_4, and their average. [5.4]

27. Approximate the value of $\int_1^3 \sqrt{x^2 + 3x}\, dx$ by finding M_4. [5.4]

28. Approximate the value of $\int_1^4 \ln(x^3 + 2)\, dx$ by finding T_6. [5.4]

29. Approximate the value of $\int_2^4 \sqrt[3]{x^2 + 1}\, dx$ by finding S_6. [5.4]

30. Consider the definite integral $\int_{-1}^1 (x^2 + x)\, dx$. [5.4]
  a) Evaluate the integral using an antiderivative.
  b) Approximate its value by finding T_4.
  c) Find the percent error in the approximation of part (b).

31. Total distance. Eduardo runs for 10 minutes. His pedometer tracks his speed, $v(t)$, in miles per hour, over 1-minute intervals. Use left- and right-rectangles and their average to approximate the total distance Eduardo runs. (*Hint:* 1 min = $\frac{1}{60}$ hr.) [5.4]

1-minute intervals, t	1	2	3	4	5	6	7	8	9	10
Speed, $v(t)$, in mi/hr	9	10	10	8	8	7	6	6	5	3

32. Business: resodding a green. The following diagram shows the widths (in feet) of a golf green, measured at 5-foot intervals. If sod costs $4.25 per square foot, find the approximate cost of resodding this green.

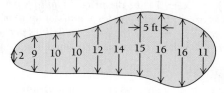

Find the volume generated by rotating about the x-axis the region bounded by the graphs of the given equations. [5.5]

33. $y = x^3, x = 1, x = 2$

34. $y = \frac{1}{x + 2}, x = 0, x = 1$

35. A water tank is formed by revolving about the x-axis the graph of the parabola $y = 16 - x^2$, where $-3 \le x \le 3$ and x and y are measured in feet. Find the volume of the water tank. [5.5]

36. A decorative container is formed by revolving about the x-axis the graph of $y = \sqrt{0.5x}$, where $2 \le x \le 18$ and x and y are measured in inches. Find the volume of the container. [5.5]

SYNTHESIS

Determine whether each improper integral is convergent or divergent, and calculate its value if it is convergent. [5.3]

37. $\int_{-\infty}^0 x^4 e^{-x^5}\, dx$

38. $\int_0^\infty \frac{dx}{(x + 1)^{4/3}}$

TECHNOLOGY CONNECTION

Approximate each integral.

39. $\int_1^\infty \frac{\ln x}{x^2}\, dx.$ [5.3]

40. $\int_0^\infty \frac{1}{x^3 + 1}\, dx.$ [5.3]

Let $D(x) = (x - 7)^2$ be the price, in dollars per unit, that consumers are willing to pay for x units of an item, and let $S(x) = x^2 + x + 4$ be the price, in dollars per unit, that producers are willing to accept for x units. Find:

1. The equilibrium point

2. The consumer surplus at the equilibrium point

3. The producer surplus at the equilibrium point

4. Business: future value. Find the future value of $12,000 invested for 10 yr at an annual percentage rate of 4.1%, compounded continuously.

5. Business: future value of a continuous income stream. Find the accumulated future value of $8000 per year, at an interest rate of 4.88%, compounded continuously, for 6 yr.

6. Physical science: demand for potash. In 2010 $(t = 0)$, the world production of potash was approximately 33 million metric tons, and demand was increasing at the rate of 9.9% a year. (*Source:* U.S. Energy Information Administration.) If the demand continues to grow at this rate, how much potash will be produced from 2010 to 2020?

7. Physical science: depletion of potash. See Exercise 6. The world reserves of potash in 2010 were approximately 9.5 billion metric tons. (*Source:* U.S. Geological Survey.) Assuming the demand for potash continues to grow at the rate of 9.9% per year and no new reserves are discovered, when will the world reserves be depleted?

8. Business: accumulated present value of a continuous income stream. Bruce Kent wants to have $25,000 in 5 yr for a down payment on a house. Find the amount he needs to save, at $R(t)$ dollars per year, at 6.125%, compounded continuously, to achieve the desired future value.

9. Business: contract buyout. Guy Laplace signs a 6-yr contract to play professional hockey at a salary of $475,000 per year. After 2 yr, his team offers to buy out the remainder of his contract. What is the least amount Guy should accept, if the going interest rate is 7.1%, compounded continuously?

10. Business: future value of a noncontinuous income stream. Stan signs a contract that will pay him an income given by $R(t) = 100,000 + 10,000t$, where t is in years and $0 \leq t \leq 8$. If he invests this money at 5%, compounded continuously, what is the future value of the income stream?

Determine whether each improper integral is convergent or divergent, and calculate its value if it is convergent.

11. $\displaystyle\int_1^\infty \frac{dx}{x^5}$

12. $\displaystyle\int_0^\infty \frac{4}{1 + 3x}\,dx$

13. Approximate the value of $\int_1^2 e^{(x-1)^2}\,dx$ by finding M_4.

14. Approximate the value of $\int_1^4 x^{\sqrt{x}}\,dx$ by finding T_6.

15. Approximate the value of $\int_0^2 \ln(1 + e^x)\,dx$ by finding S_4.

16. Consider the definite integral $\int_0^3 (3x^2 + 2x)\,dx$.
    a) Evaluate the integral using an antiderivative.
    b) Approximate its value by finding T_6.
    c) Find the percent error in the approximation of part (b).

17. General interest: treadmill. Tobias runs on a treadmill for 8 min. The calories per hour, $c(t)$, that he burns over 1-minute intervals are given below. Use left- and right-rectangles and their average to approximate the total number of calories Tobias expends on the treadmill.

1-minute intervals, t	1	2	3	4	5	6	7	8
Calories per hour, $c(t)$	40	48	50	50	45	43	40	35

Find the volume generated by rotating about the x-axis the region bounded by each of the following.

18. $y = \dfrac{1}{\sqrt{x}}, \quad x = 1, x = 5$

19. $y = \sqrt{2 + x}, \quad x = 0, x = 1$

SYNTHESIS

20. Determine whether the following improper integral is convergent or divergent, and calculate its value if it is convergent:
$$\int_{-\infty}^0 x^3 e^{-x^4}\,dx.$$

TECHNOLOGY CONNECTION

21. Approximate the integral
$$\int_{-\infty}^\infty \frac{1}{1 + x^2}\,dx.$$

Curve Fitting and Volumes of Containers

Consider the urn or vase shown at the right. How could we estimate the volume? One way would be to simply fill the container with a liquid and then pour the liquid into a measuring device.

Another way, using calculus and the curve-fitting or REGRESSION feature of a graphing calculator, would be to turn the urn on its side, as shown below, take a series of vertical measurements from the center to the top, use REGRESSION, and then integrate (either by hand or with the aid of the calculator).

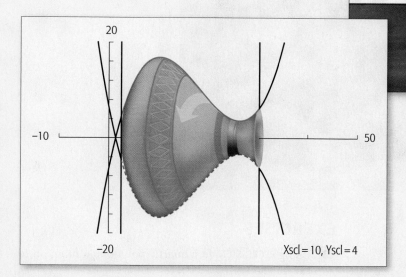

The following table is a table of values for the red curve.

x (in centimeters)	y (in centimeters)
3	4
5	10
10	17
15	16
20	10
25	5
28	3.5
34	7

EXERCISES

1. Using REGRESSION, fit a cubic polynomial function to the data.
2. Using the function found in Exercise 1, integrate over the interval $[3, 34]$ to find the volume of the urn. (*Hint:* If the function in Exercise 1 is Y1, find the volume by using the **VARS** key to enter πY1^2 as Y2. Then use the CALC option to integrate.)

Now consider the bottle shown at the right. To find the bottle's volume in a similar manner, we turn it on its side, use a measuring device to take vertical measurements, and proceed as we did with the urn.

The table of measurements is as follows.

x (in inches)	y (in inches)
1	1.125
2	1.275
3	1.250
4	1.275
5	1.275
6	1.125
7	1.000
8	0.875
9	0.750
10	0.500
11	0.500

EXERCISES

3. Using REGRESSION, fit a quartic polynomial function to the data.

4. Using the function found in Exercise 3, integrate to find the volume of the bottle. Your answer will be in cubic inches. Convert it to fluid ounces using the fact that $1 \text{ in}^3 = 0.55424$ fluid ounce.

5. The bottle in question holds 20 oz. How good was our curve-fitting procedure for making the volume estimate?

6. Find a curve that gives a better estimate of the volume. What is the curve and what is the estimated volume?

Functions of Several Variables

Chapter Snapshot

What You'll Learn

6.1 Functions of Several Variables
6.2 Partial Derivatives
6.3 Maximum–Minimum Problems
6.4 An Application: The Least-Squares Technique
6.5 Constrained Optimization
6.6 Double Integrals

Why It's Important

Functions that have more than one input are called *functions of several variables*. We introduce these functions in this chapter and learn to differentiate them to find *partial derivatives*. Then we use such functions and their partial derivatives to find regression lines and solve maximum–minimum problems. Finally, we consider the integration of functions of several variables.

Where It's Used

PREDICTING THE MINIMUM WAGE

The minimum hourly wage in the United States has grown over the years, as shown in the table. Find the regression line, and use it to predict the minimum hourly wage in 2015 and 2020.

This problem appears as Exercise 5 in Section 6.4.

INCREASE IN THE MINIMUM WAGE

NUMBER OF YEARS, x, SINCE 1990	MINIMUM HOURLY WAGE
0	3.80
1	4.25
6	4.75
7	5.15
17	5.85
18	6.55
19	7.25

(*Source*: www.workworld.org.)

525

Functions of Several Variables

Suppose that a one-product firm produces x units of its product at a profit of $4 per unit. Then its total profit P is given by

$$P(x) = 4x.$$

This is a function of one variable.

Suppose that a two-product firm produces x units of one product at a profit of $4 per unit and y units of a second product at a profit of $6 per unit. Then its total profit P is a function of the *two* variables x and y, and is given by

$$P(x, y) = 4x + 6y.$$

This function assigns to the input pair (x, y) a unique output number, $4x + 6y$.

OBJECTIVE

• Find a function value for a function of several variables.

> **DEFINITION**
>
> A **function of two variables** assigns to each input pair, (x, y), exactly one output number, $f(x, y)$.

We can regard a function of two variables as a machine that has two inputs. Thus, the domain is a set of pairs (x, y) in the plane. When such a function is given by a formula, the domain normally consists of all ordered pairs (x, y) that are meaningful replacements in the formula.

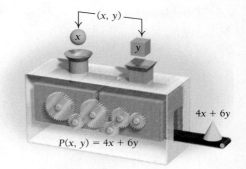

■ **EXAMPLE 1**   For the above profit function, $P(x, y) = 4x + 6y$, find $P(25, 10)$.

Solution $P(25, 10)$ is defined to be the value of the function found by substituting 25 for x and 10 for y:

$$P(25, 10) = 4 \cdot 25 + 6 \cdot 10$$
$$= 100 + 60$$
$$= \$160.$$

This result means that by selling 25 units of the first product and 10 of the second, the two-product firm will make a profit of $160.

❰ Quick Check 1

❭ **Quick Check 1**

A company's cost function is given by

$$C(x, y) = 6.5x + 7.25y.$$

Find $C(10, 15)$.

The following are examples of **functions of several variables**, that is, functions of two or more variables. If there are n variables, then there are n inputs for such a function.

■ **EXAMPLE 2**   **Business: Monthly Payment on an Amortized Loan.**   Large purchases are often financed with an amortized loan. Borrowers like to know how much they can expect to pay per month for every thousand dollars borrowed. The monthly payment P depends on the annual percentage rate (APR) i and the term of the loan t (in years). The function P of the two variables i and t is given by

$$P(i, t) = \frac{1000i\left(1 + \dfrac{i}{12}\right)^{12t}}{12\left(1 + \dfrac{i}{12}\right)^{12t} - 12}.$$

How much per month can a borrower expect to pay per thousand dollars borrowed at an APR of 6.5% for a 6-yr term?

Solution   We let $i = 0.065$ and $t = 6$ and evaluate $P(0.065, 6)$:

$$P(0.065, 6) = \frac{1000(0.065)\left(1 + \dfrac{0.065}{12}\right)^{12(6)}}{12\left(1 + \dfrac{0.065}{12}\right)^{12(6)} - 12} = \$16.81.$$

The monthly payment is $16.81 per thousand dollars borrowed.

> **Quick Check 2**
>
> Determine the monthly payment per thousand dollars borrowed at an APR of 7.25% for a term of 8 yr.

❬ Quick Check 2

:▪ **EXAMPLE 3**   **Business: Payment Tables.**   The formula in Example 2 is used to generate a table of payments that allows borrowers to easily judge the combined effects of the APR and the term. The table below shows the monthly payments per thousand dollars borrowed at various APRs and terms.

	Term, t (in years)				
Annual Percentage Rate, i	4	5	6	7	8
0.05	$23.03	$18.87	$16.10	$14.13	$12.66
0.055	$23.26	$19.10	$16.34	$14.37	$12.90
0.06	$23.49	$19.33	$16.57	$14.61	$13.14
0.065	$23.71	$19.57	$16.81	$14.85	$13.39
0.07	$23.95	$19.80	$17.05	$15.09	$13.63
0.075	$24.18	$20.04	$17.29	$15.34	$13.88

a) What can a borrower expect to pay per month at an APR of 5.5% for a 7-yr term?

b) What can a borrower expect to pay per month at the same APR as in part (a) but for 6-yr term?

Solution   The monthly payments are read directly from the table.

a) We see that $P(0.055, 7) = \$14.37$ per month.

b) From the table, $P(0.055, 6) = \$16.34$ per month.

The borrower pays more per month with a 6-yr term, but less overall than at the same rate for a 7-yr term: $16.34 for 72 months, for a total payment of $1,176.48 for the 6-yr term, or $14.37 for 84 months, for a total of $1,207.08 for the 7-yr term. Tables like the one above allow us to see the behavior of a multivariable function at a glance.

> **Quick Check 3**
>
> **a)** What is the monthly payment per thousand dollars borrowed at an APR of 6.5% for a term of 5 yr?
>
> **b)** How much less per month would the payment be with the same APR but for a term of 6 yr?

❬ Quick Check 3

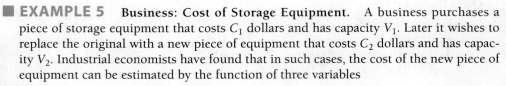

■ **EXAMPLE 4** **Business: Total Cost.** The total cost to a company, in thousands of dollars, of producing its goods is given by

$$C(x, y, z, w) = 4x^2 + 5y + z - \ln(w + 1),$$

where x dollars are spent for labor, y dollars for raw materials, z dollars for advertising, and w dollars for machinery. This is a function of four variables (all in thousands of dollars). Find $C(3, 2, 0, 10)$.

Solution We substitute 3 for x, 2 for y, 0 for z, and 10 for w:

$$\begin{aligned} C(3, 2, 0, 10) &= 4 \cdot 3^2 + 5 \cdot 2 + 0 - \ln(10 + 1) \\ &= 4 \cdot 9 + 10 + 0 - 2.397895 \\ &\approx \$43.6 \text{ thousand, or } \$43,600. \end{aligned}$$

■ **EXAMPLE 5** **Business: Cost of Storage Equipment.** A business purchases a piece of storage equipment that costs C_1 dollars and has capacity V_1. Later it wishes to replace the original with a new piece of equipment that costs C_2 dollars and has capacity V_2. Industrial economists have found that in such cases, the cost of the new piece of equipment can be estimated by the function of three variables

$$C_2 = \left(\frac{V_2}{V_1}\right)^{0.6} C_1.$$

For \$45,000, a beverage company buys a manufacturing tank that has a capacity of 10,000 gallons. Later it decides to buy a tank with double the capacity of the original. Estimate the cost of the new tank.

Solution We substitute 20,000 for V_2, 10,000 for V_1, and 45,000 for C_1:

$$\begin{aligned} C_2 &= \left(\frac{20,000}{10,000}\right)^{0.6} (45,000) \\ &= 2^{0.6}(45,000) \\ &\approx \$68,207.25. \end{aligned}$$

Note that a 100% increase in capacity was achieved by about a 52% increase in cost. This is independent of any increase in the costs of labor, management, or other equipment resulting from the purchase of the tank.

❬ Quick Check 4

> **Quick Check 4**
>
> **a)** Repeat Example 5 assuming that the company buys a tank with a capacity of 2.75 times that of the original.
>
> **b)** What is the percentage increase in cost for this tank compared to the cost of the original tank?

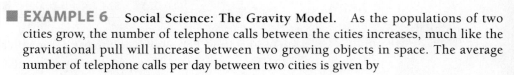

■ **EXAMPLE 6** **Social Science: The Gravity Model.** As the populations of two cities grow, the number of telephone calls between the cities increases, much like the gravitational pull will increase between two growing objects in space. The average number of telephone calls per day between two cities is given by

$$N(d, P_1, P_2) = \frac{2.8 P_1 P_2}{d^{2.4}},$$

where d is the distance, in miles, between the cities and P_1 and P_2 are their populations. The cities of Dallas and Fort Worth are 30 mi apart and have populations of 1,279,910 and 720,250, respectively. (*Sources:* Population Division, U.S. Census Bureau, 2009 estimates, and Rand McNally.) Find the average number of calls per day between the two cities.

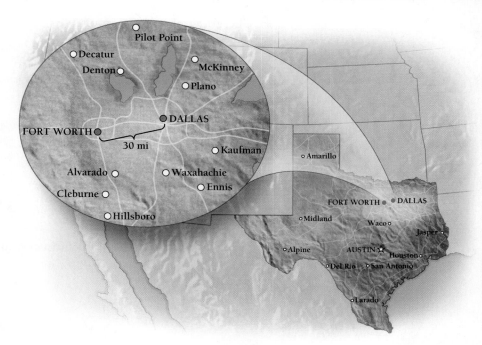

> **Quick Check 5**
>
> Find the average number of calls per day between Phoenix, Arizona (population 1,552,300) and Tucson, Arizona (population 541,800), given that the distance between the two cities is 120 mi. (*Source*: www.census.gov.)

Solution   We evaluate the function with the aid of a calculator:

$$N(30, 1{,}279{,}910, 720{,}250) = \frac{2.8(1{,}279{,}910)(720{,}250)}{30^{2.4}}$$

$$\approx 735{,}749{,}066.$$

❰ Quick Check 5

Geometric Interpretations

Visually, a function of two variables,

$$z = f(x, y),$$

can be thought of as matching a point (x_1, y_1) in the xy-plane with the number z_1 on a number line. Thus, to graph a function of two variables, we need a three-dimensional coordinate system. The axes are generally placed as shown to the left. The line z, called the z-axis, is placed perpendicular to the xy-plane at the origin.

To help visualize this, think of looking into the corner of a room, where the floor is the xy-plane and the z-axis is the intersection of the two walls. To plot a point (x_1, y_1, z_1), we locate the point (x_1, y_1) in the xy-plane and move up or down in space according to the value of z_1.

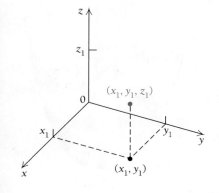

■ **EXAMPLE 7**   Plot these points:

$$P_1(2, 3, 5),$$
$$P_2(2, -2, -4),$$
$$P_3(0, 5, 2),$$
and $$P_4(2, 3, 0).$$

Solution   The solution is shown at the right.

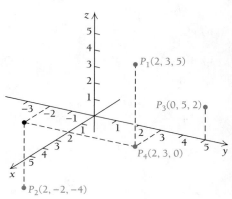

The *graph* of a function of two variables,

$$z = f(x, y),$$

consists of ordered triples (x_1, y_1, z_1), where $z_1 = f(x_1, y_1)$. This graph takes the form of a **surface**. The **domain** of a two-variable function is the set of points in the xy-plane for which f is defined.

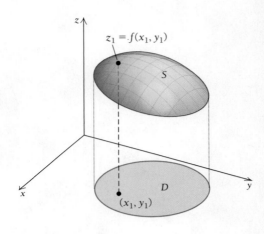

Elliptic paraboloid: $z = x^2 + y^2$

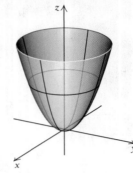

Hemisphere: $z = \sqrt{1 - x^2 - y^2}$

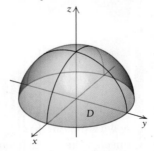

> **Quick Check 6**
>
> Determine the domain of each two-variable function.
>
> **a)** $f(x, y) = \dfrac{x + y}{x - y}$
>
> **b)** $g(x, y) = \dfrac{1}{x - 2} + \dfrac{2}{3 + y}$
>
> **c)** $h(x, y) = \ln(y - x^3)$

■ **EXAMPLE 8**   Determine the domain of each two-variable function.

a) $f(x, y) = x^2 + y^2$

b) $g(x, y) = \sqrt{1 - x^2 - y^2}$

c) $h(x, y) = x^2 + y^2 + \dfrac{1}{x^2 + y^2}$

Solution

a) Since we can square any real number and sum any two squares, the function f is defined for all x and all y. Therefore, the domain for f is

$$D = \left\{ (x, y) \,|\, -\infty < x < \infty, \; -\infty < y < \infty \right\}.$$

The graph of f is a surface called an *elliptic paraboloid*. Satellite dishes are elliptic paraboloids: the weak incoming signals bounce off the interior surface of the paraboloid and collect at a single point, called the *focus*, thus amplifying the signal.

b) The expression within the radical must be nonnegative. Therefore, $1 - x^2 - y^2 \geq 0$, which simplifies to $x^2 + y^2 \leq 1$. The domain for g is

$$D = \left\{ (x, y) \,|\, x^2 + y^2 \leq 1 \right\}.$$

The graph of g is a surface called a *hemisphere,* of radius 1. Its domain is a filled-in circle of radius 1. We can think of the domain of g as the "shadow" it casts on the xy-plane.

c) Since zero cannot be in the denominator, we must have $x^2 + y^2 \neq 0$. Therefore, x and y cannot be 0 simultaneously. The domain of h is

$$D = \left\{ (x, y) \,|\, (x, y) \neq (0, 0) \right\}.$$

The graph of h is shown at right.

❰ Quick Check 6

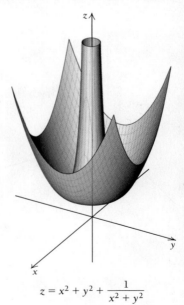

$$z = x^2 + y^2 + \frac{1}{x^2 + y^2}$$

Exploratory

Another useful and inexpensive app for the iPhone and iPod Touch is Quick Graph, a graphing calculator that creates visually appealing 3D graphs of functions of two variables. It has full graphing interactivity, with touch-based zoom and scroll features.

Some functions and their graphs are presented here as examples.

EXAMPLE 1 Graph: $\left(1 - \sqrt{x^2 + y^2}\right)^2 + z^2 = 0.2$. This is entered as follows:

(1-sqrt(x^2+y^2))^2+z^2=0.2

The graph is shown at the right.

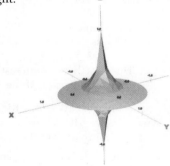

EXAMPLE 2 Graph: $\left|(2x^2 + 2y^2)^{0.25}\right| + \sqrt{|z|} = 1$. This is entered as follows:

abs((2x^2+2y^2)^0.25)+(abs(z))^0.5=1

The graph is shown at the right.

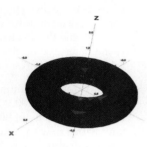

EXAMPLE 3 Graph: $z = e^{-4(x^2+y^2)}$. This is entered as follows:

z=e^(-4(x^2+y^2))

The graph is shown at the right.

EXAMPLE 4 Graph: $(xy)^2 + (yz)^2 + (zx)^2 = xyz$. This is entered as follows:

(xy)^2+(yz)^2+(xz)^2=xyz

The graph is shown at the right.

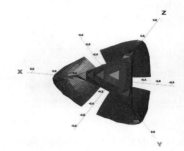

EXAMPLE 5 Graph: $4x^2 + 2y^2 + z^2 = 1$. This is entered as follows:

4x^2+2y^2+z^2=1

The graph is shown at the right.

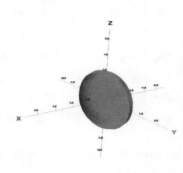

EXAMPLE 6 Graph: $z = -8xe^{-4(x^2+y^2)}$. This is entered as follows:

z=−8xe^(-4(x^2+y^2))

The graph is shown at the right.

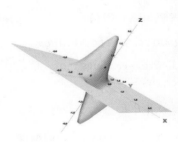

EXERCISE

Use Quick Graph or another 3D graphing utility to graph the functions in Exercises 1–12 on p. 532.

Section Summary

- A *function of two variables* assigns to each input pair, (x, y), exactly one output number, $f(x, y)$.
- A function of two variables generates points (x, y, z), where $z = f(x, y)$.

- The graph of a function of two variables is a *surface* and requires a three-dimensional coordinate system.
- The *domain* of a function of two variables is the set of points in the xy-plane for which the function is defined.

EXERCISE SET
6.1

1. For $f(x, y) = x^2 - 3xy$, find $f(0, -2)$, $f(2, 3)$, and $f(10, -5)$.

2. For $f(x, y) = (y^2 + 2xy)^3$, find $f(-2, 0)$, $f(3, 2)$, and $f(-5, 10)$.

3. For $f(x, y) = 3^x + 7xy$, find $f(0, -2)$, $f(-2, 1)$, and $f(2, 1)$.

4. For $f(x, y) = \log_{10}(x + y) + 3x^2$, find $f(3, 7)$, $f(1, 99)$, and $f(2, -1)$.

5. For $f(x, y) = \ln x + y^3$, find $f(e, 2)$, $f(e^2, 4)$, and $f(e^3, 5)$.

6. For $f(x, y) = 2^x - 3^y$, find $f(0, 2)$, $f(3, 1)$, and $f(2, 3)$.

7. For $f(x, y, z) = x^2 - y^2 + z^2$, find $f(-1, 2, 3)$ and $f(2, -1, 3)$.

8. For $f(x, y, z) = 2^x + 5zy - x$, find $f(0, 1, -3)$ and $f(1, 0, -3)$.

In Exercises 9–12, determine the domain of each function of two variables.

9. $f(x, y) = \sqrt{y - 3x}$

10. $g(x, y) = \dfrac{1}{y + x^2}$

11. $h(x, y) = xe^{\sqrt{y}}$

12. $k(x, y) = \dfrac{1}{x} + \dfrac{y}{x - 1}$

APPLICATIONS

Business and Economics

13. *Price–earnings ratio.* The *price–earnings ratio* of a stock is given by

$$R(P, E) = \frac{P}{E},$$

where P is the price of the stock and E is the earnings per share. The price per share of Hewlett-Packard stock was $32.03, and the earnings per share were $1.25. (*Source:* yahoo.finance.com.) Find the price–earnings ratio. Use decimal notation rounded to the nearest hundredth.

14. *Yield.* The *yield* of a stock is given by

$$Y(D, P) = \frac{D}{P},$$

where D is the dividend per share of stock and P is the price per share. The price per share of Texas Instruments stock was $30, and the dividend per share was $0.12. (*Source:* yahoo.finance.com.) Find the yield. Use percent notation rounded to the nearest hundredth of a percent.

15. *Cost of storage equipment.* Consider the cost model in Example 5. For $100,000, a company buys a storage tank that has a capacity of 80,000 gal. Later it replaces the tank with a new tank that has double the capacity of the original. Estimate the cost of the new tank.

16. *Savings and interest.* A sum of $1000 is deposited in a savings account for which interest is compounded monthly. The future value A is a function of the annual percentage rate i and the term t, in months, and is given by

$$A(i, t) = 1000 \left(1 + \frac{i}{12}\right)^{12t}.$$

a) Determine $A(0.05, 10)$.

b) What is the interest earned for the rate and term in part (a)?

c) How much more interest can be earned over the same term as in part (a) if the APR is increased to 5.75%?

17. *Monthly car payments.* Kim is shopping for a car. She will finance $10,000 through a lender. Use the table in Example 3 to answer the following questions.

a) One lender offers Kim an APR of 6% for a 6-yr term. What would Kim's monthly payment be?

b) A competing lender offers an APR of 5.5% but for a 7-yr term. What would Kim's monthly payment be?

c) Assume that Kim makes the minimum payment each month for the entire term of the loan. Calculate her total payments for both options described in parts (a) and (b). Which option costs Kim less overall?

Life and Physical Sciences

18. *Poiseuille's Law.* The speed of blood in a vessel is given by

$$V(L, p, R, r, v) = \frac{p}{4Lv}(R^2 - r^2),$$

where R is the radius of the vessel, r is the distance of the blood from the center of the vessel, L is the length of the blood vessel, p is the pressure, and v is the viscosity. Find $V(1, 100, 0.0075, 0.0025, 0.05)$.

19. Wind speed of a tornado. Under certain conditions, the *wind speed S*, in miles per hour, of a tornado at a distance *d* feet from its center can be approximated by the function

$$S(a, d, V) = \frac{aV}{0.51d^2},$$

where *a* is a constant that depends on certain atmospheric conditions and *V* is the approximate volume of the tornado, in cubic feet. Approximate the wind speed 100 ft from the center of a tornado when its volume is 1,600,000 ft^3 and *a* = 0.78.

20. Body surface area. The Mosteller formula for approximating the surface area *S*, in square meters (m^2), of a human is given by

$$S(h, w) = \frac{\sqrt{hw}}{60},$$

where *h* is the person's height in centimeters and *w* is the person's weight in kilograms. (*Source:* www.halls.md.) Use the Mosteller approximation to estimate the surface area of a person whose height is 165 cm and whose weight is 80 kg.

21. Body surface area. The Haycock formula for approximating the surface area *S*, in square meters (m^2), of a human is given by

$$S(h, w) = 0.024265h^{0.3964}w^{0.5378},$$

where *h* is the person's height in centimeters and *w* is the person's weight in kilograms. (*Source:* www.halls.md.) Use the Haycock approximation to estimate the surface area of a person whose height is 165 cm and whose weight is 80 kg.

General Interest

22. Goals against average. A hockey goaltender's goals against average *A* is a function of the number of goals *g* allowed and the number *m* of minutes played and is given by the formula

$$A(g, m) = \frac{60g}{m}.$$

a) Determine the goals against average of a goaltender who allows 35 goals while playing 820 min. Round *A* to the nearest hundredth.

b) A goaltender gave up 124 goals during the season and had a goals against average of 3.75. How many minutes did he play? (Round to the nearest integer.)

c) State the domain for *A*.

23. Dewpoint. The *dewpoint* is the temperature at which moisture in the air condenses into liquid (dew). It is a function of air temperature *t* and relative humidity *h*. The table below shows the dewpoints for select values of *t* and *h*.

	Relative Humidity (%)				
	20	40	60	80	100
70	29	44	55	63	70
80	35	53	65	73	80
90	43	62	74	83	90
100	52	71	84	93	100

Air Temperature (degrees Fahrenheit)

a) What is the dewpoint when the air temperature is 80°F with a relative humidity of 60%?

b) What is the dewpoint when the air temperature is 90°F with a relative humidity of 40%?

c) The air feels humid when the dewpoint reaches about 60. If the air temperature is 100°F, at what approximate relative humidity will the air feel humid?

d) Explain why the dewpoint is equal to the air temperature when the relative humidity is 100%.

SYNTHESIS

24. For the tornado described in Exercise 19, if the wind speed measures 200 mph, how far from the center was the measurement taken?

25. According to the Mosteller formula in Exercise 20, if a person's weight drops 19%, by what percentage does his or her surface area change?

26. Explain the difference between a function of two variables and a function of one variable.

27. Find some examples of functions of several variables not considered in the text, even some that may not have formulas.

TECHNOLOGY CONNECTION

General Interest

Wind chill temperature. *Because wind speed enhances the loss of heat from the skin, we feel colder when there is wind than when there is not. The* **wind chill temperature** *is what the temperature would have to be with no wind in order to give the same chilling effect. The wind chill temperature, W, is given by*

$$W(v, T) = 91.4 - \frac{(10.45 + 6.68\sqrt{v} - 0.447v)(457 - 5T)}{110},$$

where T is the actual temperature measured by a thermometer, in degrees Fahrenheit, and v is the speed of the wind, in miles per hour. Find the wind chill temperature in each case. Round to the nearest degree.

28. $T = 30°F, v = 25$ mph

29. $T = 20°F, v = 20$ mph

30. $T = 20°F, v = 40$ mph

31. $T = -10°F, v = 30$ mph

32. Use a computer graphics program such as *Maple* or *Mathematica*, an Internet site such as www.wolframalpha.com, a graphing calculator, or an iPhone app such as Quick Graph to view the graph of each function given in Exercises 1–8.

Use a 3D graphics program to generate the graph of each function.

33. $f(x, y) = y^2$

34. $f(x, y) = x^2 + y^2$

35. $f(x, y) = (x^4 - 16x^2)e^{-y^2}$

36. $f(x, y) = 4(x^2 + y^2) - (x^2 + y^2)^2$

37. $f(x, y) = x^3 - 3xy^2$

38. $f(x, y) = \dfrac{1}{x^2 + 4y^2}$

Answers to Quick Checks

1. $173.75   **2.** $13.76   **3. (a)** $19.57   **(b)** $2.76

4. (a) $82,568.07   **(b)** 83.5% increase

5. 24,095,597 calls/day   **6. (a)** $D = \{(x, y) | y \neq x\}$

   (b) $D = \{(x, y) | x \neq 2, y \neq -3\}$ **(c)** $D = \{(x, y) | y > x^3\}$

6.2   Partial Derivatives

OBJECTIVES

- Find the partial derivatives of a given function.
- Evaluate partial derivatives.
- Find the four second-order partial derivatives of a function in two variables.

Finding Partial Derivatives

Consider the function f given by

$$z = f(x, y) = x^2y^3 + xy + 4y^2.$$

Suppose for the moment that we fix y at 3. Then

$$f(x, 3) = x^2(3^3) + x(3) + 4(3^2) = 27x^2 + 3x + 36.$$

Note that we now have a function of only one variable. Taking the first derivative with respect to x, we have

$$54x + 3.$$

   In general, without replacing y with a specific number, we can consider y fixed. Then f becomes a function of x alone, and we can calculate its derivative with respect to x. This derivative is called the *partial derivative of f with respect to x*. Notation for this partial derivative is

$$\frac{\partial f}{\partial x} \quad \text{or} \quad \frac{\partial z}{\partial x}.$$

   Now, let's again consider the function

$$z = f(x, y) = x^2y^3 + xy + 4y^2.$$

The color blue indicates the variable x when we fix y and treat it as a constant. The expressions y^3, y, and y^2 are then also treated as constants. We have

$$\frac{\partial f}{\partial x} = \frac{\partial z}{\partial x} = 2xy^3 + y.$$

Similarly, we find $\partial f/\partial y$ or $\partial z/\partial y$ by fixing x (treating it as a constant) and calculating the derivative with respect to y. From

$$z = f(x, y) = x^2y^3 + xy + 4y^2, \qquad \text{The color blue indicates the variable.}$$

we get

$$\frac{\partial f}{\partial y} = \frac{\partial z}{\partial y} = 3x^2y^2 + x + 8y.$$

A definition of partial derivatives is as follows.

DEFINITION

For $z = f(x, y)$, the **partial derivatives with respect to x and y** are

$$\frac{\partial z}{\partial x} = \lim_{h \to 0} \frac{f(x + h, y) - f(x, y)}{h} \quad \text{and} \quad \frac{\partial z}{\partial y} = \lim_{h \to 0} \frac{f(x, y + h) - f(x, y)}{h}.$$

We can find partial derivatives of functions of any number of variables. Since we can apply the theorems for finding derivatives presented earlier, we will rarely need to use the definition to find a partial derivative.

■ **EXAMPLE 1** For $w = x^2 - xy + y^2 + 2yz + 2z^2 + z$, find

$$\frac{\partial w}{\partial x}, \quad \frac{\partial w}{\partial y}, \quad \text{and} \quad \frac{\partial w}{\partial z}.$$

Solution In order to find $\partial w/\partial x$, we regard x as the variable and treat y and z as constants. From

$$w = x^2 - xy + y^2 + 2yz + 2z^2 + z,$$

we get

$$\frac{\partial w}{\partial x} = 2x - y.$$

To find $\partial w/\partial y$, we regard y as the variable and treat x and z as constants. We get

$$\frac{\partial w}{\partial y} = -x + 2y + 2z;$$

To find $\partial w/\partial z$, we regard z as the variable and treat x and y as constants. We get

$$\frac{\partial w}{\partial z} = 2y + 4z + 1.$$

❰ Quick Check 1

❱ **Quick Check 1**

For $u = x^2y^3z^4$, find

$$\frac{\partial u}{\partial x}, \frac{\partial u}{\partial y}, \text{ and } \frac{\partial u}{\partial z}.$$

We will often make use of a simpler notation: f_x for the partial derivative of f with respect to x and f_y for the partial derivative of f with respect to y. Similarly, if $z = f(x, y)$, then z_x represents the partial derivative of z with respect to x, and z_y represents the partial derivative of z with respect to y.

■ **EXAMPLE 2**   For $f(x, y) = 3x^2y + xy$, find f_x and f_y.

Solution   We have

$$f_x = 6xy + y, \qquad \text{Treating } y \text{ as a constant}$$
$$f_y = 3x^2 + x. \qquad \text{Treating } x^2 \text{ and } x \text{ as constants}$$

❰ Quick Check 2

For the function in Example 2, let's evaluate f_x at $(2, -3)$:

$$f_x(2, -3) = 6 \cdot 2 \cdot (-3) + (-3)$$
$$= -39.$$

If we use the notation $\partial f / \partial x = 6xy + y$, where $f = 3x^2y + xy$, the value of the partial derivative at $(2, -3)$ is given by

$$\left. \frac{\partial f}{\partial x} \right|_{(2, -3)} = 6 \cdot 2 \cdot (-3) + (-3)$$
$$= -39.$$

However, this notation is not quite as convenient as $f_x(2, -3)$.

■ **EXAMPLE 3**   For $f(x, y) = e^{xy} + y \ln x$, find f_x and f_y.

Solution

$$f_x = y \cdot e^{xy} + y \cdot \frac{1}{x}$$
$$= ye^{xy} + \frac{y}{x},$$
$$f_y = x \cdot e^{xy} + 1 \cdot \ln x$$
$$= xe^{xy} + \ln x$$

The Geometric Interpretation of Partial Derivatives

The graph of a function of two variables $z = f(x, y)$ is a surface S, which might have a graph similar to the one shown to the right, where each input pair (x, y) in the domain D has only one output, $z = f(x, y)$.

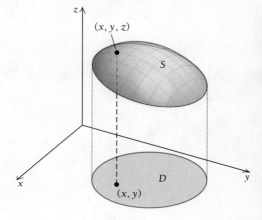

Now suppose that we hold x fixed at the value a. The set of all points for which $x = a$ is a plane parallel to the yz-plane; thus, when x is fixed at a, y and z vary along that plane, as shown to the right. The plane in the figure cuts the surface along the curve C_1. The partial derivative f_y gives the slope of tangent lines to this curve, in the positive y-direction.

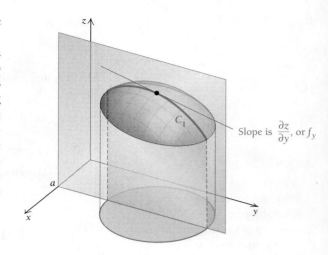

Similarly, if we hold y fixed at the value b, we obtain a curve C_2, as shown to the right. The partial derivative f_x gives the slope of tangent lines to this curve, in the positive x-direction.

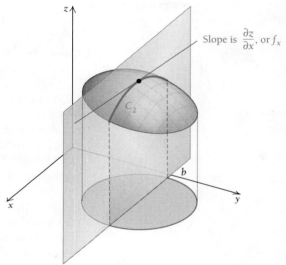

An Economics Application: The Cobb–Douglas Production Function

One model of production that is frequently considered in business and economics is the *Cobb–Douglas production function:*

$$p(x, y) = Ax^a y^{1-a}, \quad \text{for} \quad A > 0 \quad \text{and} \quad 0 < a < 1,$$

where p is the number of units produced with x units of labor and y units of capital. (Capital is the cost of machinery, buildings, tools, and other supplies.) The partial derivatives

$$\frac{\partial p}{\partial x} \quad \text{and} \quad \frac{\partial p}{\partial y}$$

are called, respectively, the *marginal productivity of labor* and the *marginal productivity of capital.*

■ **EXAMPLE 4**   A cellular phone company has the following production function for a smart phone:

$$p(x, y) = 50x^{2/3}y^{1/3},$$

where p is the number of units produced with x units of labor and y units of capital.

a) Find the number of units produced with 125 units of labor and 64 units of capital.

b) Find the marginal productivities.

c) Evaluate the marginal productivities at $x = 125$ and $y = 64$.

Solution

a) $p(125, 64) = 50(125)^{2/3}(64)^{1/3} = 50(25)(4) = 5000$ units

b) Marginal productivity of labor $= \dfrac{\partial p}{\partial x} = p_x = 50\left(\dfrac{2}{3}\right)x^{-1/3}y^{1/3} = \dfrac{100y^{1/3}}{3x^{1/3}}$

Marginal productivity of capital $= \dfrac{\partial p}{\partial y} = p_y = 50\left(\dfrac{1}{3}\right)x^{2/3}y^{-2/3} = \dfrac{50x^{2/3}}{3y^{2/3}}$

c) For 125 units of labor and 64 units of capital, we have

Marginal productivity of labor $= p_x(125, 64)$

$$= \frac{100(64)^{1/3}}{3(125)^{1/3}} = \frac{100(4)}{3(5)} = 26\tfrac{2}{3},$$

Marginal productivity of capital $= p_y(125, 64)$

$$= \frac{50(125)^{2/3}}{3(64)^{2/3}} = \frac{50(25)}{3(16)} = 26\tfrac{1}{24}.$$

> **Quick Check 3**
>
> A publisher's production function for textbooks is given by $p(x, y) = 72x^{0.8}y^{0.2}$, where p is the number of books produced, x is units of labor, and y is units of capital. Determine the marginal productivities at $x = 90$ and $y = 50$.

❰ Quick Check 3

Let's interpret the marginal productivities of Example 4. To visualize the marginal productivity of labor, suppose that capital is fixed at 64 units. Then a one-unit change in labor, from 125 to 126, will cause production to increase by about $26\tfrac{2}{3}$ units. To visualize the marginal productivity of capital, suppose that the amount of labor is fixed at 125 units. Then a one-unit change in capital from 64 to 65 will cause production to increase by about $26\tfrac{1}{24}$ units.

A Cobb–Douglas production function is consistent with the law of diminishing returns. That is, if one input (either labor or capital) is held fixed while the other increases infinitely, then production will eventually increase at a decreasing rate. With such functions, it also turns out that if a certain maximum production is possible, then the expense of more labor, for example, may be required for that maximum output to be attainable.

Higher-Order Partial Derivatives

Consider

$$z = f(x, y) = 3xy^2 + 2xy + x^2.$$

Then $\dfrac{\partial z}{\partial x} = \dfrac{\partial f}{\partial x} = 3y^2 + 2y + 2x.$

Suppose that we continue and find the first partial derivative of $\partial z/\partial x$ with respect to y. This will be a **second-order partial derivative** of the original function z. Its notation is as follows:

$$\frac{\partial}{\partial y}\left(\frac{\partial z}{\partial x}\right) = \frac{\partial}{\partial y}\left(\frac{\partial f}{\partial x}\right) = \frac{\partial}{\partial y}(3y^2 + 2y + 2x) = 6y + 2.$$

The notation $\dfrac{\partial}{\partial y}\left(\dfrac{\partial z}{\partial x}\right)$ is often expressed as

$$\frac{\partial^2 z}{\partial y\, \partial x} \quad \text{or} \quad \frac{\partial^2 f}{\partial y\, \partial x}.$$

We could also denote the preceding partial derivative using the notation f_{xy}:

$$f_{xy} = 6y + 2.$$

Note that in the notation f_{xy}, x and y are in the order (left to right) in which the differentiation is done, but in

$$\frac{\partial^2 f}{\partial y\, \partial x},$$

the order of x and y is reversed. In each case, the differentiation with respect to x is done first, followed by differentiation with respect to y.

Notation for the four second-order partial derivatives is as follows.

DEFINITION   Second-Order Partial Derivatives

1. $\dfrac{\partial^2 z}{\partial x\, \partial x} = \dfrac{\partial^2 f}{\partial x\, \partial x} = \dfrac{\partial^2 z}{\partial x^2} = \dfrac{\partial^2 f}{\partial x^2} = f_{xx}$ Take the partial with respect to x, and then with respect to x again.

2. $\dfrac{\partial^2 z}{\partial y\, \partial x} = \dfrac{\partial^2 f}{\partial y\, \partial x} = f_{xy}$ Take the partial with respect to x, and then with respect to y.

3. $\dfrac{\partial^2 z}{\partial x\, \partial y} = \dfrac{\partial^2 f}{\partial x\, \partial y} = f_{yx}$ Take the partial with respect to y, and then with respect to x.

4. $\dfrac{\partial^2 z}{\partial y\, \partial y} = \dfrac{\partial^2 f}{\partial y\, \partial y} = \dfrac{\partial^2 z}{\partial y^2} = \dfrac{\partial^2 f}{\partial y^2} = f_{yy}$ Take the partial with respect to y, and then with respect to y again.

■ **EXAMPLE 5**   For

$$z = f(x, y) = x^2 y^3 + x^4 y + x e^y,$$

find the four second-order partial derivatives.

Solution

a) $\dfrac{\partial^2 f}{\partial x^2} = f_{xx} = \dfrac{\partial}{\partial x}(2xy^3 + 4x^3 y + e^y)$ Differentiate twice with respect to x.

$= 2y^3 + 12x^2 y$

b) $\dfrac{\partial^2 f}{\partial y\, \partial x} = f_{xy} = \dfrac{\partial}{\partial y}(2xy^3 + 4x^3 y + e^y)$ Differentiate with respect to x and then with respect to y.

$= 6xy^2 + 4x^3 + e^y$

c) $\dfrac{\partial^2 f}{\partial x\, \partial y} = f_{yx} = \dfrac{\partial}{\partial x}(3x^2 y^2 + x^4 + x e^y)$ Differentiate with respect to y and then with respect to x.

$= 6xy^2 + 4x^3 + e^y$

d) $\dfrac{\partial^2 f}{\partial y^2} = f_{yy} = \dfrac{\partial}{\partial y}(3x^2 y^2 + x^4 + x e^y)$ Differentiate twice with respect to y.

$= 6x^2 y + x e^y$

⟨ Quick Check 4

> **Quick Check 4**
>
> For
> $$z = g(x, y)$$
> $$= 6x^2 + 3xy^4 - y^2,$$
> find the four second-order partial derivatives.

We see by comparing parts (b) and (c) of Example 5 that

$$\frac{\partial^2 f}{\partial y\, \partial x} = \frac{\partial^2 f}{\partial x\, \partial y} \quad \text{and} \quad f_{xy} = f_{yx}.$$

Although this will be true for virtually all functions that we consider in this text, it is *not* true for all functions. One function for which it is not true is given in Exercise 69.

In Section 6.3, we will see how higher-order partial derivatives are used in applications to find extrema for functions of two variables.

Section Summary

- For $z = f(x, y)$, the *partial derivatives with respect to x and y* are, respectively:

$$\frac{\partial z}{\partial x} = \lim_{h \to 0} \frac{f(x + h, y) - f(x, y)}{h} \quad \text{and}$$

$$\frac{\partial z}{\partial y} = \lim_{h \to 0} \frac{f(x, y + h) - f(x, y)}{h}.$$

- Simpler notations for partial derivatives are f_x and z_x for $\dfrac{\partial z}{\partial x}$

  and f_y and z_y for $\dfrac{\partial z}{\partial y}$.

- For a surface $z = f(x, y)$ and a point (x_0, y_0, z_0) on this surface, the partial derivative of f with respect to x gives the slope of the tangent line at (x_0, y_0, z_0) in the positive x-direction. Similarly, the partial derivative of f with respect to y gives the slope of the tangent line at (x_0, y_0, z_0) in the positive y-direction.

- For $z = f(x, y)$, the second-order partial derivatives are

$$f_{xx} = \frac{\partial^2 f}{\partial x^2}, f_{xy} = \frac{\partial^2 f}{\partial y\, \partial x}, f_{yx} = \frac{\partial^2 f}{\partial x\, \partial y}, \text{ and } f_{yy} = \frac{\partial^2 f}{\partial y^2}.$$

  Often (but not always), $f_{xy} = f_{yx}$.

EXERCISE SET
6.2

Find $\dfrac{\partial z}{\partial x}, \dfrac{\partial z}{\partial y}, \dfrac{\partial z}{\partial x}\bigg|_{(-2, -3)}$, and $\dfrac{\partial z}{\partial y}\bigg|_{(0, -5)}$.

1. $z = 2x - 3y$ **2.** $z = 7x - 5y$

3. $z = 3x^2 - 2xy + y$ **4.** $z = 2x^3 + 3xy - x$

Find $f_x, f_y, f_x(-2, 4)$, and $f_y(4, -3)$.

5. $f(x, y) = 2x - 5xy$ **6.** $f(x, y) = 5x + 7y$

Find $f_x, f_y, f_x(-2, 1)$, and $f_y(-3, -2)$.

7. $f(x, y) = \sqrt{x^2 + y^2}$

8. $f(x, y) = \sqrt{x^2 - y^2}$

Find f_x and f_y.

9. $f(x, y) = e^{2x-y}$ **10.** $f(x, y) = e^{3x-2y}$

11. $f(x, y) = e^{xy}$ **12.** $f(x, y) = e^{2xy}$

13. $f(x, y) = y \ln(x + 2y)$

14. $f(x, y) = x \ln(x - y)$

15. $f(x, y) = x \ln(xy)$ **16.** $f(x, y) = y \ln(xy)$

17. $f(x, y) = \dfrac{x}{y} - \dfrac{y}{3x}$ **18.** $f(x, y) = \dfrac{x}{y} + \dfrac{y}{5x}$

19. $f(x, y) = 3(2x + y - 5)^2$

20. $f(x, y) = 4(3x + y - 8)^2$

Find $\dfrac{\partial f}{\partial b}$ and $\dfrac{\partial f}{\partial m}$.

21. $f(b, m) = m^3 + 4m^2 b - b^2 + (2m + b - 5)^2$
  $+ (3m + b - 6)^2$

22. $f(b, m) = 5m^2 - mb^2 - 3b + (2m + b - 8)^2$
  $+ (3m + b - 9)^2$

Find f_x, f_y, and f_λ. (The symbol λ is the Greek letter lambda.)

23. $f(x, y, \lambda) = 5xy - \lambda(2x + y - 8)$

24. $f(x, y, \lambda) = 9xy - \lambda(3x - y + 7)$

25. $f(x, y, \lambda) = x^2 + y^2 - \lambda(10x + 2y - 4)$

26. $f(x, y, \lambda) = x^2 - y^2 - \lambda(4x - 7y - 10)$

Find the four second-order partial derivatives.

27. $f(x, y) = 5xy$

28. $f(x, y) = 2xy$

29. $f(x, y) = 7xy^2 + 5xy - 2y$

30. $f(x, y) = 3x^2 y - 2xy + 4y$

31. $f(x, y) = x^5 y^4 + x^3 y^2$

32. $f(x, y) = x^4 y^3 - x^2 y^3$

Find f_{xx}, f_{xy}, f_{yx}, and f_{yy}. (Remember, f_{yx} means to differentiate with respect to y and then with respect to x.)

33. $f(x, y) = 2x - 3y$ **34.** $f(x, y) = 3x + 5y$

35. $f(x, y) = e^{2xy}$ **36.** $f(x, y) = e^{xy}$

37. $f(x, y) = x + e^y$ **38.** $f(x, y) = y - e^x$

39. $f(x, y) = y \ln x$ **40.** $f(x, y) = x \ln y$

APPLICATIONS

Business and Economics

41. The Cobb–Douglas model. Lincolnville Sporting Goods has the following production function for a certain product:

$$p(x, y) = 2400x^{2/5} y^{3/5},$$

where p is the number of units produced with x units of labor and y units of capital.

a) Find the number of units produced with 32 units of labor and 1024 units of capital.
b) Find the marginal productivities.
c) Evaluate the marginal productivities at $x = 32$ and $y = 1024$.
d) Interpret the meanings of the marginal productivities found in part (c).

42. The Cobb–Douglas model. Riverside Appliances has the following production function for a certain product:

$$p(x, y) = 1800x^{0.621}y^{0.379},$$

where p is the number of units produced with x units of labor and y units of capital.

a) Find the number of units produced with 2500 units of labor and 1700 units of capital.
b) Find the marginal productivities.
c) Evaluate the marginal productivities at $x = 2500$ and $y = 1700$.
d) Interpret the meanings of the marginal productivities found in part (c).

Nursing facilities. *A study of Texas nursing homes found that the annual profit P (in dollars) of profit-seeking, independent nursing homes in urban locations is modeled by the function*

$$P(w, r, s, t) = 0.007955w^{-0.638}\, r^{1.038}\, s^{0.873}\, t^{2.468}.$$

In this function, w is the average hourly wage of nurses and aides (in dollars), r is the occupancy rate (as a percentage), s is the total square footage of the facility, and t is the Texas Index of Level of Effort (TILE), a number between 1 and 11 that measures state Medicaid reimbursement. (Source: K. J. Knox, E. C. Blankmeyer, and J. R. Stutzman, "Relative Economic Efficiency in Texas Nursing Facilities," Journal of Economics and Finance, Vol. 23, 199–213 (1999).) Use the preceding information for Exercises 43 and 44.

43. A profit-seeking, independent Texas nursing home in an urban setting has nurses and aides with an average hourly wage of $20 an hour, a TILE of 8, an occupancy rate of 70%, and 400,000 ft² of space.

a) Estimate the nursing home's annual profit.
b) Find the four partial derivatives of P.
c) Interpret the meaning of the partial derivatives found in part (b).

44. The change in P due to a change in w when the other variables are held constant is approximately

$$\Delta P \approx \frac{\partial P}{\partial w}\Delta w.$$

Use the values of w, r, s, and t in Exercise 43 and assume that the nursing home gives its nurses and aides a small raise so that the average hourly wage is now $20.25 an hour. By approximately how much does the profit change?

Life and Physical Sciences

Temperature–humidity heat index. *In the summer, humidity interacts with the outdoor temperature, making a person feel hotter because of reduced heat loss from the skin caused by* higher humidity. *The temperature–humidity index, T_h, is what the temperature would have to be with no humidity in order to give the same heat effect. One index often used is given by*

$$T_h = 1.98T - 1.09(1 - H)(T - 58) - 56.9,$$

where T is the air temperature, in degrees Fahrenheit, and H is the relative humidity, expressed as a decimal. Find the temperature–humidity index in each case. Round to the nearest tenth of a degree.

45. $T = 85°F$ and $H = 60\%$

46. $T = 90°F$ and $H = 90\%$

47. $T = 90°F$ and $H = 100\%$

48. $T = 78°F$ and $H = 100\%$

49. Find $\dfrac{\partial T_h}{\partial H}$, and interpret its meaning.

50. Find $\dfrac{\partial T_h}{\partial T}$, and interpret its meaning.

51. Body surface area. The Mosteller formula for approximating the surface area, S, in m², of a human is given by

$$S = \frac{\sqrt{hw}}{60},$$

where h is the person's height in centimeters and w is the person's weight in kilograms. (*Source:* www.halls.md.)

a) Compute $\dfrac{\partial S}{\partial h}$.

b) Compute $\dfrac{\partial S}{\partial w}$.

c) The change in S due to a change in w when h is constant is approximately

$$\Delta S \approx \frac{\partial S}{\partial w}\Delta w.$$

Use this formula to approximate the change in someone's surface area given that the person is 170 cm tall, weighs 80 kg, and loses 2 kg.

52. Body surface area. The Haycock formula for approximating the surface area, S, in m², of a human is given by

$$S = 0.024265h^{0.3964}w^{0.5378},$$

where h is the person's height in centimeters and w is the person's weight in kilograms. (*Source:* www.halls.md.)

a) Compute $\dfrac{\partial S}{\partial h}$.

b) Compute $\dfrac{\partial S}{\partial w}$.

c) The change in S due to a change in w when h is constant is approximately

$$\Delta S \approx \frac{\partial S}{\partial w}\Delta w.$$

Use this formula to approximate the change in someone's surface area given that the person is 170 cm tall, weighs 80 kg, and loses 2 kg.

Social Sciences

Reading ease. *The following formula is used by psychologists and educators to predict the reading ease, E, of a passage of words:*

$$E = 206.835 - 0.846w - 1.015s,$$

where w is the number of syllables in a 100-word section and s is the average number of words per sentence. Find the reading ease in each case.

53. $w = 146$ and $s = 5$

54. $w = 180$ and $s = 6$

55. Find $\dfrac{\partial E}{\partial w}$.

56. Find $\dfrac{\partial E}{\partial s}$.

SYNTHESIS

Find f_x and f_t.

57. $f(x, t) = \dfrac{x^2 + t^2}{x^2 - t^2}$

58. $f(x, t) = \dfrac{x^2 - t}{x^3 + t}$

59. $f(x, t) = \dfrac{2\sqrt{x} - 2\sqrt{t}}{1 + 2\sqrt{t}}$

60. $f(x, t) = \sqrt[4]{x^3 t^5}$

61. $f(x, t) = 6x^{2/3} - 8x^{1/4}t^{1/2} - 12x^{-1/2}t^{3/2}$

62. $f(x, t) = \left(\dfrac{x^2 + t^2}{x^2 - t^2}\right)^5$

Find f_{xx}, f_{xy}, f_{yx}, and f_{yy}.

63. $f(x, y) = \dfrac{x}{y^2} - \dfrac{y}{x^2}$

64. $f(x, y) = \dfrac{xy}{x - y}$

65. Do some research on the Cobb–Douglas production function, and explain how it was developed.

66. Explain the meaning of the first partial derivatives of a function of two variables in terms of slopes of tangent lines.

67. Consider $f(x, y) = \ln(x^2 + y^2)$. Show that f is a solution to the partial differential equation

$$\frac{\partial^2 f}{\partial x^2} + \frac{\partial^2 f}{\partial y^2} = 0.$$

68. Consider $f(x, y) = x^3 - 5xy^2$. Show that f is a solution to the partial differential equation

$$xf_{xy} - f_y = 0.$$

69. Consider the function f defined as follows:

$$f(x, y) = \begin{cases} \dfrac{xy(x^2 - y^2)}{x^2 + y^2}, & \text{for } (x, y) \neq (0, 0), \\ 0, & \text{for } (x, y) = (0, 0). \end{cases}$$

a) Find $f_x(0, y)$ by evaluating the limit

$$\lim_{h \to 0} \frac{f(h, y) - f(0, y)}{h}.$$

b) Find $f_y(x, 0)$ by evaluating the limit

$$\lim_{h \to 0} \frac{f(x, h) - f(x, 0)}{h}.$$

c) Now find and compare $f_{yx}(0, 0)$ and $f_{xy}(0, 0)$.

Answers to Quick Checks

1. $\dfrac{\partial u}{\partial x} = 2xy^3z^4$, $\dfrac{\partial u}{\partial y} = 3x^2y^2z^4$, $\dfrac{\partial u}{\partial z} = 4x^2y^3z^3$

2. $f_x = 21x^2y^2 - \dfrac{1}{y}$, $f_y = 14x^3y + \dfrac{x}{y^2}$

3. $p_x(90, 50) = 51.21$ textbooks/unit of labor, $p_y(90, 50) = 23.05$ textbooks/unit of capital

4. $g_{xx} = 12$, $g_{yy} = 36xy^2 - 2$, $g_{xy} = 12y^3$, $g_{yx} = 12y^3$

6.3

OBJECTIVE

- Find relative extrema of a function of two variables.

Maximum–Minimum Problems

We will now find maximum and minimum values of functions of two variables.

> **DEFINITION**
>
> A function f of two variables:
> 1. has a **relative maximum** at (a, b) if
> $$f(x, y) \leq f(a, b)$$
> for all points (x, y) in a region containing (a, b);
> 2. has a **relative minimum** at (a, b) if
> $$f(x, y) \geq f(a, b)$$
> for all points (x, y) in a region containing (a, b).

This definition is illustrated in Figs. 1 and 2. A relative maximum (or minimum) may not be an "absolute" maximum (or minimum), as illustrated in Fig. 3.

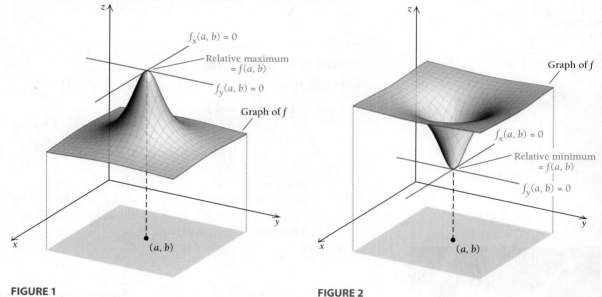

FIGURE 1

FIGURE 2

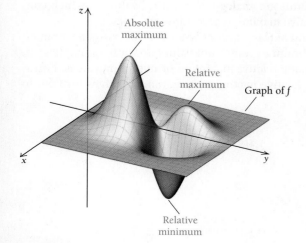

FIGURE 3

Determining Maximum and Minimum Values

Suppose that a function f has a relative maximum or minimum value at some point (a, b) inside its domain. (We assume that f and its partial derivatives exist and are "continuous" inside its domain, though we will not formally define continuity.) If we fix y at the value b, then $f(x, b)$ can be regarded as a function of x. Because a relative maximum or minimum occurs at (a, b), we know that $f(x, b)$ achieves a maximum or minimum at (a, b) and $f(x) = 0$. Similarly, if we fix x at a, then $f(a, y)$ can be regarded as a function of y that achieves a relative extremum at (a, b), and thus $f_y = 0$. In short, since an extremum exists at (a, b), we must have

$$f_x(a, b) = 0 \quad \text{and} \quad f_y(a, b) = 0. \tag{1}$$

We call a point (a, b) at which both partial derivatives are 0 a **critical point**. This concept of a critical value is comparable to that for functions of one variable. Thus, one strategy for finding relative maximum or minimum values is to solve a system of equations like (1) to find critical points. Just as for functions of one variable, this strategy does *not* guarantee that we will have a relative maximum or minimum value. We have argued only that *if f* has a maximum or minimum value at (a, b), *then* both its partial derivatives must be 0 at that point. Look back at Figs. 1 and 2. Then note Fig. 4, which illustrates a case in which the partial derivatives are 0 but the function does not have a relative maximum or minimum value at (a, b).

Considering Fig. 4, suppose that we fix y at a value b. Then $f(x, b)$, considered as the output of a function of one variable x, has a minimum at a, but f does not. Similarly, if we fix x at a, then $f(a, y)$, considered as the output of a function of one variable y, has a maximum at b, but f does not. The point $f(a, b)$ is called a **saddle point**. In other words, $f_x(a, b) = 0$ and $f_y(a, b) = 0$ [the point (a, b) is a critical point], but f does not attain a relative maximum or minimum value at (a, b).

A test for finding relative maximum and minimum values that involves the use of first- and second-order partial derivatives is stated below. We will not prove this theorem.

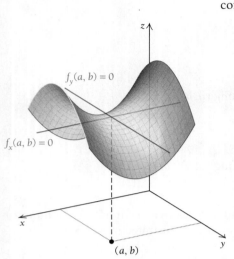

$f_y(a, b) = 0$

$f_x(a, b) = 0$

(a, b)

FIGURE 4

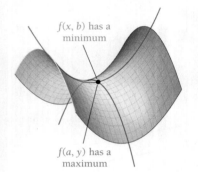

$f(x, b)$ has a minimum

$f(a, y)$ has a maximum

THEOREM 1   The *D*-Test

To find the relative maximum and minimum values of f:

1. Find f_x, f_y, f_{xx}, f_{yy}, and f_{xy}.
2. Solve the system of equations $f_x = 0$, $f_y = 0$. Let (a, b) represent a solution.
3. Evaluate D, where $D = f_{xx}(a, b) \cdot f_{yy}(a, b) - [f_{xy}(a, b)]^2$.
4. Then
   a) f has a maximum at (a, b) if $D > 0$ and $f_{xx}(a, b) < 0$.
   b) f has a minimum at (a, b) if $D > 0$ and $f_{xx}(a, b) > 0$.
   c) f has neither a maximum nor a minimum at (a, b) if $D < 0$. The function has a *saddle point* at (a, b). See Fig. 4.
   d) This test is not applicable if $D = 0$.

The *D*-test is somewhat analogous to the Second Derivative Test (Section 2.2) for functions of one variable. Saddle points are analogous to critical values at which concavity changes and there are no relative maximum or minimum values.

A relative maximum or minimum *may or may not be an absolute maximum or minimum value*. Tests for absolute maximum or minimum values are rather complicated. We will restrict our attention to finding *relative* maximum or minimum values. Fortunately, in most of our applications, relative maximum or minimum values turn out to be absolute as well.

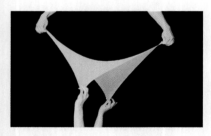

The shape of a perfect tent. *To give a tent roof the maximum strength possible, designers draw the fabric into a series of three-dimensional shapes that, viewed in profile, resemble a horse's saddle and that mathematicians call an anticlastic curve. Two people with a stretchy piece of fabric such as Spandex can duplicate the shape, as shown above. One person pulls up and out on two diagonal corners; the other person pulls down and out on the other two corners. The opposing tensions draw each point of the fabric's surface into rigid equilibrium. The more pronounced the curve, the stiffer the surface.*

■ **EXAMPLE 1** Find the relative maximum and minimum values of

$$f(x, y) = x^2 + xy + y^2 - 3x.$$

Solution

1. Find f_x, f_y, f_{xx}, f_{yy}, and f_{xy}:

$$f_x = 2x + y - 3, \qquad f_y = x + 2y,$$
$$f_{xx} = 2; \qquad\qquad f_{yy} = 2;$$
$$f_{xy} = 1.$$

2. Solve the system of equations $f_x = 0, f_y = 0$:

$$2x + y - 3 = 0, \tag{1}$$
$$x + 2y = 0. \tag{2}$$

Solving equation (2) for x, we get $x = -2y$. Substituting $-2y$ for x in equation (1) and solving, we get

$$2(-2y) + y - 3 = 0$$
$$-4y + y - 3 = 0$$
$$-3y = 3$$
$$y = -1.$$

To find x when $y = -1$, we substitute -1 for y in equation (1) or equation (2). We choose equation (2):

$$x + 2(-1) = 0$$
$$x = 2.$$

Thus, $(2, -1)$ is the only critical point, and $f(2, -1)$ is our candidate for a maximum or minimum value.

3. We must check to see whether $f(2, -1)$ is a maximum or minimum value:

$$D = f_{xx}(2, -1) \cdot f_{yy}(2, -1) - [f_{xy}(2, -1)]^2$$
$$= 2 \cdot 2 - [1]^2 \quad \text{Using step 1}$$
$$= 3.$$

4. Thus, $D = 3$ and $f_{xx}(2, -1) = 2$. Since $D > 0$ and $f_{xx}(2, -1) > 0$, it follows from the D-test that f has a relative minimum at $(2, -1)$. That minimum value is found as follows:

$$f(2, -1) = 2^2 + 2(-1) + (-1)^2 - 3 \cdot 2$$
$$= 4 - 2 + 1 - 6$$
$$= -3. \quad \text{This is the relative minimum.}$$

❯ **Quick Check 1**

Find the relative maximum and minimum values of

$$f(x, y) = x^2 + xy + 2y^2 - 7x.$$

❮ Quick Check 1

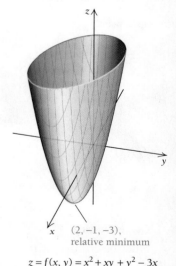

$(2, -1, -3)$, relative minimum

$z = f(x, y) = x^2 + xy + y^2 - 3x$

■ **EXAMPLE 2** Find the relative maximum and minimum values of

$$f(x, y) = xy - x^3 - y^2.$$

Solution

1. Find f_x, f_y, f_{xx}, f_{yy}, and f_{xy}:

$$f_x = y - 3x^2, \qquad f_y = x - 2y,$$
$$f_{xx} = -6x; \qquad f_{yy} = -2;$$
$$f_{xy} = 1.$$

2. Solve the system of equations $f_x = 0, f_y = 0$:

$$y - 3x^2 = 0, \tag{1}$$

$$x - 2y = 0. \tag{2}$$

Solving equation (1) for y, we get $y = 3x^2$. Substituting $3x^2$ for y in equation (2) and solving, we get

$$x - 2(3x^2) = 0$$
$$x - 6x^2 = 0$$
$$x(1 - 6x) = 0. \qquad \text{Factoring}$$

Setting each factor equal to 0 and solving, we have

$$x = 0 \quad \text{ or } \quad 1 - 6x = 0$$
$$x = 0 \quad \text{ or } \quad x = \tfrac{1}{6}.$$

To find y when $x = 0$, we substitute 0 for x in equation (1) or equation (2). We choose equation (2):

$$0 - 2y = 0$$
$$-2y = 0$$
$$y = 0.$$

Thus, $(0, 0)$ is a critical point, and $f(0, 0)$ is one candidate for a maximum or minimum value. To find the other, we substitute $\tfrac{1}{6}$ for x in either equation (1) or equation (2). We choose equation (2):

$$\tfrac{1}{6} - 2y = 0$$
$$-2y = -\tfrac{1}{6}$$
$$y = \tfrac{1}{12}.$$

Thus, $\left(\tfrac{1}{6}, \tfrac{1}{12}\right)$ is another critical point, and $f\left(\tfrac{1}{6}, \tfrac{1}{12}\right)$ is another candidate for a maximum or minimum value.

3–4. We must check both $(0, 0)$ and $\left(\tfrac{1}{6}, \tfrac{1}{12}\right)$ to see whether they yield maximum or minimum values.

For $(0, 0)$: $\quad D = f_{xx}(0, 0) \cdot f_{yy}(0, 0) - \left[f_{xy}(0, 0)\right]^2$
$$= (-6 \cdot 0) \cdot (-2) - [1]^2 \qquad \text{Using step 1}$$
$$= -1.$$

Since $D < 0$, it follows that $f(0, 0)$ is neither a maximum nor a minimum value, but a saddle point.

For $\left(\tfrac{1}{6}, \tfrac{1}{12}\right)$: $D = f_{xx}\left(\tfrac{1}{6}, \tfrac{1}{12}\right) \cdot f_{yy}\left(\tfrac{1}{6}, \tfrac{1}{12}\right) - \left[f_{xy}\left(\tfrac{1}{6}, \tfrac{1}{12}\right)\right]^2$
$$= \left(-6 \cdot \tfrac{1}{6}\right) \cdot (-2) - [1]^2$$
$$= -1(-2) - 1 \qquad \text{Using step 1}$$
$$= 1.$$

Thus, $D = 1$ and $f_{xx}\left(\tfrac{1}{6}, \tfrac{1}{12}\right) = -1$. Since $D > 0$ and $f_{xx}\left(\tfrac{1}{6}, \tfrac{1}{12}\right) < 0$, it follows that f has a relative maximum at $\left(\tfrac{1}{6}, \tfrac{1}{12}\right)$; that maximum value is

$$f\left(\tfrac{1}{6}, \tfrac{1}{12}\right) = \tfrac{1}{6} \cdot \tfrac{1}{12} - \left(\tfrac{1}{6}\right)^3 - \left(\tfrac{1}{12}\right)^2$$
$$= \tfrac{1}{72} - \tfrac{1}{216} - \tfrac{1}{144} = \tfrac{1}{432}. \qquad \text{This is the relative maximum.}$$

$(0, 0, 0)$,
saddle point

$\left(\tfrac{1}{6}, \tfrac{1}{12}, \tfrac{1}{432}\right)$,
relative maximum

$z = f(x, y) = xy - x^3 - y^2$

》 Quick Check 2

Find the critical points of

$$g(x, y) = x^3 + y^2 - 3x - 4y + 3.$$

Then use the *D*-test to classify each point as a relative maximum, a relative minimum, or a saddle point.

《 Quick Check 2

■ **EXAMPLE 3** **Business: Maximizing Profit.** A firm produces two kinds of golf ball, one that sells for $3 and one priced at $2. The total revenue, in thousands of dollars, from the sale of x thousand balls at $3 each and y thousand at $2 each is given by

$$R(x, y) = 3x + 2y.$$

The company determines that the total cost, in thousands of dollars, of producing x thousand of the $3 ball and y thousand of the $2 ball is given by

$$C(x, y) = 2x^2 - 2xy + y^2 - 9x + 6y + 7.$$

How many balls of each type must be produced and sold in order to maximize profit?

Solution The total profit $P(x, y)$ is given by

$$
\begin{aligned}
P(x, y) &= R(x, y) - C(x, y) \\
&= 3x + 2y - (2x^2 - 2xy + y^2 - 9x + 6y + 7) \\
P(x, y) &= -2x^2 + 2xy - y^2 + 12x - 4y - 7.
\end{aligned}
$$

1. Find P_x, P_y, P_{xx}, P_{yy}, and P_{xy}:

$$
\begin{array}{ll}
P_x = -4x + 2y + 12, & P_y = 2x - 2y - 4, \\
P_{xx} = -4; & P_{yy} = -2; \\
& P_{xy} = 2.
\end{array}
$$

2. Solve the system of equations $P_x = 0, P_y = 0$:

$$-4x + 2y + 12 = 0, \tag{1}$$
$$2x - 2y - 4 = 0. \tag{2}$$

Adding these equations, we get

$$-2x + 8 = 0.$$

Then

$$-2x = -8$$
$$x = 4.$$

To find y when $x = 4$, we substitute 4 for x in equation (1) or equation (2). We choose equation (2):

$$
\begin{aligned}
2 \cdot 4 - 2y - 4 &= 0 \\
-2y + 4 &= 0 \\
-2y &= -4 \\
y &= 2.
\end{aligned}
$$

Thus, $(4, 2)$ is the only critical point, and $P(4, 2)$ is a candidate for a maximum or minimum value.

3. We must check to see whether $P(4, 2)$ is a maximum or minimum value:

$$
\begin{aligned}
D &= P_{xx}(4, 2) \cdot P_{yy}(4, 2) - [P_{xy}(4, 2)]^2 \\
&= (-4)(-2) - 2^2 \qquad \text{Using step 1} \\
&= 4.
\end{aligned}
$$

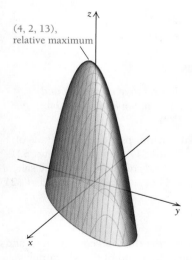

$(4, 2, 13)$,
relative maximum

$z = P(x, y) = -2x^2 + 2xy - y^2 + 12x - 4y - 7$

4. Thus, $D = 4$ and $P_{xx}(4, 2) = -4$. Since $D > 0$ and $P_{xx}(4, 2) < 0$, it follows that P has a relative maximum at $(4, 2)$. So, in order to maximize profit, the company must produce and sell 4 thousand of the $3 golf balls and 2 thousand of the $2 golf balls. The maximum profit will be

$$P(4, 2) = -2 \cdot 4^2 + 2 \cdot 4 \cdot 2 - 2^2 + 12 \cdot 4 - 4 \cdot 2 - 7 = 13,$$

or $13 thousand.

❮ Quick Check 3

Section Summary

- A two-variable function f has a *relative maximum* at (a, b) if $f(x, y) \leq f(a, b)$ for all points in a region containing (a, b) and has a *relative minimum* at (a, b) if $f(x, y) \geq f(a, b)$ for all points in a region containing (a, b).

- The *D-test* is used to classify a *critical point* as a relative minimum, a relative maximum, or a *saddle point*.

EXERCISE SET
6.3

Find the relative maximum and minimum values.

1. $f(x, y) = x^2 + xy + y^2 - y$

2. $f(x, y) = x^2 + xy + y^2 - 5y$

3. $f(x, y) = 2xy - x^3 - y^2$

4. $f(x, y) = 4xy - x^3 - y^2$

5. $f(x, y) = x^3 + y^3 - 3xy$

6. $f(x, y) = x^3 + y^3 - 6xy$

7. $f(x, y) = x^2 + y^2 - 2x + 4y - 2$

8. $f(x, y) = x^2 + 2xy + 2y^2 - 6y + 2$

9. $f(x, y) = x^2 + y^2 + 2x - 4y$

10. $f(x, y) = 4y + 6x - x^2 - y^2$

11. $f(x, y) = 4x^2 - y^2$

12. $f(x, y) = x^2 - y^2$

13. $f(x, y) = e^{x^2 + y^2 + 1}$

14. $f(x, y) = e^{x^2 - 2x + y^2 - 4y + 2}$

APPLICATIONS

Business and Economics

In Exercises 15–22, assume that relative maximum and minimum values are absolute maximum and minimum values.

15. Maximizing profit. Safe Shades produces two kinds of sunglasses; one kind sells for $17, and the other for $21. The total revenue in thousands of dollars from the sale of x thousand sunglasses at $17 each and y thousand at $21 each is given by

$$R(x, y) = 17x + 21y.$$

The company determines that the total cost, in thousands of dollars, of producing x thousand of the $17 sunglasses and y thousand of the $21 sunglasses is given by

$$C(x, y) = 4x^2 - 4xy + 2y^2 - 11x + 25y - 3.$$

Find the number of each type of sunglasses that must be produced and sold in order to maximize profit.

16. Maximizing profit. A concert promoter produces two kinds of souvenir shirt; one kind sells for $18, and the other for $25. The total revenue from the sale of x

thousand shirts at \$18 each and y thousand at \$25 each is given by

$$R(x, y) = 18x + 25y.$$

The company determines that the total cost, in thousands of dollars, of producing x thousand of the \$18 shirt and y thousand of the \$25 shirt is given by

$$C(x, y) = 4x^2 - 6xy + 3y^2 + 20x + 19y - 12.$$

How many of each type of shirt must be produced and sold in order to maximize profit?

17. Maximizing profit. A one-product company finds that its profit, P, in millions of dollars, is given by

$$P(a, p) = 2ap + 80p - 15p^2 - \tfrac{1}{10}a^2p - 80,$$

where a is the amount spent on advertising, in millions of dollars, and p is the price charged per item of the product, in dollars. Find the maximum value of P and the values of a and p at which it is attained.

18. Maximizing profit. A one-product company finds that its profit, P, in millions of dollars, is given by

$$P(a, n) = -5a^2 - 3n^2 + 48a - 4n + 2an + 290,$$

where a is the amount spent on advertising, in millions of dollars, and n is the number of items sold, in thousands. Find the maximum value of P and the values of a and n at which it is attained.

19. Minimizing the cost of a container. A trash company is designing an open-top, rectangular container that will have a volume of 320 ft^3. The cost of making the bottom of the container is \$5 per square foot, and the cost of the sides is \$4 per square foot. Find the dimensions of the container that will minimize total cost. (*Hint:* Make a substitution using the formula for volume.)

20. Two-variable revenue maximization. Boxowitz, Inc., a computer firm, markets two kinds of calculator that compete with one another. Their demand functions are expressed by the following relationships:

$$q_1 = 78 - 6p_1 - 3p_2, \tag{1}$$
$$q_2 = 66 - 3p_1 - 6p_2, \tag{2}$$

where p_1 and p_2 are the prices of the calculators, in multiples of \$10, and q_1 and q_2 are the quantities of the calculators demanded, in hundreds of units.

   a) Find a formula for the total-revenue function, R, in terms of the variables p_1 and p_2. [*Hint:* $R = p_1q_1 + p_2q_2$; then substitute expressions from equations (1) and (2) to find $R(p_1, p_2)$.]
   b) What prices p_1 and p_2 should be charged for each product in order to maximize total revenue?
   c) How many units will be demanded?
   d) What is the maximum total revenue?

21. Two-variable revenue maximization. Repeat Exercise 20, using

$$q_1 = 64 - 4p_1 - 2p_2$$

and

$$q_2 = 56 - 2p_1 - 4p_2.$$

Life and Physical Sciences

22. Temperature. A flat metal plate is located on a coordinate plane. The temperature of the plate, in degrees Fahrenheit, at point (x, y) is given by

$$T(x, y) = x^2 + 2y^2 - 8x + 4y.$$

Find the minimum temperature and where it occurs. Is there a maximum temperature?

SYNTHESIS

Find the relative maximum and minimum values and the saddle points.

23. $f(x, y) = e^x + e^y - e^{x+y}$

24. $f(x, y) = xy + \dfrac{2}{x} + \dfrac{4}{y}$

25. $f(x, y) = 2y^2 + x^2 - x^2y$

26. $S(b, m) = (m + b - 72)^2 + (2m + b - 73)^2 + (3m + b - 75)^2$

27. Is a cross-section of an anticlastic curve always a parabola? Why or why not?

28. Explain the difference between a relative minimum and an absolute minimum of a function of two variables.

TECHNOLOGY CONNECTION

Use a 3D graphics program to graph each of the following functions. Then estimate any relative extrema.

29. $f(x, y) = \dfrac{-5}{x^2 + 2y^2 + 1}$

30. $f(x, y) = x^3 + y^3 + 3xy$

31. $f(x, y) = \dfrac{3xy(x^2 - y^2)}{x^2 + y^2}$

32. $f(x, y) = \dfrac{y + x^2y^2 - 8x}{xy}$

Answers to Quick Checks

1. $(4, -1, -14)$, relative minimum   2. $(1, 2, -3)$, relative minimum; $(-1, 2, 1)$, saddle point   3. Maximum profit is \$17.031 thousand when $x = $ \$4.625 thousand and $y = $ \$3 thousand

6.4

An Application: The Least-Squares Technique

We have made frequent use in this book of a graphing calculator to perform regression. The purpose of this section is to develop an understanding of the process of regression by using the method for finding the minimum value for a function of two variables developed in the preceding section. We first considered regression in Section R.6. An equation found by regression provides a model of the phenomenon that the data measure, from which predictions can be made. For example, in business, one might want to predict future sales on the basis of past data. In ecology, one might want to predict future demand for natural gas on the basis of past usage. Suppose that we wish to find a linear equation,

$$y = mx + b,$$

to fit some data. To determine this equation is to determine the values of m and b. But how? Let's consider some factual data.

Suppose that a car rental company that offers hybrid (gas–electric) vehicles charts its revenue as shown in Fig. 1 and the accompanying table. How best could we predict the company's revenue for the year 2016?

YEARLY REVENUE OF SKY BLUE CAR RENTALS

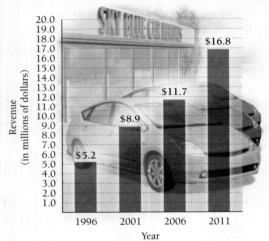

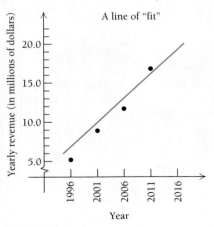

FIGURE 1

Year, x	1996	2001	2006	2011	2016
Yearly Revenue, y (in millions of dollars)	5.2	8.9	11.7	16.8	?

Suppose that we plot these points and try to draw a line through them that fits. Note that there are several ways in which this might be done (see Figs. 2 and 3). Each would give a different estimate of the company's total revenue for 2016.

Note that the years for which revenue is given follow 5-yr increments. Thus, computations can be simplified if we use the data points $(1, 5.2)$, $(2, 8.9)$, $(3, 11.7)$, and $(4, 16.8)$, as plotted in Fig. 3, where each horizontal unit represents 5 years and $x = 1$ is 1996.

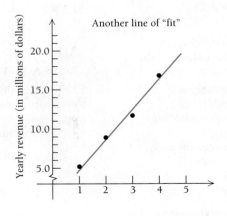

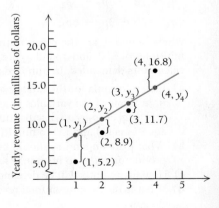

FIGURE 2

FIGURE 3

FIGURE 4

To determine the equation of the line that "best" fits the data, we note that for each data point there will be a deviation, or error, between the y-value at that point and the y-value of the point on the line that is directly above or below the point. Those deviations, in this case, $y_1 - 5.2, y_2 - 8.9, y_3 - 11.7$, and $y_4 - 16.8$, will be positive or negative, depending on the location of the line (see Fig. 4).

We wish to fit these data points with a line,

$$y = mx + b,$$

that uses values of m and b that, somehow, minimize the y-deviations in order to have a good fit. One way of minimizing the deviations is based on the *least-squares assumption*.

The Least-Squares Assumption

The line of best fit is the line for which the sum of the squares of the y-deviations is a minimum. This is called the **regression line**.

Note that squaring each y-deviation gives us a series of nonnegative terms that we can sum. Were we to simply add the y-deviations, positive and negative deviations would cancel each other out.

Using the least-squares assumption with the yearly revenue data, we want to minimize

$$(y_1 - 5.2)^2 + (y_2 - 8.9)^2 + (y_3 - 11.7)^2 + (y_4 - 16.8)^2. \tag{1}$$

Also, since the points $(1, y_1)$, $(2, y_2)$, $(3, y_3)$, and $(4, y_4)$ must be solutions of $y = mx + b$, it follows that

$$y_1 = m(1) + b = m + b,$$
$$y_2 = m(2) + b = 2m + b,$$
$$y_3 = m(3) + b = 3m + b,$$
$$y_4 = m(4) + b = 4m + b.$$

Substituting $m + b$ for y_1, $2m + b$ for y_2, $3m + b$ for y_3, and $4m + b$ for y_4 in equation (1), we now have a function of two variables:

$$S(m, b) = (m + b - 5.2)^2 + (2m + b - 8.9)^2 + (3m + b - 11.7)^2 + (4m + b - 16.8)^2.$$

Thus, to find the regression line for the given set of data, we must find the values of m and b that minimize the function S given by the sum in this last equation.

To apply the D-test, we first find the partial derivatives $\partial S/\partial b$ and $\partial S/\partial m$:

$$\frac{\partial S}{\partial b} = 2(m + b - 5.2) + 2(2m + b - 8.9) + 2(3m + b - 11.7) + 2(4m + b - 16.8)$$

$$= 20m + 8b - 85.2,$$

and

$$\frac{\partial S}{\partial m} = 2(m + b - 5.2) + 2(2m + b - 8.9)2 + 2(3m + b - 11.7)3 + 2(4m + b - 16.8)4$$

$$= 60m + 20b - 250.6.$$

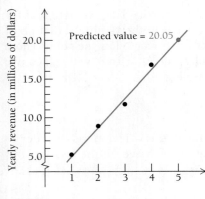

FIGURE 5

We set these derivatives equal to 0 and solve the resulting system:

$$20m + 8b - 85.2 = 0, \qquad\qquad 5m + 2b = 21.3,$$
$$\text{or}$$
$$60m + 20b - 250.6 = 0; \qquad 15m + 5b = 62.65.$$

It can be shown that the solution of this system is

$$b = 1.25, \qquad m = 3.76. \qquad \text{(See the Technology Connection below.)}$$

We leave it to the student to complete the *D*-test to verify that $(1.25, 3.76)$ does, in fact, yield the minimum of *S*. There is no need to compute $S(1.25, 3.76)$.

The values of *m* and *b* are all we need to determine $y = mx + b$. The regression line is

$$y = 3.76x + 1.25. \qquad \text{Substituting for } m \text{ and } b$$

The graph of this "best-fit" regression line together with the data points is shown in Fig. 5. Compare it to Figs. 2, 3, and 4.

TECHNOLOGY CONNECTION

Solving Linear Systems Using Matrices

In this Technology Connection, we explore a method of solution called *reduced row echelon form* (rref). We can use this method to solve the system of equations discussed above:

$$5m + 2b = 21.3,$$
$$15m + 5b = 62.65.$$

From this system, we can write the matrix

$$\begin{bmatrix} 5 & 2 & 21.3 \\ 15 & 5 & 62.65 \end{bmatrix}.$$

The first column is called the *m*-column (because the entries are the coefficients of the variable *m*), the second column is called the *b*-column, and the final column is the constants column. Before entering the numbers of the system of equations into a matrix, it is crucial that the *m* and *b* terms are in the correct positions to the left of the equal signs and the constants are to the right of the equal signs.

When a matrix is in reduced row echelon form, it has the following appearance:

$$\begin{bmatrix} 1 & 0 & a \\ 0 & 1 & b \end{bmatrix}.$$

With this form, the system is considered solved, as we can rewrite this matrix as the system

$$1x + 0y = a,$$
$$0x + 1y = b.$$

Therefore,

$$x = a \text{ and } y = b.$$

On your calculator, select MATRIX (it may be a 2nd function on some models). Under EDIT, select [A]. With this setting, matrix [A] has 2 rows and 3 columns, so it is of size 2×3. Enter these values, pressing **ENTER** after each one. Then enter the values of the matrix into the matrix field, pressing **ENTER** after each one. After you have entered the matrix values, press **2ND** and QUIT to exit. Matrix [A] is now stored in the calculator's memory.

To convert matrix [A] into reduced row echelon form, press MATRIX, and under MATH, scroll down to rref. Press **ENTER**. Now press MATRIX once again, and under NAMES, select [A], and press **ENTER**. The result will be the reduced row echelon form equivalent to the original matrix [A]:

Therefore, $m = 3.76$ and $b = 1.25$.

EXERCISES

Use the reduced row echelon form of a matrix to solve the following systems of equations with your calculator.

1. $2x + 6y = 14$
   $x - 5y = -17$

2. $3x + y = 7$
   $10x + 3y = 11$

3. $2x + y + 7 = 0$
   $x = 6 - y$

We can now extrapolate from the data to predict the car rental company's yearly revenue in 2016:

$$y = 3.76(5) + 1.25 = 20.05.$$

The yearly revenue in 2016 is predicted to be about $20.05 million. How might you check this prediction?

The method of least squares is a statistical process illustrated here with only four data points in order to simplify the explanation. Most statistical researchers would warn that many more than four data points should be used to get a "good" regression line. Furthermore, making predictions too far in the future from any mathematical model may not be valid. The further into the future a prediction is made, the more dubious one should be about the prediction.

⟩ Quick Check 1

Use the method of least squares to determine the regression line for the data points $(1, 25)$, $(2, 48)$, $(3, 76.7)$, and $(4, 104.8)$.

⟨ Quick Check 1

Exploratory

As we have seen in Section R.6 and in other parts of the book, graphing calculators can perform linear regression, as well as quadratic, exponential, and logarithmic regression. Use such a calculator now to fit a linear equation to the yearly revenue data for the car rental company.

With some calculators, you will also obtain a number r, called the **coefficient of correlation**. Although we cannot develop that concept in detail in this text, keep in mind that r is used to describe the strength of the linear relationship between x and y. The closer $|r|$ is to 1, the better the correlation.

For the yearly revenue data, $r \approx 0.993$, which indicates a fairly good linear relationship. Keep in mind that a high linear correlation does not necessarily indicate a "cause-and-effect" connection between the variables.

*The Regression Line for an Arbitrary Collection of Data Points (c_1, d_1), (c_2, d_2), ... , (c_n, d_n)

Look again at the regression line

$$y = 3.76x + 1.25$$

for the data points $(1, 5.2)$, $(2, 8.9)$, $(3, 11.7)$, and $(4, 16.8)$. Let's consider the arithmetic averages, or means, of the x-coordinates, denoted $\bar{x}$, and of the y-coordinates, denoted $\bar{y}$:

$$\bar{x} = \frac{1 + 2 + 3 + 4}{4} = 2.5,$$

$$\bar{y} = \frac{5.2 + 8.9 + 11.7 + 16.8}{4} = 10.65.$$

It turns out that the point $(\bar{x}, \bar{y})$, or $(2.5, 10.65)$, is on the regression line since

$$10.65 = 3.76(2.5) + 1.25.$$

Thus, the equation for the regression line can be written

$$y - \bar{y} = m(x - \bar{x}),$$

or, in this case,

$$y - 10.65 = m(x - 2.5).$$

All that remains, in general, is to determine m.

Suppose that we want to find the regression line for an arbitrary number of points (c_1, d_1), (c_2, d_2), ..., (c_n, d_n). To do so, we find the values m and b that minimize the function S given by

$$S(b, m) = (y_1 - d_1)^2 + (y_2 - d_2)^2 + \cdots + (y_n - d_n)^2 = \sum_{i=1}^{n} (y_i - d_i)^2,$$

where $y_i = mc_i + b$.

---

*This subsection is considered optional and can be omitted without loss of continuity.

Using a procedure like the one we used earlier to minimize S, we can show that $y = mx + b$ takes the form

$$y - \bar{y} = m(x - \bar{x}),$$

where $\quad \bar{x} = \dfrac{\sum\limits_{i=1}^{n} c_i}{n}, \bar{y} = \dfrac{\sum\limits_{i=1}^{n} d_i}{n}, \quad$ and $\quad m = \dfrac{\sum\limits_{i=1}^{n}(c_i - \bar{x})(d_i - \bar{y})}{\sum\limits_{i=1}^{n}(c_i - \bar{x})^2}.$

Let's see how this works out for the yearly revenue data from our earlier example.

c_i	d_i	$c_i - \bar{x}$	$(c_i - \bar{x})^2$	$(d_i - \bar{y})$	$(c_i - \bar{x})(d_i - \bar{y})$
1	5.2	−1.5	2.25	−5.45	8.175
2	8.9	−0.5	0.25	−1.75	0.875
3	11.7	0.5	0.25	1.05	0.525
4	16.8	1.5	2.25	6.15	9.225

$$\sum_{i=1}^{4} c_i = 10 \qquad \sum_{i=1}^{4} d_i = 42.6 \qquad \sum_{i=1}^{4}(c_i - \bar{x})^2 = 5 \qquad \sum_{i=1}^{4}(c_i - \bar{x})(d_i - \bar{y}) = 18.8$$

$$\bar{x} = 2.5 \qquad \bar{y} = 10.65 \qquad\qquad m = \frac{18.8}{5} = 3.76$$

Thus, the regression line is

$$y - 10.65 = 3.76(x - 2.5),$$

which simplifies to

$$y = 3.76x + 1.25.$$

Section Summary

- Regression is a technique for determining a continuous function that "best fits" a set of data points.

- For linear regression, the method of least squares uses calculus on a function of two variables to find the values m and b that determine the *regression line* $y = mx + b$, the line of best fit.

EXERCISE SET
6.4

For each data set, find the regression line without using a calculator.

1.

x	1	2	4	5
y	1	3	3	4

2.

x	1	3	5
y	2	4	7

3.

x	1	2	3	5
y	0	1	3	4

4.

x	1	2	4
y	3	5	8

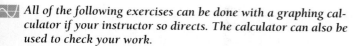

All of the following exercises can be done with a graphing calculator if your instructor so directs. The calculator can also be
used to check your work.

APPLICATIONS

Business and Economics

5. Labor force. The minimum hourly wage in the United
States has grown over the years, as shown in the table
below.

Number of Years, x, since 1990	Minimum Hourly Wage
0	$3.80
1	4.25
6	4.75
7	5.15
17	5.85
18	6.55
19	7.25

(*Source:* www.workworld.org.)

a) For the data in the table, find the regression line,
$y = mx + b$.
b) Use the regression line to predict the minimum
hourly wage in 2015 and 2020.

6. Football ticket prices. Ticket prices for NFL football
games have experienced steady growth, as shown in the
following table.

Number of Years, x, since 2005 Season	Average Ticket Price (dollars)
0	$58.95
1	62.38
2	67.11
3	72.20
4	73.18
5	76.47

(*Source:* Team Marketing Report.)

a) Find the regression line, $y = mx + b$.
b) Use the regression line to predict the average ticket
price for an NFL game in 2015 and 2020.

Life and Physical Sciences

7. Life expectancy of women. Consider the data in the following table showing the average life expectancy of women
in various years. Note that x represents the actual year.

Year, x	Life Expectancy of Women, y (years)
1950	71.1
1960	73.1
1970	74.7
1980	77.4
1990	78.8
2000	79.5
2003	80.1
2007	80.4

(*Source:* www.ssa.gov.)

a) Find the regression line, $y = mx + b$.
b) Use the regression line to predict the life expectancy
of women in 2015 and 2020.

8. Life expectancy of men. Consider the following data
showing the average life expectancy of men in various
years. Note that x represents the actual year.

Year, x	Life Expectancy of Men, y (years)
1950	65.6
1960	66.6
1970	67.1
1980	70.0
1990	71.8
2000	74.1
2003	74.8
2007	75.4

(*Source:* www.ssa.gov.)

a) Find the regression line, $y = mx + b$.
b) Use the regression line to predict the life expectancy
of men in 2015 and 2020.

General Interest

9. Grade predictions. A professor wants to predict students' final examination scores on the basis of their
midterm test scores. An equation was determined on the
basis of data on the scores of three students who took
the same course with the same instructor the previous
semester (see the following table).

Midterm Score, x	Final Exam Score, y
70%	75%
60	62
85	89

a) Find the regression line, $y = mx + b$. (*Hint*: The y-deviations are $70m + b - 75$, $60m + b - 62$, and so on.)

b) The midterm score of a student was 81%. Use the regression line to predict the student's final exam score.

10. Predicting the world record in the high jump. It has been established that most world records in track and field can be modeled by a linear function. The table below shows world high-jump records for various years. Note that x represents the actual year.

Year, x	World Record in High Jump, y (in inches)
1912 (George Horme)	78.0
1956 (Charles Dumas)	84.5
1973 (Dwight Stones)	90.5
1989 (Javier Sotomayer)	96.0
1993 (Javier Sotomayer)	96.5

(*Source*: www.wikipedia.org.)

a) Find the regression line, $y = mx + b$.

b) Use the regression line to predict the world record in the high jump in 2010 and in 2050.

c) Does your answer in part (b) for 2050 seem realistic? Explain why extrapolating so far into the future could be a problem.

11. How would you explain the concept of linear regression to a friend?

12. Discuss the idea of linear regression with a professor from another discipline in which regression is used. Explain how it is used in that field.

TECHNOLOGY CONNECTION

13. General interest: predicting the world record for running the mile. Note that x represents the actual year in following table.

Year, x	World Record, y (in minutes:seconds)
1875 (Walter Slade)	4:24.5
1894 (Fred Bacon)	4:18.2
1923 (Paavo Nurmi)	4:10.4
1937 (Sidney Wooderson)	4:06.4
1942 (Gunder Hägg)	4:06.2
1945 (Gunder Hägg)	4:01.4
1954 (Roger Bannister)	3:59.6
1964 (Peter Snell)	3:54.1
1967 (Jim Ryun)	3:51.1
1975 (John Walker)	3:49.4
1979 (Sebastian Coe)	3:49.0
1980 (Steve Ovett)	3:48.40
1985 (Steve Cram)	3:46.31
1993 (Noureddine Morceli)	3:44.39

(*Source*: USA Track & Field and infoplease.com.)

a) Find the regression line, $y = mx + b$, that fits the data in the table. (*Hint*: Convert each time to decimal notation; for instance, $4{:}24.5 = 4\frac{24.5}{60} = 4.4083$.)

b) Use the regression line to predict the world record in the mile in 2010 and in 2015.

c) In July 1999, Hicham El Guerrouj set the current (as of December 2006) world record of 3:43.13 for the mile. (*Source*: USA Track & Field and infoplease. com.) How does this compare with what is predicted by the regression line?

Answer to Quick Check

1. $y = 26.81x - 3.4$

Constrained Optimization

6.5

In Section 6.3, we discussed a method for determining maximum and minimum values on a surface represented by a two-variable function $z = f(x, y)$. If restrictions are placed on the input variables x and y, we can determine the maximum and minimum values on the surface subject to the restrictions. This process is called **constrained optimization**.

OBJECTIVES

• Find maximum and minimum values using Lagrange multipliers.

• Solve constrained optimization problems involving Lagrange multipliers.

Path Constraints: Lagrange Multipliers

Imagine that you are hiking up a mountain. If there are no constraints on your movement, you may seek out the mountain's summit—its "maximum point." The figure at the right shows a relief map of a mountaintop; its unconstrained maximum point occurs at the •, labeled with a spot elevation of 6903 ft. A hiking trail, marked as a black dashed line, bypasses the summit. If you were constrained to this hiking path, you could not reach the summit.

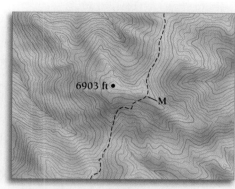

6903 ft •

M

(*Source*: USGS maps at www.mytopo.com.)

You could, however, achieve a maximum elevation along the path. This constrained maximum point is approximated at *M*.

In many applications modeled by two-variable functions, constraints on the input variables are necessary. If the input variables are related to one another by an equation, it is called a **constraint**.

Let's return to a problem we considered in Chapter 2: A hobby store has 20 ft of fencing to fence off a rectangular electric-train area in one corner of its display room. The two sides up against the wall require no fence. What dimensions of the rectangle will maximize the area?

We maximize the function

$$A = xy$$

subject to the condition, or *constraint*, $x + y = 20$. Note that A is a function of two variables.

When we solved this earlier, we first solved the constraint for y:

$$y = 20 - x.$$

We then substituted $20 - x$ for y to obtain

$$A(x, y) = x(20 - x)$$
$$= 20x - x^2,$$

which is a function of one variable. Next, we found a maximum value using Maximum–Minimum Principle 1 (see Section 2.4). By itself, the function of two variables

$$A(x, y) = xy$$

$20 - x$

x

has no maximum value. This can be checked using the *D*-test. With the constraint $x + y = 20$, however, the function does have a maximum. We see this in the following graph.

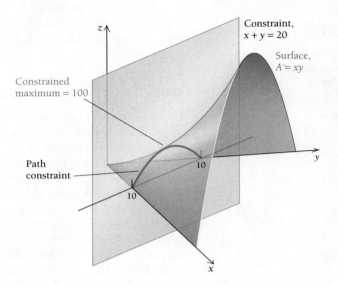

It may be quite difficult to solve a constraint for one variable. The method outlined below allows us to proceed without doing so.

The Method of Lagrange Multipliers

To find a maximum or minimum value of a function $f(x, y)$ subject to the constraint $g(x, y) = 0$:

1. Form a new function, called the **Lagrange function**:

$$F(x, y, \lambda) = f(x, y) - \lambda g(x, y).$$

The variable λ (lambda) is called a **Lagrange multiplier**.

2. Find the first partial derivatives F_x, F_y, and F_λ.

3. Solve the system

$$F_x = 0, \qquad F_y = 0, \quad \text{and} \quad F_\lambda = 0.$$

Let (a, b, λ) represent a solution of this system. We normally must determine whether (a, b) yields a maximum or minimum of the function f. For the problems in this text, we will specify that a maximum or minimum exists.

The method of Lagrange multipliers can be extended to functions of three (or more) variables.

We can illustrate the method of Lagrange multipliers by resolving the electric-train area problem.

■ EXAMPLE 1 Find the maximum value of

$$A(x, y) = xy$$

subject to the constraint $x + y = 20$.

Solution Note first that $x + y = 20$ is equivalent to $x + y - 20 = 0$.

1. We form the Lagrange function F, given by

$$F(x, y, \lambda) = xy - \lambda \cdot (x + y - 20).$$

2. We find the first partial derivatives:

$$F_x = y - \lambda,$$
$$F_y = x - \lambda,$$
$$F_\lambda = -(x + y - 20).$$

3. We set each derivative equal to 0 and solve the resulting system:

$$y - \lambda = 0, \tag{1}$$
$$x - \lambda = 0, \tag{2}$$
$$-(x + y - 20) = 0, \quad \text{or} \quad x + y - 20 = 0. \tag{3}$$

From equations (1) and (2), it follows that

$$x = y = \lambda.$$

Substituting x for y in equation (3), we get

$$x + x - 20 = 0$$
$$2x = 20$$
$$x = 10.$$

Thus, $y = x = 10$. The maximum value of A subject to the constraint occurs at $(10, 10)$ and is

$$A(10, 10) = 10 \cdot 10 = 100.$$

⟩ **Quick Check 1**

Find the maximum value of $A(x, y) = xy$ subject to the constraint $x + 2y = 30$.

❰ **Quick Check 1**

▪ **EXAMPLE 2**   Find the maximum value of

$$f(x, y) = 3xy$$

subject to the constraint

$$2x + y = 8.$$

Note: f might be interpreted, for example, as a production function with a budget constraint $2x + y = 8$.

Solution   Note that first we express $2x + y = 8$ as $2x + y - 8 = 0$.

1. We form the Lagrange function F, given by

$$F(x, y, \lambda) = 3xy - \lambda(2x + y - 8).$$

2. We find the first partial derivatives:

$$F_x = 3y - 2\lambda,$$
$$F_y = 3x - \lambda,$$
$$F_\lambda = -(2x + y - 8).$$

3. We set each derivative equal to 0 and solve the resulting system:

$$3y - 2\lambda = 0, \tag{1}$$
$$3x - \lambda = 0, \tag{2}$$
$$-(2x + y - 8) = 0, \quad \text{or} \quad 2x + y - 8 = 0. \tag{3}$$

Solving equation (2) for λ, we get

$$\lambda = 3x.$$

〉 Quick Check 2

Find the minimum value of $g(x, y) = x^2 + y^2$ subject to the constraint $3x - y = 1$.

Substituting in equation (1) for λ, we get

$$3y - 2 \cdot 3x = 0, \quad \text{or} \quad 3y = 6x, \quad \text{or} \quad y = 2x. \tag{4}$$

Substituting $2x$ for y in equation (3), we get

$$2x + 2x - 8 = 0$$
$$4x = 8$$
$$x = 2.$$

Then, using equation (4), we have

$$y = 2 \cdot 2 = 4.$$

The maximum value of f subject to the constraint occurs at $(2, 4)$ and is

$$f(2, 4) = 3 \cdot 2 \cdot 4 = 24.$$

〈 Quick Check 2

■ EXAMPLE 3 Business: The Beverage-Can Problem.   The standard beverage can holds 12 fl. oz, or has a volume of 21.66 in^3. What dimensions yield the minimum surface area? Find the minimum surface area. (Assume that the shape of the can is a right circular cylinder.)

Solution   We want to minimize the function s, given by

$$s(h, r) = 2\pi rh + 2\pi r^2$$

subject to the volume constraint

$$\pi r^2 h = 21.66,$$

or $\pi r^2 h - 21.66 = 0.$

Note that s does not have a minimum without the constraint.

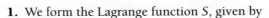

1. We form the Lagrange function S, given by

$$S(h, r, \lambda) = 2\pi rh + 2\pi r^2 - \lambda(\pi r^2 h - 21.66).$$

2. We find the first partial derivatives:

$$\frac{\partial S}{\partial h} = 2\pi r - \lambda \pi r^2,$$

$$\frac{\partial S}{\partial r} = 2\pi h + 4\pi r - 2\lambda \pi rh,$$

$$\frac{\partial S}{\partial \lambda} = -(\pi r^2 h - 21.66).$$

3. We set each derivative equal to 0 and solve the resulting system:

$$2\pi r - \lambda \pi r^2 = 0, \tag{1}$$
$$2\pi h + 4\pi r - 2\lambda \pi rh = 0, \tag{2}$$
$$-(\pi r^2 h - 21.66) = 0, \quad \text{or} \quad \pi r^2 h - 21.66 = 0. \tag{3}$$

Note that, since π is a constant, we can solve equation (1) for r:

$$\pi r(2 - \lambda r) = 0$$
$$\pi r = 0 \quad \text{or} \quad 2 - \lambda r = 0$$
$$r = 0 \quad \text{or} \qquad r = \frac{2}{\lambda}. \qquad \text{We assume } \lambda \neq 0.$$

Since $r = 0$ cannot be a solution to the original problem, we continue by substituting $2/\lambda$ for r in equation (2):

$$2\pi h + 4\pi \cdot \frac{2}{\lambda} - 2\lambda\pi \cdot \frac{2}{\lambda} \cdot h = 0$$

$$2\pi h + \frac{8\pi}{\lambda} - 4\pi h = 0$$

$$\frac{8\pi}{\lambda} - 2\pi h = 0$$

$$-2\pi h = -\frac{8\pi}{\lambda},$$

so

$$h = \frac{4}{\lambda}.$$

Since $h = 4/\lambda$ and $r = 2/\lambda$, it follows that $h = 2r$. Substituting $2r$ for h in equation (3) yields

$$\pi r^2 (2r) - 21.66 = 0$$

$$2\pi r^3 - 21.66 = 0$$

$$2\pi r^3 = 21.66$$

$$\pi r^3 = 10.83$$

$$r^3 = \frac{10.83}{\pi}$$

$$r = \sqrt[3]{\frac{10.83}{\pi}} \approx 1.51 \text{ in.}$$

Thus, when $r = 1.51$ in., we have $h = 3.02$ in. The surface area is then a minimum and is approximately

$$2\pi(1.51)(3.02) + 2\pi(1.51)^2, \quad \text{or about } 42.98 \text{ in}^2.$$

> **Quick Check 3**
>
> Repeat Example 3 for a right circular cylinder with a volume of 500 mL. (*Hint:* 1 mL = 1 cm³.) (This was Example 3 in Section 2.5.)

❰ Quick Check 3

The actual dimensions of a standard-sized 12-oz beverage can are $r = 1.25$ in. and $h = 4.875$ in. A natural question arising from the solution of Example 3 is, "Why don't beverage companies make cans using the dimensions found in that example?" To do this would mean an enormous cost for retooling. New can-making machines and new beverage-filling machines would have to be designed and purchased. Vending machines would no longer be the correct size. A partial response to the desire to save aluminum has been found in recycling and in manufacturing cans with bevelled edges. These cans require less aluminum. As a result of many engineering advances, the amount of aluminum required to make 1000 cans has been reduced over the years from 36.5 lb to 28.1 lb. Consumer preference is another very important factor affecting the shape of the can. Market research has shown that a can with the dimensions found in Example 3 is not as comfortable to hold and might not be accepted by consumers.

Closed and Bounded Regions: The Extreme-Value Theorem

In Examples 1, 2, and 3, all the constraints were given as equations. Constraints may also be stated as inequalities. If there are multiple constraints on the input variables x and y, these may be plotted on the xy-plane to form a *region of feasibility*, which contains the x and y values that satisfy all the constraints simultaneously. If the constraints form a closed and bounded region (*closed* meaning it includes the boundaries, and *bounded* meaning it has finite area, with no portions tending to infinity), then the Extreme-Value Theorem can be adapted for the two-variable function.

> **THEOREM**   **Extreme-Value Theorem for Two-Variable Functions**
>
> If $f(x, y)$ is continuous for all (x, y) within a region of feasibility that is closed and bounded, then f is guaranteed to have both an absolute maximum value and an absolute minimum value.

Critical points may occur at a vertex, along a boundary, or in the interior. Therefore, all these parts of a region must be checked for critical points.

■ **EXAMPLE 4**   **Business: Maximizing Revenue.**   Kim likes to create stylish tee shirts, one style with a script x on the front and another with a script y on the front. She sells them to her math students as a fundraiser for her favorite charity. Kim determines that her weekly revenue is modeled by the two-variable function

$$R(x, y) = -x^2 - xy - y^2 + 20x + 22y - 25,$$

where x is the number of x-shirts sold, and y is the number of y-shirts sold. Kim spends 2 hr working on each x-shirt and 4 hr working on each y-shirt, and she works no more than 40 hr per week on this project. How many of each style should she produce in order to maximize her weekly revenue? Assume $x \geq 0$ and $y \geq 0$; in other words, she cannot produce negative quantities of the tee shirts.

Solution   The number of hours Kim works per week is a constraint: $2x + 4y \leq 40$. We write the inequality with a less-than-or-equal-to sign since she may not work the full 40 hr. Along with the constraints $x \geq 0$ and $y \geq 0$, this constraint allows us to sketch the region of feasibility. This is a closed and bounded region. Since the revenue function R

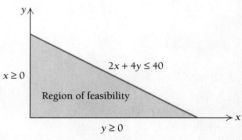

is continuous for all x and all y, the Extreme-Value Theorem guarantees an absolute minimum and an absolute maximum point. In this example, we are interested in the absolute maximum (revenue).

We determine that the three vertex points of the region are $(0, 0, -25)$, $(20, 0, -25)$, and $(0, 10, 95)$. These are all critical points.

Next, we check the interior of the region. We find the partial derivatives of R with respect to x and with respect to y:

$$R_x = -2x - y + 20,$$
$$R_y = -x - 2y + 22.$$

Setting these expressions equal to 0, we solve the system for x and y:

$$-2x - y + 20 = 0,$$
$$-x - 2y + 22 = 0;$$

or $2x + y = 20,$

$x + 2y = 22.$ After simplification

The system is solved when $x = 6$ and $y = 8$. However, this point is outside the region of feasibility; Kim would have to work $2(6) + 4(8) = 44$ hr, which is not allowed under the given constraint. Therefore, this solution must be ignored. (We address this issue at the end of this example.)

The boundaries of the region must also be checked for possible critical points:

• To check along the y-axis, we substitute $x = 0$ into the revenue function:

$$R(0, y) = -y^2 + 22y - 25.$$

The derivative is $R_y = -2y + 22$. Setting this expression equal to 0, we obtain $y = 11$. However, this is outside the region of feasibility and is ignored.

- To check along the *x*-axis, we substitute $y = 0$ into the revenue function:

$$R(x, 0) = -x^2 + 20x - 25.$$

The derivative is $R_x = -2x + 20$. Setting this expression equal to 0, we get $x = 10$. This is a feasible solution, and thus is a critical value. The critical point is $(10, 0, 75)$.

- To check along the line $2x + 4y = 40$, we use the method of Lagrange multipliers to determine possible critical values. The constraint is written as $2x + 4y - 40 = 0$, and the Lagrange function is formed:

$$L(x, y, \lambda) = -x^2 - xy - y^2 + 20x + 22y - 25 - \lambda(2x + 4y - 40).$$

Its first partial derivatives are as follows:

$$L_x = -2x - y + 20 - 2\lambda,$$
$$L_y = -x - 2y + 22 - 4\lambda,$$
$$L_\lambda = -2x - 4y + 40.$$

We set each partial derivative equal to 0:

$$-2x - y + 20 - 2\lambda = 0, \qquad (1)$$
$$-x - 2y + 22 - 4\lambda = 0, \qquad (2)$$
$$-2x - 4y + 40 = 0. \qquad (3)$$

We solve equations (1) and (2) for λ:

$$\lambda = -x - \tfrac{1}{2}y + 10 \quad \text{and} \quad \lambda = -\tfrac{1}{4}x - \tfrac{1}{2}y + \tfrac{11}{2}.$$

Equating the right-hand sides of these two equations gives us a single equation in terms of *x* and *y*. Note that the $-\tfrac{1}{2}y$ terms cancel (sum to zero):

$$-x - \tfrac{1}{2}y + 10 = -\tfrac{1}{4}x - \tfrac{1}{2}y + \tfrac{11}{2}$$
$$-\tfrac{3}{4}x = -\tfrac{9}{2}$$
$$x = 6.$$

We now substitute $x = 6$ into the constraint, $2x + 4y = 40$, to determine *y*:

$$2(6) + 4y = 40$$
$$12 + 4y = 40$$
$$4y = 28$$
$$y = 7.$$

This is a feasible solution. Therefore, $x = 6$ and $y = 7$ yield a critical point: $(6, 7, 122)$. In the graph to the right, all critical points (with their revenue values) are plotted on the region of feasibility. Therefore, Kim should produce 6 of the *x*-shirts and 7 of the *y*-shirts to maximize her weekly revenue at $122. If there were no constraints, the maximum weekly revenue would occur at $x = 6$ and $y = 8$, for a total of $123. Kim might think that working an extra 4 hr for one more dollar of revenue is not worth it.

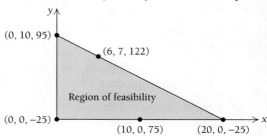

〉 Quick Check 4

Repeat Example 4, using the same revenue function but assuming that each *x*-shirt requires 4 hr to create, each *y*-shirt requires 2 hr to create, and Kim is willing to work 36 hr per week at most. How many of each style of shirt should Kim produce to maximize her weekly revenue?

❰ Quick Check 4

Section Summary

- If input variables x and y for a function $f(x, y)$ are related by another equation, that equation is a *constraint*.
- *Constrained optimization* is a method of determining maximum and minimum points on a surface represented by $z = f(x, y)$, subject to given restrictions (constraints) on the input variables x and y.
- The *method of Lagrange multipliers* allows us to find a maximum or minimum value of a function $f(x, y)$ subject to the constraint $g(x, y) = 0$.

- If the constraints are inequalities, the set of points that satisfy all the constraints simultaneously is called the *region of feasibility*.
- If the region of feasibility is closed and bounded and the surface $z = f(x, y)$ is continuous over the region, then the *Extreme-Value Theorem* guarantees that f will have both an absolute maximum and an absolute minimum value.
- Critical points may be located at vertices, along a boundary, or in the interior of a region of feasibility.

EXERCISE SET
6.5

Find the maximum value of f subject to the given constraint.

1. $f(x, y) = xy;$ $\quad 3x + y = 10$

2. $f(x, y) = 2xy;$ $\quad 4x + y = 16$

3. $f(x, y) = 4 - x^2 - y^2;$ $\quad x + 2y = 10$

4. $f(x, y) = 3 - x^2 - y^2;$ $\quad x + 6y = 37$

Find the minimum value of f subject to the given constraint.

5. $f(x, y) = x^2 + y^2;$ $\quad 2x + y = 10$

6. $f(x, y) = x^2 + y^2;$ $\quad x + 4y = 17$

7. $f(x, y) = 2y^2 - 6x^2;$ $\quad 2x + y = 4$

8. $f(x, y) = 2x^2 + y^2 - xy;$ $\quad x + y = 8$

9. $f(x, y, z) = x^2 + y^2 + z^2;$ $\quad y + 2x - z = 3$

10. $f(x, y, z) = x^2 + y^2 + z^2;$ $\quad x + y + z = 2$

Use the method of Lagrange multipliers to solve each of the following.

11. Of all numbers whose sum is 50, find the two that have the maximum product.

12. Of all numbers whose sum is 70, find the two that have the maximum product.

13. Of all numbers whose difference is 6, find the two that have the minimum product.

14. Of all numbers whose difference is 4, find the two that have the minimum product.

15. Of all points (x, y, z) that satisfy $x + 2y + 3z = 13$, find the one that minimizes
$$(x - 1)^2 + (y - 1)^2 + (z - 1)^2.$$

16. Of all points (x, y, z) that satisfy $3x + 4y + 2z = 52$, find the one that minimizes
$$(x - 1)^2 + (y - 4)^2 + (z - 2)^2.$$

APPLICATIONS

Business and Economics

17. *Maximizing typing area.* A standard piece of printer paper has a perimeter of 39 in. Find the dimensions of the paper that will give the most area. What is that area? Does standard $8\frac{1}{2} \times 11$ in. paper have maximum area?

18. *Maximizing room area.* A carpenter is building a rectangular room with a fixed perimeter of 80 ft. What are the dimensions of the largest room that can be built? What is its area?

19. *Minimizing surface area.* An oil drum of standard size has a volume of 200 gal, or 27 ft³. What dimensions yield the minimum surface area? Find the minimum surface area.

Do these drums appear to be made in such a way as to minimize surface area?

20. *Juice-can problem.* A standard-sized juice can has a volume of 99 in³. What dimensions yield the minimum surface area? Find the minimum surface area.

21. *Maximizing total sales.* The total sales, S, of a one-product firm are given by
$$S(L, M) = ML - L^2,$$

where M is the cost of materials and L is the cost of labor. Find the maximum value of this function subject to the budget constraint

$$M + L = 90.$$

22. Maximizing total sales. The total sales, S, of a one-product firm are given by

$$S(L, M) = ML - L^2,$$

where M is the cost of materials and L is the cost of labor. Find the maximum value of this function subject to the budget constraint

$$M + L = 70.$$

23. Minimizing construction costs. A company is planning to construct a warehouse whose interior volume is to be 252,000 ft^3. Construction costs per square foot are estimated to be as follows:

Walls: $3.00
Floor: $4.00
Ceiling: $3.00

a) The total cost of the building is a function $C(x, y, z)$, where x is the length, y is the width, and z is the height. Find a formula for $C(x, y, z)$.
b) What dimensions of the building will minimize the total cost? What is the minimum cost?

24. Minimizing the costs of container construction. A container company is going to construct a shipping crate of volume 12 ft^3 with a square bottom and top. The cost of the top and the sides is $2 per square foot, and the cost for the bottom is $3 per square foot. What dimensions will minimize the cost of the crate?

25. Minimizing total cost. Each unit of a product can be made on either machine A or machine B. The nature of the machines makes their cost functions differ:

Machine A: $C(x) = 10 + \dfrac{x^2}{6}$,

Machine B: $C(y) = 200 + \dfrac{y^3}{9}$.

Total cost is given by $C(x, y) = C(x) + C(y)$. How many units should be made on each machine in order to minimize total costs if $x + y = 10{,}100$ units are required?

In Exercises 26–29, find the absolute maximum and minimum values of each function, subject to the given constraints.

26. $f(x, y) = x^2 + y^2 - 2x - 2y$; $x \geq 0, y \geq 0, x \leq 4$, and $y \leq 3$

27. $g(x, y) = x^2 + 2y^2$; $-1 \leq x \leq 1$ and $-1 \leq y \leq 2$

28. $h(x, y) = x^2 + y^2 - 4x - 2y + 1$; $x \geq 0, y \geq 0$, and $x + 2y \leq 5$

29. $k(x, y) = -x^2 - y^2 + 4x + 4y$; $0 \leq x \leq 3, y \geq 0$, and $x + y \leq 6$

30. Business: maximizing profits with constraints. A manufacturer of decorative end tables produces two models, basic and large. Its weekly profit function is modeled by

$$P(x, y) = -x^2 - 2y^2 - xy + 140x + 210y - 4300,$$

where x is the number of basic models sold each week and y is the number of large models sold each week. The warehouse can hold at most 90 tables. Assume that x and y must be nonnegative. How many of each model of end table should be produced to maximize the weekly profit, and what will the maximum profit be?

31. Business: maximizing profits with constraints. A farmer has 300 acres on which to plant two crops, celery and lettuce. Each acre of celery costs $250 to plant and tend, and each acre of lettuce costs $300 to plant and tend. The farmer has $81,000 available to cover these costs.

a) Suppose the farmer makes a profit of $45 per acre of celery and $50 per acre of lettuce. Write the profit function, determine how many acres of celery and lettuce he should plant to maximize profit, and state the maximum profit. (*Hint:* Since the graph of the profit function is a plane, you will not need to check the interior for possible critical points.)
b) Suppose the farmer's profit function is instead $P(x, y) = -x^2 - y^2 + 600y - 75{,}000$. Assuming the same constraints, how many acres of celery and lettuce should he plant to maximize profit, and what is that maximum profit?

SYNTHESIS

Find the indicated maximum or minimum values of f subject to the given constraint.

32. Minimum: $f(x, y) = xy$; $x^2 + y^2 = 9$

33. Minimum: $f(x, y) = 2x^2 + y^2 + 2xy + 3x + 2y$; $y^2 = x + 1$

34. Maximum: $f(x, y, z) = x + y + z$; $x^2 + y^2 + z^2 = 1$

35. Maximum: $f(x, y, z) = x^2y^2z^2$; $x^2 + y^2 + z^2 = 2$

36. Maximum: $f(x, y, z) = x + 2y - 2z$; $x^2 + y^2 + z^2 = 4$

37. Maximum: $f(x, y, z, t) = x + y + z + t$; $x^2 + y^2 + z^2 + t^2 = 1$

38. Minimum: $f(x, y, z) = x^2 + y^2 + z^2$; $x - 2y + 5z = 1$

39. Economics: the Law of Equimarginal Productivity. Suppose that $p(x, y)$ represents the production of a two-product firm. The company produces x units of the first product at a cost of c_1 each and y units of the second product at a cost of c_2 each. The budget constraint, B, is a constant given by

$$B = c_1 x + c_2 y.$$

Use the method of Lagrange multipliers to find the value of λ in terms of p_x, p_y, c_1, and c_2. The resulting equation holds for any production function p and is called the *Law of Equimarginal Productivity*.

40. Business: maximizing production. A computer company has the following Cobb–Douglas production function for a certain product:

$$p(x, y) = 800x^{3/4}y^{1/4},$$

where x is the labor, measured in dollars, and y is the capital, measured in dollars. Suppose that the company can make a total investment in labor and capital of $1,000,000. How should it allocate the investment between labor and capital in order to maximize production?

41. Discuss the difference between solving a maximum–minimum problem using the method of Lagrange multipliers and the method of Section 6.3.

42. Write a brief report on the life and work of the mathematician Joseph Louis Lagrange (1736–1813).

TECHNOLOGY CONNECTION

43–50. Use a 3D graphics program to graph both equations in each of Exercises 1–8. Then visually check the results that you found analytically.

Answers to Quick Checks

1. $A = \dfrac{225}{2}$ at $x = 15, y = \dfrac{15}{2}$

2. $g = \dfrac{1}{10}$ at $x = \dfrac{3}{10}, y = -\dfrac{1}{10}$

3. $r \approx 4.3$ cm, $h \approx 8.6$ cm, $s \approx 348.73$ cm^2

4. $x = 5, y = 8$, maximum revenue $= \$122$

6.6

Double Integrals

OBJECTIVE

- Evaluate a double integral.

So far in this chapter, we have discussed functions of two variables and their partial derivatives. In this section, we consider integration of a function of two variables, in a process called *iterated integration*.

The following is an example of a *double integral*:

$$\int_3^6 \int_{-1}^2 10xy^2 \, dx \, dy, \quad \text{or} \quad \int_3^6 \left(\int_{-1}^2 10xy^2 \, dx \right) dy.$$

Evaluating a double integral is somewhat similar to "undoing" a second partial derivative. We first evaluate the inside integral, indicated by the innermost differential (here dx), and treat the other variable(s) (here y) as constant(s):

$$\int_{-1}^2 10xy^2 \, dx = 10y^2 \left[\frac{x^2}{2} \right]_{-1}^2 = 5y^2[x^2]_{-1}^2 = 5y^2[2^2 - (-1)^2] = 15y^2.$$

Color indicates the variable. All else is constant.

Then we evaluate the outside integral, associated with the differential dy:

$$\int_3^6 15y^2 \, dy = 15 \left[\frac{y^3}{3} \right]_3^6$$
$$= 5[y^3]_3^6$$
$$= 5(6^3 - 3^3)$$
$$= 945.$$

More precisely, the given double integral is called a **double iterated integral**. The word "iterate" means "to do again."

If dx and dy, as well as the limits of integration, are interchanged, we have

$$\int_{-1}^2 \int_3^6 10xy^2 \, dy \, dx.$$

We first evaluate the inside, y-integral, treating x as a constant:

$$\int_3^6 10xy^2 \, dy = 10x\left[\frac{y^3}{3}\right]_3^6$$

$$= \frac{10x}{3}\left[y^3\right]_3^6$$

$$= \frac{10}{3}x(6^3 - 3^3) = 630x.$$

Then we evaluate the outside, x-integral:

$$\int_{-1}^2 630x \, dx = 630\left[\frac{x^2}{2}\right]_{-1}^2$$

$$= 315\left[x^2\right]_{-1}^2$$

$$= 315\left[2^2 - (-1)^2\right] = 945.$$

Note that we get the same result.

DEFINITION

If $f(x, y)$ is defined over the rectangular region R bounded by $a \le x \le b$ and $c \le y \le d$, then the **double integral** of $f(x, y)$ over R is given by

$$\int_c^d \int_a^b f(x, y) \, dx \, dy \quad \text{or} \quad \int_a^b \int_c^d f(x, y) \, dy \, dx.$$

In a more technical definition of the double integral, Riemann sums are used. However, for the functions in this text, the above definition is sufficient.

Sometimes double integrals are defined over a nonrectangular region, in which case the bounds of integration may contain variables.

■ EXAMPLE 1 Evaluate

$$\int_0^1 \int_{x^2}^x xy^2 \, dy \, dx.$$

Solution We first evaluate the inside integral with respect to y, treating x as a constant:

$$\int_{x^2}^x xy^2 \, dy = x\left[\frac{y^3}{3}\right]_{x^2}^x$$

$$= \frac{1}{3}x\left[x^3 - (x^2)^3\right]$$

$$= \frac{1}{3}(x^4 - x^7).$$

Then we evaluate the outside integral:

$$\frac{1}{3}\int_0^1 (x^4 - x^7) \, dx = \frac{1}{3}\left[\frac{x^5}{5} - \frac{x^8}{8}\right]_0^1$$

$$= \frac{1}{3}\left[\left(\frac{1^5}{5} - \frac{1^8}{8}\right) - \left(\frac{0^5}{5} - \frac{0^8}{8}\right)\right] = \frac{1}{40}.$$

> **Quick Check 1**
>
> Evaluate
>
> $$\int_0^4 \int_{\frac{1}{2}x}^{\sqrt{x}} 2xy \, dy \, dx.$$

Thus, $\displaystyle\int_0^1 \int_{x^2}^x xy^2 \, dy \, dx = \frac{1}{40}.$

❮ Quick Check 1

The Geometric Interpretation of Multiple Integrals

Suppose that the region D in the xy-plane is bounded by the functions $y_1 = g(x)$ and $y_2 = h(x)$ and the lines $x_1 = a$ and $x_2 = b$. We want the volume, V, of the solid above D and under the surface $z = f(x, y)$. We can think of the solid as composed of many vertical columns, one of which is shown in Fig. 1 in red. The volume of this column can be thought of as $l \cdot w \cdot h$, or $z \cdot \Delta y \cdot \Delta x$. Integrating such columns in the y-direction, we obtain

$$\int_{y_1}^{y_2} z \, dy, \quad \text{so} \left[\int_{y_1}^{y_2} z \, dy \right] \Delta x$$

can be pictured as a "slab," or slice. Then integrating such slices in the x-direction, we obtain the entire volume:

$$V = \int_a^b \left[\int_{y_1}^{y_2} z \, dy \right] dx,$$

This can be thought of as a collection of slices that fills the volume.

or $$V = \int_a^b \int_{g(x)}^{h(x)} z \, dy \, dx,$$

where $z = f(x, y)$.

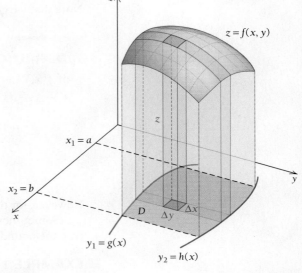

FIGURE 1

In Example 1, the region of integration D is the plane region between the graphs of $y = x^2$ and $y = x$, as shown in Figs. 2 and 3.

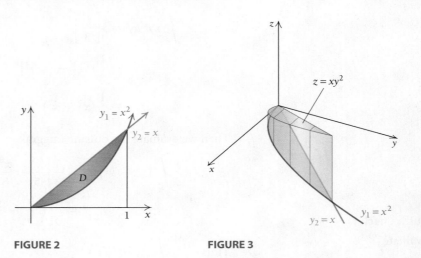

FIGURE 2   **FIGURE 3**

When we evaluated the double integral in Example 1, we found the volume of the solid based on D and capped by the surface $z = xy^2$, as shown in Fig. 3.

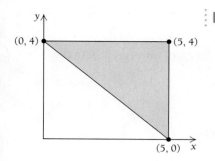

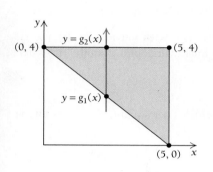

EXAMPLE 2   Business: Demographics and Vehicle Ownership.   The density of privately owned vehicles in a city is given by the two-variable function $p(x, y) = \frac{1}{8}xy$, where x is miles in the east–west direction, y is miles in the north–south direction, and p is the number of privately owned vehicles per square mile, in thousands. If the city limits are as shown in the figure to the left, what is the total number of privately owned vehicles in the city?

Solution   We must decide on the order of integration. If we decide to integrate with respect to y first, then x second, the iterated integral is

$$\int_a^b \int_{g_1(x)}^{g_2(x)} \frac{1}{8}xy \, dy \, dx.$$

Since we are integrating with respect to y first, a helpful visual method for determining the bounds of integration is to draw an arrow in the positive y-direction, intersecting the shaded region representing the city. The arrow enters the region at the diagonal boundary first; this boundary is $g_1(x)$. The arrow exits the region through the horizontal boundary, and this boundary is $g_2(x)$. The bounds for the outer integral, with respect to x, are constants: the region extends from $x = 0$ to $x = 5$.

The diagonal boundary is a line with slope $-\frac{4}{5}$ and a y-intercept of 4. Therefore, $g_1(x) = 4 - \frac{4}{5}x$. The horizontal boundary is $g_2(x) = 4$. The integral is thus

$$\int_0^5 \int_{4-\frac{4}{5}x}^4 \frac{1}{8}xy \, dy \, dx.$$

We integrate the inside integral first, with respect to y:

$$\int_{4-\frac{4}{5}x}^4 \frac{1}{8}xy \, dy = \frac{1}{8}x\left[\frac{1}{2}y^2\right]_{4-\frac{4}{5}x}^4$$

$$= \frac{1}{8}x\left[\frac{1}{2}(4)^2 - \frac{1}{2}\left(4 - \frac{4}{5}x\right)^2\right]$$

$$= \frac{1}{8}x\left(\frac{16}{5}x - \frac{8}{25}x^2\right) \qquad \text{After simplification}$$

$$= \frac{2}{5}x^2 - \frac{1}{25}x^3.$$

We now integrate with respect to x:

$$\int_0^5 \left(\frac{2}{5}x^2 - \frac{1}{25}x^3\right) dx = \left[\frac{2}{15}x^3 - \frac{1}{100}x^4\right]_0^5$$

$$= \left(\frac{2}{15}(5)^3 - \frac{1}{100}(5)^4\right) - 0$$

$$= \frac{125}{12} \approx 10.417.$$

Therefore, the city has about 10,417 privately owned vehicles.

❰ Quick Check 2

❱ **Quick Check 2**

Redo Example 2, integrating with respect to x first, then y second. (*Hint*: Draw the arrow in the positive x-direction, and define the boundaries as functions of y.)

Average Value of a Multivariable Function

Recall from Section 4.4 that the average value of a continuous single-variable function $y = f(x)$ over an interval $[a, b]$ is given by

$$y_{\text{av}} = \frac{1}{b - a}\int_a^b f(x) \, dx.$$

The average value of a continuous two-variable function $z = f(x, y)$ can be found in a similar way.

DEFINITION

Let $z = f(x, y)$ be a continuous function over a region of integration R in the xy-plane. The **average value**, z_{av}, over R is given by

$$z_{av} = \frac{1}{A(R)} \iint_R f(x, y)\, dy\, dx,$$

where $A(R)$ is the area of the region of integration.

■ **EXAMPLE 3**   **Business: Average Vehicle Ownership.**   Find the average number of privately owned vehicles per square mile in the city in Example 2.

Solution   Since the region of integration R is a triangle with vertices $(5, 0)$, $(5, 4)$, and $(0, 4)$, its area can be found using the formula for the area of a triangle, $A = \frac{1}{2}bh$. Thus, we have

$$A(R) = \tfrac{1}{2}(5)(4) = 10.$$

Therefore, the average number of privately owned vehicles per square mile in the city is

$$p_{av} = \frac{1}{A(R)} \int_0^5 \int_{4 - \frac{4}{5}x}^4 \tfrac{1}{8}xy\, dy\, dx$$

$$\approx \frac{1}{10} \cdot 10.417 \quad \text{Using the result from Example 2}$$

$$= 1.042$$

Thus, the city has an average of 1042 privately owned vehicles per square mile.

❰ Quick Check 3

> ❭ **Quick Check 3**
>
> Find the average value of $g(x, y) = 2x + y$, where $0 \le x \le 4$ and $-2 \le y \le 1$.

Section Summary

- The *double integral* of a two-variable function $f(x, y)$ over a rectangular region R bounded by $a \le x \le b$ and $c \le y \le d$ is written

  $$\int_c^d \int_a^b f(x, y)\, dx\, dy \quad \text{or} \quad \int_a^b \int_c^d f(x, y)\, dy\, dx.$$

- If the region of integration is not rectangular, the double integral may have variables in its bounds.
- The average value of a continuous two-variable function $z = f(x, y)$ over a region R is given by

  $$z_{av} = \frac{1}{A(R)} \iint_R f(x, y)\, dy\, dx,$$

  where $A(R)$ is the area of the region of integration.

**EXERCISE SET
6.6**

Evaluate

1. $\displaystyle \int_0^3 \int_0^1 2y\, dx\, dy$

2. $\displaystyle \int_0^1 \int_0^4 3x\, dx\, dy$

5. $\displaystyle \int_0^5 \int_{-2}^{-1} (3x + y)\, dx\, dy$

6. $\displaystyle \int_{-4}^{-1} \int_1^3 (x + 5y)\, dx\, dy$

3. $\displaystyle \int_{-1}^3 \int_1^2 x^2 y\, dy\, dx$

4. $\displaystyle \int_1^4 \int_{-2}^1 x^3 y\, dy\, dx$

7. $\displaystyle \int_{-1}^1 \int_x^1 xy\, dy\, dx$

8. $\displaystyle \int_{-1}^1 \int_x^2 (x + y)\, dy\, dx$

9. $\int_0^1 \int_{x^2}^x (x + y) \, dy \, dx$

10. $\int_0^2 \int_0^x e^{x+y} \, dy \, dx$

11. $\int_0^1 \int_1^{e^x} \frac{1}{y} \, dy \, dx$

12. $\int_0^1 \int_{-1}^x (x^2 + y^2) \, dy \, dx$

13. $\int_0^2 \int_0^x (x + y^2) \, dy \, dx$

14. $\int_1^3 \int_0^x 2e^{x^2} \, dy \, dx$

15. Find the volume of the solid capped by the surface $z = 1 - y - x^2$ over the region bounded on the xy-plane by $y = 1 - x^2, y = 0, x = 0$, and $x = 1$, by evaluating the integral

$$\int_0^1 \int_0^{1-x^2} (1 - y - x^2) \, dy \, dx.$$

16. Find the volume of the solid capped by the surface $z = x + y$ over the region bounded on the xy-plane by $y = 1 - x, y = 0, x = 0$, and $x = 1$, by evaluating the integral

$$\int_0^1 \int_0^{1-x} (x + y) \, dy \, dx.$$

17. Find the average value of $f(x, y) = 2x - y$, where $0 \le x \le 2$ and $2 \le y \le 3$.

18. Find the average value of $g(x, y) = 4 - x - y$, where $-1 \le x \le 1$ and $-2 \le y \le 3$.

19. Find the average value of $f(x, y) = x^2 y$, where the region of integration is a triangle with vertices $(0, 0), (6, 0)$, and $(6, 3)$.

20. Find the average value of $g(x, y) = 2 + xy$, where the region of integration is a triangle with vertices $(0, 0), (4, 4)$, and $(4, -4)$.

21. Life sciences: population. The population density of fireflies in a field is given by $p(x, y) = \frac{1}{100}x^2 y$, where $0 \le x \le 30$ and $0 \le y \le 20$, x and y are in feet, and p is the number of fireflies per square foot.

a) Determine the total population of fireflies in this field.

b) Determine the average number of fireflies per square foot of the field.

22. Life sciences: population. The population density of a city is given by $p(x, y) = 2x^2 + 5y$, where x and y are in miles and p is the number of people per square mile, in hundreds. The city limits are as shown in the graph below.

a) Determine the city's population.

b) Determine the average number of people per square mile of the city.

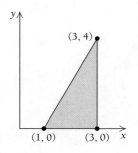

SYNTHESIS

A triple iterated integral such as

$$\int_r^s \int_c^d \int_a^b f(x, y, z) \, dx \, dy \, dz$$

is evaluated in much the same way as a double iterated integral. We first evaluate the inside x-integral, treating y and z as constants. Then we evaluate the middle y-integral, treating z as a constant. Finally, we evaluate the outside z-integral. Evaluate these triple integrals.

23. $\int_0^1 \int_1^3 \int_{-1}^2 (2x + 3y - z) \, dx \, dy \, dz$

24. $\int_0^2 \int_1^4 \int_{-1}^2 (8x - 2y + z) \, dx \, dy \, dz$

25. $\int_0^1 \int_0^{1-x} \int_0^{2-x} xyz \, dz \, dy \, dx$

26. $\int_0^2 \int_{2-y}^{6-2y} \int_0^{\sqrt{4-y^2}} z \, dz \, dx \, dy$

27. Describe the geometric meaning of the double integral of a function of two variables.

28. Explain how Exercise 1 can be answered without finding any antiderivatives.

TECHNOLOGY CONNECTION

29. Use a calculator that does multiple integration to evaluate some double integrals found in this exercise set.

Answers to Quick Checks

1. $\frac{16}{3}$ **2.** $\int_0^4 \int_{5-\frac{5}{4}y}^5 \frac{1}{8}xy \, dx \, dy = 10.417$, or 10,417 vehicles

3. $\frac{7}{2}$

KEY TERMS AND CONCEPTS

EXAMPLES

SECTION 6.1

A **function of two variables** assigns to each input pair, (x, y), exactly one output number, $f(x, y)$.

Business. A company produces two products. The first product costs $5.25 per unit to produce, and the second costs $7.50 per unit to produce. If x is the number of units of the first product and y is the number of units of the second product, the cost function C is given by

$$C(x, y) = 5.25x + 7.50y.$$

If the company produces 30 units of the first product and 45 units of the second product, the total cost of producing these products is

$$C(30, 45) = 5.25(30) + 7.50(45) = \$495$$

The graph of a two-variable function is a **surface**; graphing such a function requires a three-dimensional coordinate system. Points on the surface are expressed as ordered triples (x, y, z), where $z = f(x, y)$.

The function $g(x, y) = \sqrt{4 - x^2 - y^2}$ is a hemisphere of radius 2. Examples of points on the surface of this hemisphere are $(0, 0, 2)$, $(2, 0, 0)$, and $(0, 2, 0)$.

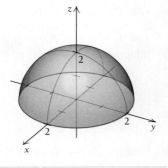

The **domain** of a two-variable function is the set of points in the xy-plane for which f is defined.

The function $f(x, y) = x^3 + y^3 - 3x - 27y - 2$ is defined for all x and for all y. Therefore, the domain of f is

$$D = \{(x, y) | -\infty < x < \infty, -\infty < y < \infty \}.$$

The function $g(x, y) = \sqrt{4 - x^2 - y^2}$ is defined as long as the expression inside the radical is nonnegative. We have $4 - x^2 - y^2 \geq 0$, which simplifies to $x^2 + y^2 \leq 4$. Therefore, the domain of g is

$$D = \{(x, y) | x^2 + y^2 \leq 4 \}.$$

For the cost function $C(x, y) = 5.25x + 7.50y$, the variables x and y represent quantities of products. Thus, they cannot be negative. Therefore, the domain for C is

$$D = \{(x, y) | x \geq 0, y \geq 0 \}.$$

SECTION 6.2

Let f be a function of two variables, x and y. The **partial derivative of f with respect to x** is defined as

$$\frac{\partial f}{\partial x} = \lim_{h \to 0} \frac{f(x + h, y) - f(x, y)}{h}.$$

The variable y is treated as a constant during the differentiation steps.

Let $f(x, y) = x^2 + 2xy^3 + \sqrt{y}$.

The partial derivative of f with respect to x is

$$\frac{\partial f}{\partial x} = 2x + 2y^3.$$

The partial derivative of f with respect to y is

$$\frac{\partial f}{\partial y} = 6xy^2 + \frac{1}{2\sqrt{y}}.$$

KEY TERMS AND CONCEPTS	EXAMPLES

The **partial derivative of f with respect to y** is defined as

$$\frac{\partial f}{\partial y} = \lim_{h \to 0} \frac{f(x, y + h) - f(x, y)}{h}.$$

The variable x is treated as a constant during the differentiation steps.

Other common notations for partial derivatives are f_x for the partial derivative of f with respect to x and f_y for the partial derivative of f with respect to y.

The partial derivative $\partial f/\partial x$ is interpreted as the slope of the tangent line at a point (x, y, z) on the surface representing the graph of f in the positive x-direction. Similarly, $\partial f/\partial y$ is interpreted as the slope of the tangent line at a point (x, y, z) on the surface in the positive y-direction.

When $x = 2$ and $y = 1$, the slope of the tangent line at $(2, 1, 13)$ on the surface representing the graph of f in the positive x-direction is

$$\left.\frac{\partial f}{\partial x}\right|_{(2,1)} = 2(2) + 2(1)^3 = 6.$$

The slope of the tangent line at that point on the surface in the positive y-direction is

$$\left.\frac{\partial f}{\partial y}\right|_{(2,1)} = 6(2)(1)^2 + \frac{1}{2\sqrt{(1)}} = 12.5.$$

For functions of many variables, the partial derivative with respect to one of the variables is found by treating all the other variables as constants and differentiating using normal techniques.

Let $w(x, y, z) = 3x^2y^3z^7$. The partial derivatives of w are

$$w_x = 6xy^3z^7,$$
$$w_y = 9x^2y^2z^7,$$
$$w_z = 21x^2y^3z^6.$$

Let f be a function of two variables, x and y. Its **second-order partial derivatives** are

$$f_{xx} = \frac{\partial^2 f}{\partial x^2}, \quad f_{xy} = \frac{\partial^2 f}{\partial y\,\partial x},$$

$$f_{yx} = \frac{\partial^2 f}{\partial x\,\partial y}, \quad \text{and} \quad f_{yy} = \frac{\partial^2 f}{\partial y^2}.$$

Often, $f_{xy} = f_{yx}$.

Let $f(x, y) = x^2 + 2xy^3 + \sqrt{y}$. Its first partial derivatives are

$$f_x = \frac{\partial f}{\partial x} = 2x + 2y^3 \quad \text{and} \quad f_y = \frac{\partial f}{\partial y} = 6xy^2 + \frac{1}{2\sqrt{y}}.$$

Its second-order partial derivatives are

$$f_{xx} = 2, \quad f_{xy} = 6y^2, \quad f_{yx} = 6y^2, \quad \text{and} \quad f_{yy} = 12xy - \frac{1}{4\sqrt{y^3}}.$$

SECTION 6.3

If f is a function of two variables, x and y, it has a **relative maximum** at (a, b) if

$$f(x, y) \le f(a, b)$$

for all points (x, y) in a region containing (a, b). Similarly, f has a **relative minimum** at (a, b) if

$$f(x, y) \ge f(a, b)$$

for all points (x, y) in a region containing (a, b).

A **critical point** occurs at (a, b) if both partial derivatives of f at (a, b) are 0. That is, $f_x(a, b) = 0$ and $f_y(a, b) = 0$.

Let $f(x, y) = x^3 + y^3 - 3x - 27y - 2$.

The first partial derivatives are $f_x(x, y) = 3x^2 - 3$ and $f_y(x, y) = 3y^2 - 27$. When $f_x = 0$, we have $x = \pm 1$. Similarly, when $f_y = 0$, we have $y = \pm 3$. There are four critical points:

$$(1, 3, -58), \text{ where } f(1, 3) = -58,$$
$$(1, -3, 50), \text{ where } f(1, -3) = 50,$$
$$(-1, 3, -54), \text{ where } f(-1, 3) = -54,$$

and $(-1, -3, 54)$, where $f(-1, -3) = 54$.

The four second-order partial derivatives are $f_{xx} = 6x$, $f_{yy} = 6y$, and $f_{xy} = f_{yx} = 0$. Therefore, $D = (6x)(6y) - 0^2 = 36xy$.

(continued)

KEY TERMS AND CONCEPTS	EXAMPLES

SECTION 6.3 (continued)

The **D-test** is used to determine whether critical points are relative maxima or minima. If (a, b) is a critical point, then

$$D = f_{xx}(a, b) \cdot f_{yy}(a, b) - [f_{xy}(a, b)]^2.$$

And:

1. If $D > 0$ and $f_{xx}(a, b) < 0$, then f has a **maximum** at (a, b).
2. If $D > 0$ and $f_{xx}(a, b) > 0$, then f has a **minimum** at (a, b).
3. If $D < 0$, then f has a **saddle point** at (a, b).
4. The D-test is not applicable if $D = 0$.

At $(1, 3)$, we have $D = 108 > 0$, and $f_{xx}(1, 3) = 6 > 0$. Therefore, $(1, 3, -58)$ is a relative minimum.

At $(1, -3)$ and at $(-1, 3)$, we have $D = -108 < 0$. Therefore, $(1, -3, 50)$ and $(-1, 3, -54)$ are saddle points.

At $(-1, -3)$, we have $D = 108 > 0$, and $f_{xx}(-1, -3) = -6 < 0$. Therefore, $(-1, -3, 54)$ is a relative maximum.

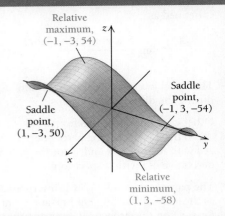

SECTION 6.4

A line of best fit for a set of data points is called a **regression line**. The method of least squares uses partial derivatives to determine this line.

A researcher obtains the data points $(1, 3), (4, 5)$, and $(6, 7)$. Find the regression line.

Using the method of least squares, we must minimize

$$(y_1 - 3)^2 + (y_2 - 5)^2 + (y_3 - 7)^2,$$

where $y_1 = m(1) + b = m + b,$
$y_2 = m(4) + b = 4m + b,$
$y_3 = m(6) + b = 6m + b.$

After substitution, we have a two-variable function:

$$S(m, b) = (m + b - 3)^2 + (4m + b - 5)^2 + (6m + b - 7)^2.$$

The partial derivatives are (after simplification):

$$S_m = 106m + 22b - 130,$$
$$S_b = 22m + 6b - 30.$$

Setting the partial derivatives equal to 0 and solving the system, we get

$$m = \frac{15}{19}, \quad b = \frac{40}{19}.$$

Therefore, the regression line is $y = \frac{15}{19}x + \frac{40}{19}$.

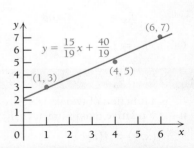

SECTION 6.5

If the input variables x and y of a two-variable function $f(x, y)$ are themselves related by an equation $g(x, y) = 0$, then $g(x, y) = 0$ is a **constraint**. The process of determining maximum and minimum values of f subject to the constraint g is called **constrained optimization**.

Maximize $f(x, y) = xy$, subject to the constraint $x + 2y = 1$.

We can substitute $x = 1 - 2y$ into f:

$$f(1 - 2y, y) = (1 - 2y)y = y - 2y^2.$$

Differentiating f with respect to y, we get $f_y = 1 - 4y$. Setting this equal to 0, we get $y = \frac{1}{4}$. Therefore, $x = 1 - 2\left(\frac{1}{4}\right) = \frac{1}{2}$. The function f has a maximum value of $\frac{1}{8}$ at $x = \frac{1}{2}$ and $y = \frac{1}{4}$, which lies on the line given by the constraint.

KEY TERMS AND CONCEPTS	EXAMPLES

The method of **Lagrange multipliers** is one way to determine a maximum or minimum value of a function f subject to a constraint g.

1. Form the **Lagrange function**
$$F(x, y, \lambda) = f(x, y) - \lambda g(x, y).$$
The variable λ is a *Lagrange multiplier*.

2. Find the first partial derivatives F_x, F_y, and F_λ.

3. Solve the system
$$F_x = 0, \quad F_y = 0, \quad \text{and} \quad F_\lambda = 0.$$

A suggested method of solution is to isolate λ in the equations $F_x = 0$ and $F_y = 0$, substitute to cancel out λ, and simplify the resulting equation in terms of x and y. Make another substitution into $F_\lambda = 0$, and determine the values of x and y.

Maximize $f(x, y) = xy$, subject to the constraint $x + 2y = 1$.

We write the constraint as $x + 2y - 1 = 0$ and form the Lagrange function:
$$F(x, y, \lambda) = xy - \lambda(x + 2y - 1).$$
Differentiating F with respect to its three input variables, we have
$$F_x = y - \lambda, \quad F_y = x - 2\lambda, \quad \text{and} \quad F_\lambda = -x - 2y + 1.$$
We set all three expressions equal to 0. From $F_x = 0$ and $F_y = 0$, we isolate λ:
$$\lambda = y \text{ and } \lambda = \tfrac{1}{2}x.$$
By substitution, we have $y = \tfrac{1}{2}x$. Substituting for y in F_λ and setting the expression equal to 0 gives
$$-x - 2\left(\tfrac{1}{2}x\right) + 1 = 0.$$
Solving for x, we have $x = \tfrac{1}{2}$. Therefore, $y = \tfrac{1}{2}\left(\tfrac{1}{2}\right) = \tfrac{1}{4}$, and f has a constrained maximum value of $\tfrac{1}{8}$ at $\left(\tfrac{1}{2}, \tfrac{1}{4}\right)$.

If $f(x, y)$ is continuous over a region of feasibility that is closed and bounded, then the **Extreme-Value Theorem** guarantees the existence of an absolute maximum value and an absolute minimum value of f.

Find the absolute maximum and minimum values of $f(x, y) = x^2 + y^2 - 2x - 2y$, subject to the constraints $0 \le x \le 2$ and $0 \le y \le 3$.

Since f is continuous on $0 \le x \le 2$ and on $0 \le y \le 3$, we check for critical points at all vertices and boundaries and in the interior. We find that f has an absolute maximum at both $(0, 3, 3)$ and $(2, 3, 3)$ and an absolute minimum at $(1, 1, -2)$.

SECTION 6.6

If $f(x, y)$ is defined over a rectangular region R bounded by $a \le x \le b$ and $c \le x \le d$, then the **double integral** of $f(x, y)$ over R is
$$\int_c^d \int_a^b f(x, y) \, dx \, dy$$
or
$$\int_a^b \int_c^d f(x, y) \, dy \, dx.$$

If the region is not rectangular, then the bounds of integration may contain variables.

Iterated integrals are evaluated by first integrating the inside integral, indicated by the innermost differential, and then integrating the outer integral.

Evaluate
$$\int_0^2 \int_1^3 x^2 y \, dy \, dx.$$

The inside integral is integrated first. We integrate with respect to y, treating x as a constant:
$$\int_1^3 x^2 y \, dy = x^2 \left[\tfrac{1}{2}y^2\right]_1^3 = x^2\left(\tfrac{9}{2} - \tfrac{1}{2}\right) = 4x^2.$$
We integrate the result with respect to x:
$$\int_0^2 4x^2 \, dx = \left[\tfrac{4}{3}x^3\right]_0^2 = \tfrac{4}{3}(8 - 0) = \tfrac{32}{3}.$$

The **average value** of a continuous two-variable function $z = f(x, y)$ over a region of integration R is given by
$$z_{av} = \frac{1}{A(R)} \iint_R f(x, y) \, dy \, dx,$$
where $A(R)$ is the area of the region of integration.

The average value of $f(x, y) = 2xy$ over a region R that is a triangle with vertices $(-2, 0)$, $(2, 0)$, and $(2, 8)$ is
$$z_{av} = \frac{1}{A(R)} \int_{-2}^2 \int_0^{2x+4} 2xy \, dy \, dx$$
$$= \tfrac{1}{16} \cdot \tfrac{256}{3} = \tfrac{16}{3}.$$

These review exercises are for test preparation. They can also be used as a practice test. Answers are at the back of the book. The blue bracketed section references tell you what part(s) of the chapter to restudy if your answer is incorrect.

CONCEPT REINFORCEMENT

Match each expression in column A with an equivalent expression in column B. Assume that $z = f(x, y)$. [6.2, 6.6]

Column A	Column B
1. $\dfrac{\partial z}{\partial x}$	a) $\displaystyle\int_2^3 \tfrac{1}{2} x \, dx$
2. $\dfrac{\partial z}{\partial y}$	b) f_{yx}
3. $\dfrac{\partial}{\partial x}(5x^3 y^7)$	c) f_{xy}
4. $\dfrac{\partial}{\partial y}(5x^3 y^7)$	d) $\displaystyle\int_2^3 y^3 \, dy$
5. $\dfrac{\partial^2 z}{\partial x \, \partial y}$	e) f_x
6. $\dfrac{\partial^2 z}{\partial y \, \partial x}$	f) $15x^2 y^7$
7. $\displaystyle\int_2^3 \int_0^1 2xy^3 \, dx \, dy$	g) $35x^3 y^6$
8. $\displaystyle\int_2^3 \int_0^1 2xy^3 \, dy \, dx$	h) f_y

REVIEW EXERCISES

Given $f(x, y) = e^y + 3xy^3 + 2y$, find each of the following. [6.1, 6.2]

9. $f(2, 0)$ 10. f_x 11. f_y

12. f_{xy} 13. f_{yx} 14. f_{xx}

15. f_{yy}

16. State the domain of $f(x, y) = \dfrac{2}{x - 1} + \sqrt{y - 2}$. [6.1]

Given $z = 2x^3 \ln y + xy^2$, find each of the following. [6.2]

17. $\dfrac{\partial z}{\partial x}$ 18. $\dfrac{\partial z}{\partial y}$ 19. $\dfrac{\partial^2 z}{\partial x \, \partial y}$

20. $\dfrac{\partial^2 z}{\partial y \partial x}$ 21. $\dfrac{\partial^2 z}{\partial x^2}$ 22. $\dfrac{\partial^2 z}{\partial y^2}$

Find the relative maximum and minimum values. [6.3]

23. $f(x, y) = x^3 - 6xy + y^2 + 6x + 3y - \tfrac{1}{5}$

24. $f(x, y) = x^2 - xy + y^2 - 2x + 4y$

25. $f(x, y) = 3x - 6y - x^2 - y^2$

26. $f(x, y) = x^4 + y^4 + 4x - 32y + 80$

27. Consider the data in the following table regarding enrollment in colleges and universities during a recent 3-year period. [6.4]

Year, x	Enrollment, y (in millions)
1	7.2
2	8.0
3	8.4

a) Find the regression line, $y = mx + b$.

b) Use the regression line to predict enrollment in the fourth year.

28. Consider the data in the table below regarding workers' average monthly out-of-pocket premium for health insurance for a family. [6.4]

a) Find the regression line, $y = mx + b$.

b) Use the regression line to predict workers' average monthly out-of-pocket premium for health insurance for a family in 2012.

Year, x	Workers' Average Monthly Out-of-Pocket Premium for Health Insurance for a Family
0 (1999)	$129
2 (2001)	149
4 (2003)	201
5 (2004)	222
6 (2005)	226

(*Source*: Kaiser Family Foundation and *The New York Times*, 10/23/05.)

29. Find the minimum value of

$$f(x, y) = x^2 - 2xy + 2y^2 + 20$$

subject to the constraint $2x - 6y = 15$. [6.5]

30. Find the maximum value of $f(x, y) = 6xy$ subject to the constraint $2x + y = 20$. [6.5]

31. Find the absolute maximum and minimum values of $f(x, y) = x^2 - y^2$ subject to the constraints $-1 \leq x \leq 3$ and $-1 \leq y \leq 2$. [6.5]

Evaluate. [6.6]

32. $\displaystyle\int_0^1 \int_1^2 x^2 y^3 \, dy \, dx$ 33. $\displaystyle\int_0^1 \int_{x^2}^x (x - y) \, dy \, dx$

34. *Business: demographics.* The density of students living near a university is modeled by

$$p(x, y) = 9 - x^2 - y^2,$$

where x and y are in miles and p is the number of students per square mile, in hundreds. Assume the university is located at $(0, 0)$. [6.6]

a) Find the number of students who live in the shaded region shown below.

b) Find the average number of students per square mile.

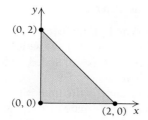

SYNTHESIS

35. Evaluate

$$\int_0^2 \int_{1-2x}^{1-x} \int_0^{\sqrt{2-x^2}} z \, dz \, dy \, dx. [6.6]$$

36. *Business: minimizing surface area.* Suppose that beverages could be packaged in either a cylindrical container or a rectangular container with a square top and bottom. Each container is designed to have the minimum surface area for its shape. If we assume a volume of 26 in³, which container would have the smaller surface area? [6.3, 6.5]

TECHNOLOGY CONNECTION

37. Use a 3D graphics program to graph $f(x, y) = x^2 + 4y^2$. [6.1]

CHAPTER 6
TEST

Given $f(x, y) = e^x + 2x^3y + y$, find each of the following.

1. $f(-1, 2)$ **2.** $\dfrac{\partial f}{\partial x}$ **3.** $\dfrac{\partial f}{\partial y}$

4. $\dfrac{\partial^2 f}{\partial x^2}$ **5.** $\dfrac{\partial^2 f}{\partial x \, \partial y}$ **6.** $\dfrac{\partial^2 f}{\partial y \, \partial x}$

7. $\dfrac{\partial^2 f}{\partial y^2}$

Find the relative maximum and minimum values.

8. $f(x, y) = x^2 - xy + y^3 - x$

9. $f(x, y) = 4y^2 - x^2$

10. *Business: predicting total sales.* Consider the data in the following table regarding the total sales of a company during the first three years of operation.

Year, x	Sales, y (in millions)
1	$10
2	15
3	19

a) Find the regression line, $y = mx + b$.

b) Use the regression line to predict sales in the fourth year.

11. Find the maximum value of

$$f(x, y) = 6xy - 4x^2 - 3y^2$$

subject to the constraint $x + 3y = 19$.

12. Evaluate

$$\int_0^3 \int_1^3 4x^3y^2 \, dx \, dy.$$

SYNTHESIS

13. *Business: maximizing production.* Southwest Appliances has the following Cobb–Douglas production function for a certain product:

$$p(x, y) = 50x^{2/3}y^{1/3},$$

where x is labor, measured in dollars, and y is capital, measured in dollars. Suppose that Southwest can make a total investment in labor and capital of $600,000. How should it allocate the investment between labor and capital in order to maximize production?

14. Find f_x and f_t:

$$f(x, t) = \frac{x^2 - 2t}{x^3 + 2t}.$$

TECHNOLOGY CONNECTION

15. Use a 3D graphics program to graph $f(x, y) = x - \frac{1}{2}y^2 - \frac{1}{3}x^3$.

Minimizing Employees' Travel Time in a Building

If employees spend considerable time moving between offices, designing a building to minimize travel time can reap enormous savings for a company.

For a multilevel building with a square base, one design concern is minimizing travel time between the most remote points. We will make use of Lagrange multipliers to help design such a building.

Let's assume that each floor has a square grid of hallways, as shown in the figure at the lower right. Suppose that you are standing at point P in the top northeast corner of the twelfth floor of this building. How long will it take to reach the most remote point at the southwest corner on the first floor—that is, point Q?

Let's call the time t. We find a formula for t in two steps:

1. You are to go from the twelfth floor to the first floor. This is a move in a vertical direction.

2. You need to cross horizontally from one corner of the building to the other.

The vertical time is h, the height of point P from the ground, divided by a, the speed at which you can travel in a vertical direction (elevator speed). Thus, vertical time is given by h/a.

The horizontal time is the time it takes to go across one level, by way of the square grid of hallways (from R to Q in the figure). If each floor is a square with side of length k, then the distance from R to Q is $2k$. If the walking speed is b, then the horizontal time is given by $2k/b$.

Thus, the time it will take to go from P to Q is a function of two variables, h and k, given by

$$t(h, k) = \text{vertical time} + \text{horizontal time}$$
$$= \frac{h}{a} + \frac{2k}{b},$$

where a and b, the elevator speed and walking speed, are constants.

What happens if we must choose between two (or more) building plans with the same floor area, but with different dimensions?

Will the travel time be the same? Or will it be different for the two buildings? First, what is the total floor area of a given building? Suppose that the building has n floors, each a square of side k. Then the total floor area is given by

$$A = nk^2.$$

Note that the area of the roof is not included.

If h is the height of point P and c is the height of each floor—that is, the distance from the carpeting on one floor to the carpeting on the floor above—then $n = 1 + h/c$, with

$$A = (1 + h/c)k^2.$$

Let's return to the problem of two buildings with the same total floor area, but with different dimensions, and see what happens to $t(h, k)$.

EXERCISES

1. Use the TABLE feature on your calculator or spreadsheet software to complete the table below. For each case in the table, let the elevator speed $a = 10$ ft/sec, the walking speed $b = 4$ ft/sec, and the height of each floor $c = 15$ ft. Each case in the table covers two situations, though the floor area stays essentially the same for a particular case.

2. Do different dimensions, with a fixed floor area, yield different travel times?

In Exercises 3–5, assume that you are finding the dimensions of a multilevel building with a square base that will minimize travel time t between the most remote points in the building. Each floor has a square grid of hallways. The height of point P is h, and the length of a side of each floor is k. The elevator speed is 10 ft/sec and the average speed of a person walking is 4 ft/sec. The total floor area of the building is 40,000 ft^2. The height of each floor is 12 ft.

3. Use the information given to find a formula for the function $t(h, k)$.

4. Find a formula for the constraint.

5. Use the method of Lagrange multipliers to find the dimensions of the building that will minimize travel time t between the most remote points in the building.

6. Use a 3D graphics program to graph both equations in Exercises 3 and 4. Then visually check the results you found analytically.

CASE	BUILDING	n	k	A	h	$t(h, k)$
1	B1	2	40	3200	15	21.5
	B2	3	32.66	3200	30	19.4
2	B1	2	60	7200		
	B2	3	48.99			
3	B1	4	40			
	B2	5	35.777			
4	B1	5	60			
	B2	10	42.426			
5	B1	5	150			
	B2	10	106.066			
6	B1	10	40			
	B2	17	30.679			
7	B1	10	80			
	B2	17	61.357			
8	B1	17	40			
	B2	26	32.344			
9	B1	17	50			
	B2	26	40.43			
10	B1	26	77			
	B2	50	55.525			

Trigonometric Functions

Chapter Snapshot

What You'll Learn

Why It's Important

Any function for which the graph repeats a pattern is said to be *periodic*. Of special importance in mathematics are the periodic *trigonometric functions*. In this chapter, we will learn how to differentiate and integrate these functions and to solve related real-world problems.

Where It's Used

MODELING PRODUCTION

The Ski Emporium's production of skis is modeled by the function

$$S(t) = 7 - 7\cos\left(\frac{\pi}{6}t\right),$$ where $S(t)$ is

the rate of production (in thousands of pairs) t months after August 1. Production, highest during winter and lowest during summer, follows this periodic pattern each year. How can trigonometry be used to model production in a specified month, total production during a year, and average production over time?

This problem appears as Example 8 in Section 7.3

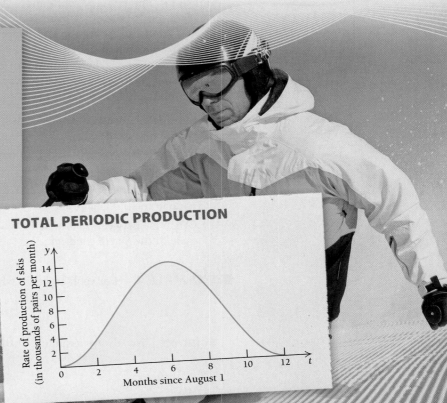

TOTAL PERIODIC PRODUCTION

Rate of production of skis (in thousands of pairs per month) vs. Months since August 1

7.1

OBJECTIVES

• Determine the quadrant in which the terminal side of an angle lies.

• Convert between radian and degree measures of an angle.

• Use symmetry to find reference angles.

• Find values of trigonometric functions.

• Find the period of a trigonometric function.

• Verify trigonometric identities.

Basics of Trigonometry

The motion of a Ferris wheel is periodic: each rider moves up and down at a consistent rate. The seasons are also periodic, as are waves. Trigonometry is the branch of mathematics that models periodic motion. This section introduces the trigonometric functions, beginning with a discussion of angles.

Angles

Consider a rotating ray, with its endpoint at the origin in the *xy*-plane and its starting position on the positive *x*-axis. We will regard a counterclockwise rotation of such a ray as positive and a clockwise rotation as negative. Note that the rotating ray and the positive *x*-axis form an angle. The rotating ray is called the *terminal side* of the angle, and the positive *x*-axis is called the *initial side* of the angle.

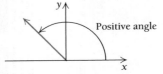

A counterclockwise rotation results in a positive angle.

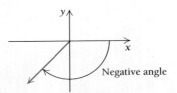

A clockwise rotation results in a negative angle.

Degree Measurement

The size, or measure, of an angle may be given in *degrees*, denoted with the symbol °. A complete revolution of a rotating ray yields an angle with a measure of 360°; half of a revolution yields a measure of 180°; a quarter of a revolution yields a measure of 90°, and so on. The ray may rotate more than one complete revolution, as with a 720° angle, or it may rotate clockwise, as with a −250° angle. Note that angles with measures 0°, 360°, and 720° share the same terminal side, as do angles of 270° and −90°. Angles that share a terminal side are called *coterminal angles*.

The *xy*-plane is divided into four quadrants. Assuming that an angle has its initial side on the positive *x*-axis and this ray is rotated counterclockwise, the four quadrants are defined as follows:

• An angle between 0° and 90° has its terminal side in the *first quadrant*.

• An angle between 90° and 180° has its terminal side in the *second quadrant*.

• An angle between 180° and 270° has its terminal side in the *third quadrant*.

• An angle between 270° and 360° has its terminal side in the *fourth quadrant*.

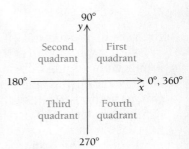

■ **EXAMPLE 1** For each angle, in which quadrant does the terminal side lie?

a) 47°   b) 212°

c) −135°   d) 740°

Solution We always assume that the initial side lies on the positive *x*-axis.

a) Since 47° is greater than 0° and less than 90°, its terminal side lies in the first quadrant.

b) The angle 212° is greater than 180° and less than 270°, so its terminal side lies in the third quadrant.

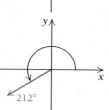

c) Since 360° − 135° = 225°, the angle −135° is coterminal with 225°, and its terminal side lies in the third quadrant.

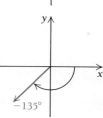

d) The angle 740° is coterminal with 20° (two complete revolutions, or 2 · 360°, plus an extra 20°). Its terminal side is in the first quadrant.

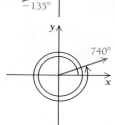

❭ **Quick Check 1**

In which quadrant does the terminal side of each angle lie?

a) 175°   **b)** −290°

c) 915°

❬ Quick Check 1

Radian Measurement

Another useful unit for measuring angles is the *radian*, which is based on the geometrical properties of the *unit circle*, a circle of radius 1 that is centered at the origin. Let angle *t* be formed by two rays that meet at the center of the unit circle. The rays intersect the circle at two points, forming an *arc*. The radian measure of angle *t* is equal to the distance along the arc of the unit circle from the initial side to the terminal side of the angle. Since the circumference of the unit circle is $C = 2\pi(1)$, or 2π, a complete revolution of the ray (360°) has a measure of 2π radians, and a half-revolution (180°) has a measure of π radians. From this, we can develop two equations relating degrees to radians:

> The distance *t* along an arc of a unit circle is the same as the radian measure *t* of the angle that forms the arc.

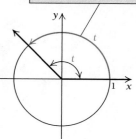

$$1° = \frac{\pi}{180} \text{ radians} \quad \text{and} \quad 1 \text{ radian} = \left(\frac{180}{\pi}\right)°.$$

In general, we can convert angle measures using the following:

- To convert an angle measure from degrees to radians, multiply the degree measure by $\dfrac{\pi \text{ radians}}{180 \text{ degrees}}$.

- To convert an angle measure from radians to degrees, multiply the radian measure by $\dfrac{180 \text{ degrees}}{\pi \text{ radians}}$.

The degree symbol (°) is not used with radian measures. So, an angle measure given simply as 20 is understood to be 20 radians.

■ **EXAMPLE 2**   Convert the measure of each angle.

a) 135° to radians

b) $\pi/4$ radians to degrees

Solution

a) We multiply 135° by $\dfrac{\pi \text{ radians}}{180 \text{ degrees}}$ and simplify:

$$135° \cdot \left(\frac{\pi \text{ radians}}{180 \text{ degrees}} \right) = \frac{135\pi}{180} \text{ radians} = \frac{3\pi}{4} \text{ radians}.$$

Thus, 135° is equivalent to $\dfrac{3\pi}{4}$ radians.

b) We multiply $\pi/4$ radians by $\dfrac{180 \text{ degrees}}{\pi \text{ radians}}$ and simplify:

$$\frac{\pi}{4} \text{ radians} \cdot \left(\frac{180 \text{ degrees}}{\pi \text{ radians}} \right) = \frac{180}{4} \text{ degrees} = 45°.$$

Thus, $\pi/4$ radians is equivalent to 45°.

❬ Quick Check 2

> **Quick Check 2**
>
> Convert the measure of each angle.
>
> **a)** 120° to radians
> **b)** $5\pi/3$ radians to degrees

Trigonometric Functions

Angles are important to *trigonometric functions*. Consider an angle t, measured in radians on the unit circle. Let the terminal side of the angle intersect the circle at point $P(x, y)$. This leads us to the definition of *cosine* and *sine*:

DEFINITION

The terminal side of an angle of t radians, whose initial side lies along the positive x-axis, intersects the unit circle at the point $P(x, y)$, where

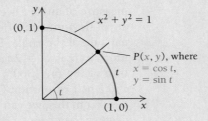

$x = $ **cosine** of t, written $\cos t$, is the first coordinate of P, and
$y = $ **sine** of t, written $\sin t$, is the second coordinate of P.

The cosine and sine values of angles whose terminal sides lie along the x-axis or y-axis can be found by noting the coordinate at which the unit circle intersects each axis.

When $t = 0$ (or $0°$), the terminal side of the angle is on the positive x-axis, and P is $(1, 0)$. Thus, $\cos 0 = 1$ and $\sin 0 = 0$.

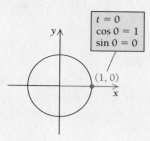

When $t = \dfrac{\pi}{2}$ (or $90°$), the terminal side of the angle is on the positive y-axis, and P is $(0, 1)$. Thus, $\cos \dfrac{\pi}{2} = 0$ and $\sin \dfrac{\pi}{2} = 1$.

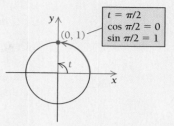

When $t = \pi$ (or $180°$), the terminal side of the angle is on the negative x-axis, and P is $(-1, 0)$. Thus, $\cos \pi = -1$ and $\sin \pi = 0$.

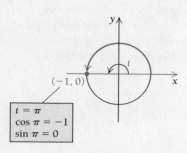

When $t = \dfrac{3\pi}{2}$ (or $270°$), the terminal side of the angle is on the negative y-axis, and P is $(0, -1)$. Thus, $\cos \dfrac{3\pi}{2} = 0$ and $\sin \dfrac{3\pi}{2} = -1$.

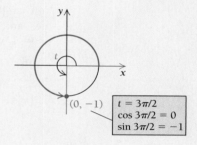

TECHNOLOGY CONNECTION

Exploratory

Use a calculator to verify that $\cos \dfrac{\pi}{5} \approx 0.809$ and $\sin \dfrac{\pi}{5} \approx 0.588$. Select any other angle t and find $x = \cos t$ and $y = \sin t$. Verify that these values satisfy $x^2 + y^2 = 1$.

A calculator can be used to determine the values of the cosine and the sine of an angle. For example, an angle of $t = \dfrac{\pi}{5}$ (or $36°$) intersects the unit circle at $P(x, y)$, where $x = \cos \dfrac{\pi}{5} \approx 0.809$ and $y = \sin \dfrac{\pi}{5} \approx 0.588$. Since this point lies on the unit circle, its coordinates satisfy $x^2 + y^2 = 1$: $(0.809)^2 + (0.588)^2 \approx 1$.

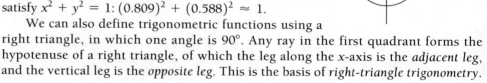

We can also define trigonometric functions using a right triangle, in which one angle is $90°$. Any ray in the first quadrant forms the hypotenuse of a right triangle, of which the leg along the x-axis is the *adjacent leg*, and the vertical leg is the *opposite leg*. This is the basis of *right-triangle trigonometry*.

DEFINITION   **Right-Triangle Trigonometry**

$$\sin t = \frac{\text{length of opposite leg}}{\text{length of hypotenuse}} = \frac{\text{opp}}{\text{hyp}}$$

$$\cos t = \frac{\text{length of adjacent leg}}{\text{length of hypotenuse}} = \frac{\text{adj}}{\text{hyp}}$$

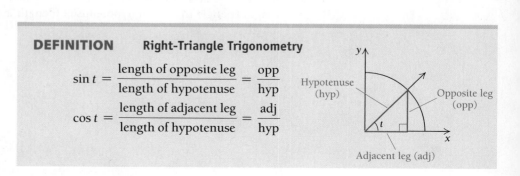

Note that for a hypotenuse of length 1, these definitions are equivalent to our earlier definitions of sine and cosine. For a hypotenuse of any other length, the adjacent and opposite legs are proportionally scaled, so the definitions are equivalent. We can use right triangles to determine the values of the sine and cosine functions for certain angles.

Function Values for Special Angles

Using the properties of right triangles, we can find values of the sine and cosine functions for the angles 45°, 30° and 60°. First, recall the Pythagorean Theorem, which states that for any right triangle, $a^2 + b^2 = c^2$, where c is the length of the hypotenuse and a and b are the lengths of the legs.

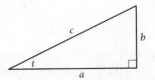

Consider the angle $t = \pi/4$ radians $= 45°$. Because the sum of all angles in a triangle is 180°, a right triangle with a 45° angle has two 45° angles. Thus, such a triangle is isosceles, and the legs are the same length, that is, $a = b$. Assuming $c = 1$, we have

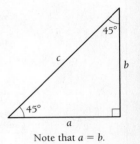

Note that $a = b$.

$$a^2 + a^2 = 1^2 \qquad \text{Using } a^2 + b^2 = c^2, \text{ with } c = 1$$
$$2a^2 = 1$$
$$a^2 = \frac{1}{2}$$
$$a = \sqrt{\frac{1}{2}} = \frac{\sqrt{2}}{2}.$$

Since $a = b$, we also have $b = \dfrac{\sqrt{2}}{2}$. Thus, for $t = \pi/4 = 45°$, we have the following.

Sine and Cosine of a 45° (or $\pi/4$ radian) Angle

$$\sin \frac{\pi}{4} = \sin 45° = \frac{\sqrt{2}}{2} \qquad \text{and} \qquad \cos \frac{\pi}{4} = \cos 45° = \frac{\sqrt{2}}{2}.$$

We can also find values of the sine and cosine functions for 30° and 60°. A right triangle with 30° and 60° angles is half of an equilateral triangle. Thus, if we choose an equilateral triangle whose sides have length 2 and take half of it, we get a right triangle that has a hypotenuse of length 2 and a leg of length 1.

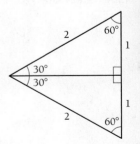

The longer leg has length a, where

$$a^2 + 1^2 = 2^2 \quad \text{Using the Pythagorean Theorem}$$
$$a^2 = 3 \quad 2^2 - 1^2 = 3$$
$$a = \sqrt{3}.$$

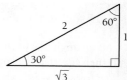

From this triangle, we can find the values of the sine and cosine for $t = \pi/6 = 30°$ and for $t = \pi/3 = 60°$. For $t = \pi/6 = 30°$, we have the following.

Sine and Cosine of a 30° (or $\pi/6$ radian) Angle

$$\sin \frac{\pi}{6} = \sin 30° = \frac{1}{2} \quad \text{and} \quad \cos \frac{\pi}{6} = \cos 30° = \frac{\sqrt{3}}{2}.$$

And for $t = \pi/3 = 60°$, we have the following.

Sine and Cosine of a 60° (or $\pi/3$ radian) Angle

$$\sin \frac{\pi}{3} = \sin 60° = \frac{\sqrt{3}}{2} \quad \text{and} \quad \cos \frac{\pi}{3} = \cos 60° = \frac{1}{2}.$$

The following table summarizes these important values for the cosine and sine functions in the first quadrant.

t (radians)	t (degrees)	$\cos t$	$\sin t$
0	0°	1	0
$\pi/6$	30°	$\dfrac{\sqrt{3}}{2}$	$\dfrac{1}{2}$
$\pi/4$	45°	$\dfrac{\sqrt{2}}{2}$	$\dfrac{\sqrt{2}}{2}$
$\pi/3$	60°	$\dfrac{1}{2}$	$\dfrac{\sqrt{3}}{2}$
$\pi/2$	90°	0	1

It is important to remember or be able to quickly derive these values.

Reference Angles and Symmetry

For any angle in the first quadrant, there are angles in each of the other three quadrants that are symmetric to that angle. The angle in the first quadrant is called the *reference angle* for the three angles formed by reflecting the terminal ray across the x-axis, across the y-axis, and across the origin.

Reflection across the *y*-axis	Reflection across the origin	Reflection across the *x*-axis
If an angle *t* is in the second quadrant, then its reference angle is symmetric across the *y*-axis and measures $\pi - t$ radians, or $(180 - t)°$.	If an angle *t* is in the third quadrant, then its reference angle is symmetric across the origin and measures $t - \pi$ radians, or $(t - 180)°$.	If an angle *t* is in the fourth quadrant, then its reference angle is symmetric across the *x*-axis and measures $2\pi - t$ radians, or $(360 - t)°$.
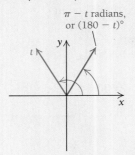	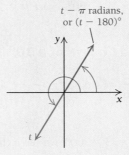	
Example: If $t = \dfrac{5\pi}{6}$ (or 150°), then its reference angle is $\pi - \dfrac{5\pi}{6} = \dfrac{\pi}{6}$, or $180° - 150° = 30°$.	**Example:** If $t = \dfrac{5\pi}{4}$ (or 225°), then its reference angle is $\dfrac{5\pi}{4} - \pi = \dfrac{\pi}{4}$, or $225° - 180° = 45°$.	**Example:** If $t = \dfrac{5\pi}{3}$ (or 300°), then its reference angle is $2\pi - \dfrac{5\pi}{3} = \dfrac{\pi}{3}$, or $360° - 300° = 60°$.

■ **EXAMPLE 3** Find $\cos \dfrac{4\pi}{3}$ and $\sin \dfrac{4\pi}{3}$, using reference angles.

Solution Since $\dfrac{4\pi}{3}$ radians is in the third quadrant, the cosine and the sine are both negative. The reference angle is $\dfrac{4\pi}{3} - \pi = \dfrac{\pi}{3}$ radians. We have

$$\cos \frac{4\pi}{3} = -\cos \frac{\pi}{3} = -\frac{1}{2} \qquad \text{The cosine is negative in the third quadrant.}$$

$$\sin \frac{4\pi}{3} = -\sin \frac{\pi}{3} = -\frac{\sqrt{3}}{2} \qquad \text{The sine is negative in the third quadrant.}$$

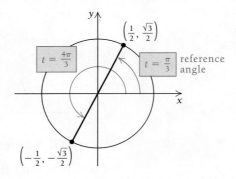

> **Quick Check 3**
>
> Without using a calculator, find $\cos \dfrac{7\pi}{4}$ and $\sin \dfrac{7\pi}{4}$.

⟨ Quick Check 3

Graphs of $f(t) = \cos t$ and $f(t) = \sin t$

Since $\cos t$ and $\sin t$ are defined for all real numbers *t*, large values of *t* (positive and negative) "wrap around" the circle many times before reaching the terminal point *P*. Nevertheless, *P* still has one first coordinate, $\cos t$, and one second coordinate, $\sin t$. Plotting points previously obtained (and finding others with a calculator), we graph the cosine and sine functions as follows. We will assume that *t* is measured in radians.

t (radians)	$\cos t$	$\sin t$
0	1	0
$\pi/6$	$\sqrt{3}/2 \approx 0.866$	$1/2 = 0.5$
$\pi/4$	$\sqrt{2}/2 \approx 0.707$	$\sqrt{2}/2 \approx 0.707$
$\pi/3$	$1/2 = 0.5$	$\sqrt{3}/2 \approx 0.866$
$\pi/2$	0	1
$2\pi/3$	$-1/2 = -0.5$	$\sqrt{3}/2 \approx 0.866$
$3\pi/4$	$-\sqrt{2}/2 \approx -0.707$	$\sqrt{2}/2 \approx 0.707$
$5\pi/6$	$-\sqrt{3}/2 \approx -0.866$	$1/2 = 0.5$
π	-1	0
$7\pi/6$	$-\sqrt{3}/2 \approx -0.866$	$-1/2 = -0.5$
$5\pi/4$	$-\sqrt{2}/2 \approx -0.707$	$-\sqrt{2}/2 \approx -0.707$
$4\pi/3$	$-1/2 = -0.5$	$-\sqrt{3}/2 \approx -0.866$
$3\pi/2$	0	-1
$5\pi/3$	$1/2 = 0.5$	$-\sqrt{3}/2 \approx -0.866$
$7\pi/4$	$\sqrt{2}/2 \approx 0.707$	$-\sqrt{2}/2 \approx -0.707$
$11\pi/6$	$\sqrt{3}/2 \approx 0.866$	$-1/2 = -0.5$
2π	1	0

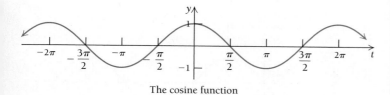

The cosine function

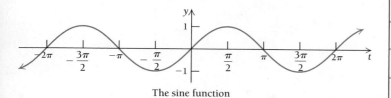

The sine function

At the origin, t is 0. Moving to the right on the graphs corresponds to rotating the terminal side of the angle t counterclockwise, and moving to the left corresponds to rotating the terminal side clockwise. Note that $\sin t = 0$ for $t = 0, \pm\pi, \pm 2\pi, \pm 3\pi$, and so on, and $\cos t = 0$ for $t = \pm\pi/2, \pm 3\pi/2, \pm 5\pi/2$, and so on. The curves repeat as the terminal side makes successive revolutions.

DEFINITION

A function f is **periodic** if there exists a positive number p such that

$$f(t + p) = f(t)$$

for every t in the domain of f. Thus, adding p to an input does not change the output. The smallest such number p is called the **period** of the function.

On the unit circle, the ray completes one cycle between $t = 0$ and $t = 2\pi$. The cycle then repeats. Thus, for any t,

$$\cos (t + 2\pi) = \cos t \quad \text{and} \quad \sin (t + 2\pi) = \sin t,$$

and the cosine and sine functions both have a period of 2π.

■ **EXAMPLE 4** What is the period of the following function?

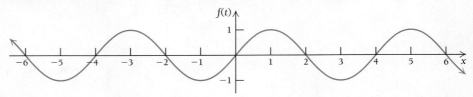

Solution This function has a period $p = 4$ units. We can see this by determining the length from peak to peak or from valley to valley:

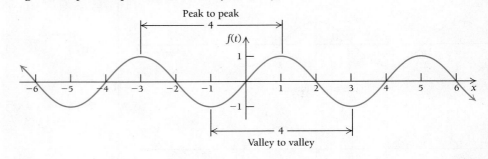

❯ **Quick Check 4**

What is the period of the following function?

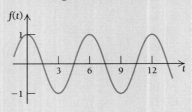

❮ Quick Check 4

Other Trigonometric Functions

The functions $\cos t$ and $\sin t$ are the basic trigonometric functions, but there are four others—the tangent, cotangent, secant and cosecant functions. These four functions are defined in terms of $\cos t$ and $\sin t$ and also by using the legs and hypotenuse of a right triangle, as follows.

> **DEFINITION**
>
> The **tangent** of t, written $\tan t$, is defined as
>
> $$\tan t = \frac{\sin t}{\cos t} = \frac{\text{opp}}{\text{adj}}.$$
>
> The **cotangent** of t, written $\cot t$, is defined as
>
> $$\cot t = \frac{\cot t}{\sin t} = \frac{1}{\tan t} = \frac{\text{adj}}{\text{opp}}.$$
>
> The **secant** of t, written $\sec t$, is defined as
>
> $$\sec t = \frac{1}{\cos t} = \frac{\text{hyp}}{\text{adj}}.$$
>
> The **cosecant** of t, written $\csc t$, is defined as
>
> $$\csc t = \frac{1}{\sin t} = \frac{\text{hyp}}{\text{opp}}.$$
>
> These functions are defined for all t such that the denominator is not zero.

Since $\cot t = \dfrac{1}{\tan t}$, we also have $\tan t = \dfrac{1}{\cot t}$.

■ **EXAMPLE 5**   Find the following values. Do not use a calculator.

a) $\tan \dfrac{\pi}{6}$ 　　　　　　　　　　　　　　　　　**b)** $\sec 45°$

Solution

a) $\tan \dfrac{\pi}{6} = \dfrac{\sin (\pi/6)}{\cos (\pi/6)} = \dfrac{1/2}{\sqrt{3}/2} = \dfrac{1}{\sqrt{3}} = \dfrac{\sqrt{3}}{3}.$

b) $\sec 45° = \dfrac{1}{\cos 45°} = \dfrac{1}{\sqrt{2}/2} = \dfrac{2}{\sqrt{2}} = \sqrt{2}.$

❮ Quick Check 5

> ❯ **Quick Check 5**
>
> Find the following values. Do not use a calculator.
>
> **a)** $\cot \dfrac{\pi}{3}$ 　　**b)** $\csc 30°$

The graph of $y = \tan t$ is shown below. It includes vertical asymptotes at values of t at which $\cos t = 0$, because $\tan t = \dfrac{\sin t}{\cos t}$ is undefined at those values. These are

$$t = \pm \frac{\pi}{2}, \pm \frac{3\pi}{2}, \pm \frac{5\pi}{2}, \ldots, \text{ or } t = \frac{\pi}{2} + k\pi, \text{ where } k \text{ is an integer. Thus, the domain}$$

of $y = \tan t$ is all real numbers t such that $t \neq \dfrac{\pi}{2} + k\pi$, where k is an integer.

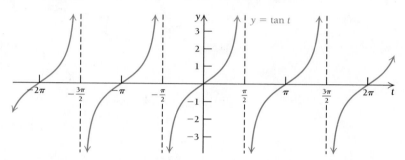

Since the graph repeats itself every π units, $\tan t$ has a period of π.

Trigonometric Identities

An *identity* is an equation that holds true for all meaningful replacements of its variables by real numbers. For example, $(t + 1)^2 = t^2 + 2t + 1$ is true for all real numbers. Another identity, $(t^2 - 4)/(t - 2) = t + 2$, is true for all real numbers except 2, which is *not* a meaningful replacement for t in the expression on the left side of the equation. The equations $\cos (t + 2\pi) = \cos t$ and $\sin (t + 2\pi) = \sin t$ are true for all real numbers t. They are examples of *trigonometric identities*.

Recall that a point $P(x, y)$ that lies on the unit circle must satisfy $x^2 + y^2 = 1$. Since $x = \cos t$ and $y = \sin t$, we have the following identities.

THEOREM 1   The Pythagorean Identities

For all real numbers t,

$$\cos^2 t + \sin^2 t = 1,$$

where $\cos^2 t = (\cos t)^2$ and $\sin^2 t = (\sin t)^2$.

Dividing $\cos^2 t + \sin^2 t = 1$ by $\cos^2 t$, we have

$$1 + \tan^2 t = \sec^2 t,$$

for all t such that $\cos t \neq 0$.

Dividing $\cos^2 t + \sin^2 t = 1$ by $\sin^2 t$, we have

$$\cot^2 t + 1 = \csc^2 t,$$

for all t such that $\sin t \neq 0$.

■ **EXAMPLE 6**   Given $\sin t = \frac{1}{4}$ with t in the first quadrant, use trigonometric identities to find $\cos t$ and $\tan t$.

Solution   We use the Pythagorean identity $\cos^2 t + \sin^2 t = 1$ to find $\cos t$ as follows:

$$\cos^2 t + \left(\frac{1}{4}\right)^2 = 1 \qquad \text{Substituting}$$

$$\cos^2 t = 1 - \frac{1}{16} \qquad \text{Noting that } \left(\frac{1}{4}\right)^2 = \frac{1}{16} \text{ and subtracting}$$

$$\cos^2 t = \frac{15}{16}$$

$$\cos t = \frac{\sqrt{15}}{4}. \qquad \text{Taking the positive root since } \cos t \text{ is positive in the first quadrant}$$

To find $\tan t$, we note that $\tan t = \dfrac{\sin t}{\cos t}$. Thus,

$$\tan t = \frac{1/4}{\sqrt{15}/4} = \frac{1}{\sqrt{15}}, \text{ or } \frac{\sqrt{15}}{15}.$$

❮ Quick Check 6

〉 **Quick Check 6**

For $\cos t = \frac{2}{3}$ with t in the first quadrant, use trigonometric identities to find $\sin t$ and $\tan t$.

We can also add or subtract two angles. However, the sine of the sum (or difference) of two angles is not the sum (or difference) of the sine of the angles: $\sin(a \pm b) \neq \sin a \pm \sin b$. The same is true for the cosine.

THEOREM 2   The Sum-Difference Identities

If t and u are two angles, then

$$\cos(t + u) = \cos t \cos u - \sin t \sin u,$$

$$\sin(t + u) = \sin t \cos u + \cos t \sin u,$$

$$\cos(t - u) = \cos t \cos u + \sin t \sin u,$$

$$\sin(t - u) = \sin t \cos u - \cos t \sin u.$$

The proof of $\cos(t - u) = \cos t \cos u + \sin t \sin u$ is outlined in Exercise 85, while the proofs of the other three identities are Exercises 86–88.

■ **EXAMPLE 7** Use a sum-difference identity to find $\cos 15°$.

Solution If we think of $15°$ as $45° - 30°$, then

$$\cos 15° = \cos(45° - 30°)$$
$$= \cos 45° \cos 30° + \sin 45° \sin 30°$$
$$= \frac{\sqrt{2}}{2} \cdot \frac{\sqrt{3}}{2} + \frac{\sqrt{2}}{2} \cdot \frac{1}{2}$$
$$= \frac{\sqrt{6} + \sqrt{2}}{4}.$$

❯ **Quick Check 7**

Use a sum-difference identity (not a calculator) to find $\cos 105°$.

❮ Quick Check 7

The sum-difference identities can be used to prove the *double-angle identities*. For example, if we let $t = u$, then

$$\sin 2t = \sin(t + t) \qquad \text{Replacing } u \text{ with } t$$
$$= \sin t \cos t + \cos t \sin t$$
$$= 2 \sin t \cos t.$$

In a similar way, we can show that $\cos 2t = \cos^2 t - \sin^2 t$.

THEOREM 3 **The Double-Angle Identities**

$$\sin 2t = 2 \sin t \cos t \qquad \text{and} \qquad \cos 2t = \cos^2 t - \sin^2 t.$$

■ **EXAMPLE 8** If $\sin t = \frac{1}{4}$ and t is in the first quadrant, find $\sin 2t$.

Solution From Example 6, we have $\cos t = \frac{\sqrt{15}}{4}$. Thus,

$$\sin 2t = 2 \sin t \cos t$$
$$= 2\left(\frac{1}{4}\right)\left(\frac{\sqrt{15}}{4}\right) \qquad \text{Substituting}$$
$$= \frac{\sqrt{15}}{8}.$$

❯ **Quick Check 8**

If $\cos t = \frac{2}{3}$ and t is in the first quadrant (see Quick Check 6), find $\cos 2t$.

❮ Quick Check 8

Section Summary

- An *angle* is formed by a ray rotating from the positive x-axis. The positive x-axis is the *initial side*, and the ray is the *terminal side*. Two angles that have the same terminal side are said to be *coterminal*.
- Angles are measured in *degrees*, denoted °, or in *radians*. They are related by $1° = \dfrac{\pi}{180}$ radians and $1 \text{ radian} = \dfrac{180}{\pi}$ degrees. To convert an angle from degrees to radians, multiply the degree measure by $\dfrac{\pi \text{ radians}}{180 \text{ degrees}}$. To convert an angle from radians to degrees, multiply the radian measure by $\dfrac{180 \text{ degrees}}{\pi \text{ radians}}$.
- If a ray at an angle t with the positive x-axis intersects the unit circle at a point $P(x, y)$, the first coordinate of P is the *cosine* of t, and the second coordinate of P is the *sine* of t, or $x = \cos t$ and $y = \sin t$.
- A function is *periodic* if there exists a positive value p such that $f(t + p) = f(t)$ for all t in the domain of f. The smallest such p is called the *period*.
- The *tangent function* is $\tan t = \dfrac{\sin t}{\cos t} = \dfrac{1}{\cot t}$, the *cotangent function* is $\cot t = \dfrac{\cos t}{\sin t} = \dfrac{1}{\tan t}$, the *secant*

function is $\sec t = \dfrac{1}{\cos t}$, and the *cosecant function* is $\csc t = \dfrac{1}{\sin t}$.

- The functions $\sin t$ and $\cos t$ have a period of 2π, while the function $\tan t$ has a period of π.
- Important trigonometric identities include the following.

  o The *Pythagorean identities*:
  $$\cos^2 t + \sin^2 t = 1,$$
  $$1 + \tan^2 t = \sec^2 t,$$
  $$\cot^2 t + 1 = \csc^2 t.$$

  o The *sum-difference identities*:
  $$\cos(t + u) = \cos t \cos u - \sin t \sin u,$$
  $$\sin(t + u) = \sin t \cos u + \cos t \sin u,$$
  $$\cos(t - u) = \cos t \cos u + \sin t \sin u,$$
  $$\sin(t - u) = \sin t \cos u - \cos t \sin u.$$

  o The *double-angle identities*:
  $$\sin 2t = 2 \sin t \cos t,$$
  $$\cos 2t = \cos^2 t - \sin^2 t.$$

EXERCISE SET
7.1

For Exercises 1–4, **(a)** identify the quadrant in which the terminal side of each angle lies and **(b)** find two coterminal angles (answers may vary).

1. $34°$

2. $320°$

3. $\dfrac{5\pi}{8}$

4. $\dfrac{7\pi}{6}$

Convert the following angles to radian measures, expressed in terms of π.

5. $15°$ **6.** $200°$ **7.** $75°$ **8.** $300°$

9. $-135°$ **10.** $-210°$ **11.** $-128°$ **12.** $-305°$

Convert the following angles to degree measures, rounded to three decimal places.

13. $\dfrac{3\pi}{2}$ **14.** $\dfrac{5\pi}{4}$ **15.** $-\dfrac{\pi}{4}$ **16.** $-\dfrac{\pi}{12}$

17. 8π **18.** 9π **19.** -5π **20.** -14π

21. 1 **22.** 2

Find each of the following values. Do not use a calculator.

23. $\sin \dfrac{9\pi}{4}$ **24.** $\sin \dfrac{7\pi}{6}$ **25.** $\cos \dfrac{11\pi}{6}$

26. $\cos 3\pi$ **27.** $\tan 4\pi$ **28.** $\tan \dfrac{\pi}{4}$

29. $\tan\left(-\dfrac{\pi}{3}\right)$ **30.** $\cot\left(-\dfrac{\pi}{6}\right)$ **31.** $\sec \dfrac{\pi}{4}$

32. $\sec \pi$ **33.** $\csc\left(-\dfrac{\pi}{6}\right)$ **34.** $\csc\left(-\dfrac{3\pi}{2}\right)$

For Exercises 35–40, a ray is drawn on the unit circle, at the given angle t. Assume that the ray intersects the circle at point $P(x, y)$. Find x and y. Do not use a calculator.

35. $t = \dfrac{4\pi}{3}$ **36.** $t = \dfrac{3\pi}{4}$ **37.** $t = -\dfrac{3\pi}{4}$

38. $t = -\dfrac{5\pi}{3}$ **39.** $t = 5\pi$ **40.** $t = -4\pi$

For Exercises 41–50, a ray is drawn on a unit circle, at the given angle t. Assume that the ray intersects the circle at point $P(x, y)$. Use a calculator to find x and y to three decimal places, and be sure the calculator is in the correct mode.

41. $t = \pi/7$

42. $t = 3\pi/8$

43. $t = 1$

44. $t = 5$

45. $t = 32°$

46. $t = 127°$

47. $t = 218°$

48. $t = 296°$

49. $t = -27°$

50. $t = -176°$

Most calculators do not have keys for the cotangent, secant, and cosecant functions. Use a calculator's sin, cos, and tan keys and an appropriate trigonometric identity to find the following values to three decimal places.

51. $\sec 54°$

52. $\sec 123°$

53. $\cot \dfrac{3\pi}{7}$

54. $\csc \dfrac{\pi}{10}$

55. $\csc 211°$

56. $\cot 19°$

For Exercises 57–70, use various trigonometric identities to find values of the given functions. Do not use a calculator. For Exercises 57–64, assume that t is in the first quadrant.

57. For $\sin t = \frac{1}{6}$, find $\cos t$ and $\tan t$.

58. For $\cos t = \frac{2}{5}$, find $\sin t$ and $\tan t$.

59. For $\cos t = \frac{3}{7}$, find $\sin t$ and $\sec t$.

60. For $\sin t = \frac{4}{9}$, find $\cos t$ and $\csc t$.

61. For $\sin t = \frac{1}{8}$, find $\sin 2t$.

62. For $\cos t = \frac{2}{7}$, find $\sin 2t$.

63. For $\cos t = \frac{3}{4}$, find $\cos 2t$.

64. For $\sin t = \frac{5}{6}$, find $\cos 2t$.

65. Find $\cos 75°$.

66. Find $\sin 15°$.

67. Find $\sin 105°$.

68. Find $\sin 75°$.

69. Find $\tan 15°$.

70. Find $\tan 75°$.

For Exercises 71 and 72, use the Pythagorean Theorem to find sin t, cos t, and tan t. Do not use a calculator.

71.

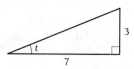

72.

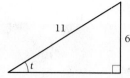

For Exercises 73–76, use a calculator to find the indicated lengths to three decimal places.

73. B and C

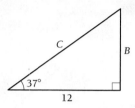

74. A and C

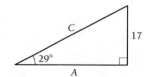

75. A and B

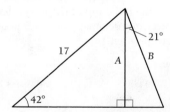

76. A and B

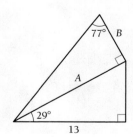

APPLICATIONS

General Interest

77. The sides of the Great Pyramid of Giza rise at an angle of 51.84°. The distance from the center of one side of the pyramid's base to the point directly below its apex is 378 ft. What is the height h of the pyramid, to the nearest foot? (*Source:* Wikipedia.com.)

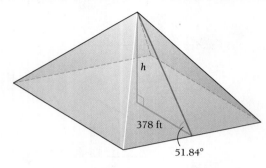

78. The Leaning Tower of Pisa in Italy is 56.3 m tall and leans at an angle of 3.97° from true vertical. Find the distance d by which the top of the tower deviates from the vertical, to three decimal places. (*Source:* www.leaningtowerofpisa.net.)

79. Business: fluctuating sales. The number of units sold (in thousands) as a function of time (in months) for Lola Summer Wear, Inc., is shown in the following graph. What is the period of this sales function?

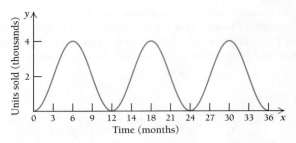

80. Tides. The tides at Fisherman's Beach, as measured from a lifeguard stand, fluctuate according to the graph below. What is the period of the tidal function?

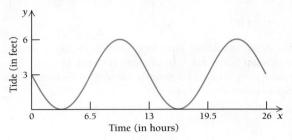

81. Electrocardiogram. The graph below shows a typical electrocardiogram of a human heartbeat. What is the period of the heartbeat? (*Source:* www.cnx.org.)

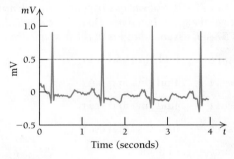

82. Violin tone. A graph representing a tone produced on a violin is shown below. What is the period of this sound? (*Source:* www.knowledgerush.com.)

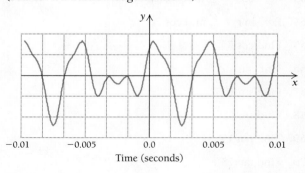

SYNTHESIS

For Exercises 83 and 84, a line passes through the origin and the given point. What is the smallest angle that this line makes with the x-axis?

83. $(3, 5)$

84. $(4, -2)$

85. To prove that $\cos(t - u) = \cos t \cos u + \sin t \sin u$, we consider two rays drawn from the origin of a unit circle at angles t and u, where $t > u$. Note that the angle between these two rays is $t - u$. In the graph on the right, the angle $t - u$ has been rotated so that its initial side lies on the positive x-axis.

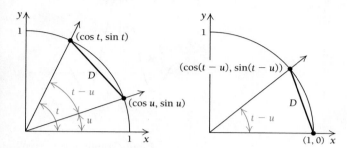

a) Recall that the distance between two points (x_1, y_1) and (x_2, y_2) is given by the Pythagorean formula, $D^2 = (x_2 - x_1)^2 + (y_2 - y_1)^2$. Use this formula to find the distance between $(\cos t, \sin t)$ and $(\cos u, \sin u)$. Simplify the expression using trigonometric identities.

b) Use the distance formula from part (a) to find the distance between $(\cos(t - u), \sin(t - u))$ and $(1, 0)$. Simplify the expression using trigonometric identities.

c) Set the two expressions from parts (a) and (b) equal, and simplify.

For Exercises 86–88, use the identity $\cos(t - u) = \cos t \cos u + \sin t \sin u$, *the symmetry identities* $\sin(-t) = -\sin t$ *and* $\cos(-t) = \cos t$, *and the shift identities* $\sin\left(t + \dfrac{\pi}{2}\right) = \cos t$ *and* $\cos\left(t - \dfrac{\pi}{2}\right) = \sin t$ *to prove the other three sum-difference identities.*

86. Prove that $\cos(t + u) = \cos t \cos u - \sin t \sin u$. (*Hint:* Replace u with $-u$.)

87. Prove that $\sin(t - u) = \sin t \cos u - \cos t \sin u$. (*Hint:* Replace t with $t - \frac{\pi}{2}$.)

88. Prove that $\sin(t + u) = \sin t \cos u + \cos t \sin u$. (*Hint:* Use the identity $\sin(t - u) = \sin t \cos u - \cos t \sin u$ and replace u with $-u$.)

89. Suppose a ray is drawn from the origin, passing through the unit circle at point P, as shown.

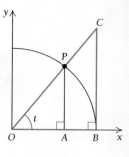

a) What are the lengths of OA and AP in terms of the angle t?

b) Use part (a) and similar triangles to prove that BC has the length $\tan t$.

c) Use part (b) and the Pythagorean Theorem to prove that OC has the length $\sec t$.

Pythagorean triples. *A Pythagorean triple consists of three integers that satisfy the Pythagorean Theorem. For example, $(3, 4, 5)$ is a Pythagorean triple since $3^2 + 4^2 = 5^2$. Other Pythagorean triples are $(5, 12, 13)$, $(8, 15, 17)$, and $(7, 24, 25)$. There are infinitely many Pythagorean triples, and the trigonometric identities can be used to generate them. For example, if $\tan t = \frac{1}{5}$, we can assume a right triangle in the first quadrant with opposite leg of length 1, adjacent leg of length 5, and hypotenuse of length $\sqrt{26}$. Using* $\sin 2t = 2 \sin t \cos t$, *we have* $2\left(\dfrac{1}{\sqrt{26}}\right)\left(\dfrac{5}{\sqrt{26}}\right) = \dfrac{5}{13}$, *and using* $\cos^2 2t + \sin^2 2t = 1$, *we have* $\cos^2 2t = 1 - \sin^2 2t = 1 - \left(\frac{5}{13}\right)^2 = \frac{144}{169}$. *Thus,* $\cos 2t = \sqrt{\frac{144}{169}} = \frac{12}{13}$, *and we have generated the Pythagorean triple $(5, 12, 13)$. Generate a Pythagorean triple starting with each of the following values.*

90. $\tan t = \frac{2}{7}$

91. $\tan t = \frac{1}{9}$

92. $\tan t = \frac{3}{8}$

93. $\tan t = \frac{5}{11}$

Answers to Quick Checks

1. (a) second; (b) first; (c) third **2.** (a) $2\pi/3$; (b) $300°$ **3.** $\cos(7\pi/4) = \sqrt{2}/2$, $\sin(7\pi/4) = -\sqrt{2}/2$ **4.** Period $= 6$ **5.** (a) $\sqrt{3}/3$; (b) 2 **6.** $\sin t = \sqrt{5}/3$, $\tan t = \sqrt{5}/2$ **7.** $\dfrac{\sqrt{2} - \sqrt{6}}{4}$ **8.** $\frac{1}{9}$

7.2

Derivatives of Trigonometric Functions

OBJECTIVES

- Differentiate trigonometric functions.

- Solve applied problems involving derivatives of trigonometric functions.

Derivatives of sin *x* and cos *x*

For consistency with our earlier work, we now regard the cosine and sine functions as functions of the variable x: $f(x) = \sin x$ and $f(x) = \cos x$. To differentiate each function, we begin by finding the value of each derivative at $x = 0$ and extend this to the general formula. The derivatives at $x = 0$ are given by the following limits:

$$\text{For } f(x) = \sin x, f'(0) = \lim_{h \to 0}\left(\frac{\sin(0 + h) - \sin 0}{h}\right) = \lim_{h \to 0}\left(\frac{\sin h}{h}\right),$$

$$\text{For } f(x) = \cos x, f'(0) = \lim_{h \to 0}\left(\frac{\cos(0 + h) - \cos 0}{h}\right) = \lim_{h \to 0}\left(\frac{\cos h - 1}{h}\right).$$

These limits are determined numerically and graphically:

Limit Numerically

h (radians)	$\sin h$	$\dfrac{\sin h}{h}$	
0.1	0.09983	0.99833	
0.01	0.0099998	0.999983	
0.001	0.000999...	0.9999998...	$\to 1$
−0.1	−0.09983	0.99833	
−0.01	−0.0099998	0.999983	
−0.001	−0.000999...	0.9999998...	$\to 1$

Right-hand limit, $h \to 0^+$ (rows with positive h)

Left-hand limit, $h \to 0^-$ (rows with negative h)

Limit Graphically

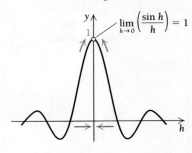

$$\lim_{h \to 0}\left(\frac{\sin h}{h}\right) = 1$$

Limit Numerically

h (radians)	$\cos h$	$\dfrac{\cos h - 1}{h}$	
0.1	0.995	−0.05	
0.01	0.99995	−0.00499996	
0.001	0.9999995	−0.00049999...	$\to 0$
−0.1	0.995	0.05	
−0.01	0.99995	0.00499996	
−0.001	0.9999995	0.00049999...	$\to 0$

Right-hand limit, $h \to 0^+$ (rows with positive h)

Left-hand limit, $h \to 0^-$ (rows with negative h)

Limit Graphically

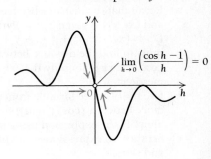

$$\lim_{h \to 0}\left(\frac{\cos h - 1}{h}\right) = 0$$

Thus, the following limits can be inferred:

$$\lim_{h \to 0}\left(\frac{\sin h}{h}\right) = 1 \quad \text{and} \quad \lim_{h \to 0}\left(\frac{\cos h - 1}{h}\right) = 0. \tag{1}$$

A geometric proof of the $\dfrac{\sin h}{h}$ case is presented in Exercise 66.

Now let's consider the general derivatives. For $f(x) = \sin x$, we have

$$\frac{d}{dx}\sin x = \lim_{h \to 0}\left(\frac{\sin(x+h) - \sin x}{h}\right).$$

$$= \lim_{h \to 0}\left(\frac{\sin x \cos h + \cos x \sin h - \sin x}{h}\right) \qquad \text{Using a sum-difference identity}$$

$$= \lim_{h \to 0}\left(\frac{\sin x \cos h - \sin x}{h} + \frac{\cos x \sin h}{h}\right) \qquad \text{Regrouping}$$

$$= \lim_{h \to 0}\left[\sin x\left(\frac{\cos h - 1}{h}\right) + \cos x\left(\frac{\sin h}{h}\right)\right] \qquad \text{Factoring}$$

$$= \sin x \cdot \lim_{h \to 0}\left(\frac{\cos h - 1}{h}\right) + \cos x \cdot \lim_{h \to 0}\left(\frac{\sin h}{h}\right) \qquad \text{Using a property of limits}$$

$$= \sin x \cdot 0 + \cos x \cdot 1 \qquad \text{Using the limits in equation (1)}$$

$$= \cos x.$$

A similar derivation (Exercise 64) shows that the derivative of $f(x) = \cos x$ is $\dfrac{d}{dx} \cos x = -\sin x$. The derivatives of the sine and cosine functions are summarized below.

THEOREM 4   **Derivatives of sin x and cos x**

$$\frac{d}{dx} \sin x = \cos x \qquad \text{and} \qquad \frac{d}{dx} \cos x = -\sin x.$$

The derivatives of the remaining trigonometric functions are found using the Quotient Rule and/or the Chain Rule. For example, since $\tan x = \dfrac{\sin x}{\cos x}$, we have

$$\frac{d}{dx} \tan x = \frac{d}{dx}\left(\frac{\sin x}{\cos x}\right)$$

$$= \frac{\cos x \cdot \cos x - \sin x \cdot (-\sin x)}{\cos^2 x}$$

$$= \frac{\cos^2 x + \sin^2 x}{\cos^2 x}$$

$$= \frac{1}{\cos^2 x}$$

$$= \sec^2 x.$$

Finding the derivatives for the other trigonometric functions is left to Exercises 61–63; the results are summarized below.

THEOREM 5   **Derivatives of the Other Trigonometric Functions**

$$\frac{d}{dx} \tan x = \sec^2 x,$$

$$\frac{d}{dx} \cot x = -\csc^2 x,$$

$$\frac{d}{dx} \sec x = \sec x \tan x,$$

$$\frac{d}{dx} \csc x = -\cot x \csc x.$$

■ **EXAMPLE 1**   Find the derivative of $y = \sin 3x^2$.

Solution   We let $u = 3x^2$, so that $y = \sin u$. To differentiate, we use the Chain Rule:

$$y' = \frac{dy}{du} \cdot \frac{du}{dx}$$

$$= \cos u \cdot 6x \qquad \frac{dy}{du} \sin u = \cos u, \frac{du}{dx}(3x^2) = 6x$$

$$= 6x \cos 3x^2 \qquad \text{Simplifying}$$

⟩ Quick Check 1

Find the derivative of $y = \cos 5x^3$.

⟨ Quick Check 1

■ **EXAMPLE 2**   Find the derivative of $y = \sec^3 x$.

Solution   Recall that $\sec^3 x = (\sec x)^3$. Thus, we use the Chain Rule:

$$y' = 3\sec^2 x \cdot \frac{d}{dx}\sec x \qquad \text{Using the Chain Rule}$$

$$= 3\sec^2 x \cdot \sec x \tan x$$

$$= 3\sec^3 x \tan x. \qquad \text{Simplifying}$$

> **Quick Check 2**
>
> Find the derivative of
> $y = \tan^4 x$.

⟨ Quick Check 2

■ **EXAMPLE 3**   Differentiate $y = x^2 \sin e^{4x}$.

Solution   We use the Product Rule and the Chain Rule:

$$y' = x^2 \cdot \frac{d}{dx}\sin e^{4x} + \sin (e^{4x}) \cdot \frac{d}{dx}x^2 \qquad \text{Using the Product Rule}$$

$$= x^2 \cdot \cos (e^{4x}) \cdot \frac{d}{dx}(e^{4x}) + \sin (e^{4x}) \cdot 2x \qquad \text{Using the Chain Rule}$$

$$= x^2 \cdot \cos (e^{4x}) \cdot e^{4x} \cdot 4 + \sin (e^{4x}) \cdot 2x$$

$$= 4x^2 e^{4x} \cos e^{4x} + 2x \sin e^{4x}. \qquad \text{Simplifying}$$

> **Quick Check 3**
>
> Find the derivative of
> $y = \dfrac{\cos x^2}{e^{2x}}$.

⟨ Quick Check 3

Sinusoidal Functions

Functions of the form $y = A \sin (Bx - C) + D$ and $y = A \cos (Bx - C) + D$ are called *sinusoidal functions*. These functions have many applications to periodic phenomena. To graph the function $y = A \sin (Bx - C) + D$, we first rewrite it as

$$y = A \sin \left[B\left(x - \frac{C}{B} \right) \right] + D.$$

The numbers represented by A, B, C and D play important roles in graphing such an equation: A determines the vertical stretching or shrinking of the graph of $y = \sin x$ (or $y = \cos x$), B determines the horizontal stretching or shrinking, C/B determines the horizontal shift, and D determines the vertical shift. The following names are given to these expressions:

$$\text{Midline: } y = D$$

$$\text{Amplitude: } |A|$$

$$\text{Period: } \frac{2\pi}{B}$$

$$\text{Phase shift: } \frac{C}{B}$$

Since B stretches or shrinks the graph horizontally, it affects the function's period. Also, we often place the midline value first to avoid possible confusion. For example, $y = 2 \sin 6x + 3$ is rewritten as $y = 3 + 2 \sin 6x$.

TECHNOLOGY CONNECTION

Exploratory Exercises

1. To see how A affects the graph of a sinusoidal function, let $y_1 = \sin x$, $y_2 = 2 \sin x$, and $y_3 = -3 \sin x$, and compare the graphs.

2. To see how B affects the graph of a sinusoidal function, let $y_1 = \sin x$, $y_2 = \sin 2x$, and $y_3 = \sin \dfrac{x}{3}$, and compare the graphs.

3. To see how C affects the graph of a sinusoidal function, let $y_1 = \cos x$, $y_2 = \cos (x + 1)$, and $y_3 = \cos (x - 2)$, and compare the graphs.

4. To see how D affects the graph of a sinusoidal function, let $y_1 = \cos x$, $y_2 = 1 + \cos x$, and $y_3 = -2 + \cos x$, and compare the graphs.

5. Let $y_1 = \sin x$. Graph $y_2 = -1 + 4 \sin 2x$. Explain how the numbers 4, 2, and -1 affect the graph of y_1.

■ **EXAMPLE 4**   Let $y = 1 + 2 \cos 4x$.

a) Find the midline, amplitude, period, and phase shift.

b) Draw two periods of the function.

Solution

a) The midline is $y = 1$; the amplitude is $|2| = 2$; the period is $2\pi/4 = \pi/2$; and the phase shift is 0.

b) Before drawing the curve, we lightly draw the midline at $y = 1$. Then we move up and down 2 units (representing the amplitude) from the midline and lightly draw top and bottom guidelines.

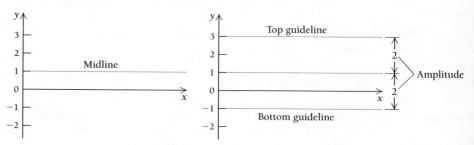

Next, we locate $x = \pi/2$ on the x-axis and divide the interval $[0, \pi/2]$ into fourths, at $x = \pi/8, \pi/4, 3\pi/8$, and $\pi/2$, representing the four quadrants of the unit circle.

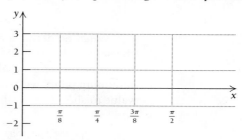

The graph of the cosine function starts at the top guideline at $x = 0$, crosses the midline at $x = \pi/8$, and intersects the bottom guideline at $x = \pi/4$. It then returns to the midline at $x = 3\pi/8$ and continues to the top guideline at $x = \pi/2$, which completes one period of the function.

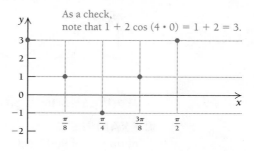

As a check, note that $1 + 2 \cos (4 \cdot 0) = 1 + 2 = 3$.

Finally, we draw a smooth curve (no discontinuities or corners) through these five points and repeat the same shape for a second cycle, or period.

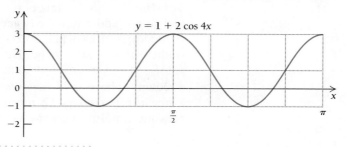

$y = 1 + 2 \cos 4x$

TECHNOLOGY CONNECTION

Graphing Sinusoidal Functions

When graphing sinusoidal functions, we must take care in setting up the window. Most of the work, whether graphing by hand or with a calculator, is in identifying the midline, amplitude, period, and phase shift. Remember, graphs of sinusoidal functions requires the input variable to be measured in radians.

To check the result of Example 4, enter Y1=1+2cos(4x) and then select WINDOW. Using the midline, amplitude, and period, we have

$$Xmin = 0,$$
$$Xmax = period,$$
$$Xscl = period/4,$$
$$Ymin = midline - amplitude,$$
and $$Ymax = midline + amplitude.$$

For $y = 1 + 2 \cos 4x$, we have Xmin $= 0$, Xmax $= \pi/2$ ($\sim 1.57079...$), Xscl $= \pi/8$ ($\sim 0.392699...$), Ymin $= -1$ and Ymax $= 3$. Now select GRAPH. One period of the function is shown in the viewing window. To see more than one period, adjust Xmin or Xmax accordingly.

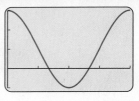

If the graph has a phase shift, there is no need to adjust the window. However, the graph will appear "shifted" within the window.

EXERCISES

Graph each function. Set the viewing window so that it frames one period of the function.

1. $y = -2 + 3 \sin 2x$ **2.** $y = 1 - 2 \sin 3x$

3. $y = -3 + 5 \cos 6x$ **4.** $y = 7 - 4 \cos 5x$

5. Graph $y_1 = 2 - 4 \cos 3x$ and $y_2 = 2 - 4 \cos (3x + 1)$. The first graph shows the function with no phase shift, and the second graph shows the function with a phase shift. What is the phase shift?

6. Graph $y = -1 + 2 \sin (4x + 3)$. What is the phase shift?

When modeling a real-world situation with a periodic function, whether we use the sine or the cosine is usually determined by the y-intercept. If the y-intercept is on the midline, we generally use the sine function. If the y-intercept is a maximum or minimum point, we generally use the cosine function. If necessary, we use a negative A value to show a vertical reflection. The four basic models are shown below.

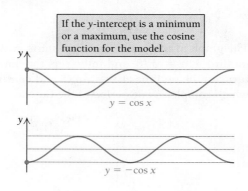

If the y-intercept is a minimum or a maximum, use the cosine function for the model.

$y = \cos x$

$y = -\cos x$

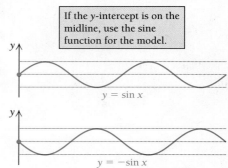

If the y-intercept is on the midline, use the sine function for the model.

$y = \sin x$

$y = -\sin x$

Applications of Sinusoidal Models

■ **EXAMPLE 5** **Business: Sales.** The Snow Hut manufactures ski jackets at a maximum rate of 9000 jackets per month on October 1 and a minimum rate of 1000 jackets per month on April 1. Assume this pattern repeats every year. Find a sinusoidal model for $P(t)$, the monthly rate of production of jackets, t months after October 1.

Solution Since the rate of production is maximized every October 1 ($t = 0, 12, 24$, and so on) and minimized every April 1 ($t = 6, 18$, and so on), we plot the points $(0, 9000)$, $(6, 1000)$, $(12, 9000)$, $(18, 1000)$, $(24, 9000)$, and so on. We can see the periodic pattern taking shape. The midline is the mean of the minimum and maximum values: $D = \frac{1000 + 9000}{2} = 5000$. Note that the maximum and minimum values are 4000 units above and below the midline, respectively, so the amplitude is $A = 4000$. Since the period is 12 months, we have $B = 2\pi/12 = \pi/6$. Furthermore, a maximum point is on the y-axis, which suggests that we use a cosine function

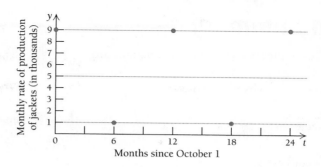

for our model. Thus, the model is $P(t) = 5000 + 4000 \cos\left(\dfrac{\pi}{6}t\right)$. Two periods of the function are shown below. As a check, note that on October 1, we have $P(0) = 5000 + 4000 \cos\left(\dfrac{\pi}{6} \cdot 0\right) = 5000 + 4000 \cdot 1 = 9000$, and on April 1, we have $P(6) = 5000 + 4000 \cos\left(\dfrac{\pi}{6} \cdot 6\right) = 5000 + 4000 \cdot (-1) = 1000$, as expected.

> **Quick Check 4**

The town of Quartzsite, Arizona, has a maximum population of 250,000 every January 1 ($t = 0$) and a minimum population of 3400 every July 1 ($t = 6$). Find a sinusoidal function that models the population of Quartzsite t months after January 1. (*Source:* www. census.gov and Arizona Department of Commerce.)

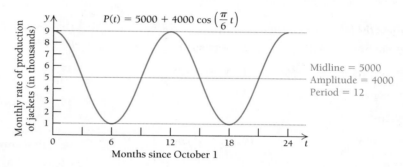

❰ Quick Check 4

■ **EXAMPLE 6** **Business: Sales.** The Snow Hut manufactures ski jackets. Its rate of production is modeled by $P(t) = 5000 + 4000 \cos\left(\dfrac{\pi}{6}t\right)$, where $P(t)$ is the number of jackets manufactured per month, t months after October 1.

a) Find the derivative of P.

b) How fast is the production rate changing on December 1?

Solution

a) The derivative is

$$P'(t) = \frac{d}{dt}\left[5000 + 4000 \cos\left(\frac{\pi}{6}t\right)\right]$$

$$= 4000\left[-\sin\left(\frac{\pi}{6}t\right) \cdot \frac{\pi}{6}\right] \qquad \text{Differentiating and using the Chain Rule}$$

$$= -2094 \sin\left(\frac{\pi}{6}t\right) \qquad\qquad 4000\left(\frac{\pi}{6}\right) \approx 2094$$

> **Quick Check 5**

Using the model in Example 6, find the following:

a) the rate of production on June 1;

b) how fast the production rate is changing on June 1.

b) Since December 1 is 2 months after October 1, we evaluate the derivative at $t = 2$:

$$P'(2) = -2094 \sin\frac{\pi}{6} \cdot 2 \approx -1813.$$

On December 1, production is changing at a rate of about -1813 jackets per month.

❰ Quick Check 5

Section Summary

- The derivatives of the six trigonometric functions are

  $\dfrac{d}{dx}\sin x = \cos x,\quad \dfrac{d}{dx}\cos x = -\sin x,\quad \dfrac{d}{dx}\tan x = \sec^2 x,$

  $\dfrac{d}{dx}\sec x = \sec x \tan x,\quad \dfrac{d}{dx}\cot x = -\csc^2 x,$

  $\dfrac{d}{dx}\csc x = -\cot x \csc x.$

- A *sinusoidal function* can be written in the form

  $y = A \sin (Bx - C) + D \quad \text{or} \quad y = A \cos (Bx - C) + D.$

- For any sinusoidal function, the *midline* is $y = D$, the *amplitude* is $|A|$, the *period* is $2\pi/B$, and the *phase shift* is C/B.

EXERCISE SET
7.2

Differentiate the following functions.

1. $y = \sin 6x$

2. $y = \sin 4x^5$

3. $y = \cos 2x$

4. $y = \cos 3x^8$

5. $y = \tan 2x^7$

6. $y = \sec 6x^2$

7. $y = \cos^2 x$

8. $y = \sin^3 x$

9. $y = x \sin x$

10. $y = x^2 \cos x$

11. $y = e^x \sin x$

12. $y = e^x \cos x$

13. $y = \dfrac{\sin x}{x}$

14. $y = \dfrac{\cos x}{x}$

15. $y = \sin x \cos x$

16. $y = \sin x + \cos x$

17. $y = \sin (x^2 + 3x + 2)$

18. $y = \cos (x^2 + 5x - 1)$

19. $y = \dfrac{e^{2x}}{\sin 2x}$

20. $y = \dfrac{e^{3x}}{\cos 4x}$

21. $y = \sqrt{\sin x}$

22. $y = \sqrt{\cos x}$

23. $y = \tan^2 x$

24. $y = \sec^2 x$

25. $y = x \sec x^2$

26. $y = x^2 \tan x$

27. $y = \cot x + \csc x$

28. $y = \tan x + \sec x$

29. $y = \dfrac{x}{\cot x}$

30. $y = \dfrac{x^2}{\csc x}$

31. $y = e^{3x} \csc x$

32. $y = e^{5x} \cot x$

33. $y = \dfrac{1 + \tan x}{\sec x}$

34. $y = \dfrac{\cot x}{1 + \csc x}$

35. $y = \ln (\tan x),$
for $0 < x < \pi/2$

36. $y = \ln (\sec x),$
for $-\pi/2 < x < \pi/2$

Differentiate. Note that certain identities may simplify the problem.

37. $y = \dfrac{\cos x}{\sec x}$

38. $y = \sin^2 x + \cos^2 x$

39. $y = \cos x \tan x$

40. $y = (\sin x + \cos x)^2$

In Exercises 41–44, find the midline, amplitude, period, and phase shift; then draw a graph showing two periods of each function.

41. $y = -5 + 2 \sin 3x$

42. $y = 1 - 3 \cos 2x$

43. $y = -2 + 4 \cos \left(\dfrac{\pi}{4} x + \dfrac{\pi}{4} \right)$

44. $y = 10 - 6 \sin \left(\dfrac{\pi}{12} x - \dfrac{\pi}{24} \right)$

45–48. *Use a graphing calculator to graph two periods of each function in Exercises 41–44.*

In Exercises 49–52, find a sinusoidal model for each situation, without a phase shift.

49. The maximum value is 10 and occurs when $t = 0, 8, 16$, and so on, and the minimum value is 4 and occurs when $t = 4, 12, 20$, and so on.

50. The minimum value is 200 and occurs when $t = 0, 12, 24$, and so on, and the maximum value is 400 and occurs when $t = 6, 18, 30$, and so on.

51. The minimum value is -8 and occurs when $t = 2, 10, 18$, and so on, and the maximum value is 0 and occurs when $t = -2, 6, 14$, and so on.

52. The maximum value is 6 and occurs when $t = 4, 20, 36$, and so on, and the minimum value is -8 and occurs when $t = -4, 12, 28$, and so on.

APPLICATIONS

Business and Economics

53. Fluctuating sales. Sal's Dune Buggy Rentals rents dune buggies at a minimum rate of 20 vehicles per month on January 1 and a maximum rate of 140 vehicles per month on July 1. Assume this pattern repeats every year.

a) Find a sinusoidal model for $f(t)$, the rate of dune buggy rentals t months after January 1.
b) Find $f'(t)$.
c) How fast is the rental rate changing on April 1?
d) How fast is the rental rate changing on June 15?

54. Fluctuating sales. The Gingerbread House sells cookies at a maximum rate of 8500 cookies per month on December 1 and a minimum rate of 1500 cookies per month on June 1. Assume this pattern repeats every year.

a) Find a sinusoidal model for $f(t)$, the rate of sales of cookies per month, t months after December 1.
b) Find $f'(t)$.
c) How fast is the rate of sales changing on August 1?
d) How fast is the rate of sales changing on September 15?

55. Gas prices. For the 3-month period between July 1 and October 1, 2011, the average price of a gallon of gasoline in the United States could be modeled by

$$P(t) = 3.64 - 0.08 \cos\left(\frac{2\pi}{5}t\right),$$ where $P(t)$ is the price

in dollars t weeks after July 1. (*Source:* www.gasbuddy.com.)

a) Find $P'(t)$.
b) Find $P'(5)$, and explain what this number represents.
c) Find the period of the price function.
d) How many weeks after July 1 was the price the highest, and what was that price?

56. Price of electricity. The average price of electricity for U.S. households since 2008 can be modeled by

$$P(t) = 11.25 - 0.75 \cos\left(\frac{\pi}{6}t\right),$$ where $P(t)$ is the price

in cents per kilowatt-hour, t months after January 1, 2008. (*Source:* www.eia.gov.)

a) Find $P'(t)$.
b) Find $P'(7)$, and explain what this number represents.
c) Find the period of the price function.
d) What is the price per kilowatt-hour when the price is minimized?

General Interest

57. Tides. The tides in the Bay of Fundy in Nova Scotia can be modeled by $f(t) = 28 - 28 \cos\left(\frac{\pi}{6.22}t\right)$, where low tide

is assumed to occur at time $t = 0$ and $f(t)$ is the height (in feet) of the water above low tide after t hours. (*Source:* Canadian Hydrological Service.)

a) Find the period, that is, the time between two high tides or two low tides.
b) Find $f'(t)$.
c) Find $f'(3)$, and explain what this number represents.
d) How many hours after low tide is high tide, and how much higher is the water?

58. Periodic motion. A weight attached to a spring hanging from the ceiling bobs up and down. The weight's distance from the ceiling is modeled by $f(t) = 2 + 0.8 \sin 3.6t$, where $f(t)$ is measured in feet and t in seconds.

a) Find the period.
b) Find $f'(t)$.
c) Find $f'(2)$, and explain what this number represents.
d) Assume the weight starts bobbing at $t = 0$. At what time is the weight farthest from the ceiling, and how far away is it at that time?

59. Temperature. The average daily high temperature in Chicago can be modeled by

$$T(t) = 57 - 27 \cos\left(\frac{2\pi}{365}t\right),$$

where $T(t)$ is the temperature in degrees Fahrenheit, t days after January 1. (*Source:* www.wunderground.com.)

a) Find $T'(t)$.
b) Find $T'(90)$, and explain what this number represents.
c) How many days after January 1 does the temperature reach its maximum, and what is the maximum temperature?

60. Average daylight. The amount of daylight in Atlanta can be modeled by $H(t) = 13 - 2 \cos\left(\frac{2\pi}{365}t\right)$, where $H(t)$ is the hours of daylight t days after December 21. (*Source:* www.gaisma.com.)

a) Find $H'(t)$.
b) Find $H'(210)$, and explain what this number represents.
c) How many days after December 21 is the amount of daylight maximized, and how many hours of daylight are there at maximum?

SYNTHESIS

61. Prove that $\dfrac{d}{dx}\cot x = -\csc^2 x$

62. Prove that $\dfrac{d}{dx}\csc x = -\cot x \csc x$

63. Prove that $\dfrac{d}{dx}\sec x = \sec x \tan x$

64. Prove that $\dfrac{d}{dx}\cos x = -\sin x$ using a method similar to the one used on page 598 to show that $\dfrac{d}{dx}\sin x = \cos x$.

65. Show that $\dfrac{d}{dx}\sin 2x = 2\cos 2x$ in two ways:

a) Using the Chain Rule
b) Using a double-angle identity

66. For a geometric proof of $\displaystyle\lim_{h\to 0}\frac{\sin h}{h}=1$, consider two right triangles OAB and OCD and a circular sector COB, as shown below. Here, h is the angle in radians, with $0 < h < \pi/2$.

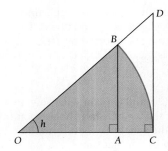

Let O be the center (origin), and assume lengths $OB = 1$ and $OC = 1$ (the radius of the circle).

a) Using trigonometry, what are the lengths of OA and AB?
b) What is the length of CD? (*Hint: CD is to 1 as AB is to OA.*)
c) Using the answers from part (a), what is the area of the triangle OAB?
d) What is the area of the circular sector COB? (*Hint: The area of a sector is to the area of the full circle as the angle h is to 2π.*)
e) Using the answer from part (b), what is the area of the triangle OCD?
f) Use the answers from parts (c), (d), and (e) to write a compound inequality that compares the three areas.
g) Since $\sin h$ is positive, divide each expression in the compound inequality by $\sin h$ to form an equivalent compound inequality.
h) Recall that if $a < b$ and a and b are positive, then $\dfrac{1}{a} > \dfrac{1}{b}$. Use this fact to rewrite the compound inequality in part (g).

i) The expression $\dfrac{\sin h}{h}$ is "trapped" between two expressions. Find the limit of these two expressions as h approaches zero.

j) Based on the answer to part (i), what must $\dfrac{\sin h}{h}$ approach as h approaches zero?

k) The preceding steps show that $\displaystyle\lim_{h\to 0^+}\frac{\sin h}{h}=1$, since $h > 0$. What adjustments would you need to make to those steps to show that $\displaystyle\lim_{h\to 0^-}\frac{\sin h}{h}=1$?

67. In Example 6, the Snow Hut's monthly production of ski jackets is modeled by $P(t) = 5000 + 4000\cos\left(\dfrac{\pi}{6}t\right)$, and we found that $P'(2) \approx -1814$, meaning that on December 1 ($t = 2$), production was decreasing at the rate of about 1814 units per month.

a) Calculate $\dfrac{P(2) - P(1)}{2 - 1}$, and explain what this number represents.
b) Calculate $\dfrac{P(3) - (2)}{3 - 2}$, and explain what this number represents.
c) Explain why the expressions in parts (a) and (b) and the expression $P'(2)$ give valid information about the change in production on December 1, even though all the values are different.

68. Using an identity, $f(x) = \sin x \cdot \cot x$ can be rewritten as $f(x) = \sin x \cdot \dfrac{\cos x}{\sin x} = \cos x$, so that $f'(x) = -\sin x$. However, this derivative is not defined for certain values of x. State the values of x for which $y' = f'(x)$ is defined, and explain why this restriction is required.

TECHNOLOGY CONNECTION

Use a graphing calculator or Graphicus to graph each function. State each function's period.

69. $y = \sin 2x + \sin 3x$

70. $y = \sin 2x + \sin 4x$

71. $y = \sin 2x + \cos 2x$

72. $y = \sin x + \sin 2x + \sin 3x$

Answers to Quick Checks

1. $y' = -15x^2 \sin 5x^3$ **2.** $y' = 4\tan^3 x \sec^2 x$

3. $y' = \dfrac{-2\left(x\sin x^2 + \cos x^2\right)}{e^{2x}}$

4. $P(t) = 126{,}700 + 123{,}300\cos\left(\dfrac{\pi}{6}t\right)$

5. (a) 3000 units; **(b)** 1813 units/month

Integration of Trigonometric Functions

OBJECTIVES

- Integrate trigonometric functions.
- Solve applied problems involving integration of trigonometric functions.

Each differentiation formula in Section 7.2 yields an integration formula. For the sine and cosine functions, we have

$$\int \sin x \, dx = -\cos x + C \qquad \text{and} \qquad \int \cos x \, dx = \sin x + C.$$

Other common trigonometric integration formulas are

$$\int \sec^2 x \, dx = \tan x + C,$$

$$\int \sec x \tan x \, dx = \sec x + C,$$

$$\int \csc^2 x \, dx = -\cot x + C,$$

$$\int \cot x \csc x \, dx = -\csc x + C.$$

We can use substitution to find antiderivatives of certain trigonometric functions.

■ **EXAMPLE 1** Find $\int \sin 3x \, dx$.

Solution Let $u = 3x$ and $du = 3 \, dx$. We multiply the integral by 1, using $3 \cdot \frac{1}{3}$, which allows us to do the substitution.

$$\int \sin 3x \, dx = \frac{1}{3} \int \sin 3x \cdot 3 \cdot dx$$

$$= \frac{1}{3} \int \sin u \cdot du \qquad u = 3x, du = 3 \, dx$$

$$= \frac{1}{3}(-\cos u) + C \qquad \text{Integrating}$$

$$= -\frac{1}{3} \cos 3x + C.$$

Check: As always, we differentiate the result to check our answer:

$$\frac{d}{dx}\left(-\frac{1}{3} \cos 3x\right) = -\frac{1}{3} \cdot (-\sin 3x) \cdot 3 \qquad \text{Using the Chain Rule: } \frac{d}{dx}(3x) = 3$$

$$= \sin 3x. \qquad \text{Simplifying}$$

〉 Quick Check 1

Find $\int x \cos x^2 \, dx$.

〈 Quick Check 1

Example 1 suggests formulas for integrals of the form $\int \sin kx \, dx$ and $\int \cos kx \, dx$. Using the substitution $u = kx$, we can show that

$$\int \sin kx \, dx = -\frac{1}{k} \cos kx + C \qquad \text{and} \qquad \int \cos kx \, dx = \frac{1}{k} \sin kx + C.$$

■ **EXAMPLE 2** Find $\int \sin^2 x \cos x \, dx$.

Solution Since $\frac{d}{dx} \sin x = \cos x$, we let $u = \sin x$, and so $du = \cos x \, dx$. We have

$$\int \sin^2 x \cos x \, dx = \int u^2 \, du \qquad \text{Substituting}$$

$$= \frac{u^3}{3} + C \qquad \text{Integrating}$$

$$= \frac{1}{3} \sin^3 x + C.$$

Check: We differentiate to check our answer:

$$\frac{d}{dx}\left(\frac{1}{3}\sin^3 x\right) = \frac{1}{3} \cdot 3 \sin^2 x \cdot \cos x \qquad \text{Using the Chain Rule: } \frac{d}{dx}\sin x = \cos x$$

$$= \sin^2 x \cdot \cos x. \qquad \text{Simplifying}$$

❯ **Quick Check 2**

Find $\int \tan^3 x \sec^2 x \, dx$.

❮ **Quick Check 2**

Formulas for $\int \tan x \, dx$ and $\int \sec x \, dx$ can be developed using substitution. We develop $\int \tan x \, dx$ as an example, and the development of $\int \sec x \, dx$ is left to Exercise 45.

■ **EXAMPLE 3** Find $\int \tan x \, dx$.

Solution Since $\tan x = \dfrac{\sin x}{\cos x}$, we have $\int \tan x \, dx = \int \dfrac{\sin x}{\cos x} \, dx$. We use the substitution $u = \cos x$. Therefore, $du = -\sin x \, dx$, and we have

$$\int \frac{\sin x}{\cos x} \, dx = -\int \frac{-\sin x}{\cos x} \, dx \qquad \text{Preparing for the substitution}$$

$$= -\int \frac{du}{u} \qquad u = \cos x, \, du = -\sin x \, dx$$

$$= -\ln |u| + C \qquad \text{Integrating}$$

$$= -\ln |\cos x| + C. \qquad \begin{array}{l}\text{Since } \cos x \text{ can be negative, we must use} \\ \text{absolute value.}\end{array}$$

Check: We differentiate the result to check our answer:

$$\frac{d}{dx}(-\ln |\cos x|) = -\frac{1}{\cos x} \cdot (-\sin x) \qquad \text{Using the Chain Rule: } \frac{d}{dx}\cos x = -\sin x$$

$$= \frac{\sin x}{\cos x} \qquad \text{Simplifying}$$

$$= \tan x.$$

❯ **Quick Check 3**

Find $\int \cot t \, dt$.

❮ **Quick Check 3**

Identities may be used to rewrite an integrand in simplified form.

■ **EXAMPLE 4** Find $\int (\cos x + \sin x)^2 \, dx$.

Solution When we expand the integrand, we can group the terms, forming two identities:

$$(\cos x + \sin x)^2 = (\cos x + \sin x)(\cos x + \sin x)$$

$$= \cos^2 x + 2 \cos x \sin x + \sin^2 x$$

$$= 1 + \sin 2x.$$

❯ **Quick Check 4**

Find $\int \tan^2 x \, dx$. (*Hint:* Use a Pythagorean identity.)

Thus, we have $\int (\cos x + \sin x)^2 \, dx = \int (1 + \sin 2x) \, dx = x - \dfrac{\cos 2x}{2} + C$.

❮ **Quick Check 4**

For a definite trigonometric integral, the bounds are always expressed in radians.

■ **EXAMPLE 5** Find the area under the graph of $y = \cos x$ over the interval $[0, \pi/2]$.

Solution We have

$$\int_0^{\pi/2} \cos x\, dx = [\sin x]_0^{\pi/2}$$

$$= \sin \frac{\pi}{2} - \sin 0$$

$$= 1 - 0 = 1.$$

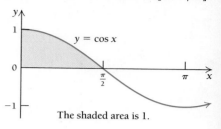

The shaded area is 1.

> **Quick Check 5**

Find the area under the graph of $y = \sin x$ over the interval $[0, \pi]$.

❰ Quick Check 5

■ **EXAMPLE 6** Find $\displaystyle\int_0^{\pi} \sin 2x\, dx$.

Solution Since $\displaystyle\int \sin kx\, dx = -\frac{1}{k}\cos kx + C$, we have $\displaystyle\int \sin 2x\, dx = -\frac{1}{2}\cos 2x + C$. We evaluate this antiderivative at the limits of integration, and simplify:

$$\int_0^{\pi} \sin 2x\, dx = \left[-\tfrac{1}{2}\cos 2x\right]_0^{\pi}$$

$$= \left(-\tfrac{1}{2}\cos(2\cdot\pi)\right) - \left(-\tfrac{1}{2}\cos(2\cdot 0)\right) \quad \text{Evaluating the limits}$$

$$= -\tfrac{1}{2}\cdot 1 + \tfrac{1}{2}\cdot 1 \quad \text{Using } \cos 2\pi = 1 \text{ and } \cos 0 = 1$$

$$= 0.$$

> **Quick Check 6**

Find $\displaystyle\int_0^{\pi/2} \cos 4x\, dx$.

❰ Quick Check 6

The graph of $y = \sin 2x$, for $0 \le x \le \pi$, shows equal areas above and below the x-axis. Thus, by the additive property of integration, the sum of the areas is zero. Note that the period of $\sin 2x$ is π. Integrating functions of the form $y = A \sin Bx$ or $y = A \cos Bx$ over one period will always result in a value of zero.

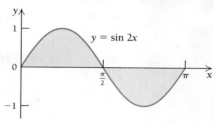

TECHNOLOGY CONNECTION

Exploratory

Verify the claim that integrating a function of the form $y = A \sin Bx$ or $y = A \cos Bx$ over one period results in a value of zero by graphing and finding the area under the graph over one period for $y = \sin 3x$, $y = 2 \sin 3x$, and $y = -4 \cos 5x$. Experiment with other functions of the same form.

Applications

Sinusoidal functions are used to model periodic situations. If the model is a rate function, we can integrate the function to find the total change. We also use integration to find the average value of a sinusoidal function over an interval.

■ **EXAMPLE 7** **General Interest: Bag-Valve Mask Respirators.** A bag-valve mask (BVM) respirator is used by a medical technician to assist a patient's breathing. Suppose the rate at which oxygen passes through the one-way valve, in milliliters (mL) of oxygen per second, is modeled by $f(t) = 120 - 120 \cos\left(\dfrac{2\pi}{5}t\right)$. Find the total amount of oxygen delivered to the patient over one period (one squeeze and release). (*Source*: emedicine.medscape.com.)

Solution   The period is $\dfrac{2\pi}{(2\pi/5)} = 5$ seconds. The total amount of oxygen delivered to the patient over one period is given by

$$\int_0^5 \left(120 - 120\cos\left(\frac{2\pi}{5}t\right)\right) dt.$$

By Theorem 3 of Section 4.1, we may write the integral as

$$\int_0^5 \left(120 - 120\cos\left(\frac{2\pi}{5}t\right)\right) dt = \int_0^5 120\, dt - \int_0^5 120\cos\left(\frac{2\pi}{5}t\right) dt.$$

Since $\displaystyle\int_0^5 120\cos\left(\frac{2\pi}{5}t\right) dt$ is integrated over one period, its value is zero. We have

$$\int_0^5 \left(120 - 120\cos\left(\frac{2\pi}{5}t\right)\right) dt = \int_0^5 120\, dt - \int_0^5 120\cos\left(\frac{2\pi}{5}t\right) dt$$

$$= \int_0^5 120\, dt - 0$$

$$= \left[120t\right]_0^5 \qquad \text{Integrating}$$

$$= 120 \cdot 5 - 120 \cdot 0 \quad \text{Evaluating}$$

$$= 600.$$

Thus, the bag-valve mask delivers 600 mL of oxygen to the patient every 5 seconds.

⟨ Quick Check 7

> **Quick Check 7**

Find

$$\int_0^3 \left(50 + 25\sin\left(\frac{2\pi}{3}t\right)\right) dt.$$

■ **EXAMPLE 8**   **Business: Total Sales.**   The Ski Emporium's monthly production of skis is modeled by $S(t) = 7 - 7\cos\left(\dfrac{\pi}{6}t\right)$, where $S(t)$ is the rate of production of skis, in thousands of pairs per month, t months after August 1.

a) Find the total production for the 1-year period from August 1 to the following August 1.

b) Find the total production from August 1 to January 1.

c) Find the average monthly production from August 1 to January 1.

Solution

a) The total production from August 1 ($t = 0$) to the following August 1 ($t = 12$) is given by $\displaystyle\int_0^{12} \left(7 - 7\cos\left(\frac{\pi}{6}t\right)\right) dt$. Noting that 12 months is one period of the function $y = \cos\left(\dfrac{\pi}{6}t\right)$, we have

$$\int_0^{12}\left(7 - 7\cos\left(\frac{\pi}{6}t\right)\right)dt = \int_0^{12} 7\, dt - \int_0^{12} 7\cos\left(\frac{\pi}{6}t\right) dt$$

$$= \int_0^{12} 7\, dt - 0 \quad \text{Since the period of } y = 7\cos\left(\frac{\pi}{6}t\right) \text{ is 12}$$

$$= \left[7t\right]_0^{12} \qquad\qquad \text{Integrating}$$

$$= 7 \cdot 12 - 7 \cdot 0 \qquad \text{Evaluating the limits}$$

$$= 84.$$

Thus, the Ski Emporium manufactures about 84,000 pairs of skis during the year.

Rate of production of skis (in thousands of pairs per month) vs. Months since August 1. Total = 84,000 pairs

b) We use $t = 5$ to represent January 1. Thus, integrating from $t = 0$ to $t = 5$ represents the total production from August 1 to January 1.

$$\int_0^5 \left(7 - 7\cos\left(\frac{\pi}{6}t\right)\right) dt = \left[7t - 7 \cdot \frac{6}{\pi} \cdot \sin\left(\frac{\pi}{6}t\right)\right]_0^5$$

$$\approx \left[7t - 13.37 \sin\left(\frac{\pi}{6}t\right)\right]_0^5$$

$$\approx \left(7 \cdot 5 - 13.37 \sin\left(\frac{\pi}{6} \cdot 5\right)\right) - \left(7 \cdot 0 - 13.37 \sin\left(\frac{\pi}{6} \cdot 0\right)\right)$$

$$\approx 35 - 6.685 \qquad 13.37 \sin\left(\frac{\pi}{6} \cdot 5\right) \approx 6.685 \text{ and } \sin 0 = 0$$

$$\approx 28.315.$$

The Ski Emporium manufactures about 28,315 pairs of skis from August 1 to January 1.

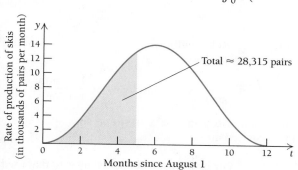

Months since August 1

Total ≈ 28,315 pairs

c) Recall from Section 4.4 that the average value of $f(x)$ over $[a, b]$ is given by

$$y_{av} = \frac{1}{b-a} \int_a^b f(x)\, dx.$$ Thus, the average monthly production of skis from August 1 to January 1 is given by

$$y_{av} = \frac{1}{5-0} \int_0^5 \left(7 - 7\cos\left(\frac{\pi}{6}t\right)\right) dt$$

$$= \tfrac{1}{5} \cdot 28.315 \qquad \text{Using the result from part (b)}$$

$$= 5.663.$$

The Ski Emporium's average monthly production of skis from August 1 to January 1 is about 5663 pairs of skis per month.

> **Quick Check 8**
>
> Using the model in Example 8, find the total number of pairs of skis manufactured between February 1 and August 1.

❮ Quick Check 8

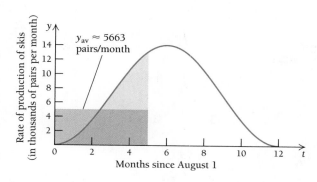

$y_{av} \approx 5663$ pairs/month

Months since August 1

Section Summary

- Some of the common *trigonometric integration formulas* are

$$\int \sin x\, dx = -\cos x + C, \qquad \int \cos x\, dx = \sin x + C,$$

$$\int \sin kx\, dx = -\frac{1}{k}\cos kx + C, \qquad \int \cos kx\, dx = \frac{1}{k}\sin kx + C$$

$$\int \tan x\, dx = -\ln|\cos x| + C, \quad \int \sec^2 x\, dx = \tan x + C,$$

$$\int \sec x \tan x\, dx = \sec x + C, \qquad \int \csc^2 x\, dx = -\cot x + C,$$

$$\int \cot x \csc x\, dx = -\csc x + C, \qquad \int \cot x\, dx = \ln|\sin x| + C.$$

- Identities and substitutions can be used to evaluate other trigonometric integrals.

EXERCISE SET
7.3

Solve.

1. $\int \sin^3 x \cos x \, dx$

2. $\int \cos^2 x \sin x \, dx$

3. $\int \cos 6x \, dx$

4. $\int \sin 4x \, dx$

5. $\int \sin \dfrac{\pi}{2} x \, dx$

6. $\int \cos \dfrac{\pi}{4} x \, dx$

7. $\int 3 \cos 2\pi x \, dx$

8. $\int -15 \sin \pi x \, dx$

9. $\int x \sin 3x^2 \, dx$

10. $\int 2x \cos 5x^2 \, dx$

11. $\int (x - 1) \cos (x^2 - 2x) \, dx$

12. $\int (2x + 1) \sin (3x^2 + 3x + 2) \, dx$

13. $\int e^{2x} \sin e^{2x} \, dx$

14. $\int 2e^{-3x} \cos 3e^{-3x} \, dx$

15. $\int \tan^2 x \sec^2 x \, dx$

16. $\int \tan^4 x \sec^2 x \, dx$

17. $\int \sec^2 10x \, dx$

18. $\int 3 \sec^2 (2x + 1) \, dx$

19. $\int \sec 5x \tan 5x \, dx$

20. $\int x \sec x^2 \tan x^2 \, dx$

21. $\int \csc^2 3x \, dx$

22. $\int x \csc^2 4x^2 \, dx$

23. $\int \cot 6x \csc 6x \, dx$

24. $\int \cot \frac{1}{2} x \csc \frac{1}{2} x \, dx$

25. $\int (x^2 + 3x + \tan 2x) \, dx$

26. $\int (e^{2x} + \sqrt{x} - \cot 4x) \, dx$

Evaluate.

27. $\displaystyle\int_0^{\pi/2} \sin x \, dx$

28. $\displaystyle\int_0^{\pi/6} \cos x \, dx$

29. $\displaystyle\int_0^{\pi/4} 2 \cos x \, dx$

30. $\displaystyle\int_0^{\pi/3} 4 \sin x \, dx$

31. $\displaystyle\int_{-\pi/6}^{\pi/2} \sin x \, dx$

32. $\displaystyle\int_{-\pi/4}^{\pi} \sin x \, dx$

33. $\displaystyle\int_{-\pi/6}^{0} 5 \cos x \, dx$

34. $\displaystyle\int_{-\pi/2}^{\pi/2} 3 \sin x \, dx$

35. $\displaystyle\int_0^{\pi/6} \sec^2 x \, dx$

36. $\displaystyle\int_{\pi/6}^{\pi/3} \sec^2 x \, dx$

37. $\displaystyle\int_{\pi/6}^{\pi/4} \csc^2 x \, dx$

38. $\displaystyle\int_{\pi/4}^{\pi/2} \csc^2 x \, dx$

39. $\displaystyle\int_0^{\pi/4} \tan^2 x \sec^2 x \, dx$

40. $\displaystyle\int_{-\pi/6}^{\pi/4} \tan x \sec x \, dx$

41. $\displaystyle\int_0^{2\pi} 5 \sin x \, dx$

42. $\displaystyle\int_0^{\pi} -3 \cos 2x \, dx$

43. $\displaystyle\int_0^{\pi} (2 + 4 \cos 2x) \, dx$

44. $\displaystyle\int_0^{\pi/2} (6 - 3 \sin 4x) \, dx$

45. Find $\int \sec x \, dx$ by multiplying by $\dfrac{\sec x + \tan x}{\sec x + \tan x}$ and making a substitution.

46. Find $\displaystyle\int \dfrac{1}{1 - \sin x} \, dx$ by multiplying by $\dfrac{1 + \sin x}{1 + \sin x}$.

47. Find $\int \sin^3 x \, dx$ using $\sin^3 x = \sin^2 x \cdot \sin x = (1 - \cos^2 x)(\sin x)$.

48. Find $\int \cos^3 x \, dx$ by using an identity similar to that used in Exercise 47.

49. a) Find $\int \sin x \cos x \, dx$ using the substitution $u = \sin x$.
    b) Find the integral using the substitution $u = \cos x$.
    c) Show that the two results are equal.

50. a) Find $\int \tan x \sec^2 x \, dx$ using the substitution $u = \tan x$.
    b) Find the integral using the substitution $u = \sec x$.
    c) Show that the two results are equal.

51. Find the average value of $y = \sin x$ over the interval $[0, \pi]$.

52. Find the average value of $y = \cos x$ over the interval $[0, \pi/4]$.

APPLICATIONS

Business and Economics

53. *Total sales.* Rentals at Sal's Dune Buggy Rentals are modeled by $f(t) = 80 - 60 \cos \left(\dfrac{\pi}{6} t \right)$, where $f(t)$ is the monthly rental rate, t months after January 1.

    a) What is the total number of rentals between January 1 and April 1?
    b) What is the average number of rentals per month between January 1 and April 1?
    c) What is the total number of rentals during the whole year?
    d) What is the average number of rentals per month for the whole year?

54. *Total sales.* The Gingerbread House sells cookies, and its sales are modeled by $f(t) = 5000 + 3500 \cos \left(\dfrac{\pi}{6} t \right)$, where $f(t)$ is the rate at which cookies are selling per month, t months after December 1.

    a) What is the total number of cookies sold between December 1 and April 1?
    b) What is the average number of cookies sold per month between December 1 and April 1?
    c) What is the total number of cookies sold during the whole year?
    d) What is the average number of cookies sold per month for the whole year?

55. Average gas price. For the 3-month period between July 1 and October 1, 2011, the weekly price of a gallon of gasoline in the United States could be modeled by

$$P(t) = 3.64 - 0.08 \cos\left(\frac{2\pi}{5} t\right),$$ where $P(t)$ is the price,

in dollars, t weeks after July 1. (*Source:* www.gasbuddy. com.) What was the average price of gas during the first 3 weeks ($t = 0$ to 3) of this period?

56. Average price of electricity. The price of electricity for U.S. households since 2008 can be modeled by

$$P(t) = 11.25 - 0.75 \cos\left(\frac{\pi}{6} t\right),$$

where $P(t)$ is the price, in cents per kilowatt-hour, t months after January 1, 2008 (*Source:* www.eia.gov.) What was the average price of electricity from the start of 2008 ($t = 0$) to the start of April 2009 ($t = 15$)?

General Interest

57. Average temperature. The daily high temperature in Chicago can be modeled by the function

$$T(t) = 57 - 27 \cos\left(\frac{2\pi}{365} t\right),$$

where $T(t)$ is the temperature, in degrees Fahrenheit, t days after January 1. Find the average high temperature in Chicago during the first 120 days of the year. (*Source:* www.wunderground.com.)

58. Average daylight. The amount of daylight in Atlanta can be modeled by $H(t) = 13 - 2 \cos\left(\frac{2\pi}{365} t\right),$ where $H(t)$ is the number of hours of daylight t days after December 21. Find the average amount of daylight during summer ($183 \le t \le 275$). (*Source:* www.gaisma.com.)

59. Bag-valve mask respirator. The rate at which oxygen passes through the one-way valve of a bag-valve mask for a child, in milliliters of oxygen per second, is modeled by

$$f(t) = 135 - 135 \cos\left(\frac{2\pi}{3} t\right).$$ Find the total amount of

oxygen delivered to the patient over one period.

60. Bellows. The rate at which air passes through a valve in a set of bellows, in milliliters per second, is modeled by

$$f(t) = 200 - 200 \cos\left(\frac{\pi}{6} t\right).$$ Find the total amount of

air that passes through the valve over one period.

SYNTHESIS

61. Suppose $\int_0^b \sin 2x\, dx = 0.$ What are some possible values of b?

62. Suppose $\int_0^{\pi/4} \sin Bx\, dx = 0.$ What are some possible values of B?

Use integration by parts to solve the following integrals.

63. $\int x \sin x\, dx$ **64.** $\int x \cos x\, dx$

65. $\int x \cos 2x\, dx$ **66.** $\int x \sin 3x\, dx$

67. $\int 2x \sin 4x\, dx$ **68.** $\int 3x \cos 5x\, dx$

69. $\int_0^{\pi/2} x \sin 2x\, dx$ **70.** $\int_0^{\pi/6} x \cos 3x\, dx$

71. Consider $\int_{\pi/2}^{\pi} \frac{\sin x}{2x}\, dx.$

a) Use the Trapezoidal Rule and $n = 4$ subdivisions to approximate the value of this integral. Round to three decimal places.

b) Use a graphing calculator or Graphicus to evaluate the integral.

72. Consider $\int_0^1 \cos x^2\, dx.$

a) Use the Trapezoidal Rule and $n = 4$ subdivisions to approximate the value of this integral. Round to three decimal places.

b) Use a graphing calculator or Graphicus to evaluate the integral.

73. Consider $\int_0^{\infty} \sin x\, dx.$

a) Explain why $\int_0^{2n\pi} \sin x\, dx = 0$ for all positive integers n.

b) Does the result from part (a) show that $\int_0^{\infty} \sin x\, dx = 0$? Why or why not?

74. If p is the period of $A \sin (Bt + C) + D$, explain why $\int_0^p (A \sin (Bt + C) + D)\, dt = p \cdot D.$

Answers to Quick Checks

1. $\frac{1}{2} \sin (x^2) + C$ **2.** $\frac{1}{4} \tan^4 x + C$ **3.** $\ln |\sin x| + C$
4. $x + \tan x + C$ **5.** 2 **6.** 0 **7.** 150 **8.** 42,000 pairs

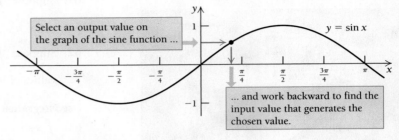

7.4 Inverse Trigonometric Functions and Applications

Inverse Trigonometric Functions

The outputs of the sine function range from −1 to 1. Suppose we work backward from an output to an input.

Select an output value on the graph of the sine function ...

$y = \sin x$

... and work backward to find the input value that generates the chosen value.

More generally, suppose that $-1 \leq x \leq 1$ and we want a number y such that $-\pi/2 \leq y \leq \pi/2$ and $\sin y = x$. This determines a function, called the *inverse sine function*, written as

$$y = \sin^{-1} x.$$

Here, -1 means "inverse of" (not "reciprocal of"), and the function $y = \sin^{-1} x$ is an *inverse trigonometric function*.

■ **EXAMPLE 1**   Find $\sin^{-1} \dfrac{\sqrt{3}}{2}$.

Solution   This problem is the same as asking, "For what number between $-\pi/2$ and $\pi/2$ is the sine equal to $\sqrt{3}/2$?"

From our earlier work, we know that number is $\pi/3$. Thus, $\sin^{-1} \dfrac{\sqrt{3}}{2} = \dfrac{\pi}{3}$ since $\sin \dfrac{\pi}{3} = \dfrac{\sqrt{3}}{2}$.

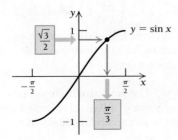

> **Quick Check 1**
>
> Find $\sin^{-1} \dfrac{\sqrt{2}}{2}$.

❮ Quick Check 1

Exploratory

To graphically represent the value of $\sin^{-1}\dfrac{\sqrt{3}}{2}$ using radian mode, let Y1=sin(X) and Y2=√(3)/2 and graph both in the viewing window $[-1.57, 1.57, -1, 1]$. Press **2ND** and CALC and select 5:intersect. Follow the directions to select the two functions, and the point of intersection will be highlighted.

The answers are given in decimal form, where $\pi/3 \approx 1.0471976\ldots$.

Use a graphing calculator to verify the answer from Quick Check 1.

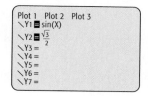
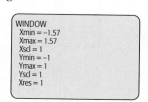
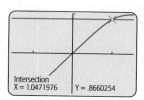

Because the sine, cosine, and tangent functions are periodic, their output values repeat for different input values. For example, $\sin x = 0$ for $x = 0$, $\pm\pi$, $\pm2\pi$, and so on. If we look for a value of $\sin^{-1} 0$, there are many potential answers: $\sin^{-1} 0 = 0$, $\pm\pi$, $\pm2\pi$, and so on. For the inverse sine to be a function, it must have precisely one output for any given input x where $-1 \le x \le 1$. To assure this, when we write $y = \sin^{-1} x$, we restrict the output y so that $-\dfrac{\pi}{2} \le y \le \dfrac{\pi}{2}$. In a similar way, we restrict the outputs for the inverse cosine and inverse tangent functions.

DEFINITIONS   The Inverse Trigonometric Functions

The **inverse sine** function:

$$y = \sin^{-1} x, \text{ where } x = \sin y \text{ and } -1 \le x \le 1 \text{ and } -\frac{\pi}{2} \le y \le \frac{\pi}{2}.$$

The **inverse cosine** function:

$$y = \cos^{-1} x \text{ where } x = \cos y \text{ and } -1 \le x \le 1 \text{ and } 0 \le y \le \pi.$$

The **inverse tangent** function:

$$y = \tan^{-1} x, \text{ where } x = \tan y \text{ and } -\infty < x < \infty \text{ and } -\frac{\pi}{2} < y < \frac{\pi}{2}.$$

The graphs of the inverse sine, inverse cosine, and inverse tangent functions are shown below. By restricting the outputs, each graph passes the vertical-line test (Section R.2).

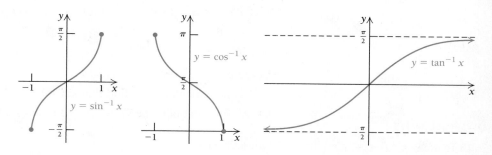

■ **EXAMPLE 2**   Evaluate. Do not use a calculator.

a) $\cos^{-1}\dfrac{1}{2}$ **b)** $\cos^{-1}\left(-\dfrac{\sqrt{2}}{2}\right)$ **c)** $\tan^{-1} 1$ **d)** $\tan^{-1}\dfrac{\sqrt{3}}{3}$

Solution

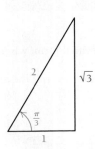

a) Since $\cos \dfrac{\pi}{3} = \dfrac{1}{2}$, we have $\cos^{-1} \dfrac{1}{2} = \dfrac{\pi}{3}$.

b) Since $\cos \dfrac{3\pi}{4} = -\dfrac{\sqrt{2}}{2}$, we have $\cos^{-1}\left(-\dfrac{\sqrt{2}}{2}\right) = \dfrac{3\pi}{4}$.

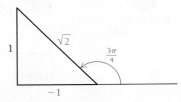

Note that $\frac{3\pi}{4}$ is in the second quadrant and that $-\dfrac{1}{\sqrt{2}} = -\dfrac{\sqrt{2}}{2}$.

c) Since $\tan \dfrac{\pi}{4} = 1$, we have $\tan^{-1} 1 = \dfrac{\pi}{4}$.

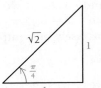 Note that $\dfrac{1}{1} = 1$.

d) Since $\tan \dfrac{\pi}{6} = \dfrac{\sqrt{3}}{3}$, we have $\tan^{-1} \dfrac{\sqrt{3}}{3} = \dfrac{\pi}{6}$.

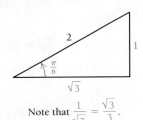

Note that $\dfrac{1}{\sqrt{3}} = \dfrac{\sqrt{3}}{3}$.

⟨ Quick Check 2

⟩ **Quick Check 2**

Evaluate $\tan^{-1}\left(-\sqrt{3}\right)$ without using a calculator.

For most input values with inverse trigonometric functions, a calculator is required.

■ **EXAMPLE 3**   Evaluate. Round each answer to three decimal places.

a) $\sin^{-1} 0.7$
b) $\tan^{-1}(-5)$

Solution

a) Using a calculator, we find that $\sin^{-1} 0.7 = 0.775$ radians (about $44.427°$).

b) We find that $\tan^{-1}(-5) = -1.373$ radians (about $-78.690°$).

⟨ Quick Check 3

⟩ **Quick Check 3**

Evaluate $\cos^{-1}(-0.45)$.

We can use the inverse trigonometric functions to solve trigonometric equations.

■ **EXAMPLE 4**   Solve for x such that x is within the indicated interval.

a) $2 - 4\cos 3x = 0$, $[0, \pi/2]$

b) $5\sin^2 x - 3\sin x = 0$, $[0, \pi/2]$

Solution

a) Recalling that $f^{-1}(f(x)) = x$, we use the inverse cosine to solve for the variable:

$$-4\cos 3x = -2$$
$$\cos 3x = \tfrac{1}{2}$$
$$\cos^{-1}(\cos 3x) = \cos^{-1}\tfrac{1}{2} \qquad \text{Finding the inverse cosine of both sides}$$
$$3x = \cos^{-1}\tfrac{1}{2}$$
$$x = \frac{\cos^{-1}\tfrac{1}{2}}{3} \qquad \text{Dividing by 3}$$
$$= \frac{\pi/3}{3} \qquad \cos^{-1}\tfrac{1}{2} = \pi/3$$
$$= \frac{\pi}{9} \approx 0.349. \qquad \text{Simplifying}$$

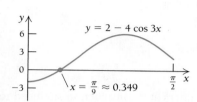

$y = 2 - 4\cos 3x$

$x = \frac{\pi}{9} \approx 0.349$

b) We first solve for $\sin x$ by factoring:

$$5\sin^2 x - 3\sin x = 0$$
$$(\sin x)(5\sin x - 3) = 0. \qquad \text{Factoring}$$

We then set each factor equal to zero and solve. For the factor $\sin x$, we have

$$\sin x = 0$$
$$\sin^{-1}(\sin x) = \sin^{-1} 0 \qquad \text{Finding the inverse sine of both sides}$$
$$x = \sin^{-1} 0$$
$$x = 0. \qquad \text{Simplifying}$$

For the factor $5\sin x - 3$, we have

$$5\sin x - 3 = 0$$
$$\sin x = \tfrac{3}{5}$$
$$\sin^{-1}(\sin x) = \sin^{-1}\tfrac{3}{5} \qquad \text{Finding the inverse sine of both sides}$$
$$x = \sin^{-1}\tfrac{3}{5}$$
$$\approx 0.644. \qquad \text{Using a calculator}$$

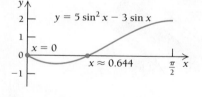

$y = 5\sin^2 x - 3\sin x$

$x = 0$

$x \approx 0.644$

Thus, the solutions of $5\sin^2 x - 3\sin x = 0$ are $x = 0$ and $x \approx 0.644$.

> **Quick Check 4**
>
> Solve for x such that x is within the indicated interval:
>
> $10\cos^2 x - 3\cos x = 0$, $[0, \pi]$. 《 Quick Check 4

Applications

■ **EXAMPLE 5**   **Business: Sales.**   Weekly sales of *The Gardener's Almanac* are modeled by $S(t) = 10 + 8\sin\left(\dfrac{\pi}{26}t\right)$, where $S(t)$ is thousands of copies sold t weeks after February 1. When will sales first reach 17,000 copies per week?

Solution   We let $S = 17$ to represent 17,000 copies sold, and solve for t:

$$17 = 10 + 8\sin\left(\frac{\pi}{26}t\right) \qquad \text{Substituting}$$

$$\sin^{-1}\frac{7}{8} = \frac{\pi}{26}t.$$

Sales reach 17,000 copies per week about 9 weeks after February 1.

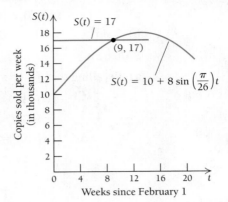

> **Quick Check 5**
>
> Using the model in Example 5, when will sales of the almanac first reach 12,000 copies per week?

❮ Quick Check 5

■ **EXAMPLE 6** **Business: Maximizing Flow.** Builders at a construction site need to build a trough through which concrete will flow. The trough will be built from a sheet of steel measuring 3 ft by 10 ft. The sheet will be bent lengthwise in thirds, and the two side flaps will be bent up to form the trough. At what angle should the sides be bent in order to maximize flow through the trough?

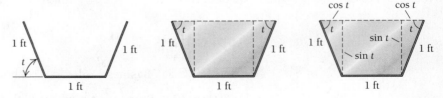

Solution Flow is maximized when the cross-sectional area of the trough is maximized. We let t represent the angle at which each side flap will be bent. Thus, we can view the cross section as two triangles and one rectangle and use trigonometry to find the unknown lengths.

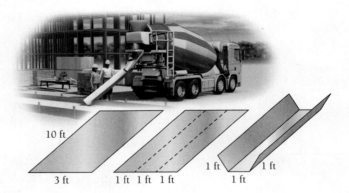

Thus, the cross-sectional area is

$$A(t) = \text{area of rectangle} + \text{area of two triangles}$$
$$= (1)(\sin t) + 2\left(\tfrac{1}{2}\right)(\cos t)(\sin t)$$
$$= \sin t + \cos t \sin t. \qquad\qquad 2 \cdot \left(\tfrac{1}{2}\right) = 1$$

To maximize $A(t)$, we differentiate:

$$A'(t) = \cos t + \cos t \cos t + (-\sin t) \sin t \qquad \text{Using the Product Rule}$$
$$= \cos t + \cos^2 t - \sin^2 t$$
$$= \cos t + \cos^2 t - (1 - \cos^2 t) \qquad \text{Using } \sin^2 t + \cos^2 t = 1$$
$$= 2 \cos^2 t + \cos t - 1. \qquad \text{Simplifying}$$

This expression is quadratic in terms of $\cos t$. We set it equal to zero and solve for t:

$$2 \cos^2 t + \cos t - 1 = 0$$
$$(2 \cos t - 1)(\cos t + 1) = 0 \qquad \text{Factoring}$$
$$2 \cos t - 1 = 0 \quad \text{or} \quad \cos t + 1 = 0$$
$$\cos t = \tfrac{1}{2} \qquad\qquad \cos t = -1$$
$$t = \cos^{-1} \tfrac{1}{2}. \qquad\qquad t = \cos^{-1}(-1).$$

The critical values are $\cos^{-1} \dfrac{1}{2} = \dfrac{\pi}{3}$, or $60°$, and $\cos^{-1}(-1) = \pi$, or $180°$. The second derivative is used to show whether the critical values minimize or maximize $A(t)$:

$$A'(t) = 2 \cos^2 t + \cos t - 1$$
$$A''(t) = -4 \cos t \sin t - \sin t$$
$$A''\left(\frac{\pi}{3}\right) = -4 \cos \frac{\pi}{3} \sin \frac{\pi}{3} - \sin \frac{\pi}{3} \qquad \text{Substituting}$$
$$= -4 \cdot \frac{1}{2} \cdot \frac{\sqrt{3}}{2} - \frac{\sqrt{3}}{2}$$
$$\approx -2.6 < 0.$$

Since $A''(\pi/3) < 0$, we conclude that $A(t)$ is maximized when $t = \pi/3$. The other critical value, $t = \pi$, minimizes the cross-sectional area. It essentially folds the side flaps onto the middle section of the sheet, creating a cross-sectional area of zero. Thus, the workers should bend each side flap at an angle of $60°$ to maximize the cross-sectional area.

Related Rates

Recall from Section 2.7 that a *related rate* means that the rates of change (represented as derivatives) of two variables are related to each other.

■ **EXAMPLE 7** **Related Rates.** Carlos is on the ground, looking through binoculars at an airplane flying toward him. The airplane is traveling at 200 mph, 1 mile above the ground, and will pass directly overhead. When the airplane is above a landmark 2 miles away from Carlos, how fast is his viewing angle increasing?

Solution Let x represent the horizontal distance from Carlos to a point directly below the airplane, and let u represent Carlos's viewing angle (angle of elevation to the airplane). When the airplane is above the landmark, we have $x = 2$ mi and $dx/dt = -200$ mph, where the rate of travel is negative since the airplane is coming toward Carlos and the distance x is decreasing.

$dx/dt = -200$ mph

This quantity does not change: it is constant.

1 mile

As the angle u changes, its rate of change, du/dt, is related to the rate of change of x, represented by dx/dt.

Carlos's position

u

x miles

Thus, the angle u and the distance x are related by

$$\tan u = \frac{1}{x}. \qquad \tan u = \frac{\text{opp}}{\text{adj}}, \text{ where opp} = 1, \text{ adj} = x$$

Differentiating implicitly with respect to t (in hours) gives us an equation relating the two rates du/dt and dx/dt. We then solve for du/dt, the rate of change of the viewing angle.

$$\frac{d}{dt}\tan u = \frac{d}{dt}\left[\frac{1}{x}\right] \qquad \text{Differentiating both sides with respect to } t$$

$$\sec^2 u \cdot \frac{du}{dt} = -\frac{1}{x^2}\cdot\frac{dx}{dt} \qquad \text{Using the Chain Rule, } \frac{d}{dx}\left(\frac{1}{x}\right) = \frac{d}{dx}x^{-1} = -x^{-2}$$

$$\frac{du}{dt} = \frac{1}{\sec^2 u}\cdot\left(-\frac{1}{x^2}\cdot\frac{dx}{dt}\right) \qquad \text{Multiplying both sides by } \frac{1}{\sec^2 u}$$

$$\frac{du}{dt} = \cos^2 u \cdot\left(-\frac{1}{x^2}\cdot\frac{dx}{dt}\right) \qquad \frac{1}{\sec^2 u} = \left(\frac{1}{\sec u}\right)^2 = (\cos u)^2 = \cos^2 u$$

We now evaluate du/dt when $x = 2$. Note that when $x = 2$, we have a right triangle with adjacent leg of length 2 mi and hypotenuse (the straight-line distance between Carlos and the airplane) of length $\sqrt{5}$ mi, as shown in the figure. Thus, when $x = 2$, we have $\cos^2 u = \left(\dfrac{2}{\sqrt{5}}\right)^2 = \dfrac{4}{5}$. Therefore,

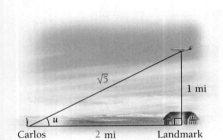

$$\frac{du}{dt} = \cos^2 u \cdot\left(-\frac{1}{x^2}\cdot\frac{dx}{dt}\right)$$

$$\frac{du}{dt} = \frac{4}{5}\cdot\left(-\frac{1}{2^2}\cdot(-200)\right) \qquad \text{Substituting}$$

$$\frac{du}{dt} = 40 \;\frac{\text{radians}}{\text{hour}}.$$

How fast is 40 radians/hour? We can convert 40 radians to degrees by multiplying it by $\dfrac{180 \text{ degrees}}{\pi \text{ radians}}$; we find that 40 radians is about $40\cdot\dfrac{180}{\pi} \approx 2292$ degrees. Since 1 hour is 3600 seconds, we have

$$2292 \;\frac{\text{degrees}}{\text{hour}}\cdot\left(\frac{1 \text{ hour}}{3600 \text{ seconds}}\right) \approx 0.64 \;\frac{\text{degrees}}{\text{second}}.$$

Carlos is raising his binoculars by about 0.6 degrees per second when the airplane is above a landmark 2 miles away.

> **Quick Check 6**
>
> Find the rate of change of Carlos's viewing angle, in degrees per second, when the airplane in Example 7 is above a landmark 1 mile away and traveling toward him.

❮ Quick Check 6

Section Summary

- The *inverse sine* function is $y = \sin^{-1} x$, where $x = \sin y$ and $-1 \le x \le 1$ and $-\dfrac{\pi}{2} \le y \le \dfrac{\pi}{2}$.

- The *inverse cosine* function is $y = \cos^{-1} x$, where $x = \cos y$ and $-1 \le x \le 1$ and $0 \le y \le \pi$.

- The *inverse tangent* function is $y = \tan^{-1} x$, where $x = \tan y$ and $-\infty < x < \infty$ and $-\dfrac{\pi}{2} < y < \dfrac{\pi}{2}$.

EXERCISE SET
7.4

Find each of the following. Do not use a calculator. State each answer in both radians and degrees.

1. $\sin^{-1}\dfrac{1}{2}$

2. $\tan^{-1}\sqrt{3}$

3. $\cos^{-1}\dfrac{\sqrt{2}}{2}$

4. $\sin^{-1}\left(-\dfrac{\sqrt{3}}{2}\right)$

5. $\cos^{-1}0$

6. $\tan^{-1}0$

7. $\tan^{-1}1$

8. $\sin^{-1}\left(-\dfrac{1}{2}\right)$

9. $\cos^{-1}(-1)$

10. $\cos^{-1}\dfrac{\sqrt{3}}{2}$

Use a calculator to find each of the following. Round all answers to three decimal places. State each answer in both radians and degrees.

11. $\sin^{-1}0.3$

12. $\cos^{-1}0.25$

13. $\tan^{-1}0.78$

14. $\tan^{-1}(-3)$

15. $\cos^{-1}(-0.17)$

16. $\sin^{-1}(-0.82)$

In Exercises 17–20, use a calculator to find the measures of angles t and u (in radians and degrees), rounded to three decimal places.

17.

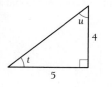

18.

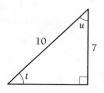

19.

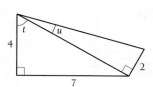

20.

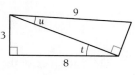

In Exercises 21–26, solve for x, in radians, so that x is within the indicated interval. Do not use a calculator.

21. $-1 + 2\sin x = 0,\quad [0, \pi/2]$

22. $1 + 4\cos x = 3,\quad [0, \pi]$

23. $\sin 3x = \sqrt{2}/2,\quad [0, \pi/6]$

24. $\cos 4x = -\sqrt{3}/2,\quad [0, \pi/4]$

25. $5\tan 2x = 5,\quad [0, \pi/4)$

26. $\tan(3x) = \sqrt{3},\quad [0, \pi/6]$

In Exercises 27–36, use a calculator to solve for x, in radians, so that x is within the indicated interval. Find each answer to three decimal places.

27. $-4 + 6\sin x = 0,\quad [0, \pi/2]$

28. $-2 - 5\cos x = 0,\quad [0, \pi]$

29. $-\frac{1}{4} + \cos(3x + 1) = 0,\quad [0, \pi/4]$

30. $-\frac{1}{3} + 2\sin(2 - 4x) = 0,\quad [0, \pi/4]$

31. $\tan(6x) = 3,\quad [0, \pi/12]$

32. $3\tan(5x) = -2,\quad (-\pi/10, 0]$

33. $3\sin^2 x - \sin x = 0,\quad (0, \pi/2]$

34. $4\cos^2 x - 3\cos x = 0,\quad [0, \pi]$

35. $2\sin^2 x - \sin x - 1 = 0,\quad [-\pi/2, \pi/2]$

36. $2\cos^2 x + 3\cos x + 1 = 0,\quad [0, \pi]$

APPLICATIONS

Business and Economics

37. Sales. Rentals at Sal's Dune Buggy Rentals are modeled by $f(t) = 80 - 60\cos\left(\dfrac{\pi}{6}t\right)$, where $f(t)$ is the rate of rentals per month, t months after January 1. When does the rate of rentals first meet or exceed 110 rentals per month?

38. Sales. The sales of cookies by The Gingerbread House are modeled by $f(t) = 5000 + 3500\cos\left(\dfrac{\pi}{6}t\right)$, where $f(t)$ is the rate of cookies sold per month, t months after December 1. When does the rate of cookie sales drop below 6750 cookies per month for the first time?

39. Optimization. Chandler Dairies has a feed trough made from a rectangular piece of sheet metal 4 ft wide and 20 ft long. The sheet is bent lengthwise into a V-shape, forming the trough. At what angle t should the sheet metal be bent in order to maximize the volume of feed it can hold?

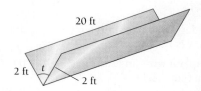

40. Optimization. A rain gutter is to be formed by attaching a strip of sheet metal to a wall and then bending the strip out from the wall at angle t. What angle t maximizes the cross-sectional area of the gutter?

General Interest

41. Temperature. The average daily high temperature in Chicago can be modeled by $T(t) = 57 - 27 \cos\left(\dfrac{2\pi}{365} t\right)$, where $T(t)$ is the average high temperature, in degrees Fahrenheit, t days after January 1. How many days after January 1 will the average daily high temperature reach 60°F? Round your answer to the nearest integer. (*Source:* www.wunderground.com.)

42. Daylight. The amount of daylight in Atlanta can be modeled by $H(t) = 13 - 2 \cos\left(\dfrac{2\pi}{365} t\right)$, where $H(t)$ is the number of hours of daylight t days after December 21. How many days after December 21 will the amount of daylight reach 11.5 hr? Round your answer to the nearest integer. (*Source:* www.gaisma.com.)

43. Related rates. An airplane is flying 2 mi above the ground at 450 mph, approaching Nicole and will pass her directly overhead. How fast is Nicole's viewing angle increasing, in degrees per second, when the airplane is above a landmark 2 mi away?

44. Related rates. A weather balloon is released and ascends straight upward at a constant rate of 30 ft/sec. Armando, standing 300 ft away, watches the balloon through a scope. How fast is Armando's viewing angle increasing, in degrees per second, 20 sec after the balloon is released?

SYNTHESIS

45. Consider the expressions $\sin\left(\sin^{-1} x\right)$ and $\sin^{-1}\left(\sin x\right)$, where x is in radians.

a) Explain why $\sin\left(\sin^{-1} 0.9\right) = 0.9$ but $\sin\left(\sin^{-1} 1.1\right) \neq 1.1$.

b) Explain why $\sin^{-1}\left(\sin \dfrac{\pi}{2}\right) = \dfrac{\pi}{2}$, but $\sin^{-1}\left(\sin \pi\right) \neq \pi$.

c) In general, for what values of x does $\sin\left(\sin^{-1} x\right) = x$?

d) In general, for what values of x does $\sin^{-1}\left(\sin x\right) = x$?

46. For what values of x does $\cos^{-1}\left(\cos x\right) = x$?

47. Suppose a right triangle has legs of length a and b. Show that $\tan^{-1} \dfrac{a}{b} + \tan^{-1} \dfrac{b}{a} = \dfrac{\pi}{2}$.

48. Show that $\cos\left(\sin^{-1} x\right) = \sqrt{1 - x^2}$ for $-1 \leq x \leq 1$.

49. Consider the equation $\sin^2 x + 4 \sin x - 2 = 0$.

a) Let $y = \sin x$, so that $y^2 + 4y - 2 = 0$. Using the quadratic formula, find two solutions of this equation. Denote these solutions as y_1 and y_2.

b) Find $x_1 = \sin^{-1} y_1$ and $x_2 = \sin^{-1} y_2$. Explain why it is possible to find one solution using the inverse sine, but not the other.

c) Confirm your solution from part (b) by using a calculator to find the x-intercepts of $f(x) = \sin^2 x + 4 \sin x - 2$, in the interval $[0, 2\pi]$.

d) Explain how the other x-intercept can be found using algebra and trigonometry.

In Exercises 50 and 51, use a method similar to that in Exercise 49.

50. Optimization. A rectangular piece of sheet metal 4 ft wide is to be bent so that the two side flaps are 1 ft wide, leaving 2 ft as the width of the center section. At what angle should the flaps be bent so that the cross-sectional area is maximized?

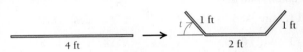

51. Optimization. A rain gutter is to be formed by attaching a 1-ft-wide piece of sheet metal to a wall at a right angle, then bending the sheet lengthwise, so that the outer flap is 4 in. wide and is bent upward, as shown in the figure. At what angle should the flap be bent so that the cross-sectional area of the gutter is maximized?

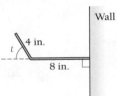

52. The range of the inverse sine function, $y = \sin^{-1} x$, is $-\dfrac{\pi}{2} \leq y \leq \dfrac{\pi}{2}$, but the range of the inverse cosine function, $y = \cos^{-1} x$, is $0 \leq y \leq \pi$. Explain why the two ranges are different.

53. Explain why the ranges of $y = \sin^{-1} x$ and $y = \tan^{-1} x$ differ.

54. Evaluate the following limits.

a) $\displaystyle \lim_{x \to \infty} \tan^{-1} x$ **b)** $\displaystyle \lim_{x \to -\infty} \tan^{-1} x$

Answers to Quick Checks

1. $\pi/4$ **2.** $-\pi/3$ **3.** 2.038 radian (about 116.7°)
4. $x = 1.266, x = 1.571 \ (\pi/2)$ **5.** 2 weeks
6. 1.59°/sec

KEY TERMS AND CONCEPTS

EXAMPLES

SECTION 7.1

A ray with its endpoint at the origin starts on the positive *x*-axis and rotates, forming an **angle**. The positive *x*-axis is the angle's **initial side**, and the ray itself is the **terminal side**.

Counterclockwise rotations are considered positive, and clockwise rotations are considered negative.

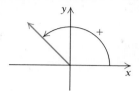 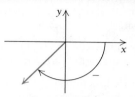

A counterclockwise rotation results in a positive angle.

A clockwise rotation results in a negative angle.

Angles are measured in **degrees** (°) or in **radians**. A complete rotation is 360°, or 2π radians. Thus, $1° = \dfrac{\pi}{180}$ radian, and $1 \text{ radian} = \left(\dfrac{180}{\pi}\right)°$.

Angles that have the same terminal side are called **coterminal**.

- $\pi/3$ radians is equivalent to 60°.
- Examples of angles that are coterminal to $\pi/3$, or 60°, are $-5\pi/3$, $7\pi/3$, and $13\pi/3$, or $-300°$, $420°$, and $780°$.

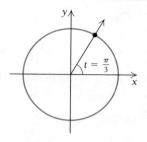

The **unit circle** is a circle of radius 1, centered at the origin, and its equation is $x^2 + y^2 = 1$. A ray at angle *t* with the *x*-axis intersects the unit circle at $P(x, y)$. The first coordinate is defined as the **cosine** of *t*, written **cos *t***, and the second coordinate is defined as the **sine** of *t*, written **sin *t***. That is, $x = \cos t$, $y = \sin t$.

The cosine and sine functions are examples of **trigonometric functions**.

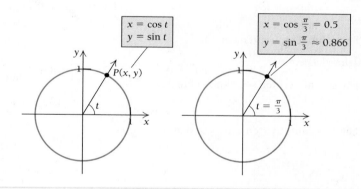

If *t* is an angle in a right triangle, the leg on the *x*-axis is the **adjacent leg**, the vertical leg is the **opposite leg**, and the **hypotenuse** is the side opposite the right angle. Therefore,

$$\sin t = \frac{\text{length of opposite leg}}{\text{length of hypotenuse}} = \frac{\text{opp}}{\text{hyp}},$$

$$\cos t = \frac{\text{length of adjacent leg}}{\text{length of hypotenuse}} = \frac{\text{adj}}{\text{hyp}}.$$

This is the basis of **right-triangle trigonometry**.

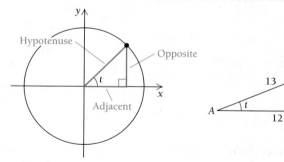

In triangle *ABC*, we have

$$\sin t = \frac{\text{opp}}{\text{hyp}} = \frac{5}{13}$$

and

$$\cos t = \frac{\text{adj}}{\text{hyp}} = \frac{12}{13}.$$

(continued)

KEY TERMS AND CONCEPTS	EXAMPLES

SECTION 7.1 *(continued)*

A function is **periodic** if there exists a positive number p such that $f(x + p) = f(x)$, for all x in the domain of f. The smallest such p is the **period** of the function.

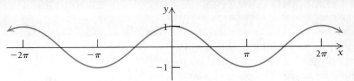

The cosine function
$y = \cos x$

The graphs of the sine and cosine functions are shown at the right. Both have period 2π.

Trigonometric functions require that the input values be in radians.

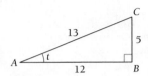

The sine function
$y = \sin x$

The other four trigonometric functions are

- The **tangent**, written **tan** t and defined as $\tan t = \dfrac{\sin t}{\cos t}$. Using the legs of a right triangle, $\tan t = \dfrac{\text{opposite leg}}{\text{adjacent leg}} = \dfrac{\text{opp}}{\text{adj}}$.
- The **cotangent**, written **cot** t and defined as $\cot t = \dfrac{1}{\tan t} = \dfrac{\cos t}{\sin t}$.
- The **secant**, written **sec** t and defined as $\sec t = \dfrac{1}{\cos t}$.
- The **cosecant**, written **csc** t and defined as $\csc t = \dfrac{1}{\sin t}$.

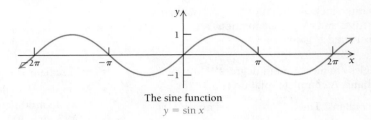

For the given triangle, we have

$$\tan t = \frac{\text{opp}}{\text{adj}} = \frac{5}{12};$$

$$\cot t = \frac{1}{\tan t} = \frac{12}{5};$$

$$\sec t = \frac{1}{\cos t} = \frac{13}{12};$$

$$\csc t = \frac{1}{\sin t} = \frac{13}{5}.$$

The graph of the tangent function is shown at the right.

The period of the tangent function is π.

Note that the tangent function is not defined for $x = \pi/2 \pm k\pi$, where k is an integer.

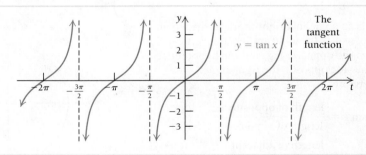

The tangent function

$y = \tan x$

A **trigonometric identity** is an equation involving trigonometric functions that is true for all meaningful replacements of the variables by real numbers.

Pythagorean Identities:
$\cos^2 t + \sin^2 t = 1$
$1 + \tan^2 t = \sec^2 t$
$\cot^2 t + 1 = \csc^2 t$

Double-Angle Identities:
$\sin 2t = 2 \sin t \cos t$
$\cos 2t = \cos^2 t - \sin^2 t$

Sum-Difference Identities:
$\cos (t + u) = \cos t \cos u - \sin t \sin u$
$\cos (t - u) = \cos t \cos u + \sin t \sin u$
$\sin (t + u) = \sin t \cos u + \cos t \sin u$
$\sin (t - u) = \sin t \cos u - \cos t \sin u$

KEY TERMS AND CONCEPTS	EXAMPLES

SECTION 7.2

The **derivatives of the trigonometric functions** are

$$\frac{d}{dx}\sin x = \cos x, \qquad \frac{d}{dx}\cos x = -\sin x,$$

$$\frac{d}{dx}\tan x = \sec^2 x, \qquad \frac{d}{dx}\sec x = \sec x \tan x,$$

$$\frac{d}{dx}\cot x = -\csc^2 x, \qquad \frac{d}{dx}\csc x = -\csc x \cot x.$$

The Product, Quotient, and Chain Rules can be used when differentiating trigonometric functions.

$$\frac{d}{dx}\sin^4 x = 4\sin^3 x \cdot \cos x.$$

$$\frac{d}{dx}\tan x^3 = \sec^2 x^3 \cdot 3x^2 = 3x^2 \sec^2 x^3.$$

$$\frac{d}{dx}e^{2x}\cos x = -e^{2x}\sin x + 2e^{2x}\cos x.$$

A **sinusoidal** function is of the form

$$y = A\sin(Bx - C) + D,$$

or

$$y = A\cos(Bx - C) + D.$$

The graph of a sinusoidal function has the following features:

- **midline**, $y = D$,
- **amplitude**, $|A|$,
- **period**, $\dfrac{2\pi}{B}$,
- **phase shift**, $\dfrac{C}{B}$.

The function $y = -3\cos(2x - 1) + 4$ is sinusoidal. The midline is 4, the amplitude is 3, the period is $2\pi/2 = \pi$, and the phase shift is $\frac{1}{2}$. Two periods of the function are shown below.

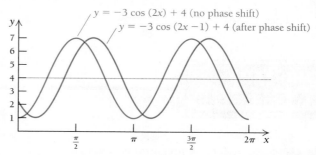

SECTION 7.3

Common trigonometric integration formulas:

$$\int \sin x \, dx = -\cos x + C,$$

$$\int \cos x \, dx = \sin x + C,$$

$$\int \sin kx \, dx = -\frac{1}{k}\cos kx + C,$$

$$\int \cos kx \, dx = \frac{1}{k}\sin kx + C,$$

$$\int \sec^2 x \, dx = \tan x + C,$$

$$\int \tan x \, dx = -\ln|\cos x| + C,$$

$$\int \cot x \, dx = \ln|\sin x| + C,$$

$$\int \sec x \tan x \, dx = \sec x + C,$$

$$\int \csc^2 x \, dx = -\cot x + C,$$

$$\int \cot x \csc x \, dx = -\csc x + C.$$

$$\int \cos 3x \, dx = \tfrac{1}{3}\sin 3x + C.$$

$$\int \sec^2 5x \, dx = \tfrac{1}{5}\tan 5x + C.$$

$$\int_0^{\pi/6} \cos 4x \, dx = \left[\tfrac{1}{4}\sin 4x\right]_0^{\pi/6}$$

$$= \frac{1}{4}\sin\left(4\cdot\frac{\pi}{6}\right) - \frac{1}{4}\sin(4\cdot 0)$$

$$= \frac{1}{4}\cdot\frac{\sqrt{3}}{2} - \frac{1}{4}\cdot 0 = \frac{\sqrt{3}}{8}.$$

(*continued*)

KEY TERMS AND CONCEPTS	EXAMPLES

SECTION 7.4

The three principal inverse trigonometric functions are as follows:

- **Inverse sine**, written $y = \sin^{-1} x$, where $\sin y = x$ and $-\pi/2 \leq y \leq \pi/2$ and $-1 \leq x \leq 1$.
- **Inverse cosine**, written $y = \cos^{-1} x$, where $\cos y = x$ and $0 \leq y \leq \pi$ and $-1 \leq x \leq 1$.
- **Inverse tangent**, written $y = \tan^{-1} x$, where $\tan y = x$ and $-\pi/2 < y < \pi/2$ and x is any real number.

$\sin^{-1} \dfrac{1}{2} = \dfrac{\pi}{6}$, since $\sin \dfrac{\pi}{6} = \dfrac{1}{2}$.

$\cos^{-1}\left(-\dfrac{1}{2}\right) = \dfrac{2\pi}{3}$, since $\cos \dfrac{2\pi}{3} = -\dfrac{1}{2}$.

$\tan^{-1} 1 = \dfrac{\pi}{4}$, since $\tan \dfrac{\pi}{4} = 1$.

CHAPTER 7
REVIEW EXERCISES

These review exercises are for test preparation. They can also be used as a practice test. Answers are in the back of the book. The blue bracketed section references tell you what part(s) of the chapter to restudy if your answer is incorrect.

CONCEPT REINFORCEMENT

In Exercises 1–10, classify each statement as true or false.

1. The unit circle is any circle with radius 1. [7.1]

2. The rotation of a ray in a counterclockwise direction is considered a positive rotation. [7.1]

3. A ray at an angle t with the x-axis intersects the unit circle at point $P(x, y)$. The first coordinate of P is given by $x = \cos t$, and the second coordinate by $y = \sin t$. [7.1]

4. An angle measure of $120°$ is equivalent to $\dfrac{2\pi}{3}$ radians. [7.1]

5. If $\sin t = \frac{3}{7}$, then $\sin 2t = \frac{6}{7}$. [7.1]

6. The derivative of $y = \sin^3 x$ is $y' = 3 \sin^2 x$. [7.2]

7. The amplitude of $y = 3 + \sin 2x$ is 3. [7.2]

8. The period of $y = 3 + \sin 2x$ is π. [7.2]

9. The antiderivative of $\cos x$ is $\sin x$. [7.3]

10. $\sin^{-1} \pi = 0$. [7.4]

REVIEW EXERCISES

11. Give the measures (in degrees) of two angles that are coterminal with an angle of $110°$. [7.1]

12. Convert $240°$ to radian measure. [7.1]

13. Convert $4\pi/9$ to degree measure. [7.1]

14. State the period of the function in the graph below. [7.1]

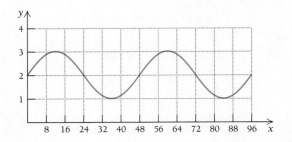

15. Find the point at which a ray at an angle of $145°$ with the x-axis intersects the unit circle. Round each coordinate to three decimal places. [7.1]

16. A right triangle is drawn in the first quadrant, with $\sin t = \frac{3}{5}$. Find $\cos t$. Do not use a calculator. [7.1]

17. A right triangle is drawn in the first quadrant, with $\cos t = \frac{1}{4}$ and the adjacent leg along the positive x-axis. Find $\tan t$. Do not use a calculator. [7.1]

18. A right triangle is drawn in the first quadrant, with $\sin t = \frac{2}{3}$. Find $\sin 2t$. Do not use a calculator. [7.1]

19. Use a calculator to find the lengths of B and C in the triangle shown below, to three decimal places. [7.1]

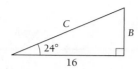

Differentiate. [7.2]

20. $y = \sin^3 4x$

21. $y = e^{2x} \cos 3x$

22. $y = \tan^2 (2x + 1)$

23. $y = \dfrac{1 + \sin 2x}{\cos 3x}$

24. $y = \sqrt{\sec x}$

25. $y = (\sin 2x + \cos 2x)^2$

26. Business: sales. The Sassy Swimsuit Store sells swimsuits at a minimum rate of 120 swimsuits per month on December 1 and at a maximum rate of 880 swimsuits per month on June 1. Assume this pattern repeats every year. Find a sinusoidal model for $S(t)$, the monthly rate of sales of swimsuits, t months after December 1. Use a model that has no phase shift. [7.2]

27. Business: sales. The rate of sales of sunglasses at The Sunglass Shack is modeled by $S(t) = 900 - 325$ $\cos \left(\dfrac{\pi}{6} t \right)$, where $S(t)$ is the monthly rate of sales of sunglasses, t months after January 1. [7.2, 7.4]

a) Find $S'(3)$, and explain what this number represents.

b) In what month does the rate of sales first reach 1100 pairs per month?

28. Weather. The rate of rainfall in a coastal community is modeled by $A(t) = 4 + 3 \sin \left(\dfrac{2\pi}{365} t \right)$, where $A(t)$ is the rate of rainfall, in millimeters (mm) per day, t days after January 1.

a) How many days will pass, from January 1, before the rate of rainfall is maximized? Round to the nearest integer. [7.2, 7.4]

b) How many days will pass, from January 1, before the rate of rainfall will reach 5 mm per day? Round to the nearest integer. [7.4]

Find each of the following. [7.3]

29. $\displaystyle\int \cos 12x \, dx$

30. $\displaystyle\int x \sin 4x^2 \, dx$

31. $\displaystyle\int \sec^2 \tfrac{1}{2} x \, dx$

32. $\displaystyle\int (\cos 2x + \sin 2x)^2 \, dx$

33. $\displaystyle\int (2 + 3 \sin 5x) \, dx$

34. $\displaystyle\int_0^{\pi/3} \sin 2x \, dx$

35. $\displaystyle\int_0^{\pi/6} \tan x \, dx$

36. $\displaystyle\int_{-\pi/4}^{\pi/4} (1 + \sin x) \, dx$

37. $\displaystyle\int_0^{12} \left(5 - 2 \cos \dfrac{\pi}{6} t \right) dt$

38. Business: total sales and average sales. The rate of sales of sunglasses at The Sunglass Shack is modeled by $S(t) = 900 - 325 \cos \left(\dfrac{\pi}{6} t \right)$, where $S(t)$ is the rate of pairs of sunglasses sold per month, t months after January 1 (as in Exercise 27). Find the total sales of sunglasses over 1 yr. [7.3]

39. Weather: total rainfall. The rate of rainfall (in millimeters per day) that falls in a coastal community is modeled by $A(t) = 4 + 3 \sin \left(\dfrac{2\pi}{365} t \right)$. Find the total rainfall during the first 6 months (January 1 to July 1, or 181 days, assuming no leap day). [7.3]

40. Without using a calculator, evaluate $\sin^{-1} \left(-\dfrac{\sqrt{2}}{2} \right)$. [7.4]

41. Find the angles t and u in the triangle shown below. State each angle in radians and in degrees to three decimal places. [7.4]

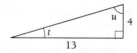

42. Maximizing cross-sectional area. Metal sheeting 4 in. wide will be bent length-wise into a parallelogram to create a wire conduit, with each side 1 in. wide. What is the angle t that maximizes the cross-sectional area? [7.4]

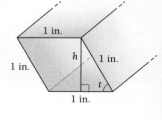

43. Related rates. A sailboat is moving at 15 mph parallel to the shore and 1 mi from shore. When the sailboat is 2 mi from an observer on the shore, how fast is the observer's viewing angle u changing as the boat moves closer to point A, as shown in the figure? State the answer in degrees per second, and let x be the distance from the sailboat to point A. [7.4]

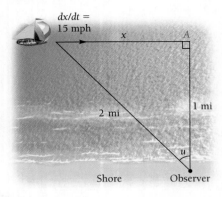

SYNTHESIS

44. Find $\int 4x \cos(4x)\, dx$. [7.4]

45. A rectangular piece of sheet metal 7 ft wide is to be bent so that the two outer flaps are 2 ft wide, leaving 3 ft as the width of the center section, as shown below. At what angle should the flaps be bent to maximize the cross-sectional area? [7.4]

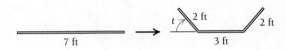

46. Solve $\sin^2 x + 3 \sin x - 1 = 0$, so that x is in $[0, 2\pi]$. Round each solution to three decimal places. [7.4]

47. Evaluate $\int_0^\pi \sqrt{\sin x}\, dx$ to three decimal places. [7.3]

48. Evaluate $\int_0^1 \ln(1 + \cos x)\, dx$ to three decimal places. [7.3]

CHAPTER 7
TEST

1. Convert $160°$ to radian measure. Leave the answer in terms of π.

2. Convert $5\pi/8$ to degree measure.

3. Find $\sin \dfrac{2\pi}{3}$ without using a calculator.

4. Suppose $\cos t = \frac{2}{9}$ and t is in the first quadrant. Find $\sin t$. Do not use a calculator.

5. Suppose $\sin t = \frac{1}{6}$ and t is in the first quadrant. Find $\sin 2t$. Do not use a calculator.

6. Find the point at which a ray at an angle of $238°$ with the x-axis intersects the unit circle. Round each coordinate to three decimal places.

7. In the triangle below, find the lengths of A and C to three decimal places.

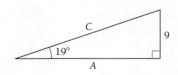

Differentiate.

8. $y = x^4 \sin 3x$

9. $y = \tan^3 2x$

10. $y = \dfrac{\cos x}{\tan 4x}$

11. $y = \sqrt{\sin x^2}$

12. **Business: optimization.** The rate of sales of souvenir football jerseys at Pinewoods College is modeled by

$$S(t) = 300 + 260 \cos\left(\frac{\pi}{6}t\right),$$ where $S(t)$ is the rate of

jerseys sold per month, t months after October 1. Assume this pattern repeats every year.

  a) In what month does the rate of sales first drop to 110 units per month?
  b) How fast is the rate of sales changing on January 1?

Find each of the following.

13. $\int (x + 1) \cos (x^2 + 2x + 4)\, dx$

14. $\int \sin^8 3x \cos 3x\, dx$

15. $\int_{-\pi/2}^{\pi/4} \cos x\, dx$

16. Use a calculator to find $\cos^{-1}(-0.73)$ in both radians and degrees. Round each answer to three decimal places.

17. Find $\tan^{-1} \frac{2}{5} + \tan^{-1} \frac{5}{2}$.

18. Find $2 \sin^{-1} 1$.

19. Find the measures of angles t and u in the triangle below. State each angle in radians and degrees to three decimal places.

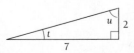

20. **Business: sales.** The rate of sales of souvenir football jerseys at Pinewoods College is modeled by

$$S(t) = 300 + 260 \cos\left(\frac{\pi}{6}t\right),$$ where $S(t)$ is the monthly

rate at which jerseys sold, t months after October 1. Use this model to find the following.

a) What is the average rate of sales per month during the first half of the year (October 1 to April 1)?

b) What are the total sales during the whole year (October 1 to the following October 1)?

21. Weather. The average monthly rainfall rate in Sandersville is modeled by $A(t) = 3 - 2 \cos\left(\dfrac{\pi}{6} t\right)$, where $A(t)$ is the rate of rainfall, in inches per month, t months after January 1. What is the total rainfall during the first 3 months of the year?

22. Related rates. A vehicle is traveling at 425 mph toward point A on a track at the Bonneville Salt Flats in Utah. Mike is standing 0.5 mi from point A and is watching the vehicle through a pair of binoculars. How fast is his viewing angle u changing (in degrees per second) at the moment when the vehicle is 2 mi away from Mike (along a straight line)? Let x represent the vehicle's distance to point A as shown in the figure.

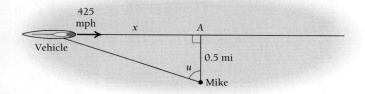

SYNTHESIS

23. Find $\displaystyle\int 2x \sin 5x \, dx$.

24. Optimization. Consider $f(x) = \sin x + \cos x$ over the interval $[0, 2\pi]$.

a) Find the absolute minimum and absolute maximum values of $f(x)$ without using a calculator.

b) Use a calculator to confirm the results from part (a).

TECHNOLOGY CONNECTION

25. A rain gutter is to be formed by attaching a 1.5-ft-wide piece of sheet metal at a right angle to a wall, then bending the sheet lengthwise so that the outer flap is 1 ft wide and is bent up as shown in the figure. At what angle should the flap be bent to maximize the cross-sectional area of the gutter?

26. Solve $\cos^2 x + 3 \cos x - 1 = 0$, so that x is in $[0, 2\pi]$. Round each solution to three decimal places.

Extended Technology Application

Parametric Curves

Recall from Section 7.1 that for any point on the unit circle $x = \cos t$ and $y = \sin t$, where t is the angle of the ray with the positive x-axis. Suppose we view t as a variable and x and y as functions of t, so $x(t) = \cos t$ and $y(t) = \sin t$, for $0 \leq t \leq 2\pi$. A graph composed of points $(x(t), y(t))$ is called a *parametric graph*, and t is the *parameter variable*.

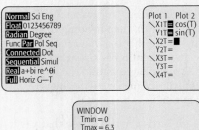

To graph parametric equations on a calculator, press MODE, and then select Radian and Par. Next, press Y= and enter the two functions as X₁ₜ and Y₁ₜ. Press WINDOW to set the bounds for t and the frame dimensions in terms of x and y. For the unit circle, set Tmin = 0, Tmax = 6.3 (rounding 2π up to the nearest tenth), and Tsₜₑₚ = 0.1. For the viewing window, set Xmin = −1, Xmax = 1, Ymin = −1 and Ymax = 1.

Select ZOOM and 5:ZSquare to make the scales consistent on both x- and y-axes, as shown below on the right.

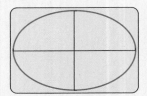 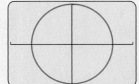

EXERCISES

For Exercises 1–5, graph the parametric equations.

1. $x(t) = \sin t, y(t) = \cos t$, for $0 \leq t \leq 2\pi$. This is a unit circle but with a different starting position and direction of travel for the ray. This illustrates that there can be more than one way to describe a shape parametrically.

2. $x(t) = 2 \cos t, y(t) = 3 \sin t$, for $0 \leq t \leq 2\pi$. This is an ellipse with horizontal width 4 and vertical length 6. Adjust the x and y window settings appropriately.

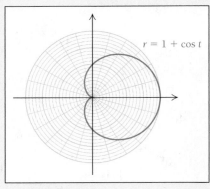

Key patterns in the 8th-century Celtic masterpiece, The Lindisfarne Gospels. See Exercise 3.

3. $x(t) = \cos 3t, y(t) = \sin 4t$, for $0 \leq t \leq 2\pi$. This is called a *key pattern*. As long as the coefficients of t have 1 as their only common divisor, the complete pattern will appear. You may need to decrease the Tsₜₑₚ value. Use 0.05 or 0.01 for a smoother curve.

4. $x(t) = (1 - \sin t) \cos t, y(t) = (1 - \sin t) \sin t$, for $0 \leq t \leq 2\pi$. This is a *cardioid*, named for its heart-shaped appearance.

$r = 1 + \cos t$

A type of cardioid.

5. $x(t) = \cos t, y(t) = 1 - \sin t$, for $0 \leq t \leq 2\pi$. This is a circle offset from the origin.

6. Based on the result of Exercise 5, what is a parametric equation for a circle of radius 1 centered at $(2, -1)$? What is a parametric equation for a circle of radius 0.5 centered at $(2, -1)$?

Many of the interesting shapes in Exercises 1–6 are difficult to express in terms of y as a function of x. By defining both x and y parametrically in terms of t, we can plot these types of graphs. We can also use calculus to study these curves. In general, if a curve is defined parametrically by $(x(t), y(t))$, then

- The derivative is given by $\dfrac{dy}{dx} = \dfrac{dy/dt}{dx/dt}$.

- The definite integral is given by $\displaystyle\int_{t_0}^{t_1} y(t) \cdot x'(t)\, dt$.

For the unit circle $x(t) = \cos t$, $y(t) = \sin t$, for $0 \le t \le 2\pi$, we have

$$\frac{dy}{dx} = \frac{dy/dt}{dx/dt} = \frac{\dfrac{d}{dt}\sin t}{\dfrac{d}{dt}\cos t} = -\frac{\cos t}{\sin t}.$$

Note that the derivative is defined in terms of the parameter variable t. For example, when $t = \pi/4$, the point on the unit circle is $\left(\cos\left(\dfrac{\pi}{4}\right), \sin\left(\dfrac{\pi}{4}\right)\right) = \left(\dfrac{\sqrt{2}}{2}, \dfrac{\sqrt{2}}{2}\right)$, and the slope of the line tangent to the unit circle at this point is $\dfrac{dy}{dx} = -\dfrac{\cos(\pi/4)}{\sin(\pi/4)} = -1$. This is shown in the graph below.

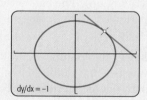

The slope of the tangent line is horizontal when the derivative dy/dx is zero, that is, when $\cos t = 0$. This occurs when $t = \pi/2$ and $3\pi/2$ and gives the points $(0, 1)$ and $(0, -1)$, respectively. The tangent line is vertical when the derivative is undefined, that is, when $\sin t = 0$, which occurs when $t = 0, \pi,$ and 2π and gives the points $(1, 0)$ and $(-1, 0)$.

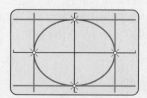

EXERCISES

Find dy/dx and all values of t and the corresponding points on the graph where there is (a) a horizontal tangent line and (b) a vertical tangent line.

7. $x(t) = 2\cos t, y(t) = 3\sin t,$ for $0 \le t \le 2\pi$. (This is the ellipse of Exercise 2.)

8. $x(t) = (1 - \sin t)\cos t, y(t) = (1 - \sin t)\sin t,$ for $0 \le t \le 2\pi$. (This is the cardioid of Exercise 4.)

To find the area within the unit circle using integration, we note that the top half of the unit circle is defined by $y = \sqrt{1 - x^2}$ and the area under the graph is given by $\displaystyle\int_{-1}^{1} \sqrt{1 - x^2}\, dx$. However, finding an antiderivative for this integrand is very difficult and is outside the scope of this book. Instead, we can use the parameterized integral form $\displaystyle\int_{t_0}^{t_1} y(t) \cdot x'(t)\, dt$. We integrate from $t_0 = 0$ to $t_1 = 2\pi$ and use substitutions. Two identities often used are the *half-angle identities*:

$$\sin^2 t = \tfrac{1}{2}(1 - \cos 2t)$$

and

$$\cos^2 t = \tfrac{1}{2}(1 + \cos 2t).$$

We start by substituting $y(t) = \sin t$ and $x'(t) = -\sin t\, dt$:

$$\int_{t_0}^{t_1} y(t) \cdot x'(t)\, dt = \int_0^{2\pi} \sin t \cdot (-\sin t\, dt)$$

$$= -\int_0^{2\pi} \sin^2 t\, dt \qquad \text{Simplifying}$$

Now we use the identity $\sin^2 t = \tfrac{1}{2}(1 - \cos 2t)$:

$$-\int_0^{2\pi} \sin^2 t\, dt = -\frac{1}{2}\int_0^{2\pi} (1 - \cos 2t)\, dt$$

$$= \tfrac{1}{2}\big[t - \tfrac{1}{2}\sin 2t\big]_0^{2\pi} \qquad \text{Integrating}$$

$$= \tfrac{1}{2}\big(2\pi - \tfrac{1}{2}\sin(2 \cdot 2\pi)\big) - \tfrac{1}{2}\big(0 - \tfrac{1}{2}\sin 0\big)$$

$$= \pi.$$

EXERCISES

9. Find the area enclosed within the ellipse of Exercise 2.

10. Find the area enclosed within the cardioid of Exercise 4.

A Famous Curve

Consider a circle of radius 1 with center at $(0, 1)$. A point P on the circle is placed at the origin, as shown below.

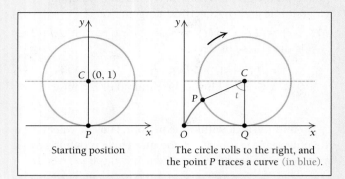

Starting position   The circle rolls to the right, and the point P traces a curve (in blue).

If the circle rolls to the right, point P rotates along with the circle and traces a curve. Let t represent the angle between segment CP and segment CQ, where Q is the point of contact between the circle and the x-axis after the circle has rolled. To define point P's coordinates in terms of the rotation t, let O be the origin and note that segments CP and CQ are radii, both with length 1.

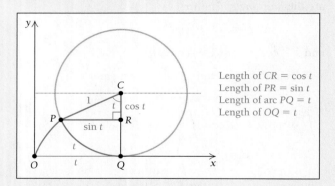

Length of $CR = \cos t$
Length of $PR = \sin t$
Length of arc $PQ = t$
Length of $OQ = t$

In the diagram above, PR is perpendicular to QC, so triangle PRC is a right triangle, with angle t at vertex C. Since CP is a radius with length 1, the length of PR is $\sin t$, and the length of CR is $\cos t$. Furthermore, the circular arc PQ has length t (recall that on a unit circle, the radian angle measure t is the same as the length of the corresponding arc). Since the circle rolls along the x-axis, the length of OQ is also t. Thus, we have the following:

- The x-coordinate of P is $OQ - PR$: $x(t) = t - \sin t$.
- The y-coordinate of P is $CQ - CR$: $y(t) = 1 - \cos t$.

This curve, $x(t) = t - \sin t$, $y(t) = 1 - \cos t$, for $0 \le t \le 2\pi$, is called a *cycloid*. One period of its graph is

shown below, with $\mathtt{Tmin} = 0$, $\mathtt{Tmax} = 6.3$, $\mathtt{Tstep} = 0.1$, $\mathtt{Xmin} = 0$, $\mathtt{Xmax} = 6.3$, $\mathtt{Ymin} = 0$, and $\mathtt{Ymax} = 2$.

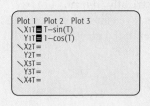
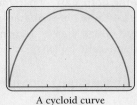

A cycloid curve

EXERCISES

11. Find the slope of the line tangent to the cycloid when the circle has rolled one quarter of a complete turn.

12. Find the area under the graph for one period of the cycloid.

The cycloid is very useful in engineering and physics. Two famous applications use cycloids:

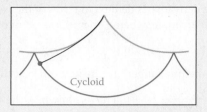

Cycloid

- The *tautochrone*. A ball is released at the lip of an inverted cycloid (facing up). The ball will roll back and forth in equal periods of time (*tauto-* means "same," *chrone* means "time").

- The *brachistochrone*. Two points, A and B, are situated so that A is above and to the side of B. The shape of the path (a path that is not necessarily the shortest distance) on which a ball under the influence of gravity alone will take the least time to travel from A to B is called a *brachistochrone curve* (*brachi-* means "shortest"). It is a half-cycloid oriented sideways so that it is momentarily vertical at A before curving toward B.

EXERCISE

13. **Other Famous Curves.** Explore the definitions of and uses for the following curves. Plot each one using a graphing calculator or software.

    a) tractrix (also known as the *pull curve*)
    b) astroid
    c) epicycloid
    d) witch of Agnesi

Differential Equations

8

Chapter Snapshot

What You'll Learn

8.1 Differential Equations
8.2 Separable Differential Equations
8.3 Applications: Inhibited Growth Models
8.4 First-Order Linear Differential Equations
8.5 Higher-Order Differential Equations and a Trigonometry Connection

Why It's Important

The rate at which a bank account grows is proportional to the amount of money in the account. This is one of many situations that can be modeled by a *differential equation*—an equation that includes a derivative and has a function as a solution. The rate of change of one quantity is often proportional to the amount of that quantity present, and differential equations can be used to model such situations in the sciences and in business.

Where It's Used

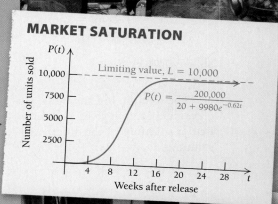

VIDEO GAME SALES

A new Roxtar video game is released in a test market with 10,000 potential customers. Sales increase as more people buy the game, and the rate of sales is proportional to both the number of units sold and the remaining room for sales in the test market. When will sales reach 70% of the test market? When are sales increasing the fastest? We use differential equations to answer questions such as these.

This problem appears as Example 1 in Section 8.3.

MARKET SATURATION

Limiting value, $L = 10{,}000$

$$P(t) = \frac{200{,}000}{20 + 9980e^{-0.62t}}$$

Number of units sold

$P(t)$
10,000
7500
5000
2500

4   8   12   16   20   24   28 t

Weeks after release

Differential Equations

In Chapter 3, we studied a very important equation:

$$\frac{d}{dt}P(t) = k \cdot P(t) \qquad \text{or} \qquad \frac{dP}{dt} = kP.$$

We saw that the solution of this equation is

$$P(t) = P_0 e^{kt},$$

where P_0 is the initial quantity at time $t = 0$. Equations such as $dP/dt = kP$ are called *differential equations.*

OBJECTIVES

• Determine general and particular solutions of a differential equation.

• Verify that a function is a solution of a differential equation.

• Solve differential equations using the uninhibited growth model.

> **DEFINITION**
>
> A **differential equation** is an equation that includes a derivative and has a function as a solution.

The motion of waves can be represented by differential equations.

In differential equations, P and $P(t)$ are often used interchangeably, as are dP/dt and $P'(t)$.

Solving Certain Differential Equations

In this chapter, we will frequently use the notation y' for a derivative. That is, if $y = f(x)$, then

$$y' = \frac{dy}{dx} = f'(x).$$

We find solutions of certain differential equations when we find their antiderivatives, which are indefinite integrals. For example, the differential equation

$$\frac{dy}{dx} = f(x), \qquad \text{or} \qquad y' = f(x),$$

has the solution

$$y = \int f(x)\, dx = F(x) + C, \text{ where } \frac{d}{dx}F(x) = f(x).$$

Recall from Chapter 4 that an indefinite integral results in a set of functions, so there are infinitely many solutions. The set of all functions that solve a differential equation is called the **general solution**.

■ **EXAMPLE 1** Find the general solution of $y' = 2x$.

Solution We find the solution by integrating both sides of the equation:

$$y = \int 2x\, dx = x^2 + C.$$

❬ Quick Check 1

> 〉**Quick Check 1**
>
> Find the general solution of $y' = 3x^2 - x$.

The solution in Example 1, $y = x^2 + C$, is a general solution because substituting *all* values of C in it gives *all* of the solutions. Substituting specific values of C gives **particular solutions** of the differential equation. For example, the following are particular solutions of $y' = 2x$:

$$y = x^2 + 3, \quad y = x^2, \quad y = x^2 - 3.$$

The graph shows the curves of these few particular solutions. The general solution can be regarded as the set of all particular solutions.

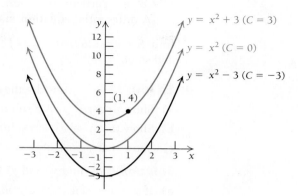

Exploratory

For $y' = 3x^2$, write the general solution. Then use a graphing calculator, iPlot, or Graphicus to graph the particular solutions for $C = -2$, $C = 0$, and $C = 1$.

Knowing the value of the function at a particular point may allow us to find a particular solution from the general solution. For example, if we are told that the solution in Example 1 must include the ordered pair $(1, 4)$, then a particular solution is $y = x^2 + 3$, since $4 = 1^2 + 3$. The requirement that the ordered pair $(1, 4)$ must be included is called an **initial condition**.

Verifying Solutions

To verify that a function is a solution of a differential equation, we find the necessary derivatives and substitute. For example, we can verify that $P(t) = P_0 e^{kt}$ is a solution of the differential equation $dP/dt = kP$. To do so, we find the derivative of P, which is $P'(t) = P_0 k e^{kt}$. We then substitute $P_0 k e^{kt}$ for dP/dt and $P_0 e^{kt}$ for P. Since $P_0 k e^{kt} = k \cdot P_0 e^{kt}$ is true, we know that $P(t) = P_0 e^{kt}$ is a solution of $dP/dt = kP$.

The solution of the differential equation $\dfrac{dP}{dt} = kP$ is

$$P(t) = P_0 e^{kt}. \tag{1}$$

■ **EXAMPLE 2** Consider the differential equation $y' = 2xy$.

a) Show that $y = e^{x^2}$ is a solution of this differential equation.

b) Show that $y = Ce^{x^2}$ is a solution, where C is a constant.

Solution For both parts (a) and (b), we find y' and then substitute.

a) If $y = e^{x^2}$, then $y' = 2xe^{x^2}$. Substituting, we have

$$\frac{y' - 2xy \overset{?}{=} 0}{2xe^{x^2} - 2x \cdot e^{x^2} \quad \Big|\quad 0}$$
$$0 \;\Big|\; 0 \quad \text{TRUE}$$

Thus, $y = e^{x^2}$ is a solution of $y' - 2xy = 0$.

b) If $y = Ce^{x^2}$, then $y' = 2Cxe^{x^2}$. Substituting, we have

$$\frac{y' - 2xy \overset{?}{=} 0}{2Cxe^{x^2} - 2x \cdot Ce^{x^2} \quad \Big|\quad 0}$$
$$0 \;\Big|\; 0 \quad \text{TRUE}$$

❭ **Quick Check 2**

Let $y' - 2y = 0$.

a) Show that $y = e^{2x}$ is a solution of this differential equation.

b) Show that $y = Ce^{2x}$ is a solution.

❮ Quick Check 2

In Example 2, $y = Ce^{x^2}$ is the general solution of $y' - 2xy = 0$, meaning that all solutions of this differential equation can be written in the form $y = Ce^{x^2}$. The solution $y = e^{x^2}$ is a particular solution, where $C = 1$.

A differential equation may include second- or higher-order derivatives. Recall from Section 1.8 that if $y = f(x)$, then $y' = \dfrac{d}{dx}f(x) = f'(x)$, $y'' = \dfrac{d}{dx}f'(x) = f''(x)$, and so on.

■ **EXAMPLE 3**   Consider the differential equation $y'' - 4y' + 3y = 0$.

a) Show that $y = e^x$ is a solution.

b) Show that $y = e^{3x}$ is a solution.

c) Show that $y = C_1 e^x + C_2 e^{3x}$ is a solution, where C_1 and C_2 are constants.

Solution   We find y' and y'' for each function and then substitute.

a) If $y = e^x$, then $y' = e^x$ and $y'' = e^x$. Substituting, we have

$$
\begin{array}{c|cc}
y'' - 4y' + 3y \overset{?}{=} 0 & \\
\hline
e^x - 4 \cdot e^x + 3 \cdot e^x & 0 \\
e^x(1 - 4 + 3) & 0 \\
0 & 0 & \text{TRUE}
\end{array}
$$

b) If $y = e^{3x}$, then $y' = 3e^{3x}$ and $y'' = 9e^{3x}$. Substituting, we have

$$
\begin{array}{c|cc}
y'' - 4y' + 3y = 0^? & \\
\hline
9e^{3x} - 4 \cdot 3e^{3x} + 3 \cdot e^{3x} & 0 \\
e^{3x}(9 - 12 + 3) & 0 \\
0 & 0 & \text{TRUE}
\end{array}
$$

c) If $y = C_1 e^x + C_2 e^{3x}$, then $y' = C_1 e^x + 3C_2 e^{3x}$ and $y'' = C_1 e^x + 9C_2 e^{3x}$. Substituting, we have

$$
\begin{array}{c|cc}
y'' - 4y' + 3y \overset{?}{=} 0 & \\
\hline
(C_1 e^x + 9C_2 e^{3x}) - 4(C_1 e^x + 3C_2 e^{3x}) + 3(C_1 e^x + C_2 e^{3x}) & 0 \\
C_1 e^x + 9C_2 e^{3x} - 4C_1 e^x - 12C_2 e^{3x} + 3C_1 e^x + 3C_2 e^{3x} & 0 \\
C_1 e^x(1 - 4 + 3) + C_2 e^{3x}(9 - 12 + 3) & 0 \\
0 & 0 & \text{TRUE}
\end{array}
$$

❯ **Quick Check 3**

Show that $y = e^{-8x}$ and $y = e^{2x}$ are solutions of $y'' + 6y' - 16y = 0$. Then show that $y = C_1 e^{-8x} + C_2 e^{2x}$ is a solution of $y'' + 6y' - 16y = 0$.

❮ Quick Check 3

An Application: The Uninhibited Growth Model

Recall from Section 3.3 that $dP/dt = kP$, with $k > 0$, is called the **uninhibited growth model** and is used for situations in which growth continues forever, unaffected by outside constraints. The equation $dP/dt = kP$ is read "the rate of change of P is directly proportional to the amount P." Thus, as P increases, the rate of change dP/dt also increases. In simple terms, the larger P is, the faster P grows. The **constant of proportionality** k is also referred to as the **continuous growth rate**.

■ EXAMPLE 4   **Business: Growth of a Savings Account.**   Suppose Donald deposits $1000 into a savings account that earns interest at a rate of 4.5%, compounded continuously. Find the particular solution of the differential equation that describes the growth in value of this account.

Solution   We let $A(t)$ represent the value of Donald's account after t years. The interest rate of 4.5% is expressed as $k = 0.045$, and at $t = 0$, we have $A_0 = \$1000$. We have

$$\frac{dA}{dt} = kA$$

$$A(t) = A_0 e^{kt} \qquad \text{Using equation (1)}$$

$$A(t) = 1000 e^{0.045t} \qquad \text{Substituting}$$

Thus, the particular solution is $A(t) = 1000 e^{0.045t}$. The graph of $A(t)$ shows the relationship between the value of A at time t and the corresponding rate of change of A. Note that as $A(t)$ increases in value, so does the rate $A'(t)$.

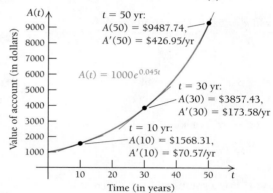

TECHNOLOGY CONNECTION 〰

Exploratory

Divide the value of $A'(t)$ by the value of $A(t)$ for each labeled point in the graph with Example 4. What is the result of each division?

❯ Quick Check 4

Find the particular solution to the differential equation $dA/dt = kA$, when $k = 0.032$ and $A_0 = \$2000$.

❮ Quick Check 4

Section Summary

- A *differential equation* is an equation that includes a derivative and has a function as a solution.
- The *general solution* of a differential equation of the form $y' = f(x)$ is a set of functions $F(x) + C$, where $\dfrac{d}{dx} F(x) = f(x)$. For many differential equations, the general solution is of the form $y = Cf(x)$ where C is any constant.
- A requirement that a solution of a differential equation include a specific ordered pair is called an *initial condition*. If an initial condition is given, then a *particular solution* in which the value of C is determined may be found.

- The differential equation $dP/dt = kP$ is called the *uninhibited growth model*, and its general solution is $P(t) = P_0 e^{kt}$, where P_0 is the initial quantity at time $t = 0$. The value k is the *constant of proportionality*, also known as the *continuous growth rate*.

EXERCISE SET
8.1

In Exercises 1–6, find the general solution and three particular solutions.

1. $y' = 5x^4$

2. $y' = 6x^5$

3. $y' = e^{2x} + x$

4. $y' = e^{4x} - x + 2$

5. $y' = \dfrac{8}{x} - x^2 + x^5$

6. $y' = \dfrac{3}{x} + x^2 - x^4$

7. Show that $y = x \ln x + 3x - 2$ is a solution of

$$y'' - \frac{1}{x} = 0.$$

8. Show that $y = x \ln x - 5x + 7$ is a solution of

$$y'' - \frac{1}{x} = 0.$$

9. Show that $y = e^x + 3xe^x$ is a solution of

$$y'' - 2y' + y = 0.$$

10. Show that $y = -2e^x + xe^x$ is a solution of

$$y'' - 2y' + y = 0.$$

11. Let $y' + 4y = 0$.
   a) Show that $y = e^{-4x}$ is a solution of this differential equation.
   b) Show that $y = Ce^{-4x}$ is a solution, where C is a constant.

12. Let $y' - 3x^2 y = 0$.
   a) Show that $y = e^{x^3}$ is a solution of this differential equation.
   b) Show that $y = Ce^{x^3}$ is a solution, where C is a constant.

13. Let $y'' - y' - 30y = 0$.
   a) Show that $y = e^{6x}$ is a solution of this differential equation.
   b) Show that $y = e^{-5x}$ is a solution.
   c) Show that $y = C_1 e^{6x} + C_2 e^{-5x}$ is a solution, where C_1 and C_2 are constants.

14. Let $y'' - 7y' - 44y = 0$.
   a) Show that $y = e^{11x}$ is a solution of this differential equation.
   b) Show that $y = e^{-4x}$ is a solution.
   c) Show that $y = C_1 e^{11x} + C_2 e^{-4x}$ is a solution, where C_1 and C_2 are constants.

In Exercises 15–22, (a) find the general solution of each differential equation, and (b) check the solution by substituting into the differential equation.

15. $\dfrac{dM}{dt} = 0.05M$

16. $\dfrac{dC}{dt} = 0.66C$

17. $\dfrac{dR}{dt} = 0.35R$

18. $\dfrac{dV}{dt} = 1.33V$

19. $\dfrac{dG}{dt} = 0.005G$

20. $\dfrac{dh}{dt} = 0.023h$

21. $\dfrac{dR}{dt} = R$

22. $\dfrac{dQ}{dt} = 2Q$

In Exercises 23–34, (a) find the particular solution of each differential equation as determined by the initial condition, and (b) check the solution by substituting into the differential equation.

23. $y' = x^2 + 2x - 3;$ $y = 4$ when $x = 0$

24. $y' = 3x^2 - x + 5;$ $y = 6$ when $x = 0$

25. $f'(x) = x^{2/3} - x;$ $f(1) = -6$

26. $f'(x) = x^{2/5} + x;$ $f(1) = -7$

27. $\dfrac{dB}{dt} = 0.03B$, where $B(0) = 500$

28. $\dfrac{dG}{dt} = 0.75G$, where $G(0) = 2000$

29. $\dfrac{dS}{dt} = 0.12S$, where $S = 750$ when $t = 0$

30. $\dfrac{dL}{dt} = 0.68L$, where $L = 1200$ when $t = 0$

31. $\dfrac{dT}{dt} = 0.015T$, where $T = 50$ when $t = 0$

32. $\dfrac{dP}{dt} = 0.024P$, where $P = 32$ when $t = 0$

33. $\dfrac{dM}{dt} = M$, where $M = 6$ when $t = 0$

34. $\dfrac{dN}{dt} = 3N$, where $N = 3.5$ when $t = 0$

APPLICATIONS

Business and Economics

35. Growth of an account. Debra deposits $A_0 = \$500$ into an account that earns interest at a rate of 3.75%, compounded continuously.
   a) Write the differential equation that represents $A(t)$, the value of Debra's account after t years.
   b) Find the particular solution of the differential equation from part (a).
   c) Find $A(5)$ and $A'(5)$.
   d) Find $A'(5)/A(5)$, and explain what this number represents.

36. Growth of an account. Jennifer deposits $A_0 = \$1200$ into an account that earns 4.2% compounded continuously.

a) Write the differential equation that represents $A(t)$, the value of Jennifer's account after t years.

b) Find the particular solution of the differential equation from part (a).

c) Find $A(7)$ and $A'(7)$.

d) Find $A'(7)/A(7)$, and explain what this number represents.

37. Growth of an account. Lily deposits A_0 into an account that earns 2.8% compounded continuously.

a) Find the particular solution, $A(t)$, the value of Lily's account after t years, in terms of A_0.

b) Find $A(4)$ and $A'(4)$.

c) Find $A'(4)/A(4)$, and explain what this number represents.

d) What happens to A_0 in part (c)? Based on part (c), does the initial quantity A_0 have an effect on the continuous growth rate k?

38. Growth of an account. Dex deposits A_0 into an account that earns 2.25% compounded continuously.

a) Find the particular solution, $A(t)$, the value of Dex's account after t years, in terms of A_0.

b) Find $A(6)$ and $A'(6)$.

c) Find $A'(6)/A(6)$, and explain what this number represents.

d) What happens to A_0 in part (c)? Based on part (c), does the initial quantity A_0 have an effect on the continuous growth rate k?

General Interest

39. Growth of a population. The city of New River had a population of 17,000 in 2002 ($t = 0$) with a continuous growth rate of 1.75% per year.

a) Write the differential equation that represents $P(t)$, the population of New River after t years.

b) Find the particular solution of the differential equation from part (a).

c) Find $P(10)$ and $P'(10)$.

d) Find $P'(10)/P(10)$, and explain what this number represents.

40. Growth of a population. An initial population of 70 bacteria is growing continuously at a rate of 2.5% per hour.

a) Write the differential equation that represents $P(t)$, the population of bacteria after t hours.

b) Find the particular solution of the differential equation from part (a).

c) Find $P(24)$ and $P'(24)$.

d) Find $P'(24)/P(24)$, and explain what this number represents.

41. Growth of a population. Before 1859, rabbits did not exist in Australia. That year, a settler released 24 rabbits into the wild. Without natural predators, the growth of the Australian rabbit population can be modeled by the uninhibited growth model $dP/dt = kP$, where $P(t)$ is the population of rabbits t years after 1859. (*Source:* www .dpi.vic.gov.au/agriculture.)

a) When the rabbit population was estimated to be 8900, its rate of growth was about 2630 rabbits per year. Use this information to find k, and then find the particular solution of the differential equation.

b) Find the rabbit population in 1900 ($t = 41$) and the rate at which the rabbit population was increasing in that year.

c) Without using a calculator, find $P'(41)/P(41)$.

42. Growth of a population. Suppose 30 sparrows are released into a region where they have no natural predators. The growth of the region's sparrow population can be modeled by the uninhibited growth model $dP/dt = kP$, where $P(t)$ is the population of sparrows t years after their initial release.

a) When the sparrow population was estimated to be 12,500, its rate of growth was about 1325 sparrows per year. Use this information to find k, and then find the particular solution of the differential equation.

b) Find the number of sparrows after 70 yr.

c) Without using a calculator, find $P'(70)/P(70)$.

SYNTHESIS

43. The amount of money, $A(t)$, in John's savings account after t years is modeled by the differential equation $dA/dt = 0.0325A$.

a) What is the continuous growth rate?

b) Find the particular solution, $A(t)$, if John's account was worth \$2582.58 after 1 yr.

c) Find the amount that John deposited initially.

44. The amount of money, $A(t)$, in Ina's savings account after t years is modeled by the differential equation $dA/dt = 0.0418A$.

a) What is the continuous growth rate?

b) Find the particular solution, $A(t)$, if Ina's account was worth \$3479.02 after 2 yr.

c) Find the amount that Ina deposited initially.

45. Explain the difference between a constant rate of growth and a constant percentage rate of growth.

46. What function is also its own derivative? Write a differential equation for which this function is a solution. Are there any other solutions to this differential equation? Why or why not?

TECHNOLOGY CONNECTION

47. Charlie deposited a sum of money in a savings account. After 1 yr, the account was worth $4467.90, and after 3 yr, the account was worth $4937.80.

    a) Use regression to find an exponential function of the form $y = ae^{bt}$ that models this situation.

    b) Write a differential equation in the form $dA/dt = kA$, including the initial condition at time $t = 0$, to model this situation.

8.2 Separable Differential Equations

OBJECTIVES

- Solve differential equations using separation of variables.
- Solve applied problems using separation of variables.

Consider the differential equation

$$\frac{dy}{dx} = 2xy. \tag{1}$$

We treat dy/dx as a quotient, as in Sections 2.6 and 4.5. Multiplying equation (1) by dx and then by $1/y$, we get

$$\frac{1}{y}\,dy = 2x\,dx, \quad \text{or} \quad \frac{dy}{y} = 2x\,dx, \quad y \neq 0. \tag{2}$$

We have separated the variables in equation (2), meaning that all the expressions involving one variable are on one side of the equation and all the expressions involving the other variable are on the other. A differential equation in which the variables can be separated is called a *separable differential equation*.

DEFINITION

A **separable differential equation** is a differential equation that can be written in the form

$$f(y)\,dy = g(x)\,dx.$$

    Equation (1) is an example of a separable differential equation, and equation (2) is equivalent to equation (1), but with the variables separated. To find a solution, we integrate both sides of equation (2):

$$\int \frac{dy}{y} = \int 2x\,dx$$

$$\ln|y| = x^2 + C.$$

We use only one constant because the two antiderivatives differ by, at most, a constant. Recall that the definition of logarithms says that if $\log_a b = t$, then $b = a^t$. Since $\ln|y| = \log_e|y| = x^2 + C$, we have

$$|y| = e^{x^2 + C}, \quad \text{or} \quad y = \pm e^{x^2} \cdot e^C.$$

At the 1968 Olympic Games in Mexico City, Bob Beamon made a miraculous long jump of 29 ft, $2\frac{1}{2}$ in. Many believed that the jump's record length was due to the altitude, which was 7400 ft. Using differential equations for analysis, M. N. Bearley refuted the altitude theory in "The Long Jump Miracle of Mexico City" (Mathematics Magazine, Vol. 45, 241–246 (November 1972)). Bearley argues that the world-record jump was a result of Beamon's exceptional speed (9.5 sec in the 100-yd dash) and the fact that he hit the take-off board in perfect position.

Thus, the solution of equation (1) is

$$y = C_1 e^{x^2}, \quad \text{where } C_1 = \pm e^C.$$

Because C_1 is also an arbitrary constant, we can drop the subscript:

$$y = C e^{x^2}.$$

As a check, we have $y' = \dfrac{dy}{dx} = 2Cxe^{x^2}$. Substituting y' and y into equation (1), we have

$$\dfrac{dy}{dx} \overset{?}{=} 2xy$$

$2Cxe^{x^2}$	$2x \cdot Ce^{x^2}$
$2Cxe^{x^2}$	$2Cxe^{x^2}$

TRUE

■ **EXAMPLE 1**   Find the particular solution of $3y^2 \dfrac{dy}{dx} + x = 0$, if $y = 5$ when $x = 0$.

Solution   We first separate the variables:

$$3y^2 \dfrac{dy}{dx} = -x \qquad \text{Adding } -x \text{ to both sides}$$

$$3y^2 \, dy = -x \, dx. \qquad \text{Multiplying both sides by } dx$$

We then integrate both sides:

$$\int 3y^2 \, dy = \int -x \, dx$$

$$y^3 = -\dfrac{x^2}{2} + C$$

$$y^3 = C - \dfrac{x^2}{2}. \tag{3}$$

We are given that $y = 5$ when $x = 0$, so we substitute to find C:

$$(5)^3 = C - \dfrac{(0)^2}{2} \qquad \text{Substituting}$$

$$125 = C.$$

Using this value for C and then taking the cube root of both sides of equation (3), we have the particular solution:

$$y = \sqrt[3]{125 - \dfrac{x^2}{2}}.$$

❭ **Quick Check 1**

Solve $2\sqrt{y}\,\dfrac{dy}{dx} - 3x = 0$, if $y = 9$ when $x = 0$.

❬ Quick Check 1

■ **EXAMPLE 2**   Find the general solution of $\dfrac{dy}{dx} = \dfrac{x}{y}$.

Solution   We first separate the variables:

$$y \dfrac{dy}{dx} = x \qquad \text{Multiplying both sides by } y$$

$$y \, dy = x \, dx. \qquad \text{Multiplying both sides by } dx$$

We then integrate both sides:

$$\int y \, dy = \int x \, dx$$

$$\frac{y^2}{2} = \frac{x^2}{2} + C_1$$

$$y^2 = x^2 + 2C_1$$

$$y^2 = x^2 + C. \qquad C = 2C_1$$

There are two general solutions:

$$y = \sqrt{x^2 + C} \quad \text{or} \quad y = -\sqrt{x^2 + C}.$$

Depending on any given initial condition, we choose one or the other of these solutions. For example, if $y = -1$ when $x = 0$, then we choose $y = -\sqrt{x^2 + C}$, since this general solution allows y to be negative. In this case, the particular solution is $y = -\sqrt{x^2 + 1}$.

> **Quick Check 2**
>
> Solve $y' = x^2 y$.

❰ Quick Check 2

■ **EXAMPLE 3**   Solve: $y' = x - xy$.

Solution   Before separating variables, we replace y' with dy/dx:

$$\frac{dy}{dx} = x - xy.$$

Then we separate the variables:

$$dy = (x - xy) \, dx$$

$$dy = x(1 - y) \, dx$$

$$\frac{dy}{1 - y} = x \, dx.$$

Next, we integrate both sides:

$$\int \frac{dy}{1 - y} = \int x \, dx$$

$$-\ln|1 - y| = \frac{x^2}{2} + C_1$$

$$\ln|1 - y| = -\frac{x^2}{2} - C_1$$

$$|1 - y| = e^{-x^2/2 - C_1} \qquad \text{Writing an equivalent exponential equation}$$

$$1 - y = \pm e^{-x^2/2 - C_1}$$

$$1 - y = \pm e^{-x^2/2} e^{-C_1} \qquad \text{Recalling that } a^{x+y} = a^x a^y$$

Since C_1 is an arbitrary constant, $\pm e^{-C_1}$ is also an arbitrary constant. Thus, we can replace $\pm e^{-C_1}$ with C_2. Since $\pm e^{-C_1}$ is never zero, C_2 cannot be zero.

$$1 - y = C_2 e^{-x^2/2}$$

$$-y = C_2 e^{-x^2/2} - 1$$

$$y = 1 - C_2 e^{-x^2/2}. \qquad \text{Multiplying both sides by } -1$$

If we replace $-C_2$ with C, we can replace the subtraction with addition. The general solution is

> **Quick Check 3**
>
> Solve
> $$(1 + x^2)y' = xy.$$

$$y = 1 + Ce^{-x^2/2}, \quad \text{where } C \neq 0.$$

❰ Quick Check 3

Exploratory

Solutions to Differential Equations

At www.wolframalpha.com, we can solve differential equations and graph families of solutions. When given an initial condition, we can find the particular solution. For example, to solve $y' = x - xy$ (Example 3), we key in y'=x-xy and press **ENTER**. Initial conditions are entered in the form y(a)=b and separated from the differential equation by a comma.

An Application to Economics: Elasticity

■ **EXAMPLE 4**   Suppose that for a certain product, the elasticity of demand (see Section 3.6) is 1 for all prices $x > 0$. That is, $E(x) = 1$ for $x > 0$. Find the demand function, $q = D(x)$.

Solution   Recall that elasticity of demand is given by

$$E(x) = -\frac{xD'(x)}{D(x)}.$$

Since $E(x) = 1$ for all $x > 0$,

$$1 = -\frac{xD'(x)}{D(x)} \qquad \text{Setting } E(x) \text{ equal to 1}$$

$$1 = -\frac{x}{q} \cdot \frac{dq}{dx} \qquad \text{Letting } D(x) = q \text{ and } D'(x) = \frac{dq}{dx}$$

$$-\frac{q}{x} = \frac{dq}{dx} \qquad \text{Multiplying both sides by } -\frac{q}{x}$$

$$\frac{dx}{x} = -\frac{dq}{q} \qquad \text{Separating variables}$$

$$\int \frac{dx}{x} = -\int \frac{dq}{q} \qquad \text{Integrating both sides}$$

$$\ln x = -\ln q + C_1 \qquad \text{Both the price, } x, \text{ and the quantity, } q, \text{ are assumed to be positive.}$$

$$\ln x + \ln q = C_1$$

$$\ln(xq) = C_1 \qquad \text{Using a property of logarithms}$$

$$xq = e^{C_1}. \qquad \text{Rewriting as an exponential equation}$$

We let $C = e^{C_1}$ so that $xq = C$:

$$q = \frac{C}{x} \quad \text{and} \quad x = \frac{C}{q}.$$

This result characterizes demand functions for which the elasticity is always 1.

❰ Quick Check 4

❱ **Quick Check 4**

Find the demand function $q = D(x)$ if the elasticity of demand is $E(x) = x$.

An Application to Psychology: Reaction to a Stimulus

In psychology, one model of stimulus-response asserts that the rate of change dR/dS of the reaction R with respect to a stimulus S is inversely proportional to the intensity of the stimulus. That is,

$$\frac{dR}{dS} = \frac{k}{S},$$

where k is a positive constant. This is known as the *Weber-Fechner Law*. To solve this equation, we separate the variables and then integrate both sides:

$$dR = k \cdot \frac{dS}{S}$$

$$\int dR = k \int \frac{dS}{S}$$

$$R = k \ln S + C. \qquad \text{We assume } S > 0.$$

(4)

Now suppose that we let S_0 be the lowest level of the stimulus that can be detected. This is the *threshold value*, or the *detection threshold*. For example, the detection threshold for light is the flame of a candle 30 miles away on a clear, dark night. If S_0 is the lowest level of the stimulus that can be detected, we can assume that $R(S_0) = 0$. Substituting this condition into equation (4), we have

$$0 = k \ln S_0 + C, \quad \text{or} \quad -k \ln S_0 = C.$$

Replacing C in equation (4) with $-k \ln S_0$ gives us

$$R = k \ln S - k \ln S_0 \qquad \text{As a check, note that } \frac{dR}{dS} = \frac{k}{S}.$$
$$= k(\ln S - \ln S_0).$$

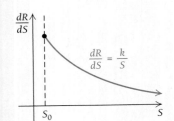

Using a property of logarithms, we have

$$R = k \ln \frac{S}{S_0}.$$

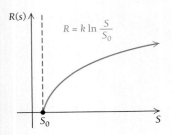

Look at the graphs of dR/dS and R. Note that as the stimulus and the response get larger, the rate of change, dR/dS, decreases. For example, suppose that a lamp has a 50-watt bulb in it. If the bulb were suddenly changed to a 100-watt one, you would probably be very aware of the difference. That is, your response would be strong. If the bulb were then changed to a 150-watt one, your response would increase, but not as dramatically as it did for the change from 50 to 100 watts. A change from a 150-watt to a 200-watt bulb would cause even less increase in response, and so on.

The following table presents some other detection thresholds you may find interesting.

Stimulus	Detection Threshold
Sound	The tick of a watch from 20 feet away in a quiet room
Taste	Water diluted with sugar in the ratio of 1 teaspoon to 2 gallons
Smell	One drop of perfume diffused into the volume of three average-size rooms
Touch	The wing of a bee dropped on your cheek at a distance of 1 centimeter (about $\frac{3}{8}$ of an inch)

Section Summary

- A *separable differential equation* is a differential equation for which the variables can be separated to opposite sides of the equals sign. That is, a separable differential equation can be written as $f(y)\,dy = g(x)\,dx$.

- To solve a separable differential equation, the variables are separated, then each side is integrated, and, if possible, the result is solved for one of the variables.

EXERCISE SET
8.2

Solve by separating variables.

1. $\dfrac{dy}{dx} = 4x^3 y$

2. $\dfrac{dy}{dx} = 5x^4 y$

3. $3y^2 \dfrac{dy}{dx} = 8x$

4. $3y^2 \dfrac{dy}{dx} = 5x$

5. $\dfrac{dy}{dx} = \dfrac{2x}{y}$

6. $\dfrac{dy}{dx} = \dfrac{x}{2y}$

7. $\dfrac{dy}{dx} = \dfrac{6}{y}$

8. $\dfrac{dy}{dx} = \dfrac{7}{y^2}$

Solve for y.

9. $y' = 3x + xy;$ $y = 5$ when $x = 0$

10. $y' = 2x - xy;$ $y = 9$ when $x = 0$

11. $y' = 5y^{-2};$ $y = 3$ when $x = 2$

12. $y' = 7y^{-2};$ $y = 3$ when $x = 1$

In Exercises 13–18, (a) write a differential equation that models the situation, and (b) find the general solution. If an initial condition is given, find the particular solution. Recall that when y is directly proportional to x, we have $y = kx$, and when y is inversely proportional to x, we have $y = k/x$, where k is the constant of proportionality. In these exercises, let $k = 1$.

13. The rate of change of y with respect to x is directly proportional to the square of y.

14. The rate of change of y with respect to x is directly proportional to the cube of y.

15. The rate of change of y with respect to x is inversely proportional to the cube of y.

16. The rate of change of y with respect to x is inversely proportional to the square root of y.

17. The rate of change of y with respect to x is directly proportional to the product of x and y, and $y = 3$ when $x = 2$.

18. The rate of change of y with respect to x is directly proportional to the quotient of x divided by y, and $y = 2$ when $x = -1$.

APPLICATIONS

Business and Economics

19. Capital expansion. *Domar's capital expansion model is*

$$\frac{dI}{dt} = hkI.$$

where I is the investment, h is the investment productivity (constant), k is the marginal productivity to the consumer (constant), and t is the time.

a) Use separation of variables to solve the differential equation.

b) Rewrite the solution in terms of the condition $I_0 = I(0)$.

20. Total profit from marginal profit. Hanna's Hat Company's marginal profit, P, as a function of its total cost, C, is given by

$$\frac{dP}{dC} = \frac{-200}{(C + 3)^{3/2}}.$$

a) Find the profit function, $P(C)$, if $P = \$10$ when $C = \$61$.

b) At what cost will the firm break even $(P = 0)$?

21. Stock growth. The value of a share of stock of Leslie's Designs, Inc., is modeled by

$$\frac{dV}{dt} = k(L - V),$$

where V is the value of the stock, in dollars, after t months; k is a constant; $L = \$24.81$, the *limiting value* of the stock; and $V(0) = 20$. Find the solution of the differential equation in terms of t and k.

22. Utility. The reaction R in pleasure units by a consumer receiving S units of a product can be modeled by the differential equation

$$\frac{dR}{dS} = \frac{k}{S + 1},$$

where k is a positive constant.

a) Use separation of variables to solve the differential equation.

b) Rewrite the solution in terms of the initial condition $R(0) = 0$.

c) Explain why the condition $R(0) = 0$ is reasonable.

Elasticity. *Find the demand function $q = D(x)$, given each set of elasticity conditions.*

23. $E(x) = \dfrac{4}{x};$ $q = 2$ when $x = 4$

24. $E(x) = \dfrac{x}{200 - x};$ $q = 190$ when $x = 10$

25. $E(x) = 2,$ for all $x > 0$

26. $E(x) = n,$ for some constant n and all $x > 0$

Life and Physical Sciences

27. Exponential growth.
a) Use separation of variables to solve the differential-equation model of uninhibited growth,

$$\frac{dP}{dt} = kP.$$

b) Rewrite the solution of part (a) in terms of the condition $P_0 = P(0)$.

Social Sciences

28. The Brentano-Stevens Law. The validity of the Weber–Fechner Law has been the subject of great debate among psychologists. An alternative model,

$$\frac{dR}{dS} = k \cdot \frac{R}{S},$$

where k is a positive constant, has been proposed. Find the general solution of this equation. (This model has also been referred to as the *Power Law of Stimulus–Response*.)

SYNTHESIS

Solve.

29. $\dfrac{dy}{dx} = 5x^4y^2 + x^3y^2$

30. $e^{-1/x} \cdot \dfrac{dy}{dx} = x^{-2} \cdot y^2$

31. What do you feel is the most important use of integration in an application found in this chapter? Why?

32. In Example 3 of this section, it is stated, "Since C is an arbitary constant, $\pm e^{-C}$ is an arbitrary constant." Explain why the $\pm$ sign is necessary.

TECHNOLOGY CONNECTION

33. Solve $dy/dx = 5/y$. Graph the particular solutions for $C_1 = 5, C_2 = -200,$ and $C_3 = 100$.

34. Solve $dy/dx = 2/y^2$. Graph the particular solutions for $C_1 = 0, C_2 = 4,$ and $C_3 = -10$.

> **Answers to Quick Checks**
> **1.** $y = \left(\frac{9}{8}x^2 + 27\right)^{2/3}$ **2.** $y = Ce^{x^3/3}$ **3.** $y = C\sqrt{1 + x^2}$
> **4.** $q = Ce^{-x}$

8.3 Applications: Inhibited Growth Models

The Inhibited Growth Model $dP/dt = kP(L - P)$

OBJECTIVES

- Solve the inhibited growth model $\dfrac{dP}{dt} = kP(L - P)$.
- Solve the inhibited growth model $\dfrac{dP}{dt} = k(L - P)$.
- Solve applied problems involving inhibited growth.

The uninhibited growth model $dP/dt = kP$ (Section 8.1) assumes that no outside factors inhibit the growth of P. A more realistic model for some growth situations takes into account factors other than the size of P. For example, the population of a herd of deer on a small island will be limited by the size of the island and the available food. Similarly, a product's sales, P, may grow quickly at first, but after the market is saturated, growth will be inhibited as fewer people purchase the product. In both of these situations, there is a limiting value, L, of P, where

$$\lim_{t \to \infty} P(t) = L.$$

In such cases, the rate of change dP/dt is directly proportional to both the quantity P and the remaining room for growth, $L - P$. Thus, if P is the value of a quantity at time t and L is the limiting value of the quantity, then the **inhibited growth model** is given by the differential equation

$$\frac{dP}{dt} = kP(L - P),$$

where $k > 0$. The solution of this differential equation is called a **logistic growth model**, an effective model for studying population growth, the spread of an epidemic, market saturation, and other phenomena in the sciences and business.

To determine a general solution of the differential equation $dP/dt = kP(L - P)$, we separate variables, noting that L is a constant:

$$\frac{dP}{dt} = kP(L - P)$$

$$dP = kP(L - P)\,dt \qquad \text{Multiplying both sides by } dt$$

$$\frac{dP}{P(L - P)} = k\,dt \qquad \text{Multiplying both sides by } \frac{1}{P(L - P)}$$

$$\int \frac{dP}{P(L - P)} = k \cdot \int dt \qquad \text{Integrating both sides}$$

$$\frac{1}{L}\ln\left(\frac{P}{L - P}\right) = kt + C. \qquad \text{Using a table of integrals (see Exercise 29)}$$

To solve for P, we multiply both sides by L and isolate $\dfrac{P}{L - P}$:

$$\ln\left(\frac{P}{L - P}\right) = Lkt + LC$$

$$\frac{P}{L - P} = e^{Lkt + LC} \qquad \text{Writing an equivalent exponential equation}$$

$$\frac{P}{L - P} = e^{Lkt}e^{LC} \qquad a^{x+y} = a^x a^y$$

$$\frac{P}{L - P} = C_1 e^{Lkt}. \qquad \text{Letting } C_1 = e^{LC}$$

We now multiply both sides by $L - P$ to clear the fraction and solve for P:

$$P = C_1 e^{Lkt}(L - P)$$

$$P = C_1 L e^{Lkt} - C_1 P e^{Lkt} \qquad \text{Using the distributive law}$$

$$P + C_1 P e^{Lkt} = C_1 L e^{Lkt}$$

$$P(1 + C_1 e^{Lkt}) = C_1 L e^{Lkt} \qquad \text{Factoring}$$

$$P = \frac{C_1 L e^{Lkt}}{1 + C_1 e^{Lkt}}.$$

We can replace C_1 with C. Thus, one form of the general solution of the inhibited growth model is

$$P(t) = \frac{CLe^{Lkt}}{1 + Ce^{Lkt}}. \tag{1}$$

This function is recognizable by the S shape of its graph and is called a *logistic growth model* or an *s-curve*.

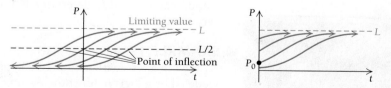

A logistic growth model P has the following characteristics:

- P is an increasing function, where $0 < P(t) < L$.
- P has a horizontal asymptote at L as $t \to \infty$; that is,

$$\lim_{t \to \infty} P(t) = L,$$

where L is the limiting value of P.

- P has a single *point of inflection* at $t = t_i$ such that $P(t_i) = \frac{1}{2}L$ (see Exercise 30). Recall from Section 2.2 that a point of inflection is where the concavity of a function changes. The point of inflection is also called the *point of diminishing returns*, since it is the point where the rate of change of P begins to decrease to zero.

Furthermore, assuming an initial population of P_0 at $t = 0$, we have

$$P_0 = \frac{CLe^{Lk0}}{1 + Ce^{Lk0}} = \frac{CL}{1 + C}. \qquad \text{Recalling that } e^{Lk0} = 1$$

Solving for C, we have

$$P_0(1 + C) = CL \qquad \text{Multiplying both sides by } 1 + C$$
$$P_0 + CP_0 = CL$$
$$P_0 = CL - CP_0 \qquad \text{Subtracting } CP_0 \text{ from both sides}$$
$$P_0 = C(L - P_0) \qquad \text{Factoring}$$
$$C = \frac{P_0}{L - P_0}.$$

Substituting this expression for C in equation (1), we have

$$P(t) = \frac{\left(\dfrac{P_0}{L - P_0}\right)Le^{Lkt}}{1 + \left(\dfrac{P_0}{L - P_0}\right)e^{Lkt}}$$

$$P(t) = \frac{P_0Le^{Lkt}}{(L - P_0) + P_0e^{Lkt}} \qquad \text{Multiplying by } \frac{L - P_0}{L - P_0}$$

$$P(t) = \frac{P_0L}{(L - P_0)e^{-Lkt} + P_0}. \qquad \text{Multiplying by } \frac{e^{-Lkt}}{e^{-Lkt}} \qquad (2)$$

Using a commutative law in the denominator, we have the following:

The general solution of the inhibited growth model $dP/dt = kP(L - P)$ is the logistic growth model

$$P(t) = \frac{P_0L}{P_0 + (L - P_0)e^{-Lkt}},$$

where $P(t)$ is the quantity at time t, L is the limiting value, and P_0 is the initial quantity at time $t = 0$.

The value of k is often stated. However, if an initial condition is known, this information can be used to determine k. For most applications, the value of k tends to be very small.

■ **EXAMPLE 1** **Business: Market Saturation.** Suppose a new Roxtar video game is released in a test market with 10,000 potential customers and the growth in its sales is modeled by a logistic function. Assume that 20 units are sold during the initial release ($t = 0$) and the point of inflection for sales occurs in week 10, when 5000 units have been sold.

a) Find the value of k and the particular solution for $P(t)$, the number of units sold after t weeks.

b) After how many weeks will 8000 units of the Roxtar video game have been sold?

Solution We assume logistic growth with $P(t)$ representing the number of games sold in week t.

a) From the information given, we have $P_0 = 20$, $L = 10,000$, and the point of inflection $(10, 5000)$. These values are substituted into the logistical growth model to solve for k:

$$P(t) = \frac{P_0 L}{P_0 + (L - P_0)e^{-Lkt}}$$ The logistic growth model

$$5000 = \frac{20 \cdot 10,000}{20 + (10,000 - 20)e^{-10,000 \cdot k \cdot 10}}$$ Substituting

$$5000 = \frac{200,000}{20 + 9980e^{-100,000k}}$$ Simplifying

$$5000(20 + 9980e^{-100,000k}) = 200,000$$

$$20 + 9980e^{-100,000k} = 40$$ Dividing both sides by 5000

$$9980e^{-100,000k} = 20$$

$$e^{-100,000k} = \frac{20}{9980}$$

$$-100,000k = \ln\left(\frac{20}{9980}\right)$$ Taking the natural logarithm of both sides

$$k = \frac{\ln(20/9980)}{-100,000} \approx 0.000061.$$

Thus, the particular solution is

$$P(t) = \frac{200,000}{20 + 9980e^{(-10,000)(0.000061)t}} = \frac{200,000}{20 + 9980e^{-0.61t}}.$$

b) To find when 8000 units have been sold, we set $P(t) = 8000$ and solve for t:

$$8000 = \frac{200,000}{20 + 9980e^{-0.61t}}$$

$$8000(20 + 9980e^{-0.61t}) = 200,000$$

$$20 + 9980e^{-0.61t} = 25$$ Dividing both sides by 8000

$$9980e^{-0.61t} = 5$$

$$e^{-0.61t} = \frac{5}{9980}$$

$$-0.61t = \ln\left(\frac{5}{9980}\right)$$ Taking the natural logarithm of both sides

$$t = \frac{\ln(5/9980)}{-0.61} \approx 12.457 \text{ weeks.}$$

❭ **Quick Check 1**

The Basketball Compendium sells 200 copies on its initial release. Its growth in sales is modeled by a logistic function. The point of inflection of sales occurs after 24 weeks, when 30,000 copies have been sold. Use this information to find the logistic growth model for this book's sales, and then find when sales will exceed 50,000 copies.

❮ Quick Check 1

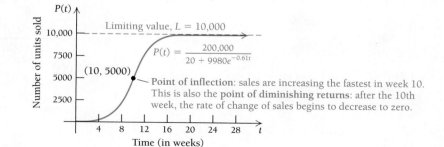

Limiting value, $L = 10,000$

$P(t) = \dfrac{200,000}{20 + 9980e^{-0.61t}}$

$(10, 5000)$

Point of inflection: sales are increasing the fastest in week 10. This is also the **point of diminishing returns**: after the 10th week, the rate of change of sales begins to decrease to zero.

■ **EXAMPLE 2**   **Medicine: Epidemic.**   Thirty students in a dormitory are initially afflicted with a highly communicable cold. The population of the dormitory is 1000, and the sick students remain in the dormitory so that the cold does not spread to the outside. After 1 hr, the number of students afflicted is 38. Assume that the spread of the cold is modeled by a logistic function.

a) Write the differential equation that models the spread of the cold.

b) Find the particular solution for $P(t)$, the number of students afflicted after t hours.

c) When will the cold be spreading the fastest?

Solution

a) We use the inhibited growth model $dP/dt = kP(L - P)$. Since $L = 1000$, the differential equation that models the spread of the cold is

$$\frac{dP}{dt} = kP(1000 - P).$$

b) To find the particular solution, we substitute $P_0 = 30$ and $L = 1000$ into the logistic growth model:

$$P(t) = \frac{P_0 L}{P_0 + (L - P_0)e^{-Lkt}} \qquad \text{The logistic growth model is the solution of the inhibited growth model.}$$

$$= \frac{30 \cdot 1000}{30 + (1000 - 30)e^{-1000kt}} \qquad \text{Substituting}$$

$$= \frac{30{,}000}{30 + 970e^{-1000kt}}.$$

After 1 hr, 38 students are afflicted. We substitute these values and solve for k:

$$38 = \frac{30{,}000}{30 + 970e^{-1000k(1)}} \qquad \text{Substituting}$$

$$38(30 + 970e^{-1000k}) = 30{,}000$$

$$30 + 970e^{-1000k} = \frac{30{,}000}{38} \qquad \text{Dividing both sides by 38}$$

$$970e^{-1000k} = \frac{30{,}000}{38} - 30$$

$$e^{-1000k} = \frac{\left(\dfrac{30{,}000}{38} - 30\right)}{970}$$

$$-1000k = \ln\left[\frac{\left(\dfrac{30{,}000}{38} - 30\right)}{970}\right] \qquad \text{Taking the natural logarithm of both sides}$$

$$k = \frac{\ln\left[\dfrac{\left(\dfrac{30{,}000}{38} - 30\right)}{970}\right]}{-1000}$$

$$\approx 0.000245. \qquad \text{Using a calculator}$$

Therefore, the logistic growth model is

$$P(t) = \frac{30{,}000}{30 + 970e^{-0.245t}}.$$

c) The cold is spreading at its fastest rate at the point of inflection. Since $L = 1000$, the point of inflection occurs when $P = 500$. Thus, we need to determine t such that $P(t) = 500$:

$$500 = \frac{30{,}000}{30 + 970e^{-0.245t}} \qquad \text{Setting } P(t) = 500$$

$$500(30 + 970e^{-0.245t}) = 30{,}000$$

$$30 + 970e^{-0.245t} = 60 \qquad \frac{30{,}000}{500} = 60$$

$$970e^{-0.245t} = 30$$

$$e^{-0.245t} = \frac{30}{970}$$

$$-0.245t = \ln\left(\frac{30}{970}\right) \qquad \text{Taking the natural logarithm of both sides}$$

$$t = \frac{\ln(30/970)}{-0.245} \approx 14.2 \text{ hr.}$$

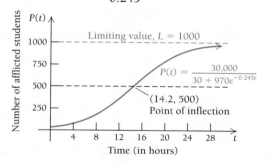

$P(t)$

Number of afflicted students

Limiting value, $L = 1000$

$$P(t) = \frac{30{,}000}{30 + 970e^{-0.245t}}$$

$(14.2, 500)$
Point of inflection

Time (in hours)

⟨ Quick Check 2

The Inhibited Growth Model $dP/dt = k(L - P)$

The inhibited growth model $dP/dt = k(L - P)$ assumes that the growth rate dP/dt is directly proportional only to the remaining room for growth, $L - P$, and furthermore, that $P(0) = 0$. The general solution is determined by separating variables:

$$\frac{dP}{dt} = k(L - P)$$

$$dP = k(L - P)\,dt$$

$$\frac{dP}{L - P} = k\,dt$$

$$\int \frac{dP}{(L - P)} = k \cdot \int dt \qquad \text{Integrating both sides}$$

$$-\ln(L - P) = kt + C \qquad \text{Since } L > P, \text{ using the absolute value is unnecessary.}$$

$$\ln(L - P) = -kt - C$$

$$L - P = e^{-kt-C} \qquad \text{Writing an equivalent exponential equation}$$

$$L - P = e^{-C} \cdot e^{-kt}$$

$$L - P = C_1 e^{-kt}. \qquad \text{Replacing } e^{-C} \text{ with } C_1$$

Solving for P, we have

$$P = L - C_1 e^{-kt}. \tag{3}$$

Since $P(0) = 0$, we can determine C_1:

$$0 = L - C_1 e^{-k(0)}$$
$$0 = L - C_1$$
$$C_1 = L.$$

We substitute L for C_1 in equation (3):

$$P(t) = L - Le^{-kt}, \quad \text{or} \quad P(t) = L(1 - e^{-kt}).$$

The general solution of the inhibited growth model $dP/dt = k(L - P)$ is

$$P(t) = L(1 - e^{-kt}),$$

where $P(t)$ is the quantity at time t and L is the limiting quantity.

The following graph shows growth from the initial point (the origin), with growth leveling off toward L as a horizontal asymptote. There is no point of inflection. Although $P(t)$ increases as t increases, the derivative dP/dt decreases to zero as t increases.

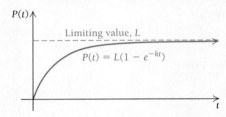

In many applications, growth starts at a point other than the origin. However, we may shift the model so that the initial point is at the origin.

■ **EXAMPLE 3**   **Business: Stock Growth.**   A share of stock for Hanson Vending Machines, Inc., is currently valued at $18. Analysts expect the value to rise but level off at $25 above the starting value. Thus, the rise in value above $18 is modeled by $dV/dt = k(25 - V)$, where t is measured in weeks and $V(0) = 0$, meaning that a share of the stock has not yet risen in value at the time $t = 0$.

a) Find the general solution of this differential equation.

b) Find the particular solution if a share of the stock has increased in value by $22.50 after 6 weeks.

c) When will the stock reach a value of $42 per share?

Solution

a) The general solution is $V(t) = 25(1 - e^{-kt})$, where $L = \$25$ is the limiting value of the rise in the price. Thus, the value of a share will rise no higher than $18 + \$25 = \$43$.

b) We substitute $t = 6$ and $V = 22.50$, and then solve for k:

$$22.50 = 25\left(1 - e^{-k(6)}\right)$$
$$0.9 = 1 - e^{-6k} \qquad 22.50/25 = 0.9$$
$$0.1 = e^{-6k}$$
$$\ln(0.1) = -6k \qquad \text{Taking the natural logarithm of both sides}$$
$$k = \frac{\ln(0.1)}{-6} \approx 0.384.$$

Thus, the particular solution is $V(t) = 25(1 - e^{-0.384t})$.

c) When the value of a share is \$42, it has increased in value by \$24, so we substitute 24 for V and solve for t:

$$24 = 25(1 - e^{-0.384t})$$
$$0.96 = 1 - e^{-0.384t} \qquad \tfrac{24}{25} = 0.96$$
$$0.04 = e^{-0.384t}$$
$$\ln(0.04) = -0.384t \qquad \text{Taking the natural logarithm of both sides}$$
$$t = \frac{\ln(0.04)}{-0.384} \approx 8.38 \text{ weeks.}$$

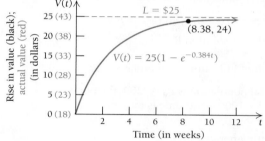

> **Quick Check 3**
>
> A stock's initial share value is \$7, and 6 months later, a share is worth \$12. The value of a share is not expected to rise more than \$10 above the \$7 starting value. Using the inhibited growth model $dV/dt = k(10 - V)$, where t is measured in months, find the particular solution, and then determine when the value of a share will be \$15 (\$8 above its starting value).

❮ Quick Check 3

Section Summary

- The *inhibited growth model* $dP/dt = kP(L - P)$ assumes that the rate of change of a population is directly proportional to both the population P and the room for growth, $L - P$, where L is the limiting population of P as $t \to \infty$.
- The general solution of $dP/dt = kP(L - P)$ is the *logistic growth model*,

$$P(t) = \frac{P_0 L}{P_0 + (L - P_0)e^{-Lkt}},$$

where P_0 is the quantity at $t = 0$. The graph of this function is also called an *s-curve*. The function P is positive

and increasing, with a horizontal asymptote at L, and a point of inflection at $(t_i, P(t_i))$ with $P(t_i) = \dfrac{1}{2}L$.

- Another inhibited growth model, $dP/dt = k(L - P)$, has a general solution

$$P(t) = L(1 - e^{-kt}).$$

This model assumes that the rate of change is proportional only to the room for growth, $L - P$, and that $P(0) = 0$.

For Exercises 1–8, use the inhibited growth model $dP/dt = kP(L - P)$. Give the values of t to three decimal places.

1. Let $L = 5000$, $k = 0.0002$, and $P_0 = 100$.

    a) Find the particular solution P.
    b) Find t when $P = 3500$.
    c) Find the point of inflection as an ordered pair.

2. Let $L = 2000$, $k = 0.0006$, and $P_0 = 300$.

    a) Find the particular solution P.
    b) Find t when $P = 1700$.
    c) Find the point of inflection as an ordered pair.

3. Let $L = 1500$, $k = 0.0002$, and $P_0 = 50$.

    a) Find the particular solution P.
    b) Find t when $P = 1350$.
    c) Find the point of inflection as an ordered pair.

4. Let $L = 10{,}000$, $k = 0.000075$, and $P_0 = 600$.

    a) Find the particular solution P.
    b) Find t when $P = 7500$.
    c) Find the point of inflection as an ordered pair.

5. Suppose the initial population is $P_0 = 200$, the point of inflection is $(15, 1000)$, and time is measured in months.

    a) Find the limiting population L.
    b) Use the result of part (a) to find k; then find the particular solution P.
    c) When will the population reach 900?

6. Suppose the initial population is $P_0 = 1500$, the point of inflection is $(20, 4000)$, and time is measured in weeks.

    a) Find the limiting population L.
    b) Use the result of part (a) to find k; then find the particular solution P.
    c) When will the population reach 7200?

7. Suppose the initial population is $P_0 = 60$, the point of inflection is $(6, 160)$, and time is measured in years.

    a) Find the limiting population L.
    b) Use the result of part (a) to find k; then find the particular solution P.
    c) When will the population reach 300?

8. Suppose the initial population is $P_0 = 78$, the point of inflection is $(22, 575)$, and time is measured in months.

    a) Find the limiting population L.
    b) Use the result of part (a) to find k; then find the particular solution P.
    c) When will the population reach 950?

For Exercises 9–12, use the inhibited growth model $dP/dt = k(L - P)$. Assume that $P(0) = 0$, and give the values of t to three decimal places.

9. Let $L = 3000$, and assume that P passes through the point $(20, 450)$.

    a) Find the particular solution P.
    b) When will $P(t) = 0.9L$?

10. Let $L = 500$, and assume that P passes through the point $(10, 300)$.

    a) Find the particular solution P.
    b) When will $P(t) = 0.7L$?

11. Let $L = 80$, and assume that P passes through the point $(12, 40)$.

    a) Find the particular solution P.
    b) When will $P(t) = 0.8L$?

12. Let $L = 12{,}000$, and assume that P passes through the point $(60, 7000)$.

    a) Find the particular solution P.
    b) When will $P(t) = 0.6L$?

APPLICATIONS

Life and Physical Sciences

13. Inhibited population growth. The population of rabbits on an island is modeled by $dP/dt = 0.000007P(4500 - P)$, where t is measured in months. Suppose there were 60 rabbits at $t = 0$.

    a) Find the particular solution P.
    b) What is the limiting population of rabbits?
    c) When does the population reach 90% of the limiting population?

14. Inhibited population growth. The population of fireflies in a small field is modeled by $dP/dt = 0.000022P(12{,}500 - P)$, where t is measured in weeks. Suppose there were 50 fireflies at $t = 0$.

    a) Find the particular solution P.
    b) What is the limiting population of fireflies?
    c) When does the population reach 90% of the limiting population?

15. Inhibited population growth: human population. In 1790 ($t = 0$), 17 adults from the British ship HMS *Bounty* came ashore on the uninhabited island of Pitcairn, in the South Pacific. By 1856, the island's human population was slightly less than 200. Let t represent years since 1790, and assume that the island's limiting population is 200. (Interestingly, the island has experienced mass emigration over the years; its current population is about 50.) (*Source:* www.government.pn.)

a) If the population was 66 in 1807, find the particular solution for $P(t)$, assuming that $dP/dt = kP(L - P)$ and $L = 200$.

b) According to the model used in part (a), in what year after 1790 was the island's population growing the fastest?

16. Inhibited population growth: tortoise population. A safe carrying capacity (limiting population) of desert tortoises in the Mojave Desert of Nevada and California is about 150 tortoises per square kilometer. (*Source:* www. deserttortoise.org.)

a) Suppose 20 tortoises are released into an unpopulated parcel of land that is 1 square kilometer in size, and 1 year later, the parcel's population is 24 tortoises. Find the particular solution for $P(t)$, using the model $dP/dt = kP(L - P)$.

b) According to this model, when will the tortoise population be growing the fastest?

17. Memory. When trying to memorize large amounts of information, we tend to memorize a lot fairly readily and then struggle to memorize the remaining information. This process can be modeled by $dM/dt = k(L - M)$, where $M(t)$ is the amount of information memorized at time t and L is the limiting amount of information that can be memorized. Suppose Omar is attempting to memorize 200 integral forms. He knows none at first (hence, $M(0) = 0$), and after 3 hr of study, he has memorized 25 of the forms.

a) Find the particular solution for $M(t)$, the number of integral forms Omar has memorized after t hours.

b) After how many hours will Omar have memorized 120 of the forms?

18. Memory. Ernest is attempting to memorize 100 words for a Spanish vocabulary test. Assume that he knows none of the words when he starts and, after 4 hr, has memorized 40 of the words. Use the same of model of growth as in Exercise 17.

a) Find the particular solution for $M(t)$, the number of vocabulary words Ernest has memorized after t hours.

b) How many hours should Ernest study in order to memorize 90 words and earn an A on the test?

19. Epidemiology. On a cruise ship carrying 2000 passengers, 5 are afflicted with a disease, and 6 hours later, the number afflicted is 25. Use the model $dP/dt = kP(L - P)$.

a) Find the particular solution for $P(t)$, the number of afflicted passengers after t days.

b) When will 80% of the passengers be afflicted?

20. Spread of a rumor. A high school has 1200 students. One day, 8 students start a rumor, and after 10 days, 300 students have heard this rumor. Use the model $dP/dt = kP(L - P)$.

a) Find the particular solution for $P(t)$, the number of students who have heard the rumor after t days.

b) When will half of the students have heard the rumor?

Business and Economics

21. Advertising. A G5 cell phone is released into a test market with 20,000 potential customers. The number of people who have heard of the new product follows the model $dP/dt = kP(L - P)$.

a) If 20 people hear of this phone initially ($t = 0$), and 12 weeks later, 500 people have heard of it, find the particular solution for $P(t)$, the number of people who have heard of the phone after t weeks of advertising.

b) When will 70% of the population have heard of the G5 cell phone?

22. Advertising. In a simplified form of the model $dP/dt = kP(L - P)$, we let $L = 1$ represent 100% of the population.

a) Assume that 2% of a population is aware of the new BodyToner Exercise Bench initially, and after 4 weeks of advertising, that figure is 18%. Find the particular solution for $P(t)$, the percentage of people who have heard of this product after t weeks of advertising.

b) When will 95% of the population have heard of the BodyToner Exercise Bench?

23. Market saturation. Suppose a math-pod is released, and initial sales are 2000 units. After 25 weeks, sales have reached 50,000 units. Use the model $dP/dt = kP(L - P)$, and assume there are 4,000,000 potential customers.

a) Find the particular solution for $P(t)$, the number of people who own a math-pod after t weeks.

b) When are sales of the math-pod increasing the fastest?

c) When will sales reach 3,000,000 units?

24. Market saturation. Initially, suppose that 0.2% of a population has purchased a copy of *Cats Are Funny*. After 6 months, 1.7% of the population has purchased a copy of the book. Let $L = 1$ represent 100% of the population, and use the model $dP/dt = kP(L - P)$.

a) Find the particular solution for $P(t)$, the percentage of people who have purchased the book after t months.

b) When are sales increasing the fastest?

c) When will 75% of the population have purchased a copy of the book?

25. Value of a stock. Shares of Carlson Industries are currently \$5. Analysts expect the stock to rise in price to no more than \$15 per share. Suppose the stock's price is \$6.50 per share after 2 months. Use the model $dP/dt = k(L - P)$.

a) Find the particular solution for $P(t)$, the price per share of the stock after t months.

b) When will a share of the stock reach a price of \$12?

26. Value of a stock. Shares of Oglethorpe Equipment are currently \$20. Analysts expect the stock to rise in price to no more than \$30 per share. Suppose the stock's price is \$24 per share after 5 months. Use the model $dP/dt = k(L - P)$.

a) Find the particular solution for $P(t)$, the price per share of the stock after t months.

b) When will a share of the stock reach a price of \$28?

SYNTHESIS

27. The number of states, $S(t)$, in the United States as a function of time t, in years since 1776, can be modeled by the logistic function $S(t) = \dfrac{50}{1 + 2.85e^{-0.022t}}$. Use this model to answer the following questions.

a) How many states were in the United States in 1836? (The actual number was 25.)

b) How many states were in the United States in 1876? (The actual number was 38.)

c) How many states were in the United States in 1912? (The actual number was 48.)

d) What is the limiting number of states?

e) Suppose a new state is added in the year 2050. Should a new model be found, or would this model still be reasonably accurate?

28. Arizona has 15 counties, and the number of counties as a function of time t, in years since 1864 (the year Arizona became a territory) can be modeled by the logistic function $C(t) = \dfrac{15}{1 + 3.35e^{-0.087t}}$. Use this model to answer the following questions.

a) How many counties did Arizona have in 1893? (The actual number was 12.)

b) How many counties did Arizona have in 1912, the year it became a state? (The actual number was 14.)

c) What is the limiting number of counties?

29. To derive the general solution of the inhibited growth model $dP/dt = kP(L - P)$, we used a table of integrals to justify $\displaystyle\int \frac{dP}{P(L - P)} = \frac{1}{L}\ln\left(\frac{P}{L - P}\right)$. To justify this equation without a table, complete the following steps.

a) Show that
$$\frac{1}{P(L - P)} = \frac{1}{L}\left(\frac{1}{P} + \frac{1}{L - P}\right).$$

b) Integrate both sides of the equation in part (a) with respect to P.

30. Show that the point of inflection for the logistic growth model
$$P(t) = \frac{P_0 L}{P_0 + (L - P_0)e^{-Lkt}}$$
occurs where $P(t) = \frac{1}{2}L$.

TECHNOLOGY CONNECTION

31. The Math Club at San Dimas Technical College starts with 10 members. Two weeks later, there are 20 members, and after 5 weeks, the membership numbers 30.

a) Use a calculator or a spreadsheet's regression feature to find a logistic function that models the growth of the Math Club membership.

b) What is the limiting size of the Math Club?

c) The Math Club meets in a room that holds 50 people. During the 7th week, 10 new people join the club. In what week since the club started will it need a larger room in which to meet? Should a new growth model be determined? Why or why not?

Answers to Quick Checks

1. $P(t) = \dfrac{12{,}000{,}000}{200 + 59800e^{-0.238t}}$; sales reach 50,000 units at $t \approx 31$ weeks   **2. (a)** $P(t) = \dfrac{300{,}000}{50 + 5950e^{-0.18t}}$; **(b)** $t \approx 26.5$ weeks   **3. (a)** $V(t) = 10(1 - e^{-0.0372t})$; **(b)** 43.3 months

8.4

First-Order Linear Differential Equations

Most of the differential equations we have seen thus far are *first order*, meaning that they include a first derivative but no higher-order derivatives. We now consider a specific type of first-order differential equation called *linear*.

> **DEFINITION**
>
> A **first-order linear differential equation** can be written in the form
> $$a_0(x)y' + a_1(x)y = b(x), \tag{1}$$
> where $a_0(x)$, $a_1(x)$, and $b(x)$ are continuous functions of x over some interval $[c, d]$. If $b(x) = 0$, then the differential equation is said to be **homogeneous**.

Equation (1) is a first-order linear differential equation because the highest derivative of y in the equation is its first derivative (first order), y and y' are in separate terms, and both y and y' are raised to the first power (linear). For example, the differential equation $2y' + 3xy = -3$ is both first order and linear. However, the differential equation $(y')^2 + 5y = x^2$ is first order but not linear since the derivative is raised to the second power.

Assuming that $a_0(x) \neq 0$, we can divide by $a_0(x)$ and write equation (1) as

$$y' + p(x)y = q(x),$$

where $p(x) = \dfrac{a_1(x)}{a_0(x)}$ and $q(x) = \dfrac{b(x)}{a_0(x)}$. The equation $y' + p(x)y = q(x)$ is written in *standard form*. When $b(x) = 0$, we have $q(x) = 0$. The resulting equation, $y' + p(x)y = 0$, can be rewritten as $y' = -p(x)y$, and separation of variables can be used to determine y. However, the example that follows shows a faster method for the case where $p(x)$ is constant.

■ **EXAMPLE 1** Find the general solution of $y' - 3y = 0$.

Solution Let's assume $y = Ce^{rx}$ and, therefore, $y' = Cre^{rx}$. We substitute these into the differential equation:

$$Cre^{rx} - 3Ce^{rx} = 0 \qquad \text{Substituting}$$
$$r - 3 = 0. \qquad \text{Dividing both sides by } Ce^{rx}, \text{ where } C \neq 0$$

Solving for r, we have $r = 3$, and the general solution is $y = Ce^{3x}$. We check by substitution:

$$\begin{array}{c|c}
\multicolumn{2}{c}{y' - 3y \overset{?}{=} 0} \\
\hline
3Ce^{3x} - 3(Ce^{3x}) & 0 \\
0 & 0 \quad \text{TRUE}
\end{array}$$

> **Quick Check 1**
>
> Find the general solution of $y' + \frac{1}{2}y = 0$.

❮ Quick Check 1

Integration Factors

When $q(x)$ is not zero, $y' + p(x)y = q(x)$ is **nonhomogeneous**. Unfortunately, separation of variables does not always work for solving nonhomogeneous differential equations. Instead, we can apply another method that uses an **integration factor**, $m(x)$, with the assumption that $m(x) > 0$ over an interval $[c, d]$. We start with the standard form

$$y' + p(x)y = q(x),$$

and multiply both sides by the integration factor $m(x)$:

$$m(x)y' + m(x)p(x)y = m(x)q(x).$$

To determine $m(x)$, note that if $m(x)p(x) = m'(x)$, then the left side of the above equation has the form of the derivative of the product $m(x)y$, since $[m(x)y]' = m(x)y' + m'(x)y$. The requirement that $m'(x) = m(x)p(x)$ can be satisfied by separating the variables:

$$m'(x) = m(x)p(x)$$

$$\frac{m'(x)}{m(x)} = p(x)$$

$$\int \frac{m'(x)}{m(x)} dx = \int p(x) \, dx$$

$$\ln m(x) = \int p(x) \, dx \qquad \text{Assuming } m(x) > 0$$

$$m(x) = e^{\int p(x) \, dx} \qquad \text{Disregarding any constants of integration for now}$$

With $m(x)$ known, we can solve the nonhomogeneous differential equation. Recall that for $m(x)$ as defined above, $m(x)y' + m(x)p(x)y = m(x)q(x)$ can be written as

$$[m(x)y]' = m(x)q(x).$$

Integrating both sides, we have

$$\int [m(x)y]' \, dx = \int m(x)q(x) \, dx.$$

The left side is $m(x)y$. Dividing both sides by $m(x)$ gives us the following result.

> The general solution of $y' + p(x)y = q(x)$ is
>
> $$y = \frac{\int m(x)q(x) \, dx + C}{m(x)}, \quad \text{with } m(x) = e^{\int p(x) \, dx}.$$

The constant of integration C is included in the above result. When an initial condition is given, a particular solution can be found in which a value for C is determined.

■ **EXAMPLE 2** Consider the equation $y' + \dfrac{y}{x} = x^2$, where $x > 0$.

a) Find the general solution.

b) Find the particular solution such that $y = 10$ when $x = 1$.

Solution

a) Writing the equation in the form $y' + p(x)y = q(x)$, we have $y' + \dfrac{1}{x}y = x^2$.

Using $p(x) = \dfrac{1}{x}$, we determine the integrating factor $m(x)$:

$$m(x) = e^{\int p(x) \, dx}$$
$$m(x) = e^{\int \frac{1}{x} dx} \qquad \text{Substituting}$$
$$= e^{\ln x} \qquad \text{Assuming } x > 0$$
$$= x.$$

Therefore, $m(x) = x$. We can now determine the general solution, y:

$$y = \frac{\int m(x)q(x)\,dx + C}{m(x)}$$

$$= \frac{\int x \cdot x^2\,dx + C}{x} \qquad \text{Substituting } m(x) = x \text{ and } q(x) = x^2$$

$$= \frac{\int x^3\,dx + C}{x}$$

$$= \frac{\left(\frac{1}{4}x^4 + C\right)}{x}$$

$$= \frac{x^3}{4} + \frac{C}{x}.$$

The general solution is $y = \dfrac{x^3}{4} + \dfrac{C}{x}$.

b) To find the particular solution for $y = 10$ when $x = 1$, we substitute and solve for C:

$$y = \frac{x^3}{4} + \frac{C}{x}$$

$$10 = \frac{1^3}{4} + \frac{C}{1} \qquad \text{Substituting}$$

$$10 = \frac{1}{4} + C$$

$$\frac{39}{4} = C$$

The particular solution is $y = \dfrac{x^3}{4} + \dfrac{39}{4x}$.

> **Quick Check 2**
>
> Use an integration factor to determine the general solution of $y' + xy = x$.

❰ Quick Check 2

Applications

First-order linear differential equations arise in many applications. For example, *Newton's Law of Cooling* states that if a heated object is placed in a large room with a constant temperature, the object will cool at a rate proportional to the difference between its temperature and the room's temperature. If we let $T(t)$ represent the heated object's temperature at time t and let R represent the constant room temperature, then the temperature of the object can be modeled by the differential equation

$$\frac{dT}{dt} = -k(T - R), \quad k > 0.$$

This is similar to the second form of the inhibited growth model in Section 8.3. However, in this situation, the object's temperature T is decreasing toward R. The factor $-k$ indicates that this cooling process is decay, as opposed to growth.

■ **EXAMPLE 3** **Physics: Newton's Law of Cooling.** A cup of hot coffee with a temperature of 80°C is placed in a room with a constant temperature of 25°C. After 30 min, the coffee's temperature has dropped to 48°C.

a) Determine the particular solution T for the coffee's temperature at time t.

b) When will the coffee's temperature be 30°C?

Solution

a) The room temperature is $R = 25$. Therefore, the differential equation is

$$\frac{dT}{dt} = -k(T - 25). \qquad \text{Newton's Law of Cooling}$$

Writing T' for dT/dt and rearranging terms, we can rewrite this equation as

$$T' + kT = 25k. \qquad k \text{ is a constant.}$$

The equation is now of the form $y' + p(t)y = q(t)$, where $p(t) = k$ and $q(t) = 25k$. Thus, the integration factor is $m(t) = e^{\int k\,dt} = e^{kt}$. We have

$$
\begin{aligned}
T(t) &= \frac{\int m(t)q(t)\,dt + C}{m(t)} \\[2mm]
&= \frac{\int e^{kt} \cdot 25k\,dt + C}{e^{kt}} \qquad \text{Substituting} \\[2mm]
&= \frac{25e^{kt} + C}{e^{kt}} \qquad \text{Note that } \int 25ke^{kt}\,dt = 25\int ke^{kt}\,dt = 25e^{kt}. \\[2mm]
&= \frac{25e^{kt}}{e^{kt}} + \frac{C}{e^{kt}} \qquad \text{Recall that } \frac{A + B}{D} = \frac{A}{D} + \frac{B}{D}. \\[2mm]
&= 25 + Ce^{-kt}
\end{aligned}
$$

To determine the constant of integration C, we use the fact that at time $t = 0$, the coffee's temperature is $80°$:

$$
\begin{aligned}
80 &= 25 + Ce^{-k(0)} \qquad \text{Substituting} \\
80 &= 25 + C \\
55 &= C.
\end{aligned}
$$

We now have $T(t) = 25 + 55e^{-kt}$. To determine k, we use the fact that at $t = 30$ min, the coffee's temperature is $48°C$:

$$
\begin{aligned}
48 &= 25 + 55e^{-k(30)} \qquad \text{Substituting} \\
23 &= 55e^{-30k} \\[2mm]
\frac{23}{55} &= e^{-30k} \qquad \text{Dividing both sides by 55} \\[2mm]
k &= \frac{\ln(23/55)}{-30} \approx 0.029. \qquad \begin{array}{l}\text{Taking the natural logarithm} \\ \text{of both sides and solving for } k\end{array}
\end{aligned}
$$

Therefore, the particular solution that models this coffee's temperature is

$$T(t) = 25 + 55e^{-0.029t}, \quad t \geq 0.$$

b) To determine when the coffee's temperature is $30°C$, we substitute 30 for T and then solve for t:

$$
\begin{aligned}
30 &= 25 + 55e^{-0.029t} \qquad \text{Substituting} \\
5 &= 55e^{-0.029t} \\[2mm]
\frac{1}{11} &= e^{-0.029t} \qquad \text{Dividing both sides by 55} \\[2mm]
t &= \frac{\ln(1/11)}{-0.029} \approx 82.7 \text{ min.} \qquad \begin{array}{l}\text{Taking the natural logarithm} \\ \text{of both sides and solving for } t\end{array}
\end{aligned}
$$

The graph of T is shown on the next page, along with the line $R = 25$, which is the horizontal asymptote. This is the limiting temperature for the cup of coffee.

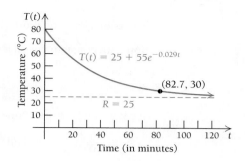

⟨ Quick Check 3

Suppose a tank is filled with brine (water and salt) to be used to make pickles. A solution of brine at a different concentration is pumped into the tank, the two solutions mix uniformly, and the mixture leaves the tank in such a way that the volume in the tank stays constant while the concentration of the brine within the tank changes. This is a *mixture problem*, and the salt content of the brine can be modeled by a differential equation.

■ **EXAMPLE 4** **Mixture Problem.** A tank contains a brine solution consisting of 300 lb of salt dissolved in 1000 gal of water. A second brine solution, containing 2 lb of salt per gallon of water, enters the tank at the rate of 4 gal per minute. Simultaneously, at the other end, the brine mixture exits at the same rate. Find the amount of salt in the tank after 6 hr.

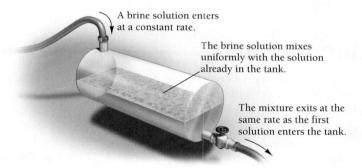

A brine solution enters at a constant rate.

The brine solution mixes uniformly with the solution already in the tank.

The mixture exits at the same rate as the first solution enters the tank.

Solution Let $A(t)$ represent the amount of salt, in pounds, in the tank after t minutes. We have the initial condition $A(0) = 300$, and we want to find $A(360)$, the amount of salt after 6 hr (360 min). Note that dA/dt represents the rate of change of salt in the tank at time t. The rate at which salt enters the tank (rate in) is

$$\text{Rate in} = \left[\frac{2 \text{ lb salt}}{1 \text{ gal}}\right] \cdot \left[\frac{4 \text{ gal}}{1 \text{ min}}\right] = 8 \text{ lb of salt per minute.}$$

The rate at which salt leaves the tank (rate out) is

$$\text{Rate out} = \left[\frac{A(t) \text{ lb salt}}{1000 \text{ gal}}\right] \cdot \left[\frac{4 \text{ gal}}{1 \text{ min}}\right] = \frac{1}{250}A(t) \text{ lb of salt per minute.}$$

The rate of change dA/dt of the amount of salt in the tank is the difference between the two rates:

$$\frac{dA}{dt} = \text{Rate in} - \text{Rate out}$$

$$A'(t) = 8 - \frac{1}{250}A(t) \qquad \text{Writing } A'(t) \text{ for } \frac{dA}{dt}$$

The equivalent equation, $A'(t) + \dfrac{1}{250}A(t) = 8$, is a first-order linear differential equation with constant coefficients. Since it is nonhomogeneous, we use an integration factor:

$$m(x) = e^{\int \frac{1}{250}dt} \qquad \text{Using } m(t) = e^{\int p(t)\,dt}$$
$$= e^{t/250}.$$

We have $m(t) = e^{t/250}$ and $q(t) = 8$, so

$$A(t) = \frac{\int m(t)q(t)\,dt + C}{m(t)}$$

$$= \frac{\int 8e^{t/250} + C}{e^{t/250}} \qquad \text{Substituting}$$

$$= \frac{2000e^{t/250} + C}{e^{t/250}}$$

$$= 2000 + Ce^{-t/250}. \qquad \text{Dividing both terms in the numerator by } e^{t/250}$$

To find C, we use the initial condition $A(0) = 300$:

$$300 = 2000 + Ce^{-0/250} \qquad \text{Substituting}$$
$$-1700 = C. \qquad e^{-0/250} = 1$$

Therefore, the particular solution is $A(t) = 2000 - 1700e^{-t/250}$. After 6 hr ($t = 360$ min), there will be $A(360) = 2000 - 1700e^{-360/250} \approx 1597.2$ lb of salt in the tank.

❮ Quick Check 4

> **❯ Quick Check 4**
>
> Repeat Example 4, assuming that the tank originally contained 1500 gal of pure water. Find the amount of salt in the tank after 10 hr.

Section Summary

- A *first-order linear differential equation* has the form $a_0(x)y' + a_1(x)y = b(x)$, where $a_0(x)$, $a_1(x)$, and $b(x)$ are continuous functions of x over some interval $[c, d]$.
  - If $b(x) = 0$, then $q(x) = 0$, and the differential equation is *homogeneous*. It is solved by separation of variables.
  - If $q(x) \neq 0$, then the differential equation is *nonhomogeneous* and solving it by separation of variables does not always work.
- Assuming that $a_0(x) \neq 0$, the *standard form* of a first-order linear differential equation is $y' + p(x)y = q(x)$, where $p(x) = \dfrac{a_1(x)}{a_0(x)}$ and $q(x) = \dfrac{b(x)}{a_0(x)}$.

- The general solution of $y' + p(x)y = q(x)$ is

$$y = \frac{\int m(x)q(x)\,dx + C}{m(x)},$$

  where the *integration factor* is $m(x) = e^{\int p(x)\,dx}$.
- *Newton's Law of Cooling*: If a heated object is placed in a large room with a constant temperature, the object will cool at a rate proportional to the difference between its temperature and the room's temperature. If $T(t)$ is the heated object's temperature at time t and if R represents the constant room temperature, then the temperature of the object can be modeled by the differential equation $dT/dt = -k(T - R)$, for $k > 0$. This equation can be solved using an integration factor.

EXERCISE SET
8.4

Find the general solution of each differential equation using an integration factor.

1. $y' + 5y = 0$

2. $y' - 10y = 0$

3. $y' + 2y = 1$

4. $y' - 6y = 3$

5. $2y' + 3y = 1$

6. $3y' + 4y = -2$

7. $y' - 2xy = x$

8. $y' + 3x^2 y = x^2$

9. $y' + \dfrac{2y}{x} = 1$

10. $y' - \dfrac{3y}{x} = 2$

Find the particular solution of each differential equation.

11. $y' + \dfrac{5y}{x} = x, \quad y(1) = 1$

12. $y' - \dfrac{4y}{x} = 2x, \quad y(2) = 3$

13. $y' + 2xy = x, \quad y(0) = 4$

14. $y' + x^2y = x^2, \quad y(0) = 2$

15. $x^2y' + xy = x^4, \quad y(2) = 5$

16. $xy' + y = x, \quad y(-1) = 6$

APPLICATIONS

Life and Physical Sciences

17. Newton's Law of Cooling. A ceramic bowl leaves a kiln with a temperature of 200°C and is placed in a room with a constant temperature of 22°C. After 45 min, the bowl's temperature is 176°C.

  a) Write an equation that gives the bowl's temperature T after t minutes.

  b) When will the bowl's temperature be 100°C?

18. Newton's Law of Cooling. A brick leaves a drying oven with a temperature of 750°F and is placed in a room with a constant temperature of 72°F. After 60 min, the brick's temperature is 625°F.

  a) Write an equation that gives the brick's temperature T after t minutes.

  b) When will the brick's temperature be 90° Fahrenheit?

19. Mixture problem. A tank contains a brine solution consisting of 100 lb of salt in 500 gal of water. Another brine solution consisting of 2 lb of salt per gallon of water is flowing in at a rate of 3 gal per minute. The mixture exits the tank at a rate of 3 gal per minute at the other end.

  a) Write an equation that gives the amount of salt A in the tank after t minutes.

  b) How much salt will be in the tank after 2 hr?

20. Mixture problem. A tank contains 1000 gal of pure water. A sugar solution containing 1 lb of sugar per gallon is flowing into the tank at a rate of 10 gal per minute. The mixture exits the tank at a rate of 10 gal per minute at the other end.

  a) Write an equation that gives the amount of sugar A in the tank after t minutes.

  b) How much sugar will be in the tank after 5 hr?

Business and Economics

21. Depreciation. The value of Peggy's convertible can be modeled by $dV/dt = -0.2(V - 12{,}000)$, where $V(t)$ is the value (in dollars) after t years. Assume that the original value of the car is $22,500.

  a) Use an integration factor to find the particular solution V.

  b) When will Peggy's convertible be worth $14,000?

  c) Explain what the number 12,000 represents in the differential equation.

22. Depreciation. The value of Paul's computer can be modeled by $dV/dt = -0.03(V - 400)$, where $V(t)$ is the value (in dollars) after t months. Assume that the original value of the computer was $4000.

  a) Use an integration factor to find the particular solution V.

  b) When will the computer have half of its original value?

  c) Explain what the number 400 represents in the differential equation.

23. Mixture: demographics. The city of Wittman has 100,000 residents, including 15,000 senior citizens. Each month, 50 people move into Wittman, of which 20 are senior citizens, and 50 people leave, so the city's population stays constant.

  a) Find a function S that models the number of senior citizens in the city after t months.

  b) How many senior citizens live in Wittman after 1 yr?

  c) When will the number of senior citizens in Wittman reach 20,000?

24. Mixture: counterfeit coins. A vending machine begins the month with 400 quarters, of which none are counterfeit. Each day, 100 quarters are deposited into the vending machine, of which 5 are counterfeit. At the end of each day, a worker randomly removes 100 coins from the vending machine.

  a) Find a function A that models the number of counterfeit coins in the vending machine after t days.

  b) How many counterfeit coins are in the vending machine after 7 days?

  c) What is the limiting number of counterfeit coins in the vending machine?

SYNTHESIS

25. Consider $y' + 4y = 0$.

  a) Find its general solution using the procedure in Example 1.

  b) Find its general solution using an integration factor.

  c) Which method is more efficient? Why?

26. Consider $y' + 4xy = 0$.

  a) Explain why this differential equation cannot be solved using the procedure in Example 1.

  b) Find its general solution using an integration factor.

  c) Find its general solution using separation of variables.

  d) Which method is more efficient? Why?

Exact equations. *A differential equation of the form* $f(x, y)\, dx + g(x, y)\, dy = 0$ *is exact if* $f_y = g_x$, *and its general solution is* $F(x, y) = C$, *where* $F_x = f$ *and* $F_y = g$. *Recall from Section 6.2 that* F_y *represents the partial derivative of* F *with respect to* y, *and so on.*

27. Consider $2xy\, dx + x^2\, dy = 0$.

  a) Find the general solution of the form $F(x, y) = C$.

  b) Rewrite the equation from part (a) in the form $y' + p(x)y = 0$, and solve it using an integration factor.

  c) Show that the solutions found in parts (a) and (b) are the same.

28. Consider $y' = -\dfrac{y}{x}$.

   a) Find its general solution in the form $F(x, y) = C$.

   b) Find its general solution using separation of variables.

   c) Rewrite the general solution in the form

$$y' + p(x)y = 0,$$

     and solve it using an integration factor.

   d) Show that the solutions found in parts (a)–(c) are the same.

29. Consider $y' = \dfrac{2xy}{-x^2 - 6y}$.

   a) Find its general solution in the form $F(x, y) = C$.

   b) Explain why this differential equation cannot be solved using separation of variables or an integration factor.

   c) Use algebra software to graph the solution when $C = 3$.

30. Consider $ye^{xy}\, dx + xe^{xy}\, dy = 0$.

   a) Show that $e^{xy} = C$ is a solution.

   b) Rewrite the differential equation in the form $dy/dx = h(x, y)$, and solve it using separation of variables.

   c) Show that the solutions found in parts (a) and (b) are the same.

In Exercises 31–34, use an integration factor to solve each differential equation. Where appropriate, use an integration technique, or refer to Table 1 in Section 4.7.

31. $y' + y = x$ **32.** $y' + 3y = 2x$

33. $(x^2 + 1)y' + 2xy = 1$

34. $(x^2 - 1)y' + 2y = 1, x \neq \pm 1$

8.5   Higher-Order Differential Equations and a Trigonometry Connection

OBJECTIVES

- Solve a higher-order homogeneous linear differential equation with constant coefficients.

- Solve a higher-order nonhomogeneous linear differential equation using the method of undetermined coefficients.

- Use trigonometric functions to solve certain higher-order linear differential equations.

Homogeneous Higher-Order Linear Differential Equations

Some differential equations contain higher-order derivatives. These higher-order linear differential equations are defined as follows.

> **DEFINITION**
>
> An **nth-order linear differential equation** is any equation that can be written in the form
>
> $$a_0(x)y^{(n)} + a_1(x)y^{(n-1)} + \cdots + a_{n-1}(x)y' + a_n(x)y = b(x),$$
>
> where $a_0(x), a_1(x), \ldots, a_n(x)$ are continuous functions of x over some interval $[c, d]$, and $y^{(n)}$ is the nth derivative of y, with $n \geq 1$.

For example, a second-order linear differential equation has the form $a_0(x)y'' + a_1(x)y' + a_2(x)y = b(x)$, and so on.

    Although we do not prove it here, all first-order linear differential equations can be solved using the techniques discussed in Sections 8.2 and 8.4. Unfortunately, not every higher-order linear differential equation can be solved explicitly, but some techniques for solving first-order equations can be extended to higher-order equations. Let's consider a homogeneous second-order linear differential equation of the form

$$a_0y'' + a_1y' + a_2y = 0, \tag{1}$$

where a_0, a_1 and a_2 are constants.

    In Section 8.4, we learned that the general solution of the homogeneous first-order linear differential equation $y' + ky = 0$ is $y = Ce^{-kx}$. Let's try to solve a homogeneous second-order linear differential equation similarly. If we let $y = e^{rx}$, then $y' = re^{rx}$ and $y'' = r^2e^{rx}$. We substitute these expressions into equation (1):

$$a_0 \cdot r^2e^{rx} + a_1 \cdot re^{rx} + a_2 \cdot e^{rx} = 0.$$

Since e^{rx} is never zero, we can divide both sides by e^{rx}:

$$a_0 r^2 + a_1 r + a_2 = 0.$$

This is called the **auxiliary equation** of the differential equation. Factoring the left side or using the quadratic formula will yield up to two values for r.

■ **EXAMPLE 1** Find the general solution of $y'' - 5y' + 6y = 0$.

Solution The given equation is of the form $a_0 y'' + a_1 y' + a_2 y = 0$, with $a_0 = 1$, $a_1 = -5$, and $a_2 = 6$. The auxiliary equation is

$$r^2 - 5r + 6 = 0. \qquad \text{Using } a_0 r^2 + a_1 r + a_2 = 0$$

The left side of this equation can be factored:

$$(r - 3)(r - 2) = 0.$$

The solutions of the auxiliary equation are $r = 2$ and $r = 3$; thus, a solution of the differential equation is $y = e^{2x} + e^{3x}$. The general solution is $y = C_1 e^{2x} + C_2 e^{3x}$, where C_1 and C_2 are constants.

Check We have $y = C_1 e^{2x} + C_2 e^{3x}$, $y' = 2C_1 e^{2x} + 3C_2 e^{3x}$, and $y'' = 4C_1 e^{2x} + 9C_2 e^{3x}$. Substituting, we obtain

$$
\begin{array}{c|cc}
y'' - 5y' + 6y \overset{?}{=} 0 & \\
\hline
(4C_1 e^{2x} + 9C_2 e^{3x}) - 5(2C_1 e^{2x} + 3C_2 e^{3x}) + 6(C_1 e^{2x} + C_2 e^{3x}) & 0 \\
4C_1 e^{2x} + 9C_2 e^{3x} - 10C_1 e^{2x} - 15C_2 e^{3x} + 6C_1 e^{2x} + 6C_2 e^{3x} & 0 \\
C_1 e^{2x}(4 - 10 + 6) + C_2 e^{3x}(9 - 15 + 6) & 0 \\
0 & 0 \quad \text{TRUE}
\end{array}
$$

《 Quick Check 1

> **Quick Check 1**
>
> Find the general solution of $y'' + y' - 12y = 0$.

In Example 1, we used the fact that if f and g are solutions of a homogeneous linear differential equation, then so is $y = C_1 f(x) + C_2 g(x)$. This fact can be stated as a theorem.

THEOREM 1   Principle of Superposition

If $f_1, f_2, f_3, \dots, f_n$ are solutions of a homogeneous linear differential equation, then so too is the sum

$$y = C_1 f_1(x) + C_2 f_2(x) + C_3 f_3(x) + \cdots + C_n f_n(x),$$

where $C_1, C_2, C_3, \dots, C_n$ are constants. This sum is called a **linear combination** of the solutions $f_1, f_2, f_3, \dots, f_n$.

A proof of this theorem is outlined in Exercise 53.

When a general solution is expressed as a linear combination, the solutions $f_1, f_2, f_3, \dots, f_n$ should be *independent* of one another. That is, none of them can be expressed as a linear combination of the other solutions. For example, the solutions of Example 1, e^{2x} and e^{3x}, are independent since neither can be written as a multiple of the other.

To find a particular solution of the differential equation in Example 1, we need two initial conditions since there are two constants to be determined. Often, one initial condition is placed on y and another on the derivative y'.

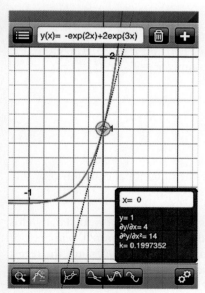

A plot by Graphicus of the particular solution $y = -e^{2x} + 2e^{3x}$, where $y(0) = 1$ and $y'(0) = 4$.

■ **EXAMPLE 2** Given that $y = C_1e^{2x} + C_2e^{3x}$ is the general solution of $y'' - 5y' + 6y = 0$, find a particular solution where $y(0) = 1$ and $y'(0) = 4$.

Solution We write y and its derivative, y':

$$y = C_1e^{2x} + C_2e^{3x} \quad \text{and} \quad y' = 2C_1e^{2x} + 3C_2e^{3x}.$$

For $x = 0$, we have $y = 1$ and $y' = 4$:

$$1 = C_1e^{2(0)} + C_2e^{3(0)} \quad \text{and} \quad 4 = 2C_1e^{2(0)} + 3C_2e^{3(0)}.$$

Since $e^0 = 1$, these equations simplify to

$$1 = C_1 + C_2 \quad \text{and} \quad 4 = 2C_1 + 3C_2.$$

> **Quick Check 2**
>
> Find the particular solution of $y'' - 3y' - 28y = 0$, where $y(0) = 5$ and $y'(0) = 13$. Use a graphing calculator or Graphicus to view the particular solution.

This is a system of two unknowns. Solving the first equation for C_1 gives $C_1 = 1 - C_2$. We substitute this expression for C_1 into the second equation:

$$4 = 2(1 - C_2) + 3C_2. \qquad \text{Substituting}$$

Simplifying, we get $C_2 = 2$. Therefore, $C_1 = 1 - 2 = -1$, and the particular solution is

$$y = -e^{2x} + 2e^{3x}.$$

❮ Quick Check 2

The procedure used in Example 1 can be extended for third- and higher-order linear homogeneous differential equations.

■ **EXAMPLE 3** Find the general solution of $y''' - 2y'' - 15y' = 0$.

Solution This is a third-order homogeneous linear differential equation. Its auxiliary equation is

$$r^3 - 2r^2 - 15r = 0.$$

Factoring, we get

$$r(r^2 - 2r - 15) = 0$$
$$r(r - 5)(r + 3) = 0.$$

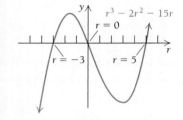

> **Quick Check 3**
>
> Find the general solution of $y''' - 4y' = 0$.

The solutions of the auxiliary equation are $r = 0$, $r = 5$, and $r = -3$. Therefore, the general solution is $y = C_1e^{0x} + C_2e^{5x} + C_3e^{-3x}$, or simply $y = C_1 + C_2e^{5x} + C_3e^{-3x}$.

❮ Quick Check 3

Repeated Solutions

Suppose we use the technique in Example 1 to solve the differential equation

$$y'' - 2y' + y = 0.$$

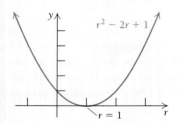

Its auxiliary equation, $r^2 - 2r + 1 = 0$, factors as $(r - 1)(r - 1) = 0$. This yields one solution, $r = 1$ with a multiplicity of 2, meaning that 1 is a *repeated solution*. Unfortunately, we cannot state the general solution as $y = C_1e^x + C_2e^x$, since the terms are multiples of one another and thus are not linearly independent. Instead, repeated real solutions are handled in a special way, stated as a theorem.

THEOREM 2

If r is a real solution with multiplicity n of an auxiliary equation of a homogeneous linear differential equation, then

$$y = C_1 e^{rx} + C_2 x e^{rx} + C_3 x^2 e^{rx} + \cdots + C_n x^{n-1} e^{rx},$$

where $C_1, C_2, C_3, \ldots, C_n$ are constants, is a solution of the differential equation.

The proof for the case of a solution of an auxiliary equation with multiplicity 2 is outlined in Exercises 67 and 68. Let's revisit $y'' - 2y' + y = 0$ as an example.

■ **EXAMPLE 4** Find the general solution of $y'' - 2y' + y = 0$.

Solution We saw above that the auxiliary equation has a solution $r = 1$ with multiplicity 2. Therefore, its general solution is $y = C_1 e^x + C_2 x e^x$.

Check We have $y' = \dfrac{d}{dx}(C_1 e^x + C_2 x e^x) = C_1 e^x + C_2(x e^x + e^x)$ and $y'' = \dfrac{d}{dx} y' = C_1 e^x + C_2(x e^x + 2e^x)$. Substituting, we obtain

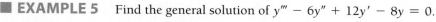

$$
\begin{array}{r|l}
 & y'' - 2y' + y \overset{?}{=} 0 \\
\hline
[C_1 e^x + C_2(x e^x + 2e^x)] - 2[C_1 e^x + C_2(x e^x + e^x)] + [C_1 e^x + C_2 x e^x] & 0 \\
C_1 e^x + C_2 x e^x + 2C_2 e^x - 2C_1 e^x - 2C_2 x e^x - 2C_2 e^x + C_1 e^x + C_2 x e^x & 0 \\
e^x(C_1 + 2C_2 - 2C_1 - 2C_2 + C_1) + x e^x(C_2 - 2C_2 + C_2) & 0 \\
e^x(2C_1 + 2C_2 - 2C_1 - 2C_2) + x e^x(2C_2 - 2C_2) & 0 \\
0 & 0 \quad \text{TRUE}
\end{array}
$$

> **Quick Check 4**

Find the general solution of $y'' - 6y' + 9y = 0$.

❮ Quick Check 4

■ **EXAMPLE 5** Find the general solution of $y''' - 6y'' + 12y' - 8y = 0$.

Solution The auxiliary equation, $r^3 - 6r^2 + 12r - 8 = 0$, can be solved by factoring:

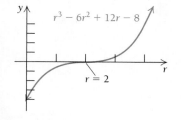

$$
\begin{aligned}
r^3 - 6r^2 + 12r - 8 &= 0 \\
r^3 - 8 - 6r^2 + 12r &= 0 && \text{Regrouping} \\
(r - 2)(r^2 + 2r + 4) - 6r(r - 2) &= 0 && \text{Factoring} \\
(r - 2)(r^2 + 2r + 4 - 6r) &= 0 && (r - 2) \text{ is a common factor} \\
(r - 2)(r^2 - 4r + 4) &= 0 \\
(r - 2)(r - 2)(r - 2) &= 0.
\end{aligned}
$$

Thus, $r = 2$ is a solution, with multiplicity 3. The general solution of the differential equation is $y = C_1 e^{2x} + C_2 x e^{2x} + C_3 x^2 e^{2x}$.

> **Quick Check 5**

Find the general solution of $y''' + 3y'' - 4y = 0$. (*Hint:* $r = 1$ is one solution of the auxiliary equation.)

❮ Quick Check 5

For a higher-order differential equation, determining the solutions of its auxiliary equation may be a real challenge, requiring careful factoring skills and graphical methods. Solutions may also be found using a graphing utility (see the Technology Connection on pages 54–55 to review the ZERO feature on the TI-83/84 calculator). The multiplicities can be found by noting how the graph passes through the horizontal axis. The graphs for solutions with multiplicity 1, 2, and 3 are shown below.

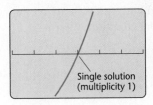

Single solution
(multiplicity 1)

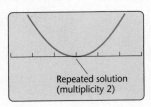

Repeated solution
(multiplicity 2)

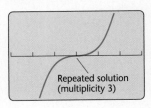

Repeated solution
(multiplicity 3)

The Method of Undetermined Coefficients

Suppose a higher-order differential equation is not homogeneous. Let's consider the case where $b(x)$ is a polynomial. Solutions for such equations are approached in two steps: the solution of the homogeneous case is determined first, and a specific solution of the nonhomogeneous case is determined second. The general solution is the sum of the homogeneous and specific solutions. We use the **method of undetermined coefficients** to find a specific solution for the nonhomogeneous case.

■ **EXAMPLE 6**   Find the general solution of $y'' - 9y = 3x^2 + 1$.

Solution   We first determine the solution of the homogeneous case, $y'' - 9y = 0$. Its auxiliary equation is $r^2 - 9 = 0$, with solutions $r = 3$ and $r = -3$. Thus, the solution of the homogeneous case is

$$y = C_1 e^{3x} + C_2 e^{-3x}.$$

To determine a solution of the nonhomogeneous case, we note that $3x^2 + 1$ is of degree 2, and so we assume a specific solution of the form $y = Ax^2 + Bx + C$. Its derivatives are $y' = 2Ax + B$ and $y'' = 2A$. These are substituted into the differential equation, and similar terms are combined:

$$y'' - 9y = 3x^2 + 1$$
$$(2A) - 9(Ax^2 + Bx + C) = 3x^2 + 1 \qquad \text{Substituting}$$
$$2A - 9Ax^2 - 9Bx - 9C = 3x^2 + 1 \qquad \text{Distributing}$$
$$-9Ax^2 - 9Bx + (2A - 9C) = 3x^2 + 1. \qquad \text{Regrouping by descending powers of } x$$

Now we equate coefficients according to powers of x:

$$-9A = 3 \qquad \text{Equating coefficients of } x^2$$
$$-9B = 0 \qquad \text{Equating coefficients of } x$$
$$2A - 9C = 1. \qquad \text{Equating constant terms}$$

From the equation $-9A = 3$, we get $A = -\frac{1}{3}$. From the equation $-9B = 0$, we get $B = 0$. Substituting $A = -\frac{1}{3}$ into $2A - 9C = 1$, we get $C = -\frac{5}{27}$. Thus, the specific solution is

$$y = Ax^2 + Bx + C$$
$$y = -\tfrac{1}{3}x^2 - \tfrac{5}{27}. \qquad \text{Substituting for } A, B, \text{ and } C$$

Therefore, the general solution is the sum of the homogeneous and the specific solutions:

$$y = C_1 e^{3x} + C_2 e^{-3x} - \tfrac{1}{3}x^2 - \tfrac{5}{27}.$$

❬ Quick Check 6

⟩ **Quick Check 6**

Find the general solution of $y'' + 3y' - 10y = 5x^2 + 7x + 16$.

⊕ A Trigonometry Connection: Trigonometric Solutions

Consider $y'' + y = 0$. Because its auxiliary equation, $r^2 + 1 = 0$, has no real solutions, let's try $y = \sin x$ as a solution. Its derivatives are $y' = \cos x$ and $y'' = -\sin x$. Substituting, we have

$$\frac{y'' + y \stackrel{?}{=} 0}{(-\sin x) + (\sin x) \quad \Big|\quad 0}$$
$$0 \quad \Big|\quad 0 \quad \text{TRUE}$$

We conclude that $y = \sin x$ is a solution of $y'' + y = 0$. The student can confirm that $y = \cos x$ is also a solution. Thus, the general solution of $y'' + y = 0$ is $y = C_1 \sin x + C_2 \cos x$.

Non-real solutions of an auxiliary equation can always be written as complex numbers of the form $a \pm bi$, where $b \neq 0$. We may use such solutions to generate trigonometric solutions. This is stated as a theorem.

THEOREM 3

If $r = a \pm bi\ (b \neq 0)$ are complex solutions of an auxiliary equation of a homogeneous linear differential equation, then a solution of the differential equation is

$$y = C_1 e^{ax} \sin(bx) + C_2 e^{ax} \cos(bx).$$

This formula is discussed in Exercise Set 9.6. Usually, complex solutions arise when the quadratic formula is used and the discriminant (radicand) is negative.

■ **EXAMPLE 7** Find the general solution of $y'' - 4y' + 9y = 0$.

Solution The auxiliary equation, $r^2 - 4r + 9 = 0$, is solved using the quadratic formula:

$$r = \frac{-(-4) \pm \sqrt{(-4)^2 - 4(1)(9)}}{2(1)} \qquad \text{Using the quadratic formula}$$

$$= \frac{4 \pm \sqrt{-20}}{2} \qquad (-4)^2 - 4(1)(9) = -20$$

$$= \frac{4 \pm 2i\sqrt{5}}{2} \qquad \sqrt{-20} = \sqrt{-1 \cdot 4 \cdot 5} = 2i\sqrt{5}$$

$$= \frac{2(2 \pm i\sqrt{5})}{2} \qquad \text{Factoring the numerator}$$

$$= 2 \pm i\sqrt{5}.$$

Thus, we have $a = 2$ and $b = \sqrt{5}$. Therefore, the general solution is

$$y = C_1 e^{2x} \sin(\sqrt{5} \cdot x) + C_2 e^{2x} \cos(\sqrt{5} \cdot x).$$

❯ **Quick Check 7**

Find the general solution of $y'' - y' + y = 0$.

❰ Quick Check 7

■ **EXAMPLE 8** Find the general solution of $y'' + y' + 3y = 2x + 3$.

Solution The solution of the homogeneous case is determined first. The auxiliary equation is $r^2 + r + 3 = 0$, and the quadratic formula gives $r = -\dfrac{1}{2} \pm i\dfrac{\sqrt{11}}{2}$ as the two solutions. Since these are non-real, the solution of the homogeneous case is

$$y = C_1 e^{(-1/2)x} \sin\left(\frac{\sqrt{11}}{2} \cdot x\right) + C_2 e^{(-1/2)x} \cos\left(\frac{\sqrt{11}}{2} \cdot x\right).$$

For the specific solution of the nonhomogeneous case, we let $y = Ax + B$ since $2x + 3$ is of degree 1. The derivatives are $y' = A$ and $y'' = 0$:

$$y'' + y' + 3y = 2x + 3$$
$$(0) + (A) + 3(Ax + B) = 2x + 3 \quad\quad \text{Substituting}$$
$$3Ax + (A + 3B) = 2x + 3. \quad\quad \text{Regrouping by descending powers of } x$$

Matching the coefficients of x, we have $3A = 2$ and $A + 3B = 3$, or $A = \frac{2}{3}$ and $B = \frac{7}{9}$. Thus, the specific solution is

$$y = Ax + B$$
$$y = \tfrac{2}{3}x + \tfrac{7}{9}. \quad\quad \text{Substituting}$$

The general solution is

$$y = C_1 e^{(-1/2)x} \sin\left(\frac{\sqrt{11}}{2} \cdot x\right) + C_2 e^{(-1/2)x} \cos\left(\frac{\sqrt{11}}{2} \cdot x\right) + \frac{2}{3}x + \frac{7}{9}.$$

> **Quick Check 8**
>
> Find the general solution of $y'' - y' + y = x^2 - 5$.

❰ Quick Check 8

Section Summary

- An *nth-order linear differential equation* is any equation that can be written in the form $a_0(x)y^{(n)} + a_1(x)y^{(n-1)} + \cdots + a_{n-1}(x)y' + a_n(x)y = b(x)$, where $a_0(x), a_1(x), \ldots, a_n(x)$ are continuous functions of x over some interval $[c, d]$ and $y^{(n)}$ is the nth derivative of y, with $n \geq 1$. If $b(x) = 0$, the equation is said to be *homogeneous*.
- The *auxiliary equation* of the homogeneous form $a_0 y'' + a_1 y' + a_2 y = 0$, where a_0, a_1, and a_2 are constants, is $a_0 r^2 + a_1 r + a_2 = 0$. If r is a real solution of the auxiliary equation, then $y = Ce^{rt}$ is a solution of the differential equation.
- If $f_1, f_2, f_3, \ldots, f_n$ are solutions of a linear homogeneous differential equation, then so too is the sum

$$y = C_1 f_1(x) + C_2 f_2(x) + C_3 f_3(x) + \cdots + C_n f_n(x),$$

where $C_1, C_2, C_3, \ldots, C_n$ are constants. This sum is called a *linear combination* of the functions $f_1, \ldots f_n$.

- If r is a real solution with multiplicity n of an auxiliary equation of a homogeneous linear differential equation, then $y = C_1 e^{rx} + C_2 x e^{rx} + C_3 x^2 e^{rx} + \cdots + C_n x^{n-1} e^{rx}$ is a solution of the differential equation.
- The *method of undetermined coefficients* is used to find a specific solution of a nonhomogeneous linear differential equation with constant coefficients. The general solution is a sum of the homogeneous solution and the specific solution.
- *Trigonometry Connection:* If $r = a \pm bi$ ($b \neq 0$) are complex solutions of an auxiliary equation of a homogeneous linear differential equation, then a *trigonometric solution* of the differential equation is

$$y = C_1 e^{ax} \sin bx + C_2 e^{ax} \cos bx.$$

EXERCISE SET
8.5

Find the general solution of each differential equation.

1. $y'' + y' - 20y = 0$

2. $y'' + 5y' + 6y = 0$

3. $y'' + 5y' - 24y = 0$

4. $y'' - 2y' - 24y = 0$

5. $y'' - 25y = 0$

6. $y'' - 100y = 0$

7. $y'' - 10y' = 0$

8. $y'' + 8y' = 0$

9. $2y'' - 3y' - 2y = 0$

10. $3y'' - y' - 2y = 0$

11. $y''' + 3y'' + 2y' = 0$

12. $y''' - 5y'' - 36y' = 0$

13. $y''' - 36y' = 0$

14. $y''' - 49y' = 0$

15. $y'' + 8y' + 16y = 0$

16. $y'' - 10y' + 25y = 0$

17. $y'' - 16y' + 64y = 0$

18. $y'' + 20y' + 100y = 0$

Find the particular solution of each differential equation.

19. $y'' + 4y' + 3y = 0$; $y(0) = 1, y'(0) = 3$

20. $y'' - 2y' - 15y = 0$; $y(0) = -2, y'(0) = -14$

21. $y'' - 6y' = 0$; $y(0) = 3, y'(0) = 12$

22. $y'' - 4y' = 0$; $y(0) = -2, y'(0) = 4$

Use a graphing utility to solve the auxiliary equation, and then write the general solution of each differential equation. (When applicable, approximate the exponents to three decimal places.)

23. $y''' + 3y'' - 13y' - 15y = 0$

24. $y''' + 8y'' + y' - 42y = 0$

25. $y''' - 3y'' - 16y' + 48y = 0$

26. $y''' + y'' - 36y' - 36y = 0$

27. $y''' - 3y' - 2y = 0$

28. $y''' - 7y'' + 16y' - 12y = 0$

29. $y'''' - 26y'' + 25y = 0$

30. $y'''' - 25y'' + 144y = 0$

31. $y''' - 2y'' - 4y' + 3y = 0$

32. $y''' + 7y'' + 6y' - 20y = 0$

Find the general solution of each nonhomogeneous differential equation.

33. $y'' + 3y' + 2y = x^2 + 1$

34. $y'' - 5y' + 6y = 2x^2 - 3x$

35. $y'' - 36y = 5x - 7$

36. $y'' - 81y = x^2$

37. $y'' - 9y' = x + 5$

38. $y'' + 7y' = 1 - 3x$

TRIGONOMETRY CONNECTION ··············· ⊕

Find the general solution of each differential equation.

39. $y'' + 4y = 0$

40. $y'' + 36y = 0$

41. $y'' + 2y = 0$

42. $y'' + 13y = 0$

43. $y'' + 3y' + 7y = 0$

44. $y'' - y' + 4y = 0$

Find the particular solution of each differential equation.

45. $y'' + 16y = 0$; $y(0) = 1, y'(0) = 2$

46. $y'' + 49y = 0$; $y(0) = -2, y'(0) = 5$

Find the general solution of each nonhomogeneous differential equation.

47. $y'' + 9y = x^2 + 8x - 2$

48. $y'' + 100y = 3x^2 + 5x - 1$

49. $y'' + y' + 6y = 3x$

50. $y'' + 2y' + 10y = 1 - 6x$

APPLICATIONS

Business and Economics

51. Price of a commodity. Over a period of 1 yr, the price P of copper follows the model $50P'' + 5P' - P = 0$, where P is in dollars per ounce and t is in months. Assume that $P(0) = \$5$ and $P'(0) = \$0.20$ per month.

a) Find the particular solution P.
b) What is the price of copper in the 6th month?
c) What is the rate of change in the price of copper in the 6th month?

52. Revenue. Over a period of 1 yr, monthly revenue R at Jason's Pool Supplies follows the model $10R'' + 3R' - R = 0$, where R is in thousands of dollars and t is in months. Assume that $R(0) = 28$ and $R'(0) = \$0$ per month.

a) Find the particular solution R.
b) What is the company's revenue in the 8th month?
c) What is the rate of change of the company's revenue in the 8th month?

SYNTHESIS

53. Suppose $y_1 = f_1(x)$ and $y_2 = f_2(x)$ are two solutions of the homogeneous linear differential equation $a_0(x)y'' + a_1(x)y' + a_2(x)y = 0$. Show that their linear combination, $y = C_1 f_1(x) + C_2 f_2(x)$, is also a solution by substituting y'', y', and y into $a_0(x)y'' + a_1(x)y' + a_2(x)y = 0$ and then rearranging terms.

54. The general solution of a nonhomogeneous differential equation is the sum of the homogeneous solution and the specific solution for the nonhomogeneous case. Explain why the general solution can include the homogeneous solution, but the homogeneous solution by itself is not a solution.

55. Let $y_1 = e^{3x}$ and $y_2 = e^{-7x}$.

a) Find a homogeneous linear differential equation with constant coefficients such that y_1 and y_2 are solutions.
b) Are there other possible homogeneous linear differential equations with constant coefficients that have these functions as solutions? Why or why not?

56. Let $y = x^2 e^{3x}$.

a) Find a homogeneous linear differential equation with constant coefficients such that y is a solution.
b) What are some other possible solutions to the differential equation in part (a)?

For each of the following higher-order differential equations use any applicable technique from this section, including graphing software, to find the general solution.

57. $y^{(4)} + y''' - 7y'' - y' + 6y = 0$

58. $y^{(4)} + 5y''' - 10y'' - 20y' + 24y = 0$

59. $y^{(4)} - 2y''' - 8y'' + 18y' - 9y = 0$

60. $y^{(4)} + 4y''' - 12y'' - 64y' - 64y = 0$

61. $y^{(4)} + y''' - 18y'' - 52y' - 40y = 0$

62. $y^{(4)} - 5y''' + 6y'' + 4y' - 8y = 0$

63. $y^{(4)} + y''' - 8y'' - y' + 7y = 0$

64. $y^{(4)} + y''' - 21y'' - 16y' + 80y = 0$

65. $y^{(4)} + 3y''' - 13y'' - 21y' + 54y = 0$

66. $y^{(4)} + 3y''' - 16y'' - 25y' + 77y = 0$

Repeated solutions: reduction of order. *If a homogeneous second-order linear differential equation with constant coefficients has r as the repeated solution of its auxiliary equation, then the general solution is* $y = C_1 e^{rx} + C_2 x e^{rx}$. *Exercises 67 and 68 outline why this is true.*

67. Let $y'' - 2y' + y = 0$. We know from Example 4 that $r = 1$ is a solution with multiplicity 2 of the auxiliary equation, so $y = e^x$ is one solution of the differential equation.

a) Let $y = v(x)e^x$ represent another possible solution of the differential equation. Using the product rule, find y' and y''.

b) Substitute y'', y', and y into the differential equation and simplify.

c) Since $e^x > 0$, divide both sides by e^x and simplify.

d) Integrate the result from part (c) twice, letting the constant of integration be 1 in the first integration and 0 in the second integration.

e) Using the results of parts (a)–(d), give the second solution and then the general solution.

68. In general, if r is a repeated solution with multiplicity 2 of the auxiliary equation of a homogeneous second-order linear differential equation with constant coefficients, then the differential equation has the form $y'' - 2ry' + r^2 y = 0$. Repeat the steps outlined in Exercise 67, starting with $y = v(x)e^{rx}$ in part (a).

Method of undetermined coefficients: nonpolynomial forms. *The method of undetermined coefficients can be used for nonpolynomial functions. For example, we can solve for a specific solution of* $y'' - 9y = 4e^{2x}$ *by setting* $y = Ae^{2x}$, *making the substitutions for y and y'', and solving for A.*

69. $y'' - 4y' - 32y = 3e^{-x}$ (*Hint:* Let $y = Ae^{-x}$.)

70. $y'' - y' - 42y = -e^{2x}$ (*Hint:* Let $y = Ae^{2x}$.)

71. $y'' + y = \cos 2x$ (*Hint:* Let $y = A \sin 2x + B \cos 2x$.)

72. $y'' + 4y = 2 \sin x$ (*Hint:* Let $y = A \sin x + B \cos x$.)

Answers to Quick Checks

1. $y = C_1 e^{-4x} + C_2 e^{3x}$ **2.** $y = 2e^{-4x} + 3e^{7x}$
3. $y = C_1 + C_2 e^{2x} + C_3 e^{-2x}$ **4.** $y = C_1 e^{3x} + C_2 x e^{3x}$
5. $y = C_1 e^x + C_2 e^{-2x} + C_3 x e^{-2x}$
6. $y = C_1 e^{2x} + C_2 e^{-5x} - \frac{1}{2}x^2 - x - 2$
7. $y = C_1 e^{x/2} \cos\left(\dfrac{\sqrt{3}}{2}x\right) + C_2 e^{x/2} \sin\left(\dfrac{\sqrt{3}}{2}x\right)$
8. $y = C_1 e^{x/2} \cos\left(\dfrac{\sqrt{3}}{2}x\right) + C_2 e^{x/2} \sin\left(\dfrac{\sqrt{3}}{2}x\right) + x^2 + 2x - 5$

KEY TERMS AND CONCEPTS

EXAMPLES

SECTION 8.1

A **differential equation** is an equation that includes a derivative and has a function as a solution.

- $y' = x^2 + 3$

- $\dfrac{dy}{dx} = 3x^2y \ (y > 0)$

- $(1 + 3x)\,dx + y^2\,dy = 0$

The **general solution** of a differential equation of the form $y' = f(x)$ is a set of functions $F(x) + C$, where $\dfrac{d}{dx}F(x) = f(x)$. For many differential equations, the general solution is of the form $y = Cf(x)$ where C is any constant.

- The general solution of $y' = x^2 + 3$ is $y = \frac{1}{3}x^3 + 3x + C$.

- The general solution of $\dfrac{dy}{dx} = 3x^2y$ is $y = Ce^{x^3}$.

  Check: Since $y = Ce^{x^3}$ and $y' = 3Cx^2e^{x^3}$, we have

$$\dfrac{dy}{dx} \overset{?}{=} 3x^2y$$

$$\underline{3Cx^2e^{x^3} \ \big|\ 3x^2(Ce^{x^3})} \quad \text{TRUE}$$

A requirement that a solution of a differential equation include a specific ordered pair is called an **initial condition**. If an initial condition is given, then a **particular solution** in which the value of C is determined may be found.

The general solution of $y' - 3y = 0$ is $y = Ce^{3x}$. If the solution includes $(0, 4)$, then C can be determined:

$$4 = Ce^{3(0)}$$
$$4 = C \cdot 1.$$

Thus, the particular solution is $y = 4e^{3x}$.

The differential equation $dP/dt = kP$ is called the **uninhibited growth model**. Its general solution is $P(t) = P_0e^{kt}$, where P_0 is the initial quantity at time $t = 0$. The value k is the **constant of proportionality**, also known as the *continuous growth rate*.

If Jay deposits \$1000 into a savings account that earns interest at a rate of 4.75%, compounded continuously, then the amount A in his account is modeled by the differential equation $dA/dt = 0.0475A$. The general solution is $A(t) = A_0e^{0.0475t}$, and the particular solution is $A(t) = 1000e^{0.0475t}$.

SECTION 8.2

A **separable differential equation** is a differential equation for which the variables can be separated to opposite sides of the equals sign. That is, it can be written in the form $f(y)dy = g(x)dx$.

To solve a separable differential equation, the variables are separated, each side is integrated, and, if possible, the result is solved for one of the variables.

The differential equation $dy/dx = 3x^2y \ (y > 0)$ is separable. Its solution is found by first separating variables, then integrating and solving for y.

$$\dfrac{dy}{y} = 3x^2\,dx$$

$$\int \dfrac{dy}{y} = \int 3x^2\,dx$$

$$\ln y = x^3 + C_1$$
$$y = e^{x^3 + C_1}$$
$$y = Ce^{x^3}, \text{ where } C = e^{C_1}.$$

SECTION 8.3

One form of the **inhibited growth model** assumes that the rate of change of P is directly proportional to P and to the remaining room for growth, $L - P$, where L is the limiting value of P. This model is expressed by the differential equation

$$\dfrac{dP}{dt} = kP(L - P).$$

The Cuddly Cathy doll sells 500 units initially in a test market with 50,000 potential customers. Its sales are modeled by

$$\dfrac{dP}{dt} = 0.000012P(50{,}000 - P),$$

where P is the number of units sold after t months.

(continued)

KEY TERMS AND CONCEPTS	**EXAMPLES**

SECTION 8.3 *(continued)*

Its particular solution is

$$P(t) = \frac{P_0 L}{P_0 + (L - P_0)e^{-Lkt}},$$

where P_0 is the initial population at time $t = 0$. The value k is usually given. The solution $P(t)$ is called a **logistic growth model**, and its graph is sometimes called an **s-curve**.

A logistic function has one point of inflection $(t_i, P(t_i))$, at which $P(t_i) = \frac{1}{2} L$, and this is the location where P is increasing the fastest. It is also called the *point of diminishing returns*, since the rate of change in P decreases to zero after this point.

The particular solution is

$$P(t) = \frac{25,000,000}{500 + (49,500)e^{-0.6t}}$$

The point of inflection occurs when $P = 25,000$. Solving for t, we get $t = 7.7$ months. Sales are increasing the fastest at this time. After that, sales will increase but at a rate that is decreasing to zero.

A second inhibited growth model assumes that P is directly proportional only to the remaining room for growth, $L - P$. This model is expressed by the differential equation

$$\frac{dP}{dt} = k(L - P),$$

and its general solution is

$$y = L(1 - e^{-kt}),$$

which assumes $P_0 = 0$.

A stock's value is currently \$12 per share and is not expected to exceed \$18 per share. The growth in share price can be modeled by $dP/dt = k(6 - P)$. If an initial condition is given, we can solve for k and determine the particular solution.

SECTION 8.4

A **first-order linear differential equation** includes y and its first derivative y', and can be written in the form $a_0(x)y' + a_1(x)y = b(x)$, where $a_0(x), a_1(x)$, and $b(x)$ are continuous functions of x over some interval $[c, d]$. If $b(x) = 0$, then the differential equation is **homogeneous**. Assuming that $a_0(x) \neq 0$, we can divide the differential equation by $a_0(x)$ and write it in *standard form*: $y' + p(x)y = q(x)$.

- $y' \cdot y = 3$ is not linear, since y and its derivative are multiplied together.
- $x^2 y' + 4y = 0$ is linear and homogeneous. Assuming that $x \neq 0$, we can rewrite this equation in standard form:

$$y' + \frac{4}{x^2} y = 0.$$

If $b(x) = 0$, then $q(x) = 0$, and the homogenous differential equation can be solved by separation of variables.

The differential equation $y' + 3xy = 0$ is linear and homogeneous. It can be solved by separation of variables (Section 8.2).

If $q(x) \neq 0$, then the differential equation is **nonhomogeneous** and its general solution is

$$y = \frac{\displaystyle\int m(x)q(x)\, dx + C}{m(x)},$$

where $m(x) = e^{\int p(x)\, dx}$ is an **integration factor**.

$y' + \dfrac{1}{x} y = 4x$, where $x > 0$, is linear and nonhomogeneous. It is solved using the integration factor

$$m(x) = e^{\int \frac{1}{x} dx} = e^{\ln x} = x.$$

The general solution is

$$y = \frac{\displaystyle\int x \cdot 4x\, dx + C}{x} = \frac{\displaystyle\int 4x^2\, dx + C}{x} = \frac{\left(\dfrac{4}{3} x^3 + C\right)}{x} = \frac{4}{3} x^2 + \frac{C}{x}.$$

KEY TERMS AND CONCEPTS	EXAMPLES

SECTION 8.5

An **nth-order linear differential equation** is any equation that can be written in the form

$$a_0(x)y^{(n)} + a_1(x)y^{(n-1)} + \cdots + a_{n-1}(x)y' + a_n(x)y = b(x),$$

where $a_0(x), a_1(x), \ldots, a_n(x)$ are continuous functions of x over some interval $[c, d]$ and $y^{(n)}$ is the nth derivative of y, with $n \geq 1$.

- $y'' + 2y' + y = 0$ is a homogenous second-order linear differential equation.
- $y''' - 9y' = 0$ is a homogenous third-order linear differential equation.

In the homogenous case, if the coefficients of y and its derivatives are constants, we may solve the **auxiliary equation**. If r is a real solution of the auxiliary equation, then $y = e^{rx}$ is a solution of the differential equation.

If $f_1, f_2, f_3, \ldots, f_n$ are solutions of a homogenous linear differential equation, then their **linear combination**

$$y = C_1 f_1(x) + C_2 f_2(x) + C_3 f_3(x) + \cdots + C_n f_n(x),$$

where $C_1, C_2, C_3, \ldots, C_n$ are constants, is also a solution of the differential equation.

For $y'' - y' - 20y = 0$, the auxiliary equation is $r^2 - r - 20 = 0$, which factors to $(r - 5)(r + 4) = 0$. The solutions are $r = 5$ and $r = -4$. Two solutions of the differential equation are e^{5x} and e^{-4x}. They are independent since neither can be written as a multiple of the other. Therefore, the general solution of the differential equation is

$$y = C_1 e^{5x} + C_2 e^{-4x}.$$

If r is a real solution with multiplicity n of a homogeneous linear differential equation, then a solution of the differential equation is

$$y = C_1 e^{rx} + C_2 x e^{rx} + C_3 x^2 e^{rx} + \cdots + C_n x^{n-1} e^{rx},$$

where $C_1, C_2, C_3, \ldots, C_n$ are constants.

For $y'' + 2y' + y = 0$, the auxiliary equation is $r^2 + 2r + 1 = 0$, which factors to $(r + 1)^2 = 0$. The repeated solution is $r = -1$ with multiplicity 2. Therefore, a solution is

$$y = C_1 e^{-x} + C_2 x e^{-x}.$$

The **method of undetermined coefficients** is used to determine solutions of nonhomogeneous higher-order differential equations, assuming that $b(x)$ is a polynomial and the coefficients of y and its derivatives are constant. The general solution is the homogenous solution plus the specific solution.

For $y'' - y' - 2y = x^2 + x$, the homogeneous solution is $y = C_1 e^{2x} + C_2 e^{-x}$. We assume a specific solution has the form $y = Ax^2 + Bx + C$. After taking derivatives and substituting, we equate coefficients and solve for A, B, and C, obtaining $A = -\frac{1}{2}, B = 0$, and $C = -\frac{1}{2}$. The specific solution is $y = -\frac{1}{2}x^2 - \frac{1}{2}$, and the general solution is

$$y = C_1 e^{2x} + C_2 e^{-x} - \tfrac{1}{2}x^2 - \tfrac{1}{2}.$$

Trigonometry Connection: If the solutions to the auxiliary equation are complex numbers of the form $r = a \pm bi\ (b \neq 0)$, then a solution of the differential equation is

$$y = C_1 e^{ax} \sin bx + C_2 e^{ax} \cos bx.$$

For $y'' + y' + 5y = 0$, the auxiliary equation is $r^2 + r + 5 = 0$. The quadratic formula gives $r = \dfrac{-1 \pm i\sqrt{19}}{2}$ as its solutions. A solution of the differential equation is

$$y = C_1 e^{(-1/2)x} \sin\left(\frac{\sqrt{19}}{2}x\right) + C_2 e^{(-1/2)x} \cos\left(\frac{\sqrt{19}}{2}x\right).$$

CHAPTER 8
REVIEW EXERCISES

These review exercises are for test preparation. They can also be used as a practice test. Answers are at the back of the book. The blue bracketed section references tell you what part(s) of the chapter to restudy if your answer is incorrect.

CONCEPT REINFORCEMENT

Match each description in column A with the appropriate equation or term in column B.

Column A

1. The equation $y' + xy = 0$ is this type of differential equation.

2. This differential equation models uninhibited growth.

3. This differential equation models Newton's Law of Cooling.

4. The differential equation $dy/dx = 4y^2x$ can be solved using this method.

5. This differential equation models growth when the rate of change is proportional only to the available room for growth.

Column B

a) $\dfrac{dP}{dt} = kP$ [8.1]

b) First-order linear differential equation [8.4]

c) Separation of variables [8.2]

d) $\dfrac{dP}{dt} = k(L - P), k > 0$ [8.3]

e) $\dfrac{dT}{dt} = -k(T - R), k > 0$ [8.4]

REVIEW EXERCISES

In Exercises 6–12, find the general solution of each differential equation. If an initial condition is given, find the particular solution.

6. $y' = 6x$ [8.1]

7. $y' = x^2$ [8.1]

8. $y' = 2x + 1, y(1) = 5$ [8.1]

9. $y' = x^3, y(3) = 0$ [8.1]

10. $y' = x^3y$ [8.2]

11. $y' = x + 2xy$ [8.2]

12. $y' = 6y, y(0) = -2$ [8.1, 8.2]

13. Verify that $y = e^{0.05x}$ is a solution of $dy/dx = 0.05y$. [8.1]

14. Verify that $y = e^{2x} - 3e^{5x}$ is a solution of $y'' - 7y' + 10y = 0$. [8.1]

15. Find the general solution of the differential equation described by this sentence: "The rate of change of y is directly proportional to the fourth power of y." Let the constant of proportionality be $k = 1$. [8.2]

16. Business: account balance. Hunter deposits $700 in a savings account. The value of the account is modeled by the differential equation $dA/dt = 0.05A$. [8.1]

   a) Find the particular solution $A(t)$, the value of Hunter's savings account, after t years.

   b) Find $A(5)$ and $A'(5)$, and explain what these numbers represent.

   c) Find $\dfrac{A'(5)}{A(5)}$, and explain what this number represents.

17. Assume that for the inhibited growth model $dP/dt = kP(L - P)$, the point of inflection of the graph of P occurs at $(25, 525)$. [8.3]

   a) Find the value of L.

   b) Find the particular solution P, where $P_0 = 200$.

18. Business: market saturation. The TouchTablet is released in a test market, and 150 units are sold initially. Sales are modeled by $dP/dt = 0.0000016P(L - P)$, where $L = 360,000$ potential customers and t is in months. [8.3]

   a) Find the particular solution $P(t)$, the number of units that have been sold, after t months.

   b) After how many months are sales increasing the fastest?

   c) When will 80% of the customers in the test market have purchased a TouchTablet?

19. Business: value of a stock. Stock in BES Industries is currently worth $15 per share, and the value of a share is modeled by $dV/dt = 0.017(22 - V)$. [8.3]

   a) Find the particular solution V, the value of a share of the stock in dollars, after t weeks.

   b) Explain the significance of the number 22 in the model.

   c) When will the stock's share value be $20?

20. Newton's Law of Cooling. A cup of hot water is removed from a microwave and set on a counter in a room with a constant temperature of 25°C. The water's initial temperature is 90°C, and after 10 min, it is 78°C. [8.4]

   a) Find the particular solution $T(t)$, the temperature of the water, after t minutes.

   b) When will the water's temperature be 30°C?

21. Mixture problem   A tank holds 2500 gal of water, in which is dissolved 200 lb of corn syrup. A liquid containing 2 lb of corn syrup per gallon of water is pumped into the tank at a rate of 5 gal per minute. The liquids mix evenly, and the mixture leaves the tank at a rate of 5 gal per minute at the other end. [8.4]

   a) Find the particular solution $A(t)$, the amount of corn syrup in the tank after t minutes.

   b) How much corn syrup is in the tank after 3 hr?

In Exercises 22–27, find the general solution of each first-order linear differential equation. If an initial condition is given, also find the particular solution.

22. $y' + 7y = 0$ [8.4]

23. $y' - 12y = 0, y(0) = \frac{1}{2}$ [8.4]

24. $y' + 3xy = x$ [8.4]

25. $y' - \dfrac{9}{x}y = 3x^2, y(1) = -1$ [8.4]

26. $y' + \dfrac{y}{2x} = 3$ [8.4]

27. $y' + 4x^3y = x^3$ [8.4]

In Exercises 28–30, determine the general solution of each higher-order linear differential equation. If an initial condition is given, also find the particular solution.

28. $y'' + 3y' - 70y = 0$ [8.5]

29. $y'' - 2y' - 99y = 0, y(0) = 1, y'(0) = 1$ [8.5]

30. $y''' - 6y'' + 9y' = 0$ [8.5]

31. Use the method of undetermined coefficients to determine the general solution of $y'' + 7y' + 10y = 3x - 2$. [8.5]

 32. Suppose $y = 3xe^{2x}$ is a solution of a homogeneous differential equation with constant coefficients. Explain why $y = 3e^{2x}$ must also be a solution to the differential equation. Identify another possible solution. [8.4]

TRIGONOMETRY CONNECTION

The differential equations in Exercises 33–35 have trigonometric solutions. Determine the general solution of each differential equation. If an initial condition is given, also find the particular solution.

33. $y'' + 10y = 0$ [8.5]

34. $y'' + y' + 6y = 0$ [8.5]

35. $y'' + y = 0, y(0) = 3, y'(0) = 1$ [8.5]

36. Use the method of undetermined coefficients to determine the general solution of $y'' - 2y' + 5y = 5x^2 - 9x - 31$. [8.5]

TECHNOLOGY CONNECTION

Find the general solution of each differential equation. [8.5]

37. $y''' - 6y'' - 31y' + 36y = 0$

38. $y^{(4)} - 6y''' + 22y' + 15y = 0$

39. $y^{(4)} - 5y''' - 24y'' - 4y' + 32y = 0$

40. $y^{(4)} - 5y'' + 4y = 0$

CHAPTER 8
TEST

1. Solve the differential equation $y' = x^2 + 5$, where $y = 3$ when $x = 1$.

2. Find the general solution of $y' = 6x^5y$.

3. Find the general solution of $\dfrac{dy}{dx} = \dfrac{9x}{y}$.

4. Find the particular solution of $y' - \dfrac{3}{x}y = 2$, where $y = 2$ when $x = 1$.

5. For the inhibited growth model $\dfrac{dP}{dt} = kP(L - P)$, suppose $P = 1200$ at the point of inflection. What is the limiting value, L?

6. Verify that $y = \dfrac{5x}{2} - \dfrac{5}{4} + Ce^{-2x}$ is a solution of $y' + 2y = 5x$.

7. Business: account balance.   Elaine deposits $2500 in a savings account. The value of the account is modeled by $dA/dt = 0.0325A$.

   a) Find the particular solution $A(t)$, the value of Elaine's savings account, after t years.

   b) Find $A(3)$ and $A'(3)$, and explain what these numbers represent.

   c) Find $\dfrac{A'(3)}{A(3)}$, and explain what this number represents.

8. Business: market saturation.   Sales of a new calculus text are modeled by $\dfrac{dP}{dt} = 0.000000125P(300{,}000 - P)$. Initially, 600 books are sold ($t = 0$).

   a) Find the particular solution $P(t)$, the sales of the book after t weeks on the market.

   b) After how many weeks on the market will sales increase the fastest?

9. Business: value of a stock. Stocks of Bradley Publishing are currently worth $7 per share, and the value of the stock is modeled by $\dfrac{dV}{dt} = 0.0088(13.25 - V)$.

   a) Find the particular solution $V(t)$, the value of the stock, in dollars, after t weeks.

   b) When will the stock's value be $12?

10. Newton's Law of Cooling. Water at 70°F is placed into a freezer with a constant temperature of 22°F. After 15 minutes, the water's temperature is 45°F.

   a) Find the particular solution $T(t)$, the temperature of the water, after t minutes in the freezer.

   b) When will the water begin to freeze (32°F)?

11. Mixture problem. A pond contains 600,000 gal of fresh-water. A stream whose water has 1 lb of contaminant per gallon flows into the pond at a rate of 10 gal per minute, the waters mix uniformly, and the mixture leaves the pond at the other end at a rate of 10 gal per minute.

   a) Find the particular solution $A(t)$, the amount of con-taminant in the pond after t minutes.

   b) How much contaminant is in the pond after 1 day (1440 min)?

12. Find the general solution of $y'' - 8y' - 33y = 0$.

13. Find the particular solution of $y'' - 25y = 0$, where $y(0) = 3$ and $y'(0) = 2.5$.

14. Find the general solution of $y''' + 2y'' + y' = 0$.

15. Find the general solution of $y''' - 100y' = 0$.

16. Use an integration factor to find the general solution of
$$y' + 4y = x.$$

17. Use the method of undetermined coefficients to find the general solution of $y'' - y' - 12y = 12x^2 + 26x + 12$.

TRIGONOMETRY CONNECTION

18. Find the general solution of $y'' + 3y' + 9y = 0$.

19. Find the general solution of $y'' + y = 0$, where $y(0) = 5$ and $y'(0) = 3$.

20. Find the general solution of $y''' - 2y'' - 25y' + 50y = 0$.

21. Suppose $y = 3x + 1$ and $y = x^2 - 3$ are solutions of a homogeneous differential equation with constant coefficients. Explain why $y = x^2 + 3x - 2$ is also a solution of the differential equation.

Extended Technology Application

Visual (Slope Fields) and Numerical (Euler's Method) Solutions of Differential Equations

Slope Fields

The set of solution curves for a differential equation can sometimes be determined visually. Consider the first-order differential equation $y' = x + y$. We can choose values for x and y and evaluate the slope of the derivative y' at the point (x, y). For example, the slope of the tangent line at $(-2, 1)$ is $y' = -2 + 1 = -1$. In the figure at the right, three sample slope lines are drawn for the differential equation $y' = x + y$.

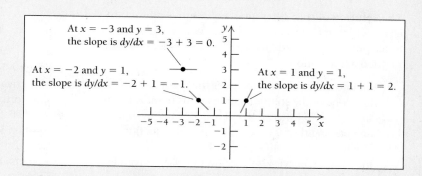

At $x = -3$ and $y = 3$, the slope is $dy/dx = -3 + 3 = 0$.

At $x = -2$ and $y = 1$, the slope is $dy/dx = -2 + 1 = -1$.

At $x = 1$ and $y = 1$, the slope is $dy/dx = 1 + 1 = 2$.

Since drawing slope fields by hand can be time-consuming, it is instead carried out with computer software (for example, GeoGebra). Free online applets that will produce slope fields are also available. For example, the plot of $y' = x + y$ shown on the left below was created with an applet by Professor Marek Rychlik, available at http://alamos.math.arizona.edu/ODEApplet/JOdeApplet.html.

Each solution curve of the differential equation $y' = x + y$ can be visualized by using these slope lines. In the figure at the right below, four possible solution curves are shown, each passing through a particular point (and thus satisfying an initial condition). Note how the solution curves "flow" with the slope lines.

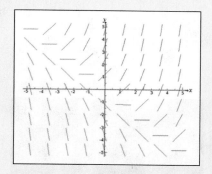

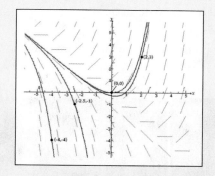

EXERCISES

1. Let $y' = x - y$. Find the slopes for all points with coordinates that are integers, with $-2 \le x \le 2$ and $-2 \le y \le 2$. Draw the slope field on a graph.

2. Use software or an online applet to draw the slope field and three particular solution curves of $y' = x - y$, one each for the initial conditions $(0, 0)$, $(-1, 1)$ and $(0, -1.5)$.

Euler's Method

A solution of a differential equation can be found numerically, using a method attributed to Leonhard Euler. Although *Euler's method* requires considerable numerical calculation, this can be performed using a spreadsheet. We start with a differential equation of the form

$$y' = f(x, y), \text{ with initial condition } (x_0, y_0).$$

The solution of this differential equation is a function $\varphi(x)$, which may or may not be known (φ is the Greek letter phi, pronounced "fee" or "fi"). The steps of Euler's method are as follows:

1. We know the solution includes the point (x_0, y_0), and the slope of the solution curve at this point is given as $y' = f(x_0, y_0)$. Recall that if a slope m is given for a line passing through (x_0, y_0), then the equation of that line is $y - y_0 = m(x - x_0)$. Thus, the line tangent to the solution curve $y = \varphi(x)$ at (x_0, y_0) is given by

$$y = y_0 + f(x_0, y_0)(x - x_0).$$

2. The tangent line in step 1 closely approximates the solution curve $y = \varphi(x)$ for x-values near x_0. Thus, we choose a new x-value, x_1, that is "close" to x_0, and from it, determine y_1:

$$y_1 = y_0 + f(x_0, y_0)(x_1 - x_0).$$

3. We now have a new point, (x_1, y_1), which approximates the solution $y = \varphi(x)$. The derivative at this point is $y' = f(x_1, y_1)$, and we repeat steps 1 and 2: From step 1, we have $y = y_1 + f(x_1, y_1)(x - x_1)$, and for step 2, we choose a new x-value, x_2, that is "close" to x_1, and from it, determine y_2:

$$y_2 = y_1 + f(x_1, y_1)(x_2 - x_1).$$

This gives a new point, (x_2, y_2), which approximates $y = \varphi(x)$.

For the third pass through these steps, the derivative is $y' = f(x_2, y_2)$, and we choose a new x-value, x_3, "close" to x_2. This gives $y_3 = y_2 + f(x_2, y_2)(x_3 - x_2)$, thus generating a new point, (x_3, y_3), and so on.

In general, each point (x_n, y_n) that approximates the true solution, $y = \varphi(x)$, is found from the formula

$$y_n = y_{n-1} + f(x_{n-1}, y_{n-1})(x_n - x_{n-1}).$$

If we let $h = x_n - x_{n-1}$, then we have

$$y_n = y_{n-1} + f(x_{n-1}, y_{n-1})h,$$

where h is called the *step size*. As long as h is small, the numerical approximation is reasonably close to the true solution, $y = \varphi(x)$.

Let's use this method to numerically solve the differential equation $y' = x + y$ with the initial condition $y(0) = 1$. We have $x_0 = 0$ and $y_0 = 1$, and the slope of the solution curve at this point is $y' = f(x_0, y_0) = f(0, 1) = 0 + 1 = 1$. Therefore, we have

$$y = y_0 + f(x_0, y_0)(x - x_0)$$
$$y = 1 + 1(x - 0) \qquad \text{Substituting}$$
$$y = x + 1. \qquad \text{Simplifying}$$

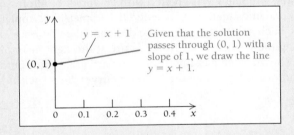

Given that the solution passes through $(0, 1)$ with a slope of 1, we draw the line $y = x + 1$.

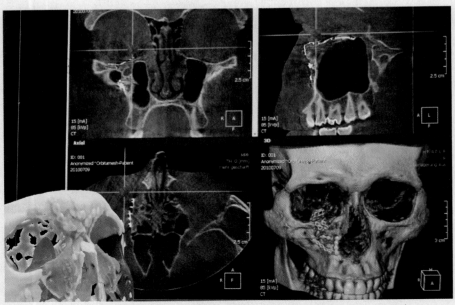

Numerical methods are sometimes used to solve differential equations in 3D image reconstruction.

Now, we select a new x-value "close" to $x_0 = 0$ and find y_1. If we choose a step size of $h = 0.1$, then for $x_1 = 0.1$, we have $y_1 = 1 + 0.1 = 1.1$.

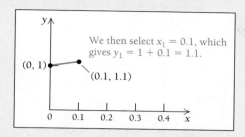

We repeat the process. We find the slope at $(0.1, 1.1)$, which is $y' = f(x_1, y_1) = f(0.1, 1.1) = 0.1 + 1.1 = 1.2$. The point and the slope define a new line:

$$y = y_1 + f(x_1, y_1)(x - x_1)$$
$$y = 1.1 + 1.2(x - 0.1) \qquad \text{Substituting}$$
$$y = 1.2x + 0.98. \qquad \text{Simplifying}$$

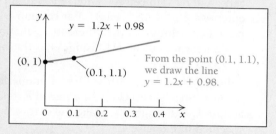

Keeping this same step size of $h = 0.1$, we next choose $x_2 = 0.2$, and using the equation $y = 1.2x + 0.98$, we get $y_2 = 1.22$.

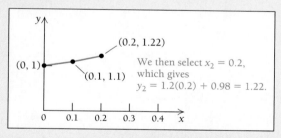

Next, we use both x_2 and y_2 to find $y' = f(x_2, y_2)$. We use this result to write an equation to generate y_3, and so on. In this way, we slowly build a solution for $y' = x + y$, with $y(0) = 1$, one point at a time.

EXERCISE

3. Let $y' = xy$, with $y(1) = 3$.

   a) Use Euler's method to find y_1, y_2, y_3, y_4, and y_5. Assume a step size of $h = 0.1$.
   b) Find the particular solution $y = \varphi(x)$ of $y' = xy$, with $y(1) = 3$.
   c) Compare the actual solutions φ_1 through φ_5 with the approximations found in part (a).

Using an Excel™ Spreadsheet

Euler's method can be adapted so that we can use a spreadsheet. We open a blank spreadsheet and enter the following field headers in the cells:

	A	B	C
1	Step-size =		
2			
3	x	dy/dx	y

For this demonstration, we again start with the differential equation $y' = x + y$, with $y(0) = 1$. We enter the following:

1. In cell B1, enter a value, for example, 0.1. This is the step size.
2. In cell A4, enter 0 (this is x_0).
3. In cell C4, enter 1 (this is y_0).
4. In cell B4, enter =A4+C4.
5. In cell A5, enter =A4+$B$1. Then copy cell A5 and paste it in cells A6 through A10. This generates the values for x_1, x_2, x_3, and so on.
6. Copy cell B4 and paste it in cells B5 through B10. This generates the slopes $f(x_n, y_n)$.
7. In cell C5, enter =C4+B4*$B$1. Copy cell C5 and paste it in cells C6 through C10. This generates the values y_1, y_2, y_3, and so on.

	A	B	C
1	Step-size =	0.1	
2			
3	x	dy/dx	y
4	0	1	1
5	0.1	1.2	1.1
6	0.2	1.42	1.22
7	0.3	1.662	1.362
8	0.4	1.9282	1.5282
9	0.5	2.22102	1.72102
10	0.6	2.543122	1.943122

Columns A and C form ordered pairs that approximate the solution of $y' = x + y$, with $y(0) = 1$. That is, the actual solution $y = \varphi(x)$ is approximated by $(0, 1)$, $(0.1, 1.1)$, $(0.2, 1.22)$, $(0.3, 1.362)$, and so on. Remember, for some differential equations, it may not be possible to find a solution $y = \varphi(x)$ directly. A numerical method such as the one described here may be the only way to solve some differential equations.

For the differential equation $y' = x + y$, with $y(0) = 1$, we can find a particular solution. Using an integration factor (Section 8.4), we have $y = \varphi(x) = -x - 1 + 2e^x$. To see how close the numerical approximations are to the actual solution, enter the following on the spreadsheet:

8. In cell D3, enter Actual y as a header.

9. In cell D4, enter =–A4–1+2*EXP(A4). Copy cell D4 and paste it in cells D5 through D10. This generates the actual y-values of the solution $y = \varphi(x)$.

We can compare the approximations of y in column C with the actual values for y in column D:

	A	B	C	D
1	Step-size =	0.1		
2				
3	x	dy/dx	y	Actual y
4	0	1	1	1
5	0.1	1.2	1.1	1.110341836
6	0.2	1.42	1.22	1.242805516
7	0.3	1.662	1.362	1.399717615
8	0.4	1.9282	1.5282	1.583649395
9	0.5	2.22102	1.72102	1.797442541
10	0.6	2.543122	1.943122	2.044237601

For small step sizes, the approximations are reasonably accurate. You may experiment with different step sizes. If you need more rows, copy cells A10 through D10 and paste them in cells A11 through D20 (or even more cells).

Graphically, the actual solution and the approximation match very well over a small interval near $x = 0$, as shown below.

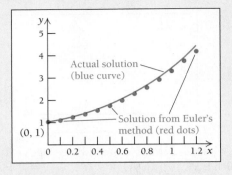

EXERCISES

4. Solve $y' = x^2 - 2y$, with $y(2) = -1$, numerically, using a spreadsheet. Use a step size of $h = 0.1$, and find y_1 through y_{10}. Make the following entries in cells: enter 2 into cell A4, enter -1 into cell C4, and enter =A4^2–2*C4 into cell B4. For column D, find the particular solution of the differential equation and type its form into cell D4, using A4 as the variable (see step 9 above).

5. Solve $y' = xy + x$, with $y(1) = 3$, numerically, using a spreadsheet. Use a step size of $h = 0.1$, and find y_1 through y_{10}. Make the entries in cells specified in Exercise 4.

Sequences and Series

9

Chapter Snapshot

What You'll Learn

9.1 Arithmetic Sequences and Series
9.2 Geometric Sequences and Series
9.3 Simple and Compound Interest
9.4 Annuities and Amortization
9.5 Power Series and Linearization
9.6 Taylor Series and a Trigonometry Connection

Why It's Important

An ordered set of numbers is called a *sequence*, and their sum is called a *series*. Sequences and series have many useful properties, including applications in business and finance. We can also express certain functions as a series of polynomial terms.

Where It's Used

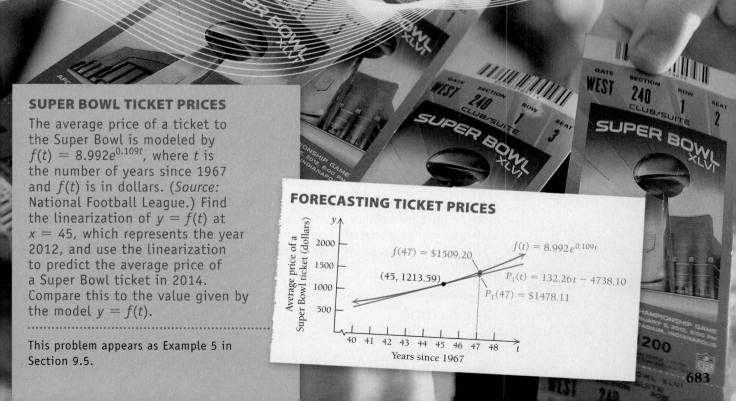

SUPER BOWL TICKET PRICES

The average price of a ticket to the Super Bowl is modeled by $f(t) = 8.992e^{0.109t}$, where t is the number of years since 1967 and $f(t)$ is in dollars. (*Source:* National Football League.) Find the linearization of $y = f(t)$ at $x = 45$, which represents the year 2012, and use the linearization to predict the average price of a Super Bowl ticket in 2014. Compare this to the value given by the model $y = f(t)$.

This problem appears as Example 5 in Section 9.5.

FORECASTING TICKET PRICES

$f(47) = \$1509.20$

$f(t) = 8.992e^{0.109t}$

$(45, 1213.59)$

$P_1(t) = 132.26t - 4738.10$

$P_1(47) = \$1478.11$

Average price of a Super Bowl ticket (dollars)

2000
1500
1000
500

40 41 42 43 44 45 46 47 48 t

Years since 1967

9.1 Arithmetic Sequences and Series

A **sequence** is any ordered set of numbers such as the following:

$$2, 5, 8, 11, 14, 17, 20, \ldots.$$

A motion picture is formed by a sequence of still images.

The three dots indicate that more terms follow. Each number is a **term** of the sequence. In the above sequence, the first term is 2, the second term is 5, the third term is 8, and so on. We can describe the terms of this sequence as follows:

$$a_1 = 2, a_2 = 5, a_3 = 8, \text{ and so on.}$$

This sequence follows a pattern we can describe with the formula $a_n = 3n - 1$. Here, a_n is called the **general *n*th term** of the sequence. It can be regarded as a function whose domain is a set of consecutive whole numbers, usually starting with $n = 1$. If the sequence does not end, it is called an *infinite sequence*. Otherwise, it is called a *finite sequence*.

■ **EXAMPLE 1**   For each general *n*th term, find the first four terms of the sequence and the 20th term.

a) $a_n = 5n + 4$ **b)** $b_n = n^2 + n$

Solution   We find the terms of each sequence by substituting $n = 1, 2, 3, \ldots$:

a) For the sequence given by $a_n = 5n + 4$, we have

$$a_1 = 5 \cdot 1 + 4 = 9,$$
$$a_2 = 5 \cdot 2 + 4 = 14,$$
$$a_3 = 5 \cdot 3 + 4 = 19,$$
$$a_4 = 5 \cdot 4 + 4 = 24.$$

The 20th term is $a_{20} = 5 \cdot 20 + 4 = 104$.

b) For the sequence given by $b_n = n^2 + n$ we have

$$b_1 = 1^2 + 1 = 2,$$
$$b_2 = 2^2 + 2 = 6,$$
$$b_3 = 3^2 + 3 = 12,$$
$$b_4 = 4^2 + 4 = 20.$$

The 20th term is $b_{20} = 20^2 + 20 = 420.$

❰ Quick Check 1

❱ **Quick Check 1**

Find the first four terms and the 30th term of the sequence defined by the general *n*th term $c_n = 9n + 8$.

Arithmetic Sequences

Consider the sequence

$$2, 5, 8, 11, 14, 17, 20, \ldots.$$

Because the same number is added to each term to obtain the next term, this is called an *arithmetic* (pronounced "a-rith-*meh*-tik") *sequence*. The number d (in the example above, $d = 3$) that is added to a term in order to obtain the next term is called the *common difference*.

> **DEFINITION**
>
> A sequence is **arithmetic** if there exists a number d, called the **common difference**, such that $a_{n+1} = a_n + d$, or $a_{n+1} - a_n = d$.

Arithmetic sequences are also called *linear sequences*, and the common difference d can be thought of as the sequence's slope, or rate of change.

■ **EXAMPLE 2** Find the common difference in each arithmetic sequence.

a) $6, 13, 20, 27, 34, \ldots$ **b)** $52, 43, 34, 25, 16, \ldots$

Solution

a) Since $13 - 6 = 7$ and $20 - 13 = 7$, and so on, the common difference is $d = 7$.

b) Since $43 - 52 = -9$ and $34 - 43 = -9$, and so on, the common difference is $d = -9$.

❬ Quick Check 2

❭ **Quick Check 2**

Find the common difference of the arithmetic sequence $1, 3.5, 6, 8.5, \ldots$.

We can generalize any arithmetic sequence in a manner that allows us to find its general nth term. Suppose a_1 is the first term of an arithmetic sequence, and d is the common difference. Then, we have the following:

$$a_2 = a_1 + d,$$
$$a_3 = a_1 + d + d = a_1 + 2d,$$
$$a_4 = a_1 + d + d + d = a_1 + 3d,$$

and so on.

This result leads to a formula for the general nth term of an arithmetic sequence.

> **THEOREM 1**
>
> The **general nth term** of an arithmetic sequence with a first term a_1 and a common difference d is given by
>
> $$a_n = a_1 + (n - 1)d.$$

■ **EXAMPLE 3** For the following arithmetic sequence, find the general nth term.

$$11, 17, 23, 29, 35, 41, \ldots.$$

Solution The common difference is $d = 6$, and the first term is $a_1 = 11$. Therefore, we have

$$a_n = a_1 + (n - 1)d \qquad \text{Recalling the formula for the general } n\text{th term of an arithmetic sequence}$$
$$= 11 + (n - 1)6 \qquad \text{Substituting}$$
$$= 6n + 5. \qquad \text{Using the distributive law and simplifying}$$

We check this result by substituting $n = 1, 2, 3, \ldots$ into $a_n = 6n + 5$:

$$a_1 = 6 \cdot 1 + 5 = 11, a_2 = 6 \cdot 2 + 5 = 17, a_3 = 6 \cdot 3 + 5 = 23, \text{ and so on.}$$

❬ Quick Check 3

❭ **Quick Check 3**

Find the general nth term of the arithmetic sequence $30, 26, 22, 18, 14, \ldots$.

Application of Arithmetic Sequences

We can now find the general nth term of arithmetic sequences in everyday situations.

■ **EXAMPLE 4** **Business: Customer Traffic.** Marion's Gym opened a new location. On the 5th day of business, the gym had 64 members, and on the 9th day, it had 88 members. Assume membership grows linearly during the first few weeks of business.

a) Find a formula for m_n, the number of members on the nth day of business.

b) How many members does Marion's Gym have on the 15th day?

Solution

a) Since the growth in membership is linear, we can model it with an arithmetic sequence. We have $m_5 = 64$ and $m_9 = 88$. To find the common difference, we note that there are $9 - 5$, or 4, terms between m_9 and m_5. Thus,

$$4d = 88 - 64$$
$$d = \frac{24}{4}$$
$$= 6.$$

We can write the general nth term as $m_n = m_1 + (n - 1)6$, where m_1 is unknown. To find m_1, we substitute one of the given points. Let's use $m_5 = 64$:

$$m_n = m_1 + (n - 1)6$$
$$64 = m_1 + (5 - 1)6 \qquad \text{Substituting}$$
$$64 = m_1 + 24$$
$$m_1 = 40.$$

Therefore, the general nth term is $m_n = 40 + (n - 1)6$, or $m_n = 34 + 6n$. This checks since we have $m_5 = 34 + 6 \cdot 5 = 64$ and $m_9 = 34 + 6 \cdot 9 = 88$.

b) On the 15th day, Marion's Gym has $m_{15} = 34 + 6 \cdot 15 = 124$ members.

⟨ Quick Check 4

> **Quick Check 4**
>
> Andrea withdraws funds daily for her living expenses while at college. On the 10th day, Andrea's account has $600, and on the 16th day, her account has $480. Assume that Andrea withdraws the same amount each day. Find the general nth term, a_n, the amount of money in her account on the nth day, and then determine the amount in her account on the 23rd day.

Arithmetic Series

The sum of the terms of any sequence is called a **series**. In general, the sum of the first n terms of a series is called the **nth partial sum** of the series and is denoted by S_n. For an arithmetic series, we can develop a formula to find the nth partial sum directly, without writing out and adding all of the terms. We illustrate the development of this formula using an example.

Suppose a recital hall has 30 curved rows of seats, with 12 seats in the first row, 14 seats in the second row, and so on; that is, each row has 2 more seats than the previous row. How many seats are in this recital hall?

We note that the number of seats a_n in the nth row is given by $a_n = 10 + 2n$. Since the recital hall has 30 rows, there are $a_{30} = 10 + 2 \cdot 30 = 70$ seats in the 30th row. The number of seats in each of the preceding rows can be found in a similar way. Thus, the total number of seats is given by

$$S_{30} = 12 + 14 + 16 + 18 + \cdots + 64 + 66 + 68 + 70. \qquad (1)$$

Let's write the series again, but in reverse order:

$$S_{30} = 70 + 68 + 66 + 64 + \cdots + 18 + 16 + 14 + 12. \qquad (2)$$

Adding equations (1) and (2), we obtain

$$\begin{aligned} S_{30} &= 12 + 14 + 16 + 18 + \cdots + 64 + 66 + 68 + 70 \\ + \; S_{30} &= 70 + 68 + 66 + 64 + \cdots + 18 + 16 + 14 + 12 \\ \hline 2S_{30} &= 82 + 82 + 82 + 82 + \cdots + 82 + 82 + 82 + 82. \end{aligned}$$

Each column on the right-hand side adds to 82, and there are 30 such columns. Thus, the value of the right-hand side is $30 \cdot 82$. The left-hand side is $S_{30} + S_{30} = 2S_{30}$. Therefore, we have

$$2S_{30} = 30 \cdot 82$$
$$S_{30} = \tfrac{30}{2} \cdot 82 = 1230 \text{ seats.} \qquad \text{Multiplying both sides by } \tfrac{1}{2}$$

Note that 30 represents the number of terms in the series, and $82 = 12 + 70$ is the sum of the first and last terms of the series. This is generalized in a formula for the nth partial sum of an arithmetic series.

THEOREM 2

The **nth partial sum** of an arithmetic series is

$$S_n = \frac{n}{2}(a_1 + a_n).$$

A more formal development of this result is outlined in Exercise 51. Recall from Section 4.2 that a sum can also be written using summation notation. Thus, we may write S_n as

$$S_n = \sum_{k=1}^{n} a_k = a_1 + a_2 + a_3 + \cdots + a_n.$$

■ **EXAMPLE 5** Find the specified partial sum for each arithmetic series.

a) The sum of the first 100 natural numbers, $1 + 2 + 3 + 4 + \cdots + 100$

b) $\displaystyle\sum_{n=1}^{50}(4n + 7)$

c) The sum of the first 200 terms of the series $3 + 12 + 21 + 30 + 39 + \cdots$

Solution

a) The first term is $a_1 = 1$, and the 100th term is $a_{100} = 100$. We have $n = 100$ terms, so the sum is

$$S_{100} = \frac{100}{2}(1 + 100) = 50(101) = 5050.$$

b) The first term is $a_1 = 7 + 4 \cdot 1 = 11$, and the 50th term is $a_{50} = 7 + 4 \cdot 50 = 207$. The sum of the first 50 terms is

$$S_{50} = \frac{50}{2}(11 + 207) = 25(218) = 5450.$$

c) The first term is $a_1 = 3$. Since we do not know a_{200}, we find it by first determining that $d = 9$ and so $a_n = 3 + (n - 1)9$, or $a_n = -6 + 9n$. Thus, the 200th term of the sequence is $a_{200} = -6 + 9 \cdot 200 = 1794$, and the sum of the first 200 terms of the series is

$$S_{200} = \frac{200}{2}(3 + 1794) = 100(1797) = 179{,}700.$$

> **Quick Check 5**
>
> Find the sum of the first 75 terms of the arithmetic series $9 + 17 + 25 + 33 + 41 + \cdots$.

❮ Quick Check 5

As a schoolboy, the great German mathematician Carl Friedrich Gauss (1777–1855) and his classmates were ordered to find the sum of $1 + 2 + 3 + \cdots + 100$, presumably as an activity to keep them occupied for a while. To his teacher's amazement, young Gauss came up with the solution in seconds by using the method in Example 5, while his classmates were laboriously adding the terms individually.

Applications of Arithmetic Series

■ **EXAMPLE 6**   **Business: Customer Traffic.**   Petrak Clothiers opens a new store that has 75 customers on its first day. Each day for the first few weeks, 20 more customers visit Petrak's than the day before.

a) How many customers visit Petrak's on the 10th day of business?

b) How many customers in total visit the store during the first 10 days?

Solution

a) We have an arithmetic sequence, 75, 95, 115, 135, … . The first term is $a_1 = 75$, and the common difference is $d = 20$. Therefore, the general nth term is $a_n = 75 + (n - 1)20$, or $a_n = 55 + 20n$. On the 10th day, there are $a_{10} = 55 + 20 \cdot 10 = 255$ customers at Petrak's Clothiers.

b) There are a total of $S_{10} = \dfrac{10}{2}(75 + 255) = 1650$ customers during the store's first 10 days of business.

❰ Quick Check 6

> ❱ **Quick Check 6**
>
> Dana saves $1.25 in her piggy bank on the 1st day of January, and each day after that, she saves $0.25 more than on the preceding day. How much money will she put in her piggy bank on the 31st day of January, and how much will she save in total over the 31 days?

> ❱ **Quick Check 7**
>
> There are 50 boxes of cookies in the bottom row of a display. Each row above that has 3 fewer boxes than the row below it, and there are 14 rows altogether.
>
> **a)** How many boxes are in the top (14th) row?
>
> **b)** How many boxes are in the whole display?

■ **EXAMPLE 7**   **Business: Product Display.**   A grocery store creates a pyramid of cans in which each row has one fewer can than the row below it. Suppose the bottom row has 25 cans, and the top row has a single can. How many cans are in the pyramid?

Solution   This is an arithmetic series: $1 + 2 + 3 + 4 + \cdots + 25$. The first term is $a_1 = 1$, and the common difference is $d = 1$. Therefore, the display has

$$S_{25} = \frac{25}{2}(1 + 25) = 325 \text{ cans.}$$

❰ Quick Check 7

Section Summary

- A *sequence* is any ordered set of numbers. Each number is a *term* of the sequence.
- Many sequences can be described by a *general nth term*, a_n. The domain of a sequence is a set of whole numbers.
- An *arithmetic sequence* is a sequence for which the same number d is added to each term to obtain the next term:
$$a_{n+1} = a_n + d, \text{ or } d = a_{n+1} - a_n.$$

- The number d is called the *common difference*, and the general nth term of an arithmetic sequence is $a_n = a_1 + (n - 1)d$.
- A *series* is the sum of the terms of a sequence. The *nth partial sum*, S_n, of a series is the sum of its first n terms.
- The nth partial sum of an arithmetic series is given by
$$S_n = \frac{n}{2}(a_1 + a_n).$$

EXERCISE SET
9.1

For Exercises 1–10, the general nth term of an arithmetic sequence is given. Find (a) the first four terms of each sequence, (b) the 20th term of the sequence, and (c) the 25th partial sum of the series.

1. $a_n = 3n + 8$

2. $b_n = 2n - 5$

3. $a_n = n + 7$

4. $b_n = 10 + n$

5. $c_n = 1 - 2n$

6. $d_n = -4n + 5$

7. $c_n = \frac{1}{2}n + \frac{3}{2}$

8. $d_n = \frac{3}{4}n + \frac{1}{2}$

9. $p_n = 1.6n + 2.1$

10. $q_n = 4.3n - 1.9$

For Exercises 11–20, find (a) the common difference, (b) the general nth term, and (c) the 30th partial sum.

11. $7, 19, 31, 43, 55, \ldots$

12. $17, 23, 29, 35, 41, \ldots$

13. $100, 104, 108, 112, 116, \ldots$

14. $250, 500, 750, 1000, 1250, \ldots$

15. $80, 78, 76, 74, 72, \ldots$

16. $57, 51, 45, 39, 33, \ldots$

17. $8, 11.5, 15, 18.5, 22, \ldots$

18. $13.1, 16.4, 19.7, 23, 26.3, \ldots$

19. $4, 3.85, 3.7, 3.55, 3.4, \ldots$

20. $20.4, 19.15, 17.9, 16.65, 15.4, \ldots$

In Exercises 21–30, find (a) the general nth term of the sequence, (b) the first five terms and the 50th term, and (c) the 50th partial sum of the series.

21. A sequence has $a_1 = 4$, and each term is found by adding 3 to the preceding term.

22. A sequence has $a_1 = 15$, and each term is found by adding 8 to the preceding term.

23. A sequence has $a_1 = 200$, and each term is found by adding -9 to the preceding term.

24. A sequence has $a_1 = 512$, and each term is found by adding -4 to the preceding term.

25. An arithmetic sequence contains the terms $a_3 = 20$ and $a_{10} = 62$.

26. An arithmetic sequence contains the terms $a_4 = 26$ and $a_9 = 61$.

27. An arithmetic sequence contains the terms $a_{11} = 91$ and $a_{27} = 43$.

28. An arithmetic sequence contains the terms $a_9 = 166$ and $a_{15} = 100$.

29. An arithmetic sequence contains the terms $a_6 = 7$ and $a_{12} = 10$.

30. An arithmetic sequence contains the terms $a_3 = 9$ and $a_{12} = 12$.

31. Find the sum of the first 40 terms of $3 + 7 + 11 + 15 + 19 + \cdots$.

32. Find the sum of the first 50 terms of $11 + 23 + 35 + 47 + 59 + \cdots$.

33. Find $\displaystyle\sum_{n=1}^{100} (56 - 4n)$.

34. Find $\displaystyle\sum_{n=1}^{50} (20.1 - 1.1n)$.

35. Find $\displaystyle\sum_{n=1}^{100} (6n - 4)$.

36. Find $\displaystyle\sum_{n=1}^{100} (52.5 - 2.5n)$.

APPLICATIONS

Business and Economics

37. Adding clients. Bantam Accountants had 100 clients at the first week of the year. Each week the firm added 7 new clients.

a) Find c_n, the number of clients the firm has in the nth week.

b) How many clients does Bantam Accountants have in the 25th week?

c) After how many weeks will Bantam Accountants have 240 clients?

38. Depleting a bank account. Tim's bank account contained $2000 on the first day. Every day after that, he withdraws $15.

a) Find a_n, the amount in his account on the nth day.

b) How much is in his account after 3 weeks?

c) After how many days will the amount in the account drop below $1000?

39. Library fines. The fine for returning a book to the Wagner Public Library after the due date is $0.50 for the first day the book is late, plus an extra $0.10 every day after that.

a) Find f_n, the amount of the fine on the nth day after the due date.

b) What is the fine for a book that is 15 days late?

c) After how many days will the fine be $5?

40. Taxi fare. Wills' Taxi Cabs charges $2.75 for the first mile and $1.95 for every mile (or part of a mile) after that.

a) Find m_n, the total fare for a trip of n miles.

b) What is the fare for a 10-mi trip?

c) After how many miles will the fare first exceed $16?

If an object loses the same amount of value each year, the value v_n of the object n years after purchase may be modeled by the straight-line depreciation formula:

$$v_n = C - n\left(\frac{C - S}{N}\right), \quad \text{for } n = 0, 1, 2, 3, \dots, N,$$

where C is the initial cost of the item, S is the end value (or salvage value), and N is the expected life of the item, in years. (See Exercise 68, Section R.4.) Note that the sequence starts at n = 0, which corresponds to the time when the object is brand-new. Use this formula in Exercises 41 and 42.

41. Straight-line depreciation. Lucas Mining purchases a copier for its office at an initial cost of $7500. The copier is expected to be used for 6 yr, at the end of which time its value will be $1500. Assume that the copier's value declines by the same amount each year.

   a) Find an expression for v_n, the value of the copier at the start of the nth year.
   b) Show that $v_0 = 7500$.
   c) What is the copier's value 4 yr after it was purchased?
   d) Write the sequence of values v_n, for $n = 0, 1, 2, \dots, 6$, which give the object's value over its 6-yr life.

42. Straight-line depreciation. Larson's Pumpkin Farm purchases a tiller for $23,750. The tiller is expected to last 8 years at which time its salvage value will be $5500. Assume the tiller's value declines the same amount each year.

   a) Find an expression for v_n, the value of the tiller after n years.
   b) Show that $v_0 = 23,750$.
   c) What is the tiller's value 6 yr after it was purchased?
   d) Write the sequence of values v_n for $n = 0, 1, 2, \dots, 8$, which give the tiller's value over its 8-yr life.

SYNTHESIS

43. Suppose the first term of an arithmetic series is $a_1 = 5$ and the 75th partial sum of the series is 5925. Find the 20th partial sum.

44. Suppose the first term of an arithmetic series is $b_1 = 3$ and the 90th partial sum of the series is 3073.5. Find the 100th partial sum.

45. The sum of the first n odd integers is $1 + 3 + 5 + 7 + 9 + 11 + \dots + (2n - 1)$.

   a) Find the first five partial sums.
   b) Find the formula for S_n, the sum of the first n odd integers.
   c) Find S_{75}.

46. The sum of the first n even integers is $2 + 4 + 6 + 8 + 10 + \dots + 2n$.

   a) Find the first five partial sums.
   b) Find the formula for S_n, the sum of the first n even integers.
   c) Find S_{250}.

47. The sum of the first n positive integers is $1 + 2 + 3 + 4 + \dots + n$.

   a) Write a formula for T_n, the sum of the first n integers. These sums $(1, 3, 6, 10, 15, \dots)$ are called *triangular numbers*, as illustrated in the diagram.

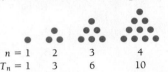

$$n = 1 \quad 2 \quad 3 \quad 4$$
$$T_n = 1 \quad 3 \quad 6 \quad 10$$

   b) A proof "by picture" for this formula is as follows:
   (i) Draw an $n \times n$ grid of dots. How many dots are there?
   (ii) Erase one diagonal line of dots, from the top left corner to the bottom right corner. How many dots remain, in terms of n?
   (iii) Erase the dots above and to the right of the diagonal. What remains is a triangle of dots. How many dots does it contain, in terms of n?

48. The handshake problem. Suppose five people are in a room, and everyone shakes everyone else's hand once.

   a) How many handshakes are possible?
   b) Explain how the formula from Exercise 47(a) can be adjusted and used to show how many handshakes are possible among n people, when everyone shakes everyone else's hand once.
   c) How many handshakes are possible among 45 people, assuming everyone shakes everyone else's hand once?

49. Show that the nth partial sum of an arithmetic series can also be written as $S_n = \frac{n}{2}(2a_1 + (n - 1)d)$.

50. How many terms are needed to make the partial sum of the arithmetic series $2 + 10 + 18 + 26 + 34 + \dots$ greater than 10,000? (*Hint:* Use the formula in Exercise 49.)

51. The following is an outline of a proof that the nth partial sum of an arithmetic sequence with a first term a_1 and a common difference d is $S_n = \frac{n}{2}(a_1 + a_n)$, where $a_n = a_1 + (n - 1)d$.

   a) Find expressions for a_2 and a_3 in terms of a_1 and d.
   b) Find expressions for a_{n-2} and a_{n-1} in terms of a_1 and d.

c) Write S_n from a_1 to a_n and then from a_n to a_1, and add the two equations. Then find and simplify an expression for $a_1 + a_n$. Do the same for $a_2 + a_{n-1}$ and $a_3 + a_{n-2}$. Simplify this sum into a product by noting there are n columns. Then solve for S_n.

d) Show that the result from part (c) is equal to

$$S_n = \frac{n}{2}(a_1 + a_n). \quad (\textit{Hint: See Exercise 49.})$$

Answers to Quick Checks

1. $c_1 = 17, c_2 = 26, c_3 = 35, c_4 = 44, c_{30} = 278$
2. $d = 2.5$ **3.** $a_n = -4n + 34$ **4.** $a_n = -20n + 800$,
$a_{23} = \$340$ **5.** $S_{75} = 22{,}875$ **6.** $a_{31} = \$8.75, S_{31} = \$155.00$ **7.** **(a)** 11 boxes; **(b)** 427 boxes

9.2 Geometric Sequences and Series

Geometric Sequences

OBJECTIVES

- Identify the common ratio of a geometric sequence.
- Find the general nth term of a geometric sequence.
- Find terms of a geometric sequence given the general nth term.
- Find the nth partial sum of a geometric series and, when it exists, the sum of an infinite geometric sequence.
- Solve applied problems involving geometric sequences and series.

Consider the sequence

$$2, 6, 18, 54, 162, \ldots .$$

Note that each term is multiplied by 3 to obtain the next term. Sequences in which each term is multiplied by a fixed number to obtain the next term are called *geometric sequences*, and the number that multiplies one term to get the next is called the *common ratio*, r.

DEFINITION

A sequence is **geometric** if there exists a number r ($r \neq 0$ and $r \neq 1$), called the **common ratio**, such that

$$a_{n+1} = a_n \cdot r \quad \text{or} \quad \frac{a_{n+1}}{a_n} = r.$$

Thus, we can find the common ratio by dividing any term of a geometric sequence (other than the first term) by the term before it.

■ EXAMPLE 1 Find the common ratio for each of the following geometric sequences.
a) $3, 6, 12, 24, 48, 96, \ldots$
b) $5, -20, 80, -320, 1280, \ldots$
c) $10, 10(1.05), 10(1.05)^2, 10(1.05)^3, \ldots$
d) $\frac{3}{5}, \frac{2}{5}, \frac{4}{15}, \frac{8}{45}, \frac{16}{135}, \ldots$

Solution

a) For the sequence $3, 6, 12, 24, 48, 96, \ldots$, the common ratio is $r = 2$, since $\frac{6}{3} = 2$ and $\frac{12}{6} = 2$, and so on.

b) For the sequence $5, -20, 80, -320, 1280, \ldots$, the common ratio is $r = -4$, since $\frac{-20}{5} = -4$ and $\frac{80}{-20} = -4$, and so on.

c) For the sequence $10, 10(1.05), 10(1.05)^2, 10(1.05)^3, \ldots$, the common ratio is $r = 1.05$, since $\frac{10(1.05)}{10} = 1.05$ and $\frac{10(1.05)^2}{10(1.05)} = 1.05$, and so on.

> **Quick Check 1**
>
> Find the common ratio of the geometric sequence $16, 8, 4, 2, 1, \ldots$.

d) For the sequence $\frac{3}{5}, \frac{2}{5}, \frac{4}{15}, \frac{8}{45}, \frac{16}{135}, \ldots$, the common ratio is $r = \frac{2}{3}$, since $\frac{2/5}{3/5} = \frac{2}{5} \cdot \frac{5}{3} = \frac{2}{3}$ and $\frac{4/15}{2/5} = \frac{4}{15} \cdot \frac{5}{2} = \frac{2}{3}$, and so on.

❰ Quick Check 1

For any geometric sequence, if a_1 is the first term and r is the common ratio, then $a_2 = a_1 \cdot r$. We also have

$$a_3 = a_1 \cdot r \cdot r = a_1 \cdot r^2,$$
$$a_4 = a_1 \cdot r \cdot r \cdot r = a_1 \cdot r^3,$$

and so on.

This result leads us to a formula for the general nth term of a geometric sequence.

THEOREM 3

The general nth term of a geometric sequence is given by

$$a_n = a_1 \cdot r^{n-1}, \quad \text{for all } n \geq 1.$$

Let's find the general nth terms of the sequences in Example 1.

■ **EXAMPLE 2**   For each geometric sequence, write the expression for the general nth term and then find the 10th term.

a) $3, 6, 12, 24, 48, 96, \ldots$

b) $5, -20, 80, -320, 1280, \ldots$

c) $10, 10(1.05), 10(1.05)^2, 10(1.05)^3, \ldots$

d) $\frac{3}{5}, \frac{2}{5}, \frac{4}{15}, \frac{8}{45}, \frac{16}{135}, \ldots$

Solution

a) For $3, 6, 12, 24, 48, 96, \ldots$, the first term is $a_1 = 3$, and the common ratio is $r = 2$. Thus, the general nth term is $a_n = 3 \cdot 2^{n-1}$, and the 10th term is $a_{10} = 3 \cdot 2^{10-1} = 3 \cdot 2^9 = 1536$.

b) For $5, -20, 80, -320, 1280, \ldots$, the first term is $a_1 = 5$, and the common ratio is $r = -4$. Thus, the general nth term is $a_n = 5 \cdot (-4)^{n-1}$, and the 10th term is $a_{10} = 5 \cdot (-4)^{10-1} = 5 \cdot (-4)^9 = -1{,}310{,}720$.

c) For $10, 10(1.05), 10(1.05)^2, 10(1.05)^3, \ldots$, the first term is $a_1 = 10$, and the common ratio is $r = 1.05$. Thus, the general nth term is $a_n = 10 \cdot 1.05^{n-1}$, and the 10th term is $a_{10} = 10 \cdot 1.05^{10-1} = 10 \cdot 1.05^9 \approx 15.513$.

d) For $\frac{3}{5}, \frac{2}{5}, \frac{4}{15}, \frac{8}{45}, \frac{16}{135}, \ldots$, the first term is $a_1 = \frac{3}{5}$, and the common ratio is $r = \frac{2}{3}$. Thus, the general nth term is $a_n = \frac{3}{5} \cdot \left(\frac{2}{3}\right)^{n-1}$, and the 10th term is

$$a_{10} = \frac{3}{5} \cdot \left(\frac{2}{3}\right)^{10-1} = \frac{3}{5} \cdot \left(\frac{2}{3}\right)^9 \approx 0.0156.$$

❭ **Quick Check 2**

Write the general nth term of the geometric sequence $16, 8, 4, 2, 1, \ldots$. Then find the 10th term of the sequence.

❬ Quick Check 2

Application: Depreciation

If an object loses the same percentage of its value from one year to the next, we can use a geometric sequence to model its depreciation over time. In many applications, we can let $n = 0$ represent the initial value or initial quantity, so that the general nth term is written $a_n = a_0 r^n$, for $n \geq 0$.

■ **EXAMPLE 3** **Business: Depreciation.** The Normans purchase a computer system for their home business. The computer system originally cost $9,500, and every year, it loses 15% of the value it had the previous year.

a) Find a_n, the value of the computer system n years after it was purchased.

b) What is the value of the computer system 10 yr after it was purchased?

Solution

a) The computer system loses 15% of its value each year. Thus, the value of the computer system each year is 85% of its value the previous year. The computer system's value is given by $a_n = 9500(0.85)^n$. Note that $a_0 = 9500(0.85)^0 = 9500 \cdot 1 = \$9500$, which represents the computer system's value when it was purchased brand-new.

b) The computer system's value 10 years after it was purchased is

$$a_{10} = 9500(0.85)^{10} = \$1870.31. \qquad \text{Using a calculator}$$

❭ **Quick Check 3**

What is the value of the computer system in Example 3 5 yr after it was purchased?

❬ Quick Check 3

Geometric Series

Recall that a series is the sum of the terms of a sequence. A *geometric series* is the sum of the terms of a geometric sequence, and we now develop a formula for the nth partial sum of any geometric series.

The first n terms of a geometric sequence are $a_1, a_1 r, a_1 r^2, a_1 r^3, \ldots, a_1 r^{n-1}$. Therefore, the nth partial sum is

$$S_n = a_1 + a_1 r + a_1 r^2 + a_1 r^3 + \cdots + a_1 r^{n-1}. \tag{1}$$

If we multiply both sides of equation (1) by r, we have

$$rS_n = r(a_1 + a_1 r + a_1 r^2 + a_1 r^3 + \cdots + a_1 r^{n-1}) \qquad \text{Multiplying both sides of (1) by } r$$

$$= a_1 r + a_1 r^2 + a_1 r^3 + a_1 r^4 + \cdots + a_1 r^n. \tag{2}$$

Subtracting the left and right sides of Equation (1) from the left and right sides of Equation (2), we obtain

$$rS_n - S_n = a_1 r^n - a_1 \qquad \text{The middle terms on the right sides of equations (1) and (2) sum to 0.}$$

$$S_n(r - 1) = a_1(r^n - 1). \qquad \text{Factoring}$$

Dividing both sides of this equation by $r - 1$ leads us to a formula for the nth partial sum of a geometric series.

THEOREM 4

The nth partial sum of a geometric series is given by

$$S_n = \frac{a_1(r^n - 1)}{r - 1},$$

where a_1 is the first term and r is the common ratio ($r \neq 0$ and $r \neq 1$).

■ **EXAMPLE 4**   Find the sum of the first 10 terms of the series $4 + 12 + 36 + 108 + \cdots$.

Solution   This series is geometric, with the first term $a_1 = 4$ and the common ratio $r = 3$. Thus, the sum of the first 10 terms of this series is

$$S_{10} = \frac{4(3^{10} - 1)}{3 - 1} = 118{,}096.$$

❰ Quick Check 4

❱ **Quick Check 4**

Find the 12th partial sum of the series $2 + 10 + 50 + 250 + \cdots$.

Infinite Geometric Series

If the number of terms in a series is finite, the series is called a *finite series*. If we allow the number of terms to increase without bound, we have an *infinite series*. Consider the infinite series $2 + 6 + 18 + 54 + 162 + \cdots$. We have the following partial sums:

$$S_1 = 2,$$
$$S_2 = 2 + 6 = 8,$$
$$S_3 = 2 + 6 + 18 = 26,$$
$$S_4 = 2 + 6 + 18 + 54 = 80.$$

If we add more terms, this series grows in value, without bound. However, consider the infinite series $\frac{1}{2} + \frac{1}{4} + \frac{1}{8} + \frac{1}{16} + \frac{1}{32} + \cdots$. Let's write its first few partial sums:

$$S_1 = \frac{1}{2},$$
$$S_2 = \frac{1}{2} + \frac{1}{4} = \frac{3}{4},$$
$$S_3 = \frac{1}{2} + \frac{1}{4} + \frac{1}{8} = \frac{7}{8},$$
$$S_4 = \frac{1}{2} + \frac{1}{4} + \frac{1}{8} + \frac{1}{16} = \frac{15}{16},$$

and so on.

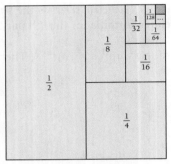

A visual "proof" illustrating that $\frac{1}{2} + \frac{1}{4} + \frac{1}{8} + \frac{1}{16} + \frac{1}{32} + \cdots$ approaches 1.

Using the formula for the nth partial sum of a geometric series, we can show that $S_{10} \approx 0.999023$ and $S_{20} \approx 0.999999046$. This geometric series grows in value but it is bounded. In fact, its partial sum is approaching the value 1.

To find a formula for the sum of an infinite geometric series, we examine the formula for the nth partial sum of a geometric series:

$$S_n = \frac{a_1(r^n - 1)}{r - 1}.$$

Note that for $|r| < 1$, as n gets large, the expression r^n approaches zero. Thus, as $n \to \infty$, we have

$$\lim_{n \to \infty} S_n = \lim_{n \to \infty} \frac{a_1(r^n - 1)}{r - 1}$$

$$= \frac{a_1(0 - 1)}{r - 1} \qquad r^n \to 0 \text{ as } n \to \infty$$

$$= \frac{a_1(-1)}{-(1 - r)} \qquad r - 1 = -(1 - r)$$

$$= \frac{a_1}{1 - r}. \qquad \frac{-1}{-1} = 1$$

The sum of an infinite geometric series is represented by S.

TECHNOLOGY CONNECTION

Exploratory

Calculate partial sums of the series $\frac{1}{2} + \frac{1}{4} + \frac{1}{8} + \frac{1}{16} + \frac{1}{32} + \cdots$. Is it possible to add enough terms that the partial sum will exceed 1? How many terms are needed to bring the partial sum to within 0.001 of 1?

THEOREM 5

If the common ratio is $|r| < 1$, then the sum of an infinite geometric series is

$$S = \sum_{n=1}^{\infty} a_1 \cdot r^{n-1} = \frac{a_1}{1 - r}.$$

Thus, for $|r| < 1$, every infinite geometric series has a finite sum, and we say that the series *converges* to the value S. If a series is not convergent, we say that it *diverges*.

■ **EXAMPLE 5** Find the sum of each infinite geometric series, if it exists.

a) $4 + 1 + \frac{1}{4} + \frac{1}{16} + \frac{1}{64} + \cdots$

b) $\displaystyle\sum_{n=1}^{\infty} \left(\frac{2}{5}\right) \cdot \left(-\frac{3}{4}\right)^{n-1}$

c) $1 + 5 + 25 + 125 + 625 + \cdots$

Solution

a) The first term is $a_1 = 4$, and the common ratio is $r = \frac{1}{4}$. Since $|r| < 1$, this infinite geometric series converges to

$$S = \frac{4}{1 - \frac{1}{4}} = \frac{4}{3/4} = 4 \cdot \frac{4}{3} = \frac{16}{3}, \text{ or } 5\frac{1}{3}.$$

b) The first term is $a_1 = \frac{2}{5}$, and the common ratio is $r = -\frac{3}{4}$. Since $|r| < 1$, this infinite geometric series converges to

$$S = \frac{\frac{2}{5}}{1 - \left(-\frac{3}{4}\right)} = \frac{2/5}{7/4} = \frac{2}{5} \cdot \frac{4}{7} = \frac{8}{35}.$$

c) The common ratio is $r = 5$. Since the common ratio does not meet the requirement $|r| < 1$, this infinite geometric series diverges.

❮ Quick Check 5

> **Quick Check 5**
>
> Find the sum of each infinite geometric series or state that it does not exist.
>
> a) $\frac{3}{4} - \frac{1}{4} + \frac{1}{12} - \frac{1}{36} + \frac{1}{108} - \cdots$
>
> b) $\displaystyle\sum_{n=1}^{\infty} 9 \cdot \left(\frac{1}{3}\right)^{n-1}$
>
> c) $1 + 3 + 9 + 27 + 81 + \cdots$

■ **EXAMPLE 6** Use an infinite series to show that $0.3333333\ldots = \frac{1}{3}$.

Solution Note that $0.3333333\ldots$ is an infinite geometric series that can be expressed as

$$0.3333333\ldots = 0.3 + 0.03 + 0.003 + 0.0003 + \cdots$$

$$= \frac{3}{10} + \frac{3}{100} + \frac{3}{1000} + \frac{3}{10,000} + \cdots,$$

where $a_1 = \frac{3}{10}$ and $r = \frac{3/100}{3/10} = \frac{3}{100} \cdot \frac{10}{3} = \frac{1}{10}$. Since $|r| < 1$, this infinite geometric series converges to

$$S = \frac{3/10}{1 - 1/10} = \frac{3/10}{9/10} = \frac{3}{10} \cdot \frac{10}{9} = \frac{1}{3}.$$

❮ Quick Check 6

> **Quick Check 6**
>
> Use an infinite series to show that $0.999999\ldots = 1$.

Applications

When money is spent, a percentage of that money is spent again, and so on. This percentage is called a *fiscal multiplier*, or a *Keynesian multiplier*. It can be a rough gauge of the health of an economy, since money is more likely to be spent multiple times in a healthy economy than in an unhealthy economy.

■ **EXAMPLE 7**   **Business: Fiscal Multiplier.**   The Book Exchange spends $1,000,000 to construct a new outlet. A portion of that $1,000,000 is spent on salaries, payments to subcontractors, and so on. The recipients of these salaries and payments spend some of what they receive on goods and services. This pattern is repeated again and again. Assume that 75% of the money received is spent at each stage. How much total spending is attributed to the original $1,000,000 expenditure?

Solution   The total spending due to the original $1,000,000 expenditure can be modeled as an infinite geometric series:

$$1{,}000{,}000 + 1{,}000{,}000(0.75) + 1{,}000{,}000(0.75)^2 + 1{,}000{,}000(0.75)^3 + \cdots .$$

We have $a_1 = 1{,}000{,}000$ and $r = 0.75$. The infinite sum is

$$S = \frac{1{,}000{,}000}{1 - 0.75} = \frac{1{,}000{,}000}{0.25} = \$4{,}000{,}000.$$

The original $1,000,000 expenditure is responsible for $4,000,000 in total spending.

❬ Quick Check 7

> **❭ Quick Check 7**
>
> Find the overall spending due to an original expenditure of $100,000 if 45% of the money is spent at every stage.

Multilevel marketing is a method of direct sales in which a salesperson recruits others to work as salespeople below him or her and to also recruit others to work as salespeople below them, forming a multilevel hierarchy. Each salesperson earns a commission on his or her own sales and a percentage of the sales made by the "downline" sales force—the salespeople whom they have recruited and so on. Those who enter the sales force early and have many downline salespeople working for them can earn considerable sums of money. However, those who enter the sales force later find that it is very difficult to find recruits to continue the hierarchy. The size of such a sales force can be modeled by a geometric series.

First level
Second level
Third level

■ **EXAMPLE 8**   **Business: Multilevel Marketing.**   HomeWares, Inc., is a direct-sales company that sells housewares to customers through an extensive sales force. Suppose each salesperson at HomeWares must recruit three people to work as downline salespeople, and each of these recruits must in turn recruit three other people to work as downline salespeople, and so on.

a) Assume that each salesperson is able to recruit three others to work as downline salespeople. How big is the total sales force after 10 levels have been established?

b) How many levels would have to be established for the sales force of HomeWares to exceed the world's population of 7,000,000,000 (7 billion)?

Solution   The first level is the original salesperson. The second level consists of the three people recruited by that salesperson. The third level consists of the nine people working below those in the second level, and so on. Thus, we have a geometric sequence of terms, $1, 3, 9, 27, 81, \ldots$, where $a_n = 3^{n-1}$ gives the number of salespeople at the nth level. The total number of salespeople up to and including the nth level is given by the nth partial sum of the geometric series $1 + 3 + 9 + 27 + 81 + \cdots$. Note that the first term is $a_1 = 1$, and the common ratio is $r = 3$.

a) The total number of salespeople up to and including the 10th level is

$$S_{10} = \frac{1 \cdot \left(3^{10} - 1\right)}{3 - 1} = 29{,}524.$$

b) To find n such that $S_n \geq 7,000,000,000$, we substitute and solve:

$$\frac{1 \cdot (3^n - 1)}{3 - 1} \geq 7,000,000,000 \qquad \text{Using } S_n = \frac{1 \cdot (3^n - 1)}{3 - 1}.$$

$$\frac{3^n - 1}{2} \geq 7,000,000,000$$

$$3^n - 1 \geq 14,000,000,000 \qquad \text{Multiplying both sides by 2}$$

$$3^n \geq 14,000,000,001$$

$$\ln 3^n \geq \ln(14,000,000,001) \qquad \text{Taking the natural logarithm of both sides}$$

$$n \ln 3 \geq \ln(14,000,000,001) \qquad \text{Using a property of logarithms}$$

$$n \geq \frac{\ln(14,000,000,001)}{\ln 3}$$

$$n \geq 21.27. \qquad \text{Using a calculator}$$

Thus, with 22 levels of salespeople, the sales force of HomeWares, Inc., will exceed the world's population. The student can verify that $S_{21} = 5,230,176,601$ people and $S_{22} = 15,690,529,803$ people, more than twice the current world's population.

Consider the challenge of working within a very large multilevel marketing system: everyone would be a salesperson as well as a potential client. On a small scale, these systems can work, but they are fundamentally unsound on larger scales.

Section Summary

- A sequence is *geometric* if there exists a number r, called the *common ratio*, such that
$$\frac{a_{n+1}}{a_n} = r \quad \text{for } r \neq 0 \text{ and } r \neq 1.$$

- The *general nth term* of a geometric sequence is given by $a_n = a_1 r^{n-1}$, for all $n \geq 1$. In many applications, we can let $n = 0$ represent the initial value or initial quantity and then the general nth term is written $a_n = a_0 r^n$, or $n \geq 0$.

- A *geometric series* is the sum of the terms of a geometric sequence.

- The sum of the first n terms of a geometric series is given by
$$S_n = \frac{a_1(r^n - 1)}{r - 1}.$$

- Any series whose number of terms is finite is called a *finite series*. If the number of terms is infinite, the series is called an *infinite series*.

- If $|r| < 1$, then the sum of an infinite geometric series is given by
$$S = \sum_{n=1}^{\infty} a_1 \cdot r^{n-1} = \frac{a_1}{1 - r}.$$

EXERCISE SET
9.2

The sequences in Exercises 1–10 are geometric. For each one, assume that $n \geq 1$ and find (a) the common ratio r, (b) the general nth term, (c) the 10th term of the sequence, and (d) the sum of the first 15 terms.

1. $3, 12, 48, 192, 384, \ldots$

2. $6, 12, 24, 48, 96, \ldots$

3. $7, 21, 63, 189, 567, \ldots$

4. $3, 15, 75, 375, 1875, \ldots$

5. $2, -6, 18, -54, 162, \ldots$

6. $3, -6, 12, -24, 48, \ldots$

7. $\frac{1}{4}, \frac{3}{20}, \frac{9}{100}, \frac{27}{500}, \frac{81}{2500}, \ldots$

8. $2, \frac{4}{3}, \frac{8}{9}, \frac{16}{27}, \frac{32}{81}, \ldots$

9. $\frac{1}{8}, -\frac{1}{12}, \frac{1}{18}, -\frac{1}{27}, \frac{2}{81}, \ldots$

10. $-\frac{2}{5}, \frac{1}{10}, -\frac{1}{40}, \frac{1}{160}, -\frac{1}{640}, \ldots$

11. The first term of a geometric sequence is $a_1 = 1$, and the common ratio is 5.

a) Find the general nth term.
b) Find the 8th term.
c) Find the sum of the first 10 terms.

12. The first term of a geometric sequence is $a_1 = 2$, and the common ratio is 4.

   a) Find general nth term.
   b) Find the 10th term.
   c) Find the sum of the first 10 terms.

13. The first term of a geometric sequence is $a_1 = 6$, and the common ratio is $-\frac{1}{2}$.

   a) Find general nth term.
   b) Find the 9th term.
   c) Find the sum of the first 10 terms.

14. The first term of a geometric sequence is $a_1 = 3$, and the common ratio is -7.

   a) Find general nth term.
   b) Find the 12th term.
   c) Find the sum of the first 10 terms.

15. Find the sum of the first 10 terms of the geometric series $4 + 8 + 16 + 32 + \cdots$.

16. Find the sum of the first 12 terms of the geometric series $2 + 10 + 50 + 250 + \cdots$.

17. Find $\displaystyle\sum_{n=1}^{12} 10 \cdot (1.2)^{n-1}$.

18. Find $\displaystyle\sum_{n=1}^{8} 100 \cdot (-0.3)^{n-1}$.

19. Consider the following series: $7 + 9.1 + 11.83 + 15.379 + 19.9927 + \cdots$.

   a) What is S_{14}?
   b) What is the minimum number of terms needed to make the partial sum at least 5000?

20. Consider the following series: $10 + 9 + 8.1 + 7.29 + 6.561 + \cdots$.

   a) What is S_{20}?
   b) What is the minimum number of terms needed to make the partial sum at least 99?

In Exercises 21–30, find the sum of each infinite geometric series. Assume that $n \geq 1$.

21. $\frac{1}{3} + \frac{1}{9} + \frac{1}{27} + \frac{1}{81} + \frac{1}{243} + \cdots$

22. $\frac{2}{7} + \frac{2}{35} + \frac{2}{175} + \frac{2}{875} + \cdots$

23. $\frac{1}{4} - \frac{1}{8} + \frac{1}{16} - \frac{1}{32} + \frac{1}{64} - \frac{1}{128} + \cdots$

24. $\frac{3}{8} - \frac{3}{32} + \frac{3}{128} - \frac{3}{512} + \frac{3}{2048} - \cdots$

25. $10 + 5 + \frac{5}{2} + \frac{5}{4} + \frac{5}{8} + \cdots$

26. $12 + 3 + \frac{3}{4} + \frac{3}{16} + \frac{3}{64} + \cdots$

27. $\displaystyle\sum_{n=1}^{\infty} 10 \cdot (-0.2)^{n-1}$ **28.** $\displaystyle\sum_{n=1}^{\infty} 6 \cdot \left(-\frac{2}{3}\right)^{n-1}$

29. $\displaystyle\sum_{n=1}^{\infty} 5 \cdot (0.9)^{n-1}$ **30.** $\displaystyle\sum_{n=1}^{\infty} 3 \cdot (0.8)^{n-1}$

31–40. Check each result of Exercises 21–30 by calculating the 5th, 10th, and 15th partial sums.

In Exercises 41–48, write each value as the sum of an infinite geometric series. Express each answer as a simplified fraction.

41. $0.4444444\ldots$ **42.** 0.7777777

43. $0.12121212\ldots$ $\left(Hint: 0.12121212\ldots = \frac{12}{100} + \frac{12}{10,000} + \frac{12}{1,000,000} + \cdots.\right)$

44. 0.73737373

45. $0.145145145\ldots$

46. $0.573573573\ldots$

47. $1.18333333\ldots$ $\left(Hint: 1.1833333\ldots = \frac{118}{100} + \left(\frac{3}{1000} + \frac{3}{10,000} + \frac{3}{100,000} + \cdots\right).\right)$

48. $2.3575757\ldots$

APPLICATIONS

Business and Economics

49. Depreciation. A tablet computer costs $1700 new and loses 18% of its value every year.

   a) Find v_n, the value of the tablet computer n years after it was purchased.
   b) Write the terms of the sequence v_n, for $n = 0, 1, 2, \ldots, 5$, which express the computer's yearly values from its purchase to 5 yr after it was purchased.
   c) Explain the meaning of v_0.

50. Depreciation. A motorcycle costs $13,500 new and loses 9% of its value every year.

   a) Find v_n, the value of the motorcycle n years after it was purchased.
   b) Write the terms of the sequence v_n, for $n = 0, 1, 2, \ldots, 6$, which express the motorcycle's yearly values from its purchase to 6 yr after it was purchased.
   c) Explain the meaning of v_0.

51. Depreciation. Shelley purchases a new car for $20,000. Its value decreases by 15% every year.

   a) Find v_n, the value of the car n years after it was purchased.
   b) Write the terms of the sequence v_n, for $n = 0, 1, 2, \ldots, 5$, which express the car's yearly values from its purchase to 5 yr after it was purchased.
   c) Explain the meaning of v_0.
   d) When will the value of Shelley's car be less than $9000?

52. Depreciation. A new public address system costs $4400. Its value decreases by 12% every year.

   a) Find v_n, the value of the system n years after it was purchased.
   b) Write the terms of the sequence v_n, for $n = 0, 1, 2, \ldots, 7$, which express the system's yearly values from its purchase to 7 yr after it was purchased.
   c) Explain the meaning of v_0.
   d) When will the system's value drop below $450?

53. *Fiscal multiplier.* The Gamer's Videogame Palace plans to spend $1,000,000 to build a new store. An analyst determines that this expenditure will generate $300,000 in spending, which in turn will create $90,000 in spending, and so on. How much total spending will be attributable to the original $1,000,000 expenditure?

54. *Fiscal multiplier.* Pete's Old-Style Hot Dog Restaurant takes in $200,000 in monthly revenue. It spends $120,000 of that amount on salaries and supplies. In turn, $72,000 of the $120,000 is spent by its employees and suppliers, and so on. How much total spending will be attributable to the original $200,000?

55. *Multilevel marketing.* Boyd's Direct Sales employs salespeople who are expected to recruit four other salespeople to work under them, and so on. The original salesperson represents the first level ($n = 1$) of the multilevel system.

 a) How many salespeople work for Boyd's in the first 10 levels?
 b) Assume that the population of the United States is 300,000,000. After how many levels will Boyd's total sales force exceed this number?

56. *Multilevel marketing.* Gaia Cosmetics employs salespeople who recruit two other salespeople to work under them, and so on. The original salesperson represents the first level ($n = 1$).

 a) How many salespeople work for Gaia in the first 12 levels?
 b) Assume that the population of the United States is 300,000,000. After how many levels will Gaia's total sales force exceed this number?

Life and Physical Sciences

57. *Elemental decay.* The half-life of the radioactive element radon is 3.8 days. That is, every 3.8 days, half of any quantity of this element will decay. Suppose 15 mg of radon is present originally ($a_0 = 15$) and n is measured in half-lives. (*Source: periodictable.com.*)

 a) Find a_n, the amount of radon after n half-lives.
 b) Write the terms of the sequence a_n, for $n = 0, 1, 2, \ldots, 5$, which express the amounts of randon after n half-lives.
 c) After how many days will the original quantity of radon have decayed to 5% or less of its original mass?

58. *Elemental decay.* The half-life of the element astatine is about 8 hr. Suppose 5 g of astatine is present originally ($a_0 = 5$) and n is measured in half-lives. (*Source: periodictable.com.*)

 a) Find a_n, the amount of astatine after n half-lives.
 b) Write the terms of the sequence a_n, for $n = 0, 1, 2, \ldots, 5$, which express the amounts of astatine after n half-lives.
 c) After how many hours will the original quantity of astatine have decayed to 2% or less of its original mass?

59. *Population decay.* Detroit's population has been declining by about 1.4% per year. In 1950, the city's population was 1.9 million people ($p_0 = 1,900,000$). Let n be the number of years since 1950. (*Source: www.city-data.com.*)

 a) Find p_n, the population of Detroit n years after 1950.
 b) Write the terms of the sequence p_n, for $n = 0, 10, 20, \ldots, 50$, which express the population of Detroit at 10-yr intervals, starting with 1950.
 c) When did Detroit's population drop below 1,000,000?

60. *Population growth.* In 2004, Blue Canyon City's population was 5400, and it has been growing by 2.4% per year since then. Let $p_0 = 5400$.

 a) Find p_n, the population of Blue Canyon City n years after 2004.
 b) Write the terms of the sequence p_n, for $n = 0, 2, 4, \ldots, 10$, which express the population of Blue Canyon City at 2-yr intervals, starting with 2004.
 c) When will the city's population exceed 8000 people?

SYNTHESIS

61. *Geometry.* Squares are nested inside one another in such a way that each square's corners are at the midpoints of the sides of the square containing it. Let A_1 represent the outermost square, with side length 1. The squares A_1, A_2, and A_3 are shown in the diagram.

 a) The area of A_2 is _____ of the area of A_1. (*Hint:* Divide A_1 into four equal-sized smaller squares and observe how A_2 fits into this pattern.)
 b) In general, the area of A_n is _____ of the area of A_{n-1}.
 c) Using the answers to parts (a) and (b), find the general nth term for the area of the nth nested square.
 d) What is the area of A_{10}?

62. *Geometry.* Equilateral triangles are nested inside one another in such a way that the vertices of one triangle are at the midpoints of the sides of the triangle containing it. Let T_1 represent the outermost triangle, with area 1. The triangles T_1, T_2, and T_3 are shown in the diagram.

 a) The area of T_2 is _____ of the area of T_1.
 b) In general, the area of T_n is _____ of the area of T_{n-1}.
 c) Using the answers to parts (a) and (b), find the general nth term for the area of the nth nested triangle.
 d) What is the area of T_7?

63. Find $\displaystyle\sum_{n=1}^{\infty} \frac{2^n + 1}{3^n}$. $\left(Hint: \displaystyle\sum_{n=1}^{\infty} \frac{2^n + 1}{3^n} = \sum_{n=1}^{\infty} \frac{2^n}{3^n} + \sum_{n=1}^{\infty} \frac{1}{3^n}. \right)$

64. Find $\displaystyle\sum_{n=1}^{\infty} \frac{3^n + 2^n + 1}{4^n}$.

65. Suppose $\displaystyle\sum_{n=1}^{\infty} \frac{a}{4^n} = 1$. Find a.

66. Suppose $\displaystyle\sum_{n=1}^{\infty} r^n = \frac{2}{5}$. Find r.

67. An infinite geometric series has $a_1 = 2$ and the sum is $S = 5$. Find r.

68. An infinite geometric series has $a_1 = 1$ and the sum is $S = \frac{3}{4}$. Find r.

69. Show that $\dfrac{1}{p} + \dfrac{1}{p^2} + \dfrac{1}{p^3} + \dfrac{1}{p^4} + \dfrac{1}{p^5} + \cdots = \dfrac{1}{p-1}$ for all $p > 1$.

70. Show that $\dfrac{1}{p} - \dfrac{1}{p^2} + \dfrac{1}{p^3} - \dfrac{1}{p^4} + \dfrac{1}{p^5} - \cdots = \dfrac{1}{p+1}$ for all $p > 1$.

71. The nth partial sum of a geometric series is
$$S_n = \frac{a_1(r^n - 1)}{r - 1},$$ with the restrictions that $r \neq 0$ and $r \neq 1$.
  a) Explain why r cannot equal 1.
  b) Show that this formula actually works for $r = 0$; then explain why we normally do not consider the case where $r = 0$.

72. When an infinite geometric series converges, it converges to a finite value. Does divergence mean that the sum of the geometric series is infinite? Why or why not?

Fractal geometry. A fractal is an object that is similar to itself at all levels of magnification. These objects are not effectively described by Euclidean, or straight-line, geometry. For example, the shape of a cloud, the distribution of trees in a forest, the path of water through a porous medium (percolation), and the motion of a dust particle in air are better described using fractal geometry. A fractal object can be formed using an iterated rule, meaning one that is applied over and over again. The fractal object is the shape that results when the number of iterations approaches infinity.

73. Sierpinski gasket. Starting with a square B_0 of side length 1, remove a square that is $\frac{1}{9}$ of the area of B_1 from it, giving B_1; next, remove squares that are $\frac{1}{9}$ of the 8 smaller squares around the hole in B_1, giving B_2. When this process is repeated an infinite number of times, the resulting object is called a *Sierpinski gasket*, denoted B_∞.

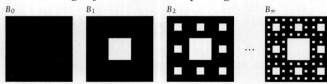

  a) In the first iteration, an area of $\frac{1}{9}$ is removed. In the second iteration, an area of $\frac{8}{81}$ is removed. In the third iteration, there will be 64 smaller squares, and an area of $\frac{1}{729}$ is removed from each, for a total area

removed of $\frac{64}{729}$. These areas form a geometric series: $\frac{1}{9} + \frac{8}{81} + \frac{64}{729} + \cdots$. Determine the total area removed from the original square as the number of iterations approaches infinity.
  b) What is the area that remains in the Sierpinski gasket, B_∞?
  c) Objects such as sponges and aerogels are mostly empty space. In what setting might this type of object be most useful? Does the concept of a two- or three-dimensional object occupying virtually no space seem plausible?

74. Square fractal. Starting with a square A_0 of side length 1, subdivide the square into ninths and remove the middle square along each edge, giving A_1. When this process is repeated an infinite number of times, the shape that remains is called the *square fractal*, denoted A_∞. Show that the total area of A_∞ is 0.

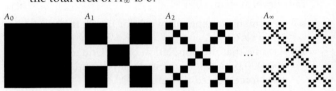

75. Fractal objects are common in everyday settings. Research how a fractal can be used to model the following items.
  a) Capillary systems such as blood vessels and tree-root structures
  b) The air-sacs (alveoli) in your lungs
  c) Your cell phone's internal antenna

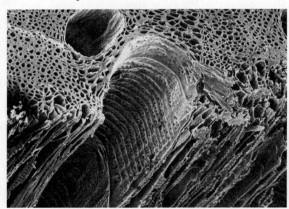

Activated charcoal is used as an absorption medium, for example, as a means of suppressing odors. A grain of activated charcoal has a diameter of about 0.1 mm, but its surface area can be several square meters. The surface of activated charcoal is highly porous on all levels—essentially, it is a fractal object. The image is an electron micrograph of a grain of activated charcoal, showing the remarkable ruggedness of its surface. Imagine the surface area of a typical bedroom (about 10 m²) "crumpled" to fit onto an object the size of a grain of sand.

Answers to Quick Checks

1. $r = \frac{1}{2}$ **2.** $a_n = 16\left(\frac{1}{2}\right)^{n-1}, a_{10} = \frac{1}{32}$ **3.** \$4215.20
4. $S_{12} = 122{,}070{,}312$ **5.** (a) $S = \frac{9}{16}$; (b) $S = \frac{27}{2}$;
(c) does not exist **6.** $a_1 = 0.9, r = 0.1, S = \dfrac{0.9}{1 - 0.1} = \dfrac{0.9}{0.9} = 1$ **7.** $S = \dfrac{100{,}000}{1 - 0.45} \approx \$181{,}818.18$

9.3

- Find the simple interest and the simple interest future value for a given principal, interest rate, and period of time.
- Find the simple interest present value for a given future value, interest rate, and period of time.
- Find the compound interest future value for a given principal, interest rate, compounding frequency, and period of time.
- Use sequences to solve real-world problems involving interest.
- Compare the growth that results from simple interest and from compound interest.

Simple and Compound Interest

When money is loaned by a bank, the borrower pays back the amount of the loan plus *interest*. In this way, the bank earns money on the loans it makes, and the bank's account holders earn money on their savings accounts, certificates of deposit, and so on. Money in an account can grow with *simple interest* (linearly) or *compound interest* (exponentially). These situations can be modeled using arithmetic and geometric sequences and series.

Simple Interest and Simple Interest Future Value

Suppose $100 is deposited in a savings account with an annual interest rate of 4%. After 1 yr, the money has earned 4% of $100 in interest, or

$$100 \cdot 0.04 \cdot 1 = \$4.$$

This interest is then added to the original deposit. Thus, the account's future value after 1 yr is

$$\$100 + \$4 = \$104.$$

We call the $100 originally deposited the *principal*, denoted P; the 4%, the *annual interest rate*, denoted i; and 1 year, the *time*, denoted t. This leads us to two definitions.

DEFINITION

For a principal P invested at an annual interest rate i for a time t, in years, the **simple interest**, denoted I, is given by

$$I = Pit,$$

and the **simple interest future value**, denoted A, is given by

$$A = P + I, \text{ or } A = P(1 + it).$$

Note that we use a lowercase i to represent the annual interest rate and a capital I to represent the total simple interest. The interest rate is always written as a decimal. The use of i in this context should not be confused with the imaginary number $i = \sqrt{-1}$.

Simple interest is calculated only once on the principal, and the annual interest rate is applied proportionally to the specified period of time. Furthermore, since we have $A = P + I$, we can also write $I = A - P$, which allows us to find the simple interest when we know the future value and the principal.

■ **EXAMPLE 1** **Business: Simple Interest.** Daisy deposits $300 into an account that earns simple interest at an annual rate of 5%. She leaves the money in the account for 9 months.

a) Find the simple interest.

b) Find the future value of Daisy's account.

Solution

a) We note that 9 months is $t = 0.75$ yr. Thus, we have the following:
   The simple interest is

$$I = Pit = \$300 \cdot 0.05 \cdot 0.75 = \$11.25.$$

b) The future value is the sum of the original deposit and the interest:

$$A = P + I = \$300 + \$11.25, \text{ or } \$311.25.$$

Note that we can also find the future value of Daisy's deposit directly by evaluating $A = P(1 + it) = 300(1 + 0.05 \cdot 0.75) = \$311.25$. Thus, the interest is $\$311.25 - \$300 = \$11.25$.

❰ Quick Check 1

> **Quick Check 1**
>
> Jeff pays $4500 for a savings bond that has an annual interest rate of 3%. Find the future value of his bond after 8 months and the interest earned.

Simple interest is sometimes used for short-term loans. The interest is calculated once and added to the principal, and the borrower repays the total amount over a period of time. This is called *add-on interest*. It is used to protect the loaner, since the full interest amount is guaranteed. The borrower does not have the option to reduce the interest by paying off the loan early, for example.

■ **EXAMPLE 2**   **Business: Add-on Interest.**   Rudy buys a set of stereo speakers from Bopper's Audio Outlet. The speakers cost $400, and Rudy elects to pay half at the time of the purchase, and the remainder over 6 months through a 10% add-on interest loan. Find Rudy's monthly payment.

Solution   Rudy pays half, or $200, right away and will pay the other half back with 10% add-on interest over 6 months. The simple interest future value of $200 at 10% for 6 months ($t = 0.5$ yr) is

$$A = \$200(1 + 0.1 \cdot 0.5) = \$210. \quad \text{Substituting}$$

Thus, Rudy will need to make 6 monthly payments of

$$\frac{\$210}{6} = \$35.$$

❰ Quick Check 2

> **Quick Check 2**
>
> Cheryl buys a dinette set for $950. She pays $400 down and the remainder over 8 months through a 9% add-on interest loan. Find Cheryl's monthly payment.

Simple Interest Present Value

Sometimes, a future value is known, and we want to find the principal that will grow to this future value over a specified time period and at a specified rate. We call this unknown principal the **present value.**

■ **EXAMPLE 3**   **Business: Simple Interest Present Value.**   A government bond pays an annual simple interest rate of 4.2% and can be redeemed for $1000 after 1 yr. How much does the bond cost initially?

Solution   We know that the simple interest future value is $A = \$1000$, and we want to know the principal P that will grow to $1000 at 4.2% simple interest in 1 yr. We have

$$A = P(1 + it) \qquad \text{Simple interest future value}$$
$$\$1000 = P(1 + 0.042 \cdot 1) \qquad \text{Substituting}$$
$$\$1000 = 1.042P$$
$$P = \frac{1000}{1.042} = \$959.69. \qquad \text{Dividing both sides by 1.042}$$

Thus, the $1000 bond costs $959.69; this is the bond's present value. Furthermore, the bond will earn $1000 - \$959.69 = \$40.31$ interest.

❰ Quick Check 3

> **Quick Check 3**
>
> A bond that pays 5.5% in simple interest will be worth $3000 in 6 months. Find the bond's original cost (the present value). How much interest will the bond earn?

Compound Interest

Suppose $100 is deposited in a savings account that pays an annual interest rate of 4%. Instead of being calculated once, interest is calculated once *every year* and added to the account each year. Each year, the interest is calculated on the current total value of the account, not just the original principal. Interest that is earned on both the principal and the previously accumulated interest is called *compound interest*. Let's find the future value of this account after 3 yr.

After 1 yr, the account has

$$A_1 = P(1 + it) \qquad \text{Simple interest future value}$$
$$= \$100(1 + 0.04 \cdot 1) \qquad \text{Substituting}$$
$$= \$100(1.04) = \$104.$$

After the 2 yr, the account has

$$A_2 = \$100(1.04)(1.04)$$
$$= \$100(1.04)^2 = \$108.16.$$

After 3 yr, the account has

$$A_3 = \$100(1.04)(1.04)(1.04)$$
$$= \$100(1.04)^3 = \$112.49 \text{ (rounded)}.$$

For the account just considered, interest is calculated once per year. If it were calculated monthly, the monthly interest rate would be $\frac{1}{12}$ of the annual rate, or, $\frac{0.04}{12}$.

If c represents the **compounding frequency** (per year), then $\frac{i}{c}$ is the *periodic rate*.

Furthermore, the total number of periods is the yearly compounding frequency c multiplied by the time t, in years. This suggests a formula for the future value of a principal that earns compound interest.

Common Compounding Frequencies

Biennially:
$c = \frac{1}{2}$ (once every 2 yr)

Annually: $c = 1$

Semiannually:
$c = 2$ (twice a year)

Quarterly: $c = 4$

Monthly: $c = 12$

Weekly: $c = 52$

Daily: $c = 365$

DEFINITION

For a given principal P, an annual interest rate i, a compounding frequency of c times per year, and a time t in years, the **compound interest future value** A is

$$A = P\left(1 + \frac{i}{c}\right)^{ct}.$$

In financial bookkeeping, future value amounts are often rounded down to the nearest cent. However, in this text we will use the common practice of rounding to the nearest cent.

We note an important distinction between simple interest and compound interest:

• Simple interest is linear growth. The interest is calculated on the original principal only.

• Compound interest is exponential growth. The interest is calculated on the total amount in the account after each period.

Applications Involving Sequences, Simple Interest, and Compound Interest

Since simple interest results in linear growth, we can describe this growth using an arithmetic sequence, and since compound interest results in exponential growth, we can describe this growth using a geometric sequence.

■ **EXAMPLE 4** **Business: Compound Interest Future Value.** Carol deposits $500 in a savings account with an annual interest rate of 3.75%, compounded monthly. Find the future value of her account after 1 month, 2 months, 3 months, and 1 year.

Solution Here, interest is calculated monthly, so $c = 12$ and the periodic monthly rate is $\frac{0.0375}{12}$. Since time t in the future value formula is in years, we treat 1 month as $t = \frac{1}{12}$ yr, 2 months as $t = \frac{2}{12}$ yr, and so on. After 1, 2, and 3 months, Carol's savings account has the following values:

$$A_1 = 500\left(1 + \frac{0.0375}{12}\right)^{12(1/12)} = 500\left(1 + \frac{0.0375}{12}\right)^{1} = \$501.56,$$

$$A_2 = 500\left(1 + \frac{0.0375}{12}\right)^{12(2/12)} = 500\left(1 + \frac{0.0375}{12}\right)^{2} = \$503.13,$$

$$A_3 = 500\left(1 + \frac{0.0375}{12}\right)^{12(3/12)} = 500\left(1 + \frac{0.0375}{12}\right)^{3} = \$504.70.$$

After 1 yr, the value of Carol's account is

$$A_{12} = 500\left(1 + \frac{0.0375}{12}\right)^{12(1)}$$

$$= 500\left(1 + \frac{0.0375}{12}\right)^{12}$$

$$= \$519.08.$$

These values form a geometric sequence: $500, \$501.56, \$503.13, \$504.70, \ldots, \$519.08, \ldots$, where the 500 is A_0, Carol's initial deposit.

> **Quick Check 4**

Bonnie deposits $1200 in an account that has an annual interest rate of 4.25%, compounded quarterly. Find the future value of her account after 3 months (a quarter), 6 months, 9 months, and 1 year.

❮ Quick Check 4

■ **EXAMPLE 5** **Business: Compound Interest.** An initial deposit of $1200 grew to $1654.61 after 6 yr of annual compounding.

a) Find the annual interest rate.

b) Write a sequence from $A_0 = 1200$ to $A_6 = 1654.61$, representing the value of the account each year from the initial deposit to the 6th year.

Solution

a) Since annual compounding is used, we have $c = 1$. We make the substitutions and solve for R:

$$1654.61 = 1200\left(1 + \frac{i}{1}\right)^{1 \cdot 6} \qquad \text{Substituting}$$

$$1654.61 = 1200(1 + i)^6 \qquad \text{Simplifying}$$

$$\frac{1654.61}{1200} = (1 + i)^6 \qquad \text{Dividing both sides by 1200}$$

$$\left(\frac{1654.61}{1200}\right)^{1/6} = 1 + i \qquad \text{Taking the positive } \tfrac{1}{6}\text{th power of each side}$$

$$i = \left(\frac{1654.61}{1200}\right)^{1/6} - 1 = 0.054999\ldots \qquad \text{Using a calculator}$$

We have $i = 0.055$, or an annual interest rate of 5.5%.

TECHNOLOGY CONNECTION

Exploratory

In Examples 4 and 5, select any consecutive pair of terms in each sequence, and divide the second by the first. What numbers do you obtain, and what do they represent?

b) The value of the account after n years is a geometric sequence whose nth term is $A_n = 1200(1.055)^n$. Thus, we have the following sequence:

$$A_0 = 1200(1.055)^0 = \$1200, \qquad A_4 = 1200(1.055)^4 = \$1486.59,$$
$$A_1 = 1200(1.055)^1 = \$1266, \qquad A_5 = 1200(1.055)^5 = \$1568.35,$$
$$A_2 = 1200(1.055)^2 = \$1335.63, \qquad A_6 = 1200(1.055)^6 = \$1654.61.$$
$$A_3 = 1200(1.055)^3 = \$1409.09,$$

⟨ Quick Check 5

■ **EXAMPLE 6** **Business: Compound Interest Present Value.** Renee wants to have $30,000 in 4 yr for a down payment on a house. Her bank offers a savings account with an annual interest rate of 3.9%, compounded monthly. What is the present value of the $30,000? That is, what initial deposit (principal) will grow to $30,000 in 4 yr at 3.9%, compounded monthly?

Solution We have $A = \$30,000$, $i = 0.039$, $c = 12$, and $t = 4$. We want to find P, the present value:

$$\$30,000 = P\left(1 + \frac{0.039}{12}\right)^{48} \qquad \text{Substituting, after noting that } 12 \cdot 4 = 48$$

$$P = \frac{\$30,000}{\left(1 + \dfrac{0.039}{12}\right)^{48}} \approx \$25,673.27.$$

Renee needs to deposit $25,673.27 at 3.9%, compounded monthly, in order to have $30,000 in 4 yr.

⟨ Quick Check 6

Simple and Compound Interest: A Comparison

The following example compares simple interest (linear) and compound interest (exponential) growth. Over a period of time, the difference in growth can be significant.

■ **EXAMPLE 7** You are offered a unique job that will extend over a 30-month period, with two payment choices:

- Option 1: You receive $5000 for the first month and monthly raises of $500 per month after that.

- Option 2: You receive $1 the first month, $2 the second month, $4 the third month, and so on, with each month's pay being twice the previous month's pay.

Which option pays more over the 30 months?

Solution The first option pays you $5000 for the 1st month, $5500 for the 2nd month, $6000 for the 3rd month, and so forth. Each month's pay is a simple interest calculation ($5000, plus 10% of $5000 times the number of months after the first month), and the monthly payments can be represented as an arithmetic sequence: $A_n = 4500 + 500n$. Thus, $A_{30} = \$19,500$ would be your pay for the 30th month. Therefore, the *total* payment over the 30 months is the 30th partial sum of the arithmetic series, or

$$S_{30} = \tfrac{30}{2}(5000 + 19{,}500) = \$367{,}500.$$

The second option represents geometric growth, since each month's pay is double the previous month's pay, with $B_n = 2^{n-1}$. Over 30 months, the *total* payment is the 30th partial sum of a geometric series, or

$$S_{30} = \frac{1(2^{30} - 1)}{2 - 1} = \$1,073,741,823,$$

which is over a billion dollars! Clearly, option 2 pays significantly more than option 1.

Example 7 may not be realistic, but it illustrates the long-term difference between simple interest (linear growth) and compound interest (exponential growth). Recognizing the short- and long-term advantages of these options can help you make better decisions.

Section Summary

- Let P be the principal, i be the annual interest rate, t be the time in years, and c be the *compounding frequency* (per year).
  - Simple interest: $I = Pit$.
  - Simple interest future value: $A = P + I = P(1 + it)$.
  - Compound interest future value: $A = P\left(1 + \dfrac{i}{c}\right)^{ct}$.
  - If the future value A and the principal P are known, the interest is $I = A - P$.
  - If the future value A is known, then P is the *present value* of A.

- Simple interest is calculated only on the principal. It yields linear growth, which can be modeled using an arithmetic sequence.
- Compound interest is calculated more than once and is earned on the principal and the accrued interest, yielding exponential growth. This growth can be modeled using a geometric sequence.

EXERCISE SET
9.3

In Exercises 1–10, find (a) the simple interest and (b) the simple interest future value using the given principal P, interest rate i, and time t.

1. $P = \$300, i = 5\%, t = 1$ yr

2. $P = \$800, i = 4\%, t = 1$ yr

3. $P = \$1200, i = 1.4\%, t = 2$ yr

4. $P = \$1500, i = 2.2\%, t = 4$ yr

5. $P = \$500, i = 3.1\%, t = 20$ months

6. $P = \$900, i = 4.2\%, t = 18$ months

7. $P = \$2000, i = 2.5\%, t = 9$ months

8. $P = \$2400, i = 2.7\%, t = 6$ months

9. $P = \$1300, i = 1.95\%, t = 25$ weeks

10. $P = \$400, i = 2.13\%, t = 42$ weeks

11. A principal of \$200 earned \$40 in interest in 2 yr. Find the simple interest rate.

12. A principal of \$500 earned \$67.50 in interest in 3 yr. Find the simple interest rate.

13. A principal of \$350 grew to \$400 in 5 yr. Find the simple interest rate.

14. A principal of \$1275 grew to \$1530 in 4 yr. Find the simple interest rate.

15. Find the time needed for \$500 to earn \$15 at a simple interest rate of 3%.

16. Find the time needed for \$1100 to earn \$69.30 at a simple interest rate of 4.2%.

17. What principal will earn \$20 in simple interest at 4% in 1 yr?

18. What principal will earn \$50 in simple interest at 3% in 6 months?

19. What is the present value of \$500 that earns 2.5% simple interest for 18 months?

20. What is the present value of \$2200 that earns 3.25% simple interest for 2 yr?

21. What is the present value of \$10,000 that earns 1.78% simple interest for 30 months?

22. What is the present value of \$7500 that earns 2.13% simple interest for 42 months?

In Exercises 23–32, use the given principal, interest rate, time, and compounding frequency to find (a) the compound interest future value, (b) the interest earned, and (c) the first five terms of the sequence A_n, representing the value of the account after n compounding periods.

23. $P = \$600$; $i = 4\%$; $t = 5$ yr, compounded monthly

24. $P = \$1200$; $i = 5\%$; $t = 4$ yr, compounded monthly

25. $P = \$500$; $i = 2\%$; $t = 7$ yr, compounded quarterly

26. $P = \$900$; $i = 3\%$; $t = 3$ yr, compounded quarterly

27. $P = \$450$; $i = 2.25\%$; $t = 6$ yr, compounded annually

28. $P = \$630$; $i = 1.85\%$; $t = 8$ yr, compounded annually

29. $P = \$2500$; $i = 2.12\%$; $t = 4$ yr, compounded weekly

30. $P = \$3200$; $i = 3.08\%$; $t = 5$ yr, compounded weekly

31. $P = \$1750$; $i = 4.13\%$; $t = 6$ yr, compounded semiannually

32. $P = \$2940$; $i = 2.92\%$; $t = 10$ yr, compounded biennially

In Exercises 33–40, use the given principal, interest rate, time, and compounding frequency to find (a) the present value, (b) the interest earned, and (c) the first five terms of the sequence A_n, representing the value of the account after n compounding periods.

33. $A = \$10,000$; $i = 4\%$; $t = 3$ yr, compounded monthly

34. $A = \$12,000$; $i = 5\%$; $t = 4$ yr, compounded monthly

35. $A = \$4000$; $i = 4.2\%$; $t = 2$ yr, compounded quarterly

36. $A = \$7000$; $i = 3.7\%$; $t = 5$ yr, compounded quarterly

37. $A = \$2500$; $i = 3.15\%$; $t = 18$ months, compounded monthly

38. $A = \$3350$; $i = 5.05\%$; $t = 42$ months, compounded quarterly

39. $A = \$17,250$; $i = 2.25\%$; $t = 10$ yr, compounded biennially

40. $A = \$35,000$; $i = 3.1\%$; $t = 20$ yr, compounded biennially

APPLICATIONS

Business and Economics

41. Simple interest. A deposit of $1000 grows in value to $1035 after 1 yr. Assume simple interest.

a) Find the simple interest rate.

b) Write the first five terms, A_1 through A_5, of the sequence representing the value of the deposit after n years.

42. Simple interest. A deposit of $2500 grows in value to $2555 after 1 yr. Assume simple interest.

a) Find the simple interest rate.

b) Write the first five terms, A_1 through A_5, of the value of the deposit after n years.

43. Library fines. The fine on an overdue book at the Birchwood Public Library is 0.5% of the value of the book multiplied by the number of days the book is overdue. Suppose Glen's overdue library book is valued at $25.00.

a) Find a formula for F_n, the amount of Glen's fine when the book has been overdue for n days.

b) Write the first five terms, F_1 through F_5, of the sequence representing the amount of Glen's fine after n days.

44. Credit card late fee. For overdue payments on its credit card accounts, Robertson's Department Store charges a penalty of 1.2% of the finance charge multiplied by the number of days the payment is overdue. Suppose Hilda's finance charge is $30.

a) Find a formula for F_n, the penalty applied to Hilda's payment when it is n days overdue.

b) Write the first five terms, F_1 through F_5, of the sequence representing the penalty after n days.

45. Government bonds. A treasury bond is purchased, and 3 months later, it is redeemed for $500. Assume an annual simple interest rate of 2%.

a) Find the present value of the bond (its purchase price).

b) Find the interest earned.

46. Government bonds. A treasury bond is purchased, and 2 yr later, it is redeemed for $1200. Assume an annual simple interest rate of 1.45%.

a) Find the present value of the bond.

b) Find the interest earned.

47. Add-on interest. Paul's Appliance Mart sells refurbished appliances. It advertises a refrigerator for $600, with half down and the other half to be paid in 8 monthly payments. Assume Paul's charges an annual interest rate of 5%. Find the simple interest and the monthly payment.

48. Add-on interest. Kevin agrees to sell a motorcycle to his friend Sam for $600. Sam pays $200 down, and Kevin agrees to finance the balance for 10 months at an annual simple interest rate of 4.25%. Find the simple interest and Sam's monthly payment.

49. Compound interest future value. Gina deposits $3000 in a savings account at an annual interest rate of 4.5%, compounded monthly.

a) Find a formula for A_n, the value of Gina's account after n years.

b) Write the first five terms, A_1 through A_5, of the sequence representing the value of Gina's account after n years.

50. Compound interest future value. Dana deposits $11,200 in a savings account at an annual interest rate of 3.7%, compounded monthly.

a) Find a formula for A_n, the amount in Dana's account after n years.

b) Write the first five terms, A_1 through A_5, of the sequence representing the value of Dana's account after n years.

51. Present value. Yvonne wants to have $5000 in 2 yr. She opens an account that has an annual interest rate of 4.8%, compounded monthly. Find the present value (the initial lump-sum deposit) of her account and the interest her deposit will earn.

52. Present value. Ted wants to have $2000 in 3 yr. He opens an account that has an annual interest rate of 3.7% compounded quarterly. Find the present value (the initial lump-sum deposit) of his account and the interest his deposit will earn.

53. Compound interest. An initial deposit of $1500 grows to $1690.91 after 3 yr. Assume monthly compounding.

   a) What is the annual interest rate?
   b) Write the first three terms, A_1 through A_3, of the sequence representing the value of the account after n years.

54. Compound interest. An initial deposit of $2000 grows to $2398.57 after 5 yr. Assume monthly compounding.

   a) What is the annual interest rate?
   b) Write the first five terms, A_1 through A_5, of the sequence representing the value of the account after n years.

55. Comparison. Greg wants to deposit $10,000 for 3 yr. First Federal offers a certificate of deposit that will earn 4% simple interest for a 3-yr term, while Valley View Bank offers a savings account that will earn 3.8%, compounded monthly, over the 3-yr period.

   a) Find the future value for both options.
   b) Find the interest earned for both options.
   c) Which option is better, and why?

56. Comparison. Paulette wants to deposit $2000 for 4 yr. Mutual Savings offers a certificate of deposit that will earn 3.5% simple interest for a 4-yr term, while Perkins Savings offers a savings account that will earn 3.37%, compounded weekly, over the 4-yr period.

   a) Find the future value for both options.
   b) Find the interest earned for both options.
   c) Which option is better, and why?

57. Comparison. Clara wants to have $3000 for a trip to Europe in 2 yr. She has two options: a simple-interest account at an annual interest rate of 4.2% and a compound-interest account at 4%, compounded monthly.

   a) Find the present value (the amount she needs to deposit now) for both options.
   b) Find the interest earned for both options.
   c) Which option is better, and why?

58. Comparison. George wants to have $7500 in 3 yr for college expenses. He has two options: a simple-interest account at an annual interest rate of 3.75% and a compound-interest account at 3.67%, compounded quarterly.

   a) Find the present value (the amount he needs to deposit now) for both options.
   b) Find the interest earned for both options.
   c) Which option is better, and why?

SYNTHESIS

Annual yield. *The annual interest rate i, when compounded more than once per year, results in a slightly higher yearly interest rate; this is called the annual (or effective) yield and denoted as Y. For example, $1000 deposited at 5%, compounded monthly for 1 yr (12 months), has a future value of $A = 1000\left(1 + \frac{0.05}{12}\right)^{12} = \$1051.16$. The interest earned is $51.16/$1000, or 0.05116, which is 5.116% of the original deposit. Thus, we say this account has a yield of $Y = 0.05116$, or 5.116%. The formula for annual yield depends on the annual interest rate i and the compounding frequency c:*

$$Y = \left(1 + \frac{i}{c}\right)^c - 1.$$

For Exercises 59–62, find the annual yield as a percentage, to two decimal places, given the annual interest rate and the compounding frequency.

59. Annual interest rate of 5.3%, compounded monthly

60. Annual interest rate of 4.1%, compounded quarterly

61. Annual interest rate of 3.75%, compounded weekly

62. Annual interest rate of 4%, compounded daily

63. Lena is considering two savings accounts: Western Bank offers 4.5%, compounded annually, on saving accounts, while Commonwealth Savings offers 4.43%, compounded monthly.

   a) Find the annual yield for both accounts.
   b) Which account has the higher annual yield?

64. Chris is considering two savings accounts: Sierra Savings offers 5%, compounded annually, on savings accounts, while Foothill Bank offers 4.88%, compounded weekly.

   a) Find the annual yield for both accounts.
   b) Which account has the higher annual yield?

65. Stockman's Bank will pay 4.2%, compounded annually, on a saving account. A competitor, Mesalands Savings, offers monthly compounding on savings accounts. What is the minimum annual interest rate that Mesalands needs to pay to make its annual yield exceed that of Stockman's?

66. Belltown Bank offers a certificate of deposit at 3.75%, compounded annually. Shea Savings offers savings accounts with interest compounded quarterly. What is the minimum annual interest rate that Shea needs to pay to make its annual yield exceed that of Belltown?

TECHNOLOGY CONNECTION

67. In Example 7, the monthly pay with option 1 is given by $A_n = 4500 + 500n$, for $1 \le n \le 30$, and the monthly pay with option 2 is given by $B_n = 2^{n-1}$, for $1 \le n \le 30$.

   a) In what month does the monthly pay with option 2 exceed the monthly pay with option 1?
   b) In what month does the total amount earned with option 2 exceed the total amount earned with option 1?

68. Suppose $1000 is deposited in a savings account that earns 3.25%, compounded monthly.

   a) In what month is the amount in the account twice the original deposit?
   b) In what month is the amount in the account three times the original deposit?
   c) Repeat parts (a) and (b) for a principal of $5000.
   d) Does the doubling time or tripling time for a savings account depend on the principal? Why or why not?

Answers to Quick Checks

1. $A = \$4590, I = \$90$ **2.** $72.88 (rounded)
3. $P = \$2919.71, I = \$80.29$ **4.** $1212.75, $1225.64,
$1238.66, $1251.82   **5.** $i = 3; A_0 = \$3000,$
$A_1 = \$3090, A_2 = \$3182.70, A_3 = \$3278.18$
6. $P = \$6423.39$

Annuities and Amortization

Annuities

Suppose $1000 is deposited each year into an account with an annual interest rate of 5%, compounded annually. What will be the future value of the account at the start of the 4th year? The table below shows the growth of each deposit. Assume that each deposit is made at the start of the year. Note that each $1000 deposit grows according to the compound interest future value formula, so its value after t years is given by $1000(1.05)^t$.

	Value at the start of the first year	Value at the start of the second year	Value at the start of the third year	Value at the start of the fourth year
First deposit	$1000	$1000(1.05)	$1000(1.05)^2$	$1000(1.05)^3$
Second deposit		$1000	$1000(1.05)	$1000(1.05)^2$
Third deposit			$1000	$1000(1.05)
Fourth deposit				$1000
Total	$1000	$2050	$3152.50	$4310.13

Adding the values in the last column, we obtain the future value of the account at the start of the 4th year:

$$A = 1000 + 1000(1.05) + 1000(1.05)^2 + 1000(1.05)^3, \text{ or } \$4310.13.$$

This is a geometric series, so we can use the formula for the nth partial sum of such a series to find the sum. We have the first term, $a_1 = 1000$, and the common ratio, $r = 1.05$, and there are $n = 4$ terms, so

$$S_n = A = \frac{a_1(r^n - 1)}{r - 1} \qquad \text{Finding the sum of a geometric series}$$

$$= \frac{1000[(1.05)^4 - 1]}{(1.05) - 1} = \$4310.13. \qquad \text{Substituting}$$

A savings account into which equal-sized deposits (called *payments*) are made on a regular basis is called an **annuity**. Annuities are popular ways to save money over long periods of time, as these accounts can grow very large. For example, many people use annuities to save for retirement or college.

If we let p represent the amount paid into the annuity on a regular basis, i the annual interest rate, c the compounding frequency, and t the time in years, we can develop a formula for the future value of an annuity. The first term is $a_1 = p$, the common ratio is $r = \left(1 + \dfrac{i}{c}\right)$, and the total number of payments is ct. We assume that each payment is made at the start of a compounding period.

$$S_n = \frac{a_1(r^n - 1)}{r - 1} \qquad \text{Sum of a geometric series}$$

$$A = \frac{p\left[\left(1 + \dfrac{i}{c}\right)^{ct} - 1\right]}{\left(1 + \dfrac{i}{c}\right) - 1}. \qquad \text{Substituting}$$

The denominator simplifies to i/c. This gives us the following formula.

THEOREM 6

If we assume an annual interest rate i compounded c times per year for t years and equal payments p into the account made with the same frequency c at the start of each compounding period, then the **future value of an annuity** is given by

$$A = \frac{p\left[\left(1 + \dfrac{i}{c}\right)^{ct} - 1\right]}{\dfrac{i}{c}}.$$

The total deposits (called the *personal contribution*) made into an annuity are equal to $p \cdot c \cdot t$, and the interest earned is given by $A - p \cdot c \cdot t$.

Note that the payment p is different from the principal P. Instead of one deposit, P, an annuity receives regular deposits, each equal to p, that coincide with the compounding periods.

■ **EXAMPLE 1**   **Business: Annuity.**   Javier is going to deposit $100 into an annuity every month for 20 yr. This account pays an annual interest rate of 4.75%, compounded monthly.

a) Find the future value of Javier's annuity.

b) Find Javier's personal contribution to the annuity.

c) Find the interest earned.

Solution

a) Javier will make $c = 12$ payments per year for $t = 20$ yr, for a total of $ct = 12 \cdot 20 = 240$ payments. Thus, the future value of Javier's annuity is

$$A = \frac{p\left[\left(1 + \dfrac{i}{c}\right)^{ct} - 1\right]}{\dfrac{i}{c}} \qquad \text{Using the formula for an annuity}$$

$$= \frac{100\left[\left(1 + \dfrac{0.0475}{12}\right)^{240} - 1\right]}{\dfrac{0.0475}{12}} \qquad \text{Substituting}$$

$$= \$39{,}937.65.$$

b) Javier will make 240 payments of $100 each, for a personal contribution of $24,000.

c) The interest earned is $\$39{,}937.65 - \$24{,}000 = \$15{,}937.65$.

❮ Quick Check 1

TECHNOLOGY CONNECTION 〰

Exploratory

The calculation made in Example 1(a) can be entered as a single expression:

100((1+0.0475/12)^240-1)/(0.0475/12),

followed by **ENTER**. Be sure to include parentheses for the denominator. Entering the entire equation at once reduces the potential for rounding error when calculating the equation in intermediate steps.

❯ **Quick Check 1**

Sharlene deposits $500 every 3 months (quarterly) into an annuity with an annual interest rate of 5.1%, compounded quarterly. Find the future value of her annuity after 15 yr, her personal contribution, and the interest earned.

Sinking Funds

When an annuity is used to save a specific amount by some particular point in the future, we call this type of savings account a **sinking fund**.

Suppose the Geigers want to have $50,000 in 18 yr to help pay for college for their new baby. They decide to open a college fund with an annual interest rate of 4.5%, compounded monthly. What is the monthly payment they need to make for the sinking fund to grow as desired? To determine the payment size that allows the fund to achieve the desired future value, we use the formula for the future value of an annuity:

$$A = \dfrac{p\left[\left(1 + \dfrac{i}{c}\right)^{ct} - 1\right]}{\dfrac{i}{c}}$$

Using the formula for an annuity

$$\$50{,}000 = \dfrac{p\left[\left(1 + \dfrac{0.045}{12}\right)^{216} - 1\right]}{\dfrac{0.045}{12}}$$

Substituting, after noting that $18 \cdot 12 = 216$

$$50{,}000 \cdot \dfrac{0.045}{12} = p\left[\left(1 + \dfrac{0.045}{12}\right)^{216} - 1\right]$$

$$p = \dfrac{187.5}{\left[\left(1 + \dfrac{0.045}{12}\right)^{216} - 1\right]} = \$150.66$$

Solving for p

Thus, the Geigers' monthly payment needs to be $150.66. If they deposit this amount every month for 18 yr (216 payments), they will achieve their goal of $50,000. They will personally contribute $150.66 \cdot 216 = \$32{,}542.56$. The rest, $\$50{,}000 - \$32{,}542.56 = \$17{,}457.44$, is interest.

■ **EXAMPLE 2** **Sinking Fund versus Lump-Sum Deposit.** Kerrie wants to have $10,000 in 4 yr for graduate school expenses. At a local bank, she can open a savings account with an annual interest rate of 3.85%, compounded quarterly. She has two options:

- Option 1: Make quarterly deposits (payments) into this account (a sinking fund).
- Option 2: Make a one-time lump-sum deposit into this account.

Find Kerrie's quarterly payment for option 1 and the lump-sum deposit for option 2. Compare the two options.

Solution   We have $A = \$10{,}000$, $i = 0.0385$, $c = 4$, and $t = 4$. Note that $c \cdot t = 4 \cdot 4 = 16$. For option 1, Kerrie's quarterly payment is

$$A = \dfrac{p\left[\left(1 + \dfrac{i}{c}\right)^{ct} - 1\right)}{\dfrac{i}{c}}$$

Using the formula for an annuity

$$\$10{,}000 = \dfrac{p\left[\left(1 + \dfrac{0.0385}{4}\right)^{16} - 1\right)}{\dfrac{0.0385}{4}}$$

Substituting

$$10{,}000 = p(17.20853974)$$

Simplifying

$$p = \dfrac{10{,}000}{17.20853974} = \$581.11.$$

Solving for p

For option 2, we need to find the present value P of $10,000, and since Kerrie makes just one payment, we use the compound interest future value formula:

$$A = P\left(1 + \frac{i}{c}\right)^{ct}$$ Using the formula for compound interest future value

$$\$10,000 = P\left(1 + \frac{0.0385}{4}\right)^{16}$$ Substituting

$$P = \frac{10,000}{\left(1 + \dfrac{0.0385}{4}\right)^{16}} = \$8579.04.$$ Solving for P

If Kerrie chooses option 1, she will make 16 quarterly payments of $581.11, for a personal contribution of $9297.76, and the rest of the account's final value, $10,000 − $9297.76 = $702.24, is interest. If Kerrie chooses option 2, her personal contribution is $8579.04, meaning that the rest, $10,000 − $8579.04 = $1420.96, is interest. More interest is earned with option 2, but Kerrie may not have that amount of money available to make the initial lump-sum deposit. Thus, option 1, with the smaller quarterly payments, may be her best choice.

⟨ Quick Check 2

> **Quick Check 2**

Sarah wants to have $1000 in 1 yr for her holiday shopping. She opens an account that has an annual interest rate of 4%, compounded monthly. Find (a) her monthly payment if she chooses to use the account as a sinking fund and (b) the one-time lump-sum deposit she needs to make if she chooses that option. Compare the interest earned with both options.

Amortization

Most large purchases, such as a house or a car, are made with an amortized loan. The borrower pays off such a loan in equal payments. Unlike an add-on interest loan (Section 9.3), for which the total interest is added to the principal at the start, an amortized loan allows the borrower to pay less total interest by paying more than the minimum required payment at any time. The process by which such a loan is paid off is called **amortization**; it is represented graphically below.

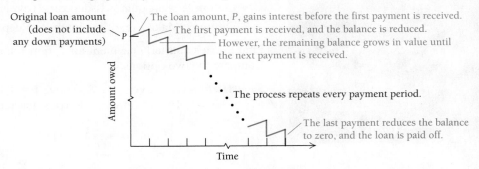

Suppose you buy a new car. After making a down payment, you elect to finance the remaining balance through a loan on which you will make monthly payments. However, the money borrowed accrues interest as long as the loan is active. We let P represent the loan amount and p represent the minimum regular payment needed to pay off (*amortize*) the loan. The formula for amortization equates the formulas for compound interest future value (to represent the growth of the loan amount P) and annuity future value (which includes the payment amount p).

DEFINITION

A loan amount P is financed at an annual interest rate i compounded c times per year for t years. The payment amount p must satisfy the *amortization formula*:

$$P\left(1 + \frac{i}{c}\right)^{ct} = \frac{p\left[\left(1 + \dfrac{i}{c}\right)^{ct} - 1\right]}{\dfrac{i}{c}}.$$

■ **EXAMPLE 3** **Amortization: Car Loan.** Henri buys a new car for $20,000, with 30% down and the rest financed through an amortized loan at 7%, compounded monthly for 5 yr.

a) Find his monthly loan payment.

b) Assuming that Henri makes every payment for the life of the loan, find the total amount he will pay.

Solution

a) Henri pays 30% of $20,000, or $6000, as a down payment and finances the rest of the car's cost through an amortized loan. Since 70% of $20,000 is $14,000, we have $P = \$14,000$, $c = 12, t = 5$, and $i = 0.07$. These are substituted into the amortization formula:

$$P\left(1 + \frac{i}{c}\right)^{ct} = \frac{p\left[\left(1 + \frac{i}{c}\right)^{ct} - 1\right]}{\frac{i}{c}}$$ Using the amortization formula

$$14{,}000\left(1 + \frac{0.07}{12}\right)^{60} = \frac{p\left[\left(1 + \frac{0.07}{12}\right)^{60} - 1\right]}{\frac{0.07}{12}}$$ Substituting, after noting that $ct = 5 \cdot 12 = 60$

$$19{,}846.75363 = p(71.59290165)$$

$$\left. p = \frac{19{,}846.75363}{71.59290165} = \$277.22. \right\}$$ Solving for p

Thus, Henri's monthly car payment is $277.22.

b) If Henri makes the minimum monthly payment every month for 60 months, he will pay a total of $\$277.22 \cdot 60 = \$16{,}633.20$ to cover the loan.

❰ Quick Check 3

> **Quick Check 3**
>
> The Sharmas purchase a new home for $275,000. Suppose they put 40% down and finance the rest through a mortgage (an amortized home loan) at an annual rate of 4.1%, compounded monthly for 30 yr.
>
> a) Find their monthly payment.
> b) Find the total interest they will pay on their mortgage, assuming that they make all payments.

A person may have a desired or maximum amount for the regular payment and want to use that to find loan amount P that is affordable.

■ **EXAMPLE 4** **Business: Car Loan.** Chelsea wants to buy a new car and can afford to make payments of $300 per month. She prequalifies for a car loan at 5%, compounded monthly for 6 years. What is the largest loan amount she can afford?

Solution We use the amortization formula and make the substitutions:

$$P\left(1 + \frac{i}{c}\right)^{ct} = \frac{p\left[\left(1 + \frac{i}{c}\right)^{ct} - 1\right]}{\frac{i}{c}}$$ Using the amortization formula

$$P\left(1 + \frac{0.05}{12}\right)^{72} = \frac{300\left[\left(1 + \frac{0.05}{12}\right)^{72} - 1\right]}{\frac{0.05}{12}}$$ Substituting

$$P(1.349017744) = 25129.27758$$

$$\left. P = \frac{25129.27758}{1.349017744} \right.$$ Solving for P

$$= \$18{,}627.83.$$

Chelsea can afford a loan of up to $18,627.83.

> **Quick Check 4**
>
> Mr. and Mrs. Urrutia qualify for a 30-yr mortgage at 4.5%, compounded monthly. They are able to make payments of $2000 per month. What is the largest loan amount they can afford?

❰ Quick Check 4

Amortization Schedules

To see the effect each payment has on a loan, we can set up an *amortization schedule*.

■ **EXAMPLE 5**   **Amortization Schedule.**   From Example 3, Henri's car loan amount is $P = \$14,000$ and his monthly payment is $277.22. Fill in the first two rows of the amortization schedule shown below.

Balance	Payment	Portion of payment applied to interest	Portion of payment applied to principal	New balance
$14,000	$277.22			

Solution   During the first month, interest is calculated once. Therefore, we use the simple interest formula to determine 1 month's interest on $14,000 at the annual interest rate of $i = 7\% = 0.07$, with $t = \frac{1}{12}$ representing 1 month:

$$I = Pit$$
$$= \$14,000 \cdot 0.07 \cdot \tfrac{1}{12}$$
$$= \$81.67.$$

Henri's payment is split into two parts: $81.67 pays off the interest for the month, and the remainder, $277.22 - \$81.67 = \$195.55$, is applied toward the principal. The new balance is $14,000 - \$195.55 = \$13,804.45$.

Balance	Payment	Portion of payment applied to interest	Portion of payment applied to principal	New balance
$14,000	$277.22	$81.67	$195.55	$13,804.45
$13,804.45	$277.22			

We repeat the process: The next month's interest is calculated based on $P = \$13,804.45$. We obtain $I = \$13,804.45 \cdot 0.07 \cdot \frac{1}{12} = \$80.53$. The second payment of $277.22 is split in two parts: $80.53 pays off the interest for the month, and the rest, $277.22 - \$80.53 = \$196.69$, is applied to the principal. The new balance is $13,804.45 - \$196.69 = \$13,607.76$.

Balance	Payment	Portion of payment applied to interest	Portion of payment applied to principal	New balance
$14,000	$277.22	$81.67	$195.55	$13,804.45
$13,804.45	$277.22	$80.53	$196.69	$13,607.76

Slowly, the portion of the monthly payment applied to interest decreases and the portion applied to principal increases. Eventually, very little interest will accrue on the balance, and nearly all of the payment will be applied to the principal.

❰ Quick Check 5

> **Quick Check 5**
>
> Complete the first two rows of an amortization schedule for the Sharmas' mortgage in Quick Check 3.

TECHNOLOGY CONNECTION

Amortization Schedules on a Spreadsheet

An amortization schedule like that in Example 5 can be created with a spreadsheet. In cells A1 through E1, enter the headings shown below. In cell A2, enter 14000, and in cell B2, enter 277.22. Highlight the whole sheet and click on the $ button to convert all cells to dollar format:

	A	B	C	D	E
1	Balance Forward	Payment	Portion to Interest	Portion to Principal	New Balance
2	$ 14,000.00	$ 277.22			
3					

Next, in cell C2, enter = A2*0.07/12; in cell D2, enter =B2-C2; in cell E2, enter =A2-D2; and in cell A3, enter =E2. Then copy cells B2 through E2 and paste them into cells B3 through E3. Finally, copy cells A3 through E3 and paste them into cells A4 through E61. This process generates the amortization schedule for all 60 months of the loan period.

Note that the monthly portion devoted to interest decreases steadily. The final payment reduces the balance to zero (due to rounding error, some small fluctuations in the last row usually occur).

	A	B	C	D	E
1	Balance Forward	Payment	Portion to Interest	Portion to Principal	New Balance
2	$ 14,000.00	$ 277.22	$ 81.67	$ 195.55	$ 13,804.45
3	$ 13,804.45	$ 277.22	$ 80.53	$ 196.69	$ 13,607.75
4	$ 13,607.75	$ 277.22	$ 79.38	$ 197.84	$ 13,409.91
5	$ 13,409.91	$ 277.22	$ 78.22	$ 199.00	$ 13,210.92
6	$ 13,210.92	$ 277.22	$ 77.06	$ 200.16	$ 13,010.76
7	$ 13,010.76	$ 277.22	$ 75.90	$ 201.32	$ 12,809.44
8	$ 12,809.44	$ 277.22	$ 74.72	$ 202.50	$ 12,606.94
9	$ 12,606.94	$ 277.22	$ 73.54	$ 203.68	$ 12,403.26
10	$ 12,403.26	$ 277.22	$ 72.35	$ 204.87	$ 12,198.39
11	$ 12,198.39	$ 277.22	$ 71.16	$ 206.06	$ 11,992.33
12	$ 11,992.33	$ 277.22	$ 69.96	$ 207.26	$ 11,785.06
13	$ 11,785.06	$ 277.22	$ 68.75	$ 208.47	$ 11,576.59
49	$ 3,460.68	$ 277.22	$ 20.19	$ 257.03	$ 3,203.65
50	$ 3,203.65	$ 277.22	$ 18.69	$ 258.53	$ 2,945.12
51	$ 2,945.12	$ 277.22	$ 17.18	$ 260.04	$ 2,685.08
52	$ 2,685.08	$ 277.22	$ 15.66	$ 261.56	$ 2,423.52
53	$ 2,423.52	$ 277.22	$ 14.14	$ 263.08	$ 2,160.44
54	$ 2,160.44	$ 277.22	$ 12.60	$ 264.62	$ 1,895.82
55	$ 1,895.82	$ 277.22	$ 11.06	$ 266.16	$ 1,629.66
56	$ 1,629.66	$ 277.22	$ 9.51	$ 267.71	$ 1,361.95
57	$ 1,361.95	$ 277.22	$ 7.94	$ 269.28	$ 1,092.67
58	$ 1,092.67	$ 277.22	$ 6.37	$ 270.85	$ 821.82
59	$ 821.82	$ 277.22	$ 4.79	$ 272.43	$ 549.40
60	$ 549.40	$ 277.22	$ 3.20	$ 274.02	$ 275.38
61	$ 275.38	$ 277.22	$ 1.61	$ 275.61	$ (0.23)

EXERCISES

1. Suppose Henri decides to pay $300 per month. Make the necessary changes to the amortization schedule. By how many months is the length of the loan reduced, and how much interest does he save?

2. Suppose Henri decides to pay $350 per month. Make the necessary changes to the amortization schedule. By how many months is the length of the loan reduced, and how much interest does he save?

Section Summary

- An *annuity* is a savings account in which equal-sized deposits, or payments, are made at regular intervals. The *future value of an annuity* is given by

$$A = \frac{p\left[\left(1 + \dfrac{i}{c}\right)^{ct} - 1\right]}{\dfrac{i}{c}},$$

where p is the regular payment amount, i is the annual interest rate, c is the compounding (and payment) frequency, and t is the time, in years.

- A *sinking fund* is an annuity for which the future value A is known and the regular payment amount p needed to reach that future value is then determined.
- *Amortization* is a process in which a loan amount P at an annual interest rate i compounded c times per year for t years is paid off (*amortized*) in equal-sized payments p, according to the formula

$$P\left(1 + \frac{i}{c}\right)^{ct} = \frac{p\left[\left(1 + \dfrac{i}{c}\right)^{ct} - 1\right]}{\dfrac{i}{c}}.$$

EXERCISE SET
9.4

In Exercises 1–10, find the future value of each annuity using the given payment amount, interest rate, time, and compounding frequency.

1. Monthly payment of $200; $i = 5\%$; $t = 12$ yr, compounded monthly

2. Monthly payment of $150; $i = 4\%$; $t = 10$ yr, compounded monthly

3. Quarterly payments of $250; $i = 4.5\%$; $t = 20$ yr, compounded quarterly

4. Quarterly payments of $175; $i = 4.8\%$; $t = 15$ yr, compounded quarterly

5. Semiannual payments of $500; $i = 5.3\%$; $t = 8$ yr, compounded semiannually

6. Semiannual payments of $225; $i = 3.7\%$; $t = 14$ yr, compounded semiannually

7. Annual payments of $1500; $i = 3.25\%$; $t = 25$ yr, compounded annually

8. Annual payments of $1750; $i = 3.55\%$; $t = 30$ yr, compounded annually

9. Weekly payments of $10; $i = 4.75\%$; $t = 2$ yr, compounded weekly

10. Weekly payments of $25; $i = 3.67\%$; $t = 3$ yr, compounded weekly

In Exercises 11–16, find the sinking-fund payment needed to achieve the given future value. Assume that the payment frequency is the same as the compounding frequency.

11. $A = \$5000$; $i = 4\%$; $t = 6$ yr, compounded monthly

12. $A = \$10{,}000$; $i = 5.5\%$; $t = 10$ yr, compounded monthly

13. $A = \$12{,}000$; $i = 4.25\%$; $t = 9$ yr, compounded annually.

14. $A = \$8000$; $i = 4.56\%$; $t = 14$ yr, compounded annually

15. $A = \$15{,}000$; $i = 3.175\%$; $t = 15$ yr, compounded quarterly

16. $A = \$12{,}500$; $i = 3.667\%$; $t = 5$ yr, compounded quarterly

In Exercises 17–26, find the payment needed to amortize the given loan amount. Assume that the payment frequency is the same as the compounding frequency.

17. $P = \$7000$; $i = 6\%$; $t = 5$ yr, compounded monthly

18. $P = \$8000$; $i = 5\%$; $t = 8$ yr, compounded monthly

19. $P = \$12{,}000$; $i = 5.7\%$; $t = 6$ yr, compounded quarterly

20. $P = \$20{,}000$; $i = 6.2\%$; $t = 7$ yr, compounded quarterly

21. $P = \$500$; $i = 4.1\%$; $t = 1$ yr, compounded monthly

22. $P = \$800$; $i = 3.8\%$; $t = 2$ yr, compounded monthly

23. $P = \$150{,}000$; $i = 5.15\%$; $t = 30$ yr, compounded semiannually

24. $P = \$235{,}000$; $i = 4.35\%$; $t = 30$ yr, compounded semiannually

25. $P = \$75{,}000$; $i = 8\%$; $t = 15$ yr, compounded annually

26. $P = \$90{,}000$; $i = 7\%$; $t = 12$ yr, compounded annually

In Exercises 27–30, find (a) the future value of the annuity, (b) the personal contribution and (c) the interest earned.

27. Tim deposits $125 every month for 8 yr into a savings account that has an annual interest rate of 4.5%, compounded monthly.

28. Tom deposits $500 every 3 months for 5 yr into a savings account that has an annual interest rate of 5.2%, compounded quarterly.

29. Cindy deposits $2000 every year for 15 yr into a savings account that has an annual interest rate of 7%, compounded annually.

30. Julie deposits $50 weekly for 1 yr into a savings account that has an annual interest rate of 4.45%, compounded weekly.

In Exercises 31–34, find (a) the payment size, (b) the personal contribution and (c) the interest earned. Assume that each payment is made at the start of the compounding period.

31. Gayla wants to have $5000 in 2 yr for a down payment on a car. She establishes a sinking fund at an annual interest rate of 4.35%, compounded monthly.

32. Ricardo wants to have $25,000 in 5 yr for college expenses. He establishes a sinking fund at an annual interest rate of 5.1%, compounded quarterly.

33. The Miyokawas want to have $200,000 in 25 yr for their retirement. They start a sinking fund at an annual interest rate of 5.4%, compounded annually.

34. Klaus wants to have $30,000 in 6 yr for a down payment on a house. He starts a sinking fund at an annual interest rate of 3.9%, compounded monthly.

APPLICATIONS

Business and Economics

35. Investment strategies. The Monroes want to have $10,000 in 3 yr for a vacation trip to the Amazon. They consider two options: (1) a one-time lump-sum deposit into an account that has an annual interest rate of 4.5%, compounded monthly, and (2) monthly payments into a sinking fund that has an annual interest rate of 4.35%, compounded monthly.

    a) Find the lump-sum deposit needed for option 1 and the interest earned.
    b) Find the monthly payments needed for option 2 and the interest earned.
    c) Which option earns more interest, and how much more?

36. Investment strategies. Gerald wants to have $2000 in 1 yr for a climbing trip to the Grand Teton. He considers two options: (1) a one-time lump-sum deposit into an account that has an annual interest rate of 5%, compounded monthly, and (2) quarterly payments into a sinking fund that has an annual interest rate of 4.7%, compounded quarterly.

a) Find the lump-sum deposit needed for option 1 and the interest earned.

b) Find the quarterly payments needed for option 2 and the interest earned.

c) Which option earns more interest, and how much more?

37. Retirement. For 12 yr, Janice deposits $1500 every 3 months into a retirement account that has an annual interest rate of 5.25%, compounded quarterly.

a) Find the future value of Janice's account.

b) Janice then allows the total amount in the account to accrue interest at the same interest rate and compounding frequency, without making further deposits, for another 20 yr. Find the future value of her account after that time has elapsed.

c) Find the total interest Janice's account earns over the 32-yr period.

38. Retirement. For 10 yr, Briana deposits $100 every month into a retirement account that has an annual interest rate of 4.3%, compounded monthly.

a) Find the future value of Briana's account.

b) Briana then allows the total amount in the account to accrue interest at the same interest rate and compounding frequency, without making further deposits, for another 15 yr. Find the future value of her account after that time has elapsed.

c) Find the total interest Briana's account earns over the 25-yr period.

39. Car loans. Todd purchases a new Honda Accord LX for $22,150. He makes a $4000 down payment and finances the remainder through an amortized loan at an annual interest rate of 6.5%, compounded monthly for 5 yr.

a) Find Todd's monthly car payment.

b) Assume that Todd makes every payment for the life of the loan. Find his total payments.

c) How much interest does Todd pay?

40. Car loans. Katie purchases a new Jeep Wrangler Sport for $23,000. She makes a $5000 down payment and finances the remainder through an amortized loan at an annual interest rate of 5.7%, compounded monthly for 7 yr.

a) Find Katie's monthly car payment.

b) Assume that Katie makes every payment for the life of the loan. Find her total payments.

c) How much interest does Katie pay?

41. Home mortgages. The Hogansons purchase a new home for $195,000. They make a 25% down payment and finance the remainder with a 30-yr mortgage at an annual interest rate of 5.2%, compounded monthly.

a) Find the Hogansons' monthly mortgage payment.

b) Assume that the Hogansons make every payment for the life of the loan. Find their total payments.

c) How much interest do the Hogansons pay?

42. Home mortgages. Andre purchases an office building for $450,000. He makes a 30% down payment and finances the remainder through a 15-yr mortgage at an annual interest rate of 4.15%, compounded monthly.

a) Find Andre's monthly mortgage payment.

b) Assume that Andre makes every payment for the life of the loan. Find his total payments.

c) How much interest does Andre pay?

43. Credit cards. Joanna uses her credit card to finance a $500 purchase. Her card charges an annual interest rate of 22.75%, compounded monthly, and assumes a 10-yr term. Assume that Joanna makes no further purchases on her credit card.

a) Find Joanna's monthly credit card payment.

b) Assume that Joanna makes every payment for the life of the loan. Find her total payments.

c) How much interest does Joanna pay?

44. Credit cards. Isaac uses his credit card to finance a $1200 purchase. His card charges an annual interest rate of 19.25%, compounded monthly, and assumes a 10-yr term. Assume that Isaac makes no further purchases on his credit card.

a) Find Isaac's monthly credit card payment.

b) Assume that Isaac makes every payment for the life of the loan. Find his total payments.

c) How much interest does Isaac pay?

45. Loans. Western Sky Financial will loan a borrower $9925 at an annual interest rate of 89.68%, compounded monthly for 7 yr. (*Source*: www.westernsky.com.)

a) Find the monthly payment.

b) Assume that the borrower makes every payment for the life of the loan. Find the total payments.

c) How much interest will the borrower pay?

46. Payday loans. Advance America will loan a borrower $83.50 at an annual interest rate of 456.25%, compounded weekly for a 2-week term. (*Source*: www.advanceamerica.com.)

a) Find the weekly payment.

b) Assume the borrower makes both weekly payments. Find the total payments.

c) How much interest will the borrower pay?

In Exercises 47–52, complete the first two lines of an amortization schedule for each situation, using a table as shown below.

Balance	Payment	Portion of payment applied to interest	Portion of payment applied to principal	New balance

47. Todd's car loan in Exercise 39

48. Katie's car loan in Exercise 40

49. The Hogansons' home loan in Exercise 41

50. Andre's office building loan in Exercise 42

51. Joanna's credit card loan in Exercise 43

52. Isaac's credit card loan in Exercise 44

53. Maximum loan amount. Desmond plans to purchase a new car. He qualifies for a loan at an annual interest rate of 5.8%, compounded monthly for 6 yr. He is willing to pay up to $300 per month. What is the largest loan he can afford?

54. Maximum loan amount. Curtis plans to purchase a new car. He qualifies for a loan at an annual interest rate of 7%, compounded monthly for 5 yr. He is willing to pay up to $200 per month. What is the largest loan he can afford?

55. Maximum loan amount. The Daleys plan to purchase a new home. They qualify for a mortgage at an annual interest rate of 4.15%, compounded monthly for 30 yr. They are willing to pay up to $1800 per month. What is the largest loan they can afford?

56. Maximum loan amount. Martina plans to purchase a new home. She qualifies for a mortgage at an annual interest rate of 4.45%, compounded monthly for 20 yr. She is willing to pay up to $2000 per month. What is the largest loan she can afford?

57. Comparing loan options. Kathy plans to finance $12,000 for a new car through an amortized loan. The lender offers two options: (1) a 5-yr term at an annual interest rate of 5.2%, compounded monthly, and (2) a 6-yr term at an annual interest rate of 5%, compounded monthly.

  a) Find the monthly payments for options 1 and 2.
  b) Assume that Kathy makes every payment for the life of the loan. Find her total payments for options 1 and 2.
  c) Assume that Kathy intends to make every payment for the life of either loan. Which option will result in less interest paid, and how much less?

58. Comparing loan options. The Aubrys plan to finance a new home through an amortized loan of $275,000. The lender offers two options: (1) a 30-yr term at an annual interest rate of 4%, compounded monthly, and (2) a 20-yr term at an annual interest rate of 5%, compounded monthly.

  a) Find the monthly payments for options 1 and 2.
  b) Assume that the Aubrys make every payment for the life of the loan. Find their total payments for options 1 and 2.
  c) Assume that the Aubrys intend to make every payment for the life of each loan. Which option will result in less interest paid, and by how much?

59. Comparing rates. Darnell plans to finance $200,000 for a new home through a 30-yr mortgage.

  a) Find his monthly payment assuming an annual interest rate of 5%, compounded monthly.
  b) Find his monthly payment assuming an annual interest rate of 6%, compounded monthly.
  c) Assume that Darnell makes every payment for the life of the loan. How much will he save in interest over 30 yr if he accepts the 5% rate?

60. Comparing rates. The Salazars plan to finance $300,000 for a new home through a 20-yr mortgage.

  a) Find their monthly payment assuming an annual interest rate of 4.25%, compounded monthly.
  b) Find their monthly payment assuming an annual interest rate of 4.75%, compounded monthly.
  c) Assume that the Salazars make every payment for the life of the loan. How much will they save in interest over 30 yr if they accept the 4.25% rate?

61. Multiple approaches. Dwight is 25 years old. He plans to retire at age 60 and wants to have saved a sum of money by then that will allow him to withdraw $500 per month for 25 more years (until age 85), at which time the sum will be depleted (amortized). Assume a fixed annual interest rate of 4.5%, compounded monthly for the entire duration.

  a) Find the amount of money Dwight will need at age 60.
  b) Find the monthly deposit Dwight needs to make in order to reach this sum, assuming that he starts saving immediately.

62. Multiple approaches. Kenna is 30 years old. She plans to retire at age 55 and wants to have saved a sum of money by then that will allow her to withdraw $700 per month for 25 more years (until age 80), at which time the sum will be depleted (amortized). Assume a fixed annual interest rate of 5%, compounded monthly for the entire duration.

  a) Find the amount of money Kenna will need at age 55.
  b) Find the monthly deposit Kenna needs to make in order to reach this sum, assuming that she starts saving immediately.

SYNTHESIS

63. Lottery winnings. Suppose you win $5,000,000 (after taxes) in a lottery. Instead of a one-time payment, you accept a structured plan of annual payments over 20 yr. Under this plan, you do not receive $5,000,000 divided by 20, or $250,000 per year. Instead, you receive yearly payments W that assume an annual interest rate of 5% and that will accrue to $5,000,000 over 20 yr. Find the annual payment W you receive under this plan.

64. Structured settlement. Suppose you won a $700,000 (after taxes) settlement in court, to be paid in quarterly payments at an annual interest rate of 4%, compounded quarterly for 5 yr. Find your quarterly payment.

Shortening the life of a loan. *Amortization gives the borrower an advantage: by paying more than the minimum required payment, the borrower can pay off (amortize) the principal faster and save money in interest. Assume P is the loan amount or principal, i is the interest rate, c the compounding frequency, and t is the time in years. The amortization formula is used to determine the payment amount p that will amortize the loan in t years. However, if the borrower consistently pays Q, where Q > p, the formula for the time needed to amortize the loan amount P is*

$$t = \frac{\ln(cQ) - \ln(cQ - Pi)}{c \ln\left(1 + \dfrac{i}{c}\right)}.$$

65. Business: home loan. The Begays finance $200,000 for a 30-yr home mortgage at an annual interest rate of 5%, compounded monthly.

   a) Find the monthly payment needed to amortize this loan in 30 yr.

   b) Assuming that the Begays make the payment found in part (a) every month for 30 yr, find the total interest they will pay.

   c) Suppose the Begays pay an extra 15% every month (thus, $Q = 1.15 \cdot p$). Find the time needed to amortize the $200,000 loan.

   d) About how much total interest will the Begays pay if they pay Q every month?

   e) About how much will the Begays save on interest if they pay Q every month?

66. Business: car loan. Gwen finances a $15,000 car loan at an annual interest rate of 6%, compounded monthly over a 5-yr term.

   a) Find the payment needed to amortize this loan in 5 yr.

   b) Assuming that Gwen makes the payment found in part (a) every month for 5 yr, find the total interest she will pay.

   c) Suppose Gwen pays an extra 20% every month. Find the time needed to amortize the $15,000 loan.

   d) About how much total interest will Gwen pay if she pays Q every month?

   e) About how much will Gwen save on interest if she pays Q every month?

67. Derive the formula for t given above.

68. If you are shopping for a car, you will likely be offered loans at different interest rates and/or for different terms. For example, a longer term usually results in a lower monthly payment. Is increasing the term of the loan a good long-term strategy? In general, what is always a good strategy when choosing among various loan options?

TECHNOLOGY CONNECTION · · · · · · · · · · · · ·

69–74. Use a spreadsheet to complete the first twelve lines of an amortization schedule for each loan in Exercises 47–52.

Answers to Quick Checks

1. $A = \$44,652.05$, personal contribution $= \$30,000$, $I = \$14,652.05$ **2.** (a) $81.82 per month, $I = \$18.16$; (b) $960.85, $I = \$39.15$ **3.** (a) $797.28; (b) $122,020.80 **4.** $394,722.32 **5.** First row: $165,000, $797.28, $563.75, $233.53, $164,766.47; second row: $164,766.47, $797.28, $562.95, $234.33, $164,532.14

9.5

Power Series and Linearization

Power Series

Many exponential, logarithmic, radical, rational, and trigonometric functions can be expressed as an infinite series of terms called a *power series*.

DEFINITION

A **power series** is an infinite series of the form

$$\sum_{n=0}^{\infty} c_n x^n = c_0 + c_1 x + c_2 x^2 + c_3 x^3 + c_4 x^4 + c_5 x^5 + \cdots,$$

where $c_0, c_1, c_2, \ldots$ are real coefficients.

For example, when all coefficients are 1, we have the following power series:

$$1 + x + x^2 + x^3 + x^4 + x^5 + \cdots.$$

In this special case, we have an infinite geometric power series with first term $a_1 = 1$ and common ratio $r = x$. For $|x| < 1$, we know from Section 9.2 that this series converges and its sum is given by

$$S = \frac{1}{1 - x}, \qquad \text{for } -1 < x < 1.$$

Thus, for $|x| < 1$, we can equate $f(x) = \dfrac{1}{1 - x}$ and $1 + x + x^2 + x^3 + x^4 + x^5 + \cdots$.

The power series of $f(x) = \dfrac{1}{1 - x}$ is

$$\frac{1}{1 - x} = 1 + x + x^2 + x^3 + x^4 + x^5 + \cdots, \qquad \text{for } -1 < x < 1.$$

We call the interval $-1 < x < 1$ the **interval of convergence**. Since $x = 0$ is centered within this interval, we call $x = 0$ the **center of convergence**. All power series of the form

$$c_0 + c_1 x + c_2 x^2 + c_3 x^3 + c_4 x^4 + c_5 x^5 + \cdots$$

are centered at $x = 0$. If a power series is shifted k units horizontally, we say that the power series is centered at $x = k$ and is of the form

$$c_0 + c_1(x - k) + c_2(x - k)^2 + c_3(x - k)^3 + c_4(x - k)^4 + c_5(x - k)^5 + \cdots.$$

Depending on the power series, the interval of convergence can vary: the power series may converge for all real numbers, over an interval, or only at a single point. The center of convergence k is the center of the interval of convergence (a, b): $k = \dfrac{a + b}{2}$.

If the interval of convergence is all real numbers, then its center is agreed to be $k = 0$.

For any power series, we define the **nth approximating polynomial**, denoted $P_n(x)$, as a partial sum of the power series, where n is the degree of the approximation. Thus, we have

$$P_1(x) = c_0 + c_1(x - k),$$ First approximating polynomial

$$P_2(x) = c_0 + c_1(x - k) + c_2(x - k)^2,$$ Second approximating polynomial

$$P_3(x) = c_0 + c_1(x - k) + c_2(x - k)^2 + c_3(x - k)^3,$$ Third approximating polynomial

and so on.

As n increases, the approximating polynomials approach the function as a limit, for all x within the interval of convergence $a < x < b$. That is,

$$\lim_{n \to \infty} P_n(x) = f(x), \qquad \text{for } a < x < b.$$

For $f(x) = \dfrac{1}{1 - x}$, whose power series is $1 + x + x^2 + x^3 + x^4 + x^5 + \cdots$, for $-1 < x < 1$, the first approximating polynomial is $P_1(x) = 1 + x$, the second approximating polynomial is $P_2(x) = 1 + x + x^2$, and so on. The first approximating polynomial gives the equation of the tangent line to $f(x) = \dfrac{1}{1 - x}$ at $(0, 1)$.

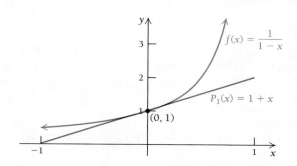

Although we do not prove it here, the first approximating polynomial $P_1(x)$ is always the equation of the tangent line of f at $x = k$.

The second approximating polynomial is a parabola that better approximates $f(x) = \dfrac{1}{1 - x}$ at $(0, 1)$.

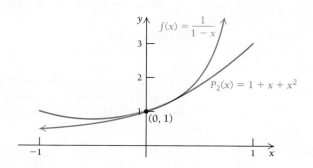

The approximating polynomials with more terms agree more closely with $f(x) = \dfrac{1}{1 - x}$.

■ **EXAMPLE 1**   Let $f(x) = \dfrac{2}{3 + x}$.

a) Use the power series for $\dfrac{1}{1 - x}$ to find a power series for f.

b) Find the interval of convergence.

Solution

a) To find a power series for $f(x) = \dfrac{2}{3 + x}$, we rewrite it in the form $c \cdot \dfrac{1}{1 - r}$.

$$\frac{2}{3 + x} = \frac{2}{3} \cdot \left(\frac{1}{1 + x/3} \right) \qquad \text{Factoring}$$

$$= \frac{2}{3} \cdot \left(\frac{1}{1 - (-x/3)} \right) \qquad \text{Writing in the form } c \cdot \frac{1}{1 - r}$$

The expression $\dfrac{1}{1 - (-x/3)}$ can be viewed as the sum of a geometric series with first term $a_1 = 1$ and common ratio $r = -x/3$. Thus, for $|-x/3| < 1$, we have

$$f(x) = \frac{2}{3} \left(\frac{1}{1 - (-x/3)} \right)$$

$$= \frac{2}{3} \left(1 + \left(-\frac{x}{3} \right) + \left(-\frac{x}{3} \right)^2 + \left(-\frac{x}{3} \right)^3 + \left(-\frac{x}{3} \right)^4 + \cdots \right)$$

$$\text{Using } \frac{1}{1 - r} = 1 + r + r^2 + r^3 + \cdots$$

$$= \frac{2}{3} \left(1 - \frac{x}{3} + \frac{x^2}{9} - \frac{x^3}{27} + \frac{x^4}{81} - \cdots \right) \qquad \text{Simplifying}$$

$$= \frac{2}{3} - \frac{2x}{9} + \frac{2x^2}{27} - \frac{2x^3}{81} + \frac{2x^4}{243} - \cdots. \qquad \text{Multiplying}$$

b) This series converges for $\left| -\dfrac{x}{3} \right| < 1$:

$$\left| \frac{x}{3} \right| < 1 \qquad \text{Noting that } |-a| = |a|$$

$$-1 < \frac{x}{3} < 1 \qquad \text{Recalling that } |x| < a \text{ is equivalent to } -a < x < a$$

$$-3 < x < 3. \qquad \text{Multiplying}$$

Thus, the interval of convergence is $-3 < x < 3$.

❮ Quick Check 1

TECHNOLOGY CONNECTION

Exploratory

Graph Y1=2/(3+X),
Y2=2/3-2X/9+(2X^2)/27, and
Y3=2/3-2X/9+(2X^2)/27-(2X^3)/81
in the window $[-3, 3, 0, 3]$ to
confirm that the approximating
polynomials y_2 and y_3 closely
agree with y_1. What approximating polynomial would agree even
more closely with y_1?

❯ **Quick Check 1**

Write the first six terms
of the power series for
$g(x) = \dfrac{4}{2 - x}$, and give the
interval of convergence.

The figure shows the graph of f from Example 1 (in blue) along with the graphs of two approximating polynomials. The graphs closely agree with that of the function f, serving as evidence that our calculations are probably correct.

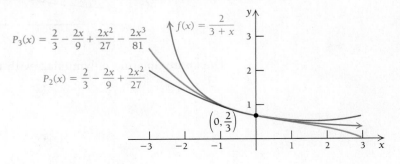

$$P_3(x) = \frac{2}{3} - \frac{2x}{9} + \frac{2x^2}{27} - \frac{2x^3}{81}$$

$$P_2(x) = \frac{2}{3} - \frac{2x}{9} + \frac{2x^2}{27}$$

$$f(x) = \frac{2}{3 + x}$$

Differentiation and antidifferentiation can be used to develop additional power series. Although we do not prove it here, if a power series is convergent over an interval (a, b), then its derivatives and antiderivatives are also convergent over that interval.

■ **EXAMPLE 2** Let $g(x) = \dfrac{1}{1 + x}$.

a) Use the power series for $\dfrac{1}{1 - x}$ to find the power series for g.

b) Find the power series for g'.

c) Find the power series for G, where $G(x) = \displaystyle\int g(x)\, dx$.

Solution

a) Knowing that $\dfrac{1}{1 - x} = 1 + x + x^2 + x^3 + x^4 + x^5 + \cdots$, we express $\dfrac{1}{1 + x}$ as

$$\frac{1}{1 - (-x)}.$$

$$
\begin{aligned}
\frac{1}{1 + x} &= \frac{1}{1 - (-x)} \\
&= 1 + (-x) + (-x)^2 + (-x)^3 + (-x)^4 + (-x)^5 + \cdots && \text{Substituting} \\
&= 1 - x + x^2 - x^3 + x^4 - x^5 + \cdots. && \text{Simplifying}
\end{aligned}
$$

Thus, we have $g(x) = \dfrac{1}{1 + x} = 1 - x + x^2 - x^3 + x^4 - x^5 + \cdots$. This series is convergent for $|-x| < 1$, or $-1 < x < 1$.

b) To find a power series for g', we differentiate both sides of the result of part (a):

$$\frac{d}{dx}\left[\frac{1}{1 + x}\right] = \frac{d}{dx}[1 - x + x^2 - x^3 + x^4 - x^5 + \cdots] \qquad \begin{array}{l}\text{Differentiating}\\ \text{both sides}\end{array}$$

$$-\frac{1}{(1 + x)^2} = -1 + 2x - 3x^2 + 4x^3 - 5x^4 + \cdots. \qquad \frac{d}{dx}\left[\frac{1}{1 + x}\right] = -\frac{1}{(1 + x)^2}$$

Thus, we have $g'(x) = -\dfrac{1}{(1 + x)^2} = -1 + 2x - 3x^2 + 4x^3 - 5x^4 + \cdots$. Since the power series for g converges over $-1 < x < 1$, this series is also convergent over that interval.

c) To find a power series for G, where $G(x) = \displaystyle\int g(x)\, dx$, we integrate the power series for g term by term:

$$\int \left(\frac{1}{1 + x}\right) dx = \int (1 - x + x^2 - x^3 + x^4 - x^5 + \cdots)\, dx$$

$$\ln|1 + x| = x - \tfrac{1}{2}x^2 + \tfrac{1}{3}x^3 - \tfrac{1}{4}x^4 + \tfrac{1}{5}x^5 - \tfrac{1}{6}x^6 + \cdots + C$$

$$\int \left(\frac{1}{1 + x}\right) dx = \ln|1 + x|$$

$$\ln|1 + 0| = 0 - \tfrac{1}{2}\left(0^2\right) + \tfrac{1}{3}\left(0^3\right) - \tfrac{1}{4}\left(0^4\right) + \tfrac{1}{5}\left(0^5\right) - \tfrac{1}{6}\left(0^6\right) + \cdots + C \qquad \text{Letting } x = 0$$

$$0 = C. \qquad \ln 1 = 0$$

Thus, we have $G(x) = \ln|1 + x| = x - \tfrac{1}{2}x^2 + \tfrac{1}{3}x^3 - \tfrac{1}{4}x^4 + \tfrac{1}{5}x^5 - \tfrac{1}{6}x^6 + \cdots$. Since the power series for g converges over $-1 < x < 1$, this series also converges over that interval; however, it is also convergent for $x = 1$. (See the Extended Technology Application, page 742).

> **Quick Check 2**

Given $g(x) = \dfrac{4}{2 - x}$, find the power series for g'.

❰ Quick Check 2

In the next example, we complete the square to find a power series that is not centered at $x = 0$. For example, a function of the form $f(x) = \dfrac{1}{1 - (x - a)^2}$ has a power series centered at $x = a$.

■ **EXAMPLE 3**   Find the power series for $h(x) = \dfrac{1}{x^2 + 4x + 5}$, its interval of convergence, and its center of convergence.

Solution   To write h in the form $\dfrac{1}{1 - r}$, we first complete the square in the denominator:

$$
\begin{aligned}
x^2 + 4x + 5 &= x^2 + 4x + 4 + 5 - 4 & &\text{Adding and subtracting 4; } 4 = \left(\tfrac{4}{2}\right)^2 \\
&= (x^2 + 4x + 4) + (5 - 4) & &\text{Grouping} \\
&= (x + 2)^2 + 1. & &\text{Factoring and simplifying}
\end{aligned}
$$

Therefore, we have

$$
\begin{aligned}
h(x) &= \frac{1}{x^2 + 4x + 5} \\
&= \frac{1}{(x + 2)^2 + 1} & &\text{Using the above result} \\
&= \frac{1}{1 - \left[-(x + 2)^2\right]}. & &\text{Rewriting in the form } \frac{1}{1 - r}
\end{aligned}
$$

Thus, the first term is $a_1 = 1$, and the common ratio is $r = -(x + 2)^2$. The power series is

$$
\begin{aligned}
\frac{1}{x^2 + 4x + 5} &= \frac{1}{1 - \left[-(x + 2)^2\right]} \\
&= 1 + \left[-(x + 2)^2\right] + \left[-(x + 2)^2\right]^2 \\
&\quad + \left[-(x + 2)^2\right]^3 + \left[-(x + 2)^2\right]^4 + \cdots & &\text{Substituting} \\
&= 1 - (x + 2)^2 + (x + 2)^4 - (x + 2)^6 + (x + 2)^8 - \cdots.
\end{aligned}
$$

The power series converges for $|r| < 1$. Since $r = -(x + 2)^2$, the interval of convergence is

$$
\begin{aligned}
\left|(x + 2)^2\right| &< 1 & &\text{The negative sign is unnecessary inside the absolute-value bars.} \\
(x + 2)^2 &< 1 & &\text{Since } (x + 2) \text{ is squared, we do not need to take the absolute value.} \\
-1 < x + 2 &< 1 & &\text{Noting that if } a^2 < 1, \text{ then } -1 < a < 1 \\
-3 < x &< -1. & &\text{Subtracting 2 to isolate } x
\end{aligned}
$$

This series converges over $(-3, -1)$ and is centered at $x = -2$. The figure shows the graph of h (in blue) along with its approximating polynomial P_8 (in red).

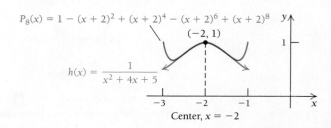

$P_8(x) = 1 - (x + 2)^2 + (x + 2)^4 - (x + 2)^6 + (x + 2)^8$

$(-2, 1)$

$h(x) = \dfrac{1}{x^2 + 4x + 5}$

Center, $x = -2$

> **Quick Check 3**
>
> Find the power series for
> $h(x) = \dfrac{1}{x^2 - 2x + 2}$, its
> interval of convergence, and
> its center of convergence.

❮ Quick Check 3

As mentioned earlier, the first approximating polynomial is the equation of the tangent line to the graph of the function f at $x = k$. We explore this further as we discuss linearization.

Linearization

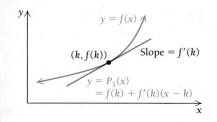

The equation of a tangent line to the graph of $y = f(x)$ at $(k, f(k))$ can be found using the point–slope equation:

$$y - y_1 = m(x - x_1) \qquad \text{Recalling the point–slope equation of a line}$$
$$y - f(k) = f'(k)(x - k) \qquad \text{Substituting}$$
$$y = f(k) + f'(k)(x - k). \qquad \text{Solving for } y$$

The equation $y = f(k) + f'(k)(x - k)$ is called the **linearization** of $y = f(x)$ centered at $x = k$. It is the same as the first approximating polynomial, $P_1(x) = c_0 + c_1(x - k)$, with $c_0 = f(k)$ and $c_1 = f'(k)$.

■ **EXAMPLE 4** Consider the power series of $f(x) = \dfrac{2}{3 + x}$ centered at $x = 0$.

a) Find the linearization of $y = f(x)$ at $x = 0$.

b) Use the linearization from part (a) to estimate $f(0.15)$.

Solution From Example 1, the power series for $f(x) = \dfrac{2}{3 + x}$ centered at $x = 0$ is

$$f(x) = \frac{2}{3 + x} = \frac{2}{3} - \frac{2x}{9} + \frac{2x^2}{27} - \frac{2x^3}{81} + \frac{2x^4}{243} - \cdots.$$

a) The linearization of $y = f(x)$ at $x = 0$ is

$$P_1(x) = \frac{2}{3} - \frac{2x}{9}.$$

b) Since $x = 0.15$ is near the center, $x = 0$, we expect a good estimate:

$$f(0.15) \approx P_1(0.15)$$
$$= \frac{2}{3} - \frac{2(0.15)}{9} = 0.6333\ldots \qquad \text{Using a calculator}$$

A calculator shows that $f(0.15) = \dfrac{2}{3 + 0.15} = 0.63492\ldots$. The estimate, $P_1(0.15) = 0.6333\ldots$, is close to the actual value of $f(0.15)$.

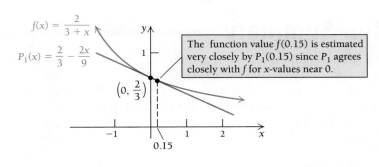

The function value $f(0.15)$ is estimated very closely by $P_1(0.15)$ since P_1 agrees closely with f for x-values near 0.

❬ Quick Check 4

❭ **Quick Check 4**

Find the linearization of $f(x) = \dfrac{4}{2 - x}$ at $x = 0$, and use it to approximate a value for $f(0.07)$.

Application: Ticket Prices

■ **EXAMPLE 5**   **Business: Super Bowl Ticket Prices.**   The average price of a ticket to the Super Bowl is modeled by $f(t) = 8.992e^{0.109t}$, where t is the number of years since 1967 and $f(t)$ is in dollars. (*Source*: National Football League.)

a) Find the linearization of $y = f(t)$ at $x = 45$, representing the year 2012.

b) Use the linearization to predict the average price of a Super Bowl ticket in 2014. Compare this to the value given by the model $y = f(t)$.

Solution

a) To find the linearization, we use $P_1(t) = f(k) + f'(k)(t - k)$, with $k = 45$. Note that $f(45) = 8.992e^{0.109(45)} = 1213.59$ and $f'(t) = 0.980e^{0.109t}$, so $f'(45) = 0.980e^{0.109(45)} = 132.26$. We have

$$P_1(t) = f(k) + f'(k)(t - k)$$
$$= 1213.59 + 132.26(t - 45) \qquad \text{Substituting}$$
$$= 132.26t - 4738.11. \qquad \text{Simplifying}$$

b) Using the linearization from part (a) to estimate the ticket price in 2014, we set $t = 47$:

$$P_1(47) = 132.26(47) - 4738.11 \qquad \text{Substituting}$$
$$= \$1478.11.$$

The model gives $f(47) = 8.992e^{0.109(47)} = \$1509.20$. The two values are relatively close to each another.

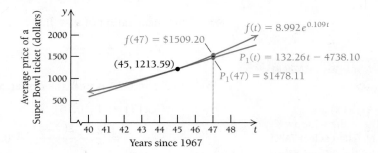

> **Quick Check 5**

Using the linearization found in Example 5, estimate the price of a Super Bowl ticket in 2016.

❮ Quick Check 5

Section Summary

- A *power series* is an infinite series of the form
  $c_0 + c_1x + c_2x^2 + c_3x^3 + c_4x^4 + c_5x^5 + \cdots$, where
  $c_1, c_2, \ldots, c_n, \ldots$ are real coefficients.
- A power series converges to $y = f(x)$ over some interval $a < x < b$, called the *interval of convergence*. Its *center of convergence* k is the average of a and b:

  $$k = \frac{a + b}{2}.$$

  A power series centered at $x = k$ is written
  $c_0 + c_1(x - k) + c_2(x - k)^2 + c_3(x - k)^3 + c_4(x - k)^4 + c_5(x - k)^5 + \cdots$.

- Suppose a power series converges to $y = f(x)$ over some interval $a < x < b$. The *nth approximating polynomial*, denoted $P_n(x)$, is a partial sum of the power series, where n is the degree of the polynomial. As n increases, the approximating polynomials approach f as a limit, for all x in the interval of convergence.
- Given a power series for $y = f(x)$, the power series for $\frac{d}{dx}[f(x)]$ or $F(x) = \int f(x)\, dx$ may be found by differentiating or integrating the terms of the power series of f.
- The *linearization* of a differentiable function $y = f(x)$ at $x = k$ is

  $$P_1(x) = f(k) + f'(k)(x - k).$$

EXERCISE SET
9.5

In Exercises 1–6, find (a) the first five terms of the power series for each function and (b) the interval and center of convergence.

1. $f(x) = \dfrac{1}{1 + 2x}$

2. $f(x) = \dfrac{1}{1 - 3x}$

3. $f(x) = \dfrac{1}{1 - 5x}$

4. $f(x) = \dfrac{1}{1 + 4x}$

5. $g(x) = \dfrac{1}{1 - x^2}$

6. $g(x) = \dfrac{1}{1 + x^2}$

In Exercises 7–10, use the factorization $\dfrac{a}{b + cx} =$

$\dfrac{a}{b}\left(\dfrac{1}{1 - (-cx/b)}\right)$ *to find (a) the first five terms of the power series for each function and (b) the interval and center of convergence.*

7. $f(x) = \dfrac{2}{5 - 3x}$

8. $g(x) = \dfrac{3}{4 + 5x}$

9. $h(x) = \dfrac{5}{3 + 6x}$

10. $m(x) = \dfrac{10}{4 - x}$

11–20. Graph each of the functions in Exercises 1–10 along with P_2 and P_4 over the interval of convergence.

In Exercises 21–24, (a) write the first five terms of the power series for f', then (b) do the same for F, where $F(x) = \int f(x)\, dx$ and $F(0) = 0$.

21. $f(x) = \dfrac{1}{1 + 2x}$, for $-\frac{1}{2} < x < \frac{1}{2}$. (See Exercise 1.)

22. $f(x) = \dfrac{1}{1 - 3x}$, for $-\frac{1}{3} < x < \frac{1}{3}$. (See Exercise 2.)

23. $f(x) = \dfrac{1}{1 - 5x}$, for $-\frac{1}{5} < x < \frac{1}{5}$. (See Exercise 3.)

24. $f(x) = \dfrac{1}{1 + 4x}$, for $-\frac{1}{4} < x < \frac{1}{4}$. (See Exercise 4.)

In Exercises 25–28, (a) write the first five terms of the power series by first completing the square in the denominator, and (b) find the interval and center of convergence.

25. $f(x) = \dfrac{1}{x^2 + 2x + 2}$

26. $f(x) = \dfrac{1}{x^2 - 6x + 10}$

27. $g(x) = \dfrac{1}{x^2 - 4x + 13}$

28. $g(x) = \dfrac{1}{x^2 + 10x + 41}$

29. Let $f(x) = x^2 + 3x + 2$.

a) Find the linearization $P_1(x)$ of f at $x = 4$.
b) Graph f and P_1.
c) Evaluate $P_1(4.1)$ and compare this value to $f(4.1)$.

30. Let $f(x) = 4x^3 + 5x - 1$.

a) Find the linearization $P_1(x)$ of f at $x = 2$.
b) Graph f and P_1.
c) Evaluate $P_1(1.95)$ and compare this value to $f(1.95)$.

31. Let $g(x) = e^{3x}$.

a) Find the linearization $P_1(x)$ of g at $x = 0$.
b) Graph g and P_1.
c) Evaluate $P_1(0.08)$ and compare this value to $g(0.08)$.

32. Let $g(x) = \ln x$.

a) Find the linearization $P_1(x)$ of g at $x = 1$.
b) Graph g and P_1.
c) Evaluate $P_1(1.12)$ and compare this value to $g(1.12)$.

APPLICATIONS

Business and Economics

33. Supply. The supply function for a new rollerball pen is given by $S(p) = 0.007p^3 - 0.5p^2 + 150p$, where p is the price in dollars.

a) Find the linearization $P_1(p)$ of S at $p = 20$.
b) Use the linearization from part (a) to estimate the supply when the price is raised to \$21, and compare this value to $S(21)$.

34. Supply. The supply function for a backpack is given by $S(p) = -0.05p^3 - 0.33p^2 + 120p$, where p is the price in dollars.

a) Find the linearization $P_1(p)$ of S at $p = 12$.
b) Use the linearization from part (a) to estimate the supply when the price is raised to \$13, and compare this value to $S(13)$.

35. Ticket prices. The average price of a ticket to a National Hockey League game can be modeled by $p(t) = 0.35t^2 - 2.75t + 47.95$, where $p(t)$ is the average price of a ticket t years after 2000. (*Source:* www.teammarketing.com.)

a) Find the linearization $P_1(t)$ of p at $t = 13$.

b) Use the linearization from part (a) to estimate the price of a ticket in 2014, and compare this to the value given by the model.

36. Baseball game attendance. The total yearly attendance for all Major League baseball games since 1998 can be modeled by $A(t) = 3.87t^2 + 213.47t + 70{,}656$, where $A(t)$ is the attendance in thousands, t years after 1998. (*Source*: www.ballparksofbaseball.com/attendance.htm.)

a) Find the linearization $P_1(t)$ of A at $t = 16$.
b) Use the linearization from part (a) to estimate the total attendance in 2015, and compare this to the value given by the model.

37. Toll charges. Since 1975, the Golden Gate Bridge in San Francisco collects tolls from southbound vehicles only. The toll for a southbound passenger vehicle on a weekday can be modeled by $C(t) = 0.002t^2 + 0.117t + 0.761$, where $C(t)$ is the toll in dollars t years after 1975. (*Source*: www.goldengate.org.)

a) Find the linearization $P_1(t)$ of C at $t = 34$ (representing 2009).
b) Use the linearization from part (a) to estimate the toll for a passenger vehicle in 2014, and compare this to the value given by the model.

38. Food prices. Since 1990, the price of 1 lb of Oreos™ cookies can be modeled by $C(t) = 0.001t^3 - 0.031t^2 + 0.313t + 2.669$, where $C(t)$ is the price in dollars t years after 1990. (*Source*: www.foodtimeline.org.)

a) Find the linearization $P_1(t)$ of C at $t = 22$ (representing 2012).
b) Use the linearization from part (a) to estimate the price of a pound of Oreos in 2015, and compare this to the value given by the model.

SYNTHESIS

39. From Example 2(c), we have $\ln(1 + x) = x - \frac{1}{2}x^2 + \frac{1}{3}x^3 - \frac{1}{4}x^4 + \frac{1}{5}x^5 - \frac{1}{6}x^6 + \cdots$, for $-1 < x < 1$. Write a series that approximates $\ln 2$ to three decimal places. (*Hint*: Let $x = -\frac{1}{2}$ and use a property of logarithms.)

40. Let $f(x) = \dfrac{1}{x}$.

a) Show that $\dfrac{1}{x} = 1 + (1 - x) + (1 - x)^2 + (1 - x)^3 + \cdots$. (*Hint*: $x = 1 - (1 - x)$.)
b) What is the interval of convergence?
c) Graph f along with P_2 and P_4.
d) Use P_2 and P_4 to approximate $f(0.8)$.

41. Is it possible for a geometric power series to be convergent for all x? At only a single value of x? Why or why not?

42. Is the derivative or the antiderivative of a geometric power series also geometric? Why or why not?

Answers to Quick Checks

1. $h(x) = 2 + x + \frac{1}{2}x^2 + \frac{1}{4}x^3 + \frac{1}{8}x^4 + \frac{1}{16}x^5 + \cdots$, for $-2 < x < 2$ **2.** $g'(x) = 1 + x + \frac{3}{4}x^2 + \frac{1}{2}x^3 + \frac{5}{16}x^4 + \cdots$
3. $h(x) = 1 - (x - 1)^2 + (x - 1)^4 - (x - 1)^6 + (x - 1)^8 - \cdots$, for $0 < x < 2$. The center of convergence is $x = 1$.   **4.** $P_1(x) = 2 + x$; $P(0.07) = 2.07$
5. \$1742.64

9.6 Taylor Series and a Trigonometry Connection

Taylor Series

A *Taylor series* is a power series for which a specific method is used to find each of the coefficients. Taylor series can be used to approximate nonpolynomial functions such as exponential, logarithmic, radical, and trigonometric functions.

OBJECTIVES

- Find the Taylor series for certain functions.
- Given a Taylor series for some function, use calculus to find other Taylor series.

Taylor series are named for the English mathematician Brook Taylor (1685–1731). A Taylor series that is centered at $x = 0$ is sometimes called a Maclaurin series, for the Scottish mathematician Colin Maclaurin (1698–1746), a contemporary of Taylor.

DEFINITION

Assume that all derivatives of f exist in some open interval that includes $x = k$. The **Taylor series** of f centered at $x = k$ is a power series of the form

$$f(x) = c_0 + c_1(x - k) + c_2(x - k)^2 + c_3(x - k)^3 + c_4(x - k)^4 + c_5(x - k)^5 + \cdots,$$

where each coefficient is given by the formula

$$c_n = \frac{f^{(n)}(k)}{n!}, \qquad \text{for } n = 0, 1, 2, 3, \ldots.$$

Factorial Values	
$0! = 1$	$4! = 24$
$1! = 1$	$5! = 120$
$2! = 2$	$6! = 720$
$3! = 6$	$7! = 5040$

We represent the nth derivative of f by $f^{(n)}$, where $f^{(0)}$ represents the function f. In the above formula, $n!$ is read "**n factorial**" and is defined as $n! = 1 \cdot 2 \cdot 3 \cdot 4 \cdot \cdots \cdot n$, with the special case that $0! = 1$. The factorial values for n from 0 to 7 are given in the table in the margin.

To see why the coefficients are given by the above formula, we express f as a power series centered at $x = k$:

$$f(x) = c_0 + c_1(x - k) + c_2(x - k)^2 + c_3(x - k)^3 + c_4(x - k)^4 + c_5(x - k)^5 + \cdots.$$

Taking derivatives, we have

$$f'(x) = c_1 + 2c_2(x - k) + 3c_3(x - k)^2 + 4c_4(x - k)^3 + 5c_5(x - k)^4 + \cdots,$$
$$f''(x) = 2c_2 + 6c_3(x - k) + 12c_4(x - k)^2 + 20c_5(x - k)^3 + \cdots,$$
$$f^{(3)}(x) = 6c_3 + 24c_4(x - k) + 60c_5(x - k)^2 + \cdots,$$
$$f^{(4)}(x) = 24c_4 + 120c_5(x - k) + \cdots,$$
$$f^{(5)}(x) = 120c_5 + \cdots,$$

and so on.

We then evaluate each derivative at $x = k$:

$$f(k) = c_0 + c_1(k - k) + c_2(k - k)^2 + c_3(k - k)^3 + c_4(k - k)^4 + c_5(k - k)^5 + \cdots,$$
$$f'(k) = c_1 + 2c_2(k - k) + 3c_3(k - k)^2 + 4c_4(k - k)^3 + 5c_5(k - k)^4 + \cdots,$$
$$f''(k) = 2c_2 + 6c_3(k - k) + 12c_4(k - k)^2 + 20c_5(k - k)^3 + \cdots,$$
$$f^{(3)}(k) = 6c_3 + 24c_4(k - k) + 60c_5(k - k)^2 + \cdots,$$
$$f^{(4)}(k) = 24c_4 + 120c_5(k - k) + \cdots,$$
$$f^{(5)}(k) = 120c_5 + \cdots,$$

and so on.

Since $k - k = 0$, we have $f(k) = c_0$, $f'(k) = c_1$, $f''(k) = 2c_2$, $f^{(3)}(k) = 6c_3$, $f^{(4)}(k) = 24c_4$, $f^{(5)}(k) = 120c_5$, and so on. Thus, we can write the coefficients in terms of f and its derivatives evaluated at k:

$$c_0 = f(k), c_1 = f'(k), c_2 = \frac{1}{2}f''(k), c_3 = \frac{1}{6}f^{(3)}(k), c_4 = \frac{1}{24}f^{(4)}(k),$$

$$c_5 = \frac{1}{120}f^{(5)}(k), \text{ and so on.}$$

Using factorial notation, we have

$$c_0 = f(k), c_1 = f'(k), c_2 = \frac{f''(k)}{2!}, c_3 = \frac{f^{(3)}(k)}{3!},$$

$$c_4 = \frac{f^{(4)}(k)}{4!}, c_5 = \frac{f^{(5)}(k)}{5!}, \text{ and so on.}$$

Thus, the Taylor series for f is

$$f(x) = c_0 + c_1(x - k) + c_2(x - k)^2 + c_3(x - k)^3 + c_4(x - k)^4 + c_5(x - k)^5 + \cdots$$

$$= f(k) + f'(k)(x - k) + \frac{f''(k)}{2!}(x - k)^2 + \frac{f^{(3)}(k)}{3!}(x - k)^3 + \frac{f^{(4)}(k)}{4!}(x - k)^4 + \frac{f^{(5)}(k)}{5!}(x - k)^5 + \cdots.$$

Substituting

■ **EXAMPLE 1** Find the Taylor series for $f(x) = e^x$ centered at $x = 0$.

Solution We set $f(x) = e^x$ equal to a power series centered at $x = 0$:

$$e^x = c_0 + c_1(x - 0) + c_2(x - 0)^2 + c_3(x - 0)^3 + c_4(x - 0)^4 + c_5(x - 0)^5 + \cdots$$

$$= c_0 + c_1 x + c_2 x^2 + c_3 x^3 + c_4 x^4 + c_5 x^5 + \cdots$$

$$= f(0) + \frac{f'(0)}{1!}x + \frac{f''(0)}{2!}x^2 + \frac{f^{(3)}(0)}{3!}x^3 + \frac{f^{(4)}(0)}{4!}x^4 + \frac{f^{(5)}(0)}{5!}x^5 + \cdots.$$

Note that $f'(x) = e^x$, $f''(x) = e^x$, $f^{(3)}(x) = e^x$, and so on. Thus, $f^{(n)}(0) = e^0 = 1$ for all n. Thus, the Taylor series for e^x centered at $x = 0$ is

$$e^x = 1 + x + \frac{1}{2!}x^2 + \frac{1}{3!}x^3 + \frac{1}{4!}x^4 + \frac{1}{5!}x^5 + \cdots.$$

The figure shows the graph of $f(x) = e^x$ in blue, with three of its approximating polynomials for comparison. Note that the more terms the approximating polynomials include, the more closely they agree with f.

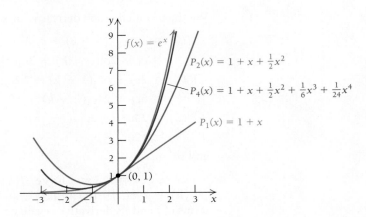

> **Quick Check 1**

Find the Taylor series for $f(x) = e^{-x}$ centered at $x = 0$.

❮ Quick Check 1

It is possible to show that the Taylor series for $f(x) = e^x$ is convergent for all x. A method for doing this is presented in the Synthesis section of the exercise set.

■ **EXAMPLE 2** Use the Taylor series for $f(x) = e^x$ to approximate e to three decimal places.

Solution As shown in Example 1, the Taylor series for $f(x) = e^x$ is

$$e^x = 1 + x + \frac{1}{2!}x^2 + \frac{1}{3!}x^3 + \frac{1}{4!}x^4 + \frac{1}{5!}x^5 + \cdots.$$

For $x = 1$, we have

$$e^1 = 1 + (1) + \frac{1}{2!}(1)^2 + \frac{1}{3!}(1)^3 + \frac{1}{4!}(1)^4 + \frac{1}{5!}(1)^5 + \cdots \qquad \text{Substituting}$$

$$\approx 1 + 1 + \frac{1}{2!} + \frac{1}{3!} + \frac{1}{4!} + \frac{1}{5!} \qquad \text{Evaluating } P_5(1)$$

$$\approx 1 + 1 + \frac{1}{2} + \frac{1}{6} + \frac{1}{24} + \frac{1}{120} = 2.716666\ldots.$$

To ensure accuracy to three decimal places, we extend the pattern until the number in the third decimal place no longer changes:

$$1 + 1 + \frac{1}{2!} + \frac{1}{3!} + \frac{1}{4!} + \frac{1}{5!} + \frac{1}{6!} = 2.7180555\ldots,$$

$$1 + 1 + \frac{1}{2!} + \frac{1}{3!} + \frac{1}{4!} + \frac{1}{5!} + \frac{1}{6!} + \frac{1}{7!} = 2.718254\ldots.$$

Rounding to three decimal places, we have $e = 2.718$.

⟨ Quick Check 2

> **Quick Check 2**
>
> Use the Taylor series for $f(x) = e^x$ to approximate the value of $e^{0.5}$ to three decimal places.

■ **EXAMPLE 3** Find the Taylor series for $f(x) = \sqrt{x}$ centered at $x = 1$.

Solution The Taylor series for $f(x) = \sqrt{x}$ centered at $x = 1$ has the following form:

$$\sqrt{x} = c_0 + c_1(x - 1) + c_2(x - 1)^2 + c_3(x - 1)^3 + c_4(x - 1)^4 + \cdots.$$

To find the coefficients, we use the formula $c_n = \dfrac{f^{(n)}(k)}{n!}$. Since $k = 1$, we have

$$c_0 = \frac{f^{(0)}(1)}{0!} = \frac{\sqrt{1}}{1} = 1, \qquad\qquad f^{(0)}(x) = f(x) = \sqrt{x}$$

$$c_1 = \frac{f^{(1)}(1)}{1!} = \frac{\left(\frac{1}{2}(1)^{-1/2}\right)}{1} = \frac{1}{2}, \qquad f^{(1)}(x) = f'(x) = \tfrac{1}{2}x^{-1/2}$$

$$c_2 = \frac{f^{(2)}(1)}{2!} = \frac{\left(-\frac{1}{4}(1)^{-3/2}\right)}{2} = -\frac{1}{8}, \qquad f^{(2)}(x) = f''(x) = -\tfrac{1}{4}x^{-3/2}$$

$$c_3 = \frac{f^{(3)}(1)}{3!} = \frac{\left(\frac{3}{8}(1)^{-5/2}\right)}{6} = \frac{1}{16}, \qquad f^{(3)}(x) = \tfrac{3}{8}x^{-5/2}$$

$$c_4 = \frac{f^{(4)}(1)}{4!} = \frac{\left(-\frac{15}{16}(1)^{-7/2}\right)}{24} = -\frac{5}{128}, \qquad f^{(4)}(x) = -\tfrac{15}{16}x^{-7/2}$$

and so on.

Thus, the Taylor series for $f(x) = \sqrt{x}$ centered at $x = 1$ is

$$\sqrt{x} = 1 + \tfrac{1}{2}(x - 1) - \tfrac{1}{8}(x - 1)^2 + \tfrac{1}{16}(x - 1)^3 - \tfrac{5}{128}(x - 1)^4 + \cdots.$$

The figure shows the graph of $f(x) = \sqrt{x}$ in blue and the graph of its fourth approximating polynomial $P_4(x)$ in red. We see that over the interval $(0, 2)$, the graphs are nearly identical.

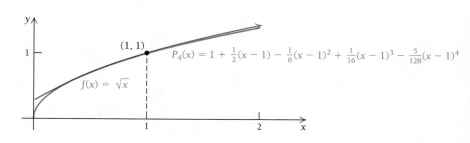

> **Quick Check 3**
>
> Write the first five terms of the Taylor series for $f(x) = \sqrt{x}$ centered at $x = 4$.

⟨ Quick Check 3

Taylor series offer a way to evaluate integrals when antiderivatives are not easily found.

■ **EXAMPLE 4** Use the Taylor series for e^x to evaluate $\int_0^1 e^{-x^2}\, dx$ to three decimal places.

Solution Centered at $x = 0$, the Taylor series for $f(x) = e^{-x^2}$ can be found by replacing x in the Taylor series for $f(x) = e^x$ with $-x^2$:

$$e^x = 1 + x + \tfrac{1}{2}x^2 + \tfrac{1}{6}x^3 + \tfrac{1}{24}x^4 + \tfrac{1}{120}x^5 + \cdots \qquad \text{The Taylor series for } e^x$$

$$e^{-x^2} = 1 + (-x^2) + \tfrac{1}{2}(-x^2)^2 + \tfrac{1}{6}(-x^2)^3 + \tfrac{1}{24}(-x^2)^4 + \tfrac{1}{120}(-x^2)^5 + \cdots \quad \text{Substituting}$$

$$= 1 - x^2 + \tfrac{1}{2}x^4 - \tfrac{1}{6}x^6 + \tfrac{1}{24}x^8 - \tfrac{1}{120}x^{10} + \cdots . \qquad \text{Simplifying}$$

We now find $\int_0^1 e^{-x^2}\, dx$ using the Taylor series for $f(x) = e^{-x^2}$:

$$\int_0^1 e^{-x^2}\, dx = \int_0^1 \left(1 - x^2 + \tfrac{1}{2}x^4 - \tfrac{1}{6}x^6 + \tfrac{1}{24}x^8 - \tfrac{1}{120}x^{10} + \cdots\right) dx$$

$$= \left[x - \tfrac{1}{3}x^3 + \tfrac{1}{10}x^5 - \tfrac{1}{42}x^7 + \tfrac{1}{216}x^9 - \tfrac{1}{1320}x^{11} + \cdots\right]_0^1 \qquad \text{Integrating}$$

$$= 1 - \tfrac{1}{3} + \tfrac{1}{10} - \tfrac{1}{42} + \tfrac{1}{216} - \tfrac{1}{1320} + \cdots - 0 \approx 0.747.$$

> **Quick Check 4**
>
> Use the first five terms of the Taylor series for $g(x) = e^{-x^3}$ to evaluate $\int_0^{0.5} e^{-x^3}\, dx$ to three decimal places.

❮ Quick Check 4

⊕ A Trigonometry Connection: Taylor Series for Sinusoidal Functions

The trigonometric functions $y = \sin x$ and $y = \cos x$ can be expressed as Taylor series. The development of the Taylor series for $f(x) = \sin x$ centered at $x = 0$ follows.

■ **EXAMPLE 5** Find the Taylor series for $f(x) = \sin x$ centered at $x = 0$.

Solution We use the formula $c_n = \dfrac{f^{(n)}(k)}{n!}$ to find the coefficients. Recall that $\sin 0 = 0$ and $\cos 0 = 1$.

$$c_0 = \frac{f^{(0)}(0)}{0!} = \frac{\sin 0}{1} = 0, \qquad f^{(0)}(x) = f(x) = \sin x$$

$$c_1 = \frac{f^{(1)}(0)}{1!} = \frac{\cos 0}{1} = 1, \qquad f^{(1)}(x) = f'(x) = \cos x$$

$$c_2 = \frac{f^{(2)}(0)}{2!} = \frac{-\sin 0}{2} = 0, \qquad f^{(2)}(x) = f''(x) = -\sin x$$

$$c_3 = \frac{f^{(3)}(0)}{3!} = \frac{-\cos 0}{6} = -\frac{1}{6}, \qquad f^{(3)}(x) = -\cos x$$

$$c_4 = \frac{f^{(4)}(0)}{4!} = \frac{\sin 0}{24} = 0, \qquad f^{(4)}(x) = \sin x$$

$$c_5 = \frac{f^{(5)}(0)}{5!} = \frac{\cos 0}{120} = \frac{1}{120}, \qquad f^{(5)}(x) = \cos x$$

and so on.

We see patterns: the even coefficients, $c_0, c_2, c_4, \ldots$, are all zero. The other coefficients are of the form $1/n!$, with alternating signs. Thus, we can conclude that $c_6 = 0$ and $c_7 = -\dfrac{1}{7!} = -\dfrac{1}{5040}$, and so on. The Taylor series for $f(x) = \sin x$, centered at $x = 0$, is

$$\sin x = x - \frac{1}{6}x^3 + \frac{1}{120}x^5 - \frac{1}{5040}x^7 + \cdots$$

$$= x - \frac{1}{3!}x^3 + \frac{1}{5!}x^5 - \frac{1}{7!}x^7 + \cdots. \qquad \text{Using factorial notation}$$

Although we do not show it here, this Taylor series is convergent for all x.

The figure shows the graph of $f(x) = \sin x$ in blue, with its first, third, fifth, and seventh approximating polynomials in red. Note that the more terms the approximating polynomials have, the more closely they agree with f.

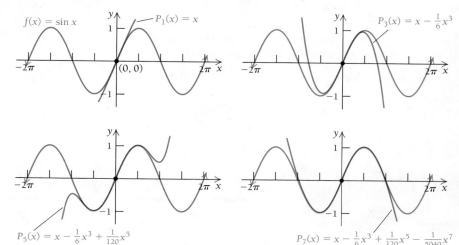

Since $\dfrac{d}{dx}\sin x = \cos x$, we can use calculus to find the Taylor series for $y = \cos x$.

■ EXAMPLE 6   Find the Taylor series for $y = \cos x$ by differentiating the Taylor series for $\sin x$.

Solution   We have

$$\sin x = x - \frac{1}{3!}x^3 + \frac{1}{5!}x^5 - \frac{1}{7!}x^7 + \cdots \qquad \begin{array}{l}\text{Using the result from}\\ \text{Example 5}\end{array}$$

$$\frac{d}{dx}[\sin x] = \frac{d}{dx}\left[x - \frac{1}{3!}x^3 + \frac{1}{5!}x^5 - \frac{1}{7!}x^7 + \cdots\right]$$

$$\cos x = 1 - \frac{1}{3!}\cdot 3x^2 + \frac{1}{5!}\cdot 5x^4 - \frac{1}{7!}\cdot 7x^6 + \cdots \qquad \text{Differentiating}$$

Note that $\dfrac{1}{3!}\cdot 3 = \dfrac{3}{1\cdot 2\cdot 3} = \dfrac{1}{2!}$ and $\dfrac{1}{5!}\cdot 5 = \dfrac{5}{1\cdot 2\cdot 3\cdot 4\cdot 5} = \dfrac{1}{4!}$, and so on. Thus, we have

$$\cos x = 1 - \frac{1}{2!}x^2 + \frac{1}{4!}x^4 - \frac{1}{6!}x^6 + \frac{1}{8!}x^8 - \cdots.$$

This Taylor series is also convergent for all x.

❮ Quick Check 5

> **❯ Quick Check 5**
>
> Show that $\displaystyle\int_0^x \cos t\, dt = \sin x$ by integrating each term of the Taylor series for $\cos t$ and evaluating the bounds.

Calculators and computers use Taylor series to evaluate nonpolynomial functions. Expressing a function as a Taylor series simplifies the evaluation of expressions such as e^2 and $\sin\dfrac{\pi}{3}$ to a sequence of basic arithmetic operations—addition, subtraction, multiplication, and division—that a calculator or computer can perform very quickly.

Section Summary

- Assuming that all derivatives of f exist in some open interval that includes $x = k$, the *Taylor series of f centered at $x = k$* is a power series of the form

$$f(x) = c_0 + c_1(x - k) + c_2(x - k)^2 + c_3(x - k)^3 \\ + c_4(x - k)^4 + c_5(x - k)^5 + \cdots,$$

where $c_n = \dfrac{f^{(n)}(k)}{n!}$, for $n = 0, 1, 2, 3, \ldots$.

- The expression $n!$ is read as n *factorial* and is defined as $n! = 1 \cdot 2 \cdot 3 \cdot \cdots \cdot n$, with the special case that $0! = 1$.

- The Taylor series for $f(x) = e^x$, centered at $x = 0$, is

$$e^x = 1 + x + \frac{1}{2!}x^2 + \frac{1}{3!}x^3 + \frac{1}{4!}x^4 + \frac{1}{5!}x^5 + \cdots,$$

for $-\infty < x < \infty$.

- *Trigonometry Connection*: The Taylor series for $f(x) = \sin x$, centered at $x = 0$, is

$$\sin x = x - \frac{1}{3!}x^3 + \frac{1}{5!}x^5 - \frac{1}{7!}x^7 + \cdots,$$

for $-\infty < x < \infty$.

The Taylor series for $f(x) = \cos x$, centered at $x = 0$, is

$$\cos x = 1 - \frac{1}{2!}x^2 + \frac{1}{4!}x^4 - \frac{1}{6!}x^6 + \frac{1}{8!}x^8 - \cdots,$$

for $-\infty < x < \infty$.

EXERCISE SET
9.6

In Exercises 1–6, (a) find the first five terms of the Taylor series for each function, and (b) graph each function along with the specified approximating polynomials.

1. $f(x) = \sqrt[3]{x}$, centered at $x = 1$; P_1 and P_3

2. $f(x) = \sqrt[3]{x}$, centered at $x = 8$; P_1 and P_3

3. $g(x) = \sqrt[4]{x}$, centered at $x = 1$; P_1 and P_3

4. $g(x) = \sqrt[5]{x}$, centered at $x = 1$; P_1 and P_3

5. $h(x) = e^{x^2}$, centered at $x = 0$; P_2 and P_4

6. $h(x) = e^{x^3}$, centered at $x = 0$; P_3 and P_6

In Exercises 7–10, use the Taylor series for $f(x) = e^x$ centered at $x = 0$ to approximate each expression to four decimal places.

7. $\sqrt{e}$

8. $e^{-0.25}$

9. $e^{0.3}$

10. $\dfrac{1}{\sqrt{e}}$

11. Use the Taylor series for $f(x) = \sqrt{x}$, centered at $x = 1$, to approximate $\sqrt{1.1}$ to four decimal places.

12. Use the Taylor series for $f(x) = \sqrt{x}$, centered at $x = 4$, to approximate $\sqrt{4.5}$ to four decimal places.

13. Use the Taylor series for $f(x) = \sqrt[3]{x}$, centered at $x = 8$, to approximate $\sqrt[3]{9}$ to four decimal places.

14. Use the Taylor series for $f(x) = \sqrt[4]{x}$, centered at $x = 1$, to approximate $\sqrt[4]{0.9}$ to four decimal places.

15. Use a Taylor series to approximate $\displaystyle\int_0^1 e^{x^3}\, dx$ to three decimal places.

16. Use a Taylor series to approximate $\displaystyle\int_0^{0.5} e^{-x^2}\, dx$ to three decimal places.

APPLICATIONS

Business and Economics

17. **Ticket prices.** From Example 5 in Section 9.5, the average price for a ticket to the Super Bowl is modeled by $f(t) = 8.992e^{0.109t}$, where $f(t)$ is the price t years after 1967. (*Source*: National Football League.)

    a) Find the second approximating polynomial, $P_2(t)$, centered at $t = 20$, and use it to predict the price of a Super Bowl ticket in 1985.

    b) Find the fourth approximating polynomial, $P_3(t)$, centered at $t = 20$, and use it to predict the price of a Super Bowl ticket in 1985.

    c) Compare the answers from parts (a) and (b) with the price given by the model for 1985.

    d) Would P_2 or P_3 in parts (a) and (b) accurately predict the price of a Super Bowl ticket in 2012? Why or why not?

18. **Batman comic book.** From Section 3.3, Example 7, the value of an original Batman comic book is modeled by $V(t) = 0.10e^{0.228t}$, where $V(t)$ is the value in dollars t years after 1939, when the comic book was published.

    a) Find the second approximating polynomial, $P_2(t)$, centered at $t = 0$, and use it to predict the price of an original Batman comic book in 2010. The actual price paid at auction in 2010 was $1.075 million. (*Source:* Heritage Auction Galleries.)

b) Find the second approximating polynomial, $P_2(t)$, centered at $t = 61$ (representing 2000), and use it to predict the price of an original Batman comic book in 2010.

c) How does the prediction of part (b) compare to that of part (a) and to the actual value in 2010? How can you get a better approximation?

TRIGONOMETRY CONNECTION ·············· ⊕

19. Find the first five terms of the Taylor series for $g(x) = \sin x$ centered at $x = \pi/4$.

20. Find the first five terms of the Taylor series for $g(x) = \cos x$ centered at $x = \pi/3$.

21. Use the Taylor series for $f(x) = \sin x$ centered at $x = 0$ to approximate $\sin 0.3$ to four decimal places.

22. Use the Taylor series for $f(x) = \cos x$ centered at $x = 0$ to approximate $\cos(-0.1)$ to four decimal places.

23. Use a Taylor series to approximate $\int_0^{0.5} \sin x^2 \, dx$ to four decimal places.

24. Use a Taylor series to approximate $\int_0^1 \cos \sqrt{x} \, dx$ to four decimal places.

SYNTHESIS

The ratio test: intervals of convergence. *To find the interval of convergence of a Taylor series, we use the ratio test. Let a_n be the general nth term of the Taylor series. We form the ratio $|a_{n+1}/a_n|$ and evaluate its limit as $n \to \infty$:*

$$\lim_{n \to \infty} \left| \frac{a_{n+1}}{a_n} \right| = L.$$

If $L < 1$, then the series converges; if $L > 1$, the series diverges; and if $L = 1$, there is no conclusion. For example, the Taylor series for e^x centered at $x = 0$ is

$$e^x = 1 + x + \frac{1}{2!}x^2 + \frac{1}{3!}x^3 + \frac{1}{4!}x^4 + \frac{1}{5!}x^5 + \cdots = \sum_{n=0}^{\infty} \frac{x^n}{n!}.$$

The general nth term is $a_n = x^n/n!$, so

$$\left| \frac{a_{n+1}}{a_n} \right| = \left| \frac{x^{n+1}/(n+1)!}{x^n/n!} \right| = \left| \frac{x^{n+1} \cdot n!}{x^n \cdot (n+1)!} \right| = \left| \frac{x}{n+1} \right|.$$

Now the limit is taken as $n \to \infty$. Note that the variable x can be moved to the front, since the limit depends only on n. Thus,

$$\lim_{n \to \infty} \left| \frac{a_{n+1}}{a_n} \right| = \lim_{n \to \infty} \left| \frac{x}{n+1} \right| = |x| \cdot \lim_{n \to \infty} \left(\frac{1}{n+1} \right) = |x| \cdot 0 = 0.$$

Since $L = 0$, which is less than 1, the series converges regardless of the value of x, and the series converges for all x.
   Use the ratio test to determine the interval of convergence for each Taylor series.

25. $y = \sin x = \sum_{n=0}^{\infty} (-1)^n \frac{x^{2n+1}}{(2n+1)!}.$

26. $y = \cos x = \sum_{n=0}^{\infty} (-1)^n \frac{x^{2n}}{(2n)!}.$

Euler's formula. *The functions e^x, $\sin x$, and $\cos x$ are related by the formula $e^{ix} = \cos x + i \sin x$. This is Euler's formula, named for the Swiss mathematician Leonhard Euler. To derive this formula, we use the Taylor series for each function.*

27. Use the fact that the pattern $i^1 = i$, $i^2 = -1$, $i^3 = -1$, and $i^4 = 1$ repeats for higher powers of i to derive Euler's formula through the following steps.

a) Find and simplify the Taylor series of e^{ix}.

b) Find and simplify the Taylor series of $\sin(ix)$.

c) Use the results from parts (a) and (b) to show that $e^{ix} = \cos x + i \sin x$.

28. Use Euler's formula to show that $e^{-i\pi} = -1$.

29. Use Euler's formula to show that $i^i = e^{-\pi/2}$.

In Example 7 from Section 8.5, we learned that the differential equation $y'' - 4y' + 9y = 0$ has the auxiliary equation $r^2 - 4r + 9 = 0$. One solution of this auxiliary equation is $r = 2 + i\sqrt{5}$. Thus, a solution of the differential equation is

$$
\begin{aligned}
y &= Ce^{(2+i\sqrt{5})x} = Ce^{2x+i\sqrt{5}x} = Ce^{2x}e^{i\sqrt{5}x} \\
&= Ce^{2x}(\cos(\sqrt{5} \cdot x) + i \sin(\sqrt{5} \cdot x)) \\
&\qquad\qquad\qquad\qquad \text{Using Euler's formula} \\
&= Ce^{2x}\cos(\sqrt{5} \cdot x) + Cie^{2x}\sin(\sqrt{5} \cdot x) \quad \text{Distributing} \\
&= C_1 e^{2x}\cos(\sqrt{5} \cdot x) + C_2 e^{2x}\sin(\sqrt{5} \cdot x). \\
&\qquad\qquad\qquad\qquad \text{Replacing } Ci \text{ with } C_2
\end{aligned}
$$

The other solution of the auxiliary equation, $r = 2 - i\sqrt{5}$, produces a similar result in which only the leading constants may differ. Thus, the general solution is simply

$$y = C_1 e^{2x} \cos \sqrt{5}x + C_2 e^{2x} \sin \sqrt{5}x.$$

30. Use Euler's formula to find the general solution of $y'' + y = 0$.

31. Use Euler's formula to find the general solution of $y'' + y' + 3y = 0$.

Answers to Quick Checks

1. $e^{-x} = 1 - x + \frac{1}{2!}x^2 - \frac{1}{3!}x^3 + \frac{1}{4!}x^4 - \frac{1}{5!}x^5 + \cdots$

2. $e^{0.5} = 1 + (0.5) + \frac{1}{2!}(0.5)^2 + \frac{1}{3!}(0.5)^3 + \frac{1}{4!}(0.5)^4$
$+ \frac{1}{5!}(0.5)^5 + \frac{1}{6!}(0.5)^6 \approx 1.6487$ **3.** $2 + \frac{1}{4}(x-4)$
$- \frac{1}{64}(x-4)^2 + \frac{1}{512}(x-4)^3 - \frac{5}{16{,}384}(x-4)^4 + \cdots$

4. $e^{-x^3} = 1 - x^3 + \frac{1}{2}x^6 - \frac{1}{6}x^9 + \frac{1}{24}x^{12} - \cdots;$ $\int_0^{0.5} e^{-x^3} \, dx \approx 0.883$

5. $\int_0^x \cos t \, dt = \int_0^x \left(1 - \frac{1}{2!}t^2 + \frac{1}{4!}t^4 \right.$
$\left. - \frac{1}{6!}t^6 + \cdots \right) dt = \left[t - \frac{1}{2!}\left(\frac{1}{3}t^3\right) + \frac{1}{4!}\left(\frac{1}{5}t^5\right) \right.$
$\left. - \frac{1}{6!}\left(\frac{1}{7}t^7\right) + \cdots \right]_0^x = \left(x - \frac{1}{2!}\left(\frac{1}{3}x^3\right) + \frac{1}{4!}\left(\frac{1}{5}x^5\right) \right.$
$\left. - \frac{1}{6!}\left(\frac{1}{7}x^7\right) + \cdots \right) - 0 = x - \frac{1}{3!}x^3 + \frac{1}{5!}x^5 - \frac{1}{7!}x^7 + \cdots$

KEY TERMS AND CONCEPTS	EXAMPLES

SECTION 9.1

A **sequence** is any ordered set of numbers, which are the **terms** of the sequence. Many sequences can be defined by a **general nth term**, denoted a_n where the domain of n is any consecutive set of whole numbers.

The numbers $3, 10, 17, 24, 31, \ldots$ form a sequence. It is described by $a_n = 7n - 4$. As a check, note that $a_1 = 3$, $a_2 = 10$, $a_3 = 17$, and so on.

An **arithmetic sequence** is one in which the same number d is added to each term to obtain the next term. The number d is called the **common difference**, and the **general nth term** is $a_n = a_1 + (n - 1)d$, where a_1 is the first term of the sequence. Arithmetic sequences are sometimes called *linear sequences*.

The sequence $3, 10, 17, 24, 31, \ldots$ is arithmetic. Its common difference is $d = 7$, and its general nth term is

$$a_n = 3 + (n - 1)7,$$

which simplifies to $a_n = 7n - 4$.

A **series** is the sum of the terms of a sequence. The **nth partial sum**, written S_n, is the sum of the first n terms of a series. For any arithmetic series, the nth partial sum is

$$S_n = \frac{n}{2}(a_1 + a_n),$$ where a_1 is the first term

of the sequence and a_n is the nth term.

The sum of the first 100 terms of the series $3 + 10 + 17 + 24 + \cdots$ is

$$S_{100} = \frac{100}{2}(3 + 696) = 34{,}950,$$

where $a_{100} = 7 \cdot 100 - 4 = 696$.

SECTION 9.2

A **geometric sequence** is one in which each term is multiplied by the same number r ($r \neq 0$ and $r \neq 1$) to obtain the next term. The number r is called the **common ratio**. In any geometric sequence, $r = \frac{a_{n+1}}{a_n}$. The **general nth term** of a geometric sequence is $a_n = a_1 r^{n-1}$, where a_1 is the first term of the sequence.

The sequence $3, 6, 12, 24, 48, \ldots$ is geometric. Its common ratio is $r = 2$. Its nth term is $a_n = 3 \cdot 2^{n-1}$. As a check, note that

$$a_1 = 3 \cdot 2^{1-1} = 3 \cdot 2^0 = 3,$$
$$a_2 = 3 \cdot 2^{2-1} = 3 \cdot 2^1 = 6,$$
$$a_3 = 3 \cdot 2^{3-1} = 3 \cdot 2^2 = 12,$$
$$a_4 = 3 \cdot 2^{4-1} = 3 \cdot 2^3 = 24,$$

and so on.

The sum of the first n terms of a geometric series is given by

$$S_n = \sum_{n=1}^{\infty} a_1 \cdot r^{n-1} = \frac{a_1(r^n - 1)}{r - 1}.$$

If $|r| < 1$, then the sum of an infinite geometric series is given by $S = \frac{a}{1 - r}$, and we say that the series is *convergent*. An infinite series that does not converge is *divergent*.

- The sum of the first 10 terms of the series $3 + 6 + 12 + 24 + 48 + \cdots$ is

$$S_{10} = \frac{3(2^{10} - 1)}{2 - 1} = 3069.$$

- The infinite series $\frac{3}{7} + \frac{1}{7} + \frac{1}{21} + \frac{1}{63} + \cdots$ has $r = \frac{1}{3}$. Since $|r| < 1$, the infinite series converges, and the sum is

$$S = \frac{3/7}{1 - 1/3} = \frac{9}{14}.$$

SECTION 9.3

Let P be the principal, i be the annual interest rate, t be the time in years, and c be the **compounding frequency** (per year). Then,

- The **simple interest** is $I = Pit$.
- The **simple interest future value** is $A = P + I$, or $A = P(1 + it)$.

Let $P = \$250$, $i = 4\%$, and $t = 3$ yr. Then the simple interest is

$$I = 250 \cdot 0.04 \cdot 3 = \$30.$$

The simple interest future value is

$$A = \$250 + \$30 = \$280,$$

or

$$A = 250(1 + 0.04 \cdot 3) = \$280.$$

KEY TERMS AND CONCEPTS

EXAMPLES

- The **compound interest future value** is
$$A = P\left(1 + \frac{i}{c}\right)^{ct}.$$

In general, if the future value A and the principal P are known, then the interest is $I = A - P$.

Simple interest yields linear growth, which can be modeled by an arithmetic sequence. Compound interest yields exponential growth, which can be modeled by a geometric sequence.

If interest is compounded monthly, then the compound interest future value is

$$A = 250\left(1 + \frac{0.04}{12}\right)^{36} = \$281.82,$$

and the interest earned is $\$281.82 - \$250 = \$31.32$.

If a future value A is known, then the **present value** of A is the principal P that if invested now will grow to A at some specified future date.

If the Hill Elementary School PTA wants $A = \$1000$ in 2 yr and can earn simple interest at an annual rate of 2.5%, then the present value is

$$1000 = P(1 + 0.025 \cdot 2)$$

$$P = \frac{1000}{1 + 0.025 \cdot 2} = \$952.38.$$

If we assume the interest is compounded quarterly, then the present value is

$$1000 = P\left(1 + \frac{0.025}{4}\right)^{8}$$

$$P = \frac{1000}{\left(1 + \frac{0.025}{4}\right)^{8}} = \$951.38.$$

SECTION 9.4

An **annuity** is a savings account into which equal-sized deposits called *payments* (denoted p) are made on a regular basis. The future value A of an annuity paying an interest rate i compounded c times per year for t years is

$$A = \frac{p\left[\left(1 + \frac{i}{c}\right)^{ct} - 1\right]}{\frac{i}{c}}.$$

The total contribution is $p \cdot c \cdot t$, and the interest earned is

$$I = A - p \cdot c \cdot t.$$

The payments (deposits) are made at the beginning of each compounding period.

Bart deposits $\$200$ at the start of each month into an annuity that has an annual interest rate of 4.5%, compounded monthly for 20 yr. The future value of his annuity is

$$A = \frac{200\left[\left(1 + \frac{0.045}{12}\right)^{240} - 1\right]}{\frac{0.045}{12}} = \$77,624.87.$$

His personal contribution will be
$$200 \cdot 12 \cdot 20 = \$48,000.$$

His annuity will earn
$$\$77,624.87 - \$48,000 = \$29,624.87$$
in interest.

A **sinking fund** is an annuity for which the future value is known and the regular payment amount needed to reach that future value is then determined using the formula for the future value of an annuity.

Twyla wants to have $\$2500$ in 2 yr. She opens a savings account at an annual interest rate of 3.8%, compounded monthly. Twyla's monthly payment (deposit) is

$$2500 = \frac{p\left[\left(1 + \frac{0.038}{12}\right)^{24} - 1\right]}{\frac{0.038}{12}}$$

$$2500 = p(24.89463796)$$

$$p = \frac{2500}{24.89463796} = \$100.42.$$

KEY TERMS AND CONCEPTS	EXAMPLES

SECTION 9.4   *(continued)*

Amortization is a process in which a loan amount P is paid off in $c \cdot t$ equal payments over a period of t years. The loan amount P and the payment amount p are related by the formula

$$P\left(1 + \frac{i}{c}\right)^{ct} = \frac{p\left[\left(1 + \frac{i}{c}\right)^{ct} - 1\right]}{\frac{i}{c}}.$$

Andy buys a 2012 Nissan Sentra for \$25,000. He pays \$5000 and finances the rest through a 5-yr loan at an annual interest rate of 6%. His monthly car payment is

$$20{,}000\left(1 + \frac{0.06}{12}\right)^{60} = \frac{p\left[\left(1 + \frac{0.06}{12}\right)^{60} - 1\right]}{\frac{0.06}{12}}$$

$$26977.00305 = p(69.77003051)$$

$$p = \frac{26977.00305}{69.77003051} = \$386.66.$$

SECTION 9.5

A **power series** centered at $x = 0$ has the form $c_0 + c_1 x + c_2 x^2 + c_3 x^3 + \cdots$, where $c_0, c_1, c_2, \ldots$ are real coefficients. A power series has an **interval of convergence** $a < x < b$ and a **center of convergence** k, where $k = \dfrac{a + b}{2}$. Depending on the power series, the interval of convergence may be a single point, an interval, or all real numbers.

The power series for $f(x) = \dfrac{3}{1 + 4x}$ is

$$3 - 12x + 48x^2 - 192x^3 + 768x^4 - \cdots .$$

It is convergent for $|4x| < 1$, or

$$-\frac{1}{4} < x < \frac{1}{4}.$$

Its center of convergence is $x = 0$.

Suppose $y = f(x)$ has a power series over an interval of convergence $a < x < b$. The **nth approximating polynomial** is a partial sum of the power series, where n is the degree of the polynomial. As n increases, the approximating polynomials more closely agree with f within the interval of convergence; that is

$$\lim_{n \to \infty} P_n(x) = f(x), \qquad \text{for } a < x < b.$$

The **linearization** of $y = f(x)$ at $x = k$ is

$$y = f(k) + f'(k)(x - k).$$

It is the equation of the line tangent to the graph of f at $x = k$ and is the same as the first approximating polynomial of f at $x = k$.

The first approximating polynomial of $f(x) = \dfrac{3}{1 + 4x}$, centered at $x = 0$, is
$$P_1(x) = 3 - 12x.$$
This is also the linearization of f at $x = 0$.
The second approximating polynomial of $f(x) = \dfrac{3}{1 + 4x}$, centered at $x = 0$, is

$$P_2(x) = 3 - 12x + 48x^2.$$

The figure shows the graph of f in blue and the graph of its third approximating polynomial, $P_3(x) = 3 - 12x + 48x^2 - 192x^3$, in red.

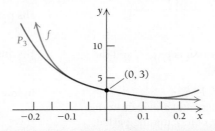

SECTION 9.6

The expression $n!$ is read "n **factorial**" and is defined as

$$n! = 1 \cdot 2 \cdot 3 \cdot \cdots \cdot n,$$

with the special case that $0! = 1$.

$0! = 1$, $1! = 1$, $2! = 1 \cdot 2 = 2$, $3! = 1 \cdot 2 \cdot 3 = 6$, $4! = 1 \cdot 2 \cdot 3 \cdot 4 = 24$, and so on.

KEY TERMS AND CONCEPTS

Assuming that all nth derivatives of f exist in some open interval that includes $x = k$, the **Taylor series** of f centered at $x = k$ is a power series of the form

$$f(x) = c_0 + c_1(x - k) +$$
$$c_2(x - k)^2 + c_3(x - k)^3 +$$
$$c_4(x - k)^4 + \cdots,$$

where each coefficient is given by

$$c_n = \frac{f^{(n)}(k)}{n!}, \qquad \text{for } n = 0, 1, 2, 3, \ldots$$

EXAMPLES

The Taylor series for $f(x) = e^x$ entered at $x = 0$ is

$$e^x = 1 + x + \frac{1}{2!}x^2 + \frac{1}{3!}x^3 + \frac{1}{4!}x^4 + \frac{1}{5!}x^5$$
$$+ \cdots, \text{ for } -\infty < x < \infty.$$

Trigonometry Connection:

The Taylor series for $f(x) = \sin x$ centered at $x = 0$ is

$$\sin x = x - \frac{1}{3!}x^3 + \frac{1}{5!}x^5 - \frac{1}{7!}x^7 + \cdots, \text{ for } -\infty < x < \infty.$$

The Taylor series for $f(x) = \cos x$ centered at $x = 0$ is

$$\cos x = 1 - \frac{1}{2!}x^2 + \frac{1}{4!}x^4 - \frac{1}{6!}x^6 + \frac{1}{8!}x^8 - \cdots, \text{ for} -\infty < x < \infty.$$

CHAPTER 9
REVIEW EXERCISES

These review exercises are for test preparation. They can also be used as a practice test. Answers are in the back of the book. The blue bracketed section references tell you what part(s) of the chapter to restudy if your answer is incorrect.

CONCEPT REINFORCEMENT

In Exercises 1–10, classify each statement as true or false.

1. The sequence $1, 2, 4, 5, 7, 8, \ldots$ is arithmetic. [9.1]

2. The sequence $7, 4, 1, -2, -5, -8, \ldots$ has a common difference of 3. [9.1]

3. Every arithmetic sequence has a common difference. [9.1]

4. Every geometric series has a common ratio. [9.2]

5. The sequence $2, -8, 32, -64, 128, \ldots$ has a common ratio of 4. [9.2]

6. The infinite geometric series $\frac{1}{4} + \frac{1}{8} + \frac{1}{16} + \frac{1}{32} + \frac{1}{64} + \cdots$
converges to $\dfrac{1/4}{1 - (1/2)} = \dfrac{1}{2}$. [9.2]

7. The simple interest earned on $200 in 2 yr at an annual interest rate of 6% is $24. [9.3]

8. Making a one-time deposit and allowing the money to accrue interest over time is how one sets up an annuity. [9.4]

9. The power series for $f(x) = \dfrac{1}{1 - x}$ is convergent for $-1 < x < 1$. [9.5]

10. If a power series is convergent over the interval $1 < x < 5$, then the center of convergence is 4. [9.5]

REVIEW EXERCISES

11. Consider the sequence $11, 20, 29, 38, 47, \ldots$ [9.1]
    a) Find the general nth term.
    b) Find the 200th term of this sequence.
    c) Find the sum of the first 50 terms.

12. An arithmetic sequence has $a_4 = 13$ and $a_7 = 25$. [9.1]
    a) Find the general nth term.
    b) Find the sum of the first 200 terms.

13. **Business: straight-line depreciation.** Meili purchases a new computer for her business at an initial cost of $3500. She expects to keep the computer for 5 yr, at which time its value is expected to be $1000. Assume that the computer drops in value by the same amount each year, and let its value when brand-new be $a_0 = 3500$. [9.1]
    a) Find a_n, the value of Meili's computer n years after it was purchased.
    b) What is the value of her computer after 3 yr?
    c) Write a sequence of terms a_n, for $n = 0, 1, 2, \ldots, 5$, representing the value of Meili's computer n years after it was purchased.

14. **Business: depleting an account.** Ryan deposits $4000 in an account. At the start of each week after that, he withdraws $120 for expenses. [9.1]
    a) Find a_n, the amount in Ryan's account n weeks after the $4000 was deposited. Assume that $a_0 = \$4000$, the amount of the original deposit.
    b) How much money is in Ryan's account after 16 weeks?
    c) Write a sequence of terms a_n, for $n = 0, 1, 2, \ldots, 5$, representing the value of Ryan's account n weeks after the $4000 was deposited.

15. *Business: fiscal multiplier.* Suppose for every $1 spent in a local economy, 37% is saved and the remainder is spent again, the process repeating itself at every stage. How much total spending will be attributable to the original $1 spent? [9.2]

16. Determine the sum of the following infinite series: $\frac{1}{4} + \frac{3}{20} + \frac{9}{100} + \frac{27}{500} + \cdots$. [9.2]

17. Write the decimal $0.5656565656\ldots$ as a fraction in lowest terms. [9.2]

18. *Business: add-on interest.* Linda buys a washing machine for $1200, paying half down and the rest in 8 monthly payments through an add-on interest loan at an annual interest rate of 10%. Find Linda's monthly payment. [9.3]

19. *Business: present value.* Dennis wants to have $2500 in 2 yr. He opens a savings account that has an annual interest rate of 3.82%, compounded quarterly. [9.3]

   a) What lump-sum deposit does Dennis need to make to meet his goal?
   b) How much interest will Dennis earn?

20. *Business: sinking fund.* Wayne is saving $5000 for a new motorcycle. He starts a sinking fund at an annual interest rate of 3.75%, compounded semiannually for 3 yr. [9.4]

   a) What semiannual payment does Wayne need to make to meet his goal?
   b) What will be Wayne's personal contribution?
   c) How much interest will Wayne earn?

21. *Business: annuity.* Patrice deposits $50 into a savings account every month at an annual interest rate of 4.7%, compounded monthly. [9.4]

   a) How much will her account contain after 8 yr, assuming she makes no withdrawals?
   b) How much interest will she earn?

22. *Business: retirement.* Lorraine opens a savings account that earns an annual interest rate of 3.5%, compounded monthly. She deposits $100 every month for 10 yr. Then, she lets the account continue to grow, at the same annual interest rate and compounding frequency, for another 15 yr. Find the future value in Lorraine's account at the end of the 25 years. [9.4]

23. *Business: car loan.* Glenda buys a used Subaru Outback for $13,000. She pays 25% down and finances the rest at an annual interest rate of 6.5%, compounded monthly for 5 yr. [9.4]

   a) Find Glenda's monthly car payment.
   b) If Glenda makes every payment for the life of the loan, find the total interest she will pay.

24. *Business: home mortgage.* The Savards qualify for a 30-yr mortgage at an annual interest rate of 4.25%, compounded monthly. They are willing to pay up to $2500 per month as their mortgage payment. What is the most they can afford to borrow? [9.4]

25. *Business: credit card.* Vicki uses her credit card to buy $1200 in goods. Her credit card has an annual interest rate of 20.75%, compounded monthly over a 10-yr term. Assume that Vicki makes no additional purchases with this credit card. [9.4]

   a) Find Vicki's monthly payment.
   b) Assume that Vicki pays the amount found in part (a). In the first month, what will she pay in interest, and what will she pay toward the principal?

26. Let $f(x) = \dfrac{4}{1-x}$. [9.5]

   a) Find the first five terms of the power series for f.
   b) Find the interval of convergence and the center of convergence of the power series.

27. Let $g(x) = \dfrac{2}{1+6x}$. Find the first five terms of the power series for the first derivative, g'. [9.5]

28. *Business: estimating average cost.* The average cost, in dollars, of producing x units of a plastic storage container is given by $A(x) = 8 + \dfrac{20}{x} + \dfrac{x^2}{100}$. Find $P_1(x)$ at $x = 20$ units, and use $P_1(x)$ to estimate the average cost of producing 21 units. [9.5]

29. Use a power series and a substitution to evaluate
$$\int_0^1 e^{-x^6}\, dx$$
to three decimal places. [9.6]

TRIGONOMETRY CONNECTION · · · · · · · · · · · · ·

30. Let $h(x) = \sin x^4$. [9.6]

   a) Use a substitution and find the first five terms of the Taylor series for h.
   b) Find the first five terms of the power series of h'.

31. Use a Taylor series and a substitution to evaluate
$$\int_0^1 \cos x^4\, dx$$
to four decimal places. [9.6]

1. Consider the sequence 6, 10, 14, 18, 22, ...

  a) Find the common difference.
  b) Find the general nth term.
  c) Find the 50th term.
  d) Find the sum of the first 100 terms.

2. Consider the sequence 81, 27, 9, 3, 1, ...

  a) Find the common ratio.
  b) Find the general nth term.
  c) Find the 9th term.
  d) Find the sum of the first 12 terms.

3. Find S_{80} of the arithmetic sequence in which $a_4 = 29$ and $a_8 = 61$.

4. Find the sum of the first 45 terms of the following series:
$7 + 17 + 27 + 37 + 47 + \cdots$.

5. Find the sum of the following infinite series:
$2 - \frac{1}{4} + \frac{1}{32} - \frac{1}{256} + \frac{1}{2048} - \cdots$.

6. Find the simple interest future value of $2000 at an annual interest rate of 3.89% for 4 yr.

7. Business: government bond. Heath purchases a bond and then redeems it 6 months later for $500. Assume that the bond grew in value at an annual simple interest rate of 2.5%. What did Heath pay for the bond?

8. Business: depleting a bank account. Tania's business account has $3000 on the first day of the month. She withdraws $90 per day for expenses. Let a_n be the amount in Tania's account after n days, and $a_0 = 3000$.

  a) Find the general nth term.
  b) How much is in Tania's account after 12 days?

9. Business: add-on interest. Zahra buys a new bedroom set for $1200. She pays $400 down and finances the rest with an add-on interest loan at an annual interest rate of 15% for 8 months. Find her monthly payment.

10. Business: fiscal multiplier. Find the total effect of a $100,000 expenditure if 52% of the money is spent and then the same percentage of that is spent, and so on.

11. Business: compound interest future value. Henry deposits $1350 into an account with an annual interest rate of 4.2% compounded monthly for 6 yr.

  a) Find the future value of Henry's deposit.
  b) Find the interest earned.

12. Business: present value. Gena wants to have $4000 in 3 yr. She opens a savings account that has an annual interest rate of 3.75%, compounded weekly.

  a) Find the lump-sum deposit that Gena needs to make to meet her goal.
  b) How much interest will Gena's deposit earn?

13. Business: annuity. Jill deposits $20 per week into her account for holiday shopping. The account has an annual interest rate of 4.3%, compounded weekly.

  a) Find the future value of Jill's account after 1 yr (52 weeks).
  b) Find Jill's personal contribution.
  c) Find the interest earned after 1 year.

14. Business: retirement. Craig wants to have $1,000,000 when he retires. He starts a sinking fund in an account that has an annual interest rate of 5.25%, compounded quarterly for 25 yr. Find the quarterly payment Craig will need to make in order to reach his goal.

15. Business: amortized loan. The Langways purchase a new home for $450,000. They pay 20% down and finance the rest through a 30-yr mortgage at an annual interest rate of 3.75%, compounded monthly.

  a) Find the Langways' monthly mortgage payment.
  b) Assuming that the Langways make the monthly payment found in part (a) for the life of the loan, how much will they pay in total?
  c) How much interest will they pay?

16. Business: car loan. Carl qualifies for a car loan at an annual interest rate of 5.75%, compounded monthly for 5 yr. He is willing to pay up to $400 per month in car payments. What is the maximum loan amount he can afford?

17. Let $h(x) = \dfrac{1}{3 - x}$.

  a) Find the first five terms of the power series for h.
  b) Find the first five terms of the power series for the first derivative, h'.

18. Use a power series and a substitution to evaluate the integral $\displaystyle\int_{0}^{0.5} e^{x^4}\, dx$ to three decimal places.

19. Business: estimating average cost. The average cost, in dollars, of producing x units of a bookshelf is given by $A(x) = 16 + \dfrac{12}{x} + \dfrac{x^2}{70}$. Find $P_1(x)$ at $x = 30$ units, and use $P_1(x)$ to estimate the average unit cost for 31 units.

TRIGONOMETRY CONNECTION

20. Let $h(x) = \cos \sqrt[3]{x}$.

  a) Use a substitution and find the first six terms of the Taylor series of h.
  b) Find the first five terms of the Taylor series of h'.

21. Use a Taylor series and a substitution to evaluate $\displaystyle\int_{0}^{1} \sin x^5\, dx$ to four decimal places.

Sequences and Series on a Spreadsheet

A spreadsheet can be a powerful tool for finding the terms of a sequence and the partial sums of a series. Hundreds or even thousands of terms can be determined very quickly this way.

For example, suppose we use a spreadsheet to explore the sequence given by $a_n = \dfrac{1}{2^n}$, and the series $\frac{1}{2} + \frac{1}{4} + \frac{1}{8} + \frac{1}{16} + \cdots$. First, we generate a list of index values:

1. Enter 1 in cell A1.

2. Enter the expression =1+A1 in cell A2. Be sure to include the equals sign.

3. Select cell A2 and copy it.

4. Select cells A3 through A100 and then paste. The numbers 1 through 100 now occupy cells A1 through A100. These are the index values.

Now we need to enter the terms of the sequence into column B:

5. Enter =1/(2^A1) in cell B1.

6. Select cell B1 and copy it.

7. Select cells B2 through B100 and then paste. These are the terms (in decimal form) of the sequence $\frac{1}{2} + \frac{1}{4} + \frac{1}{8} + \frac{1}{16} + \cdots$.

The partial sums are entered into column C:

8. Enter =SUM($B$1:$B1) in cell C1. The dollar signs "lock" the initial cell.

9. Select cell C1 and copy it.

10. Select cells C2 through C100 and then paste. The partial sums of the series $\frac{1}{2} + \frac{1}{4} + \frac{1}{8} + \frac{1}{16} + \cdots$ are listed in cells C2 through C100.

Experiment by increasing the column widths for columns B and C to 30 and hitting the increase-decimal tab: . This allows more decimal places to appear in each cell. Below is an image of the first 30 terms (column B) of the sequence $\frac{1}{2}, \frac{1}{4}, \frac{1}{8}, \frac{1}{16}, \ldots$, and the first 30 partial sums (column C) for the series $\frac{1}{2} + \frac{1}{4} + \frac{1}{8} + \frac{1}{16} + \cdots$. We see that the sequence of partial sums converges toward 1.

	A	B	C
1	1	0.500000000000000000000000000	0.500000000000000000000000000
2	2	0.250000000000000000000000000	0.750000000000000000000000000
3	3	0.125000000000000000000000000	0.875000000000000000000000000
4	4	0.062500000000000000000000000	0.937500000000000000000000000
5	5	0.031250000000000000000000000	0.968750000000000000000000000
6	6	0.015625000000000000000000000	0.984375000000000000000000000
7	7	0.007812500000000000000000000	0.992187500000000000000000000
8	8	0.003906250000000000000000000	0.996093750000000000000000000
9	9	0.001953125000000000000000000	0.998046875000000000000000000
10	10	0.000976562500000000000000000	0.999023437500000000000000000
11	11	0.000488281250000000000000000	0.999511718750000000000000000
12	12	0.000244140625000000000000000	0.999755859375000000000000000
13	13	0.000122070312500000000000000	0.999877929687500000000000000
14	14	0.000061035156250000000000000	0.999938964843750000000000000
15	15	0.000030517578125000000000000	0.999969482421875000000000000
16	16	0.000015258789062500000000000	0.999984741210937000000000000
17	17	0.000007629394531250000000000	0.999992370605468000000000000
18	18	0.000003814697265620000000000	0.999996185302734000000000000
19	19	0.000001907348632812500000000	0.999998092651367000000000000
20	20	0.000000953674316406250000000	0.999999046325683000000000000
21	21	0.000000476837158203125000000	0.999999523162841000000000000
22	22	0.000000238418579101562000000	0.999999761581420000000000000
23	23	0.000000119209289550781000000	0.999999880790710000000000000
24	24	0.000000059604644775390600000	0.999999940395355000000000000
25	25	0.000000029802322387695300000	0.999999970197677000000000000
26	26	0.000000014901161193847700000	0.999999985098838000000000000
27	27	0.000000007450580596923830000	0.999999992549419000000000000
28	28	0.000000003725290298461910000	0.999999996274709000000000000
29	29	0.000000001862645149230960000	0.999999998137354000000000000
30	30	0.000000000931322574615479000	0.999999999068677000000000000

EXERCISES

In the following exercises, you need to change the expression used for the general nth term of the sequence in cell B1, then copy and paste it to cells B2 through B100. The contents in column C will change automatically.

1. Use a spreadsheet to find the first 100 terms of the geometric sequence $\frac{1}{3}, \frac{1}{9}, \frac{1}{27}, \frac{1}{81}, \dots$, and the first 100 partial sums of the series $\frac{1}{3} + \frac{1}{9} + \frac{1}{27} + \frac{1}{81} + \cdots$. What is the sum of this infinite series?

2. Use a spreadsheet to find the first 100 terms of the geometric sequence $\frac{1}{4}, \frac{1}{16}, \frac{1}{64}, \frac{1}{256}, \dots$, and the first 100 partial sums of the series $\frac{1}{4} + \frac{1}{16} + \frac{1}{64} + \frac{1}{256} + \cdots$. What is the sum of this infinite series?

3. Use a spreadsheet to find the first 100 terms of the geometric sequence $\frac{2}{5}, \frac{3}{10}, \frac{9}{40}, \frac{27}{160}, \dots$, and the first 100 partial sums of the series $\frac{2}{5} + \frac{3}{10} + \frac{9}{40} + \frac{27}{160} + \cdots$. What is the sum of this infinite series?

4. A ball is released from a height of 10 ft. After each bounce, the ball reaches a height that is 60% of the height of the previous bounce. Use a spreadsheet to determine the total vertical distance the ball will travel before coming to a stop. Note that the ball will travel up and down on each bounce, except for the initial drop.

The Harmonic Sequence and Series

The sequence $1, \frac{1}{2}, \frac{1}{3}, \frac{1}{4}, \frac{1}{5}, \frac{1}{6}, \frac{1}{7}, \frac{1}{8}, \dots$ is called the *harmonic sequence*, and its general nth term is given by $a_n = \frac{1}{n}$. The *harmonic series* is the sum of these terms:

$$1 + \frac{1}{2} + \frac{1}{3} + \frac{1}{4} + \frac{1}{5} + \frac{1}{6} + \frac{1}{7} + \frac{1}{8} + \frac{1}{9} + \frac{1}{10} + \cdots.$$

This series diverges very slowly. On a spreadsheet, we adjust column B by entering =1/A1 in cell B1; then we copy and paste that cell to cells B2 through B100. We then examine the sequence of partial sums in column C.

In the image below, the first 20 partial sums are shown in column C, as well as the 40th, 60th, 80th, and 100th partial sums. For more terms, select cells A100 through C100, then paste them into cells A101 through C500. More terms can be created as needed.

Note that the partial sums continue to grow. The harmonic series exceeds the value 5 after 83 terms and 6 after 227 terms. This growth continues forever without bound.

	A	B	C
1	1	1.000000000000000000000000000	1.000000000000000000000000000
2	2	0.500000000000000000000000000	1.500000000000000000000000000
3	3	0.333333333333333300000000000	1.833333333333333300000000000
4	4	0.250000000000000000000000000	2.083333333333333300000000000
5	5	0.200000000000000000000000000	2.283333333333333300000000000
6	6	0.166666666666666700000000000	2.450000000000000000000000000
7	7	0.142857142857143000000000000	2.592857142857143000000000000
8	8	0.125000000000000000000000000	2.717857142857143000000000000
9	9	0.111111111111111100000000000	2.828968253968253500000000000
10	10	0.100000000000000000000000000	2.928968253968253500000000000
11	11	0.090909090909090900000000000	3.019877344877344400000000000
12	12	0.083333333333333330000000000	3.103210678210678800000000000
13	13	0.076923076923076900000000000	3.180133755133755700000000000
14	14	0.071428571428571400000000000	3.251562326562233300000000000
15	15	0.066666666666666670000000000	3.318228993228999900000000000
16	16	0.062500000000000000000000000	3.380728993228999900000000000
17	17	0.058823529411764700000000000	3.439552522264076000000000000
18	18	0.055555555555555560000000000	3.495108078196310000000000000
19	19	0.052631578947368400000000000	3.547739657143680000000000000
20	20	0.050000000000000000000000000	3.597739657143680000000000000
40	40	0.025000000000000000000000000	4.278543038963638000000000000
60	60	0.016666666666666700000000000	4.679870412951740000000000000
80	80	0.012500000000000000000000000	4.965479278945520000000000000
100	100	0.010000000000000000000000000	5.187377517639620000000000000

EXERCISES

5. How many terms are needed for the harmonic series to exceed the value 7?

6. Consider ordered pairs of the form $(x, h(x))$, where $h(x)$ is the number of terms needed for the harmonic series to exceed the value x. For example, we have $h(2) = 4$, since 4 terms are needed for the harmonic series to exceed the value 2. This is the ordered pair $(2, 4)$.

  a) Find ordered pairs for $x = 3$ through $x = 7$.

  b) Using regression, find an exponential function that fits the ordered pairs $(x, h(x))$, for $x = 2, 3, 4, 5, 6, 7$.

  c) Use the function from part (b) to estimate $h(10)$, the number of terms needed for the harmonic series to exceed the value 10. Then find $h(10)$ using the spreadsheet. How close is the estimate to the actual value?

7. The *alternating harmonic series* is

$$1 - \frac{1}{2} + \frac{1}{3} - \frac{1}{4} + \frac{1}{5} - \frac{1}{6} + \frac{1}{7} - \frac{1}{8} + \cdots.$$

This series converges very slowly to ln 2.

  a) In cell B1, enter =(-1)^(A1-1)* (1/A1). Copy this cell and paste it into cells B2 through the bottom of column B. What is the 100th partial sum? 200th partial sum? 500th partial sum?

  b) The value of ln 2 is 0.693147.... For which partial sum does the value of the alternating harmonic series "lock in" to two decimal places? That is, at what index value n does the partial sum stay at 0.69...?

  c) For which partial sum does the value of the alternating harmonic series lock in to three decimal places, at 0.693...?

8. The alternating series $4 - \frac{4}{3} + \frac{4}{5} - \frac{4}{7} + \frac{4}{9} - \cdots$ converges to π. In cell B1, enter =(-1)^(A1-1)*(4/(2*A1-1)). Copy this cell and paste it to the bottom of column B. How many terms are needed to lock the series in at 3.141? (This series is one form of the *Madhava-Leibniz series* and was one of the first nongeometric methods used to calculate a value of π. Madhava was an Indian mathematician of the late 14th-century.)

A Business Application:

The Coupon Collector Problem

In business, a common way to increase sales is to entice customers with a set of free prizes and a challenge to collect them all. For example, suppose a cereal manufacturer randomly includes one of ten small toys in each box. On average, how many boxes of cereal would a customer have to purchase to collect all ten toys? This is known as the *coupon collector problem*, and it can be shown that the average number of boxes one needs to purchase is given by the sum

$$10 \cdot \left(1 + \tfrac{1}{2} + \tfrac{1}{3} + \tfrac{1}{4} + \tfrac{1}{5} + \tfrac{1}{6} + \tfrac{1}{7} + \tfrac{1}{8} + \tfrac{1}{9} + \tfrac{1}{10}\right) \approx 29.28$$

In this type of situation, we always round up. Thus, on average, purchasing 30 boxes is usually sufficient to collect all ten toys. Note that there is no upper bound that guarantees the collection of all ten toys: it's possible to purchase 100 boxes and not complete the set. However, this is very unlikely to happen.

In general, if n objects (such as toys or coupons) are randomly distributed among the units of a product being sold, a customer needs to purchase

$$C(n) = n \cdot \left(1 + \tfrac{1}{2} + \tfrac{1}{3} + \tfrac{1}{4} + \tfrac{1}{5} + \cdots + \frac{1}{n}\right) \text{ units, on}$$

average, in order to collect all n objects. Note that the series in the parentheses is the harmonic series.

EXERCISE

9. In 1979, the band Led Zeppelin released its album *In Through the Out Door*. Each vinyl LP came in one of six randomly selected album covers, which were wrapped in plain brown paper so that the purchaser could not identify the cover until after the purchase. On average, how many LPs would one need to purchase in order to collect all six variations?

Recursive Sequences

A sequence in which each term is defined by one or more preceding terms is said to be *recursive*. For example, the sequence $a_{n+1} = a_n + 2$ is recursive. If we define $a_1 = 3$, then $a_2 = a_1 + 2 = 3 + 2 = 5$, $a_3 = 5 + 2 = 7$, $a_4 = 7 + 2 = 9$, and so on.

The most famous recursive sequence is the *Fibonacci sequence*, in which each term is defined by $F_n = F_{n-1} + F_{n-2}$, with $F_1 = 1$ and $F_2 = 1$. Thus, each term after F_2 is the sum of the two terms that come before it: $F_3 = 1 + 1 = 2$, $F_4 = 2 + 1 = 3$, $F_5 = 3 + 2 = 5$, and so on. The first 12 terms of the Fibonacci sequence are 1, 1, 2, 3, 5, 8, 13, 21, 34, 55, 89, 144, ...

On a spreadsheet, generate 100 index values in column A. In cells B1 and B2, enter 1. In cell B3, enter =B1+B2. Copy this cell and paste it into cells B4 through B100. The first 100 terms of the Fibonacci sequence will appear in column B.

The Fibonacci sequence has many remarkable properties. Perhaps the best known is its relationship to the *golden ratio*, $\varphi = \dfrac{\sqrt{5} - 1}{2} \approx 0.6180339\ldots$, where φ is the Greek letter phi. This number can be shown to be the limit of the quotients of the terms of the Fibonacci sequence:

$$\lim_{n \to \infty} \left(\frac{F_n}{F_{n+1}} \right) = \varphi.$$

The Fibonacci sequence and the golden ratio are found in many fields of study, including geometry, architecture, and art and, intriguingly, in many natural proportions of the human body. A fantastic resource is the site www.goldennumber.net; you are encouraged to learn more about this fascinating number.

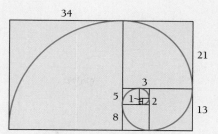

The golden spiral

EXERCISES

10. On the spreadsheet, enter 1 in cell C1 and =B1/B2 in cell C2. Then copy cell C2 and paste it into cells C3 through C100. What does this ratio tend toward as *n* increases?

11. The limit given above works for all initial values of F_1 and F_2 (as long as neither is zero). In cells B1 and B2, enter any two nonzero integers. The remaining cells should update automatically. What do you believe is the limit of the quotients as *n* increases?

12. What happens if you take the reciprocal of the quotient? In cell C2, enter =B2/B1. Copy this cell and paste it into cells C3 through C100. What is the new limit?

The golden ratio is often used in art and architecture since it yields dimensions that are pleasing to the eye.

Probability Distributions

Chapter Snapshot

What You'll Learn

Why It's Important

Probability is often used to make decisions in our daily lives and in the sciences and business. In this chapter, we develop the foundations of probability and probability distributions in a manner that ultimately relies on calculus.

Where It's Used

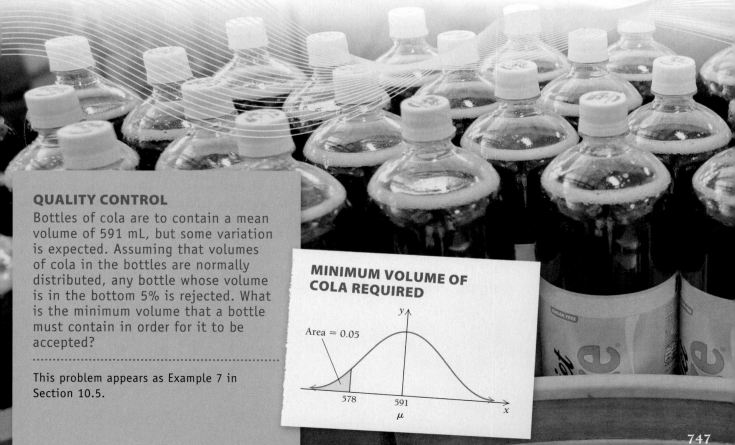

QUALITY CONTROL

Bottles of cola are to contain a mean volume of 591 mL, but some variation is expected. Assuming that volumes of cola in the bottles are normally distributed, any bottle whose volume is in the bottom 5% is rejected. What is the minimum volume that a bottle must contain in order for it to be accepted?

This problem appears as Example 7 in Section 10.5.

MINIMUM VOLUME OF COLA REQUIRED

Area = 0.05

578   591
       μ

10.1

OBJECTIVES

- Use and interpret the empty set and a universal set in various situations.

- Determine when one set is a subset of another set and when two sets are equal.

- Find the complement of a set, the intersection of two sets, and the union of two sets.

- Find the cardinality of a set.

A Review of Sets

We often consider collections of objects in everyday life. For example, the students in your class, the letters of the alphabet, the members of your family, and the positive even integers are all examples of *sets*. The objects within a set are the **elements** or the **members** of the set.

Recall from Section R.3 that a set may be written in two ways: with the *roster method*, in which the elements are listed, or with *set-builder notation*, in which the elements are described. For example, the set of chess pieces C can be written in two ways:

- Roster method: $C = \{$ king, queen, knight, bishop, rook, pawn $\}$
- Set-builder notation: $C = \{ x \mid x$ is a piece used in the game of chess $\}$

The elements of a set are enclosed within braces, $\{\ \}$, and the vertical slash, $\mid$, used in set-builder notation is read "such that." The symbol $\in$ means "is an element of," and $\notin$ means "is not an element of." Thus, we can write "king $\in C$," since the king is an element of C, the set of chess pieces.

Empty Set and Universal Set

How can we describe the set of people who have been to Mars or the set of states in the United States that begin with the letter Z? These sets have no elements, which suggests a need for a specially defined set called the *empty set*.

> **DEFINITION**
>
> The **empty set**, denoted by $\varnothing$, is the set that contains no elements.

We also need a set of "all" elements that exist within the context of a given situation. The set of all letters in the alphabet is an example of a *universal set*.

> **DEFINITION**
>
> A **universal set** (also called a **universe**), denoted by U, is the set of all elements within the context of a situation.

When we work with equations, we try to find the set of all solutions to that equation, known as the *solution set*. Some equations have no solution, and some may be true for all possible values of x, as shown in the following example.

■ **EXAMPLE 1**   Describe the elements in each set:

a) The solution set of $x = x + 1$

b) The solution set of $(x + 1)^2 = x^2 + 2x + 1$

Solution

a) The equation $x = x + 1$ has no solution because no matter what real number we substitute for x, we get a false statement. The solution set of this equation is the empty set, $\varnothing$.

b) The equation $(x + 1)^2 = x^2 + 2x + 1$ is true for all real numbers. Thus, its solution set is the set of all real numbers. We can write this set using set-builder notation: $\{x \mid x \text{ is a real number}\}$. Here, and throughout most of algebra and calculus, the universe is assumed to be the set of real numbers (denoted $\mathbb{R}$).

❬ Quick Check 1

Subsets and Equality

We are often interested in subgroups of elements from within a set. For example, from the universal set of the 50 states of the United States, we can form a subgroup of the states that border Canada or a subgroup of the states with ten or more representatives in the House of Representatives. These subgroups are called *subsets*.

DEFINITION

A set A is a **subset** of another set B if all elements of A are elements of B. The symbol for "is a subset of" is $\subseteq$, and we write $A \subseteq B$.

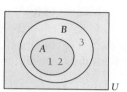

For example, the set $A = \{1, 2\}$ is a subset of $B = \{1, 2, 3\}$, and we write $A \subseteq B$. Also, every set is a subset of itself; that is, $A \subseteq A$ is true for any set A. By definition, the empty set is a subset of all sets. Note that the subset symbol is used when comparing two sets, while the element symbol is used to indicate that an element is a member of a set.

◼ **EXAMPLE 2**   Classify each statement as true or false. If false, explain why, and provide a correct statement.

a) $\{3\} \in \{1, 2, 3, 4, 5\}$ **b)** $\{2, 3\} \subseteq \{1, 2, 3, 4, 5\}$

Solution

a) Since $\{3\}$ is a subset, not an element, of the set $\{1, 2, 3, 4, 5\}$, the statement $\{3\} \in \{1, 2, 3, 4, 5\}$ is *false*. The correct statement is $\{3\} \subseteq \{1, 2, 3, 4, 5\}$. Alternatively, we can remove the braces from $\{3\}$ and write $3 \in \{1, 2, 3, 4, 5\}$, which indicates that 3 is an element of the set $\{1, 2, 3, 4, 5\}$.

b) Since every element in the set $\{2, 3\}$ is also an element in $\{1, 2, 3, 4, 5\}$, the statement $\{2, 3\} \subseteq \{1, 2, 3, 4, 5\}$ is *true*.

❬ Quick Check 2

We can also compare two sets to see if they contain the same elements. Consider set $A = \{x \mid x \text{ is a positive odd integer less than or equal to } 9\}$ and set $B = \{1, 3, 5, 7, 9\}$. By comparing the elements in sets A and B, we see that each set contains the same elements, and we conclude that the two sets are *equal*.

DEFINITION

Sets A and B are **equal**, written $A = B$, if they contain the same elements. The order in which elements are listed does not matter, and repeated elements are ignored.

For example, $\{a, b, c\}$ is equal to $\{c, a, b\}$ and to $\{a, b, b, a, c, b, a, b\}$. Also, every set is equal to itself.

The Three Stooges: the order in which the characters appear does not affect the set itself. Both images show the same set of characters.

■ **EXAMPLE 3** Let $A = \{1, 2, 3, 4, 5\}$, $B = \{1, 2, 3\}$, and $C = \{3, 3, 2, 1, 2, 1, 2\}$.

a) Which of the sets are equal to one another?

b) Which sets are subsets of any of the other sets?

Solution

a) Sets B and C contain the same elements: 1, 2, and 3. We disregard repeated elements and the order of elements, so $B = C$. Also, each set is equal to itself, so $A = A$, $B = B$, and $C = C$.

b) Since all elements of B are in A, we have $B \subseteq A$, and similarly, $C \subseteq A$. Since $B = C$, we have $B \subseteq C$ and $C \subseteq B$. And since every set is a subset of itself, we have $A \subseteq A$, $B \subseteq B$, and $C \subseteq C$.

❰ Quick Check 3

Complements, Unions and Intersections

Suppose a college has a soccer team and we consider the team as a set whose elements are the soccer players. A second set, consisting of all of the college's students who are *not* on the soccer team, is called the *complement* of the first set.

> **DEFINITION**
>
> Given a universal set U and a set A such that $A \subseteq U$, the **complement** of A, denoted A', is the set of all elements in U that are *not* in A:
>
> $$A' = \{x \mid x \notin A \text{ and } x \in U\}.$$

■ **EXAMPLE 4** Let $U = \{1, 2, 3, 4, 5, 6, 7, 8, 9, 10\}$ and $A = \{1, 2, 3, 5, 7, 9\}$. Find A'.

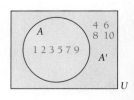

Solution The complement of A are those elements in U that are not in A. Thus, $A' = \{4, 6, 8, 10\}$.

❰ Quick Check 4

Note that A'' is the complement of A', and so $A'' = \{1, 2, 3, 5, 7, 9\}$ for the set in Example 4. In general, $A'' = A$ for any set A.

Suppose the college also has a basketball team, and we consider this team as a set whose elements are the basketball players. Then, we can compare the set of players on the soccer team and the set of players on the basketball team in two ways:

- The set of students who play soccer *or* basketball are members of the *union* of the sets of soccer players and basketball players. Here, the word *or* means "and/or," so this set includes students who play on one or both teams.

- The set of students who play soccer *and* basketball are members of the *intersection* of the sets of soccer players and basketball players.

DEFINITION

The **union** of A and B, denoted $A \cup B$, is the set of elements in A *or* B:

$$A \cup B = \{x \mid x \in A \text{ or } x \in B\}.$$

The **intersection** of A and B, denoted $A \cap B$, is the set of elements in A *and* B:

$$A \cap B = \{x \mid x \in A \text{ and } x \in B\}.$$

If $A \cap B = \varnothing$, then A and B are *mutually exclusive*.

■ **EXAMPLE 5**   Let $A = \{1, 2, 3, 5, 7, 9\}$, $B = \{1, 4, 5, 8, 9\}$, and $C = \{4, 6, 10\}$. Find the following sets.

a) $A \cup B$ **b)** $A \cap B$ **c)** $A \cap C$

Solution

a) The union of A and B contains those elements in A or B or in both A and B. Thus, $A \cup B = \{1, 2, 3, 4, 5, 7, 8, 9\}$.

b) The intersection of A and B are those elements in both A and B. Thus, $A \cap B = \{1, 5, 9\}$.

c) There are no elements that are in both A and C. Thus, the intersection is empty, $A \cap C = \varnothing$, and we conclude that A and C are mutually exclusive.

❮ Quick Check 5

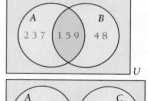

> **Quick Check 5**
>
> Let $M = \{4, 5, 6, 7, 8, 9\}$ and $N = \{1, 2, 3, 4, 5\}$. Find $M \cup N$ and $M \cap N$.

A Business Application

■ **EXAMPLE 6**   **Business: Sandwich Topping Choices.** Subway™ sandwich shops offer the following topping options for sandwiches: bell pepper (b), cucumber (c), jalapeno pepper (j), lettuce (l), olive (o), pickles (p), red onion (r), and tomato (t). (*Source:* www.subway.com.) Let the universe be the set of these toppings, $U = \{b, c, j, l, o, p, r, t\}$. Customers may choose any number of these toppings, or none at all. Suppose Jeff chooses bell pepper, cucumber, lettuce, and tomato for his sandwich, while Katie chooses bell pepper, jalapeno, lettuce, red onion, and tomato for her sandwich. Thus, $J = \{b, c, l, t\}$ and $K = \{b, j, l, r, t\}$.

a) Express the set J' in set notation, and state its meaning in words.

b) Express the set $J \cup K$ in set notation, and state its meaning in words.

c) Express the set $J \cap K$ in set notation, and state its meaning in words.

d) What is the empty set in this situation?

Solution

a) Set J', or the complement of J, is the set of elements in U and not in J. Thus, $J' = \{j, o, p, r\}$. These are the toppings that Jeff did *not* choose.

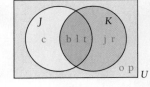

b) Set $J \cup K$, the union of J and K, is $\{b, c, j, l, r, t\}$. These are the toppings that were chosen by Jeff *or* Katie.

c) Set $J \cap K$, the intersection of J and K, is $\{b, l, t\}$. These are the toppings that were chosen by Jeff *and* Katie.

d) The empty set represents the case where a customer does not choose any toppings.

❮ Quick Check 6

> **Quick Check 6**
>
> For the situation in Example 6, express the set K' in set notation, and state its meaning in words.

Cardinality

The set of chess pieces, $\{\text{king, queen, knight, bishop, rook, pawn}\}$, consists of six different elements, so we say that the *cardinality* of this set is 6.

> **DEFINITION**
>
> The **cardinality** of a set is the number of elements contained in the set. If A is a set, then its cardinality is denoted by $n(A)$.*

For example, let $A = \{m, a, t, h\}$. This set contains 4 elements, so we say that the cardinality of set A is 4, or $n(A) = 4$. However, be careful to note repeated elements when determining a set's cardinality. For example, if $B = \{e, l, e, m, e, n, t\}$, then the cardinality of B is $n(B) = 5$, since the repeated e's are ignored.

The cardinality of the empty set is zero: $n(\varnothing) = 0$. Here we see a relationship between the empty set and the number 0, since the empty set has zero elements in it. However, the set containing the number 0, written $\{0\}$, has one element in it, so it is not the same as the empty set.

For $A = \{m, a, t, h\}$ and $B = \{e, l, e, m, e, n, t\}$, is it true that $n(A) + n(B) = n(A \cup B)$? To answer this, note that $n(A) = 4$, $n(B) = 5$, and $A \cup B = \{m, a, t, h, e, l, n\}$. Since $n(A \cup B) = 7$, we see that $n(A) + n(B) \neq n(A \cup B)$. Note that $A \cap B = \{m, t\}$, so to find $n(A \cup B)$, we can add $n(A)$ and $n(B)$, but we have to subtract $n(A \cap B)$, since the elements in $A \cap B$ would otherwise be counted twice.

In this image, there are three circles and three green shapes. However, there are five objects that are green or circular. Note that the green circle is counted once.

> **THEOREM 1   Cardinality of the Union**
>
> The number of elements in the union of A and B is
> $$n(A \cup B) = n(A) + n(B) - n(A \cap B).$$
>
> This is also known as the *Inclusion-Exclusion Principle*.

■ **EXAMPLE 7**   Let A and B be two sets such that $n(A) = 21, n(B) = 30$, and $n(A \cap B) = 9$. How many elements are in $A \cup B$?

Solution   We use the formula for the cardinality of the union:

$$n(A \cup B) = n(A) + n(B) - n(A \cap B)$$
$$= 21 + 30 - 9 \qquad \text{Substituting}$$
$$= 42.$$

The union of A and B has 42 elements.

❮ Quick Check 7

> **Quick Check 7**
>
> Let A and B be two sets such that $n(A) = 84, n(B) = 65$, and $n(A \cap B) = 33$. Find $n(A \cup B)$.

---

*The notation $|A|$ is also used to represent the cardinality of a set.

■ **EXAMPLE 8**   **Business: Sandwich Topping Choices.**   Recall the sandwich toppings in Example 6: the universe of toppings is $U = \{b, c, j, l, o, p, r, t\}$, and Jeff chose $J = \{b, c, l, t\}$, while Katie chose $K = \{b, j, l, r, t\}$.

a) How many toppings were chosen by Jeff *or* Katie?

b) How many toppings were *not* chosen by Jeff or Katie?

Solution

a) Four toppings were chosen by Jeff and five by Katie, so we have $n(J) = 4$ and $n(K) = 5$. Three toppings were chosen by both Jeff and Katie, so we have $n(J \cap K) = 3$. Thus,

$$n(J \cup K) = n(J) + n(K) - n(J \cap K)$$
$$= 4 + 5 - 3 \qquad \text{Substituting}$$
$$= 6.$$

Therefore, 6 toppings were chosen by Jeff *or* Katie.

b) The universe has $n(U) = 8$ elements, and we know from part (a) that $n(J \cup K) = 6$. Thus, $n((J \cup K)') = 2$, and 2 toppings were *not* chosen by Jeff or Katie.

❯ **Quick Check 8**

From Example 8, find $n(K')$, and describe what it represents.

❮ Quick Check 8

Section Summary

- A *set* is a collection of objects, and each object in a set is an *element* or a *member*.
- The *empty set* contains no elements and is denoted by $\varnothing$.
- A *universal set*, also called a *universe*, contains all elements in the situation being considered and is denoted by U.
- Set A is a *subset* of set B, denoted $A \subseteq B$, if every element of A is an element of B. The empty set is a subset of all sets.
- Two sets are *equal* if they contain the same elements, where repeats and different orderings are ignored.
- The *complement* of a set A is $A' = \{x \mid x \notin A \text{ and } x \in U\}$.

- The *union* of two sets A and B is the set of elements in A or B:
$$A \cup B = \{x \mid x \in A \text{ or } x \in B\}.$$
- The *intersection* of two sets A and B is the set of elements in A *and* B:
$$A \cap B = \{x \mid x \in A \text{ and } x \in B\}.$$
- The *cardinality* of a set A is the number of elements in A and is denoted $n(A)$. The empty set has a cardinality of zero: $n(\varnothing) = 0$.
- The cardinality of the union of two sets is
$$n(A \cup B) = n(A) + n(B) - n(A \cap B).$$

EXERCISE SET
10.1

In Exercises 1–14, (a) use the roster method to identify the elements in each set, and (b) state the cardinality of each set.

1. The set of positive integers less than or equal to 12

2. The set of positive odd integers less than 19

3. The set of even integers greater than 20 and less than or equal to 40

4. The set of positive integer multiples of 4 that are less than or equal to 40

5. The set of the seasons of the year

6. The set of the letters of the alphabet

7. $\{x \mid x$ is a positive integer multiple of 3, less than or equal to 42$\}$

8. $\{x \mid x$ is a positive integer multiple of 10, less than or equal to 150$\}$

9. The solution set of $3x + 1 = 5$

10. The solution set of $4x = 2x + 1$

11. The solution set of $2x = 2(x - 3)$

12. The solution set of $x^2 = x^2 - 6$

13. The solution set of $(x + 2)^2 = x^2 + 4x + 4$

14. The solution set of $x^2 - 1 = (x + 1)(x - 1)$

Express each set in Exercises 15–20 using set-builder notation.

15. $\{2, 4, 6, 8, 10, 12, 14\}$

16. $\{11, 22, 33, 44, 55, \ldots, 99\}$

17. $\{$Alabama, Alaska, Arizona, Arkansas, California, Colorado, $\ldots$, Wyoming$\}$

18. $\{$January, February, March, $\ldots$, December$\}$

19. $\{$Sunday, Monday, Tuesday, $\ldots$, Saturday$\}$

20. $\{$Europe, Asia, Africa, North America, South America, Australia, Antarctica$\}$

In Exercises 21–28, let the universal set be $U = \{1, 2, 3, 4, 5, 6, 7, 8, 9, 10\}$, $A = \{1, 2, 3, 4, 5\}$, $B = \{6, 7, 8, 9, 10\}$, $C = \{1, 3, 5, 7, 9\}$, and $D = \{6, 7, 8\}$. Classify each statement as true or false. If false, explain why and provide a correct statement.

21. $3 \in A$ **22.** $7 \in A$

23. $\{4\} \in A$ **24.** $\{6, 7\} \in D$

25. $B \subseteq D$ **26.** $D \subseteq B$

27. $D = \{6, 7, 8, 7, 6\}$ **28.** $B = \{10, 9, 8, 7, 6\}$

Use the sets from Exercises 21–28 to identify the following sets. Describe each set using the roster method and state its cardinality.

29. $B \cup C$ **30.** $A \cup C$

31. $B \cap C$ **32.** $A \cap B$

33. A' **34.** D'

35. $(C \cup D)'$ **36.** $(C \cap D)'$

37. A and B are sets such that $n(A) = 14, n(B) = 22$, and $n(A \cap B) = 7$. Find $n(A \cup B)$.

38. A and B are sets such that $n(A) = 75, n(B) = 128$, and $n(A \cap B) = 33$. Find $n(A \cup B)$.

39. A and B are sets such that $n(A) = 6, n(B) = 11$, and $n(A \cup B) = 14$. Find $n(A \cap B)$.

40. A and B are sets such that $n(A) = 38, n(B) = 27$, and $n(A \cup B) = 50$. Find $n(A \cap B)$.

41. Let the universe U be defined such that $n(U) = 100$, with A and B subsets of U such that $n(A) = 45$ and $n(B) = 27$. Find $n(A')$ and $n(B')$.

42. Let the universe U be defined such that $n(U) = 570$, with A and B subsets of U such that $n(A) = 230$ and $n(B) = 400$. Find $n(A')$ and $n(B')$.

43. Let $n(U) = 30$, with A and B subsets of U such that $n(A) = 20$ and $n(B) = 13$. What is the minimum value of $n(A \cap B)$?

44. Let $n(U) = 1000$, with A and B subsets of U such that $n(A) = 250$ and $n(B) = 855$. What is the minimum value of $n(A \cap B)$?

45. A student states that the solution of $x^2 = -9$ is $x = \varnothing$. Explain why this is an incorrect statement.

46. A student states that the solution of $3x = 0$ is $x = \varnothing$. Explain why this is an incorrect statement.

APPLICATIONS

General

47. Survey results. A survey of 100 people has the following results: 25 own a car, 62 own a bicycle, and 13 own both a car and a bicycle. How many people own a car or a bicycle?

48. Survey results. A survey of 60 students shows that 34 are taking a math class, 27 are taking an English class, and 12 are taking both classes. How many students are taking math or English?

49. Weather. In September 2011 (30 days), the temperature in Phoenix, Arizona, reached 100°F on 23 days. Also, measurable rain fell at the airport on 5 days, and on 3 days, both events occurred. (*Source*: www.wunderground.com.)

  a) On how many days did at least one of these events occur?

  b) On how many days did the temperature reach 100°F without measurable rain?

  c) On how many days did the temperature not reach 100°F and rain did not fall?

50. Professional basketball. During the 1980s (10 seasons), the Los Angeles Lakers played in the National Basketball Association's (NBA) championship eight times, the Boston Celtics played five times, and the teams played each other three times. (*Source*: NBA.com.)

  a) How many times did the Lakers or the Celtics play in the NBA championship?

  b) How many times did the Lakers play in the NBA championship, but not against the Celtics?

  c) How many times did neither team play in the championships?

Business and Economics

51. Shopping habits. The Middlesex Shopping Center includes a grocery store and a drug store. One afternoon, a total of 200 people visited the shopping center, and 125 shopped at the grocery store, 94 shopped at the drug store, and 31 shopped at both stores.

  a) How many people shopped at the grocery store or the drug store?

  b) How many people shopped at neither the grocery store nor the drug store?

52. Shopping habits. The Coffee Bean Company had 95 customers one morning, with 70 purchasing coffee, 54 purchasing a pastry, and 37 purchasing both coffee and a pastry.

a) How many customers purchased coffee or a pastry?
b) How many customers purchased neither coffee nor a pastry?

53. Travel. A survey of 50 travelers showed that, during the previous year, 20 had been to Los Angeles, 32 had been to New York, and 11 had been to both cities.

a) How many travelers had been to Los Angeles or New York in the previous year?
b) How many travelers had been to New York but not to Los Angles in the previous year?
c) How many travelers had not been to either city in the previous year?

54. Tire conditions. A random inspection of 100 vehicles showed that 65 had tires with low air-pressure and 42 had tires with a bald tread. Furthermore, 17 had tires with both low pressure and bald tread.

a) How many vehicles had tires with low pressure or bald tread?
b) How many vehicles had tires with good tread?
c) How many vehicles had tires with good tread and good air pressure?

SYNTHESIS

55. What must be true about A and B for $n(A) + n(B) = n(A \cup B)$ to be true?

56. What must be true about A and B for $n(A) + n(B) = n(B)$ to be true?

57. If $A \subseteq B$ and $B \subseteq A$, then what must be true about A and B?

58. What is another name for U'?

deMorgan's Laws. *Given two sets A and B, deMorgan's Laws are* $(A \cup B)' = A' \cap B'$ *and* $(A \cap B)' = A' \cup B'$. *These laws allow us to "distribute" the complement to the sets within the parentheses, switching* $\cup$ *for* $\cap$, *and vice versa. Assume that* $U = \{1, 2, 3, 4, 5, 6, 7, 8, 9, 10\}$, $A = \{1, 3, 5, 7, 9\}$, $B = \{2, 3, 4, 5, 6, 7\}$ *and* $C = \{4, 5, 7, 8, 10\}$.

59. Verify $(A \cup B)' = A' \cap B'$ by finding $(A \cup B)'$ and $A' \cap B'$ and comparing the results.

60. Verify $(A \cap B)' = A' \cup B'$ by finding $(A \cap B)'$ and $A' \cup B'$ and comparing the results.

61. Find $(A \cup B \cup C)'$.

62. Find $(A \cap B \cap C)'$.

63. Find $[(A' \cap B)' \cup C']'$.

64. Find $[B' \cap (A \cup C')']'$.

65. Business: travel. One hundred tourists visiting the Navajo Nation were surveyed. Three of the top tourist attractions in the area are Four Corners, Betatakin Ruins, and Monument Valley, and the data from the survey are displayed in the Venn diagram, which shows the number of people who had visited each site (or a combination of sites):

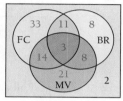

a) How many people visited at least one of the tourist attractions?
b) How many people visited the Betatakin Ruins and Monument Valley but not Four Corners?
c) How many people visited Monument Valley or Four Corners and the Betatakin Ruins?
d) How many people visited none of the three locations?

Four Corners, where Arizona, New Mexico, Colorado, and Utah meet at a single point.

Answers to Quick Checks

1. (a) $\{1\}$; **(b)** $\varnothing$; **(c)** $\{x \mid x \text{ is a real number}\}$ **2. (a)** true; **(b)** false, $\{2\} \subseteq \{2, 3, 4, 5, 6\}$ or $2 \in \{2, 3, 4, 5, 6\}$ **3. (a)** $L = N, K = K, L = L, M = M, N = N$; **(b)** $L \subseteq K$, $M \subseteq L, M \subseteq K, M \subseteq N, N \subseteq L, N \subseteq K, K \subseteq K, L \subseteq L, M \subseteq M$, $N \subseteq N$. **4.** $C' = \{1, 2, 3, 7, 8, 9, 10\}$ **5.** $M \cup N = \{1, 2, 3, 4, 5, 6, 7, 8, 9\}, M \cap N = \{4, 5\}$ **6.** $K' = \{c, o, p\}$; this is the set of toppings Katie did not choose—cucumber, olives, and pickles. **7.** $n(A \cup B) = 116$ **8.** $n(K') = 3$; this indicates that Katie did not choose 3 toppings from the set of all possible choices.

10.2

OBJECTIVES

- Find the sample space of an experiment.

- Find the theoretical probability of an event, the probability of the complement of an event, and the probability of the union of two events, using mathematical means only.

- Find the experimental probability of an event using data and frequency tables.

- Solve applied problems involving probability.

Probability

Probability is the branch of mathematics devoted to the calculation of the likelihood of events. It had its origins in the mid-17th century, in the efforts of mathematicians such as Blaise Pascal, Pierre Fermat, and Jakob Bernoulli. The earliest questions in probability theory focused on betting or wagering, as gamblers tried to understand the likelihood of various outcomes in games of chance. Despite its humble beginnings, probability has important applications in the sciences, industry, and business.

Preliminary Definitions

In probability, an **experiment** can be any action, and its results are called **outcomes**. Each performance of an experiment is known as a **trial**. The table below shows some possible outcomes for various experiments.

Experiment	Possible Outcomes
Toss a coin	Tails, heads
Roll a single die	1, 2, 3, 4, 5, 6
Draw a card from a standard deck of playing cards	Hearts (♥): ace, 2, 3, 4, 5, 6, 7, 8, 9, 10, jack, queen, king; Diamonds (♦): ace, 2, 3, 4, 5, 6, 7, 8, 9, 10, jack, queen, king; Clubs (♣): ace, 2, 3, 4, 5, 6, 7, 8, 9, 10, jack, queen, king; Spades (♠): ace, 2, 3, 4, 5, 6, 7, 8, 9, 10, jack, queen, king
One spin of a roulette wheel	0, 00, 1, 2, 3, 4, 5, 6, 7, . . . , 36
Shooting free throws in basketball	Shot is made; shot is missed.

In *experimental probability*, an experiment is performed many times, and the probabilities of the outcomes are generated using actual data from these trials. For example, the probability of a basketball player making a free throw attempt is based on his or her past shooting performance. If a player has been successful on 200 of the last 250 free throw attempts, we can conclude that the player has a probability of 80% (200 divided by 250) of making the next free throw he or she attempts.

In *theoretical probability*, the probabilities are determined by mathematical reasoning. We can calculate the probability of a tossed coin showing heads, or the probability of rolling two dice and getting two sixes, using mathematical reasoning. In some situations, we can verify theoretical probabilities by performing the experiment.

For any experiment, the set of all possible outcomes is called the **sample space**, denoted S. Any subset of the sample space is called an **event** (or *event space*). Thus, we can think of the sample space as being the universe of possible outcomes for an experiment. However, we must be careful when defining the sample space for an experiment. For example, for rolling two dice, we could define the sample space to be the sums of two dice, $S = \{2, 3, 4, 5, 6, 7, 8, 9, 10, 11, 12\}$, but from experience, we know that some sums are more likely to occur than others. Whenever possible, we define the sample space so that all outcomes are equally likely. Thus, a preferable sample space for rolling two dice is a list of the possible outcomes as ordered pairs: $S = \{(1,1), (1,2), (1,3), (1,4), (1,5), (1,6), (2,1), (2,2), (2,3), \ldots, (6,6)\}$ (see Example 2).

Theoretical Probability: A Mathematical Approach

What is the probability of a tossed coin showing heads? A coin has two sides, and we assume that each side is equally likely to be facing up after the coin has landed. Since heads is one of two possible sides, we conclude that the probability of a coin showing heads is 1 of the 2 possibilities, or $\frac{1}{2}$, or 50%. This leads us to a definition of *theoretical probability*.

DEFINITION

In an experiment, the **theoretical probability** that event E occurs is

$$P(E) = \frac{n(E)}{n(S)},$$

where $n(E)$ is the number of outcomes (elements) for the event E and $n(S)$ is the number of outcomes (elements) in the sample space S.

Recall that the outcomes contained in the sample space S should be equally likely. For example, when a coin is tossed once, its sample space is $S = \{T, H\}$, where T represents the outcome of a tail and H represents the outcome of a head. The word *fair* means that no outcome in the sample space is more likely to occur than another when a trial is performed. Thus, when tossing a fair coin, we assume that the coin shows a head or a tail as equally likely outcomes.

■ **EXAMPLE 1**   A single fair die is rolled. What is the probability of rolling a 2?

Solution   Each outcome of rolling a fair die is equally likely. The sample space is $S = \{1, 2, 3, 4, 5, 6\}$, and the event space for rolling a 2 is $E = \{2\}$. Thus, the probability of rolling a 2 is

$$P(E) = \frac{n(E)}{n(S)} = \frac{1}{6}.$$

> **Quick Check 1**
>
> A single fair die is rolled. What is the probability of rolling an even number?

❬ Quick Check 1

■ **EXAMPLE 2**   Two fair dice are rolled. Let event E be a sum of 12 for the two dice and event F be a sum of 7. Find the following.

a) $P(E)$, the probability that the sum of the two dice is 12

b) $P(F)$, the probability that the sum of the two dice is 7

Solution   To find the sample space for this experiment, we list every possible combined roll of two dice as ordered pairs, with the first number representing the number on the first die and the second number representing the number on the second die:

$$S = \left\{ \begin{array}{cccccc}
(1,1), & (2,1), & (3,1), & (4,1), & (5,1), & (6,1), \\
(1,2), & (2,2), & (3,2), & (4,2), & (5,2), & (6,2), \\
(1,3), & (2,3), & (3,3), & (4,3), & (5,3), & (6,3), \\
(1,4), & (2,4), & (3,4), & (4,4), & (5,4), & (6,4), \\
(1,5), & (2,5), & (3,5), & (4,5), & (5,5), & (6,5), \\
(1,6), & (2,6), & (3,6), & (4,6), & (5,6), & (6,6)
\end{array} \right\}.$$

There are $n(S) = 36$ possible outcomes.

❭ **Quick Check 2**

Using the sample space in Example 2, find the probability of rolling (a) a sum of 3 and (b) a sum of 5. (c) Is it more likely to roll a sum of 4 than a sum of 10?

a) Event $E = \{(6, 6)\}$, so we have $n(E) = 1$. Therefore, the probability of rolling a sum of 12 is $P(E) = \dfrac{n(E)}{n(S)} = \dfrac{1}{36}$.

b) Event $F = \{(1, 6), (2, 5), (3, 4), (4, 3), (5, 2), (6, 1)\}$, so we have $n(F) = 6$. Therefore, the probability of rolling a sum of 7 is $P(F) = \dfrac{n(F)}{n(S)} = \dfrac{6}{36}$, or $\dfrac{1}{6}$.

❬ Quick Check 2

The theoretical probabilities found in Examples 1 and 2 can be demonstrated by performing the actual experiments. If dice are not available, a random-number generator can be used to simulate these (and other) experiments.

TECHNOLOGY CONNECTION

Exploratory

Generating Random Numbers

On the TI-83/84, press **MATH** and then select PRB and option 1: rand. Press **ENTER**, and a random number between 0 and 1 is generated. Press **ENTER** again, and another random number is generated, as shown below.

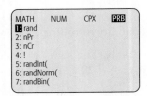
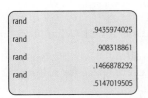

We can use random numbers to simulate certain experiments. For example, the single die experiment in Example 1 can be simulated using the command int(6*rand)+1, which randomly generates the integers 1, 2, 3, 4, 5, and 6.

Generate 30 random integers between 1 and 6. Let n be the number of times a 2 is generated. How close is $n/30$ to $1/6$? What happens if you generate 100 random numbers, and compare $n/100$ to $1/6$? How might you use the random number generator to simulate the experiment in Example 2?

Random numbers can also be generated with a spreadsheet such as Excel™. This is demonstrated in the Extended Technology Application at the end of this chapter.

Properties of Probability

If S is a sample space and E and F are two events, then the following properties are true:

> **THEOREM 2   Properties of Probability**
>
> **P1.** For any event E, $0 \le P(E) \le 1$.
>
> **P2.** For any event E, $P(E) + P(E') = 1$.
>
> **P3.** For any events E and F, $P(E \cup F) = P(E) + P(F) - P(E \cap F)$.

Property P1 states that all probabilities are nonnegative values between 0 and 1. If $P(E) = 0$, then E is an *impossible* event, and if $P(E) = 1$, then E is a *certain* event.

Property P2 states that the probability of an event E or its complement E' is certain; that is, event E will occur, or it will not occur. Property P2 is often written $P(E) = 1 - P(E')$. The proofs of Properties P1 and P2 are left as Exercises 43 and 44.

Property P3 is the probability of the union of two events. Recall the cardinality of the union formula from Section 10.1:

$$n(E \cup F) = n(E) + n(F) - n(E \cap F).$$

Multiplying both sides by $\dfrac{1}{n(S)}$, we have

$$\frac{n(E \cup F)}{n(S)} = \frac{n(E)}{n(S)} + \frac{n(F)}{n(S)} - \frac{n(E \cap F)}{n(S)},$$

from which Property P3 follows.

■ **EXAMPLE 3** A single fair die is rolled. Let event E be that the die shows a 3 and event F be that the die shows a 7. Find the following probabilities.

a) $P(E)$ and $P(E')$ **b)** $P(F)$ and $P(F')$

Solution

a) Event $E = \{3\}$, so $P(E) = \frac{1}{6}$. Its complement is $E' = \{1, 2, 4, 5, 6\}$, so $P(E') = \frac{5}{6}$. Note that $P(E) + P(E') = 1$: it is certain that a die rolled once will show a 3 or will not show a 3.

b) Since 7 is not an element of the sample space S, the event F is empty. It is impossible to get a 7 on a single roll of one die, so $P(F) = 0$. However, it is certain that a roll of a die will *not* show a 7. Thus, $P(F') = 1$.

> **Quick Check 3**

Two fair dice are rolled. Let event E be that the sum is 12 and event F be that the sum is 20. Find $P(E)$, $P(E')$, $P(F)$, and $P(F')$.

❰ Quick Check 3

■ **EXAMPLE 4** A jar contains 7 black balls, 6 yellow balls, 4 green balls, and 3 red balls, all the same size and weight. The jar is shaken well, and then 1 ball is randomly selected. Let event B be that the ball is black, event Y be that the ball is yellow, event G be that the ball is green, and event R be that the ball is red.

a) What is $P(R)$, the probability that the ball is red?

b) What is $P(R')$, the probability that the ball is not red?

c) What is $P(Y \cup B)$, the probability that the ball is yellow or black?

Solution

a) There are 20 balls altogether and of these 3 are red, so the probability that the ball is red is $P(R) = \frac{3}{20}$.

b) Since the probability of drawing a red ball is $P(R) = \frac{3}{20}$, the probability that the ball is not red is $P(R') = 1 - P(R) = 1 - \frac{3}{20} = \frac{17}{20}$.

c) The probability of drawing a yellow ball is $P(Y) = \frac{6}{20}$, and the probability of drawing a black ball is $P(B) = \frac{7}{20}$. Note that it is impossible to draw a ball that is yellow and black simultaneously, so that these two events are mutually exclusive, or $P(Y \cap B) = 0$. Thus, the probability that the ball is yellow or black is $P(Y \cup B) = P(Y) + P(B) - P(Y \cap B) = \frac{6}{20} + \frac{7}{20} - 0 = \frac{13}{20}$.

> **Quick Check 4**

Repeat the experiment in Example 4.

a) What is $P(B \cup G)$, the probability that the ball selected at random is black or green?

b) What is $P(G')$, the probability that the ball selected at random is not green?

❰ Quick Check 4

■ **EXAMPLE 5** Two fair dice are rolled. Let event E be that the sum is 5 and event F be that at least one die shows a 1. Find $P(E \cup F)$.

Solution We have

$$E = \{(1, 4), (2, 3), (3, 2), (4, 1)\},$$
$$F = \{(1, 1), (1, 2), (1, 3), (1, 4), (1, 5), (1, 6), (2, 1), (3, 1), (4, 1), (5, 1), (6, 1)\},$$

and

$$E \cap F = \{(1, 4), (4, 1)\}.$$

Thus, $n(E) = 4, n(F) = 11$, and $n(E \cap F) = 2$. Therefore, the probability that a roll of two dice results in a sum of 5 or that at least one die shows a 1 is

$$P(E \cup F) = P(E) + P(F) - P(E \cap F)$$

$$= \tfrac{4}{36} + \tfrac{11}{36} - \tfrac{2}{36} \qquad \text{Substituting}$$

$$= \tfrac{13}{36}.$$

The following display of the sample space shows that 13 elements are contained within the union of events E and F:

$$S = \left\{ \begin{array}{cccccc}
(1,1), & (2,1), & (3,1), & (4,1), & (5,1), & (6,1), \\
(1,2), & (2,2), & (3,2), & (4,2), & (5,2), & (6,2), \\
(1,3), & (2,3), & (3,3), & (4,3), & (5,3), & (6,3), \\
(1,4), & (2,4), & (3,4), & (4,4), & (5,4), & (6,4), \\
(1,5), & (2,5), & (3,5), & (4,5), & (5,5), & (6,5), \\
(1,6), & (2,6), & (3,6), & (4,6), & (5,6), & (6,6)
\end{array} \right\}.$$

> **Quick Check 5**
>
> In Example 5, let event G be that the sum is 7 and find $P(F \cup G)$.

❬ Quick Check 5

Experimental Probabilities: Frequency Tables

Often, probabilities are found by collecting data. The data can be presented in a **frequency table**, from which the probabilities can be determined. For example, suppose each of the 20 students in a calculus class is asked how many courses he or she is taking. If 5 students are taking one course, 4 are taking two courses, 8 are taking three courses, and 3 are taking four courses, we can create a frequency table in which the top row is the number of courses a student is taking (1, 2, 3, or 4) and the bottom row is the number of students enrolled in that number of courses:

Number of courses a student is taking	1	2	3	4
Number of students (frequency)	5	4	8	3

For a student chosen randomly from this class, we can find probabilities based on the data. For example, a randomly chosen student from this class has a probability of $\tfrac{8}{20}$, or 0.4, of being enrolled in three courses.

■ **EXAMPLE 6**   **Business: Visitation Frequency.**   According to a poll conducted by the Pew Internet Group, the numbers of people who made a given number of visits to a fast-food restaurant within a 1-month period are as shown in the frequency table below. A total of 2512 people were polled. (Source: www.pewinternet.org.)

Number of visits	0	1	2	3	4	5	6
Number of people (frequency)	799	324	354	201	196	163	475

A person is selected at random.

a) Find the probability that the person visited a fast-food restaurant 3 times in the 1-month period.

b) Find the probability the person visited a fast-food restaurant at most 3 times in the 1-month period.

Solution

a) Since 201 people visited a fast-food restaurant 3 times in the 1-month period, the probability that a randomly selected person visited a fast-food restaurant 3 times is $\frac{201}{2512} \approx 0.08$.

b) To find the probability that a randomly selected person visited a fast-food restaurant at most 3 times, we add the probabilities that the person visited a fast-food restaurant 0, 1, 2, or 3 times:

$$P(\text{at most 3 visits}) = P(0 \text{ visits}) + P(1 \text{ visit}) + P(2 \text{ visits}) + P(3 \text{ visits})$$
$$= \frac{799}{2512} + \frac{324}{2512} + \frac{354}{2512} + \frac{201}{2512}$$
$$= \frac{1678}{2512} \approx 0.668.$$

Thus, there is about a 66.8% probability that a randomly selected person visited a fast-food restaurant at most 3 times in a 1-month period.

❬ Quick Check 6

> **❭ Quick Check 6**
>
> Using the data in Example 6, find **(a)** the probability that a randomly selected person visited a fast-food restaurant 2 times in the 1-month period and **(b)** the probability that a randomly selected person visited a fast-food restaurant at most 2 times in the 1-month period.

Section Summary

- An *experiment* is any action whose results are called *outcomes*. Each time the experiment is performed is a *trial*. The set of all possible outcomes of an experiment is the *sample space*, denoted S. An *event* is any subset of S.
- The *theoretical probability* of an event E is

$$P(E) = \frac{n(E)}{n(S)}.$$

- For any event E, $0 \leq P(E) \leq 1$. The probability of an *impossible* event is 0, and the probability of a *certain* event is 1.

- Given an event E and its complement E', the sum of their probabilities is $P(E) + P(E') = 1$.
- The probability of the union of events E and F is $P(E \cup F) = P(E) + P(F) - P(E \cap F)$.
- For experimental probabilities, the data are often presented in *frequency tables*. A frequency value divided by the total number of data points is interpreted as a probability.

EXERCISE SET
10.2

In Exercises 1–18, state all probabilities as reduced fractions.

1. A spinner marked off in four equal-sized regions, numbered 1 through 4, is spun once. Let event E be that the result is an even number and event F be that the result is 3, 2, or 1.

a) Find $P(E)$ and $P(E')$. **b)** Find $P(F)$ and $P(F')$.

2. A dartboard is divided into 9 equal-sized sectors, numbered 1 through 9. A player throws a dart randomly at the dartboard, so that any point on the surface of the dartboard is equally likely to be hit. Let event E be that the dart hits an odd-numbered sector and event F be that the dart hits the sector 4.

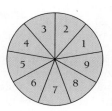

a) Find $P(E)$ and $P(E')$. **b)** Find $P(F)$ and $P(F')$.

3. One card is selected at random from a deck of well-shuffled playing cards. Let event E be that the card is a face card (king, queen, or jack) and event F be that the card is a club. (For a description of a deck of cards, see page 756.)

   a) Find $P(E)$ and $P(E')$.
   b) Find $P(F)$ and $P(F')$.
   c) Find $P(E \cup F)$.

4. One card is selected at random from a deck of well-shuffled playing cards. Let event E be that the card is an ace and event F be that the card is a diamond.

   a) Find $P(E)$ and $P(E')$.
   b) Find $P(F)$ and $P(F')$.
   c) Find $P(E \cup F)$.

5. Two fair dice are rolled. Let event E be that the sum is 10 and event F be that at least one die shows a 5.

   a) Find $P(E)$ and $P(E')$.
   b) Find $P(F)$ and $P(F')$.
   c) Find $P(E \cup F)$.

6. Two fair dice are rolled. Let event E be that the first die shows a 1 and event F be that the second die shows a 3.

   a) Find $P(E)$ and $P(E')$.
   b) Find $P(F)$ and $P(F')$.
   c) Find $P(E \cup F)$.

7. A jar contains 10 red, 8 blue, 7 green, and 5 yellow candies. A candy is randomly selected from the jar. Let event R be that the candy selected is red, event B be that it is blue, event G be that it is green, and event Y be that it is yellow.

   a) Find $P(R)$ and $P(R')$.
   b) Find $P(R \cap Y)$ and $P(R \cup Y)$.
   c) Find $P(R \cap G \cap B)$ and $P(R \cup G \cup B)$.

8. A wallet contains five \$1 bills, two \$5 bills, and one \$10 bill. A bill is randomly selected. Let event E be that a \$1 bill is selected, event F be that a \$5 bill is selected, and event G be that a \$10 bill is selected.

   a) Find $P(E)$ and $P(E')$.
   b) Find $P(E \cap F)$ and $P(E \cup F)$.
   c) Find $P(E \cap F \cap G)$ and $P(E \cup F \cup G)$.

9. Zach is one of five men in his office. Seven women also work in the office. One person is randomly selected to plan the upcoming holiday party. Let event Z be that Zach is selected, event M be that a man is selected, and event W be that a woman is selected.

   a) Find $P(Z)$ and $P(Z')$.
   b) Find $P(M \cap Z)$ and $P(M \cup Z)$.
   c) Find $P(M \cap W)$ and $P(M \cup W)$.

10. Gina is one of 15 women students attending a math lecture. There are 32 students in the lecture room. The professor randomly selects one student to write a solution on the board. Let event G be that Gina is selected, event M be that a man is selected, and event W be that a woman is selected.

   a) Find $P(G)$ and $P(G')$.
   b) Find $P(W \cap G)$ and $P(W \cup G)$.
   c) Find $P(M \cap G)$ and $P(M \cup G)$.

11. A college has 1125 freshmen, 1170 sophomores, 1281 juniors, 1263 seniors, and 290 graduate students. One student is randomly selected to serve on the school's curriculum committee. Let event A be that a freshman is selected, event B be that a sophomore is selected, event C be that a junior is selected, event D be that a senior is selected, and event E be that a graduate student is selected.

   a) Find $P(A)$ and $P(A')$.
   b) Find $P(E)$ and $P(E')$.
   c) Find $P(A \cap B \cap C)$ and $P(A \cup B \cup C)$.

12. Bart is one of 12,000 people at the baseball game. He is sitting in section K, which has 250 people. One person at the game will be chosen at random to receive two free tickets for next Saturday's boat show at the civic center. Let event B be that Bart is selected and event K be that someone from section K is selected.

   a) Find $P(K)$ and $P(K')$.
   b) Find $P(B)$ and $P(B')$.
   c) Find $P(B \cap K)$ and $P(B \cup K)$

13. Two fair coins are tossed simultaneously. Let event E be that at least one tail appears and event F be that both coins show the same side.

   a) Write the sample space S. Use TH to represent "tail-head," and so on.
   b) Find $P(E)$ and $P(E')$.
   c) Find $P(F)$ and $P(F')$.
   d) Find $P(E \cup F)$.

14. Three fair coins are tossed simultaneously. Let event E be that exactly two coins show heads and event F be that at least one coin shows a tail.

   a) Write the sample space S. Use TTH to represent "tail-tail-head," and so on.
   b) Find $P(E)$ and $P(E')$.
   c) Find $P(F)$ and $P(F')$.
   d) Find $P(E \cup F)$.

15. Four fair coins are tossed simultaneously. Let event E be that all four coins show the same side and event F be that at least one coin shows a head.

   a) Write the sample space S. Use TTHT to represent "tail-tail-head-tail," and so on.
   b) Find $P(E)$ and $P(E')$.
   c) Find $P(F)$ and $P(F')$.
   d) Find $P(E \cup F)$.

16. A spinner marked off in five equal-sized regions, numbered 1 through 5, is spun twice. Let event E be that both spins give 1 and event F be that the sum of the spins is 6.

   a) Write the sample space S. Use ordered pairs such as (2, 3), and so on.
   b) Find $P(E)$ and $P(E')$.
   c) Find $P(F)$ and $P(F')$.
   d) Find $P(E \cup F)$.

17. A spinner marked off in three equal-sized regions, numbered 1 through 3, is spun three times. Let event E be that at least one spin gives 2 and event F be that the sum of the spins is 7.

a) Write the sample space S. Use ordered triples such as $(1, 3, 2)$, and so on.
b) Find $P(E)$ and $P(E')$.
c) Find $P(F)$ and $P(F')$.
d) Find $P(E \cup F)$.

18. A spinner marked off in four equal-sized regions, numbered 1 through 4, is spun three times. Let event E be that exactly one spin gives 3 and event F be that the sum of the spins is 12.

a) Write the sample space S. Use ordered triples such as $(4, 1, 3)$, and so on.
b) Find $P(E)$ and $P(E')$.
c) Find $P(F)$ and $P(F')$.
d) Find $P(E \cup F)$.

APPLICATIONS

Business and Economics

19. Souvenir sales. Paul applies to sell souvenirs at a sporting event. He may be assigned to any of sections A through E, or he may be rejected. The table below shows the probabilities of his being assigned to a particular section or being rejected.

Section	Probability
A	0.1
B	0.3
C	0.2
D	0.1
E	0.1
Rejected	0.2

a) What is the probability that Paul is assigned to section B?
b) What is the probability that Paul is not assigned to section B?
c) What is the probability that Paul's application is not rejected?
d) What is the probability that Paul is assigned to sections A, B, or C?

20. Revenue. Cathy is applying for a booth at the local art fair. The table below shows the probabilities of her being assigned to certain locations or being rejected.

Location	Main Street	Baker Street	City Park	Rejected
Probability	0.33	0.21	0.19	0.27

a) What is the probability that her application is accepted?
b) What is the probability that she gets a booth on Main Street or in City Park?
c) What is the probability that she is accepted, but not assigned to Main Street?

21. Revenue. For a week, Contempo Women's Fashions tracks the amounts spent by its customers, with the following results:

Amount spent	Frequency
$0–$39.99	33
$40–$79.99	59
$80–$119.99	92
$120–$159.99	97
$160–$199.99	40
$200 or more	16

a) What is the probability that a randomly chosen customer spent $120 or more?
b) What is the probability that a randomly chosen customer did not spend less than $80?
c) What is the probability that a randomly chosen customer spent between $40 and $159.99?

22. Gas prices. The table below summarizes the price per gallon charged for unleaded gasoline at a number of gas stations around Las Cruces, New Mexico. (*Source:* www.newmexicogasprices.com, April 19, 2012.)

Price	Frequency
$3.50–$3.59	9
$3.60–$3.69	8
$3.70–$3.79	11
$3.80–$3.89	11

a) What is the probability that a randomly selected gas station around Las Cruces, New Mexico, sells unleaded gasoline for at most $3.69 per gallon?
b) What is the probability that a randomly selected gas station sells unleaded gasoline for at least $3.60 per gallon?
c) What is the probability that a randomly selected gas station sells unleaded gasoline for between $3.60 and $3.79 per gallon?

General Interest

23. Two unfair coins, both favoring heads, are tossed simultaneously for many trials, with the following results:

Outcome	Two heads	One head, one tail	Two tails
Frequency	61	103	36

The two coins are tossed again.

a) What is the probability that both coins show heads?
b) What is the probability that at least one of the coins shows heads?
c) What is the probability that the two coins show the same side?

24. Three unfair coins, all favoring heads, are tossed simultaneously for many trials, with the following results:

Outcome	All heads	Two heads	One head	No heads
Frequency	69	113	52	16

The three coins are tossed again.

a) What is the probability that at least one head shows?
b) What is the probability that at least one tail shows?
c) What is the probability of all three coins not showing the same side?

25. Cami plays softball. The following table shows the frequency of the number of hits she has per game.

Number of hits	0	1	2	3	4
Frequency	12	18	11	7	2

a) For a randomly selected game, what is the probability that she gets exactly 2 hits?
b) What is the probability that she gets at most two hits?
c) What is the probability that she gets at least one hit?

26. Devin plays goalkeeper for his soccer team. The following table shows the frequency with which he has given up various numbers of goals per game.

Goals allowed	0	1	2	3	4
Frequency	7	10	12	8	1

a) For a randomly selected game, what is the probability that he does not give up any goals?
b) What is the probability that he gives up at least one goal?
c) What is the probability that he gives up at most two goals?

27. An article contains 1307 words. The frequencies of words of various lengths are given in the table below.

Length (number of letters)	Frequency
1	36
2	224
3	285
4	169
5	174
6	124
7	82
8	73
9	70
10	34
11	17
12	8
13	7
14	0
15	4

A word in the article is randomly chosen. Using the information in the table, find the following probabilities.

a) The probability that the word has 4 letters
b) The probability that the word has 5 or fewer letters
c) The probability that the word has 8 or more letters

28. A randomly chosen paragraph written in French consists of 679 letters, with the following frequencies for the vowels, including Y, in either upper or lower case. (*Source*: Benoit Clairoux, *Les Nordiques de Quebec*, Les Editions de l'Homme, 2001.)

Vowel	A	E	I	O	U	Y
Frequency	55	115	40	33	37	0

A letter is randomly selected from the paragraph. Using the information in the table, find the following probabilities.

a) The probability that the letter E is selected
b) The probability that a vowel (A, E, I, O, U, or Y) is not selected
c) The probability that O, U, or Y is selected
d) Since Y has a frequency of 0 in the table, does this suggest that Y is not used in French?

29. Stanley Cup finals. The National Hockey League's Stanley Cup finals are a best-of-seven series, meaning that the first team to win four games is the champion. The following table shows the numbers of series that lasted for 4, 5, 6, or 7 games since the seven-game series format was adopted, through 2012. (*Source:* www.championshiphistory.com.)

Number of games	4	5	6	7
Frequency	20	17	19	16

a) What is the probability that a series will last exactly 5 games?

b) What is the probability that a series will not last 7 games?

c) What is the probability that one team will not sweep the series (win four games in a row with no losses)?

d) Is this table a reliable indicator of the number of games in future Stanley Cup series? What other factors should be considered when calculating the probability that a series will last a certain number of games?

30. Super Bowl margins of victory. The table below shows the margins of victory (point differentials of the final scores) between the winning and losing teams for all the Super Bowls played through 2012. (*Source:* www.championshiphistory.com.)

Margin of victory	Frequency
1–5	12
6–10	8
11–15	7
16–20	8
21–25	4
26–30	3
31–35	2
36–40	1
41–45	1

a) What is the probability that a game will be decided by 15 or fewer points?

b) What is the probability that a game will be decided by 21 or more points?

c) What is the probability that a game will be decided by between 6 and 25 points?

d) Is this table a reliable indicator of future Super Bowl margins of victory? What other factors should be considered when calculating the probability of a certain margin of victory?

In Exercises 31 and 32, state the probabilities as reduced fractions and as percentages to three decimal places.

31. Mortality tables. The following mortality table shows the number of deaths in various age groups for males born in the United States (out of 100,000 live births, all races), based on mortality rates as of 2007. (*Source:* www.ssa.gov.)

The age groups are given as intervals of the form $[a, b)$, where, for example, $[0, 1)$ means that the child did not survive to his first birthday.

Age group	Number of deaths
$[0, 1)$	738
$[1, 2)$	49
$[2, 3)$	31
$[3, 4)$	24
$[4, 5)$	20
$[5, 6)$	18
$[6, 7)$	16
$[7, 8)$	15
$[8, 9)$	14
$[9, 10)$	10

For a randomly chosen male born in 2007, find the following probabilities:

a) The child survived to his first birthday.

b) The child survived to his third birthday.

c) The child survived to his fifth birthday.

32. Mortality tables. Sir Edmond Halley (1656–1742) constructed a mortality table for the city of Breslau (now Wroclaw, Poland) for the year 1693, based on a sample of 1238 live births. (*Source:* N. Bacaer, *A Short History of Mathematical Population Dynamics*, Springer-Verlag, 2011.)

Age group	Number of deaths
$[0, 1)$	238
$[1, 2)$	145
$[2, 3)$	57
$[3, 4)$	38
$[4, 5)$	28
$[5, 6)$	22
$[6, 7)$	18
$[7, 8)$	12
$[8, 9)$	10
$[9, 10)$	9

For a randomly chosen baby born in 1693 in Breslau, find the following probabilities:

a) The child survived to his or her first birthday.

b) The child survived to his or her fifth birthday.

c) The child survived to his or her tenth birthday.

SYNTHESIS

Probabilities as areas.   *A regulation dartboard has a radius of 225 mm. The central bull's-eye (red circle) has a radius of 6.35 mm, and the secondary bull (green ring) has a radius of 15.9 mm. The outermost red and green band (the double ring) has an outer radius of 170 mm and is 8 mm wide. The inner red and green band (the treble ring) has an outer radius of 107 mm and is 8 mm wide. (Source: www.dartboards.com.)*

Double ring,
outer radius,
r = 170 mm

Outer radius,
r = 225 mm

Treble ring,
outer radius, r = 107 mm

Bull's-eye (innermost red region), r = 6.35 mm
Bull (secondary green band), r = 15.9 mm

For a randomly thrown dart, the probability of hitting a certain region is that region's area divided by the area of the whole dartboard. Assume a dart is thrown randomly at the dartboard, so that any point on the surface is equally likely to be hit, and ignore any instances where the dart misses the dartboard completely. Find the following probabilities to four nonzero decimal places:

33. What is the probability that the dart hits the bull's-eye?

34. What is the probability that the dart hits the bull, but not the bull's-eye?

35. What is the probability that the dart hits within the treble ring?

36. In reality, the bull's-eye is hit more often than the theoretical probability given in Exercise 33. What factors might account for the higher probability at which the bull's-eye is actually hit?

37. Two numbers, x and y, are randomly chosen from the interval $(0, 1)$.

    a) On a square with corners at $(0, 0)$, $(1, 0)$, $(0, 1)$, and $(1, 1)$, shade the region in which the x- and y-coordinates add to 1.5 or more. Based on the area of this shaded region, what is the probability that two randomly selected numbers chosen from the interval $(0, 1)$ add to 1.5 or greater?

    b) Using a calculator as a random-number generator, create 20 pairs of numbers (40 numbers total). How many of the pairs add to 1.5 or more? What is the probability that the numbers add to 1.5 or more?

38. Three numbers, x, y, and z, between 0 and 1 are randomly chosen. What is the probability that the three numbers add to 1 or more?

39. A number between 0 and 1 is chosen at random. What is the probability that the square of this number is less than or equal to 0.4?

40. A number between 0 and 1 is chosen at random. What is the probability that the cube of this number is greater than or equal to 0.65?

41. You toss a coin 10 times, and it shows heads 7 times. Is this enough to claim that the coin is unfair? Suppose you get 70 heads in 100 tosses, or 700 heads in 1000 tosses. When do you decide that the coin is unfair?

42. You roll a pair of dice 360 times, and the sum of 12 appears 8 times. Does this show that the dice are unfair? How would you determine whether the dice are unfair?

43. Prove that for any event E, $0 \le P(E) \le 1$. (*Hint:* Recall that $E \subseteq S$, so that $n(E) \le n(S)$.)

44. Prove that for any event E, $P(E) + P(E') = 1$. (*Hint:* Recall that $E \cup E' = S$.)

Answers to Quick Checks

1. $\frac{3}{6}$, or $\frac{1}{2}$ **2. (a)** $\frac{2}{36}$, or $\frac{1}{18}$; **(b)** $\frac{4}{36}$ or $\frac{1}{9}$; **(c)** they are equally likely events.   **3.** $P(E) = \frac{1}{36}$, $P(E') = \frac{35}{36}$, $P(F) = 0$, $P(F') = 1$
4. (a) $P(B \cup G) = \frac{11}{20}$; **(b)** $P(G') = \frac{16}{20}$ **5.** $P(F \cup G) = \frac{15}{36}$
6. (a) $\frac{354}{2512} = 0.141$; **(b)** $\frac{1477}{2512} = 0.588$

Discrete Probability Distributions

The figure on the left below shows a machine into which coins are dropped through a slot and allowed to fall randomly. Each coin dropped into the machine is considered one trial of the experiment. Over many trials, more coins collect in the middle than on the ends, suggesting a higher probability that a coin will land in a slot toward the middle. As the coins stack up along the bottom, they form a *probability distribution*, which can also be represented by a graph of the probability of an outcome as a function of that outcome. Such a graph is a bar chart called a *histogram*. Here, the coins themselves form bars; the higher the bar, the more likely a coin is to land in that slot.

Discrete Random Variables

We graph probability distributions the same way we graph points. However, in some experiments, the outcomes are not numerical. For example, when a fair coin is tossed once, the outcomes are $S = \{T, H\}$, where T is tails and H is heads. These outcomes are not numbers, so they cannot be plotted easily on a graph. To address this difficulty, we introduce a *random variable*.

> ### DEFINITION
>
> For any sample space S, a **random variable** is a function $X(s)$ that assigns a numerical value to each element s in the sample space S. We write $X(s) = x$, where x is the numerical value assigned to the element s.
>
> If the elements of the sample space can be listed individually, then the sample space is said to be **discrete**, and the random variable is called a **discrete random variable**.

A random variable is a measurement used in an experiment with random outcomes that allows us to read (or "measure") the outcomes of the experiment. For convenience, the notation for a random variable is often shortened to X, instead of $X(s)$. The set of outputs is called the **range** of the random variable.

For example, we can define a random variable X to be the number of heads that show after one toss of a fair coin. We have $X = 0$ (if the coin shows a tail) or $X = 1$ (if the coin shows a head), and the range of X is $\{0, 1\}$. This is illustrated in the figure, along with examples of random variables defined for other experiments.

X: number of heads after one toss of a coin

X: number of heads after one toss of two coin

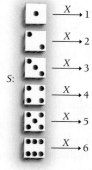

X: number of dots on the top face of a die

■ **EXAMPLE 1** Three fair coins are tossed together. The sample space is $S = \{$TTT, TTH, THT, THH, HTT, HTH, HHT, HHH$\}$. Let the discrete random variable X be the number of heads that show. List the value of X for each element in the sample space, and state the range of X.

Solution   We examine each outcome and count the number of heads.

- The outcome TTT has no heads. Thus, $X(\text{TTT}) = 0$.
- The outcomes TTH, THT, and HTT each have one head. Thus, $X(\text{TTH}) = 1, X(\text{THT}) = 1$, and $X(\text{HTT}) = 1$.
- The outcomes THH, HTH, and HHT each have two heads. Thus, $X(\text{THH}) = 2$, $X(\text{HTH}) = 2$, and $X(\text{HHT}) = 2$.
- The outcome HHH has three heads. Thus, $X(\text{HHH}) = 3$.

The range of X is $\{0, 1, 2, 3\}$.

❮ Quick Check 1

❯ **Quick Check 1**

For each outcome in the sample space S of Example 1, let the discrete random variable W be the number of tails that show. List the value of W for each element in the sample space and state the range of W.

Probability Mass Functions and Distributions

Once a random variable for an experiment is defined, we can match the output values of the random variable with their corresponding probabilities. We do so by creating a function of the form $f(x) = P(X = x)$ called a *probability mass function*. Note that the output values (the range) of the random variable X become the input values (the domain) for the probability mass function f.

> **DEFINITION**
>
> Let X be a discrete random variable. A **probability mass function** is a function f that meets these conditions:
>
> 1. $f(x) \geq 0$, for all x.   (Probabilities are never negative.)
> 2. $f(x) = P(X = x)$.   ($f(x)$ is the probability that the event denoted by x occurs.)
> 3. The sum of all $f(x)$ is 1.   (The sum of all the probabilities is 1.)

Probability mass functions are often written as a set of ordered pairs.

■ **EXAMPLE 2**   Consider the three-coin experiment described in Example 1. Find the probability mass function $f(x) = P(X = x)$, where the discrete random variable X is the number of heads that show.

Solution   Since each coin is fair, each outcome in the sample space, $S = \{\text{TTT, TTH,}$ $\text{THT, THH, HTT, HTH, HHT, HHH}\}$, is equally likely. There are 8 possible outcomes, so each outcome has a probability of $\frac{1}{8}$. Recall that the range of X is $\{0, 1, 2, 3\}$.

- One outcome (TTT) has no heads, so we have $f(0) = P(X = 0) = \frac{1}{8}$.
- Three outcomes (TTH, THT, HTT) have one head, so we have $f(1) = P(X = 1) = \frac{3}{8}$.
- Three outcomes (THH, HTH, HHT) have two heads, so we have $f(2) = P(X = 2) = \frac{3}{8}$.
- One outcome (HHH) has three heads, so we have $f(3) = P(X = 3) = \frac{1}{8}$.

Thus, the probability mass function f can be written as the following set of ordered pairs:

$$f = \left\{\left(0, \tfrac{1}{8}\right), \left(1, \tfrac{3}{8}\right), \left(2, \tfrac{3}{8}\right), \left(3, \tfrac{1}{8}\right)\right\}.$$

The function f is a probability mass function because the function values are never negative, the function values represent probabilities, and the sum of the function values is 1.

❮ Quick Check 2

❯ **Quick Check 2**

Repeat Example 2 assuming that two fair coins are tossed and the discrete random variable X is defined as the number of tails that show. Write the probability mass function as a set of ordered pairs.

The domain of the probability mass function f in Example 2 is all real numbers x. However, for any x-value not in the range of the random variable X, we have $f(x) = 0$.

For example, in Example 2, we have $f(4) = 0$ since $x = 4$ is not in the range of X. We interpret this situation as being impossible (probability of 0): four heads cannot occur when three coins are tossed.

› Quick Check 3

Explain why $g(x) = \dfrac{5 - x}{11}$, for $x = -1, 2, 3, 4, 6$, is not a probability mass function.

■ **EXAMPLE 3** Explain why $f(x) = \frac{1}{10}x$, for $x = 1, 2, 5$, is not a probability mass function.

Solution The given function f is not a probability mass function because the sum of the function values is not 1. We have $f(1) + f(2) + f(5) = \frac{1}{10} + \frac{2}{10} + \frac{5}{10} = \frac{8}{10} \neq 1$.

❰ Quick Check 3

■ **EXAMPLE 4** Find c such that each function is a probability mass function, and rewrite the probability mass function to include the value of c.
  a) $f = \{(-3, 0.4), (-2, 0.2), (-1, 0.1), (0, c)\}$
  b) $g(x) = cx^2$, for $x = 1, 2, 3, 4$

Solution

a) The sum of the function values must equal 1:

$$0.4 + 0.2 + 0.1 + c = 1$$
$$0.7 + c = 1$$
$$c = 0.3. \qquad \text{Solving for } c$$

The probability mass function is $f = \{(-3, 0.4), (-2, 0.2), (-1, 0.1), (0, 0.3)\}$.

b) The sum of the function values must equal 1:

$$g(1) + g(2) + g(3) + g(4) = 1$$
$$c \cdot 1^2 + c \cdot 2^2 + c \cdot 3^2 + c \cdot 4^2 = 1 \qquad \text{Evaluating}$$
$$c + 4c + 9c + 16c = 1$$
$$30c = 1$$
$$c = \tfrac{1}{30}. \qquad \text{Solving for } c$$

› Quick Check 4

Find c such that the function $f = \{(1, c), (2, 2c), (3, 4c)\}$ is a probability mass function.

Note that for $g(x) = \frac{1}{30}x^2$, we have $g(x) \geq 0$ for all x. The probability mass function can be written as $g(x) = \frac{1}{30}x^2$, for $x = 1, 2, 3, 4$, or as a set of ordered pairs, $g = \{(1, \frac{1}{30}), (2, \frac{4}{30}), (3, \frac{9}{30}), (4, \frac{16}{30})\}$.

❰ Quick Check 4

The graph of a probability mass function is represented as a bar chart called a **histogram**. The probability mass function together with its histogram represents a **probability distribution**. A histogram allows us to visualize the probabilities of an experiment and, along with the probability mass function, will vary, depending on the definition of the random variable and the experiment.

■ **EXAMPLE 5** A fair die is rolled. Let the random variable X be the number that ends up on the top face. Define a probability mass function for this experiment, and draw the probability distribution histogram.

› Quick Check 5

A spinner marked off in five equal-sized regions numbered 1 through 5 is spun once. Let the random variable X be the number of the region where the pointer stops. Find the probability mass function, and draw its histogram.

Solution The range of the random variable is $\{1, 2, 3, 4, 5, 6\}$. Since the die is fair, each outcome has an equal probability of $\frac{1}{6}$. Thus, the probability mass function is $f(x) = \frac{1}{6}$, for $x = 1, 2, 3, 4, 5, 6$. Its histogram is as follows:

❰ Quick Check 5

When, as in Example 5, each outcome has the same probability, we say that the distribution is *uniform*. Its histogram has bars of equal height.

■ **EXAMPLE 6**　Two fair dice are rolled. Let the random variable X be the sum of the numbers shown.

a) Write the probability mass function $f(x) = P(X = x)$ for this experiment as a set of ordered pairs, and draw its histogram.

b) Find $P(X \leq 5)$, and interpret its meaning.

Solution　For reference, the sample space S is shown below as ordered pairs representing the numbers on the two dice.

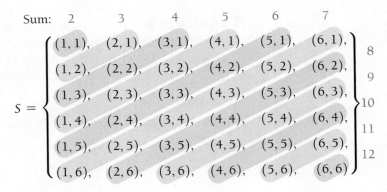

$$\text{Sum:} \quad 2 \qquad 3 \qquad 4 \qquad 5 \qquad 6 \qquad 7$$

$$S = \left\{ \begin{array}{llllll} (1,1), & (2,1), & (3,1), & (4,1), & (5,1), & (6,1), \\ (1,2), & (2,2), & (3,2), & (4,2), & (5,2), & (6,2), \\ (1,3), & (2,3), & (3,3), & (4,3), & (5,3), & (6,3), \\ (1,4), & (2,4), & (3,4), & (4,4), & (5,4), & (6,4), \\ (1,5), & (2,5), & (3,5), & (4,5), & (5,5), & (6,5), \\ (1,6), & (2,6), & (3,6), & (4,6), & (5,6), & (6,6) \end{array} \right\} \begin{array}{l} 8 \\ 9 \\ 10 \\ 11 \\ 12 \end{array}$$

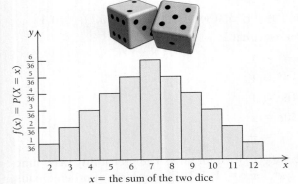

x = the sum of the two dice

a) The range of the random variable X is $\{2, 3, 4, 5, 6, 7, 8, 9, 10, 11, 12\}$. There is one way to roll a sum of 2, so $f(2) = P(X = 2) = \frac{1}{36}$; there are two ways to roll a sum of 3, so $f(3) = P(X = 3) = \frac{2}{36}$, and so on. The probability mass function f, written as a set of ordered pairs, is shown below, with its histogram to the left:

$$f = \left\{ \left(2, \tfrac{1}{36}\right), \left(3, \tfrac{2}{36}\right), \left(4, \tfrac{3}{36}\right), \left(5, \tfrac{4}{36}\right), \left(6, \tfrac{5}{36}\right), \left(7, \tfrac{6}{36}\right), \left(8, \tfrac{5}{36}\right), \left(9, \tfrac{4}{36}\right), \right.$$
$$\left. \left(10, \tfrac{3}{36}\right), \left(11, \tfrac{2}{36}\right), \left(12, \tfrac{1}{36}\right) \right\}.$$

b) The expression $P(X \leq 5)$ represents the probability that the sum of the dice is 5 or less. Thus,

$$P(X \leq 5) = f(2) + f(3) + f(4) + f(5)$$
$$= \tfrac{1}{36} + \tfrac{2}{36} + \tfrac{3}{36} + \tfrac{4}{36} \qquad \text{Using the definitions of } f \text{ from part (a)}$$
$$= \tfrac{10}{36}, \text{ or } \tfrac{5}{18}.$$

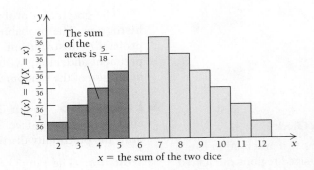

x = the sum of the two dice

The probability of rolling a sum of 5 or less is $\frac{5}{18}$. Note that we are representing the probability as the sum of the areas of the shaded bars.

⟨ Quick Check 6

⟩ **Quick Check 6**

Use the data in Example 6 to find:

a) $P(X \leq 7)$;

b) $P(X \geq 4)$;

c) $P(5 \leq X \leq 10)$.

Cumulative Probability Functions

In Example 6(b), the expression $P(X \leq 5)$ is an example of a *cumulative probability*.

> **DEFINITION**
>
> For a discrete random variable X and a probability mass function $f(x) = P(X = x)$, the **cumulative probability function** of f, labeled F, is defined as
>
> $$F(x) = P(X \leq x).$$
>
> Cumulative probability functions are defined for all x and are increasing functions with a minimum value of 0 and a maximum value of 1.

The following example illustrates how a probability mass function is constructed, and from it, its cumulative probability function.

■ **EXAMPLE 7**   **Business: Visitation Frequency.**   According to a poll conducted by the Pew Internet Group (see Example 6, Section 10.2), the numbers of people who made a given number of visits to a fast-food restaurant within a 1-month period are as shown in the table below. A total of 2512 people were polled. (*Source:* www.pewinternet.org.)

Number of visits, x	0	1	2	3	4	5	6
Number of respondents (frequency)	799	324	354	201	196	163	475

Let the random variable X be the number of visits to a fast-food restaurant in the 1-month period by a randomly selected person who participated in the poll.

a) Find the probability mass function $f(x) = P(X = x)$.

b) Draw the probability distribution histogram.

c) Find the cumulative probability function, and draw its graph.

d) Find the probability that the person visited a fast-food restaurant at most 3 times.

Solution

a) Since 799 respondents did not visit a fast-food restaurant, we have $f(0) = P(X = 0) = \frac{799}{2512} \approx 0.318$. Similarly, $f(1) = P(X = 1) = \frac{324}{2512} \approx 0.129$, and so on. The student can confirm that

$$f = \{(0, 0.318), (1, 0.129), (2, 0.141), (3, 0.08), (4, 0.078), (5, 0.065), (6, 0.189)\}.$$

b) The probability distribution histogram is as follows:

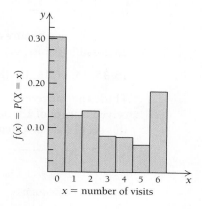

x = number of visits

c) We calculate the cumulative probability $F(x) = P(X \leq x)$ for each x. For example, when $x = 0$, we have $F(0) = P(X \leq 0) = f(0) = 0.318$; when $x = 1$, we have $F(1) = P(X \leq 1) = f(0) + f(1) = 0.318 + 0.129 = 0.447$, and so on. The cumulative probability function, written as a piece-wise function, along with its graph, is shown below.

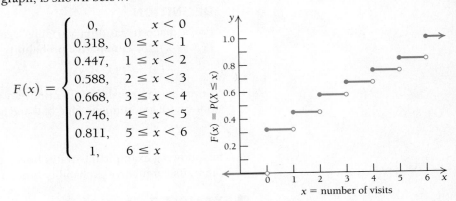

$$F(x) = \begin{cases} 0, & x < 0 \\ 0.318, & 0 \leq x < 1 \\ 0.447, & 1 \leq x < 2 \\ 0.588, & 2 \leq x < 3 \\ 0.668, & 3 \leq x < 4 \\ 0.746, & 4 \leq x < 5 \\ 0.811, & 5 \leq x < 6 \\ 1, & 6 \leq x \end{cases}$$

d) The probability that a randomly selected respondent visited a fast-food restaurant at most 3 times is $F(3) = 0.668$. This is illustrated below in two ways: as the sum of areas of the bars in the probability distribution histogram and as a point on the cumulative probability graph.

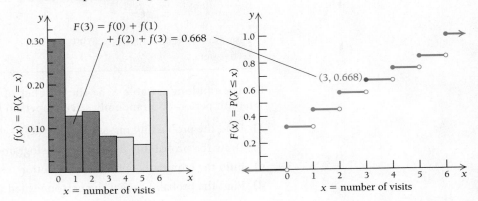

$F(3) = f(0) + f(1)$
$+ f(2) + f(3) = 0.668$

$(3, 0.668)$

> **Quick Check 7**
>
> For the data in Example 7, find and interpret the meaning of $P(X \leq 5)$.

❮ Quick Check 7

The Mean of a Discrete Random Variable

Recall that the *arithmetic mean* of a set of numbers is the sum of the numbers, divided by the number of values in the set. For example, the mean of 1, 3, 5, and 8 is

$$\mu = \frac{1 + 3 + 5 + 8}{4} = 4.25,$$ where the Greek letter μ (mu) represents the mean.

Suppose a botany class took a quiz; let the random variable X be the score (out of 20 possible points) each student received. The scores on the quiz are

$13, 15, 15, 17, 16, 17, 18, 20, 20, 13, 15, 14, 16, 17, 17, 13, 16, 18, 19, 20, 20, 13, 16, 17, 15.$

Thus, the range of X is $\{13, 14, 15, 16, 17, 18, 19, 20\}$. Note that four students received a score of 13, one student received a score of 14, and so on. We can summarize this set of data in a frequency table:

x	13	14	15	16	17	18	19	20
Frequency	4	1	4	4	5	2	1	4

The probabilities $P(X = x)$ are found by dividing the frequency of each data point by 25, the number of data points (students in the class). We have

x	13	14	15	16	17	18	19	20
$P(X = x)$	$\frac{4}{25}$	$\frac{1}{25}$	$\frac{4}{25}$	$\frac{4}{25}$	$\frac{5}{25}$	$\frac{2}{25}$	$\frac{1}{25}$	$\frac{4}{25}$

Therefore, the probability mass function is

$$f = \left\{ \left(13, \tfrac{4}{25}\right), \left(14, \tfrac{1}{25}\right), \left(15, \tfrac{4}{25}\right), \left(16, \tfrac{4}{25}\right), \left(17, \tfrac{5}{25}\right), \left(18, \tfrac{2}{25}\right), \left(19, \tfrac{1}{25}\right), \left(20, \tfrac{4}{25}\right) \right\}.$$

The mean score on the quiz can be calculated as follows:

$$\mu = \frac{13 \cdot 4 + 14 \cdot 1 + 15 \cdot 4 + 16 \cdot 4 + 17 \cdot 5 + 18 \cdot 2 + 19 \cdot 1 + 20 \cdot 4}{25}$$

$$= 13 \cdot \tfrac{4}{25} + 14 \cdot \tfrac{1}{25} + 15 \cdot \tfrac{4}{25} + 16 \cdot \tfrac{4}{25} + 17 \cdot \tfrac{5}{25} + 18 \cdot \tfrac{2}{25} + 19 \cdot \tfrac{1}{25} + 20 \cdot \tfrac{4}{25}$$

$$= 16.4.$$

In the second line of the above calculation, $\frac{4}{25}$ is the probability that a randomly selected student scored 13 on the quiz, $\frac{1}{25}$ is the probability that a randomly selected student scored 14 on the quiz, and so on. Thus, the mean score is found by multiplying each score x by its probability $f(x)$, and adding these products.

DEFINITION

For a discrete random variable X and a probability mass function f, the **mean** μ of X is given by

$$\mu = x_1 \cdot f(x_1) + x_2 \cdot f(x_2) + x_3 \cdot f(x_3) + \cdots + x_n \cdot f(x_n)$$

$$= \sum_{i=1}^{n} x_i \cdot f(x_i).$$

Note that the mean does not have to be in the range of the random variable. In the botany class, it is impossible (probability of 0) for any single student to score exactly 16.4 on the quiz. However, we expect that, on average, students in the botany class scored about 16.4 on the quiz.

■ **EXAMPLE 8**   Find the mean of the random variable in Example 7.

Solution   Recall that the probability mass function is

$$f = \left\{ (0, 0.318), (1, 0.129), (2, 0.141), (3, 0.08), (4, 0.078), (5, 0.065), (6, 0.189) \right\}.$$

Therefore, the mean is

$$\mu = 0 \cdot 0.318 + 1 \cdot 0.129 + 2 \cdot 0.141 + 3 \cdot 0.08 + 4 \cdot 0.078 + 5 \cdot 0.065 + 6 \cdot 0.189.$$

$$= 2.422.$$

Thus, on average, a person who participated in the poll visited a fast-food restaurant about 2.422 times in the 1-month period.

❭ **Quick Check 8**

A random variable X has the probability mass function

$$f(x) = \frac{x}{19}, \text{ for } x = 3, 7, 9.$$

Find the mean of X.

❰ Quick Check 8

Section Summary

- A *random variable* is a function $X(s)$ that assigns a numerical value x to each element s in a sample space S. If the elements in S can be listed, then the sample space is *discrete*, and the random variable is called a *discrete random variable*. The set of output values x is the *range* of the random variable.
- For a discrete random variable X, a *probability mass function* is a function f that meets the following conditions:
  1. $f(x) \geq 0$, for all x.
  2. $f(x) = P(X = x)$.
  3. The sum of all $f(x)$ is 1.
- The graph of a probability mass function is usually presented as a bar chart called a *histogram*. A probability

mass function along with its graph represents a *probability distribution*.
- For a discrete random variable X and a probability mass function $f(x) = P(X = x)$, the *cumulative probability function F* is defined as $F(x) = P(X \leq x)$. It is an increasing function with a minimum value of 0 and a maximum value of 1.
- For a discrete random variable X and a probability mass function f, the *mean of X* is

$$\mu = x_1 \cdot f(x_1) + x_2 \cdot f(x_2) + x_3 \cdot f(x_3) + \cdots + x_n \cdot f(x_n)$$

$$= \sum_{i=1}^{n} x_i \cdot f(x_i).$$

EXERCISE SET
10.3

1. A spinner marked off in four equal-sized regions, numbered 1, 2, 3, and 4, is spun once. Let the random variable X be the number of the region where the pointer stops, and define a probability mass function $f(x) = P(X = x)$.

   a) State the range of X.
   b) Write f as a set of ordered pairs.
   c) Draw the probability distribution as a histogram.
   d) Find $f(3)$, and explain what this number represents.
   e) Find μ, and explain what this number represents.

2. A spinner marked off in four equal-sized regions, numbered 1, 2, 3, and 4 is spun twice. Let the random variable X be the sum of the numbers of the regions where the pointer stops, and define a probability mass function $f(x) = P(X = x)$.

   a) State the range of X.
   b) Write f as a set of ordered pairs.
   c) Draw the probability distribution as a histogram.
   d) Find $f(5)$, and explain what this number represents.
   e) Find μ, and explain what this number represents.

3. A spinner marked off in four equal-sized regions, numbered 1, 2, 3, and 4 is spun twice. Let the random variable X be the positive difference between the numbers of the regions where the pointer stops, and define a probability mass function $g(x) = P(X = x)$.

   a) State the range of X.
   b) Write g as a set of ordered pairs.
   c) Draw the probability distribution as a histogram.
   d) Find $g(1)$, and explain what this number represents.
   e) Find μ, and explain what this number represents.

4. Two fair dice are thrown. Let the random variable X be the positive difference between the values shown on the two dice, and define a probability mass function $g(x) = P(X = x)$.

   a) State the range of X.
   b) Write g as a set of ordered pairs.
   c) Draw the probability distribution as a histogram.
   d) Find $g(4)$, and explain what this number represents.
   e) Find μ, and explain what this number represents.

In Exercises 5–18, state whether each function is a probability mass function or not. If not, explain why not.

5. $f = \{(0, 0.5), (2, 0.3), (3, 0.1), (5, 0.1)\}$

6. $g = \{(1, 0.4), (2, 0.25), (6, 0.15), (10, 0.2)\}$

7. $f = \{(-3, 0.3), (-2, 0.4), (0, 0.3)\}$

8. $g = \{(-4, 0.1), (-3, 0.2), (-2, 0.3), (-1, 0.1)\}$

9. $p = \{(1, 0.4), (4, 0.2), (5, 0.2), (7, 0.3)\}$

10. $q = \{(0, \frac{1}{2}), (1, \frac{1}{3}), (2, \frac{1}{4})\}$

11. $p = \{(10, 0.7), (11, 0.4), (12, -0.1)\}$

12. $q = \{(-5, 0.4), (-4, 0.4), (-3, 0.4), (-2, -0.2)\}$

13. $f(x) = \frac{1}{10}x$, for $x = 1, 2, 3, 4$

14. $g(x) = \frac{1}{17}x$, for $x = 2, 3, 5, 7$

15. $f(x) = \frac{1}{6}x^2$, for $x = -1, 0, 1, 2$

16. $g(x) = \frac{1}{153}x^3$, for $x = 1, 3, 5$

17. $h(x) = \frac{1}{5}x$, for $x = -1, 1, 2, 3$

18. $f(x) = \frac{1}{20}x$, for $x = 1, 3, 6, 8$

In Exercises 19–30, determine the value of c such that the function is a probability mass function.

19. $f = \{(1, 0.2), (2, 0.15), (3, 0.38), (4, c)\}$

20. $g = \{(3, 0.1), (4, c), (5, 0.45), (6, 0.19)\}$

21. $f = \{(-2, c), (-1, 0.25), (0, 3c), (1, 0.4)\}$

22. $g = \{(15, 0.19), (17, 0.5c), (19, 0.3c), (21, 0.27)\}$

23. $f = \{(-1, c), (0, c), (1, 2c)\}$

24. $g = \{(4, 3c), (6, 2c), (8, c), (10, c)\}$

25. $g(x) = cx, \quad x = 2, 3, 4$

26. $h(x) = cx^2, \quad x = -1, 1, 2, 3, 4$

27. $g(x) = c\sqrt{x}, \quad x = 1, 4, 9, 16$

28. $f(x) = c(x - 1), \quad x = 2, 3, 4, 5, 6$

29. $h(x) = cx(x - 1), \quad x = 2, 4, 6, 8$

30. $f(x) = cx^2(x - 1), \quad x = 2, 3, 4, 5$

31. Let $f(x) = \dfrac{c}{x}$, for $x = 1, 3, 5, 7$, be a probability mass function.

   a) Find c.   **b)** Find $f(3)$.
   c) Find $f(4)$.   **d)** Find μ.

32. Let $f(x) = \dfrac{c}{x^2}$, for $x = 1, 2, 3, 4$, be a probability mass function.

   a) Find c.   **b)** Find $f(2)$.
   c) Find $f(5)$.   **d)** Find μ.

33. Let $f(x) = \frac{1}{15}x^2$, for $x = -2, -1, 1, 3$, be a probability mass function.

   a) Find the cumulative probability function F, and draw its graph.

   b) Is it true that $F(2) = 0$? Why or why not?

34. Let $g(x) = \frac{1}{6}\sqrt{x}$, for $x = 1, 4, 9$, be a probability mass function.

   a) Find the cumulative probability function G, and draw its graph.

   b) Is it true that $G(10) = 0$? Why or why not?

35. A spinner marked off in four equal-sized regions, numbered 1, 2, 3, and 4, is spun twice. Let the random variable X be the higher value resulting from the two spins (or the common value if both spins have the same result).

   a) Write the probability mass function $f(x) = P(X = x)$ as a set of ordered pairs.

   b) Find the cumulative probability function F, and draw its graph.

   c) Find $f(2)$ and $F(2)$. What does each number represent?

   d) Find $1 - f(2)$ and $1 - F(2)$. What does each number represent?

36. A spinner marked off in four equal-sized regions, numbered 1, 2, 3, and 4, is spun twice. Let the random variable X be the product of the outcomes of the two spins.

   a) Write the probability mass function $g(x) = P(X = x)$ as a set of ordered pairs.

   b) Find the cumulative probability function G, and draw its graph.

   c) Find $g(2)$ and $G(2)$. What does each number represent?

   d) Find $1 - g(2)$ and $1 - G(2)$. What does each number represent?

APPLICATIONS

Business and Economics

37. Price of an app. Fifty people were asked how much they paid for the last iPhone app they purchased. The following data were collected:

Price, x	Number of people
$1.00	18
$2.00	17
$3.00	9
$4.00	3
$5.00	2
$6.00	0
$7.00	1

Let the random variable X be the price paid for the app, and let $f(x) = P(X = x)$ be a probability mass function based on the result of this survey.

   a) Write f as a set of ordered pairs, and draw the probability distribution as a histogram.

   b) Find the cumulative probability function F.

   c) Find $f(2)$ and $F(2)$, and explain what each number represents.

   d) Find $1 - f(2)$ and $1 - F(2)$, and explain what each number represents.

   e) Find μ, and explain what this number represents.

38. Transportation. Steve's Airport Shuttle runs a van that holds up to 9 passengers and leaves the civic center every hour for the airport. The table below shows the frequencies of the various numbers of passengers for the last 50 trips. Let the random variable X be the number of passengers on the shuttle, with the probability mass function $f(x) = P(X = x)$.

Passengers, x	Frequency
0	2
1	5
2	5
3	12
4	8
5	7
6	4
7	3
8	2
9	2

a) Write f as a set of ordered pairs, and draw the probability distribution as a histogram.
b) Find the cumulative probability function F.
c) Find $f(5)$ and $F(5)$, and explain what each number represents.
d) Find $1 - f(5)$ and $1 - F(5)$, and explain what each number represents.
e) Find μ, and explain what this number represents.

39. Credit rating. An individual's credit rating is a score that ranges up to 850. The distribution of these scores for the United States as of April 2010 is given in the following table, where each score x represents the midpoint of a 50-point subdivision. For example, 675 represents all scores from 650 to 699. (*Source:* www.money-zine.com.)

Score, x	Percentage
475	6.9%
525	9.0%
575	9.6%
625	9.5%
675	11.9%
725	15.7%
775	19.5%
825	17.9%

Let the random variable X be the credit score, and let $g(x) = P(X = x)$ be a probability mass function based on these data.

a) Find $g(675)$, and explain what this number represents.
b) Find the cumulative probability function G.
c) Find $G(675)$, and explain what this number represents.
d) Find μ, and explain what this number represents.

40. Transportation. A ferry transports vehicles across the Franklin River once each hour. Let the random variable X be the number of vehicles on the ferry during a passage, where the ferry can carry a minimum of zero vehicles and a maximum of five. A traffic engineer calculates the probability mass function to be $f(x) = \frac{1}{21}(x + 1)$, for $x = 0, 1, 2, 3, 4, 5$.

a) Find $f(3)$, and explain what this number represents.
b) Find the cumulative probability function F.
c) Find the probability that 4 or fewer cars are on the ferry during a passage.
d) Find μ, and explain what this number represents.

41. Visitation frequency. The Pew Internet Group poll (see Example 7) asked the same 2512 people how often they visited a coffee shop in the 1-month period. The poll results are shown in the following table. (*Source:* www. pewinternet.org.)

Number of visits, x	Number of respondents (frequency)
0	1367
1	208
2	246
3	136
4	168
5	83
6	304

Let the random variable X be the number of monthly visits to a coffee shop made by a randomly selected respondent, and let $h(x) = P(X = x)$ be a probability mass function based on these poll results.

a) Find $h(4)$, and explain what this number represents.
b) Find $H(4)$, and explain what this number represents.
c) Find $h(1.5)$ and $H(1.5)$, and explain what these numbers represents.
d) Find μ, and explain what this number represents.

42. Visitation frequency. The Pew Internet Group poll (see Example 7) asked the same 2512 people how often they visited a church or temple in the 1-month period. The poll results are shown in the following table. (*Source:* www.pewinternet.org.)

Number of visits, x	Number of respondents (frequency)
0	1168
1	173
2	173
3	113
4	482
5	108
6	295

Let the random variable X be the number of monthly visits to a church or temple made by a randomly selected respondent, and let $k(x) = P(X = x)$ be a probability mass function based on these poll results.

a) Find $k(2)$, and explain what this number represents.
b) Find $K(2)$, and explain what this number represents.
c) Find μ, and explain what this number represents.
d) What might explain the significantly higher number of respondents who attended a church or temple 4 times during the month compared to 3 or 5 times?

SYNTHESIS

Descriptive mean: the expected value. **Consider an experiment with a sample space S of outcomes. If $E_1, E_2, \ldots, E_n$ are mutually exclusive events in S (subsets of S) whose union is S, then the expected value of the random variable X can be defined descriptively by**

$$E(X) = P(E_1) \cdot (\text{value of } E_1) + P(E_2) \cdot (\text{value of } E_2) + \cdots + P(E_n) \cdot (\text{value of } E_n).$$

This value is the same as the mean of X. For example, suppose 100 raffle tickets are sold at $1 each, and the holder of the winning ticket wins a $50 jackpot. Here, the random variable X represents the value of a ticket, where $X = \$49$ if the ticket is the winner ($50 minus the $1 cost) or $X = -\$1$ if the ticket is not the winner. The probability of holding the winning ticket, $\frac{1}{100}$, is multiplied by the value of that ticket, $49, and added to the probability of holding a losing ticket, $\frac{99}{100}$, multiplied by its value, $-$1. Each lottery ticket's expected value is thus

$$E(X) = \frac{1}{100} \cdot \$49 + \frac{99}{100} \cdot (-\$1) = \frac{-\$50}{100} = -\$0.50.$$

Thus, the average loss on a ticket is 50 cents. If you purchased all the tickets, you would win the $50 jackpot, but because you spent $100 to purchase all the tickets, you would lose $50 overall. Thus, each ticket has an expected value of $-\$0.50$.

43. Business: expected profit or loss. Olsen's Tailoring plans to open a new store at an initial cost of $250,000. Olsen's forecasts a 10% chance that in its first year the new store will generate $400,000 in revenue, a 55% chance that it will generate $300,000 in revenue, and a 35% chance that it will generate $150,000 in revenue. What is the expected profit (or loss) for this new store during its first year?

44. Business: insurance premiums. American Express offers single-trip flight insurance for $4.99 per policy, with a $2,000,000 payout in the event of death or dismemberment. The probability of an individual being in an airline accident is about 1 in 5.4 million. What is American Express's expected profit per policy sold? (*Sources*: www.americanexpress.com and www.planecrashinfo.com.)

45. Education: standardized testing. A test has many multiple-choice items, each with five answer choices. A student receives 1 point for every correct answer, and each incorrect answer brings a penalty (loss) of $\frac{1}{4}$ of a point. If Chris guesses on every question, what score can he expect on the test?

46. Expected value. On her daily commute to work, Joelle discovers a short-cut through a construction site. Using this shortcut, she saves 5 min on 80% of workdays. The rest of the days she gets caught in construction traffic and has to wait 1 hr to get through. In the long run, is taking this short-cut saving Joelle time? How much time does she save (or waste) on average?

The dreidel game. **A dreidel is a four-sided wooden top, with one of four letters of the Hebrew alphabet on each face: nun, gimel, hei, or shin. In the dreidel game, players compete for a pot of tokens. A player spins the top and takes one of the following actions, depending on which letter faces up.**

Nun

Gimel

Hei

Shin

- *If nun faces up, the player neither adds nor subtracts tokens from the pot.*
- *If gimel faces up, the player wins the entire pot of tokens.*
- *If hei faces up, the player wins half of the tokens in the pot, rounding up if the number is odd.*
- *If shin faces up, the player adds a token to the pot.*

Assume that each face of the dreidel is equally likely to face upward and that the pot holds 20 tokens. Let the random variable X be the amount of tokens won by a player. Thus, the range of X is $\{-1, 0, 10, 20\}$. Let the probability mass function be $f(x) = P(X = x)$ and the cumulative probability function be $F(x) = P(X \leq x)$.

47. Find $f(0)$ and $F(0)$.

48. Find $f(5)$ and $F(5)$.

49. Find $f(20)$ and $F(20)$.

50. Find μ, and explain what this number represents.

TECHNOLOGY CONNECTION

51. Find c such that $f = \{(1, c), (2, c^2)\}$ is a probability mass function.

52. Find c such that $f = \{(0, c), (1, c^2), (2, c^3)\}$ is a probability mass function.

Answers to Quick Checks

1. $W(TTT) = 3$, $W(TTH) = 2$, $W(THT) = 2$, $W(HTT) = 2$, $W(THH) = 1$, $W(HTH) = 1$, $W(HHT) = 1$, $W(HHH) = 0$. The range of W is $\{0, 1, 2, 3\}$. **2.** $f = \{(0, \frac{1}{4}), (1, \frac{1}{2}), (2, \frac{1}{4})\}$
3. $g(6) = -\frac{1}{11}$, and function values cannot be negative.
4. $c = \frac{1}{7}$ **5.** $f(x) = \frac{1}{5}$, for $x = 1, 2, 3, 4, 5$;

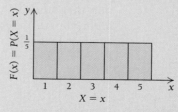

6. **(a)** $\frac{21}{36}$, or $\frac{7}{12}$; **(b)** $\frac{33}{36}$, or $\frac{11}{12}$; **(c)** $\frac{27}{36}$, or $\frac{3}{4}$ **7.** $F(5) = 0.811$; 81.1% of people visit a fast-food restaurant at most 5 times in a month. **8.** $\mu = 7.32$

10.4

Continuous Probability Distributions

In Section 10.3, we considered experiments whose outcomes were *discrete*; that is, each outcome could be listed individually. However, it is often reasonable to define a *continuous* random variable when the outcomes of an experiment are defined over some interval of the real numbers. Using calculus and definite integrals, we can go beyond discrete outcomes and consider probabilities and probability distributions that are continuous in nature.

OBJECTIVES

- Construct a probability density function based on a continuous random variable.

- Verify certain properties of probability density functions.

- Find the cumulative density function given a probability density function.

- Solve applied problems involving probability density functions and cumulative density functions, including uniform and exponential density functions.

Continuous Random Variables and Probability Density Functions

Suppose the wait time between trains at a subway station ranges from 5 min to 15 min. These wait times vary *continuously* over the interval $[5, 15]$. If we let x represent a wait time in $[5, 15]$, then x is a quantity that can be measured, and since x ranges over an interval of real numbers, we call x a **continuous random variable**.

Now suppose the wait time between trains at this subway station is uniformly distributed: the next train may appear at any time between 5 min and 15 min with equal probability. From Section 10.3, Example 5, a uniform distribution is one in which all outcomes have the same probability. Thus, we could model the wait times for this situation with a *continuous* function, $f(x) = \frac{1}{10}$, for $5 \le x \le 15$, where the random variable x is in minutes. Note that f is defined over an interval of real numbers, not for a discrete list of numbers, and that the area under the graph of this function is $\frac{1}{10} \cdot 10 = 1$.

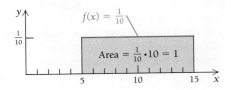

We can use this model to find probabilities. For example, the probability that a wait time is between 5 min and 10 min is 0.5, since the area under $f(x) = \frac{1}{10}$ between $x = 5$ and $x = 10$ is 0.5.

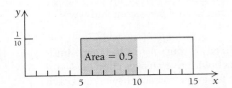

Using probability notation, we express this as $P([5, 10]) = P(5 \le x \le 10) = 0.5$.

In situations where the random variable is defined over an interval of the real numbers, we define probabilities as areas under the graph of a continuous function that models the situation. Thus, for the subway example, $P(9) = 0$, since the probability that the wait time is exactly 9.000. . . min is 0. However, as the student can confirm, there is a 5% chance that the wait time will be within 15 sec of 9 min; that is, $P([8.75, 9.25]) = 0.05$.

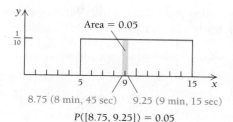

8.75 (8 min, 45 sec)   9.25 (9 min, 15 sec)

$P([8.75, 9.25]) = 0.05$

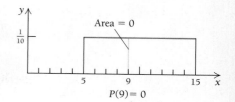

As the interval surrounding $t = 9$ min shrinks, the probability of the wait time being *exactly* 9 min approaches zero.

$P(9) = 0$

Suppose the wait times are not uniform but instead are modeled by $g(x) = \frac{1}{50}x - \frac{1}{10}$, for $5 \le x \le 15$, where the random variable x is in minutes. Note that the area below g from $5 \le x \le 15$ is 1.

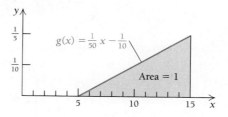

Viewing the graph of $g(x)$, we see that the wait times tend to be longer rather than shorter. Using this model, the probability that a wait time is between 10 min and 15 min is 0.75, since the area under g between $x = 10$ and $x = 15$ is 0.75:

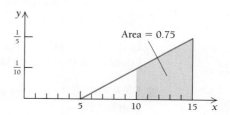

Thus,

$$P([10, 15]) = P(10 \le x \le 15) = 0.75.$$

Note that this result can be found using a definite integral:

$$P(10 \le x \le 15) = \int_{10}^{15} \left(\frac{1}{50}x - \frac{1}{10} \right) dx$$
$$= \left[\frac{1}{100}x^2 - \frac{1}{10}x \right]_{10}^{15}$$
$$= \left(\frac{1}{100}(15)^2 - \frac{1}{10}(15) \right) - \left(\frac{1}{100}(10)^2 - \frac{1}{10}(10) \right)$$
$$= 0.75 - 0$$
$$= 0.75.$$

Certain nonnegative functions, called *probability density functions*, can be used to model probability situations in which x is a continuous random variable.

DEFINITION

Let x be a continuous random variable. A function f is said to be a **probability density function** of x if:

1. For all x in the domain of f, we have $f(x) \geq 0$.
2. The area under the graph of f is 1 (see Fig. 1).
3. For any subinterval $[c, d]$ in the domain of f (see Fig. 2), the probability that x will be in that subinterval is given by

$$P([c, d]) = \int_c^d f(x)\,dx.$$

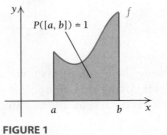

FIGURE 1   **FIGURE 2**

In this book, we do not use capital X to represent a continuous random variable but instead use lowercase x since it is already defined over an interval of real numbers.

■ **EXAMPLE 1**   Verify that Property 2 of the definition of a probability density function holds for

$$f(x) = \frac{3}{117}x^2, \quad \text{for } 2 \leq x \leq 5.$$

Solution

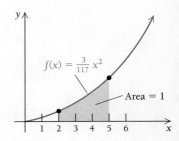

$$\int_2^5 \frac{3}{117}x^2 \, dx = \frac{3}{117}\left[\frac{1}{3}x^3\right]_2^5$$

$$= \frac{1}{117}\left[x^3\right]_2^5$$

$$= \frac{1}{117}\left(5^3 - 2^3\right) \quad \text{Substituting}$$

$$= \frac{117}{117} = 1.$$

❯ **Quick Check 1**

Assume that x is a continuous random variable. Verify that $g(x) = \frac{3}{14}\sqrt{x}$, for $1 \leq x \leq 4$, is a probability density function.

❮ Quick Check 1

■ **EXAMPLE 2** **Business: Life of a Product.** A company that produces compact fluorescent bulbs determines that the life t of a bulb is from 3 to 6 yr and that the probability density function for t is given by

$$f(t) = \frac{24}{t^3}, \quad \text{for } 3 \leq t \leq 6.$$

a) Verify Property 2 of the definition of a probability density function.

b) Find the probability that a bulb will last no more than 4 yr.

c) Find the probability that a bulb will last at least 4 yr and at most 5 yr.

Solution

a) We want to show that $\int_3^6 f(t)\, dt = 1$. We have

$$\int_3^6 \frac{24}{t^3}\, dt = -12\left[\frac{1}{t^2}\right]_3^6 \qquad \text{Integrating: } 24\left[\frac{t^{-2}}{-2}\right] = -12\left[\frac{1}{t^2}\right]$$

$$= -12\left(\frac{1}{6^2} - \frac{1}{3^2}\right) \qquad \text{Substituting}$$

$$= -12\left(-\frac{3}{36}\right) = 1. \qquad \text{Simplifying}$$

b) The probability that a bulb will last no more than 4 yr is

$$P(3 \leq t \leq 4) = \int_3^4 \frac{24}{t^3}\, dt$$

$$= -12\left[\frac{1}{t^2}\right]_3^4$$

$$= -12\left(\frac{1}{4^2} - \frac{1}{3^2}\right) \qquad \text{Substituting}$$

$$= -12\left(-\frac{7}{144}\right) = \frac{7}{12} \approx 0.58. \qquad \text{Simplifying}$$

c) The probability that a bulb will last at least 4 yr and at most 5 yr is

$$P(4 \leq t \leq 5) = \int_4^5 \frac{24}{t^3}\, dt$$

$$= -12\left[\frac{1}{t^2}\right]_4^5$$

$$= -12\left(\frac{1}{5^2} - \frac{1}{4^2}\right) \qquad \text{Substituting}$$

$$= -12\left(-\frac{9}{400}\right)$$

$$= \frac{27}{100} = 0.27. \qquad \text{Simplifying}$$

❮ Quick Check 2

Exploratory

Graph the function

$$f(x) = \tfrac{3}{117} x^2$$

using a viewing window of $[0, 5, 0, 1]$. Then successively evaluate each of the following integrals, shading the appropriate area if possible:

$$\int_2^3 \tfrac{3}{117} x^2\, dx,$$

$$\int_3^4 \tfrac{3}{117} x^2\, dx,$$

and

$$\int_4^5 \tfrac{3}{117} x^2\, dx.$$

Add your results, and explain the meaning of the total.

❯ **Quick Check 2**

The time between arrivals of buses at a station is modeled by the probability density function

$$h(x) = \frac{10}{x^2}, \quad \text{for } 5 \leq x \leq 10,$$

where x is in minutes. Find the probability that:

a) the time between buses is between 5 and 7 min;

b) the time between buses is between 8 and 10 min.

Cumulative Density Functions

Consider the probability density function in Example 2, $f(t) = \dfrac{24}{t^3}$, for $3 \leq t \leq 6$. Let's explore what happens when we evaluate the definite integral of f by keeping the lower bound $a = 3$ and allowing the upper bound t to vary such that $3 \leq t \leq 6$. Thus, we have a function F written as a definite integral, with the variable as the top bound of the integral. When we write a function in integral form and the variable (in this case, t) is a bound of the integral, we must use a different letter (called a *dummy variable*) for the variable in the integrand. Any letter other than that used for the bound can be the dummy variable. Let's use s. We then have

$$F(t) = \int_3^t \frac{24}{s^3}\, ds, \qquad 3 \leq t \leq 6.$$

When the integrand (in the temporary dummy variable s) is integrated and evaluated at the bounds t and 3, the result is a function F in the variable t. As t increases continuously from 3 to 6, the value of $F(t) = \int_3^t \dfrac{24}{s^3}\, ds$ increases from 0 to 1, as shown in the figure.

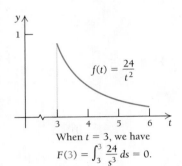

When $t = 3$, we have
$F(3) = \int_3^3 \frac{24}{s^3}\, ds = 0.$

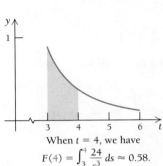

When $t = 4$, we have
$F(4) = \int_3^4 \frac{24}{s^3}\, ds \approx 0.58.$

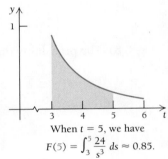

When $t = 5$, we have
$F(5) = \int_3^5 \frac{24}{s^3}\, ds \approx 0.85.$

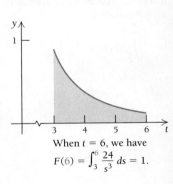

When $t = 6$, we have
$F(6) = \int_3^6 \frac{24}{s^3}\, ds = 1.$

The function F is called a *cumulative density function*.

DEFINITION

Let x be a continuous random variable, and let f be a probability density function of x on $[a, b]$. The **cumulative density function** of f is defined by

$$F(x) = \int_a^x f(s)\, ds.$$

Note that the lower bound of the integral is a, while the upper bound x varies such that $a \leq x \leq b$. The cumulative density function F is an increasing function with $F(a) = 0$ and $F(b) = 1$. Furthermore, we have $F(x) = 0$ for all $x \leq a$ and $F(x) = 1$ for all $x \geq b$. A probability is read as a point on the graph of F, which corresponds to the area between the lower bound a and the variable x on the graph of f.

■ **EXAMPLE 3**   **Business: Wait Times.** Customers at Cornelius Pharmacy experience a wait time modeled by the probability density function $f(x) = \frac{1}{8}x - \frac{1}{4}$, for $2 \leq x \leq 6$, where the continuous random variable x is measured in minutes.

a) Find the cumulative density function F.

b) Find the probability that a randomly chosen customer waits at most 4 min.

c) Find the probability that a randomly chosen customer waits between 3 and 5 min.

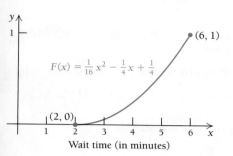

F(x) = $\frac{1}{16} x^2 - \frac{1}{4} x + \frac{1}{4}$

(6, 1)

(2, 0)

Wait time (in minutes)

Solution

a) The cumulative density function is given by

$$F(x) = \int_2^x \left(\tfrac{1}{8} s - \tfrac{1}{4} \right) ds \qquad \text{Here } s \text{ is a temporary dummy variable.}$$
$$= \left(\tfrac{1}{16} s^2 - \tfrac{1}{4} s \right)_2^x$$
$$= \left(\tfrac{1}{16} x^2 - \tfrac{1}{4} x \right) - \left(\tfrac{1}{16} \cdot 2^2 - \tfrac{1}{4} \cdot 2 \right)$$
$$= \tfrac{1}{16} x^2 - \tfrac{1}{4} x + \tfrac{1}{4}, \quad \text{for } 2 \le x \le 6.$$

The student can confirm that $F(x) = \frac{1}{16}x^2 - \frac{1}{4}x + \frac{1}{4}$, for $2 \le x \le 6$, is an increasing function with $F(2) = 0$ and $F(6) = 1$.

b) Using the cumulative density function F, the probability that a customer waits at most 4 min is

$$F(4) = \tfrac{1}{16} (4)^2 - \tfrac{1}{4} (4) + \tfrac{1}{4} = 0.25.$$

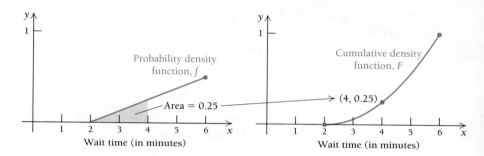

Probability density function, f

Area = 0.25

Wait time (in minutes)

Cumulative density function, F

(4, 0.25)

Wait time (in minutes)

c) Using the cumulative density function F, the probability that a customer waits between 3 and 5 min is

$$F(5) - F(3) = \left(\tfrac{1}{16} (5)^2 - \tfrac{1}{4} (5) + \tfrac{1}{4} \right) - \left(\tfrac{1}{16} (3)^2 - \tfrac{1}{4} (3) + \tfrac{1}{4} \right)$$
$$= 0.5625 - 0.0625 = 0.5.$$

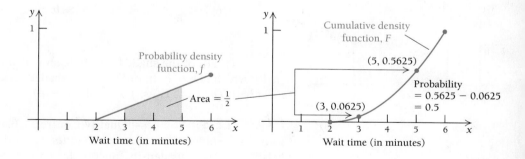

Probability density function, f

Area = $\frac{1}{2}$

Wait time (in minutes)

Cumulative density function, F

(5, 0.5625)

Probability
= 0.5625 − 0.0625
= 0.5

(3, 0.0625)

Wait time (in minutes)

Constructing Probability Density Functions

Consider an arbitrary nonnegative function $f(x)$ whose definite integral over some interval $[a, b]$ is K. Then

$$\int_a^b f(x) \, dx = K.$$

Multiplying on both sides by $1/K$ gives us

$$\frac{1}{K} \int_a^b f(x) \, dx = \frac{1}{K} \cdot K = 1, \quad \text{or} \quad \int_a^b \frac{1}{K} \cdot f(x) \, dx = 1.$$

Thus, when we multiply $f(x)$ by $1/K$, we create a function whose area over the given interval is 1. Such a function satisfies the definition of a probability density function.

■ **EXAMPLE 4**   Find k such that

$$f(x) = kx^2$$

is a probability density function over the interval $[1, 4]$. Then write the probability density function and the cumulative density function.

Solution   Note that for $k \geq 0$, we have $kx^2 \geq 0$. For f to be a probability density function, we must also have $\int_1^4 kx^2 \, dx = 1$. Since $\int_1^4 kx^2 \, dx = k\left[\frac{1}{3}x^3\right]_1^4 = k\left[\frac{64}{3} - \frac{1}{3}\right] = k \cdot 21$, we have $21k = 1$, so $k = \frac{1}{21}$. Thus, the probability density function is

$$f(x) = \frac{1}{21}x^2, \quad \text{for } 1 \leq x \leq 4.$$

The cumulative density function is

$$F(x) = \int_1^x \frac{1}{21}s^2 \, ds = \frac{1}{21}\left[\frac{s^3}{3}\right]_1^x = \frac{x^3}{63} - \frac{1}{63}, \text{ for } 1 \leq x \leq 4.$$

Check: We note that $F(1) = 0$ and $F(4) = 1$.

❮ Quick Check 3

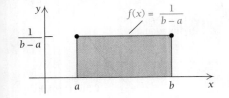

> **Quick Check 3**

Find k such that $g(x) = \dfrac{k}{x}$ is a probability density function over $[2, 7]$. Then write the probability density function g, and the cumulative density function G.

Uniform Distributions

Consider the probability density function $f(x) = \frac{1}{10}$, for $5 \leq x \leq 15$, which describes the wait times at a subway station, discussed earlier in this section. The wait times are uniform, and the graph of f is a horizontal line that has an area of 1 below it over $[5, 15]$.

DEFINITION

A continuous random variable x is said to be uniformly distributed over an interval $[a, b]$ if it has a probability density function f given by

$$f(x) = \frac{1}{b - a}, \quad \text{for } a \leq x \leq b.$$

The length of the shaded rectangle shown at the left is the length of the interval $[a, b]$, which is $b - a$. In order for the shaded area to be 1, the height of the rectangle must be $1/(b - a)$. Thus, $f(x) = 1/(b - a)$.

■ **EXAMPLE 5**   **Business: Quality Control.**   BlastOut Inc. produces sirens used for tornado warnings. The maximum loudness, L, of the sirens ranges from 70 to 100 decibels. The probability density function for L is

$$f(L) = \frac{1}{30}, \quad \text{for } 70 \leq L \leq 100.$$

A siren is selected at random off the assembly line. Find the probability that its maximum loudness is from 70 to 92 decibels.

Solution   The probability is

$$P(70 \leq L \leq 92) = \int_{70}^{92} \frac{1}{30} \, dL = \frac{1}{30}\left[L\right]_{70}^{92}$$

$$= \frac{1}{30}(92 - 70) = \frac{22}{30} = \frac{11}{15} \approx 0.73.$$

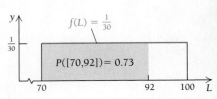

> **Quick Check 4**

Business: Quality Control.
The probability density function for the weight x, in pounds, of bags of feed sold at a feedstore is

$$f(x) = \frac{1}{8}, \quad \text{for } 45 \leq x \leq 53.$$

A bag is selected at random. Find the probability the bag weighs between 47.5 and 50.25 lb.

❮ Quick Check 4

Exponential Distributions

The duration of a phone call, the distance between successive cars on a highway, and the amount of time required to learn a task are all examples of exponentially distributed random variables. That is, their probability density functions are exponential.

> **DEFINITION**
>
> A continuous random variable is exponentially distributed if it has a probability density function of the form
>
> $$f(x) = ke^{-kx}, \qquad \text{over the interval } [0, \infty).$$

To see that $f(x) = 2e^{-2x}$ is such a probability density function, note that

$$\int_0^\infty 2e^{-2x}\, dx = \lim_{b \to \infty} \int_0^b 2e^{-2x}\, dx = \lim_{b \to \infty} \left[-e^{-2x}\right]_0^b = \lim_{b \to \infty} \left(\frac{-1}{e^{2b}} - (-1)\right) = 1.$$

The general case,

$$\int_0^\infty ke^{-kx}\, dx = 1,$$

can be verified in a similar way.

Why is it reasonable to assume that the distance between cars is exponentially distributed? Part of the reason is that there are many more cases in which traffic is highly congested and involves many cars. Similarly with the duration of phone calls; that is, there are many more short calls than long ones.

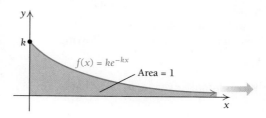

EXAMPLE 6 **Business: Transportation Planning.** The distance x, in feet, between successive cars on a certain stretch of highway has a probability density function

$$f(x) = ke^{-kx}, \quad \text{for } 0 \le x < \infty,$$

where $k = 1/a$ and a is the average distance between successive cars over some period of time. A transportation planner determines that the average distance between cars on a certain stretch of highway is 166 ft. Find the following:

a) The cumulative density function

b) The probability that the distance between two successive cars, chosen at random, is 50 ft or less

c) The probability that the distance between two successive cars, chosen at random, is between 30 and 100 ft

A transportation planner can determine the probabilities that cars are certain distances apart.

Solution We first determine k:

$$k = \tfrac{1}{166} \approx 0.006024.$$

The probability density function for x is given by

$$f(x) = 0.006024e^{-0.006024x}, \quad \text{for } 0 \le x < \infty.$$

a) The cumulative density function is given by

$$F(x) = \int_0^x 0.006024 e^{-0.006024s} \, ds = \left[-e^{-0.006024s}\right]_0^x$$

$$= \left(-e^{-0.006024x}\right) - \left(-e^{-0.006024(0)}\right)$$

$$= 1 - e^{-0.006024x}.$$

> **Quick Check 5**
>
> The response time for a paramedic unit has the probability density function
>
> $$f(t) = 0.05e^{-0.05t}$$
> for $0 \le t < \infty$,
>
> where t is in minutes. Find the probability of each response time:
>
> **a)** between 5 and 10 min;
> **b)** between 8 and 20 min;
> **c)** between 1 and 45 min;
> **d)** more than 60 min.

b) The probability that the distance between the cars is 50 ft or less is given by

$$P(0 \le x \le 50) = F(50)$$
$$= \left(1 - e^{-0.006024(50)}\right)$$
$$= 0.26.$$

c) The probability that the distance between the cars is between 30 ft and 100 ft is given by

$$P(30 \le x \le 100) = F(100) - F(30)$$
$$= \left(1 - e^{-0.006024(100)}\right) - \left(1 - e^{-0.006024(30)}\right)$$
$$= 0.2872.$$

❮ Quick Check 5

Section Summary

- A *continuous random variable* is a quantity that can be observed (or measured) and whose possible values comprise an interval of real numbers.
- If x is a continuous random variable, then function f is a *probability density function* of x if it meets the following criteria:
  1. For all x in $[a, b]$, $f(x) \ge 0$.
  2. The area under the graph of f over $[a, b]$ is 1; that is,
  $$\int_a^b f(x) \, dx = 1.$$
  3. The probability that x is within the subinterval $[c, d]$ is given by $P([c, d]) = \int_c^d f(x) \, dx.$

- If f is a probability density function over $[a, b]$, then its *cumulative density function* is
  $$F(x) = \int_a^x f(s) \, ds.$$
- A continuous random variable x is *uniformly distributed* over an interval $[a, b]$ if its probability density function has the form $f(x) = \dfrac{1}{b-a}$ over the interval.
- A continuous random variable x is *exponentially distributed* over $[0, \infty)$ if its probability density function has the form $f(x) = ke^{-kx}$.

EXERCISE SET
10.4

In Exercises 1–12, verify Property 2 of the definition of a probability density function over the given interval.

1. $f(x) = \frac{1}{4}x$, $[1, 3]$

2. $f(x) = 2x$, $[0, 1]$

3. $f(x) = 3$, $\left[0, \frac{1}{3}\right]$

4. $f(x) = \frac{1}{5}$, $[3, 8]$

5. $f(x) = \frac{3}{64}x^2$, $[0, 4]$

6. $f(x) = \frac{3}{26}x^2$, $[1, 3]$

7. $f(x) = \frac{1}{x}$, $[1, e]$

8. $f(x) = \frac{1}{e-1}e^x$, $[0, 1]$

9. $f(x) = \frac{3}{2}x^2$, $[-1, 1]$

10. $f(x) = \frac{1}{3}x^2$, $[-2, 1]$

11. $f(x) = 3e^{-3x}$, $[0, \infty)$

12. $f(x) = 4e^{-4x}$, $[0, \infty)$

Find k such that each function is a probability density function over the given interval. Then write the probability density function.

13. $f(x) = kx$, $[2, 5]$

14. $f(x) = kx$, $[1, 4]$

15. $f(x) = kx^2$, $[-1, 1]$

16. $f(x) = kx^2$, $[-2, 2]$

17. $f(x) = k$, $[1, 7]$

18. $f(x) = k$, $[3, 9]$

19. $f(x) = k(2 - x)$, $[0, 2]$

20. $f(x) = k(4 - x)$, $[0, 4]$

21. $f(x) = \dfrac{k}{x}$, $[1, 3]$ **22.** $f(x) = \dfrac{k}{x}$, $[1, 2]$

23. $f(x) = ke^x$, $[0, 3]$ **24.** $f(x) = ke^x$, $[0, 2]$

25–48. Find the cumulative density function of each probability density function in Exercises 1–24.

49. A dart is thrown at a number line in such a way that it always lands in the interval $[0, 10]$. Let x represent the number that the dart hits. Suppose that the probability density function for x is given by

$$f(x) = \tfrac{1}{50}x, \quad \text{for } 0 \le x \le 10.$$

Find $P(2 \le x \le 6)$, the probability that the dart lands in $[2, 6]$.

50. In Exercise 49, suppose that the dart always lands in the interval $[0, 5]$ and that the probability density function for x is given by

$$f(x) = \tfrac{3}{125}x^2, \quad \text{for } 0 \le x \le 5.$$

Find $P(1 \le x \le 4)$, the probability that the dart lands in $[1, 4]$.

51. A number x is selected at random from the interval $[4, 20]$. The probability density function for x is given by

$$f(x) = \tfrac{1}{16}, \quad \text{for } 4 \le x \le 20.$$

Find the probability that a number selected is between 9 and 20.

52. A number x is selected at random from the interval $[5, 29]$. The probability density function for x is given by

$$f(x) = \tfrac{1}{24}, \quad \text{for } 5 \le x \le 29.$$

Find the probability that a number selected is between 14 and 29.

APPLICATIONS

Business and Economics

53. Transportation planning. Refer to Example 6. A transportation planner determines that the average distance between cars on a certain highway is 100 ft. What is the probability that the distance between two successive cars, chosen at random, is 40 ft or less?

54. Transportation planning. Refer to Example 6. A transportation planner determines that the average distance between cars on a certain highway is 200 ft. What is the probability that the distance between two successive cars, chosen at random, is 10 ft or less?

55. Duration of a phone call. A telephone company determines that the duration t, in minutes, of a phone call is an exponentially distributed random variable with a probability density function

$$f(t) = 2e^{-2t}, \quad 0 \le t < \infty.$$

Find the probability that a phone call will last no more than 5 min.

56. Duration of a phone cell. Referring to Exercise 55, find the probability that a phone call will last no more than 2 min.

57. Time to failure. The *time to failure*, t, in hours, of a machine is often exponentially distributed with a probability density function

$$f(t) = ke^{-kt}, \quad 0 \le t < \infty,$$

where $k = 1/a$ and a is the average amount of time that will pass before a failure occurs. Suppose that the average amount of time that will pass before an arcade game breaks down is 100 hr. What is the probability that an arcade game fails in 50 hr or less?

58. Reliability of a machine. The *reliability* of the game (the probability that it will work) in Exercise 57 is defined as

$$R(T) = 1 - \int_0^T 0.01e^{-0.01t}\, dt,$$

where $R(T)$ is the reliability at time T. Write $R(T)$ without using an integral.

Life and Physical Sciences

59. Wait time for 911 calls. The wait time before a 911 call is answered in the state of California has a probability density function $f(t) = 0.23e^{-0.23t}$, for $0 \le t < \infty$, where t is in seconds. (*Source*: California Government Code.)

   a) The state standard is that 90% of 911 calls are to be answered within 10 sec. Verify that this standard is met using the probability density function f.

   b) What is the probability that a 911 call is answered within 15 to 25 sec after being made?

60. Emergency room wait times. The wait time at an emergency room has the probability density function $f(t) = 0.116e^{-0.116t}$, for $0 \le t < \infty$, where t is in hours. (*Source*: www.pressganey.com.)

   a) Find the probability that a wait time is at most 1 hr.

   b) In 2009, half of all emergency room patients waited up to 6 hr. Verify this using the probability density function f.

Social Sciences

61. Time in a maze. In a psychology experiment, the time t, in seconds, that it takes a rat to find its way through a maze is an exponentially distributed random variable with the probability density function

$$f(t) = 0.02e^{-0.02t}, \quad 0 \le t < \infty.$$

Find the probability that a rat will find its way through a maze in 150 sec or less.

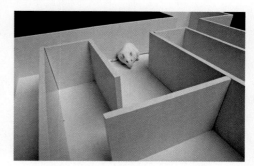

62. Time in a maze. Use the equation in Exercise 61 to find the following probabilities.

  a) The probability that a rat will find its way through the maze in more than 150 sec

  b) The probability that a rat will find its way through the maze in exactly 150 sec

SYNTHESIS

63. The function $f(x) = x$ is a probability density function over $[0, b]$. Find b.

64. The function $f(x) = 12x^2$ is a probability density function over $[-a, a]$. Find a.

In Exercises 65 and 66, find the error in each statement, and write a correct statement.

65. For the probability density function $f(x) = \frac{1}{2}x - \frac{1}{2}$, for $1 \le x \le 3$, it follows that $F(x) = \frac{1}{4}x^2$.

66. For the probability density function $f(x) = 0.04e^{-0.04x}$, for $0 \le x < \infty$, it follows that $F(x) = e^{-0.04x}$, for $0 \le x < \infty$.

67. Show that if $f(x) = ke^{-kx}$, for $0 \le x < \infty$, then $F(x) = 1 - e^{-kx}$, for $0 \le x < \infty$.

68. Show that if $f(x) = \dfrac{1}{b - a}$, for $a \le x \le b$, then $F(x) = \dfrac{x - a}{b - a}$, for $a \le x \le b$.

69. The wait times at an urgent care center are exponentially distributed. There is a 30% probability that a patient will have to wait up to 1 hr to see a doctor.

  a) Find k, and then write the probability density function f. (*Hint*: Use the result from Exercise 67.)

  b) Find the probability that a patient will have to wait between 90 min and 3 hr for a doctor.

70. The elapsed time between the arrivals of cars at a rural intersection is exponentially distributed. There is a 20% probability that 10 min will pass between the arrivals of cars at the intersection.

  a) Find k, and then write the probability density function f.

  b) Find the probability that two cars will come to the intersection within 5 min of one another.

71. The graph of f is a probability density function.

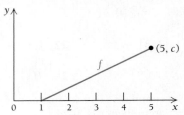

  a) Find c.
  b) Find f.
  c) Find the cumulative density function F.
  d) Find $P(2 \le x \le 3)$.

72. The graph of f is a probability density function.

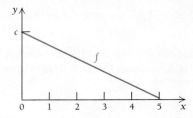

  a) Find c.
  b) Find f.
  c) Find the cumulative density function F.
  d) Find $P(1 \le x \le 4)$.

73. Find c such that $f(x) = cxe^{2x}$, for $1 \le x \le 2$, is a probability density function.

74. Find c such that $f(x) = cx\sqrt{1 + x}$, for $0 \le x \le 1$, is a probability density function.

TECHNOLOGY CONNECTION

75–86. Verify Property 2 of the definition of a probability density function for each of the functions in Exercises 1–12.

Answers to Quick Checks

1. $\displaystyle\int_1^4 \frac{3}{14}\sqrt{x}\,dx = \frac{3}{14}\left[\frac{2}{3}x^{3/2}\right]_1^4 = \frac{3}{14}\left(\frac{16}{3} - \frac{2}{3}\right) = 1;$

$g(x) \ge 0$ on $[1, 4]$ **2.** **(a)** 0.57; **(b)** 0.25

3. $k = 0.7982$; $g(x) = \dfrac{0.7982}{x}$, for $2 \le x \le 7$;

$G(x) = 0.7982 \ln x - 0.5533$, for $2 \le x \le 7$

4. 0.34375   **5.** **(a)** 0.172; **(b)** 0.302; **(c)** 0.846; **(d)** 0.050

10.5

Mean, Variance, Standard Deviation, and the Normal Distribution

Mean of a Continuous Random Variable

In Section 10.4, we considered a situation in which wait times for subway trains were uniformly distributed and modeled by the probability density function $f(x) = \frac{1}{10}$, for $5 \le x \le 15$, where the continuous random variable x is in minutes.

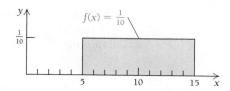

What is the mean wait time between trains? For the moment, let's view the wait times as a discrete distribution, $f(x) = \frac{1}{10}$, for $x = 5.5, 6.5, 7.5, 8.5, 9.5, 10.5, 11.5, 12.5, 13.5, 14.5$, where each value of x is the midpoint of a 1-min interval. That is, $x = 5.5$ represents the interval between 5 min and 6 min, and so on. This discrete distribution is shown below as a histogram:

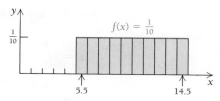

Therefore, the mean wait time is

$$\mu = 5.5 \cdot \tfrac{1}{10} + 6.5 \cdot \tfrac{1}{10} + 7.5 \cdot \tfrac{1}{10} + 8.5 \cdot \tfrac{1}{10} + 9.5 \cdot \tfrac{1}{10} + 10.5 \cdot \tfrac{1}{10} + 11.5 \cdot \tfrac{1}{10} + 12.5 \cdot \tfrac{1}{10} + 13.5 \cdot \tfrac{1}{10} + 14.5 \cdot \tfrac{1}{10}$$
$$= \tfrac{1}{10}(5.5 + 6.5 + 7.5 + 8.5 + 9.5 + 10.5 + 11.5 + 12.5 + 13.5 + 14.5)$$
$$= \tfrac{1}{10}(100) = 10 \text{ min.}$$

If we subdivide $[5, 15]$ into smaller subintervals and let x be the midpoint of each subinterval, we again obtain a mean of $\mu = 10$ min. A sum like the one used here to obtain the mean is of the form $\sum x \cdot f(x)$, suggesting that we use integration to determine the mean wait time:

$$\mu = \int_5^{15} x \cdot f(x)\, dx \qquad \text{The integrand is a product of a wait time and its probability.}$$

$$= \int_5^{15} x \cdot \frac{1}{10}\, dx \qquad \text{Substituting } f(x) = \tfrac{1}{10}$$

$$= \frac{1}{10} \int_5^{15} x\, dx$$

$$= \frac{1}{10}\left[\frac{1}{2}x^2\right]_5^{15} \qquad \text{Integrating}$$

$$= \tfrac{1}{20}\left[(15)^2 - (5)^2\right] \qquad \text{Evaluating the bounds; } \tfrac{1}{10} \cdot \tfrac{1}{2} = \tfrac{1}{20}$$
$$= \tfrac{1}{20}\left[225 - 25\right] \qquad \text{Simplifying}$$
$$= \tfrac{1}{20}\left[200\right]$$
$$= 10 \text{ min.} \qquad \text{In this uniform distribution, } \mu \text{ is also the midpoint of } [5, 15].$$

This result suggests a way to define the mean of a continuous random variable.

DEFINITION

Let x be a continuous random variable over the interval $[a, b]$ with probability density function f. The **mean**, μ, of x is defined by

$$\mu = \int_a^b x \cdot f(x)\, dx.$$

■ **EXAMPLE 1**   Suppose wait times between trains are given by the probability density function $g(x) = \frac{1}{50}x - \frac{1}{10}$, for $5 \le x \le 15$, where the continuous random variable x is in minutes. Find the mean of x, the wait time between trains.

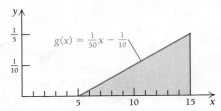

Solution   We evaluate $\displaystyle\int_5^{15} x \cdot g(x)\, dx$:

$$\mu = \int_5^{15} x \cdot \left(\frac{1}{50}x - \frac{1}{10} \right) dx \qquad \text{Substituting } g(x) = \frac{1}{50}x - \frac{1}{10}$$

$$= \int_5^{15} \left(\frac{1}{50}x^2 - \frac{1}{10}x \right) dx \qquad \text{Distributing}$$

$$= \left[\frac{1}{150}x^3 - \frac{1}{20}x^2 \right]_5^{15} \qquad \text{Integrating}$$

$$= \left(\frac{1}{150}(15)^3 - \frac{1}{20}(15)^2 \right) - \left(\frac{1}{150}(5)^3 - \frac{1}{20}(5)^2 \right) \qquad \text{Evaluating}$$

$$= \frac{45}{4} - \left(-\frac{5}{12} \right)$$

$$= \frac{35}{3} = 11.6666\ldots \text{ min.}$$

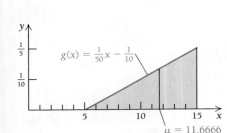

For this distribution, the mean wait time is $\mu = 11.6666\ldots$ min. Since wait times in this distribution tend to be longer, it makes sense that the mean wait time is larger than the midpoint value, 10.

> **Quick Check 1**
>
> Find the mean of the random variable x, where the probability density function is $f(x) = \frac{1}{2}x$, for $0 \le x \le 2$.

❮ Quick Check 1

We can get a physical idea of the mean of a random variable by pasting the graph of a probability density function on cardboard and cutting out the area under the curve over the interval $[a, b]$. Then we try to find a balance point on the x-axis. That balance point is the mean, μ.

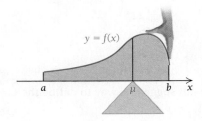
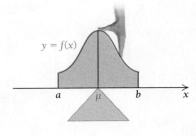

Variance and Standard Deviation of a Continuous Random Variable

Because two very different distributions can have the same mean, it is useful to have a second statistic that serves as a measure of how the data in a distribution are spread out. Statistics that provide such a measure are the *variance* and (especially) the *standard deviation* of a distribution. A full derivation of how these statistics were developed is beyond the scope of this book.

DEFINITION

The **variance**, σ^2, of a continuous random variable x, defined on $[a, b]$, with probability density function f, is

$$\sigma^2 = \int_a^b x^2 f(x)\, dx - \left[\int_a^b x f(x)\, dx\right]^2 = \int_a^b x^2 f(x)\, dx - \mu^2.$$

The **standard deviation**, σ, of a continuous random variable is defined as

$$\sigma = \sqrt{\text{variance}}.$$

(The symbol σ is the lowercase Greek letter *sigma*.)

■ **EXAMPLE 2** Consider the probability density function $g(x) = \frac{1}{50}x - \frac{1}{10}$, for $5 \le x \le 15$, where the random variable x is the number of minutes between buses at a depot.

a) Find the variance and the standard deviation of x.

b) Find the probability that the time between buses is within one standard deviation of the mean.

Solution

a) To calculate σ^2, we need to evaluate $\displaystyle\int_5^{15} x^2 g(x)\, dx$ and find μ, or $\displaystyle\int_5^{15} x g(x)\, dx$.

From Example 1, we have $\mu = \frac{35}{3}$. We evaluate $\displaystyle\int_5^{15} x^2 g(x)\, dx$:

$$
\begin{aligned}
\int_5^{15} x^2 g(x)\, dx &= \int_5^{15} x^2 \left(\frac{1}{50}x - \frac{1}{10}\right) dx \qquad \text{Substituting } g(x) = \tfrac{1}{50}x - \tfrac{1}{10} \\
&= \int_5^{15} \left(\frac{1}{50}x^3 - \frac{1}{10}x^2\right) dx \\
&= \left[\tfrac{1}{200}x^4 - \tfrac{1}{30}x^3\right]_5^{15} \\
&= \left(\tfrac{1}{200}(15)^4 - \tfrac{1}{30}(15)^3\right) - \left(\tfrac{1}{200}(5)^4 - \tfrac{1}{30}(5)^3\right) \qquad \text{Evaluating} \\
&= \tfrac{425}{3}.
\end{aligned}
$$

Therefore, the variance is

$$
\begin{aligned}
\sigma^2 &= \int_5^{15} x^2 g(x)\, dx - \left[\int_5^{15} x g(x)\, dx\right]^2 \\
&= \tfrac{425}{3} - \left(\tfrac{35}{3}\right)^2 \qquad \text{Substituting} \\
&= \tfrac{50}{9}.
\end{aligned}
$$

The standard deviation is $\sigma = \sqrt{50/9} \approx 2.357$ min.

b) To find the probability that the wait time is within one standard deviation of the mean, we first find the limits of integration, using $\mu = 1.666$ and $\sigma = 2.357$. Since $\mu - \sigma = 11.666 - 2.357 = 9.309$ and $\mu + \sigma = 11.666 + 2.357 = 14.023$, we have

$$P(9.309 \le x \le 14.023) = \int_{9.309}^{14.023} \left(\frac{1}{50}x - \frac{1}{10} \right) dx \qquad \text{Substituting the limits of integration}$$

$$= \left[\frac{1}{100}x^2 - \frac{1}{10}x \right]_{9.309}^{14.023}$$

$$= \left(\frac{1}{100}(14.023)^2 - \frac{1}{10}(14.023) \right) - \left(\frac{1}{100}(9.309)^2 - \frac{1}{10}(9.309) \right)$$

$$\approx 0.628. \qquad \text{Using a calculator}$$

Therefore, the probability that the wait time is within one standard deviation of the mean is about 0.628.

> **Quick Check 2**
>
> Given a continuous random variable x with a probability density function $f(x) = \frac{1}{2}x$, for $0 \le x \le 2$, find:
>
> **a)** the variance and the standard deviation of x;
>
> **b)** the probability that x is within one standard deviation of the mean.

❬ Quick Check 2

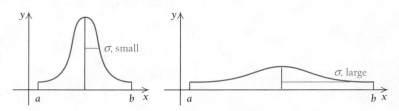

Loosely speaking, standard deviation is a measure of how closely bunched the graph of f is, or how far the points on f are, on average, from the line $x = \mu$.

The Normal Distribution

Suppose that the average score on a test is 70. Usually there are about as many scores above the average as there are below the average; and the farther away from the average a particular score is, the fewer people there are who get that score. On this test, it is probable that more people scored in the 80s than in the 90s and more people scored in the 60s than in the 50s. Test scores, heights of human beings, and weights of human beings are all examples of random variables that are often *normally* distributed.

Consider the function

$$g(x) = e^{-x^2/2}, \quad \text{over the interval } (-\infty, \infty).$$

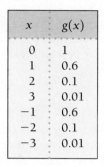

Exploratory

Use a graphing calculator or iPlot to confirm that

$$\int_{-\infty}^{\infty} e^{-x^2/2} \, dx = \sqrt{2\pi},$$

letting $y_1 = e^{-x^2/2}$.

1. Set the limits of integration at -2 and 2.
2. Repeat, setting the limits at -3 and 3.
3. Compare your results with $\sqrt{2\pi} \approx 2.507$.

This function has the entire set of real numbers as its domain. Its graph is the bell-shaped curve shown below. We can find function values by using a calculator:

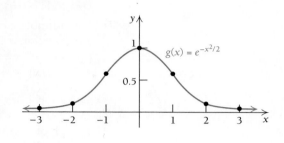

x	$g(x)$
0	1
1	0.6
2	0.1
3	0.01
-1	0.6
-2	0.1
-3	0.01

This function has an antiderivative, but that antiderivative has no basic integration formula. Nevertheless, it can be shown that the improper integral converges over the interval $(-\infty, \infty)$ to a number given by

$$\int_{-\infty}^{\infty} e^{-x^2/2} \, dx = \sqrt{2\pi}.$$

That is, although an antiderivative cannot be easily found, there is a numerical value for the improper integral evaluated over the set of real numbers. Since the area is not 1, the function g is not a probability density function. However, the function given by $f(x) = g(x)/\sqrt{2\pi}$ is

$$f(x) = \frac{1}{\sqrt{2\pi}} e^{-x^2/2}, \quad \text{over } (-\infty, \infty).$$

 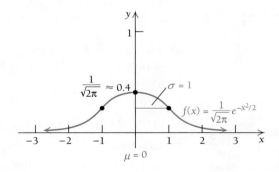

Exploratory

Use a graphing calculator or iPlot to approximate

$$\int_{-b}^{b} \frac{1}{\sqrt{2\pi}} e^{-x^2/2} \, dx,$$

for $b = 10$, 100, and 1000. What does this suggest about

$$\int_{-\infty}^{\infty} \frac{1}{\sqrt{2\pi}} e^{-x^2/2} \, dx?$$

This is a way to verify part of the assertion that

$$f(x) = \frac{1}{\sqrt{2\pi}} e^{-x^2/2}$$

is a probability density function. Use a similar approximation procedure to show that the mean is 0 and the standard deviation is 1.

DEFINITION

A continuous random variable x has a **standard normal distribution** if its probability density function is

$$f(x) = \frac{1}{\sqrt{2\pi}} e^{-x^2/2}, \quad \text{over } (-\infty, \infty).$$

The standard normal distribution has a mean of 0 and a standard deviation of 1. Its graph follows.

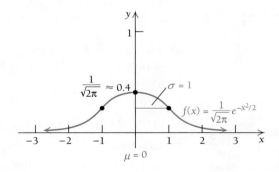

The general case is defined as follows.

> **DEFINITION**
>
> A continuous random variable x is **normally distributed** with mean μ and standard deviation σ if its probability density function is given by
>
> $$f(x) = \frac{1}{\sigma\sqrt{2\pi}} e^{-(1/2)[(x-\mu)/\sigma]^2}, \quad \text{over } (-\infty, \infty).$$

The graph of any normal distribution is a transformation of the graph of the standard normal distribution. This can be shown by translating the graph of a normal distribution along the x-axis and adjusting how tightly clustered the graph is about the mean. Some examples follow.

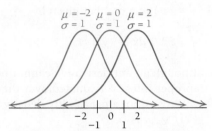

Normal distributions with same standard deviations but different means

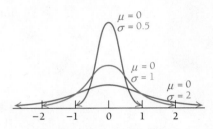

Normal distributions with same means but different standard deviations

Probability and the Normal Distribution

The normal distribution is extremely important in statistics; it underlies much of the research in the behavioral and social sciences. Because of this, tables of approximate values of the definite integral of the standard normal distribution have been prepared using numerical approximation methods like the Trapezoidal Rule given in Section 5.4. Table A at the back of the book (p. 841) is such a table. It contains values of

$$P(0 \le x \le z) = \int_0^z \frac{1}{\sqrt{2\pi}} e^{-x^2/2}\, dx.$$

The symmetry of the graph of this function about the mean allows many types of probabilities to be computed from the table. Some involve addition or subtraction of areas.

■ **EXAMPLE 3** Let x be a continuous random variable with a standard normal distribution. Using Table A at the back of the book, find each of the following.

a) $P(0 \le x \le 1.68)$ **b)** $P(-0.97 \le x \le 0)$

c) $P(-2.43 \le x \le 1.01)$ **d)** $P(1.90 \le x \le 2.74)$

e) $P(-2.98 \le x \le -0.42)$ **f)** $P(x \ge 0.61)$

Solution

a) $P(0 \le x \le 1.68)$ is the area bounded by the standard normal curve and the lines $x = 0$ and $x = 1.68$. We look this up in Table A by going down the left column to 1.6, then moving to the right to the column headed 0.08. There we read 0.4535. Thus,

$$P(0 \le x \le 1.68) = 0.4535.$$

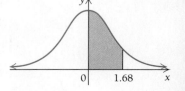

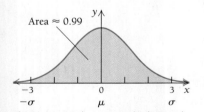

The 68-95-99 Rule: In a standard normal distribution, about 68% of the data is within one standard deviation of the mean, about 95% of the data is within two standard deviations of the mean, and about 99% of the data is within three standard deviations of the mean.

b) Because of the symmetry of the graph,

$$P(-0.97 \leq x \leq 0)$$
$$= P(0 \leq x \leq 0.97) \quad \text{Using symmetry}$$
$$= 0.3340.$$

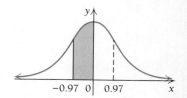

c) $P(-2.43 \leq x \leq 1.01)$
$$= P(-2.43 \leq x \leq 0) + P(0 \leq x \leq 1.01)$$
$$= P(0 \leq x \leq 2.43) + P(0 \leq x \leq 1.01)$$
$$= 0.4925 + 0.3438$$
$$= 0.8363$$

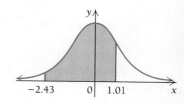

d) $P(1.90 \leq x \leq 2.74)$
$$= P(0 \leq x \leq 2.74) - P(0 \leq x \leq 1.90)$$
$$= 0.4969 - 0.4713$$
$$= 0.0256$$

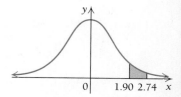

e) $P(-2.98 \leq x \leq -0.42)$
$$= P(0.42 \leq x \leq 2.98) \quad \text{Using symmetry}$$
$$= P(0 \leq x \leq 2.98) - P(0 \leq x \leq 0.42)$$
$$= 0.4986 - 0.1628$$
$$= 0.3358$$

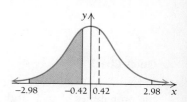

〉 Quick Check 3

Let x be a continuous random variable with a standard normal distribution. Using Table A, find each of the following.

a) $P(-1 \leq x \leq \frac{1}{2})$

b) $P(x \leq -0.77)$

c) $P(x \geq \frac{2}{5})$

f) $P(x \geq 0.61)$
$$= P(x \geq 0) - P(0 \leq x \leq 0.61)$$
$$= 0.5000 - 0.2291 \quad \text{Because of the}$$
$$= 0.2709 \qquad\qquad \text{symmetry about}$$
the line $x = 0$, half the area is on each side of the line, and since the entire area is 1, we have $P(x \geq 0) = 0.5000$.

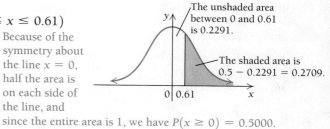

The unshaded area between 0 and 0.61 is 0.2291.

The shaded area is $0.5 - 0.2291 = 0.2709$.

❮ Quick Check 3

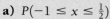

For most normal distributions, $\mu \neq 0$ and $\sigma \neq 1$. It would be a hopeless task to make tables for all values of the mean μ and the standard deviation σ. For any normal distribution, the transformation

$$z = \frac{x - \mu}{\sigma} \qquad \text{This is used to convert } x\text{-values to } z\text{-values.}$$

standardizes the distribution, since $(x - \mu)/\sigma$ is a measure of how many standard deviations x is from μ. Subtracting μ from all x-values and then dividing by σ preserves the order of the x-values while permitting the use of Table A. Such converted values are called *z-scores*, or *z-values*.

$$P(a \leq x \leq b) = P\left(\frac{a - \mu}{\sigma} \leq z \leq \frac{b - \mu}{\sigma}\right).$$

and this last probability can be found using Table A.

■ **EXAMPLE 4** The weights, w, of the members of a high-school cross-country track team are normally distributed with a mean, μ, of 150 lb and a standard deviation, σ, of 25 lb. Find the probability that the weight of a randomly selected member of the team is between 160 lb and 180 lb.

Solution We first standardize the weights:

$$180 \text{ is standardized to } \frac{b - \mu}{\sigma} = \frac{180 - 150}{25} = 1.2;$$

$$160 \text{ is standardized to } \frac{a - \mu}{\sigma} = \frac{160 - 150}{25} = 0.4.$$

These z-values measure the distance of w from μ in terms of σ.

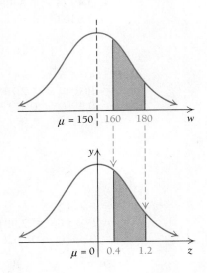

> **Quick Check 4**
>
> Referring to Example 4, find the probability of each of the following:
>
> **a)** The weight of a randomly selected member of the team is below 165 lb.
>
> **b)** The weight of a randomly selected member of the team is between 135 lb and 155 lb.
>
> **c)** The weight of a randomly selected member of the team is above 175 lb.

Then we have

$$
\begin{aligned}
P(160 \leq w \leq 180) &= P(0.4 \leq z \leq 1.2) \\
&= P(0 \leq z \leq 1.2) - P(0 \leq z \leq 0.4) \\
&= 0.3849 - 0.1554 \\
&= 0.2295.
\end{aligned}
$$

Now we can use Table A.

Thus, the probability that the weight of a randomly selected member of the team is 160 lb to 180 lb is 0.2295. That is, about 23% of the cross-country track team weighs between 160 lb and 180 lb.

❮ Quick Check 4

TECHNOLOGY CONNECTION

Statistics on a Calculator

It is possible to use a T1-83/84 Plus to make an approximation of the probability in Example 4 without performing a standard conversion, using Table A , or entering the normal probability density function. We first select an appropriate window, $[0, 300, 0, 0.02]$, with Xscl = 50 and Yscl = 0.01. Next, we use the ShadeNorm command, which we find by pressing **2ND** DISTR ▷ **ENTER**. We enter the values as shown and press **ENTER**. (If necessary, the ClearDraw option on the DRAW menu can be used to clear the graph.)

Left endpoint of interval   Right endpoint of interval

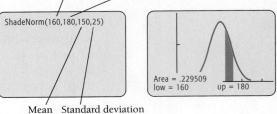

Mean   Standard deviation

The area is shaded and given as 0.229509, or about 23%.

(continued)

Alternatively, the probability can be calculated directly by pressing **2ND** and DISTR and selecting normalcdf. The values are entered as shown:

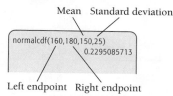

Mean   Standard deviation

normalcdf(160,180,150,25)
                0.2295085713

Left endpoint   Right endpoint

Open-ended intervals can be handled by selecting a "distant" value for the open end. For example, if we want to know the probability that a member of the cross-country team weighs less than 145 lb, we can use 0 as the left endpoint:

normalcdf(0,145,150,25)
                0.4207403112

If we want to know the probability that a member of the cross-country team weighs more than 160 lb, we can use 300 as the right endpoint:

normalcdf(160,300,150,25)
                0.3445783019

For cases involving an open-ended (infinite) bound, a rule of thumb is to set that bound at a value approximately five standard deviations below or above the mean. This will give areas that are accurate to over six decimal places.

EXERCISES

1. The annual rainfall in Crosleyville is normally distributed with mean $\mu = 150$ mm and standard deviation $\sigma = 25$ mm.

   a) What is the probability that the annual rainfall is between 125 mm and 170 mm?
   b) What is the probability that the annual rainfall is greater than 200 mm?

2. *Test scores.* All the high school students in a state take a standardized math test, and the scores are normally distributed with $\mu = 68$ and $\sigma = 5.3$.

   a) What is the probability that a randomly selected student scored between 65 and 75?
   b) What is the probability that the student scored above 70?

Percentiles

Suppose you take an exam and score better than 85% of all the students taking that exam. We say that your score is in the 85th *percentile*. For the standard normal distribution, the **percentile** for each z-value is the area under the standard normal curve to the left of z, multiplied by 100.

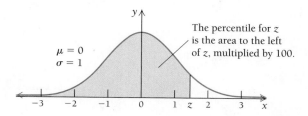

The percentile for z is the area to the left of z, multiplied by 100.

■ **EXAMPLE 5**   For the standard normal distribution, with $\mu = 0$ and $\sigma = 1$, determine the percentile corresponding to each of the following z-values.

   a) $z = 0$ b) $z = 2.25$ c) $z = -1.75$

Solution

a) The percentile corresponding to $z = 0$ is the area under the curve shown at the right from $-\infty$ to 0 (that is, to the left of 0). This is half of the total area of the standard normal distribution. Thus, a z-value of 0 corresponds to the 50th percentile: a score exactly at the mean is higher than 50% of all other scores.

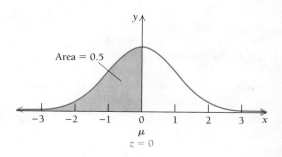

Area = 0.5

b) Table A shows that the area from 0 to 2.25 is 0.4878. We add this to 0.5. so the total area from $-\infty$ to 2.25 is 0.9878. A z-value of 2.25 corresponds to the 98.78th percentile.

c) We use Table A to determine the area from 0 to 1.75, which is 0.4599. By symmetry, the area between -1.75 and 0 is also 0.4599. Since the area from $-\infty$ to 0 is 0.5, to get the area from $-\infty$ to -1.75, we subtract: $0.5 - 0.4599 = 0.0401$. Therefore, a z-value of -1.75 corresponds to the 4th percentile (rounded).

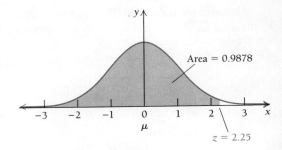

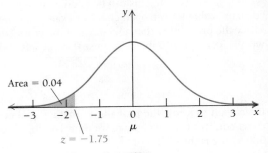

> **Quick Check 5**
>
> For the standard normal distribution, with $\mu = 0$ and $\sigma = 1$, determine the percentile corresponding to each of the following z-values.
>
> **a)** $z = -1$
>
> **b)** $z = 0.25$
>
> **c)** $z = 2.8$

❮ **Quick Check 5**

⬛ **EXAMPLE 6**   A large class takes an exam, and the students' scores are normally distributed: the mean score is $\mu = 72$, and the standard deviation is $\sigma = 4.5$. The professor curves the grading scale so that anyone who scored in the top 10% receives an A. What is the minimum score needed to get an A?

Solution   The top 10% corresponds to the 90th percentile. We need to determine the z-value that corresponds to an area of 0.9. From Table A, we see that an area of 0.4 is achieved when $z = 1.28$ (with rounding). Therefore, when the area of 0.5 for the left half of the distribution is included, we see that a z-value of 1.28 corresponds to the 90th percentile.

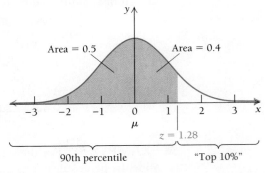

We use the transformation formula to determine the x-value (test score) that corresponds to $z = 1.28$:

$$1.28 = \frac{x - 72}{4.5}$$

$$5.76 = x - 72 \qquad \text{Multiplying both sides by 4.5}$$

$$x = 77.76. \qquad \text{Adding 72 to both sides}$$

Therefore, a score of 78 (rounded) is the minimum score needed to get an A.

> **Quick Check 6**
>
> Speeds along a stretch of state highway are normally distributed and have a mean of $\mu = 59$ mph with a standard deviation of $\sigma = 8$. The state police will ticket any driver whose speed is in the top 2% of this distribution. What is the minimum speed that will get a driver a citation?

❮ **Quick Check 6**

■ **EXAMPLE 7** **Business: Quality Control.** Bottles of cola are to contain a mean volume of 591 mL, but some variation is expected. Any bottle at or below the 5th percentile of the volume distribution is rejected. Suppose we know that a bottle that contains 593 mL of cola is in the 60th percentile. What is the minimum volume that will be accepted? Assume that the volumes are normally distributed.

Solution We are not given the standard deviation, but we can determine it from the given information. Since we know that a bottle with 593 mL is in the 60th percentile, we can find the z-value that corresponds to the 60th percentile and use that to find σ. Table A shows that the area from 0 to 0.253 (interpolated) is 0.10, which we add to the 0.5 from the left half of the distribution. Therefore, $z = 0.253$. We use the transformation formula to solve for σ, with $x = 593$ and $\mu = 591$:

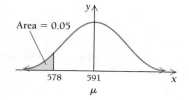

$$0.253 = \frac{593 - 591}{\sigma} \qquad \text{It is important to remember that } z = \frac{x - \mu}{\sigma}.$$

$$0.253 = \frac{2}{\sigma}$$

$$\sigma = \frac{2}{0.253} = 7.91. \qquad \text{Solving for } \sigma$$

We now determine the z-value that corresponds to the 5th percentile. Table A shows that the area from 0 to 1.645 is 0.45. So, by symmetry, the area from -1.645 to 0 is also 0.45. Therefore, the area to the left of -1.645 is 0.05. The 5th percentile corresponds to a z-value of -1.645. We now solve for the x-value that corresponds to this z-value:

$$-1.645 = \frac{x - 591}{7.91}$$

$$-13.01 = x - 591 \qquad \text{Multiplying both sides by 7.91}$$

$$x = 578. \qquad \text{Adding 591 and rounding to the nearest integer}$$

Rounding, we conclude that any bottle containing less than 578 mL is rejected and any bottle containing more than 578 mL is accepted.

❭ **Quick Check 7**

In Example 7, suppose any bottle at or above the 95th percentile is rejected, since its contents may overflow. What is the maximum volume that will be accepted?

❮ Quick Check 7

TECHNOLOGY CONNECTION

Percentiles on a Calculator

On the TI-83 Plus and TI-84 Plus calculators, z-values can be determined for percentiles of the standard normal distribution ($\mu = 0$ and $\sigma = 1$). Press **2ND** and DISTR and select InvNorm. Enter the percentile as a decimal between 0 and 1, and press **ENTER**. The result is the z-value that corresponds to the given percentile. For example, the z-value that corresponds to the 60th percentile is 0.253.

Percentile

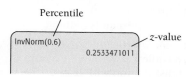

z-value

For any mean and standard deviation, the x-value can be found directly by entering the percentile, followed by the given mean and standard deviation, separated by commas. For example, if the mean is $\mu = 591$ and the standard

deviation is $\sigma = 7.91$, then the x-value that corresponds to the 5th percentile is 577.990.

Percentile   Mean   Standard deviation

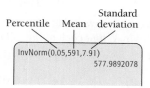

EXERCISES

Use InvNorm to find the x-values for the given percentiles.

1. 50th percentile, $\mu = 0$, $\sigma = 1$

2. 25th percentile, $\mu = 0$, $\sigma = 1$

3. 95th percentile, $\mu = 5$, $\sigma = 3$

4. 3rd percentile, $\mu = 10$, $\sigma = 4$

5. 58.45th percentile, $\mu = 100$, $\sigma = 15$

Section Summary

Assume that x is a continuous random variable and f is a probability density function of x on an interval $[a, b]$.

- The *mean* of x is given by $\mu = \displaystyle\int_a^b x f(x)\, dx$.

- The *variance* of x is $\sigma^2 = \displaystyle\int_a^b x^2 f(x)\, dx - \mu^2$.

- The *standard deviation* of x is $\sigma = \sqrt{\text{variance}}$. It is used to describe the "spread" of the data.
- The *standard normal distribution* of x is defined by the probability density function

$$f(x) = \frac{1}{\sqrt{2\pi}} e^{-x^2/2},$$

over $(-\infty, \infty)$, where $\mu = 0$ and $\sigma = 1$.

- The general case of a normally distributed random variable x with mean μ and standard deviation σ has the probability density function

$$f(x) = \frac{1}{\sigma\sqrt{2\pi}} e^{-(1/2)[(x-\mu)/\sigma]^2},$$

over the interval $(-\infty, \infty)$.
- For any normal distribution, the data, or x-values, are converted into z-values using the transformation formula:

$$z = \frac{x - \mu}{\sigma}.$$

- For the standard normal distribution, the *percentile* for each z-value is the area under the standard normal curve to the left of z, multiplied by 100.

EXERCISE SET
10.5

For each probability density function, over the given interval, find the mean, the variance, and the standard deviation.

1. $f(x) = \frac{1}{4}$, $[3, 7]$

2. $f(x) = \frac{1}{5}$, $[3, 8]$

3. $f(x) = \frac{1}{8}x$, $[0, 4]$

4. $f(x) = \frac{2}{9}x$, $[0, 3]$

5. $f(x) = \frac{1}{4}x$, $[1, 3]$

6. $f(x) = \frac{2}{3}x$, $[1, 2]$

For each probability density function, on the given interval, find the mean, the standard deviation, and the area under the function within one standard deviation of the mean.

7. $g(x) = \frac{1}{8}$, $[2, 10]$

8. $h(x) = \frac{1}{12}$, $[6, 18]$

9. $f(x) = \frac{1}{8}x$, $[0, 4]$

10. $f(x) = \frac{1}{8}x - \frac{1}{8}$, $[1, 5]$

Let x be a continuous random variable with a standard normal distribution. Using Table A, find each of the following.

11. $P(0 \leq x \leq 2.13)$

12. $P(0 \leq x \leq 0.36)$

13. $P(-1.37 \leq x \leq 0)$

14. $P(-2.01 \leq x \leq 0)$

15. $P(-1.89 \leq x \leq 0.45)$

16. $P(-2.94 \leq x \leq 2.00)$

17. $P(1.35 \leq x \leq 1.45)$

18. $P(0.76 \leq x \leq 1.45)$

19. $P(-1.27 \leq x \leq -0.58)$

20. $P(-2.45 \leq x \leq -1.24)$

21. $P(x \geq 3.01)$

22. $P(x \geq 1.01)$

23. **a)** $P(-1 \leq x \leq 1)$
b) What percentage of the area is from -1 to 1?

24. **a)** $P(-2 \leq x \leq 2)$
b) What percentage of the area is from -2 to 2?

Let x be a continuous random variable that is normally distributed with mean $\mu = 22$ and standard deviation $\sigma = 5$. Using Table A, find each of the following.

25. $P(24 \leq x \leq 30)$

26. $P(22 \leq x \leq 27)$

27. $P(19 \leq x \leq 25)$

28. $P(18 \leq x \leq 26)$

29–46. Use a graphing calculator to do Exercises 11–28.

47. Find the following percentiles for a standard normal distribution.

a) 30th percentile
b) 50th percentile
c) 95th percentile

48. In a normal distribution with $\mu = 60$ and $\sigma = 7$, find the x-value that corresponds to the

a) 75th percentile
b) 35th percentile

49. In a normal distribution with $\mu = -15$ and $\sigma = 0.4$, find the x-value that corresponds to the

a) 92nd percentile
b) 46th percentile

50. In a normal distribution with $\mu = 0$ and $\sigma = 4$, find the x-value that corresponds to the

a) 50th percentile
b) 84th percentile

APPLICATIONS

Business and Economics

51. Cereal box weights. The weights of boxes of Sugarpow cereal are uniformly distributed between 17.5 oz and 18.7 oz.

a) Find a probability density function f that models this situation.
b) Find the mean, the variance, and the standard deviation.
c) What is the probability that the weight of a randomly chosen box of Sugarpow is within one standard deviation of the mean?

52. Wait times. Customers at Hazen's Clothiers wait between 0 and 4 minutes, with the wait times uniformly distributed, before being helped by a salesperson.

a) Find a probability density function f that models this situation.
b) Find the mean, the variance, and the standard deviation.
c) What is the probability that a randomly chosen customer will have a wait time that is within one standard deviation of the mean?

53. Wait times. Patients at the Pearlview Medical Clinic have a wait time modeled by $f(x) = -\frac{1}{450}x + \frac{1}{15}$, for $0 \le x \le 30$, where the random variable x is in minutes.

a) Find the mean, the variance, and the standard deviation.
b) Find the probability that a randomly chosen patient will have a wait time within one standard deviation of the mean.

54. Wait times. The time between television commercial breaks is modeled by $g(x) = \frac{2}{9}x - \frac{14}{9}$, for $7 \le x \le 10$, where the random variable x is in minutes.

a) Find the mean, the variance, and the standard deviation.
b) Find the probability that the time between commercial breaks is within one standard deviation of the mean.

55. Mail orders. The number of orders, N, received daily by an online vendor of used Blu-Ray DVDs is normally distributed with mean 250 and standard deviation 20. The company has to hire extra help or pay overtime on days when over 300 orders are received. On what percentage of days will the company have to hire extra help or pay overtime?

56. Bread baking. The number of loaves of bread, N, baked each day by Fireside Bakers is normally distributed

with mean 1000 and standard deviation 50. The bakery pays bonuses on days when at least 1100 loaves are baked. On what percentage of days will the bakery pay a bonus?

57. Auto body welding. The processing time for the robogate welding station has a normal distribution with mean 38.6 sec and standard deviation 1.729 sec. Find the probability that the next operation of the robogate welding station will take 40 sec or less.

58. Piercing. The processing time for an automatic piercing station has a normal distribution with mean 36.2 sec and standard deviation 2.108 sec. Find the probability that the next operation of the piercing station will take between 35 and 40 sec.

59. Test score distribution. In 2011, combined SAT reading and math scores were normally distributed with mean 1011 and standard deviation 100. Find the SAT scores that correspond to these percentiles. (*Source:* www. collegeboard.com.)

a) 35th percentile
b) 60th percentile
c) 92nd percentile

General Interest

60. Test score distribution. The scores on a biology test are normally distributed with mean 65 and standard deviation 20. A score from 80 to 89 is a B. What is the probability of getting a B?

61. Test score distribution. In a large class, students test scores had a mean of $\mu = 76$ and a standard deviation $\sigma = 7$.

a) The top 12% of students got an A. Find the minimum score needed to get an A (round to the appropriate integer).
b) Scores in the 70th percentile earned a B. Find the minimum score needed to get a B (round to the appropriate integer).

62. Average temperature. Las Vegas, Nevada, has an average daily high temperature of 104 degrees in July, with a standard deviation of 6 degrees. (*Source:* www.wunderground.com.)

a) In what percentile is a temperature of 112 degrees?
b) What temperature would be at the 67th percentile?
c) What temperature would be in the top 0.5% of all July temperatures for Las Vegas?

63. Heights of basketball players. Players in the National Basketball Association have a mean height of 79 in. (6 ft 7 in.). (*Source:* www.apbr.org.) If a basketball player who is 7 ft 2 in. tall is in the top 1% of players by height, in what percentile is a 6 ft 11 in. player?

64. Bowling scores. At the time this book was written, the bowling scores, S, of author Marv Bittinger (shown below) were normally distributed with mean 201 and standard deviation 23.

a) Find the probability that a score is from 185 to 215, and interpret your results.
b) Find the probability that a score is from 160 to 175, and interpret your results.
c) Find the probability that a score is greater than 200, and interpret your results.

SYNTHESIS

For each probability density function, over the given interval, find the mean, the variance, and the standard deviation.

65. $f(x) = \dfrac{1}{b-a}$, over $[a, b]$

66. $f(x) = \dfrac{3a^3}{x^4}$, over $[a, \infty)$

Median. *Let x be a continuous random variable over* $[a, b]$ *with probability density function f. The* **median** *of the x-values is that number m for which*

$$\int_a^m f(x)\, dx = \frac{1}{2}.$$

Find the median.

67. $f(x) = \frac{1}{2}x$, $[0, 2]$

68. $f(x) = \frac{3}{2}x^2$, $[-1, 1]$

69. $f(x) = ke^{-kx}$, $[0, \infty)$

70. Are the median and the mean the same? Why or why not?

71. Business: coffee production. Suppose that the amount of coffee beans loaded into a vacuum-packed bag has a mean weight of μ ounces, which can be adjusted on the filling machine. Suppose that the amount dispensed is normally distributed with $\sigma = 0.2$ oz. What should μ be set at to ensure that only 1 bag in 50 will have less than 16 oz?

72. Business: vending machines. Suppose that the mean amount of cappuccino, μ, dispensed by a vending machine can be set. If a cup holds 8.5 oz and the amount dispensed is normally distributed with $\sigma = 0.3$ oz, what should μ be set at to ensure that only 1 cup in 100 will overflow?

73. Explain why a normal distribution may not apply if you were analyzing the distribution of weights of students in a classroom.

74. A professor gives an easy test worth 100 points. The mean is 94, and the standard deviation is 5. Is it possible to apply a normal distribution to this situation? Why or why not?

TECHNOLOGY CONNECTION

75. Approximate the integral $\displaystyle\int_{-\infty}^{\infty} e^{-x^2}\, dx$.

Answers to Quick Checks
1. $\mu = \frac{4}{3}$ **2. (a)** $\sigma^2 = \frac{2}{9}, \sigma = \frac{1}{3}\sqrt{2} \approx 0.47$; **(b)** 0.63
3. (a) 0.533; **(b)** 0.279; **(c)** 0.345
4. (a) 0.726; **(b)** 0.305; **(c)** 0.159
5. (a) 15.9th percentile; **(b)** about 59.9th percentile;
(c) 99.7th percentile **6.** About 75.4 mph
7. 604 mL

KEY TERMS AND CONCEPTS	EXAMPLES

SECTION 10.1

A **set** is a collection of objects. Each object is an **element** or a **member** of the set.	$V = \{a, e, i, o, u\}$ is the set of vowels. The element e is a member of set V, so we write $e \in V$. The element f is not a member of set V, so we write $f \notin V$.
The **empty set** contains no elements. It is denoted $\varnothing$.	The set of real-number solutions to the equation $x^2 = -1$ is empty. Its solution set is $\varnothing$.
The **universal set** (also called the **universe**) is the set of all elements within the context of a situation. It is written U.	The set of possible solutions to a linear equation comes from the universal set of real numbers.
Two sets are **equal** if they contain the same elements, ignoring repeated elements and orderings.	$\{a, e, i, o, u\} = \{e, o, i, a, a, a, e, u\} = \{u, o, i, e, a\}$
A is a **subset** of B if all elements of A are elements of B. We write $A \subseteq B$. The empty set is a subset of all sets.	$\{a, e, i\} \subseteq \{a, e, i, o, u\}$ $\{a, e, i, o, u\} \subseteq \{a, e, i, o, u\}$ $\varnothing \subseteq \{a, e, i, o, u\}$
The **complement** of A is the set of all elements in U and not in A: $$A' = \{x \mid x \in U \text{ and } x \notin A\}.$$	If $U = \{a, b, e, i, k, o, t, u, z\}$ and $V = \{a, e, i, o, u\}$, then $V' = \{b, k, t, z\}$.
The **intersection** of A and B is the set of all elements in A and B: $$A \cap B = \{x \mid x \in A \text{ and } x \in B\}.$$ If $A \cap B = \varnothing$, then A and B are **mutually exclusive**.	$\{a, e, i\} \cap \{a, e, o, u\} = \{a, e\}$ $\{a, e, i\} \cap \{o, u\} = \varnothing$
The **union** of A and B is the set of all elements in A or B: $$A \cup B = \{x \mid x \in A \text{ or } x \in B\}.$$	$\{a, e, i\} \cup \{e, o, u\} = \{a, e, i, o, u\}$
The **cardinality** of a set is the number of elements in the set and is denoted by $n(A)$. The empty set has a cardinality of zero: $n(\varnothing) = 0$	If $V = \{a, e, i, o, u\}$, then $n(V) = 5$.
The cardinality of the union of two sets is $$n(A \cup B) = n(A) + n(B) - n(A \cap B).$$	Let $A = \{a, b, c, d\}$ and $B = \{c, d, e, f, g\}$. Then $n(A \cup B) = 4 + 5 - 2 = 7$.

SECTION 10.2

| An **experiment** produces **outcomes**. Each performance of an experiment is called a **trial**. The set of all possible outcomes is called the **sample space** S. An **event** (or **event space**) E is any subset of the sample space. | Experiment: toss a coin once

 Outcomes: tails (T) or heads (H)

 Sample space: $S = \{T, H\}$

 Event: the coin showing heads is $E = \{H\}$ |

(continued)

KEY TERMS AND CONCEPTS	EXAMPLES

A *probability* is a number that represents the likelihood of an event occurring. Probabilities are **experimental**, meaning that an experiment must be performed in order to determine them, or **theoretical**, meaning that they can be determined by mathematical reasoning.

A baseball player's batting average is a probability that indicates the likelihood of a hit each time at bat. It is determined experimentally, based on past performance.

The probability of rolling a sum of 10 on one throw of two dice can be found mathematically.

The **theoretical probability** of an event E occurring is the number of elements in E divided by the number of elements in S:

$$P(E) = \frac{n(E)}{n(S)}.$$

The probability of a fair coin landing heads-up when tossed is

$$P(E) = \frac{n(E)}{n(S)} = \frac{1}{2}, \text{ or } 50\%.$$

For any event E, $0 \le P(E) \le 1$.

A probability of 0 indicates an *impossible* event.

A probability of 1 indicates a *certain* event.

Roll two dice and let event E be that the sum is 12. Then, $P(E) = \frac{1}{36}$.

The probability of rolling two dice and getting a sum of 13 is 0, meaning it is impossible.

The probability of rolling two dice and getting a sum that is not 13 is 1, meaning it is certain.

The sum of the probability of an event E and its complementary event E' is 1:

$$P(E) + P(E') = 1.$$

Therefore, $P(E') = 1 - P(E)$.

The probability of rolling two dice and getting a sum that is not 12 is $P(E') = 1 - \frac{1}{36} = \frac{35}{36}$.

The probability of the union of two events is

$$P(E \cup F) = P(E) + P(F) - P(E \cap F).$$

Roll two fair dice and let event E be that the sum is 4 and event F be that at least one die shows a 3. Then

$$P(E) = \frac{3}{36}, \quad P(F) = \frac{11}{36}, \quad \text{and} \quad P(E \cap F) = \frac{2}{36}.$$

Therefore,

$$P(E \cup F) = \frac{3}{36} + \frac{11}{36} - \frac{2}{36} = \frac{12}{36} = \frac{1}{3}.$$

A **frequency table** can be used to find experimental probabilities. A frequency value divided by the total number of data points is interpreted as a probability.

Thirty people are surveyed, and asked how many pets they own. The data are shown in the following frequency table:

Number of pets, x	0	1	2	3
Frequency	11	13	5	1

A randomly chosen respondent from this survey has a probability of $\frac{13}{30}$ of owning one pet.

KEY TERMS AND CONCEPTS	**EXAMPLES**

SECTION 10.3

A **random variable** is a function $X(s)$ that assigns a numerical value to each element s in the sample space S. Often, a random variable is denoted by X, instead of $X(s)$. If the elements of S can be listed, then the sample space is **discrete** and the random variable is a **discrete random variable**.

Three fair dice are rolled, and the random variable X is the number of times a 2 shows on a die. Thus, a roll of 3-2-4 gives $X(3\text{-}2\text{-}4) = 1$. A roll of 2-2-6 gives $X(2\text{-}2\text{-}6) = 2$, and so on. Since each outcome can be listed, this is considered a discrete sample space, and X is a discrete random variable.

If X is a discrete random variable, then a **probability mass function** is a function f that meets these conditions:

1. $f(x) \geq 0$ for all x.
2. $f(x) = P(X = x)$.
3. The sum of all $f(x)$ is 1.

The function $f(x) = \frac{1}{10}x$, for $x = 1, 2, 3, 4$, is a probability mass function since it meets all three conditions of the definition:

1. $f(1) = \frac{1}{10}, f(2) = \frac{2}{10}, f(3) = \frac{3}{10}$, and $f(4) = \frac{4}{10}$ are all nonnegative.
2. The probability that $x = 2$ is $f(2) = P(X = 2) = \frac{2}{10}$, for example.
3. $f(1) + f(2) + f(3) + f(4) = 1$.

The graph of a probability mass function is represented as a bar chart called a **histogram**. The probability mass function and the histogram together represent a **probability distribution**.

The function $f(x) = \frac{1}{10}x$, for $x = 1, 2, 3, 4$, is a probability mass function. Its histogram is as follows:

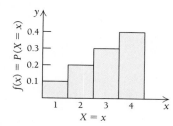

Together, they form a probability distribution.

If X is a discrete random variable with a probability mass function $f(x) = P(X = x)$, then the **cumulative probability function** of f is

$$F(x) = P(X \leq x),$$

which is an increasing function with a minimum value of 0 and a maximum value of 1.

The cumulative probability function of $f(x) = \frac{1}{10}x$, for $x = 1, 2, 3, 4$, is

$$F = \begin{cases} 0, & x < 1, \\ \frac{1}{10}, & 1 \leq x < 2, \\ \frac{3}{10}, & 2 \leq x < 3, \\ \frac{6}{10}, & 3 \leq x < 4, \\ 1, & 4 \leq x. \end{cases}$$

If X is a discrete random variable with a probability mass function f, then the **mean** of X is given by

$$\mu = x_1 \cdot f(x_1) + x_2 \cdot f(x_2) + \cdots + x_n \cdot f(x_n)$$
$$= \sum_{i=1}^{n} x_i \cdot f(x_i)$$

The mean of X, which has a probability mass function $f(x) = \frac{1}{10}x$, for $x = 1, 2, 3, 4$, is

$$\mu = 1 \cdot \frac{1}{10} + 2 \cdot \frac{2}{10} + 3 \cdot \frac{3}{10} + 4 \cdot \frac{4}{10} = 3.$$

KEY TERMS AND CONCEPTS	**EXAMPLES**

SECTION 10.4

In probability, a **continuous random variable** is a quantity that can be observed (or measured) repeatedly and whose possible values comprise an interval of real numbers.

A function f is a **probability density function** of a continuous random variable x if it meets the following conditions:

1. For all x in its domain, $f(x) \geq 0$.
2. The area under the graph of f is 1.
3. For any subinterval $[c, d]$ in the domain of f, the probability that x will be in that subinterval is $P([c, d]) = \int_c^d f(x)\, dx$.

A probability density function is always stated with its domain.

The function $f(x) = \frac{2}{9}x$, for $0 \leq x \leq 3$, is a probability density function since

- $f(x) \geq 0$ for all x in $[0, 3]$.

- $\int_0^3 \frac{2}{9}x\, dx = \left[\frac{1}{9}x^2\right]_0^3 = \frac{3^2}{9} - 0 = 1.$

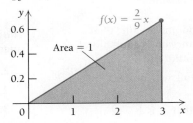

The probability that x is between 1.5 and 2.3 is

$$\int_{1.5}^{2.3} \frac{2}{9}x\, dx = \left[\frac{1}{9}x^2\right]_{1.5}^{2.3} = \frac{(2.3)^2}{9} - \frac{(1.5)^2}{9} \approx 0.338.$$

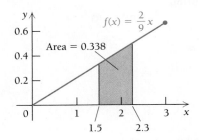

If X is a continuous random variable and f is a probability density function of x on $[a, b]$, then the **cumulative density function** of f is given by

$$F(x) = \int_a^x f(s)\, ds.$$

A cumulative density function is an increasing function such that $0 \leq F(x) \leq 1$, with $F(a) = 0$ and $F(b) = 1$.

The function $f(x) = \frac{2}{9}x$, for $0 \leq x \leq 3$, is a probability density function. Its cumulative density function is

$$F(x) = \int_0^x \frac{2}{9}s\, ds = \left[\frac{1}{9}s^2\right]_0^x = \frac{1}{9}x^2 - 0 = \frac{1}{9}x^2, \quad \text{for } 0 \leq x \leq 3.$$

A continuous random variable is **uniformly distributed** over an interval $[a, b]$ if it has a probability density function f given by

$$f(x) = \frac{1}{b - a}, \quad \text{for } a \leq x \leq b.$$

Helicopter tours over Hoover Dam last from 45 to 55 min. with the times uniformly distributed. If x is the time that a tour lasts, the probability density function f is given by

$$f(x) = \tfrac{1}{10}, \quad \text{for } 45 \leq x \leq 55.$$

The probability that a flight lasts between 48 and 53.5 min is

$$\int_{48}^{53.5} \frac{1}{10}\, dx = \left[\frac{1}{10}x\right]_{48}^{53.5} = \frac{1}{10}(53.5 - 48) = 0.55.$$

KEY TERMS AND CONCEPTS	**EXAMPLES**
A continuous random variable is **exponentially distributed** if it has a probability density function f of the form $$f(x) = ke^{-kx}, \quad \text{over the interval } [0, \infty).$$	The time x (in minutes) between shoppers entering a store is modeled by the probability density function $$f(x) = 3e^{-3x}, \quad \text{for } 0 \leq x < \infty.$$ The probability that the time between shoppers is 2 min or less is $$\int_0^2 3e^{-3x}\,dx = [-e^{-3x}]_0^2 = (-e^{-6} - (-1)) = 0.9975.$$

SECTION 10.5

Let x be a continuous random variable over $[a, b]$ with probability density function f. Then the **mean**, μ, of x is $$\mu = \int_a^b x \cdot f(x)\,dx.$$	Consider the probability density function $f(x) = \frac{2}{9}x$ over the interval $[0, 3]$. Its mean is $$\mu = \int_0^3 x \cdot \frac{2}{9}x\,dx = \int_0^3 \frac{2}{9}x^2\,dx = 2.$$
The **variance**, σ^2, of x is $$\sigma^2 = \int_a^b x^2 \cdot f(x)\,dx - \left[\int_a^b x \cdot f(x)\,dx\right]^2$$ $$= \int_a^b x^2 \cdot f(x)\,dx - \mu^2.$$	Its variance is $$\sigma^2 = \left[\int_0^3 x^2 \cdot \frac{2}{9}x\,dx\right] - \mu^2 = 4.5 - (2)^2 = 0.5.$$
The **standard deviation**, σ, of x is the square root of the variance: $$\sigma = \sqrt{\text{variance}}.$$	Its standard deviation is $$\sigma = \sqrt{0.5} \approx 0.71.$$
A continuous random variable x has a **standard normal distribution** if it has a probability density function f given by $$f(x) = \frac{1}{\sqrt{2\pi}}e^{-x^2/2}, \quad \text{over } (-\infty, \infty),$$ with $\mu = 0$ and $\sigma = 1$. Tables or calculators are used to determine areas within the standard normal distribution. To convert an x-value into a z-value for use with the standard normal distribution, we use the transformation formula $$z = \frac{x - \mu}{\sigma}.$$	Weights of packages of ground coffee are normally distributed with mean $\mu = 3$ oz and standard deviation $\sigma = 0.5$. The probability that a packet of ground coffee has a weight between 2.75 oz and 3.15 oz is $$P\left(\frac{2.75 - 3}{0.5} \leq x \leq \frac{3.15 - 3}{0.5}\right) = P(-0.5 \leq z \leq 0.3)$$ $$= 0.309 = 30.9\%.$$ 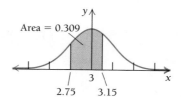
In a standard normal distribution, the **percentile** for a z-value is the area under the curve to the left of z multiplied by 100.	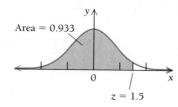 In the standard normal distribution, the area to the left of $z = 1.5$ is 0.933. Therefore, a z-value of 1.5 corresponds to the 93.3rd percentile.

These review exercises are for test preparation. They can also be used as a practice test. Answers are in the back of the book . The blue bracketed section references tell you what part(s) of the chapter to restudy if your answer is incorrect.

CONCEPT REINFORCEMENT

In Exercises 1–10, classify each statement as true or false.

1. If $A = \{d, e, f, g\}$ and $B = \{e, g, h\}$, then $A \cap B = \{e, g\}$. [10.1]

2. If $G = \{1, 2, 4, 5, 1, 2, 4, 5\}$, then $n(G) = 8$. [10.1]

3. If an event has a probability of 0.4, then its complementary event has a probability of 0.6. [10.2]

4. An event that is certain to happen has a probability of 1. [10.2]

5. If $f(x) = \frac{1}{15}x$, for $x = 1, 2, 3, 4, 5$, is a probability mass function, then $f(8) = \frac{8}{15}$. [10.3]

6. Two fair coins are tossed, and the random variable X is the number of heads that show. Therefore, $P(X = 2) = \frac{1}{4}$. [10.3]

7. If $f(x) = \frac{1}{2}x$, for $0 \le x \le 2$, is a probability mass function, then its cumulative density function is $F(x) = \frac{1}{4}x^2$, for $0 \le x \le 2$. [10.4]

8. The function $g(x) = \frac{1}{12}$, for $20 \le x \le 32$, is a probability density function. [10.4]

9. The area under the graph of a probability density function is always 1. [10.4]

10. If f is a probability density function and $x = 1.5$ corresponds to the 45th percentile, then the area under the graph of f to the left of $x = 1.5$ is 45. [10.5]

REVIEW EXERCISES

11. Let $U = \{1, 2, 3, 4, 5, 6, 7, 8, 9, 10\}$, $A = \{3, 5, 6, 8, 9\}$, $B = \{1, 4, 7\}$ and $C = \{5, 6, 7, 8, 9, 10\}$. [10.1]
    a) Find $A \cup C$.
    b) Find $A \cap B$.
    c) Find C'.
    d) Find $n(A \cup B)$.
    e) Find $n(B \cup C)$.

12. A survey of 80 patrons at Big Roy's Rib Joint shows that 55 like beef brisket, 40 like pulled pork, and 31 like both. How many patrons like neither item? [10.1]

13. Three fair coins are tossed. Determine the probability that exactly two show tails. [10.2]

14. Two fair dice are rolled. Let event E be that at least one die shows a 5 and event F be that the sum of the two dice is 7. [10.2]
    a) Find $P(E)$ and $P(E')$.
    b) Find $P(F)$ and $P(F')$.
    c) Find $P(E \cup F)$.

15. Determine k such that $\{(0, 0.2), (1, 0.05), (2, 0.31), (4, k)\}$ is a probability mass function. [10.3]

16. Let $f(x) = \frac{c}{x}$, for $x = 2, 3, 5$, be a probability mass function. [10.3]
    a) Find c.
    b) Draw the probability distribution histogram.
    c) Find the cumulative probability function $F(x)$, and draw its graph.

17. The table below shows the amounts of cash won by contestants on the game show *Cash Bonanza* over a 2-month period. [10.3]

Amount	Frequency
$0	3
$1000	8
$5000	12
$10,000	17
$25,000	19
$50,000	6
$100,000	2

Let the random variable X represent the cash amount won by a randomly selected contestant.

a) Find $P(X = 5000)$.
b) Find $P(X \ge 10,000)$.
c) Find μ, and explain what this number represents.

18. Five cards are sequentially numbered 1 through 5, and two cards are randomly selected from the set without replacement, meaning that it is not possible to select the same card twice (and, therefore, impossible to get the same number twice). Let the random variable X represent the sum of the numbers on the two cards selected. [10.3]
    a) Write the sample space as a set of ordered pairs.
    b) Find $P(X = 4)$.
    c) Find $P(X \le 4)$.
    d) Find μ, and explain what this number represents.

19. Consider the experiment and sample space in Exercise 18. Let the random variable Y represent the positive difference of the numbers on the two cards selected. [10.3]

   a) Find $P(Y = 2)$.
   b) Find $P(Y \leq 2)$.
   c) Find μ, and explain what this number represents.

20. Let $g(x) = cx$, for $2 \leq x \leq 4$, be a probability density function. [10.4, 10.5]

   a) Find c.
   b) Find $P([2.5, 3.5])$.
   c) Find the mean μ and the standard deviation σ.
   d) Find the cumulative density function G.

21. Business: wait times. A customer at Eastwood's Furniture Showroom waits between 0 and 3 min, with the wait times uniformly distributed, before being helped by a salesperson. [10.4, 10.5]

   a) What is the probability that a customer's wait time is between 30 sec and 2 min?
   b) What is the probability that a customer waits exactly 2 min?
   c) What is the mean wait time?
   d) Find the variance and the standard deviation.
   e) What is the probability that a customer's wait time is within one standard deviation of the mean?

22. Business: wait times. The wait time between customers at Smalley's Lumber is exponentially distributed with a probability density function $f(x) = 1.5e^{-1.5x}$, for $0 \leq x < \infty$. [10.4, 10.5]

   a) Find the probability that two customers show up within 2 min of one another.
   b) Find $F(4)$, and explain what this number represents.

Assuming a standard normal distribution, use Table A to find each of the following, rounded to four decimal places. [10.5]

23. $P(0 \leq z \leq 1.85)$

24. $P(-1.74 \leq z \leq 1.43)$

25. $P(-2.08 \leq z \leq -1.18)$

26. $P(z \geq 0)$

27. Determine the percentile corresponding to $z = 1.52$.

28. For a normal distribution with mean $\mu = 17$ and standard deviation $\sigma = 3.5$, find each of the following, rounded to four decimal places. [10.5]

   a) $P(15 \leq x \leq 17.5)$
   b) The percentile corresponding to $x = 19$

29. Business: pizza sales. The number of pizzas sold daily at Benito's Pizzeria is normally distributed with mean $\mu = 90$ and standard deviation $\sigma = 20$. What is the probability that at least 100 pizzas are sold during a day? [10.5]

30. Business: distribution of revenue. Benito's Pizzeria has daily mean revenues that are normally distributed, with $\mu = \$5500$ and standard deviation $\sigma = \$425$. What is the lowest amount in the top 5% of the daily revenues? [10.5]

SYNTHESIS

31. Find c such that the function $f(x) = cxe^x$, for $0 \leq x \leq 1$, is a probability density function.

TECHNOLOGY CONNECTION

32. Find c such that the function $f = \{(0, c), (1, c^2), (2, c^4)\}$ is a probability mass function.

33. Using a random-number generator, generate 50 random numbers between 0 and 1, then square each random number. Based on the resulting data, what is the probability that a randomly chosen number between 0 and 1 has a square that is less than or equal to 0.5?

CHAPTER 10 TEST

1. Let $U = \{a, b, c, d, e, f, g, h\}$, $A = \{a, b, c, d\}$, $B = \{c, d, f, g, h\}$, and $C = \{e, g, h\}$.

   a) Find A'.
   b) Find $B \cup C$.
   c) Find $A \cap C$.
   d) Find $n(B \cap C)$.
   e) Find $n(B \cup C)$.

2. Two fair dice are thrown. Let event E be that at least one die shows a 3 and event F be that the sum of the two dice is 5.

   a) Find $P(E)$ and $P(E')$.
   b) Find $P(F)$ and $P(F')$.
   c) Find $P(E \cup F)$.

3. A spinner marked off in five equal-sized regions, numbered 1 through 5, is spun twice. Let the random variable X represent the sum of the numbers from the two spins.

   a) Find $P(X = 3)$
   b) Find $P(X \leq 5)$.

4. In Exercise 3, let the random variable Y represent the product of the numbers from the two spins.

   a) Find $P(Y = 7)$
   b) Find $P(Y \leq 10)$.

5. Let $f(x) = cx$, for $x = 1, 3, 5, 7$, be a probability mass function.

  a) Determine c.
  b) Draw the probability distribution histogram.
  c) Find $F(x)$, and draw its graph.
  d) Find μ.

6. Explain why $f(x) = \frac{1}{41}x^2$, for $x = 1, 4, 5$, is not a probability mass function.

7. Let $g(x) = cx^2$ over the interval $[1, 3]$ be a probability density function.

  a) Find $P([1.2, 2.1])$.
  b) Find the cumulative density function G.
  c) Find μ.

8. Dan's commute to work takes between 25 and 35 min, with the commute times uniformly distributed. Let $f(x) = \frac{1}{10}$, for $25 \le x \le 35$, be a probability distribution function for Dan's commute times, where x is in minutes.

  a) Find the probability that Dan's commute takes between 29 and 34.5 mins.
  b) Find the probability that his commute takes exactly 27 mins.
  c) Find the mean, the variance, and the standard deviation of Dan's commute times.
  d) What is the probability that Dan's commute time is within one standard deviation of the mean?

9. General interest: highway traffic. The time between vehicles passing a certain checkpoint on a highway is exponentially distributed, with a probability density function $g(x) = 1.35e^{-1.35x}$, for $0 \le x < \infty$, where x is measured in minutes.

  a) Find the probability that the time between two vehicles is between 1 and 2 mins.
  b) Find the probability that the time between two vehicles is greater than 3 mins.
  c) Find the mean time between two vehicles.

10. Business: sales. Stan the Hot Dog Man tracks the number of hot dogs he sells to each customer. The table below shows how many customers bought 0, 1, 2, 3, or 4 hot dogs during a busy Wednesday. (Some customers buy only a beverage.)

Number of hot dogs sold	0	1	2	3	4
Frequency	18	59	53	18	6

Let the random variable X represent the number of hot dogs purchased by a randomly selected customer.

  a) Find $P(X = 2)$.
  b) Find $P(X \le 2)$.
  c) Find μ, and explain what this number represents.

11. Let the random variable x be normally distributed, with mean $\mu = 0$ and standard deviation $\sigma = 1$. Use Table A to determine the following probabilities.

  a) $P(-1.33 \le x \le 0)$
  b) $P(x \ge 1.75)$
  c) $P(2.14 \le x \le 3)$

12. Business: gas mileage efficiency. Advertisements for a new vehicle claim that it will get 22 miles per gallon (mpg), but in reality the vehicles have a mean gas consumption of $\mu = 22$ mpg, with a standard deviation of $\sigma = 2.5$ mpg. Assuming a normal distribution, what is the probability that one of these new vehicles will get between 20 and 23 mpg?

13. New students at Bonita College take a math placement test. The scores are normally distributed with a mean of $\mu = 40$ and a standard deviation of $\sigma = 5$. Any student who scores in the bottom 25% must take remedial algebra. What is the highest score (rounded to the appropriate integer) a student can get and still be required to take remedial algebra?

14. A large field of cross-country runners is competing to advance to the state finals. Suppose the mean time to run the course is $\mu = 115$ min with a standard deviation of $\sigma = 12$ min and the times are distributed normally. The top 15% of runners will advance to the state finals. Determine the maximum time (rounded to the nearest tenth of a minute) needed to advance to the state finals.

SYNTHESIS

15. Find c such that $f(x) = cx \ln x$, for $1 \le x \le 2$, is a probability density function.

TECHNOLOGY CONNECTION

16. A number is randomly chosen between 0 and 1. Use a random-number generator to find the probability that the cube of the number is less than or equal to 0.7.

Extended Technology Application

Experimental Probability and the Law of Large Numbers

Theoretical probability can answer many questions, but not all. There are some situations in which probabilities can only be determined experimentally. For example, probabilities in weather reports are based on sophisticated computer models that are run many times in succession. If 30% of the runs predict rainfall, then the weather report will say "30% chance of rain." Some situations can be examined using both experimental and theoretical probability. Many of the theoretical probability examples presented in this chapter can be performed as experiments. It is interesting to see how well the results of many experimental trials agree with the theoretical results.

There is no rule that says how many trials of an experiment need to be performed to obtain an experimental probability that is close to the theoretical probability. If the experiment involves many possible outcomes, then more trials may be necessary. For example, it may be sufficient to toss a fair coin 50 times to see that half the tosses result in heads. However, to verify how often a roll of two fair dice results in a sum of 12 may require many hundreds of rolls. In both cases, after many trials, the experimental probability will be very close to the theoretical probability, and, most importantly, it will *stay* close as more trials are performed.

We know from Example 1 in Section 10.2 that when a fair die is rolled, the theoretical probability that it will show a 2 is $\frac{1}{6} = 0.167$. How do the outcomes of rolling a real die many times compare to this theoretical figure? Students at Arizona State University performed this experiment: each student rolled a die 25 times and tallied the number of times a 2 came up. Their results were collected (in no particular order) and tallied on a spreadsheet. A total of 82 students participated, for a total of 2050 trials. The running tally was as follows:

Assuming that the theoretical probability of an event has been accurately determined, the *Law of Large Numbers* states that the experimental probability of the same event will get very close to the theoretical probability, and stay close, after many trials. For example, if you toss a fair coin 10 times, it's possible that 7 of the 10 tosses will result in heads. However, after 100 tosses, the occurrence of heads will tend toward 50, and the more tosses you make, the stronger the trend will be toward half the tosses resulting in heads. Streaks and other forms of "luck" may skew the outcomes for a small number of trials, but these effects become negligible after more trials.

Number of rolls	Frequency of a 2	Experimental probability
100	17	0.17
200	38	0.19
300	52	0.173
400	68	0.17
500	83	0.166
1000	166	0.166
2000	338	0.169
2050	347	0.169

After 500 rolls, the experimental probability of rolling a 2 was very close to the theoretical probability of $\frac{1}{6}$. More importantly, the experimental probability *stayed* close to the theoretical probability. After 2050 rolls, we would expect $2050/6 \approx 342$ occurrences of a 2. The experimental frequency of 347 is very close to 342, suggesting that the die was indeed fair.

EXERCISES

Each exercise is designed for whole-class participation, with the data from all individuals pooled and examined as a whole.

1. **Toss a fair coin.** The theoretical probability of a fair coin landing heads-up on one toss is $\frac{1}{2}$. To verify this experimentally, each student should use a coin of the same type.

   a) Each student tosses his or her coin 10 times (or more, depending on how many students there are) and tallies the outcomes (for example, 6 heads and 4 tails). The instructor collects these data.

   b) How many trials were performed in all?

   c) How many times did the coin land heads-up?

   d) What is the class's experimental probability for a coin landing heads-up?

2. **Toss two fair coins.** The theoretical probability of two fair coins both showing heads when tossed simultaneously is $\frac{1}{4}$. Each student should use two identical coins.

   a) Each student tosses his or her pair of coins 20 times (or more, depending on how many students there are) and tallies the outcomes (for example, 4 TT, 7 HT, 6 TH, 3 HH). The instructor collects these data.

   b) How many trials were performed in all?

   c) How many times did both coins land heads-up?

   d) What is the class's experimental probability for both coins landing heads-up?

3. **Roll two fair dice.** The theoretical probability of getting a sum of 12 on one roll of two fair dice is $\frac{1}{36}$. Each student should use two dice.

   a) Each student rolls his or her pair of dice 20 times (or more, depending on how many students there are) and tallies the outcomes. The instructor collects these data.

   b) How many trials were performed in all?

   c) How many times did the dice show a sum of 12?

   d) What is the class's experimental probability for getting a sum of 12 on one roll of two dice?

Using a Spreadsheet

Random numbers can also be generated on a spreadsheet (such as Excel™). This allows a student working alone to simulate an experiment with many hundreds or thousands of trials. For example, the coin-toss experiment can be simulated by letting any number less than 0.5 be tails and any number greater than 0.5 be heads.

To do this, open a spreadsheet, and in cell A1, enter =RAND(). This generates a random number between 0 and 1. Copy this cell; then paste it into cells A2 through A1000. In cell Bl, enter =IF(A1 < 0.5,1,0). This will return a value of 1 if the number in the cell to its left is less than 0.5 and a value of 0 if the number is greater than 0.5. Copy cell B1, and paste it into cells B2 through B1000. In cell C1, enter =SUM(B$1:B1). Copy cell C1, and paste it into cells C2 through C1000. This generates a running sum of the cells in column B.

Below is an example of a simulation of the coin-toss experiment using an Excel spreadsheet, which shows results for trials 1 through 20 (left) and up through trials 100 and 1000 (right). We can see that the frequency of heads is trending very closely to the expected theoretical probability of 0.5.

	A	B	C		A	B	C
1	0.027201	1	1	90	0.735026	0	48
2	0.671373	0	1	91	0.537605	0	48
3	0.927818	0	1	92	0.29239	1	49
4	0.482646	1	2	93	0.685191	0	49
5	0.109804	1	3	94	0.653409	0	49
6	0.589723	0	3	95	0.869636	0	49
7	0.16161	1	4	96	0.096449	1	50
8	0.139522	1	5	97	0.060718	1	51
9	0.127048	1	6	98	0.164192	1	52
10	0.843891	0	6	99	0.208989	1	53
11	0.320001	1	7	100	0.223315	1	54
12	0.956926	0	7	992	0.215703	1	499
13	0.338328	1	8	993	0.453159	1	500
14	0.756018	0	8	994	0.046426	1	501
15	0.198861	1	9	995	0.212272	1	502
16	0.320295	1	10	996	0.481376	1	503
17	0.262637	1	11	997	0.189144	1	504
18	0.832752	0	11	998	0.075167	1	505
19	0.967296	0	11	999	0.409387	1	506
20	0.274474	1	12	1000	0.044764	1	507

Here's a useful short-cut: to regenerate all the random numbers in the Excel spreadsheet, go to any unused cell and type in a space, then press **ENTER**. All of the random numbers will regenerate at once. You can perform 1000 trials at once, as often as you like.

How could a spreadsheet be used to emulate the roll of two dice? The tossing of two coins? There are many possible correct answers, and a discussion with your classmates may come up with some clever ways to emulate the experiments.

EXERCISES

4. Repeat the coin-toss experiment (Exercise 1) using a random-number generator. Track your experimental probabilities after 10 trials, 50 trials, 100 trials, 500 trials, and 1000 trials. How close are your results to the theoretical probability of getting a tail on one toss of a coin?

5. Repeat the coin-toss experiment with two coins (Exercise 2) using a random-number generator. Let any values less than 0.25 represent TT, values between 0.25 and 0.5 represent TH, values between 0.5 and 0.75 represent HT, and values above 0.75 represent HH. Thus, if we are interested in the occurrences of HH, we can track the frequency of values above 0.75. On the Excel spreadsheet, change cell B1 to =IF(A1 > 0.75,1,0). Copy cell B1 and paste into cells B2 through B1000. Track your experimental probabilities after 10 trials, 50 trials, 100 trials, 500 trials, and 1000 trials. How close are your results to the theoretical probability of getting two heads on the toss of two coins?

6. Repeat the experiment of rolling two fair dice (Exercise 3) using a random-number generator. The theoretical probability of rolling a sum of 12 is $\frac{1}{36}$,

or about 0.0278. Thus, you can change cell B1 to =IF(A1 < 0.0278,1,0) to emulate the occurrence of a sum of 12. Copy cell B1, and paste it into cells B2 through B1000. Track your experimental probabilities after 10 trials, 50 trials, 100 trials, 500 trials, and 1000 trials. How close are your results to the theoretical probability of getting a sum of 12 when rolling two dice?

7. A man has two children, and the oldest is a boy. A woman has two children, at least one of which is a boy.

    a) Are the probabilities that the woman and the man have two sons the same? Why or why not?

    b) Use a random-number generator to simulate the man's two children (ignoring any instance when the first is a girl) and the woman's two children. Does this support (or contradict) your answer to part (a)?

8. Consider some other experiment discussed in this chapter or think of one of your own. How can you simulate the experiment using a random-number generator? How close are the randomly generated probabilities to the theoretical probability?

Cumulative Review

1. Write an equation of the line with slope -4 and containing the point $(-7, 1)$.

2. For $f(x) = x^2 - 5$, find $f(x + h)$.

3. **a)** Graph:
$$f(x) = \begin{cases} 5 - x, & \text{for } x \neq 2, \\ -3, & \text{for } x = 2. \end{cases}$$
   b) Find $\lim\limits_{x \to 2} f(x)$.
   c) Find $f(2)$.
   d) Is f continuous at 2?

Find each limit, if it exists. If a limit does not exist, state that fact.

4. $\lim\limits_{x \to -4} \dfrac{x^2 - 16}{x + 4}$

5. $\lim\limits_{x \to 1} \sqrt{x^3 + 8}$

6. $\lim\limits_{x \to 3} \dfrac{4}{x - 3}$

7. $\lim\limits_{x \to \infty} \dfrac{12x - 7}{3x + 2}$

8. $\lim\limits_{x \to \infty} \dfrac{2x^3 - x}{8x^5 - x^2 + 1}$

9. If $f(x) = x^2 + 3$, find $f'(x)$ by determining
$$\lim\limits_{h \to 0} \dfrac{f(x + h) - f(x)}{h}.$$

For exercises 10–12, refer to the following graph of $y = h(x)$.

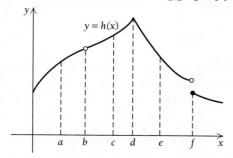

10. Identify the input values for which h has no limit.

11. Identify the input values for which h is discontinuous.

12. Identify the input values for which the derivative of h does not exist.

Differentiate.

13. $y = -9x + 3$

14. $y = x^2 - 7x + 3$

15. $y = x^{1/4}$

16. $f(x) = x^{-6}$

17. $f(x) = \sqrt[3]{2x^5 - 8}$

18. $f(x) = \dfrac{5x^3 + 4}{2x - 1}$

19. $y = \ln(x^2 + 5)$

20. $y = e^{\ln x}$

21. $y = e^{3x} + x^2$

22. $y = e^{\sqrt{x - 3}}$

23. $f(x) = \ln(e^x - 4)$

24. For $y = x^2 - \dfrac{2}{x}$, find d^2y/dx^2.

25. Business: average cost. Doubletake Clothing finds that the cost, in dollars, of producing x pairs of jeans is given by $C(x) = 320 + 9\sqrt{x}$. Find the rate at which the average cost is changing when 100 pairs of jeans have been produced.

26. Differentiate implicitly to find dy/dx if $x^3 + x/y = 7$.

27. Find an equation of the tangent line to the graph of $y = e^x - x^2 - 3$ at the point $(0, -2)$.

28. Find the x-value(s) at which the tangent lines to $f(x) = x^3 - 2x^2$ have a slope of -1.

Sketch the graph of each function. List and label the coordinates of any extrema and points of inflection. State where the function is increasing or decreasing, where it is concave up or concave down, and where any asymptotes occur.

29. $f(x) = x^3 - 3x + 1$

30. $f(x) = 2x^2 - x^4 - 3$

31. $f(x) = \dfrac{8x}{x^2 + 1}$

32. $f(x) = \dfrac{8}{x^2 - 4}$

Find the absolute maximum and minimum values, if they exist, over the indicated interval. If no interval is indicated, consider the entire real number line.

33. $f(x) = 3x^2 - 6x - 4$

34. $f(x) = -5x + 1$

35. $f(x) = \frac{1}{3}x^3 - x^2 - 3x + 5; [-2, 0]$

36. Business: maximizing profit. For custom sweatshirts, Detailed Clothing's total revenue and total cost, in dollars, are given by

$$R(x) = 4x^2 + 11x + 110,$$

$$C(x) = 4.2x^2 + 5x + 10.$$

Find the number of sweatshirts, x, that must be produced and sold in order to maximize profit.

37. Business: minimizing inventory costs. An appliance store sells 450 MP3 players each year. It costs $4 to store a player for a year. When placing an order, there is a fixed cost of $1 plus $0.75 for each player. How many times per year should the store reorder MP3 players, and in what lot size, in order to minimize inventory costs?

38. Let $y = 3x^2 - 2x + 1$. Use differentials to find the approximate change in y when $x = 2$ and $\Delta x = 0.05$.

39. Business: exponential growth. A national frozen yogurt firm is experiencing growth of 10% per year in the number, N, of franchises that it owns; that is,

$$\frac{dN}{dt} = 0.1N,$$

where N is the number of franchises and t is the time, in years, from 2001.

   a) Given that there were 8000 franchises in 2001, find the solution of the equation, assuming that $N_0 = 8000$ and $k = 0.1$.
   b) How many franchises were there in 2009?
   c) What is the doubling time of the number of franchises?

40. Economics: elasticity of demand. Consider the demand function

$$q = D(x) = 240 - 20x,$$

where q is the quantity of coffee mugs demanded at a price of x dollars.

   a) Find the elasticity.
   b) Find the elasticity at $x = \$2$, and state whether the demand is elastic or inelastic.
   c) Find the elasticity at $x = \$9$, and state whether the demand is elastic or inelastic.
   d) At a price of $2, will a small increase in price cause total revenue to increase or decrease?
   e) Find the value of x for which the total revenue is a maximum.

41. Business: approximating cost overage. A large square plot of ground measures 75 ft by 75 ft, with a tolerance of ± 4 in. Landscapers are going to cover the plot with grass sod. Each square of sod costs $8 and measures 3 ft by 3 ft.

   a) Use differentials to estimate the change in area when the measurement tolerance is taken into acount.
   b) How many extra squares of sod should the landscapers bring to the job, and how much extra will this cost?

Evaluate.

42. $\displaystyle\int 3x^5 \, dx$

43. $\displaystyle\int_{-1}^{0} (2e^x + 1) \, dx$

44. $\displaystyle\int \frac{x}{(7 - 3x)^2} \, dx$ (Use Table 1 on pp. 454–455.)

45. $\displaystyle\int x^3 e^{x^4} \, dx$ (Do not use Table 1.)

46. $\displaystyle\int (x + 3) \ln x \, dx$

47. $\displaystyle\int \frac{75}{x} \, dx$

48. $\displaystyle\int_{0}^{1} 3\sqrt{x} \, dx$

49. Find the area under the graph of $y = x^2 + 3x$ over the interval $[1, 5]$.

50. Business: present value. Find the present value of $250,000 due in 30 yr at 6%, compounded continuously.

51. Business: value of a fund. Leigh Ann wants to have $50,000 saved in 10 yr.

   a) She could make a one-time deposit at an APR of 5.45%, compounded continuously. Find the amount she should deposit.
   b) She could instead make an investment that would yield a constant revenue stream of $R(t)$ dollars per year, at 5.45% compounded continuously. Find $R(t)$.
   c) Calculate the interest earned in part (a) and in part (b).

52. Business: contract buyout. An executive works under an 8-yr contract that pays him $200,000 per year. He invests the money at an APR of 4.85%, compounded continuously. After 5 yr, the company offers him a buyout of the contract. What is the lowest amount he should accept, if the continuously compounded APR is the same?

53. Determine whether the following improper integral is convergent or divergent, and calculate its value if it is convergent:

$$\int_{3}^{\infty} \frac{1}{x^7} \, dx.$$

54. Economics: supply and demand. Demand and supply functions are given by

$$p = D(x) = (x - 20)^2$$

and

$$p = S(x) = x^2 + 10x + 50,$$

where p is the price per unit, in dollars, when x units are sold. Find the equilibrium point and the consumer's surplus.

55. The table below shows Julia's speeds (in mph) at 30-sec intervals during the first 10 min of her morning run. Approximate the total distance Julia traveled (in miles) by finding L_{10} and R_{10} and their average. (*Hint:* 1 hr = 3600 sec.)

Time (30-sec intervals)	Speed (mph)
0	0
1	8
2	9
3	9
4	10
5	9.5
6	8.5
7	6
8	6
9	7
10	8

56. Approximate the value of $\int_0^3 \sqrt{e^x + 1}\, dx$ to three decimal places by finding T_6.

57. Find the volume of the solid of revolution generated by rotating the region under the graph of

$y = e^{-x}$, from $x = 0$ to $x = 5$,

about the x-axis.

58. Let $f(x, y) = \sqrt{16 - x^2 - y^2}$.

a) Evaluate $f(3, 1)$.
b) State the domain of f.

59. Consider the data in the following table.

Age of business (in years)	1	3	5
Profit (in tens of thousands of dollars)	4	7	9

a) Find the regression line, $y = mx + b$.
b) Use the regression line to predict the profit when the business is 10 years old.

Given $f(x, y) = e^y + 4x^2y^2 + 3x$. find each of the following.

60. f_x **61.** f_{yy}

62. Find the relative maximum and minimum values of $f(x, y) = 8x^2 - y^2$.

63. Maximize $f(x, y) = 4x + 2y - x^2 - y^2 + 4$, subject to the constraint $x + 2y = 9$.

64. Evaluate

$$\int_0^3 \int_{-1}^2 e^x \, dy \, dx.$$

65. Business: demographics. The number of shoppers, in hundreds per square mile, who frequent a mall is modeled by the two-variable function $f(x, y) = 10 - x - y^2$, where x is miles from the mall toward the east and y is miles from the mall toward the north. The graph below shows a shaded region to the northeast of the mall, which is at $(0, 0)$. Find the total number of frequent mall shoppers in the region.

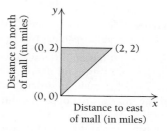

66. Differentiate.

a) $f(x) = 2 \sin 3x$
b) $g(x) = x \cos(x^2)$
c) $h(x) = \dfrac{x}{\tan(4x)}$

67. The temperature on a spring day in Seattle is modeled by $T(t) = 59 - 11 \sin\left(\dfrac{\pi}{12}t\right)$, where T is the temperature (in°F) t hours after midnight.

a) Find the midline, amplitude, period, and phase shift.
b) What is the temperature at 2 p.m.?
c) Find $T'(15)$, and explain what this number represents.
d) Find the minimum and maximum temperatures and the times at which they occur.

68. Business: total Sales. Sales of overcoats at a men's clothier are modeled by $S(t) = 300 + 210 \cos\left(\dfrac{\pi}{6}t\right)$, where $S(t)$ is the rate at which overcoats are sold (in units per month) t months after January 1. Find the total number of overcoats sold for the first 9 months of the year.

69. Solve $3 + 8 \cos(2x - 5) = 7$ for x in the interval $[0, \pi]$.

70. Solve the differential equation $dy/dx = xy$.

71. Solve the differential equation $y' + 4xy = 3x$, where $y = 3$ when $x = -1$.

72. Business: profit. The revenue of Rumsey's Roadhouse Restaurant over a 1-yr period is modeled by $5R'' + 4R' - R = 0$, where $R(t)$ is the revenue (in thousands of dollars) t months after January 1.

a) Find the general solution for $R(t)$.
b) Find the particular solution where $R(0) = 35$ and $R'(0) = -11$.
c) How much revenue does Rumsey's Roadhouse Restaurant receive in April?

73. Consider the following sequence: $15, 22, 29, 36, 43, \ldots$.

    a) Find the general nth term.
    b) Find the 100th term of the sequence.
    c) Find S_{100}.

74. Business: fiscal multiplier. Find the cumulative effect in an economy of an initial expenditure of $1,000,000, if 62% of the money is spent at every subsequent level.

75. Business: savings. Gail wants to save $10,000 in 2 years for a trip to Europe. She considers two options.

    a) A savings account has an interest rate of 3.5%, compounded monthly. Find the one-time deposit Gail would need to make in order to reach her goal and the interest that deposit would earn.
    b) A sinking fund account has an interest rate of 2.95%, compounded quarterly. Find the quarterly payments Gail would need to make to reach her goal, her total contribution, and the interest earned.

76. Use a series to approximate the value of $\int_0^1 e^{-x^4}\, dx$ to four decimal places.

77. A jar of candy contains 12 yellow, 15 red, and 20 blue candies. You reach in and select one candy at random. Let event Y be selecting a yellow candy, event R be selecting a red candy, and event B be selecting a blue candy. Find the following probabilities:

    a) $P(Y)$
    b) $P(Y')$
    c) $P(Y \cup B)$
    d) $P(B \cap R)$

78. The table below shows the numbers of students who achieved certain scores on a history quiz.

Score	14	15	16	17	18	19	20
Number of students (frequency)	2	5	3	1	4	3	6

Let the random variable X be the score of a randomly selected student.

    a) Find $P(X = 18)$, and explain what this number represents.
    b) Find $P(X \leq 18)$, and explain what this number represents.
    c) Find μ, and explain what this number represents.

79. Let $f(x) = \dfrac{c}{x^2}$ be a probability density function over the interval $[1, 3]$.

    a) Find c.
    b) Find the mean μ and the standard deviation σ.
    c) Find the cumulative density function F.

80. Business: distribution of salaries. The salaries paid by a large corporation are normally distributed with a mean $\mu = \$45{,}000$ and a standard deviation $\sigma = \$6{,}000$.

    a) Find the probability that a randomly selected employee earns between $42,000 and $55,000 per year.
    b) An executive of the corporation earns $60,000 per year. In what percentile of the salaries does this salary place him?
    c) A new employee insists on a salary that is in the top 2% of salaries. What is the minimum salary that this employee would accept?

APPENDIX A

Review of Basic Algebra

This appendix covers most of the algebraic topics essential to a study of calculus. It might be used in conjunction with Chapter R or as the need for certain skills arises throughout the book.

OBJECTIVES

- Manipulate exponential expressions.
- Multiply and factor algebraic expressions.
- Solve equations, inequalities, and applied problems.

Exponential Notation

Let's review the meaning of the expression

$$a^n,$$

where a is any real number and n is an integer; that is, n is a number in the set $\{\ldots, -3, -2, -1, 0, 1, 2, 3, \ldots\}$. The number a is called the **base** and n is called the **exponent**. If n is greater than 1, then

$$a^n = \underbrace{a \cdot a \cdot a \cdots a}_{n \text{ factors}}.$$

In other words, a^n is the product of n **factors**, each of which is a.

■ **EXAMPLE 1**   Express each of the following without exponents:

a) 4^3 **b)** $(-2)^5$ **c)** $(-2)^4$ **d)** -2^4 **e)** $(1.08)^2$ **f)** $\left(\dfrac{1}{2}\right)^3$

Solution

a) $4^3 = 4 \cdot 4 \cdot 4 = 64$

b) $(-2)^5 = (-2)(-2)(-2)(-2)(-2) = -32$

c) $(-2)^4 = (-2)(-2)(-2)(-2) = 16$

d) $-2^4 = -(2^4) = -(2)(2)(2)(2) = -16$ The base is 2, not −2.

e) $(1.08)^2 = 1.08 \times 1.08 = 1.1664$

f) $\left(\dfrac{1}{2}\right)^3 = \dfrac{1}{2} \cdot \dfrac{1}{2} \cdot \dfrac{1}{2} = \dfrac{1}{8}$

We define an exponent of 1 as follows:

$$a^1 = a, \quad \text{for any real number } a.$$

In other words, any real number to the first power is that number itself.

We define an exponent of 0 as follows:

$$a^0 = 1, \quad \text{for any nonzero real number } a.$$

That is, any nonzero real number a to the zero power is 1.

■ **EXAMPLE 2**   Express without exponents:

a) $(-2x)^0$ **b)** $(-2x)^1$ **c)** $\left(\frac{1}{2}\right)^0$ **d)** e^0 **e)** e^1 **f)** $\left(\frac{1}{2}\right)^1$

Solution

a) $(-2x)^0 = 1$ **b)** $(-2x)^1 = -2x$ **c)** $\left(\frac{1}{2}\right)^0 = 1$

d) $e^0 = 1$ **e)** $e^1 = e$ **f)** $\left(\frac{1}{2}\right)^1 = \frac{1}{2}$

The meaning of a negative integer as an exponent is as follows:

$$a^{-n} = \frac{1}{a^n} = \left(\frac{1}{a}\right)^n, \quad \text{for any nonzero real number } a.$$

That is, any nonzero real number a to the $-n$ power is the reciprocal of a^n, or equivalently, (the reciprocal of a)^n.

■ **EXAMPLE 3**   Express without negative exponents:

a) 2^{-5} **b)** 10^{-3} **c)** $\left(\frac{1}{4}\right)^{-2}$ **d)** x^{-5} **e)** e^{-k} **f)** t^{-1}

Solution

a) $2^{-5} = \frac{1}{2^5} = \frac{1}{2 \cdot 2 \cdot 2 \cdot 2 \cdot 2} = \frac{1}{32}$ **b)** $10^{-3} = \frac{1}{10^3} = \frac{1}{10 \cdot 10 \cdot 10} = \frac{1}{1000}$, or 0.001

c) $\left(\frac{1}{4}\right)^{-2} = \left(\frac{4}{1}\right)^2 = 4^2 = 16$ **d)** $x^{-5} = \frac{1}{x^5}$

e) $e^{-k} = \frac{1}{e^k}$ **f)** $t^{-1} = \frac{1}{t^1} = \frac{1}{t}$

Properties of Exponents

Note the following:

$$b^5 \cdot b^{-3} = (b \cdot b \cdot b \cdot b \cdot b) \cdot \frac{1}{b \cdot b \cdot b}$$

$$= \frac{b \cdot b \cdot b}{b \cdot b \cdot b} \cdot b \cdot b$$

$$= 1 \cdot b \cdot b = b^2.$$

We can obtain the same result by adding the exponents. This is true in general.

THEOREM 1

For any nonzero real number a and any integers n and m,

$$a^n \cdot a^m = a^{n+m}.$$

(To multiply when the bases are the same, add the exponents.)

■ **EXAMPLE 4**   Multiply:

a) $x^5 \cdot x^6$ **b)** $x^{-5} \cdot x^6$ **c)** $2x^{-3} \cdot 5x^{-4}$ **d)** $r^2 \cdot r$

Solution

a) $x^5 \cdot x^6 = x^{5+6} = x^{11}$ **b)** $x^{-5} \cdot x^6 = x^{-5+6} = x$

c) $2x^{-3} \cdot 5x^{-4} = 10x^{-3+(-4)} = 10x^{-7}, \quad \text{or} \quad \dfrac{10}{x^7}$ **d)** $r^2 \cdot r = r^{2+1} = r^3$

Note the following:

$$b^5 \div b^2 = \frac{b^5}{b^2} = \frac{b \cdot b \cdot b \cdot b \cdot b}{b \cdot b}$$

$$= \frac{b \cdot b}{b \cdot b} \cdot b \cdot b \cdot b$$

$$= 1 \cdot b \cdot b \cdot b = b^3.$$

We can obtain the same result by subtracting the exponents. This is true in general.

THEOREM 2

For any nonzero real number a and any integers n and m,

$$\frac{a^n}{a^m} = a^{n-m}.$$

(To divide when the bases are the same, subtract the exponent in the denominator from the exponent in the numerator.)

■ **EXAMPLE 5**   Divide:

a) $\dfrac{a^3}{a^2}$ **b)** $\dfrac{x^7}{x^7}$ **c)** $\dfrac{e^3}{e^{-4}}$ **d)** $\dfrac{e^{-4}}{e^{-1}}$

Solution

a) $\dfrac{a^3}{a^2} = a^{3-2} = a^1 = a$ **b)** $\dfrac{x^7}{x^7} = x^{7-7} = x^0 = 1$

c) $\dfrac{e^3}{e^{-4}} = e^{3-(-4)} = e^{3+4} = e^7$ **d)** $\dfrac{e^{-4}}{e^{-1}} = e^{-4-(-1)} = e^{-4+1} = e^{-3}, \quad \text{or} \quad \dfrac{1}{e^3}$

Note the following:

$$(b^2)^3 = b^2 \cdot b^2 \cdot b^2 = b^{2+2+2} = b^6.$$

We can obtain the same result by multiplying the exponents. The other results in Theorem 3 can be similarly motivated.

THEOREM 3

For any nonzero real numbers a and b, and any integers n and m,

$$(a^n)^m = a^{nm}, \qquad (ab)^n = a^n b^n, \qquad \text{and} \qquad \left(\frac{a}{b}\right)^n = \frac{a^n}{b^n}.$$

■ **EXAMPLE 6**   Simplify:

a) $(x^{-2})^3$ **b)** $(e^x)^2$ **c)** $(2x^4y^{-5}z^3)^{-3}$ **d)** $\left(\dfrac{x^2}{p^4q^5}\right)^3$

Solution

a) $(x^{-2})^3 = x^{-2\cdot3} = x^{-6}$,   or $\dfrac{1}{x^6}$

b) $(e^x)^2 = e^{2x}$

c) $(2x^4y^{-5}z^3)^{-3} = 2^{-3}(x^4)^{-3}(y^{-5})^{-3}(z^3)^{-3}$
$= \dfrac{1}{2^3}x^{-12}y^{15}z^{-9}$,   or $\dfrac{y^{15}}{8x^{12}z^9}$

d) $\left(\dfrac{x^2}{p^4q^5}\right)^3 = \dfrac{(x^2)^3}{(p^4q^5)^3} = \dfrac{x^6}{(p^4)^3(q^5)^3}$
$= \dfrac{x^6}{p^{12}q^{15}}$

Multiplication

The distributive law is important when multiplying. This law is as follows.

The Distributive Law

For any numbers A, B, and C,

$$A(B + C) = AB + AC.$$

Because subtraction can be regarded as addition of an additive inverse, it follows that

$$A(B - C) = AB - AC.$$

■ **EXAMPLE 7**   Multiply:

a) $3(x - 5)$ **b)** $P(1 + i)$ **c)** $(x - 5)(x + 3)$ **d)** $(a + b)(a + b)$

Solution

a) $3(x - 5) = 3\cdot x - 3\cdot5 = 3x - 15$

b) $P(1 + i) = P\cdot1 + P\cdot i = P + Pi$

c) $(x - 5)(x + 3) = (x - 5)x + (x - 5)3$
$= x\cdot x - 5x + 3x - 5\cdot3$
$= x^2 - 2x - 15$

d) $(a + b)(a + b) = (a + b)a + (a + b)b$
$= a\cdot a + ba + ab + b\cdot b$
$= a^2 + 2ab + b^2$

The following formulas, which are obtained using the distributive law, are also useful when multiplying. All three are used in Example 8, which follows.

$$(A + B)^2 = A^2 + 2AB + B^2$$
$$(A - B)^2 = A^2 - 2AB + B^2$$
$$(A - B)(A + B) = A^2 - B^2$$

■ **EXAMPLE 8**   Multiply:

a) $(x + h)^2$ **b)** $(2x - t)^2$ **c)** $(3c + d)(3c - d)$

Solution

a) $(x + h)^2 = x^2 + 2xh + h^2$

b) $(2x - t)^2 = (2x)^2 - 2(2x)t + t^2 = 4x^2 - 4xt + t^2$

c) $(3c + d)(3c - d) = (3c)^2 - d^2 = 9c^2 - d^2$

Factoring

Factoring is the reverse of multiplication. That is, to factor an expression, we find an equivalent expression that is a product. Always remember to look first for a common factor.

■ **EXAMPLE 9**   Factor:

a) $P + Pi$ **b)** $2xh + h^2$ **c)** $x^2 - 6xy + 9y^2$
d) $x^2 - 5x - 14$ **e)** $6x^2 + 7x - 5$ **f)** $x^2 - 9t^2$

Solution

a) $P + Pi = P \cdot 1 + P \cdot i = P(1 + i)$ We used the distributive law.
b) $2xh + h^2 = h(2x + h)$
c) $x^2 - 6xy + 9y^2 = (x - 3y)^2$
d) $x^2 - 5x - 14 = (x - 7)(x + 2)$ We looked for factors of -14 whose sum is -5.
e) $6x^2 + 7x - 5 = (2x - 1)(3x + 5)$ We first considered ways of factoring the first coefficient—for example, $(2x \quad)(3x \quad)$. Then we looked for factors of -5 such that when we multiply, we obtain the given expression.

f) $x^2 - 9t^2 = (x - 3t)(x + 3t)$ We used the formula $(A - B)(A + B) = A^2 - B^2$.

Some expressions with four terms can be factored by first looking for a common binomial factor. This is called **factoring by grouping**.

■ **EXAMPLE 10**   Factor:

a) $t^3 + 6t^2 - 2t - 12$ **b)** $x^3 - 7x^2 - 4x + 28$

Solution

a) $t^3 + 6t^2 - 2t - 12 = t^2(t + 6) - 2(t + 6)$ Factoring the first two terms and then the second two terms

$= (t^2 - 2)(t + 6)$ Factoring out the common binomial factor, $t + 6$

b) $x^3 - 7x^2 - 4x + 28 = x^2(x - 7) - 4(x - 7)$ Factoring the first two terms and then the second two terms

$= (x - 7)(x^2 - 4)$ Factoring out the common binomial factor, $x - 7$

$= (x - 7)(x - 2)(x + 2)$ Using $(A - B)(A + B) = A^2 - B^2$

Solving Equations

Basic to the solution of many equations are the *Addition Principle* and the *Multiplication Principle*. We can add (or subtract) the same number on both sides of an equation and obtain an equivalent equation, that is, a new equation that has the same solutions as the original equation. We can also multiply (or divide) by a nonzero number on both sides of an equation and obtain an equivalent equation.

The Addition Principle

For any real numbers $a, b,$ and $c,$

$a = b$ is equivalent to $a + c = b + c$.

The Multiplication Principle

For any real numbers $a, b,$ and $c,$ with $c \neq 0,$

$a = b$ is equivalent to $a \cdot c = b \cdot c$.

When solving a linear equation, we use these principles and other properties of real numbers to get the variable alone on one side. Then it is easy to determine the solution.

■ **EXAMPLE 11**   Solve: $-\frac{5}{6}x + 10 = \frac{1}{2}x + 2$.

Solution   We first multiply by 6 on both sides to clear the fractions:

$$6\left(-\frac{5}{6}x + 10\right) = 6\left(\frac{1}{2}x + 2\right) \qquad \text{Using the Multiplication Principle}$$
$$6\left(-\frac{5}{6}x\right) + 6 \cdot 10 = 6\left(\frac{1}{2}x\right) + 6 \cdot 2 \qquad \text{Using the distributive law}$$
$$-5x + 60 = 3x + 12 \qquad \text{Simplifying}$$
$$60 = 8x + 12 \qquad \text{Using the Addition Principle: We add } 5x \text{ on both sides.}$$
$$48 = 8x \qquad \text{Adding } -12 \text{ on both sides}$$
$$\tfrac{1}{8} \cdot 48 = \tfrac{1}{8} \cdot 8x \qquad \text{Multiplying by } \tfrac{1}{8} \text{ on both sides}$$
$$6 = x.$$

The variable is now alone on one side, and we see that 6 is the solution. We can check by substituting 6 into the original equation.

The third principle for solving equations is the *Principle of Zero Products*.

The Principle of Zero Products

For any numbers a and b, if $ab = 0$, then $a = 0$ or $b = 0$; and if $a = 0$ or $b = 0$, then $ab = 0$.

To solve an equation using this principle, we must have a 0 on one side and a product on the other. The solutions are then obtained by setting each factor equal to 0 and solving the resulting equations.

■ **EXAMPLE 12**   Solve: $3x(x - 2)(5x + 4) = 0$.

Solution   We have

$$3x(x - 2)(5x + 4) = 0$$
$$3x = 0 \quad \text{or} \quad x - 2 = 0 \quad \text{or} \quad 5x + 4 = 0 \qquad \text{Using the Principle of Zero Products}$$
$$\tfrac{1}{3} \cdot 3x = \tfrac{1}{3} \cdot 0 \quad \text{or} \quad x = 2 \quad \text{or} \quad 5x = -4 \qquad \text{Solving each equation separately}$$
$$x = 0 \quad \text{or} \quad x = 2 \quad \text{or} \quad x = -\tfrac{4}{5}.$$

The solutions are 0, 2, and $-\frac{4}{5}$.

Note that the Principle of Zero Products applies *only* when a product is 0. For example, although we may know that $ab = 8$, *we do not know* that $a = 8$ or $b = 8$.

■ **EXAMPLE 13**   Solve: $4x^3 = x$.

Solution   We have
$$4x^3 = x$$
$$4x^3 - x = 0 \qquad \text{Adding } -x \text{ to both sides}$$
$$x(4x^2 - 1) = 0$$
$$x(2x - 1)(2x + 1) = 0 \qquad \text{Factoring}$$
$$x = 0 \quad \text{or} \quad 2x - 1 = 0 \quad \text{or} \quad 2x + 1 = 0 \qquad \text{Using the Principle of Zero Products}$$
$$x = 0 \quad \text{or} \quad 2x = 1 \quad \text{or} \quad 2x = -1$$
$$x = 0 \quad \text{or} \quad x = \tfrac{1}{2} \quad \text{or} \quad x = -\tfrac{1}{2}.$$

The solutions are 0, $\frac{1}{2}$, and $-\frac{1}{2}$.

Rational Equations

Expressions like the following are polynomials in one variable:

$$x^2 - 4, \qquad x^3 + 7x^2 - 8x + 9, \qquad t - 19.$$

The **least common multiple, LCM,** of two polynomials is found by factoring and using each factor the greatest number of times that it occurs in any one factorization.

■ **EXAMPLE 14** Find the LCM: $x^2 + 2x + 1$, $5x^2 - 5x$, and $x^2 - 1$.

Solution

$$\left.\begin{aligned} x^2 + 2x + 1 &= (x + 1)(x + 1); \\ 5x^2 - 5x &= 5x(x - 1); \\ x^2 - 1 &= (x + 1)(x - 1) \end{aligned}\right\} \quad \text{Factoring}$$

$$\text{LCM} = 5x(x + 1)(x + 1)(x - 1)$$

A **rational expression** is a ratio of polynomials. Each of the following is a rational expression:

$$\frac{x^2 - 6x + 9}{x^2 - 4}, \quad \frac{x - 2}{x - 3}, \quad \frac{a + 7}{a^2 - 16}, \quad \frac{5}{5t - 15}.$$

A **rational equation** is an equation containing one or more rational expressions. Here are some examples:

$$\frac{2}{3} - \frac{5}{6} = \frac{1}{x}, \quad x + \frac{6}{x} = 5, \quad \frac{2x}{x - 3} - \frac{6}{x} = \frac{18}{x^2 - 3x}.$$

To solve a rational equation, we first clear the equation of fractions by multiplying on both sides by the LCM of all the denominators. The resulting equation might have solutions that are *not* solutions of the original equation. Thus, we must check all possible solutions in the original equation.

■ **EXAMPLE 15** Solve: $\dfrac{2x}{x - 3} - \dfrac{6}{x} = \dfrac{18}{x^2 - 3x}$.

Solution Note that $x^2 - 3x = x(x - 3)$. The LCM of the denominators is $x(x - 3)$. We multiply by $x(x - 3)$.

$$x(x - 3)\left(\frac{2x}{x - 3} - \frac{6}{x}\right) = x(x - 3)\left(\frac{18}{x^2 - 3x}\right) \qquad \text{Multiplying by the LCM on both sides}$$

$$x(x - 3) \cdot \frac{2x}{x - 3} - x(x - 3) \cdot \frac{6}{x} = x(x - 3)\left(\frac{18}{x^2 - 3x}\right) \qquad \text{Using the distributive law}$$

$$2x^2 - 6(x - 3) = 18 \qquad \text{Simplifying}$$

$$2x^2 - 6x + 18 = 18$$

$$2x^2 - 6x = 0$$

$$2x(x - 3) = 0$$

$$2x = 0 \quad or \quad x - 3 = 0$$

$$x = 0 \quad or \quad x = 3$$

The numbers 0 and 3 are possible solutions. We look at the original equation and see that each makes a denominator 0. We can also carry out a check, as follows.

Check

For 0:

$$\frac{2x}{x - 3} - \frac{6}{x} = \frac{18}{x^2 - 3x}$$

$$\frac{2(0)}{0 - 3} - \frac{6}{0} \overset{?}{\underset{|}{}} \frac{18}{0^2 - 3(0)}$$

$$0 - \frac{6}{0} \; \Big|\; \frac{18}{0} \qquad \text{UNDEFINED; FALSE}$$

For 3:

$$\frac{2x}{x - 3} - \frac{6}{x} = \frac{18}{x^2 - 3x}$$

$$\frac{2(3)}{3 - 3} - \frac{6}{3} \overset{?}{\underset{|}{}} \frac{18}{3^2 - 3(3)}$$

$$\frac{6}{0} - 2 \; \Big|\; \frac{18}{0} \qquad \text{UNDEFINED; FALSE}$$

The equation has *no solution*.

■ **EXAMPLE 16**   Solve: $\dfrac{x^2}{x-2} = \dfrac{4}{x-2}$.

Solution   The LCM of the denominators is $x-2$. We multiply by $x-2$.

$$(x-2) \cdot \frac{x^2}{x-2} = (x-2) \cdot \frac{4}{x-2}$$

$$x^2 = 4 \qquad \text{Simplifying}$$

$$x^2 - 4 = 0$$

$$(x+2)(x-2) = 0$$

$$x = -2 \quad or \quad x = 2 \qquad \text{Using the Principle of Zero Products}$$

Check

For 2:

$$\frac{x^2}{x-2} = \frac{4}{x-2}$$

$$\frac{2^2}{2-2} \overset{?}{\;\vert\;} \frac{4}{2-2}$$

$$\frac{4}{0} \;\bigg\vert\; \frac{4}{0} \qquad \text{Undefined; False}$$

For -2:

$$\frac{x^2}{x-2} = \frac{4}{x-2}$$

$$\frac{(-2)^2}{-2-2} \overset{?}{\;\vert\;} \frac{4}{-2-2}$$

$$\frac{4}{-4} \;\bigg\vert\; \frac{4}{-4}$$

$$-1 \;\bigg\vert\; -1 \qquad \text{True}$$

The number -2 is a solution, but 2 is not (it results in division by 0).

Solving Inequalities

Two inequalities are **equivalent** if they have the same solutions. For example, the inequalities $x > 4$ and $4 < x$ are equivalent. Principles for solving inequalities are similar to those for solving equations. We can add the same number to both sides of an inequality. We can also multiply on both sides by the same nonzero number, but if that number is negative, we must reverse the inequality sign. The following are the inequality-solving principles.

The Inequality-Solving Principles

For any real numbers a, b, and c,

$$a < b \quad \text{is equivalent to} \quad a + c < b + c.$$

For any real numbers a, b, and any *positive* number c,

$$a < b \quad \text{is equivalent to} \quad ac < bc.$$

For any real numbers a, b, and any *negative* number c,

$$a < b \quad \text{is equivalent to} \quad ac > bc.$$

Similar statements hold for $\leq$ and $\geq$.

■ **EXAMPLE 17**   Solve: $17 - 8x \geq 5x - 4$.

Solution   We have

$$17 - 8x \geq 5x - 4$$
$$-8x \geq 5x - 21 \qquad \text{Adding } -17 \text{ to both sides}$$
$$-13x \geq -21 \qquad \text{Adding } -5x \text{ to both sides}$$
$$-\tfrac{1}{13}(-13x) \leq -\tfrac{1}{13}(-21) \qquad \begin{array}{l}\text{Multiplying both sides by } -\tfrac{1}{13} \text{ and} \\ \text{reversing the inequality sign}\end{array}$$
$$x \leq \tfrac{21}{13}.$$

Any number less than or equal to $\frac{21}{13}$ is a solution.

Applications

To solve applied problems, we first translate to mathematical language, usually an equation. Then we solve the equation and check to see whether the solution to the equation is a solution to the problem.

■ **EXAMPLE 18**   **Life Science: Weight Gain.**   After a 5% gain in weight, a grizzly bear weighs 693 lb. What was its original weight?

Solution   We first translate to an equation:

$$\underbrace{(Original\ weight)}_{w} + 5\%\underbrace{(Original\ weight)}_{w} = 693$$
$$w + 5\% \qquad\qquad w \qquad = 693.$$

Now we solve the equation:

$$w + 5\%w = 693$$
$$1 \cdot w + 0.05w = 693$$
$$(1 + 0.05)w = 693$$
$$1.05w = 693$$
$$w = \frac{693}{1.05} = 660.$$

Check $660 + 5\% \cdot 660 = 660 + 0.05 \cdot 660 = 660 + 33 = 693.$

The original weight of the bear was 660 lb.

■ **EXAMPLE 19**   **Business: Total Sales.**   Raggs, Ltd., a clothing firm, determines that its total revenue, in dollars, from the sale of x suits is given by

$$200x + 50.$$

Determine the number of suits that the firm must sell to ensure that its total revenue will be more than $70,050.

Solution   We translate to an inequality and solve:

$$200x + 50 > 70{,}050$$
$$200x > 70{,}000 \qquad \text{Adding } -50 \text{ to both sides}$$
$$x > 350. \qquad \text{Multiplying both sides by } \tfrac{1}{200}$$

Thus the company's total revenue will exceed $70,050 when it sells more than 350 suits.

EXERCISE SET
A

Express as an equivalent expression without exponents.

1. 5^3

2. 7^2

3. $(-7)^2$

4. $(-5)^3$

5. $(1.01)^2$

6. $(1.01)^3$

7. $\left(\dfrac{1}{2}\right)^4$

8. $\left(\dfrac{1}{4}\right)^3$

9. $(6x)^0$

10. $(6x)^1$

11. t^1

12. t^0

13. $\left(\dfrac{1}{3}\right)^0$

14. $\left(\dfrac{1}{3}\right)^1$

Express as an equivalent expression without negative exponents.

15. 3^{-2}

16. 4^{-2}

17. $\left(\dfrac{1}{2}\right)^{-3}$

18. $\left(\dfrac{1}{2}\right)^{-2}$

19. 10^{-1}

20. 10^{-4}

21. e^{-b}

22. t^{-k}

23. b^{-1}

24. h^{-1}

Multiply.

25. $x^2 \cdot x^3$

26. $t^3 \cdot t^4$

27. $x^{-7} \cdot x$

28. $x^5 \cdot x$

29. $5x^2 \cdot 7x^3$

30. $4t^3 \cdot 2t^4$

31. $x^{-4} \cdot x^7 \cdot x$

32. $x^{-3} \cdot x \cdot x^3$

33. $e^{-t} \cdot e^t$

34. $e^k \cdot e^{-k}$

Divide.

35. $\dfrac{x^5}{x^2}$

36. $\dfrac{x^7}{x^3}$

37. $\dfrac{x^2}{x^5}$

38. $\dfrac{x^3}{x^7}$

39. $\dfrac{e^k}{e^k}$

40. $\dfrac{t^k}{t^k}$

41. $\dfrac{e^t}{e^4}$

42. $\dfrac{e^k}{e^3}$

43. $\dfrac{t^6}{t^{-8}}$

44. $\dfrac{t^5}{t^{-7}}$

45. $\dfrac{t^{-9}}{t^{-11}}$

46. $\dfrac{t^{-11}}{t^{-7}}$

47. $\dfrac{ab(a^2b)^3}{ab^{-1}}$

48. $\dfrac{x^2y^3(xy^3)^2}{x^{-3}y^2}$

Simplify.

49. $(t^{-2})^3$

50. $(t^{-3})^4$

51. $(e^x)^4$

52. $(e^x)^5$

53. $(2x^2y^4)^3$

54. $(2x^2y^4)^5$

55. $(3x^{-2}y^{-5}z^4)^{-4}$

56. $(5x^3y^{-7}z^{-5})^{-3}$

57. $(-3x^{-8}y^7z^2)^2$

58. $(-5x^4y^{-5}z^{-3})^4$

59. $\left(\dfrac{cd^3}{2q^2}\right)^4$

60. $\left(\dfrac{4x^2y}{a^3b^3}\right)^3$

Multiply.

61. $5(x - 7)$

62. $x(1 + t)$

63. $(x - 5)(x - 2)$

64. $(x - 4)(x - 3)$

65. $(a - b)(a^2 + ab + b^2)$

66. $(x^2 - xy + y^2)(x + y)$

67. $(2x + 5)(x - 1)$

68. $(3x + 4)(x - 1)$

69. $(a - 2)(a + 2)$

70. $(3x - 1)(3x + 1)$

71. $(5x + 2)(5x - 2)$

72. $(t - 1)(t + 1)$

73. $(a - h)^2$

74. $(a + h)^2$

75. $(5x + t)^2$

76. $(7a - c)^2$

77. $5x(x^2 + 3)^2$

78. $-3x^2(x^2 - 4)(x^2 + 4)$

Use the following equation for Exercises 79–82.

$$\begin{aligned}
(x + h)^3 &= (x + h)(x + h)^2 \\
&= (x + h)(x^2 + 2xh + h^2) \\
&= (x + h)x^2 + (x + h)2xh + (x + h)h^2 \\
&= x^3 + x^2h + 2x^2h + 2xh^2 + xh^2 + h^3 \\
&= x^3 + 3x^2h + 3xh^2 + h^3
\end{aligned}$$

79. $(a + b)^3$

80. $(a - b)^3$

81. $(x - 5)^3$

82. $(2x + 3)^3$

Factor.

83. $x - xt$

84. $x + xh$

85. $x^2 + 6xy + 9y^2$

86. $x^2 - 10xy + 25y^2$

87. $x^2 - 2x - 15$

88. $x^2 + 8x + 15$

89. $x^2 - x - 20$

90. $x^2 - 9x - 10$

91. $49x^2 - t^2$

92. $9x^2 - b^2$

93. $36t^2 - 16m^2$

94. $25y^2 - 9z^2$

95. $a^3b - 16ab^3$

96. $2x^4 - 32$

97. $a^8 - b^8$

98. $36y^2 + 12y - 35$

99. $10a^2x - 40b^2x$

100. $x^3y - 25xy^3$

101. $2 - 32x^4$

102. $2xy^2 - 50x$

103. $9x^2 + 17x - 2$

104. $6x^2 - 23x + 20$

105. $x^3 + 8$ (*Hint:* See Exercise 66.)

106. $a^3 - 27$ (*Hint:* See Exercise 65.)

107. $y^3 - 64t^3$

108. $m^3 + 1000p^3$

109. $3x^3 - 6x^2 - x + 2$

110. $5y^3 + 2y^2 - 10y - 4$

111. $x^3 - 5x^2 - 9x + 45$ **112.** $t^3 + 3t^2 - 25t - 75$

Solve.

113. $-7x + 10 = 5x - 11$ **114.** $-8x + 9 = 4x - 70$

115. $5x - 17 - 2x = 6x - 1 - x$

116. $5x - 2 + 3x = 2x + 6 - 4x$

117. $x + 0.8x = 216$ **118.** $x + 0.5x = 210$

119. $x + 0.08x = 216$ **120.** $x + 0.05x = 210$

121. $2x(x + 3)(5x - 4) = 0$

122. $7x(x - 2)(2x + 3) = 0$

123. $x^2 + 1 = 2x + 1$ **124.** $2t^2 = 9 + t^2$

125. $t^2 - 2t = t$ **126.** $6x - x^2 = x$

127. $6x - x^2 = -x$ **128.** $2x - x^2 = -x$

129. $9x^3 = x$ **130.** $16x^3 = x$

131. $(x - 3)^2 = x^2 + 2x + 1$ **132.** $(x - 5)^2 = x^2 + x + 3$

133. $\dfrac{4x}{x + 5} + \dfrac{20}{x} = \dfrac{100}{x^2 + 5x}$

134. $\dfrac{x}{x + 1} + \dfrac{3x + 5}{x^2 + 4x + 3} = \dfrac{2}{x + 3}$

135. $\dfrac{50}{x} - \dfrac{50}{x - 2} = \dfrac{4}{x}$ **136.** $\dfrac{60}{x} = \dfrac{60}{x - 5} + \dfrac{2}{x}$

137. $0 = 2x - \dfrac{250}{x^2}$ **138.** $5 - \dfrac{35}{x^2} = 0$

139. $3 - x \le 4x + 7$ **140.** $x + 6 \le 5x - 6$

141. $5x - 5 + x > 2 - 6x - 8$

142. $3x - 3 + 3x > 1 - 7x - 9$

143. $-7x < 4$ **144.** $-5x \ge 6$

145. $5x + 2x \le -21$ **146.** $9x + 3x \ge -24$

147. $2x - 7 < 5x - 9$ **148.** $10x - 3 \ge 13x - 8$

149. $8x - 9 < 3x - 11$ **150.** $11x - 2 \ge 15x - 7$

151. $8 < 3x + 2 < 14$ **152.** $2 < 5x - 8 \le 12$

153. $3 \le 4x - 3 \le 19$ **154.** $9 \le 5x + 3 < 19$

155. $-7 \le 5x - 2 \le 12$ **156.** $-11 \le 2x - 1 < -5$

APPLICATIONS

Business and Economics

157. Investment increase. An investment is made at $8\frac{1}{2}\%$, compounded annually. It grows to $705.25 at the end of 1 yr. How much was invested originally?

158. Investment increase. An investment is made at 7%, compounded annually. It grows to $856 at the end of 1 yr. How much was invested originally?

159. Total revenue. Sunshine Products determines that the total revenue, in dollars, from the sale of x flowerpots is $3x + 1000$. Determine the number of flowerpots that must be sold so that the total revenue will be more than $22,000.

160. Total revenue. Beeswax Inc. determines that the total revenue, in dollars, from the sale of x candles is $5x + 1000$. Determine the number of candles that must be sold so that the total revenue will be more than $22,000.

Life and Physical Sciences

161. Weight gain. After a 6% gain in weight, an elk weighs 508.8 lb. What was its original weight?

162. Weight gain. After a 7% gain in weight, a deer weighs 363.8 lb. What was its original weight?

Social Sciences

163. Population increase. After a 2% increase, the population of a city is 826,200. What was the former population?

164. Population increase. After a 3% increase, the population of a city is 741,600. What was the former population?

General Interest

165. Grade average. To get a B in a course, a student's average must be greater than or equal to 80% (at least 80%) and less than 90%. On the first three tests, Claudia scores 78%, 90%, and 92%. Determine the scores on the fourth test that will guarantee her a B.

166. Grade average. To get a C in a course, a student's average must be greater than or equal to 70% and less than 80%. On the first three tests, Horace scores 65%, 83%, and 82%. Determine the scores on the fourth test that will guarantee him a C.

Indeterminate Forms and l'Hôpital's Rule

Indeterminate Forms 0/0 and ∞/∞ and l'Hôpital's Rule

In Example 1 of Section 1.1, we showed that

$$\lim_{x \to 1}\left(\frac{x^2 - 1}{x - 1}\right) = 2.$$

Note that as x approaches 1, both the numerator and the denominator approach 0. The expression $0/0$ is an *indeterminate form*. Another common indeterminate form, ∞/∞, arises when x approaches some value a and both the numerator and the denominator increase (or decrease) without bound.

When we attempt to evaluate a limit by substitution and get an indeterminate form such as $0/0$ or ∞/∞, the limit may exist, but we must use a method other than direct evaluation to find it. In Chapter 1, we used numerical methods. Here, we use differentiation to find the limit (if it exists) using l'Hôpital's Rule.

OBJECTIVES

- Find the limit of the indeterminate form 0/0 or ∞/∞ using l'Hôpital's Rule.
- Find the limit of the indeterminate form 0^0, 1^∞, or ∞^0 using l'Hôpital's Rule.
- Find the limit of the indeterminate form ∞ − ∞ using l'Hôpital's Rule.

L'Hôpital's Rule is named for the French mathematician and author Guillaume de l'Hôpital (1661–1704). The name is pronounced "low-pi-tall," with the accent on the last syllable.

> **THEOREM**   **l'Hôpital's Rule**
>
> Let f and g be differentiable over an open interval containing $x = a$ (although not necessarily at a itself). If
>
> $$\lim_{x \to a}\left(\frac{f(x)}{g(x)}\right) = \frac{0}{0} \quad \text{or} \quad \lim_{x \to a}\left(\frac{f(x)}{g(x)}\right) = \frac{\pm\infty}{\pm\infty},$$
>
> and if $\lim\limits_{x \to a}\left(\dfrac{f'(x)}{g'(x)}\right)$ exists, then
>
> $$\lim_{x \to a}\left(\frac{f(x)}{g(x)}\right) = \lim_{x \to a}\left(\frac{f'(x)}{g'(x)}\right).$$

A simplified proof of l'Hôpital's Rule for the case of the indeterminate form $0/0$ is outlined in Exercise 37.

When using l'Hôpital's Rule, we first check whether $f(x)/g(x)$ gives an indeterminate form by substituting $x = a$. If we obtain an indeterminate form, we differentiate the numerator and the denominator of the original expression separately. We then evaluate $\lim\limits_{x \to a}[f'(x)/g'(x)]$.

■ **EXAMPLE 1**   Evaluate

$$\lim_{x \to -3}\left(\frac{x^2 - x - 12}{x + 3}\right).$$

(This is Example 4 in Section 1.2, on page 112.)

Solution   We attempt to evaluate the limit by substituting $x = -3$ into the given expression:

$$\lim_{x \to -3}\left(\frac{x^2 - x - 12}{x + 3}\right) = \frac{(-3)^2 - (-3) - 12}{(-3) + 3}$$

$$= \frac{9 + 3 - 12}{-3 + 3} \qquad \text{Simplifying}$$

$$= \frac{0}{0}. \qquad \text{This is an indeterminate form.}$$

Using l'Hôpital's Rule, we differentiate the numerator and the denominator, and then find the limit:

$$\lim_{x \to -3}\left(\frac{x^2 - x - 12}{x + 3}\right) = \lim_{x \to -3}\frac{2x - 1}{1} \qquad \begin{array}{l}\text{Differentiating numerator} \\ \text{and denominator}\end{array}$$

$$= \frac{2(-3) - 1}{1} \qquad \text{Substituting}$$

$$= -7.$$

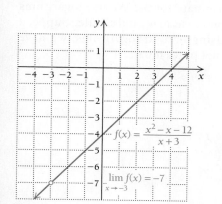

$f(x) = \dfrac{x^2 - x - 12}{x + 3}$

$\lim_{x \to -3} f(x) = -7$

〉 **Quick Check 1**

Evaluate $\lim_{x \to 2}\left(\dfrac{x^3 - 8}{x - 2}\right)$.

〈 Quick Check 1

■ **EXAMPLE 2**   Evaluate $\lim_{x \to \infty}\left(\dfrac{x^2}{e^{3x}}\right)$.

Solution   As x approaches ∞, the expression x^2/e^{3x} approaches the indeterminate form ∞/∞. To use l'Hôpital's Rule, we differentiate the numerator and the denominator of the expression, and then evaluate the limit:

$$\lim_{x \to \infty}\left(\frac{x^2}{e^{3x}}\right) = \lim_{x \to \infty}\left(\frac{2x}{3e^{3x}}\right) \qquad \text{Differentiating numerator and denominator}$$

$$= \frac{\infty}{\infty}. \qquad \lim_{x \to \infty} 2x = \infty \text{ and } \lim_{x \to \infty} 3e^{3x} = \infty$$

Since we obtain the same indeterminate form, ∞/∞, we use l'Hôpital's Rule again:

$$\lim_{x \to \infty}\left(\frac{2x}{3e^{3x}}\right) = \lim_{x \to \infty}\left(\frac{2}{9e^{3x}}\right) \qquad \text{Differentiating numerator and denominator}$$

$$= \frac{2}{\infty} \qquad \lim_{x \to \infty} 2 = 2 \text{ and } \lim_{x \to \infty} 9e^{3x} = \infty$$

$$= 0. \qquad \text{Using } \lim_{x \to \infty}\left(\frac{1}{x}\right) = 0 \text{ (from Example 5 in Section 1.1)}$$

Thus, as x approaches ∞, the expression x^2/e^{3x} approaches 0. This indicates that the graph of $y = x^2/e^{3x}$ has a horizontal asymptote at $y = 0$ (the x-axis).

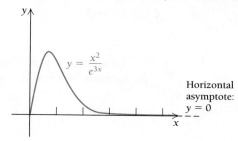

$y = \dfrac{x^2}{e^{3x}}$

Horizontal asymptote: $y = 0$

〉 **Quick Check 2**

Evaluate $\lim_{x \to \infty}\left(\dfrac{e^{4x}}{x^4}\right)$.

〈 Quick Check 2

Indeterminate Forms 0^0, 1^∞, and ∞^0

The expressions 0^0, 1^∞, and ∞^0 are also indeterminate forms. We use logarithms to rewrite such a form as a ratio so that we may use l'Hôpital's Rule. Suppose $y = f(x)^{g(x)}$ and $\lim_{x \to a} f(x)^{g(x)}$ is indeterminate. Using natural logarithms, we have $\ln y = \ln f(x)^{g(x)} = g(x) \ln f(x)$. Although we do not prove it here, if

$$\lim_{x \to a} [g(x) \ln f(x)] = L,$$

then

$$\lim_{x \to a} f(x)^{g(x)} = e^L.$$

■ EXAMPLE 3   Evaluate

$$\lim_{h \to 0} (1 + h)^{1/h}.$$

Solution   As h approaches 0, the base $1 + h$ approaches 1, and the exponent $1/h$ approaches ∞. Thus, we have the indeterminate form 1^∞. We let $y = (1 + h)^{1/h}$ and take the natural logarithm of both sides:

$$y = (1 + h)^{1/h}$$
$$\ln y = \ln(1 + h)^{1/h}$$
$$\ln y = \frac{1}{h} \ln(1 + h) \qquad \text{Using a property of logarithms}$$
$$\ln y = \frac{\ln(1 + h)}{h}.$$

As h approaches 0, we have for the numerator, $\ln(1 + h) = \ln(1 + 0) = \ln 1 = 0$, and for the denominator, $h = 0$. Thus, the expression $\ln(1 + h)/h$ approaches the indeterminate form $0/0$, and we use l'Hôpital's Rule:

$$\lim_{h \to 0} (\ln y) = \lim_{h \to 0} \left(\frac{\ln(1 + h)}{h} \right)$$
$$= \lim_{h \to 0} \left(\frac{\frac{1}{1 + h}}{1} \right) \qquad \text{Differentiating numerator and denominator}$$
$$= \lim_{h \to 0} \left(\frac{1}{1 + h} \right) \qquad \text{Simplifying}$$
$$= \frac{1}{1 + 0} = 1. \qquad \text{Evaluating the limit}$$

Since $\lim_{h \to 0} (\ln y) = 1$, we have $\lim_{h \to 0} y = e^1 = e$. Therefore,

$$\lim_{h \to 0} (1 + h)^{1/h} = e.$$

❭ **Quick Check 3**

Evaluate $\lim_{x \to \infty} \left(1 + \dfrac{2}{x} \right)^x$.

❬ Quick Check 3

In Section 3.3, we used the limit found in Example 3 as part of the derivation of the formula for continuous exponential growth, $P(t) = P_0 e^{kt}$.

Indeterminate Form ∞ − ∞

L'Hôpital's Rule may also be applied in cases where the limit is of the indeterminate form $\infty - \infty$. The expression is first rewritten as a ratio, and then, if possible, l'Hôpital's Rule is applied.

■ **EXAMPLE 4**   Evaluate

$$\lim_{x \to \infty} (\sqrt{x^2 + x} - x).$$

Solution   As x approaches ∞, the expression $\sqrt{x^2 + x} - x$ approaches $\infty - \infty$, which is an indeterminate form. Before we can use l'Hôpital's Rule, we must rewrite $\sqrt{x^2 + x} - x$ as a ratio:

$$\sqrt{x^2 + x} - x = \sqrt{x^2 \left(1 + \frac{1}{x}\right)} - x \qquad \text{Factoring}$$

$$= x\sqrt{1 + \frac{1}{x}} - x \qquad \text{Assuming } x > 0$$

$$= x\left(\sqrt{1 + \frac{1}{x}} - 1\right) \qquad \text{Factoring}$$

$$= \frac{\sqrt{1 + \frac{1}{x}} - 1}{\frac{1}{x}} \qquad \text{Rewriting } x \text{ as } 1/(1/x)$$

The expression is now a ratio. Note that as x approaches ∞, both the numerator and the denominator approach 0. Thus, the expression approaches the indeterminate form $0/0$, and we can use l'Hôpital's Rule:

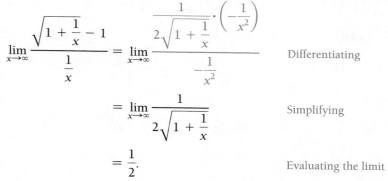

$$\lim_{x \to \infty} \frac{\sqrt{1 + \frac{1}{x}} - 1}{\frac{1}{x}} = \lim_{x \to \infty} \frac{\frac{1}{2\sqrt{1 + \frac{1}{x}}} \cdot \left(-\frac{1}{x^2}\right)}{-\frac{1}{x^2}} \qquad \text{Differentiating}$$

$$= \lim_{x \to \infty} \frac{1}{2\sqrt{1 + \frac{1}{x}}} \qquad \text{Simplifying}$$

$$= \frac{1}{2}. \qquad \text{Evaluating the limit}$$

Thus, $\lim_{x \to \infty} (\sqrt{x^2 + x} - x) = \frac{1}{2}$. This indicates that the graph of $y = \sqrt{x^2 + x} - x$ has a horizontal asymptote at $y = \frac{1}{2}$.

‹ Quick Check 4

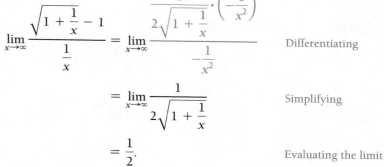

Horizontal asymptote: $y = \frac{1}{2}$

$y = \sqrt{x^2 + x} - x$

> **Quick Check 4**

Evaluate
$$\lim_{x \to \infty} (\sqrt{x^2 + 3x} - x).$$

EXERCISE SET
B

Evaluate each limit. Use l'Hôpital's Rule when necessary.

1. $\lim\limits_{x \to 5}\left(\dfrac{x^2 - 25}{2x - 10}\right)$

2. $\lim\limits_{x \to -2}\left(\dfrac{x^2 - 4}{x + 2}\right)$

3. $\lim\limits_{x \to 1}\left(\dfrac{x^3 + 2x - 3}{x^2 - 1}\right)$

4. $\lim\limits_{x \to 3}\left(\dfrac{x^3 - x - 24}{x^2 - 9}\right)$

5. $\lim\limits_{x \to -3}\left(\dfrac{x^2 - 9}{x - 3}\right)$

6. $\lim\limits_{x \to -4}\left(\dfrac{x^2 + x - 12}{x - 2}\right)$

7. $\lim\limits_{x \to 2}\left(\dfrac{x^3 + 5x + 1}{2x^2 - 6}\right)$

8. $\lim\limits_{x \to 10}\left(\dfrac{x^2 + x - 120}{x + 10}\right)$

9. $\lim\limits_{x \to \infty}\left(\dfrac{4x^2 + x - 3}{2x^2 + 1}\right)$

10. $\lim\limits_{x \to -\infty}\left(\dfrac{3x^3 + x + 11}{6x^3 + x + 2}\right)$

11. $\lim\limits_{x \to 0}\left(\dfrac{e^{2x} - 1}{3x}\right)$

12. $\lim\limits_{x \to 0}\left(\dfrac{e^{4x} + x - 1}{5x}\right)$

13. $\lim\limits_{x \to 0}\left(\dfrac{2e^{-3x} + x - 2}{4x}\right)$

14. $\lim\limits_{x \to 0}\left(\dfrac{5e^{-2x} + x - 5}{2x}\right)$

15. $\lim\limits_{x \to 0}\left(\dfrac{x^2 + x}{e^{3x} - 1}\right)$

16. $\lim\limits_{x \to 0}\left(\dfrac{4x}{e^{2x} - 1}\right)$

17. $\lim\limits_{x \to 2}\left(\dfrac{x - 2}{\ln(x - 1)}\right)$

18. $\lim\limits_{x \to -3}\left(\dfrac{x + 3}{\ln(x + 4)}\right)$

19. $\lim\limits_{x \to 0}(1 + 2x)^{1/x}$

20. $\lim\limits_{x \to 0}(1 - 3x)^{1/x}$

21. $\lim\limits_{x \to \infty}\left(1 - \dfrac{4}{x}\right)^{x}$

22. $\lim\limits_{x \to \infty}\left(1 + \dfrac{5}{x}\right)^{x}$

23. $\lim\limits_{x \to 0} x^{x}$

24. $\lim\limits_{x \to \infty} \sqrt[x]{x}$

25. $\lim\limits_{x \to \infty}(\sqrt{x^2 - x} - x)$

26. $\lim\limits_{x \to \infty}(\sqrt{x^2 + 5x} - x)$

27. $\lim\limits_{x \to \infty}(\sqrt[3]{x^3 + x^2} - x)$

28. $\lim\limits_{x \to \infty}(\sqrt[3]{8x^3 + x^2} - 2x)$

29. Show that for all $k, \lim\limits_{x \to \infty}\left(1 + \dfrac{k}{x}\right)^{x} = e^{k}$.

30. Show that for all $k > 0, \lim\limits_{x \to \infty}(\sqrt{x^2 + kx} - x) = \dfrac{k}{2}$.

31. Evaluate $\lim\limits_{x \to \infty}\left(\dfrac{1}{\sqrt{x^2 + 4x} - x}\right)$.

32. Evaluate $\lim\limits_{x \to \infty}\left(\dfrac{1}{\sqrt{x^2 - 6x} - x}\right)$.

33. Evaluate $\lim\limits_{x \to \infty}\left(\dfrac{\sqrt{x^2 - 4x} - x}{\sqrt{x^2 + 10x} - x}\right)$.

34. Evaluate $\lim\limits_{x \to \infty}\left(\dfrac{\sqrt{x^2 + 7x} - x}{\sqrt{x^2 - 3x} - x}\right)$.

SYNTHESIS

35. Consider $\lim\limits_{x \to \infty}\left(\dfrac{x^n}{e^x}\right)$.

   a) Evaluate this limit for $n = 3, 4, 5$, and 6.

   b) Predict the limit when $n = 100$, and explain how you would demonstrate this using l'Hôpital's Rule.

   c) Is there a positive integer n for which this limit is not zero? Why or why not?

36. Consider $\lim\limits_{x \to \infty}\left(\dfrac{e^{2x}}{e^{3x}}\right)$.

   a) Use algebra to simplify the expression; then evaluate the limit.

   b) Explain why l'Hôpital's Rule is not needed to find this limit.

37. The proof of l'Hôpital's Rule for the indeterminate form $0/0$ is outlined below. Assume that f and g are differentiable at $x = a$ and that $f(a) = 0$ and $g(a) = 0$, and recall that the derivative of f at $x = a$ can be defined as

$$f'(a) = \lim\limits_{x \to a}\left(\dfrac{f(x) - f(a)}{x - a}\right).$$ Give a reason that justifies each step.

   a) $\lim\limits_{x \to a}\dfrac{f(x)}{g(x)} = \lim\limits_{x \to a}\dfrac{f(x) - f(a)}{g(x) - g(a)}$

   _____

   b) $\lim\limits_{x \to a}\dfrac{f(x) - f(a)}{g(x) - g(a)} = \lim\limits_{x \to a}\dfrac{\left(\dfrac{f(x) - f(a)}{x - a}\right)}{\left(\dfrac{g(x) - g(a)}{x - a}\right)}$

   _____

   c) $\lim\limits_{x \to a}\dfrac{\left(\dfrac{f(x) - f(a)}{x - a}\right)}{\left(\dfrac{g(x) - g(a)}{x - a}\right)} = \dfrac{\lim\limits_{x \to a}\left(\dfrac{f(x) - f(a)}{x - a}\right)}{\lim\limits_{x \to a}\left(\dfrac{g(x) - g(a)}{x - a}\right)}$

   _____

   d) $\dfrac{\lim\limits_{x \to a}\left(\dfrac{f(x) - f(a)}{x - a}\right)}{\lim\limits_{x \to a}\left(\dfrac{g(x) - g(a)}{x - a}\right)} = \dfrac{f'(a)}{g'(a)}$

   _____

   e) $\dfrac{f'(a)}{g'(a)} = \lim\limits_{x \to a}\dfrac{f'(x)}{g'(x)}$

   _____

38. Find and correct the error in the following limit calculation.

$$\lim_{x \to 2}\left(\frac{x^2 - 4}{x + 2}\right) = \lim_{x \to 2}\left(\frac{2x}{1}\right) = 4.$$

TRIGONOMETRY CONNECTION

39. Evaluate $\lim_{x \to 0}\left(\dfrac{\sin x}{x}\right)$.

40. Evaluate $\lim_{x \to 0}\left(\dfrac{\cos x - 1}{x}\right)$.

41. Evaluate $\lim_{x \to 0}\left(\dfrac{\sin 5x}{\sin 2x}\right)$.

42. Evaluate $\lim_{x \to 0}\left(\dfrac{\sin 7x}{\sin 4x}\right)$.

Answers to Quick Checks
1. 12   **2.** ∞ **3.** e^2 **4.** $\frac{3}{2}$

Regression and Microsoft Excel

APPENDIX C

Using Excel 2007

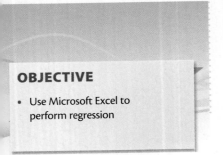

OBJECTIVE

- Use Microsoft Excel to perform regression

We can use Microsoft Excel to enter and plot data and to find lines of best fit using regression. Suppose we are given the following data:

x	0	2	4	5	6
y	3	4.7	6	6.8	8

Step 1: We enter the data into two columns, as shown in Fig. 1.

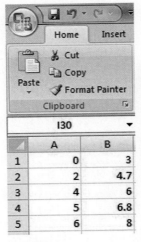

FIGURE 1

FIGURE 2

Step 2: We highlight the columns of data (see Fig. 2). Then we go to the Insert tab and select Scatter. Next, we choose the first option, with the markers shown as distinct points.

Step 3: A graph of the data points appears, as shown in Fig. 3. Under the Layout tab, we select Trendline and then choose More Trendline Options.

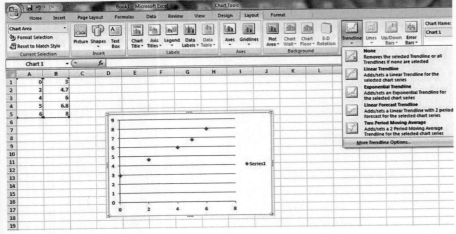

FIGURE 3

836

FIGURE 4

Step 4: A box opens, with Linear preselected as the type of regression to be used; see Fig. 4. At the bottom, we check the box next to Display Equation on chart. As an option, we can also check the box next to Display R-squared value on chart.

Step 5: The line of best fit is now displayed on the graph, along with the R^2 value, as shown in Fig. 5. We can visually inspect how closely the line models the data. The R^2 value is called the *squared correlation coefficient*: an R^2 value close to 1 indicates that the line fits the data well, or, equivalently, that the data have a strong linear trend.

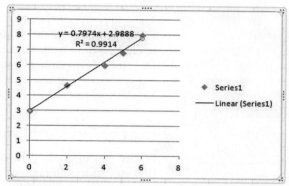

FIGURE 5

In this book, regression is also used to fit exponential and polynomial functions to data. The student can experiment with the various regression options in Excel.

Using Excel for Mac 2008

The steps for finding the line of best fit using Excel for Mac 2008 are given below. (*Note:* There are three toolbar levels in Excel for Mac, the primary toolbar along the top of the screen, the secondary set of options directly below the primary toolbar, and a tertiary toolbar that is connected to the spreadsheet cells directly, along the top.)

Step 1: We enter the data into two columns.

Step 2: Under the Charts option (tertiary toolbar), we select X Y (Scatter), then click on the first choice. A scatterplot appears on the screen.

Step 3: To add a trendline, we click on one of the points on the scatterplot to "activate" them (they will appear as X shapes).

Step 4: We go to the Chart option on the primary toolbar, and select Add Trendline from the drop-down menu. A Format Trendline window appears. We click on Options.

Step 5: We check the boxes next to Display equation on chart and Display R-squared value on chart. These will appear on the scatterplot, along with the trendline. Other types of regression can be viewed by clicking on Type in the Format Trendline window.

EXERCISE SET

C

Use Excel to find the line of best fit for the following data.

x	4	6	8	10	12
y	15	22	27	33	44

2. Use Excel to find the line of best fit for the following data.

x	−5	−1	3	8
y	2	10	19	35

3. Use Excel to find a quadratic function (polynomial of power 2) that best fits the data in Exercise 2.

MathPrint Operating System for TI-84 and TI-84 Plus Silver Edition

OBJECTIVE

• Upgrade a TI calculator with MathPrint

The graphing calculator screens in this text display math in the format of the TI Math-Print operating system. With MathPrint, the math looks more like that seen in a printed book. You can obtain MathPrint and install it by following the instructions given below. Only the TI-84 family of graphing calculators can be updated with the MathPrint operating system. If you own a TI-83 graphing calculator, you can use this brief appendix to help you "translate" what you see in the Classic mode shown on your calculator.

How to Get MathPrint Mode on a TI Graphing Calculator

Before you upgrade your operating system, you must **archive all items** (programs, lists etc.) stored in random access memory (RAM); otherwise, they will be lost. Follow these steps to accomplish this task:

1. Press **2ND** and Mem. Select 2:Mem Mgmt/Del and then 1:All on the next screen.
2. Press **ENTER** with the cursor next to any item you wish to archive (it will be marked with an asterisk).
3. Upgrade the operating system (see method 1 or method 2 below).
4. After you upgrade the operating system, repeat steps 1 and 2 to move items out of the archive back into RAM. Unarchived items no longer have an asterisk next to them.

There are several ways to upgrade a TI-84 calculator to the latest operating system, which includes MathPrint mode. Two of these methods are presented here.

Method 1: Using TI-Connect

You can install the latest operating system by following these steps:

1. Launch TI-Connect software.
2. Connect your calculator to your computer using the Silver USB cable.
3. On a PC, click the Update button on the TI-Connect main menu page. On a Macintosh, double-click the TI Software Update button and wait for the software to recognize the device. Click ▾ to reveal the contents of your calculator, scroll down and check the box next to TI-84 Plus family Operating System, and click the Update button at the top of the window.
4. Follow the prompts to upgrade the operating system. Make sure that your calculator's batteries are fresh.

Method 2: Transfering the MathPrint Operating System from Another Graphing Calculator

Press **2ND** and Mem and then **ENTER** on the sending calculator and check that the latest version of the operating system (2.53 MP or higher) is installed. Then follow these steps:

1. Connect the two graphing calculators with a unit-to-unit link cable.
2. On the receiving calculator, press **2ND** and LINK, followed by ▶ and **ENTER**. The screen will display Waiting....

3. On the sending calculator, press **2ND** and LINK, scroll down and highlight G:SendOS, and then press **ENTER**.

4. Follow the prompts to upgrade the operating system.

You must also install version 1.1 of the application CatalogHelp on your graphing calculator. This application can be downloaded at no charge from education.ti.com and transferred from a computer to your calculator using TI-Connect software or from another calculator using the preceding steps.

Switching between MathPrint Mode and Classic Mode

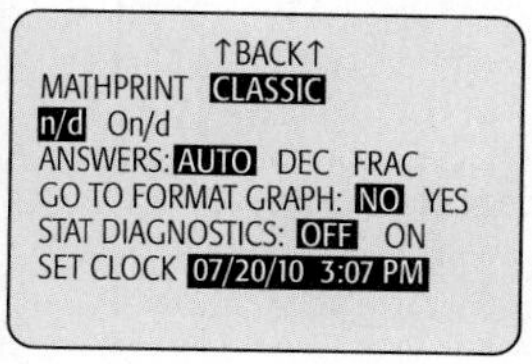

A TI-84 graphing calculator with MathPrint can be switched from MathPrint to Classic mode by changing the mode settings. To do this, press the [MODE] key and scroll to the second page, shown at the left. Use the arrow keys to highlight MATHPRINT or CLASSIC and press **ENTER**.

A TI-84 graphing calculator loaded with the MathPrint operating system and running in Classic mode will show many of the MathPrint features. Below are two examples

Feature	MathPrint	MathPrint in Classic Mode
Improper fractions	$\frac{2}{5} - \frac{1}{3}$ $\frac{1}{15}$	2/5 – 1/3 1/15
Logarithms	$\log_2(32)$ 5	logBASE(32,2) 5

Translating between MathPrint Mode and Classic Mode

The following table compares displays of several types in MathPrint mode and Classic mode (on a calculator without MathPrint installed).

Feature	MathPrint	Classic (MathPrint not installed)
Improper fractions	$\frac{2}{5} - \frac{1}{3}$ $\frac{1}{15}$	2/5–1/3▶Frac 1/15 Enter an expression and press **MATH** and select 1:Frac.
Mixed fractions	$2\frac{1}{5} * \left(3\frac{2}{3}\right)$ $\frac{121}{15}$	Not supported
Absolute values	|10 – 15| 5 Press **ALPHA** and F2 and select 1:abs(.	abs(10–15) 5 Press **MATH** and ▶ and select 1:abs(.

(continued)

Feature	MathPrint	Classic (MathPrint not installed)
Summation	$\sum\limits_{1=1}^{10}(I^2)$ 385 Press **ALPHA** and F2 and select $2:\sum($.	sum(seq(I²,I,1,10) 385 Press **2ND** and LIST and then ▶ and select 5:seq(for seq and press **2ND** and LIST and then ▶ twice and select 5:sum(for sum.
Numerical derivatives	$\frac{d}{dX}(X^2)\big|_{X=3}$ 6 Press **ALPHA** and F2 and select 3:nDeriv(.	nDeriv(X²,X,3) 6 Press **MATH** and select 8:nDeriv(.
Numerical values of integrals	$\int_1^5(X^2)dX$ 41.33333333 Press **ALPHA** and F2 and select 4:fnInt(.	fnInt(X²,X,1,5) 41.33333333 Press **MATH** and select 9:fnInt(.
Logarithms	$\log_2(32)$ 5 Press **ALPHA** and F2 and select 5:logBASE(.	Evaluating logs with bases other than 10 or e cannot be done on a graphing calculator if the Math-Print operating system is not installed. To evaluate $\log_2 32$, use the change-of-base formula: log(32)/log(2) 5

The [Y=] Editor

MathPrint features can be accessed from the [Y=] editor as well as from the home screen. The following table shows examples that illustrate differences between MathPrint in the [Y=] editor and Classic mode.

Feature	MathPrint	Classic Mode (MathPrint not installed)
Graphing the derivative of $y = x^2$	Plot 1 Plot 2 Plot 3 \Y₁ ■ $\frac{d}{dX}(X^2)\big|_{X=X}$ \Y₂ =	Plot 1 Plot 2 Plot 3 \Y₁ ■ nDeriv(X²,X,X) \Y₂ =
Graphing an antiderivative of $y = x^2$	Plot 1 Plot 2 Plot 3 \Y₁ ■ $\int_0^x(X^2)dX$ \Y₂ =	Plot 1 Plot 2 Plot 3 \Y₁ ■ fnInt(X²,X,0,X) \Y₂ =

Areas for a Standard Normal Distribution

Entries in the table represent area under the curve between $z = 0$ and a positive value of z. Because of the symmetry of the curve, area under the curve between $z = 0$ and a negative value of z are found in a similar manner.

Area = Probability
$$= P(0 \le x \le z)$$
$$= \int_0^z \frac{1}{\sqrt{2\pi}} e^{-x^2/2}\, dx$$

z	0.00	0.01	0.02	0.03	0.04	0.05	0.06	0.07	0.08	0.09
0.0	.0000	.0040	.0080	.0120	.0160	.0199	.0239	.0279	.0319	.0359
0.1	.0398	.0438	.0478	.0517	.0557	.0596	.0636	.0675	.0714	.0753
0.2	.0793	.0832	.0871	.0910	.0948	.0987	.1026	.1064	.1103	.1141
0.3	.1179	.1217	.1255	.1293	.1331	.1368	.1406	.1443	.1480	.1517
0.4	.1554	.1591	.1628	.1664	.1700	.1736	.1772	.1808	.1844	.1879
0.5	.1915	.1950	.1985	.2019	.2054	.2088	.2123	.2157	.2190	.2224
0.6	.2257	.2291	.2324	.2357	.2389	.2422	.2454	.2486	.2517	.2549
0.7	.2580	.2611	.2642	.2673	.2704	.2734	.2764	.2794	.2823	.2852
0.8	.2881	.2910	.2939	.2967	.2995	.3023	.3051	.3078	.3106	.3133
0.9	.3159	.3186	.3212	.3238	.3264	.3289	.3315	.3340	.3365	.3389
1.0	.3413	.3438	.3461	.3485	.3508	.3531	.3554	.3577	.3599	.3621
1.1	.3643	.3665	.3686	.3708	.3729	.3749	.3770	.3790	.3810	.3830
1.2	.3849	.3869	.3888	.3907	.3925	.3944	.3962	.3980	.3997	.4015
1.3	.4032	.4049	.4066	.4082	.4099	.4115	.4131	.4147	.4162	.4177
1.4	.4192	.4207	.4222	.4236	.4251	.4265	.4279	.4292	.4306	.4319
1.5	.4332	.4345	.4357	.4370	.4382	.4394	.4406	.4418	.4429	.4441
1.6	.4452	.4463	.4474	.4484	.4495	.4505	.4515	.4525	.4535	.4545
1.7	.4554	.4564	.4573	.4582	.4591	.4599	.4608	.4616	.4625	.4633
1.8	.4641	.4649	.4656	.4664	.4671	.4678	.4686	.4693	.4699	.4706
1.9	.4713	.4719	.4726	.4732	.4738	.4744	.4750	.4756	.4761	.4767
2.0	.4772	.4778	.4783	.4788	.4793	.4798	.4803	.4808	.4812	.4817
2.1	.4821	.4826	.4830	.4834	.4838	.4842	.4846	.4850	.4854	.4857
2.2	.4861	.4864	.4868	.4871	.4875	.4878	.4881	.4884	.4887	.4890
2.3	.4893	.4896	.4898	.4901	.4904	.4906	.4909	.4911	.4913	.4916
2.4	.4918	.4920	.4922	.4925	.4927	.4929	.4931	.4932	.4934	.4936
2.5	.4938	.4940	.4941	.4943	.4945	.4946	.4948	.4949	.4951	.4952
2.6	.4953	.4955	.4956	.4957	.4959	.4960	.4961	.4962	.4963	.4964
2.7	.4965	.4966	.4967	.4968	.4969	.4970	.4971	.4972	.4973	.4974
2.8	.4974	.4975	.4976	.4977	.4977	.4978	.4979	.4979	.4980	.4981
2.9	.4981	.4982	.4982	.4983	.4984	.4984	.4985	.4985	.4986	.4986
3.0	.4987	.4987	.4987	.4988	.4988	.4989	.4989	.4989	.4990	.4990

Photo Credits

p. 1: Howard Berman/Iconica/Getty Images. p. 3: Howard Berman/Iconica/Getty Images. p. 11: (upper) Kyodo/AP Images; (lower) Mike Stotts/WENN/Newscom. p. 36: Rainprel/Shutterstock. p. 39: Pincasso/Shutterstock. p. 42: Scott Surgent. p. 48: Olly/Shutterstock. p. 62: Lightpoet/Shutterstock.

p. 73: William Ju/Shutterstock. p. 75: Howard Berman/Iconica/Getty Images. p. 90: Stephen Coburn/Shutterstock. p. 93: Mike Kemp/Getty Images. p. 113: Xavier Pironet/Shutterstock. p. 121: Rob Pitman/Shutterstock. p. 133: Pulen/Shutterstock. p. 156: Silver-John/Shutterstock. p. 169: Brian Spurlock. p. 172: Mike Kemp/Getty Images. p. 194: Todd Taulman/Shutterstock. p. 196: NY Daily News/Getty Images. p. 197: API/Alamy. p. 214: NASA. p. 232: Glow Images/Getty. p. 249: Getty Images. p. 260: Dimitriadi Kharlampiy/Shutterstock. p. 261: Scott Surgent. p. 265: API/Alamy.

p. 269: Jochen Sand/Digital Vision/Jupiter Images. p. 276: Annieannie/Dreamstime. p. 291: Dmitrijs Dmitrijevs/Shutterstock. p. 293: (left) Intrepix/Dreamstime; (right) Thomas Deerinck/NCMIR/Photo Researchers, Inc. p. 305: Guy Sagi/Dreamstime. p. 306: (left) Alain/Dreamstime; (lower right) Twildlife/Dreamstime; (upper right) Jack Schiffer/Dreamstime. p. 307: Heritage Auctions/UPI/Newscom.

p. 320: Donvictorio/Shutterstock. p. 338: Maigi/Shutterstock. p. 340: David Lichtneker/Alamy. p. 341: Vera Kailova/Dreamstime. p. 342: Heritage Auctions/UPI/Newscom. p. 344: A. T. Willett/Alamy. p. 348: (left) SE/Shutterstock; (right) *Boy with a Pipe* (1905), Pablo Picasso. Oil on canvas, 39.4 in × 32.0 in (100 cm × 81.3 cm). Private collection. Copyright © 2011 Estate of Pablo Picasso/Artists Rights Society (ARS)/Ezio Petersen/UPI/Newscom. p. 349: Janis Rozentals/Shutterstock.

p. 350: (left) Chuck Crow/The Plain Dealer/Landov; (right) Paul Knowles/Shutterstock. p. 351: VVO/Shutterstock. p. 355: Debbie Hill/UPI/Newscom. p. 361: Jeffrey M. Frank/Shutterstock. p. 365: Macs Peter/Shutterstock. p. 370: Landov. p. 372: Exactostock/Superstock. p. 386: John Phillips/PA Photos/Landov. p. 387: Alcon Entertainment/Album/Newscom. p. 388: Henri Lee/PictureGroup/AP Images. p. 389: Alon Brik/Dreamstime. p. 397: Bojan Pavlukovic/Dreamstime. p. 398: Monkeybusiness/Dreamstime. p. 399: Radius Images/Alamy. p. 402: Jordan Tan/Shutterstock. p. 424: Rechitan Sorin/Shutterstock. p. 468: NASA. p. 470: Hunta/Shutterstock. p. 471: (left) Ilene MacDonald/Alamy; (middle) Ken Welsh/Alamy; (right) GoodMood Photo/Shutterstock. p. 472: Henrik Winther Andersen/Shutterstock. p. 473: Melinda Nagy/Dreamstime. p. 480: Melinda Nagy/Dreamstime. p. 488: Worldpics/Shutterstock. p. 492: Jim Parkin/Shutterstock. p. 509: (upper left) www.pgagolfd.com and www.yourgolftravel.com; (middle left) www.mapquest.com and www.durbinbrothers.net.

p. 511: Ekaterina Pokrovsky/Shutterstock. p. 512: Steven Hockney/Shutterstock. p. 513: Mirec/Fotolia. p. 524: Anna Jurkovska/Shutterstock. p. 525: Mangostock/Shutterstock. p. 528: Gozzoli/Dreamstime. p. 533: James Thew/Shutterstock. p. 534: Vladyslav Morozov/Shutterstock.

p. 541: Helder Almeida/Shutterstock. p. 544: Rick Hanston/Latent Images. p. 555: Mangostock/Shutterstock. p. 556: Gray Mortimore/Getty Images Sport/Getty Images. p. 561: Fotosearch/SuperStock.

p. 564: Andre Mueller/Shutterstock. p. 578: Anna Jurkovska/Shutterstock. p. 581: Ilja Mašík/Shutterstock. p. 582: (left) Vitalliy/Shutterstock; (middle) Ingrid Prats/Shutterstock; (right) Chad McDermott/Fotolia. p. 596: (upper left) Brandelet Didier/Fotolia; (lower left) Galina Mikhalishina/Fotolia; (middle right) www.cnx.org; (lower right) www.knowledgerush.com. p. 605: (left) Raymond Gehman/National Geographic Image Collection/Glow Images; (right) Graça Victoria/Shutterstock.

p. 610: Ilja Mašík/Shutterstock. p. 613: Dean Bertoncelj/Shutterstock. p. 622: Ryan McGinnis/Alamy. p. 630: Agnieszka Barbara/Shutterstock. p. 633: Imagebroker.net/SuperStock. p. 634: Epic-StockMedia/Fotolia. p. 640: Sport The Library SportsChrome/Newscom. p. 646: Accent/Fotolia.

p. 648: Imagebroker.net/SuperStock. p. 652: Ford Prefect/Shutterstock. p. 655: Rafael Ben-Ari/Alamy.

p. 665: Graphicus screenshot reprinted by permission of Serafim Chekalkin. p. 671: Terekhov Igor/Shutterstock. p. 679: Applet screenshot reprinted by permission of Dr. Marek Rychlik. p. 680: Sigrid Gombert/Cultura/Glow Images. p. 683: European Pressphoto Agency (EPA)/Alamy. p. 684: Reha Mark/Shutterstock. p. 688: Juice Images/Glow Images. p. 700: Biophoto Associates/Photo Researchers, Inc. p. 711: Zimmytws/Fotolia. p. 716: Moritz Buchty/Fotolia. p. 726: European Pressphoto Agency (EPA)/Alamy. p. 727: AP Images. p. 745: (left) AGE Fotostock/SuperStock; (right) *Portrait of Lisa Gherardini* ("Mona Lisa") (1503–1506), Leonardo da Vinci. Oil on poplar wood, 77 × 53 cm. Collection of the Louvre Museum; Purchased by François I in 1518 [Inv. 779]/SuperStock. p. 747: David Gordon/Alamy. p. 748: Africa Studio/Shutterstock. p. 750: (left) Pictorial Press Ltd/Alamy; (right) Hulton Archive/Getty Images. p. 751: Jeff Greenberg/Alamy. p. 755: Hero/Corbis/Glow Images. p. 755: Scott Surgent. p. 757: Piai/Fotolia. p. 775: Alex Segre/Alamy.

p. 777: Zee/Fotolia. p. 785: Tim Roberts Photography/Shutterstock. p. 787: Matt Meadows/Getty Images. p. 799: David Gordon/Alamy. p. 802: Brian Spurlock. p. 811: (upper) Mark and Audrey Gibson/Glow Images; (lower) Nikkytok/Fotolia.

Answers

Chapter R

Technology Connection, p. 6

1–20. Left to the student

Exercise Set R.1, p. 10

1.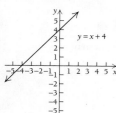

3.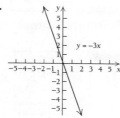

5.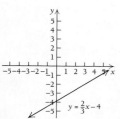

7.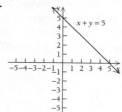

9.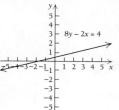

11.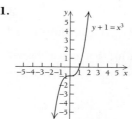

13.

15.

17.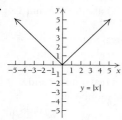

19.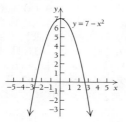

21.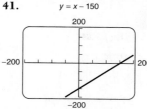

23. 3.98 min (1954), 3.66 min (2008), 3.64 min (2012) **25.** About 27.25 mi/hr, or mph **27. (a)** 1.8 million, 3.7 million, 4.4 million, 4.5 million; **(b)** 44 and 70; **(c)** about 58; **(d)** ✎

29. (a) \$102,800.00; **(b)** \$102,819.60; **(c)** \$102,829.54; **(d)** \$102,839.46; **(e)** \$102,839.56 **31. (a)** \$31,200.00; **(b)** \$31,212.00; **(c)** \$31,218.12; **(d)** \$31,224.25; **(e)** \$31,224.32 **33.** \$550.86 **35.** \$97,881.97 **37. (a)** 1996–2000, 2002; **(b)** 1987, 1990; **(c)** 1999; **(d)** 1987, 1990 **39. (a)** \$206,780.16; **(b)** \$42,000; \$164,780.16

41.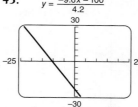

43.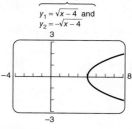

45.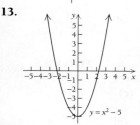

47.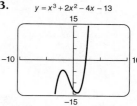

Technology Connection, p. 16

1. 951; 42,701 **2.** 21.813

Technology Connection, p. 17

1. 6; 3.99; 150; $-1.\overline{5}$, or $-\frac{14}{9}$ **2.** -21.3; -18.39; -117.3; $3.2\overline{5}$, or $\frac{293}{90}$ **3.** -75; -65.466; -420.6; $1.6\overline{8}$, or $\frac{76}{45}$

Technology Connection, p. 20

1–3. Left to the student

Exercise Set R.2, p. 21

1. Yes **3.** Yes **5.** Yes **7.** Yes **9.** Yes **11.** Yes
13. Yes **15.** No **17.** Yes
19. (a)

x	5.1	5.01	5.001	5
$f(x)$	17.4	17.04	17.004	17

A ✎ indicates that the exercise asks for a written interpretation or explanation; answers will vary.

(b) $f(4) = 13, f(3) = 9, f(-2) = -11, f(k) = 4k - 3,$
$f(1 + t) = 4t + 1, f(x + h) = 4x + 4h - 3$
21. $g(-1) = -2, g(0) = -3, g(1) = -2, g(5) = 22,$
$g(u) = u^2 - 3, g(a + h) = a^2 + 2ah + h^2 - 3,$ and
$\dfrac{g(a + h) - g(a)}{h} = 2a + h, h \neq 0$ **23. (a)** $f(4) = \dfrac{1}{49}, f(-3)$

is undefined, $f(0) = \dfrac{1}{9}, f(a) = \dfrac{1}{(a + 3)^2}, f(t + 4) = \dfrac{1}{(t + 7)^2},$

$f(x + h) = \dfrac{1}{(x + h + 3)^2},$ and $\dfrac{f(x + h) - f(x)}{h} =$

$\dfrac{-2x - h - 6}{(x + h + 3)^2(x + 3)^2}, h \neq 0$ **(b)** Take an input, square it, add

six times the input, add 9, and then take the reciprocal of the result.

25. 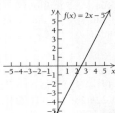   **27.**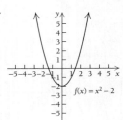

29. 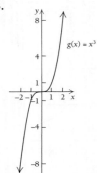   **31.**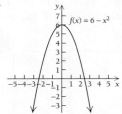

33. 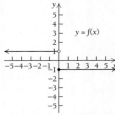   **35.** Yes   **37.** Yes   **39.** No
41. No   **43.** Yes   **45.** Yes
47. (a) 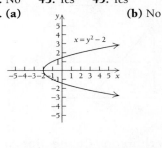   **(b)** No

49. $\dfrac{f(x + h) - f(x)}{h} = 2x + h - 3, h \neq 0$

51. $f(-1) = 3, f(1) = -2$ **53.** $f(0) = 17, f(10) = 6$
55. 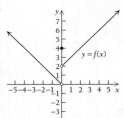   **57.**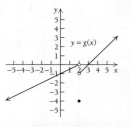

59.   **61.**

63. 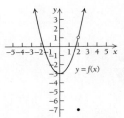   **65.** \$563.25   **67. (a)** 1.818 m²;
(b) 2.173 m²; **(c)** 1.537 m²
69. (a) Yes; a unique "scale of
impact" number is assigned to
each event. **(b)** The inputs are
the events; the outputs are the
scale of impact numbers.

71. $y = \pm\sqrt{\dfrac{x + 5}{2}}$; this is *not* a function.
73. $y = 2 \cdot \sqrt[3]{x}$; this is a function.   **75.**
77.

X	Y1	
−3	.6	
−2	ERR:	
−1	−1	
0	−.75	
1	−1	
2	ERR:	
3	.6	
X = −3		

79. Left to the student

Technology Connection, p. 28

1.

X	Y1	
−3	.2	
−2.5	.4444	
−2	ERR:	
−1.5	−.5714	
−1	−.3333	
−.5	−.2667	
0	−.25	
X = −3		

2. Answers may vary.

Technology Connection, p. 29

1. Domain $= \mathbb{R}$; range $= [-4, \infty)$
2. Domain $= \mathbb{R}$; range $= \mathbb{R}$
3. Domain $= \{x \mid x \text{ is a real number and } x \neq 0\}$;
range $= \{x \mid x \text{ is a real number and } x \neq 0\}$
4. Domain $= \mathbb{R}$; range $= [-8, \infty)$
5. Domain $= [-4, \infty)$; range $= [0, \infty)$
6. Domain $= [-3, 3]$; range $= [0, 3]$
7. Domain $= [-3, 3]$; range $= [-3, 0]$
8. Domain $= \mathbb{R}$; range $= \mathbb{R}$

Exercise Set R.3, p. 31

1. $[-2, 4]$ **3.** $(0, 5)$ **5.** $[-9, -4)$ **7.** $[x, x + h]$
9. (p, ∞) **11.** $[-2, 2]$
13. $[-4, -1)$
15. $(-\infty, -2]$
17. $(-2, 3]$
19. $(-\infty, 12.5)$ **21. (a)** 3;
(b) $\{-3, -1, 1, 3, 5\}$; **(c)** 3; **(d)** $\{-2, 0, 2, 3, 4\}$
23. (a) 4; **(b)** $\{-5, -3, 1, 2, 3, 4, 5\}$; **(c)** $\{-5, -3, 4\}$;
(d) $\{-3, 2, 4, 5\}$ **25. (a)** -1; **(b)** $[-2, 4]$; **(c)** 3;
(d) $[-3, 3]$ **27. (a)** -2; **(b)** $[-4, 2]$; **(c)** -2;
(d) $[-3, 3]$ **29. (a)** 3; **(b)** $[-3, 3]$; **(c)** about -1.4 and 1.4;
(d) $[-5, 4]$ **31. (a)** 1; **(b)** $[-5, 5)$; **(c)** $[3, 5)$;
(d) $\{-2, -1, 0, 1, 2\}$ **33.** $\{x \mid x \text{ is a real number and } x \neq 2\}$
35. $\{x \mid x \geq 0\}$ **37.** $\mathbb{R}$ **39.** $\{x \mid x \text{ is a real number and } x \neq 2\}$
41. $\mathbb{R}$ **43.** $\{x \mid x \text{ is a real number and } x \neq 3.5\}$
45. $\{x \mid x \geq -\frac{4}{5}\}$ **47.** $\mathbb{R}$ **49.** $\{x \mid x \text{ is a real number and } x \neq 5,$
$x \neq -5\}$ **51.** $\mathbb{R}$ **53.** $\{x \mid x \text{ is a real number and } x \neq 5,$
$x \neq 1\}$ **55.** $[-1, 2]$ **57. (a)** $A(t) = 5000\left(1 + \dfrac{0.08}{2}\right)^{2t}$;
(b) $\{t \mid t \geq 0\}$ **59. (a)** $[0, 84.7]$; **(b)** $[0, 4,600,000]$; **(c)**
61. (a) $[0, 70]$; **(b)** $[8, 75]$ **63.**   **65.**
67. $(-\infty, 0) \cup (0, \infty)$; $[0, \infty)$; $\mathbb{R}$; $[1, \infty)$; $\mathbb{R}$

Technology Connection, p. 34

1. The line will slant up from left to right, will intersect the y-axis at $(0, 1)$, and will be steeper than $y = 10x + 1$.
2. The line will slant up from left to right, will pass through the origin, and will be less steep than $y = \dfrac{2}{31}x$.   **3.** The line will slant down from left to right, will pass through the origin, and will be steeper than $y = -10x$ **4.** The line will slant down from left to right, will intersect the y-axis at $(0, -1)$, and will be less steep than $y = -\frac{5}{32}x - 1$.

Technology Connection, p. 37

1. The graph of y_2 is a shift 3 units up of the graph of y_1, and y_2 has y-intercept $(0, 3)$. The graph of y_3 is a shift 4 units down of the graph of y_1, and y_3 has y-intercept $(0, -4)$. The graph of $y = x - 5$ is a shift 5 units down of the graph of $y = x$, and $y = x - 5$ has y-intercept $(0, -5)$. All lines are parallel.
2. For any x-value, the y_2-value is 3 more than the y_1-value and the y_3-value is 4 less than the y_1-value.

Exercise Set R.4, p. 45

1.
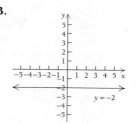

3.
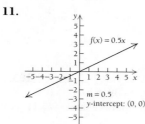

5.
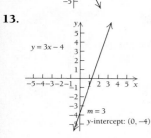

7.
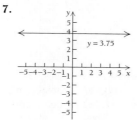

9.
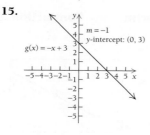

11.

13.

15.

17.
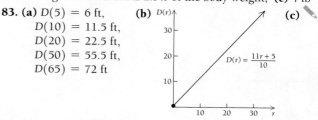

19. $m = 3$, y-intercept: $(0, 6)$
21. $m = -2$, y-intercept: $(0, 3)$
23. $m = -1$, y-intercept: $(0, -4)$
25. $m = \frac{1}{3}$, y-intercept: $(0, -\frac{7}{3})$
27. $y + 3 = -5(x + 2)$, or $y = -5x - 13$
29. $y - 3 = -2(x - 2)$, or $y = -2x + 7$
31. $y - 0 = 2(x - 3)$, or $y = 2x - 6$ **33.** $y + 6 = \frac{1}{2}(x - 0)$, or $y = \frac{1}{2}x - 6$ **35.** $y - 3 = 0 \cdot (x - 2)$, or $y = 3$ **37.** $-\frac{4}{7}$
39. $\frac{1}{3}$ **41.** Undefined slope **43.** $-\frac{34}{3}$ **45.** 0 **47.** 3
49. 2 **51.** $y = -\frac{4}{7}x - \frac{1}{7}$ **53.** $y = \frac{1}{3}x - \frac{11}{3}$ **55.** $x = 3$
57. $y = -\frac{34}{3}x + \frac{91}{15}$ **59.** $y = 3$ **61.** $\frac{2}{5}$ **63.** 3.5%
65. (a) $I(s) = 0.0057s$; (b) 18 cartridges
67. (a) $C(x) = 80x + 45{,}000$; (b) $R(x) = 225x$; (c) $P(x) = 175x - 45{,}000$; (d) profit of \$480,000; (e) 258 pairs
69. (a) $C(x) = 4x + 250$; (b) $R(x) = C(x) + P(x) = 13x$, so Jimmy charges \$13 per lawn; (c) 28 lawns
71. $V(3) = \$25{,}200$ **73.** About 91% (don't round up, for legal reasons!) **75.** \$200 per year **77.** $-\$1717.50$ per year
79. $t \approx 0.02$ sec **81.** (a) $B(W) = 0.025W$; (b) $0.025 = 2.5\%$, so the weight of the brain is 2.5% of the body weight; (c) 4 lb
83. (a) $D(5) = 6$ ft,
$D(10) = 11.5$ ft,
$D(20) = 22.5$ ft,
$D(50) = 55.5$ ft,
$D(65) = 72$ ft
(b) (c)

85. (a) Approximately $y = \frac{20}{9}x - 4372.44$; (b) approximately 94.7%; (c) 2012; (d)   **87.** (a) $N = 1.02P$;
(b) $N = 204{,}000$ people; (c) $P = 360{,}000$ people   **89.**
91. (a) Graph III; (b) graph IV; (c) graph I; (d) graph II
93. Answers may vary.

Technology Connection, p. 51

1–2. Left to the student

Technology Connection, p. 53

1. (a) 2 and 4; (b) 2 and -5; (c) and (d) left to the student

Technology Connection, p. 54

1. -5 and 2   **2.** -4 and 6   **3.** -2 and 1
4. 0, -1.414, and 1.414 (approx.)   **5.** 0 and 700
6. -2.079, 0.463, and 3.116 (approx.)   **7.** -3.096, -0.646, 0.646, and 3.096 (approx.)   **8.** -1 and 1   **9.** -0.387 and 1.721   **10.** 6.133   **11.** -2, -1.414, 1, and 1.414
12. -3, -1, 2, and 3

Technology Connection, p. 55

Left to the student

Technology Connection, p. 58

1. Left to the student   **2.** For $x = 3$, y_1 is ERR and y_2 is 6.

Technology Connection, p. 59

1–8. Left to the student

Technology Connection, p. 60

1–2. Left to the student

Technology Connection, p. 64

1. Approximately $(14, 266)$

Exercise Set R.5, p. 65

1.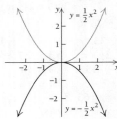

3.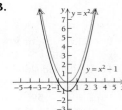

5.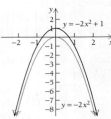

7.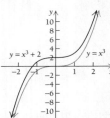

9.

11.

13. Parabola with vertex at $(-2, -11)$ **15.** Not a parabola

17.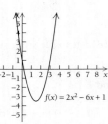

19.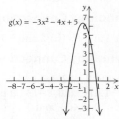

21.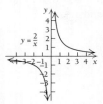

23.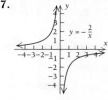

25.

27.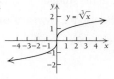

29.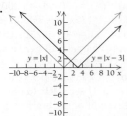

31.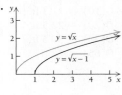

33.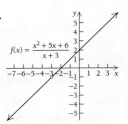

35.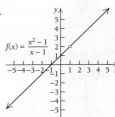

37. $1 \pm \sqrt{3}$, or $-0.732, 2.732$ **39.** $-3 \pm \sqrt{10}$, or -6.162, 0.162 **41.** $\dfrac{1 \pm \sqrt{2}}{2}$, or $-0.207, 1.207$ **43.** $\dfrac{-4 \pm \sqrt{10}}{3}$, or $-2.387, -0.279$ **45.** $\dfrac{-7 \pm \sqrt{13}}{2}$, or $-5.303, -1.697$

47. $x^{3/2}$ **49.** $a^{3/5}$ **51.** $t^{1/7}$ **53.** x^3 **55.** $t^{-5/2}$

57. $(x^2 + 7)^{-1/2}$ **59.** $\sqrt[5]{x}$ **61.** $\sqrt[3]{y^2}$ **63.** $\dfrac{1}{\sqrt[5]{t^2}}$ **65.** $\dfrac{1}{\sqrt[3]{b}}$

67. $\dfrac{1}{\sqrt[6]{e^{17}}}$ **69.** $\dfrac{1}{\sqrt{x^2 - 3}}$ **71.** $\dfrac{1}{\sqrt[3]{t^2}}$ **73.** 27 **75.** 16

77. 8 **79.** $\{x \mid x \neq 5\}$ **81.** $\{x \mid x \neq 2, x \neq 3\}$

83. $\{x \mid x \geq -\frac{4}{5}\}$ **85.** $\{x \mid x \leq 7\}$ **87.** $(50, 500)$; $x = \$50$, $q = 500$ items   **89.** $(5, 1)$; price is $500, and quantity is 1000.

91. $(1, 4)$; price is $1, and quantity is 400.   **93.** $(2, 3)$; price is $2000, and quantity is 3000.   **95.** $140.90 per share

97. (a) 166 mi, **(b)** 176 mi, 184 mi

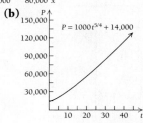

99. (a) 99,130 particles/cm^3, 108,347 particles/cm^3, 127,322 particles/cm^3 **(b)**

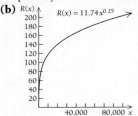

101. 16 cities; 30 cities   **103.**   **105.** $-1.831, -0.856, 3.188$
107. $1.489, 5.673$ **109.** $-2, 3$ **111.** $[-1, 2]$
113. Approximately $(75.11, 7893)$; produce 7893 units at a price of $75.11 each.

Technology Connection, p. 72

1. (a) $y = 2.7x + 63.8$ **(b)** 93.5; **(c)**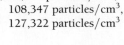

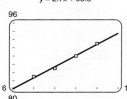

Technology Connection, p. 74

1. (a) $y = -0.00005368295x^4 + 0.037566680x^3$
$- 3.4791715x^2 + 105.81080x - 916.68952$ **(b)**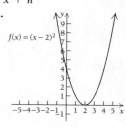

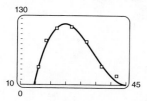

2. (a) $y = -62.8327x^2 + 5417.8404x - 57264.7856$

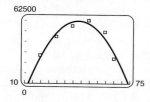

(b) $y = -1.6519x^3 + 145.6606x^2 - 2658.3088x + 36491.7730$

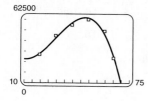

(c) $y = -0.0771x^4 + 11.3952x^3 - 639.2276x^2 + 17037.1915x - 135483.9938$

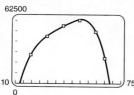

(d) quartic function;   **(e)** \$38,853, \$58,887

3. $y = 93.2857x^2 - 1336x + 5460.8286$

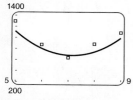

Exercise Set R.6, p. 76

1. Linear   **3.** Quadratic, $a < 0$ **5.** Linear
7. Polynomial, neither linear nor quadratic   **9.** Linear
11. (a) $y = \frac{2}{9}x + 3.6$; **(b)** \$6.3 million; \$8.0 million; **(c)** 2024
13. (a) $y = 0.144x^2 - 4.63x + 60$; **(b)** 188.5 ft; **(c)**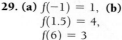
15. Answers will vary.   **17.**   **19.**
21. (a) $y = -0.224x + 6.5414$; **(b)** 2.51% **(c)** The regression
answer seems more plausible; it uses all the data.
(d) $y = -0.009856x^3 + 0.1993x^2 - 1.3563x + 8.103$;
-9.217%. **(e)**

Chapter Review Exercises, p. 85

1. (d)   **2.** (b)   **3.** (f)   **4.** (a)   **5.** (e)   **6.** (c)   **7.** True
8. False   **9.** True   **10.** True   **11.** False   **12.** False
13. True   **14.** True   **15. (a)** About 56 per 1000 women;

(b) 18, 30; **(c)** [15, 45]; this interval covers typical human child-bearing ages.   **16.** \$1340.24   **17.** \$5017.60
18. Not a function. One input, Richard, has three outputs.
19. (a) $f(3) = -6$; **(b)** $f(-5) = -30$; **(c)** $f(a) = -a^2 + a$;
(d) $f(x + h) = -x^2 - 2xh - h^2 + x + h$

20. 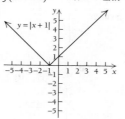   **21.**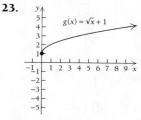

22. 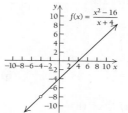   **23.**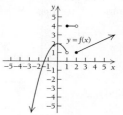

24. Not a function   **25.** Function   **26.** Function
27. Not a function   **28. (a)** $f(2) = 1$;   **(b)** $[-4, 4]$;
(c) $x = -3$; **(d)** $[-1, 3]$
29. (a) $f(-1) = 1$, **(b)**
$f(1.5) = 4$,
$f(6) = 3$

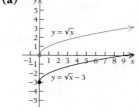

30. (a) $[-2, 5]$; **(b)** $(-1, 3]$; **(c)** $(-\infty, a)$
31. (a) $[-4, 5)$;

(b) $(2, \infty)$;   **32. (a)** $f(-3) = -2$;
(b) $\{-3, -2, -1, 0, 1, 2, 3\}$; **(c)** $-1, 3$; **(d)** $\{-2, 1, 2, 3, 4\}$
33. (a) $(-\infty, 5) \cup (5, \infty)$; **(b)** $[-6, \infty)$ **34.** Slope, -3;
y-intercept, 2   **35.** $y + 5 = \frac{1}{4}(x - 8)$, or $y = \frac{1}{4}x - 7$ **36.** -3

37. $-\$350$ per year   **38.** 75 pages per day   **39.** $A = \dfrac{7}{200}V$
40. (a) $C(x) = 0.50x + 4000$; **(b)** $R(x) = 10x$;
(c) $P(x) = 9.5x - 4000$ **(d)** 422 CDs

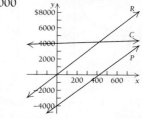

41. (a)   **(b)**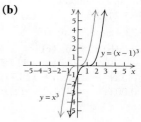

42. (a)

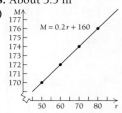

$f(x) = x^2 - 6x + 8$

Vertex at $(3, -1)$

(b)

$g(x) = \sqrt[3]{x} + 2$

(c)

$y = -\dfrac{1}{x}$

(d)

$y = \dfrac{x^2 + x - 6}{x - 2}$

43. (a) $x = 1, x = 3$; **(b)** $x = \dfrac{2 \pm \sqrt{10}}{2}$ **44. (a)** $x^{4/5}$; **(b)** t^4;

(c) $m^{-2/3}$; **(d)** $(x^2 - 9)^{-1/2}$ **45. (a)** $\sqrt[5]{x^2}$ **(b)** $\dfrac{1}{\sqrt[5]{m^3}}$;

(c) $\sqrt{x^2 - 5}$; **(d)** $\sqrt[3]{t}$ **46.** $\left[\dfrac{9}{2}, \infty\right)$ **47.** $(3, 16)$; price $= \$3$,
quantity $= 1600$ units   **48.** About 3.3 hr

49. (a) $y = 0.2x + 160$ **(b)**
(c) 173.4 beats/min

$M = 0.2r + 160$

50. (a)

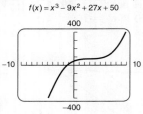

(b) Data fit a quadratic function.
(c) $y = 2.0x^2 - 89.8\overline{4}x + 870$;
(d) About $-\$25.33$; **(e)**

51. (a) 525,375 lb; **(b)** \$4.31/lb

52. $f(x) = x^3 - 9x^2 + 27x + 50$ **53.** $y = \sqrt[3]{|4 - x^2|} + 1$

Zero: $x = -1.25$;
domain: $\mathbb{R}$; range: $\mathbb{R}$

Zero: none; domain: $\mathbb{R}$;
range: $[1, \infty)$

54. $(-1.21, 2.36)$ **55. (a)** $y = 0.2x + 160$;
(b) 173.4 beats/min; **(c)**   **56. (a)** $y = 1.86x^2 - 84.18x +$
943.86; **(b)** \$95.46; **(c)**   **57. (a)** $y = 37.58x + 294.48$;
$y = -0.59x^2 + 74.61x - 117.72$;
$y = 0.02x^3 - 2.60x^2 + 125.71x - 439.65$;
$y = 0.003x^4 - 0.324x^3 + 11.46x^2 - 88.51x + 507.84$

(b) $y_1 = 37.58x + 294.48$
$y_2 = -0.59x^2 + 74.61x - 117.72$
$y_3 = 0.02x^3 - 2.60x^2 + 125.71x - 439.65$
$y_4 = 0.003x^4 - 0.324x^3 + 11.46x^2 - 88.51x + 507.84$

(c)

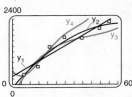

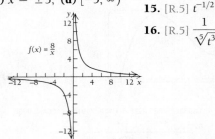

Chapter R Test, p. 88

1. [R.1] \$750   **2.** [R.2] **(a)** $f(-3) = -4$;
(b) $f(a + h) = -a^2 - 2ah - h^2 + 5$ **3.** [R.4] Slope, $\frac{4}{5}$;
y-intercept, $-\frac{2}{3}$ **4.** [R.4] $y - 7 = \frac{1}{4}(x + 3)$, or $y = \frac{1}{4}x + \frac{31}{4}$
5. [R.4] $-\frac{1}{2}$ **6.** [R.4] $-\$700$/yr   **7.** [R.4] $\frac{1}{2}$ lb/bag
8. [R.4] $F = \frac{2}{3}W$ **9.** [R.4] **(a)** $C(x) = 0.08x + 8000$;
(b) $R(x) = 0.50x$; **(c)** $P(x) = 0.42x - 8000$; **(d)** 19,048
cards   **10.** [R.5] $(3, 25)$; $x = \$3$, $q = 25$ thousand units
11. [R.2] Yes   **12.** [R.2] No   **13.** [R.3] **(a)** $f(1) = -4$;
(b) $\mathbb{R}$ **(c)** $x = \pm 3$; **(d)** $[-5, \infty)$
14. [R.5]

$f(x) = \dfrac{8}{x}$

15. [R.5] $t^{-1/2}$

16. [R.5] $\dfrac{1}{\sqrt[5]{t^3}}$

17. [R.5]

$f(x) = \dfrac{x^2 - 1}{x + 1}$

18. [R.5] $(-\infty, -7) \cup (-7, 2) \cup (2, \infty)$ **19.** [R.5] $(-2, \infty)$
20. [R.3] $[c, d)$
21. [R.2]

$f(x) = \begin{cases} x^2 + 2, & \text{for } x \geq 0 \\ x^2 - 2, & \text{for } x < 0 \end{cases}$

22. [R.6] **(a)**

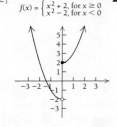

(b) yes; **(c)** $y = -1.94x^2 + 102.74x + 1253.49$; **(d)** 2589.9
calories; **(e)**   **23.** [R.5] $\frac{1}{16}$ **24.** [R.5] Domain: $\left(-\infty, \frac{5}{3}\right]$;
zero: $x = -798\frac{2}{3}$ **25.** [R.5] Answers will vary. One possibility
is $(x + 3)(x - 1)(x - 4) = 0$.   **26.** [R.4] $\frac{51}{7}$
27. [R.5] Zeros: $\pm \sqrt{8} \approx \pm 2.828$,
$\pm \sqrt{10} \approx \pm 3.162$;
domain: $\mathbb{R}$; range: $[-1, \infty)$

$y = \sqrt[3]{|9 - x^2|} - 1$

28. [R.6] **(a)** $y = -1.51x^2 + 79.98x + 1436.93$;
(b) 2480.4 calories; **(c)**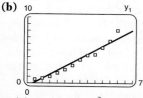

Extended Technology Application, p. 90

1. (a) $y = 0.121x - 0.465x$
(b) 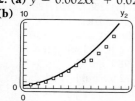 **(c)** $7.04, $8.01. Not reasonable estimates. The data are increasing, but these prices seem lower than the trend of the curve indicates. **(d)** 2119; seems too far in the future.

2. (a) $y = 0.002xx^2 + 0.024x + 0.420$
(b) 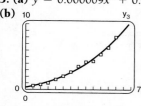 **(c)** $9.60, $11.90. Yes, the curve seems to better follow the trend in the data. **(d)** 2043; seems more reasonable.

3. (a) $y = 0.000009x^3 + 0.00082x^2 + 0.0426x + 0.347$
(b) 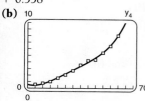 **(c)** $8.29, $10.43. Yes, the curve seems to follow the trend in the data, but the resulting ticket prices are virtually the same as those found with the quadratic function. **(d)** 2066; seems reasonable.

4. (a) $y = 0.0000012x^4 - 0.00023x^3 + 0.0098x^2 - 0.067x + 0.558$
(b) **(c)** $8.81, $13.02. Yes, but the estimates are higher than those found using the quadratic or cubic function. **(d)** 2028; seems too soon.

5. (a) If the scatterplot and the linear function are graphed on the same axes, the predicted prices seem lower than the trend of the curve indicates.
(b) Graphing the scatterplot, the linear function, and the quadratic function, we see that the quadratic function seems to better follow the trend of the data.

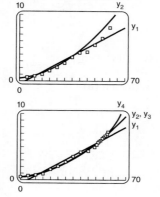

(c) Look at the leading coefficient of the cubic function. Note that it is virtually 0, so the term can be deleted, making a quadratic. Plus, the graphs of the quadratic and cubic functions are virtually identical. Using a higher-order polynomial such as a quartic allows results to show very erratic increases, and since the quadratic does just as good a job as the cubic, the researcher might reject the quartic.

Chapter 1

Technology Connection, p. 96

1. 5   **2.** -4   **3.** $g(x) = 327, 456.95, 475.24, 492.1, 493.81, 494.19, 495.9, 513.24, 573.9, 685.17$ **4.** 494   **5.** -1

Exercise Set 1.1, p. 106

1. 11   **3.** -2   **5.** The limit, as x approaches 4, of $f(x)$
7. The limit, as x approaches 5 from the left, of $F(x)$ **9.** $\lim\limits_{x \to 2^+}$
11. 2   **13.** -3   **15.** Does not exist   **17.** 3   **19.** 4
21. -1   **23.** -1   **25.** 0   **27.** 5   **29.** Does not exist
31. 2   **33.** 2   **35.** 1   **37.** 4   **39.** Does not exist   **41.** 0
43. 0   **45.** 1   **47.** 4   **49.** Does not exist   **51.** 1   **53.** 1
55. Does not exist   **57.** 0   **59.** 3   **61.** 1

63.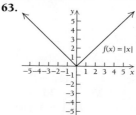
$\lim\limits_{x \to 0} f(x) = 0$;
$\lim\limits_{x \to -2} f(x) = 2$

65.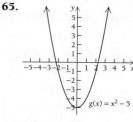
$\lim\limits_{x \to 0} g(x) = -5$;
$\lim\limits_{x \to -1} g(x) = -4$

67.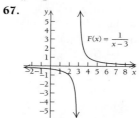
$\lim\limits_{x \to 3} F(x)$ does not exist;
$\lim\limits_{x \to 4} F(x) = 1$

69.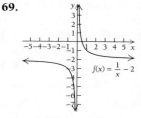
$\lim\limits_{x \to \infty} f(x) = -2$;
$\lim\limits_{x \to 0} f(x)$ does not exist

71.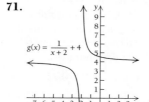
$\lim\limits_{x \to \infty} g(x) = 4$;
$\lim\limits_{x \to -2} g(x)$ does not exist

73.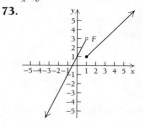
$\lim\limits_{x \to 1^-} F(x) = 3$; $\lim\limits_{x \to 1^+} F(x) = 1$;
$\lim\limits_{x \to 1} F(x)$ does not exist

75.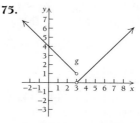
$\lim\limits_{x \to 3^-} g(x) = 1$; $\lim\limits_{x \to 3^+} g(x) = 0$;
$\lim\limits_{x \to 3} g(x)$ does not exist

77.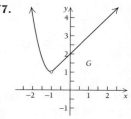
$\lim\limits_{x \to -1} G(x) = 1$

79.

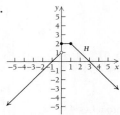

$\lim\limits_{x\to 0} H(x)$ does not exist;
$\lim\limits_{x\to 1} H(x) = 2$

81. \$3.30; \$3.30; \$3.30   **83.** \$3.70; \$4.10; limit does not exist
85. \$1.10; \$1.30; limit does not exist   **87.** Limit does not exist.
89. 1100 deer; 1200 deer; limit does not exist   **91.** ✎
93. 35 bears; 34 bears; limit does not exist   **95.** 3   **97.** −1
99. Limit does not exist; 2   **101.** Limit does not exist; limit
does not exist.

Technology Connection, p. 113

1. 53   **2.** 2.8284   **3.** 0.25   **4.** 0.16667

Exercise Set 1.2, p. 117

1. True   **3.** False   **5.** True   **7.** False   **9.** 5   **11.** −3
13. 4   **15.** 15   **17.** 1   **19.** 6   **21.** $\frac{7}{2}$ **23.** $\frac{13}{4}$ **25.** 3
27. $\frac{1}{10}$ **29.** Limit does not exist.   **31.** $\sqrt{7}$ **33.** Limit does
not exist.   **35.** 0   **37.** Not continuous   **39.** Not continuous
41. Not continuous   **43. (a)** −2, −2, −2; **(b)** −2 **(c)** yes,
$\lim\limits_{x\to 1} g(x) = g(1)$; **(d)** limit does not exist; **(e)** −3
(f) no, the limit does not exist   **45. (a)** 2; **(b)** 2; **(c)** yes,
$\lim\limits_{x\to 1} h(x) = h(1)$; **(d)** 0; **(e)** 0; **(f)** yes, $\lim\limits_{x\to -2} h(x) = h(-2)$
47. (a) 3; **(b)** 1; **(c)** limit does not exist; **(d)** 1; **(e)** no,
$\lim\limits_{x\to 3} G(x)$ does not exist; **(f)** yes, $\lim\limits_{x\to 0} G(x) = G(0)$; **(g)** yes,
$\lim\limits_{x\to 2.9} G(x) = G(2.9)$ **49.** Yes; $\lim\limits_{x\to 5} f(x) = f(5)$ **51.** No;
$\lim\limits_{x\to 0} G(x)$ does not exist and $G(0)$ does not exist.   **53.** Yes;
$\lim\limits_{x\to 3} g(x) = g(3)$ **55.** No; $\lim\limits_{x\to 3} F(x)$ does not exist.   **57.** Yes;
$\lim\limits_{x\to 3} f(x) = f(3)$ **59.** No; $\lim\limits_{x\to 2} G(x)$ does not equal $G(2)$.   **61.** Yes;
$\lim\limits_{x\to 5} f(x) = f(5)$ **63.** No; $g(5)$ does not exist and $\lim\limits_{x\to 5} g(x)$ does
not exist.   **65.** Yes; $\lim\limits_{x\to 4} F(x) = F(4)$ **67.** Yes; $g(x)$ is
continuous at each point on $(-4, 4)$.   **69.** No; $f(x)$ is not
continuous at $x = 0$.   **71.** Yes, since $g(x)$ is not continuous at
each real number   **73. (a)** $k = 5$;   **(b)** so that the Candy
Factory does not lose revenue   **75.** Limit does not exist.
77. 6   **79.** −0.2887, or $-\dfrac{1}{2\sqrt{3}}$ **81.** 0.75   **83.** 0.25

Technology Connection, p. 125

1–2. Left to the student

Exercise Set 1.3, p. 128

1. (a) $8x + 4h$;   **(b)** 48, 44, 40.4, 40.04   **3. (a)** $-8x - 4h$;
(b) −48, −44, −40.4, −40.04   **5. (a)** $2x + h + 1$;   **(b)** 13,
12, 11.1, 11.01   **7. (a)** $\dfrac{-2}{x\cdot(x+h)}$;   **(b)** $-\frac{2}{35}, -\frac{1}{15}, -\frac{4}{51}, -\frac{40}{501}$
9. (a) −2;   **(b)** −2, −2, −2, −2   **11. (a)** $-3x^2 - 3xh - h^2$;
(b) −109, −91, −76.51, 75.1501   **13. (a)** $2x + h - 3$;
(b) 9, 8, 7.1, 7.01   **15. (a)** $2x + h + 4$; **(b)** 16, 15, 14.1, 14.01
17. About 0.3% per yr; about −0.5% per yr; about −0.07% per yr
19. About 0.35% per yr; about −0.56% per yr; about −0.05% per yr
21. About 3.7% per yr; about 2.6% per yr; about 3.25% per yr
23. About 0.97% per yr; about 3% per yr; about 1.9% per yr
25. 1.045 quadrillion BTUs/yr; 0.638 quadrillion BTUs/yr;
−0.089 quadrillion BTUs/yr

27. (a) 70 pleasure units/unit of product, 39 pleasure units/unit
of product, 29 pleasure units/unit of product, 23 pleasure units/unit
of product; **(b)** ✎   **29. (a)** \$11.35; **(b)** \$26.82; **(c)** \$15.47;
(d) \$1.19, the average price of a ticket increases by \$1.19/yr.
31. \$909.72 is the annual increase in the debt from the 2nd to the
3rd year.   **33.** \$19.95 is the cost to produce the 301st unit.
35. (a) 1.0 lb/month; **(b)** 0.54 lb/month; **(c)** 0.77 lb/month;
(d) 0.67 lb/month; **(e)** growth rate is greatest in the first 3 months.
37. (a) Approximately 1.49 hectares/g; **(b)** 1.09 represents the
average growth rate, in hectares/g, of home range with respect to
body weight when the mammal grows from 200 to 300 g
39. (a) 1.25 words/min, 1.25 words/min, 0.625 word/min, 0
words/min, 0 words/min; **(b)** ✎   **41. (a)** 256 ft; **(b)** 128 ft/sec
43. (a) 125 million people/yr for both countries; **(b)** ✎
(c) A: 290 million people/yr, −40 million people/yr, −50
million people/yr, 300 million people/yr, B: 125 million people/
yr in all intervals; **(d)** ✎   **45. (a)** 1985–86; **(b)** 1975–76,
2003–04, and 2004–05; **(c)** about \$2472 for public and about
\$5356 for private   **47.** $2ax + b + ah$
49. $4x^3 + 6x^2h + 4xh^2 + h^3$ **51.** $5ax^4 + 10ax^3h + 10ax^2h^2 +$
$5axh^3 + ah^4 + 4bx^3 + 6bx^2h + 4bxh^2 + bh^3$
53. $\dfrac{1}{(1 - x - h)(1 - x)}$ **55.** $\dfrac{2}{\sqrt{2x + 2h + 1} + \sqrt{2x + 1}}$

Technology Connection, p. 138

1. $f'(x) = -\dfrac{3}{x^2}; f'(-2) = -\frac{3}{4}; f'(-\frac{1}{2}) = -12$

2. $y = -\frac{3}{4}x - 3; y = -12x - 12$ **3.** Left to the student

Technology Connection, p. 140

1–8. Left to the student

Exercise Set 1.4, p. 141

1. (a) and **(b)**

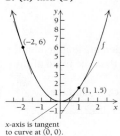

x-axis is tangent
to curve at (0, 0).

3. (a) and **(b)**

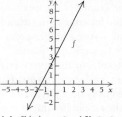

x-axis is tangent
to curve at (0, 0).

(c) $f'(x) = 3x$;
(d) −6, 0, 3

(c) $f'(x) = -4x$;
(d) 8, 0, −4

5. (a) and **(b)**

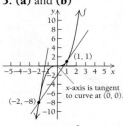

x-axis is tangent
to curve at (0, 0).

(c) $f'(x) = 3x^2$;
(d) 12, 0, 3

7. (a) and **(b)** All tangent
lines are identical to the graph
of the original function.

(c) $f'(x) = 2$; **(d)** 2, 2, 2

9. (a) and **(b)** All tangent lines are identical to the graph of the original function.

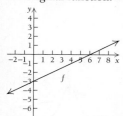

(c) $f'(x) = \frac{1}{2}$; **(d)** $\frac{1}{2}, \frac{1}{2}, \frac{1}{2}$

11. (a) and **(b)**

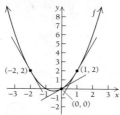

(c) $f'(x) = 2x + 1$;
(d) $-3, 1, 3$

13. (a) and **(b)**

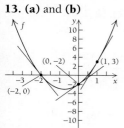

(c) $f'(x) = 4x + 3$;
(d) $-5, 3, 7$

15. (a) and **(b)** There is no tangent line for $x = 0$.

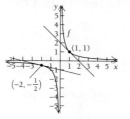

(c) $f'(x) = -\dfrac{1}{x^2}$;

(d) $-\frac{1}{4}$, does not exist, -1

17. (a) $y = 6x - 9$; **(b)** $y = -2x - 1$; **(c)** $y = 20x - 100$
19. (a) $y = -2x + 4$; **(b)** $y = -2x - 4$;
(c) $y = -0.0002x + 0.04$ **21. (a)** $y = 2x + 5$; **(b)** $y = 4$;
(c) $y = -10x + 29$ **23.** $f'(x) = m$ **25.** $x_0, x_3, x_4, x_6, x_{12}$
27. x_1, x_2, x_3, x_4 **29–34.** Answers will vary.
35. $x = 1, 2, 3, 4, 5, 6, 7, 8, 9, 10, 11, 12$ **37.** $x \le 0$ **39.** No
41. **43.** $f'(x) = \dfrac{1}{(1-x)^2}$ **45.** $f'(x) = -\dfrac{2}{x^3}$

47. $f'(x) = \dfrac{1}{\sqrt{2x+1}}$ **49. (a)** f' is not defined at $x = -3$;

(b) 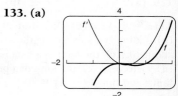 ... **51. (a)** $x = 3$; **(b)** $h'(0) = -1, h'(1) = -1,$
$h'(4) = 1, h'(10) = 1$ **53.** $f'(x)$ is not defined at $x = -1$.
55. (a) $\lim\limits_{x \to 2} F(x) = 5, F(2) = 5$; therefore, $\lim\limits_{x \to 2} F(x) = F(2)$
(b) no, the graph has a corner there **57.** $m = 11, b = -18$
58–63. Left to the student **65.** $f'(x)$ does not exist for $x = 5$.

Technology Connection, p. 147

1. $152, -76, -100, -180$ **2.** $36, 0, 12, 0, 43.47$
3. $-2.31, 3.69, 0.81$

Technology Connection, p. 153

1. Tangent line is horizontal at $\left(2, \frac{8}{3}\right)$.

Exercise Set 1.5, p. 154

1. $7x^6$ **3.** -3 **5.** 0 **7.** $30x^{14}$ **9.** $-6x^{-7}$ **11.** $-8x^{-3}$
13. $3x^2 + 6x$ **15.** $\dfrac{4}{\sqrt{x}}$ **17.** $0.9x^{-0.1}$ **19.** $\frac{2}{5}x^{-1/5}$

21. $-\dfrac{21}{x^4}$ **23.** $\frac{4}{5}$ **25.** $\dfrac{1}{4\sqrt[4]{x^3}} + \dfrac{3}{x^2}$ **27.** $\dfrac{1}{2\sqrt{x}} + \dfrac{1}{x\sqrt{x}}$

29. $-\frac{10}{3}\sqrt[3]{x^2}$ **31.** $10x - 7$ **33.** $0.9x^{0.5}$ **35.** $\frac{2}{3}$ **37.** $-\dfrac{12}{7x^4}$

39. $-\dfrac{5}{x^2} - \dfrac{2}{3}x^{-1/3}$ **41.** 4 **43.** $\frac{1}{3}x^{1/3}$ **45.** $-0.02x - 0.5$

47. $-2x^{-5/3} + \dfrac{3}{4}x^{-1/4} + \dfrac{6}{5}x^{1/5} - \dfrac{24}{x^4}$ **49.** $-\dfrac{2}{x^2} - \dfrac{1}{2}$

51. 24 **53.** 1 **55.** 14 **57.** $\frac{4}{3}$ **59. (a)** $y = 10x - 15$;
(b) $y = x + 3$; **(c)** $y = -2x + 1$ **61. (a)** $y = -2x + 3$;
(b) $y = -\frac{2}{27}x + \frac{1}{3}$; **(c)** $y = \frac{1}{4}x + \frac{3}{4}$ **63.** $(0, -3)$
65. $(0, 1)$ **67.** $\left(\frac{5}{6}, \frac{23}{12}\right)$ **69.** $(-25, 76.25)$ **71.** None
73. The tangent line is horizontal at *all* points of the graph.
75. $(-1, -4), \left(\frac{5}{3}, 5\frac{13}{27}\right)$ **77.** $\left(\sqrt{3}, 2 - 2\sqrt{3}\right)$,
or approximately $(1.73, -1.46)$; $\left(-\sqrt{3}, 2 + 2\sqrt{3}\right)$,
or approximately $(-1.73, 5.46)$ **79.** $(0, -2), \left(-1, -\frac{11}{6}\right)$
81. $(9.5, 99.75)$ **83.** $(60, 150)$ **85.** $\left(-2 + \sqrt{3}, \frac{4}{3} - \sqrt{3}\right)$,
or approximately $(-0.27, -0.40)$; $\left(-2 - \sqrt{3}, \frac{4}{3} + \sqrt{3}\right)$, or
approximately $(-3.73, 3.07)$ **87. (a)** $A'(r) = 6.28r$; **(b)**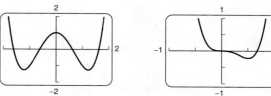
89. (a) $w'(t) = 1.82 - 0.1192t + 0.002274t^2$; **(b)** about 21 lb;
(c) about 0.86 lb/month **91. (a)** $R'(v) = -\dfrac{6000}{v^2}$;

(b) 75 beats/min; **(c)** -0.94 beat/min per mL

93. (a) $\dfrac{dP}{dt} = 4000t$; **(b)** 300,000 people; **(c)** 40,000 people/yr;
(d) 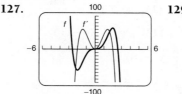 **95. (a)** $V' = \dfrac{0.61}{\sqrt{h}}$; **(b)** 244 mi; **(c)** 0.0031 mi/ft;
(d) 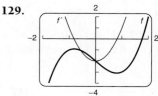 **97.** $(2, \infty)$ **99.** $(-\infty, -1)$ and $(3, \infty)$
101. $(0, -2), \left(\sqrt{\frac{1}{3}}, -\frac{55}{27}\right), \left(-\sqrt{\frac{1}{3}}, -\frac{55}{27}\right)$
103. Always increasing **105.** Always increasing on $(0, \infty)$

107. $2x + 1$ **109.** $3x^2 - 1$ **111.** $3x^2 - \dfrac{1}{x^2}$ **113.** $-192x^2$

115. $\dfrac{2}{3\sqrt[3]{x^2}}$ **117.** $1 - \dfrac{1}{x^2}$ **119.** $3x^2 + 6x + 3$ **121.**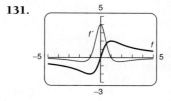

123.

$y = x^4 - 3x^2 + 1$

$(0, 1), (-1.225, -1.25),$
$(1.225, -1.25)$

125.

$y = 10.2x^4 - 6.9x^3$

$(0, 0), (0.507, -0.225)$

127.

$f'(1) = 45$

129.

$f'(1) = 1$

131.

$f'(1) = 0$

133. (a)

(b) $f'(x) = 3x^2 - x; x = 0, x = \frac{2}{3}$;
(c) $0 < x < \frac{2}{3}$;
(d) no, all x such that $f'(x) = 0$ were found in part (b).

Technology Connection, p. 161

1. (c)   **2–5.** Left to the student

Technology Connection, p. 163

1–2. Left to the student

Exercise Set 1.6, p. 163

1. $11x^{10}$ **3.** $12x + 7$ **5.** $20x^4 + 60x^2$ **7.** $\frac{15}{2}x^{3/2} + 4x$

9. $24x^2 + 12x + 11$ **11.** $\frac{9\sqrt{t}}{2} - \frac{1}{2\sqrt{t}} + 2$

13. $4x^3$, for $x \neq 0$ **15.** $8x^3 + 1$, for $x \neq 0$
17. $8x + 2$, for $x \neq \frac{1}{2}$ **19.** 1, for $t \neq -4$

21. $48x^3 + 3x^2 + 22x + 17$ **23.** $\dfrac{-2x(5x^3 - 3x - 15)}{(2x^3 + 3)^2}$

25. $120x^2 + \frac{25}{2}x^{3/2} + \frac{3}{2}x^{-1/2} + 24$ **27.** $\dfrac{3}{(3 - t)^2} + 15t^2$

29. $2x + 6$ **31.** $2x(x^2 - 4)(3x^2 - 4)$

33. $5 - 100x^{-3} + 30x^{-4}$ **35.** $3t^2 - 1 + \dfrac{6}{t^2}$

37. $\dfrac{-x^4 - 3x^2 - 2x}{(x^3 - 1)^2} - 10x$

39. $\dfrac{(x^{1/2} + 3)(\frac{1}{3}x^{-2/3}) - (x^{1/3} - 7)(\frac{1}{2}x^{-1/2})}{(\sqrt{x} + 3)^2}$, or

$\dfrac{6 - \sqrt{x} + 21x^{1/6}}{6x^{2/3}(\sqrt{x} + 3)^2}$ **41.** $\dfrac{2x^{-1} + 1}{(x^{-1} + 1)^2}$, or $\dfrac{x(x + 2)}{(x + 1)^2}$, for $x \neq 0$

43. $\dfrac{-1}{(t - 4)^2}$ **45.** $\dfrac{-2(x^2 - 3x - 1)}{(x^2 + 1)^2}$

47. $\dfrac{(t^2 - 2t + 4)(-2t + 3) - (-t^2 + 3t + 5)(2t - 2)}{(t^2 - 2t + 4)^2}$, or

$\dfrac{-t^2 - 18t + 22}{(t^2 - 2t + 4)^2}$ **49–96.** Left to the student

97. **(a)** $y = 2$; **(b)** $y = \frac{1}{2}x + 2$ **99.** **(a)** $y = x + 5$;
(b) $y = \frac{21}{4}x - \frac{21}{4}$ **101.** $-\$0.006875/\text{jacket}$
103. $-\$0.0053/\text{jacket}$ **105.** About $\$0.0016/\text{jacket}$
107. $\$1.64/\text{vase}$ **109.** **(a)** $P'(t) = 57.6t^{0.6} - 104$;
(b) $\$461.4\,\text{billion/yr}$; **(c)** ✎   **111.** **(a)** $T'(t) = \dfrac{-4(t^2 - 1)}{(t^2 + 1)^2}$

(b) $100.2°\text{F}$; **(c)** $-0.48°\text{F/hr}$ **113.** $30t^2 + 10t - 15$

115. $3x^2\left(\dfrac{x^2 + 1}{x^2 - 1}\right) + \dfrac{-4x}{(x^2 - 1)^2}(x^3 - 8)$, or

$\dfrac{3x^6 - 4x^4 - 3x^2 + 32x}{(x^2 - 1)^2}$ **117.** $\dfrac{-x^6 + 4x^3 - 24x^2}{(x^4 - 3x^3 - 5)^2}$

119. **(a)** $f'(x) = \dfrac{-2x}{(x^2 - 1)^2}$; **(b)** $g'(x) = \dfrac{-2x}{(x^2 - 1)^2}$; **(c)** ✎
121. ✎   **123.** **(a)** Definition of a derivative;
(b) adding and subtracting the same quantity is the same as
adding 0; **(c)** the limit of a sum is the sum of the limits;
(d) factoring; **(e)** the limit of a product is the product of the
limits and $\lim\limits_{h \to 0} f(x + h) = f(x)$; **(f)** definition of a derivative;
(g) using Leibniz notation
125. $\$2.58/\text{jacket}$; $\$0.014/\text{jacket at } x = 184$

127. 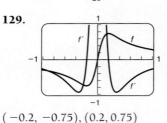   There are no points at
which the tangent line is
horizontal.

129. 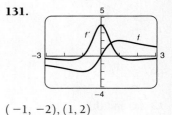   **131.**

$(-0.2, -0.75), (0.2, 0.75)$ $(-1, -2), (1, 2)$

Exercise Set 1.7, p. 173

1. $8x + 4$ **3.** $-55(7 - x)^{54}$ **5.** $\dfrac{4}{\sqrt{1 + 8x}}$

7. $\dfrac{3x}{\sqrt{3x^2 - 4}}$ **9.** $-640x(8x^2 - 6)^{-41}$

11. $4(x - 4)^7(2x + 3)^5(7x - 6)$ **13.** $\dfrac{-6}{(3x + 8)^3}$

15. $\dfrac{4x(5x + 14)}{(7 - 5x)^4}$ **17.** $9x^2(1 + x^3)^2 - 32x^7(2 + x^8)^3$

19. $4x - 400$ **21.** $\dfrac{1}{2\sqrt{x}} + 3(x - 3)^2$

23. $-5(2x - 3)^3(10x - 3)$ **25.** $(3x - 1)^6(2x + 1)^4$

27. $\dfrac{2x(5x - 1)}{\sqrt{4x - 1}}$ **29.** $\dfrac{5x^4 + 6}{3\sqrt[3]{(x^5 + 6x)^2}}$, or $\dfrac{5x^4 + 6}{3x^{2/3}(x^4 + 6)^{2/3}}$

31. $\dfrac{44(3x - 1)^3}{(5x + 2)^5}$ **33.** $\dfrac{-7}{2(x + 3)^{3/2}(4 - x)^{1/2}}$

35. $200(2x^3 - 3x^2 + 4x + 1)^{99}(3x^2 - 3x + 2)$

37. $\dfrac{68(5x - 1)^3}{(2x + 3)^5}$ **39.** $\dfrac{-1}{(x - 1)^{3/2}(x + 1)^{1/2}}$

41. $\dfrac{-(2x + 3)^3(6x + 61)}{(3x - 2)^6}$ **43.** $\dfrac{(3x - 4)^{1/4}(138x - 19)}{(2x + 1)^{1/3}}$

45. $\dfrac{1}{2\sqrt{u}}, 2x, \dfrac{x}{\sqrt{x^2 - 1}}$ **47.** $50u^{49}, 12x^2 - 4x$,

$50(4x^3 - 2x^2)^{49}(12x^2 - 4x)$ **49.** $2u + 1, 3x^2 - 2$,
$(2x^3 - 4x + 1)(3x^2 - 2)$ **51.** $3x^2(10x^3 + 13)$

53. $\dfrac{2(2x - 1)}{3(2x^2 - 2x + 5)^{2/3}}$ **55.** $\dfrac{3(-6t - 11)}{(5 + 3t)^2(6 + 3t)^2}$, or

$\dfrac{-6t - 11}{3(t + 2)^2(3t + 5)^2}$ **57.** $y = \dfrac{5}{4}x + \dfrac{3}{4}$ **59.** $y = 4x - 3$

61. **(a)** $\dfrac{2x - 3x^2}{(1 + x)^6}$; **(b)** $\dfrac{2x - 3x^2}{(1 + x)^6}$; **(c)** They are the same.

63. $f(x) = x^5, g(x) = 3x^2 - 7$ **65.** $f(x) = \dfrac{x + 1}{x - 1}, g(x) = x^3$

67. -216 **69.** $4(13)^{-2/3}$, or about 0.72
71. $f'(x) = 6[2x^3 + (4x - 5)^2]^5[6x^2 + 8(4x - 5)]$

73. $f'(x) = \dfrac{1}{2\sqrt{x^2 + \sqrt{1 - 3x}}}\left(2x - \dfrac{3}{2\sqrt{1 - 3x}}\right)$

75. $\$1,000,000/\text{item}$ **77.** $P'(x) = \dfrac{500(2x - 0.1)}{\sqrt{x^2 - 0.1x}} - \dfrac{4000x}{3(x^2 + 2)^{2/}}$

79. (a) $0.84x^3 - 17.76x^2 + 101.06x - 18.92$;
(b) 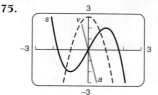 **(c)** \$336 billion/yr   **81. (a)** $dA/di = 3000(1 + i)^2$;
(b) 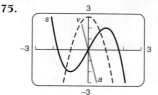   **83. (a)** $D(t) = \dfrac{80,000}{1.6t + 9}$; **(b)** -4.482 units/day

85. (a) $D(c) = 4.25c + 106.25, c(w) = \dfrac{95w}{43.2} \approx 2.199w$
(b) 4.25 mg/unit of creatine clearance; **(c)** 2.199 units of creatine clearance/kg; **(d)** 9.35 mg/kg; **(e)**

87. $1 + \dfrac{1}{2\sqrt{x}} + \dfrac{1}{2\sqrt{x + \sqrt{x}}} \cdot \left(1 + \dfrac{1}{2\sqrt{x}}\right)$ **89.** $\frac{1}{27}x^{-26/27}$

91. $\dfrac{6x^7 + 32x^5 + 5x^4}{(x^3 + 6x + 1)^{2/3}}$ **93.** $\dfrac{3x^2(x - 2)}{2(x - 1)^{5/2}}$

95. $\dfrac{1}{(1 - x)\sqrt{1 - x^2}}$ **97.** $\dfrac{3(x^2 - x - 1)^2(x^2 + 4x - 1)}{(x^2 + 1)^4}$

99. $\dfrac{6\sqrt{t} + 1}{4\sqrt{t}\sqrt{3t + \sqrt{t}}}$ **101.**

103.
$(-2.14476, -7.728),$
$(2.14476, 7.728)$

105. $\dfrac{4 - 2x^2}{\sqrt{4 - x^2}}$

107. $\dfrac{5(\sqrt{2x - 1} + x^3)^4(3x^2\sqrt{2x - 1} + 1)}{\sqrt{2x - 1}}$,
or $5(\sqrt{2x - 1} + x^3)^4[(2x - 1)^{-1/2} + 3x^2]$

Exercise Set 1.8, p. 182

1. $20x^3$ **3.** $24x^2$ **5.** 8   **7.** 0   **9.** $\dfrac{6}{x^4}$ **11.** $\dfrac{-1}{4x^{3/2}}$

13. $12x^2 + \dfrac{6}{x^3}$ **15.** $\dfrac{-4}{25x^{9/5}}$ **17.** $\dfrac{48}{x^5}$

19. $14(x^2 + 3x)^5(13x^2 + 39x + 27)$
21. $10(2x^2 - 3x + 1)^8(152x^2 - 228x + 85)$
23. $\dfrac{3(x^2 + 2)}{4(x^2 + 1)^{5/4}}$ **25.** $\dfrac{-2}{9x^{4/3}}$ **27.** $\dfrac{45x^4 - 54x^2 - 3}{16(x^3 - x)^{5/4}}$
29. $\frac{5}{8}x^{-3/4} - \frac{1}{4}x^{-3/2}$ **31.** $24x^{-5} + 6x^{-4}$ **33.** $24x - 2$
35. $\dfrac{44}{(2x - 3)^3}$ **37.** 24   **39.** $720x$ **41.** $120x^{-6} + \dfrac{15}{16}x^{-7/2}$
43. 0   **45. (a)** $v(t) = 3t^2 + 1$; **(b)** $a(t) = 6t$;
(c) $v(4) = 49$ ft/sec, $a(4) = 24$ ft/sec^2 **47. (a)** $v(t) = 3$;
(b) $a(t) = 0$; **(c)** $v(2) = 3$ mi/hr, $a(2) = 0$ mi/hr^2; **(d)**
49. (a) 144 ft; **(b)** 96 ft/sec; **(c)** 32 ft/sec^2
51. $v(2) = 19.62$ m/sec, $a(2) = 9.81$ m/sec^2 **53. (a)** The velocity at $t = 20$ sec is greater, since the slope of a tangent line is greater there. **(b)** The acceleration is positive, since the velocity (slope of a tangent line) is increasing over time.
55. (a) \$146,000/month, \$84,000/month, $-\$4000$/month;
(b) $-\$68,000$/month^2, $-\$56,000$/month^2, $-\$32,000$/month^2;
(c)   **57. (a)** 11.34, 1.98, 0.665; **(b)** $-0.789, -0.0577,$
-0.0112; **(c)**   **59.** $\dfrac{6}{(1 - x)^4}$ **61.** $\dfrac{-15}{(2x - 1)^{7/2}}$ $(72x + 11)$
63. $\dfrac{3x^{1/2} - 1}{2x^{3/2}(x^{1/2} - 1)^3}$ **65.** $k(k - 1)(k - 2)(k - 3)(k - 4)x^{k-5}$

67. $f'(x) = \dfrac{3}{(x + 2)^2}, f''(x) = \dfrac{-6}{(x + 2)^3}, f'''(x) = \dfrac{18}{(x + 2)^4}$,
$f^4(x) = \dfrac{-72}{(x + 2)^5}$ **69.** 2.29 sec   **71. (a)** 3.24 m;
(b) 3.24 m/sec; **(c)** 1.62 m/sec^2; **(d)** It is the gravitational constant for the moon.   **73.** 42.33 ft/sec
75.

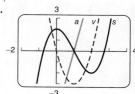

77.

$v(t)$ switches at $t = 0$. $v(t)$ switches at $t = 1$.

Chapter Review Exercises, p. 190

1. False   **2.** False   **3.** True   **4.** False   **5.** True
6. True   **7.** False   **8.** True   **9.** (e)   **10.** (c)
11. (a)   **12.** (f)   **13.** (b)   **14.** (d)
15. (a)

$x \to -7^-$	$f(x)$
-8	-11
-7.5	-10.5
-7.1	-10.1
-7.01	-10.01
-7.001	-10.001
-7.0001	-10.0001

$x \to -7^+$	$f(x)$
-6	-9
-6.5	-9.5
-6.9	-9.9
-6.99	-9.99
-6.999	-9.999
-6.9999	-9.9999

(b) $\lim\limits_{x \to -7^-} f(x) = -10$; $\lim\limits_{x \to -7^+} f(x) = -10$; $\lim\limits_{x \to -7} f(x) = -10$
16.

17. $\lim\limits_{x \to -7} \dfrac{x^2 + 4x - 21}{x + 7} = \lim\limits_{x \to -7} \dfrac{(x + 7)(x - 3)}{x + 7} =$
$\lim\limits_{x \to -7} (x - 3) = -10$ **18.** -4 **19.** 10   **20.** -12
21. 3   **22.** Not continuous, since $\lim\limits_{x \to -2} g(x)$ does not exist
23. Continuous   **24.** -4 **25.** -4
26. Continuous, since $\lim\limits_{x \to 1} g(x) = g(1)$ **27.** Does not exist
28. -2 **29.** Not continuous, since $\lim\limits_{x \to -2} g(x)$ does not exist
30. 2   **31.** -3 **32.** $4x + 2h$ **33.** $y = x - 1$ **34.** $(4, 5)$
35. $(5, -108)$ **36.** $45x^4$ **37.** $\dfrac{8}{3}x^{-2/3}$ **38.** $\dfrac{24}{x^9}$ **39.** $6x^{-3/5}$
40. $0.7x^6 - 12x^3 - 3x^2$ **41.** $\frac{5}{2}x^5 + 32x^3 - 2$ **42.** $2x, x \neq 0$
43. $\dfrac{-x^2 + 16x + 8}{(8 - x)^2}$ **44.** $2(5 - x)(2x - 1)^4(-7x + 26)$
45. $35x^4(x^5 - 3)^6$ **46.** $\dfrac{x(11x + 4)}{(4x + 2)^{1/4}}$ **47.** $-48x^{-5}$
48. $3x^5 - 60x + 26$ **49. (a)** $v(t) = 1 + 4t^3$; **(b)** $a(t) = 12t^2$;
(c) $v(2) = 33$ ft/sec, $a(2) = 48$ ft/sec^2

50. (a) $A_C(x) = 5x^{-1/2} + 100x^{-1}, A_R(x) = 40,$
$A_P(x) = 40 - 5x^{-1/2} - 100x^{-1};$ **(b)** average cost is dropping at
approximately $1.33 per item.   **51. (a)** $P'(t) = 100t;$
(b) 30,000; **(c)** 2000/yr   **52.** $(f \circ g)(x) = 4x^2 - 4x + 6;$
$(g \circ f)(x) = -2x^2 - 9$ **53.** $\dfrac{-9x^4 - 4x^3 + 9x + 2}{2\sqrt{1+3x}(1+x^3)^2}$

54. -0.25 **55.** $\frac{1}{6}$ **56.**

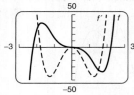

$(-1.7137, 37.445), (0, 0), (1.7137, -37.445)$

Chapter 1 Test, p. 192

1. [1.1] **(a)**

$x \to 6^-$	$f(x)$
5	11
5.7	11.7
5.9	11.9
5.99	11.99
5.999	11.999
5.9999	11.9999

$x \to 6^+$	$f(x)$
7	13
6.5	12.5
6.1	12.1
6.01	12.01
6.001	12.001
6.0001	12.0001

(b) $\lim\limits_{x \to 6^-} f(x) = 12; \lim\limits_{x \to 6^+} f(x) = 12; \lim\limits_{x \to 6} f(x) = 12$
2. [1.1]

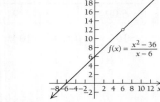

3. [1.2] $\lim\limits_{x \to 6} \dfrac{x^2 - 36}{x - 6} = \lim\limits_{x \to 6} \dfrac{(x+6)(x-6)}{x-6} = \lim\limits_{x \to 6} (x+6) = 12$
4. [1.1] Does not exist   **5.** [1.1] 0   **6.** [1.1] Does not exist
7. [1.1] 2   **8.** [1.1] 4   **9.** [1.1] 1   **10.** [1.1] 1
11. [1.1] 1   **12.** [1.2] Continuous   **13.** [1.2] Not continuous,
since $\lim\limits_{x \to 3} f(x)$ does not exist   **14.** [1.1, 1.2] Does not exist
15. [1.1, 1.2] 1   **16.** [1.1, 1.2] No   **17.** [1.1, 1.2] 3
18. [1.1, 1.2] 3   **19.** [1.1, 1.2] Yes   **20.** [1.1, 1.2] 6
21. [1.1, 1.2] $\frac{1}{8}$ **22.** [1.1, 1.2] Does not exist, since

$\lim\limits_{x \to 0^-} \dfrac{1}{x} \neq \lim\limits_{x \to 0^+} \dfrac{1}{x}$ **23.** [1.3] $4x + 3 + 2h$

24. [1.4] $y = \frac{3}{4}x + 2$

25. [1.5] $(0, 0), (2, -4)$ **26.** [1.5] $23x^{22}$

27. [1.5] $\frac{4}{3}x^{-2/3} + \frac{5}{2}x^{-1/2}$ **28.** [1.5] $\dfrac{10}{x^2}$ **29.** [1.5] $\frac{5}{4}x^{1/4}$

30. [1.5] $-1.0x + 0.61$ **31.** [1.5] $x^2 - 2x + 2$

32. [1.6] $\dfrac{-6(x-2)}{x^4}$ **33.** [1.6] $\dfrac{5}{(5-x)^2}$

34. [1.7] $(x+3)^3(7-x)^4(-9x + 13)$

35. [1.7] $-5(x^5 - 4x^3 + x)^{-6}(5x^4 - 12x^2 + 1)$

36. [1.6, 1.7] $\dfrac{2x^2 + 5}{\sqrt{x^2 + 5}}$ **37.** [1.8] $24x$

38. [1.6] **(a)** $A_R = 50, A_C = x^{-1/3} + 750x^{-1},$
$A_P = 50 - x^{-1/3} - 750x^{-1};$ **(b)** average cost is dropping at
approximately $11.74 per item.   **39.** [1.5] **(a)** $M'(t) =$
$-0.003t^2 + 0.2t;$ **(b)** 9; **(c)** 1.7 words/min
40. [1.7] $(f \circ g)(x) = 4x^6 - 2x^3; (g \circ f)(x) = 2(x^2 - x)^3$
41. [1.6, 1.7] $\dfrac{-1 - 9x}{2(1-3x)^{2/3}(1+3x)^{5/6}}$ **42.** [1.2] 27
43. [1.5] 50

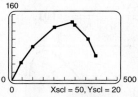

(1.0836, 25.1029) and
(2.9503, 8.6247)
44. [1.5] 0.5

Extended Technology Application, p. 195

1.

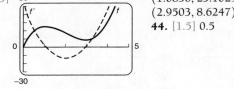

2. (a) $y = -0.0000045x^3 + 0.000204x^2 + 0.7806x + 4.6048;$
(b)
(c) acceptable fit;
(d) about 441 ft;

(e) $dy/dx = -0.0000135x^2 + 0.000408x + 0.7806;$
(f) approximately $(256, 142);$ at about 256 ft from home plate,
the ball reached its maximum height of approximately 142 ft.
3. (a) $y = -0.0000000024x^4 - 0.0000026x^3 - 0.00026x^2$
$+ 0.8150x + 4.3026;$
(b)
(c) acceptable fit;
(d) about 440 ft;

(e) $dy/dx = -0.0000000096x^3 - 0.0000078x^2 -$
$0.00053x + 0.815;$ **(f)** approximately $(257, 142);$ at about
257 ft from home plate, the ball reached its maximum height of
approximately 142 ft.
4. (a)

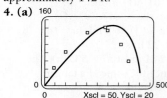

(b) 450 ft;
(c) $\dfrac{dy}{dx} = \dfrac{303.75 - 0.003x^2}{\sqrt{202,500 - x^2}};$
(d) approximately $(318, 152);$
at about 318 ft from home
plate, the ball reached its
maximum height of
approximately 152 ft.
5. The two models are very similar. The main difference seems to
be that the maximum height is reached further from home plate
with the model in Exercise 4.   **6.** 466 ft, 442 ft, 430 ft
7. The estimate of the reporters is way off. Even with a low
trajectory, the ball would at best have traveled a horizontal
distance of about 526 ft.   **8. (a)** 523 ft; **(b)** 430 ft;
(c) 464 ft (rounded)

Chapter 2

Technology Connection, p. 208

1. $f(x) = 2 - (x-1)^{2/3}$

2. $f'(x) = -\frac{2}{3}(x-1)^{-1/3}$

The derivative is not defined at $(1, 2)$.

Technology Connection, p. 210

1. Relative maximum at $(-1, 19)$; relative minimum at $(2, -8)$

Technology Connection, p. 212

1–8. Left to the student

Exercise Set 2.1, p. 212

1. Relative minimum at $(-2, 1)$

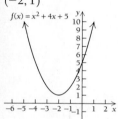

3. Relative maximum at $\left(-\frac{1}{2}, \frac{21}{4}\right)$

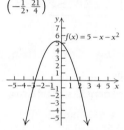

5. Relative minimum at $(-1, -2)$

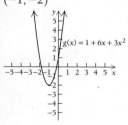

7. Relative minimum at $(1, 1)$; relative maximum at $\left(-\frac{1}{3}, \frac{59}{27}\right)$

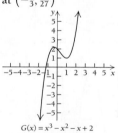

9. Relative minimum at $(1, 4)$; relative maximum at $(-1, 8)$

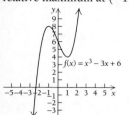

11. Relative minimum at $(0, 0)$; relative maximum at $(-1, 1)$

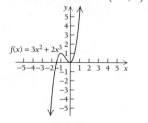

13. No relative extrema exist.

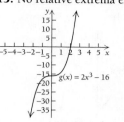

15. Relative minimum at $(4, -22)$; relative maximum at $(0, 10)$

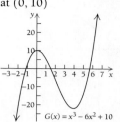

17. Relative maximum at $\left(\frac{3}{4}, \frac{27}{256}\right)$

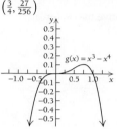

19. No relative extrema exist.

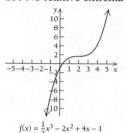

21. Relative minima at $\left(-\sqrt{5}, -32\right)$ and $\left(\sqrt{5}, -32\right)$; relative maximum at $(0, 18)$

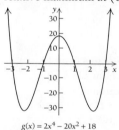

23. No relative extrema exist.

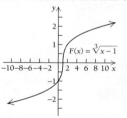

25. Relative maximum at $(0, 1)$

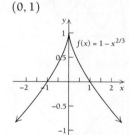

27. Relative minimum at $(0, -8)$

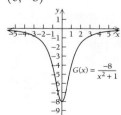

29. Relative minimum at $(-1, -2)$; relative maximum at $(1, 2)$

31. No relative extrema exist.

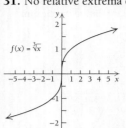

$f(x) = \sqrt[3]{x}$

33. Relative minimum at $(-1, 2)$

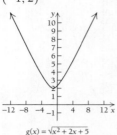

$g(x) = \sqrt{x^2 + 2x + 5}$

35–83. Left to the student.
87. Relative minimum at $(1.94, 15{,}882)$; relative maximum at $(7.05, 17{,}773)$

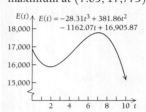

$E(t) = -28.31t^3 + 381.86t^2 - 1162.07t + 16{,}905.87$

85. ✎

89. Relative maximum at $(6, 102.2)$

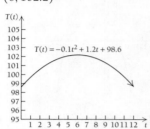

$T(t) = -0.1t^2 + 1.2t + 98.6$

91. Increasing on $(-1, \infty)$, decreasing on $(-\infty, -1)$; relative minimum at $x = -1$.
93. Increasing on $(-\infty, 1)$, decreasing on $(1, \infty)$; relative maximum at $x = 1$.
95. Increasing on $(-4, 2)$, decreasing on $(-\infty, -4)$ and $(2, \infty)$; relative minimum at $x = -4$, relative maximum at $x = 2$.
97.

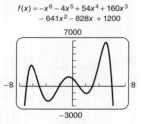

$f(x) = -x^6 - 4x^5 + 54x^4 + 160x^3 - 641x^2 - 828x + 1200$

Relative minima at $(-3.683, -2288.03)$ and $(2.116, -1083.08)$; relative maxima at $(-6.262, 3213.8)$ and $(-0.559, 1440.06)$ and $(5.054, 6674.12)$

99.

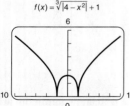

$f(x) = \sqrt[3]{|4 - x^2|} + 1$

Relative minima at $(-2, 1)$ and $(2, 1)$; relative maximum at $(0, 2.587)$

101.

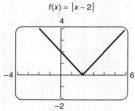

$f(x) = |x - 2|$

Relative minimum at $(2, 0)$; increasing on $(2, \infty)$; decreasing on $(-\infty, 2)$; f' does not exist at $x = 2$

103.

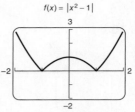

$f(x) = |x^2 - 1|$

Relative maximum at $(0, 1)$; relative minima at $(-1, 0)$ and $(1, 0)$; increasing on $(-1, 0)$ and $(1, \infty)$; decreasing on $(-\infty, -1)$ and $(0, 1)$; f' does not exist at $x = -1$ and $x = 1$

105.

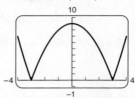

$f(x) = |9 - x^2|$

Relative maximum at $(0, 9)$; relative minima at $(-3, 0)$ and $(3, 0)$; increasing on $(-3, 0)$ and $(3, \infty)$; decreasing on $(-\infty, -3)$ and $(0, 3)$; f' does not exist at $x = -3$ and $x = 3$

107.

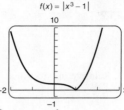

$f(x) = |x^3 - 1|$

Relative minimum at $(1, 0)$; increasing on $(1, \infty)$; decreasing on $(-\infty, 1)$; f' does not exist at $x = 1$

109. ✎ **111.** ✎

Technology Connection, p. 224

Left to the student

Technology Connection, p. 226

1. Relative minimum at $(1, -1)$; inflection points at $(0, 0)$, $(0.553, -0.512)$, $(1.447, -0.512)$, and $(2, 0)$

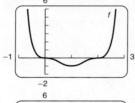
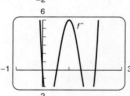

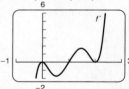

Technology Connection, p. 230

1–8. Left to the student

Technology Connection, p. 231

1. Critical values: $-1, 0,$ and 1

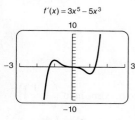

$f'(x) = 3x^5 - 5x^3$

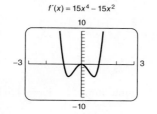

$f'(x) = 15x^4 - 15x^2$

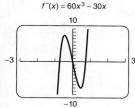

$f''(x) = 60x^3 - 30x$

2. Inflection points at $-0.707, 0,$ and 0.707

Exercise Set 2.2, p. 231

1. Relative maximum is $f(0) = 5$.

3. Relative minimum is $f\left(\frac{1}{2}\right) = -\frac{1}{4}$.

5. Relative maximum is $f\left(\frac{4}{5}\right) = -\frac{19}{5}$.

7. Relative minimum is $f\left(\frac{1}{2}\right) = -1$; relative maximum is $f\left(-\frac{1}{2}\right) = 3$.

9.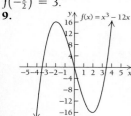

$f(x) = x^3 - 12x$

Relative minimum at $(2, -16)$, relative maximum at $(-2, 16)$; inflection point at $(0, 0)$; increasing on $(-\infty, -2)$ and $(2, \infty)$, decreasing on $(-2, 2)$, concave down on $(-\infty, 0)$, concave up on $(0, \infty)$

11.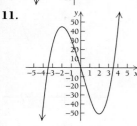

$f(x) = 3x^3 - 36x - 3$

Relative minimum at $(2, -51)$, relative maximum at $(-2, 45)$; inflection point at $(0, -3)$; increasing on $(-\infty, -2)$ and $(2, \infty)$, decreasing on $(-2, 2)$; concave down on $(-\infty, 0)$, concave up on $(0, \infty)$

13.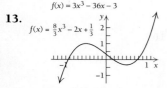

$f(x) = \frac{8}{3}x^3 - 2x + \frac{1}{3}$

Relative minimum at $\left(\frac{1}{2}, -\frac{1}{3}\right)$, relative maximum at $\left(-\frac{1}{2}, 1\right)$; inflection point at $\left(0, \frac{1}{3}\right)$; increasing on $\left(-\infty, -\frac{1}{2}\right)$ and $\left(\frac{1}{2}, \infty\right)$, decreasing on $\left(-\frac{1}{2}, \frac{1}{2}\right)$; concave down on $(-\infty, 0)$, concave up on $(0, \infty)$

15.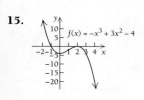

$f(x) = -x^3 + 3x^2 - 4$

Relative minimum at $(0, -4)$, relative maximum at $(2, 0)$; inflection point at $(1, -2)$; increasing on $(0, 2)$, decreasing on $(-\infty, 0)$ and $(2, \infty)$; concave up on $(-\infty, 1)$, concave down on $(1, \infty)$

17.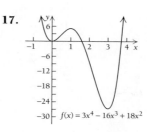

$f(x) = 3x^4 - 16x^3 + 18x^2$

Relative minima at $(0, 0)$ and $(3, -27)$, relative maximum at $(1, 5)$; inflection points at $(0.451, 2.321)$ and $(2.215, -13.358)$; increasing on $(0, 1)$ and $(3, \infty)$, decreasing on $(-\infty, 0)$ and $(1, 3)$; concave up on $(-\infty, 0.451)$ and $(2.215, \infty)$, concave down on $(0.451, 2.215)$

19. $f(x) = x^4 - 6x^2$

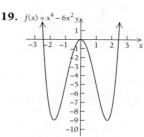

Relative minima at $\left(-\sqrt{3}, -9\right)$ and $\left(\sqrt{3}, -9\right)$, relative maximum at $(0, 0)$; inflection points at $(-1, -5)$ and $(1, -5)$; increasing on $\left(-\sqrt{3}, 0\right)$ and $\left(\sqrt{3}, \infty\right)$, decreasing on $\left(-\infty, -\sqrt{3}\right)$ and $\left(0, \sqrt{3}\right)$; concave up on $(-\infty, -1)$ and $(1, \infty)$, concave down on $(-1, 1)$

21. $f(x) = x^3 - 2x^2 - 4x + 3$

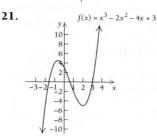

Relative minimum at $(2, -5)$, relative maximum at $\left(-\frac{2}{3}, \frac{121}{27}\right)$; inflection point at $\left(\frac{2}{3}, -\frac{7}{27}\right)$; increasing on $\left(-\infty, -\frac{2}{3}\right)$ and $(2, \infty)$, decreasing on $\left(-\frac{2}{3}, 2\right)$; concave down on $\left(-\infty, \frac{2}{3}\right)$, concave up on $\left(\frac{2}{3}, \infty\right)$

23.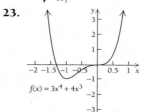

$f(x) = 3x^4 + 4x^3$

Relative minimum at $(-1, -1)$; inflection points at $\left(-\frac{2}{3}, -\frac{16}{27}\right)$ and $(0, 0)$; increasing on $(-1, \infty)$, decreasing on $(-\infty, -1)$; concave up on $\left(-\infty, -\frac{2}{3}\right)$ and $(0, \infty)$, concave down on $\left(-\frac{2}{3}, 0\right)$

25.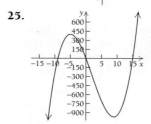

$f(x) = x^3 - 6x^2 - 135x$

Relative minimum at $(9, -972)$, relative maximum at $(-5, 400)$; inflection point at $(2, -286)$; increasing on $(-\infty, -5)$ and $(9, \infty)$, decreasing on $(-5, 9)$; concave down on $(-\infty, 2)$, concave up on $(2, \infty)$

27.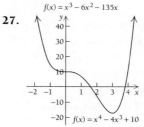

$f(x) = x^4 - 4x^3 + 10$

Relative minimum at $(3, -17)$; inflection points at $(0, 10)$ and $(2, -6)$; increasing on $(3, \infty)$, decreasing on $(-\infty, 3)$; concave down on $(0, 2)$, concave up on $(-\infty, 0)$ and $(2, \infty)$

29.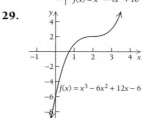

$f(x) = x^3 - 6x^2 + 12x - 6$

No relative extrema; inflection point at $(2, 2)$; increasing on $(-\infty, \infty)$; concave down on $(-\infty, 2)$, concave up on $(2, \infty)$

31.

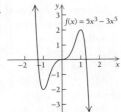

Relative minimum at $(-1, -2)$, relative maximum at $(1, 2)$; inflection points at $(-0.707, -1.237)$, $(0, 0)$, and $(0.707, 1.237)$; increasing on $(-1, 1)$, decreasing on $(-\infty, -1)$ and $(1, \infty)$; concave down on $(-0.707, 0)$ and $(0.707, \infty)$, concave up on $(-\infty, -0.707)$ and $(0, 0.707)$

33.

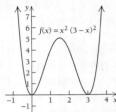

Relative minima at $(0, 0)$ and $(3, 0)$, relative maximum at $\left(\frac{3}{2}, \frac{81}{16}\right)$; inflection points at $(0.634, 2.25)$ and $(2.366, 2.25)$; increasing on $\left(0, \frac{3}{2}\right)$ and $(3, \infty)$, decreasing on $(-\infty, 0)$ and $\left(\frac{3}{2}, 3\right)$; concave down on $(0.634, 2.366)$, concave up on $(-\infty, 0.634)$ and $(2.366, \infty)$

35.

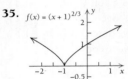

Relative minimum at $(-1, 0)$; no inflection points; increasing on $(-1, \infty)$, decreasing on $(-\infty, -1)$; concave down on $(-\infty, -1)$ and $(-1, \infty)$

37.

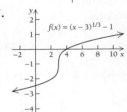

No relative extrema; inflection point at $(3, -1)$; increasing on $(-\infty, \infty)$; concave up on $(-\infty, 3)$, concave down on $(3, \infty)$

39.

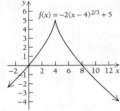

Relative maximum at $(4, 5)$; no inflection points; increasing on $(-\infty, 4)$, decreasing on $(4, \infty)$; concave up on $(-\infty, 4)$ and $(4, \infty)$

41.

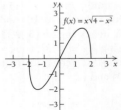

Relative minimum at $\left(-\sqrt{2}, -2\right)$, relative maximum at $\left(\sqrt{2}, 2\right)$; inflection point at $(0, 0)$; increasing on $\left(-\sqrt{2}, \sqrt{2}\right)$, decreasing on $\left(-2, -\sqrt{2}\right)$ and $\left(\sqrt{2}, 2\right)$; concave up on $(-2, 0)$, concave down on $(0, 2)$

43.

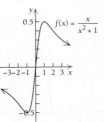

Relative minimum at $\left(-1, -\frac{1}{2}\right)$, relative maximum at $\left(1, \frac{1}{2}\right)$; inflection points at $\left(-\sqrt{3}, -\frac{\sqrt{3}}{4}\right)$ and $(0, 0)$ and $\left(\sqrt{3}, \frac{\sqrt{3}}{4}\right)$; increasing on $(-1, 1)$, decreasing on $(-\infty, -1)$ and $(1, \infty)$; concave up on $\left(-\sqrt{3}, 0\right)$ and $\left(\sqrt{3}, \infty\right)$; concave down on $\left(-\infty, -\sqrt{3}\right)$ and $\left(0, \sqrt{3}\right)$

45.

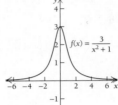

Relative maximum at $(0, 3)$; inflection points at $\left(-\sqrt{\frac{1}{3}}, \frac{9}{4}\right)$ and $\left(\sqrt{\frac{1}{3}}, \frac{9}{4}\right)$; increasing on $(-\infty, 0)$, decreasing on $(0, \infty)$; concave up on $\left(-\infty, -\sqrt{\frac{1}{3}}\right)$ and $\left(\sqrt{\frac{1}{3}}, \infty\right)$, concave down on $\left(-\sqrt{\frac{1}{3}}, \sqrt{\frac{1}{3}}\right)$

47–101. Left to the student

103.

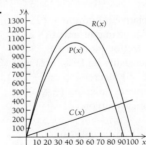

105.

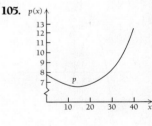

107. Radius $= \frac{40}{3}$, or $13\frac{1}{3}$ mm   **109.** ✏   **111.** ✏

113. Left to the student   **115.** False   **117.** True

119. True   **121.** True

123.

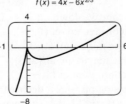

Relative maximum at $(0, 0)$; relative minimum at $(1, -2)$

125.

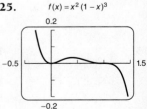

Relative minimum at $(0, 0)$; relative maximum at $(0.4, 0.035)$

127.

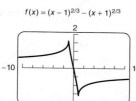

$f(x) = (x - 1)^{2/3} - (x + 1)^{2/3}$

Relative maximum at $(-1, 1.587)$; relative minimum at $(1, -1.587)$

Technology Connection, p. 237

1. Vertical asymptotes: $x = -7$ and $x = 4$ **2.** Vertical asymptotes: $x = 0, x = 3$, and $x = -2$

Technology Connection, p. 238

1. and 2. Left to the student   **3.** 2   **4.** Left to the student

Technology Connection, p. 239 (top)

1. and 2. Left to the student

Technology Connection, p. 239 (bottom)

1. Horizontal asymptote: $y = 0$ **2.** Horizontal asymptote: $y = 3$
3. Horizontal asymptote: $y = 0$ **4.** Horizontal asymptote: $y = \frac{1}{2}$

Technology Connection, p. 240 (top)

1. $y = 3x - 1$ **2.** $y = 5x$

Technology Connection, p. 240 (bottom)

1. x-intercepts: $(0, 0)$, $(3, 0)$, and $(-5, 0)$; y-intercept: $(0, 0)$
2. x-intercepts: $(0, 0)$, $(1, 0)$, and $(-3, 0)$; y-intercept: $(0, 0)$

Exercise Set 2.3, p. 247

1. $x = 5$ **3.** $x = -3$ and $x = 3$ **5.** $x = 0, x = 2$, and $x = 4$
7. $x = -1$ **9.** No vertical asymptotes   **11.** $y = \frac{3}{4}$ **13.** $y = 0$
15. $y = 5$ **17.** No horizontal asymptotes   **19.** $y = 0$
21. $y = \frac{1}{2}$

23.

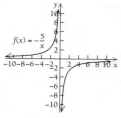

Increasing on $(-\infty, 0)$ and $(0, \infty)$
No relative extrema
Asymptotes: $x = 0$ and $y = 0$
Concave up on $(-\infty, 0)$;
concave down on $(0, \infty)$ No intercepts

25.

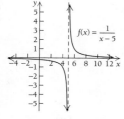

Decreasing on $(-\infty, 5)$ and $(5, \infty)$
No relative extrema
Asymptotes: $x = 5$ and $y = 0$
Concave down on $(-\infty, 5)$;
concave up on $(5, \infty)$
y-intercept: $\left(0, -\frac{1}{5}\right)$

27.

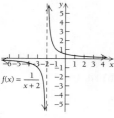

Decreasing on $(-\infty, -2)$ and $(-2, \infty)$
No relative extrema
Asymptotes: $x = -2$ and $y = 0$
Concave down on $(-\infty, -2)$;
concave up on $(-2, \infty)$
y-intercept: $\left(0, \frac{1}{2}\right)$

29.

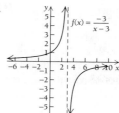

Increasing on $(-\infty, 3)$ and $(3, \infty)$
No relative extrema
Asymptotes: $x = 3$ and $y = 0$
Concave up on $(-\infty, 3)$; concave down on $(3, \infty)$
y-intercept: $(0, 1)$

31.

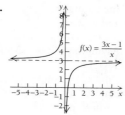

Increasing on $(-\infty, 0)$ and $(0, \infty)$
No relative extrema
Asymptotes: $x = 0$ and $y = 3$
Concave up on $(-\infty, 0)$; concave down on $(0, \infty)$
x-intercept: $\left(\frac{1}{3}, 0\right)$

33.

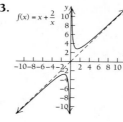

Increasing on $\left(-\infty, -\sqrt{2}\right)$ and $\left(\sqrt{2}, \infty\right)$; decreasing on $\left(-\sqrt{2}, 0\right)$ and $\left(0, \sqrt{2}\right)$
Relative minimum at $\left(\sqrt{2}, 2\sqrt{2}\right)$; relative maximum at $\left(-\sqrt{2}, -2\sqrt{2}\right)$;
Asymptotes: $x = 0$ and $y = x$
Concave down on $(-\infty, 0)$; concave up on $(0, \infty)$
No intercepts

35.

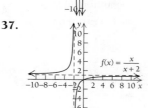

Decreasing on $(-\infty, 0)$; increasing on $(0, \infty)$
No relative extrema
Asymptotes: $x = 0$ and $y = 0$
Concave down on $(-\infty, 0)$ and $(0, \infty)$
No intercepts

37.

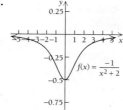

Increasing on $(-\infty, -2)$ and $(-2, \infty)$
No relative extrema
Asymptotes: $x = -2$ and $y = 1$
Concave up on $(-\infty, -2)$; concave down on $(-2, \infty)$
x- and y-intercept: $(0, 0)$

39.

Decreasing on $(-\infty, 0)$; increasing on $(0, \infty)$
Relative minimum at $\left(0, -\frac{1}{2}\right)$
Asymptote: $y = 0$
Concave up on $\left(-\sqrt{\frac{2}{3}}, \sqrt{\frac{2}{3}}\right)$; concave down on $\left(-\infty, -\sqrt{\frac{2}{3}}\right)$ and $\left(\sqrt{\frac{2}{3}}, \infty\right)$
Inflection points: $\left(-\sqrt{\frac{2}{3}}, -\frac{3}{8}\right)$ and $\left(\sqrt{\frac{2}{3}}, -\frac{3}{8}\right)$
y-intercept: $\left(0, -\frac{1}{2}\right)$

41. 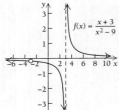 $f(x) = \dfrac{x+3}{x^2-9}$ Decreasing on $(-\infty, -3)$, $(-3, 3)$, and $(3, \infty)$
No relative extrema
Asymptotes: $x = 3$ and $y = 0$
Concave down on $(-\infty, -3)$ and $(-3, 3)$; concave up on $(3, \infty)$
y-intercept: $\left(0, -\dfrac{1}{3}\right)$

43. 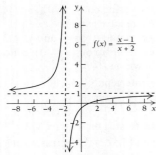 $f(x) = \dfrac{x-1}{x+2}$ Increasing on $(-\infty, -2)$ and $(-2, \infty)$
No relative extrema
Asymptotes: $x = -2$ and $y = 1$
Concave up on $(-\infty, -2)$; concave down on $(-2, \infty)$
x-intercept: $(1, 0)$; y-intercept: $\left(0, -\dfrac{1}{2}\right)$

45. 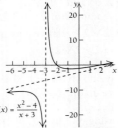 $f(x) = \dfrac{x^2-4}{x+3}$

Increasing on $\left(-\infty, -3 - \sqrt{5}\right)$ and $\left(-3 + \sqrt{5}, \infty\right)$, or approximately $(-\infty, -5.236)$ and $(-0.764, \infty)$; decreasing on $\left(-3 - \sqrt{5}, 3\right)$ and $\left(-3, -3 + \sqrt{5}\right)$, or approximately $(-5.236, -3)$ and $(-3, -0.764)$
Relative maximum at $\left(-3 - \sqrt{5}, -6 - 2\sqrt{5}\right)$ or approximately $(-5.236, -10.472)$; relative minimum at $\left(-3 + \sqrt{5}, -6 + 2\sqrt{5}\right)$, or approximately $(-0.764, -1.528)$
Asymptotes: $x = -3$ and $y = x - 3$
Concave down on $(-\infty, -3)$; concave up on $(-3, \infty)$
x-intercepts: $(-2, 0)$, $(2, 0)$; y-intercept: $\left(0, -\dfrac{4}{3}\right)$

47. 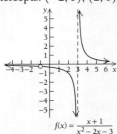 $f(x) = \dfrac{x+1}{x^2-2x-3}$ Decreasing on $(-\infty, -1)$, $(-1, 3)$, and $(3, \infty)$
No relative extrema
Asymptotes: $x = 3$ and $y = 0$
Concave down on $(-\infty, -1)$ and $(-1, 3)$; concave up on $(3, \infty)$
y-intercept: $\left(0, -\dfrac{1}{3}\right)$

49. 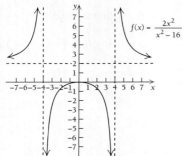 $f(x) = \dfrac{2x^2}{x^2-16}$

Increasing on $(-\infty, -4)$ and $(-4, 0)$; decreasing on $(0, 4)$ and $(4, \infty)$
Relative maximum at $(0, 0)$
Asymptotes: $x = -4$, $x = 4$, and $y = 2$
Concave up on $(-\infty, -4)$ and $(4, \infty)$; concave down on $(-4, 4)$
x- and y-intercept: $(0, 0)$

51. 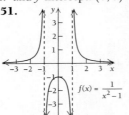 $f(x) = \dfrac{1}{x^2-1}$ Increasing on $(-\infty, -1)$ and $(-1, 0)$; decreasing on $(0, 1)$ and $(1, \infty)$
Relative maximum at $(0, -1)$
Asymptotes: $x = -1$, $x = 1$, and $y = 0$
Concave up on $(-\infty, -1)$ and $(1, \infty)$; concave down on $(-1, 1)$
y-intercept: $(0, -1)$

53. 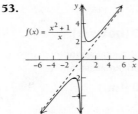 $f(x) = \dfrac{x^2+1}{x}$

Increasing on $(-\infty, -1)$ and $(1, \infty)$; decreasing on $(-1, 0)$ and $(0, 1)$
Relative maximum at $(-1, -2)$; relative minimum at $(1, 2)$
Asymptotes: $x = 0$ and $y = x$
Concave down on $(-\infty, 0)$; concave up on $(0, \infty)$
No intercepts

55. 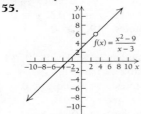 $f(x) = \dfrac{x^2-9}{x-3}$

Increasing on $(-\infty, 3)$ and $(3, \infty)$
No relative extrema
No asymptotes
No concavity
x-intercept: $(-3, 0)$; y-intercept: $(0, 3)$

57. $f(x) = \dfrac{-2x}{x-2}$ **59.** $g(x) = \dfrac{x^2-2}{x^2-1}$

61. $h(x) = \dfrac{x-9}{x^2+x-6}$

63. (a) \$50, \$37.24, \$32.64, \$26.37;
(b) maximum = 50 at $t = 0$;
(c) 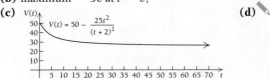 $V(t) = 50 - \dfrac{25t^2}{(t+2)^2}$ **(d)**

65. (a) \$480, \$600, \$2400, \$4800;   **(b)** $[0, 100)$

(c)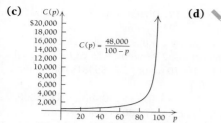

$$C(p) = \frac{48,000}{100 - p}$$

(d)

67. (a) $1.22, $0.79, $0.47;   **(b)** 36.8 yr after 1970;   **(c)** 0

69. (a)

n	9	6	3	1	2/3	1/3
E	4.00	6.00	12.00	36.00	54.00	108.00

(b) $\lim_{n \to 0} E(n) = \infty$. The pitcher gives up one or more runs but gets no one out (0 innings pitched).   **(c)** $E = 2.00$; pitcher gave up an average of 2 earned runs per game (9 innings).   **71.**

73. Does not exist   **75.** $-\infty$ **77.** $\frac{3}{2}$ **79.** $-\infty$

81.

$$f(x) = \frac{x}{\sqrt{x^2 + 1}}$$

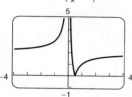

83.

$$f(x) = \frac{x^3 + 2x^2 - 15x}{x^2 - 5x - 14}$$

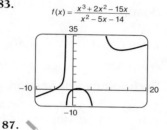

85.

$$f(x) = \left| \frac{1}{x} - 2 \right|$$

87.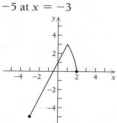

89. (a) 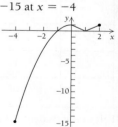   **(b)** $f'(x) = \dfrac{x^2 - 10x + 1}{(x^2 + x - 6)^2}$; critical values: ~0.101 and ~9.899;   **(c)** ;   **(d)** ;   **(e)**

Technology Connection, p. 253

1. On $[-2, 1]$, absolute minimum is -8 at $x = -2$, and absolute maximum is 2.185 at $x = -0.333$; on $[-1, 2]$, absolute minimum is 1 at $x = -1$ and $x = 1$, and absolute maximum is 4 at $x = 2$

Technology Connection, p. 254

1. Absolute minimum: -4 at $x = 2$; no absolute maximum

Technology Connection, p. 256

1. No absolute maximum; absolute minimum: 6.325 at $x = 0.316$

Exercise Set 2.4, p. 257

1. (a) 55 mph   **(b)** 5 mph   **(c)** 25 mpg   **3.** Absolute maximum: $5\frac{1}{4}$ at $x = \frac{1}{2}$; absolute minimum: 3 at $x = 2$
5. Absolute maximum: 4 at $x = 2$; absolute minimum: 1 at $x = -1$ and $x = 1$ **7.** Absolute maximum: $\frac{86}{27}$ at $x = -\frac{1}{3}$; absolute minimum: 2 at $x = -1$ **9.** Absolute maximum: 8 at $x = 3$; absolute minimum: -17 at $x = -2$
11. Absolute maximum: 15 at $x = -2$; absolute minimum: -13 at $x = 5$ **13.** Absolute maximum: -5 for $-1 \le x \le 1$; absolute minimum: -5 for $-1 \le x \le 1$

15. Absolute maximum: 4 at $x = -1$; absolute minimum: -12 at $x = 3$ **17.** Absolute maximum: $\frac{16}{5}$ at $x = -\frac{1}{5}$; absolute minimum: -48 at $x = 3$ **19.** Absolute maximum: 50 at $x = 5$; absolute minimum: -4 at $x = 2$ **21.** Absolute maximum: 2 at $x = -1$; absolute minimum: -110 at $x = -5$ **23.** Absolute maximum: 513 at $x = -8$; absolute minimum: -511 at $x = 8$
25. Absolute maximum: 17 at $x = 1$; absolute minimum: -15 at $x = -3$ **27.** Absolute maximum: 32 at $x = -2$; absolute minimum: $-\frac{27}{16}$ at $x = \frac{3}{2}$ **29.** Absolute maximum: 13 at $x = -2$ and $x = 2$; absolute minimum: 4 at $x = -1$ and $x = 1$
31. Absolute maximum: -1 at $x = 5$; absolute minimum: -5 at $x = -3$ **33.** Absolute maximum: $20\frac{1}{20}$ at $x = 20$; absolute minimum: 2 at $x = 1$ **35.** Absolute maximum: $\frac{4}{5}$ at $x = -2$ and $x = 2$; absolute minimum: 0 at $x = 0$ **37.** Absolute maximum: 3 at $x = 26$; absolute minimum: -1 at $x = -2$
39–47. Left to the student   **49.** Absolute maximum: 36 at $x = 6$ **51.** Absolute maximum: 70 at $x = 10$ **53.** Absolute maximum: $\frac{1}{3}$ at $x = \frac{1}{2}$ **55.** Absolute maximum: 900 at $x = 30$
57. Absolute maximum: $2\sqrt{3}$ at $x = -\sqrt{3}$; absolute minimum: $-2\sqrt{3}$ at $x = \sqrt{3}$ **59.** Absolute maximum: 5700 at $x = 2400$
61. Absolute minimum: $-55\frac{1}{3}$ at $x = 1$ **63.** Absolute maximum: 2000 at $x = 20$; absolute minimum: 0 at $x = 0$ and $x = 30$ **65.** Absolute minimum: 24 at $x = 6$
67. Absolute minimum: 108 at $x = 6$ **69.** Absolute maximum: 3 at $x = -1$; absolute minimum: $-\frac{3}{8}$ at $x = \frac{1}{2}$ **71.** Absolute maximum: 2 at $x = 8$; absolute minimum: 0 at $x = 0$
73. No absolute maximum or minimum   **75.** Absolute maximum: -1 at $x = 1$; absolute minimum: -5 at $x = -1$
77. No absolute maximum; absolute minimum: -5 at $x = -1$ **79.** Absolute maximum: 1 at $x = -1$ and $x = 1$; absolute minimum: 0 at $x = 0$ **81.** No absolute maximum or minimum
83. Absolute maximum: $-\frac{10}{3} + 2\sqrt{3}$ at $x = 2 - \sqrt{3}$; absolute minimum: $-\frac{10}{3} - 2\sqrt{3}$ at $x = 2 + \sqrt{3}$ **85.** No absolute maximum; absolute minimum: -1 at $x = -1$ and $x = 1$
87–95. Left to the student   **97.** 1430 units; 25 yr of service
99. 1986   **101.** 1999; 37.4 billion barrels
103. (a) $P(x) = -\frac{1}{2}x^2 + 400x - 5000$;   **(b)** 400 items
105. About 1.26 at $x = \frac{1}{9}$ cc, or about 0.11 cc
107. Absolute maximum: 3 at $x = 1$; absolute minimum: -5 at $x = -3$

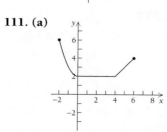

109. Absolute maxima: 1 at $x = 0$ and 1 at $x = 2$; absolute minimum: -15 at $x = -4$

111. (a)

(b) Absolute maximum: 6 at $x = -2$;   **(c)** $f(x) = 2$ over $[0, 4]$
113. Absolute maximum: $3\sqrt{6}$ at $x = 3$; absolute minimum: -2 at $x = -2$
115. Minimum: $20,000 at $x = 7$ "quality units"

117. ✎ **119. (a)** 2.755 billion barrels in 1981; **(b)** 0.0765 billion barrels/yr; 0.1014 billion barrels/yr **121.** No absolute maximum; absolute minimum: 0 at $x = 1$
123. (a) $P(t) = t + 8.857$; $P(7) = 15.857$ mm Hg;
(b) $P(t) = 0.117t^4 - 1.520t^3 + 6.193t^2 - 7.018t + 10.009$; $P(7) = 24.86$ mm Hg; $P(0.765) = 7.62$ mm Hg is the smallest contraction

Technology Connection, p. 263

1.

x	$y = 20 - x$	$A = x(20 - x)$
0	20	0
4	16	64
6.5	13.5	87.75
8	12	96
10	10	100
12	8	96
13.2	6.8	89.76
20	0	0

2. Left to the student   **3.** Maximum: 100 at $x = 10$

Technology Connection, p. 264

1.

x	$8 - 2x$	$4x^3 - 32x^2 + 64x$
0	8	0
0.5	7	24.5
1.0	6	36
1.5	5	37.5
2.0	4	32
2.5	3	22.5
3.0	2	12
3.5	1	3.5
4.0	0	0

2. Left to the student   **3.** Maximum: about 37.9 at $x \approx 1.33$

Technology Connection, p. 271

1. Left to the student   **2.** Minimum: $23,500 at $x = 100$, yes

Exercise Set 2.5, p. 273

1. Maximum $Q = 625$; $x = 25$, $y = 25$ **3.** ✎
5. Minimum product $= -4$; $x = 2$, $y = -2$

7. Maximum $Q = \frac{1}{4}$; $x = \frac{1}{2}$, $y = \sqrt{\frac{1}{2}}$ **9.** Minimum $Q = 30$; $x = 3$, $y = 2$ **11.** Maximum $Q = 21\frac{1}{3}$; $x = 2$, $y = 10\frac{2}{3}$
13. Maximum area $= 4050$ yd^2; width is 45 yd, and length (parallel to shoreline) is 90 yd   **15.** $x = 13.5$ ft, $y = 13.5$ ft;

maximum area $= 182.25$ ft^2 **17.** Dimensions: $33\frac{1}{3}$ cm by $33\frac{1}{3}$ cm by $8\frac{1}{3}$ cm; maximum volume $= 9259\frac{7}{27}$ cm^3
19. Dimensions: 5 in. by 5 in. by 2.5 in.; minimum surface area $= 75$ in^2 **21.** Dimensions: 2.08 yd by 4.16 yd by 1.387 yd
23. $1048; 46 units   **25.** $19; 70 units   **27.** $5481; 1667 units
29. (a) $R(x) = x(150 - 0.5x)$;
(b) $P(x) = -0.75x^2 + 150x - 4000$; **(c)** 100 suits; **(d)** $3500;
(e) $100/suit   **31.** $12.75/ticket; 57,500 people
33. 25 trees/acre   **35. (a)** $q(x) = 3.13 - 0.04x$; **(b)** $39.13
37. 4 ft by 4 ft by 20 ft   **39.** Order 5 times/yr; lot size is 20.
41. Order 12 times/yr; lot size is 60.   **43.** Order 8 times/yr; lot size is 32.   **45.** $r \approx 3.414$ in., $h \approx 6.828$ in.
47. $r \approx 2.879$ in., $h \approx 9.598$ in.   **49.** 14 in. by 14 in. by 28 in.
51. $x \approx 3.36$ ft, $y \approx 3.36$ ft   **53.** $\sqrt[3]{0.1}$, or approximately 0.4642   **55.** 9%   **57.** S is 3.25 mi downshore from A.
59. $x = \dfrac{bp}{a + b}$ **61. (a)** $A'(x) = \dfrac{x' \cdot C'(x) - C(x)}{x^2}$;
(b) $A'(x_0) = 0 = \dfrac{x_0 \cdot C'(x_0) - C(x_0)}{x_0^2}$; solving for $C'(x_0)$,
we get $C'(x_0) = \dfrac{C(x_0)}{x_0} = A(x_0)$.   **63.** $x = -\sqrt{2}$, $y = 0$,
$Q = -3\sqrt{2} \approx -4.24$ **65.** Order 25 times; lot size: 100 units

Technology Connection, p. 279 (top)

1. $P(x) = -3 + 40x - 0.5x^2$; $R(40) = 1200$, $C(40) = 403$, $P(40) = 797$, $R'(40) = 10$, $C'(40) = 10$, $P'(40) = 0$; marginal cost is constant.

Technology Connection, p. 279 (bottom)

1. $P'(50) = \$140/unit$; $P(51) - P(50) = \$217/unit$

Exercise Set 2.6, p. 285

1. (a) $P(x) = -0.001x^2 + 3.8x - 60$;
(b) $R(100) = \$500$, $C(100) = \$190$, $P(100) = \$310$;
(c) $R'(x) = 5$; $C'(x) = 0.002x + 1.2$; $P'(x) = -0.002x + 3.8$;
(d) $R'(100) = \$5$, $C'(100) = \$1.40$, $P'(100) = \$3.60$; **(e)** ✎
3. (a) $1234.38; **(b)** $24.52; **(c)** $24.38; **(d)** $48.75;
(e) $1283.13   **5. (a)** $1799; **(b)** $235.88; **(c)** $75.40;
(d) $R(71) = \$1874.40$, $R(72) = \$1949.80$, $R(73) = \$2025.20$
7. (a) $4572.78; **(b)** $594.03; **(c)** $593.63; **(d)** $5166.41
9. If the price increases from $1000 to $1001, sales will decrease by 100 units.   **11.** $2.01; $2.00   **13.** $2; $2
15. (a) $P(x) = -0.01x^2 + 1.4x - 30$; **(b)** $-\$0.01$; $0
17. (a) $dS/dp = 0.021p^2 - p + 150$; **(b)** 3547 units;
(c) ✎ **(d)** ✎   **19.** $-\$0.01$ **21.** $491.03 billion   **23.** ✎
25. About $0.21 paid in taxes per dollar earned   **27.** 0.0401; 0.04
29. 0.2816; 0.28   **31.** -0.556; -1 **33.** 6; 6   **35.** 5.1   **37.** 10.1
39. 10.017   **41.** $\dfrac{1}{2\sqrt{x + 1}}\,dx$ **43.** $9x^2\sqrt{2x^3 + 1}\,dx$
45. $\dfrac{1}{5(x + 27)^{4/5}}\,dx$ **47.** $(4x^3 - 6x^2 + 10x + 3)\,dx$ **49.** 3.1
51. 7.2   **53.** 657.00   **55.** -0.01345 m^2 **57.** The concentration changes more from 1 hr to 1.1 hr   **59.** $\dfrac{10}{2\pi} = 1.59$ ft
61. (a) $dA = 628$ ft^2 **(b)** 3 extra cans   **(c)** $90
63. $R'(x) = 100 - \dfrac{3\sqrt{x}}{2}$ **65.** $R'(x) = 500 - 2x$
67. $R'(x) = 5$ **69.** ✎

Exercise Set 2.7, p. 292

1. $\dfrac{-x^2}{2y^2}$; -2 **3.** $\dfrac{4x}{9y^2}$; $-\dfrac{8}{9}$ **5.** $\dfrac{x}{y}$; $\sqrt{\dfrac{3}{2}}$ **7.** $\dfrac{-y}{2x}$; $\dfrac{1}{4}$

9. $\dfrac{3x - 2y^2}{2xy}$; $-\dfrac{1}{12}$ **11.** $\dfrac{1 - y}{x + 2}$; $-\dfrac{1}{9}$ **13.** $\dfrac{6x^2 - 2xy}{x^2 - 3y^2}$; $-\dfrac{36}{23}$

15. $\dfrac{-y}{x}$ **17.** $\dfrac{x}{y}$ **19.** $\dfrac{3x^2}{5y^4}$ **21.** $\dfrac{-3xy^2 - 2y}{4x^2y + 3x}$ **23.** $\dfrac{3}{3p^2 + 1}$

25. $\dfrac{-p}{3x}$ **27.** $\dfrac{2 - p}{x - 2}$ **29.** $\dfrac{-p - 4}{x + 3}$ **31.** $-\dfrac{3}{4}$

33. \$400/day, \$80/day, \$320/day

35. \$16/day, \$8/day, \$8/day   **37.** -1.18 sales/day

39. $-17{,}915 \text{ mi}^2/\text{yr}$ **41.** Decreasing by $0.0256 \text{ m}^2/\text{month}$

43. (a) $\dfrac{dV}{dt} = 952.38R\,\dfrac{dR}{dt}$;   **(b)** 0.0143 mm/sec^2

45. $-2\dfrac{1}{12}\text{ft/sec}$ **47.** $494.8 \text{ cm}^3/\text{week}$ **49.** $\dfrac{-y^3}{x^3}$

51. $\dfrac{2x}{y(x^2 + 1)^2}$, or $\dfrac{x(1 - y^2)}{y(1 + x^2)}$

53. $\dfrac{5x^4 - 3(x - y)^2 - 3(x + y)^2}{3(x + y)^2 - 3(x - y)^2 - 5y^4}$, or $\dfrac{6y^2 - 5x^4 + 6x^2}{y(5y^3 - 12x)}$

55. $\dfrac{-6(y^2 - xy + x^2)}{(2y - x)^3}$ **57.** $\dfrac{2x(y^3 - x^3)}{y^5}$ **59.**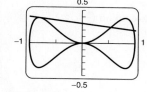

61.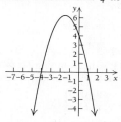

$x^4 = y^2 + x^6$

63.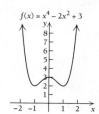

$x^3 = y^2(2 - x)$

Chapter Review Exercises, p. 301

1. (g)   **2.** (e)   **3.** (f)   **4.** (a)   **5.** (b)   **6.** (d)   **7.** (c)

8. False   **9.** False   **10.** True   **11.** False   **12.** False

13. False   **14.** Relative maximum: $\dfrac{25}{4}$ at $x = -\dfrac{3}{2}$

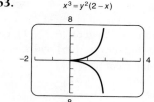

15. Relative minima: 2 at $x = -1$ and 2 at $x = 1$; relative maximum: 3 at $x = 0$

$f(x) = x^4 - 2x^2 + 3$

16. Relative minimum: -4 at $x = 1$; relative maximum: 4 at $x = -1$

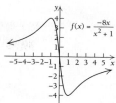

$f(x) = \dfrac{-8x}{x^2 + 1}$

17. No relative extrema

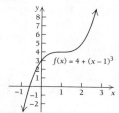

$f(x) = 4 + (x - 1)^3$

18. Relative minimum: $\dfrac{76}{27}$ at $x = \dfrac{1}{3}$; relative maximum: 4 at $x = -1$

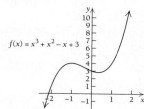

$f(x) = x^3 + x^2 - x + 3$

19. Relative minimum: 0 at $x = 0$

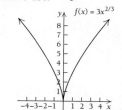

$f(x) = 3x^{2/3}$

20. Relative maximum: 17 at $x = -1$; relative minimum: -10 at $x = 2$

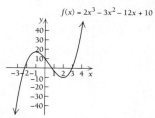

$f(x) = 2x^3 - 3x^2 - 12x + 10$

21. Relative maximum: 4 at $x = -1$; relative minimum: 0 at $x = 1$

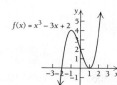

$f(x) = x^3 - 3x + 2$

22.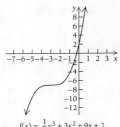

$f(x) = \frac{1}{3}x^3 + 3x^2 + 9x + 2$

No relative extrema
Inflection point at $(-3, -7)$
Increasing on $(-\infty, \infty)$
Concave down on $(-\infty, -3)$;
concave up on $(-3, \infty)$

23.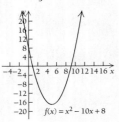

$f(x) = x^2 - 10x + 8$

Relative minimum: -17 at $x = 5$
Decreasing on $(-\infty, 5)$; increasing
on $(5, \infty)$
Concave up on $(-\infty, \infty)$

24.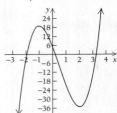

$f(x) = 4x^3 - 6x^2 - 24x + 5$

Relative minimum: -35 at $x = 2$;
relative maximum: 19 at $x = -1$
Inflection point at $\left(\frac{1}{2}, -8\right)$
Increasing on $(-\infty, -1)$ and
$(2, \infty)$; decreasing on $(-1, 2)$
Concave down on $\left(-\infty, \frac{1}{2}\right)$;
concave up on $\left(\frac{1}{2}, \infty\right)$

25.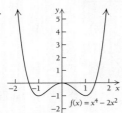

$f(x) = x^4 - 2x^2$

Relative minima: -1 at $x = -1$ and -1 at $x = 1$;
relative maximum: 0 at $x = 0$
Inflection points at $\left(-\sqrt{\frac{1}{3}}, -\frac{5}{9}\right)$ and $\left(\sqrt{\frac{1}{3}}, -\frac{5}{9}\right)$
Increasing on $(-1, 0)$ and $(1, \infty)$;
decreasing on $(-\infty, -1)$ and $(0, 1)$
Concave up on $\left(-\infty, -\sqrt{\frac{1}{3}}\right)$ and $\left(\sqrt{\frac{1}{3}}, \infty\right)$;
concave down on $\left(-\sqrt{\frac{1}{3}}, \sqrt{\frac{1}{3}}\right)$

26.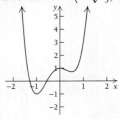

$f(x) = 3x^4 + 2x^3 - 3x^2 + 1$

Relative minima: -1 at $x = -1$ and $\frac{11}{16}$ at $x = \frac{1}{2}$;
relative maximum: 1 at $x = 0$
Inflection points at $(-0.608, -0.147)$ and $(0.274, 0.833)$
Increasing on $(-1, 0)$ and $\left(\frac{1}{2}, \infty\right)$;
decreasing on $(-\infty, -1)$ and $\left(0, \frac{1}{2}\right)$
Concave down on $(-0.608, 0.274)$; concave up on
$(-\infty, -0.608)$ and $(0.274, \infty)$

27.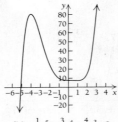

$f(x) = \frac{1}{5}x^5 + \frac{3}{4}x^4 - \frac{4}{3}x^3 + 8$

Relative minimum: $\frac{457}{60}$ at $x = 1$;
relative maximum: $\frac{1208}{15}$ at $x = -4$
Inflection points at $(-2.932, 53.701)$, $(0, 8)$, and $(0.682, 7.769)$
Increasing on $(-\infty, -4)$ and $(1, \infty)$; decreasing on $(-4, 1)$
Concave down on $(-\infty, -2.932)$ and $(0, 0.682)$; concave up on
$(-2.932, 0)$ and $(0.682, \infty)$

28.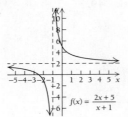

$f(x) = \frac{2x + 5}{x + 1}$

No relative extrema
Decreasing on $(-\infty, -1)$ and $(-1, \infty)$
Concave down on $(-\infty, -1)$; concave up on $(-1, \infty)$
Asymptotes: $x = -1$ and $y = 2$
x-intercept: $\left(-\frac{5}{2}, 0\right)$;
y-intercept: $(0, 5)$

29.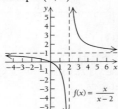

$f(x) = \frac{x}{x - 2}$

No relative extrema
Decreasing on $(-\infty, 2)$ and $(2, \infty)$
Concave down on $(-\infty, 2)$; concave up on $(2, \infty)$
Asymptotes: $x = 2$ and $y = 1$
x-intercept: $(0, 0)$; y-intercept: $(0, 0)$

30.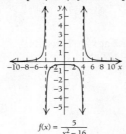

$f(x) = \frac{5}{x^2 - 16}$

Relative maximum at $\left(0, -\frac{5}{16}\right)$
Decreasing on $(0, 4)$ and $(4, \infty)$;
increasing on $(-\infty, -4)$ and $(-4, 0)$
Concave down on $(-4, 4)$; concave up on $(-\infty, -4)$ and $(4, \infty)$
Asymptotes: $x = -4$, $x = 4$, and $y = 0$
y-intercept: $\left(0, -\frac{5}{16}\right)$

31.

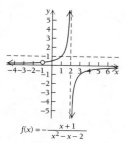

$$f(x) = -\frac{x+1}{x^2 - x - 2}$$

No relative extrema
Increasing on $(-\infty, -1)$, $(-1, 2)$, and $(2, \infty)$
Concave up on $(-\infty, -1)$ and $(-1, 2)$; concave down on $(2, \infty)$
Asymptotes: $x = 2$ and $y = 0$
y-intercept: $(0, \frac{1}{2})$

32.

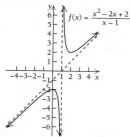

$$f(x) = \frac{x^2 - 2x + 2}{x - 1}$$

Relative minimum at $(2, 2)$; relative maximum at $(0, -2)$
Decreasing on $(0, 1)$ and $(1, 2)$; increasing on $(-\infty, 0)$ and $(2, \infty)$
Concave down on $(-\infty, 1)$; concave up on $(1, \infty)$
Asymptotes: $x = 1$ and $y = x - 1$
y-intercept: $(0, -2)$

33.

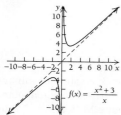

$$f(x) = \frac{x^2 + 3}{x}$$

Relative minimum at $(\sqrt{3}, 2\sqrt{3})$; relative maximum at $(-\sqrt{3}, -2\sqrt{3})$
Decreasing on $(-\sqrt{3}, 0)$ and $(0, \sqrt{3})$; increasing on $(-\infty, -\sqrt{3})$ and $(\sqrt{3}, \infty)$
Concave down on $(-\infty, 0)$; concave up on $(0, \infty)$
Asymptotes: $x = 0$ and $y = x$
No intercepts

34. Absolute maximum: 66 at $x = 3$; absolute minimum: 2 at $x = 1$ **35.** Absolute maximum: $75\frac{23}{27}$ at $x = \frac{16}{3}$; absolute minima: 0 at $x = 0$ and $x = 8$ **36.** No absolute maxima; absolute minimum: $10\sqrt{2}$ at $x = 5\sqrt{2}$
37. No absolute maxima; absolute minima: 0 at $x = -1$ and $x = 1$ **38.** 30 and 30   **39.** $Q = -1$ when $x = -1$ and $y = -1$
40. Maximum profit is \$451 when 30 units are produced and sold.   **41.** 10 ft by 10 ft by 25 ft   **42.** Order 12 times per year with a lot size of 30   **43. (a)** \$108;   **(b)** \$1/dinner;   **(c)** \$109
44. $\Delta y = -0.335$, $dy = -0.35$ **45. (a)** $(6x^2 + 1)\, dx$;

(b) 0.25   **46.** 9.111   **47.** $dV = \pm 240{,}000 \text{ ft}^3$ **48.** $\dfrac{-3y - 2x^2}{2y^2 + 3x}$, $\dfrac{4}{5}$

49. -1.75 ft/sec   **50.** \$600/day, \$450/day, \$150/day

51. No maximum; absolute minimum: 0 at $x = 3$
52. Absolute maxima: 4 at $x = 2$ and $x = 6$; absolute minimum: -2 at $x = -2$ **53.** $\dfrac{3x^5 - 2(x - y)^3 - 2(x + y)^3}{2(x + y)^3 - 2(x - y)^3 - 3y^5}$
54. Relative maximum at $(0, 0)$; relative minima at $(-9, -9477)$ and $(15, -37{,}125)$ **55.** $f(x) = \dfrac{3x + 3}{x + 2}$ (answers may vary)

56. Relative maxima at $(-1.714, 37.445)$; relative minimum at $(1.714, -37.445)$ **57.** Relative maximum at $(0, 1.08)$; relative minima at $(-3, -1)$ and $(3, -1)$
58. (a) Linear: $y = 6.998187602x - 124.6183581$
Quadratic: $y = 0.0439274846x^2 + 2.881202838x - 53.51475166$
Cubic: $y = -0.0033441547x^3 + 0.4795643605x^2 - 11.35931622x + 5.276985809$
Quartic: $y = -0.00005539834x^4 + 0.0067192294x^3 - 0.0996735857x^2 - 0.8409991942x - 0.246072967$

(b) The quartic function best fits the data.   **(c)** The domain is $[26, 102]$. Very few women outside of the age range from 26 to 102 years old develop breast cancer.   **(d)** Maximum: 466 per 100,000 women at $x = 79.0$ years old

Chapter 2 Test, p. 303

1. [2.1, 2.2] Relative minimum: -9 at $x = 2$
Decreasing on $(-\infty, 2)$ increasing on $(2, \infty)$

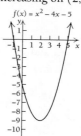

$$f(x) = x^2 - 4x - 5$$

2. [2.1, 2.2] Relative minimum: 2 at $x = -1$; relative maximum: 6 at $x = 1$
Decreasing on $(-\infty, -1)$ and $(1, \infty)$; increasing on $(-1, 1)$

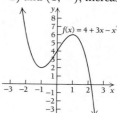

$$f(x) = 4 + 3x - x^3$$

3. [2.1, 2.2] Relative minimum: -4 at $x = 2$
Decreasing on $(-\infty, 2)$; increasing on $(2, \infty)$

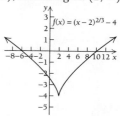

$$f(x) = (x - 2)^{2/3} - 4$$

4. [2.1, 2.2] Relative maximum: 4 at $x = 0$
Increasing on $(-\infty, 0)$; decreasing on $(0, \infty)$

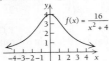

5. [2.3]

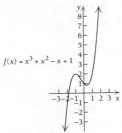

Relative maximum: 2 at $x = -1$; relative minimum: $\frac{22}{27}$ at $x = \frac{1}{3}$
Inflection point: $\left(-\frac{1}{3}, \frac{38}{27}\right)$

6. [2.3]

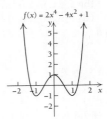

Relative maximum: 1 at $x = 0$;
relative minima: -1 at $x = -1$ and $x = 1$
Inflection points: $\left(-\sqrt{\frac{1}{3}}, -\frac{1}{9}\right)$ and $\left(\sqrt{\frac{1}{3}}, -\frac{1}{9}\right)$

7. [2.3]

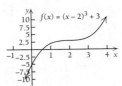

No relative extrema
Inflection point: $(2, 3)$

8. [2.3]

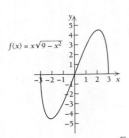

Relative maximum: $\frac{9}{2}$ at $x = \sqrt{\frac{9}{2}}$;
relative minimum: $-\frac{9}{2}$ at $x = -\sqrt{\frac{9}{2}}$
Inflection point: $(0, 0)$

9. [2.3]

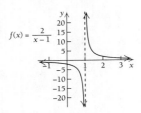

No relative extrema
Asymptotes: $x = 1$ and $y = 0$

10. [2.3]

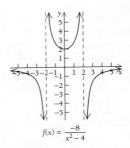

Relative minimum: 2 at $x = 0$
Asymptotes: $x = -2, x = 2$, and $y = 0$

11. [2.3]

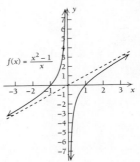

No relative extrema
Asymptotes: $x = 0$ and $y = x$

12. [2.3]

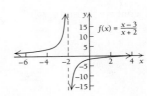

No relative extrema
Asymptotes: $x = -2$ and $y = 1$
13. [2.4] Absolute maximum: 9 at $x = 3$; no absolute minimum
14. [2.4] Absolute maximum: 2 at $x = -1$; absolute minimum: -1
at $x = -2$ **15.** [2.4] Absolute maximum: 28.49 at $x = 4.3$; no
absolute minimum   **16.** [2.4] Absolute maximum: 7 at $x = -1$;
absolute minimum: 3 at $x = 1$ **17.** [2.4] There are no absolute
extrema.   **18.** [2.4] Absolute minimum: $-\frac{13}{12}$ at $x = \frac{1}{6}$
19. [2.4] Absolute minimum: 48 at $x = 4$ **20.** [2.5] 4 and -4
21. [2.5] $Q = 50$ for $x = 5$ and $y = -5$ **22.** [2.5] Maximum
profit: \$24,980; 500 units   **23.** [2.5] Dimensions: 40 in. by
40 in. by 10 in; maximum volume: 16,000 in^3 **24.** [2.5] Order
35 times per year; lot size, 35   **25.** [2.6] $\Delta y = 1.01$; $f'(x)\Delta x = 1$

26. [2.6] 7.0714   **27.** [2.6]   (a) $\dfrac{x}{\sqrt{x^2 + 3}} \, dx$;   (b) 0.00756

28. [2.7] $\dfrac{-x^2}{y^2}$; $-\dfrac{1}{4}$ **29.** [2.6] $dV = \pm 1413$ cm^3

30. [2.7] -0.96 ft/sec   **31.** [2.4] Absolute maximum:
$\dfrac{2^{2/3}}{3} \approx 0.529$ at $x = \sqrt[3]{2}$; absolute minimum: 0 at $x = 0$

32. [2.5] 10,000 units   **33.** [2.4] Absolute minimum:
0 at $x = 0$; relative maximum: 25.103 at $x = 1.084$; relative
minimum: 8.625 at $x = 2.95$ **34.** [2.4] Relative minimum:
-0.186 at $x = 0.775$; relative maximum: 0.186 at $x = -0.775$
35. [2.1, 2.2]   (a) Linear: $y = -0.7707142857x + 12691.60714$
Quadratic: $y = -0.9998904762x^2 + 299.1964286x$
$+ 192.9761905$
Cubic: $y = 0.000084x^3 - 1.037690476x^2 + 303.3964286x$
$+ 129.9761905$

Quartic: $y = -0.000001966061x^4 + 0.0012636364x^3$
        $-1.256063636x^2 + 315.8247403x + 66.78138528$

(b) Since the number of bowling balls sold cannot be negative, the domain is $[0, 300]$. This is supported by both the quadratic model and the raw data. The cubic and quartic models can also be used but are more complicated. **(c)** Based on the quadratic function, the maximum value is 22,575 bowling balls. The company should spend $150,000 on advertising.

Extended Technology Application, p. 306

1. (a) 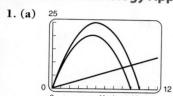   **(b)** 4500;   **(c)** 20,250

2. (a) 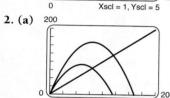   **(b)** 60,000;   **(c)** 90,000

3. (a) 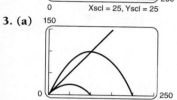   **(b)** 50,000;   **(c)** 25,000

4. (a) 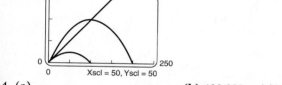   **(b)** 400,000;   **(c)** 400,000

5. (a) 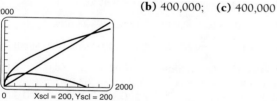   **(b)** 30,513;   **(c)** 205,923

6. (a) $y = -0.0011P^3 + 0.0715P^2 - 0.0338P + 4$
(b) 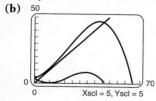   **(c)** 33,841

Chapter 3

Technology Connection, p. 308

1. 156.993   **2.** 16.242   **3.** 0.064   **4.** 0.000114

Technology Connection, p. 312

Left to the student

Exercise Set 3.1, p. 319

1.   **3.**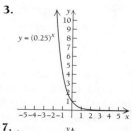

5. 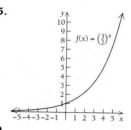   **7.**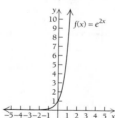

9.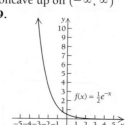

11. $-e^{-x}$ **13.** $3e^{3x}$ **15.** $6e^x$
17. $-7e^{-7x}$ **19.** $8e^{4x}$
21. $3e^{-x}$ **23.** $-\frac{5}{2}e^{-5x}$
25. $-\frac{4x}{3} \cdot e^{x^2}$ **27.** $15e^{5x}$
29. $5x^4 - 12e^{6x}$
31. $5x^4 \cdot e^{2x} + 2x^5 \cdot e^{2x}$

33. $\dfrac{2e^{2x}(x-2)}{x^5}$ **35.** $e^x(x^2 + 5x - 6)$ **37.** $\dfrac{e^x(x-4)}{x^5}$

39. $(-2x + 7)e^{-x^2+7x}$ **41.** $-xe^{-x^2/2}$ **43.** $\dfrac{e^{\sqrt{x-7}}}{2\sqrt{x-7}}$

45. $\dfrac{e^x}{2\sqrt{e^x - 1}}$ **47.** $-2xe^{-2x} + e^{-2x} - e^{-x} + 3x^2$ **49.** e^{-x}

51. ke^{-kx} **53.** $(4x^2 + 3x)e^{x^2-7x}(2x - 7) + (8x + 3)e^{x^2-7x}$, or
$(8x^3 - 22x^2 - 13x + 3)e^{x^2-7x}$

55.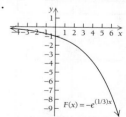

No critical values
No inflection points
Increasing on $(-\infty, \infty)$
Concave up on $(-\infty, \infty)$

57.

No critical values
No inflection points
Increasing on $(-\infty, \infty)$
Concave up on $(-\infty, \infty)$

59.

No critical values
No inflection points
Decreasing on $(-\infty, \infty)$
Concave up on $(-\infty, \infty)$

61.

No critical values
No inflection points
Decreasing on $(-\infty, \infty)$
Concave down on $(-\infty, \infty)$

63.

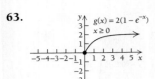

No critical values on $[0, \infty)$
No inflection points
Increasing on $[0, \infty)$
Concave down on $[0, \infty)$

65–73. Left to the student **75.** 1 **77.** $y = -x + 1$
79. Left to the student **81.** (a) \$1.6 billion, \$2.7 billion; (b) 15 yr
83. (a) $C'(t) = 50e^{-t}$; (b) \$50 million/yr; (c) \$916,000/yr;
(d) 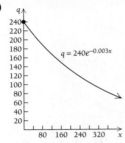 **85.** (a) 113,000; (b)
(c) $q'(x) = -0.72e^{-0.003x}$;
(d)

$q = 240e^{-0.003x}$

87. (a) 0 ppm, 3.7 ppm, 5.4 ppm, 4.5 ppm, 0.05 ppm;
(b) 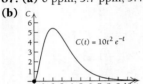 (c) $C'(t) = 10te^{-t}(2 - t)$;
(d) 5.4 ppm at $t = 2$ hr
$C(t) = 10t^2 e^{-t}$ (e)

89. $15e^{3x}(e^{3x} + 1)^4$ **91.** $-e^{-t} - 3e^{3t}$ **93.** $\dfrac{(x^2 - 2x + 1)e^x}{(x^2 + 1)^2}$
95. $\dfrac{e^{\sqrt{x}}}{2\sqrt{x}} + \dfrac{1}{2}e^{x/2}$ **97.** $e^{x/2}\left(\dfrac{x}{2\sqrt{x - 1}}\right)$ **99.** $\dfrac{4}{(e^x + e^{-x})^2}$
101. 2; 2.25; 2.48832; 2.59374; 2.71692
103. $4e^{-2} \approx 0.5413$, for $x = 2$ **105.** Left to the student
107. $f(x) = x^2 e^{-x}$ **109.** $f(x) = f'(x) = f''(x) = e^x$

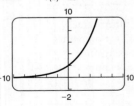

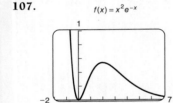

Relative minimum at $(0, 0)$;
relative maximum at $(2, 0.5413)$
111. $f(x) = 2e^{0.3x}$ $f'(x) = 0.6e^{0.3x}$

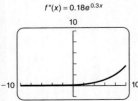

$f''(x) = 0.18e^{0.3x}$ **113.** $f(x) = \left(1 + \dfrac{1}{x}\right)^x$

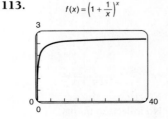

Technology Connection, p. 323

Left to the student

Technology Connection, p. 325

1. Graph is left to the student; function values are 1000, 5, 0.699, 3.
2. Left to the student; graph is obtained by entering $y = \log x/\log 2$.

Technology Connection, p. 327

1. $t = 6.9$ **2.** $x = -4.1$ **3.** $t = 74.9$ **4.** $x = 46.2$
5. $x = 38.7$

Technology Connection, p. 328

Left to the student

Exercise Set 3.2, p. 334

1. $2^3 = 8$ **3.** $8^{1/3} = 2$ **5.** $a^J = K$ **7.** $10^{-p} = h$
9. $\ln b = M$ **11.** $\log_{10} 100 = 2$ **13.** $\log_{10} 0.1 = -1$
15. $\log_M V = p$ **17.** 0.51 **19.** 2.708 **21.** 2.609
23. 2.9957 **25.** 0.2231 **27.** 2.6094 **29.** 3 **31.** -1.3863
33. -0.6094 **35.** 8.681690 **37.** -4.006334 **39.** 8.999619
41. $t \approx 4.382$ **43.** $t \approx 3.454$ **45.** $t \approx 2.303$
47. $t \approx 140.671$ **49.** $-\dfrac{8}{x}$ **51.** $x^3 + 4(\ln x)x^3 - x$
53. $\dfrac{1}{x}$ **55.** $x + 2x \ln (7x)$ **57.** $\dfrac{1 - 4 \ln x}{x^5}$ **59.** $\dfrac{2}{x}$
61. $\dfrac{2(3x + 1)}{3x^2 + 2x - 1}$ **63.** $\dfrac{x^2 + 7}{x(x^2 - 7)}$ **65.** $\dfrac{2e^x}{x} + 2e^x \ln x$
67. $\dfrac{e^x}{e^x + 1}$ **69.** $\dfrac{4(\ln x)^3}{x}$ **71.** $\dfrac{1}{x \ln (8x)}$
73. $\dfrac{\ln (5x) + \ln (3x)}{x}$ **75.** $y = 8.455x - 11.94$
77. $y = 0.732x - 0.990$ **79.** (a) 2000 units;
(b) $N'(a) = \dfrac{500}{a}$, $N'(10) = 50$ units per \$1000 spent on
advertising; (c) minimum is 2000 units; (d) **81.** 58 days
83. (a) \$58.69, \$78.00; (b) $V'(t) = 63.8e^{-1.1t}$;
(c) 2.7 months; (d) **85.** (a) $P'(x) = 1.7 - 0.3 \ln x$;
(b) (c) 289.069 **87.** (a) 68%; (b) 35.8%; (c) 3.6%;
(d) 5.3%; (e) $S'(t) = \dfrac{-20}{t + 1}$; (f) maximum = 68%, and
minimum approaches 0%; (g) **89.** (a) 2.4 ft/s; (b) 3.4 ft/s;
(c) $v'(p) = \dfrac{0.37}{p}$; (d) **91.** $t = \dfrac{\ln (P/P_0)}{k}$ **93.** $\dfrac{7(2t - 1)}{t(t - 1)}$
95. $\dfrac{1}{x \ln (3x) \cdot \ln (\ln (3x))}$ **97.** $\dfrac{-1}{1 - t} - \dfrac{1}{1 + t}$,
or $\dfrac{-2}{(1 - t)(1 + t)}$ **99.** $\dfrac{1}{x \ln 5}$ **101.** $\dfrac{x}{x^2 + 5}$
103. $x^4 \ln x$ **105.** $\dfrac{1}{\sqrt{x}(1 - \sqrt{x})(1 + \sqrt{x})}$, or $\dfrac{1}{\sqrt{x}(1 - x)}$
107. Definition of logarithm; Product Rule for exponents;
definition of logarithm; substitution **109.** Definition of
logarithm; if $a = b$, then $a^c = b^c$; Power Rule for exponents;
definition of logarithm; substitution and the Commutative Law
for Multiplication **111.** 1 **113.** e^π **115.** 0
117. Left to the student **119.** Left to the student
121. Minimum: $-e^{-1} \approx -0.368$

Technology Connection, p. 341

1. 8.159 billion   **2.** 12.98 billion   **3.** 15.152 billion
4. 20.648 billion   **5.** $y = 9689.991(1.0347623)^x = 9689.991e^{0.0341717x}$ **6.** $16,741; $18,548; $39,336

Exercise Set 3.3, p. 347

1. $f(x) = ce^{4x}$ **3.** $A(t) = ce^{-9t}$ **5.** $Q(t) = ce^{kt}$
7. (a) $N(t) = 112,000e^{0.046t}$;   **(b)** $N(40) = 705,212$;
(c) 15.1 yr   **9. (a)** $P(t) = P_0e^{0.059t}$;   **(b)** $1060.78, $1125.24;
(c) 11.7 yr   **11. (a)** $G(t) = 4.7e^{0.093t}$;   **(b)** 48.07 billion gallons;
(c) 7.5 yr   **13.** 4.62%   **15.** 6.9 yr after 2006
17. 11.2 yr; $102,256.88   **19.** $7500; 8.3 yr
21. (a) $k = 0.151$, or 15.1%, $V(t) = 30,000e^{0.151t}$;
(b) $549,188,702;   **(c)** 4.6 yr;   **(d)** 69 yr
23. (a) $E(t) = 1.031e^{0.047101t}$;   **(b)** $3.347 billion;
(c) after about 48.3 yr, or in 2038
25. (a) $y = 136.3939183 \cdot 1.071842825^x$,
$y = 136.3939183e^{0.0693794334x}$, and exponential
growth rate $= 0.069$, or 6.9%;   **(b)** 444 million, 628 million;
(c) 18.7 yr;   **(d)** 10 yr   **27.** Approximately $8.6 billion
29. (a) $S(t) = 4e^{0.0494t}$;   **(b)** 4.94%/yr;   **(c)** 50¢, 58¢, 67¢
(d) For the years 2011–2021, the total cost of Forever
Stamps is $11 \times 4500, or $49,500. For the years 2011–2012,
the cost of regular first-class stamps is $2 \times 4500, or $9000.
If the price of a regular postage stamp increases to 50¢ in 2013,
the cost of postage for the years 2013–2015 would be
$3 \times $0.50 \times 10,000$, or $15,000. If the price increases to
58¢ in 2016, the cost for the years 2016–2018 would be
$3 \times $0.58 \times 10,000$, or $17,400. If the price increases to
67¢ in 2019, the cost for the years 2019–2021 would be
$3 \times $0.67 \times 10,000$, or $20,100. Thus, the total cost of
regular first-class stamps for the years 2011–2021 would be
$9000 + $15,000 + $17,400 + $20,100 = $61,500.
Thus, by buying Forever Stamps, the firm would save
$61,500 - $49,500 = $12,000.   **(e)**
31. (a) 2%;   **(b)** 3.8%, 7%,   **(d)**
21.6%, 50.2%, 93.1%, 98%;
(c) $P'(x) = \dfrac{637e^{-0.13x}}{(1 + 49e^{-0.13x})^2}$

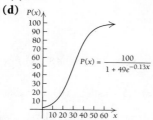

$P(x) = \dfrac{100}{1 + 49e^{-0.13x}}$

33. (a) $V(t) = 0.10e^{0.224t}$;   **(b)** $9,486,828;   **(c)** 3.09 yr;
(d) after 87.1 yr, or in 2025   **35.** 2019   **37.** 1%/yr
39. 4%/yr   **41.** $B(t) = 190e^{0.035t}$; 886 bears
43. (a) 1000, 1375, 1836,   **(c)**
3510, 5315, 5771;
(b) $P'(t) = \dfrac{11,051.36e^{-0.4t}}{(1 + 4.78e^{-0.4t})^2}$

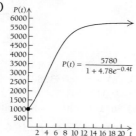

$P(t) = \dfrac{5780}{1 + 4.78e^{-0.4t}}$

45. $N(t) = 48,869e^{0.0378t}$, where $t_0 = 1930$; exponential
growth rate $= 3.78\%$

47. (a) 0%, 33%, 55%, 70%,   **(c)**
86%, 99.2%, 99.8%;
(b) at 7 months, the
percentage of doctors who
are prescribing the
medication is growing
by 2.4% per month

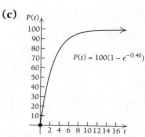

$P(t) = 100(1 - e^{-0.4t})$

49. (a) $N(t) = \dfrac{29.47232081}{1 + 79.56767122e^{-0.809743969t}}$;
(b) 29 students;   **(c)** $N(t) = \dfrac{29.47232081}{1 + 79.56767122e^{-0.809743969t}}$

(d) $N'(t) = \dfrac{1898.885181e^{-0.809743969t}}{(1 + 79.56767122e^{-0.809743969t})^2}$;   **(e)**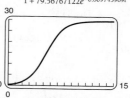
51–55.   **57.** $\ln 4 = kT_4$ **59.** 2 yr   **61.** 7.57%   **63.** 9%
65. $k = \dfrac{\ln (y_2/y_1)}{t_2 - t_1}$ **67.**   **69. (a)** $R(0) = $2 million; this
represents the initial revenue of the corporation at its inception.
(b) $\lim_{t \to \infty} R(t) = $4000 million $= R_{max}$; this represents the upper
limit of the revenue of the company over all time. It is never
actually attained.   **(c)** $t = 24$

Exercise Set 3.4, p. 360

1. (a) $N(t) = N_0e^{-0.096t}$;   **(b)** 341 g;   **(c)** 7.2 days
3. (a) $A(t) = A_0e^{-kt}$;   **(b)** 11 hr   **5.** 23.1%/min   **7.** 22 yr
9. 42.9 g   **11.** 4223 yr   **13.** 25 days   **15.** 3965 yr
17. $13,858.23   **19.** $6,393,134   **21.** $42,863.76
23. (a) $40,000;   **(b)** $5413.41;   **(c)**
25. (a) 0.022, 0.031, 0.069;   **(b)**

$Q(t) = (Q_0 - 0.00055)e^{0.163t} + 0.00055$

27. (a) $N(t) = 5,650,000e^{-0.018t}$;   **(b)** 1,953,564 farms,
1,753,573 farms;   **(c)** about 2046   **29. (a)** $B(t) = 64.6e^{-0.0068t}$;
(b) 58.3 lb;   **(c)** 2172   **31. (a)** $P(t) = 51.9e^{-0.0085224t}$;
(b) 43.77 million;   **(c)** after 463 yr, or in 2458   **33. (a)** 27;
(b) 0.05878;   **(c)** 83°;   **(d)** 28.7 min;   **(e)**   **35.** The murder
was committed at 7 P.M.   **37. (a)** 145 lb;   **(b)** -1.2 lb/day
39. (a) 11.2 W;   **(b)** 173 days;   **(c)** 402 days;   **(d)** 50 W;   **(e)**
41. (c)   **43.** (e)   **45.** (f)   **47.** (d)   **49.** (a)
51. $x = $166.16, $q = 292$ printers   **53. and 55.**

Exercise Set 3.5, p. 368

1. $(\ln 7)7^x$ **3.** $(\ln 8)8^x$ **5.** $x^3 \cdot \ln 5.4 \cdot (5.4)^x + 3x^2(5.4)^x$

7. $(\ln 7) \cdot 7^{x^4+2} \cdot 4x^3$ **9.** $8e^{8x}$ **11.** $(\ln 3) \cdot 3^{x^4+1} \cdot (4x^3)$

13. $\dfrac{1}{x \cdot \ln 4}$ **15.** $\dfrac{1}{x \cdot \ln 17}$ **17.** $\dfrac{5}{(5x+1)\ln 6}$

19. $\dfrac{6}{(6x-7)\ln 10}$ **21.** $\dfrac{3x^2+1}{(x^3+x)\ln 8}$ **23.** $\dfrac{2}{(x-2\sqrt{x})\ln 7}$

25. $\dfrac{6^x}{x \cdot \ln 7} + 6^x \cdot \ln 6 \cdot \log_7 x$ **27.** $5(\log_{12} x)^4\left(\dfrac{1}{x\ln 12}\right)$

29. $\dfrac{(4x+1)7^x \cdot \ln 7 - 4 \cdot 7^x}{(4x+1)^2}$

31. $\dfrac{6 \cdot 5^{2x^3-1}}{(6x+5)(\ln 10)} + (\ln 5)5^{2x^3-1} \cdot 6x^2 \cdot \log(6x+5)$

33. $(\ln 7)7^x \cdot (\log_4 x)^9 + \dfrac{7^x \cdot 9 \cdot (\log_4 x)^8}{x \cdot \ln 4}$

35. $5(3x^5+x)^4(15x^4+1) \cdot (\log_3 x) + \dfrac{(3x^5+x)^5}{\ln 3 \cdot x}$

37. (a) $V'(t) = 5200(\ln 0.80)(0.80)^t$;   (b)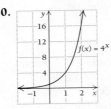
39. (a) \$19.84 trillion;   (b) \$905 billion/yr;   (c)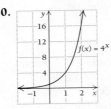
41. (a) 0.82;   (b) -0.015/yr;   (c) 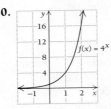   **43.** 8.8
45. (a) $I = I_0 10^{10}$;   (b) $I = I_0 10$;   (c) the power
mower is 10^9 times louder than a just audible sound;
(d) $dI/dL = I_0 10^{0.1L}(\ln 10)(0.1)$;   (e)

47. (a) $\dfrac{dL}{dI} = \dfrac{10}{(\ln 10)I}$;   (b)   **49.** $\ln 3 \approx 1.0986$

51. $(\ln 2)2^{x^4} \cdot 4x^3$ **53.** $\dfrac{1}{\ln 3 \cdot \log x \cdot \ln 10 \cdot x}$

55. $\ln a \cdot a^{f(x)} \cdot f'(x)$

57. $\left(\dfrac{g(x) \cdot f'(x)}{f(x)} + g'(x) \cdot \ln (f(x))\right) \cdot [f(x)]^{g(x)}$ **59.**

Technology Connection, p. 373

1. $E(x) = \dfrac{x}{300-x}$; $R(x) = 300x - x^2$ **2.** Left to the student

3. \$150

Exercise Set 3.6, p. 376

1. (a) $E(x) = \dfrac{x}{400-x}$;   (b) $\dfrac{5}{11}$, inelastic;   (c) \$200

3. (a) $E(x) = \dfrac{x}{50-x}$;   (b) 11.5, elastic;   (c) \$25

5. (a) $E(x) = 1$;   (b) 1, unit elasticity;   (c) total revenue is
independent of x.

7. (a) $E(x) = \dfrac{x}{2(600-x)}$;   (b) 0.10, inelastic;   (c) \$400

9. (a) $E(x) = 0.25x$;   (b) 2.5, elastic;   (c) \$4

11. (a) $E(x) = \dfrac{2x}{x+3}$;   (b) 0.5, inelastic;   (c) \$3

13. (a) $E(x) = \dfrac{25x}{967-25x}$;   (b) approximately 19¢;
(c) prices greater than 19¢;   (d) prices less than 19¢;
(e) approximately 19¢;   (f) decrease

15. (a) $E(x) = \dfrac{3x^3}{2(200-x^3)}$;   (b) $\dfrac{81}{346}$;   (c) increase

17. (a) $E(x) = n$;   (b) no;   (c) yes, at $n = 1$
19. $E(x) = (-x)L'(x)$ **21.**

Chapter Review Exercises, p. 382

1. (b)   **2.** (e)   **3.** (f)   **4.** (c)   **5.** (a)   **6.** (d)
7. False   **8.** True   **9.** True   **10.** False   **11.** True

12. False   **13.** True   **14.** False   **15.** True   **16.** $\dfrac{1}{x}$

17. e^x **18.** $\dfrac{4x^3}{x^4+5}$ **19.** $\dfrac{e^{2\sqrt{x}}}{\sqrt{x}}$ **20.** $\dfrac{1}{2x}$ **21.** $3x^4e^{3x} + 4x^3e^{3x}$

22. $\dfrac{1-3\ln x}{x^4}$ **23.** $2xe^{x^2}(\ln 4x) + \dfrac{e^{x^2}}{x}$ **24.** $4e^{4x} - \dfrac{1}{x}$

25. $8x^7 - \dfrac{8}{x}$ **26.** $\dfrac{1-x}{e^x}$ **27.** $(\ln 9)9^x$ **28.** $\dfrac{1}{(\ln 2)x}$

29. $3^x(\ln 3)(\log_4 (2x+1)) + \dfrac{(3^x)2}{(2x+1)(\ln 4)}$

30.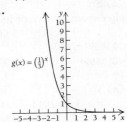
$f(x) = 4^x$

31.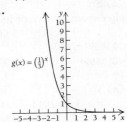
$g(x) = \left(\tfrac{1}{3}\right)^x$

32. 6.93   **33.** -3.2698 **34.** 8.7601   **35.** 3.2698
36. 2.54995   **37.** -3.6602 **38.** $Q(t) = 25e^{7t}$ **39.** 4.3%
40. 10.2 yr   **41.** (a) $C(t) = 15.81e^{0.024t}$;   (b) \$29.51, \$35.75
42. (a) $N(t) = 60e^{0.12t}$;   (b) 123 franchises;   (c) 5.8 yr after 2007
43. 5.3 yr   **44.** 18.2%   **45.** (a) $A(t) = 800e^{-0.07t}$;
(b) 197 g;   (c) 9.9 days   **46.** (a) 0.50, 0.75, 0.97, 0.999, 0.9999;
(b) $p'(t) = 0.7e^{-0.7t}$;   (c)   (d)

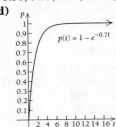
$p(t) = 1 - e^{-0.7t}$

47. \$186,373.98   **48.** (a) $E(x) = \dfrac{2x}{x+4}$;   (b) 0.4, inelastic;
(c) 1.5, elastic;   (d) decrease;   (e) \$4   **49.** $\dfrac{-8}{(e^{2x}-e^{-2x})^2}$
50. $-\dfrac{1}{1024e} \approx 0$ **51.**
$f(x) = \dfrac{e^{1/x}}{(1+e^{1/x})^2}$ **52.** 0

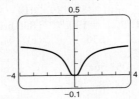

53. (a) $y = 9.033880177 \cdot 1.431864118^x$,
$y = 9.033880177e^{0.3589771744x}$, 0.3589771744;
(b) \$671.0 billion, \$24.3 trillion;   (c) 10.56 yr;   (d) 1.93 yr

Chapter 3 Test, p. 384

1. [3.1] $6e^{3x}$ **2.** [3.2] $\dfrac{4(\ln x)^3}{x}$ **3.** [3.1] $-2xe^{-x^2}$

4. [3.2] $\dfrac{1}{x}$ **5.** [3.1] $e^x - 15x^2$ **6.** [3.1, 3.2] $\dfrac{3e^x}{x} + 3e^x \cdot \ln x$

7. [3.5] $(\ln 7)7^x + (\ln 3)3^x$ **8.** [3.5] $\dfrac{1}{(\ln 14)x}$ **9.** [3.2] 1.0674

10. [3.2] 0.5554   **11.** [3.2] 0.4057   **12.** [3.3] $M(t) = 2e^{6t}$
13. [3.3] 23.1%   **14.** [3.3] 10.0 yr
15. [3.3]   **(a)** $C(t) = 3.22e^{0.021t}$;   **(b)** $3.65, $4.14
16. [3.4]   **(a)** $A(t) = 3e^{-0.1t}$;   **(b)** 1.1 cc;   **(c)** 6.9 hr
17. [3.4] About 16.47 centuries, or 1647 yr   **18.** [3.4] 4.0773%/sec
19. [3.3]   **(a)** 4%;   **(b)** 5.2%, 14.5%, 40.7%, 73.5%, 91.8%,
99.5%, 99.9%;   **(c)** $P'(t) = \dfrac{672e^{-0.28t}}{(1 + 24e^{-0.28t})^2}$;   **(d)** ✎;
(e)

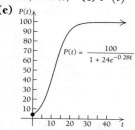

20. [3.4] $10,043,738   **21.** [3.6]   **(a)** $E(x) = 0.2x$;
(b) 0.6, inelastic;   **(c)** 3.6, elastic;   **(d)** increase;   **(e)** $5
22. [3.2] $(\ln x)^2$ **23.** [3.1] Maximum is $\dfrac{256}{e^4} \approx 4.689$;
minimum is 0   **24.** [3.1]

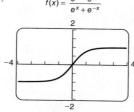

$f(x) = \dfrac{e^x - e^{-x}}{e^x + e^{-x}}$

25. [3.1] 0   **26.** [3.3]   **(a)** $y = 686996.582 \cdot 1.086747476^x$,
$y = 686996.582 \cdot e^{0.0831892684x}$.   **(b)** $4.28 million, $5.50 million;
(c) 87.5 yr;   **(d)** 8.3 yr;   **(e)** ✎

Extended Technology Application, p. 387

1. Linear: $R(t) = -5.2444t + 52.3333$
Quadratic: $R(t) = 0.756x^2 - 13.5603x + 68.965$
Cubic: $R(t) = -0.2107x^3 + 4.2327x^2 - 29.5954x + 87.044$
Exponential: $R(t) = 68.4552(0.7868)^x$
The linear and cubic functions both reach $R = 0$ too fast. The value from the quadratic function decreases as we might expect over time, but then it makes a dramatic rise. Most movies do not have this revenue pattern. The exponential function shows a steady decrease and approaches 0 as a limit, but never reaches it. It is a reasonable assumption that G gets smaller and smaller over time. Eventually, box office revenue reaches 0. Gross revenue thereafter comes from DVD rentals, TV rights, and electronic outlets, such as iTunes.   **2.** $G = $4.90, $3.85, $3.03, $2.39, $1.88, $1.48, $1.16, $0.91, all in millions   **3.** $R = $239.79, $243.64, $246.67, $249.06, $250.94, $252.43, $253.58, $254.49, all in millions. There are costs, such as marketing and shipping costs, associated with distributing a movie to theaters. Eventually, movie executives want the jump in revenue that comes with DVD and electronic rentals.
4. $R(t) = \dfrac{251.1}{1 + 3.4687e^{-0.4183t}}$
5. $R'(t) = \dfrac{364.3354}{(1 + 3.4687e^{-0.4183t})^2}$, which represents the rate of change of the total revenue; $\lim\limits_{t\to\infty} R'(t) = 0$, which means that eventually the total revenue does not change. From the logistic function, it would be about $251.1 million, but the table shows about $254 million.

6. From the logistic function, it would be about $745 million, but the table shows about $762 million.

Chapter 4

Technology Connection, p. 396

1. **(a)** Left to the student;   **(b)** 400;   **(c)** the area is the square of x;
(d) $A(x) = x^2$;   **(e)** $A(x)$ is the antiderivative of $f(x)$.
2. **(a)** Left to the student;   **(b)** 60;   **(c)** the area is 3 times x;
(d) $A(x) = 3x$;   **(e)** $A(x)$ is the antiderivative of $f(x)$.
3. **(a)** Left to the student;   **(b)** 800;   **(c)** the area is the cube of x;
(d) $A(x) = x^3$;   **(e)** $A(x)$ is the antiderivative of $f(x)$.

Exercise Set 4.1, p. 396

1. $\dfrac{x^7}{7} + C$ **3.** $2x + C$ **5.** $\frac{4}{5}x^{5/4} + C$ **7.** $\frac{1}{3}x^3 + \frac{1}{2}x^2 - x + C$
9. $\frac{2}{3}t^3 + \frac{5}{2}t^2 - 3t + C$ **11.** $-\dfrac{x^{-2}}{2} + C$ **13.** $\frac{3}{4}x^{4/3} + C$
15. $\frac{2}{7}x^{7/2} + C$ **17.** $-\dfrac{x^{-3}}{3} + C$ **19.** $\ln x + C$
21. $3\ln x - 5x^{-1} + C$ **23.** $-21x^{1/3} + C$ **25.** $e^{2x} + C$
27. $\frac{1}{3}e^{3x} + C$ **29.** $\frac{1}{7}e^{7x} + C$ **31.** $\frac{5}{3}e^{3x} + C$
33. $\frac{3}{4}e^{8x} + C$ **35.** $-\frac{2}{27}e^{-9x} + C$ **37.** $\frac{5}{3}x^3 - \frac{2}{7}e^{7x} + C$
39. $\dfrac{x^3}{3} - x^{3/2} - 3x^{-1/3} + C$ **41.** $3x^3 + 6x^2 + 4x + C$
43. $3\ln x - \frac{5}{2}e^{2x} + \frac{2}{9}x^{9/2} + C$ **45.** $14x^{1/2} - \frac{2}{15}e^{5x} - 8\ln x + C$
47. $f(x) = \frac{1}{2}x^2 - 3x + 13$ **49.** $f(x) = \frac{1}{3}x^3 - 4x + 7$
51. $f(x) = \frac{5}{3}x^3 + \frac{3}{2}x^2 - 7x + 9$ **53.** $f(x) = x^3 - \frac{5}{2}x^2 + x + 4$
55. $f(x) = \frac{5}{2}e^{2x} - 2$ **57.** $f(x) = 8x^{1/2} - 13$
59. $D(t) = -270.1t^3 + 865.15t^2 + 3648t + 41,267$
61. $C(x) = \dfrac{x^4}{4} - x^2 + 7000$ **63.** **(a)** $R(x) = \dfrac{x^3}{3} - 3x$;   **(b)** ✎
65. $D(x) = \dfrac{4000}{x} + 3$ **67.** **(a)** $E(t) = 32 + 30t - 5t^2$;
(b) $E(3) = 77\%, E(5) = 57\%$ **69.** **(a)** $I(t) = -3.17t^2 + 141.6t + 1408$;   **(b)** 930 people;   **(c)** 1522 people;   **(d)** 348 people
71. **(a)** $h(t) = -16t^2 + 75t + 30$;   **(b)** $h(2) = 116$ ft, $h'(2) = 11$ ft/sec;   **(c)** $t = \dfrac{75}{32} \approx 2.344$ sec;   **(d)** $h(2.344) \approx 117.89$ ft;
(e) 5.06 sec;   **(f)** $h'(5.06) = -86.92$ ft/sec
73. $f(t) = \frac{2}{3}t^{3/2} + 2t^{1/2} - \frac{28}{3}$ **75.** $\frac{25}{7}t^7 + \frac{20}{3}t^6 + \frac{16}{5}t^5 + C$
77. $\frac{2}{3}t^{3/2} - \frac{2}{5}t^{5/2} + C$ **79.** $\dfrac{x^2}{2} - 6\ln x + \frac{7}{2}x^{-2} + C$
81. $\dfrac{1}{\ln 10} \cdot \ln x + C$, or $\log x + C$ **83.** $3x^4 - \frac{8}{3}x^3 - \frac{17}{2}x^2 - 5x + C$ **85.** $\dfrac{x^2}{2} - x + C$ **87.** ✎

Exercise Set 4.2, p. 407

1. $1060   **3.** 46,800¢, or $468   **5.** $-$255,000   **7.** $8400
9. 23,302.4¢, or $233.02   **11.** $471.96   **13.** $\sum\limits_{i=1}^{6} 3i$
15. $\sum\limits_{i=1}^{4} f(x_i)$ **17.** $\sum\limits_{i=1}^{15} G(x_i)$ **19.** $2^1 + 2^2 + 2^3 + 2^4$, or 30
21. $f(x_1) + f(x_2) + f(x_3) + f(x_4) + f(x_5)$ **23.** **(a)** 1.4914;
(b) 1.1418   **25.** 3,166,250¢, or $31,662.50   **27.** 247.68
29. 124   **31.** 4   **33.** 12   **35.** $\frac{9}{2}$ **37.** 25   **39.** 8
41. **(a)** 4;   **(b)** 1;   **(c)** 2;   **(d)** $\frac{9}{2}$;   **(e)** $\frac{23}{2}$ **43.** 37.96
(Exact area is $\frac{1}{2} \cdot 25\pi$.)

Technology Connection, p. 413

1. $\dfrac{32}{3}$ **2.** $\dfrac{9}{4}$ **3.** $\dfrac{5 - \ln 6}{6} \approx 0.535$ **4.** ~ 1.59359

5. 313.24

Technology Connection, p. 418

1. 0   **2.** 13.75   **3.** 0.535   **4.** 27.972   **5.** -260

Exercise Set 4.3, p. 421

1. 8   **3.** 8   **5.** $41\frac{2}{3}$ **7.** $\frac{1}{4}$ **9.** $10\frac{2}{3}$ **11.** $e^3 - 1 \approx 19.086$
13. $3 \ln 6 \approx 5.375$ **15.** Total cost, in dollars, for t days
17. Total number of kilowatts used in t hours   **19.** Total revenue,
in dollars, for x units produced   **21.** Total amount of the drug, in
milligrams, in v cubic centimeters of blood   **23.** Total number
of words memorized in t minutes   **25.** 4   **27.** $9\frac{5}{6}$ **29.** 12
31. $e^5 - e^{-1}$, or approximately 148.045   **33.**   **35.** 0; the
area above the x-axis is the same as the area below the x-axis.
37. 0; the area above the x-axis is the same as the area below it.
39–42. Left to the student   **43.** 40   **45.** $\frac{5}{3}$ **47.** $\frac{637}{6}$
49. $e^2 - e^{-5}$, or approximately 7.382   **51.** $\dfrac{b^3 - a^3}{6}$
53. $\dfrac{e^{2b} - e^{2a}}{2}$ **55.** $\dfrac{e^2 + 1}{2}$, or approximately 4.195
57. $\frac{8}{3}$ **59.** $628.56   **61.** $29.13   **63. (a)** $2948.26;
(b) $2913.90   **65.** $7627.28 billion   **67.** 18.69 hr; 20.12 hr
69. 7 words   **71.** About 5 words   **73.** $s(t) = t^3 + 4$
75. $v(t) = 2t^2 + 20$ **77.** $s(t) = -\dfrac{t^3}{3} + 3t^2 + 6t + 10$
79. (a) 104.17 m; **(b)** 229.17 m   **81. (a)** 60 mph;
(b) $\frac{1}{8}$ mi   **83. (a)** 16.67 km/hr; **(b)** 0.1875 km
85. $s(t) = -16t^2 + v_0 t + s_0$ **87.** $\frac{1}{4}$ mi   **89.** 148 mi
91. On the 10th day   **93.** 3.5   **95.** $359\frac{7}{15}$ **97.** 6.75
99. 30   **101.** $5\frac{1}{3}$ **103.** $14\frac{2}{3}$ **105.**   **107.** 4068.789
109. 7.571   **111.** 9.524   **113.** 10.987

Technology Connection, p. 429

1. $\frac{4}{3}$ **2.** Left to the student

Technology Connection, p. 432

1.

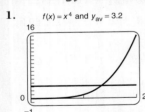

$f(x) = x^4$ and $y_{av} = 3.2$

Over the interval $[0, 2]$, the areas
under $f(x) = x^4$ and $y_{av} = 3.2$
are equal.

Exercise Set 4.4, p. 433

1. 22   **3.** $18\frac{1}{6}$ **5.** $89\frac{11}{12}$ **7.** 5   **9.** $\frac{7}{2}$ **11.** $x = -3$ and $x = 3$
13. $x = \dfrac{3 \pm \sqrt{29}}{2}$, or approximately $x = -1.193$ and $x = 4.193$
15. $x = -3$ and $x = 5$ **17.** $\frac{3}{2}$ **19.** 62.5   **21.** $\frac{1}{4}$ **23.** $4\frac{1}{2}$
25. $20\frac{5}{6}$ **27.** $4\frac{1}{2}$ **29.** $\frac{3}{10}$ **31.** $41\frac{2}{3}$ **33.** $10\frac{2}{3}$ **35.** 3
37. $85\frac{1}{3}$ **39.** $\frac{8}{3}$ **41.** $-e^{-1} + 1$, or approximately 0.632
43. $\frac{16}{3}$ **45.** $2a + 5$ **47.** $\dfrac{2^{n+1} - 1}{n + 1}$ **49. (a)** $2,201,556.58;
(b) $220,155.66   **51.** $26,534.37   **53.** $32,781.35
55. (a) Ben; **(b)** 2 more words; **(c)** 0.7 word per minute;
(d) 0.9 word per minute   **57. (a)** 90 words per minute;
(b) 96 words per minute, at $t = 1$ min; **(c)** 70 words per minute
59. (a) 42.03 μg/mL; **(b)** 22.44 μg/mL   **61. (a)** 31.7°;

(b) $-10°$; **(c)** 46.25°   **63.** $40\frac{8}{15}$ **65.** 16   **67.** 6   **69.** 4
71. 4   **73.** 5.886   **75.** 0.237

Technology Connection, p. 441

1. 4.673

Exercise Set 4.5, p. 443

1. $\frac{1}{6}(8 + x^3)^6 + C$ **3.** $\frac{1}{16}(x^2 - 6)^8 + C$ **5.** $\frac{1}{24}(3t^4 + 2)^2 + C$
7. $\ln(2x + 1) + C$ **9.** $\frac{1}{4}(\ln x)^4 + C$ **11.** $\frac{1}{3}e^{3x} + C$
13. $3e^{x/3} + C$ **15.** $\frac{1}{5}e^{x^5} + C$ **17.** $-\frac{1}{2}e^{-t^2} + C$
19. $\frac{1}{2}\ln(5 + 2x) + C$ **21.** $\frac{1}{3}\ln(12 + 3x) + C$
23. $-\ln(1 - x) + C$ **25.** $\frac{1}{12}(t^2 - 1)^6 + C$
27. $\frac{1}{8}(x^4 + x^3 + x^2)^8 + C$ **29.** $\ln(4 + e^x) + C$
31. $(\ln x)^2 + C$ **33.** $\ln(\ln x) + C$ **35.** $\dfrac{1}{3a}(ax^2 + b)^{3/2} + C$
37. $\dfrac{P_0}{k}e^{kt} + C$ **39.** $\dfrac{1}{24(2 - x^4)^6} + C$ **41.** $\frac{5}{6}(1 + 6x^2)^{6/5} + C$
43. $e - 1$ **45.** $\frac{21}{4}$ **47.** $\ln 5$ **49.** $\ln 19$ **51.** $1 - e^{-b}$
53. $1 - e^{-mb}$ **55.** $\frac{208}{3}$ **57.** $\frac{1640}{6561}$ **59.** $\frac{315}{8}$
61. Left to the student   **63.** $\frac{3}{2}x - \frac{3}{4}\ln(2x + 1) + C$
65. $x + 5\ln(x - 2) + C$
67. $\frac{1}{13}(x + 1)^{13} - \frac{1}{6}(x + 1)^{12} + \frac{1}{11}(x + 1)^{11} + C$
69. $\frac{2}{7}(x - 2)^{7/2} + \frac{8}{5}(x - 2)^{5/2} + \frac{8}{3}(x - 2)^{3/2} + C$
71. $D(x) = 2000\sqrt{25 - x^2} + 5000$ **73.** $P(x) = \dfrac{1500}{x^2 - 6x + 10}$
75. $5\frac{1}{3}$ **77.** $\dfrac{1}{a}\ln(ax + b) + C$ **79.** $2e^{\sqrt{t}} + C$
81. $\frac{1}{100}(\ln x)^{100} + C$ **83.** $\frac{1}{2}(e^t + 2)^2 + C$
85. $\frac{4}{9}(2 + t^3)^{3/4} + C$ **87.** $\frac{1}{3}(\ln x)^3 + \frac{3}{2}(\ln x)^2 + 4\ln x + C$
89. $\frac{1}{8}[\ln(t^4 + 8)]^2 + C$ **91.** $x + \dfrac{9}{x + 3} + C$
93. $t - 4 - \ln(t - 4) + C$, or $t - \ln(t - 4) + K$, where
$K = -4 + C$ **95.** $-\ln(1 + e)^{-x} + C$ **97.** $\dfrac{(\ln x)^{n+1}}{n + 1} + C$
99. $\dfrac{1}{am}\ln(1 - ae^{-mx}) + C$ **101.** $\dfrac{5}{6(n + 1)}(2x^3 - 7)^{n+1} + C$

Technology Connection, p. 449

1. 1.941

Exercise Set 4.6, p. 452

1. $xe^{4x} - \frac{1}{4}e^{4x} + C$ **3.** $\dfrac{x^6}{2} + C$ **5.** $\frac{1}{5}xe^{5x} - \frac{1}{25}e^{5x} + C$
7. $-\frac{1}{2}xe^{-2x} - \frac{1}{4}e^{-2x} + C$ **9.** $\dfrac{x^3 \ln x}{3} - \dfrac{x^3}{9} + C$
11. $\frac{1}{4}x^2 \ln x - \frac{1}{8}x^2 + C$ **13.** $(x + 5)\ln(x + 5) - x + C$
15. $\left(\dfrac{x^2}{2} + 2x\right)\ln x - \dfrac{x^2}{4} - 2x + C$
17. $\left(\dfrac{x^2}{2} - x\right)\ln x - \dfrac{x^2}{4} + x + C$
19. $\frac{2}{3}x(x + 2)^{3/2} - \frac{4}{15}(x + 2)^{5/2} + C$
21. $\dfrac{x^4}{4}\ln(2x) - \dfrac{x^4}{16} + C$, or $\dfrac{x^4 \ln 2}{4} - \dfrac{x^4 \ln x}{4} + \dfrac{x^4}{16} + C$
23. $x^2 e^x - 2xe^x + 2e^x + C$ **25.** $\frac{1}{2}x^2 e^{2x} - \frac{1}{2}xe^{2x} + \frac{1}{4}e^{2x} + C$
27. $-\frac{1}{2}x^3 e^{-2x} - \frac{3}{4}x^2 e^{-2x} - \frac{3}{4}xe^{-2x} - \frac{3}{8}e^{-2x} + C$
29. $\frac{1}{3}(x^4 + 4)e^{3x} - \frac{4}{9}x^3 e^{3x} + \frac{4}{9}x^2 e^{3x} - \frac{8}{27}xe^{3x} + \frac{8}{81}e^{3x} + C$
31. $\frac{8}{3}\ln 2 - \frac{7}{9}$ **33.** $14\ln 14 - 10\ln 10 - 4$ **35.** 1
37. $\frac{1192}{15}$ **39.** $C(x) = \frac{8}{3}x(x + 3)^{3/2} - \frac{16}{15}(x + 3)^{5/2}$
41. (a) $-10Te^{-T} - 10e^{-T} + 10$; **(b)** about 9.084 kW-h

43. $\frac{2}{125}(5x + 1)^{5/2} - \frac{2}{75}(5x + 1)^{3/2} + C$; they are the same.

45. $2\sqrt{x}e^{\sqrt{x}} - 2e^{\sqrt{x}} + C$ **47.** $\frac{2}{3}x^{3/2}\ln x - \frac{4}{9}x^{3/2} + C$

49. $2\sqrt{x}(\ln x) - 4\sqrt{x} + C$

51. $\frac{2}{7}(27x^3 + 83x - 2)(3x + 8)^{7/6}$
$- \frac{4}{91}(81x^2 + 83)(3x + 8)^{13/6} + \frac{1296}{1729}x(3x + 8)^{19/6}$
$- \frac{2592}{43,225}(3x + 8)^{25/6} + C$

53. $\frac{x^{n+1}}{n + 1}(\ln x)^2 - \frac{2x^{n+1}}{(n + 1)^2}\ln x + \frac{2x^{n+1}}{(n + 1)^3} + C$

55. Let $u = x^n$ and $dv = e^x\, dx$. Then $du = nx^{n-1}\, dx$ and $v = e^x$. Next, use integration by parts. **57.** ✏️ **59.** About 355,986

Exercise Set 4.7, p. 457

1. $-\frac{1}{9}e^{-3x}(3x + 1) + C$ **3.** $\frac{6^x}{\ln 6} + C$

5. $\frac{1}{10}\ln\left|\frac{5 + x}{5 - x}\right| + C$ **7.** $3 - x - 3\ln|3 - x| + C$

9. $\frac{1}{8(8 - x)} + \frac{1}{64}\ln\left|\frac{x}{8 - x}\right| + C$

11. $(\ln 3)x + x\ln x - x + C$ **13.** $\frac{x^5}{5}(\ln x) - \frac{x^5}{25} + C$

15. $\frac{x^4}{4}(\ln x) - \frac{x^4}{16} + C$ **17.** $\ln|x + \sqrt{x^2 + 7}| + C$

19. $\frac{2}{5 - 7x} + \frac{2}{5}\ln\left|\frac{x}{5 - 7x}\right| + C$ **21.** $-\frac{5}{4}\ln\left|\frac{x - 1/2}{x + 1/2}\right| + C$

23. $m\sqrt{m^2 + 4} + 4\ln|m + \sqrt{m^2 + 4}| + C$

25. $\frac{5}{2x^2}(\ln x) + \frac{5}{4x^2} + C$ **27.** $x^3e^x - 3x^2e^x + 6xe^x - 6e^x + C$

29. $\frac{1}{15}(3x - 1)(1 + 2x)^{3/2} + C$

31. $S(x) = 100\left[\frac{20}{20 - x} + \ln(20 - x)\right]$

33. $-4\ln\left|\frac{x}{3x - 2}\right| + C$ **35.** $\frac{-1}{2(x - 2)} + \frac{1}{4}\ln\left|\frac{x}{x - 2}\right| + C$

37. $\frac{-3}{e^{-x} - 3} + \ln|e^{-x} - 3| + C$

Chapter Review Exercises, p. 466

1. True **2.** False **3.** True **4.** False **5.** (e) **6.** (d)
7. (a) **8.** (f) **9.** (b) **10.** (c) **11.** $77,000
12. $4x^5 + C$ **13.** $3e^x + 2x + C$ **14.** $t^3 + \frac{5}{2}t^2 + \ln t + C$
15. 9 **16.** 21 **17.** Total number of words keyboarded in
t minutes **18.** Total sales in t days

19. $\frac{b^6 - a^6}{6}$ **20.** $-\frac{2}{5}$ **21.** $e - \frac{1}{2}$ **22.** $2\ln 4$, or $4\ln 2$

23. $\frac{22}{3}$ **24.** Zero **25.** Negative **26.** Positive
27. $13\frac{1}{2}$ **28.** $\frac{1}{4}e^{x^4} + C$ **29.** $\ln(4t^6 + 3) + C$
30. $\frac{1}{4}(\ln 4x)^2 + C$ **31.** $-\frac{2}{3}e^{-3x} + C$ **32.** $xe^{3x} - \frac{1}{3}e^{3x} + C$
33. $-\frac{2x}{3} + x\ln x^{2/3} + C$ **34.** $x^3\ln x - \frac{x^3}{3} + C$

35. $e^{3x}\left(\frac{1}{3}x^4 - \frac{4}{9}x^3 + \frac{4}{9}x^2 - \frac{8}{27}x + \frac{8}{81}\right) + C$

36. $\frac{1}{14}\ln\left|\frac{7 + x}{7 - x}\right| + C$ **37.** $\frac{1}{5}x^2e^{5x} - \frac{2}{25}xe^{5x} + \frac{2}{125}e^{5x} + C$

38. $\frac{1}{49} + \frac{x}{7} - \frac{1}{49}\ln|7x + 1| + C$

39. $\ln|x + \sqrt{x^2 - 36}| + C$ **40.** $x^7\left(\frac{\ln x}{7} - \frac{1}{49}\right) + C$

41. $\frac{1}{64}e^{8x}(8x - 1) + C$ **42.** About $70,666.67
43. $\frac{1}{2}(1 - 3e^{-2})$, or approximately 0.297 **44.** 80 mi
45. About $162,753.79
46. $10x^3e^{0.1x} - 300x^2e^{0.1x} + 6000xe^{0.1x} - 60,000e^{0.1x} + C$

47. $\ln|4t^3 + 7| + C$ **48.** $\frac{2}{75}(5x - 8)\sqrt{4 + 5x} + C$
49. $e^{x^5} + C$ **50.** $\ln(x + 9) + C$ **51.** $\frac{1}{96}(t^8 + 3)^{12} + C$

52. $x\ln(7x) - x + C$ **53.** $\frac{x^2}{2}\ln(8x) - \frac{x^2}{4} + C$

54. $\frac{1}{10}\left[\ln|t^5 + 3|\right]^2 + C$ **55.** $-\frac{1}{2}\ln(1 + 2e^{-x}) + C$

56. $(\ln\sqrt{x})^2 + C$, or $\frac{1}{4}(\ln x)^2 + C$

57. $x^{92}\left(\frac{\ln x}{92} - \frac{1}{8464}\right) + C$

58. $(x - 3)\ln(x - 3) - (x - 4)\ln(x - 4) + C$

59. $\frac{1}{3(\ln x)^3} + C$ **60.** $\frac{3}{7}(x + 3)^{7/3} - \frac{9}{4}(x + 3)^{4/3} + C$

61. $\frac{1}{16}(2x + 1)^2 - \frac{1}{4}(2x + 1) + \frac{1}{8}\ln(2x + 1) + C$
62. 1.343

Chapter 4 Test, p. 468

1. $[4.2]$ 95 **2.** $[4.1]\frac{2\sqrt{3}}{3}x^{3/2} + C$ **3.** $[4.1]\frac{500}{3}x^6 + C$

4. $[4.1]$ $e^x + \ln x + \frac{8}{11}x^{11/8} + C$ **5.** $[4.3]\frac{1}{6}$ **6.** $[4.3]$ $4\ln 3$
7. $[4.3]$ Total miles run in t hours **8.** $[4.3]$ 12

9. $[4.3]\frac{1 - e^{-2}}{2}$ **10.** $[4.3]$ 1 **11.** $[4.4]\frac{61}{6}$ **12.** $[4.3]$ Positive

13. $[4.5]$ $\ln(x + 12) + C$ **14.** $[4.5]$ $-2e^{-0.5x} + C$

15. $[4.5]\frac{(t^4 + 3)^{10}}{40} + C$ **16.** $[4.6]$ $\frac{1}{5}xe^{5x} - \frac{1}{25} + C$

17. $[4.6]\frac{x^4}{4}\ln x^4 - \frac{x^4}{4} + C$, or $x^4\ln|x| - \frac{x^4}{4} + C$

18. $[4.7]\frac{2^x}{\ln 2} + C$ **19.** $[4.7]\frac{1}{7}\ln\left|\frac{x}{7 - x}\right| + C$

20. $[4.4]$ 6 **21.** $[4.4]\frac{1}{3}$ **22.** $[4.4]$ $49,000 **23.** $[4.3]$ 94 words
24. $[4.3]$ 5.4 km **25.** $[4.5]\frac{6}{7}\ln(5 + 7x) + C$
26. $[4.6]$ $x^5e^x - 5x^4e^x + 20x^3e^x - 60x^2e^x + 120xe^x - 120e^x + C$
27. $[4.5]\frac{1}{6}e^{x^6} + C$ **28.** $[4.6, 4.7]\frac{2}{3}x^{3/2}(\ln x) - \frac{4}{9}x^{3/2} + C$

29. $[4.7]\frac{1}{16}\ln\left(\frac{8 + x}{8 - x}\right) + C$

30. $[4.6, 4.7]-10x^4e^{-0.1x} - 400x^3e^{-0.1x} - 12,000x^2e^{-0.1x} - 240,000xe^{-0.1x} - 2,400,000e^{-0.1x} + C$

31. $[4.6]\frac{x^2}{2}\ln(13x) - \frac{x^2}{4} + C$

32. $[4.6]\frac{1}{15}(3x^2 - 8)(x^2 + 4)^{3/2} + C$

33. $[4.5]\frac{(\ln x)^4}{4} - \frac{4}{3}(\ln x)^3 + 5\ln x + C$

34. $[4.6]$ $(x + 3)\ln(x + 3) - (x + 5)\ln(x + 5) + C$

35. $[4.6, 4.7]\frac{3}{10}(8x^3 + 10)(5x - 4)^{2/3} - \frac{108}{125}x^2(5x - 4)^{5/3} + \frac{81}{625}x(5x - 4)^{8/3} - \frac{243}{34,375}(5x - 4)^{11/3} + C$

36. $[4.6]\frac{2}{27}(3x - 2)^{3/2} + \frac{4}{9}(3x - 2)^{1/2} + C$

37. $[4.6]$ $x + 8\ln x - \frac{16}{x} + C$

38. $[4.5]\frac{1}{\ln 5}e^{(\ln 5)x} + C$, or $\frac{5^x}{\ln 5} + C$ **39.** $[4.4]$ 16

Extended Technology Application, p. 470

1. (a) 36%; (b) 33.3 **2.** (a) 16.7%; (b) 55.5
3. (a) $f(x) = x^{1.75}$, where $0 \le x \le 1$; (b) 0.272, 27.2;
(c) ~59% **4.** (a) $f(x) = x^{2.34}$, where $0 \le x \le 1$;
(b) 0.2, 20; (c) ~15.4%; (d) ~21.9%
5. Left to the student **6.** (a) $f(x) = x^{2.64}$; (b) ~20.6%
7. (a) $f(x) = x^{1.86}$; (b) ~32.9% **8.** (a) 34.3%;
(b) $f(x) = 0.0000763(11022.2)^x$, where $0 \le x \le 1$;
(c) 0.409; (d) 0.819, 81.9; (e) ~0.8%; (f) ~79.2%

Chapter 5

Technology Connection, p. 474

1. 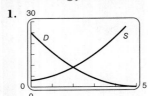 The point of intersection is $(2, 9)$; this is the equilibrium point.

Exercise Set 5.1, p. 479

1. (a) $(6, \$4)$; **(b)** \$15; **(c)** \$9 **3. (a)** $(1, \$9)$;
(b) \$3.33; **(c)** \$1.67 **5. (a)** $(3, \$9)$; **(b)** \$36; **(c)** \$18
7. (a) $(50, \$500)$; **(b)** \$12,500; **(c)** \$6250 **9. (a)** $(2, \$3)$;
(b) \$2; **(c)** \$0.35 **11. (a)** $(100, \$10)$; **(b)** \$1000;
(c) \$333.33 **13. (a)** $(0.8, \$10.24)$; **(b)** \$2.22; **(c)** \$0.98
15. (a) $(5, \$0.61)$; **(b)** \$86.36; **(c)** \$2.45 **17.**
19. (a) $(6, \$2)$; **(b)**
(c) \$7.62; **(d)** \$7.20

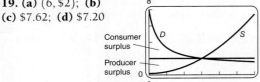

21. (a) Linear **(b)** $y = -2.5x + 22.5$;
(c) \$45; **(d)** \$24.20

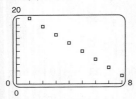

Exercise Set 5.2, p. 489

1. \$119,721.74 **3.** \$235,955.31 **5.** \$83,527.02
7. \$173,773.94 **9.** \$2,617,560 **11.** \$19,765,160
13. \$3,207,800 **15.** \$7,981,030 **17.** \$380,920
19. \$216,192 **21.** \$70,408.74 **23. (a)** \$1,321,610;
(b) \$7,995,280 **25.** \$160,777.75 **27.** A: \$598,884,
B: \$601,377; B is the better buy. **29. (a)** Crunchers:
\$2,338,910, Radars: \$2,364,760; **(b)** the difference of the
accumulated present values of the two offers, or \$25,850
31. (a) \$62,144.41; **(b)** \$4796.74 **33.** \$379,358.53
35. (a) \$6,080,740; **(b)** \$1,179,540 **37. (a)** \$2,182,290;
(b) \$538,145; **(c)** 4%: \$688,339, 6%: \$582,338, 8%: \$498,815,
10%: \$432,332 **(d)** **39.** 80.29 billion cubic meters
41. in 113.3 yr, or 2123 **43. (a)** approximately 0.0284, or 2.84%;
(b) approximately 39.76 billion barrels; **(c)** approximately
27.7 years after 2006 **45.** 16.031 lb **47.** \$535,847
49. \$732,121 **51.**

Exercise Set 5.3, p. 496

1. Convergent; $\frac{1}{2}$ **3.** Divergent **5.** Convergent; 1
7. Convergent; $\frac{1}{2}$ **9.** Divergent **11.** Convergent; 2
13. Divergent **15.** Divergent **17.** Divergent
19. Convergent; 1 **21.** Covergent; $\dfrac{1000}{\pi^{0.001}}$ **23.** Divergent
25. $\frac{1}{2}$ **27.** 1 **29.** \$51,428.57 **31.** \$6250 **33.** \$4500
35. \$62,500 **37.** \$900,000 **39.** $33,333\frac{1}{3}$ lb
41. (a) 4.20963; **(b)** 0.702858 rem; **(c)** 2.37551 rems
43. Divergent **45.** Convergent; 2 **47.** Convergent; $\frac{1}{2}$

49. $\dfrac{1}{k^2}$ is the total dose of the drug. **51.**

53. 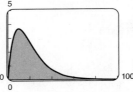 **55.** About 1.252

Exercise Set 5.4, p. 507

1. $L_4 = 18, R_4 = 34$; average $= 26$
3. $L_6 = 24.375, R_6 = 49.875$; average $= 37.125$
5. $L_8 = 0.725, R_8 = 0.663$; average $= 0.694$
7. $L_6 = 117.102, R_6 = 318.317$; average $= 217.709$
9. $L_8 = 2151.875, R_8 = 3839.875$; average $= 2995.875$
11. $M_4 = 25; -1.32\%$ **13.** $M_6 = 35.438; -1.56\%$
15. $M_8 = 0.693; 0\%$ **17.** $M_6 = 193.069; -4.05\%$
19. $M_8 = 2894.109; -1.16\%$ **21.** $T_4 = 2.977$ **23.** $T_8 = 1.325$
25. $T_5 = 25.051$ **27.** $T_4 = 1.155$ **29.** $S_4 = 5.641$
31. $S_6 = 2.160$ **33.** $S_6 = 3.142$ **35.** $S_4 = 1.432$
37. $2.958; 0.64\%$ **39.** $1.326; -0.08\%$ **41.** $25.130; -0.32\%$
43. $1.178; -1.95\%$ **45.** $5.641; 0\%$ **47.** $2.149; 0.51\%$
49. $3.142; 0\%$ **51.** $1.431; 0.07\%$ **53.** $L_8 = 2.867$ mi,
$R_8 = 3.033$ mi; average $= 2.95$ mi **55.** 55.2 ft^2
57. (a) $T_8 = 9.269$; **(b)** 37.076 **59. (a)** $M_6 = 1.397$; **(b)** $\pi/2$;
(c) -11.08% **61.** $S_6 = 7.984$ **63.** \$2106 **65. (a)** About
3455 ft^2; **(b)** about \$9500 **67. (a)** $<$; **(b)** $>$; **(c)** $>$
69. (a) $<$; **(b)** $>$; **(c)** $<$ **71.** **73.** **75.**

Exercise Set 5.5, p. 513

1. $\dfrac{\pi}{3}$, or about 1.05 **3.** $\dfrac{15\pi}{2}$, or about 23.56
5. $\dfrac{\pi}{2}(e^{10} - e^{-4})$, or about 34,599.06 **7.** $\dfrac{2\pi}{3}$, or about 2.09
9. $4\pi \ln \frac{9}{4}$, or about 10.19 **11.** 32π, or about 100.53
13. $\dfrac{32\pi}{5}$, or about 20.11 **15.** 56π, or about 175.93
17. $\dfrac{32\pi}{3}$, or about 33.51 **19.** $1,703,703.7\pi$ ft^3

21.

The graphs are semicircles. Their rotation about the x-axis creates spheres of radius 2 and r, respectively. **23.** $2\pi e^3$, or about 126.20 **25.** π

Chapter Review Exercises, p. 520

1. (c) **2.** (d) **3.** (e) **4.** (a) **5.** (b) **6.** True **7.** False
8. True **9.** False **10.** True **11.** True **12.** False
13. $(2, \$16)$ **14.** \$18.67 **15.** \$5.33 **16.** \$7195.37
17. \$6603.40 **18.** \$25,948.85 **19.** \$7919.65 per yr
20. \$639,668.38 **21.** 27.99 billion metric tons **22.** in 39.29 yr,
or 2049–2050 **23.** Convergent; 1 **24.** Divergent
25. Convergent; $\frac{1}{2}$ **26.** $L_4 = 2.591, R_4 = 4.172$;
average $= 3.381$ **27.** 6.301 **28.** 8.334 **29.** 4.300
30. (a) $\frac{2}{3}$; **(b)** $\frac{3}{4}$; 12.50% **31.** $L_9 = 1.15$ mi, $R_9 = 1.05$ mi;
average $= 1.10$ mi **32.** \$2305.63 **33.** $127\pi/7$ **34.** $\pi/6$
35. $5286\pi/5 \approx 3321.29$ ft^3 **36.** $80\pi \approx 251.33$ in^3
37. Divergent **38.** Convergent; 3 **39.** 1 **40.** 1.209

Chapter 5 Test, p. 522

1. [5.1] (3, 16)   **2.** [5.1] $45   **3.** [5.1] $22.50
4. [5.2] $18,081.81   **5.** [5.2] $55,766.35   **6.** [5.2] 563.7
million metric tons   **7.** [5.2] in 34.2 yr, or 2044
8. [5.2] $4273.39 per yr   **9.** [5.2] $1,906,391.71
10. [5.2] $1403,270.10   **11.** [5.3] Convergent; $\frac{1}{4}$
12. [5.3] Divergent   **13.** [5.4] 3.161
14. [5.4] 17.095   **15.** [5.4] 2.691   **16.** [5.4] **(a)** 36;
(b) 36.375; **(c)** -1.04% **17.** $L_7 = 5.267$ cal, $R_7 = 5.183$ cal;
average $= 5.225$ cal   **18.** [5.5] $\pi \ln 5$ **19.** [5.5] $\frac{5\pi}{2}$
20. [5.3] Convergent; $-\frac{1}{4}$ **21.** [5.3] π

Extended Technology Application, p. 524

1. $y = 0.00525488582427x^3 - 0.31949926791313x^2 + 5.2617546608767x - 8.994864719578$ **2.** 12,348.287 cm^3
3. $y = 0.00109000713314x^4 - 0.02219272885861x^3 + 0.11944088382992x^2 - 0.11585042438606x + 1.0466143607372$ **4.** 35.170535 in^3, 19.493 fl. oz
5. It seems good since the volume estimate was 19.493 fl. oz.
6. The cubic function $y = 0.00149729174531x^3 - 0.05193696325118x^2 + 0.34763221627001x + 0.68805240705899$ yields an estimated volume of 35.4635 in^3, or 19.655 fl. oz, which is a better estimate.

Chapter 6

Exercise Set 6.1, p. 532

1. $0; -14; 250$ **3.** $1; -\frac{125}{9}; 23$ **5.** $9; 66; 128$ **7.** $6; 12$
9. $\{(x, y) | y \geq 3x\}$ **11.** $\{(x, y) | y \geq 0\}$ **13.** 25.62
15. $151,571.66   **17.** **(a)** $165.70; **(b)** $143.70; **(c)** for (a), approximately $11,930.40; for (b), approximately $12,070.80; she spends less with option (a).   **19.** 244.7 mph   **21.** 1.939 m^2
23. **(a)** 65; **(b)** 62; **(c)** about 30%; **(d)**   **25.** Drops by approximately 10%   **27.**   **29.** $-10°$F   **31.** $-64°$F

33.

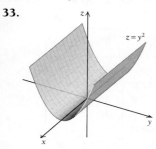

$z = y^2$

35.

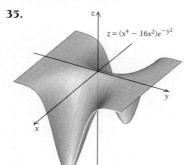

$z = (x^4 - 16x^2)e^{-y^2}$

37.

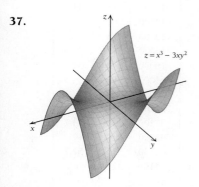

$z = x^3 - 3xy^2$

Exercise Set 6.2, p. 540

1. $2; -3; 2; -3$ **3.** $6x - 2y; -2x + 1; -6; 1$ **5.** $2 - 5y;$

$-5x; -18; -20$ **7.** $\dfrac{x}{\sqrt{x^2 + y^2}}; \dfrac{y}{\sqrt{x^2 + y^2}}; \dfrac{-2}{\sqrt{5}}; \dfrac{-2}{\sqrt{13}}$
9. $2e^{2x-y}; -e^{2x-y}$ **11.** $ye^{xy}; xe^{xy}$
13. $\dfrac{y}{x + 2y}; \dfrac{2y}{x + 2y} + \ln(x + 2y)$ **15.** $1 + \ln(xy); \dfrac{x}{y}$
17. $\dfrac{1}{y} + \dfrac{y}{3x^2}; -\dfrac{x}{y^2} - \dfrac{1}{3x}$ **19.** $12(2x + y - 5); 6(2x + y - 5)$
21. $4m^2 + 10m + 2b - 22; 3m^2 + 8mb + 10b + 26m - 56$
23. $5y - 2\lambda; 5x - \lambda; -(2x + y - 8)$
25. $2x - 10\lambda; 2y - 2\lambda; -(10x + 2y - 4)$
27. $f_{xx} = 0; f_{yy} = 0; f_{xy} = 5; f_{yx} = 5$
29. $f_{xx} = 0; f_{yy} = 14x; f_{xy} = 14y + 5; f_{yx} = 14y + 5$
31. $f_{xx} = 20x^3y^4 + 6xy^2; f_{yy} = 12x^5y^2 + 2x^3;$
$f_{xy} = 20x^4y^3 + 6x^2y; f_{yx} = 20x^4y^3 + 6x^2y$ **33.** $0; 0; 0; 0$
35. $4y^2e^{2xy}; 4xye^{2xy} + 2e^{2xy}; 4xye^{2xy} + 2e^{2xy}; 4x^2e^{2xy}$
37. $0; 0; 0; e^y$ **39.** $\dfrac{-y}{x^2}; \dfrac{1}{x}; \dfrac{1}{x}; 0$ **41.** **(a)** 614,400 units
(b) $\dfrac{\partial p}{\partial x} = 960\left(\dfrac{y}{x}\right)^{3/5}, \dfrac{\partial p}{\partial y} = 1440\left(\dfrac{x}{y}\right)^{2/5};$
(c) $\left.\dfrac{\partial p}{\partial x}\right|_{(32, 1024)} = 7680, \left.\dfrac{\partial p}{\partial y}\right|_{(32, 1024)} = 360;$ **(d)**
43. **(a)** $1.274 million;
(b) $\dfrac{\partial P}{\partial w} = -0.005075w^{-1.638}r^{1.038}s^{0.873}t^{2.468},$
$\dfrac{\partial P}{\partial r} = 0.008257w^{-0.638}r^{0.038}s^{0.873}t^{2.468},$
$\dfrac{\partial P}{\partial s} = 0.006945w^{-0.638}r^{1.038}s^{-0.127}t^{2.468},$
$\dfrac{\partial P}{\partial t} = 0.019633w^{-0.638}r^{1.038}s^{0.873}t^{1.468};$ **(c)**
45. $99.6°$F   **47.** $121.3°$F   **49.**   **51.** **(a)** $\dfrac{\sqrt{w}}{120\sqrt{h}};$
(b) $\dfrac{\sqrt{h}}{120\sqrt{w}};$ **(c)** -0.0243 m^2 **53.** 78.244   **55.** -0.846
57. $f_x = \dfrac{-4xt^2}{(x^2 - t^2)^2}; f_t = \dfrac{4x^2t}{(x^2 - t^2)^2}$
59. $f_x = \dfrac{1}{\sqrt{x}(1 + 2\sqrt{t})}; f_t = \dfrac{-1 - 2\sqrt{x}}{\sqrt{t}(1 + 2\sqrt{t})^2}$
61. $f_x = 4x^{-1/3} - 2x^{-3/4}t^{1/2} + 6x^{-3/2}t^{3/2};$
$f_t = -4x^{1/4}t^{-1/2} - 18x^{-1/2}t^{1/2}$
63. $f_{xx} = \dfrac{-6y}{x^4}; f_{xy} = \dfrac{-2}{y^3} + \dfrac{2}{x^3}; f_{yx} = \dfrac{-2}{y^3} + \dfrac{2}{x^3};$
$f_{yy} = \dfrac{6x}{y^4}$ **65.**
67. $f_{xx} = \dfrac{-2x^2 + 2y^2}{(x^2 + y^2)^2}$ and $f_{yy} = \dfrac{2x^2 - 2y^2}{(x^2 + y^2)^2}$, so $f_{xx} + f_{yy} = 0$
69. **(a)** $\lim\limits_{h \to 0} \dfrac{y(h^2 - y^2)}{h^2 + y^2} = -y;$ **(b)** $\lim\limits_{h \to 0} \dfrac{x(x^2 - h^2)}{x^2 + h^2} = x;$
(c) $f_{yx}(0, 0) = 1$ and $f_{xy}(0, 0) = -1;$ at $(0, 0)$, the mixed partial derivatives are *not* equal.

Exercise Set 6.3, p. 548

1. Relative minimum $= -\frac{1}{3}$ at $\left(-\frac{1}{3}, \frac{2}{3}\right)$
3. Relative maximum $= \frac{4}{27}$ at $\left(\frac{2}{3}, \frac{2}{3}\right)$
5. Relative minimum $= -1$ at $(1, 1)$
7. Relative minimum $= -7$ at $(1, -2)$
9. Relative minimum $= -5$ at $(-1, 2)$
11. No relative extrema   **13.** Relative minimum $= e$ at $(0, 0)$

15. 6 thousand of the $17 sunglasses and 5 thousand of the $21 sunglasses   **17.** Maximum value of $P = \$55$ million when $a = \$10$ million and $p = \$3$ **19.** The bottom measures 8 ft by 8 ft, and the height is 5 ft.

21. (a) $R = 64p_1 - 4p_1^2 - 4p_1p_2 + 56p_2 - 4p_2^2$; **(b)** $p_1 = 6$, or $\$60, p_2 = 4$, or $\$40$; **(c)** $q_1 = 32$, or 3200 units, $q_2 = 28$, or 2800 units; **(d)** \$304,000   **23.** No relative extrema; saddle point at $(0, 0)$ **25.** Relative minimum $= 0$ at $(0, 0)$; saddle points at $(2, 1)$ and $(-2, 1)$ **27.** 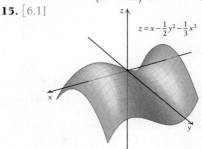   **29.** Relative minimum $= -5$ at $(0, 0)$ **31.** No relative extrema

Technology Connection p. 552

1. $x = -2, y = 3$ **2.** $x = -10, y = 37$ **3.** $x = -13, y = 19$

Exercise Set 6.4, p. 554

1. $y = 0.6x + 0.95$ **3.** $y = \frac{36}{35}x - \frac{29}{35}$ **5. (a)** $y = 0.15x + 3.95$; **(b)** \$7.70, \$8.45   **7. (a)** $y = 0.165x - 249.287$; **(b)** 83.2 yr, 84.0 yr   **9. (a)** $y = 1.07x - 1.24$; **(b)** 85%   **11.**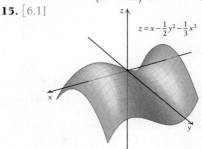
13. (a) $y = -0.0059379586x + 15.57191398$; **(b)** 3:38.2, 3:36.4; **(c)** the value predicted is 3:42.1, about a second less than the actual record.

Exercise Set 6.5, p. 564

1. Maximum $= \frac{25}{3}$ at $\left(\frac{5}{3}, 5\right)$ **3.** Maximum $= -16$ at $(2, 4)$
5. Minimum $= 20$ at $(4, 2)$ **7.** Minimum $= -96$ at $(8, -12)$
9. Minimum $= \frac{3}{2}$ at $\left(1, \frac{1}{2}, -\frac{1}{2}\right)$ **11.** 25 and 25   **13.** 3 and -3
15. $\left(\frac{3}{2}, 2, \frac{5}{2}\right)$ **17.** $9\frac{3}{4}$ in. by $9\frac{3}{4}$ in.; $95\frac{1}{16}$ in^2; no
19. $r = \sqrt[3]{\frac{27}{2\pi}} \approx 1.6$ ft, $h = 2r \approx 3.2$ ft; about 48.3 ft^2
21. Maximum value of S is 1012.5 at $L = 22.5, M = 67.5$
23. (a) $C(x, y, z) = 7xy + 6yz + 6xz$; **(b)** $x = 60$ ft, $y = 60$ ft, $z = 70$ ft; minimum cost is \$75,600   **25.** 10,000 units on A, 100 units on B   **27.** Absolute maxima at $(-1, 2, 9)$ and $(1, 2, 9)$; absolute minimum at $(0, 0, 0)$ **29.** Absolute maximum at $(2, 2, 8)$; absolute minimum at $(0, 6, -12)$
31. (a) $P(x, y) = 45x + 50y$, 180 acres of celery and 120 acres of lettuce, \$14,100 profit; **(b)** 270 acres of lettuce and zero acres of celery, \$14,100 profit   **33.** Minimum $= -\frac{155}{128}$ at $\left(-\frac{7}{16}, -\frac{3}{4}\right)$
35. Maximum $= \frac{8}{27}$ at $\left(\pm\sqrt{\frac{2}{3}}, \pm\sqrt{\frac{2}{3}}, \pm\sqrt{\frac{2}{3}}\right)$
37. Maximum $= 2$ at $\left(\frac{1}{2}, \frac{1}{2}, \frac{1}{2}, \frac{1}{2}\right)$ **39.** $\lambda = \frac{p_x}{c_1} = \frac{p_y}{c_2}$ **41.**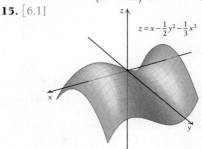
43–49. Left to the student

Exercise Set 6.6, p. 570

1. 9   **3.** 14   **5.** -10 **7.** 0   **9.** $\frac{3}{20}$ **11.** $\frac{1}{2}$ **13.** 4   **15.** $\frac{4}{15}$
17. $-\frac{1}{2}$ **19.** $\frac{108}{5}$ **21. (a)** 18,000 fireflies; **(b)** 30 fireflies/ft^2
23. 39   **25.** $\frac{13}{240}$ **27.** 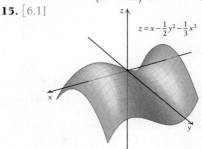   **29.** Left to the student

Chapter Review Exercises, p. 576

1. (e)   **2.** (h)   **3.** (f)   **4.** (g)   **5.** (b)   **6.** (c)   **7.** (d)   **8.** (a)
9. 1   **10.** $3y^3$ **11.** $e^y + 9xy^2 + 2$ **12.** $9y^2$ **13.** $9y^2$ **14.** 0
15. $e^y + 18xy$ **16.** $D = \{(x, y) | x \neq 1, y \geq 2\}$ **17.** $6x^2 \ln y + y^2$
18. $\frac{2x^3}{y} + 2xy$ **19.** $\frac{6x^2}{y} + 2y$ **20.** $\frac{6x^2}{y} + 2y$ **21.** $12x \ln y$
22. $\frac{-2x^3}{y^2} + 2x$ **23.** Relative minimum $= -\frac{549}{20}$ at $\left(5, \frac{27}{2}\right)$
24. Relative minimum $= -4$ at $(0, -2)$
25. Relative maximum $= \frac{45}{4}$ at $\left(\frac{3}{2}, -3\right)$
26. Relative minimum $= 29$ at $(-1, 2)$ **27. (a)** $y = \frac{3}{5}x + \frac{20}{3}$; **(b)** 9.1 million   **28. (a)** $y = 17.94x + 124.41$; **(b)** \$358
29. Minimum $= \frac{125}{4}$ at $\left(-\frac{3}{2}, -3\right)$ **30.** Maximum $= 300$ at $(5, 10)$ **31.** Absolute maximum at $(3, 0, 9)$; absolute minimum at $(0, 2, -4)$ **32.** $\frac{5}{4}$ **33.** $\frac{1}{60}$

34. (a) About 1533 students; **(b)** About 767 students/mi^2
35. 0   **36.** The cylindrical container
37.

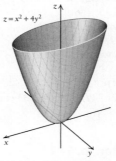

Chapter 6 Test, p. 577

1. $[6.1]$ $e^{-1} - 2$ **2.** $[6.2]$ $e^x + 6x^2y$ **3.** $[6.2]$ $2x^3 + 1$
4. $[6.2]$ $e^x + 12xy$ **5.** $[6.2]$ $6x^2$ **6.** $[6.2]$ $6x^2$ **7.** $[6.2]$ 0
8. $[6.3]$ Minimum $= -\frac{7}{16}$ at $\left(\frac{3}{4}, \frac{1}{2}\right)$ **9.** $[6.3]$ None
10. $[6.4]$ **(a)** $y = \frac{9}{2}x + \frac{17}{3}$; **(b)** \$24 million
11. $[6.5]$ Maximum $= -19$ at $(4, 5)$ **12.** $[6.6]$ 720
13. $[6.5]$ \$400,000 for labor, \$200,000 for capital
14. $[6.2]$ $f_x = \dfrac{-x^4 + 4xt + 6x^2t}{(x^3 + 2t)^2}$; $f_t = \dfrac{-2x^3 - 2x^2}{(x^3 + 2t)^2}$
15. $[6.1]$

Extended Technology Application, p. 579

1.

Case	Bldg.	n	k	A	h	$t(h, k)$
1	B1	2	40	3200	15	21.5
	B2	3	32.66	3200	30	19.4
2	B1	2	60	7200	15	31.5
	B2	3	48.99	7200	30	27.5
3	B1	4	40	6400	45	24.5
	B2	5	35.777	6400	60	23.9
4	B1	5	60	18000	60	36
	B2	10	42.426	18000	135	34.7
5	B1	5	150	112500	60	81
	B2	10	106.066	112500	135	66.5
6	B1	10	40	16000	135	33.5
	B2	17	30.679	16000	240	39.3
7	B1	10	80	64000	135	53.5
	B2	17	61.357	64000	240	54.7
8	B1	17	40	27200	240	44
	B2	26	32.344	27200	375	53.7
9	B1	17	50	42500	240	49
	B2	26	40.43	42500	375	57.7
10	B1	26	77	154154	375	76
	B2	50	55.525	154152	735	101.3

2. Yes   **3.** $t(h, k) = \dfrac{h}{10} + \dfrac{k}{2}$ **4.** $A = \left(1 + \dfrac{h}{12}\right)k^2 = 40,000$
5. About 57.7 ft by 57.7 ft by 132.2 ft (11 floors)
6. Left to the student

Chapter 7

Exercise Set 7.1, p. 594

1. Quadrant 1; $394°, 754°, -326°, -686°$, and so on
3. Quadrant 2; $21\pi/8, 37\pi/8, -11\pi/8, -27\pi/8$, and so on
5. $\pi/12$ **7.** $5\pi/12$ **9.** $-3\pi/4$ **11.** $-32\pi/45$ **13.** $270°$
15. $-45°$ **17.** $1440°$ **19.** $-900°$ **21.** About $57.296°$
23. $\sqrt{2}/2$ **25.** $\sqrt{3}/2$ **27.** 0 **29.** $-\sqrt{3}$ **31.** $\sqrt{2}$
33. -2 **35.** $(-1/2, -\sqrt{3}/2)$ **37.** $(-\sqrt{2}/2, -\sqrt{2}/2)$
39. $(-1, 0)$ **41.** $(0.901, 0.434)$ **43.** $(0.540, 0.841)$
45. $(0.848, 0.530)$ **47.** $(-0.788, -0.616)$
49. $(0.891, -0.454)$ **51.** 1.701 **53.** 0.228 **55.** -1.942
57. $\cos t = \sqrt{35}/6, \tan t = \sqrt{35}/35$
59. $\sin t = 2\sqrt{10}/7, \sec t = 7/3$ **61.** $3\sqrt{7}/32$
63. $\frac{1}{8}$ **65.** $(\sqrt{6} - \sqrt{2})/4$ **67.** $(\sqrt{6} + \sqrt{2})/4$
69. $(\sqrt{6} - \sqrt{2})/(\sqrt{6} + \sqrt{2})$ or $2 - \sqrt{3}$
71. $\sin t = 3\sqrt{58}/58, \cos t = 7\sqrt{58}/58, \tan t = 3/7$
73. $B \approx 9.043, C \approx 15.026$ **75.** $A \approx 11.375, B \approx 12.184$
77. 481 ft   **79.** 12 months   **81.** About 1 sec   **83.** $59.036°$
85. ✎   **87.** ✎   **89.** ✎   **91.** $18, 80, 82$ or $9, 40, 41$
93. $96, 110, 146$ or $48, 55, 73$

Technology Connection, p. 602

1.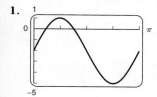
2.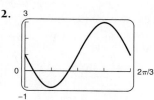
3.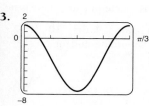
4.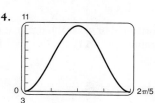
5.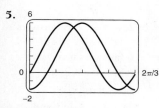
6.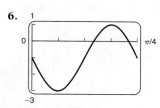

Phase shift: $-\frac{1}{3}$
(shift $\frac{1}{3}$ unit to the left)

Phase shift: $-\frac{3}{4}$
(shift $\frac{3}{4}$ unit to the left)

Exercise Set 7.2, p. 604

1. $6\cos 6x$ **3.** $-2\sin 2x$ **5.** $14x^6 \sec^2 2x^7$
7. $-2\cos x \sin x$ or $-\sin 2x$ **9.** $x\cos x + \sin x$
11. $e^x(\cos x + \sin x)$ **13.** $(x\cos x - \sin x)/x^2$
15. $\cos^2 x - \sin^2 x$ or $\cos 2x$ **17.** $(2x + 3)\cos(x^2 + 3x + 2)$
19. $2e^{2x}(\sin 2x - \cos 2x)/\sin^2 2x$ **21.** $\dfrac{\cos x}{2\sqrt{\sin x}}$, or $\frac{1}{2}\cot x \sqrt{\sin x}$
23. $2\tan x \sec^2 x$ **25.** $(\sec x^2)(1 + 2x^2 \tan x^2)$
27. $(-\csc x)(\csc x + \cot x)$ **29.** $x\sec^2 x + \tan x$
31. $e^{3x} \csc x (3 - \cot x)$ **33.** $\cos x - \sin x$
35. $\sec x \csc x$, or $1/(\cos x \sin x)$ **37.** $-2\cos x \sin x$ or $-\sin 2x$
39. $\cos x$ **41.** Midline: -5, amplitude: 2, period: $2\pi/3$, phase
shift: 0 **43.** Midline: -2, amplitude: 4, period: 8, phase shift: -1

45.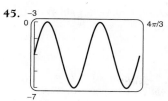
47.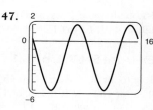

49. $y = 7 + 3\cos\left(\dfrac{\pi}{4}t\right)$ **51.** $y = -4 - 4\sin\left(\dfrac{\pi}{4}t\right)$

53. (a) $f(t) = 80 - 60\cos\left(\dfrac{\pi}{6}t\right)$; (b) $f'(t) = 31.416\sin\left(\dfrac{\pi}{6}t\right)$;
(c) $f'(3) \approx 31.4$, sales are increasing by about 31.4 rentals/
month on April 1; (d) $f'(5.5) \approx 8.13$, sales are
increasing by about 8.13 rentals/month on June 15
55. (a) $P'(t) = 0.101\sin\left(\dfrac{2\pi}{5}t\right)$; (b) $P'(5) = 0$, at the 5th
week, there is no change in price; (c) 5 weeks;
(d) 2.5 weeks, $3.64.   **57.** (a) 12.44 hr (12 hr, 26 min);
(b) $f'(t) = 14.142\sin\left(\dfrac{\pi}{6.22}t\right)$; (c) $f'(3) \approx 14.12$, tide is rising by
14.12 ft/hr; (d) 6.22 hr, 56 ft   **59.** (a) $T'(t) = 0.465\sin\left(\dfrac{2\pi}{365}t\right)$;
(b) $T'(90) \approx 0.465$, 90 days after Jan 1, the temperature
is increasing by 0.465°F/day; (c) about the 182nd or 183rd day,
84°F   **61.** $\dfrac{d}{dx}\cot x = \dfrac{d}{dx}\left(\dfrac{\cos x}{\sin x}\right) = \dfrac{-\sin^2 x - \cos^2 x}{\sin^2 x} =$
$-\dfrac{1}{\sin^2 x} = -\csc^2 x$ **63.** $\dfrac{d}{dx}\sec x = \dfrac{d}{dx}\left(\dfrac{1}{\cos x}\right) =$
$\dfrac{d}{dx}(\cos x)^{-1} = -1 \cdot (\cos x)^{-2} \cdot -\sin x = \dfrac{\sin x}{(\cos x)^2} = \tan x \sec x$
65. (a) $\dfrac{d}{dx}\sin 2x = 2\cos 2x$; (b) $\dfrac{d}{dx}\sin 2x = \dfrac{d}{dx}(2\sin x \cos x) =$
$2[\sin x \cdot (-\sin x) + \cos x(\cos x)] = 2(\cos^2 x - \sin^2 x) = 2\cos 2x$
67. (a) -1464, production decreases by about 1464 units
between November and December; (b) -2000, production
decreases by about 2000 units between December and January;
(c) ✎   **69.** Period: 2π **71.** Period: π

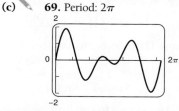
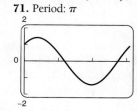

Exercise Set 7.3, p. 612

1. $\frac{1}{4}\sin^4 x + C$ **3.** $\frac{1}{6}\sin 6x + C$ **5.** $-\dfrac{2}{\pi}\cos\dfrac{\pi}{2}x + C$
7. $\dfrac{3}{2\pi}\sin 2\pi x + C$ **9.** $-\frac{1}{6}\cos 3x^2 + C$ **11.** $\frac{1}{2}\sin(x^2 - 2x) + C$
13. $-\frac{1}{2}\cos e^{2x} + C$ **15.** $\frac{1}{3}\tan^3 x + C$ **17.** $\frac{1}{10}\tan 10x + C$
19. $\frac{1}{5}\sec 5x + C$ **21.** $-\frac{1}{3}\cot 3x + C$ **23.** $-\frac{1}{6}\csc 6x + C$
25. $\frac{1}{3}x^3 + \frac{3}{2}x^2 - \frac{1}{2}\ln|\cos 2x| + C$ **27.** 1 **29.** $\sqrt{2}$ **31.** $\sqrt{3}/2$
33. $\frac{5}{2}$ **35.** $\sqrt{3}/3$ **37.** $\sqrt{3} - 1$ **39.** $\frac{1}{3}$ **41.** 0 **43.** 2π
45. $\ln|\sec x + \tan x| + C$ **47.** $-\cos x + \frac{1}{3}\cos^3 x + C$
49. (a) $\frac{1}{2}\sin^2 x + C$; (b) $-\frac{1}{2}\cos^2 x + C$; (c) $\frac{1}{2}\sin^2 x + C =$
$\frac{1}{2}(1 - \cos^2 x) + C = \frac{1}{2} - \frac{1}{2}\cos^2 x + C = -\frac{1}{2}\cos^2 x + C$
(C absorbs $\frac{1}{2}$)   **51.** $2/\pi \approx 0.637$ **53.** (a) About 125 rentals;
(b) about 42 rentals/month; (c) 960 rentals; (d) 80 rentals/month
55. About $3.65   **57.** About 45.5 °F   **59.** 405 mL
61. $0, \pi, 2\pi, 3\pi$, and so on   **63.** $-x\cos x + \sin x + C$

65. $\frac{1}{2}x \sin 2x + \frac{1}{4}\cos 2x + C$ **67.** $-\frac{1}{2}x \cos 4x + \frac{1}{8}\sin 4x + C$
69. $\pi/4$ **71.** (a) $T_4 = 1.007$; (b) 1.071 **73.** ✏

Exercise Set 7.4, p. 621

1. $\pi/6$, or $30°$ **3.** $\pi/4$, or $45°$ **5.** $\pi/2$, or $90°$ **7.** $\pi/4$, or $45°$
9. π, or $180°$ **11.** 0.305, or $17.458°$ **13.** 0.662, or $37.954°$
15. 1.742, or $99.788°$ **17.** $t \approx 0.675$, or $38.660°$; $u \approx 0.896$,
or $51.340°$ **19.** $t \approx 1.052$, or $60.255°$; $u \approx 0.251$, or $14.363°$
21. $\pi/6$ **23.** $\pi/12$ **25.** $\pi/8$ **27.** 0.730 **29.** 0.106
31. 0.208 **33.** 0.340 **35.** $-0.524 \, (-\pi/6)$, $1.571 \, (\pi/2)$
37. $t = 4$ (May)   **39.** $\pi/4$, or $45°$ **41.** 98th day
43. About $1.790°/\text{sec}$ **45.** (a) 0.9 is in the domain of
$y = \sin^{-1} x$, but 1.1 is not in the domain, so $\sin^{-1} 1.1$ is
not defined; (b) $\sin \pi/2 = 1$, which is in the domain
of $y = \sin^{-1}x$, and $\sin^{-1} 1 = \pi/2$, while $\sin \pi = 0$, but
$\sin^{-1} 0 = 0$ (π is outside the range of $y = \sin^{-1} x$);
(c) $-1 \le x \le 1$; (d) $-\pi/2 \le x \le \pi/2$ **47.** $\tan^{-1}(a/b)$ is one
angle of the right triangle and $\tan^{-1}(b/a)$ is the other angle. Since
the two angles add to $90°$, or $\pi/2$, the formula follows.
49. (a) $y = \dfrac{-4 \pm 2\sqrt{6}}{2} = -2 \pm \sqrt{6}$, so $y_1 = -2 + \sqrt{6}$
and $y_2 = -2 - \sqrt{6}$; (b) $x_1 = \sin^{-1}(-2 + \sqrt{6}) \approx 0.466$,
$x_2 = \sin^{-1}(-2 - \sqrt{6})$ is not defined since $-2 - \sqrt{6} \approx -4.45$
is outside the domain for $y = \sin^{-1} x$; (c) $x = 0.466$ and
$x = 2.675$; (d) ✏   **51.** 1.344, or about $77.006°$ **53.** ✏

Chapter Review Exercises, p. 626

1. False   **2.** True   **3.** True   **4.** True   **5.** False
6. False   **7.** False   **8.** True   **9.** True   **10.** False
11. $470°, 830°, -250°, -610°$, and so on   **12.** $4\pi/3$
13. $80°$ **14.** 48 **15.** $(-0.819, 0.574)$ **16.** $\frac{4}{5}$
17. $\sqrt{15}$ **18.** $4\sqrt{5}/9$ **19.** $B \approx 7.124$, $C \approx 17.514$
20. $y' = 12 \sin^2 4x \cos 4x$ **21.** $y' = e^{2x}(2 \cos 3x - 3 \sin 3x)$
22. $y' = 4 \tan(2x + 1) \sec^2(2x + 1)$
23. $y' = \dfrac{2 \cos 3x \cos 2x + 3 \sin 3x \, (1 + \sin 2x)}{\cos^2 3x}$
24. $y' = \frac{1}{2} \tan x \sqrt{\sec x}$ **25.** $y' = 4 \cos 4x$
26. $y = 500 - 380 \cos\left(\dfrac{\pi}{6}t\right)$ **27.** (a) $S'(3) \approx 170.2$; the
rate of sales is increasing by about 170 pairs/month on April 1;
(b) $t \approx 4.26$, in May   **28.** (a) $t = 91$ days; (b) $t = 20$ days
29. $\frac{1}{12}\sin 12x + C$ **30.** $-\frac{1}{8}\cos 4x^2 + C$ **31.** $2 \tan \frac{1}{2}x + C$
32. $x - \frac{1}{4}\cos 4x + C$ **33.** $2x - \frac{3}{5}\cos 5x + C$ **34.** 0.75
35. $-\ln \dfrac{\sqrt{3}}{2}$, or $\ln 2 - \frac{1}{2}\ln 3$ **36.** $\pi/2$ **37.** 60
38. 10,800 sunglasses   **39.** 1072.5 mm   **40.** $-\pi/4$
41. $t \approx 0.298$, or $17.103°$; $u \approx 1.272$, or $72.897°$
42. $\pi/2$, or $90°$ **43.** About $0.239°/\text{sec}$
44. $x \sin 4x + \frac{1}{4}\cos 4x + C$ **45.** $t \approx 1.131$, or $64.802°$
46. 0.308, or about $17.647°$; 2.834, or about $162.376°$
47. 2.396 **48.** 0.608

Chapter 7 Test, p. 628

1. [7.1] $8\pi/9$ **2.** [7.1] $112.5°$ **3.** [7.1] $\sqrt{3}/2$ **4.** [7.1] $\sqrt{77}/9$
5. [7.1] $\sqrt{35}/18$ **6.** [7.1] $(-0.530, -0.848)$
7. [7.1] $A \approx 26.138$, $C \approx 27.644$
8. [7.2] $y' = x^3 (3x \cos 3x + 4 \sin 3x)$
9. [7.2] $y' = 6 \tan^2 2x \sec^2 2x$
10. [7.2] $y' = -\dfrac{\tan 4x \sin x + 4 \sec^2 4x \cos x}{\tan^2 4x}$

11. [7.2] $y' = \dfrac{x \cos x^2}{\sqrt{\sin x^2}}$ **12.** [7.2] (a) $t \approx 4.565$, in February;
(b) $S'(3) \approx -136.1$, rate of sales is decreasing by about 136
jerseys/month on January 1   **13.** [7.3] $\frac{1}{2}\sin(x^2 + 2x + 4) + C$
14. [7.3] $\frac{1}{27}\sin^9 3x + C$ **15.** [7.3] $1 + \dfrac{\sqrt{2}}{2}$ **16.** [7.4] 2.389, or
about $136.886°$ **17.** [7.4] $\pi/2$ **18.** [7.4] π
19. [7.4] $t \approx 0.278$, or $15.945°$; $u \approx 1.292$, or $74.055°$
20. [7.3] (a) 300 jerseys/month; (b) 3600 jerseys
21. [7.3] About 5.180 in.   **22.** [7.4] About $12.683°/\text{sec}$
23. [7.4] $-\dfrac{2x}{5}\cos 5x + \frac{2}{25}\sin 5x + C$
24. [7.4] (a) max $= \sqrt{2}$, min $= -\sqrt{2}$; (b) left to the student
25. [7.4] $t \approx \pi/3$, or $60°$ **26.** [7.4] 1.263, or $72.365°$; 5.020,
or $287.625°$

Extended Technology Application, p. 630

1–5. Left to the student   **6.** $x = 2 + \cos t$, $y = -1 +$
$\sin t$, $0 \le t \le 2\pi$; $x = 2 + \dfrac{\sqrt{2}}{2}\cos t$, $y = -1 + \dfrac{\sqrt{2}}{2}\sin t$,
$0 \le t \le 2\pi$ **7.** (a) $t = \dfrac{\pi}{2}$, $(0, 3)$ and $t = \dfrac{3\pi}{2}$, $(0, -3)$;
(b) $t = 0$, $(2, 0)$ and $t = \pi$, $(-2, 0)$. Note that $t = 2\pi$
repeats the point $(2, 0)$.
8. (a) $t = \dfrac{\pi}{6}$, $\left(\dfrac{\sqrt{3}}{4}, \dfrac{1}{4}\right)$; $t = \dfrac{5\pi}{6}$, $\left(-\dfrac{\sqrt{3}}{4}, \dfrac{1}{4}\right)$; $t = \dfrac{3\pi}{2}$, $(0, -2)$;
(b) $t = \dfrac{7\pi}{6}$, $\left(-\dfrac{3\sqrt{3}}{4}, -\dfrac{3}{4}\right)$; $t = \dfrac{11\pi}{6}$, $\left(\dfrac{3\sqrt{3}}{4}, -\dfrac{3}{4}\right)$.
Note that at $t = \dfrac{\pi}{2}$, $\dfrac{dy}{dx} = \dfrac{0}{0}$, which is indeterminate. There is no
horizontal or vertical tangent at this point.   **9.** 6π **10.** $3\pi/2$,
about 4.712 **11.** 1 **12.** 3π **13.** Left to the student

Chapter 8

Exercise Set 8.1, p. 638

1. $y = x^5 + C$; $y = x^5 + 1$, $y = x^5 - 2$, $y = x^5 + 8$ (answers
may vary)   **3.** $y = \frac{1}{2}e^{2x} + \frac{1}{2}x^2 + C$; $y = \frac{1}{2}e^{2x} + \frac{1}{2}x^2 + 1$,
$y = \frac{1}{2}e^{2x} + \frac{1}{2}x^2 - 3$, $y = \frac{1}{2}e^{2x} + \frac{1}{2}x^2 + 6$ (answers may vary)
5. $y = 8 \ln |x| - \frac{1}{3}x^3 + \frac{1}{6}x^6 + C$; $y = 8 \ln |x| - \frac{1}{3}x^3 +$
$\frac{1}{6}x^6 - 2$, $y = 8 \ln |x| - \frac{1}{3}x^3 + \frac{1}{6}x^6 + 4$, $y = 8 \ln |x| -$
$\frac{1}{3}x^3 + \frac{1}{6}x^6 - 11$ (answers may vary)   **7.** $y' = 4 + \ln x$ and
$y'' = \dfrac{1}{x}$, so $\left(\dfrac{1}{x}\right) - \dfrac{1}{x} = 0$ **9.** $y' = 3xe^x + 4e^x$ and
$y'' = 3xe^x + 7e^x$, so $(3xe^x + 7e^x) - 2(3xe^x + 4e^x) +$
$(3xe^x + e^x) = (3xe^x - 6xe^x + 3xe^x) + (7e^x - 8e^x + e^x) =$
$xe^x(3 - 6 + 3) + e^x(7 - 8 + 1) = 0$
11. (a) $y' = -4e^{-4x}$, so $(-4e^{-4x}) + 4(e^{-4x}) = 0$;
(b) $y' = -4Ce^{-4x}$, so $(-4Ce^{-4x}) + 4(Ce^{-4x}) = 0$
13. (a) $y' = 6e^{6x}$, $y'' = 36e^{6x}$, so $(36e^{6x}) - (6e^{6x}) - 30(e^{6x}) =$
$e^{6x}(36 - 6 - 30) = 0$; (b) $y' = -5e^{-5x}$, $y'' = 25e^{-5x}$, so
$(25e^{-5x}) - (-5e^{-5x}) - 30(e^{-5x}) = e^{-5x}(25 + 5 - 30) = 0$;
(c) $y' = 6C_1e^{6x} - 5C_2e^{-5x}$, $y'' = 36C_1e^{6x} + 25C_2e^{-5x}$,
so $(36C_1e^{6x} + 25C_2e^{-5x}) - (6C_1e^{6x} - 5C_2e^{-5x}) -$
$30(C_1e^{6x} + C_2e^{-5x}) = C_1e^{6x}(36 - 6 - 30) +$
$C_2e^{-5x}(25 + 5 - 30) = 0$ **15.** (a) $M(t) = M_0e^{0.05t}$;
(b) $dM/dt = 0.05M_0e^{0.05t}$, so $0.05M_0e^{0.05t} = 0.05(M_0e^{0.05t})$

17. (a) $R(t) = R_0e^{0.35t}$; **(b)** $dR/dt = 0.35R_0e^{0.35t}$, so $0.35R_0e^{0.35t} = 0.35(R_0e^{0.35t})$ **19. (a)** $G(t) = G_0e^{0.005t}$; **(b)** $dG/dt = 0.005G_0e^{0.005t}$, so $0.005G_0e^{0.005t} = 0.005(G_0e^{0.005t})$
21. (a) $R(t) = R_0e^t$; **(b)** $dR/dt = R_0e^t$, so $R_0e^t = (R_0e^t)$
23. (a) $y = \frac{1}{3}x^3 + x^2 - 3x + 4$; **(b)** $y' = \frac{1}{3}(3x^2) + 2x - 3 = x^2 + 2x - 3$ **25. (a)** $f(x) = \frac{3}{5}x^{5/3} - \frac{1}{2}x^2 - \frac{61}{10}$; **(b)** $f'(x) = \frac{3}{5}(\frac{5}{3}x^{2/3}) - \frac{1}{2}(2x) = x^{2/3} - x$ **27. (a)** $B(t) = 500e^{0.03t}$; **(b)** $dB/dt = 15e^{0.03t}$, so $15e^{0.03t} = 0.03(500e^{0.03t})$
29. (a) $S(t) = 750e^{0.12t}$; **(b)** $dS/dt = 90e^{0.12t}$, so $90e^{0.12t} = 0.12(750e^{0.12t})$ **31. (a)** $T(t) = 50e^{0.015t}$; **(b)** $dT/dt = 0.75e^{0.015t}$, so $0.75e^{0.015t} = 0.015(50e^{0.015t})$ **33. (a)** $M(t) = 6e^t$; **(b)** $dM/dt = 6e^t$, so $6e^t = (6e^t)$ **35. (a)** $dA/dt = 0.0375A$; **(b)** $A(t) = 500e^{0.0375t}$; **(c)** $A(5) = \$603.12$, $A'(5) = \$22.62/\text{yr}$; **(d)** $22.62/603.12 = 0.0375$, the continuous growth rate **37. (a)** $A(t) = A_0e^{0.028t}$; **(b)** $A(4) = \$1.1185A_0$, $A'(4) = \$0.0313A_0/\text{yr}$; **(c)** $0.0313A_0/1.1185A_0 = 0.028$, the continuous growth rate; **(d)** $A_0/A_0 = 1$, so the initial quantity has no effect on the continuous growth rate.
39. (a) $dP/dt = 0.0175P$; **(b)** $P(t) = 17{,}000e^{0.0175t}$; **(c)** $P(10) \approx 20{,}251$ people, $P'(10) = 354.4$ people/yr; **(d)** $\frac{354.4}{20{,}251} = 0.0175$, the continuous growth rate
41. (a) $k = 0.296$, $P(t) = 24e^{0.296t}$; **(b)** $P(41) = 4{,}475{,}165$ rabbits, $P'(41) = 1{,}324{,}649$ rabbits/yr; **(c)** 0.296
43. (a) $3.25\%/\text{yr}$; **(b)** $A(t) = 2500e^{0.0325t}$; **(c)** $\$2500$
45. ✎   **47. (a)** $y = 4250e^{0.05t}$; **(b)** $dA/dt = 0.05A$, $A_0 = 4250$

Exercise Set 8.2, p. 645

1. $y = Ce^{x^4}$ **3.** $y = \sqrt[3]{4x^2 + C}$ **5.** $y = \pm\sqrt{2x^2 + C}$
7. $y = \pm\sqrt{12x + C}$ **9.** $y = 8e^{x^2/2} - 3$ **11.** $y = \sqrt[3]{15x - 3}$
13. (a) $dy/dx = y^2$, or $y' = y^2$; **(b)** $y = \dfrac{1}{C - x}$
15. (a) $dy/dx = \dfrac{1}{y^3}$, or $y' = y^{-3}$; **(b)** $y = \pm\sqrt[4]{4x + C}$
17. (a) $dy/dx = xy$, $y(2) = 3$; **(b)** $y = 0.406e^{x^2/2}$
19. (a) $I = Ce^{hkt}$; **(b)** $I = I_0e^{hkt}$ **21.** $V = -4.81e^{-kt} + 24.81$
23. $q = 2e^{(4/x)-1}$ **25.** $q = \dfrac{C}{x^2}$ **27. (a)** $P = Ce^{kt}$; **(b)** $P = P_0e^{kt}$
29. $y = -\dfrac{4}{4x^5 + x^4 + C}$ **31.** ✎   **33.** $y = \pm\sqrt{10x + C}$

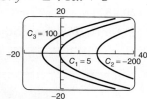

Exercise Set 8.3, p. 654

1. (a) $P(t) = \dfrac{500{,}000}{100 + 4900e^{-t}}$; **(b)** 4.739; **(c)** $(3.892, 2500)$
3. (a) $P(t) = \dfrac{75{,}000}{50 + 1450e^{-0.3t}}$; **(b)** 18.548; **(c)** $(11.224, 750)$
5. (a) 2000; **(b)** $k = 0.0000732$, $P(t) = \dfrac{400{,}000}{200 + 1800e^{-0.146t}}$;
(c) 13.7 months   **7. (a)** 320; **(b)** $k = 0.000764$,
$P(t) = \dfrac{19{,}200}{60 + 260e^{-0.244t}}$; **(c)** 17.1 yr
9. (a) $P(t) = 3000(1 - e^{-0.00813t})$; **(b)** 283.221
11. (a) $P(t) = 80(1 - e^{-0.0578t})$; **(b)** 27.845

13. (a) $P(t) = \dfrac{270{,}000}{60 + 4440e^{-0.0315t}}$; **(b)** 4500; **(c)** in 206.4 months
15. (a) $P(t) = \dfrac{3400}{17 + 183e^{-0.0982t}}$; **(b)** in 24.198 yr, or about 1814–1815 **17. (a)** $M(t) = 200(1 - e^{-0.0445t})$; **(b)** 20.59 hr
19. (a) $P(t) = \dfrac{10{,}000}{5 + 1995e^{-0.27t}}$; **(b)** 27.32 days
21. (a) $P(t) = \dfrac{400{,}000}{20 + 19{,}980e^{-0.27t}}$; **(b)** 28.719 weeks
23. (a) $P(t) = \dfrac{8{,}000{,}000{,}000}{2000 + 3{,}998{,}000e^{-0.129t}}$; **(b)** 58.918 weeks;
(c) 67.434 weeks   **25. (a)** $P(t) = 10(1 - e^{-0.0813t})$; **(b)** 14.809 months **27. (a)** About 28 states; **(b)** about 38 states; **(c)** about 44 states; **(d)** 50 states; **(e)** ✎
29. (a) $\dfrac{1}{P(L - P)} = \dfrac{A}{P} + \dfrac{B}{L - P} = \dfrac{A(L - P) + BP}{P(L - P)}$

$= \dfrac{(B - A)P + AL}{P(L - P)}$. Thus, $AL = 1$, so $A = 1/L$, and since

$B - A = 0$, $B = 1/L$. Substituting gives $\dfrac{1}{P(L - P)} =$

$\dfrac{1/L}{P} + \dfrac{1/L}{L - P} = \dfrac{1}{L}\left(\dfrac{1}{P} + \dfrac{1}{L - P}\right)$.

(b) $\displaystyle\int \dfrac{1}{P(L - P)}\,dP = \int \dfrac{1}{L}\left(\dfrac{1}{P} + \dfrac{1}{L - P}\right)dP$

$= \dfrac{1}{L}\displaystyle\int\left(\dfrac{1}{P} + \dfrac{1}{L - P}\right)dP = \dfrac{1}{L}\ln P -$

$\dfrac{1}{L}\ln(L - P) = \dfrac{1}{L}\ln\left(\dfrac{P}{L - P}\right)$

31. (a) $P(t) = \dfrac{32.889}{1 + 2.289e^{-0.634t}}$; **(b)** about 33 students; **(c)** ✎

Exercise Set 8.4, p. 662

1. $y = Ce^{-5x}$ **3.** $y = Ce^{-2x} + \frac{1}{2}$ **5.** $y = Ce^{-3x/2} + \frac{1}{3}$
7. $y = Ce^{x^2} - \frac{1}{2}$ **9.** $y = \frac{1}{3}x + \dfrac{C}{x^2}$ **11.** $y = \frac{1}{7}x^2 + \dfrac{6}{7x^5}$
13. $y = \frac{7}{2}e^{-x^2} + \frac{1}{2}$ **15.** $y = \frac{1}{4}x^3 + \dfrac{6}{x}$
17. (a) $T(t) = 22 + 178e^{-0.00322t}$; **(b)** after 256.2 min
19. (a) $A(t) = 1000 - 900e^{-0.006t}$; **(b)** about 562 lb
21. (a) $V(t) = 12{,}000 + 10{,}500e^{-0.2t}$; **(b)** 8.3 yr; **(c)** ✎
23. (a) $S(t) = 40{,}000 - 25{,}000e^{-0.0005t}$; **(b)** $15{,}150$ senior citizens; **(c)** about 446 months   **25. (a)** $y = Ce^{-4x}$; **(b)** $y = Ce^{-4x}$;
(c) ✎   **27. (a)** $x^2y = C$; **(b)** $y' + \dfrac{2}{x}y = 0$, $y = \dfrac{C}{x^2}$; **(c)** ✎
29. (a) $x^2y + 3y^2 = C$; **(b)** ✎; **(c)** left to the student
31. $y = x - 1 + Ce^{-x}$ **33.** $y = \dfrac{x + C}{x^2 + 1}$

Exercise Set 8.5, p. 670

1. $y = C_1e^{-5x} + C_2e^{4x}$ **3.** $y = C_1e^{-8x} + C_2e^{3x}$
5. $y = C_1e^{-5x} + C_2e^{5x}$ **7.** $y = C_1 + C_2e^{10x}$
9. $y = C_1e^{(-1/2)x} + C_2e^{2x}$ **11.** $y = C_1 + C_2e^{-2x} + C_3e^{-x}$
13. $y = C_1 + C_2e^{-6x} + C_3e^{6x}$ **15.** $y = C_1e^{-4x} + C_2xe^{-4x}$
17. $y = C_1e^{8x} + C_2xe^{8x}$ **19.** $y = -2e^{-3x} + 3e^{-x}$
21. $y = 2e^{6x} + 1$ **23.** $y = C_1e^{-5x} + C_2e^{-x} + C_3e^{3x}$
25. $y = C_1e^{-4x} + C_2e^{3x} + C_3e^{4x}$

27. $y = C_1e^{-x} + C_2xe^{-x} + C_3e^{2x}$
29. $y = C_1e^{-5x} + C_2e^{-x} + C_3e^x + C_4e^{5x}$
31. $y = C_1e^{-1.618x} + C_2e^{0.618x} + C_3e^{3x}$
33. $y = C_1e^{-2x} + C_2e^{-x} + \frac{1}{2}x^2 - \frac{3}{2}x + \frac{9}{4}$
35. $y = C_1e^{-6x} + C_2e^{6x} - \frac{5}{36}x + \frac{7}{36}$
37. $y = C_1 + C_2e^{9x} - \frac{1}{18}x^2 - \frac{46}{81}x$
39. $y = C_1 \sin 2x + C_2 \cos 2x$
41. $y = C_1 \sin \sqrt{2}x + C_2 \cos \sqrt{2}x$
43. $y = C_1e^{-3x/2} \sin \frac{\sqrt{19}}{2}x + C_2e^{-3x/2} \cos \frac{\sqrt{19}}{2}x$
45. $y = \frac{1}{2} \sin 4x + \cos 4x$
47. $y = C_1 \sin 3x + C_2 \cos 3x + \frac{1}{9}x^2 + \frac{8}{9}x - \frac{20}{81}$
49. $y = C_1e^{-x/2} \sin \frac{\sqrt{23}}{2}x + C_2e^{-x/2} \cos \frac{\sqrt{23}}{2}x + \frac{1}{2}x - \frac{1}{12}$
51. (a) $y = 10e^{-0.2t} + 40e^{0.1t}$; **(b)** about $75.90;
(c) about $6.69/month **53.** $y' = C_1f_1'(x) + C_2f_2'(x)$ and
$y'' = C_1f_1''(x) + C_2f_2''(x)$, so $a_0(C_1f_1''(x) + C_2f_2''(x)) + a_1(C_1f_1'(x) + C_2f_2'(x)) + a_2(C_1f_1(x) + C_2f_2(x)) = C_1(a_0f_1''(x) + a_1f_1'(x) + a_2f_1(x)) + C_2(a_0f_2''(x) + a_1f_2'(x) + a_2f_2(x)) = C_1(0) + C_2(0) = 0$
55. (a) $y'' + 4y' - 21y = 0$; **(b)**
57. $y = C_1e^{-3x} + C_2e^{-x} + C_3e^x + C_4e^{2x}$
59. $y = C_1e^{-3x} + C_2e^x + C_3xe^x + C_4e^{3x}$
61. $y = C_1e^{-2x} + C_2xe^{-2x} + C_3x^2e^{-2x} + C_4e^{5x}$
63. $y = C_1e^{-3.193x} + C_2e^{-x} + C_3e^x + C_4e^{2.193x}$
65. $y = C_1e^{-4.162x} + C_2e^{-3x} + C_3e^{2x} + C_4e^{2.162x}$
67. (a) $y' = e^x(v'(x) + v(x)), y'' = e^x(v''(x) + 2v'(x) + v(x))$; **(b)**
$v''(x)e^x = 0$; **(c)** $v''(x) = 0$; **(d)** $v'(x) = \int v''(x)dx = C$,
$v(x) = \int v'(x)dx = \int C\,dx = Cx + D$. Letting $C = 1$ and
$D = 0$, we have $v(x) = x$. **(e)** Second solution is $y = v(x)e^x = xe^x$;
general solution is $y = C_1e^x + C_2xe^x$.
69. $y = C_1e^{-4x} + C_2e^{8x} - \frac{1}{9}e^{-x}$
71. $y = C_1 \sin x + C_2 \cos x - \frac{1}{3} \cos 2x$

Chapter Review Exercises, p. 676

1. (c) **2.** (a) **3.** (f) **4.** (d) **5.** (e) **6.** $y = 3x^2 + C$
7. $y = \frac{1}{3}x^3 + C$ **8.** $y = x^2 + x + 3$ **9.** $y = \frac{1}{4}x^4 - \frac{81}{4}$
10. $y = Ce^{x^4/4}$ **11.** $y = Ce^{x^2} - \frac{1}{2}$ **12.** $y = -2e^{6t}$
13. $y' = 0.05e^{0.05x}$, so $(0.05e^{0.05x}) = 0.05(e^{0.05x})$
14. $y' = 2e^{2x} - 15e^{5x}, y'' = 4e^{2x} - 75e^{5x}$. Thus,
$(4e^{2x} - 75e^{5x}) - 7(2e^{2x} - 15e^{5x}) + 10(e^{2x} - 3e^{5x}) = e^{2x}(4 - 14 + 10) + e^{5x}(-75 + 105 - 30) = 0$.
15. $y = \frac{1}{\sqrt[3]{C - 3x}}$ **16. (a)** $A(t) = 700e^{0.05t}$;
(b) $A(5) = $898.82, A'(5) = $44.94/yr$; **(c)** 0.05,
the continuous interest rate **17. (a)** 1050;
(b) $P(t) = \dfrac{210,000}{200 + 850e^{-0.0579t}}$ **18. (a)** $P(t) = \dfrac{54,000,000}{150 + 359,850e^{-0.576t}}$; **(b)** 13.5 months; **(c)** 15.9 months
19. (a) $V(t) = 22(1 - e^{-0.017t})$ **(b)** The rise in the value of a share is not expected to exceed $22. **(c)** 15.2 weeks
20. (a) $T(t) = 25 + 65e^{-0.0204t}$; **(b)** 125.7 min
21. (a) $A(t) = 5000 - 4800e^{-0.002t}$; **(b)** 1651.2 lb
22. $y = Ce^{-7x}$ **23.** $y = \frac{1}{2}e^{12x}$ **24.** $y = Ce^{-3x^2/2} + \frac{1}{3}$

25. $y = -\frac{1}{2}x^3 - \frac{1}{2}x^9$ **26.** $y = 2x + \dfrac{C}{\sqrt{x}}$ **27.** $y = Ce^{-x^4} + \frac{1}{4}$
28. $y = C_1e^{-10x} + C_2e^{7x}$ **29.** $y = \frac{1}{2}e^{-9x} + \frac{1}{2}e^{11x}$
30. $y = C_1e^{3x} + C_2xe^{3x} + C_3$
31. $y = C_1e^{-5x} + C_2e^{-2x} + \frac{3}{10}x - \frac{41}{100}$
32. ✎ **33.** $y = C_1 \sin \sqrt{10}x + C_2 \cos \sqrt{10}x$
34. $y = C_1e^{-x/2} \sin \frac{\sqrt{23}}{2}x + C_2e^{-x/2} \cos \frac{\sqrt{23}}{2}x$
35. $y = \sin x + 3 \cos x$
36. $y = C_1e^x \sin 2x + C_2e^x \cos 2x + x^2 - x - 7$
37. $y = C_1e^{-4x} + C_2e^x + C_3e^{9x}$
38. $y = C_1e^{-x} + C_2xe^{-x} + C_3e^{3x} + C_4e^{5x}$
39. $y = C_1e^{-2x} + C_2xe^{-2x} + C_3e^x + C_4e^{8x}$
40. $y = C_1e^{-2x} + C_2e^{-x} + C_3e^x + C_4e^{2x}$

Chapter 8 Test, p. 677

1. [8.1] $y = \frac{1}{3}x^3 + 5x - \frac{7}{3}$ **2.** [8.2] $y = Ce^{x^6}$
3. [8.2] $y = \pm\sqrt{9x^2 + C}$ **4.** [8.4] $y = 3x^3 - x$
5. [8.3] 2400 **6.** [8.1] $y' = \dfrac{5}{2} - 2Ce^{-2x}$, so
$\left(\dfrac{5}{2} - 2Ce^{-2x}\right) + 2\left(\dfrac{5x}{2} - \dfrac{5}{4} + Ce^{-2x}\right) = 5x$
7. [8.1] **(a)** $A(t) = 2500e^{0.0325t}$; **(b)** $A(3) = $2576.03, A'(3) = $89.57/yr; **(c)** 0.0325, the continuous interest rate
8. [8.3] **(a)** $P(t) = \dfrac{180,000,000}{600 + 299,400e^{-0.0375t}}$; **(b)** 165.7 weeks
9. [8.3] **(a)** $V(t) = 13.25(1 - e^{-0.0088t})$; **(b)** 268.3 weeks
10. [8.4] **(a)** $T(t) = 22 + 48e^{-0.049t}$; **(b)** 32.01 min
11. [8.4] **(a)** $A(t) = 600,000 - 600,000e^{-t/60,000}$; **(b)** 14,228.6 lb
12. [8.5] $y = C_1e^{-3x} + C_2e^{11x}$ **13.** [8.5] $y = \frac{5}{4}e^{-5x} + \frac{7}{4}e^{5x}$
14. [8.5] $y = C_1 + C_2e^{-x} + C_3xe^{-x}$
15. [8.5] $y = C_1 + C_2e^{-10x} + C_3e^{10x}$
16. [8.5] $y = \dfrac{x}{4} - \dfrac{1}{16} + Ce^{-4x}$
17. [8.5] $y = C_1e^{-3x} + C_2e^{4x} - x^2 - 2x - 1$
18. [8.5] $y = C_1e^{-3x/2} \sin \frac{\sqrt{27}}{2}x + C_2e^{-3x/2} \cos \frac{\sqrt{27}}{2}x$
19. [8.5] $y = 3 \sin x + 5 \cos x$
20. [8.5] $y = C_1e^{-5x} + C_2e^{2x} + C_3e^{5x}$ **21.** [8.5] ✎

Extended Technology Application, p. 679

1.

			x		
y	-2	-1	0	1	2
-2	0	1	2	3	4
-1	-1	0	1	2	3
0	-2	-1	0	1	2
1	-3	-2	-1	0	1
2	-4	-3	-2	-1	0

Graph is left to the student.

2.

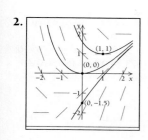

3. (a) $y_1 = 3.3, y_2 = 3.663, y_3 = 4.1026\ldots, y_4 = 4.6359\ldots,$ $y_5 = 5.2849\ldots$; **(b)** $y = 3e^{x^2/2 - 1/2}$; **(c)** $\varphi_1 = 3.3321\ldots, \varphi_2 = 3.7382\ldots, \varphi_3 = 4.2360\ldots, \varphi_4 = 4.8482\ldots, \varphi_5 = 5.6047\ldots$
4. The particular solution is $\varphi(x) = \frac{1}{2}x^2 - \frac{1}{2}x + \frac{1}{4} - \frac{9}{4}e^{4-2x}$. For a step size of 0.1, the spreadsheet looks something like this.

	A	B	C	D
1	Step size =	0.1		
2				
3	x	dy/dx	y	actual y
4	2	6	-1	-1
5	2.1	5.21	-0.4	-0.437144194
6	2.2	4.598	0.121	0.061779896
7	2.3	4.1284	0.5808	0.510173819
8	2.4	3.77272	0.99364	0.919009831
9	2.5	3.508176	1.370912	1.297271257
10	2.6	3.3165408	1.7217296	1.652313023
11	2.7	3.18323264	2.05338368	1.990156831
12	2.8	3.096586112	2.371706944	2.315732835
13	2.9	3.04726889	2.681365555	2.633077502
14	3	3.027815112	2.986092444	2.945495613

5. The particular solution is $\varphi(x) = 4e^{x^2/2 - 1/2} - 1$. For a step size of 0.1, the spreadsheet looks something like this.

	A	B	C	D
1	Step size =	0.1		
2				
3	x	dy/dx	y	actual y
4	1	4	3	3
5	1.1	4.84	3.4	3.442842441
6	1.2	5.8608	3.884	3.984306922
7	1.3	7.111104	4.47008	4.647959679
8	1.4	8.65366656	5.1811904	5.464297609
9	1.5	10.56983558	6.046557056	6.47298383
10	1.6	12.96586498	7.103540614	7.725889062
11	1.7	15.98018209	8.400107113	9.291253514
12	1.8	19.79662558	9.998125322	11.25941681
13	1.9	24.65779697	11.97778788	13.75075637
14	2	30.88713515	14.44356758	16.92675628

Chapter 9

Exercise Set 9.1, p. 689

1. (a) 11, 14, 17, 20; **(b)** 68; **(c)** 1175 **3. (a)** 8, 9, 10, 11;
(b) 27; **(c)** 500 **5. (a)** $-1, -3, -5, -7$; **(b)** -39; **(c)** -625
7. (a) $2, \frac{5}{2}, 3, \frac{7}{2}$; **(b)** $\frac{23}{2}$; **(c)** 225 **9. (a)** 3.7, 5.3, 6.9, 8.5;
(b) 34.1; **(c)** 572.5 **11. (a)** 12; **(b)** $a_n = -5 + 12n$; **(c)** 5430
13. (a) 4; **(b)** $a_n = 96 + 4n$; **(c)** 4740 **15. (a)** -2;
(b) $a_n = 82 - 2n$; **(c)** 1530 **17. (a)** 3.5; **(b)** $a_n = 4.5 + 3.5n$;
(c) 1762.5 **19. (a)** -0.15; **(b)** $a_n = 4.15 - 0.15n$; **(c)** 54.75
21. (a) $a_n = 1 + 3n$; **(b)** 4, 7, 10, 13, 16; $a_{50} = 151$; **(c)** 3875
23. (a) $a_n = 209 - 9n$; **(b)** 200, 191, 182, 173, 164; $a_{50} = -241$;
(c) -1025 **25. (a)** $a_n = 2 + 6n$; **(b)** 8, 14, 20, 26, 32; $a_{50} = 302$;
(c) 7750 **27. (a)** $a_n = 124 - 3n$; **(b)** 121, 118, 115, 112, 109;
$a_{50} = -26$; **(c)** 2375 **29. (a)** $a_n = 4 + 0.5n$; **(b)** 4.5, 5, 5.5, 6,
6.5; $a_{50} = 29$; **(c)** 837.5 **31.** 3240 **33.** $-14,600$ **35.** 29,900
37. (a) $c_n = 93 + 7n$; **(b)** 268; **(c)** in the 21st week
39. (a) $f_n = 0.4 + 0.1n$; **(b)** \$1.90; **(c)** in the 46th day
41. (a) $v_n = 7500 - 1000n$; **(b)** $v_0 = 7500 - 1000(0) = \$7500$;
(c) \$3500; **(d)** $v_0 = \$7500, v_1 = \$6500, v_2 = \$5500,$
$v_3 = \$4500, v_4 = \$3500, v_5 = \$2500, v_6 = \$1500$

43. 480 **45. (a)** 1, 4, 9, 16, 25; **(b)** $S_n = n^2$; **(c)** $S_{75} = 5625$
47. (a) $T_n = n(n + 1)/2$; **(b)** ✎ **49.** Since $a_n = a_1 + (n - 1)d$,
$$S_n = \frac{n}{2}(a_1 + a_n) = \frac{n}{2}(a_1 + a_1 + (n - 1)d) =$$
$$\frac{n}{2}(2a_1 + (n - 1)d).$$ **51. (a)** $a_2 = a_1 + d, a_3 = a_2 + 2d$;
(b) $a_{n-2} = a_1 + (n - 3)d, a_{n-1} = a_1 + (n - 2)d$;
(c) $2S_n = n(2a_1 + (n - 1)d)$. Therefore, $S_n = \frac{n}{2}(2a_1 + (n - 1)d)$;
(d) $S_n = \frac{n}{2}(2a_1 + (n - 1)d) = \frac{n}{2}(a_1 + a_1 + (n - 1)d) = \frac{n}{2}(a_1 + a_n)$

Exercise Set 9.2, p. 697

1. (a) 4; **(b)** $a_n = 3(4)^{n-1}$; **(c)** 786,432; **(d)** 1,073,741,823
3. (a) 3; **(b)** $a_n = 7(3)^{n-1}$; **(c)** 137,781; **(d)** 50,221,171
5. (a) -3; **(b)** $a_n = 2(-3)^{n-1}$; **(c)** $-39,366$; **(d)** 7,174,454
7. (a) $\frac{3}{5}$; **(b)** $a_n = \left(\frac{1}{4}\right)\left(\frac{3}{5}\right)^{n-1}$; **(c)** 19,683/7,812,500;
(d) $0.624706\ldots$ **9. (a)** $-\frac{2}{3}$; **(b)** $a_n = \left(\frac{1}{8}\right)\left(-\frac{2}{3}\right)^{n-1}$; **(c)** $-\frac{512}{157,464}$;
(d) $0.07517127\ldots$ **11. (a)** $a_n = 5^{n-1}$; **(b)** 78,125;
(c) 2,441,406 **13. (a)** $a_n = 6\left(-\frac{1}{2}\right)^{n-1}$; **(b)** $\frac{3}{128}$; **(c)** $\frac{1023}{256}$
15. 4092 **17.** $395.8050224\ldots$ **19. (a)** $895.3878233\ldots$;
(b) 21 **21.** $\frac{1}{2}$ **23.** $\frac{1}{6}$ **25.** 20 **27.** $\frac{25}{3}$ **29.** 50
31. $0.497942387\ldots, 0.49991532\ldots, 0.499999965\ldots$
33. $0.171875, 0.166503906\ldots, 0.166671753\ldots$
35. $19.375, 19.980468750\ldots, 19.999389648\ldots$
37. $8.336, 8.333332480, 8.333333334$ **39.** 20.4755,
32.566077995, 39.705443395 **41.** $\frac{4}{9}$ **43.** $\frac{12}{99}$ **45.** $\frac{145}{999}$ **47.** $\frac{71}{60}$
49. (a) $v_n = 1700(0.82)^n$; **(b)** $v_0 = \$1700, v_1 = \$1394,$
$v_2 = \$1143.08, v_3 = \$937.33, v_4 = \$768.61, v_5 = \$630.26$;
(c) $v_0 = \$1700$, the original price of the tablet
51. (a) $v_n = 20,000(0.85)^n$; **(b)** $v_0 = \$20,000, v_1 = \$17,000,$
$v_2 = \$14,450, v_3 = \$12,282.50, v_4 = \$10,440.13,$
$v_5 = \$8874.11$; **(c)** $v_0 = \$20,000$, the original price of the car;
(d) 5th year **53.** \$1,428,571.43 **55. (a)** 349,525; **(b)** 15 levels
57. (a) $a_n = 15(0.5)^n$; **(b)** $a_0 = 15$ mg, $a_1 = 7.5$ mg,
$a_2 = 3.75$ mg, $a_3 = 1.875$ mg, $a_4 = 0.9375$ mg,
$a_5 = 0.46875$ mg; **(c)** 4 half-lives, or about 15.2 days
59. (a) $p_n = 1,900,000(0.986)^n$; **(b)** $p_0 = 1,900,000,$
$p_{10} = 1,650,147, p_{20} = 1,433,151, p_{30} = 1,244,690,$
$p_{40} = 1,081,011, p_{50} = 938,857$; **(c)** about 1996 **61. (a)** $\frac{1}{2}$;
(b) $\frac{1}{2}$; **(c)** $A_n = \left(\frac{1}{2}\right)^{n-1}$; **(d)** $\frac{1}{512}$ square unit **63.** $\frac{5}{2}$ **65.** 3 **67.** $\frac{3}{5}$
69. $a_1 = \dfrac{1}{p}, r = \dfrac{1}{p}, S_\infty = \dfrac{1/p}{1 - (1/p)} = \dfrac{1/p}{(p - 1)/p} = \dfrac{1}{p - 1}$
71. ✎ **73. (a)** 1; **(b)** 0; **(c)** ✎ **75.** ✎

Exercise Set 9.3, p. 706

1. (a) \$15; **(b)** \$315 **3. (a)** \$33.60; **(b)** \$1233.60 **5. (a)** \$25.83;
(b) \$525.83 **7. (a)** \$37.50; **(b)** \$2037.50 **9. (a)** \$12.19;
(b) \$1312.19 **11.** 10% **13.** 2.86% **15.** 1 yr **17.** \$500
19. \$481.93 **21.** \$9573.96 **23. (a)** \$732.60; **(b)** \$132.60;
(c) $A_1 = \$602.00, A_2 = \$604.01, A_3 = \$606.02, A_4 = \$608.04,$
$A_5 = \$610.07$ **25. (a)** \$574.94; **(b)** \$74.94; **(c)** $A_1 = \$502.50,$
$A_2 = \$505.01, A_3 = \$507.54, A_4 = \$\$510.08, A_5 = \$512.63$
27. (a) \$514.27; **(b)** \$64.27; **(c)** $A_1 = \$460.13, A_2 = \$470.48,$
$A_3 = \$481.06, A_4 = \$491.89, A_5 = \$502.95$ **29. (a)** \$2721.20;
(b) \$221.20; **(c)** $A_1 = \$2501.02, A_2 = \$2502.04, A_3 = \$2503.06,$
$A_4 = \$2504.08, A_5 = \$2505.10$ **31. (a)** \$2236.45;
(b) \$486.45; **(c)** $A_1 = \$1786.14, A_2 = \$1823.02, A_3 = \$1860.67,$
$A_4 = \$1899.09, A_5 = \$1938.31$ **33. (a)** \$8870.97;
(b) \$1129.03; **(c)** $A_1 = \$8900.54, A_2 = \$8930.21,$
$A_3 = \$8959.98, A_4 = \$8989.54, A_5 = \$9019.81$

35. (a) $3679.34; **(b)** $320.66; **(c)** $A_1 = 3717.97, $A_2 = 3757.01, $A_3 = 3796.46, $A_4 = 3836.32, $A_5 = 3876.60 **37. (a)** $2384.77; **(b)** $115.24; **(c)** $A_1 = 2391.03, $A_2 = 2397.31, $A_3 = 2403.60, $A_4 = 2409.91, $A_5 = 2416.23 **39. (a)** $13,842.28; **(b)** $3407.72; **(c)** $A_1 = $14,465.18$, $A_2 = $15,116.12$, $A_3 = $15,796.02$, $A_4 = $16,507.18$, $A_5 = $17,250.00$
41. (a) 3.5%; **(b)** $A_1 = 1035, $A_2 = 1070, $A_3 = 1105, $A_4 = 1140, $A_5 = 1175 **43. (a)** $F_n = 0.125n$; **(b)** $F_1 = 0.13, $F_2 = 0.25, $F_3 = 0.38, $F_4 = 0.50, $F_5 = 0.63 **45. (a)** $497.51; **(b)** $2.49   **47.** $10, $38.75
49. (a) $A_n = 3000(1 + 0.045/12)^{12n}$; **(b)** $A_1 = 3137.82, $A_2 = 3281.97, $A_3 = 3432.74, $A_4 = 3590.44, $A_5 = 3755.39 **51.** $4543.19, $456.81   **53. (a)** 4%; **(b)** $A_1 = 1561.11, $A_2 = 1624.71, $A_3 = 1690.91
55. (a) First Federal: $11,200, Valley View: $11,205.50; **(b)** First Federal: $1200, Valley View: $1205.50; **(c)** ✎
57. (a) Simple: $2767.53, compound: $2769.72; **(b)** simple: $232.47, compound: $230.28; **(c)** ✎   **59.** 5.43%
61. 3.82%   **63. (a)** Western: 4.5%, Commonwealth: 4.52%; **(b)** Commonwealth   **65.** 4.1212...%   **67. (a)** 15th month; **(b)** 18th month

Technology Connection, p. 715

1. 5 months; about $243.27   **2.** 14 months; about $644.14

Exercise Set 9.4, p. 715

1. $39,352.75   **3.** $32,161.67   **5.** $9804.70   **7.** $56,519.90
9. $1090.48   **11.** $61.56   **13.** $1122.35   **15.** $196.15
17. $135.33   **19.** $593.88   **21.** $42.60   **23.** $4936.26
25. $8762.22   **27. (a)** $14,412.16; **(b)** $12,000; **(c)** $2412.16
29. (a) $50,258.04; **(b)** $30,000; **(c)** $20,258.04   **31. (a)** $199.78; **(b)** $4794.72; **(c)** $205.28   **33. (a)** $3964.69; **(b)** $99,117.25; **(c)** $100,882.75   **35. (a)** $8739.37, $1260.63; **(b)** $260.55, $620.20; **(c)** option 1, $640.43   **37. (a)** $99,420.67; **(b)** $282,175.47; **(c)** $210,175.47   **39. (a)** $355.13; **(b)** $21,307.80; **(c)** $3157.80   **41. (a)** $803.07; **(b)** $289,105.20; **(c)** $142,855.20   **43. (a)** $10.59; **(b)** $1270.80; **(c)** $770.80   **45. (a)** $743.47; **(b)** $62,451.48; **(c)** $52,526.48
47.

Balance	Payment	Portion to interest	Portion to principal	New balance
$18,150.00	$355.13	$98.31	$256.82	$17,893.18
$17,893.18	$355.13	$96.92	$258.21	$17,634.97

49.

Balance	Payment	Portion to interest	Portion to principal	New balance
$146,250.00	$803.07	$633.75	$169.32	$146,080.68
$146,080.68	$803.07	$633.02	$170.05	$145,910.63

51.

Balance	Payment	Portion to interest	Portion to principal	New balance
$500.00	$10.59	$9.48	$1.11	$498.89
$498.89	$10.59	$9.46	$1.13	$497.76

53. $18,205.38   **55.** $370,291.65   **57. (a)** $227.56, $193.26; **(b)** $13,653.60, $13,914.72; **(c)** option 1, $261.12
59. (a) $1073.64; **(b)** $1199.10; **(c)** $45,165.60   **61. (a)** $89,955.16; **(b)** $88.39   **63.** $151,212.94   **65. (a)** $1073.64; **(b)** $186,510.40; **(c)** about 22.5 yr; **(d)** about $133,366; **(e)** about $53,144

67. Start with $P\left(1 + \dfrac{i}{c}\right)^{ct} = \dfrac{p\left[\left(1 + \dfrac{i}{c}\right)^{ct} - 1\right]}{\dfrac{i}{c}}$.

Multiply both sides by i/c, distribute p to clear the parentheses, and rearrange terms so that the two terms containing $\left(1 + \dfrac{i}{c}\right)^{ct}$ are on one side of the equation. Isolate the expression $\left(1 + \dfrac{i}{c}\right)^{ct}$, and take the natural logarithm of both sides of the equation.

69.

Balance	Payment	Portion to interest	Portion to principal	New balance
$18,150.00	$355.13	$98.31	$256.82	$17,893.18
$17,893.18	$355.13	$96.92	$258.21	$17,634.97
$17,634.97	$355.13	$95.52	$259.61	$17,375.37
$17,375.37	$355.13	$94.12	$261.01	$17,114.35
$17,114.35	$355.13	$92.70	$262.43	$16,851.93
$16,851.93	$355.13	$91.28	$263.85	$16,588.08
$16,588.08	$355.13	$89.85	$265.28	$16,322.80
$16,322.80	$355.13	$88.42	$266.71	$16,056.08
$16,056.08	$355.13	$86.97	$268.16	$15,787.92
$15,787.92	$355.13	$85.52	$269.61	$15,518.31
$15,518.31	$355.13	$84.06	$271.07	$15,247.24
$15,247.24	$355.13	$82.59	$272.54	$14,974.70

71.

Balance	Payment	Portion to interest	Portion to principal	New balance
$146,250.00	$803.07	$633.75	$169.32	$146,080.68
$146,080.68	$803.07	$633.02	$170.05	$145,910.63
$145,910.63	$803.07	$632.28	$170.79	$145,739.84
$145,739.84	$803.07	$631.54	$171.53	$145,568.30
$145,568.30	$803.07	$630.80	$172.27	$145,396.03
$145,396.03	$803.07	$630.05	$173.02	$145,223.01
$145,223.01	$803.07	$629.30	$173.77	$145,049.24
$145,049.24	$803.07	$628.55	$174.52	$144,874.72
$144,874.72	$803.07	$627.79	$175.28	$144,699.44
$144,699.44	$803.07	$627.03	$176.04	$144,523.40
$144,523.40	$803.07	$626.27	$176.80	$144,346.60
$144,346.60	$803.07	$625.50	$177.57	$144,169.03

73.

Balance	Payment	Portion to interest	Portion to principal	New balance
$500.00	$10.59	$9.48	$1.11	$498.89
$498.89	$10.59	$9.46	$1.13	$497.76
$497.76	$10.59	$9.44	$1.15	$496.60
$496.60	$10.59	$9.41	$1.18	$495.43
$495.43	$10.59	$9.39	$1.20	$494.23
$494.23	$10.59	$9.37	$1.22	$493.01
$493.01	$10.59	$9.35	$1.24	$491.77
$491.77	$10.59	$9.32	$1.27	$490.50
$490.50	$10.59	$9.30	$1.29	$489.21
$489.21	$10.59	$9.27	$1.32	$487.89
$487.89	$10.59	$9.25	$1.34	$486.55
$486.55	$10.59	$9.22	$1.37	$485.19

Exercise Set 9.5, p. 727

1. (a) $1 - 2x + 4x^2 - 8x^3 + 16x^4 - \cdots$; **(b)** $\left(-\frac{1}{2}, \frac{1}{2}\right), 0$
3. (a) $1 + 5x + 25x^2 + 125x^3 + 625x^4 + \cdots$; **(b)** $\left(-\frac{1}{5}, \frac{1}{5}\right), 0$
5. (a) $1 + x^2 + x^4 + x^6 + x^8 + \cdots$; **(b)** $(-1, 1), 0$
7. (a) $\frac{2}{5} + \frac{6}{25}x + \frac{18}{125}x^2 + \frac{54}{625}x^3 + \frac{162}{3125}x^4 + \cdots$;
(b) $\left(-\frac{5}{3}, \frac{5}{3}\right), 0$ **9. (a)** $\frac{5}{3} - \frac{10}{3}x + \frac{20}{3}x^2 - \frac{40}{3}x^3 + \frac{80}{3}x^4 - \cdots$;
(b) $\left(-\frac{1}{2}, \frac{1}{2}\right), 0$

11.

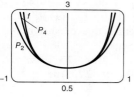

13.

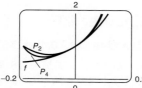

15.

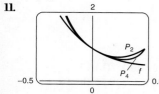

17.

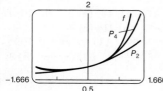

19.

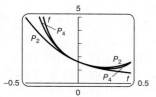

21. (a) $f'(x) = -\dfrac{2}{(1 + 2x)^2} = -2 + 8x - 24x^2 + 64x^3 - 160x^4 + \cdots$; **(b)** $F(x) = \frac{1}{2}\ln(1 + 2x) = x - x^2 + \frac{4}{3}x^3 - 2x^4 + \frac{16}{5}x^5 - \cdots$ **23. (a)** $f'(x) = \dfrac{5}{(1 - 5x)^2} = 5 + 50x + 375x^2 + 2500x^3 + 15{,}625x^4 + \cdots$;

(b) $F(x) = -\frac{1}{5}\ln(1 - 5x) = x + \frac{5}{2}x^2 + \frac{25}{3}x^3 + \frac{125}{4}x^4 + 125x^5 + \cdots$ **25. (a)** $1 - (x + 1)^2 + (x + 1)^4 - (x + 1)^6 + (x + 1)^8 - \cdots$; **(b)** $(-2, 0), -1$ **27. (a)** $\frac{1}{9} - \frac{1}{81}(x - 2)^2 + \frac{1}{729}(x - 2)^4 - \frac{1}{6561}(x - 2)^6 + \frac{1}{59{,}049}(x - 2)^8 - \cdots$;

(b) $(1, 3), 2$
29. (a) $P_1(x) = 11x - 14$; **31. (a)** $P_1(x) = 3x + 1$;
(b)

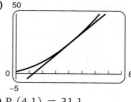

(b)

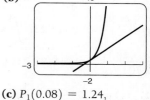

(c) $P_1(4.1) = 31.1$, **(c)** $P_1(0.08) = 1.24$,
$f(4.1) = 31.11$ $g(0.08) = 1.27\ldots$

33. (a) $P_1(p) = 138.4p + 88$; **(b)** $P_1(21) = 2994.4$ units, $D(21) = 2994.3$ units **35. (a)** $P_1(t) = 6.35t - 11.2$; **(b)** $P_1(14) = \$77.70, p(14) = \$78.05$ **37. (a)** $P_1(t) = 0.253t - 1.551$; **(b)** $P_1(39) = \$8.32, C(39) = \$8.37$
39. $\ln 2 = \frac{1}{2} + \frac{1}{8} + \frac{1}{24} + \frac{1}{64} + \frac{1}{160} + \frac{1}{384} + \frac{1}{896} + \frac{1}{2048} + \frac{1}{4608} + \frac{1}{10{,}240} + \cdots \approx 0.693$ **41.**

Exercise Set 9.6, p. 734

1. (a) $1 + \frac{1}{3}(x - 1) - \frac{1}{9}(x - 1)^2 + \frac{5}{81}(x - 1)^3 - \frac{10}{243}(x - 1)^4 + \cdots$;
(b)

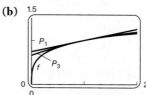

3. (a) $1 + \frac{1}{4}(x - 1) - \frac{3}{32}(x - 1)^2 + \frac{7}{128}(x - 1)^3 - \frac{77}{2048}(x - 1)^4 + \cdots$;
(b)

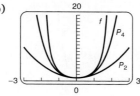

5. (a) $1 + x^2 + \frac{1}{2}x^4 + \frac{1}{6}x^6 + \frac{1}{24}x^8 + \cdots$;
(b)

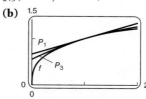

7. $e^{0.5} \approx 1 + \frac{1}{2} + \frac{1}{8} + \frac{1}{48} + \frac{1}{384} + \frac{1}{3840} + \frac{1}{46{,}080} = 1.6487\ldots$
9. $e^{0.3} \approx 1 + \frac{3}{10} + \frac{9}{200} + \frac{27}{6000} + \frac{81}{240{,}000} + \frac{243}{12{,}000{,}000} = 1.3498\ldots$
11. $\sqrt{1.1} \approx 1 + \frac{1}{20} - \frac{1}{800} + \frac{1}{16{,}000} = 1.0488\ldots$
13. $\sqrt[3]{9} \approx 2 + \frac{1}{12} - \frac{1}{288} + \frac{5}{20{,}736} - \frac{5}{248{,}832} = 2.0800\ldots$
15. $\displaystyle\int_0^1 e^{x^3}\,dx \approx 1 + \frac{1}{4} + \frac{1}{14} + \frac{1}{60} + \frac{1}{312} + \frac{1}{1920} + \frac{1}{13{,}680} = 1.342$

17. (a) $P_2(t) = 79.544 + 8.671(t - 20) + 0.473(t - 20)^2$, $P_2(18) = \$64.94$; **(b)** $P_3(t) = 79.544 + 8.671(t - 20) + 0.473(t - 20)^2 + 0.0172(t - 20)^3$, $P_3(18) = \$63.96$; **(c)** $f(18) = \$63.96$; **(d)**

19. $\frac{\sqrt{2}}{2} + \frac{\sqrt{2}}{2}\left(x - \frac{\pi}{4}\right) - \frac{\sqrt{2}}{4}\left(x - \frac{\pi}{4}\right)^2 - \frac{\sqrt{2}}{12}\left(x - \frac{\pi}{4}\right)^3 + \frac{\sqrt{2}}{48}\left(x - \frac{\pi}{4}\right)^4 + \cdots$ **21.** $\sin 0.3 \approx \frac{3}{10} - \frac{27}{6000} + \frac{243}{12,000,000} = 0.2955$ **23.** $\int_0^{0.5} \sin x^2\, dx \approx \frac{1}{24} - \frac{1}{5376} + \frac{1}{2,703,360} = 0.04148$ **25.** $(-\infty, \infty)$

27. (a) $e^{ix} = 1 + ix - \frac{1}{2!}x^2 - \frac{1}{3!}ix^3 + \frac{1}{4!}x^4 + \frac{1}{5!}ix^5 - \frac{1}{6!}x^6 - \frac{1}{7!}ix^7 + \cdots$; **(b)** $\sin(ix) = ix - \frac{1}{3!}ix^3 + \frac{1}{5!}ix^5 - \frac{1}{7!}ix^7 + \cdots = i\left(x - \frac{1}{3!}x^3 + \frac{1}{5!}x^5 - \frac{1}{7!}x^7 + \cdots\right) = i \sin x$; **(c)** all even-powered terms of the series for e^{ix} equal $\cos x$; all odd-powered terms equal $i \sin x$.   **29.** $e^{i(\pi/2)} = \cos \frac{\pi}{2} + i \sin \frac{\pi}{2} = i$, so $e^{i(\pi/2)} = i$. Therefore, $\left(e^{i(\pi/2)}\right)^i = e^{-\pi/2} = i^i$.   **31.** Since $r = -\frac{1}{2} \pm \frac{i\sqrt{11}}{2}$, we have $y = Ce\left(-\frac{1}{2} \pm \frac{i\sqrt{11}}{2}\right)x = Ce^{(-1/2)x}\left(\cos \frac{\sqrt{11}}{2}x + i \sin \frac{\sqrt{11}}{2}x\right) = e^{(-1/2)x}\left(C_1 \cos \frac{\sqrt{11}}{2}x + C_2 \sin \frac{\sqrt{11}}{2}x\right)$.

Chapter Review Exercises, p. 739

1. False   **2.** False   **3.** True   **4.** True   **5.** False   **6.** True   **7.** True   **8.** False   **9.** True   **10.** False   **11. (a)** $a_n = 9n + 2$; **(b)** 1802; **(c)** 11,575   **12. (a)** $a_n = 4n - 3$; **(b)** 79,800   **13. (a)** $a_n = 3500 - 500n$; **(b)** $2000; **(c)** $a_0 = \$3500$, $a_1 = \$3000$, $a_2 = \$2500$, $a_3 = \$2000$, $a_4 = \$1500$, $a_5 = \$1000$ **14. (a)** $a_n = 4000 - 120n$; **(b)** $2080; **(c)** $a_0 = \$4000$, $a_1 = \$3880$, $a_2 = \$3760$, $a_3 = \$3640$, $a_4 = \$3520$, $a_5 = \$3400$ **15.** $2.70   **16.** $\frac{5}{8}$ **17.** $\frac{56}{99}$ **18.** $80   **19. (a)** $2316.95; **(b)** $183.05   **20. (a)** $795.11; **(b)** $4770.66; **(c)** $229.34   **21. (a)** $5813.33; **(b)** $1013.33   **22.** $24,228.15   **23. (a)** $190.77; **(b)** $1696.20   **24.** $508,192.17   **25. (a)** $23.79; **(b)** $20.75 to interest, $3.04 to principal   **26. (a)** $4 + 4x + 4x^2 + 4x^3 + 4x^4 + \cdots$; **(b)** $(-1, 1)$, 0   **27.** $-12 + 144x - 1296x^2 + 10,368x^3 - 77,760x^4 + \cdots$ **28.** $P_1(x) = 0.35x + 6$; $P_1(21) = \$13.35$ **29.** $\int_0^1 e^{-x^6}\, dx \approx 1 - \frac{1}{7} + \frac{1}{26} - \frac{1}{114} + \frac{1}{600} - \frac{1}{3720} + \frac{1}{26,640} = 0.888$ **30. (a)** $x^4 - \frac{1}{6}x^{12} + \frac{1}{120}x^{20} - \frac{1}{5040}x^{28} + \frac{1}{362,880}x^{36} - \cdots$; **(b)** $4x^3 - 2x^{11} + \frac{1}{6}x^{19} - \frac{1}{180}x^{27} + \frac{1}{10,080}x^{35} - \cdots$ **31.** $\int_0^1 \cos x^4\, dx \approx 1 - \frac{1}{18} + \frac{1}{408} - \frac{1}{18,000} = 0.9468$

Chapter 9 Test, p. 741

1. [9.1] **(a)** 4; **(b)** $a_n = 4n + 2$; **(c)** 202; **(d)** 20,400   **2.** [9.2] **(a)** $\frac{1}{3}$; **(b)** $a_n = 81\left(\frac{1}{3}\right)^{n-1}$; **(c)** $\frac{1}{81}$; **(d)** 121.4997714   **3.** [9.1] 25,680   **4.** [9.1] 10,215   **5.** [9.2] $\frac{16}{9}$ **6.** [9.3] $2311.20   **7.** [9.3] $493.83   **8.** [9.1] **(a)** $a_n = 3000 - 90n$; **(b)** $1920   **9.** [9.3] $110   **10.** [9.2] $208,333.33

11. [9.3] **(a)** $1736.14; **(b)** $386.14   **12.** [9.3] **(a)** $3574.53; **(b)** $425.47   **13.** [9.4] **(a)** $1062.24; **(b)** $1040; **(c)** $22.24   **14.** [9.4] $4890.34   **15.** [9.4] **(a)** $1667.22; **(b)** $600,199.20; **(c)** $240,199.20   **16.** [9.4] $20,815.15   **17.** [9.5] **(a)** $\frac{1}{3} + \frac{1}{9}x + \frac{1}{27}x^2 + \frac{1}{81}x^3 + \frac{1}{243}x^4 + \cdots$; **(b)** $\frac{1}{9} + \frac{2}{27}x + \frac{1}{27}x^2 + \frac{4}{243}x^3 + \frac{5}{729}x^4 + \cdots$ **18.** [9.6] $\int_0^{0.5} e^{x^4}\, dx \approx \frac{1}{2} + \frac{1}{160} + \frac{1}{9216} = 0.506$ **19.** [9.5] $P_1(x) = 0.844x + 3.943$; $30.11 per unit   **20.** [9.6] **(a)** $1 - \frac{1}{2}x^{2/3} + \frac{1}{24}x^{4/3} - \frac{1}{720}x^2 + \frac{1}{40,320}x^{8/3} - \frac{1}{3,628,800}x^{10/3} + \cdots$; **(b)** $-\frac{1}{3}x^{-1/3} + \frac{1}{18}x^{1/3} - \frac{1}{360}x + \frac{1}{15,120}x^{5/3} - \frac{1}{1,088,640}x^{7/3} + \cdots$ **21.** [9.6] $\int_0^1 \sin x^5\, dx \approx \frac{1}{6} - \frac{1}{96} + \frac{1}{3120} - \frac{1}{181,440} = 0.1566$

Extended Technology Application, p. 743

1. Left to the student; $\frac{1}{2}$ **2.** Left to the student; $\frac{1}{3}$ **3.** Left to the student; $\frac{8}{5}$ **4.** 40 feet   **5.** 616 terms   **6. (a)** (3, 11), (4, 31), (5, 83), (6, 227), (7, 616); **(b)** $y = 0.539e^{1.00717x}$; **(c)** model: 12,754 terms, spreadsheet: 12,367 terms   **7. (a)** $S_{100} = 0.6881721793101950$, $S_{200} = 0.6906534304818240$, $S_{500} = 0.6921481805579460$; **(b)** 160; **(c)** 3398   **8.** 2455 terms   **9.** 15   **10.** 0.61803398874990 . . .   **11.** Left to the student   **12.** 1.61803398874990 . . . , or $1 + \phi$

Chapter 10

Exercise Set 10.1, p. 753

1. (a) $\{1, 2, 3, 4, 5, 6, 7, 8, 9, 10, 11, 12\}$; **(b)** 12   **3. (a)** $\{22, 24, 26, 28, 30, 32, 34, 36, 38, 40\}$; **(b)** 10   **5. (a)** $\{\text{fall, winter, spring, summer}\}$; **(b)** 4   **7. (a)** $\{3, 6, 9, 12, 15, 18, 21, 24, 27, 30, 33, 36, 39, 42\}$; **(b)** 14   **9. (a)** $\{\frac{4}{3}\}$; **(b)** 1   **11. (a)** $\varnothing$; **(b)** 0   **13. (a)** $\mathbb{R}$; **(b)** ∞ **15.** $\{x \,|\, x \text{ is a positive even integer less than or equal to } 14\}$ **17.** $\{x \,|\, x \text{ is a state in the United States}\}$ **19.** $\{x \,|\, x \text{ is a day of the week}\}$ **21.** True   **23.** False, $\{4\} \subseteq A$ or $4 \in A$ **25.** False, $D \subseteq B$ **27.** True   **29.** $\{1, 3, 5, 6, 7, 8, 9, 10\}$, 8   **31.** $\{7, 9\}$, 2   **33.** $\{6, 7, 8, 9, 10\}$, 5   **35.** $\{2, 4, 10\}$, 3   **37.** 29   **39.** 3   **41.** 55, 73   **43.** 3   **45.**   **47.** 74   **49. (a)** 25; **(b)** 20; **(c)** 5   **51. (a)** 188; **(b)** 12   **53. (a)** 41; **(b)** 21; **(c)** 9   **55.** $A \cap B = \varnothing$ **57.** $A = B$ **59.** $(A \cup B)' = \{8, 10\}$, $A' \cap B' = \{8, 10\}$ **61.** $\varnothing$ **63.** $\{4\}$ **65. (a)** 98; **(b)** 8; **(c)** 22; **(d)** 2

Exercise Set 10.2, p. 761

1. (a) $\frac{1}{2}, \frac{1}{2}$; **(b)** $\frac{3}{4}, \frac{1}{4}$ **3. (a)** $\frac{3}{13}, \frac{10}{13}$; **(b)** $\frac{1}{4}, \frac{3}{4}$; **(c)** $\frac{11}{26}$ **5. (a)** $\frac{1}{12}, \frac{11}{12}$; **(b)** $\frac{11}{36}, \frac{25}{36}$; **(c)** $\frac{13}{36}$ **7. (a)** $\frac{1}{3}, \frac{2}{3}$; **(b)** $0, \frac{1}{2}$; **(c)** $0, \frac{5}{6}$ **9. (a)** $\frac{1}{12}, \frac{11}{12}$; **(b)** $\frac{1}{12}, \frac{5}{12}$; **(c)** $0, 1$ **11. (a)** $\frac{1125}{5129}, \frac{4004}{5129}$; **(b)** $\frac{290}{5129}, \frac{4839}{5129}$; **(c)** $0, \frac{3576}{5129}$ **13. (a)** $S = \{\text{TT, TH, HT, HH}\}$; **(b)** $\frac{3}{4}, \frac{1}{4}$; **(c)** $\frac{1}{2}, \frac{1}{2}$; **(d)** 1   **15. (a)** $S = \{\text{TTTT, TTTH, TTHT, TTHH, THTT, THTH, THHT, THHH, H T̄TT, HTTH, HTHT, HTHH, HHTT, HHTH, HHHT, HHHH}\}$; **(b)** $\frac{1}{8}, \frac{7}{8}$; **(c)** $\frac{5}{16}, \frac{1}{16}$; **(d)** 1   **17. (a)** $S = \{(1, 1, 1), (1, 1, 2), (1, 1, 3), (1, 2, 1), (1, 2, 2), (1, 2, 3), (1, 3, 1), (1, 3, 2), (1, 3, 3), (2, 1, 1), (2, 1, 2), (2, 1, 3), (2, 2, 1), (2, 2, 2), (2, 2, 3), (2, 3, 1), (2, 3, 2), (2, 3, 3), (3, 1, 1), (3, 1, 2), (3, 1, 3), (3, 2, 1), (3, 2, 2), (3, 2, 3), (3, 3, 1), (3, 3, 2), (3, 3, 3)\}$; **(b)** $\frac{19}{27}, \frac{8}{27}$; **(c)** $\frac{2}{9}, \frac{7}{9}$; **(d)** $\frac{22}{27}$ **19. (a)** 0.3; **(b)** 0.7; **(c)** 0.8; **(d)** 0.6   **21. (a)** $\frac{153}{337}$; **(b)** $\frac{245}{337}$; **(c)** $\frac{248}{337}$ **23. (a)** $\frac{61}{200}$;

(b) $\frac{41}{50}$; **(c)** $\frac{97}{200}$ **25. (a)** $\frac{11}{50}$; **(b)** $\frac{41}{50}$; **(c)** $\frac{19}{25}$ **27. (a)** $\frac{169}{1307}$;
(b) $\frac{888}{1307}$; **(c)** $\frac{213}{1307}$ **29. (a)** $\frac{17}{72}$; **(b)** $\frac{7}{9}$; **(c)** $\frac{13}{18}$; **(d)**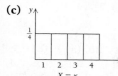
31. (a) $\frac{49,631}{50,000} \approx 99.262\%$; **(b)** $\frac{49,591}{50,000} \approx 99.182\%$;
(c) $\frac{49,569}{50,000} \approx 99.138\%$ **33.** 0.0007965   **35.** 0.03255
37. (a) 0.125; **(b)** answers will vary but be near 0.125.
39. About 0.6324   **41.** 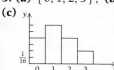   **43.**

Exercise Set 10.3, p. 774

1. (a) $\{1, 2, 3, 4\}$; **(b)** $\{(1, \frac{1}{4}), (2, \frac{1}{4}), (3, \frac{1}{4}), (4, \frac{1}{4})\}$;
(c) *[graph]*   **(d)** $f(3) = P(X = 3) = \frac{1}{4}$; the probability the spinner points to 3 is $\frac{1}{4}$; **(e)** 2.5, the average value of the spins

3. (a) $\{0, 1, 2, 3\}$; **(b)** $\{(0, \frac{4}{16}), (1, \frac{6}{16}), (2, \frac{4}{16}), (3, \frac{2}{16})\}$;
(c) *[graph]*   **(d)** $f(1) = P(X = 1) = \frac{6}{16}$, or $\frac{3}{8}$; the probability the difference of the two spins is 1 is $\frac{3}{8}$; **(e)** 1.25, the average difference of two spins

5. Yes   **7.** Yes   **9.** No, sum $\neq 1$ **11.** No, $p(12)$ is negative.
13. Yes   **15.** Yes   **17.** No, $h(-1)$ is negative.   **19.** 0.27
21. 0.0875   **23.** $\frac{1}{4}$ **25.** $\frac{1}{9}$ **27.** $\frac{1}{10}$ **29.** $\frac{1}{100}$ **31. (a)** $\frac{105}{176}$;
(b) $\frac{105}{528} \approx 0.199$; **(c)** 0; **(d)** 2.386

33. (a) $F(x) = \begin{cases} 0, & x < -2 \\ \frac{4}{15}, & -2 \le x < -1 \\ \frac{1}{3}, & -1 \le x < 1 \\ \frac{2}{5}, & 1 \le x < 3 \\ 1, & x \ge 3 \end{cases}$

[graph labeled $X \le x$]

(b) no, $F(2) = \frac{6}{15}$, or $\frac{2}{5}$ **35. (a)** $f = \{(1, \frac{1}{16}), (2, \frac{3}{16}), (3, \frac{5}{16}),$
$(4, \frac{7}{16})\}$; **(b)** $F(x) = \begin{cases} 0, & x < 1 \\ \frac{1}{16}, & 1 \le x < 2 \\ \frac{1}{4}, & 2 \le x < 3 \\ \frac{9}{16}, & 3 \le x < 4 \\ 1, & x \ge 4 \end{cases}$

[graph labeled $X \le x$]

(c) $f(2) = \frac{3}{16}$ is the probability that the highest spin is 2, and $F(2) = \frac{1}{4}$ is the probability that the highest spin is at most 2;
(d) $1 - f(2) = \frac{13}{16}$ is the probability that the highest spin is not 2, and $1 - F(2) = \frac{3}{4}$ is the probability that the highest spin is greater than 2.   **37. (a)** $f = \{(1, 0.36), (2, 0.34), (3, 0.18), (4, 0.06),$
$(5, 0.04), (7, 0.02)\}$; *[graph labeled $X = x$]*

(b) $F(x) = \begin{cases} 0, & x < 1 \\ 0.36, & 1 \le x < 2 \\ 0.70, & 2 \le x < 3 \\ 0.88, & 3 \le x < 4 \\ 0.94, & 4 \le x < 5 \\ 0.98, & 5 \le x < 7 \\ 1, & x \ge 7 \end{cases}$ **(c)** $f(2) = 0.34$ is the probability that a randomly selected person paid \$2 for an app, and $F(2) = 0.7$ is the probability that a randomly selected person paid at most \$2 for an app;

(d) $1 - f(2) = 0.66$ is the probability that a randomly selected person did not pay \$2 for an app, and $1 - F(2) = 0.3$ is the probability that a randomly selected person paid more than \$2 for an app; **(e)** \$2.16 is the average price paid for an app among the 50 people surveyed.   **39. (a)** $g(675) = 0.119$, the probability that a randomly selected person has a credit score of 675;

(b) $G(x) = \begin{cases} 0, & x < 475 \\ 0.069, & 475 \le x < 525 \\ 0.159, & 525 \le x < 575 \\ 0.255, & 575 \le x < 625 \\ 0.35, & 625 \le x < 675; \\ 0.469, & 675 \le x < 725 \\ 0.626, & 725 \le x < 775 \\ 0.821, & 775 \le x < 825 \\ 1, & x \ge 825 \end{cases}$

(c) $G(675) = 0.469$, the probability that a randomly selected person has a credit score of at most 675; **(d)** 687.9, the average credit score   **41. (a)** $h(4) = \frac{168}{2512} = 0.0669$, the probability that a randomly selected respondent visited a coffee shop 4 times during the month; **(b)** $H(4) = \frac{2125}{2512} = 0.8459$, the probability that a randomly selected respondent visited a coffee shop at most 4 times during the month; **(c)** $h(1.5) = 0$, because it is impossible to visit a coffee shop 1.5 times. $H(1.5) = H(1) = \frac{1575}{2512} = 0.627$; the probability that a respondent visited a coffee shop at most 1.5 times during the month is the same as the probability of visiting at most 1 time during the month. **(d)** 1.6, the average number of visits to a coffee shop during the month by a respondent
43. \$7500   **45.** 0   **47.** $f(0) = \frac{1}{4}$, $F(0) = \frac{1}{2}$
49. $f(20) = \frac{1}{4}$, $F(20) = 1$ **51.** $c \approx 0.618$

Exercise Set 10.4, p. 786

1. $\int_1^3 \frac{1}{4} x \, dx = \left[\frac{x^2}{8}\right]_1^3 = \frac{9}{8} - \frac{1}{8} = 1$

3. $\int_0^{1/3} 3 \, dx = [3x]_0^{1/3} = 3\left(\frac{1}{3} - 0\right) = 1$

5. $\int_0^4 \frac{3}{64} x^2 \, dx = \left[\frac{x^3}{64}\right]_0^4 = \frac{1}{64}(4^3 - 0^3) = 1$

7. $\int_1^e \frac{1}{x} \, dx = [\ln x]_1^e = \ln e - \ln 1 = 1 - 0 = 1$

9. $\int_{-1}^1 \frac{3}{2} x^2 \, dx = \left[\frac{x^3}{2}\right]_{-1}^1 = \frac{1^3}{2} - \frac{(-1)^3}{2} = \frac{1}{2} + \frac{1}{2} = 1$

11. $\int_0^\infty 3e^{-3x} \, dx = \lim_{b \to \infty} \int_0^b 3e^{-3x} \, dx = \lim_{b \to \infty} [-e^{-3x}]_0^b$
$= \lim_{b \to \infty} [-e^{-3b} - (-e^0)] = \lim_{b \to \infty} \left(-\frac{1}{e^{3b}} + 1\right) = 1$

13. $\frac{2}{21}$; $f(x) = \frac{2}{21} x$ **15.** $\frac{3}{2}$; $f(x) = \frac{3}{2} x^2$ **17.** $\frac{1}{6}$; $f(x) = \frac{1}{6}$

19. $\frac{1}{2}$; $f(x) = \frac{2 - x}{2}$ **21.** $\frac{1}{\ln 3}$; $f(x) = \frac{1}{x \ln 3}$

23. $\frac{1}{e^3 - 1}$; $f(x) = \frac{e^x}{e^3 - 1}$ **25.** $F(x) = \frac{1}{8} x^2 - \frac{1}{8}, [1, 3]$
27. $F(x) = 3x, [0, \frac{1}{3}]$ **29.** $F(x) = \frac{1}{64} x^3, [0, 4]$
31. $F(x) = \ln x, [1, e]$ **33.** $F(x) = \frac{1}{2} x^3 + \frac{1}{2}, [-1, 1]$
35. $F(x) = 1 - e^{-3x}, [0, \infty)$ **37.** $F(x) = \frac{1}{21} x^2 - \frac{4}{21}, [2, 5]$
39. $F(x) = \frac{1}{2} x^3 + \frac{1}{2}, [-1, 1]$ **41.** $F(x) = \frac{1}{6} x - \frac{1}{6}, [1, 7]$

43. $F(x) = x - \frac{1}{4}x^2, [0, 2]$ **45.** $F(x) = \frac{\ln x}{\ln 3}, [1, 3]$

47. $F(x) = \frac{e^x - 1}{e^3 - 1}, [0, 3]$ **49.** $\frac{8}{25}$, or 0.32   **51.** $\frac{11}{16}$, or 0.6875

53. 0.3297   **55.** 0.999955   **57.** 0.3935

59. (a) $\int_0^{10} 0.23e^{-0.23t}\, dt = 0.9$; **(b)** 0.0286   **61.** 0.950213

63. $\sqrt{2}$ **65.** $F(x) = \frac{1}{4}x^2 - \frac{1}{2}x + \frac{1}{4}, [1, 3]$ **67.** ✎

69. (a) $k = 0.357, f(t) = 0.357e^{-0.357t}, [0, \infty)$; **(b)** 0.2427

71. (a) $c = \frac{1}{2}$ **(b)** $f(x) = \frac{1}{8}x - \frac{1}{8}, [1, 5]$;

(c) $F(x) = \frac{1}{16}x^2 - \frac{1}{8}x + \frac{1}{16}, [1, 5]$; **(d)** 0.1875 $\left(\frac{3}{16}\right)$

73. $c = 4/(3e^4 - e^2) \approx 0.02557$ **75–85.** Left to the student.

Technology Connection, p. 797

1. (a) 0.6295, or 62.95%; **(b)** 0.0228, or 2.28%
2. (a) 0.621, or 62.1%; **(b)** 0.353, or 35.3%

Technology Connection, p. 799

1. $x = 0$ **2.** $x = -0.6745$ **3.** $x = 9.935$ **4.** $x = 2.477$
5. $x = 103.201$

Exercise Set 10.5, p. 800

1. $\mu = 5, \sigma^2 = \frac{4}{3}, \sigma = \frac{2}{\sqrt{3}}$

3. $\mu = \frac{8}{3}, \sigma^2 = \frac{8}{9}, \sigma = \frac{2\sqrt{2}}{3}$

5. $\mu = \frac{13}{6}, \sigma^2 = \frac{11}{36}, \sigma = \frac{\sqrt{11}}{6}$

7. $\mu = 6, \sigma^2 = \frac{16}{3}, \sigma \approx 2.309$; area ≈ 0.58

9. $\mu = \frac{8}{3}, \sigma^2 = \frac{8}{9}, \sigma \approx 0.943$; area ≈ 0.63 **11.** 0.4834

13. 0.4147   **15.** 0.6442   **17.** 0.0150   **19.** 0.1790   **21.** 0.0013

23. (a) 0.6826; **(b)** 68.26%   **25.** 0.2898   **27.** 0.4514

29–45. Check using the answers to Exercises 11–27.

47. (a) −0.52; **(b)** 0; **(c)** 1.645   **49. (a)** −14.44;

(b) −15.04   **51. (a)** $f(x) = \frac{1}{1.2}$, or $\frac{5}{6}, [17.5, 18.7]$;

(b) $\mu = 18.1, \sigma^2 = 0.12, \sigma \approx 0.346$; **(c)** about 0.58

53. (a) $\mu = 10, \sigma^2 = 50, \sigma = 7.071$; **(b)** about 0.63

55. 0.62%   **57.** 0.7910   **59. (a)** 972; **(b)** 1036;

(c) 1152   **61. (a)** 84; **(b)** 80   **63.** 90.8th   **65.** $\mu = \frac{b - a}{2}$,

$\sigma^2 = \frac{(b - a)^2}{12}, \sigma = \frac{b - a}{2\sqrt{3}}$ **67.** $\sqrt{2}$

69. $\frac{\ln 2}{k}$ **71.** 16.412 oz   **73.** ✎   **75.** About 1.772

Chapter Review Exercises, p. 808

1. True   **2.** False   **3.** True   **4.** True   **5.** False   **6.** True
7. True   **8.** True   **9.** True   **10.** False

11. (a) $\{3, 5, 6, 7, 8, 9, 10\}$; **(b)** $\varnothing$; **(c)** $\{1, 2, 3, 4\}$; **(d)** 8;

(e) 8   **12.** 16   **13.** $\frac{3}{8}$ **14. (a)** $\frac{11}{36}, \frac{25}{36}$; **(b)** $\frac{6}{36}, \frac{30}{36}$; **(c)** $\frac{15}{36}$

15. 0.44   **16. (a)** $\frac{30}{31}$;

(b)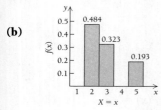

(c) $F(x) = \begin{cases} 0, & x < 2 \\ 0.484, & 2 \le x < 3 \\ 0.807, & 3 \le x < 5 \\ 1, & x \ge 5 \end{cases}$

17. (a) $\frac{12}{67}$; **(b)** $\frac{44}{67}$; c) $18,104.48, the average amount won
18. (a) $\{(1, 2), (1, 3), (1, 4), (1, 5), (2, 1), (2, 3), (2, 4),$
$(2, 5), (3, 1), (3, 2), (3, 4),\ (3, 5), (4, 1), (4, 2), (4, 3),$
$(4, 5), (5, 1), (5, 2), (5, 3), (5, 4)\}$; **(b)** $\frac{2}{20}$; **(c)** $\frac{4}{20}$; **(d)** 6,
the average sum of the two cards   **19. (a)** $\frac{6}{20}$; **(b)** $\frac{14}{20}$;
(c) 2, the average difference of the two cards   **20. (a)** $\frac{1}{6}$; **(b)** 0.5;
(c) $\mu = 3.111, \sigma = 0.566$; **(d)** $G(x) = \frac{1}{12}x^2 - \frac{1}{3}$, for $2 \le x \le 4$
21. (a) 0.5; **(b)** 0; **(c)** 1.5; **(d)** $\sigma^2 = 0.75, \sigma = 0.866$; **(e)** 0.577
22. (a) 0.9502; **(b)** 0.9975, the probability that two customers
show up within 4 min of one another   **23.** 0.4678   **24.** 0.8827
25. 0.1002   **26.** 0.5   **27.** 93.57th   **28. (a)** 0.2729; **(b)** 71.61st
29. 0.3085   **30.** $6200   **31.** 1   **32.** 0.57   **33.** About 0.71

Chapter 10 Test, p. 809

1. [10.1] **(a)** $\{e, f, g, h\}$; **(b)** $\{c, d, e, f, g, h\}$; **(c)** $\varnothing$; **(d)** 2;
(e) 6   **2.** [10.2] **(a)** $\frac{11}{36}, \frac{25}{36}$; **(b)** $\frac{4}{36}, \frac{32}{36}$; **(c)** $\frac{13}{36}$
3. [10.3] **(a)** $\frac{2}{25}$; **(b)** $\frac{10}{25}$ **4.** [10.3] **(a)** 0; **(b)** $\frac{17}{25}$

5. [10.3] **(a)** $\frac{1}{16}$; **(b)**

(c) $F(x) = \begin{cases} 0, & x < 1 \\ 0.0625, & 1 \le x < 3 \\ 0.25, & 3 \le x < 5 \\ 0.5625, & 5 \le x < 7 \\ 1, & x \ge 7 \end{cases}$

(d) 5.25   **6.** [10.3] The sum $f(1) + f(4) + f(5) = \frac{42}{41}, \ne 1$.
7. [10.4] **(a)** 0.2897; **(b)** $G(x) = \frac{1}{26}x^3 - \frac{1}{26}, 1 \le x \le 3$;
(c) 2.308   **8.** [10.4] **(a)** 0.55; [10.4] **(b)** 0; **(c)** [10.5]
$\mu = 30, \sigma^2 = 8.335, \sigma = 2.887$; **(d)** [10.5] 0.577   **9.** [10.4]
(a) 0.1920; **(b)** 0.017; **(c)** 0.741 min   **10.** [10.3] **(a)** $\frac{53}{154}$;
(b) $\frac{130}{154}$; **(c)** 1.58, average number of hot dogs sold per customer
11. [10.5] **(a)** 0.4082; **(b)** 0.0401; **(c)** 0.0148
12. [10.5] 0.4436   **13.** [10.5] 36   **14.** [10.5] 127.4 min
15. [10.4] About 1.572   **16.** [10.2] About 0.89

Extended Technology Application, p. 812

1–8. Answers will vary.

Cumulative Review, p. 815

1. [R.4] $y = -4x - 27$ **2.** [R.2] $x^2 + 2xh + h^2 - 5$
3. [1.2] **(a)**   **(b)** 3; **(c)** −3; **(d)** no

4. [1.2] -8 **5.** [1.2] 3 **6.** [1.2] Does not exist **7.** [2.3] 4
8. [2.3] 0 **9.** [1.4] $f'(x) = 2x$ **10.** [1.1] f **11.** [1.2] b, f
12. [1.4] b, d, f **13.** [1.5] -9 **14.** [1.5] $2x - 7$
15. [1.5] $\frac{1}{4}x^{-3/4}$ **16.** [1.5] $-6x^{-7}$ **17.** [1.7] $\frac{10}{3}x^4(2x^5 - 8)^{-2/3}$
18. [1.6] $\dfrac{20x^3 - 15x^2 - 8}{(2x - 1)^2}$ **19.** [3.2] $\dfrac{2x}{x^2 + 5}$
20. [3.2] 1 **21.** [3.1] $3e^{3x} + 2x$ **22.** [3.1] $\dfrac{e^{\sqrt{x-3}}}{2\sqrt{x - 3}}$
23. [3.2] $\dfrac{e^x}{e^x - 4}$ **24.** [1.8] $2 - 4x^{-3}$ **25.** [2.5] $-\$0.04$/pair
26. [2.7] $3xy^2 + \dfrac{y}{x}$ **27.** [3.1] $y = x - 2$

28. [1.5] $x = 1, x = \dfrac{1}{3}$ **29.** [2.2]
Relative maximum at $(-1, 3)$, relative
minimum at $(1, -1)$; point of inflection
at $(0, 1)$; increasing on $(-\infty, -1)$
and $(1, \infty)$, decreasing on $(-1, 1)$;
concave down on $(-\infty, 0)$, concave
up on $(0, \infty)$

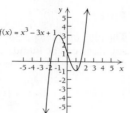

30. [2.2] Relative maxima at $(-1, -2)$
and $(1, -2)$, relative minimum at
$(0, -3)$; points of inflection at
$\left(-\dfrac{1}{\sqrt{3}}, -\dfrac{22}{9}\right)$ and $\left(\dfrac{1}{\sqrt{3}}, -\dfrac{22}{9}\right)$;
increasing on $(-\infty, -1)$ and $(0, 1)$,
decreasing on $(-1, 0)$ and $(1, \infty)$;
concave down on $\left(-\infty, -\dfrac{1}{\sqrt{3}}\right)$ and $\left(\dfrac{1}{\sqrt{3}}, \infty\right)$, concave up on
$\left(-\dfrac{1}{\sqrt{3}}, \dfrac{1}{\sqrt{3}}\right)$

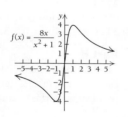

31. [2.2, 2.3] Relative maximum at
$(1, 4)$; relative minimum at $(-1, -4)$;
points of inflection at $(-\sqrt{3}, -2\sqrt{3})$ and
$(0, 0)$ and $(\sqrt{3}, 2\sqrt{3})$; decreasing on
$(-\infty, -1)$ and $(1, \infty)$, increasing on
$(-1, 1)$; concave down on $(-\infty, -\sqrt{3})$
and $(0, \sqrt{3})$, concave up on $(-\sqrt{3}, 0)$
and $(\sqrt{3}, \infty)$; horizontal asymptote at
$y = 0$

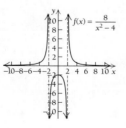

32. [2.3] Relative maximum at $(0, -2)$;
no points of inflection; vertical
asymptotes at $x = -2$ and $x = 2$;
horizontal asymptote at $y = 0$;
increasing on $(-\infty, -2)$ and $(-2, 0)$,
decreasing on $(0, 2)$ and $(2, \infty)$; concave
up on $(-\infty, -2)$ and $(2, \infty)$, concave
down on $(-2, 2)$
33. [2.4] Minimum $= -7$
at $x = 1$ **34.** [2.4] No absolute extrema **35.** [2.4]
Maximum $= 6\frac{2}{3}$ at $x = -1$; minimum $= 4\frac{1}{3}$ at $x = -2$
36. [2.5] 15 sweatshirts **37.** [2.5] 30 times; lot size of 15
38. [2.6] $\Delta y \approx 0.5$; $f'(x)\Delta x = 3$
39. [3.3] **(a)** $N(t) = 8000e^{0.1t}$; **(b)** 17,804; **(c)** 6.9 yr
40. [3.6] **(a)** $E(x) = \dfrac{x}{12 - x}$; **(b)** $E(2) = \frac{1}{5}$, inelastic;
(c) $E(9) = 3$, elastic; **(d)** increase; **(e)** $x = \$6$
41. [2.6] **(a)** $\Delta A \approx \pm 50$ ft^2; **(b)** 6 squares, $48

42. [4.2] $\frac{1}{2}x^6 + C$ **43.** [4.3] $3 - \dfrac{2}{e}$
44. [4.7] $\dfrac{7}{9(7 - 3x)} + \dfrac{1}{9}\ln|7 - 3x| + C$ **45.** [4.5] $\frac{1}{4}e^{x^4} + C$
46. [4.6] $\left(\dfrac{x^2}{2} + 3x\right)\ln x - \dfrac{x^2}{4} - 3x + C$
47. [4.2] $75\ln x + C$ **48.** [4.3] 2 **49.** [4.3] $\frac{232}{3}$, or $77\frac{1}{3}$
50. [5.2] $41,324.72 **51.** [5.2] **(a)** $28,992.08;
(b) $3832.91/yr; **(c)** $21,007.91, $11,670.90
52. [5.2] $711,632.48 **53.** [5.3] Convergent, $\dfrac{1}{4374}$
54. [5.1] $(7, \$169)$; $751.33 **55.** [5.4] $L_{10} = 0.675$ mi,
$R_{10} = 0.742$ mi, average $= 0.709$ mi **56.** [5.4] $T_6 = 7.714$
57. [5.6] $-\dfrac{\pi}{2}\left(\dfrac{1}{e^{10}} - 1\right)$, or approximately 1.571
58. [6.1] **(a)** $\sqrt{6}$; **(b)** $\{(x, y)|x^2 + y^2 \leq 4\}$
59. [6.4] **(a)** $y = \frac{5}{4}x + \frac{35}{12}$; **(b)** $154,167 **60.** [6.2] $8xy^3 + 3$
61. [6.2] $e^y + 24x^2y$ **62.** [6.3] No relative extrema
63. [6.5] Maximum $= 4$ at $(3, 3)$ **64.** [6.6] $3(e^3 - 1)$
65. [6.6] 1467 shoppers **66.** [7.2] **(a)** $f'(x) = 6\cos 3x$;
(b) $g'(x) = \cos x^2 - 2x^2 \sin x^2$;
(c) $h' = \dfrac{\tan 4x - 4x\sec^2 4x}{\tan^2 4x}$ **67.** [7.2] **(a)** Midline $= 59$,
amplitude $= 11$, period $= 24$, phase shift $= 0$; **(b)** 64.5°F;
(c) $T'(15) = 2.036$; **(d)** Minimum is 48°F at 6 a.m., and
maximum is 70°F at 6 p.m. $(t = 18)$. **68.** [7.3] About 2299 units
69. [7.4] $x = \dfrac{-\pi + 15}{6} \approx 1.976, x = \dfrac{\pi + 15}{6} \approx 3.024$
70. [8.2] $y = C_1e^{x^2/2}$, where $C_1 = e^C$ **71.** [8.4] $y = \frac{9}{4}e^{2-2x^2} + \frac{3}{4}$
72. [8.5] **(a)** $R(t) = C_1e^{0.2t} + C_2e^{-t}$; **(b)** $R(t) = 20e^{0.2t} + 15e^{-t}$;
(c) $R(4) = 44.785$, or about $44,785 **73.** [9.1] **(a)**
$a_n = 7n + 8$; **(b)** $a_{100} = 708$; **(c)** 36,150 **74.** [9.2]
$2,631,578.95 **75.** [9.3] **(a)** $9324.89, $675.11; [9.4]
(b) $1218.09, $9744.22, $255.20 **76.** [9.6]
$1 - \frac{1}{5} + \frac{1}{18} - \frac{1}{78} + \frac{1}{408} - \frac{1}{2520} + \frac{1}{18,000} \approx 0.8448$
77. [10.1] **(a)** $\frac{12}{47}$; **(b)** $\frac{35}{47}$; **(c)** $\frac{32}{47}$; **(d)** 0 **78.** [10.3] **(a)** $\frac{4}{24}$, or $\frac{1}{6}$, the
probability that a randomly selected student scored 18 on the
quiz; **(b)** $\frac{15}{24}$, or $\frac{5}{8}$, the probability that a randomly selected student
scored 18 or below on the quiz; **(c)** 17.375, the mean (average)
score on the quiz **79.** [10.4] **(a)** $\frac{3}{2}$; **(b)** $\mu = \frac{3}{2}\ln 3, \sigma = 0.5333$;
(c) $F(x) = -\dfrac{3}{2x} + \dfrac{3}{2}, 1 \leq x \leq 3$ **80.** [10.5] **(a)** 0.644;
(b) 99.4th; **(c)** $57,322.49

Appendix A

Exercise Set A, p. 828

1. $5 \cdot 5 \cdot 5$, or 125 **2.** $7 \cdot 7$, or 49 **3.** $(-7)(-7)$, or 49
4. $(-5)(-5)(-5)$, or -125 **5.** 1.0201 **6.** 1.030301 **7.** $\frac{1}{16}$
8. $\frac{1}{64}$ **9.** 1 **10.** $6x$ **11.** t **12.** 1 **13.** 1 **14.** $\frac{1}{3}$ **15.** $\dfrac{1}{3^2}$, or $\dfrac{1}{9}$
16. $\dfrac{1}{4^2}$, or $\dfrac{1}{16}$ **17.** 8 **18.** 4 **19.** 0.1 **20.** 0.0001 **21.** $\dfrac{1}{e^b}$
22. $\dfrac{1}{t^k}$ **23.** $\dfrac{1}{b}$ **24.** $\dfrac{1}{h}$ **25.** x^5 **26.** t^7 **27.** x^{-6}, or $\dfrac{1}{x^6}$ **28.** x^6
29. $35x^5$ **30.** $8t^7$ **31.** x^4 **32.** x **33.** 1 **34.** 1 **35.** x^3
36. x^4 **37.** x^{-3}, or $\dfrac{1}{x^3}$ **38.** x^{-4}, or $\dfrac{1}{x^4}$ **39.** 1 **40.** 1 **41.** e^{t-4}

42. e^{k-3}　**43.** t^{14}　**44.** t^{12}　**45.** t^2　**46.** t^{-4}, or $\frac{1}{t^4}$　**47.** a^6b^5

48. x^7y^7　**49.** t^{-6}, or $\frac{1}{t^6}$　**50.** t^{-12}, or $\frac{1}{t^{12}}$　**51.** e^{4x}　**52.** e^{5x}

53. $8x^6y^{12}$　**54.** $32x^{10}y^{20}$　**55.** $\frac{1}{81}x^8y^{20}z^{-16}$, or $\frac{x^8y^{20}}{81z^{16}}$

56. $\frac{1}{125}x^{-9}y^{21}z^{15}$, or $\frac{y^{21}z^{15}}{125x^9}$　**57.** $9x^{-16}y^{14}z^4$, or $\frac{9y^{14}z^4}{x^{16}}$

58. $625x^{16}y^{-20}z^{-12}$, or $\frac{625x^{16}}{y^{20}z^{12}}$　**59.** $\frac{c^4d^{12}}{16q^8}$　**60.** $\frac{64x^6y^3}{a^9b^9}$

61. $5x - 35$　**62.** $x + xt$　**63.** $x^2 - 7x + 10$

64. $x^2 - 7x + 12$　**65.** $a^3 - b^3$　**66.** $x^3 + y^3$

67. $2x^2 + 3x - 5$　**68.** $3x^2 + x - 4$　**69.** $a^2 - 4$

70. $9x^2 - 1$　**71.** $25x^2 - 4$　**72.** $t^2 - 1$　**73.** $a^2 - 2ah + h^2$

74. $a^2 + 2ah + h^2$　**75.** $25x^2 + 10xt + t^2$

76. $49a^2 - 14ac + c^2$　**77.** $5x^5 + 30x^3 + 45x$

78. $-3x^6 + 48x^2$　**79.** $a^3 + 3a^2b + 3ab^2 + b^3$

80. $a^3 - 3a^2b + 3ab^2 - b^3$　**81.** $x^3 - 15x^2 + 75x - 125$

82. $8x^3 + 36x^2 + 54x + 27$　**83.** $x(1 - t)$　**84.** $x(1 + h)$

85. $(x + 3y)^2$　**86.** $(x - 5y)^2$　**87.** $(x - 5)(x + 3)$

88. $(x + 5)(x + 3)$　**89.** $(x - 5)(x + 4)$

90. $(x - 10)(x + 1)$　**91.** $(7x - t)(7x + t)$

92. $(3x - b)(3x + b)$　**93.** $4(3t - 2m)(3t + 2m)$

94. $(5y - 3z)(5y + 3z)$　**95.** $ab(a + 4b)(a - 4b)$

96. $2(x^2 + 4)(x + 2)(x - 2)$

97. $(a^4 + b^4)(a^2 + b^2)(a + b)(a - b)$

98. $(6y - 5)(6y + 7)$　**99.** $10x(a + 2b)(a - 2b)$

100. $xy(x + 5y)(x - 5y)$　**101.** $2(1 + 4x^2)(1 + 2x)(1 - 2x)$

102. $2x(y + 5)(y - 5)$　**103.** $(9x - 1)(x + 2)$

104. $(3x - 4)(2x - 5)$　**105.** $(x + 2)(x^2 - 2x + 4)$

106. $(a - 3)(a^2 + 3a + 9)$　**107.** $(y - 4t)(y^2 + 4yt + 16t^2)$

108. $(m + 10p)(m^2 - 10mp + 100p^2)$

109. $(3x^2 - 1)(x - 2)$　**110.** $(y^2 - 2)(5y + 2)$

111. $(x - 3)(x + 3)(x - 5)$　**112.** $(t - 5)(t + 5)(t + 3)$

113. $\frac{7}{4}$　**114.** $\frac{79}{12}$　**115.** -8　**116.** $\frac{4}{5}$　**117.** 120　**118.** 140

119. 200　**120.** 200　**121.** $0, -3, \frac{4}{5}$　**122.** $0, 2 -\frac{3}{2}$　**123.** $0, 2$

124. $3, -3$　**125.** $0, 3$　**126.** $0, 5$　**127.** $0, 7$　**128.** $0, 3$

129. $0, \frac{1}{3}, -\frac{1}{3}$　**130.** $0, \frac{1}{4}, -\frac{1}{4}$　**131.** 1　**132.** 2

133. No solution　**134.** No solution　**135.** -23　**136.** -145

137. 5　**138.** $-\sqrt{7}, \sqrt{7}$　**139.** $x \geq -\frac{4}{5}$　**140.** $x \geq 3$

141. $x > -\frac{1}{12}$　**142.** $x > -\frac{5}{13}$　**143.** $x > -\frac{4}{7}$　**144.** $x \leq -\frac{6}{5}$

145. $x \leq -3$　**146.** $x \geq -2$　**147.** $x > \frac{2}{3}$　**148.** $x \leq \frac{5}{3}$

149. $x < -\frac{2}{5}$　**150.** $x \leq \frac{5}{4}$　**151.** $2 < x < 4$　**152.** $2 < x \leq 4$

153. $\frac{3}{2} \leq x \leq \frac{11}{2}$　**154.** $\frac{6}{5} \leq x < \frac{16}{5}$　**155.** $-1 \leq x \leq \frac{14}{5}$

156. $-5 \leq x < -2$　**157.** $\$650$　**158.** $\$800$　**159.** More than

7000 units　**160.** More than 4200 units　**161.** 480 lb

162. 340 lb　**163.** 810,000　**164.** 720,000

165. $60\% \leq x < 100\%$　**166.** $50\% \leq x < 90\%$

Appendix B

Exercise Set B, p. 834

1. 5　**2.** -4　**3.** $\frac{5}{2}$　**4.** $\frac{13}{3}$　**5.** 0　**6.** 0　**7.** $\frac{19}{2}$　**8.** $-\frac{1}{2}$　**9.** 2

10. $\frac{1}{2}$　**11.** $\frac{2}{3}$　**12.** 1　**13.** $-\frac{5}{4}$　**14.** $-\frac{9}{2}$　**15.** $\frac{1}{3}$　**16.** 2　**17.** 1

18. 1　**19.** e^2　**20.** e^{-3}　**21.** e^{-4}　**22.** e^5　**23.** 1　**24.** 1

25. $-\frac{1}{2}$　**26.** $\frac{5}{2}$　**27.** $\frac{1}{3}$　**28.** $\frac{1}{12}$　**29.** Left to the student

30. Left to the student　**31.** $\frac{1}{2}$　**32.** $-\frac{1}{3}$　**33.** $-\frac{2}{5}$　**34.** $-\frac{7}{3}$

35. (a) $0, 0, 0, 0$; **(b)** ✎; **(c)** ✎　**36. (a)** $e^{2x}/e^{3x} = 1/e^x, 0$; **(b)** ✎.

37. (a) Subtract 0 from numerator and denominator; **(b)** since

$x \neq a$, multiply numerator and denominator by $1/(x - a)$;

(c) property of limits, limit of quotient is quotient of limits;

(d) definition of derivative at $x = a$; **(e)** differentiability of f and

g at $x = a$　**38.** l'Hôpital's Rule does not apply since the first

expression is not indeterminate at $x = 3$. The correct limit is 0.

39. 1　**40.** 0　**41.** $\frac{5}{2}$　**42.** $\frac{7}{4}$

Appendix C

Exercise Set C, p. 837

1. $y = 3.45x + 0.6$　**2.** $y = 2.5283x + 13.34$

3. $y = 0.0732x^2 + 2.3x + 11.813$

Diagnostic Test, p. xxi

Part A

The blue bracketed references indicate where worked-out

solutions can be found in Appendix A: Review of Basic Algebra.

For example, [Ex.1] means that the problem is worked out in

Example 1 of the appendix.

1. 64 [Ex. 1]　**2.** -32 [Ex. 1]　**3.** $\frac{1}{8}$ [Ex. 1]　**4.** $-2x$ [Ex. 2]

5. 1 [Ex. 2]　**6.** $\frac{1}{x^5}$ [Ex. 3]　**7.** 16 [Ex. 3]　**8.** $\frac{1}{t}$ [Ex. 3]

9. x^{11} [Ex. 4]　**10.** x [Ex. 4]　**11.** $\frac{10}{x^7}$ [Ex. 4]　**12.** a [Ex. 5]

13. e^7 [Ex. 5]　**14.** $\frac{1}{x^6}$ [Ex. 6]　**15.** $\frac{y^{15}}{8x^{12}z^9}$ [Ex. 6]

16. $3x - 15$ [Ex. 7]　**17.** $x^2 - 2x - 15$ [Ex. 7]

18. $a^2 + 2ab + b^2$ [Ex. 7]　**19.** $4x^2 - 4xt + t^2$ [Ex. 8]

20. $9c^2 - d^2$ [Ex. 8]　**21.** $h(2x + h)$ [Ex. 9]

22. $(x - 3y)^2$ [Ex. 9]　**23.** $(x + 2)(x - 7)$ [Ex. 9]

24. $(2x - 1)(3x + 5)$ [Ex. 9]

25. $(x - 7)(x + 2)(x - 2)$ [Ex. 10]　**26.** $x = 6$ [Ex. 11]

27. $x = 0, 2, -\frac{4}{5}$ [Ex. 12]　**28.** $x = 0, \frac{1}{2}, -\frac{1}{2}$ [Ex. 13]

29. No solution [Ex. 15]　**30.** $x \leq \frac{21}{13}$ or $\left(-\infty, \frac{21}{13}\right]$ [Ex. 17]

31. 660 lb [Ex. 18]　**32.** 351 suits [Ex. 19]

Part B

The blue bracketed references indicate where worked-out solutions

can be found in Chapter R. For example, [Ex. R.2.5] means that the

problem is worked out in Example 5 of Section R.2.

1.

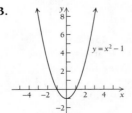

[Ex. R.1.1]

2.

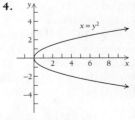

[Ex. R.1.2]

3.

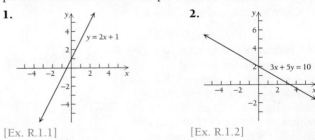

[Ex. R.1.3]

4.

[Ex. R.1.4]

5. $f(0) = 8; f(-5) = 143; f(7a) = 147a^2 - 14a + 8$ [Ex. R.2.4]

6. $\dfrac{(f(x+h)-f(x))}{h} = 1 - 2x - h$ [Ex. R.2.5]

7. 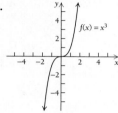 [Ex. R.2.9]

$f(x) = \begin{cases} 4, & \text{for } x \geq 0, \\ 3 - x^2, & \text{for } 0 < x \leq 2, \\ 2x - 6, & \text{for } x > 2. \end{cases}$

8. $(-4, 5)$ [Ex. R.3.1a]

9. $\left\{ x \,\middle|\, x \text{ is any real number and } x \neq \frac{5}{2} \right\}$ or $\left\{ x \,\middle|\, \left(-\infty, \frac{5}{2}\right) \cup \left(\frac{5}{2}, \infty\right) \right\}$ [Ex. R.3.4]

10. Slope $m = \frac{1}{2}$; y-intercept: $\left(0, -\frac{7}{4}\right)$ [Ex. R.4.4]

11. $y = 3x - 2$ [Ex. R.4.5]   **12.** $m = -\frac{3}{2}$ [Ex. R.4.7]

13.

$f(x) = x^2 - 2x - 3$

[Ex. R.5.1]

14.

$f(x) = x^3$

[Ex. R.5.4]

15.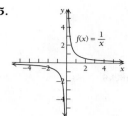

$f(x) = \dfrac{1}{x}$

[Ex. R.5.6]

16.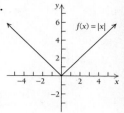

$f(x) = |x|$

[Ex. R.5.8]

17.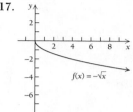

$f(x) = -\sqrt{x}$

[Ex. R.5.9]

18. \$1,166.40 [Ex. R.1.6]

Index of Applications

Index

Summary of Important Formulas for Differentiation

1. Power Rule. For any real number k, $\dfrac{d}{dx}x^k = kx^{k-1}$.

2. Derivative of a Constant Function. If $F(x) = c$, then $F'(x) = 0$.

3. Derivative of a Constant Times a Function. If $F(x) = cf(x)$,

$$\text{then } F'(x) = cf'(x).$$

4. Derivative of a Sum. If $F(x) = f(x) + g(x)$, then

$$F'(x) = f'(x) + g'(x).$$

5. Derivative of a Difference. If $F(x) = f(x) - g(x)$, then

$$F'(x) = f'(x) - g'(x).$$

6. Derivative of a Product. If $F(x) = f(x)g(x)$, then

$$F'(x) = f(x)g'(x) + g(x)f'(x).$$

7. Derivative of a Quotient. If $F(x) = \dfrac{f(x)}{g(x)}$, then

$$F'(x) = \frac{g(x)f'(x) - f(x)g'(x)}{[g(x)]^2}.$$

8. Extended Power Rule. If $F(x) = [g(x)]^k$, then

$$F'(x) = k[g(x)]^{k-1}g'(x).$$

Summary of Important Formulas for Differentiation

(*continued*)

9. Chain Rule. If $F(x) = f[g(x)]$, then $F'(x) = f'[g(x)]g'(x)$.
Or, if $y = f(u)$ and $u = g(x)$, then

$$\frac{dy}{dx} = \frac{dy}{du} \cdot \frac{du}{dx}.$$

10. $\dfrac{d}{dx} e^x = e^x$

11. $\dfrac{d}{dx} e^{f(x)} = e^{f(x)} \cdot f'(x)$

12. $\dfrac{d}{dx} \ln x = \dfrac{1}{x}, \quad x > 0$

13. $\dfrac{d}{dx} \ln f(x) = \dfrac{f'(x)}{f(x)}, \quad f(x) > 0$

14. $\dfrac{d}{dx} \ln |x| = \dfrac{1}{x}, \quad x \neq 0$

15. $\dfrac{d}{dx} \ln |f(x)| = \dfrac{f'(x)}{f(x)}, \quad f(x) \neq 0$

16. $\dfrac{d}{dx} a^x = (\ln a)a^x, \quad a > 0$

17. $\dfrac{d}{dx} \log_a |x| = \dfrac{1}{\ln a} \cdot \dfrac{1}{x}, \quad a > 0 \text{ and } x \neq 0$